AUTOMOTIVE, MECHANICAL AND ELECTRICAL ENGINEERING

PROCEEDINGS OF THE 2016 INTERNATIONAL CONFERENCE ON AUTOMOTIVE ENGINEERING, MECHANICAL AND ELECTRICAL ENGINEERING (AEMEE 2016), HONG KONG, CHINA, 9–11 DECEMBER 2016

Automotive, Mechanical and Electrical Engineering

Editor

Lin Liu
Wuhan University of Technology, Wuhan, China

CRC Press
Taylor & Francis Group
Boca Raton London New York Leiden

CRC Press is an imprint of the
Taylor & Francis Group, an **informa** business

A BALKEMA BOOK

CRC Press/Balkema is an imprint of the Taylor & Francis Group, an informa business

© 2017 Taylor & Francis Group, London, UK

Typeset by V Publishing Solutions Pvt Ltd., Chennai, India
Printed and bound in Great Britain by CPI Group (UK) Ltd, Croydon, CR0 4 YY

Published by: CRC Press/Balkema
P.O. Box 11320, 2301 EH Leiden, The Netherlands
e-mail: Pub.NL@taylorandfrancis.com
www.crcpress.com – www.taylorandfrancis.com

ISBN: 978-1-138-62951-6 (Hbk)
ISBN: 978-1-315-21044-5 (eBook)

Table of contents

Network, communications and applied information technologies

Technologies in energy and power, cells, engines, generators, electric vehicles

System test and diagnosis, monitoring and identification, video and image processing

Technologies in electrical and electronic control and automation

Industrial production, manufacturing, management and logistics

Automotive, Mechanical and Electrical Engineering – Liu (Ed.)
© *2017 Taylor & Francis Group, London, ISBN 978-1-138-62951-6*

Preface

The organizing committee of AEMEE 2016 is proud to present the proceedings of the 2016 International Conference on Automotive Engineering, Mechanical and Electrical Engineering (AEMEE 2016: http:// www.aemee.org/), held in Hong Kong, China during December 09–11, 2016.

AEMEE 2016 was a platform for presenting excellent results and new challenges facing the field of the automotive engineering, mechanical and electrical engineering. It brought together experts from industry, governments and academia, experienced in engineering, design and research.

AEMEE 2016 received 356 manuscripts, and 115 authors participated in this conference. By submitting a paper to AEMEE 2016, the authors agreed to the review process and understood that papers would undergo a peer-review process. Manuscripts were reviewed by appropriately qualified experts in the field selected by the Conference Committee, who gave detailed comments and-if the submission was accepted-the authors would submit a revised version that took into account this feedback. All papers were reviewed using a double-blind review process: authors declared their names and affiliations in the manuscript for the reviewers to see, but reviewers did not know each other's identities, nor did the authors receive information about who had reviewed their manuscript. The Committees of AEMEE 2016 invested great efforts in reviewing the papers submitted to the conference and organizing the sessions to enable the participants to gain maximum benefit.

Hopefully, all participants and other interested readers will benefit scientifically from the proceedings and also find it stimulating in the process.

With our warmest regards,
Lin Liu
Conference Organizing Chair
Wuhan, China

Automotive, Mechanical and Electrical Engineering – Liu (Ed.)
© 2017 Taylor & Francis Group, London, ISBN 978-1-138-62951-6

Committees

CONFERENCE CHAIRS

Prof. L. Liu, *Wuhan University of Technology, China*
Prof. T.S. Ma, *Wuhan University, China*

TECHNICAL COMMITTEE

Prof. L. Liu, *Wuhan University of Technology, China*
Prof. Y. Wang, *Chongqing University, China*
Prof. H. Davis, *Boya Century Publishing Ltd., Hong Kong*
Prof. W. Liu, *Huazhong University of Science and Technology, China*
Prof. S. Zhu, *Wuhan University, China*
Prof. G.S. Liu, *Hunan University, China*
A. Prof. P. Wang, *Guangxi College of Education, China*
Dr. C. Yang, *Wuhan University of Technology, China*
Dr. Z.G. Fang, *Wuhan University of Technology, China*
Dr. Z.B. You, *Wuhan University of Technology, China*
Dr. J.F. Ke, *Wuhan University of Technology, China*
Dr. Z.H. Tan, *Wuhan University of Technology, China*
Dr. H. Zhang, *GAC Automotive Engineering Institute, China*
Mr. Y. Zhou, *Dongfeng Citroen Automobile Co. Ltd., China*
Dr. J. Cai, *Dongfeng Motor Corporation Technical Center, China*
Dr. C. Yang, *Wuhan University of Technology, China*
Dr. S.F. Zhao, *Wuhan University of Science and Technology, China*
Mr. Yang Zhou, *Dongfeng Citroen Automobile Co. Ltd., China*

ORGANIZING COMMITTEE

Prof. L. Liu, *Wuhan University of Technology, China*
Dr. Z.B. You, *Wuhan University of Technology, China*
Dr. C. Zhang, *Asian Union of Information Technology, China*
Dr. D.W. Fang, *Asian Union of Information Technology, China*
Mr. C. Ma, *Wuhan Vike Technology Co. Ltd., China*
Mr. Y.Y. Liu, *Wuhan Vike Technology Co. Ltd., China*
Mr. C. Liu, *Wuhan Vike Technology Co. Ltd., China*
Ms. H.H. You, *Wuhan Zhicheng Times Cultural Development Co. Ltd., China*
Mr. X.T. Ke, *Wuhan Zhicheng Times Cultural Development Co. Ltd., China*
Mr. B. Zhou, *Wuhan Zhicheng Times Cultural Development Co. Ltd., China*
Mr. X. Yi, *Wuhan Zhicheng Times Cultural Development Co. Ltd., China*
Mr. K. Mai, *Wuhan Zhicheng Times Cultural Development Co. Ltd., China*
Ms. Y.F. Ma, *Wuhan Zhicheng Times Cultural Development Co. Ltd., China*
Mr. X.T. Ke, *Wuhan Zhicheng Times Cultural Development Co. Ltd., China*
Mr. B. Zhou, *Wuhan Zhicheng Times Cultural Development Co. Ltd., China*
Mr. X. Yi, *Wuhan Zhicheng Times Cultural Development Co. Ltd., China*
Mr. K. Mai, *Wuhan Zhicheng Times Cultural Development Co. Ltd., China*

Automotive, Mechanical and Electrical Engineering – Liu (Ed.)
© 2017 Taylor & Francis Group, London, ISBN 978-1-138-62951-6

Sponsors

Guangxi College of Education
Northeast Petroleum University
Research Center of Engineering and Science (RCES)
Asian Union of Information Technology
HuBei XinWenSheng Conference Co. Ltd.

Automotive engineering and rail transit engineering

Automotive, Mechanical and Electrical Engineering – Liu (Ed.)
© 2017 Taylor & Francis Group, London, ISBN 978-1-138-62951-6

A study of the fuzzy three-parameter shifting rules of the AMT vehicle

Qinghong Chen & Yong Wang
Chongqing College of Electronic Engineering, Chongqing, China

ABSTRACT: In order to enable the automatic transmission shifting of the AMT vehicle to reflect the changes of road conditions and vehicle conditions, and to meet the requirement of shifting smoothness of the vehicle, we studied the fuzzy three-parameter shifting control of the AMT vehicle and put forward that acceleration should be used to reflect the changes of road conditions. This paper will introduce the principle of fuzzy three-parameter shifting control at first, and then it will show the design of fuzzy three-parameter shifting controller. Finally, it will give the fuzzy three-parameter shifting simulation of a certain model of AMT car produced by Changan Automobile. In addition, we will also compare it with the fuzzy two-parameter shifting. The result shows that the fuzzy three-parameter shifting is fit with drivers' shifting experience and habits than the two-parameter shifting. And compared with solving the shifting rules with traditional calculation method, it is more convenient and easier to be implemented with stronger robustness.

Keywords: Automotive engineering; Automatic transmission; Three-parameter shifting; Fuzzy control

1 INTRODUCTION

At present, several typical automatic transmissions are (A. bastian, 1995): hydraulic mechanical Automatic Transmission (AT), metal-belt-type Continuously Variable Transmission (CVT), electronic-controlled Automatic Mechanical Transmission (AMT). These different kinds of transmissions have their own advantages and disadvantages, and research on them at home and abroad has been continuously deepened to meet people's increasing demands for comfortable, shifting smoothness, fuel economy and low emission. The realization of electronic-controlled Automatic Mechanical Transmission (AMT) is based on the manual gear transmission. Its structure is simple, and keeps the majority of assembly components of dry clutch and manual transmission, only with the gear lever of the manual operating system being replaced with automatic control mechanism. It retains the advantages of the gear of the original manual transmission, such as high transmission efficiency, low cost, simple structure, and easy production, which, as a result, brings good inheritance in production and low cost in transformation, making it suitable for the situation of China.

The key techniques of automatic transmission control of the AMT vehicle are the establishment of the shifting rules and the control of clutch. This paper mainly focuses on the establishment of the shifting rules, that is, to determine the best gear of a vehicle according to the driver's intention, the vehicle's running status, road conditions, etc., as well as the principle that some vehicle parameters are of the optimal property. Currently, the methods to determine the gear can be divided into two large categories, one is to solve the rule of shifting according to the principle of optimal index of specific performances after determining the shifting control parameters; the other one is to take advantage of the driving experience of drivers and the knowledge of experts to form the fuzzy control rules, that is, the shifting rules based on the expert system. The former is a traditional method with a complete theory and a set of complete solutions, and the given gear can ensure the vehicle's optimal performance when the operational environment and working conditions of the vehicle are consistent with the preset conditions of the solved optimal shifting rules, but the given gear could be obviously not the best, or not the best in some fields when there exist large differences between the actual operational conditions of the vehicle and the preset conditions when solving the shifting rules. In this paper, the fuzzy shifting control based on expert knowledge will be discussed. At present, the shifting rule which is comparatively mature is the two-parameter shifting rule (the speed of the vehicle, and the accelerator opening). But the two-parameter shifting rule cannot reflect the impact of the changes of such external conditions as road conditions and vehicle load on shifting. Therefore, I will adopt the three-parameter shifting in this paper, so as to consider the impact of the changes of driving conditions on shifting.

Many domestic and foreign scholars have studied the establishment of the shifting rules, A.Bastian et al. studied the fuzzy shifting rules of the AT vehicle (Huang Zongyi, 2006; Huifang Kong, 2008; J Yi, 2007). Yi Jun et al. studied the fuzzy shifting of the tracked vehicle. Li Pingkang et al. studied the fuzzy shifting of the construction vehicle (Wang Lixin, 2003; Weibo Yu, 2016; Yi Jun, 2006; Yi Jun, et al. 2008; Zhiyi Zhang, 2008). These studies had something in common, that is, the transmissions, which was the subject of these studies, all had a hydraulic torque converter in them, and when establishing the shifting rules, the rotating speed of the turbine and pump of the hydraulic torque converter were added, so that the hydraulic torque converter could serve as a buffer. However, due to a power interruption problem when shifting the gear of the AMT vehicle, the vehicles were extremely sensitive to the change of acceleration, so the way in which the AT and construction vehicles shift gears fuzzily could not be applied to the AMT vehicle directly. Huifang Kong (Zhang Zhiyi, 2005) and B Mashadi (Zhang Yong, 2003) studied the fuzzy shifting of the AMT vehicle. Since what they adopted was the two-parameter shifting, they were unable to fully consider the impact of external environmental changes on the shifting of the vehicle. In order to correctly reflect the impact of external environmental changes on shifting, in this paper I adopt the fuzzy three-parameter shifting (vehicle speed, accelerator opening, and acceleration). To some extent, acceleration reflects the changes of the driving environment, and after introducing acceleration, the three-parameter shifting can make the impact in the process of shifting smaller, thus improving riding comfort. First of all, this paper describes the principle of the fuzzy shifting of the AMT vehicle, then studies the fuzzy shifting controller, and finally the fuzzy control system is simulated. It turns out that the adoption of the fuzzy three-parameter shifting is more fit for the operation of skilled drivers than the adoption of the two-parameter shifting, because it makes shifting process more comfortable and smoother, and the vehicle has better fuel economy.

2 THE PRINCIPLE OF THE FUZZY THREE-PARAMETER SHIFTING CONTROL OF THE AMT VEHICLE

The principle of the fuzzy three-parameter shifting control of the AMT vehicle is shown in Figure 1.

In Figure 1, a is the opening of the engine's accelerator, V is the vehicle's driving speed, and Ac is the vehicle's acceleration. The core of fuzzy three-parameter shifting control is the design of

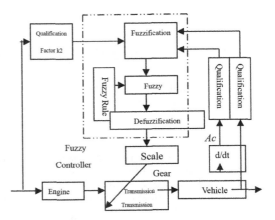

Figure 1. The principle of the fuzzy three-parameter shifting control of the AMT vehicle.

fuzzy controller (the part in the double dotted line frame). First quantify and then fuzzy process the signals of the accelerator opening, the vehicle speed and acceleration, and then do fuzzy reasoning combining with the fuzzy rule base to get the results, which can determine the transmission's gear through defuzzification and adjusting the scale factor, thus changing the transmission ratio of AMT transmission and satisfying the requirements of the vehicle for driving power and gear when running.

3 THE DESIGN OF THE FUZZY CONTROLLER

The design of the fuzzy controller includes the determination of the quantification factor and the scale factor, the fuzzy strategy, the design of fuzzy rule base, the establishment of fuzzy reasoning mechanism, and defuzzification.

When the vehicle is running, the accelerator opening reflects the driver's intention, and the speed reflects the current running conditions of the vehicle. And to a certain extent, acceleration reflects the changes of driving environment, while upshifts and downshifts are the effective methods to realize drivers' intentions according to the vehicle's conditions and the environment. Choose the accelerator opening, the vehicle's speed and acceleration as the input of the fuzzy controller and gear as the output.

3.1 *The fuzzification of controlled variables and qualification factor*

According to the driving experiences of excellent drivers and the specific parameters of vehicles in actual testing, the fuzzy subsets of the three input quantities of the fuzzy three-parameter shifting controller respectively are: "Negative Big" (NB), "Negative Medium" (NM), "Negative Small" (NS), "Medium" (M), "Positive Small" (PS), "Positive Medium" (PM), "Positive Big" (PB), which are seven fuzzy quantities in total. The accelerator opening is taken as: "Very Small" (VS), "Small" (S), "Medium" (M), "Big" (B), "Very Big" (VB), which are five fuzzy quantities in total. And the acceleration is taken as: "Very Small" (VB), "Small" (S), "Medium" (M), "Big" (B), "Very Big" (VB), which are five fuzzy quantities in total. The output gear Dy is indicated by single point, and there are five gears in total: I, II, III, IV, V.

When the total number of the elements in the discourse domain is 2–3 times than the total number of fuzzy subsets, the degree of coverage of the fuzzy subsets for the discourse domain is the best (B Mashadi, 2007). Quantify the vehicle speed, the accelerator opening and the acceleration, and then their discourse domains are taken respectively as:

{0, 1, 2, 3, 4, 5, 6, 7, 8, 9, 10, 11, 12, 13, 14, 15}

{0, 1, 2, 3, 4, 5, 6, 7, 8, 9, 10}

{0, 1, 2, 3, 4, 5, 6, 7, 8, 9, 10}

Based on the experience of drivers, when vehicle speed is less than 10 km per hour, the vehicle usually stays at the first gear, and when the vehicle speed is more than 70 km per hour, the vehicle usually stays at the fifth gear, so the physical discourse domain of vehicle speed is $Vi = [10, 70]$, and the quantification factor is $k_i = 0.25$; When the accelerator opening is more than 55%, the gear shifting is mainly based on the driving speed and acceleration, so the physical discourse domain of accelerator opening is taken as $a = [0, 50]$, and the quantification factor is $k_j = 0.2$; the range of acceleration is generally less than 4 m/s², so the physical discourse domain is taken as $Ac = [-5, 5]$, and the quantification factor is $k_l = 1.25$. The specific algorithms obtained are as follows:

$$V = \begin{cases} 15 & k_i(v-10) \geq 15 \\ k_i v & 0 < k_i(v-10) < 15 \\ 0 & k_i(v-10) \leq 0 \end{cases} \quad (1)$$

$$a = \begin{cases} 10 & k_j(a-10) \geq 10 \\ k_j a & 0 < k_j(a-10 < 10 \\ 0 & k_j(a-10) < 0 \end{cases} \quad (2)$$

$$Ac = \begin{cases} 5 & k_l A_c \geq 5 \\ k_l A_c & 0 < k_l A_c < 10 \\ -5 & k_l A_c \leq -5 \end{cases} \quad (3)$$

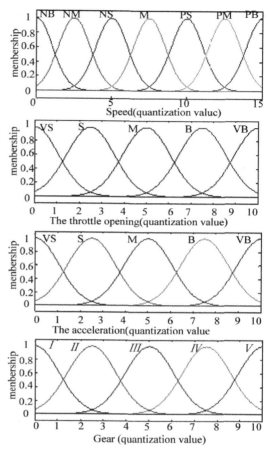

Figure 2. The fuzzy three-parameter shifting input quantity/output quantity membership functions of the AMT vehicle.

The membership functions of the input and output quantities are the type of Gaussian function, as shown in Figure 2. The Gaussian membership function can simplify the calculation designed by some fuzzy reasoning machine (B Mashadi, 2007). The three input quantities and the membership function expression of the output quantities are shown in Equation (4).

$$f(x, \sigma, c) = e^{-\frac{(x-c)^2}{2\sigma^2}} \quad (4)$$

Among them, C determines the central position of the membership function, and σ determines the width of the curve of the function.

From the figure of the membership function, it can be seen that the selected membership function meets three basic features: completeness, consistency, and interactivity.

3.2 The establishment of the fuzzy control rules

According to the shifting experience of excellent drivers, when establishing the fuzzy three-parameter shifting rules, there are 175 fuzzy rules in total according to the fuzzy subsets of the three input quantities. The three-dimensional fuzzy rules are shown in Figure 3, and each cube corresponds to one fuzzy rule described below.

IF V is NB and a is VS and Ac is VS, THEN Dy is *I*;

IF V is NM and a is S and Ac is S, THEN Dy is *II*;

IF V is M and a is M and Ac is B, THEN Dy is *III*;

IF V is PB and a is VB and Ac is VB, THEN Dy is *V*;

3.3 Fuzzy reasoning

The fuzzy relation of the *i* fuzzy rule is:

$$R_i(v, a, Ac, Dy) = u_{V_i \times a_i \times Ac_i \times Dy_i}(v, a, Ac, Dy) \quad (5)$$

The total fuzzy relation is:

$$R = \bigcup_{i=1}^{175} R_i \quad (6)$$

The compositional rule of inference adopts Mamdani (max-min) compound operation:

$$Dy' = (V' \times a' \times Ac') \circ R \quad (7)$$

O is the synthesis operator.

3.4 Defuzzification

According to the features of this research subject, defuzzification adopts the maximum membership degree law:

$$Dy = \sup\{u_{V' \times a' \times Ac'}\} \quad (8)$$

4 THE SIMULATION ANALYSIS OF THE FUZZY THREE-PARAMETER SHIFTING OF THE AMT VEHICLE

After designing the fuzzy three-parameter shifting controller, the controlling system is simulated by using MATLAB/Simulink program.

The simulation uses a Changan Null car equipped with AMT transmission as the subject. The main parameters of the car are shown in Table 1, and the simulation module diagram is shown in Figure 4. The simulation results are shown in Figures 5, 6.

In Figure 5, (a) is the fuzzy three-parameter shift when the acceleration is 3 m/s²; (b) is the fuzzy three-parameter shift when the acceleration is 1.5 m/s²; (c) is the fuzzy three-parameter shift when the acceleration is −1.5 m/s²; (d) is the fuzzy three-parameter shift when the acceleration is −3 m/s². It can be seen from the figure that with the same accelerator opening, the bigger the acceleration is, the lower the vehicle speed will be when shifting the gear, while at the same acceleration, the bigger the accelerator opening is, the higher the vehicle speed will be when shifting the gear. And it is totally the same with the gear-shifting behavior of skilled drivers, which indicates that the adoption of the fuzzy three-parameter shifting is successful, because it is able to reflect the impact of the changes of the external environment on gear-shifting decisions completely.

Figure 6 is the fuzzy three-parameter shifting when acceleration is taken as 0 m/s². Arrow lines 1, 2, 3, 4, 5 refer to the five gears respectively. The adoption of the fuzzy two-parameter shifting is equal to the fuzzy three-parameter shifting when acceleration is taken as 0 m/s² (the four cases shown in Figure 5 are unattainable). Therefore, it can be seen that when adopting the fuzzy three-parameter shifting, the impact of acceleration on the process of shifting is taken into account when shifting the gear, making gear shifting be more fit with the

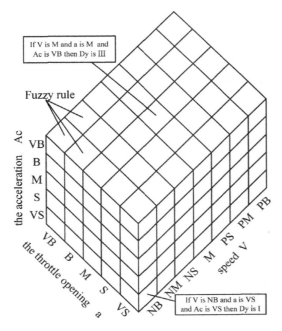

Figure 3. The three-dimensional fuzzy rule base of the fuzzy three-parameter shifting.

Table 1.   The table of main parameters of a certain testing model of Changan car.

Power kw	Idle speed r/min	Wheel diameter mm	The speed ratio of the transmission and the main decelerator						
			First gear	Second gear	Third gear	Fourth gear	Fifth gear	Reverse gear	Main decelerator
63	800	274	3.416	1.894	1.280	0.914	0.757	3.272	3.272

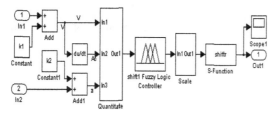

Figure 4.   The simulation model of the fuzzy three-parameter shifting of the AMT vehicle.

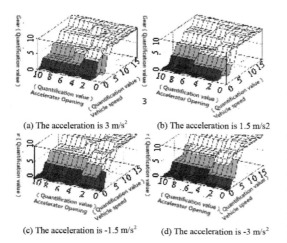

(a) The acceleration is 3 m/s²

(b) The acceleration is 1.5 m/s2

(c) The acceleration is -1.5 m/s²

(d) The acceleration is -3 m/s²

Figure 5.   The simulation of the fuzzy three-parameter shifting of the AMT vehicle.

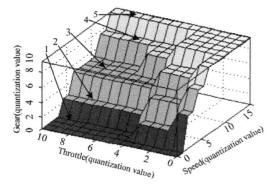

Figure 6.   The fuzzy three-parameter shifting when acceleration is taken as 0 m/s².

driving behaviors of skilled drivers and the gear-shifting decisions more reasonable. In addition, it also enhances the vehicles' abilities to adapt to the environment, which increases the economy of fuel and the smoothness of shifting in the process of gear shifting.

5   CONCLUSION

As people's requirements for the driving performance and emission of the vehicle are getting increased, in order to make the gear-shifting decision and shifting control of the AMT automatic transmission vehicle be more fit with vehicles, the environment, the drivers' intentions, this paper adopts the fuzzy three-parameter shifting method to effectively solve the problem that the relation between the conditions of the vehicle and the road and the gear-shifting decision and control cannot be reflected well, which, as a result, makes the shifting process be more fit with people's driving habits and more adapt to the road conditions. In addition, it also avoids the complex calculation in the traditional gear-shifting decision control, which is easier and more convenient with better robustness and adaptability. The fuzzy three-parameter shifting has better controlling precision than the two-parameter shifting, and the process of shifting is smoother.

REFERENCES

Bastian, A. S. Tano, T. Oyama etc, System Overview and Special Feature of FATE: Fuzzy Logic Automatic Transmission Expert System [J]. IEEE1995 0-7803-2461-7.

Huang Zongyi, Principle and Design of the Modern Transmission Automatic Cars [M], Shanghai, Tongji University Press, Sept. 2006.

Huifang Kong, Study on AMT Fuzzy shifting Strategy and Realization [J], IEEE2008 978-1-4244-2503-7.

Li Pingkang, Jin Taotao, Li Bei, Fuzzy Recursive Algorithm Based on Vehicle Shifting Strategy [J]. Journal of Beijing Jiaotong University, Vol. 32, No. 1, Feb. 2008.

Mashadi, B. A Kazemkhani, and R Baghaei Lakeh, An automatic gear-shifting strategy for manual transmissions [J], JSCE253 IMechE 2007, Proc. IMechE Vol. 221 Part I: J. systems and control engineering.

Sakaguchi, S., I. Sakai, T. Haga, Application of Fuzzy Logic to Shift Scheduling Method for Automatic Transmission [J], IEEE1993 0-7803-0614-7.

Wang Lixin, Fuzzy System and Fuzzy Control Tutorial [M], Beijing, Tsinghua University Press, June 2003.

Weibo Yu, Nan Li, Dingxuan Zhao etc., Adaptive fuzzy shift strategy in automatic transmission of construction Vehicles [J], IEEE2016 1-4244-0466-5.

Yi, J., X-L Wang, Y-J Hu etc, Modelling and simulation of a fuzzy controller of automatic transmission of a tracked vehicle in complicated driving conditions [J], JAUTO335 IMechE 2007.

Yi Jun, Wang Xuelin, Hu Yujin etc, Fuzzy Control and Simulation on Automatic Transmission of Tracked Vehicle in Complicated Driving Conditions [J]. IEEE2006 1-4244-0759-1.

Yi Jun, Xu Zhongbao, Wang Xuelin etc, Adaptive Fuzzy Control of Shift Strategy of Off-road Vehicle [J], Automotive Engineering, 2008 (Vol. 30), No. 1

Zhang Yong, Liu Jie, Lu Xintian etc, Adaptive Fuzzy Shift Strategy of Construction Vehicles [J], Journal of Tongji University, Vol. 31, No. 1, Jan. 2003.

Zhang Zhiyi, Zhao Dinxuan, Chen Ning, Research on Fuzzy Automatic Transmission Strategy of Vehicles [J], Transactions of the Chinese Society for Agricultural Machinery, Vol. 36, No. 10, Oct. 2005.

Zhiyi Zhang, Dingxuan Zhao, BeiSun, Study on Fuzzy Automatic Transmission Strategy of Vehicles [J], IEEE2008 978-4244-1674-5.

Automotive, Mechanical and Electrical Engineering – Liu (Ed.)
© 2017 Taylor & Francis Group, London, ISBN 978-1-138-62951-6

Design process and kinematic characteristics analysis of a minivan's Macpherson suspension system

Yong Wei
SAIC GM Wuling Automobile Co. Ltd., Liuzhou, China

Zhuoyu Su
Hubei Province Key Laboratory of Modern Automotive Technology, Wuhan, China
School of Automotive Engineering, Wuhan University of Technology, Wuhan, China
Hubei Collaborative Innovation Centre for Automotive Components Technology, Wuhan, China

Changye Liu
SAIC GM Wuling Automobile Co. Ltd., Liuzhou, China

Fengxiang Xu & Hao Chen
Hubei Province Key Laboratory of Modern Automotive Technology, Wuhan, China
School of Automotive Engineering, Wuhan University of Technology, Wuhan, China
Hubei Collaborative Innovation Centre for Automotive Components Technology, Wuhan, China

ABSTRACT: Taking a minivan system as the research object, a process of kinematic characteristics analysis and optimisation design is presented. The simulation of parallel wheel travel of the suspension of the minicar is carried out and analysed. The characteristics curves of four front wheel alignment parameters including toe angle, camber angle, caster angle, and kingpin inclination angle are drawn, the unreasonable alignment parameters are drawn, and the non-ideal characteristics of the wheel alignment parameters are established. The objective function is then to reduce the variation of the unreasonable alignment parameters; the design variables are given by the sensitivity analysis, and the constraint condition is the change in the coordinate values of key hard points. By optimising wheel alignment parameters, and the original and optimised simulation results, a better solution is obtained and the system performance of the suspension is improved.

Keywords: Minivan; Macpherson suspension; Kinematic characteristics; Alignment parameters; Optimisation design

1 INTRODUCTION

The vehicle suspension is a connecting device transferring force and torques from the frame to the vehicle axle, to ease the impact of load and restrain irregular vibration of the vehicle bearing system (Chen, 2012). In the research and design of vehicle suspension, the ADAMS virtual prototype test is an important means of suspension analysis and optimisation design (Chen, 2008).

Over the years, Macpherson suspension analysis and optimisation design has been widely studied. In Ren et al. (2010), the multi-body dynamics model of a minivan's Macpherson suspension was established and analysed. In Zhang et al. (2013), the sensitivity analysis and optimisation of hard points in a suspension system were researched by ADAMS. In Sun (2014), based on the mathematical relationship between suspension characteristics and the wheel movement, establish variables constraint functions and a whole vehicle dynamics model for researching the suspension characteristics. Wang et al. (2015) proposed the wheel rotation centre method to evaluate suspension performance and vehicle comfort. In Sagi et al. (2015), a multi-objective optimisation model was researched for determining the optimal parameters of a suspension system. Jamali et al. (2014) combined a simulation road with pavement power spectral density, and established a model with five degrees of freedom for evaluating the vehicle performance approximately.

To a certain extent, Macpherson suspension kinematics and dynamics characteristics were improved in previous studies. However, few researchers have studied the analysis and design process of the Macpherson suspension. In order to further research the analysis and design process, improving the whole vehicle comfort and handling stability, taking a minivan suspension system as the research object, the process of kinematic characteristics analysis and optimisation design is given.

The rest of the paper is organised as follows. The design process of the suspension system, including the general suspension design requirements, and the composition of the suspension design are developed in Section 2. The kinematic model of suspension is established in Section 3. To realise the fundamental characteristics of the research object, the kinematics characteristics and wheel alignment parameters of a minivan suspension system are analysed and confirmed in Section 4. To further improve the suspension system performance, the optimisation design of the suspension system kinematics characteristics is developed in Section 5, which is followed by the concluding remarks in Section 6.

2 DESIGN PROCESS OF THE SUSPENSION SYSTEM

Figure 1 presents the design process of a minivan's suspension system in this study. Considering the suspension system plays a decisive role in the vehicle handing stability and comfort, there are five design requirements during the research (Ding, 2010). Firstly, reasonable elastic properties of springs and damper damping characteristics are important to avoid impact between springs and dampers. Secondly, rational is necessary for transferring force and torque between frame and wheels. At the same time, the steering mechanism movement should be coordinated with the movement of the guide mechanism to avoid movement interference. It is crucial to determine the rational suspension roll centre and longitudinal metacentre to improve anti-roll and longitudinal ability during steering. Lastly, high-strength and lightweight components and parts are essential.

2.1 Design object parameters

Based on the market requirement and design objective, basic model and standard model needed to develop was choose. The suspension parameters of the basic model and standard model were measured and collected in K&C tests. Comparing with two kinds of parameters data, the design object parameters of the suspension were defined.

2.2 Suspension system hard points

The structural features of the Macpherson suspension include simple structure, small volume, and less frame stress point. According to the basic vehicle suspension hard points data and body structure, the hard points of the front suspension are roughly determined. The kinematics simulation model is established using ADAMS, the simulation results are compared with the object parameters, and then the data of the hard points are adjusted so that the simulation results coincide with the object parameters.

2.3 Components design and checking

According to the change in suspension hard points, the structure and size of the base model parts are improved accordingly. Considering the material, processing technology and connections of the parts, but also using a simulation model, the stiffness of the bushings is adjusted and determined, and the parts are optimised by finite element analysis. For the coil spring, according to the suspension stiffness and suspension structure to calculate the stiffness, and to determine the parameters of the spring. For the damper, the shape of the structure and the damping curve is determined, and the appropriate damper is selected to meet the design requirements.

2.4 Kinetic characteristic simulation

After all the parameters of the suspension system are determined, the dynamic simulation model is

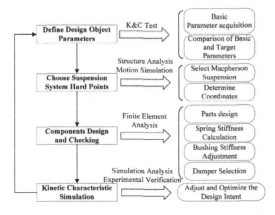

Figure 1. Design process of a minivan's Macpherson suspension.

reestablished, and the complete K&C characteristic simulation analysis is carried out. The design target is adjusted and the optimisation design is carried out. The design of the suspension system is completed by the test.

3 KINEMATIC MODELLING

Figure 2(a) shows the Macpherson suspension structure. Based on the theory of vehicle dynamics, the suspension is reasonably simplified and assumed. In addition to elastic components and rubber components, the other parts are considered rigid, regardless of its deformation. All connections between components are simplified as ball joints, sleeve connections, etc., excluding internal clearance. The frictional forces within the motion pairs between the components are negligible (Huang et al., 2010).

Figure 2(b) shows a minivan's Macpherson suspension CAD model. The relative position of the component parts is determined, the coordinate, quality, inertia and geometric characteristics of the component parts are obtained, finding out all kinds of constraints and connection point, the reference point coordinate information. The detailed parameter information for the simulation model establishment is provided.

Some key parameters of the Macpherson suspension are listed in Table 1. Figure 3 shows a multi-rigid-body dynamic model of Macpherson front suspension assembly in ADAMS/CAR. The wheel travel is generated by two-wheel co-oscillating excitation.

In this study, the technical roadmap of Macpherson suspension system kinematics analysis and optimisation is shown in Figure 4.

Table 1. The initial key point coordinate of the left Macpherson suspension components (mm).

Number	Hard points	x	y	z
P1	Top damper mounting point	−697.34	−535.60	421.27
P2	Under spring mounting point	−713.72	−562.38	151.39
P3	Steering tie rod outside point	−849.36	−605.78	−114.13
P4	Steering tie rod inside point	−848.00	−330.05	−57.90
P5	Under damper mounting point	−727.01	−584.11	−68.71
P6	Lower control arm inside front point	−759.77	−288.18	−142.18
P7	Lower control arm inside rear point	−422.00	−289.80	−133.96
P8	Lower control arm outside point	−745.10	−634.24	−214.86
P9	Steering knuckle center point	−730.40	−644.24	−138.29
P10	Wheel centre point	−725.03	−692.62	−138.29

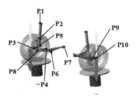

Figure 3. A multi-rigid-body dynamic model of Macpherson front suspension.

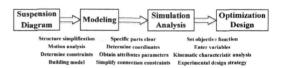

Figure 4. The technical roadmap of suspension system kinematics analysis and optimisation.

4 SUSPENSION KINEMATIC CHARACTERISTIC SIMULATION ANALYSIS

This paper investigates the kinematics simulations of the wheels on the front suspension of both sides on the same direction jumping; the simulation step is set to 100 with the jumping stroke of the wheel set as 50 mm and the rebound stroke of the wheel set as −50 mm. Figure 5 shows the varying curves of the wheel alignment parameters obtained through the kinematics simulations.

The ideal design characteristics of the variation range of the toe angle are (−0.5°~0°)/50 mm

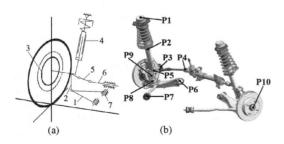

Figure 2. (a) A minivan's Macpherson suspension diagram; (b) A minivan's Macpherson suspension CAD model.
1. Lower control arm; 2. Steering knuckle; 3. Wheel hub; 4. Damper; 5. Steering tie rod; 6. Steering gear rack; 7. Frame.

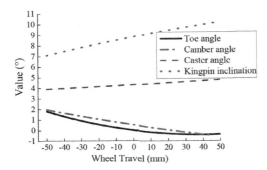

Figure 5. The wheel alignment parameters of front suspension.

Table 2 Comparison of the parameters on the front suspension.

Wheel alignment parameters	Ideal design requirements/(°)		Simulation analysis value/(°)	
	Variation range	The amount of change	Variation range	The amount of change
Toe angle	−0.5~0	0.5	−0.4119~0	0.4119
Camber angle	−1.5~1	2.5	−0.583~1.9354	2.5184
Caster angle	3~10	7	3.885~4.77	0.8850
Kingpin inclination	7~13	6	7.0459~10.1832	3.1373

during jumping stroke. It can be seen from Figure 5 that the variation range of the toe angle is (−0. 4119°~0°)/50 mm. Thus the variation range of the toe angle matches the requirement of the rules.

The initial value of the camber angle is set to 0.5°; however, the value of the camber will be changed on the basis of the initial value as the wheel travels. Therefore, the ideal range of the camber angle is from −1.5° to 1°. It is obvious that the simulation value range of the camber angle in Figure 5 varies from −0.583° to 1.9354°, exceeding the ideal range.

Figure 5 shows the variation range of the caster angle. It can be seen from the curves that the variation range of the simulation is between 3.885° and 4.77°, which is in the ideal range of changes (3°~10°).

In the actual design process, the proper range of the kingpin inclination is set to 7°~13° to keep the automatic aligning torque of the vehicle at low speed and steering lightness, and the angle should be set to a smaller value. In Figure 5, the kingpin inclination from the simulation is from 7.0459° to 10.1832°, which meets the reasonable range of values.

The results of the dynamics simulation analysis of the kinematics characteristics of the suspension system are shown in Table 2. The conclusion can be illustrated as follows:

1. For the values of the toe angle, although the range of values is in line with the ideal design requirements, the values changed greatly, and further optimisation is still required;
2. The camber angle exceeds the ideal range. It needs to be optimised to achieve the desired design value range requirements;
3. The caster angle and the kingpin inclination are within the desired value range.

Above all, the toe angle and camber angles need to make some optimisation design changes to meet the requirements of the desired design value, while the other parameters must be limited under the design requirements.

5 SUSPENSION KINEMATIC CHARACTERISTIC OPTIMISATION DESIGN

5.1 Sensitivity analysis

Allowing for dampers and other components in the Macpherson suspension, the suspension needs to be redesigned. In this paper, the parameters that have most effect on the feature of the suspension are chosen from 30 parameters corresponding to coordinates of ten hard points (every hard point with three coordinates x, y, z) by using sensitivity analysis.

Sensitivity factors are the derivative value of objective function to design variables. The more the sensitive factors, the more the corresponding design variable influences the objective function. After sensitivity analysis, the influence rate of every parameter on the wheel alignment parameters is obtained as shown in Figure 6. Six parameters, P8_x, P8_y, P8_z, P4_z, P3_z, and P3_x, are the key parameters. In the research, P3 corresponds to the position of outside endpoint of the tie rod; P4 corresponds to the position of internal endpoint of the tie rod; P8 corresponds to the outside endpoint of the lower control arm. In the next step, these six parameters are works as the only variables in optimisation design so as to improve the efficiency of optimisation.

5.2 Optimisation design

The results obtained from kinematics characteristics and sensitivity analysis are used for multi-variables and multi-objective optimisation design.

In the process of optimisation, the conflicting objectives cannot achieve most optimisation values at the same time.

In the kinematics simulation of the Macpherson suspension, the wheel toe angle and camber angle changed a lot. Hence the standard deviation value of the wheel toe angle and camber angle needed to reach the minimum. The formula and constraint condition are as follows:

$$\min \sqrt{\frac{1}{N}\sum_{i=1}^{N}(T_i - \overline{T})^2} \qquad i = 1, 2, \ldots, N \qquad (1)$$

$$\min \sqrt{\frac{1}{N}\sum_{i=1}^{N}(\alpha_j - \overline{\alpha})^2} \qquad j = 1, 2, \ldots, N \qquad (2)$$

$$P_{k\,\min} \leq P_k \leq P_{k\,\max} \qquad k = 1, 2, \ldots, 6 \qquad (3)$$

where T is wheel toe angle, α is wheel camber angle, N is the number of times of the optimisation test, P_k is design variables corresponding to outside endpoint of steering tie rod in x and z direction, internal endpoint of steering tie rod in z direction, outside endpoint of lower control arm in x, y, z direction. Considering the position of the low control arm pin and steering tie rod, the range of the six variables is −10 mm to 10 mm.

After repeated iteration optimisation calculations, the optimal values of the wheel toe angle and camber angle are shown in Table 3. The optimal values are then imported to ADAMS/CAR for dynamic simulation to analyse how alignment parameters change before and after optimisation of hard points.

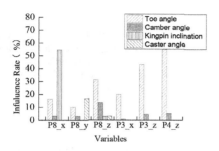

Figure 6. The influence rate of design variables.

Table 3. Comparison of hard points coordinates (mm).

Hard points	Prior to optimisation			After optimisation		
	x	y	z	x	y	z
P3	−849.36	−605.78	−114.13	−843.36	−605.78	−108.13
P4	−848	−330.05	−57.9	−848	−330.05	−67.9
P8	−745.10	−634.24	−214.86	−735.10	−644.24	−204.86

5.3 Optimisation results analysis

After iterative optimisation calculation, the theoretical optimal solution based on toe angle and camber angle is obtained (Table 3). The wheel alignment parameters optimisation results are shown in Figure 7 and Table 4.

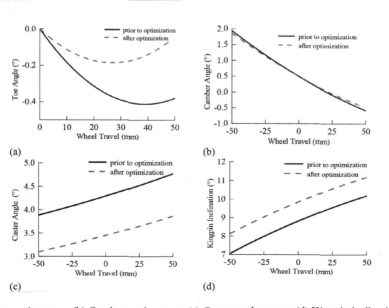

Figure 7. (a) Toe angle curves; (b) Camber angle curves; (c) Caster angle curves; (d) Kingpin inclination.

13

Table 4. Comparison of wheel alignment parameters optimisation.

Wheel alignment parameters	The change interval prior to optimisation (Change)/(°)	The change interval after optimisation (Change)/(°)	Optimisation values/(°)	Optimisation ratio/(%)
Toe angle	−0.4119~0 (0.4119)	−0.1852~0 (0.1852)	0.2267	55.0376
Camber angle	−0.583~1.9354 (2.5184)	−0.4857~1.8589 (2.3446)	0.1738	6.9012
Caster angle	3.885~4.77 (0.885)	3.0958~3.866 (0.7702)	0.1148	12.9717
Kingpin inclination	7.0459~10.1832 (3.1373)	8.1242~11.1734 (3.0492)	0.0881	2.8081

In Figure 7 and Table 4, it can be seen that wheel toe angle, camber angle and kingpin inclination change less when the wheel moves. Consequently, the kinematics characteristics of the Macpherson suspension are improved, tyre wear is reduced and the stability of the car is improved.

6 CONCLUSION

In this paper, by analysing kinematics characteristics and optimising the design of the Macpherson suspension, the following conclusions are made:

a. Obtaining the change rule of Macpherson suspension wheel alignment parameters with different wheel vibrating and determining the reasonable alignment parameters.
b. By simulation, the position of the lower control arm outside endpoint has a great effect on cam-ber angle and kingpin caster angle and the positions of the steering tie rod internal endpoint strongly influence the wheel toe angle.
c. The result shows that the kinematics characteristics of the optimised Macpherson suspension are improved significantly. In particular, the range of wheel toe angle and kingpin caster angle is more reasonable.

To conclude, the proposed optimisation design is useful in the process of kinematic modelling and simulation. It offers an effective optimisation approach that can shorten the design period. Although the research object in this paper is the Macpherson suspension, the concept and method of the proposed approach can be instructive for other suspensions.

ACKNOWLEDGEMENTS

This work was supported by the National Science Foundation for China (Grant No. 51605353) and the Joint Funds of the National Science Foundation of China (U1564202). The support from the Innovative Research Team in University (IRT13087), Research Project of State Key Laboratory of Mechanical System and Vibration (MSV201608) and the Foundation Research Funds for the Central Universities (WHUT: 2016IVA037) is also greatly appreciated.

REFERENCES

Automotive Engineering Manual Editorial Board. (2001). *Automotive engineering manual.* Beijing: China Communications Press.
Chen, J.R. (2012). *Automobile structure.* Beijing: China Communications Press.
Chen, J. (2008). *MSC.ADAMS technology and engineering analysis example.* Beijing: China Water & Power Press.
Ding, L. (2010). Design of suspension and process analysis for automobile. *Society of Automotive Engineers of China, SAE-China Congress Proceedings,* 1466–1471. Beijing: China Machine Press.
Huang, D.M., Wang, L.M., Jing, G.H., & Yin, C.H. (2010). Simulative analysis of front suspension of a light duty truck based on ADAMS/CAR. *Automobile Technology,* (9), 28–32.
Jamali, A., Shams, H., & Fasihozaman, M. (2014). Pareto multi-objective optimum design of vehicle-suspension system under random road excitations. *Proceedings of the Institution of Mechanical Engineers Part K-Journal of Multi-body Dynamics, 228*(3), 282–293.
Ren, K., Wang, J.J., & Wu, D.H. (2010). The simulation and optimization of Macpherson front suspension of a certain minibus based on ADMAS/CAR. *Machinery Design and Manufacture,* (3), 36–38.
Sagi, G., Lulic, Z., & Ilincic, P. (2015). Multi-objective optimisation model in the vehicle suspension system development process. *Tehnicki Vjesnik-technical Gazette, 22*(4), 1021–1028.
Sun, Z. (2014). Dynamic modeling, verification and optimization of full vehicle based on suspension system performance. Changsha: Hunan University.
Wang, B., Guan, H., & Lu, P. (2015). Novel evaluation method of vehicle suspension performance based on concept of wheel turn center. *Chinese Journal of Mechanical Engineering, 28*(5), 935–944.
Zhang, Y.C., Dai, J., Hou, Y., & Wu, Z.D. (2013). Optimization of Suspension Hard Points Based on ADAMS. CAD/CAM YU ZHISAOYE XINXIHUA, (12), 86–8.

Automotive, Mechanical and Electrical Engineering – Liu (Ed.)
© 2017 Taylor & Francis Group, London, ISBN 978-1-138-62951-6

Energy conservation of electric vehicles by applying multi-speed transmissions

Xiaoxiao Wu, Peng Dong & Xiangyang Xu
Beijing Key Laboratory for High Efficient Transmission and System Control of New Energy Resource Vehicle,
School of Transportation Science and Engineering, Beihang University, Beijing, China

D.A. Kupka & Yanqin Huang
Chair of Industrial and Automotive Drivetrains, Department of Mechanical Engineering,
Ruhr-University Bochum, Bochum, Germany

ABSTRACT: Thanks to good torque-to-speed characteristics of electric motors, current electric vehicles are normally equipped with single-speed transmission. However, the driving distance is limited due to the low energy capacity of the battery system. This paper tries to reduce the energy consumption and extend the driving distance of electric vehicles by applying multi-speed transmissions. An efficiency-optimised gear shift strategy is developed to keep the electric motor operating in its most efficient range. Through the forward driving simulation in the WLTC (Worldwide Harmonised Light-Duty Test-Cycle) driving cycle, it is found that a two-speed transmission can greatly reduce the energy consumption and extend the maximum driving distance of electric vehicles. The improvement becomes less significant when continuing to increase the transmission speeds. Therefore, electric vehicles are suggested to have a two-speed transmission with respect to energy conservation and driving distance extension.

Keywords: Electric vehicle; Multi-speed transmission; Shift strategy; Energy conservation

1 INTRODUCTION

A pure Electric Vehicle (EV) applies electric motor as its power source. It allows CO_2 emission to be zero; thus it is the best solution to the emission problem caused by road transportation. However, large-scale application of electric vehicles is inhibited by the high cost of the electric motor, limited energy storage capacity of the batteries and lack of charging stations. A technology breakthrough is necessary for a battery system with higher capacity and quicker charging time. In addition, the electric motor also needs to be improved with better efficiency and lower cost (Nakagawa et al., 2013).

The efficiency of the electric motor varies at different operating points. It will take a long time to improve the motor efficiency in all speed and torque ranges. Although the electric motor has a better torque-to-speed characteristic, equipped with a multi-speed transmission, the motor can work in the high efficiency range as much as possible. This will benefit the electric vehicle a lot in view of the low energy capacity of the current battery system, for the energy loss can be reduced. Correspondingly, the electric vehicle will have a

longer driving distance. Besides, a smaller motor can be selected with the help of the multi-speed transmission. Many studies (Hofman & Dai, 2010; Holdstock et al., 2012; Qin et al., 2011; Ren et al., 2009; Sorniotti et al., 2012; Wu et al., 2013; Walker et al., 2013; Wang et al., 2014) have tried to improve the efficiency of the electric vehicle by introducing a multi-speed transmission.

In this paper, we investigate how many transmission speeds is the best choice for an electric vehicle based on the forward driving simulation. Firstly, in Section 2, a detailed driving simulation model for an electric vehicle is developed and introduced. A different number of transmission speeds can be selected in the simulation model. The influence of the ancillary power loss is first considered in the model. In Section 3, a shift strategy is developed with respect to the optimum efficiency of the electric motor. This shift strategy can keep the operating points of the electric motor in the high efficiency range during the vehicle driving. Section 4 discusses the simulation results of the energy conservation brought by the multi-speed transmission based on the Worldwide Harmonised Light-Duty Test-Cycle (WLTC). Finally, a short conclusion is presented in Section 5.

2 DEVELOPMENT OF THE EV POWERTRAIN MODEL

Many simulation models are developed for the design and development of electric vehicles (Chew et al., 2014; Goncalves et al., 2009; Maia et al., 2011; McDonald, 2012; Park et al., 2014; Yang et al., 2012). Depending on the research focus, how detailed each component is modelled differs greatly. For example, the user may need a more precise model for the battery system to investigate the battery degradation (Hulsebusch et al., 2011; Serrao et al., 2005; Yaguesgoma et al., 2013). Different methods can be applied for the powertrain modelling such as the physics-based resistive companion form technique and the bond graph method (Mi et al., 2007). In this paper, a physics-based mathematical model is built up. It consists of the following sub-models.

2.1 Battery model

A lithium-ion battery system is applied in the simulation model. Its energy capacity is assumed to be 22 kWh, which is similar to the battery capacity of most of the current electric vehicles. Many battery models focus on studies of battery behaviour and electrical efficiency. These models are detailed as built up electrochemically (Ceraolo, 2000), mathematically (Salameh et al., 1992), or electrically (Chen & Rinconmora, 2006). They provide a high level of accuracy due to a low abstraction level and many input parameters (Schellenberg et al., 2014). Hu established a battery model with a moderate complexity to characterise the voltage behaviour of a lithium-ion battery for electric vehicles (Hu et al., 2011). However, such a detailed model is time consuming and not necessary in the forward driving simulation. Here, in order to simplify the model and speed up the computation, we calculate the battery State of Charge (SOC) value depending on the vehicle power requirement in each time step. It is expressed by

$$SOC(t_i) = SOC(t_{i-1}) - \frac{P_{Veh}(t_i) \cdot (t_i - t_{i-1})}{Q_{Bat} \cdot (\eta_{Bat} \cdot \eta_{Inv} \cdot \eta_{Mot} \cdot \eta_{Tra})^{\omega}}$$

(1)

where $SOC\ (t_i)$ = the state of charge in time step t_i; $P_{Veh}(t_i)$ = the vehicle power requirement in time step t_i; Q_{Bat} = the battery capacity; η_{Bat} = the battery efficiency; η_{Inv} = the inverter efficiency; η_{Mot} = the motor efficiency; η_{Tra} = the mechanical efficiency of the transmission and ω indicates the direction of power flow. When the energy is recovered from vehicle braking or coasting, ω is equal to −1. When the electric motor outputs power to the vehicle, it is equal to 1. In this way, the SOC value of the battery system and the energy consumption in a specific driving cycle can be calculated quickly. Actually,

the battery charging and discharging efficiency is greatly influenced by the temperature and SOC level of the battery since the internal resistance and the open circuit voltage of the battery depend on those parameters. In this paper, it is assumed to be constant. Although the battery aging effects and the temperature effects are neglected in the battery model, it does not influence the scope of our work.

2.2 Electric motor model

The most common type of electric motor in current electric vehicles is the Permanent Magnet Synchronous Machine (PMSM). Besides, the Separately Excited Synchronous Machine (SESM) and the Asynchronous Machine (ASM) can also be seen in some electric vehicles. PMSM is significantly more efficient than SESM and ASM at low speed (Braun, 2012). Because the electric vehicle is currently mainly designed for low-speed urban driving, the lightweight and compact PMSM with high efficiency is preferred by vehicle manufacturers.

A permanent magnet synchronous machine with peak torque 210 Nm, peak power 60 kW, and peak speed 10,500 rpm is applied in the simulation model. The efficiency map of this PMSM is shown in Figure 1, whose data were provided by our supplier. It shows that the electric motor works in a higher efficiency area when the torque is between 20 Nm and 120 Nm and the speed between 1,500 rpm and 4,500 rpm. At other operating points the efficiency is low. Especially at the operating points of low torque and low speed (dark area on the left side of Figure 1), the motor efficiency is very low and its values are not shown in Figure 1. The electric model calculates the output torque according to the accelerator pedal position from the driver model. The torque-to-speed characteristic curve at different pedal positions is acquired through interpolation between the full-load curve and the thrust characteristic curve. The efficiency of each operating point in Figure 1 is considered when calculating the motor output torque.

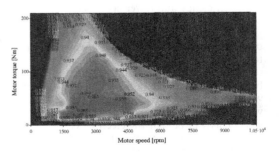

Figure 1.   Efficiency map of the electric motor applied in the simulation model.

2.3 Power loss of ancillary devices

For the electric vehicle, the power loss of ancillary devices also has a big influence on its maximum driving distance. The average power loss of ancillary devices differs with different driving cycles. At the beginning of the driving cycle, a higher power is usually required for the ancillary devices such as air conditioning or heating. In order to determine the maximum driving distance of the electric vehicle, the Chair of Mechatronics at University of Duisburg-Essen investigated the influence of ancillary power loss based on different driving cycles (Proff et al., 2012). It was found that the ancillary power loss is higher in winter than in summer in Germany because of the temperature requirement for heating. In this paper, the simulation results from the study by Proff et al. (2012) are applied as the nominal values for the calculation of the ancillary power loss. It is expressed by

$$P_{auxiliary}(t_i) = \max[P_{min\_auxiliary}, P_{auxiliary}(t_0) - kP_{auxiliary} \cdot (t_i - t_0)] \qquad (2)$$

where $P_{ancillary}(t_i)$ = the average ancillary power loss in time step t_i; $P_{min\_ancillary}$ = the minimum ancillary power loss in the driving cycle, which is a basic electric power demand for lighting and other electric devices; $P_{ancillary}(t_0)$ = the initial ancillary power loss at the beginning of the driving cycle; and $kP_{ancillary}$ = the gradient of the ancillary power loss decrease. The ancillary power loss consumes the battery energy and reduces the maximum driving distance. This equation assumes that it reaches its maximum level at the beginning of the driving cycle due to heating or air conditioning. Then it will decrease gradually as the driving time extends until it reaches the minimum value.

2.4 Definition of transmission ratio

A mini-class electric vehicle is used in the simulation model. According to the vehicle data, the speed ratio range of the transmission is predefined. Firstly, the largest ratio is determined with respect to the requirement of the vehicle climbing ability based on

$$i_{max} = \frac{r_{dyn} \cdot m_F \cdot g \cdot (f_R \cdot cos\alpha_{St} + sin\alpha_{St})}{T_{max} \cdot \eta} \qquad (3)$$

where r_{dyn} = the tyre radius, (m); m_F = the vehicle mass, (kg); f_R = the rolling resistance coefficient; α_{St} = the gradient angle; T_{max} = the maximum motor torque, (N · m); and η = the powertrain efficiency. In the simulation model, the transmission efficiency is a function of torque and speed. It is variable at different operating points and is improved as the load increases.

Secondly, the smallest ratio should satisfy the maximum vehicle speed requirement at the maximum electric motor speed. In addition, the smallest ratio should also consider the overdrive ratio, which is calculated according to the motor speed requirement at the vehicle cruising speed.

$$i_{min} = min\left(\frac{n_{max}}{v_{max}/2\pi r_{dyn}}, \frac{n_{cruising}}{v_{cruising}/2\pi r_{dyn}}\right) \qquad (4)$$

where v_{max} = the maximum vehicle speed, (m/s); n_{max} = the maximum motor speed, (rpm); $v_{cruising}$ = the vehicle cruising speed, (m/s); $n_{cruising}$ = the motor speed at the vehicle cruising speed, (rpm); and r_{dyn} = the tyre radius, (m). Table 1 shows the vehicle parameters and some requirements for this electric vehicle.

Here the vehicle cruising speed is defined as 100 km/h. The motor speed requirement at this vehicle speed is 3,200 rpm, which is in the optimum efficiency range according to Figure 1. The calculated transmission ratio is then limited in the range of 3.426 to 8.826. In the simulation model, the number of the transmission speed is firstly defined before each driving simulation. When the speed number is more than two, the intermediate gear ratio is calculated based on the geometrical gear steps. In this paper, the maximum speed number is limited to four, as shown in Table 2.

2.5 Driving resistance model

The total driving resistances of the vehicle consist of the rolling resistance, the air resistance, the acceleration resistance and the gradient resistance. In the forward driving simulation, these resistances act on the

Table 1. Vehicle parameters and requirements.

Vehicle	Value	Vehicle	Value
Class	Mini class	Max. climbing ability	55%
Weight	1,250 kg	Max. speed requirement	135 km/h
Wheel radius	0.284 m	Cruising speed	100 km/h
Vehicle cross-section	2.07 m²	Motor speed at vehicle cruising speed	3,200 rpm

Table 2. Transmission speed ratio.

Transmission	Speed ratio
Single-spped	i_1: 8.826
2-speed	i_1: 8.826, i_2: 3.426
3-speed	i_1: 8.826, i_2: 5.499, i_3: 3.426
4-speed	i_1: 8.826, i_2: 6.438, i_3: 4.700, i_4: 3.426

rolling wheels and must be balanced by the driving torque. The output load is calculated according to

$$T_L = [f_R \cdot m_F \cdot g \cdot cos(\alpha_{St}) + \frac{1}{2} \cdot \rho_L \cdot c_W \cdot A_F \cdot v_F^2$$
$$+ m_F \cdot g \cdot sin(\alpha_{St}) + \lambda_F \cdot m_F \cdot a_F] \cdot r_{dyn} \qquad (5)$$

where ρ_L = the air density; c_W = the drag coefficient; A_F = the vehicle cross-section; v_F = the vehicle speed; λ_F = the rotational inertia coefficient; and a_F = the vehicle acceleration.

2.6 Driver model

A driver model is included to calculate the accelerator pedal position and the brake pedal position according to the speed profile of the reference driving cycle. The driver model controls the pedal position through a PID controller to make the actual vehicle speed in the simulation follow the speed profile of the reference cycle closely. The change rate of the pedal position is expressed by

$$\Delta xP(t_i) = K_P \cdot \Delta v(t_i) + K_I \cdot \Delta s(t_i) + K_D \cdot \Delta a(t_i) \qquad (6)$$

where $\Delta xP(t_i)$ = the change rate of the pedal position in time step t_i; $\Delta v(t_i)$ = the difference between the actual vehicle speed and the reference vehicle speed in time step t_i; $\Delta s(t_i)$ = the difference between the actual driving distance and the reference driving distance in time step t_i; $\Delta a(t_i)$ = the difference between the actual vehicle acceleration and the reference vehicle acceleration in time step t_i; K_P = the proportional gain of the PID controller; K_I = the integral gain of the PID controller; and K_D = the derivative gain of the PID controller. Then the pedal position is calculated through iteration in

$$xP(t_i) = xP(t_{i-1}) + \Delta xP(t_i) \qquad (7)$$

where $xP(t_i)$ is the pedal position in time step t_i. When it is positive, the pedal position stands for the accelerator pedal. When it is negative, the pedal position stands for the brake pedal. In addition, the simulation model also considers the energy regeneration from the vehicle braking and coasting. When the vehicle is braking, the electric motor will provide braking forces and work as a generator to charge the battery. The regeneration is only activated when the brake pedal is applied in the simulation. It is modulated through the brake pedal. The regeneration has certain prerequisites. For example, it will not work at very low vehicle speed. The regeneration is also limited by the SOC and the charging power of the battery. In addition, the regeneration is optimised by considering the coordinated control of the wheel brake torque and the regenerative brake torque.

Different driving cycles can be selected in the forward driving simulation, e.g. the New European Driving Cycle (NEDC), which is supposed to represent the typical usage of a passenger car in Europe. In this paper, a Worldwide Harmonised Light-Duty Test-Cycle (WLTC) is applied in the simulation. The WLTC represents real driving scenarios more accurately than the NEDC. The load transients of the WLTC are more frequent. It will replace the NEDC from 2017. The WLTC is divided into three classes according to the vehicle power-weight ratio:

Class 1: low-power vehicles with power-weight ratio smaller than 22 W/kg
Class 2: medium-power vehicles with power-weight ratio between 22 W/kg and 34 W/kg
Class 3: high-power vehicles with power-weight ratio higher than 34 W/kg

In the driving cycle of each class, there are several parts designed to represent real-world vehicle operation on urban and extra-urban roads, motorways and highways (Wikipedia, 2015). According to the motor power and the vehicle mass described in the previous section, the driving cycle of class 3 is applied in the simulation. Its speed profile is shown in Figure 2. More cycle information is shown in Table 3.

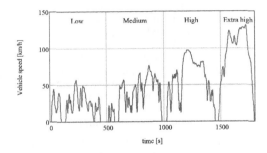

Figure 2. Speed profile of the driving cycle applied in the simulation.

Table 3. WLTC class 3 driving cycle.

Vehicle	Low	Med-ium	High	Extra high	Total
Duration [s]	589	433	455	323	1,800
Stop duration [s]	156	48	31	7	242
Distance [m]	3,095	4,756	7,158	8,254	23,262
Percentage of stops	26.5%	11.1%	6.8%	2.2%	13.4%
Max. speed [km/h]	56.5	76.6	97.4	131.3	131.3
Average speed [km/h]	18.9	39.5	56.6	92.0	46.5
Min. acceleration [m/s²]	−1.5	−1.5	−1.5	−1.2	−1.5
Max. acceleration [m/s²]	1.5	1.6	1.6	1.0	1.6

3 EFFICIENCY-OPTIMISED GEAR SHIFT STRATEGY

An efficiency-optimised gear shift strategy is developed in the simulation model. The basic idea of the shift strategy is to keep the operating points of the electric motor in its optimum efficiency range through gear shifting. The efficiency map of the electric motor can be converted to the transmission output side through the gear ratio. Then the motor efficiency can be depicted depending on the wheel speed and the wheel torque. We can choose the most suitable gear for motor efficiency for a specific operating point of the vehicle. For example, as shown in Figure 3, the motor efficiency is obviously higher in second gear than in first gear at the vehicle operating point (1,000 rpm, 300 Nm) when the two-speed transmission in Table 2 is used. Therefore, in order to develop the efficiency-optimised gear shift strategy, we must find the most suitable gear for each vehicle operating point by comparing the motor efficiency in different gears.

Figure 4 shows the suitable gear of each vehicle operating point for the two-speed transmission in Table 2. There is an obvious boundary line between first gear and second gear. In the left region of the boundary, first gear can provide the vehicle with a high driving torque and is more efficient for the electric motor. In the right area of the boundary, second gear is more suitable for a high vehicle speed because the electric motor operates more efficiently.

An inverse tangent function, which is expressed by Equation (8), is used in the simulation model to fit the upshift line according to

$$n_{1-2}(T) = A \cdot arctan\left[\left(\frac{T}{B}\right)^3\right] \cdot \frac{2}{\pi} + C \qquad (8)$$

where $n_{1-2}(T)$ = the upshift speed for a specific driving torque; and A, B, C are the function coefficients for the fitted upshift line. In the forward driving simulation, the vehicle will shift into a higher gear based on this upshift line. Figure 5 shows the fitted upshift line for the two-speed transmission in Table 2. For the three-speed and the four-speed transmission in Table 2, their upshift lines are shown in Figure 6.

In order to avoid frequent shift, the speed of the downshift line is defined to be lower than the speed of the upshift line in the shift map. Besides, the efficiency-optimised gear shift strategy sets a minimum time interval between two shifts in the driving simulation. Based on the developed shift strategy, Figure 7 shows how the area of the optimum motor efficiency expands in the characteristic map of wheel torque versus wheel speed with the help of the four transmissions in Table 2.

The maximum vehicle speed of 135 km/h approximately corresponds to the wheel rotational speed of 1,260 rpm. In this speed range, it can be seen that the efficient operating range of the electric motor expands with the help of the multi-speed transmission and the developed effi-

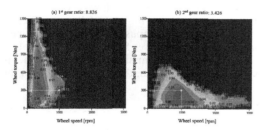

Figure 3. Motor efficiency versus wheel torque and wheel speed.

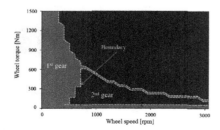

Figure 4. Suitable gear of each vehicle operating point for the two-speed transmission.

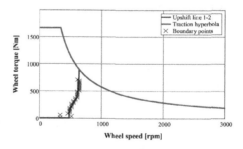

Figure 5. Upshift line of the two-speed transmission in Table 2.

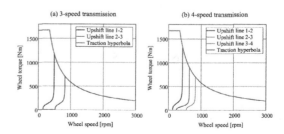

Figure 6. Upshift line of the three-speed and four-speed transmission in Table 2.

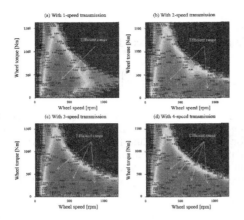

Figure 7. Motor efficiency in the characteristic map of wheel torque versus wheel speed.

ciency-optimised gear shift strategy. The efficient range is narrow with single-speed transmission. It is suitable for the vehicle in urban traffic at low driving speed. However, the motor efficiency is low when the electric vehicle runs on the highway at high driving speed. When the vehicle is equipped with the two-speed transmission in Table 2, the second gear improves the motor efficiency at high driving speed. As shown in graph (b), the red efficient range obviously expands towards the high wheel speed area. When the transmission speed number increases to three or four, the cover area of the red efficient range expands correspondingly under the control of the developed shift strategy. The total efficiency is further improved.

However, it is observed that the red efficient range mainly expands towards the small wheel torque and high wheel speed area. This is because the ratio of a higher speed becomes smaller. The red efficient range thus cannot expand towards the high wheel torque and low wheel speed area, and the efficiency of the electric motor is normally low at the operating point of its maximum torque. In addition, the efficiency on the traction hyperbola curve where the electric motor reaches its peak power cannot be improved by the multi-speed transmission and the efficiency-optimised gear shift strategy.

4 SIMULATION RESULTS AND DISCUSSION

In the forward driving simulation, the driver model calculates the accelerator pedal position and the brake pedal position according to the speed profile of the reference driving cycle. Then the electric motor outputs torque based on the pedal position.

This torque is transmitted through transmission to tyres to overcome the driving resistance and the braking torque. Shift automatically takes place according to the developed shift strategy. The actuating energy for the gear shifting is also considered in the simulation model. We assume that the shifting element is the wet multi-plate clutch which is electro-hydraulically actuated. It needs hydraulic pressure from the oil pump. Therefore, the energy consumption of the oil pump is calculated in the driving simulation. Besides, the friction loss in each shifting process is also considered. By comparing the total energy consumption with different transmissions in Table 2, we can know how many transmission speeds is the best choice for current electric vehicles.

Figure 8 shows the simulation results of the motor operating points with different transmissions. The operating points in the negative torque area mean that the electric motor works as a generator for energy regeneration. It can be seen from graph (a) that most of the operating points are in the area below 100 Nm and 5,000 rpm when equipped with the single-speed transmission. However, the motor does not make full use of its efficient range. A large area around 100 Nm with high efficiency is not covered by any operating point, and some operating points appear in the motor high speed area with low efficiency. They correspond to the high vehicle speed in the WLTC driving cycle. Therefore, the transmission with only single-speed is not efficient for the electric vehicle.

When the transmission speed number increases, the developed shift strategy allows more operating points of the electric motor to fall in its efficient range. As shown in graphs (b), (c) and (d), no operating point appears over 5,000 rpm. It is clear that the multi-speed transmission will reduce the energy consumption. However, the difference between the

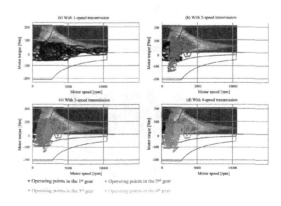

Figure 8. Motor operating points with different transmissions.

two-speed, three-speed, and four-speed transmission is very small according to Figure 8.

Table 4 compares the simulation results with the four different transmissions in Table 4. It can be seen that energy consumption is greatest with the single-speed transmission. The average motor efficiency for the whole driving cycle is also the lowest. The energy consumption is greatly reduced when the transmission speed number increases to two. About 3% battery SOC is saved in the WLTC driving cycle through one more gear ratio and the efficiency-optimised shift strategy. That means the maximum driving distance can increase by about 30 km for the electric vehicle with 22 kWh battery capacity. It has a big advantage compared with the single-speed transmission. However, continuing to increase the transmission speeds will not reduce the energy consumption significantly. A two-speed transmission is thus the best choice for the mini-sized vehicle with respect to a trade-off between the energy conservation and the increased cost for the development of multi-speed transmission.

This paper also investigates the small class vehicle and the compact class vehicle. The vehicle mass increases and certain vehicle parameters change in these two class vehicles. The gear ratio is also predefined differently depending on the vehicle parameters. The energy consumption and the average motor efficiency in the WLTC driving cycle are shown in Figure 9.

It is known from Figure 9 that the two-speed transmission can significantly reduce the energy consumption of all three class electric vehicles. However, the improvement is small when the speed number is more than two, and the transmission

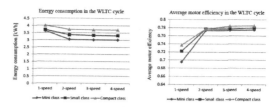

Figure 9. Simulation results of the energy consumption and the average motor efficiency.

structure will become very complex with increasing weight, space, and development cost. The weight and cost of a two-speed transmission will not increase very much compared with a single-speed transmission. Only an additional shifting element and a planetary gear set (or two transfer gear sets) are necessary, but it could help to reduce the size and cost of the electric motor. Therefore, a two-speed transmission is the best choice for current electric vehicles to save battery energy and to extend their driving distances.

5 CONCLUSION

1. In this paper, the effect of increasing transmission speed number on the energy conservation of electric vehicles was investigated.
2. A powertrain model of electric vehicle is developed for the forward driving simulation. The model takes into account the battery system, the electric motor, the multi-speed transmission, the ancillary power loss, the driving resistance, the driver actuation, and the energy regeneration.
3. An efficiency-optimised gear shift strategy is developed to control the shift in the driving simulation. Simulation results show that a two-speed transmission can greatly reduce the energy consumption of electric vehicles by keeping the electric motor working in its high efficiency range. Compared with the two-speed transmission, the further improvement of more speeds is small, but the transmission structure will become very complex with increasing weight, space, and development cost.
4. It can be concluded that a two-speed transmission is the best choice for current electric vehicles with respect to energy conservation and driving distance extension.

Table 4. Simulation results with different transmission.

Transmission	Single-speed	2-speed	3-speed	4-speed
Energy consumption [kWh]	3.631	3.04	3.008	3.012
SOC consumption	16.5%	13.8%	13.7%	13.7%
Average motor efficiency	0.695	0.773	0.774	0.775
Max. driving distance [km]	143	171	173	173
Percentage of Gear 1	100%	42.82%	16.49%	1.69%
Percentage of Gear 2	/	57.18%	46.14%	21.25%
Percentage of Gear 3	/	/	37.37%	43.12%
Percentage of Gear 4	/	/	/	33.94%

REFERENCES

Braun, L. (2012). Fahrer- und fahrsituation sabhängige Bewertung unterschiedlicher Elektromotorkonzepte. Karlsruhe, Germany: Karlsruhe Institute of Technology.

Ceraolo, M. (2000). New dynamical models of lead-acid batteries. *IEEE Transactions on Power Systems, 15*(4), 1184–1190.

Chen, M., & Rinconmora, G. (2006). Accurate electrical battery model capable of predicting runtime and I-V performance. *IEEE Transactions on Energy Conversion, 21*(2), 504–511.

Chew, K.W., Koay, C.K., & Yong, Y.R. Advisor. (2014). Simulation of electric vehicle performance on various driving cycles. *International Journal of Innovative Science, Engineering & Technology, 1*(8), 70–76.

Goncalves, G.A., Bravo, J.T., Baptista, P.C., Silva, C.M., & Farias, T.L. (2009). Monitoring and simulation of fuel cell electric vehicles. *World Electric Vehicle Journal, 3*, 1–8.

Hofman, T., & Dai, C.H. (2010). Energy efficiency analysis and comparison of transmission technologies for an electric vehicle. Proceedings of the 2010 IEEE Vehicle Power and Propulsion Conference (VPPC), Lille, France.

Holdstock, T., Sorniotti, A., Everitt, M., Fracchia, M., Bologna, S., & Bertolotto, S. (2012). Energy consumption analysis of a novel four-speed dual motor drivetrain for electric vehicles. *Proceedings of the 2012 IEEE Vehicle Power and Propulsion Conference, VPPC (6422721)*, 295–300.

Hu, X-S., Sun, F-C., & Zou, Y. (2011). Online model identification of lithium-ion battery for electric vehicles. *Journal of Central South University, 18*(5), 1525–1531.

Hulsebusch, D., Schwunk, S., Caron, S., & Propfe, B. (2011). Modeling and simulation of electric vehicles— the effect of different Li-ion battery technologies. *Journal of Automotive Safety and Energy, 2*(2), 59–67.

Maia, R., Silva, M., Araujo, R., & Nunes, U. (2011). Electric vehicle simulator for energy consumption studies in electric mobility systems. *Proceedings of IEEE Forum on Integrated and Sustainable Transportation Systems*, 227–232.

McDonald, D. (2012). Electric vehicle drive simulation with MATLAB/Simulink. Proceedings of the 2012 North-Central Section Conference, American Society for Engineering Education, 125–149.

Mi, C., Abulmasrur, M., & Gao, W. (2007). Modeling and simulation of electric and hybrid vehicles. *Proceedings of the IEEE*, 729–745.

Nakagawa, M., Schwarz, U., Özbek, M., Pfluger, J., & Andert, J. (2013). Global direction of powertrain electrification. *Proceedings of the 12th CTI Symposium, Automatic Transmissions, HEV and EV Drives, Berlin, Germany*, 25–34.

Park, C.H., Kwon, M.H., Jeong, N.T., Lee, S.G., Suh, M.W., Kim, H.S., & Hwang, S.H. (2014). Development of electric vehicle simulator for performance analysis. *Universal Journal of Mechanical Engineering, 2*(7), 231–239.

Proff, H., Schonharting, J., Schramm, D., & Ziegler, J. (2012). *Zukünftige Entwicklungen in der Mobilität* (1st ed.), 91-104. Wiesbaden, Germany: Springer Gabler.

Qin, D-T., Zhou, B-H., Hu, M-H., Hu, J-J., & Wang, X. (2011). Parameters design of powertrain system of electric vehicle with two-speed gearbox. *Journal of Chongqing University, 34*(1), 1–6.

Ren, Q., Crolla, D.A., & Morris, A. (2009). Effect of transmission design on electric vehicle (EV) performance. *Proceedings of the Vehicle Power and Propulsion Conference, Dearborn, USA*, 1260–1265.

Salameh, Z., Casacca, M., & Lynch, W.A. (1992). A mathematical model for lead-acid batteries. *IEEE Transactions on Energy Conversion, 7*(1), 93–98.

Schellenberg, S., Berndt, R., Eckhoff, D., & German, R.A. (2014). Computationally inexpensive battery model for the microscopic simulation of electric vehicles. *Proceedings of IEEE 80th Vehicular Technology Conference, Vancouver, BC*, 1–6.

Serrao, L., Chehab, Z., Guezennee, Y., & Rizzoni, G. (2005). An aging model of Ni-MH Batteries for hybrid electric vehicles. *Proceedings of Vehicle Power and Propulsion Conference*, 78–85.

Sorniotti, A., Holdstock, T., & Pilone, G.L. et al. (2012). Analysis and simulation of the gearshift methodology for a novel two-speed transmission system for electric powertrains with a central motor. *Proceedings of the Institution of Mechanical Engineers, Part D: Journal of Automobile Engineering*, 916–929.

Walker, P.D., Abdul, R.S., Zhu, B., & Zhang, N. (2013). Modelling, simulations, and optimization of electric vehicles for analysis of transmission ratio selection. *Advances in Mechanical Engineering*, 323917.

Wang, X-J., Cai, Y-C., Zhou, Y-S., & Gao, S. (2014). A study on the effects of the matching of automatic transmission on the energy consumption of electric vehicle. *Automotive Engineering, 36*(7), 871–878.

Wikipedia. (2015). Worldwide harmonized light vehicles test procedures. Accessed: 15–01–15. Available at: http://en.wikipedia.org/wiki/Worldwide_harmonized_Light_vehicles_Test_Procedures/.

Wu, G., Zhang X., & Dong Z-M. (2013). Impacts of two-speed gearbox on electric vehicle's fuel economy and performance. *Proceedings of the SAE 2013 World Congress and Exhibition, Detroit, MI, United States. SAE Technical Paper*, 2013-01-0349.

Yaguesgoma, M., Olivellarosell, P., Villafafilarobles, R., & Sumper, A. (2013). Ageing of electric vehicle battery considering mobility needs for urban areas. *Proceedings of International Conference on Renewable Energies and Power Quality*.

Yang, S-B., Li Z-Z., Xu, F., Zhang, G-G., Deng, Y-F., & Wang, J. (2012). An electric vehicle powertrain simulation and test of driving cycle based on AC electric dynamometer test bench. *Proceedings of 2012 International Conference on Mechanical Engineering and Material Science*, 273–276.

Automotive, Mechanical and Electrical Engineering – Liu (Ed.)
© 2017 Taylor & Francis Group, London, ISBN 978-1-138-62951-6

Finite element analysis and optimisation analysis of thrust rod for mining truck suspension

Shunan Hu, Qiuyue Zhen & Meilin Ji
School of Automotive Engineering, Changshu Institute of Technology, Suzhou, P.R. China

Guorui Shang
R&D Center of Shandong Pengxiang Automobile Co. Ltd., Yantai, P.R. China

ABSTRACT: The force formula of the thrust rod was deduced for the suspension of mining trucks driven on a ramp. By Finite Element Analysis (FEA), we found that the insufficient stress factor was the main cause of breaking and buckling was the main cause of the bending failure. The FEA results were consistent with the actual failure modes. The main reasons for the thrust rod failures were found and the optimal designs were conducted within a defined range. After optimal FEA, the optimal results met with the requirements of the safety factor and the reliability of the thrust rod structure design.

Keywords: Mining truck suspension; Thrust rod; FEA; Optimisation of design

1 INTRODUCTION

Several occurrences in which thrust rod heads were fractured and thrust rod shafts were bent in mining truck suspension systems using I shape thrust rod appeared. After surveying the working environment, we found that the mining car was fully loaded during uphill and was empty during downhill journeys between the mining area and the unloading yard.

Finite Element Analysis (FEA) often has been used for developing mining truck parts, but has been less used for thrust rods. Gang (2012) studied the failure cases from all work conditions of vehicles. Hu and Song (2012) considered the component force, which is caused by gravity, and the friction force between the spring and axle, and analysed their influences on the stress of the thrust rod.

Before driving the uphill, the mining trucks needed to speed up for some distances to ensure they could drive over the ramp at a certain speed. Although the road was wide, there were some potholes, so the highest speed of the mining trucks was between 30 and 40 km/h on the lever load. If the ramp was too long, the downhill speed gradually decreased and the transmission gear was shifted down to ensure the driving force. The highest load mass from the mining truck was transferred to the suspension on the uphill. The transferred forces were acted on the ground from the axle to the tyre by the thrust rod. In this process, the largest

force of the thrust rod occurred. So the forces of the thrust rods should be analysed on the uphill (Figure 1).

Through the force and moment balance equations the following formulas were deduced:

$$F_1 + F_x = F_2 + F_\mu + G \sin \alpha \tag{1}$$

$$F_1 L_1 = (F_\mu + G \sin \alpha) L_3 + F_x L_2 \tag{2}$$

From the Equations (1) and (2) we obtained:

$$F_1 = \frac{(F_\mu + G \sin \alpha) L_3 + F_x L_2}{L_1} \tag{3}$$

$$F_2 = \frac{F_x (L_2 + L_1) + (F_\mu + G \sin \alpha)(L_1 - L_3)}{L_1} \tag{4}$$

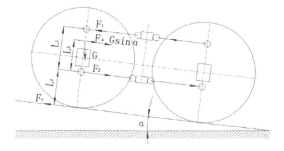

Figure 1. Force analysis of the thrust rod on the ramp.

On the uphill, the relationship between F_μ and G is:

$$F_\mu = \mu G \cos \alpha \qquad (5)$$

From the Equations (3), (4) and (5), we obtained:

$$F_1 = \frac{(\mu \cos \alpha + \sin \alpha)GL_3 + F_x L_2}{L_1} \qquad (6)$$

$$F_2 = \frac{F_x(L_2 + L_1) + G(\mu \cos \alpha + \sin \alpha)(L_1 - L_3)}{L_1} \qquad (7)$$

In the formulae:

F_x is driving force, its value was about 170700 N.
L_1 is the distance from the upper thrust rod to the nether thrust rod and its value was 0.591 m.
L_2 is the distance from the nether thrust rod to the tyre touching the ground surface and its value was 0.385 m.
L_3 is the distance from the spring seat to the nether thrust rod and its value was 0.370 m.
G is the gravity of the goods that is on the axle and its value was 450000 N.
A is the uphill angle and the maximum angle was about 15°.
μ is the friction coefficient and its value was 0.18.

Using these values in Equations (6) and (7), we obtained:

F1 = 233099 N
F2 = 209091 N

So failure happened on the uphill and the upper thrust rod force was bigger than the nether thrust rod force. Therefore, it is important to analyse the stress of the upper thrust rod.

2 FEA OF THE UPPER THRUST ROD

The thrust rod has a thin rod structure; its design not only requires the strength and stiffness to meet the use requirement, but there is also a need to analyse the stability of the compressive rod. So static structural analysis and linear bucking analysis is needed for the upper thrust rod.

The upper thrust rod was modelled in ANSYS Workbench. It was made up of three parts: two thrust rod heads and a thrust rod shaft. The material of the upper thrust rod head was 40 Cr, and its tensile yield strength was 785 MPa. The material of the upper thrust rod shaft was steel 35, and its tensile yield strength was 345 MPa. In order to obtain a high quality grid, the two thrust rod heads and the thrust rod shaft were divided as shown in Figure 2.

Figure 2.   Meshing of the upper thrust rod.

Figure 3.   Equivalent stress of the upper thrust rod.

	Mode	✔ Load Multiplier
1	1.	1.2976
2	2.	1.3135
3	3.	10.648
4	4.	11.345
5	5.	23.638

Figure 4.   The load multiplier of the upper thrust rod.

2.1 The static FEA of the upper thrust rod

According to the force analysis of the upper thrust bar, and considering the buffer action of the rubber, the dynamic load coefficient was selected as 2.0. We fixed one head of the upper thrust rod and forced the other head with the pressure value 466198 N. The static FEA showed the comprehensive stress detailed in Figure 3.

The maximum equivalent stress of the upper thrust rod was 767.8 MPa on the upper thrust rod head (Figure 3). The stress factor was 1.02 less than the safety factor of 1.5. It is easy to break down. This FEA result is consistent with the failure mode demonstrated in which the thrust rod heads were fractured.

2.2 Linear bucking analysis of the upper thrust rod

Linear bucking was based on the static FEA. Its boundary conditions were similar to the static FEA boundary conditions. After linear bucking analysis, its load multipliers were 1.30 and 1.31 in the first two steps, as seen in Figure 5. The load multipliers were smaller than the safety load multiplier of 1.5. The upper thrust rod would be bent in the up and down direction in the first step, as Figure 5 shows. The upper thrust rod would be bent in the left and right direction in the second step, as Figure 6 shows.

Figure 5. First bucking shape mode.

Figure 6. Second bucking shape mode.

Figure 7. The actual bending failure mode.

The first and second bucking analysis results (shown in Figures 5 and 6) are consistent with the actual bending failure mode, as shown in Figure 7.

3 OPTIMISATION ANALYSIS OF THE UPPER THRUST ROD

According to Jiang (2010), who studied the stability analysis of the longitudinal thrust rod, the outer diameter of the thrust rod shaft had a great effect on the stress of the thrust rod. Above, in the data, the FEA results and the actual failure mode, the most important parameters are the inner and outer diameters of the upper thrust rod head and the outer radius of its shaft, as input parameters to optimise. The static stress and the first and second load multipliers are the output parameters to optimise. The input parameters' ranges are shown in Table 1.

Exploring the design in ANSYS Workbench, a sensitivity analysis of the equivalent maximum

Table 1. Input parameters range.

Input parameters	Min. value mm	Max. value mm
Outer diameter of the upper thrust rod head	125	135
Inner diameter of the upper thrust rod head	96	103
Outer radius of the upper thrust rod shaft	60	75

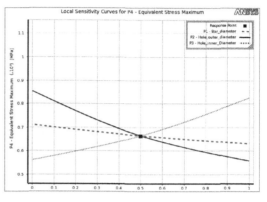

Figure 8. Maximum equivalent stress of the input parameters.

stress using the range of the input parameters in Table 1 produced the results in Figure 8.

In Figure 8, the inner and outer diameters of the head had the greatest influence on the maximum equivalent stress. The influence of the inner and outer diameter of the maximum equivalent stress was opposite. The outer radius of its shaft had a smaller influence on the maximum equivalent stress.

So, the inner diameter of the head was maintained at the same value. The outer diameter of the head and outer radius of the shaft was optimised. Through the 3-D response surface of the three parameters, the influence of the equivalent stress between the three parameters was calculated (Figure 9). In Figure 9, the optimal value of the maximum equivalent stress is near 500 MPa when the outer diameter of the head is 135 mm and the outer radius of the shaft is 30 mm. The stress factor is 1.57, which meets with the design requirements. Using the optimal values in the linear buckling analysis, the load multipliers of the first and second steps were 2.32 and 2.34. The values are bigger

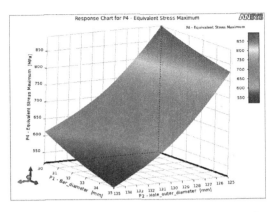

Figure 9. 3-D response surface of the maximum equivalent stress, outer diameter of the head and outer radius of the shaft.

than the safety load multiplier of 1.5. Therefore, the optimal values can be used in the design.

4 CONCLUSION

Several occurrences in which thrust rod heads were fractured and the shafts were bent originated during the uphill journeys, according to an analysis of the mining working environment. The stress factor and load multiplier of the thrust rod head was insufficient, according to static FEA and linear bulking analysis. The FEA results were confirmed in the failure mode. After using the optimisation function of ANSYS Workbench, the most important parameters were found. These could reduce the workload of the design and, at the same time, avoid the failure problem.

REFERENCES

Bu, J. et al. (2016). Stability analysis for vehicle torque rod based on the rubber bushing. *Natural Science Journal of Xiangtan University, 1*, 39–41.
Chen, N. (2015). Research on finite element modeling method for large off-highway mining dump truck frame structure. *Science Technology and Engineering, 1*, 309–314.
Hu, S. & Song, J. (2012). Design and check of thrust bar in heavy truck. *Special Purpose Vehicle, 2*, 78–79.
Jiang, H. (2010). Optimization design on longitudinal rod of air suspension guiding structure. *Machinery Design & Manufacture, 6*, 1–3.
Ju, G. (2012). Design match of thrust rod in heavy engineering truck. *Special Purpose Vehicle, 2*, 74–77.

Automotive, Mechanical and Electrical Engineering – Liu (Ed.)
© 2017 Taylor & Francis Group, London, ISBN 978-1-138-62951-6

Research on crashworthy structure for the straddled-type monorail vehicle

Xiaoxu Zhang
Chongqing College of Electronic Engineering, Chongqing, China

ABSTRACT: On the basis of the simulation theory of the crashworthy vehicle, this paper takes the Chongqing straddled-type monorail vehicle as the major object and uses the application of the non-linear finite element technique. First establish the CAD structure and import the corresponding CAE software, according to the requirements of the inspection of the grid units, to analyse and research the optimisation unit. Finally complete the straddled-type monorail vehicle body finite element model. Determine the initial speed of vehicle collision by the body and human body impact evaluation criteria and show the deformation of the vehicle collision in the process, in order to analyse the crash construction of the vehicle body in a virtual environment. The results show the effective dissipation of the crash energy and the ordered deformation of the car-body structure, so that the reasonability and feasibility of the design are verified.

Keywords: Straddled-type monorail vehicle; Frontal crash; Crashworthiness; Energy absorption device; Simulation

1 INTRODUCTION

As urban rail transit gradually becomes the mainstream of the development of urban transportation, its high level of active safety protection measures can prevent potential train collisions effectively, thus decreasing the probability of major traffic accidents. However, in the actual operating process, various forms of human error and sudden changes in the operating environment are inevitable, so accidents still happen frequently. Once an accident occurs, there will be heavy losses because there are a great number of passengers on the train. Therefore, in the development of urban rail transit, full attention must be paid to the safety of monorail vehicles. As a result, the issue of collisions gradually becomes an urgent issue to be discussed and solved for scholars all over the world.

2 SOLUTIONS FOR THE QUESTIONS OF COLLISION

Before experimenting with a real car, we should first simulate the value, which is quite important for improving and optimising the model of the experiment, using material objects. Besides, the nature of the research on the large deformation and collision of the vehicle also determines that the numerical simulation will be a main research method.

In the finite element analysis programs, we usually utilise full three-dimensional finite element models to analyse the impact of one or multi-section vehicles. Meanwhile, people use finite element analysis programs to conduct computer simulation research on trains' dynamic performance, single collision, double collisions, the safety of passengers when accidents happen, etc. In addition, through combining computer simulation with CAD/CAM technology, we can control the passive safety performance of the vehicles during the process of development. It can be said that the computer numerical simulation of the impact resistance of the body structure is more and more widely used in the research of the impact resistance of the vehicle body structure. The numerical simulation of the impact resistance of the vehicle body structure generally includes the analysis of the typical energy-absorbing components, the simulation of the whole vehicle and the structure optimisation of the impact resistance of the vehicle body. Through a large number of examples, it has proved to be an effective method in the current research of the passive safety of the vehicle body. The method of simulation calculation has a good consistency with the experimental results. The simulation experiment has good reproducibility, and the amount of information that is stored is large. Furthermore, the vehicle body can be opened along an arbitrary cross section according to the experimental requirements, so as to observe the deformation of internal components and the distribution of stress and strain.

3 THE DESIGN OF THE IMPACT RESISTANCE STRUCTURE OF THE BODY OF STRADDLED-TYPE MONORAIL VEHICLES

When the vehicle collides with the fore-end face to face, the first vehicle goes into the state of collision first. Therefore, the frontal collision simulation experiments of the first vehicle are the focus of the research of the design of impact resistance of monorail vehicles. As for the first vehicles, which are located at both ends of the straddled-type monorail vehicles, they are shared by drivers and passengers, while non-passenger zones of the vehicle are quite small, and the deformable zone decreases a lot correspondingly, which decreases the energy-absorbing effect and makes it difficult to satisfy the safety design standards. Having recognised the importance of developing the safety design standards for vehicle collision, we conduct research for the safety performance of collisions to be carried out on the design of straddled-type monorail vehicles. The results are summarised as follows:

In designing the body of the straddled-type monorail vehicles, the focus is the rational design of the collision energy-absorbing zone of the vehicles, which aims to rationally distribute the inflexibility of the end wall, side wall and underframe, and which makes the non-passenger zone become a high energy-absorbing deformation zone. The shadow zone in Figure 1 can be considered to be the deformation zone for the energy absorption of the collision of the first vehicle. The strength of the vehicle body in the deformation zone must be less than that of the passenger zone. The ideal deformation is that, firstly, the plastic deformation happens from the end of the vehicles, and in accordance with the process of progressive development that has been set, to absorb a large amount of the energy caused by impact to control the changes of the collision force and deceleration.

In the phase of "single collision", in order to keep the deceleration at a certain appropriate level

when the vehicle collides, we have to balance the force of collision and control its peak value as much as we possibly can, keeping the force of collision that transmits to the human body and the value of acceleration in an acceptable range and making sure that the impact on passengers meets the requirements of safety regulations.

For "double collisions", we should arrange the interior space of the vehicle properly and prevent the equipment installation, interior decoration, apron board and the bogie connecting equipment from dropping or being damaged, so as to achieve the aim of protecting people.

4 THE CONFIRMATION OF COLLISION SIMULATION VELOCITY

In order to simulate the collision of straddled-type monorail vehicles, certain initial velocity should firstly be given to the vehicle body so that it can collide with the specific rigid wall. As for the setting of the initial velocity, there are certain experiences and theories to follow. According to the research of Reference, we can find that the number of vehicles has limited influence on the energy absorption of the first vehicle. The energy absorption of the first vehicle can be calculated from Formula (1):

$$E_d = \frac{R_1}{2f_d} \times \frac{1}{2} M_{c1} v^2 \qquad (1)$$

In the formula, R_1 is the energy absorption coefficient of the interface, and the recommended value of the reference is 0.9; coefficient 2 indicates that the two vehicles absorb equal energy; f_d is the energy coefficient, and the recommended value is 1.2; M_{c1} is the mass of the first vehicle; V is the velocity.

Substitute each value of the straddled-type monorail vehicle into the above formula, and the energy absorption of the first vehicle is as follows:

$$E_d = \frac{0.9}{2 \times 1.2} \times \frac{1}{2} M_{c1} v^2 = 0.1875 M_{c1} v^2 \qquad (2)$$

In the references, it says that two urban rail vehicles of the same specification collide at a velocity of 20 km/h, and whether the energy at this velocity can be absorbed by energy-absorbing components when the body structure is not basically deformed is verified.

Since the rigid wall is fixed in the model of calculation, according to the theorem of momentum and the law of conservation of energy, under the situation that the plastic deformation of the vehicle body absorbs the same energy, the velocity with

The deformation part   Non deformation part

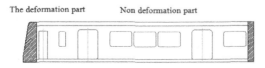

Figure 1.   Structure diagram of the crashworthy vehicle.

Table 1. Collision velocity of straddled-type monorail vehicle.

The velocity that the vehicle collides with the fixed rigid wall	The velocity that vehicle collides with another static vehicle of the same mass
7.86 m/s (28.29 km/h)	11.11 m/s (40 km/h)

which the vehicle collides with another static vehicle of the same mass and the velocity with which the vehicle collides with the fixed rigid wall have a relationship of $\sqrt{2}$. The calculation velocity used in this paper is as shown in Table 1.

5 THE ANALYSIS OF COLLISION SIMULATION RESULT

5.1 *The deformation analysis of the body frame of vehicles*

The body of the straddled-type monorail vehicle with the initial velocity 28.29 km/h makes the head-on collision with the fixed rigid wall happen. The deformation process of the body is shown in Figure 2.

From Figure 3 we can see that there is no collision between the body of the vehicle and the rigid wall in the 5 ms until collision happens. The deformation value of the vehicle head is 0. Starting from 5 ms, the vehicle head collides with the rigid wall and the deformation value increases gradually until 45 ms, as shown in Figure 2. At this moment, the maximum deformation value is 101 mm. Basically, the value does not change after that. The impact energy is dispersed through the body frame successfully. The guiding mechanism of the rail vehicle also has a large deformation. The collision causes the vehicle to swing back and forth and causes significant deformations on the vehicle's cab, the left and right side wall, roof and the front part of its underframe.

The coupler at the rear of the vehicle is deformed. The main reason for this is that when the first vehicle collides with the rigid wall, it slows down immediately while the second vehicle still moves on at initial velocity. The interaction between two vehicles produces stress on the coupler, causing it to be deformed. The second vehicle does not deform, which means that colliding with other vehicles at this velocity does not exert any influence on the vehicle behind.

5.2 *The analysis of the changes of the deformation velocity of vehicle body and time*

From Figure 4 (the velocity versus time curve), it is concluded that the velocity of the deformation

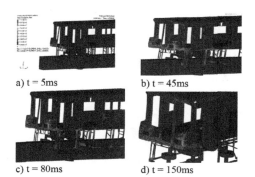

a) t = 5ms   b) t = 45ms

c) t = 80ms   d) t = 150ms

Figure 2. Course of deformation vs time.

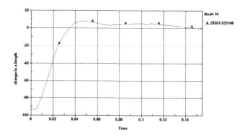

Figure 3. Course of body's deformation vs time.

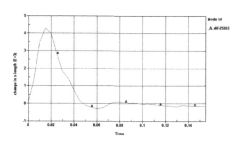

Figure 4. Curve of velocity deformation vs time.

increases gradually as the collision begins. The velocity reaches the maximum at 15 ms, about 4.3 m/s, and then drops to 0. The fluctuation of the collision velocity is not large and no second collision happens. However, the velocity drops quite quickly, indicating that the energy disperses quickly.

5.3 *The analysis of collision energy*

From Figure 5 (the total kinetic energy versus time curve), it is concluded that in this paper the initial kinetic energy of the first vehicle is 1,736 kJ, while after the collision, the final kinetic energy is 704 kJ, which indicates that in this paper, the total kinetic

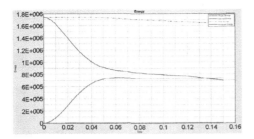

Figure 5. Curve of the vehicle structure energy vs time.

Figure 6. The major absorption component of the straddled-type monorail vehicle.

energy dissipated by the first vehicle during collision is 1,032 kJ.

The energy change is evidently shown in the collision process of the vehicles. Firstly, after the collision, the front-end of the vehicle touches the rigid wall and the deformation begins from small to large, then to the maximum, and then the rear part bounces back and separates until the movement ceases. This important rule reveals the characteristics of energy change during the process of collision, and also provides an important theoretical foundation for the identification of the collapse and deformation of the vehicle body.

In addition, the hourglass energy is only 11.6 kJ, accounting for 0.67% of the total energy, which ensures the correctness of the calculation results. The ordinary simulation requires that the ratio of hourglass energy to total energy is less than 5%. Meanwhile, the sum of the energy is larger than the initial total energy. This is because during the process of collision, some illegal energy is produced. However, as its proportion is very small, the calculation results are still convincing.

5.4 *The situation of energy absorption*

From Figure 5 (the internal energy versus time curve), it is concluded that during the collision, the plastic deformation occurs mainly on the energy-absorbing part, while the non-deformed zone only has elastic deformation. From the curve, the maximum internal energy generated from the plastic deformation of this energy-absorbing structure is 749.6 kJ, which indicates that the maximum internal energy that the vehicle absorbs is 749.6 kJ. The energy that the energy-absorbing structure absorbs accounts for 72.6% of the total dissipated kinetic energy during the collision.

The energy absorption of the vehicle body mainly relies on the deformed components. The major deformed components during collision are illustrated in Figure 6.

6 CONCLUSION

Taking the analytical method of the collision safety of the straddled-type monorail vehicle as the research objective and the Chongqing straddled-type monorail vehicles as the research subject, this paper explores the basic algorithm and theories of non-linear finite elements theoretically. We adopted the related software programs to establish the collision simulation analysis model and carried out the head-on collision simulation analysis of the vehicle. With the initial velocity of 40 km/h (the collision velocity of the monorail vehicle and the static monorail vehicle of the same mass), it is concluded that the non-passenger zone at the front of the vehicle has obvious plastic deformation and absorbs most of the collision energy, which greatly reduces the peak value of the collision force. Furthermore, it does not impact the survival room of the crew and passengers. The change to each part of the human body does not cause its functions to be affected. From the analysis of the situation of body injury, the maximum value of the injury index of each part is within the standards, indicating that people are safe when collision occurs at this initial velocity.

REFERENCES

Canjea, S., & Thornes, C. Buf. (1996). Load and crashworthiness requirements for the NJ transit low-floor light rail vehicle. *Railroad Conference, Proceedings of the 1996 ASME/IEEE Joint.*

Du, Z., & Zhang X. (2010). The bumping resistance design for straddled-type monorail vehicle. *Railway Locomotive & Car*, 30(5), 69–71.

Li, L., et al. (2008). Numerical study on crashworthy structure for urban rail vehicle. *Railway Locomotive & Car*, 28(2), 28–31.

Lu, G. (2002). Energy absorption requirement for crashworthy vehicles. *IMechE*, 216(Part F), 31–39.

Lu, G. (2006). Energy absorption requirement for crashworthy vehicles. *Foreign Rolling Stock*, 43, 8–13.

Zhang, Z., et al. (2013). Research on energy-absorbing structures for two ends of high-speed train car body. *Electric Drive for Locomotives*, 2013(1), 43–47.

Automotive, Mechanical and Electrical Engineering – Liu (Ed.)
© 2017 Taylor & Francis Group, London, ISBN 978-1-138-62951-6

Research on a hybrid electric vehicle regenerative braking system

Zhigang Fang & Can Yang
School of Automotive Engineering, Wuhan University of Technology, Wuhan, China

Xianghong Zhu
Hubei Tri-ring Special Purpose Vehicle Limited Liability Company, Wuhan, China

Lei Zhao
Shiyan Product Quality Supervision and Testing Institute, Shiyan, China

ABSTRACT: The Hybrid Electric Vehicle (HEV) has been a new orientation for vehicles as it combines the two advantages of low emissions and low fuel consumption. As a key technology for reducing energy consumption, more and more importance has been attached to regenerative braking, which has been extensively utilised by people. This thesis aims to introduce the features of regenerative braking, including structure, function and control strategy. The appraisal and development trend of the regenerative braking system is discussed.

Keywords: Hybrid Electric Vehicle; Regenerative braking; Control strategy

1 INTRODUCTION

Having both a fuel engine and motor power, hybrid electric vehicles not only have the characteristics of environmental protection like pure electric cars, but they can also make up for the deficiency of the short trip range of pure electric vehicles, which significantly improves the fuel economy and emissions of the vehicle. The common structural characteristic between hybrid electric vehicles and electric cars is the motor. When braking or decelerating (braking or downhill), the regenerative braking, which is unique to the electric vehicle, can transfer kinetic energy into electrical energy stored in the storage device, resulting in the reduction of fuel consumption and the eventual increase in the range of electric cars. Therefore, regenerative braking is a potential direction of hybrid electric vehicle research.

2 STRUCTURE AND WORKING PRINCIPLE OF REGENERATIVE BRAKING SYSTEM

One of the most important characteristics of the hybrid electric vehicle is its ability to recycle braking energy significantly. Hybrid electric vehicle motors can be controlled to run as a generator, which converts the vehicle's kinetic energy to potential energy and to electricity stored in the energy storage device, to be used again.

Structure of Regenerative Braking System. For the hybrid electric vehicle, which is shown in Figure 1, a mechanical friction braking system should exist in conjunction with the electric regenerative braking. Therefore, it is a hybrid braking system. Like the hybrid drive system, it can have a variety of structures and control strategies. The ultimate goal of the system's design and control is to ensure the vehicle's braking performance, and its ability to recycle braking energy as much as possible. The key to the problem between regenerative braking and fluid (gas) is to suppress dynamic coordination.

Moreover, it should consider the following special requirements:

In order to make the driver have a smooth feeling when braking, fluid (gas) to suppress dynamic torque should be controlled according to the change of the regenerative braking torque. In the end, make the total torque of drivers get what they want. At the same time, the regenerative braking control should not cause the impact of the brake pedal, and do not give the driver a sense of abnormal.

In order to make the vehicle brake steadily, the front and rear wheels of the braking force distribution must be well balanced. In addition, in order to prevent the car from slipping, the front and the maximum braking force should be lower than the maximum allowed.

In order to improve the electric vehicle energy recovery, the proportion of regenerative braking

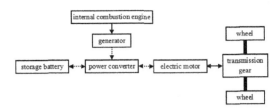

(1) SHEV

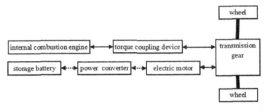

(2) PHEV

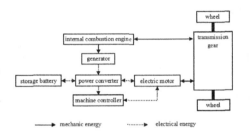

→ mechanic energy ┈┈┈► electrical energy

(3)PSHEV

Figure 1.   Hybrid electric vehicle structure.

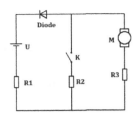

Figure 2.   Regenerate—hydraulic hybrid braking system structure.

and mechanical braking must have a good distribution. At the same time, the capacity of the motor and battery charging power should be taken into consideration.

To introduce the structure, the regenerate—hydraulic hybrid braking system is taken as an example, as shown in Figure 2. Recycling is only in the front wheel braking, whose torque is determined by the regenerative braking torque produced in the motor braking system and the friction brake torque produced in the mechanical braking system. On the brake pedal, the brake fluid pressure electric pump is needed for the braking force, braking control and motor control work, to determine the regenerative braking torque on the electric car and the hydraulic braking torque on the front and rear wheels. After using the brake pedal, the braking force produced from the brake fluid

by the electric pump makes the braking control and motor control work together to determine the regenerative braking torque on the electric car and the hydraulic braking torque on the front and rear wheels. The regenerative braking control module gets energy feedback to the storage battery, electric cars on the ABS and the role of the control valve and the traditional fuel in the car is the same, its function is to produce maximum braking force. The control valve and ABS on electric cars play the same role as traditional fuel in cars, whose function is to produce maximum braking force.

Principle of Regenerative Braking System. Motor speed is achieved by reducing the operating frequency. At the instant the frequency decreases, the synchronous speed of the motor decreases as well. But because of inertia, the motor rotor speed changes with time lag. At the moment that the actual speed is greater than the given speed, the counter electromotive force will be higher than the input voltage of the inverter dc motor. Thus the electric motor changes into a generator, working in a regenerative braking state and recycling energy instead of electric consumption. In this way, the kinetic energy can be transferred into electrical energy.

The generator voltage of the generative braking system is always lower than the battery voltage. In order to store the regenerative braking of electric energy in the storage device, the electronic brake control system must be used to make the motor power work. The basic principle of regenerative braking energy recovery is shown in Figure 3.

In Figure 3, R1 is the current-limiting resistor in the total circuit; R2 is the current-limiting resistance for braking; R3 is the resistance of the motor circuits; U is battery voltage.

The following makes E for motor induced electromotive force and L for motor armature inductance.

The motor armature drive current is disconnected at work, the armature ends are connected to the switch circuit, and its open circuit is controlled by the control unit. Because the motor is a perceptual component, induction electromotive force E

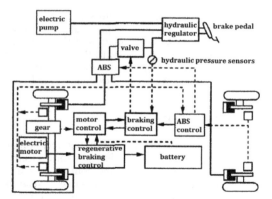

Figure 3. Basic principle of regenerative braking energy recovery.

and inductive current I along with the change of time t have the following relationship:

$$E = -L\frac{di}{dt} \qquad (1)$$

When the switch K is closed, the induced current caused by the induced electromotive force of the motor forms a loop through the switch K. The function of K is directed by IGBT (Insulated Gate Bipolar Transistor) components. The open circuit of the IGBT is provided by the control unit. Induced current is the braking current, whose value is:

$$i_1 = -\frac{E}{R1+R3} \qquad (2)$$

After the switch K is disconnected, the absolute value of di/dt rapidly increases. When the induction electromotive force is greater than the battery voltage: E > U, the current at the time of energy recovery is:

$$i_2 = \frac{E-U}{R1+R3} \qquad (3)$$

As a result, the energy generated in the process of motor regenerative braking is filled into the storage device.

3 THE PROSPECT OF TECHNOLOGY OF REGENERATION BRAKING SYSTEM OF HYBRID ELECTRIC VEHICLE

Two prominent advantages of hybrid cars are energy saving and environmental protection.

Then, compared to two other kinds of electric vehicles, the cost is just right, and they do not need to change the current infrastructure. For the current technology, which is relatively mature, it is the car with the most performance advantages at this stage. The environmental benefit and economic benefit brought by hybrid cars is self-evident, and a market potential exists in all of the countries in the world. The hybrid technology attached high attention in domestic and foreign, which pay more efforts to develop hybrid electric vehicles.

Among the key technologies of hybrid electric vehicles, the battery technology has been the research focus at home and abroad. But with a lot of manpower, financial and material resources, its bottleneck still has not been broken through. So the power of the electric car that can carry energy and can travel distances on a single charge is relatively short. Regenerative braking can prolong the travelling distance of electric vehicles with a one time charge. As a key energy saving technology, hybrid cars will have very considerable economic and social benefits, which has received increasing attention.

In the regenerative braking control strategy, there are key technical problems: brake stability, the adequacy of the braking energy recovery, braking pedal stationarity and composite braking coordination compatibility.

Braking stability: due to the electric braking torque (the regenerative braking torque) varying with the speed range, its braking process is different from the friction braking system. For later driven cars, the electric braking torque is added on the rear axle.

Energy recovery adequacy: energy recovery capacity is influenced by factors of battery, motor work characteristics, charging speed, etc. For the reason that the braking energy recycling often exists problem such as charging and fast charging, the motor and battery work auspicious is complicated. So by improving the motor control technology, the battery energy storage technology and the use of advanced energy storage as a way to improve the charging efficiency of the short time charging system, this is the key to ensuring the braking energy recycling is sufficient.

Brake pedal stationarity: as the auxiliary braking torque, the electric braking torque will affect the brake pedal to beat and driving comfort. It is necessary to optimise the size of the electric braking torque control to make the car brake feel the same as a conventional automobile brake system.

Composite braking coordination compatibility: to realise the energy under the condition of ensure the safety of the brake fully recycling, regenerative braking must be compatible with conventional braking system, ABS system good coordination.

Reasonable design electric mechanism regulated torque and the brake torque adjustment of comprehensive control strategy are needed to realise the motor feedback braking energy and ABS braking coordination control.

To overcome these problems, our scientific research ability needs to be improved in the following respects:

Regenerative braking system modelling and vehicle braking dynamics modelling ability: based on a regenerative braking and friction braking model, establish a hybrid integrated multiple operating conditions braking kinetics model. According to the requirements of regenerative braking, establish a reasonable control model of the battery energy, and the efficiency of the motor/generator control model.

Optimisation ability based on vehicle braking dynamics simulation: based on the comprehensive braking dynamics simulation of the hybrid electric vehicle, through the analysis on the battery in the process of regenerative braking, motor/generator work change rule and correlation, finish working parameters of subsystem comprehensive control based on maximum braking energy recovery rate. Through analysis and evaluation on braking performance and braking energy recovery in multiple operating conditions, optimise the distribution of the braking energy control strategy.

Regenerative braking energy management and control strategy development capabilities: according to the different conditions of hybrid cars, the road adhesion conditions and braking requirements on the condition of braking safety, determine the reasonable regenerative braking and friction braking energy distribution management model and control strategy, braking and comprehensive coordination of ABS control, and improve the braking energy recovery.

Regenerative braking system experiment, matching control and comprehensive evaluation of simulation ability: build a physical simulation environment for the hybrid regenerative braking system that can reflect the state of the vehicle, and conduct the regenerative braking system parameter matching and performance optimisation. According to the experimental simulation of the regenerative braking system, put forward the scientific and reasonable hybrid regenerative braking comprehensive performance test and evaluation methods.

4 CONCLUSION

The theory of regenerative braking technology is one of the key technologies of hybrid electric vehicles, whose research of the control system is very complex, involving different subjects such as automobiles, machinery, electrical and electronic, etc., which is a considerable breadth and depth of the subject. Because the research started late in our country, there are now many key problems to be solved. For the reason that regenerative braking has broad application prospects, it is necessary to increase the investment in research and development in our country.

REFERENCES

Ahn, J.K., Jung, K.H., Kim, D.H., Jin, H.B., Kim, H.S., & Hwang, S.H. (2009). Analysis of a regenerative braking system for hybrid electric vehicles using an electro-mechanical brake. *International Journal of Automotive Technology*, 10(2), 229–234.

Han, J.F., Tao, J., Lu, H.J., et al. (2014) Development and prospect of regenerative braking technology of electric vehicles. *Applied Mechanics and Materials*, 2808 (448).

Han, Z.L., & Wang, Y. (2010). On the study of electric vehicle regenerative braking. *Applied Mechanics and Materials*, 1016 p. (33).

Li, G.F., & Wang, H.X. (2012). Study on regenerative braking control strategy for EV based on the vehicle braking mechanics. *Advanced Materials Research*, 1700 (490).

Wang, F., & Wu, Y.H. (2013). Study of regenerative braking model and simulation of a car. *Advanced Materials Research*, 2109 (605).

Automotive, Mechanical and Electrical Engineering – Liu (Ed.)
© *2017 Taylor & Francis Group, London, ISBN 978-1-138-62951-6*

Research on seal performance of rotating articulated skirt and the underwater vehicle

Haixia Gong, Zhe Liu & Jian Zhou
School of Mechanical and Electrical Engineering, Harbin Engineering University, Harbin, China

ABSTRACT: The rotating articulated skirt and the underwater vehicle are sealed by an O-ring which is compressed by bolts pre-tightening flange, to solve the difficulty of sealing the rotating articulated skirt and the underwater vehicle in a deep water operation. In this paper, the O-ring was taken as the object of study. 3 order Yeoh model is used as the corresponding material parameters of the hyper-elastic constitutive relation curve fitting of the O-ring, using the finite element software to analyse the influence of different number of bolts pre-tightening on the contact stress and shear stress of the O-ring. Thus the dangerous section position is obtained, and the optimal number of bolts is acquired through the analysis of contact stress on the dangerous sections. The results show that: on the contact surfaces A and B, the contact stress of O-rings which are pre-tightened by 24 bolts is 14.635 MPa and 13.011 MPa, while the seal areas that meet the requirements are 52.52% and 67.14%, which is far stronger than a setting 22, 20 or 26 bolts.

Keywords: Articulated shirt; Carrier; Sealing; O-ring; Yeoh Model; non-linear finite element analysis

1 INTRODUCTION

At present, underwater docking technology is widely used in underwater space station docking, and marine life rescue (Wang Liquan, 2008; Wang Liquan, 2007). The technical parameters of the submersible seal mechanism are still not exactly defined, so the airtightness of the articulated skirt and the carrier should be studied. Research on the rotary articulated skirt both domestic and foreign is rare. However, Wang Liquan (2011) studied the airtightness of structure, from which a rotating articulated skirt kinematics model was established. Schoof (2007) studied buckling strength and the structural stability of the skirt by using the finite element. Mao Jiyu (2000) studied the seal technical requirements of the rotating articulated skirt and created the structure design of the device. The seal failure form of the fixed rescue bell was analysed by Hu Yong (2007), who studied the influence of many parameters on the airtightness of rescue bell.

There is, however, no research on airtightness of rotary articulated skirt and rescue carrier. In this paper, the numerical method of finite element is used to analyse the airtightness of the rotary articulated skirt and rescue carrier, and contact stress is used to determine the airtightness. This has resulted in the design of four kinds of assembly of articulated skirt upper flange and carrier lower flange whose bolts are respectively 20, 22, 24 and 26. Considering the non-linear contact and O-ring non-linear material, the finite element model of a combined structure of articulated skirt upper flange, carrier lower flange, bolts and O-ring is established, while the contact stress and shear stress of O-rings of the four assemblies are analysed, as a result of which the number of bolts in the assembly which meet the airtightness is established.

2 THE SEAL OF THE ROTATING ARTICULATED SKIRT AND THE UNDERWATER VEHICLE

The assembly of a rotating articulated skirt and an underwater vehicle is shown in Figure 1. In order to ensure the reliability of the seal, the sealing structure in this paper choose the structural form of the static seal of the flange end face, and the rubber seal ring which is more mature in the seal technology is used as the sealing element. Because of the sediment stirred up by the underwater carrier propeller, the sealing groove is therefore arranged in the lower flange of the carrier, which effectively prevents the accumulation of sediment in the groove thus reducing the wear of sand on the O-ring, resulting, in turn, in an improvement

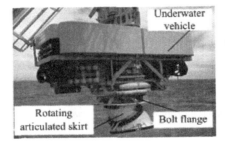

Figure 1. The assembly of articulated shirt and carrier.

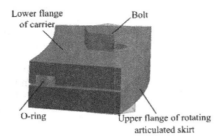

Figure 2. Partial enlarged drawing of seal.

of airtightness. The lower flange of the carrier is connected with the rotary articulated skirt through the bolt, the sealing effect is achieved by pre-tightening and compressing the O-ring, which is shown in Figure 2.

3 SUPER-ELASTIC CONSTITUTIVE EQUATION OF THE O-RING

3.1 3 order Yeoh model

The Yeoh model can be used to simulate the non-linear behaviour of carbon black reinforced natural rubber by uniaxial test data. Due to the introduction of the high order terms of the I_1, the strain energy and the test data are in good agreement, so the characteristic of the O-ring is characterised by the 3-order Yeoh model.

Strain energy function of Yeoh model (Wei Yintao, 1999; Li Xiaofang, 2005):

$$W = C_{10}(I_1 - 3) + C_{20}(I_1 - 3)^2 + C_{30}(I_1 - 3)^3 \quad (1)$$

The expressions of Cauchy stress in several common tests are as follows:
In the uniaxial tensile test:

$$\sigma_{11} = 2\left(\lambda_1^2 - \frac{1}{\lambda_1}\right)\frac{\partial W}{\partial I_1} \quad (2)$$

In the simple shear test:

$$\sigma_{12} = 2\left(\lambda_1 - \frac{1}{\lambda_1}\right)\frac{\partial W}{\partial I_1} \quad (3)$$

In the biaxial tension test:

$$\sigma_1 = 2\left(\lambda_1^2 - \frac{1}{\lambda_1^2 \lambda_2^2}\right)\frac{\partial W}{\partial I_1} \quad (4)$$

$$\sigma_2 = 2\left(\lambda_2^2 - \frac{1}{\lambda_2^2 \lambda_2^2}\right)\frac{\partial W}{\partial I_1} \quad (5)$$

where: C_{10}, C_{20}, C_{30} are the material parameters related to temperature; I_1 is the strain invariant; and λ_1, λ_2 are the main elongations.

Factoring the 3-order Yeoh strain energy function (1) into the equation (2), (3), (4), (5), the relationship between Cauchy stress and the main elongation of the uniaxial tensile test, simple shear test and biaxial tensile test can be respectively obtained.

$$\sigma_{11} = 2\left(\lambda_1^2 - \frac{1}{\lambda_1}\right)\left[C_{10} + 2C_{20} + 3C_{30}(I_1 - 3)^2\right] \quad (6)$$

$$\sigma_{12} = 2\left(\lambda_1 - \frac{1}{\lambda_1}\right)\left[C_{10} + 2C_{20} + 3C_{30}(I_1 - 3)^2\right] \quad (7)$$

$$\sigma_1 = 2\left(\lambda_1^2 - \frac{1}{\lambda_1^2 \lambda_2^2}\right)\left[C_{10} + 2C_{20} + 3C_{30}(I_1 - 3)^2\right] \quad (8)$$

$$\sigma_2 = 2\left(\lambda_2^2 - \frac{1}{\lambda_2^2 \lambda_1^2}\right)\left[C_{10} + 2C_{20} + 3C_{30}(I_1 - 3)^2\right] \quad (9)$$

Taking the stress and strain points measured in the experiment into the expression, the material parameters C_{10}, C_{20}, C_{30} can be obtained.

3.2 Super-elastic constitutive relation curve fitting material parameters

Because the rubber is a non-linear material, ANSYS fitting test data to obtain the required parameters of O-ring simulation is used.

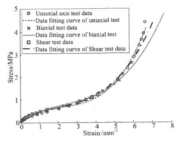

Figure 3. The curve fitted test data.

The curve fitted the above test data by using ANSYS is shown in Figure 3.

By using ANSYS, the parameters obtained by the above-mentioned curve can be reached: C_{10} = 1.29 MPa, C_{20} = −1.03 × 10⁻² MPa, C_{30} = 2.7 × 10⁻⁴ MPa.

4 FINITE ELEMENT ANALYSIS AND SIMULATION

4.1 Material properties

The material properties of the carrier lower flange and the articulated skirt upper flange and bolts are assumed to be the linear-elastic body which is isotropy. 921 A steel was chosen as the material for the carrier lower flange, the articulated shirt upper flange and the bolts, which has the features of high strength, is tough, and has good welding ability and strong resistance to the corrosion of seawater, whose modulus of elasticity is 200 GPa. Poisson's ratio is 0.3; the O-ring is compressed into super-elastic material on the pre-tightening force of bolts, the parameters of C_{10}, C_{20}, C_{30} are described in the previous section.

4.2 Finite element model

Because the bolts around the axis are uniformly distributed over 360°, the four assemblies that have bolts of 20, 22, 24 and 26 are simplified as the assembly along the circumferential direction of 18°, (180/11)°, 15°, and (180/13)°, and are analysed by ANSYS. The designed working depth of the rotary articulated skirt is 600 m, and the outside of the O-ring is subjected to hydrostatic pressure of 6 MPa. The articulated skirt is in both the hydrostatic pressure and the impact of the flow of force. Due to bad working conditions, the criterion of contact stress is 7.5 MPa (1.25 times of the hydrostatic pressure) (Lu Junfeng, 2014). Contact stress is often used as a measure of airtightness for the requirements of the seal of O-ring; the method that the hydrostatic pressure is less than the maximum contact stress is currently widely used (Ren Quanbin, 1995).

The SOLID186 solid with intermediate nodes is used in the carrier lower flange, the articulated shirt upper flange, the bolt and the O-ring, which is shown in Figure 4. The contact between the parts is simulated by the face-to-face contact element, and all contact surfaces are simulated by CONTACT174 and TARGET170; the contact of the O-ring and the carrier lower flange, the articulated shirt upper flange are rubbing contact, whose friction coefficient is 0.2; the contact of bolts and the carrier lower flange, the articulated shirt upper flange are a binding contact. The preload of bolts is simulated by a Prets179 pretension element and

Figure 4.   Finite element model of assembly.

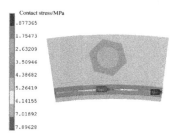

Figure 5.   Nephogram of surface contact stress of the O-ring.

load to the centre of the screw cylindrical surface; the fixed constraint is loaded onto the face of the carrier lower flange.

The specifications of bolts are M24 mm × 50 mm. In order to accurately simulate the real sealing process of the assembly, the total load step in the calculation process is divided into three steps. Bolt preload is applied at the first load step with 2 mm, which is the compression value of the O-ring; LOCK command is arranged on the second load step, which means that the compression of bolts in the first step will remain from the second load step to the third load step. Hydrostatic pressure 6 MPa is loaded to the outside of the O-ring at the third load step.

4.3 Simulation result

Due to the fact that the diameter of the O-ring is large (900 mm), there is an uneven distribution of circumferential contact stress. Analysis of the contact surface of the O-ring upper surface and the lower flange groove of the carrier will indicate the upper side of the contact stress of the O-ring (shown in Figure 5). It can be seen that the contact stress is gradually reduced from the centre to the two sides, and that both sides of the sealing ring are dangerous cross-sections. As long as both sides of the contact stress of the O-ring are greater than the hydrostatic pressure, the whole O-ring meets the requirements of the seal.

Three contact surfaces A, B and C are present on the surface of the O-ring, the lower flange of the carrier and the upper flange of the rotary articulated skirt (see Figure 6).

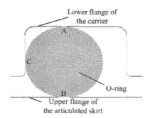

Figure 6.   Three contact surfaces of the sealing ring.

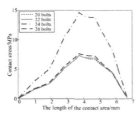

Figure 7.   Contact stress curve of A.

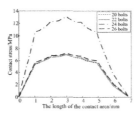

Figure 8.   Contact stress curve of B.

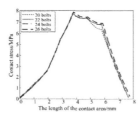

Figure 9.   Contact stress curve of C.

In order to carry out the quantitative analysis of the dangerous section, the contact stress curves of the three contact surfaces are generated, which are A, B and C, as shown in Figures 7–9.

Figure 6 shows that as long as the contact stress in the A and B contact surfaces of the O-ring is greater than the contact stress criterion (7.5 MPa), the requirements of sealing can be achieved. Figures 7–9 show that the lower flange of the carrier and the upper flange of the articulated shirt pre-tightened by 24 bolts can produce a maximum contact of 14.635 MPa and 13.011 MPa in the A and B contact surfaces of the O-ring; and

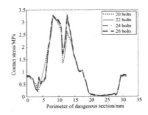

Figure 10.   Shear stress curve.

the region whose contact stress is greater than 7.5 MPa in A and B contact surfaces are 52.52% and 67.14% respectively, which is far more than the contact stress produced by the O-rings pre-tightened by 20, 22 and 26 bolts. Therefore, the structure pre-tightened by 24 bolts can meet the requirements of sealing.

Not only the sealing requirements need to be achieved, but also the failure of the O-ring should be prevented. If the shear stress of the O-ring is greater than the shear strength of the material, the O-ring will undergo wear and tear (Hu Dianyin, 2005). Therefore, the shear stress curves of 20, 22, 24 and 26 bolts corresponding to the dangerous section of the circular section, are obtained, as shown in Figure 10. It is known that the four shear stress curves are highly coincident, and the maximum shear stress is 3.278 MPa, 3.219 MPa, 3.235 MPa and 3.253 MPa, which means 20, 22, 24 and 26 bolt preload O-ring is consistent failure or not.

5   CONCLUSION

1. Different numbers of bolts preload the lower flange of the carrier and the upper flange of the articulated skirt to make the O-ring contact stress significantly different, but not much difference between the shear stress.
2. Under the condition of this paper, the sealing effect of the O-ring pre-tightened by 24 bolts is better than in those pre-tightened by 20, 22 and 26 bolts. Under a pressure of 6 MPa, the compression ratio of the O-ring is 20%, and the optimal bolt number is 24.
3. The sealing performance of the O-ring is gradually decreased with the distance from the bolt, the dangerous section appears at the most distal end of the bolt, and the contact stress is of approximately parabola distribution.

REFERENCES

Hu Dianyin, Wang Rongqiao, Ren Quanbin and Hong Jie. (2005). Finite element analysis of O-ring seal structure [J]. *Journal of Beijing University of Aeronautics and Astronautics.* 31(2), 255–260.

Hu Yong, Zhang Jin and Cui Weicheng. (2007). Sealing ability research on movable rescue bell. *Journal of Ship Mechanics*. 11(2), 221–230.

Li Xiaofang and Yang Xiaxiang. (2005). A review of elastic constitutive model for rubber materials. *China Elastomerics*. 15(1), 50–58.

Lu Junfeng and Zhao Yao. (2014). Research on sealing ability of the spherical transfer skirt at large mating angle. *Shipbuilding of China*. 61(8), 453–457.

Mao Jiyu, Zhang Xiangming, Luo Ziyei and Liu Yan. (2000). A research for the revolving on the skirt sealing technology of DSRV mating system. *Lubrication Engineering*. 25(5), 47–48.

Ren Quanbin, Chen Ruxun, and Yang Weiguo. (1995). Deformation and stress analysis of rubber O-ring. *Journal of Aerospace Power*. 10(3), 38–41.

Schoof, C., Goland, L. & Lo, D. (2007). Pressurised Rescue Module System Hull and Transfer Skirt Design and Experimental Validation. *Oceans*. 81(9), 1–8.

Wang Liquan, Tang Dedong, Wu Jianrong and Zhang Zhonglin. (2008). Research on Structure and kinematics of a novel mating skirt on the underwater vehicle. *China Mechanical Engineering*. 19(23), 2814–2818.

Wang Liquan, Tang Dedong and Zhang Zhonglin. (2007). *Research on Novel Underwater Mating Device with Articulated Skirt*. Proceedings of the 2007 IEEE International Conference on Mechatronics and Automation, Harbin, China, 769–773.

Wang Liquan, Tang Dedongi, Meng Qingxin and Zhang Zhonglin. (2011). Prototype development and test study on double articulated mating skirts of rescue vehicle. *Journal of Mechanical Engineering*. 47(9), 39–44.

Wei Yintao, Yang Tingqing and Du Xingwen. (1999). On the large deformation rubber-like materials: Constitutive laws and finite element method. *Chinese Journal of Solid Mechanics*. 20(4), 281–289.

Automotive, Mechanical and Electrical Engineering – Liu (Ed.)
© *2017 Taylor & Francis Group, London, ISBN 978-1-138-62951-6*

Research on the improvement of the pedestrians' overtaking behaviours based on a social force model in urban rail channel

Yi Xing, Xiangyong Yin & Yafei Liu
School of Traffic and Transportation, Beijing, China

ABSTRACT: In this paper, based on researching the behaviour of pedestrians' overtaking in Xizhimen and Chongwenmen subway station channel in Beijing, the characteristics of pedestrians' overtaking behaviour are analysed, when the rear pedestrians pass the front pedestrians, considering the required space and the shortest distance. Then choosing a temporary target improves the existing pedestrian social force model. In this paper, the modified pedestrians' overtaking behaviours based on the social force model program is compiled by MATLAB, and the reasonable model parameters are selected. Taking the Xizhimen and Chongwenmen subway station channel as the research background, the model simulation program is used to simulate the pedestrians' overtaking behaviours in a subway station channel. The results show the effectiveness of the proposed model.

Keywords: Urban rail transit; Pedestrians' overtaking behaviours; Social force model; Pedestrian flow simulation

1   INTRODUCTION

There are a large number of pedestrians' overtaking behaviours in the urban rail transit channel. This behaviour can easily lead to pedestrians crossing and conflict in the process of pedestrian walking, thus reducing pedestrian walking speed and passing efficiency. In order to solve the above problems, in this paper improving the pedestrians' overtaking behaviours on social force model and making the model of the pedestrians' overtaking behaviours simulation is closer to reality.

At present, domestic and foreign scholars have made some achievements in the study of pedestrians' overtaking behaviours on social force model. Helbing and Monár (1995) proposed the social force model, which attracted wide attention from scholars in the field of transportation, The micro-simulation model is a pedestrian dynamic model, which is mainly used to simulate pedestrians' walking behaviours. The model is continuous in time and space, and integrates the psychological characteristics of pedestrians, so that pedestrians can move under the action of various social forces. Helbing et al. (2005) uses pedestrian traffic experiments to further study the microscopic traffic behaviours of pedestrian flow in a bottleneck area. It is concluded that bi-directional pedestrian flow will form pedestrian zone and pedestrian self-organisation phenomena under certain density conditions, in order to predict the pedestrians'

danger and other emergencies in the studied bottleneck area. This will improve pedestrian traffic facilities, depending on the hazards.

Introducing pedestrian collision avoidance mechanism revises the behaviors of pedestrians collision (Zhang & Shi, 2013). There are two different collision avoidance mechanisms for pedestrians moving in the same direction and in different directions. The simulation results show that the improved model can reduce the number of collisions between pedestrians and improve pedestrian travel speed and travel efficiency. Hewang Sh (2014) explored the problem of excessive pedestrian contact in the social force model, and using Daniel R. Parisi and A. Sayfried's method, modified the expected speed and instantaneous velocity of pedestrians in the model.

The existing social force model mainly changes the speed direction of pedestrians by changing the expected speed direction of pedestrians, and completes the pedestrians' overtaking behaviours. Changing the direction of the expected speed, while not considering the direction of change in the other pedestrians, easily leads to conflict with other pedestrians. In this paper, after pedestrians cross the front pedestrian, according to their own required space and the shortest overtake distance, selecting the temporary target points, through updating each step of the temporary target points to complete the pedestrians' overtaking behaviours and improve the existing pedestrian overtake social force model. The model of pedestrians overtake social force is developed by using MATLAB, selecting the reasonable model parameters. Taking the research area of Xizhimen and Chongwenmen subway stations as the research background, the model simulation program is used to simulate the pedestrians' overtaking behaviours in the subway stations, to verify the validity of the model.

2 ANALYSIS OF PEDESTRIANS' OVERTAKING BEHAVIOURS

This paper investigates the pedestrians' overtaking behaviours of a one-way and bi-directional pedestrian walkway in Beijing Xizhimen and Chongwenmen subway stations, and analyses the characteristics of the pedestrians' behaviours. Figure 1 shows the Chongwenmen subway station research channel (one-way) area pedestrians' overtaking behaviours process diagram. The first picture shows the behaviour of a pedestrian (circle-marked) preparing to overtake the front pedestrians. The second picture shows the pedestrian change the direction of stride, and overtake from the right side of the front pedestrian. The third picture shows that in accordance with the direction of the target forward, the pedestrian overtakes the front pedestrians. Pedestrians choose to overtake from the right side of the front pedestrian because the right side of the front pedestrian can provide the required space.

Figure 2 shows that the research on the process of the pedestrian flow overtaking behaviours in the same direction in the Xizhimen subway research channel area. The first picture shows the rear pedestrian (red circle marker) ready to overtake the pedestrian in front. The second picture shows the pedestrian overtake the pedestrian ahead on their left side. The third picture shows after the pedes-

trian has overtaken the pedestrian ahead, and go ahead in accordance with the target direction. The reason for the pedestrian choosing to overtake the pedestrian ahead on their left side is that on the one hand, the ahead pedestrians' left sides can meet the overtaking required space. On the other hand, the pedestrians (circle-marked) on the left side of the front pedestrian can realise the overtaking behaviours in the shortest distance.

The pedestrians' crossing behaviours in the Chongwenmen and Xizhimen subway stations can be seen. First it is determined beyond the pedestrians around, whether they can meet the required space of the overtaking pedestrian. Then the pedestrians can finish the overtaking behaviours in the shortest distance.

3 IMPROVEMENT OF PEDESTRIANS' OVERTAKING BEHAVIOURS ON SOCIAL FORCE MODEL IN THE STRAIGHT LINE AREA OF URBAN RAIL CHANNEL

This section analyses the pedestrians' overtaking behaviours on social force model of the existing urban rail channel, by considering the required space and the shortest distance of the pedestrians' overtake process, selecting the temporary target point and improving the existing pedestrians' overtaking behaviours on social force model.

3.1 *The problems of the pedestrians' overtaking behaviours on social force model in the existing urban rail channel*

The description of the pedestrians' overtaking behaviours on social force model in the existing urban rail channel.

The existing pedestrians' overtaking behaviours on social force model describing the pedestrian's overtaking behaviours by changing the intended speed direction of the pedestrian self-driving force (Haijun Zhang). Figure 3 is a schematic diagram of the change of intended speed direction when pedestrians overtake other pedestrians. When the

Figure 1. The overtaking behaviours of pedestrians in research area of Chongwenmen station subway.

Figure 2. The overtaking behaviours of pedestrians in research area of subway Xizhimen station.

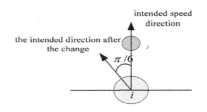

Figure 3. Schematic diagram of the change of pedestrians' intended speed direction.

distance between pedestrians is less than 0.75 m (including the radius of pedestrians), the pedestrian changes the intended speed direction according to the current intended direction with the current position as the base point, and the angle of change is $\pi/6$.

The pedestrians' speed is updated in accordance with the traditional social force model after the pedestrians find the intended direction, and describing the pedestrian transcendence, as shown in the formula 1, 2.

$$\overrightarrow{F_{desired}} = m_i \cdot \left(v_0 \vec{e}_i - \vec{v}_t\right)/\tau \tag{1}$$

where $F_{desired}$ is the pedestrian self-driving force according the intended speed direction; m_i is the pedestrian weight; v_0 is the intended speed; e_i is the direction of intended speed; v_t is the actual speed; τ controls the acceleration.

$$\vec{F} = \frac{m_i d_i(t)}{d(t)} = \overrightarrow{F_{desired}} + \overrightarrow{F_{wall}} + \overrightarrow{F_{pedestrian}} \tag{2}$$

where $\vec{F}$ is the pedestrian by the resultant force; F_{wall} is the interaction between pedestrians and the wall; $F_{pedestrian}$ is the force between pedestrians.
The model has the following deficiencies:

1. Changing the direction of pedestrians by using the intended speed may conflict with other pedestrians. Pedestrian i considered avoiding conflict with pedestrian j, but did not consider conflicting with other pedestrians. It is possible in the process of overtaking pedestrian j, that there is conflict with other pedestrians.
2. The effect of multiple forces may cause pedestrian i to deviate from the intended speed direction of the overtaking behaviour. In the traditional social force model, the position of the pedestrian i in the next step is determined by the self-driving force of pedestrians, the interaction between pedestrians and the interaction force between pedestrians and obstructions, in which the self-driving force is considered without other obstacles or pedestrian interference, advancing according to the intended speed for direction of the destination. When pedestrian i wants to overtake pedestrian j, the intended speed in the direction of the destination is changed to the intended speed direction of the overtaking behaviours, which may cause the pedestrian i to deviate from the intended speed direction of the overtaking behaviours.
3. Describe the required space

The Seyfried study found that pedestrians needed a certain distance to meet the required space of next step in the process of walking, and

the relation between the required lengths for one pedestrian to move with velocity v, where v itself is linear, at least for velocities 0.1 m/s < v < 1.0 m/s. The required length of the pedestrian to move can be determined by formula 3.

$$d = a + bv \tag{3}$$

With a = 0.36 m and b = 1.06 s.

In reality, pedestrians' walking behaviour takes place in a two-dimensional system. Therefore, on the basis of Seyfried A's research, the required space for pedestrians walking is defined as two-dimensional required space. It is defined as: under the hypothetical conditions, pedestrians along the direction of velocity v, the required space occupied by the length of d, and the width of the pedestrians' diameter (Shanshan Li), as shown in Figure 4.

In she selecting for the length and width in required space, there are the following deficiencies:

In reality, the traffic environment is different, the intended speed V_{des} of pedestrians are different, and the length d of required space is more fixed, based on hypothetical conditions selected. When pedestrians move to the next step and the step is greater than the length d, it may cause collisions with other pedestrians.

The width of the required space is equal to the pedestrian's diameter, which easily leads to the pedestrian conflicting with other pedestrians (Figure 5), pedestrians i by the force of the pedestrian j, the two person will be bounced off.

3.2 The improvement of pedestrian overtaking social force model

By analysing the existing pedestrian overtaking social force model and the shortcomings of the

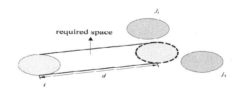

Figure 4. The required space for pedestrian i.

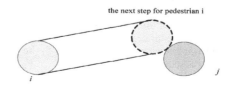

Figure 5. The contact diagram between pedestrian i and pedestrian j.

Figure 6. The schematic of pedestrians setting target point.

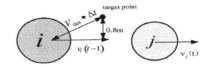

Figure 7. Schematic diagram of initial target selection.

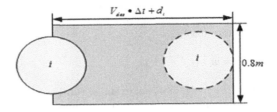

Figure 8. Required space parameters.

selection for the pedestrian required space, combined with research and analysis, the pedestrian overtaking social force model mainly made the following improvements.

1. Set the temporary target point. In each step in the progress of rear pedestrian i overtaking pedestrian j, according to the direction of a number of pedestrians in front, adjust the temporary target points, to prevent conflicting with other pedestrians.
2. The shortest overtaking distance. In the progress of overtaking behaviours, pedestrian i will first determine the temporary position. If the pedestrian j in the left front of the pedestrian i, pedestrians i will walk into the right front, so that makes the shortest distance, and vice versa, as shown in Figure 6.
3. The guarantee of required space.

The selection of temporary target points must guarantee that rear pedestrians have more dynamic and static required space to overtake the pedestrian in front. This is the most important factor to overtake the pedestrian j and to determine temporary target points. It is best to have the shortest distance from the direction of the required space, but sometimes the shortest path is on the right, and there is no required space on the right, the left side has the demand space, and pedestrians can select the temporary target point on the left.

After selecting the temporary target point, the pedestrian i adjusts the speed direction of each step. The improved model of pedestrians' overtaking behaviours consists of the following parts:

1. Selecting of initial target points when pedestrians i overtakes the front pedestrian j. According to Pelechano N's research, the scope of the psychological impact between pedestrians is 1.5 m*0.8 m. In order to ensure that pedestrian i are not affected by the front pedestrian j, when the distance between the pedestrian i and the front pedestrian j is less than 1.5, the pedestrian i is selected as the target $G_{(x1,y2)}$: V_{des} is the intended speed of pedestrian i, and Δt is the time of the simulation step size. The distance from the current position is $V_{des} \bullet \Delta t$, the distance from the pedestrian at time t in the vertical $v_i(t-1)$ direction of 0.8 m, and consider the shortest

distance to complete overtake behaviour (left or right), as shown in Figure 7.

2. Judgment of required space. In order to ensure there is no conflict with other pedestrians when pedestrian i reaches the target point, and can meet the pedestrian i static required space, the length of the required space is set to $V_{des} \bullet \Delta t + d_i$. According to the required space of a pedestrian walking in China is $0.6 \times 0.8\ m^2$ (Shanshan Li) and the radius of pedestrians are $d_i \in [0.2, 0.3]$. In order to ensure that pedestrians are not affected by other pedestrian contact, the width of the required space is determined to be 0.8 m, as shown in Figure 8.
3. Adjust the selection of target points. When the pedestrian i overtakes the pedestrian j, the required space of initial target point occupied by other pedestrians, the pedestrian i needs to re-select the target point, known as the adjustment target. First, determine the relative position of pedestrian i and pedestrian j, such as pedestrian i is on the left side of pedestrian j, taking the position $P_{(x,y).(t-1)}$ of pedestrian i as the centre of the circle and the $V_{des} \bullet \Delta t$ as the radius. Starting from the current position of the initial target point of the pedestrian i, find the target point that meets the pedestrian i required space in accordance with the clockwise direction, until the target point is connected with the pedestrian in the horizontal position. If did not find the target point, search from the direction of the intended speed of the pedestrian i and meeting the pedestrian i required space with the counterclockwise direction, until the target point is connected with the pedestrian i in the horizontal position, as shown in Figure 9. If the pedestrian i is on the right side of the pedestrian j, the target point selection first starts from the position of the initial target point, selects the target point in the counter-clockwise direction. If the target point is not found, then it is selected in the clockwise direction.

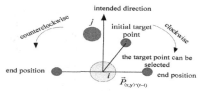

Figure 9. Adjust the selection of target points.

4. Calculation of deflection force. In order to overtake pedestrian j, pedestrian i needs to be within a certain range of psychological effects and temporary change in the direction of the target, towards the temporary target point forward. Then it is necessary to add a deflection force and change the direction of walking in order to complete overtake behaviours. The deflection force F_{tur} not only ensures the pedestrian walks towards the target point, but also guarantees the next step position of the pedestrian i, which does not exceed the required space by the pedestrians at the target point. The calculation is shown in formula 4.

$$\overrightarrow{F_{tur}} = \kappa_{tur} \times m_i \left(V_{des} \bullet \vec{e}_{op} - \overrightarrow{v_{est}(t)} \right) / \tau \qquad (4)$$

where $\overrightarrow{F_{tur}}$ is the deflection force; κ_{tur} is deflection force coefficient; m_i is the weight of pedestrian I; $v_{est}(t)$ is the actual speed at the moment; V_{des} is the intended speed; τ is the reaction time, $\tau = 0.5$; $\vec{e}_{op}$ is the unit vector for the direction of the next pedestrian's walk; $\vec{e}_{op}$ is the current position of the pedestrian points to the target point.

$$\vec{e}_{op} = \left(\overrightarrow{G_{(x1,y2)}} - \overrightarrow{P_{(x,y)}} \right) / \left\| \overrightarrow{G_{(x1,y2)}} - \overrightarrow{P_{(x,y)}} \right\| \qquad (5)$$

where $\overrightarrow{G_{(x1,y2)}}$ is the pedestrian i next temporary target location; $P_{(x, y)}$ is the pedestrian i current location.

5. Calculate the speed and the next step position for pedestrian i. After the pedestrian determines the target point, the pedestrian speed and position of the next step is determined by the formula 6, 7. Pedestrian i by the role of the deflection force in the process of overtaking instead of the self-driving force. After the end of the deflection force F_{tur} disappears, the driving force is recovered.

$$\begin{cases} \vec{F} = \dfrac{m_i d_i(t)}{d(t)} \overrightarrow{F_{tur}} + \overrightarrow{F_{wall}} + \overrightarrow{F_{pedestrian}} \\ \quad + \varphi_1 \left(\overrightarrow{F_{turning}} + \overrightarrow{F_{attractive}} \right) \\ \vec{a} = \vec{F} / m_i \\ \overrightarrow{v_i(t)} = \vec{a} \bullet \Delta t + \overrightarrow{v_i(t-1)} \end{cases} \qquad (6)$$

where $\vec{F}$ is the pedestrian by the resultant force, $\overline{F_{wall}}$ is the interaction between pedestrians and the wall, $\overline{F_{pedestrian}}$ is the force between pedestrians, $\overline{F_{turning}}$ is the gravity of the pedestrian in the corner, φ_1 is the coefficients of the piecewise function. $v_i(t)$ is the speed of the pedestrians in the step t, $v_i(t-1)$ is the speed of the pedestrians in the step t − 1. When the pedestrian is not affected by other pedestrians, $\overline{F_{pedestrian}} = 0$.

$$\overline{P_{(x,y),t}} = \overline{P_{(x,y),(t-1)}} + \overline{v_i(t)} \bullet \Delta t \qquad (7)$$

where $\overline{P_{(x,y),(t-1)}}$ is the position of the pedestrian at time (t − 1), $\overline{P_{(x,y),t}}$ is the position of the pedestrian at time t.

4 ANALYSIS ON THE IMPROVEMENT EFFECT OF PEDESTRIANS' OVERTAKING BEHAVIOURS ON SOCIAL FORCE MODEL

Determining the pedestrian parameters of the social force model, inputting the relevant data, and simulating the improved pedestrian overtaking social force model to verify the validity and feasibility of the improved model. Analysing the overtaking process and trajectory of the pedestrian.

4.1 Select the model parameters

According to research for the pedestrian social force model simulation parameters, pedestrian quality and pedestrian radius obey normal distribution. The range of pedestrian quality m is $m \in [50, 80]kg$, $N(65, 5^2)$, the range of pedestrian radius is $d \in [0.2, 0.3]m$, the initial intended speed V^0 and initial actual speed V are uniformly distributed, respectively $V^0 \in [0.8, 1.6]m/s$ and $V \in [0.5, 1.8]m/s$; This simulation takes two pedestrians which contains the front pedestrian 1 and rear pedestrian 2, respectively the intended speed is 1.0 m/s and 1.6 m/s. The initial speed is 0.8 m/s and 1.3 m/s, and the initial position is (1.8, 2.7), (0, 2.7).

In the normal pedestrian flow, pedestrian relaxation time τ takes a fixed value of 0.5 s. The parameters of pedestrian transcending social force modelca according to the reference (Shanshan Li).

4.2 Analysis on the improvement effect of pedestrians' overtaking behaviors on social force model

In Figure 10, (a) and (b) show respectively for the existing pedestrian overtaking social force model and for the improved model, the trajectory

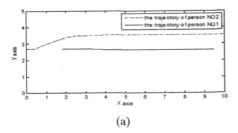

(a)

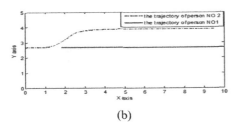

(b)

Figure 10. Comparison of existing and improved model pedestrian overtake simulation trajectories.

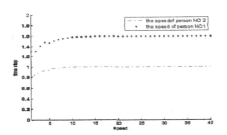

Figure 11. The speed changes of pedestrians 1 and 2 during the transition.

of the simulation when pedestrian 2 overtakes pedestrian 1.

Figure 10 shows that when the pedestrian 2 overtakes pedestrian 1 in the existing social force model, the number of steps is 12, which is about 5.85 m to complete the overtake behaviour. The deviation from the original track is 1.19 m (a). It is far away compared with the deviation from the track of 0.89 m (b) in the improvement model, and cannot reflect the phenomenon when pedestrians overtake according to the shortest distance walking.

Figure 11 shows the changes in the speed of pedestrian 1 and pedestrian 2 during the whole simulation process in the improved pedestrians' behaviours on social force model. When pedestrian 2 walked into the scope of action, pedestrian 2 in the role of the deflection force F_{tur}, changed the direction of speed and reduced the psychological impact of pedestrian 1 which caused the speed to

continue to rise gradually until pedestrian 2 completed the overtake behaviour. After the 15th step, pedestrian 2 was no longer affected by pedestrians 1 and accelerated to the desired speed of 1.6 m/s. In the simulation process, pedestrian 1 was not affected by pedestrian 2. In the role of self-driving force, the speed of pedestrian 1 increased from the initial speed 0.8 m/s to the intended speed 1 m/s.

5 SIMULATION ANALYSIS OF PEDESTRIANS' OVERTAKE BEHAVIOUR IN URBAN RAIL CHANNEL

In order to validate the improvement of pedestrian overtake social force model, this paper constructs a one-way channel 10 m × 5 m to analyse the behaviour of one-way pedestrian in pedestrian flow, as shown in Figure 12.

In this passage, the amount of pedestrian flow simulation steps is 300, each step is 0.2 s, and pedestrian arrival rate of 64 person/min.

Figure 12 illustrates the process of pedestrian 52 overtaking the front of a pedestrian. Simulation step 246 to step 254 is the process of pedestrian 52 overtaking the front of pedestrians from beginning to end. The first picture shows pedestrian 52 at 1.23 m (within the range of 1.5 m × 0.8 m) behind pedestrian 51, and the actual speed 1.2 m/s is greater than the speed (0.71) of pedestrian 51, satisfying the condition of overtake. Pedestrian 52 begins to overtake the front pedestrians. The second picture shows the process of pedestrian 52 overtaking the front pedestrians. As pedestrian 52 is on the right side of pedestrian 51, pedestrian 52 overtakes the right side of the pedestrians according to the way in which the pedestrians have the shortest distance in the improved overtaking model. The third picture shows the location of pedestrian 52 who completed the overtaking behaviour.

According to the overtaking process of pedestrian 52, the velocity variation of pedestrian 52 during the whole overtaking process is obtained, as shown in Figure 13. The speed of pedestrian 52 increased from 1.4 m/s to 1.46 m/s in the course of overtake, and the speed of pedestrian 52 gradually increased (the expected speed of pedestrian 52

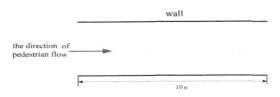

Figure 12. The design of simulation environment.

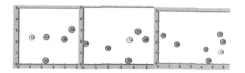

Figure 13. Pedestrian overtaking process simulation diagram.

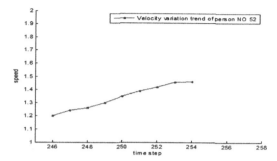

Figure 14. Speed change chart of rear pedestrians in the process of overtaking.

was 1.55 m/s), which was consistent with the speed of pedestrians overtaking process in reality.

6 CONCLUSION

This paper analyses the characteristics of pedestrians' overtaking behaviours in an urban rail channel. On the basis of existing pedestrians' overtaking social force model, the judgment of temporary target point and required space, and the pedestrians self-driving force formula replaced by deflection force F_{tur} (formula 4), can effectively describe the behaviour of pedestrians overtaking the front pedestrian in the shortest distance. In this paper,

the existing and improved models are compared by MATLAB. Based on the two-way channel of Xizhimen subway station and the one-way channel of Chongwenmen subway station, the improved pedestrians flow model is simulated. The simulation results are in agreement with the analysis of the investigation, and validates the improvement of pedestrians' overtaking behaviours on social force model.

REFERENCES

Helbing, D. and Monár, P. (1995). Social force model for pedestrian dynamic. *Physical Review, E51*(5), 4282–4286.

Helbing, D., Buzna, L., Johansson, A. and Werner, T. (2005). Self-organized pedestrian crowd dynamics: experiments, simulation, and design solution. *Transportation Science, 39*(1) 1–24.

Pelechano, N., Allbeck, J.M. and Badler, N.I. (2007). Controlling individual agents in high-density crowd simulation. *Proceedings of the 2007 ACM SIGGRAPH/Eurographics symposium on computer animation* (pp. 99–108).

Seyfried, A., Steffen, B. and Lippert, T. (2006). Basics of modeling the pedestrian flow. *Physica A, 368*(1), 232–238.

Shanshan, L. (2013). Research on the microscopic behavior models of vehicle, bicycle, pedestrian, and their interactive interference at the signalized plane intersection. Beijing Jiaotong University, Beijing.

Shi, H. (2014) *Research on behaviors and strategies of pedestrian flow through bottleneck based on social force model.* Inner Mongolia University, Hohhot.

Yu, Q. (2015) Improving pedestrian social force model based on pedestrians predicting theory in the rail transit station. Beijing Jiaotong University, Beijing.

Zhang, H. and Shi, H. (2013) Research of bi-directional pedestrian flow's model on modification of social force. *Computer Engineering and Applications, Jiangsu: School of Computer Science and Telecommunication Engineering, 16,* 236–239.

Automotive, Mechanical and Electrical Engineering – Liu (Ed.)
© 2017 Taylor & Francis Group, London, ISBN 978-1-138-62951-6

Sound quality of vehicle exhaust noise prediction by using support vector regression

Xiaoxu Zhang

Chongqing College of Electronic Engineering, Chongqing, China

ABSTRACT: The Support Vector Regression (SVR) was proposed to establish the relationship between the objective psychoacoustic parameters and the sound quality of vehicle exhaust noise. Sensory pleasantness evaluation of vehicle exhaust noise samples were obtained under the same training and test sample sets and the results were compared with that obtained through Multiple Linear Regression (MLR) prediction models. The results showed that the prediction values of SVR were close to the measured values, the mean absolute percentage error is smaller than MLR and in the 8% error range. The SVR model represented the nonlinear of sensory pleasantness and objective parameters exactly. It is suggested that SVR is an effective and powerful tool for predicting sound quality of vehicle exhaust noise.

Keywords: Highway transportation; Vehicle; Sound quality; Exhaust noise; Support vector regression

1 INTRODUCTION

With the rapid economic development of our country, vehicle ownership has been increasing rapidly, especially in some big cities. As a result, vehicle noise has become the main factor that impairs the urban environment. The exhaust noise is one of the main noise sources of vehicles. Along with the continuous improvement of the scientific and technological level, the sound pressure of vehicle noise has been under effective control. The improvement of the sound quality makes the vehicle sound adapt to the requirements of consumers better and better. And the sound of vehicles has become one of the selling point of vehicles and directly affects the sales and marketing of vehicles. However, the study finds that people's subjective feelings about noise are affected by various factors, such as personal experiences, the surrounding environment and the psychological state. These factors restrict and influence each other, causing the effect of various objective evaluation parameters on the subjective feelings to be nonlinear. So it is very difficult to establish a complete and accurate mathematical model using conventional modeling methods, but the support vector regression machine proves to be an effective solution to this problem. Shi Yan et al. use the multiple linear regression method, which has good predictive ability, to establish a predictive model between people's subjective feelings about vehicle noise and the objective evaluation parameters. Inspired by the work of Shi Yan et al., this paper uses the support vector regression machine to establish a predictive model between people's subjective feelings about vehicle noise and the objective evaluation parameters, and conducts a study by comparing it with the result of multiple linear regression models.

2 A BRIEF INTRODUCTION OF THE PRINCIPLES AND METHODS

The Support Vector Regression Machine (SVR) is a consistent machine learning method proposed by Vapnik and his coworkers based on the statistical learning theory and the structural risk minimization principle. It is a relatively good way to solve practical problems such as small samples, nonlinearity, high dimensions, and local minima and over learning. With the features of high fitting accuracy, strong generalization ability and global optimum, the method has been successfully applied to many practical fields. The basic idea of SVR is: set the sample data set as $\{(x_1, y_1), (x_2, y_2), ..., (x_m, y_m)\}$, use a nonlinear mapping φ to map x in the sample data set to high dimensional space F, and carry out linear regression in the feature space F:

$$f(x) = \sum_{i=1}^{m} (a_i - a_i^*) k(x, x_i) + b \qquad (1)$$

$k(x, x_i) = \varphi(x) \cdot \varphi(x_i)$ is the kernel function of the formula. By choosing different kernel functions, different SVR models will be generated. The papers uses the radial basis function to establish SVR regression model:

$$k(x, x_i) = \exp\left(-\gamma (x - x_i)^2\right) \qquad (2)$$

3 THE ESTABLISHMENT OF SVR REGRESSION MODEL

3.1 Data

The data employed in this paper are from Document. This data set including 28 groups of measured samples (Table 1) is a satisfaction assessment of the exhaust noise of vehicles obtained by Shi Yan et al. through the test using the method of paired comparison of the jury. By adopting the multiple linear regression model, they carried out a prediction of the satisfaction through modeling, and the specific experimental procedure and modeling process are shown in Document.

3.2 The establishment of the model

In the modeling process of SVR for satisfaction, the five parameters of loudness, sharpness, roughness, fluctuation and kurtosis are taken as input variables, and satisfaction is taken as an output variable to conduct the learning and training of modeling. In order to compare with the prediction results of the multiple linear regression method in Document, the training samples and testing samples for the conduction of modeling prediction research adopted by this paper are completely the same with those in Document.

3.3 The evaluation of the performance of the model

Mean Absolute Percentage Error (MAPE), Root Mean Square Error (RMSE) and correlation coefficient (R^2) are used to evaluate the predictive performance of the established model, which is defined as:

$$MAPE = \frac{1}{n}\sum_{j=1}^{n}\left|\frac{\hat{y}_j - y_j}{y_j}\right| \tag{3}$$

$$RMSE = \sqrt{\frac{1}{n}\sum_{j=1}^{n}(\hat{y}_j - y_j)^2} \tag{4}$$

Table 1. Measured values.

Samples SN	Loudness L/sone	Sharpness S/asum	Roughness R/asper	Fluctuition F/	Kurtosis K	Satisfaction P
1	47.9	1.71	0.42	0.13	0.06	4.2
2	46.3	1.78	0.46	0.18	0.02	5.4
3	23.4	1.75	0.95	0.13	1.79	28.9
4	22.8	1.62	0.57	0.19	2.27	28.5
5	29.3	1.72	0.14	0.14	−0.11	21.9
6	45.7	1.64	0.26	0.12	−0.52	9.4
7	39.3	1.75	1.15	0.11	0.14	13.1
8	49.4	1.64	0.34	0.15	−0.35	5.4
9	30.6	1.24	2.18	0.06	−0.94	19.8
10	27.7	1.18	1.92	0.11	−0.37	22.6
11	36.9	1.41	0.94	0.09	−0.01	14.0
12	29.0	1.30	1.14	0.10	0.61	24.4
12	29.0	1.30	1.14	0.10	0.61	24.4
13	47.2	1.90	0.29	0.11	−0.12	2.2
14	26.1	1.47	0.50	0.09	−0.49	21.3
15	25.3	1.44	1.21	0.08	−0.18	23.8
16	46.5	1.51	1.10	0.24	−0.23	11.0
17	25.5	1.81	2.09	0.08	−0.32	25.3
18	30.6	1.58	1.00	0.12	0.31	23.9
19*	41.23	1.68	0.69	0.13	0.24	11.0
20*	36.84	1.51	0.95	0.16	0.48	18.4
21*	31.37	1.76	0.53	0.09	−0.27	18.9
22	45.10	1.62	0.70	0.21	0.10	9.10
23*	24.62	1.50	1.63	0.15	−0.35	28.9
24*	29.32	1.24	1.06	0.10	1.01	24.3
25*	35.58	1.45	0.82	0.17	0.63	21.2
26*	48.60	1.19	1.20	0.16	−0.43	7.50
27*	43.93	1.80	0.21	0.19	−0.72	8.70
28*	27.87	1.47	0.92	0.12	0.11	24.6

*is a testing sample.

$$R^2 = \frac{\left(\sum_{j=1}^{n}(\hat{y}_j - \bar{\hat{y}})(y_j - \bar{y})\right)^2}{\sum_{j=1}^{n}(\hat{y}_j - \bar{\hat{y}})^2 \sum_{j=1}^{n}(y_j - \bar{y})^2} \qquad (5)$$

In the formula, n is the number of testing samples. y_j and $\hat{y}_j$ are the target value and predictive value of the testing samples respectively. $\bar{y}$ and $\bar{\hat{y}}$ are the target mean value and predictive mean value of the testing samples respectively.

4 THE ANALYSIS AND DISCUSSION OF THE RESULTS

The previous 18 groups are used as training samples to establish SVR predictive model. In order to verify the effectiveness of the SVR model, the last 10 groups are used as the predictive samples of other samples which are prepared. Through the established model, the predictive values of all the satisfaction are calculated and compared with the measured values, and the predictive values of the multiple linear regression method. Table 2 shows the psychoacoustic parameters of ten predictive samples, as well as the measured values of the sound quality satisfaction and the predictive values of two predictive models. It can be seen from Table 2 that the sample satisfaction prediction result produced by the subjective valuation predictive model for the exhaust noise of vehicles which SVR established is relatively ideal. All the MAPEs of the predictive values are lower

Table 2. Comparison between the predictive values and measured values.

Samples SN	Meas-ured	SVR prediction		Multiple regression predition	
		Predictive value	Error %	Predictive value	Error %
19	11.0	11.09	0.82	11.34	3.1
20*	18.4	19.15	4.07	16.67	−9.4
21*	18.9	18.58	−1.70	18.38	−2.8
22*	9.10	9.77	7.38	10.15	11.5
23*	28.9	27.52	−4.77	27.36	−5.3
24*	24.3	25.04	3.05	22.51	−7.4
25*	21.2	21.15	−0.23	18.11	−14.6
26*	7.50	7.500	0.00	6.73	−10.2
27*	8.70	9.23	6.15	9.46	8.8
28*	24.6	25.38	3.17	23.34	−5.2

Table 3. Comparison of two prediction performances.

Regression method	Satisfaction P		
	MAPE%	RMSE	R^2
MLR	8.83	1.99	0.96
SVR	3.13	0.67	0.99

than 8%. Among the 10 samples, the maximum error is 7.38%(26*), 4 MAPEs are lower than 2%, and the others are between 3% and 6%, with the changes being small. However, 60% of the predictive values of the multiple linear regression model have errors that are lower than 10%, with 14.6% being the biggest and 2.8% being the smallest. The deviation fluctuates in a relatively wide range, making the predictive result unstable. The result of the comparison between the predictive values and the measured values of the model shows that the predictive ability of the SVR model is superior to that of the multiple linear regression models, and the former can better reflect non-linear relationship between objective parameters and satisfaction.

Table 3 shows the statistics of MAPE, RMSE and R^2, which are calculated using two methods, about the satisfaction of the samples produced by the subjective evaluation prediction model of the exhaust noise of vehicles. Table 3 indicates that MAPE of SVR of the predictive sample is 3.13, which is much smaller than the 8.83 of MLR. RMSE is 0.67, which is also much smaller than the 1.99 of MLR. The correlation coefficient R^2 it obtained is 0.99, which is closer to 1 than 0.96 of MLR. It fully shows that the predictive model of SVR is more accurate than MLR. Its generalization ability is also stronger, and it is more effective.

5 CONCLUSION

According to the measured data set of satisfaction according to different objective psychoacoustic parameters, we used the support vector regression machine to establish the predictive model for the sound quality of the exhaust noise of vehicles, and compared it with the multiple linear regression models. The result shows that there is a non-linear relationship between objective parameters and subjective satisfaction. The impact of objective parameters on subjective satisfaction is very complex. The predictive value of the SVR model is closer to the measured value, with smaller prediction error, better prediction accuracy and stronger generalization ability.

REFERENCES

Cai C Z et al. 2003. SVM-Prot: web-based support vector machine software for functional classification of a protein from its primary sequence [J]. Nucleic Acids Research, 31: 3692–3697.

Jiao F L et al. 2006. Listener clustering and subject assessment of car interior noise quality [J]. Technical Acoustics, 25 (6): 568–572 (in Chinese).

Liao J P et al. 2010. Elementary Introduction Noise Control of Automobile [J]. Design, Computational research. 1:15–19.

Pei J F et al. 2012. Prediction on the glass transition temperature of styrenic copolymers by using SVR [J].

Journal of Macromolecular Science Part B: Physics, 51(7): 1437–1448.

Shi Y et al. 2010. Sound Quality Prediction of Vehicle Exhaust Noise Based on Neural Network. Transactions of the Chinese Society for Agricultural Machinery [J]. 41(8): 16–20.

Shin S H et al. 2009. Sound quality evaluation of the booming sensation for passenger cars [J]. Applied Acoustics, 2009, 70(2): 309–320.

Tang J L et al. 2012. Support vector regression model for direct methanol fuel cell [J]. Int J Mod PhysC, 2012, 23(7): 1250055.

Vapnik V. The nature of statistical learning theory [M]. New York: Springer, 1995.

Automotive, Mechanical and Electrical Engineering – Liu (Ed.)
© 2017 Taylor & Francis Group, London, ISBN 978-1-138-62951-6

A simulation analysis of the exhaust muffler basic unit transmission loss

Daolai Cheng
School of Railway Transportation, Shanghai Institute of Technology, Shanghai, China

Jianxiang Fan & Linzhang Ji
School of Mechanical Engineering, Shanghai Institute of Technology, Shanghai, China

ABSTRACT: The basic unit is often used to reduce the noise of a certain frequency or a certain frequency segment. In this paper, an acoustic finite element analysis is carried out on the transmission loss of the muffler, such as the expansion ratio of the muffler, the inner tube, the number of the expansion chamber and so on. HyperMesh divides the mesh of mufflers inner region, defining acoustic grid, posting treatment and calculates the muffler transmission loss. Through such methods, we can get the effect of the muffler transmission loss under the circumstance of the different basic unit of muffler.

Keywords: Muffler; Acoustic finite element; HyperMesh; Transmission loss

1 INTRODUCTION

The evaluation index of an exhaust muffler's basic unit is transmission loss, insertion loss, and sound pressure level difference. Transmission loss is mainly used to evaluate the performance of the basic unit of the muffler (Pang, 2013). With the rapid development of computers and the emergence of noise simulation software, such as LMS Vitual. Lab ANSYS, the foundation was laid for the analysis of the muffler performance. In this paper, the simulation is mainly aimed at the transmission loss of the basic unit, including the single expansion muffler, the expansion ratio of the expansion cavity, the length of the expansion cavity, the number of the expansion chamber and so on.

2 SINGLE EXPANSION MUFFLER TRANSMISSION LOSS

2.1 Single expansion muffler

The single expansion muffler is composed of a pipe which is mainly composed of an expansion chamber and two sides connected with the two sides (Pang, 2013), as shown in Figure 1.

The section areas of inlet pipe 1 and outlet pipe 3 are S_1 and S_3 respectively, and the cross section area of the middle expansion tube 2 is S_2. Due to the change of the section area of the pipe, the internal impedance of the muffler is changed. Because of meeting part of the cross-sectional area change, when the sound waves along the inlet pipe ahead, resulting in a part of sound be

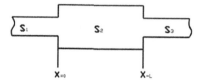

Figure 1. Single expansion muffler.

reflected back, and another part of the sound forward propagation (Pang, 2013; Beranek, 1954). When $S_1 = S_3$, we can obtain the transmission power coefficient:

$$T_w = \cfrac{4}{4+\left(\cfrac{S_1}{S_2}-\cfrac{S_2}{S_1}\right)^2 \sin^2 kl} = \cfrac{1}{1+\left(\cfrac{S_1}{S_2}-\cfrac{S_2}{S_1}\right)\sin^2 kl}$$

(1)

The transmission loss of the expanded muffler is:

$$TL = 10\lg\frac{1}{Tw} = 10\lg\left[1+\frac{1}{4}\left(\frac{S_1}{S_2}-\frac{S_2}{S_1}\right)^2 \sin^2 kl\right],$$

$$\left(\frac{S_2}{S_1} = m, m \text{ is expansion ratio}\right)$$

(2)

2.2 Establishment of physical size model of single expansion muffler

The model size of the simple expansion muffler is as follows: the diameters of inlet pipe 1 and outlet

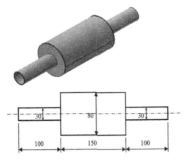

Figure 2. Physical geometry model of muffler.

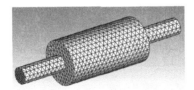

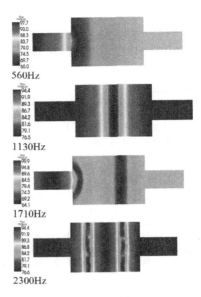

Figure 4. Sound pressure cloud image.

560Hz

1130Hz

1710Hz

2300Hz

Figure 3. Internal area grid.

pipe 3 are 30 mm and the diameter of the expansion tube 2 is 80 mm.

2.3 Acoustic finite element model of single expansion muffler

The internal construction grid of the muffler is divided by HyperMesh. For the division of the acoustic mesh, the accuracy of the acoustic computation is controlled by the majority of units. So the local mesh is too small to improve the accuracy of the calculation. However, at the same time, the grid is too coarse to produce large errors. It is generally assumed that there are 6 units in the minimum wave length. The internal area grid is shown in Figure 3.

The inner sound pressure of the muffler was calculated by the acoustic software. Its fluid material is defined as air, sound velocity 340 m/s, mass density 1.225 kg/m³.

Defining inlet boundary conditions—define unit vibration velocity-1 m/s and exit non-reflecting boundary condition, acoustic impedance 416.5 kg/(m²·s) (Zhang, 2013; Selameta, 2003). The sound pressure of the muffler is shown in Figure 4:

As is revealed in the picture, at the frequency of 560 Hz and 1710 Hz, the sound pressure at the outlet is the lowest, and the noise elimination is the highest. But at the frequencies of 1130 Hz and 2300 Hz, the sound pressure of the inlet and outlet is almost no different. So, it has no effect on noise elimination.

3 THE INFLUENCE ON THE MUFFLER BASIC UNIT TRANSMISSION LOSS

3.1 The effect of the expansion ratio m on the transmission loss

As is revealed in the picture 1, the muffler expansion ratio m = 7. On the basis of this, the influence of the expansion ratio on the transmission loss is discussed.

We can see the single expansion muffler transmission loss under different expansion ratios. As shown in Figure 5.

As is revealed in the picture, when the expansion ratio increases, the peak value of the transmission loss increases with the increase of m. At the same time, the expansion ratio has no effect on the pass frequency.

3.2 The effect of expansion chamber length on the transmission loss

By changing the length of the expansion chamber, we can learn the effect on the transmission loss, which is under the case of a certain expansion ratio (Munjal, 2008). As shown in Table 2.

We can see the effect of the expansion chamber length on the transmission loss. As shown in Figure 6.

With the increase of the length of the expansion chamber, the noise reduction frequency moves to the low frequency. But the peak value of the transmission loss is not changed. Therefore, we can conclude that the variation of the length of the muffler

Table 1. Single expansion muffler with different expansion ratios.

Number	Scheme 1	Scheme 2	Scheme 3	Scheme 4	Scheme 5	Scheme 6
1	m = 7	m = 9	m = 16	m = 25	m = 36	m = 49

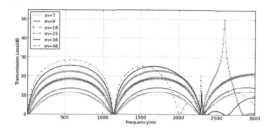

Figure 5. Different expansion ratio.

Table 2. The length of expansion chamber.

Number	Scheme 1	Scheme 2	Scheme 3	Scheme 4
1	L = 160 mm	L = 200 mm	L = 320 mm	L = 400 mm

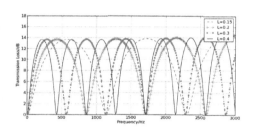

Figure 6. Different expansion chamber lengths.

only affects the noise elimination frequency and the pass frequency, but has no effect on the peak value of the transmission loss.

3.3 The effect of expansion chamber number on the transmission loss

We build two different numbers of expanding expansion chambers (Yuan X, 2009), its specific parameters are as shown in Figure 7.

Through the simulation analysis, the transmission loss curve is as shown in Figure 8.

As is revealed in the picture, the transmission loss of the double expansion chamber muffler is still higher than the single one under the condition of the same cavity volume. It can effectively inhibit the single expansion chamber muffler from producing the pass frequency.

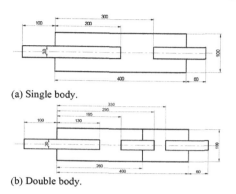

(a) Single body.

(b) Double body.

Figure 7. Two different numbers of expanding expansion chambers.

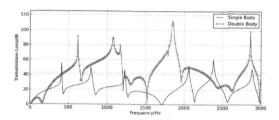

Figure 8. Different transmission loss curve.

4 CONCLUSION

With the increase of the expansion ratio, the transmission loss will increase. However, there is no effect on the pass frequency. With the increase of the length of the expansion chamber, the variation of the length of the muffler only affects the noise elimination frequency and the pass frequency, but has no effect on the peak value of the transmission loss. The double expansion chamber can not only improve the overall noise reduction of the muffler but it can also effectively inhibit the single expansion chamber muffler from producing the pass frequency. Acoustic finite element method is a kind of advanced calculation method, which is compared with the experience formula calculation in the past.

ACKNOWLEDGMENTS

This work was financially supported by the laboratory construction project of Shanghai Institute of Technology (No. 10210D150008) fund.

REFERENCES

Beranek L. (1954). Acoustics. McGraw Hill.

Munjal M.L.K (2008). Muffler Acoustics [M]. *Formulas of Acoustics*, 796–798.

Pang J., Chen, G., He, H. (2013). Automotive noise and vibration. 1st edn. Beijing: Beijing Institute of Technology Press. Reference to a chapter in an edited book.

Selameta, A., Deniab, A.J. (2003). Acoustic behavior of circular dual-chamber mufflers. *Journal of Sound and Vibration*, *265*, 967–985.

Yuan X., Liu, Z.S., Bi, R., and (2009). The contrast of simple expanded-muffler with large porosity perforated muffler. *Journal of AUTO SU-TECH*, 4, 18–21.

Zhang, F.L., Xu, J.W. (2013). Vitual. Lab Acoustics acoustic simulation from entry to master, 1st edn. Xi'an: Northwestern Polytechnical University Press.

Automotive, Mechanical and Electrical Engineering – Liu (Ed.)
© 2017 Taylor & Francis Group, London, ISBN 978-1-138-62951-6

The Weibull model of the diesel engine based on reliability data distribution fitting

Zhongzheng Liu, Guanfeng Wang, Qiang Zhang & Xueying Li
School of Mechanic Engineering, Shandong University, Jinan, China

ABSTRACT: Reliability improvement of the diesel engine is a systematic and probabilistic assessment that needs a large amount of useful reliability data for this study. Collected diesel engine failure records are taken as data samples, and then the two-parameter Weibull distribution fitting model is used for failure mileage distribution. The parameters are estimated by using the linear regression and maximum likelihood estimation methods, and the parameters are verified to come to an agreement with the collected data of the Weibull model.

Keywords: Diesel engine reliability; Weibull distribution; Parameter estimation

1 INTRODUCTION

Due to the harsh and complex working environment of the engine, the failure rate of the engine in the whole system reached 30% (Shaoliang Long, 1994). Therefore, the engine not only had good performance but also had high reliability. The automotive engine reliability theory was developed from the extension of electrical engineering reliability, but due to the compact nature and complexity of the vehicle engine design, and also the harshness and variability of the environment in which it is used, this theory is widely applied to engine products (Wang, Z., 2010).

There are several mature methods of engine reliability research. The dynamic reliability model proposed by Wang uses time as the life index. By changing the load, material and size of the part, the dynamic reliability model of the dynamic load under different strengths is obtained (Wang, Z., 2008). Another means is widely used on the basis of the reliable characteristics of parts. Parts in the structure and dynamic load have their own characteristics, and should be combined with its characteristics to design. For instance, referring to the reliable characteristics of connecting rods and cylinders, Shanxi Institute of Automotive Engine used the reliability prediction method instead of the safety factor method for design, and achieved good results (Xiang, Y., 2008). The actual value of the reliability of the engine system depends on the degree of failure association (Xie Liyang, 2004).

On the basis of the fault correlation theory, the Weibull distribution model is used to fit the data distribution of the reliability data. The maximum likelihood estimation method and the linear regression method are used to estimate and test the two parameters by the least squares method.

2 ESTABLISHMENT OF WEIBULL DISTRIBUTION

2.1 Data collation

In order to understand the probability distributions of the reliability failure data directly, the data are divided into several groups.

As shown in Table 1, the fault data are sorted into 20 groups in ascending order, and the frequency of each group of fault data are calculated, where the mileage is t [220,115486].

Secondly, the probability distribution density of the fault data is plotted on the coordinate paper by using the median of each interval as the x-axis and the calculated probability density p (t) as the y-axis.

The probability density is calculated as follows:

$$\hat{p}(t) = \frac{n_i}{n\Delta t_i} \qquad (1)$$

where n_i is the number of failures in each group, n is the total number of faults and Δt_i is group spacing, by dividing the mileage into 20 groups, the group spacing is 5,763.3 km.

As shown in Figure 1, the probability density value of the engine failure data decreases with the increase of the mileage, the trend is gradually flat, and the mileage growth in the process of use

Table 1. Faulty data packets.

Group	The median value	Quantity	Frequency
1	3101.65	596	0.199264
2	8864.95	233	0.0779
3	14628.25	257	0.085924
4	20391.55	214	0.071548
5	26154.85	190	0.063524
6	31918.15	181	0.060515
7	37681.45	141	0.047141
8	43444.75	142	0.047476
9	49208.05	136	0.04547
10	54971.35	170	0.056837
11	60734.65	117	0.039117
12	66497.95	91	0.030425
13	72261.25	95	0.031762
14	78024.55	85	0.028419
15	83787.85	78	0.026078
16	89551.15	83	0.02775
17	95314.45	92	0.030759
18	101077.8	35	0.011702
19	106841.1	28	0.009361
20	112604.4	27	0.009027

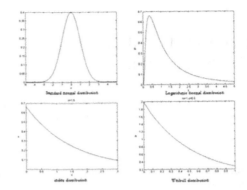

Figure 2. Probability density plots for common distributions.

resulting from component failures can be described with Weibull distribution as follows:

$$f(t) = \frac{m}{\eta}\left(\frac{t-\gamma}{\eta}\right)^{m-1} e^{-\left(\frac{t-\gamma}{\eta}\right)^m} \quad t \geq \gamma \tag{2}$$

where η is the scale parameter, and $\eta > 0$. η does not affect the starting position and shape of the Weibull probability density curve, but the abscissa of the curve is scaled, and the relative position and the degree of dispersion of the curve in the coordinate system are transformed. γ is a position parameter, and $\gamma > 0$. γ does not affect the shape of the Weibull probability density curve, only the starting point position of the abscissa of the curve in the coordinate system is changed.

In the process of studying Weibull distribution, we often make $\gamma = 0$.

The probability distribution density is:

$$f(t) = \frac{m}{\eta}\left(\frac{t}{\eta}\right)^{m-1} e^{-\left(\frac{t}{\eta}\right)^m} \tag{3}$$

The probability distribution function is:

$$F(t) = 1 - e^{-\left(\frac{t}{\eta}\right)^m} \tag{4}$$

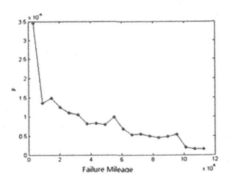

Figure 1. Probability density distribution trend of failure mileage.

is a continuous process. The probability density is roughly a continuous monotonically decreasing function of the mileage. In contrast to Figure 2, it is clear that the probability density distribution of the reliability failure mileage does not match the trend of the probability density function image of normal distribution or logarithmic normal distribution, which firstly increases and then decreases. Therefore, the distribution of failure miles is unlikely to be a normal or lognormal distribution, and is more likely to be a Weibull distribution or an exponential distribution.

2.2 Weibull distribution

The Weibull distribution consists of three parameters: position, shape and scale. A system failure

2.3 Parameter estimation

2.3.1 Linear regression method

Linear regression is a method to optimise parameters by establishing the minimum deviation between the observed value of the sample and the estimated value of the parameter to the objective function. Abstract frame

First, establish a linear equation:

$$y = ax + b \tag{5}$$

The estimated value of parameters a and b are calculated by substituting the sample observation values $(x_1, y_1), (x_2, y_2), \ldots, (x_n, y_n)$ into the Equation (5). Each (x_i, y_i) can yield a linear regression parameter estimation:

$$\hat{y}_i = \hat{a}x_i + \hat{b} \quad (i = 1, 2, \ldots, n) \tag{6}$$

The difference between the sample observation value y_i and the parameter estimation value $\hat{y}_i$ is:

$$y_i - \hat{y}_i = y_i - \hat{a}x_i - \hat{b} \quad (i = 1, 2, \ldots, n) \tag{7}$$

Equation (7) expresses the degree of deviation between the estimated value $\hat{y}_i$ and the observed value y_i. The smaller the deviation, the better the fitting degree between the estimated value and the actual value, and the more accurate the result is.

The concept of residual r is used to denote the cumulative sum of squared deviations:

$$r(a, b) = \sum_{i=1}^{n} (y_i - \hat{a}x_i - \hat{b})^2 \tag{8}$$

The least square method is adopted to obtain the optimal solution of the estimated value:

$$r(\hat{a}, \hat{b}) = \min r(a, b) \tag{9}$$

By solving the partial derivative of r with respect to a, b, and making it zero, we get:

$$\begin{cases} \dfrac{\partial r}{\partial a} = -2\sum_{i=1}^{n}(y_i - ax_i - b)x_i = 0 \\ \dfrac{\partial r}{\partial b} = -2\sum_{i=1}^{n}(y_i - ax_i - b) = 0 \end{cases} \tag{10}$$

Solution can be obtained:

$$\begin{cases} \hat{a} = \left(\sum_{i=1}^{n} x_i y_i - n\overline{xy}\right) \Big/ \left(\sum_{i=1}^{n} x_i^2 - n\overline{x}^2\right) \\ \hat{b} = \overline{y} - \overline{x}\hat{a} \end{cases} \tag{11}$$

By linearising the Weibull distribution of the two parameters, the data points can be approximated by straight line fitting. From the probability distribution function of the Weibull distribution, we get:

$$\frac{1}{1 - F(t)} = e^{\left(\frac{t}{\eta}\right)^m} \tag{12}$$

Both sides take the logarithm at the same time:

$$\ln\ln\frac{1}{1 - F(t)} = m\ln t - m\ln\eta \tag{13}$$

In the linear regression fitting calculation, we must first verify the linear relationship between the variables x and y, calculated by Equation (14):

$$\rho = \frac{\sum_{i}^{n} x_i y_i - n\overline{xy}}{\sqrt{\left(\sum_{i=1}^{n} x_i^2 - n\overline{x}^2\right)\left(\sum_{i=1}^{n} y_i^2 - n\overline{y}^2\right)}} = 0.9885 \approx 1 \tag{14}$$

Therefore, the calculation can be a = 0.8924, b = −9.5455. Then it can be drawn that m = 0.8924, $\eta = 44{,}198.08$.

2.3.2 Maximum likelihood estimation

In the experiment, the probability function of the population is p(t; θ), θ ∈ A, where θ is the vector composed of several unknown parameters, A is the vector space of the parameters, which can be obtained by θ. $x_1, \cdots, x_n$ is the observed value of the sample in the test, since the samples in the test are from the population, and when the total truth value is unknown, the value at which the probability of occurrence of the observation in the possible maximum of the possible values of θ is taken as the estimated value estimate. In this case, the probabilities of the joint samples are treated as a function of θ as a likelihood function of the sample, denoted as $L(\theta; x_1, \cdots, x_n)$.

Usually L(θ) is abbreviated:

$$L(\theta) = L(\theta, x_1, \cdots, x_n) = p(x_1; \theta) \cdot p(x_2; \theta) \cdots$$
$$p(x_n; \theta) = \prod_{i=1}^{n} p(x_i; \theta) \tag{15}$$

Then the probability of the sub-sample in the adjacent domain can be estimated as:

$$p = \prod_{i=1}^{n} p(x_i; \theta)\Delta x_i \tag{16}$$

In order to find the maximum value of $\hat{\theta} = \hat{\theta}(x_1, \cdots, x_n)$, since the change of Δx_i is not affected by the function θ, it must satisfy:

$$L(\hat{\theta}) = \max_{\theta \in A} L(\theta) \tag{17}$$

Then $\hat{\theta}$ is the maximum likelihood estimation of θ. When we need to determine the unknown parameters of the distribution model as $\theta_1, \cdots, \theta_m$, we can get the expression of maximum likelihood estimation:

$$L'\left(\hat{\theta}\right) = \prod_{i=1}^{n} p\left(x_i; \theta_1, \cdots, \theta_m\right) \qquad (18)$$

By making the Weibull distribution probability density expression into (18), we can get:

$$L'\left(x; m, \eta\right) = \prod_{i=1}^{n} p\left(x_i\right) = \prod_{i=1}^{n} \frac{m}{\eta}\left(\frac{t}{\eta}\right)^{m-1} e^{-\left(\frac{t}{\eta}\right)^m} \qquad (19)$$

Since lnx is a monotonic increasing function of x, the maximum value for L(θ) is equivalent to the maximum for lnL(θ). Therefore, it is often used to solve the maximum likelihood estimation of θ from inL(θ), and we can get the likelihood equation, that is:

$$\frac{dL\left(\theta\right)}{d\theta} = 0 \; or \; \frac{dlnL\left(\theta\right)}{d\theta} = 0 \qquad (20)$$

Take the logarithm of both sides. Then let the first partial derivative of m, η be 0, and the likelihood estimate of the parameter:

$$L\left(x; m, \eta\right) = lnL^{\prime}\left(x; m, \eta\right) = nln\left(\frac{m}{\eta}\right) +$$
$$\left(m-1\right)\sum_{i=1}^{n} \ln\frac{x_i}{\eta} - \sum_{i=1}^{n}\left(\frac{x_i}{\eta}\right)^m$$

$$\begin{cases} \dfrac{\partial L}{\partial m} = \dfrac{n}{m} + \sum_{i=1}^{n} lnx_i - \dfrac{1}{\eta}\sum_{i=1}^{n}\left[x_i{}^n lnx_i\right] = 0 \\ \dfrac{\partial L}{\partial \eta} = -\dfrac{n}{\eta} + \dfrac{1}{\eta^2}\sum_{i=1}^{n} x_i{}^m = 0 \end{cases} \qquad (21)$$

Solving the equation can obtain the likelihood estimated value of m, η. According to the maximum likelihood estimation method, two parameters can be obtained, which are m = 1.0480, η = 37068.93. The linear regression method and maximum likelihood method of the Weibull distribution parameters obtained are shown in Table 2.

2.4 Parameter test

By Glevenchenko theorem:

$$P\{\lim_{n\to\infty} sup_{-\infty}^{\infty}\left|F\left(x\right)-F_n\left(x\right)\right| = 0\} = 1 \qquad (22)$$

Table 2. Linear regression, Maximum likelihood estimation.

	m	η
Linear Regression	0.8924	44198.08
Maximum Likelihood	1.0480	37068.93

That is, $F_n(x)$ is a sufficient estimate of F(x) when the number of samples n is large enough.

Let $x_1, \cdots, x_n$ be the sample observations of the continuous distribution function $F(x)$. Assuming $H_0 : F\left(x\right) = F_0\left(x\right)$, alternative hypotheses $H_1 : F\left(x\right) \neq F_0\left(x\right)$.

Inspection procedure:
1. Assuming that the assumption $H_0 : F(x) = F_0(x)$ is accepted, the alternative hypothesis $H_1 : F(x) \neq F_0(x)$ holds;
2. Building random variables:
   The maximum deviation D_n between the theoretical distribution $F(x)$ and the empirical distribution $F_n(x)$ is taken as the difference degree:

$$D_n = sup_{-\infty}^{\infty}\left|F\left(x\right) - F_n\left(x\right)\right| = \max_{1 \leq i \leq n} \delta_i \qquad (23)$$

3. Given the significance level α, we get the rejection domain, where we take α = 0.05 $P\{D_n > D_{n,\alpha}\} = \alpha$, then the rejection domain is $D_n > D_{n,\alpha}$
4. $D_{n,\alpha}$ and D_n are compared, if $D_n > D_{n,\alpha}$ the assumption $H_0 : F\left(x\right) = F_0\left(x\right)$ is rejected, or the assumption is accepted.

The K-S test is used to estimate the parameters obtained by the two estimation methods. The results are shown in Table 3.

It can be seen that the D_n of the parameters estimated by the linear regression and maximum likelihood methods are less than $D_{(n,\alpha)}$, therefore the original hypothesis H_0 is accepted, which means that the faulty mileage data of the two methods are in line with the two parameters of Weibull distribution. At the same time, it can be obtained that the D_n of the linear regression method is smaller than that of the maximum likelihood method, and

Table 3. K-S Test of parameter estimation.

	D_n	$D_{n,\alpha}$
Linear Regression	0.0428	0.2980
Maximum Likelihood	0.1075	

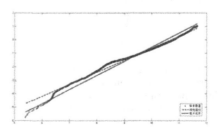

Figure 3. Graphical estimation of parameter estimation.

whether the estimated distribution in the linear regression method is closer to the distribution of the actual data sample needs further research.

The results of the two methods are tested on the Weibull probability map, and it can be found that the estimated values of the two methods can generalise well the distribution trend of the sample data, and the fitting effect of the linear regression is better.

3 CONCLUSION

We mainly focus on the failure mileage reliability modelling of the diesel engine. The first step is to establish a two-parameter Weibull distribution model. The distribution trend of the two-parameter Weibull distribution model is obtained by arranging the data, and the two-parameter Weibull distribution model is selected as the fitting model. Then the parameters of the distribution of the least squares method and the maximum likelihood method are prepared to estimate a third method of K-S test, a graph plotted by these two methods, which estimates the value of the parameters in a direct way and proves that the distribution model fits the engine failure data well.

REFERENCES

Lianglong Shao, (1994). Research on life distribution law of diesel engine, 1994, (2), 55–58.

Toscano, R., & Lyonnet, P. (2008). Online reliability prediction via dynamic failure rate model, 2008, 57(3), 452–457.

Wang, Z. (2008). *Theory and Method of Dynamic Reliability Modeling of Components and Systems*. Shenyang: Northeastern University.

Wang, Z. (2010). *Failure Rate Calculation of Parts and Systems Based on Load—Strength Interference*. Beijing: Beihang University, 35–37.

Xiang, Y., Chen, G., & Ding, L. (2008). Reliability design and experiment of engine, 2008, 33(2), 131–133.

Xie, L., & Zhou, J. (2004). System level load-strength interference based reliability modeling of auto fan system. *Reliability Engineering and System Safety*, 2001(3), 33–37.

Mechanical, manufacturing, and process engineering

Automotive, Mechanical and Electrical Engineering – Liu (Ed.)
© *2017 Taylor & Francis Group, London, ISBN 978-1-138-62951-6*

Analysis of metal structure cracks on the miter gate of the ship lock

Panpan Zhang
China Waterborne Transport Research Institute, Beijing, China
Key Laboratory of Logistics Equipment and Control Engineering, Beijing, China

Jiahai Zhou
China Waterborne Transport Research Institute, Beijing, China

ABSTRACT: With the development of water transportation, the ship lock is playing a more and more important part. The miter gate is one of the most significant facilities for the ship lock. In this paper, different kinds of miter gate for the ship lock gate were introduced, and the characteristics of the miter gate were summarised. The reasons for the metal structure crack were analysed and measures were put forward to solve the problem.

Keywords: Metal structure; Crack; Miter gate; Ship lock; Stress gauge

1 INTRODUCTION

As inland water transportation develops, ship locks are playing a more and more important role in this field. There are advantages for waterborne transportation, such as low cost and large volume. Heretofore, over 2000 ship locks have been built and put into use in China. Also, thousands of ship locks are working all around the world. Taking the Three Gorges lock as an example, the capacity of the lock was over 100 million tons in the year 2014, shown in Figure 1.

Metal structures represented by miter gates are widely used in water and hydraulic power engineering and their health will directly affect the safety of the ship lock. The miter gate plays a significant part in the navigation of inland water. A metal structure crack is the main type of failure that greatly affects the safety of the miter gate, and it is becoming more and more important to monitor the working condition of the miter gate for the ship lock. In this paper, different kinds of miter gate for the ship lock gate were introduced. Characteristics of the miter gate were summarised. The reasons for the metal structure crack were analysed and measures were put forward to solve the problem.

2 STRUCTURE OF THE MITER GATE

The miter gate is now widely used in the ship lock and there are different kinds of miter gates shown in Figure 2, which are the flat type and circular arc

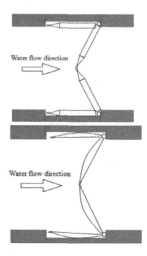

Figure 1. The Three Gorges ship lock.

Figure 2. Different structures of miter gates.

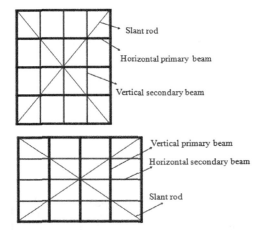

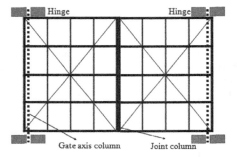

Figure 3. Different flat type miter gates.

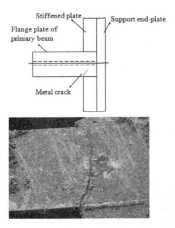

metal cracks can be found in the conjunction of the primary beam flange plate and the stiffened plate. At the same time, the crack occurs a lot in the underwater part. When cracks are found, measures such as repair welding, as shown in Figure 6, should be taken to avoid further damage to the metal structure.

In order to solve the problem, the crack has to be cleaned and re-welded. In addition, reinforcing plates, as shown in Figure 7, are welded to improve the situation of the miter gate.

However, metal cracks still emerged in the connection area between the reinforcing plate and the flange plate of the primary beam, as shown in Figure 7. According to comparison between different

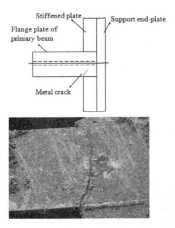

Figure 5. The crack area on the miter gate.

Figure 4. The schematic diagram of miter gates.

type. The application of the circular arc type miter gate is limited because of the disadvantages of bad gate stiffness and bigger gate recess.

According to the arrangement of the beams, there are two kinds of flat type miter gates, as shown in Figure 3, which are the horizontal primary beam gate and the vertical primary beam gate. The vertical primary beam miter gate is usually applied in a relatively wide ship lock.

The horizontal primary beam miter gate was used in the Gezhouba ship lock. As is shown in Figure 4, there are a gate axis column and a joint column on each miter gate. The miter gate can rotate around the gate axis column, and the joint column is able to seal the gate in case of water leakage.

3 ANALYSIS OF MITER GATE CRACKS

In the manufacturing of the miter gate, welding techniques are widely used. And thus, through several years of application, cracks happen in the connection of different metal plates. As is illustrated in Figure 5,

Figure 6. Repair welding of the metal crack.

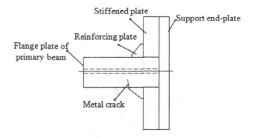

Figure 7. The reinforcement of the miter gate structure.

miter gates in different ship locks of the Gezhouba, stress concentration in the welding area was found to be the main reason for this. In the design of a miter gate, stress concentration should be taken into consideration and measures should be taken to reduce the structure stress concentration.

Metal structure cracks have a great influence on the safety of the miter gate. Therefore, it is necessary to monitor the crack growth under the water.

Different methods of metal structure monitoring, such as acoustic emission testing, are shown in Figure 8. A stress gauge, eddy current testing and so on can be used. But for underwater structure monitoring, a reliable method has to be chosen. Compared with different testing methods, the stress gauge is an easy way to monitor the condition of the miter gate.

As shown in Figure 9, there are different types of stress gauges, such as the electric resistance strain gauge and the Fibre Bragg Grating (FBG) stress gauge. According to the application condition of the miter gate, the light signal of the FBG stress gauge is preferable when applied to an underwater structure.

The fibre gratings have an extensive use in the fields of optical communication and optical sensing. As sensing cells, they can transduce physical quantities like strain, temperature, etc., and have attractive merits such as being small and light, resistant to corrosion, immune to electromagnetic interference, etc. Among fibre gratings, FBG

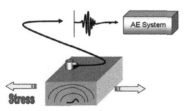

Figure 8. The schematic diagram for acoustic emission testing.

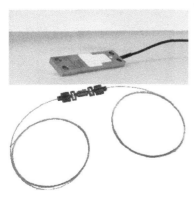

Figure 9. Stress gauges.

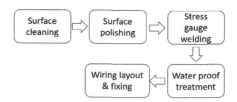

Figure 10. The procedure of stress gauge mounting.

occupies the dominant position in the sensing field and is now studied focusing mainly on the demodulation technology and novel applications.

In this paper, the fibre Bragg grating stress gauge was chosen to set up the monitoring system for the miter gate. The FBG stress gauge will be arranged along the welding seam. When the stress gauges are mounted, steps should be taken as shown in Figure 10.

The signal acquisition system is placed in the control room. Through the stress data, the condition of the miter gate can be obtained and pertinent repairs can be carried out.

4 CONCLUSION

As water transportation develops rapidly, ship locks play a huge role. The miter gate is an important element for the ship lock. Different miter gates for ship locks were introduced in this paper, and the characteristics of the flat type and the circular arc type miter gate were discussed. Taking the miter gate used in the Gezhouba, the reasons for the metal structure crack were analysed and measures were put forward to solve the problem. The FBG stress gauge can be mounted on the surface to evaluate the preliminary condition of the miter gate. It should be pointed out that the condition evaluation of the whole miter gate is a systematic project and more measures need to be taken in the future.

REFERENCES

Li, Z., et al. (2000). Fiber Bragg grating strain sensor with drift compensation for smart structures. *Proceedings of SPIE, 4223*, 131–134.

Liao, P., (2009). Review on research of lock capacity at inland waterway locks. *Hydro-science and Engineering*, 3, 34–38.

Zhang, P., & Li, H. (2015). Design of a vibration monitoring system for the hoisting machine in ship locks. *Advances in Computer Science Research*, 2164–2167.

Zhang, Y. (2013). *Safety test technology and evaluation method study on miter gates of ship lock*. Zhejiang University, 2013.

Zingoni, A. (2005). Structural health monitoring, damage detection and long-term performance. *Engineering Structures*, 27(12), 1713–1714.

Automotive, Mechanical and Electrical Engineering – Liu (Ed.)
© 2017 Taylor & Francis Group, London, ISBN 978-1-138-62951-6

Design and analysis of a wave-piercing buoy

Dong Jiang & Laihao Ma
Dalian Maritime University, Dalian, China

Zhijiang Zhang
Ningbo Navigation aid Department of Donghai Navigation Safety Administration (DNSA) MOT Zhejiang, China

Hongwei Dai
Donghai Navigation Safety Administration (DNSA) MOT Shanghai, China

Haiquan Chen
Dalian Maritime University, Dalian, China

ABSTRACT: This paper proposed the design of a wave-piercing buoy based on a columnar buoy. Six middle pontoons are used in this new type of buoy to obtain a larger wet surface area, which can increase the resistance and resilience of the buoy in the water. The buoy is made of Ultra-High Molecular Weight Polyethylene (UHMWPE) to reduce the weight, which can improve its natural frequency to avoid resonance with the waves. This paper also uses the ANSYS-Hydraulic to analyse the performance of different schemes to improve the performance and survivability of buoys.

Keywords: Wave-piercing buoy; Hydrodynamic analysis; RAO; Heave; Roll

1 INTRODUCTION

The buoy is an important facility for navigation security. It is mainly used to mark channels or obstacles to ensure the safety of navigation. In addition, the buoy is an excellent carrier as it can be equipped with sensors and a data transmission system, and work together with maritime surveillance aircraft and satellites to form a marine monitoring network. The development work of buoys in China started 1960s, where the Shandong Academy of Sciences, as the representative of scientific institutions, has developed FZF series marine data buoys and other products. However, the survivability of the existing buoys in high sea conditions is not sufficient to meet the needs of navigation and ocean monitoring. Foreign countries like America have developed NOMAD ship buoys and other mature buoys, which can work in different depths from 50–6,000 metres (WANG J CH, 2013). So there is still a gap between buoy technology in China and other foreign countries, hence there is a need for China to develop new types of buoy to ensure the viability of the buoy in hard sea conditions as well as enhance its performance.

The high stability buoy commonly used is the columnar buoy; this paper proposes the design of a wave-piercing buoy base on the aforementioned. This new type of buoy is processed with Ultra-High Molecular Weight Polyethylene (UHMWPE) to further improve its performance and survivability, thus meeting the requirements for navigability and ocean monitoring.

2 THE DESIGN OF THE STRUCTURE

2.1 The theoretical basis of the design

Buoys can be divided into various shapes: disc, column, ship and ball types. The movement of the buoy in waves involves six degrees of freedom, the main forms of motion which affect buoy performance are heave, roll and pitch. The heaving and swaying motion amplitude of the columnar buoy is smaller compared to other types of buoys as its waterline area is much smaller. This is also why the columnar buoy becomes the common high stability buoy. The Newman studied the heave inherent period of the columnar buoy and summed up the formula:

$$w_0 = \sqrt{\frac{\rho g A_0}{M}} \qquad (1)$$

In formula 1, $W'0$ is the heave natural frequency of the buoy, ρ is the density of seawater, $A0$ is the waterline area of the buoy and M is the weight of the buoy. To avoid buoy resonance with the waves, the natural frequency should be kept away from the main energy frequency of the wave (QU Shao chun, 2010). Research shows that the main energy frequency of the wave is mostly distributed in the low-frequency part, so the buoy's natural frequency should be improved to avoid resonance. According the formula 1, reducing the quality or increasing the area of the waterline, can improve the buoy's inherent period, reduce the heave motion amplitude and improve the performance significantly.

This paper puts forward the design scheme of 'wave-piercing buoy' based on the traditional columnar buoy, in which the number of intermediate pontoons will be increased, and their diameter reduced to obtain a larger wet surface area, which can increase the resistance and resilience of the buoy in the water, and reduce the swing angle.

2.2 Structural design

The difference between a wave-piercing buoy and columnar buoy is the number and size of the intermediate pontoons, therefore, the number and size of them must first be determined. This paper uses the ANSYS-Hydraulic to analyse the performance of different schemes. The main forms of motion that affect buoy performance are heave, roll and pitch; in this paper the motion response of roll and heave under the unit amplitude wave for the symmetry of structure are analysed, and the main indicator is RAO (Berteaux, 1977).

2.2.1 Determination of the number of intermediate pontoons

The number of middle pontoons of buoy is set to 4, 5, 6, 7, then the analysis was carried out on the different schemes. It was found that the value of the roll RAO of the six intermediate pontoons is only 10^{-9}, which means the buoy does not suffer violent rolling movement in the waves, while its heave RAO has a smaller amplitude which is less than one. It can be concluded therefore that there is no overlap between the buoy's natural frequency and the wave main energy frequency, so it was decided to adopt a six-pontoon design. The results of heave and roll RAO calculations are shown below:

2.2.2 Determination of the size of intermediate pontoons

In the case of the same displacement, increasing the diameter of the pontoon can improve its seakeeping, but its metacentric height will also be greater, so hydrodynamic analysis is therefore required to optimise the size of the intermediate pontoon.

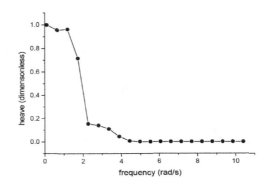

Figure 1. Heave RAO of six-pontoon design.

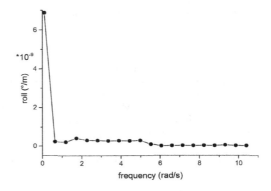

Figure 2. Roll RAO of six-pontoon design.

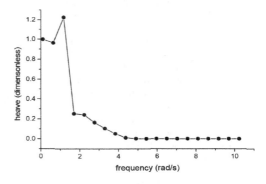

Figure 3. Heave RAO of 500 mm design.

In order to save costs and make the process easier, the diameter sizes of 500 mm, 600 mm, and 650 mm were chosen for the design scheme for hydrodynamic analysis, and the results of heave and roll RAO calculations are shown below.

The hydrodynamic analysis results of the different design schemes were compared and the data of the 600 mm scheme is excellent. Its rolling RAO

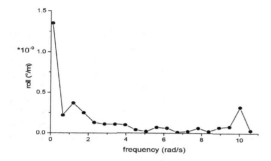

Figure 4.   Roll RAO of 500 mm design.

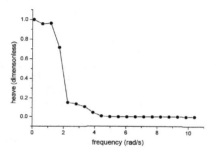

Figure 5.   Heave RAO of 600 mm design.

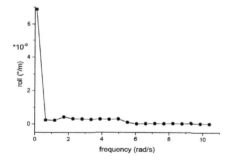

Figure 6.   Roll RAO of 600 mm design.

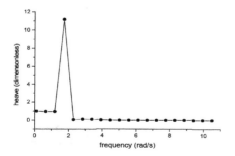

Figure 7.   Heave RAO of 650 mm design.

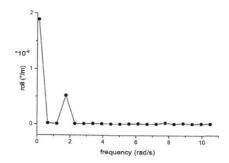

Figure 8.   Roll RAO of 650 mm design.

Table 1.   The basic parameters of the buoy.

Name	Outer diameter (mm)	Inner diameter (mm)	Length (mm)	Thickness (mm)
Middle pontoon	600	570	2600	
Terminal column	740	700	1130	
Cover plate	2400			30

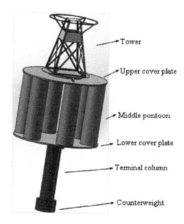

Figure 9.   The three-dimensional structure of a wave-piercing buoy.

curve is not only a small order of magnitude, only 10^{-9}, but the curve is gentler too. It was therefore decided to adopt the 600 mm design.

After determining the number and size of the intermediate pontoons, use the steel buoy HF-3.0 in the East China Sea area as a reference to determine the parameters of the wave-piercing buoy as shown in the table below:

The three-dimensional structure diagram of the wave-piercing buoy was obtained using Solidworks software.

3 THE ANALYSIS AND VERIFICATION OF THE BUOY

3.1 *Determination of parameters*

After completing the design, the basic parameters of the buoy should be calculated to ensure its reliability, including the centre of gravity, centre of buoyancy and initial stability distance.

The centre of gravity is the mass of the centre of the buoy, calculated by the product of the gravity of each component and its respective centre of gravity, then divided by the total weight of it as follows:

$$Z_G = \frac{\sum\limits_{i=1}^{n} M_i Z_i}{\sum\limits_{i=1}^{n} M_i} \qquad (2)$$

M is the gravity of the component, Z is the centre of gravity of the component.

The buoyancy of the buoy is the centroid of volume, calculated by the product of the volume of drained water and its centroid which is divided by the buoy total drainage volume as follows:

$$Z_v = \frac{\sum\limits_{i=1}^{m} V_i Z_i}{\sum\limits_{i=1}^{m} V_i} \qquad (3)$$

In formula 3, V represents the volume of the component-discharged water, and Z represents the centroid height of the discharged water.

The initial stability distance represents the distance between the transverse metacentric and the centre of gravity according to the buoy engineering, the formula for calculating the initial stability distance of the buoy is as follows:

$$\overline{Gm} = Z_v - Z_G + \frac{\pi D^4}{64V} \qquad (4)$$

In formula 4, D represents the diameter of the buoy, and V represents the total discharged volume. The parameters of the designed buoy are brought into the above formulas, and related parameters are calculated by Matlab as follows:

Table 2. The value of important parameters.

Name	Z_G	Z_V	G_m
Value(m)	2.160	2.107	0.413

3.2 *RAO response analysis*

The RAO response curves of the buoy in the six degrees of freedom are obtained by mathematical modelling and numerical simulation using the frequency domain analysis method (RyuS, 2006). The main forms of motion that affect the performance of the buoy are heave, roll and pitch, because of the buoy structure form complete symmetry in the plane of the horizon, so its rolling and pitching response is consistent. The RAO response curves of the heave and buoy of the traditional column buoy and wave-piercing buoy are given when the wave direction is 0°.

It was found from the curve that the wave-piercing buoy has an inherent period of about 5.3 seconds, which can effectively avoid the wave main energy period. The heave RAO of the wave-piercing buoy is smaller than the traditional one, the overall trend of the movement increases with decreasing wave frequency. It can be concluded that the wavy buoy has excellent performance in the heave direction when the wave period is large. Observing the roll RAO curves of the wave-piercing buoy, it is found that the magnitude of the roll direction is very small, which means there is no obvious resonance phenomenon in the rolling direction.

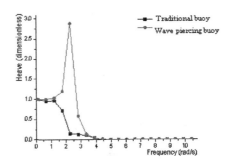

Figure 10. Heave RAO of buoy.

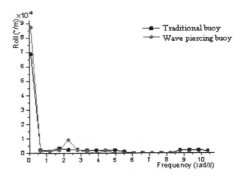

Figure 11. Roll RAO of buoy.

4 CONCLUSION

The wave-piercing buoy in this paper used multi-pontoon design, and was processed by UHMWPE, therefore its wet surface area is larger and lighter than the traditional columnar buoy. The result of hydrodynamic analysis of this new type of buoy shows that the magnitude of the RAO amplitude in the heave and roll direction is very small. It can be concluded that the performance and survivability in hard sea conditions of the wave-piercing buoy are superior to those of traditional buoys and can effectively meet the needs of navigational and ocean monitoring.

ACKNOWLEDGEMENTS

This work was supported by the Fundamental Research Funds for the Central Universities (313201 6356, 3132016345, and 3132016344) and Natural Science Foundation of Liaoning Province (20150 20132).

REFERENCES

Berteaux, Goldsmith, Schott. (1977). Heave and Roll Response of Free Floating Bodies of Cylindrical Shape. *WHOI Technical Report* 77–12.

Dai Honglei, Mou Naixia, Wand Chunyu, Tian Maoyi Development status and trend of ocean buoy in China. Meteorological, Hydrological and Marine instruments. *Draft Rep 17th IALA Conference on aids to navigation.*

Malcolm, N. *e-Navigation: the role of visual aids to navigation.*

QU Shao chun, Zheng Kun, WANG ying min. (2010). Analysis and Simulation of Spar buoy M otion. *Computer Simulation.* 006: 363–367.

Ryu, S, Duggal A.S., Heyl, C.N., et al. (2006). Prediction of deep water oil offloading buoy response and experimental validation. *International Journal of Offshore and Polar Engineering.* 16(4): 290–296.

Wang Bo, Li Min, Liu Shi xuan, Chen Shi zhel, Zhu Qinglin, Wang Hong guang. *Current status and trend of ocean data buoy observation technology applications.*

Wang J CH. (2013). Ocean data buoy: *Principles and engineering.* Beijing: Ocean Press.

Automotive, Mechanical and Electrical Engineering – Liu (Ed.)
© 2017 Taylor & Francis Group, London, ISBN 978-1-138-62951-6

Research on tenuous shaft assembly technique based on a three-dimensional pose

Kun Wei & Bingyin Ren
School of Mechatronics Engineering, Harbin Institute of Technology, Harbin, China

Liren Chai
China Academy of Aerospace Aerodynanics, Beijing, China

ABSTRACT: A tenuous shaft assembly technique based on a three-dimensional pose is proposed in this paper. Also servo feedback pose error is defined in a Cartesian space. Furthermore, a simulated model is established in MATLAB/Simulink. Simultaneously, the line tracking experiment is simulated. The result shows that the presented control method can track the desired trajectory very well. Finally, the tenuous shaft assembly experiment is carried out with a YASKAWA six-DOF Industrial manipulator, which verifies the feasibility of the proposed method. The overall process accomplishes assembly rapidly and automatically, which provides effective methods for a great variety of applications where industrial manipulators are guided by vision.

Keywords: Pose; Cartesian space; Vision guided; Industrial manipulator; Tenuous shaft

1 INTRODUCTION

1.1 Background

Nowadays, industrial manipulators guided by machine vision have been widely applied in industrial factories, such as for welding, palletising, painting and assembling, which increases manufacturing automation and working efficiency (Navarro et al., 2015) and achieves flexible manufacturing (Aviles et al., 2016). The crucial precondition for industrial manipulators guided by machine vision to complete all kinds of jobs is to estimate the pose of components and parts. Then, pose data are transmitted to robot controllers, which command end effectors to reach specific positions at a given orientation (Jia et al., 2015).

1.2 Present research

Most industrial manipulators adopting linear control methods based on joint space are independent and decoupling. Joint angles are computed through inverse kinematics after acquiring three-dimensional pose. Then, robot servo controllers require each joint to reach a desired position (Brogardh, 2007).

2 GETTING STARTED

A Cartesian space control method based on three-dimensional pose is proposed in this paper on the basis of the acquisition of the tenuous shaft pose by an orthogonal binocular vision system (Ren et al., 2017). The feedback error of this method is defined in a Cartesian space, in which pose information is integrated in the whole closed-loop control process.

The assembly task of the tenuous shaft in this research paper is to make the end slots plug in to the fixed grid plate in a certain order. The control model in the Cartesian space is set up in MATLAB/Simulink after acquiring the tenuous shaft pose. Also the line tracking experiment is simulated. Finally, the tenuous shaft assembly experiment is carried out with a YASKAWA six-DOF industrial manipulator, which verifies the feasibility of the proposed method and lays the foundation for the rapid assembly of precision shaft parts by means of industrial manipulators guided by machine vision.

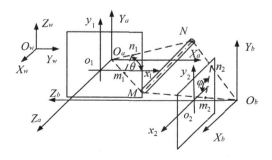

Figure 1. Model of orthogonal binocular camera for detecting the pose of shaft part.

3 POSE ESTIMATION FOR TENUOUS SHAFT

The image of the tenuous shaft end is acquired by orthogonal binocular vision in this paper. The fundamental principle of pose estimation is shown in Figure 1. The end centre point M is made the feature point. Then, the end image is processed in order to obtain the pixel position of the feature point and the orientation angle of the projection line in the image frame. Furthermore, according to the intrinsic and extrinsic parameters of the camera, the position and orientation of the point M in world frame $O_w - X_w Y_w Z_w$ are computed. The detailed calculating process can look up the literature numbered 6 in which the pose estimation methods of the tenuous shaft parts are proposed (Chai, 2016).

4 MODELLING OF MOTION CONTROL AND SIMULATION

The current pose and desired pose of the tenuous shaft gripped by the manipulator end effector are identified by the orthogonal binocular vision system. Determined pose information is fully integrated into the robot controller, which develops a Cartesian space closed-loop control system based on the three-dimensional pose. The servo error is defined in the Cartesian space. The control block diagram is shown in Figure 2.

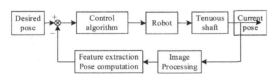

Figure 2. Control block diagram.

Related functions in the robot toolbox are encapsulated in modules and the simulation model is established in the MATLAB/Simulink environment, shown in Figure 3. Since downward plugged in grid plate belongs to linear path planning after the pose of tenuous shaft is adjusted by robot end gripper. Therefore, the simulation input is a linear path and the simulation time set up is 2 seconds. The tracking result is shown in Figure 4.

From the vertical displacement tracking curve, it is known that the Cartesian space closed-loop control system can demand the robot end effector to track the desired linear trajectory with a relative error of 0.32%.

5 EXPERIMENT AND RESULT

5.1 *Hardware configuration*

This experiment hardware system consists of the YASKAWA MOTOMAN six-DOF Vertical multi-joint industrial manipulator, two DAHENGCMOS cameras with the resolution 2592 × 1944 and a host computer, shown in Figure 5. The communication

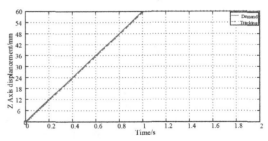

Figure 4. Vertical displacement tracking curve.

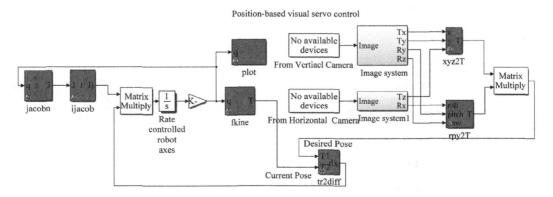

Figure 3. Simulation model in Simulink.

76

between the host computer and the robot controller is used by series port RS232. The images of the tenuous shaft end and fixed grid plate are captured by the orthogonal binocular vision system. Then, they are processed by the host computer in real time. The current pose and desired pose are computed through model match and feature extraction so as to obtain their relative pose error. Also, the Cartesian space motion instructions are planned. Then the motion instructions are transmitted into the robot controller aiming at demanding the robot to move along with a planned trajectory, namely, completing the plug-in action. Eventually, the automated assembly task of the tenuous shaft with vision guided robots based on three-dimensional pose is achieved

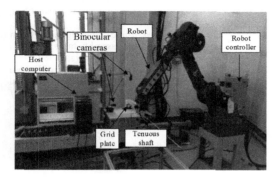

Figure 5.   Experiment hardware configuration.

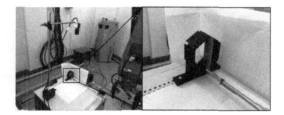

Figure 6.   Tenuous shaft plug-in process.

5.2   Assembly experiment

Camera calibration and eye to hand calibration must firstly be carried out before the assembly experiment in order to map from the image frame to the robot base frame. Then, the grid plate is fixed before the opening cameras and the target is located in the grid plate. The desired pose parameters of the grid plate are computed. During the assembly process, the tenuous shaft end pose is adjusted by the manipulator to make its slot align to the grid plate. The tenuous shaft is controlled to plug in to the grid plate downwardly, as shown in Figure 6. After the assembly experiment has been performed five times, the grid plate desired pose, tenuous shaft initial pose and final pose are given in Table 1.

The tenuous shaft assembly experiment is accomplished successfully. The maximum displacement error is controlled within 0.006 mm and the maximum angle error is controlled within 0.192°. Meanwhile, the assembly period is within 90 seconds. The overall process accomplishes assembly rapidly and automatically.

Table 1.   Measurement result in assembly process.

No	Parameter type	X_w mm	Y_w mm	Z_w mm	α °	β °	γ °
1	Desired pose	0.968	−2.166	382.484	0.128	0	1.006
	Initial pose	2.236	1.405	386.659	−0.383	0.105	0.495
	Final pose	0.964	−2.162	384.396	0.086	0.095	0.987
2	Desired pose	0.975	−2.164	382.474	0.116	0	1.035
	Initial pose	2.125	1.325	384.342	0.172	0.097	0.376
	Final pose	0.978	−2.162	383.263	0.167	0.014	1.166
3	Desired pose	0.964	−2.152	382.484	0.117	0	0.761
	Initial pose	2.174	1.424	380.542	0.121	0.346	0.542
	Final pose	0.962	−2.153	381.372	0.092	0.102	0.735
4	Desired pose	0.973	−2.153	382.422	0.998	0	1.021
	Initial pose	2.456	1.562	381.456	0.345	0.312	0.563
	Final pose	0.968	−2.156	382.419	1.098	−0.046	1.014
5	Desired pose	0.972	−2.159	382.486	1.126	0	1.012
	Initial pose	2.542	1.575	381.434	0.768	0.734	0.682
	Final pose	0.975	−2.153	382.487	0.934	0.116	1.123

6 CONCLUSION

The Cartesian space closed-loop control method is presented in this paper after the tenuous shaft parts poses are estimated by the vision system. Simulation results show that the proposed method can track the planned linear trajectory well in advance. Also, assembly experiment results reveal that the vision system has a high inspection speed and the manipulator has a high positioning accuracy, which can meet assembly requirements. In conclusion, the overall process can be applied in all kinds of industry fields where precise shaft parts are assembled automatically with industrial manipulators guided by machine vision.

REFERENCES

Aviles, J.F., et al. (2016). On-line learning of welding bead geometry in industrial robots. *The International Journal of Advanced Manufacturing Technology*, 83(1–4), 217–231.

Brogardh, T. (2007). Present and future robot control development: An industrial perspective. *Annual Reviews in Control*, 31(1), 69–79.

Chai, L.R. (2016). *Research on assembly technology of nuclear rods based on binocular vision*. Harbin: Harbin Institute of Technology.

Jia, B.X., et al. (2015). Visual trajectory tracking of industrial manipulator with iterative learning control. *Industrial Robot: An International Journal*, 42(1), 54–63.

Navarro, J.L., et al. (2015). On-line knowledge acquisition and enhancement in robotic assembly tasks. *Robotics and Computer-Integrated Manufacturing*, 78–89.

Ren, B.Y., et al. (2017). A method for measuring 3D pose of assembly end of long shaft part using orthogonal binocular stereovision. *Journal of Harbin Institute of Technology*, 49(1).

Automotive, Mechanical and Electrical Engineering – Liu (Ed.)
© 2017 Taylor & Francis Group, London, ISBN 978-1-138-62951-6

Research on the meshing characteristic of the gear with the engagement misalignment

Xiaohe Deng
School of Automotive Engineering, Wuhan University of Technology, Wuhan, China
Hubei Province Key Laboratory of Modern Automotive Technology, Wuhan University of Technology, Wuhan, China
Hubei Collaborative Innovation Center for Automotive Components Technology, Wuhan University of Technology, Wuhan, China

ABSTRACT: The meshing position of gear in the actual meshing process would deviate from the theoretical position, which is called gear engagement misalignment. The gear engagement misalignment could lead to changes in actual meshing characteristic of gear, and then influence gear's meshing capability. Therefore, it is necessary to analyse the actual meshing situation of gear in detail for gear design. This paper aims at the three typical meshing misalignment forms of gear, using finite element method to establish a gear meshing model, and analysing gear contact ratio, contact stress distribution and size of contact zone, and the contact stress of gear in different meshing misalignment values. Then the impact of the meshing misalignment on gear meshing rule is analysed.

Keywords: Engagement misalignment; Meshing characteristic of gear

1 INTRODUCTION

The system components have their own rigidities, and the gear transmission systems under certain conditions produce the corresponding deformation. The accumulation and transmission of deformation leads to dislocation of gear meshing. Due to the gear design and manufacture and installation of the system of deformation caused by the error, the gear stiffness system itself, the gear stiffness, bending and torsion deformation caused by itself, making gears mesh on the actual location of the form is varied. The meshing position of gear in the actual meshing process would deviate from the theoretical position, which is called gear engagement misalignment. The gear engagement misalignment could lead to changes in actual meshing characteristic of gear, and then influence the gear's meshing capability. Therefore, it is necessary to analyse the actual meshing situation of gear in detail for gear design.

2 TYPES OF GEAR MISALIGNMENT

Gear shaft combined deformation will happen in gear meshing force, which contains both bending deformation and torsion deformation. Due to action of forces of driving wheel and driven wheel of the same size in opposite directions, the input shaft and output shaft bending deformation must be on a plane, which causes angular misalignment of the gear centre axis in the same plane. At the same time, bending torsional deformation leads to the centre of the gear axis appear angle displacement in a different plane. Therefore, gear meshing misalignment is a variety of combinations in the form of dislocation. For the two axes in 3D space, there are three types of relative relationship. One is parallel to each other, which is the ideal situation. The second is the two lines cut or its extension. The third form is space disjoint, with two lines in a plane at this time. Therefore, in front of the two kinds of relations can also be called with the surface, behind a relationship for different surface. The ideal meshing gear axis is parallel, but in fact processing assembling error exists with the load shaft, bearings, housing and other components. The deformation makes the relative location of the two shafts under the working state change. The shaft of the gear centre line is different from the previous parallel relationship, namely the dislocation. According to the relative position of the axis of the meshing gears, the gear meshing misalignment can be divided into three categories: (1) the gear axis parallel dislocation; (2) the gear axis intersection dislocation; (3) the gear axis space angular misalignment.

3 MATHEMATICAL MODEL

Both driving and driven meshing gear have the misalignments caused by the deformation of system. Thus the amount of misalignments are the integrated for the both dislocation. Assuming that one of the meshing gear displacement quantities is zero, another gear displacement quantity for the two gear integrated dislocation. The detailed operation is as follows: transfer the amount of misalignments of a gear pair into one of the dislocation of the gear, for there is no deviation in the position of a gear benchmark position, the gear as a benchmark, another gear relative to the benchmark gear mesh displacement quantity to a certain extent, the meshing gear for integrated dislocation. Since the meshing of gear tooth surfaces touch each other, the gears will mesh in a plane. Therefore, the relative displacement quantity can be used to make a pair of meshing gears meshing displacement quantity into a gear relative to another without dislocation of the gear meshing.

The parameters of adopted meshing gear pair are shown in Table 1.

The finite element method is used to establish the corresponding model of the meshing gears. Applying load and boundary condition, the contact stress diagram of the driving gear when the pair meshes in the node positon is shown in Figure 1. Figure 1 is the case of no meshing

Table 1. Parameters of meshing gear pair.

	Driving gear	Driven gear
Number of tooth	19	19
Normal modulus (mm)	2.87	2.87
Normal pressure angle (°)	22.69°	22.69°
Tooth width	16	16
Addendum coefficient	1.002	1.002
Tip clearance coefficient	0.364	0.364

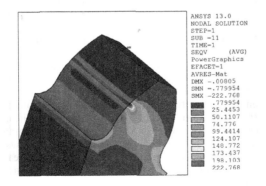

Figure 1.   The contact stress diagram (no misalignment).

misalignment. It can be seen that when there is no meshing dislocation, there is uniform load distribution on tooth width direction, and the maximum stress of the gear is at the end of gear width with a maximum of 222.768 MPa.

4 MESH MISALIGNMENT INFLUENCE LAW OF GEAR ENGAGEMENT

4.1 *Gear parallel misalignment effect on the law of gear engagement*

The parallel misalignment of meshing gear will lead to changes of the centre distance of gear shaft. In theory, the change of the centre distance will lead to slight interference of external diameter of gear, thus leading to the change of tooth profile of gear contact ratio. Due to the reduction of centre distance, the contact area will be slightly narrowed, leading to a small transverse contact ratio increase. The transverse contact ratio calculation formula is:

$$\varepsilon_a = \left[z_1(\tan\alpha_{a1} - \tan\tan\alpha_a) + z_2(\tan\alpha_{a2} - \tan\tan\alpha_a) \right] / 2$$

where, αa is the pressure angle, satisfying formula

$$\alpha_a = \arccos\,(\mathrm{a}\cos\alpha / a')$$

where, a′ is the actual centre distance, a is the theoretical centre distance. It can be seen that when cent distance is reduced, contact ratio increases. If there is a parallel misalignment of 5 μm, the reduction of centre distance will lead to tooth profile contact ratio increasing from 1.463 to 1.463, up 0.1%, so this change is very small and usually can be ignored. Centre distance can be thought of as the involute tooth profile is not sensitive, parallel misalignment have less effect on the gear meshing gears, the gear contact ratio as the meshing and slight changes in the quantity of dislocation, smaller influence on the change of the gear meshing state.

4.2 *Gear axis intersect misalignment influence law of gear engagement*

Meshing of gear centre line in the same plane displacement will lead to partial load on the gear width. This is expressed as follows. The edge on one side of the gap increase tooth width and tooth width of the edge of the other side clearance is reduced, uneven load distribution along the tooth width, to transfer the load to the tooth width edge on one side. In this case, the theory contact plane will remain the same shape and size. This form of misalignment causes major changes in the load distribution in the tooth width direction, assuming the maximum dislocation value of one gear side is 10 μm. Figure 2 is the contact stress diagram of the corresponding driving gear when the gear pair meshes in the node position. Figure 2 shows the case of 10 μm mesh misalignment. Inn the dislocation cases, there are partial loads on the width direction presented obvious, side gear on the edge of the maximum value is 284.135 MPa, and the biggest stress rises by 27% in comparison to the no dislocation cases in Figure 1. The length of the gear meshing contact line still covers the whole gear width.

If we select the maximum mesh misalignment values from 0 μm to 25 μm by 5 μm intervals, Figure 3 shows the corresponding driving wheel contact force distribution along the gear width. We can see that when the misalignment is zero, the gear tooth width direction of the distribution of contact force is symmetric, and with the amount of dislocation increasing, gear on one side of the contact force increases, and the other side of the contact force is reduced. The total contact force is constant. The gear in the inhomogeneity of tooth width direction load causes partial load phenom-

enon, and the gear is prone to stress and damage. On both sides of the gear end face, the value of the contact force is low. Although the load on one end of the gear width is reduced, along the direction of the tooth width theload is covered.

Tooth contact stress along the tooth width is shown in Figure 4. In line with the trend of the contact force, with the increment of displacement, the lines of contact stress are shown as the "seesaw" form. Among them, there is a fixed point near the centre of gear width. If there is no mesh displacement misalignment, the contact stress of gear central symmetry is slightly larger, on both sides of the end face of the contact stress. When the misalignment value is not zero, close to the maximum displacement quantity for the position of the maximum contact stress of the position, (as shown in Figure), and the meshing area is consistent.

4.3 Gear space angle misalignment effect on the law of gear engagement

The gear meshing condition is not the same when there are angle misalignments of gear axis which are not in the same plane. Due to angular misalignment, the gear outside diameter will also twist correspondingly. Assume that the maximum dislocation value is 10 μm, which is caused by gear space angle misalignment. Figure 5 is the contact stress diagram of the corresponding driving gear when the gear pair meshes in the node position. In the gear space angle misalignment cases, partial load on the width direction load is obviously presented, side gear on the edge of the maximum value is 382.007 MPa, and the biggest stress rises by 71.5%. In this situation, the contact line of the meshing gear deviates to one side of the tooth width, which cannot cover the entire gear width. It can be concluded that with the increment of dislocation, meshing area gradually shift to one side, and the effective areas decrease greatly.

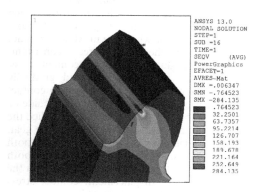

Figure 2. The contact stress diagram (10 μm gear parallel misalignment).

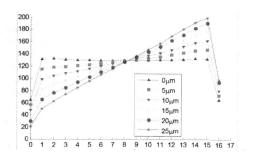

Figure 3. Contact force diagram under different gear parallel misalignment values.

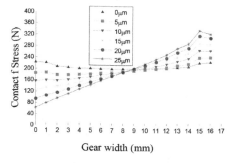

Figure 4. Contact stress diagram under different gear parallel misalignment values.

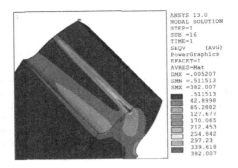

Figure 5. The contact stress diagram (10 μm gear space angle misalignment).

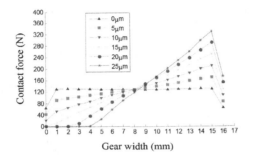

Figure 6. Contact force diagram under different gear space angle misalignments.

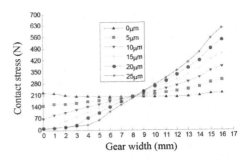

Figure 7. Contact stress diagram under different gear space angle misalignments.

If we also establish the finite element model of meshing gears by selecting the maximum mesh misalignment values from 0 μm to 25 μm by 5 μm intervals, Then Figure 6 shows the corresponding driving wheel contact force distribution along the gear width. It can be seen that when the misalignment value is zero, the gear tooth width direction of the distribution of contact force is symmetric. When the misalignment values increases, the contact force increases on one side of the gear width, and decreases on the other side of the gear width. However, total contact force is constant. Because of load uniformity of the gear in the gear width direction, there is partial load. On both sides of the gear end face, the value of the contact force is low. When the mesh is greater than 15 μm, the load on one side of the gear is close to zero. In the gear width, load does not cover the whole area. Due to a tooth bearing the total load is in the same place, the partial load is on one side of the gear width, and there will be inevitable lift of gear load on the other side. But when the gear tooth to the partial load is big enough, the gear on the left sideclose to the tooth width end position of the contact force is zero, and the trend of the gear end tooth width oriented extension centre. Such gear meshing width

decreases, gear needs not only inherit the torque, still under bending moment in the side of the gear. The distribution of gear contact force creates an extremely unreasonable state, seriously influencing the loading of the gear.

Under the condition of different meshing gear axis angle displacements, tooth contact stress is shown in Figure 7. If there is no mesh displacement misalignment, the contact stress of gear central is symmetrical, and on both sides of the end face the contact stress are slightly larger. When misalignment value is zero, the position of maximum misalignment is where the maximum contact stress located, and the meshing area is consistent, as is shown in Figure 5. Due to the contact stress, reduced tooth width side bear load gradually decreases to zero, and the corresponding contact stress value was close to zero. On the other side of the shoulder, the bulk of the load area, reached the maximum contact stress on the edge of the gear, to form the edge contact. Due to the gear tooth side out gradually, at the end of the tooth-to-tooth face contact stress also tends to zero, and with the increment of mesh displacement, contact stress non-uniform distribution of speed increases.

5 CONCLUSION

Gear parallel misalignment leads to the changes of load distribution in the direction of gear tooth width when the gear load increases, and the corresponding contact stress increases. The maximum contact stress appears near the greatest dislocation position. The load still covers the whole tooth width direction when the gear parallel misalignment happens. Gear axis space angular misalignment leads to the changes of load distribution in the direction of gear tooth width. When the gear load increases, the corresponding contact stress increases. The maximum contact stress appears at the greatest dislocation position. When the misalignment value increases to a certain amount, the no-load situation happens gradually in the tooth width side direction. Gear axis space angle displacement quantity is largest for the influence of the gear mesh, in the actual gear design, gear system should be fully considered, the deformation of the gear meshing displacement quantity control in a reasonable range, in particular to control the third form of meshing misalignment.

ACKNOWLEDGEMENTS

The research is supported by the Fundamental Research Funds for the Central Universities (WUT: 2016IVA039).

REFERENCES

Gonzalez-Perez, I. Iserte, J.L. & Fuentes, A. (2011). Implementation of Hertz theory and validation of a finite element model for stress analysis of gear drives with localized bearing contact. *Mechanism and Machine Theory*, 46, 765–783.

Huang, C., Wang, J.X., Xiao, K., Li, M. & Li, J.Y. (2013). Dynamic characteristics analysis and experimental research on a new type planetary gear apparatus with small tooth number difference. *Journal of Mechanical Science and Technology* 27(5), 1233–1244.

Litvin, F.L. & Fuentes, A. (2004). *Gear geometry and applied theory* (2nd ed.), Cambridge: Cambridge University Press.

Litvin, F.L., Fuentes, A., Gonzalez-Perez, I., Carvenal, L., Kawasaki, K. & Handschuh R.F. (2003). Modified involute helical gears: computerized design, simulation of meshing and stress analysis. *Computer Methods in Applied Mechanics and Engineering*, 192, 3619–3655.

Litvin, F.L., Lian, Q.M. & Kapelevich, A.L. (2000). Asymmetric modified spur gear drives: reduction of noise, localization of contact, simulation of meshing and stress analysis. *Computer Methods in Applied Mechanics and Engineering*, 188(1–3), 363–390.

Smith, J.D. (2003). *Gear noise and vibration* (2nd ed.). New York: Marcel Dekker.

Automotive, Mechanical and Electrical Engineering – Liu (Ed.)
© 2017 Taylor & Francis Group, London, ISBN 978-1-138-62951-6

Simulation of towing operation for tractor-aircraft system

Jingyuan Bao
Naval Military Representative Office, China Ship Development and Design Center, Wuhan, China

Qijun Wu & Junwei Xie
China Ship Development and Design Center, Wuhan, China

ABSTRACT: A framework of modelling and simulation for tractor-aircraft system is presented in this article. A dynamic model for tractor-aircraft system is established and then a controller for the driver is also established. Simulations and analyses for the tow tractor-aircraft system are conducted. The simulation model can follow the given path well. It is expected that the research is useful for tow tractor in design and manufacturing.

Keywords: Tractor-aircraft system; Dynamic; Simulation; Path

1 INTRODUCTION

In recent years, civil or military tractors have been widely used. In the maritime field, the Operating Conditions are very complex (TANG Qing, 2004; ZHANG Ning, 2001), it's inconvenience to research and design the tractor—aircraft system. Based on virtual prototyping technology (XIAO Li-jun, 2006), the tractor-aircraft system can be simulated and analyzed, which can simulate the state of the tractor-aircraft system under complex conditions. In the traction vehicle-traction operation of aircraft systems in the process of safety performance research, it's mainly through multi-body dynamics simulation technology traction operation of the security for the actual operation. Modeling of the traction process should consider two aspects of the problem:

First, the traction system is a typical complex multi-body dynamics system, the fundamental purpose of multi-body dynamics is the use of computer technology for complex mechanical system dynamics modeling and solution (Yang Yong, 2007). Multi-body system dynamics is a branch of the new disciplines that developed on the basis of classical mechanics. In the dynamics of classical rigid body systems, the dynamics of multibody systems and the dynamics of multibody systems two stages of development, has developed quite mature.

On the other hand, in the simulation calculation of traction operation, the tractor traction aircraft is required to move along a certain traction trajectory, and a driver model must be introduced to constitute a closed-loop control of the tractor-aircraft system.

From the model whether to include the driver's preview link, divided into compensation tracking model and preview tracking model. The compensation model is represented by the PID model proposed by Iguchi and the Crossover model proposed by McRure (Iguchi M, 2005). The input of the driver compensation tracking model is the deviation between the information of the expected trajectory of the current time and the state information of the vehicle. The model assumes that the driver compensates and corrects only according to the deviation and outputs the steering wheel angle. MacAdam proposed optimal preview control model (Macadam C.C., 1980), the parameters can be determined directly by the vehicle dynamics characteristics, car track to follow a very high accuracy. Reddy and other local optimization preview model to control the vehicle in the future time of the driving track and the expected deviation of the road as the goal, the application of local optimization algorithm optimization of the steering wheel angle. The relationship between the model parameters and vehicle maneuvering characteristics and driver characteristic parameters is established by the preview-follow system theory (Guo K.H., 1984) proposed by Guo Konghui. After a lot of man-vehicle-road closed-loop system simulation and actual driver test comparison, Obtained with the real skilled drivers driving a car very close to the track to follow the results.

Virtual prototyping technology based on multi-body system dynamics is an important way to shorten the development cycle of tractor-aircraft system, reduce development cost, improve product design and manufacture quality. For tractor-and-airplane systems, the size of the travel path and traction is particularly important. In order to reduce the development risk of the system, it is necessary to carry on dynamic modeling and simulation analysis before the tractor is used, and design a reasonable traction route.

2 MULTI-BODY DYNAMICS AND CONTROL MODEL OF TRACTOR-AIRCRAFT SYSTEM

2.1 *Multi-body dynamics model and its solution*

The multibody system dynamics equation can be written in the form of equation (1) (Dan N., 2009; Shabana A. A., 2005).

$$
\begin{aligned}
\dot{\mathbf{q}} &= \mathbf{v} \\
\mathbf{M}(\mathbf{q})\dot{\mathbf{v}} &= \mathbf{Q}(t,\mathbf{q},\mathbf{v}) - \mathbf{\Phi}_{\mathbf{q}}^{\mathrm{T}}(\mathbf{q},t)\lambda \\
\mathbf{0} &= \mathbf{\Phi}(\mathbf{q},t)
\end{aligned}
\tag{1}
$$

Among them, $\mathbf{q},\mathbf{v} \in R^n$ are the generalized coordinates and the generalized velocity, respectively, $\lambda \in R^m$ are the Lagrangian multipliers. $\mathbf{M}$ represents the generalized mass matrix, $\mathbf{Q}$ representing the generalized force, $\mathbf{\Phi}$ representing the complete constraint.

Because high order integration methods are limited by stiffness problems, low-order integration methods such as Newmark, HHT are often used to solve equation (1). HHT-I3 algorithm (Dan N., 2009) for solving the equation (1) the specific format as shown in equation (2) below.

$$
\begin{aligned}
\mathbf{q}_{n+1} &= \mathbf{q}_n + h\dot{\mathbf{q}}_n + \frac{h^2}{2}\left[(1-2\beta)\mathbf{a}_n + 2\beta\mathbf{a}_{n+1}\right] \\
\dot{\mathbf{q}}_{n+1} &= \dot{\mathbf{q}}_n + h\left[(1-\gamma)\mathbf{a}_n + \gamma\mathbf{a}_{n+1}\right] \\
\frac{1}{1+\alpha}&(\mathbf{M}(\mathbf{q})\mathbf{a})_{n+1} + \left(\mathbf{\Phi}_{\mathbf{q}}^T\lambda - \mathbf{Q}\right)_{n+1} - \frac{\alpha}{1+\alpha}\left(\mathbf{\Phi}_{\mathbf{q}}^T - \mathbf{Q}\right)_n = \mathbf{0} \\
\frac{1}{\beta h^2}&\mathbf{\Phi}\left(\mathbf{q}_{n+1},t_{n+1}\right) = \mathbf{0}
\end{aligned}
\tag{2}
$$

where h is the integral step size, $\mathbf{a}_{n+1}$ is the approximation $\ddot{\mathbf{q}}$, α, β and γ are parameters. Note that Eq. (2) contains a Newton iteration for $\mathbf{a}_{n+1}$ and λ_{n+1} and the corresponding Jacobian is

$$
\mathbf{J}_{HHTI3} = \begin{bmatrix} \dfrac{1}{1+\alpha}\mathbf{M} + \hat{\mathbf{P}} & \mathbf{\Phi}_{\mathbf{q}}^T \\ \mathbf{\Phi}_{\mathbf{q}} & \mathbf{0} \end{bmatrix}
\tag{3}
$$

Among them,

$$
\hat{\mathbf{P}} = \beta h^2\left(\mathbf{M}(\mathbf{q})\ddot{\mathbf{q}} + \mathbf{\Phi}_{\mathbf{q}}^T\lambda - \mathbf{Q}\right)_{\mathbf{q}} - \gamma h\,\mathbf{Q}_{\dot{\mathbf{q}}}
$$

The tire force $\mathbf{Q}$ is included in the generalized force described above. Fiala tire model is a widely used tire force model, this paper is the use of this model. Fiala model of vertical force in the formula

$$
F_z = \min\{0, F_{zk} + F_{zc}\}
\tag{4}
$$

Which F_{zk} is caused by the elastic deformation of the vertical force, F_{zc} caused by the vertical force.

The longitudinal forces in the Fiala model are calculated as follows:

$$
F_x = \begin{cases} -C_{slip} \times S_s, & |S_s| < S_{critical} \\ -sign(S_s)\left(F_{x1} - F_{x2}\right), & |S_s| > S_{critical} \end{cases}
\tag{5}
$$

where, $F_{x1} = U \times F_z$, $F_{x2} = \left|\dfrac{(U\times F_z)^2}{4\times |S_s| \times C_{slip}}\right|$, U is the friction coefficient, S_s is the longitudinal slip, $S_{critical}$ is the longitudinal slip threshold, C_{slip} is the slip rate.

The lateral force in the Fiala model is calculated by:

$$
F_y = \begin{cases} -U \times |F_z| \times (1-H^3) \times sign(\alpha), & |\alpha| < \alpha_{critical} \\ -U \times |F_z| \times sign(\alpha), & |\alpha| > \alpha_{critical} \end{cases}
\tag{6}
$$

In the formula, α is the lateral deflection angle, $\alpha_{critical}$ is the lateral deflection angle threshold value, $H = 1 - \dfrac{C_\alpha \times \tan\alpha}{3\times U\times |F_z|}$, C_α is the lateral stiffness coefficient. The positive moment in the Fiala model is calculated by:

$$
T_z = \begin{cases} 2U|F_z|R_2(1-H)H^3 sign(\alpha), & |\alpha| < \alpha_{critical} \\ 0, & |\alpha| > \alpha_{critical} \end{cases}
\tag{7}
$$

The equations (1–7) above form the multibody dynamics model of the tractor-aircraft system and the solution format. Specific modeling and solution will be completed with the help of commercial software ADAMS.

2.2 *Driver control model*

In order to move the tractor-aircraft system in accordance with the set traction trajectory, a driver model must be established to form a closed-loop control of the tractor-aircraft system. Whether the preview link is included in the model can be divided into compensation tracking model and preview tracking model.

In this paper, the driver's preview tracking model is adopted. Driver preview tracking model block diagram shown in Figure 1. In the figure, P (s), F (s) and B (s) respectively indicate the preview link, forward correction link and feedback estimation link of the driver, which are the expected

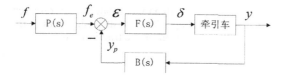

Figure 1. The driver's preview tracking model.

trajectory information. The preview link is based on the current vehicle motion state the estimated future time vehicle position information is estimated by estimating the link estimated vehicle state information at a future time, the deviation of the two estimated values, the control information applied to the vehicle, and the position of the trajectory for the vehicle. The preview tracking model is more in line with the actual handling characteristics of the driver because the driver always follows the road path in advance during the usual driving process. The driver control model is established by ADAMS SUBROUTINE. The driver model is expressed in terms of VARSUB to control the steering angle input of the vehicle model to control the direction of the vehicle.

3 NUMERICAL SIMULATION

3.1 *Kinetic model*

Modeling of Tractor-Aircraft System Based on Multi-body System Dynamics Software ADAMS. The dynamics model of the tractor consists of cab, frame, wheel, and steering mechanism and so on. Aircraft dynamics model mainly by the aircraft fuselage, wheels, steering agencies. The components are connected and simplified according to the actual physical model to form a unified tractor-aircraft dynamics model.

It is noteworthy that the ship deck of the tractor-aircraft system has independent movements to simulate the effect of waves on the hull motion. According to the theory of ship kinematics, deck movement consists of three directions, namely, roll, pitch and heave. When the deck kinematics model is established, its motion is approximated as a sinusoidal motion in three directions, namely:

$$x(t) = a_x \sin(2\pi f_x t)$$
$$y(t) = a_y \sin(2\pi f_y t) \qquad (8)$$
$$z(t) = a_z \sin(2\pi f_z t)$$

where a_x, a_y, a_z are three directions of motion amplitude, respectively, f_x, f_y, f_z are three directions of motion frequency.

The resulting tractor-aircraft-deck kinetic model is shown in Figure 2.

3.2 *Simulation conditions and results*

Select a common traction operation as a simulation case. In the case where the deck has independent movement in three directions, the traction operation shown in Fig. 3 is set and the tractor traction aircraft is moved from right to left. The simulation results of the traction operation are shown in Figure 4 and Figure 5.

Figure 4 is the default traction trajectory and the actual traction comparison results. The solid trajectory is the preset trajectory, and the broken line is the actual traction trajectory. The maximum trajectory deviation between them is 294 mm, which indicates that the trajectory tracking algorithm has enough precision.

Figure 5 shows the traction force when the traction operation is calculated. As can be seen from

Figure 2. Tractor-aircraft dynamics model.

Figure 3. Preset traction mode.

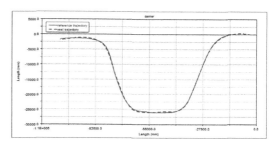

Figure 4. The default traction trajectory and the actual traction comparison.

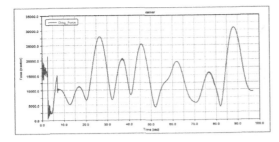

Figure 5.   Traction tension.

the figure, at t = 88 s, the traction force reaches a maximum value of 30927 N. Comparison of the actual traction maximum and traction rating, you can check the safety of traction institutions.

4   CONCLUSION

This paper presents a modeling and simulation and analysis process of tractor marine operations. Based on the multi-body dynamics theory, the tractor-aircraft dynamics model is established. Based on the modern control theory, the driver's preview control model is established. Finally, the complex tractor-aircraft dynamics and control model are established. The independent movement of the deck in three directions is used to simulate the sea surface waves, which increases the difficulty of modeling and simulation. A typical traction operation is selected as a simulation example. The simulation results verify the reliability and accuracy of the dynamic model. The research method in this paper has a positive significance for the quick verification of the traction scheme.

REFERENCES

Curvature Concept. SAE Paper 841016, 1984.
Dan N., Laurent O., Jay K., A Discussion of Low-order Numerical Integration Formulas for Rigid and Flexible Multibody Dynamics. Journal of Computational and Nonlinear Dynamics, 4 (2), 021008, 2009.
Guo K.H., Modeling a Closed-loop Directional Control System by an Optimal.
Iguchi M., A study of Manual Control. Journal of Mechanic Society of Japan, 62 (1): 112–116, 1959.
Macadam C.C., An Optimal Preview Control for Linear Systems. Journal of Dynamic Systems Measurement and Control, 102 (3): 188–190, 1980.
Reddy R.N., et al., Contribution to the Simulation of Driver-Vehicle-Road System. SAE Paper 810513.
Shabana A.A., Dynamics of Multibody dynamics, 3rd ed. Cambridge: Cambridge University Press, 2005.
Tang Qing. Application analysis of vehicle lashing and fastening system on Ro-Ro ship. Dalian Maritime University, Master's Thesis, 2004.
Xiao Li-jun. ADAMS/CAR based on the virtual prototype of vehicle suspension system design and performance analysis. Hunan University, master's degree thesis, 2006.
Yang Yong, Ren Weiqun, Chen Liping. Simulation and Analysis of Ride Comfort of Three-axle Trailer. Automobile Engineering, 29 (2): 165–169, 2007.
Zhang Ning. Mathematic model of solid dynamic simulation of vehicle lashing on Ro-Ro ship. Dalian Maritime University, Master's Thesis, 2001.

Automotive, Mechanical and Electrical Engineering – Liu (Ed.)
© *2017 Taylor & Francis Group, London, ISBN 978-1-138-62951-6*

Study of the influence of the process parameters of incremental forming of 6061 aluminium alloy

Zhifeng Liu, Zheng Shi & Tieneng Guo
Beijing Key Laboratory of Advanced Manufacturing Technology, Beijing, China

Zhijie Li
School of North China Institute of Aerospace Engineering, Langfang, China

Chao Cui
Beijing Spacecrafts, Beijing, China

ABSTRACT: The 6061 aluminium alloy is widely used in the aerospace field due to its advantages of high toughness, good forming performance and lack of deformation after forming. However, its manufacturing lacks theoretical guidance. Taking the 6061-T6 aluminium alloy with thickness of 1 mm as the research object, the stress and strain of the machining area established by the mechanical model are first analysed. The influence of the main process parameters, such as tool radius, tool depth ΔZ, half-apex angle α on the sheet thickness distribution and forming quality using Abaqus/explicit, are then analysed. The stress and strain data are obtained by numerical simulation. By comparing the numerical simulation with the theoretical value, it shows that a big tool diameter and a small tool depth make the thickness distribution more uniform, but it is necessary to consider the reasonable choice of processing efficiency. Half-apex angle has a great effect on sheet thickness distribution—the smaller the half-apex angle is, the larger the sheet thickness thins. When the forming angle exceeds the sheet metal forming limit, cracking occurs.

Keywords: Incremental sheet forming; Numerical simulation; 6061 aluminium alloy; Formability; Abaqus/explicit

1 INTRODUCTION

Sheet metal incremental forming is layered into the idea of rapid prototyping manufacturing technology. The complex 3D digital model is discretised into a series of contour layers along the height direction, which is decomposed into a series of contour lines. Finally, the sheet is formed into the desired workpiece (Zhou, 2004). Progressive forming technology is proposed by Matsuhara Shio and the forming process is described in detail. Compared with the traditional mould processing technology, progressive forming technology has an advanced nature of not needing the mould (Mo et al., 2002). Along with developments in computer technology and the finite element theory, the most effective scientific research method is the numerical simulation of the forming process.

The process parameters that affect forming performance and forming quality in the progressive forming process are tool diameter, spacing, angle, plate thickness, a reasonable choice of parameters needed in the manufacturing process, to get the best forming performance. A lot of research on the process parameters has been done at home and abroad. Ayed made a numerical simulation with square cone parts and studied the numerical method of the progressive forming process of sheet metal (Ayed et al., 2014). Hu Jianbiao analysed the effects on the sheet forming process caused by the process parameters of half angle, spacing, tool diameter with Abaqus software in the process of simulation (Hu et al., 2011). Arfa studied the influence of the process parameters and the impact pressure on the single point incremental forming of aluminium alloy (Arfa et al., 2013). Zhang Yong analysed the impact of the parameters during the simulation of sheet metal forming by the numerical analysis software Abaqus and combined with the experiments to verify (Zhang et al., 2015).

Most of the scholars have analysed the influence of the main process parameters on the forming of sheet metal by a numerical simulation method. The mechanical model of the contact area is not established and compared with the simulation results. The materials of the research board are mostly 1060 aluminium, 08 AL and other materials, but there has been less analysis of the forming of 6061 aluminium alloy sheet, which has excellent forming properties. In this paper, based on the 6061-T6 aluminium alloy sheet, firstly the mechanical model to analyse the stress and strain of the contact zone between the tool head and the plate is set out. Then the finite element model is established, analysing the effects of tool diameter, spacing, and half angle on the forming process, and making the comparison between simulation results and the theoretical analysis.

2 MECHANICAL MODEL ESTABLISHMENT

2.1 Stress zone strain analysis

As shown in Figure 1, the deformation of the plate in the area A, the contact area between the plate and the tool head, is subject to tension and bending. Hence the amount of deformation is the sum of the amount of tensile deformation and the amount of bending deformation. Taking into account the pure bending without stretching, if the bending is not too large, the surface of the middle position of the plate can be approximated as a neutral plane. By stretching the overlap effect, the original neutral plane will be offset inward. The inner surface is under tensile strain induced by tension and the compressive strain caused by bending. The outer surface is under tensile strain caused by stretching and bending. Without taking into account the bending deformation, the tensile strain caused by the tension can be expressed by the tension strain of the neutral layer. According to the volume invariance hypothesis, $\varepsilon_t + \varepsilon_\phi + \varepsilon_\theta = 0$ plane strain assumption $\varepsilon_\theta = 0$, get the equivalent strain of the regional a unit $\bar{\varepsilon}^A$ (Fanga et al., 2014).

$$\bar{\varepsilon}^A = \sqrt{\frac{2}{3}\varepsilon_{ij}\varepsilon_{ij}} = \sqrt{\frac{2}{3}\left(\varepsilon_t^2 + \varepsilon_\theta^2 + \varepsilon_\phi^2\right)} = \frac{2}{\sqrt{3}}\ln\frac{rt_0}{R_m t} \quad (1)$$

ε_θ—tangential strain
ε_ϕ—meridional strain
ε_t—thickness strain
$\bar{\varepsilon}$—equivalent strain
t_0—initial sheet thickness
t—sheet thickness
R_m—curvature radius of the mid surface of sheet

2.2 Stress analysis of the stress area

In Figure 2, the tangential force σ_θ is obtained by the equilibrium equation, the constitutive equation and the hardening criterion equation. Radial force σ_ϕ and thickness direction stress σ_t, and finally get the equivalent stress for the stress state of the region A, the same element, the equilibrium equation should be satisfied in the thickness direction (Xu & Liu, 1995):

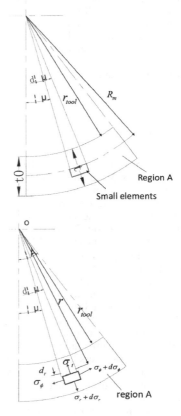

Figure 2. Analysis of stress and strain in region A.

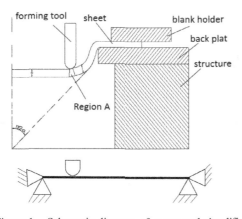

Figure 1. Schematic diagram of stress and simplified mechanical model of sheet metal forming.

$$(\sigma_t + d\sigma_t) \cdot (r + dr) \cdot d\phi - \sigma_t \cdot r \cdot d\phi - \sigma_\phi \cdot \sin\left(\frac{d\phi}{2}\right) \cdot dr$$

$$-\left(\sigma_\phi + d\sigma_\phi\right) \cdot \sin\left(\frac{d\phi}{2}\right) \cdot dr = 0 \qquad (2)$$

In area an any element in the thickness direction of stress σ_t^A can be expressed as:

$$\sigma_t^A = \frac{C_0}{n+1}\left[\left(\ln\frac{rt_0}{R_m t}\right)^n - \ln\left(\frac{r_{tool} \cdot t_0 + t \cdot t_0}{R_m t}\right)^{n+1}\right] \qquad (3)$$

The radial stress σ_ϕ^A is obtained by means of the constitutive equations:

$$\sigma_\phi^A = C_0\left(\ln\frac{rt_0}{R_m t}\right)^n + \frac{C_0}{n+1}$$

$$\left[\left(\ln\frac{rt_0}{R_m t}\right)^{n+1} - \left(\ln\frac{r_{tool} \cdot t_0 + t \cdot t_0}{R_m t}\right)^{n+1}\right] \qquad (4)$$

As the shear strain increment is 0, $d\varepsilon_\theta = 0$
$\sigma_\theta = (1/2)(\sigma_t + \sigma_\theta)$
The shear stress can be calculated as:

$$\sigma_\theta^A = \frac{C_0}{2}\left(\ln\frac{rt_0}{R_m t}\right)^n + \frac{C_0}{n+1}$$

$$\left[\left(\ln\frac{rt_0}{R_m t}\right)^{n+1} - \left(\ln\frac{r_{tool} \cdot t_0 + t \cdot t_0}{R_m t}\right)^{n+1}\right] \qquad (5)$$

By the combined Equations (2)–(5), the equivalent stress $\bar{\sigma}$ of the stressed area was obtained:

$$\bar{\sigma} = \frac{1}{\sqrt{2}}\sqrt{\left(\sigma_\phi^{AT} - \sigma_t^{AT}\right)^2 + \left(\sigma_\phi^{AT} - \sigma_\theta^{AT}\right)^2}$$

$$+\left(\sigma_T^{AT} - \sigma_\theta^{AT}\right)^2 = \frac{\sqrt{3}}{2}C_0\left(\ln\frac{rt_0}{R_m t}\right)^n \qquad (6)$$

C_0 —constant value for the given material
n —strain hardening index
r —distance from unit to the centre of rotation

3 FINITE ELEMENT ANALYSIS MODEL

Figure 3 shows the change in shape of 6061 aluminium alloy sheet metal before and after incremental forming and shape trajectory. Firstly, the plate is fixed with a clamp and the degree of freedom of the plate is fully restrained. The movement of the tool head in the process of machining is a contour line processed by a clockwise and anticlockwise direction, until the sheet material is extruded out of the desired shape result.

A finite element analysis model is established in Abaqus software, which is shown in Figure 4(a). The sheet size is $120\ mm \times 120\ mm$ 6061 aluminium alloy, with density 2.9 g/cm^3. The specific performance parameters are shown in Table 1 and the stress-strain curve (Sun et al., 2014) is shown in Figure 5. The tool head and the plate are arranged as a discrete rigid

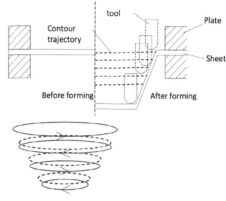

Contour trajectory

Figure 3. Schematic diagram of single point incremental forming of sheet metal.

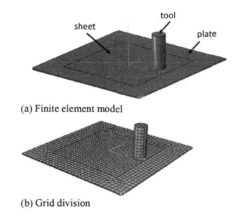

(a) Finite element model

(b) Grid division

Figure 4. Finite element analysis model.

Table 1. Performance parameters of 6061 aluminium alloy.

Material	Elastic modulus/GPa	Poisson ratio	Strength/ MPa	Yield strength/ MPa
6061-T6	68.9	0.33	343	269

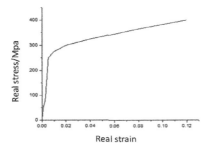

Figure 5. Stress and strain curves of 6061-T6 aluminium alloy.

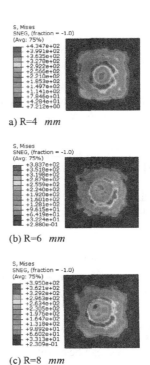

a) R=4 mm

(b) R=6 mm

(c) R=8 mm

Figure 6. Forming mises von of different tool tip radius.

body. As the shell element is suitable for simulation analysis of progressive forming of the plate (Zhuang & You, 2009), so parts are modelled using SR4 shell element. The coefficient of friction between the tool head and the sheet material is 0.1 and the contact algorithm between the tool head and the plate is a face-to-face contact, as are the plate and the sheet. All degrees of freedom of the upper and lower plate are bound when setting boundary conditions and when the blank holder force is defined, in the initial setting, the pressing edge is fixed. The upper plate dropped 0.1 mm to simulate the pressure of the edge of the plate. The tool head is rotated down in the form of contour lines, rotation direction opposite to each turn, the decrease in each cycle was ΔZ, each horizontal is feed based on half angle, and the grid partition is tetrahedral dominated. The minimum unit size is 3 mm and the total number of cells is 1,600, as shown in Figure 4 (b).

4 NUMERICAL SIMULATION RESULTS

4.1 Effect of tool head diameter

Firstly, the half angle and the press amount are both under the same amount, simulating the NC incremental forming process with different diameters of spherical tool heads. The parameters for the layer spacing ΔZ, forming half angle $\alpha = 45°$, sheet thickness 1 mm, tool head radius 4 mm, 6 mm and 8 mm. According to the simulation results, the stress and strain of the sheet are compared: the stress unit is MPa, and the strain unit is %.

As shown in Figure 6, the tool head radius is 4 mm, 6 mm and 8 mm, the simulation results show that the maximum equivalent stresses are 434.7 MPa, 383.7 MPa and 395 MPa, the maximum value of the stress is more than the yield strength 269 MPa of 6061-T6 aluminium alloy sheet. Plastic deformation occurred in the sheet. As the plastic deformation of aluminium alloy sheet is not uniform, the stress distribution of the

sheet is therefore not uniform. The residual stress in the sheet metal forming can be in the interior of the sheet. It can be seen that the maximum stress is at the end of the shape. The comparison shows that the maximum stress of R = 6 mm is less than stress value of R = 8 mm, which can be expressed by Equation (6):

$$2\bar{\sigma}/\sqrt{3}\,C_0 = \ln\left(\frac{(r_{tool} + t_1)t_0}{(r_{tool} + t)t}\right)^n \qquad (7)$$

When the equivalent stress of the inner side of the plate is analysed, the thickness of the forming plate averages t = 0.7 mm, t_0 = 1 mm, r = r_{tool}, n = 0.223. When these are put into Equation (7) and the curve is obtained as shown in Figure 7(a), it can be seen that the stress of the inner side of the plate increases with the increase of the radius of the tool head. When it changes from the inner side of the sheet to the outer side, the relationship between the stress and the tool head radius changes, as shown in Figure 7(b), when more than half of the plate thickness $t_1 > t/2$, The lateral stress gradually decreases with the increase of the radius of the tool head. The simulation results show stress value of R = 4 mm, is maximum, the rea-

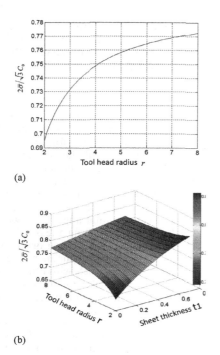

(a)

(b)

Figure 7. Relationship between the stress of the plate and the radius of the tool head.

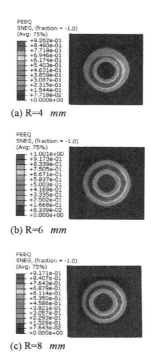

(a) R=4 *mm*

(b) R=6 *mm*

(c) R=8 *mm*

Figure 8. Strain PEEQ cloud images of different tool tip radius.

son being that the smaller the radius of the tool head, the more obvious the stress concentration, and as the stress becomes larger, the material is prone to rupture. R = 4 mm, smaller radius, the stress concentration is more obvious. When the radius of the tool head is increased, the contact area between the tool head and the sheet metal is increased, and the stress concentration of the contact between the tool head and the sheet is relieved. From Figure 8 we can see that there is a law of strain; the maximum area of strain is in the main deformation zone, and the maximum strain of the three is not large.

4.2 *Effect of layer spacing*

In the process of sheet metal forming, the two process parameters, the vertical displacement and the horizontal offset, are considered in the planning process. The drop in the vertical direction of the tool head each walk round is layer spacing. When machining the same vertical depth, the larger the layer spacing, the less time it is used. This paper in the case of other process parameters unchanged, making forming simulation of conical parts with different layer spacing. The tool head diameter is d = 6 mm, the forming depth is 20 mm, the layer spacing ΔZ is respectively selected as 0.5 mm,

1 mm and 2 mm. It can be seen from Figure 9 that when the layer spacing is 0.5 mm, the maximum equivalent stress of sheet material is 338.9 MPa; when the layer spacing is 1 mm, the maximum equivalent stress is 383.7 MPa; when the layer spacing is 2 mm, the maximum equivalent stress is 416.9 MPa. It can be seen from the above that the maximum value of the equivalent stress of the plate is proportional to the distance between the layers. As the increase of layer spacing will make the vertical feeding distance of tool head in contour processing of each circle increases, which leads to the contact area of the tool head with the sheet increases, which required the extrusion pressure of forming of sheet metal increases, the equivalent stress of the sheet increases by extrusion pressure increasing.

From Figure 10 we can see that the layer spacing respectively is 0.5 mm, 1 mm and 2 mm. The strain in the main deformation zone of sheet metal forming is basically the same. To sum up, the other process parameters are unchanged, and with an increase in the layer spacing, the equivalent stress of the plate also increases. However, with an increase in forming tensile stress, the sheet will have reaction force on the tool head, influencing the tool head service life. Smaller layer spacing can make the machining process take longer to form.

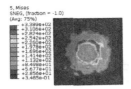

(a) $\Delta Z = 0.5\,mm$

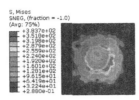

(b) $\Delta Z = 1\,mm$

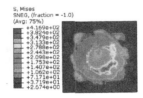

(c) $\Delta Z = 2\,mm$

Figure 9. Forming von Mises cloud images of different layers.

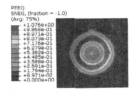

(a) $\Delta Z = 0.5\,mm$

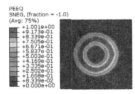

(b) $\Delta Z = 1\,mm$

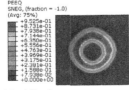

(c) $\Delta Z = 2\,mm$

Figure 10. Strain PEEQ cloud images of different layers.

Therefore, in the process of sheet metal processing, both the forming precision and the quality should be considered and to achieve processing efficiency the best layer spacing should be selected.

4.3 The effect of half angle

Among the parameters that affect the incremental sheet forming process, half angle is an important factor affecting the sheet metal forming limit. If the forming angle is too small, the sheet processing is cracks easily; if the angle is too large it is fairly easy to produce an indentation. The 6061 aluminium alloy material forming limit corresponding to the maximum angle, so after forming half angle cannot be less than the limit angle, otherwise the material will have severe thinning, the problem that the material is extruded and broken or the material strength is not up to the requirement. Here for 6061 aluminium alloy materials with other process parameters unchanged, making numerical simulation respectively on the forming process of conical parts with different forming half angle. For the wall angle α obtained by the relationship between the horizontal feed h_1 and vertical feed h_2, the formula is $\alpha = \arctan\left(h_1 / h_2\right)$, as shown in Table 2.

It can be seen from Figure 11 that the maximum equivalent stress reached 596.6 MPa, far more than the material yield pole 296 MPa. In the marked position on Figure 11(a) we can see that a sheet crack has occurred. It can be seen from the figure that, with the increase in the horizontal feed h_1, the forming radius of the bottom surface decreases, which leads to more and more serious stress of the bottom surface. When $h_1 = 1.5$ mm, the equivalent stress in the centre of the lower bottom is 421 MPa–316 MPa. When $h_1 = 1$ mm, the equivalent stress in the bottom of the cone is only 160 MPa–96 MPa. The equivalent stress distribution becomes more and more uniform with the increase of the horizontal feed in the main deformation area of the plate.

Figure 12 shows the thickness distribution as $\alpha = 45^0$ and $\alpha = 26.56^0$. As can be seen from the figure, when $\alpha = 45°$ the thickness change is more uniform, and when $\alpha = 26.56°$ the thickness change is quite different. As shown in Figure 13, the thickness distribution curve of the conical

Table 2. The amount of feed and forming half-apex angle.

Horizontal feed h_1/mm	Vertical feed h_2/mm	Half angle α
0.5	1	26.56⁰
1	1	45⁰
1.5	1	56.31⁰

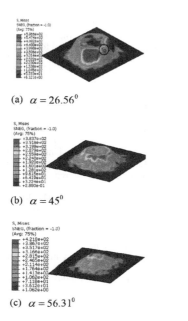

(a) $\alpha = 26.56^0$

(b) $\alpha = 45^0$

(c) $\alpha = 56.31^0$

Figure 11. Von Mises stress nephogram of different forming angles.

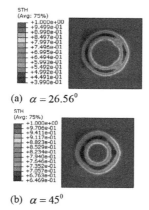

(a) $\alpha = 26.56^0$

(b) $\alpha = 45^0$

Figure 12. Forming thickness of different half angle forming cloud images.

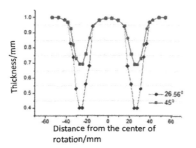

Figure 13. Comparison of the thickness under different half angles.

part along the radial compared with the half angle $\alpha = 45°$ and $\alpha = 26.56°$. From the curve we can see that the forming thickness of $\alpha = 26.56°$ is less than the thickness of $\alpha = 45°$ and thickness distribution of $\alpha = 45°$ is uniform after forming. The thickness of deformation zone of the sheet is in the range of $0.67 - 0.71\,mm$. This is consistent with the plate thickness theory of $t = t_0 \sin \alpha$ (Zhou et al., 2003; Li et al., 2011). When $\alpha = 26.56°$, the thickness of the plate forming area is larger change, the thinnest position has occurred. In the actual processing, there is need to select the right half angle which cannot exceed the material forming limit angle.

5 CONCLUSION

In this paper, the forming region of conical parts formed by 6061 aluminium alloy sheet is theoretically modelled. The stress and strain of the contact area between the tool head and the plate are analysed. Getting influence law of the process parameters such as tool radius, layer spacing, half angle on forming regional stress and strain by numerical simulation. The distribution of stress and strain under different process parameters was simulated by the finite element method. According to the theoretical and numerical methods, the following conclusions are drawn:

1. In the case of other parameters remaining unchanged, the bigger the radius of the tool head, the larger the stress on the inside of the sheet. But the stress from the inside to the outside of the sheet is constantly changing and not increasing all the time. The larger the radius of tool head so as the contact area between the plate and the plate, the stress concentration can be reduced effectively. According to the actual needs to choice tool head radius.
2. In the case of other parameters remaining unchanged, the larger the distance between the layers, the larger the tensile stress generated by extrusion, the larger the contact area between the tool head and the sheet. At the same time, the reaction force of the sheet to the tool head will increase, so that the wear of the tool is relatively large, which will affect the service life. But the larger layer spacing can shorten the forming time and improve the processing efficiency.
3. When the influence of half angle on uniformity of thickness of forming of 6061 aluminium alloy sheet is large, the half angle increases, the thickness distribution is uniform, and sheet thinning is small. When the half angle decreases, the thickness distribution is uneven, and the larger the sheet thinning is. Each material corresponds to a forming limit angle, which

cannot exceed the limit angle in the actual processing. Otherwise there may be material rupture, so the half angle can be used as an evaluation index of the sheet metal incremental forming limit.

REFERENCES

Arfa, H., Bahloul, R., & Belhadj, S.H. (2013). Finite element modelling and experimental investigation of single point incremental forming process of aluminium sheets. *Springer, 6*, 483–510.

Ayed, L.B., Robert, C., & Delamézière. (2014). A simplified numerical approach for incremental sheet metal forming process. *Engineering Structures*, 75–86.

Fanga, Y., Lu, B., Chen, J., Xu, D.K., & Ou, H. (2014). Analytical and experimental investigations on deformation mechanism and fracture behavior in single point incremental forming. *Journal of Materials Processing Technology*, 1503–1515.

Hu, J.B., Zhou, J., & Li, J.C. ect. (2011). Research on sheet metal incremental forming process based on numerical simulation. *Hot Working Process*, (15), 76–78.

Li, J.C., Zhang, X., & Pang, S.T. (2011). Experimental and numerical research on thickness variation of dieless sheet metal incremental forming. *Hot Working Technology, 40* (7), 1–4.

Mo, J.H., Ye, C.S., & Huang, S.K. (2002). Numerical control incremental forming technology of metal sheet. *Aviation Manufacturing Technology*, (12), 25–27.

Sun, L.M., Tian, N., & Cong, F.G. (2014) Study on the strain hardening exponent of 6016 aluminium alloy sheet for automotive body. *42*, 11.

Xu, B.Y., & Liu, X.S. (1995). *Application of elastic and plastic mechanics*. Tsinghua University Press.

Zhang, Y., Li, P.Y., & Wang, Q.D. (2015). Numerical analysis of single point incremental forming based on ABAQUS. *Mechanical Strength, 37*(1), 99–103.

Zhou, L.R., Xiao, X.Z., & Mo, J.H. (2003). Research on NC incremental forming process of sheet metal parts. *Journal of Plasticity Engineering*, (4), 27–29.

Zhou, L.R. (2004). *The study of forming mechanism and process of NC incremental metal sheet* (dissertation). Wuhan: Huazhong University of Science and Technology.

Zhuang, Z., & You, X.C. (2009). *Finite element analysis and application based on ABAQUS*. Tsinghua University Press.

Network, communications and applied information technologies

Automotive, Mechanical and Electrical Engineering – Liu (Ed.)
© *2017 Taylor & Francis Group, London, ISBN 978-1-138-62951-6*

Design of a single-input and dual-input reconfigurable pipelined Analogue to Digital Converter (ADC)

Xin Yi & Yongsheng Yin
Institute of VLSI Design, Hefei University of Technology, Hefei, China

Zhijian Xie
NC A & T State University, Greensboro, USA

ABSTRACT: According to the characteristics of multi-standard wireless communication systems, a new type of single-input and dual-input reconfigurable pipelined Analogue to Digital Converter (ADC) were designed. The single-input reconfigurable ADC can be achieved by shutting down sub-stages gradually, and its systematic index is 8–14 bit without deteriorating other performance aspects. It can also reduce the power consumption of the system. The dual-input ADC can acquire two input signals simultaneously and be extended to more input signals as needed. It was achieved by external control logics and algorithm adjustment in digital correction. The proposed architecture can achieve a distribution of 14-bit precision: 7b+8b and 5b+10b respectively. Both reconfigurable pipelined ADCs were designed in Simulink models and the simulation were implemented.

Keywords: wireless communication; pipeline A/D converter; reconfigurable control; precision distribution; dual-input system

1 INTRODUCTION

With the development of wireless communication technology, there is an increasing demand for speed and accuracy of A/D converters (Chen Zhenyu et al., 2012). Because different standards allow different input signal frequency ranges, bandwidth, and dynamic ranges, the required A/D converter operating mode for each protocol will not be the same. So it is essential to enable a variety of different protocol standards coexist. In order to meet the needs of the different properties in multi-protocol wireless terminal applications, multi-input ADC systems are becoming more and more necessary. A pipelined architecture ADC can realise high precision in digital to analogue conversion as well as maintain a high speed and low power consumption because of its structural characteristics (Gu Anqiang, 2008). Moreover, the design freedom and flexibility of a pipeline itself provide a broad research space for the optimisation problem of speed, power, voltage and area (Yin Yongsheng, 2006). Through the research, it is possible for a pipelined ADC to change its operating mode to achieve adjustable accuracy, thereby reducing the device's power consumption and improving

data processing speed. Research on this subject is of great value.

2 DESIGN

2.1 *The requirements of different wireless standards systems for an A/D converter*

The standards of different wireless communication systems for an A/D converter are shown in Table 1 (Stojcevski A et al., 2002). The required resolution ranges from 6 to 14 bit and the sampling rate varies from 10 to 30 MS/s. For future communication systems, one important characteristic is that there is a collection of various services in multi-standard radio terminals.

Table 1. Basic characteristics of different wireless standards.

Standards	GPS	GSM	Blue tooth	WCDMA	WLAN 802.11a	WLAN 802.11b	WLAN 802.11g
Bandwidth /MHz	2	200	1	3.84	20	22	22
Resolution/b	10	12~14	13	6~8	10~14	6~8	10~14

2.2 Design of single-input reconfigurable pipelined ADC

2.2.1 Single-input pipelined ADC using scaling down technology

Based on the analysis of existing ADC structures applied to multi-standard wireless terminals (Van de Vel H et al., 2008), it is clear that in the design of pipelined ADCs, the former stages consume a greater proportion of the power consumption than the latter ones.

For this reason, a novel reconfigurable control method—scaling down technique—was proposed. Chen Zhenyu used this method, powering off the front-stages. Although this method can reduce the power consumption in a great proportion, it does not go for the high-resolution mode. In fact, for pipeline structure, the requirements of accuracy and other performance are different in each stage. The first stage requires the highest property, which almost determines the overall system performance.

The overall reconfigurable pipelined ADC structure is shown in Figure 1.

As shown in Figure 1, the single-input reconfigurable pipelined ADC designed in this paper consists of the following components: a reconfigurable programmable controller, six precision control switches K1-K6, sample and hold circuit, 10 controllable pipeline stages (1.5 bit/stage), a delay alignment circuit and a digital correction circuit. The working state of each stage is controlled by a programmable state control unit.

The signal F_{RC} is from the system level. Vin is the analogue input signal. Figure 1 shows that the improved reconfigurable pipelined ADC is designed on the basis of 14 stages (1.5 bit/stage) pipeline ADC structure with an introduced reconfigurable programmable control module. The proposed design is able to adjust precision by means of scaling down the latter pipeline stages gradually, which can achieve a dynamic change of precision—in the bits of 8, 9, 10, 11, 12, 13, 14. Meanwhile, stages that are not involved in contributing to the resolution of the structure can be shut off, entering to sleep mode and saving power consumption.

2.2.2 How does a single-input reconfigurable pipelined ADC work

A reconfigurable entire pipelined ADC works in this way: First, the system will send F_{RC} to the reconfigurable controller to select 6 different states—a, b, c, d, e, f, which can be set to 0 or 1 respectively, controlling the on and off of the K1-K6 switches so that the resolution changes according to the different states of the switches.

2.3 Design of dual-input reconfigurable ADC

2.3.1 The architecture of dual-input reconfigurable ADC

The overall architecture of dual-input reconfigurable ADC is shown in Figure 2. The dual-input structure can be reconstructed on the basis of a 14-bit pipelined ADC by adding some control logic circuits. The analogue input Vin1 goes through the first S/H circuit, while Vin2 chooses to be sampled by the middle stage (here is the third and fifth stage). S2 is a reconfigurable control signal, it is used to select which stage Vin2 is supposed to enter. Besides, S2 also controls the algorithm in the digital correction circuit, which guarantees

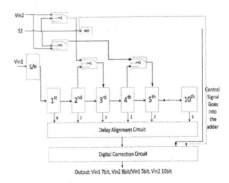

Figure 2. Overall architecture of dual-input reconfigurable pipelined ADC.

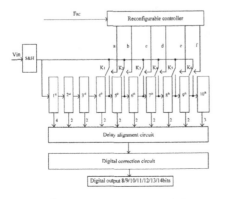

Figure 1. Overall configuration of single-input reconfigurable pipelined ADC.

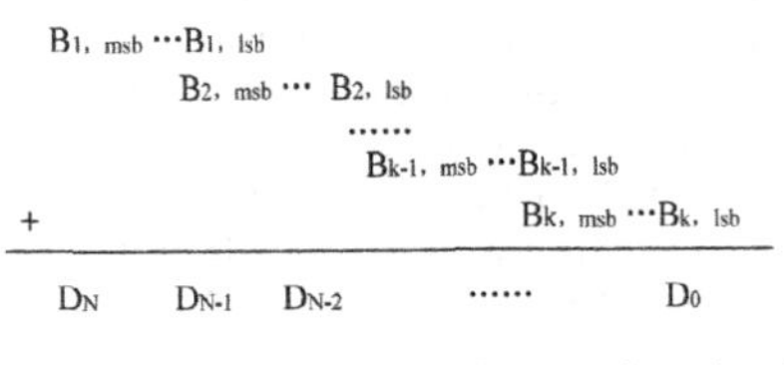

Figure 3. Pipelined ADC digital correction algorithm.

that the two outputs are converted correctly. The advantage of this circuit is that only one control signal is used to control all circumstances.

2.3.2 The algorithm and principle of a dual-input reconfigurable ADC

In the case of two inputs, we hope to achieve a precision-assignable mode. That is, the 14 bits of pipelined ADC can be distributed to X + (14-X) mode.

First, the pipelined ADC digital correction algorithm is as follows:

According to this scheme, we can achieve the correct conversion of two inputs by making changes in the digital correction algorithm. Because of the spit of a certain bit, the entire resolution will increase one bit. In terms of this dual-input system, total resolution will be 15 bit.

The whole control process is as follows:

If the control signal S2 is valid, Vin2 selects to input from the third stage, passing through the fourth, fifth level up to the tenth. In a similar way, S2 cuts off the carry-bit between the second and third stages, and outputs B20 (before algorithm) as the LSB of Vin1, and outputs B31 (after algorithm) as the MSB of Vin2. We obtain Vin1 for 5 bit and Vin2 for 10 bit precision.

If the control signal S2 is invalid, then Vin2 selects to input from the fifth stage. So we avoid the addition of B40 and B51, output B40 (before algorithm) as the LSB of Vin1, and output B51 (after algorithm) as the MSB of Vin2. In this case, Vin1 is 7 bit and Vin2 is 8 bit.

3 SIMULATION

3.1 The performance simulation of a single-input reconfigurable ADC

Both reconfigurable pipelined ADCs were designed in Simulink models and the simulations were performed.

FRC controls the switches on and off to output distinct accuracy respectively. The simulation results are as follows:

Table 2. The performance of a single-input reconfigurable ADC in different control modes.

Control signal F_{RC}	Resolution
$F_{RC} = 1$	13.8 bit
$F_{RC} = 2$	11.86 bit
$F_{RC} = 3$	10.9 bit
$F_{RC} = 4$	9.87 bit
$F_{RC} = 5$	8.87 bit
$F_{RC} = 6$	7.85 bit

From Table 2 we can see that the designed single-input reconfigurable ADC can achieve the required performance in 6 different resolution modes. Under the requirements of the aforementioned communication standards, its effective resolution almost reaches the ideal value (only less than 0.2 bit), which have met the demands successfully.

In addition, the power consumption expression of the reconfigurable pipelined ADC is as follows (Arias J, 2005):

$$P_{ADC} = nP_{block} + P_{S/H} + P_{DSP} + P_{DEC} \qquad (1)$$

where n is the resolution of the ADC output, P_{block} is the power consumed by every stage, $P_{S/H}$ is the power consumption of the S/H circuit, P_{DSP} is the power consumption of the reconfigurable controller, and P_{DEC} is the power consumption of is digital correction circuit.

The power consumption curve is shown below in Figure 4. It can be seen that the circuit has a positive effect on power reduction when the system does not need high resolution, also it minimises the number of control signals. Power consumption is close to linearly proportional to the number of bits, i.e. 8-bit mode consumes about four seventh of the power in 14-bit mode. There is no doubt that the power efficiency has been improved.

3.2 The performance simulation of a single-input reconfigurable ADC

As for dual-input reconfigurable ADC, the simulation is under different operating modes, also using Simulink models. The simulation results are shown in Figures 5 to 8.

From the simulation results, we can see that when S2 is valid, the actual precision of Vin1 is 4.97 bit and Vin2 is 9.78 bit. When S2 is invalid, Vin1 is 7.00 bit and Vin2 is 7.84 bit. Both performances are close to the ideal value.

From Figures 5 to 8 we can see that the reconfigurable ADC can achieve satisfactory performance

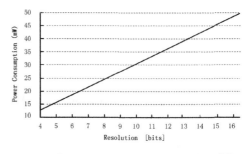

Figure 4. Power consumption curve in different resolutions.

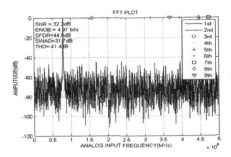

Figure 5.   The output of Vin1 when s2 is valid.

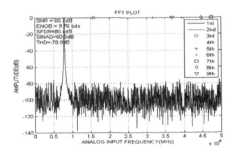

Figure 6.   The output of Vin2 when s2 is valid.

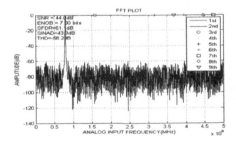

Figure 7.   The output of Vin1 when s2 is invalid.

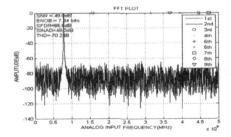

Figure 8.   The output of Vin1 when s2 is invalid.

Table 3.   Simulation of two different modes.

Control Mode	Vin1	Vin2
S2 = 1	4.97 bit	9.78 bit
S2 = 0	7.00 bit	7.84 bit

in two operating modes. Its effective resolution has met the standards presented in Table 1. It means that a novel reconfigurable pipeline architecture can be realised without increasing circuit complexity and degrading system performance.

4   CONCLUSION

In this paper, a single-input and a dual-input reconfigurable ADC are designed on a system-level and the simulation shows that both structures can achieve good performance in a variety of circumstances. The design has met the multi-mode operation required by a multi-standard wireless terminal. The former have used simply several control switches and control logics to change the number of pipeline stages dynamically. The latter have managed to achieve a precision-assignable mode according to system requirements. Both proposed structures can improve the power efficiency and save area. In this way the unnecessary power consumption can be reduced. Also, the method used in the dual-input structure can be extended to more input signals as needed. The paper is a good guide to a reconfigurable pipelined ADC design. The intensive study needs is transistor-level circuit design.

REFERENCES

Arias J. (2005). Low-Power pipeline ADC for wireless LANs. *IEEE Journal of Solid-State Circuits, 39(8)*, 1338–1340.

Chen Zhenyu, Wang Lizhi, Yang Yong. (2012). A design of a reconfigurable pipeline ADC., *35(12)*, 146–148–385, Xian: Air Force Engineering University.

Gu Anqiang. (2008). Reconfigurable pipelined ADC research suitable for multi-standard wireless terminal. Wuhan: Huazhong University of Science and Technology.

Stojcevski A, Singh A, Zayegh A. (2002). A reconfigurable ADC for UTRA-TDD mobile terminal receiver. *Proceedings of the 45th IEEE Midwest Symposium on Circuits and Systems*, Oklahoma, IEEE Solid-State Circuits Society, Pp. 613–616

Van De vel H, Buter B, Van Der ploeg H, et al. A 1.2 V 250 mW 14b 100MS/s digitally calibrated pipeline ADC in 90 nm CMOS. *Proceedings of the IEEE Symposium on VLSI Circuits*, 74–75.

Yin Yongsheng. (2006). Research on reconfigurable pipeline computing systems.. Hefei: Institute of VLSI Design, Hefei University of Technology.

Automotive, Mechanical and Electrical Engineering – Liu (Ed.)
© 2017 Taylor & Francis Group, London, ISBN 978-1-138-62951-6

A new method of particle filter location based on multi-view fusion

Zhongliang Deng, Yan Zhou, Jichao Jiao & Chenyang Zhai
Beijing University of Posts and Telecommunications, Beijing, China

ABSTRACT: With the development of wireless communication technology and the prevalence of Location-Based Service (LBS), wireless location technology has become a research hotspot. Among wireless positioning technologies, TDOA positioning technology, with its high precision and wide range of applications, has been getting considerable attention. However, the TDOA technique is also easily disturbed by noise and other interferences, and the accuracy and stability of the technology will be affected. Therefore, other positioning technologies are needed for a better result. This paper proposes a new TDOA/RSSI fusion location method based on the idea of the particle filter. Also, this paper proposes a real-time calculation of the accuracy of TDOA location technology to meet the requirements of the fusion method for real-time TDOA location and the accuracy range of the results.

Keywords: TDOA; Accuracy analysis; Fusion localisation; Particle filter

1 INTRODUCTION

With the popularity of mobile Internet and mobile devices, positioning technology has become the focus of navigation technology research. The Beidou and GPS-led satellite navigation system, under the support of an outdoor positioning navigation service, has become an important part of people's lives. Items such as Baidu map and Tak Tak map rely on location-based applications such as mushrooming after another. Common methods of indoor positioning include the TDOA positioning method, RSSI positioning method, AOA positioning method and so on. The TDOA method has the advantages of high positioning accuracy and convenient base station layout, and is widely used in indoor location technology.

However, the TDOA technology has its shortcomings. When faced with a complex indoor environment the signal, before arriving at the mobile station, often goes through the reflex, refraction process, which makes the mobile station received signal propagation time and true location of the signal transmission time inconsistent, resulting in a solution between the results and the true position of the deviation. Therefore, fusion localisation through a variety of algorithms has become the main research direction, and the particle filter, with its wide range of applications and excellent positioning accuracy, has been valued. The prerequisite for implementing the particle filter is to get the resolution accuracy of the real-time positioning method. So, how to get the resolution precision of the TDOA algorithm in real time becomes the focus of the research.

2 TDOA SOLUTION METHOD

The TDOA location method avoids the clock synchronisation requirement between the mobile station and the base station by measuring the propagation time difference between different base stations and the mobile station. When the arrival time difference between a mobile station and a neighbouring base station is known, it is possible to know that the position of the mobile station is on the hyperbola with the two base stations as the focus, and the intersection of the hyperbola for a plurality of base stations is the required mobile station location. In the process of particle filter fusion, we need to estimate the variance of the solution of TDOA, and a now commonly used method is the mobile end of the case known TDOA solution to the results of statistics, resulting in the results and real value. This method requires knowledge of the true location of the mobile terminal, while not meeting the requirements of real-time. This makes it necessary to analyse TDOA results from the TDOA solution process. In the TDOA positioning solution process, the Newton iterative method is the most commonly used solution. Using the Newton iterative method, we need to know the location of each base station received and the difference in signal propagation time of each base station received. Wherein each base station location is denoted as (x_i, y_i, z_i), and the signal propagation time difference is multiplied by the speed of light to find the signal propagation distance difference, expressed as Los_{ij}. If (x, y, z) denotes the mobile station coordinate, then relationship between them is as follows:.

$$\begin{cases} Los_{12} = \sqrt{(x-x_1)^2 + (y-y_1)^2 + (z-z_1)^2} \\ \quad -\sqrt{(x-x_2)^2 + (y-y_2)^2 + (z-z_2)^2} \\ Los_{13} = \sqrt{(x-x_1)^2 + (y-y_1)^2 + (z-z_1)^2} \\ \quad -\sqrt{(x-x_3)^2 + (y-y_3)^2 + (z-z_3)^2} \\ Los_{14} = \sqrt{(x-x_1)^2 + (y-y_1)^2 + (z-z_1)^2} \\ \quad -\sqrt{(x-x_4)^2 + (y-y_4)^2 + (z-z_4)^2} \\ \cdots \end{cases} \quad (1)$$

Take $R_i = \sqrt{(x-x_i)^2 + (y-y_i)^2 + (z-z_i)^2}$, the equation can be simplified to $Los_{ij} = R_i - R_j = t_{ij}c$, wherein $t_{ij} = t_i - t_j$, the difference in the propagation time between the base station i, j received by the mobile station.

As a recursive algorithm, the Newton iteration requires an initial estimated position as the initial point of the recursion, and then in the recursive process by TDOA measurement error of the local least-square solution, until the error is less than the set threshold when the output. Take the initial set position of the mobile station (x', y', z'). According to Taylor, by an expansion of Los_{ij}, we can get:

$$\begin{aligned} F(x, y, z) = F(x', y', y') &+ F'(x', y', z')\Delta x \\ &+ F'(x', y', z')\Delta y + F'(x', y', z')\Delta z \\ &+ o(x', y', z') \end{aligned} \quad (2)$$

Ignoring the higher-order small quantities, the formula $G\delta = b$ can be obtained. In this formula:

$$G = \begin{bmatrix} \dfrac{x_1 - x'}{R'_1} - \dfrac{x_2 - x'}{R'_2} & \dfrac{y_1 - y'}{R'_1} - \dfrac{y_2 - y'}{R'_2} & \dfrac{z_1 - z'}{R'_1} - \dfrac{z_2 - z'}{R'_2} \\ \dfrac{x_1 - x'}{R'_1} - \dfrac{x_3 - x'}{R'_3} & \dfrac{y_1 - y'}{R'_1} - \dfrac{y_3 - y'}{R'_3} & \dfrac{z_1 - z'}{R'_1} - \dfrac{z_3 - z'}{R'_3} \\ \dfrac{x_1 - x'}{R'_1} - \dfrac{x_4 - x'}{R'_4} & \dfrac{y_1 - y'}{R'_1} - \dfrac{y_4 - y'}{R'_4} & \dfrac{z_1 - z'}{R'_1} - \dfrac{z_4 - z'}{R'_4} \end{bmatrix} \quad (3)$$

$$b = \begin{bmatrix} ct_{12} - (R'_2 - R'_1) \\ ct_{13} - (R'_3 - R'_1) \\ ct_{14} - (R'_4 - R'_1) \end{bmatrix} \quad (4)$$

Can be solved by the least squares method

$$\delta = \begin{bmatrix} \Delta x \\ \Delta y \\ \Delta z \end{bmatrix} = (G^{-1}Q^{-1}G)^{-1}G^T Q^{-1}b \quad (5)$$

Q is the distance covariance matrix.

After obtaining $(\Delta x, \Delta y, \Delta z)$, the initial value (x', y', z') will be replaced with $(x'', y'', z'') = (x' + \Delta x, y' + \Delta y, z' + \Delta z)$, then repeat the entire process until $\Delta x, \Delta y, \Delta z$ is below the set threshold value and the three-dimensional coordinates of this time are the required coordinates.

3 TDOA ERROR ANALYSIS

In the actual TDOA solution, we cannot always get the arrival time difference between the signals of different base stations due to the interference of non-line-of-sight errors and the propagation condition of the channel. As a result, our signal propagation distance difference formula is corrected to
$$Los_{ij} = R_i - R_j = (t_i - t_j)c + (\varepsilon_i - \varepsilon_j)c = (t_{ij} + \varepsilon_{ij})c$$

Wherein, ε_i is the error of the arrival time of the signal of the ith road base station. From the signal propagation conditions it can be seen that ε_i is subject to a Gaussian distribution with a mean of 0 and a variance of σ_i^2.

After this correction, the original solve equation $G\delta = b$ updates to $G\delta_{new} = b_{new}$

$$G = \begin{bmatrix} \dfrac{x_1 - x'}{R'_1} - \dfrac{x_2 - x'}{R'_2} & \dfrac{y_1 - y'}{R'_1} - \dfrac{y_2 - y'}{R'_2} & \dfrac{z_1 - z'}{R'_1} - \dfrac{z_2 - z'}{R'_2} \\ \dfrac{x_1 - x'}{R'_1} - \dfrac{x_3 - x'}{R'_3} & \dfrac{y_1 - y'}{R'_1} - \dfrac{y_3 - y'}{R'_3} & \dfrac{z_1 - z'}{R'_1} - \dfrac{z_3 - z'}{R'_3} \\ \dfrac{x_1 - x'}{R'_1} - \dfrac{x_4 - x'}{R'_4} & \dfrac{y_1 - y'}{R'_1} - \dfrac{y_4 - y'}{R'_4} & \dfrac{z_1 - z'}{R'_1} - \dfrac{z_4 - z'}{R'_4} \end{bmatrix} \quad (6)$$

$$b_{new} = \begin{bmatrix} ct_{12} + c\varepsilon_{12} - (R'_2 - R'_1) \\ ct_{13} + c\varepsilon_{13} - (R'_3 - R'_1) \\ ct_{14} + c\varepsilon_{14} - (R'_4 - R'_1) \end{bmatrix} = b + \varepsilon_\rho \quad (7)$$

$$\delta_{new} = \begin{bmatrix} \Delta x \\ \Delta y \\ \Delta z \end{bmatrix} + \begin{bmatrix} \varepsilon_x \\ \varepsilon_y \\ \varepsilon_z \end{bmatrix} = (G^{-1}Q^{-1}G)^{-1}G^T Q^{-1}b_{new} \quad (8)$$

Wherein $\varepsilon_\rho = \begin{bmatrix} c\varepsilon_{12} \\ c\varepsilon_{13} \\ \cdots \\ c\varepsilon_{1i} \end{bmatrix}$ is the measurement error matrix, $\begin{bmatrix} \varepsilon_x \\ \varepsilon_y \\ \varepsilon_z \end{bmatrix}$ is the result error matrix, $\varepsilon_x, \varepsilon_y, \varepsilon_z$

are the projection vector of the error between the mobile station position coordinate and the actual position coordinate of the mobile station in three dimensions.

Due to:

$$\delta_{new} = (G^{-1}Q^{-1}G)^{-1}G^TQ^{-1}b_{new}$$
$$= (G^{-1}Q^{-1}G)^{-1}G^TQ^{-1}(b + \varepsilon_\rho) \qquad (9)$$

$$\delta = \begin{bmatrix} \Delta x \\ \Delta y \\ \Delta z \end{bmatrix} = (G^{-1}Q^{-1}G)^{-1}G^TQ^{-1}b \qquad (10)$$

We can obtain:

$$\begin{bmatrix} \varepsilon_x \\ \varepsilon_y \\ \varepsilon_z \end{bmatrix} = \delta_{new} - \begin{bmatrix} \Delta x \\ \Delta y \\ \Delta z \end{bmatrix} = (G^{-1}Q^{-1}G)^{-1}G^TQ^{-1}\varepsilon_\rho \qquad (11)$$

When the time of arrival of each signal is independent of each other, there is:

$$Q = Q^{-1} = E \qquad (12)$$

The error formula can be simplified as:

$$\begin{bmatrix} \varepsilon_x \\ \varepsilon_y \\ \varepsilon_z \end{bmatrix} = (G^{-1}G)^{-1}G^T\varepsilon_\rho \qquad (13)$$

The error covariance matrix of the arrival time difference location can be:

$$A = Cov\left(\begin{bmatrix} \varepsilon_x \\ \varepsilon_y \\ \varepsilon_z \end{bmatrix}\right) = E\left(\begin{bmatrix} \varepsilon_x \\ \varepsilon_y \\ \varepsilon_z \end{bmatrix}\begin{bmatrix} \varepsilon_x & \varepsilon_y & \varepsilon_z \end{bmatrix}\right)$$

$$= E\left((G^TG)G^T\varepsilon_\rho((G^TG)G^T\varepsilon_\rho)^T\right)$$

$$= (G^TG)^{-1}G^TE(\varepsilon_\rho\varepsilon_\rho^T)G(G^TG)^{-1}$$

$$= c^2(G^TG)^{-1}G^T\begin{bmatrix} D(\varepsilon_{12}) & 0 & 0 \\ 0 & D(\varepsilon_{13}) & 0 \\ 0 & 0 & D(\varepsilon_{14}) \end{bmatrix}$$
$$\times G(G^TG)^{-1}$$

$$= c^2(G^TG)^{-1}G^T\begin{bmatrix} \sigma_1^2 + \sigma_2^2 & 0 & 0 \\ 0 & \sigma_1^2 + \sigma_3^2 & 0 \\ 0 & 0 & \sigma_1^2 + \sigma_4^2 \end{bmatrix}$$
$$\times G(G^TG)^{-1}$$
$$(14)$$

The trace of matrix A is the variance of the final result of the TDOA localisation.

At last, we complete the derivation of the TDOA precision real-time calculation formula, and through this formula, we can calculate the variance range of the TDOA positioning result in real time by the spatial position of each signal transmitting base station and the statistical characteristics of the signal propagation time, in order to use TDOA precision as the basis for fusion of the particle filter method to lay the foundation.

4 PARTICLE FILTERING METHOD BASED ON TDOA AND RSSI RESULTS

The core of the particle algorithm is to scatter particles near the possible true result positions, and then calculate the probability density function of each particle as the correct result according to the results of the TDOA and RSSI localisation methods. Then eliminate the less likely particles, the probability of high-particle re-sampling and the total number of particles in the same conditions to increase the number of high probability particles, and then the weight of particles and particle coordinates to be weighted fusion results.

As a result of cost constraints, a smart terminal positioning of mobile devices with only low accuracy of the compass and gyroscope is commonly used, which cannot simply rely on these two sensors to locate. Therefore, the combination of the actual situation and the use of the compass as a predictor of particle position transfer equation parameters is a better choice.

The state of the particle filter can be described as:

$$\begin{cases} x_k = f(x_{k-1}) + u_{k-1} \\ y_k = h(x_k) + v_k \end{cases} \qquad (15)$$

Wherein the $f(\cdot), h(\cdot)$ are the particle filter state transition equation and observation equation, where x_k represents the system state, y_k represents observations, u_k represents process noise, v_k represents observation noise. Use:

$$\begin{cases} X_k = x_{0:k} = \{x_0, x_1, \cdots, x_k\} \\ Y_k = y_{1:k} = \{y_1, \cdots, y_k\} \end{cases} \qquad (16)$$

Respect all status and observations from 0 to k moment. In analysing the continuous positioning problem it is usually assumed that the target state transition process to obey a first-order Markov chain model that the current time status x_k only previous time the state of the relevant x_{k-1}. Another hypothesis is independent observations,

105

that the observation y_k is only with statement x_k, the state at k time. In these two assumed conditions, the importance of particle filter probability density function $q(x_{0:k} \mid y_{1:k})$ can be decomposed into:

$$q(x_{0:k} \mid y_{1:k}) = q(x_{0:k-1} \mid y_{1:k-1})q(x_k \mid x_{0:k-1}, y_{1:k}) \quad (17)$$

Let the system state is a Markov process, and the system at a given state of each independent observations, there

$$p(x_{0:k}) = p(x_0)\prod_{i=1}^{k} p(x_i \mid x_{i-1}) \quad (18)$$

$$p(y_{1:k} \mid x_{1:k}) = \prod_{i=1}^{k} p(y_i \mid x_i) \quad (19)$$

Therefore, the posterior probability density function can be expressed as a recursive form:

$$
\begin{aligned}
p(x_{0:k} \mid Y_k) &= \frac{p(y_k \mid x_{0:k}, Y_{k-1})p(x_{0:k} \mid Y_{k-1})}{p(y_k \mid Y_{k-1})} \\
&= \frac{p(y_k \mid x_{0:k}, Y_{k-1})p(x_k \mid x_{0:k-1}, Y_{k-1})p(x_{0:k-1} \mid Y_{k-1})}{p(y_k \mid Y_{k-1})} \\
&= \frac{p(y_k \mid x_k)p(x_k \mid x_{k-1})p(x_{0:k-1} \mid Y_{k-1})}{p(y_k \mid Y_{k-1})}
\end{aligned}
$$

$$(20)$$

Recursive form particle weight $w_k^{(i)}$ can be expressed as:

$$
\begin{aligned}
w_k^{(i)} &\propto \frac{p(x_{0:k}^{(i)} \mid Y_k)}{q(x_{0:k}^{(i)} \mid Y_k)} \\
&= \frac{p(y_k \mid x_k^{(i)})p(x_k^{(i)} \mid x_{k-1}^{(i)})p(x_{0:k-1}^{(i)} \mid Y_{k-1})}{q(x_k^{(i)} \mid x_{0:k-1}^{(i)}, Y_k)q(x_{0:k-1}^{(i)} \mid Y_{k-1})} \\
&= w_{k-1}^{(i)} \frac{p(y_k \mid x_k^{(i)})p(x_k^{(i)} \mid x_{k-1}^{(i)})}{q(x_k^{(i)} \mid x_{0:k-1}^{(i)}, Y_k)}
\end{aligned}
$$

$$(21)$$

In order to meet the efficiency of the calculation, the transfer probability density function of the state variable $p(x_k \mid x_{k-1})$ can be used as the approximation of the importance probability density function. At this time, the weight of the particle formula can be transformed into:

$$w_k^{(i)} = w_{k-1}^{(i)} p(y_k \mid x_k^{(i)}) \quad (22)$$

Particle filter operation steps:

1. Extract the calculation result (x_t, y_t) obtained by TDOA and the variance of results σ_t^2, the results obtained from Section 3 of the conclu-

sions, (x_t, y_t) is a position in the Cartesian coordinate system. The results of TDOA are subject to a Gaussian distribution with a mean of (x_t, y_t) and a variance of σ_t^2. Then obtain the RSSI calculation result (x_r, y_r), and from the RSSI channel conditions get σ_r^2.

2. Assume that the time is k, when $k = 1$, initialise, take two positions in the plane as the two-dimensional uniform distribution of a particle. Each particle has a weight average $w_1^i = \frac{1}{N}$ When k is not 1, go directly to the next step.

3. Since the system is mainly used for human localisation, the motion model of the particle is the same as the human walking model, and can be described by the following equation:

$$\begin{bmatrix} x^i{}_k \\ y^i{}_k \end{bmatrix} = \begin{bmatrix} x^i{}_{k-1} \\ y^i{}_{k-1} \end{bmatrix} + \begin{bmatrix} \sin(\theta + \Delta\theta) \\ \cos(\theta + \Delta\theta) \end{bmatrix} * (L + \Delta L) \quad (23)$$

Wherein L is the average set in the step length, θ is determined by the direction of movement of the compass. ΔL is the error, Poisson distribution of $\Delta L - L \sim P(L)$, $\Delta\theta$ normal distribution with mean zero. We can predict the position of the particle at k by the position of the particle at time $k-1$.

4. In step 1, we got $(x_r, y_r)\sigma_t^2\sigma_r^2$. Due to the system of observational equations:

$$Z_{kt} = \begin{bmatrix} x_k \\ y_k \end{bmatrix} + \begin{bmatrix} u_{ktx} \\ u_{kty} \end{bmatrix} \quad (24)$$

Wherein $\begin{bmatrix} u_{ktx} \\ u_{kty} \end{bmatrix}$ is noise, meet variance σ_t^2, mean 0 Gaussian distribution. Therefore, for the TDOA results, there is a backward probability density:

$$P_t(Z_{kt} \mid X_k) = \frac{1}{\sigma_t\sqrt{2\pi}}\exp\left(-\frac{(x_t - x_k)^2 + (y_t - y_k)^2}{2\sigma_t^2}\right)$$

$$(25)$$

Similarly, for the RSSI results, there is a backward probability density:

$$P_r(Z_{kr} \mid X_k) = \frac{1}{\sigma_r\sqrt{2\pi}}\exp\left(-\frac{(x_r - x_k)^2 + (y_r - y_k)^2}{2\sigma_r^2}\right)$$

$$(26)$$

Due to the TDOA result variance σ_t^2, the RSSI result variance is σ_r^2, after allocation according to the variance of the final results of the probability density:

$$P(Z_k \mid X_k) = \frac{\sigma_r^2}{\sigma_t^2 + \sigma_r^2} * P_t(Z_k \mid X_k)$$
$$+ \frac{\sigma_t^2}{\sigma_t^2 + \sigma_r^2} * P_r(Z_k \mid X_k) \tag{27}$$

When $\frac{\sigma_r^2}{\sigma_t^2 + \sigma_r^2}$ is less than the threshold value, we believe that the results from the TDOA are no longer credible, then take:

$$P(Z_k \mid X_k) = P_r(Z_k \mid X_k) \tag{28}$$

Similarly, when this is less than the threshold value, whichever:

$$P(Z_k \mid X_k) = P_t(Z_k \mid X_k) . \tag{29}$$

According to current observations (x_r, y_r), calculate the weight of each particles corresponding value:

$$(w_k^i)' = w_{k-1}^i * P(Z_k \mid X_k^i) \tag{30}$$

Then the weights of the particles are normalised:

$$(w_k^i)'' = \frac{(w_k^i)'}{\sum_{i=1}^{N}(w_k^i)'} \tag{31}$$

5. To avoid wasting a lot of computational resources on particles with very low weights, we will re-sample the particles, delete the lower-weight particles, and copy the higher-weight particles.

We choose to remove M particles that weigh below the threshold $w = \frac{1}{N}$, from the largest to start copying the weights remaining residual particles of weight median particle largest particles to ensure that the same total number of particles. When the remaining number of particles is less than, from the maximum weight of the residual particles start copying again, then repeat the process until the total number of particles reaches N.

The normalisation of particle weights is carried out again.

$$w_k^i = \frac{(w_k^i)''}{\sum_{i=1}^{N}(w_k^i)''} \tag{32}$$

6. The calculated positioning results:

$$\begin{bmatrix} x \\ y \end{bmatrix} = \sum_{i=1}^{N} w_k^i * \begin{bmatrix} x^i \\ y^i \end{bmatrix} \tag{33}$$

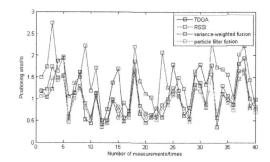

Figure 1. Comparison of accuracy of positioning results.

At the end of the current round of positioning, go back to step 1 to enter the next cycle.

5 EXPERIMENTAL RESULTS AND CONCLUSION

In the experimental environment, eight nodes are set as the beacon base station of the experiment, and the receivers are placed at different positions in the experimental environment. The TDOA location method is used separately, Using the RSSI localisation method, the localisation method based on the variance weighted fusion and the deviation of the positioning result of the four methods using the particle filter fusion localisation method and the actual position of the receiver. A total of 40 sites were tested at the site. After the positioning results were stable, the average of the ten consecutive positioning results was selected at each location, and the results were statistically analysed. At the same time the experiment also included a group of walk tests, with the experimental hand-held receiver taken in the corridor along a fixed path at uniform speed, real-time recording of the experimenter's position, TDOA positioning solution results, RSSI solution results, position variance weighted fusion methods and particle filter fusion methods. The location of the results was repeated ten times for statistical analysis.

The fixed-point test results are as follows:

The average error of the TDOA method is 1.11 m, the average error of the RSSI method is 1.37 m, the average error of the variance weighting method is 1.017 m, and the average error of the particle filter is 0.95 m. In 67.5% of cases, the result of the particle filter is better than that of the TDOA method, the result of the particle filter is better than that of the RSSI method in 90% of cases, and the result of the particle filter is better than the variance weighted fusion method in all cases.

It can be found that the result of the particle filter is better than that of the TDOA, RSSI and variance weighted fusion in the walk test.

It can be concluded that the particle filter method can improve the robustness of the localisation process compared with using the TDOA method or the RSSI method alone, so that the jitter of the positioning results is gentler and the accuracy of the positioning result is improved.

ACKNOWLEDGEMENTS

The National High Technology Research and Development Program ("863" Program) of China (No. 2015 AA124103) and (No.2015 AA124101).

REFERENCES

Deng, Z., & Dai, C. (2015). A method of positioning signal transmission based on frequency division multiplexing. *Proceedings of 2015 5th International Conference on Computer Sciences and Automation Engineering (ICCSAE 2015, 5.)*

Deng, Z., & Wang, X. (2013). The Kalman positioning algorithm based on TDOA/AOA, *Proceedings of 2013 World Congress on Industrial Materials-Applications; Products; and Technologies WCIM 2013 V739)*, 6.

Xia, J., Deng, Z., & Gao, L. (2015). The research on integrated communication networks and satellite navigation positioning technology. *Proceedings of 2015 5th International Conference on Computer Sciences and Automation Engineering (ICCSAE 2015)*, 5.

Yang, Y., & Deng, Z. (2015). Research of TDOA NLOS error compensation interpolation positioning, *Proceedings of 2015 2nd International Workshop on Materials Engineering and Computer Sciences (IWMECS 2015)*, 6.

Yin, L., Deng, Z., Xi, Y., Dong, H., Zhan, Z., & Gao, Z. (2013). A satellite selection algorithm for GNSS multi-system based on pseudorange measurement accuracy. *Proceedings of 2013 5th IEEE International Conference on Broadband Network & Multimedia Technology*, 4.

Automotive, Mechanical and Electrical Engineering – Liu (Ed.)
© 2017 Taylor & Francis Group, London, ISBN 978-1-138-62951-6

A review of DNA sequence data analysis technologies and their combination with data mining methods

Tiange Yu
Department of Biology, Hong Kong Baptist University, Hong Kong, China

Yang Chen
Life Science Institute, Central South University, Changsha, China

Bowen Zhang
Shenzhen Middle School, Shenzhen, China

ABSTRACT: Analysis of DNA sequence data is important for comprehending the meaningful biological information buried in the data set. Due to the high time and computational complexity, developing efficient and effective analysis methods for processing large-scale DNA sequence data has always been a crucial task in the post-genomic era. With the recent progresses in computer science and electrical engineering, multiple novel methods have been introduced to the biology regime and applied to the DNA sequence analysis. These methods significantly improved the efficiency and accuracy of DNA sequence data analysis. Among these methods, data mining technique, which uncovers the information hidden in the data, is regarded as one of the most promising and effective approach. This paper provides a comprehensive review of recent progress on DNA sequence analysis technologies, and discusses the use of data mining method in this area.

Keywords: DNA sequence; Bioinformatics; Data mining; Computer science; Biological sequence

1 INTRODUCTION

Nucleic acid, especially DNA, is the informational molecule in cells that carries the genetic instructions for growth, development, functions and reproduction of all known cellular organisms and many viruses. The intensively studied human genome contains around 2.85 billion nucleotides and it took 13 years to handle all the sequences in the Human Genome Project. Analysis of DNA sequences is essential for reaching a deep understanding of the functions and regulation of genes, which could result in improvements from many aspects, including personalized medicine, agriculture, forensics, etc (J. Shendure and H. Ji, 2008). Since there are different sequence features for different functional DNA fragments, DNA data mining is the key to identify the specific DNA fragments and their regulatory interactions efficiently (M. J. Zaki, 2007).

Though sequencing data could be obtained more and more easily with the development of next-generation sequencing technology, we are still facing the challenges of understanding the functions and evolution of these sequences. DNA sequence motifs are the short patterns that appear frequently in the genome which carries out biological functions, including transcriptional factor binding sites, mRNA splicing signaling, transcriptional termination signal, etc (P. D'Haeseleer, 2006). By extracting them and comprehending the functions, it would be much easier to uncover the regulatory mechanisms of genes, which could facilitate understanding the functions of new sequences without conducting laborious and costly functional assays on candidate genes individually.

Methods to identify DNA sequence motifs include both experimental techniques and computational technologies, such as SELEX (J. Hoinka, et al., 2012), CNEs (Q. Li, 2010), IMUNE (R. J. Pantazes, 2016), etc. For experimental identifications, a common approach in previous studies is to used Mass Spectrometry to identify short linear motifs that are abundant but hard to find in eukaryotic cells (T. J. Gibson, 2015). Meanwhile, for computational identification, numerous algorithms to extract motifs were developed. They could be classified into three categories, based on co-regulated genes promoter sequences, phylogenetic footprinting, and both coregulated genes promoter sequences and phylogenetic footprinting. For instance, Oligo-Analysis, which is based on coregulated-genes-promoter-sequences, could be used to identify the motifs efficiently in Saccharomyces cerevisiae. Gibbs sampler, phylogenetic-footprinting based that able to do the work within several genes, could be used to identify motifs in proteobacterial genomes. Smith-Waterman, based on both aspects, was used to identify the motifs in Archaea (M. K. Das and H. K. Dai, 2007). Given the availability of numerous methodologies, evaluation of various algorithms has becoming increasingly important since the performance of methods varies when they dealing with motifs with different features for different data sets.

In this review, we will start with introducing the fundamentals of DNA sequence data analysis, and then touch on the history and previous progresses in DNA sequence data analysis techniques and the application of data mining techniques in analyzing DNA sequence data, finally, the review will end with the current key challenges and major questions in the field.

2 FUNDAMENTALS OF DNA SEQUENCE DATA ANALYSIS TECHNIQUES

Numerous previous research has shown that DNA sequence is not a random combination of the four types of nucleotide: A (adenine), G (guanine), C (cytosine), and T (thymine). The four nucleotides forms different sections (such as genes) on the long DNA chain and executes given functions (M. K. Das and H. K. Dai, 2012). Therefore, DNA sequence with similar functions can contain sections that have strong similarities. The similarities among DNA sequences is the foundation for DNA sequence data analysis. A gold standard commonly used in Bioinformatics is that if the coherence between two arrays is over 30%, the chance that they have the same origin is high. Genes with the same origin have the same common ancestor, and will be likely to have similar functions. Therefore, it is an important task to identify the sections that have strong coherences, and then use data mining technology to dig out the information hidden behind the sequence. This would allow the function and structure of the unknown DNA sections to be predicted, the coherence between DNA sequences to be discovered, and the abnormal sequences to be detected (H. Yang, 2012).

The basic measurement of coherence in DNA sequences is the coherence between the nucleotides (L. Excoffier, 1992). There are 16 possible combinations of nucleotide pairs, resulting in 16 coherence functions:

$$\{\Gamma_{\alpha\beta}(d)\} = \{\Gamma_{AA}(d), \Gamma_{AC}(d), \Lambda, \Gamma_{GT}(d)\Gamma_{TT}(d)\})$$

$$(1)$$

Each of these coherence functions are defined as two nucleotides α and β separated by

$$d\{\Gamma_{\alpha\beta}(d)\} = P_{\alpha\beta}(d) - P_{\alpha} * P_{\beta} \qquad (2)$$

where $\Gamma_{\alpha\beta}(d)$ is the probability that α and β separated by d appears together. P_{α} and P_{β} are the density of α and β on the DNA sequence, given by

$$\begin{cases} P_{\alpha} = \sum_{\beta} P_{\alpha\beta} \\ P_{\beta} = \sum_{\alpha} P_{\alpha\beta} \end{cases} \qquad (3)$$

Therefore, there are 9 independent coherence functions (L. Excoffier, 1992). The study on the coherence based on Eq. (1) – Eq. (3) is the basis of DNA sequence analysis. As biological information, DNA sequence data has its own unique properties (Y. X. Yangyong Zhu, 2007):

1. Small symbol sets
The symbol set used to represent a DNA sequence is very small, namely, A, C, T, G.
2. Non-numerical values
The symbols use for representing a DNA sequence are character type, not numerical type.
3. Large diversion in length
Some of the DNA sequences may be very short (tens of symbols), while some can be very long (more than several megabits).
4. The existence of noise
The DNA sequence is likely to be contaminated with noise.
5. The biological meaning
The DNA sequence is not a simple array of characters; instead, the combination of the symbols represents specific biological meanings.

3 HISTORY AND RECENT PROGRESSES ON DNA SEQUENCE DATA ANALYSIS TECHNIQUES

The mathematical background and fundamental methods for DNA sequence analysis can date back to the 1960 s. In the early stage, it was difficult to obtain consecutive long DNA sequences. Therefore, most of the research was focused on the coherence study between adjacent nucleotide pairs. Recently, with the development in high-throughput DNA sequencing technology, large quantities of long DNA sequences have become available. The availability of long DNA sequences

Table 1. DNA sequence data analysis methods developed in the early years.

Methods	Features	Algorithms
Pattern enumerate method based on exhaustive search	Enumerate all possible patterns. Algorithm may run in exponential time. Suitable for small scale analysis; inefficient for large scale analysis.	MOTIF (H. O. Smith, 1990), etc.
Applying prune strategy in the search	Reduce computational complexity to fit typical DNA sequences. Allows searching for longer and more complicated patterns than simple exhaustive search	Pratt (I. Jonassen, 1997), TEIRESIAS (I. Rigoutsos and A. Floratos, 1998), etc.
Iterative heuristic method	Converge to a local maximum instead of the global optimum. May lose some useful patterns, but is almost guaranteed to find a pattern as good as possible.	Gibbs (C. E. Lawrence, 1993), COPIA (C. Liang, 2006), etc.
Constructing stochastic model	Useful to process patterns that are difficult to be described by a simple deterministic pattern and is expressed in form of stochastic model.	EM (C. E. Lawrence and A. A. Reilly, 1990), MEME (W. N. Grundy, 1997), etc.
Methods using additional information	Using information from sequence alignment, or global properties of a sequence. Suitable for analyzing sequences that contain biological meanings	EMOTIF (C. G. Nevill-Manning, 1998), etc.

shines light on the study of the correlation among genes which are separated by short- and long-spaces. The historic as well as recent analysis methods is summarized as in Table 1 (Y. X. Yangyong Zhu, 2007).

4 APPLICATION OF DATA MINING TECHNIQUES IN THE ANALYSIS OF DNA SEQUENCE DATA

Since the scale of DNA sequence data sets generated today is commonly in the order of Megabits to hundreds of Gigabits, the study and analysis on such sequences are limited by computational capabilities. The analysis methods developed in the early years (pre-1990 s) have been shown as inefficient for processing these long DNA sequences (A. Brazma, 1998). Therefore, techniques developed in computer science regime have been used to accelerate the analysis speed and improve the analysis accuracy. Some of these techniques include big data analysis, statistics, artificial intelligence, machine learning, etc. Data mining, combining the advantages in these techniques and aiming at digging the hidden information behind the symbols and sequences, has been the mainstream method for processing the DNA sequence data.

The development of efficient and effective DNA sequence data mining methods can be divided roughly into three stages, i.e. statistics-based data mining methods; general data mining methods; and specialized DNA sequences-oriented data mining methods (Y. X. Yangyong Zhu, 2007), as shown in Table 2

The main research areas of DNA sequence data mining technology includes the following five aspects:

4.1 DNA sequence pattern mining

DNA sequence patterns carries important biological meaning in the survival and evolution of species, and the identification of such patterns is a crucial task for DNA sequence data analysis. Researchers have developed sequence pattern mining algorithms which are highly efficient and suitable to analyze large data sets. Sequence pattern mining was firstly proposed by Agrawal and Srikant in 1995 (R. Agrawal and R. Srikant, 1995). Then in 1996, Srikant et al. described the Generalized Sequential Patterns mining (GSP) algorithm (R. Srikant and R. Agrawal, 1996). The GSP algorithm introduced time and conceptual constraint and searches the patterns within the dataset with a bottom-up Breadth-First-Search (BFS) method. However, it has the disadvantage that it generates a large set of patterns which reduces the overall efficiency. Regarding to this issue, Pei et al. proposed the PrefixSpan algorithm which is based on growing pattern sets (J. Han, 2001). The PrefixSpan method reduces the computational complexity by dividing the overall data set into smaller subsets and performing data mining on these individual subsets.

It is soon realized that bottom-up search strategy such as the GSP algorithm tend to generate short candidate patterns that are not frequently used. Therefore, one direction to improve the efficiency is to remove the infrequently used patterns during the searching process. As another approach, Ester et al. proposed an up-bottom algorithm named ToMMSA in 2004 (M. Ester and X. Zhang, 2004). To ensure finding all useful patterns, ToMMSA firstly perform an initial bottom-up search to find the maximum length of the frequent patterns. Then, starting from the sub-sequences that are shorter than the maximum

Table 2. Different stages of the development of DNA sequence data mining techniques.

Stages	Pros	Cons
Stage 1: statistics-based data mining methods	Effective at capturing sequential constraints present in the data	Inefficient for large-scale data set; result lacks interpretability
Stage 2: general data mining methods	Effective at the analysis of coherence between genes in large scale data sets	DNA data need to be mapped to a suitable form; accuracy is usually compromised
Stage 3: Specialized DNA sequences-oriented data mining methods	Effective at capturing sequential characteristics, and are scalable for large data sets. Especially good at processing DNA sequences that contains noise	

length, the algorithm reduces the size of the pattern set by removing infrequent sequences. The maximum number of candidate patterns generated by this method is $n*l*(l-1)/2$ which is $O(nl^2)$ where l is the length of the longest sequence in the pattern set, and n is the number of patterns. This is far less than the candidate patterns generated by a bottom-up search method, which is $O(|\Sigma|^m)$ where Σ represents the number of symbols in the DNA sequence and m represents the maximum length of the frequent patterns.

DNA sequences often contains patterns which are separated by random distances. Realizing this issue, Wang et al. proposed a two-phase algorithm which effectively analyzes DNA sequences that contains an arbitrary space (K. Wang, 2004). Because of the complexity of DNA sequence data, constraints in the analysis is comparatively complex, related work on this direction is still a hot topic under research currently.

In order to optimize the data mining results, multiple algorithms have also been proposed to reduce the redundant and low-value patterns. For example, Yan et al. proposed the CloSpan algorithm which works based on closed sequence patterns (E. Bahar, 2003). Another example is the BIDE algorithm, which mines frequent closed sequences without candidate maintenance (J. Wang and J. Han, 2004).

4.2 DNA sequence association rules mining

Mining association rules between sets of items is firstly introduced by Agrawal in 1993 (R. Agrawal, 1993), and has received wide interests thereafter. It is very useful in discovering the correlation and cooperation between DNA sequences, detecting the redundant sequences in DNA sequence sets as well as being excel in evolutional analysis.

Generally speaking, algorithms for mining DNA sequence association rules are two-step procedures: firstly generate the sequence pattern, and then generate the strong association rules based on the sequence patterns.

4.3 DNA sequence clustering analysis

DNA sequence clustering analysis is an effective approach to identify the classifications of unknown DNA sequences, and to reveal the correlation between sequences, and therefore to analyze the function of DNA sequences.

In its early stage, DNA clustering analysis methods are based on side-by-side comparisons, e.g. K-medoid method (L. Kaufman and P. J. Rousseeuw, 2009), complete-link method (W. H. Day and H. Edelsbrunner, 1984). These methods have high time and computational complexity and cannot be applied to large scale DNA sequence analysis.

Regarding this issue, many new algorithms have been proposed. For example, Wang et al. proposed the CLUSEQ algorithm (J. Yang and W. Wang, 2003), which constructs the correlation assessment method based on statistical characteristics of the sequences. Guralnik et al. (V. Guralnik and G. Karypis, 2001) described the retractable clustering method. Another approach is to combine biological knowledge and background into the clustering of DNA sequences, e.g. the BiCluster method (Y. Cheng and G. M. Church, 2000).

4.4 DNA sequence classification analysis

The purpose of DNA sequence classification is to classify unknown DNA sequences and therefore to predict its function and its correlation with other DNA sequences. The current DNA sequence classification analysis methods can be roughly categorized into four types: Comparison-based methods, statistics-based (Markov chain) methods; conventional machine learning methods; and pattern-mining-based characteristic extraction classification method. Comparison-based method is usually the easiest to implement but the computational complexity increases rapidly with the scale of data set. Statistics-based methods can extract the characteristics of the sequence but the results are more difficult to interpret as compared to machine-learning-based methods. Pattern-mining based methods is comparably satisfying in achieving a good balance among the effectiveness of characteristic extraction, the accuracy of classification, and the computational complexity.

4.5 *DNA sequence abnormality analysis*

DNA sequence abnormality analysis is often regarded as pre-processing of DNA data mining since it can effectively detect abnormal DNA sequences and exclude their negative impact on the accuracy of the data mining result. It is an area that receives relatively less attention than the other aspects of analysis; however, the interest in this area is growing rapidly. For example, Sun et al. introduced an abnormal analysis method based on Probabilistic Suffix Trees (PST) in 2006 (P. Sun, 2006), which reduced the time and memory cost in computation as well as improved the accuracy. However, there are still many remaining issues in this area, calling for more intensive research.

5 CONCLUSION

Efficient and effective analysis methods for processing huge amount of DNA sequence data has always been a crucial task in post-genomic era. Data mining technique, which digs the information hidden in the data, is regarded as the most promising and effective approach for analyzing large-scale DNA sequence data. This paper provides a comprehensive review of recent progress on DNA sequence analysis technologies, and discusses the use of data mining method in this area.

Key questions and major challenges in DNA data mining lies in multiple aspects. The first key question is to better understand the so-called "junk" DNA sequences that prevails in eukaryotic genomes. Given the revelation of diverse functions carried out by LincRNA, we expect the expedition of research in this area with the help of data mining techniques. Secondly, apart from predicting functions from sequence similarity, various case studies have shown drastic functional changes upon gene duplication, subfunctionalization and neofunctionalization. It is of great importance to use data mining techniques to compare and contrast motif changes across species in evolutionary time.

Given with the rapid development in statistical and computational methods, it is expected that interdisciplinary research combining bioinformatics, data mining and molecular biology is going to uncover more meaningful results with high-throughput, serving as a doorway to future development in human medicine and basic biological research.

REFERENCES

Agrawal, R. and R. Srikant, "Mining sequential patterns," in Data Engineering, 1995. Proceedings of the Eleventh International Conference on, 1995, pp. 3–14.

Agrawal, R., T. Imieliński, and A. Swami, "Mining association rules between sets of items in large databases," in Acm sigmod record, 1993, pp. 207–216.

Brazma, A., I. Jonassen, I. Eidhammer, and D. Gilbert, "Approaches to the automatic discovery of patterns in biosequences," J Comput Biol, vol. 5, pp. 279–305, Summer 1998.

Bahar, E., "CloSpan: Mining: Closed sequential patterns in large datasets," 2003.

C. International., Human Genome Sequencing, "Finishing the euchromatic sequence of the human genome," Nature, vol. 431, pp. 931–45, Oct 21 2004.

Cheng, Y. and G. M. Church, "Biclustering of expression data," in Ismb, 2000, pp. 93–103.

Day, W. H. and H. Edelsbrunner, "Efficient algorithms for agglomerative hierarchical clustering methods," Journal of classification, vol. 1, pp. 7–24, 1984.

Das, M. K. and H. K. Dai, "A survey of DNA motif finding algorithms," BMC Bioinformatics, vol. 8 Suppl 7, p. S21, Nov 01 2007.

D'Haeseleer, P., "What are DNA sequence motifs?," Nat Biotechnol, vol. 24, pp. 423–5, Apr 2006.

Excoffier, L., P. E. Smouse, and J. M. Quattro, "Analysis of molecular variance inferred from metric distances among DNA haplotypes: application to human mitochondrial DNA restriction data," Genetics, vol. 131, pp. 479–91, Jun 1992.

Ester, M. and X. Zhang, "A top-down method for mining most specific frequent patterns in biological sequence data," in Proc. the Fourth SIAM International Conference on Data Mining, 2004, pp. 90–101.

Gibson, T. J., H. Dinkel, K. Van Roey, and F. Diella, "Experimental detection of short regulatory motifs in eukaryotic proteins: tips for good practice as well as for bad," Cell Commun Signal, vol. 13, p. 42, Nov 18 2015.

Guralnik, V. and G. Karypis, "A scalable algorithm for clustering sequential data," in Data Mining, 2001. ICDM 2001, Proceedings IEEE International Conference on, 2001, pp. 179–186.

Grundy, W. N., T. L. Bailey, C. P. Elkan, and M. E. Baker, "Meta-MEME: motif-based hidden Markov models of protein families," Comput Appl Biosci, vol. 13, pp. 397–406, Aug 1997.

Han, J., J. Pei, B. Mortazavi-Asl, H. Pinto, Q. Chen, U. Dayal, et al., "Prefixspan: Mining sequential patterns efficiently by prefix-projected pattern growth," in proceedings of the 17th international conference on data engineering, 2001, pp. 215–224.

Hoinka, J., E. Zotenko, A. Friedman, Z. E. Sauna, and T. M. Przytycka, "Identification of sequence-structure RNA binding motifs for SELEX-derived aptamers," Bioinformatics, vol. 28, pp. i215–23, Jun 15 2012.

Jonassen, I., "Efficient discovery of conserved patterns using a pattern graph," Comput Appl Biosci, vol. 13, pp. 509–22, Oct 1997.

Kaufman, L. and P. J. Rousseeuw, Finding groups in data: an introduction to cluster analysis vol. 344: John Wiley & Sons, 2009.

Liang, C., "COPIA: A New Software for Finding Consensus Patterns in Unaligned Protein Sequences," UWSpace, 2006.

Lawrence, C. E. and A. A. Reilly, "An Expectation Maximization (EM) algorithm for the identification and characterization of common sites in unaligned biopolymer sequences," Proteins, vol. 7, pp. 41–51, 1990.

Lawrence, C. E., S. F. Altschul, M. S. Boguski, J. S. Liu, A. F. Neuwald, and J. C. Wootton, "Detecting subtle sequence signals: a Gibbs sampling strategy for multiple alignment," Science, vol. 262, pp. 208–14, Oct 8 1993.

Li, Q., D. Ritter, N. Yang, Z. Dong, H. Li, J. H. Chuang, et al., "A systematic approach to identify functional motifs within vertebrate developmental enhancers," Dev Biol, vol. 337, pp. 484–95, Jan 15 2010.

Mount, D. W., Bioinformatics sequence and genome analysis vol. 21–22. New York: Cold Spring Harbor Laboratory Press, 2001.

Nevill-Manning, C. G., T. D. Wu, and D. L. Brutlag, "Highly specific protein sequence motifs for genome analysis," Proc Natl Acad Sci U S A, vol. 95, pp. 5865–71, May 26 1998.

Pantazes, R. J., J. Reifert, J. Bozekowski, K. N. Ibsen, J. A. Murray, and P. S. Daugherty, "Identification of disease-specific motifs in the antibody specificity repertoire via next-generation sequencing," Sci Rep, vol. 6, p. 30312, Aug 02 2016.

Rigoutsos, I. and A. Floratos, "Combinatorial pattern discovery in biological sequences: The TEIRESIAS algorithm," Bioinformatics, vol. 14, pp. 55–67, 1998.

Smith, H. O., T. M. Annau, and S. Chandrasegaran, "Finding sequence motifs in groups of functionally related proteins," Proc Natl Acad Sci U S A, vol. 87, pp. 826–30, Jan 1990.

Srikant, R. and R. Agrawal, "Mining sequential patterns: Generalizations and performance improvements," in International Conference on Extending Database Technology, 1996, pp. 1–17.

Sun, P., S. Chawla, and B. Arunasalam, "Mining for Outliers in Sequential Databases," in SDM, 2006, pp. 94–105.

Shendure, J. and H. Ji, "Next-generation DNA sequencing," Nat Biotechnol, vol. 26, pp. 1135–45, Oct 2008.

Wang, J. and J. Han, "BIDE: Efficient mining of frequent closed sequences," in Data Engineering, 2004. Proceedings. 20th International Conference on, 2004, pp. 79–90.

Wang, K., Y. Xu, and J. X. Yu, "Scalable sequential pattern mining for biological sequences," in Proceedings of the thirteenth ACM international conference on Information and knowledge management, 2004, pp. 178–187.

Yang, J. and W. Wang, "CLUSEQ: Efficient and effective sequence clustering," in Data Engineering, 2003. Proceedings. 19th International Conference on, 2003, pp. 101–112.

Yang, H., "Review of biological sequence data mining techniques," Journal of Hefei University of Technology, vol. 35, pp. 1212–1216, 2012.

Yangyong Zhu, Y. X., "DNA sequence data mining technique," Journal of Software, vol. 18, pp. 2766–2781, 2007.

Zaki, M. J., G. Karypis, and J. Yang, "Data Mining in Bioinformatics (BIOKDD)," Algorithms Mol Biol, vol. 2, p. 4, Apr 11 2007.

Automotive, Mechanical and Electrical Engineering – Liu (Ed.)
© *2017 Taylor & Francis Group, London, ISBN 978-1-138-62951-6*

A routing protocol of vehicular heterogeneous network based on the Dijkstra algorithm

Hewei Yu & Zhihao Bin
School of Computer Science and Engineering, South China University of Technology, Guangzhou, China

ABSTRACT: With the development of intelligent transportation, the vehicular smart system has become a focus of the academic and industrial communities. The routing protocol is one of the most significant issues in the vehicular heterogeneous network. The paper uses the Dijkstra algorithm to obtain the optimal routing path. To evaluate the performance of the Dijkstra algorithm, it is compared with the minimum tree model based on the Prim algorithm. The simulation experiments are based on the MATLAB platform. The experimental results indicate the Dijkstra algorithm achieves the best performance.

Keywords: Intelligent transportation; Vehicular heterogeneous network; Routing protocol; Dijkstra algorithm; Minimum tree model

1 INTRODUCTION

With the development of computer and communication technology, multiple technology fusing has a grade progress in the modern society. Intelligent transportation has become a research hotspot (Cheng,, Yang, & Shen, 2015; Ghazal et al., 2015). The explosive growth in vehicles and ubiquitous information has promoted the development of intelligent transportation (Shaw & William, 2015; Schurch, Hodler, & Rodic, 2015).

The traditional vehicle communication network is used in the closed network. It cannot adapt to the demands of intelligent transportation. Due to the drawback of the network framework, the traditional communication system is only used in the local area network. The new generation intelligent transportation system includes the communication of intelligent transportation, safety information of vehicles and smart applications. It is important for vehicle networks to realise the inter-vehicle digital links and real-time communication between vehicles with outer Internet.

The Vehicle Ad Hoc Network (VANET) (Khairnar & Kotecha, 2013) can realise the communication of different moving vehicles (Dressler et al. 2014; Liu et al., 2014; Wei et al., 2014). Besides this, the vehicles need to link to other networks. The different networks together compose the vehicle heterogeneous network. The optimal routing protocol of the heterogeneous network is one of the most important issues in wireless resource sharing. The vehicles make an optimal choice by detecting the network information state in real time.

In this work, we use the Dijkstra algorithm (Chen et al., 2014; Fan et al., 2010; Ding, Chang, & S. M, 2014) to obtain the optimal routing in the vehicular heterogeneous network. The Dijkstra algorithm was proposed by E. W. Dijkstra in 1959. It uses the greedy algorithm model to calculate the shortest distance in the directed network diagram. As one of the most outstanding algorithms in obtaining the shortest distance, the key concept is selecting the nearest node beside selected node set.

The remainder of this work is organised as follows: Section 2 gives the related works of the vehicular heterogeneous network. Section 3 describes the Dijkstra algorithm. The experiments and analysis are listed in Section 4. Section 5 concludes the work.

2 RELATED WORK

The vehicular heterogeneous network (Xu et al., 2014; Li, Xu, & Ma, 2015) is a Mobile Ad Hoc Network that uses in communication of vehicles and other facilities. The vehicles need to be equipped with a wireless transceiver and computer control model. Every vehicle is not only a network node, but is also a rout. And the range of the wireless network is only in the hundreds of metres. The remote communication needs help from other middle nodes.

The main applications of the vehicular heterogeneous network focus on developing safety. Emergency and warning messages can be sent to the vehicles. If an accident happens, the network can provide real-time information to the drivers. The architecture of a vehicular heterogeneous network is described in Figure 1.

Vehicle networks began in Japan in the 1980s. Since this time, vehicular heterogeneous networks have had important functions in transportation management, detection of road conditions, vehicle tracking and transportation safety. A lot of relevant research has been applied. For example, the Germanic Network on Wheels, the Japanese Cooperative Group Driving, and the Thinking Partner of Mercedes-Benz.

The protocol of vehicular heterogeneous networks include path generation, path selection and path maintenance. Path generation is based on the distribution network and demands from the users. Based on the information of network state and service state, the best path should be selected. Path maintenance is used to maintain the selected path (Satyajeet, A. R., & S. S, 2016; Shafiee & Leung, 2012).

In our work, the path selection is mainly discussed. The business requirements, network capacity, network characteristics and operating environment would be considered in the routing protocol.

The vehicle network routing protocol can be divided into three categories: topological routing, geographic routing and map routing. Topological routing is based on the network link state, to realise the data transmission. Geographic routing uses the near nodes and destination node to develop the efficiency of routing. According to the GPS positioning system, the nodes obtain the geographic information. The positioning system and electronic map are used to obtain the routing information for the map routing. Besides this, map routing can be used for path planning.

3 DIJKSTRA ALGORITHM

The network diagram is described as $G = (V, E, W)$, the node set is defined as $V = \{v_1, v_2, \cdots, v_p\}$, and p is the number of V. E and W denote the edge and weight of the network diagram, respectively. The fundamental steps are as follows.

Step 1: Initialisation.

$$d(v_i) = \begin{cases} 0, i = s \\ w_{ij}, i \neq s \,\&\, \langle v_s, v_i \rangle \in E \\ \infty, i \neq s \,\&\, \langle v_s, v_i \rangle \notin E \end{cases}$$

The other nodes are $d(v_j) = w_{1j}(j = 1, 2, 3, \cdots, n)$, and $S = \{v_1\}$, $R = V\text{-}S = \{v_2, v_3, \cdots, v_p\}$. Where w_{ij} denotes the weight of the edge (v_i, v_j), and $d(v_j)$ is the weight of the selected nodes.

Step 2: Calculate the distance of $d(v_k)$.

Find a node v_k, and get $d(v_k) = \min_{v_j \in R}\{d(v_j)\}$. Let $S = S \bigcup \{v_k\}$, $R = V\text{-}S$.

If $R = \varnothing$, end the algorithm; else go to Step 3.

Step 3: Revise the $d(v_k)$.

Get $d(v_j) = \min\{d(v_j), d(v_k) + w_{kj}\}$, and go to Step 2.

The flow diagram is described in Figure 2.

4 EXPERIMENTS AND ANALYSIS

In this work, the vehicular heterogeneous network is simulated in MATLAB software. There are ten vehicles and one signal tower in the simulation network. The VANET is shown in Figure 3.

From Figure 3, we define the connectivity of the 11 nodes. The connectivity indicates that the conjoint nodes can communicate with each other.

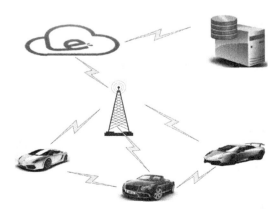

Figure 1. Architecture of a vehicular heterogeneous network.

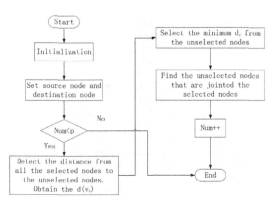

Figure 2. Flow diagram of the Dijkstra algorithm.

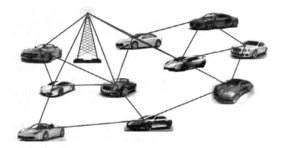

Figure 3.   Vehicular ad hoc network.

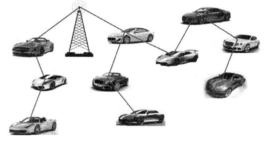

Figure 4.   Minimum path routing based on the Prim algorithm.

Table 1.   The communication of conjoint nodes.

Node	1	2	3	4	5	6	7	8	9	10	11
1	1	1	0	1	1	0	0	0	0	0	0
2	1	1	1	1	1	0	0	0	0	0	0
3	0	1	1	0	0	1	0	0	0	0	0
4	1	1	0	1	1	0	1	1	0	0	0
5	1	1	0	1	1	1	1	0	0	0	0
6	0	0	1	0	1	1	0	0	1	0	0
7	0	0	0	1	1	0	1	1	0	1	0
8	0	0	0	1	0	0	1	1	1	1	1
9	0	0	0	0	0	1	0	1	1	0	1
10	0	0	0	0	0	0	1	1	0	1	1
11	0	0	0	0	0	0	0	1	1	1	1

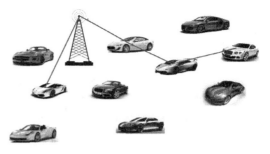

Figure 5.   Minimum path routing based on Dijkstra algorithm.

If one vehicle communicates with another vehicle, it should pass middle nodes to connect to the destination node. The communication of conjoint nodes is demonstrated in Table 1.

The Prim algorithm is a calculating method of the minimum tree model. In our simulation network model, the minimum tree model is based on the Prim algorithm, as shown in Figure 4.

In the above vehicular network, if the second car wants to communicate with the 11th car, the minimum path routing is calculated by the Dijkstra algorithm; the result is shown in Figure 5.

It is obvious that the Dijkstra algorithm is superior to the Prim algorithm in the communication cost between node 2 and node 11.

Besides this, except for the communication between the vehicles and signal tower, the information also transmits between the vehicles themselves, called 'V2V'. In the vehicular heterogeneous network, all the nodes compose the VANET, the information is transmitted by the vehicles, and they do not have to go through the signal tower. Every vehicle has a signal transceiver that can transmit the information. In this network, the information

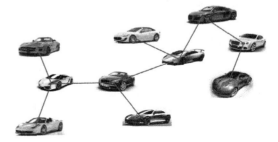

Figure 6.   Minimum path routing based on Prim algorithm (V2V).

platform is fast changing, and the information exchange is abundant.

The node information exchange is based on the connectivity of vehicles. The connectivity of all the vehicles is shown in Figure 6.

If node 2 communicates with node 10, the minimum path routing based on the Dijkstra algorithm is shown in Figure 7.

Figures 6 and 7 demonstrate that the Dijkstra algorithm is superior to the Prim algorithm. It only needs four hops from node 2 to node 10, but the

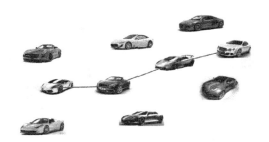

Figure 7. Minimum path routing based on Dijkstra algorithm (V2V).

minimum tree model needs five hops to achieve the same communication.

5 CONCLUSION

In this paper, we use the Dijkstra algorithm to obtain the optimal routing path in the vehicular heterogeneous network. To evaluate the performance of the Dijkstra algorithm, the Prim algorithm is compared with it. The communication between a vehicle and the signal tower, and the inter-vehicle network is compared. Simulation experiments indicate that the Dijkstra algorithm is superior to the minimum tree model, based on the Prim algorithm. In further research, a more complex vehicular heterogeneous network should be evaluated with the Dijkstra algorithm.

ACKNOWLEDGEMENTS

This study was supported by the Guangdong Science & Technology Project (Grant No.2013B010401006), the Guangzhou Science & Technology Project (Grant No. 2014 J4100019), the Key Collaborative Innovation Project among Industry, University and Institute of Guangzhou (Grant No. 201604010001), and Yuexiu Sci-ence & Technology Project (Grant No. 2015-CY-003).

REFERENCES

Chen, Y.Z., et al. (2014). Path optimization study for vehicles evacuation based on Dijkstra algorithm. *Procedia Engineering, 71*,159–165.

Cheng, X., Yang L., & Shen X. (2015). D2D for intelligent transportation systems: a feasibility study. *IEEE Transactions on Intelligent Transportation Systems 16*(4), 1784–1793.

Ding, H., Chang, D.F., & University, S.M. (2014). The courier vehicle distribution route optimization based on Dijkstra algorithm. Value Engineering.

Dressler, F., et al. (2014). Inter-vehicle communication: Quo vadis. *IEEE Communications Magazine 52*(6), 170–177.

Fan, Y.Z., et al. (2010). An improved Dijkstra algorithm used on vehicle optimization route planning. *Computer Engineering and Technology (ICCET), 2010 2nd International Conference on 2010: V3-693-V3-696.*

Ghazal, A., et al. (2015). A nonstationary wideband mimo channel model for high-mobility intelligent transportation systems. *IEEE Transactions on Intelligent Transportation Systems 16*(2), 885–897.

Khairnar, V.D. & Kotecha, K. (2013). Performance of vehicle-to-vehicle communication using IEEE 802.11p in Vehicular Ad-hoc Network environment. *International Journal of Network Security & Its Applications 5*(2),

Li, L., Xu, Y., & Ma, L. (2015). Vertical handoff strategy on achieving throughput in Vehicular Heterogeneous Network. *Vehicular Technology Conference IEEE, 2015.*

Liu, P., et al. (2014). Path loss modeling for vehicle-to-vehicle communication on a slope. *IEEE Transactions on Vehicular Technology 63*(6), 2954–2958.

Satyajeet, D., A.R., & S.S. (2016). Heterogeneous approaches for cluster based routing protocol in Vehicular Ad Hoc Network (VANET). *International Journal of Computer Applications 134.*

Schurch, B., Hodler, J., & Rodic., A.B. (2015) Advanced technologies for intelligent transportation systems. *International Journal of Radiation Oncology Biology Physics 66*(4), 1143–51.

Shafiee, K. & Leung V.C. (2012) Hybrid multi-technology routing in heterogeneous vehicular networks. *EURASIP Journal on Wireless Communications and Networking 2012*(1), 1–19.

Shaw, W.T. (2015). Towards smart cities: interaction and synergy of the smart grid and intelligent transportation systems. *Smart Grid: Networking, Data Management and Business Models*, 75–81.

Wei, S., et al. (2014). Simulation study of unmanned aerial vehicle communication networks addressing bandwidth disruptions. *SPIE Defense + Security International Society for Optics and Photonics*, 11730–41.

Xu, Y., et al. (2014). Fuzzy Q-learning based vertical handoff control for vehicular heterogeneous wireless network. *ICC 2014—2014 IEEE International Conference on Communications*, 5653–5658.

Automotive, Mechanical and Electrical Engineering – Liu (Ed.)
© 2017 Taylor & Francis Group, London, ISBN 978-1-138-62951-6

Agile C&C decision-making information combination technique based on the theory of evidence

Huayong Sun
College of Command Information Systems, PLA University of Science and Technology, Nanjing, China
Bengbu Automobile NCO Academy, Bengbu, China

Jichuan Quan, Haiyan Huang & Zhipeng Lei
College of Command Information Systems, PLA University of Science and Technology, Nanjing, China

ABSTRACT: Put forward by studying decision-making agility to improve Command and Control (C&C) agility. In the relevant problems of C&C decision-making, incomplete decision information combination problem are studied. In order to realise that the express of decision-making preference can be quickly and accurately under the condition of incomplete information, this paper proposes a mass functions construction method based on linguistic decision-making information. Then the method of Theory of Evidence can be used in incomplete decision information combination.

Keywords: C&C decision-making; Agility; Theory of Evidence; Information combination

1 INTRODUCTION

Agility is the ability to successfully affect, cope with, and/or exploit changes in circumstances (Alberts, 2011). Research to improve the agility of C&C is an effective way to deal with the complex changes of the battlefield environment. It has become one of the hot issues in the field of C&C. Because the core function of C&C is scientific decision-making, and people's participation in the decision-making process leads to an increase in uncontrollable factors, the study of decision-making agility is of great significance to enhance the agility of C&C. However, war is complex and changeable, and the decision-making process of C&C may not have "enough" information, so agile decision information combination under incomplete information is a key technical problem to be solved. Research on this problem will reduce the uncertainty of the decision result to a certain extent, improve the flexibility and adaptability of the decision-making process, so as to improve the agility of decision-making.

Theory of Evidence is an effective method to deal with incomplete information. It can process and fuse uncertain information effectively in the absence of *a priori* probability. Relying on the accumulation of evidence, the Theory of Evidence can narrow down the hypothesis set, so it can deal with the uncertainty caused by random and fuzzy. Theory of Evidence has advantages in the aspects of uncertainty representation, measurement and

combination, so it is widely used. Adopting the Theory of Evidence, this paper studied the incomplete preference information combination problems based on language expression, solving the C&C decision-making problem with the consideration of incomplete information and efficiency of decision-making, so as to enhance the agility of C&C decision-making.

2 AGILE C&C DECISION-MAKING INFORMATION COMBINATION TECHNOLOGY BASED ON THE THEORY OF EVIDENCE

2.1 *Related concepts of Theory of Evidence*

Definition 1 For a certain problem, it is possible to recognise the possible results. We call them the hypotheses of the problem. If all the hypotheses are mutually exclusive and can describe all the possible problems, we called the hypotheses set: The Frame of Discernment.

Definition 2 Assume that $P(\Theta)$ represents a set of all subsets of The Frame of Discernment Θ (called the power set of Θ). If map m: $P(\Theta) \rightarrow [0,1]$ meet: (1) $m(\varnothing) = 0$, (2) $\sum_{A \subseteq \Theta} m(A) = 1$, where m is a Basic Probability Assignment (BPA) or mass function.

If $m(A) > 0$, A is the Focal Element. In the Theory of Evidence, the evidence is a broad concept. Experts' estimates, historical information, experimental data, and simulation results can all be called evidence.

Theorem 1 Dempster-Shafer combination rule

Let $m_1, m_2, \ldots, m_n$ be n BPA in the same Frame of Discernment Θ, Focal Elements were $A_1, A_2, \ldots, A_n$. For proposition $A(A \subset \Theta)$, $m(A)$ indicates the reliability of the A after multiple evidence are aggregated.

$$m(A)$$
$$= \begin{cases} 0 & , A = \varnothing \\ \dfrac{1}{1-K} \displaystyle\sum_{A_1 \cap A_2 \ldots \cap A_n = A} m_1(A_1) m_2(A_2) \ldots m_n(A_n), A \neq \varnothing \end{cases}$$
$$(1)$$

In the formula,

$$K = \sum_{A_1 \cap A_2 \ldots \cap A_n = \varnothing} m_1(A_1) m_2(A_2) \ldots m_n(A_n) \qquad (2)$$

means conflict coefficient; the smaller the value of K the better the indication of the consistency of the evidence; the bigger the value of K the greater the indication of conflict, $K \to 1$ indicates that the evidence are highly conflicting.

2.2 A new method for constructing mass function

In order to improve the efficiency of C&C decision-making, the general choice is "language based decision-making". Experts use the linguistic evaluation values to express their preferences which are consistent with cognitive psychology. In 2000, Professor Herrera proposed the method of linguistic 2-tuple (Herrera 2000), using 2-tuple (s_i, ρ) to represent the decision-maker's linguistic evaluation information. In the use of 2-tuple method, the decision-makers can not only give the program evaluation using a language phrase, but they can also give the close degree between the evaluation and the language phrase. And the express of evaluation is more accurately fit decision-makers psychological, so it is widely adopted. Based on the above reasons, this paper proposes to define the transformation ways of the linguistic 2-tuple to the mass function, so as to obtain the mass function.

According to "Weber-Fechner's law", subjective sensation is proportional to the logarithm of the stimulus intensity (Li, 2004). That is:

$$M = c \log R \qquad (3)$$

In the formula, M is the sensation, R is the stimulus intensity, c is the constant determined depending on the sense and type of stimulus.

Here, deform the Equation (3)

$$R = a^M \qquad (4)$$

This formula shows that the relationship between the subjective sensation and the stimulus intensity. In the linguistic decision-making, a is related to the number g of the used language phrase, and the determination of the value is usually done by the experiment or the expert method, in particular, when $g = 7$, $a \approx 1.4$. This paper proposes a method of transforming linguistic 2-tuple to mass function as follows. The Frame of Discernment is $\Theta = \{support(s), notsupport(n)\}$.

When the evaluation is (s_i, ρ_i), the mass function can be obtained by the following formula:

$$\begin{cases} m(s) = \dfrac{a^{\Delta^{-1}(s_i, \rho_i)}}{a^{\Delta^{-1}(s_{g-1}, 0.5)}} \\ m(\Theta) = 1 - m(s) \end{cases} \qquad (5)$$

where, $\Delta^{-1}(s_i, \rho) = i + \rho$ is an operator that can translate the linguistic 2-tuple (s_i, ρ) to the corresponding value;

When the decision-maker does not give the evaluation, the mass function is:

$$m(\Theta) = 1 \qquad (6)$$

2.3 The steps of evidence combination based on the new mass function

Problem description: Assume that a group of decision-making experts is represented by a set $D = \{d_1, d_2, \ldots, d_P\} (P \geq 2)$, d_p indicates the pth decision-maker; An alternative set is $X = \{x_1, x_2, \ldots, x_Q\} (Q \geq 2)$, x_q indicates the qth option. The language phrase set is $S = \{s_0, s_1, \ldots s_{g-1}\}$, and the g is an even number. Experts use linguistic 2-tuple to express decision preference, and there may be some experts who did not give the evaluation information of a program. Now we need to determine the sort of the program based on the evaluation of experts.

Step one: Experts evaluate the alternatives by linguistic 2-tuple. Expert d_p evaluates the alternative x_q, the preference by 2-tuple (s_q^p, ρ_q^p) means, if the experts did not give the preference, with a null value "*" means. Expert d_p's evaluation vector can be expressed as

$$C_p = \left[(s_1^p, \rho_1^p) \quad (s_2^p, \rho_2^p) \quad \ldots \quad (s_Q^p, \rho_Q^p) \right] \text{ or}$$
$$C_p = \left[(s_1^p, \rho_1^p) \quad (s_2^p, \rho_2^p) \quad \ldots * \ldots \quad (s_Q^p, \rho_Q^p) \right].$$

Step two: By Equations (5) and (6) the expert evaluation can be converted to a mass function. The mass function of the expert d_p evaluating alternative x_q by m_q^p means;

$m_q^p = \left\{ m_q^p(\varnothing), m_q^p(s), m_q^p(n), m_q^p(\Theta) \right\}$, so

$C_p = \begin{bmatrix} m_1^p & m_2^p & \dots & m_n^p \end{bmatrix}$.

Step three: Using the evaluation of all the decision-makers, the decision matrix can be obtained:

$$C = \begin{bmatrix} m_1^1 & m_2^1 & \dots & m_Q^1 \\ m_1^2 & m_2^2 & \dots & m_Q^2 \\ \dots & \dots & \dots & \dots \\ m_1^P & m_2^P & \dots & m_Q^P \end{bmatrix}.$$

Step four: Combine all the mass functions of experts. For each alternative, using the synthesis Equations (1) and (2) to combine all the experts' mass functions and then the integrated mass function of all alternatives can be obtained as $m_1, m_2, \dots, m_Q$.

Step five: According to the practical significance of the mass function, $m_q(s)$ means the degree of support for the alternative and the order of the alternative set can be obtained.

Through the above five steps, the combination of incomplete decision-making information, which is expressed by a linguistic method, can be realised. At the same time, the linguistic method is more in line with people's cognitive psychology, with which the decision-makers can express preference as quickly as possible. In the decision-making process, combined with the relevant knowledge of psychology, the language information is translated into a mass function, so the transformation is more scientific and easy to understand. In the combination process, D-S Theory of Evidence has a strong theoretical basis, so the combined result is of high reliability, also it can deal with incomplete information. Therefore, this method, to a certain extent, can improve the agility of C&C decision-making.

3 CONCLUSION

First of all, the paper points out that the core functions of C&C are scientific decision-making, and the participation of people in the decision-making process leads to an increase of uncontrollable factors. Therefore, the study of decision-making agility is of great significance to improve the agility of C&C. In the problem of C&C decision-making, the decision information combination problem is studied under the condition of incomplete information based on Theory of Evidence.

According to "Weber-Fechner's law", put forward a method of constructing mass function based on linguistic decision information. By using the decision-making method proposed in this paper not only can we make the decision expression quickly and accurately, but also handle the incomplete decision information. This method can enhance the response and adaptability of the C&C decision-making, so it can enhance the agility of C&C to a certain extent.

ACKNOWLEDGMENTS

This paper was financially supported by National Natural Science Found of China (61174198).

REFERENCES

Alberts, D.S. (2011). The agility advance—A survival guide for complex enterprises and endeavors. CCRP Publication, 283–448.

Herrera, F. and Martinez L. (2000). A 2-tuple fuzzy linguistic representation model for computing with words. *IEEE Trans on Fuzzy Systems, 8(6)*, 746–752.

Li, W.B. (2004). History of Western Psychology. Zhejiang Education Press.

Automotive, Mechanical and Electrical Engineering – Liu (Ed.)
© 2017 Taylor & Francis Group, London, ISBN 978-1-138-62951-6

An operational amplifier based non-foster matching network for small loop antenna

Yi Ren, Juan Tang & Jia Sun
School of Electronic Engineering, Chongqing University of Posts and Telecommunications, Chongqing, P.R. China

Jianmei Zhou
School of Electronics and Information, Chongqing Institute of Engineering, Chongqing, P.R. China

Hongsheng Zhang
School of Electronic Engineering, Chongqing University of Posts and Telecommunications, Chongqing, P.R. China

ABSTRACT: This work presents a novel non-Foster matching circuit to improve the radiation of small loop antennas. The designed non-Foster circuit is achieved using grounded Negative Impedance Converters (NICs), which are based on the operational amplifier (op-amp). A passive matching network is added to offset the input's negative inductance and negative resistance of the proposed non-Foster circuit. The extracted S1P data file of the loaded small loop antenna is used to instead of the equivalent circuit of an antenna in a traditional integrated design. With the novel designed non-Foster circuit, the loop antenna matched well at 10 MHz–60 MHz frequency band and its reflection coefficient was less than −20 dB. The good agreements between simulated and measured results validate the design concept.

Keywords: Small loop antennas; Non-Foster circuit; Operational amplifiers (op-amps)

1 INTRODUCTION

Electrically small loop antennas always suffer from the problem of low radiation efficiency, especially at the HF or VHF frequencies. This is because the antenna size is too small and leads to a high radiation Quality factor (Q). As a result, impedance matching of the small loop antenna is necessary to improve the radiation efficiency (Hansen, 2011). The active matching circuit, named 'non-Foster circuit', using Negative Impedance Converters (NICs) is considered. Non-Foster circuits with negative components can cancel out the large input reactance of the small loop antenna and are not restricted to the efficiency-band trade-off (Wheeler, 1947). The main components of the NICs are active devices, such as transistors, field effect transistors and operational amplifiers (op-amps). Recently, the have transistorised NICs received more attention. In 2007, Marcellis first proposed the concept of NICs with op-amps, and also presented the simulation results of various NICs' architectures based on op-amps (Marcellis, 2007). Due to their high integration property, the NICs based on op-amps have fewer tuneable

peripheral circuit elements, which simplifies the design and simulation. This gives NICs based on op-amps an advantage in the future of antenna design.

In this paper, a grounded NIC with op-amp is proposed for a small loop antenna. In the NIC circuit, the load is the extracted S1P data file of the loop antenna rather than its equivalent circuit. The equivalent circuit is designed in the front NIC as a passive matching network to offset negative inductance and negative resistance of the loaded small loop antenna. The matched antenna can then work properly in the frequency range of interest.

2 NON-FOSTER MATCHING LOOP ANTENNAS

The basic form of the NIC based on op-amps consists of one op-amp, two resistors, and one load element (Marcellis, 2007). The NIC produces a negative impedance out of the load impedance scaled by the ratio of the two resistors. Figure 1 shows the proposed non-Foster matching of the load impedance Z_{ant}. $Z_{ant} = R_{ant} + jX_{ant}$. Z_{ant} is the input

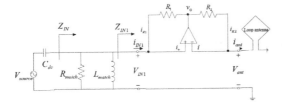

Figure 1. Proposed non-Foster matching network.

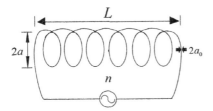

Figure 2. Multi-loop antenna.

impedance of the unmatched loop antenna, R_{ant} is the radiation resistance and X_{ant} is the antenna reactance. The circuit is composed of one stage of the NIC and the shunt-connected inductor and resistor in front of it. Assuming the ideal op-amp (Sun Xiaozi, 2002), the input impedance Z_{IN1} can be calculated as follows:

$$Z_{IN1} = -\frac{R_1}{R_2}(R_{ant} + jX_{ant}) \qquad (1)$$

The load impedance is converted to the negative impedance scaled by the ratio of R_1/R_2. The converted input reactance shows the non-Foster reactance behaviour.

The input equivalent impedance Z_{IN} can be written as follows:

$$Z_{IN} = Z_{IN1} \parallel R_{match} \parallel j\omega L_{match} \qquad (2)$$

where, Z_{IN1} and Z_{IN} are the input impedance of NIC circuit and the matched circuit, respectively. C_{dc} is the isolated DC capacitor for circuit protection.

The electrically small loop antenna is defined by its Circumference (C) being less than 1/3 working wavelength (Kraus, 2013), i.e., $C \leq 1/3\lambda$. A typical multi-loop antenna is represented in Figure 2, where a is the loop radius, a_0 is the wire radius, L is the total height and n is the number of turns.

In this work, a loop antenna with $a = 13$ mm, $a_0 = 1.25$ mm, $L = 70$ mm and $n = 6$ is fabricated as the load. In the working frequency band 10 MHz–60 MHz, the loop antenna has a large input reactance and very small input resistance. Therefore, impedance matching becomes critical for the small loop antenna.

3 SIMULATION AND MEASUREMENT RESULTS ANALYSIS

The super high speed current feedback amplifier tube AD8009 (Analog Devices Inc., 2000) is adopted as the active component in the proposed design. The simulation shows that the designed circuit performed best at $R_1 = 200$ Ω, $R_2 = 110$ Ω, $L_{match} = 1$ uH and $R_{match} = 51$ Ω. The layout and schematic simulation results show that the reflection coefficient (S11) is less than −38 dB, as shown in Figure 3. The input impedance is almost matched to 50 Ω's pure resistance in the frequency range of interest. Based on the simulation results, we tested the matching circuit with the small loop antenna.

Figure 4 shows a photograph of the fabricated non-Foster matching circuit with a loaded loop antenna. A DC power supply voltage of ±3.5 V is applied to the op-amp AD8009, and the coupling capacitors are imposed between the DC power supply and op-amp. Figures 5(a) and (b) give the comparison of the input resistance and reactance of the unmatched antenna, NIC circuit and the matched antenna. The measurement results validate the loop antenna achieving the impedance matching of 50 Ω.

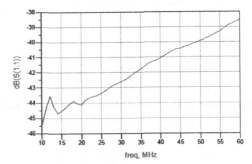

Figure 3. The simulated reflection coefficient of the matched loop antenna.

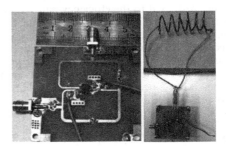

Figure 4. Photograph of the non-Foster matched loop antenna.

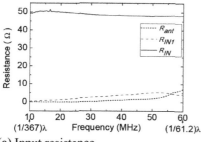

(a) Input resistance

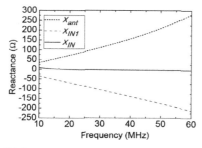

(b) Input reactance

Figure 5. Comparison of the impedances of the matched and unmatched loop antenna.

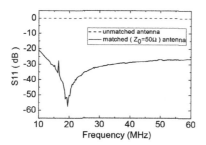

Figure 6. Comparison of the reflection coefficient of the matched and unmatched loop antenna.

The measured reflection coefficients of the matched and unmatched loop antenna are shown in Figure 6. The reflection coefficient of the matched loop antenna is less than −20 dB at 10 MHz–60 MHz whilst the unmatched one is almost 0. It should be noted that the discrepancies between the measured and simulated results of the matched antenna are strong, probably caused by the fabrication, measurement deviation and non-ideal characteristics of the commercial operational amplifier. Figure 7 shows the gain improvement of the matched loop antenna.

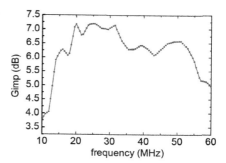

Figure 7. Gain improvement of matched loop antenna.

It shows that the maximum gain improvement is 7.2 dB at 25 MHz.

4 CONCLUSION

The measurements of the matched small loop antenna, at 10 MHz–60 MHz, show that the reflection coefficient is less than −20 dB and the input impedance almost match 50 Ω. At the same time, the maximum gain improvement can run up to 7.2 dB in the frequency range of interest. The test process is under stable circuit condition. Future works will be smaller circuits and lower operating frequency bands.

ACKNOWLEDGEMENTS

This work is supported in part by grant number NSFC 61301032, 61401051 and in part by Natural Science Foundation of Chongqing grant number cstc2013 jcyjA40037.

REFERENCES

Analog Devices Inc. (2000). "Low Distortion Amplifier," AD8009 datasheet.
Hansen R.C. and R.E. Collin (2011). Small antenna handbook. Hoboken, NJ, USA: Wiley.
Kraus, J.D., R.J. Marhefka, Translated by Wen Xun Zhang, Antennas: For All Applications Third Edition [M]. Beijing: Publishing House of Electronic Industry, 2013: 154–171.
Marcellis, A.D., G. Ferri, and V. Stornelli (2007). NIC-based capacitance multipliers for low-frequency integrated active filter applications. Proc. Res. Microelectr. Elect., 225–228.
Sun Xiaozi, Zhang Qimin (2002). Analog electronic technology foundation. Xi'an: Xidian University Press.
Wheeler, H.A., (1947). Fundamental limitations of small antennas. Proc. IRE, 35, 1479–1484.

Automotive, Mechanical and Electrical Engineering – Liu (Ed.)
© 2017 Taylor & Francis Group, London, ISBN 978-1-138-62951-6

Analysis of the precision influence factors of SDS-TWR method based on UWB

Zheng Zhang, Shaojun Li & Zejun Lu
School of Mechanical Engineering, Hubei University of Technology, Wuhan, China

ABSTRACT: With the rapid development of wireless communication technology, Ultra-Wideband (UWB) positioning technology has gained in importance. The research on indoor positioning technology in complex environments has received much attention in recent years. In this paper, the accuracy of the Two-Way Ranging (TWR) algorithm depends on the time synchronisation of the ranging nodes. Therefore, the Symmetrical Double-Sided Two-Way Ranging (SDS-TWR) method is adopted to reduce the requirement of high precision time synchronisation for ranging nodes. The influence of antenna height, pulse repetition rate, signal centre frequency and bandwidth on the accuracy of the SDS-TWR method based on UWB has all been studied by the experiment. Experimental results show that appropriate installation position, suitable height of antenna, high UWB pulse repetition rate, high UWB centre frequency and wide bandwidth conditions all lead to a high ranging accuracy, whose ranging error is less than 10 cm. The SDS-TWR method based on UWB in the complex environment shows an excellent ranging performance.

Keywords: Wireless communication; Ranging; UWB; SDS-TWR

1 INTRODUCTION

Wireless communication technology has brought great convenience to human life. For example, the Global Positioning System (GPS) is widely used in navigation, monitoring, positioning, the smart city concept and other fields. In Ware & Fulker (2014), it is shown that, with the rapid development of wireless communication technology, the relationship between people is closer, and communication is smoother. GPS has excellent positioning performance outdoors, but the satellite signal strength and quality decrease sharply in indoor and other sheltering environments, resulting in the problem of too-large positioning accuracy error and even loss of target, which creates great inconvenience for people (Deng, 2014).

Location of short-distance wireless technology based on UWB can effectively compensate for the shortcomings of the GPS technology. It has a very wide spectrum, unique advantages in penetrating ability, fine resolution, accurate location, and anti-multipath and anti-jamming, making it one of the most promising short-distance wireless indoor positioning technologies, as proposed by Hu & Zhu (2005). After several years of development, the basic technology of UWB has been well developed, and its potential of ranging and positioning in indoor complex environments also needs further excavation developed by Kristem & Molisch

(2014). UWB indoor location based on the characteristics of positioning, with a very strong practicality, will further promote the development of UWB indoor location measurement technology proposed by Zhang & Yang (2013).

2 SDS-TWR METHOD

2.1 *TWR algorithm principle and error analysis*

The TWR algorithm is based on the signal propagation velocity and time in the medium to calculate the distance between the transceiver nodes. The algorithm process is as follows, assuming that node A sends a signal and records the transmission and reception timestamp. Node A calculates the time from sending to the reception, recorded as T_A. Node B receives the signal and records the reception and transmission timestamp. Node B calculates the time from reception to sending, recorded as T_B. Then the system can calculate the distance between node A and node B by:

$$d = \frac{(T_A - T_B)}{2} \times c \qquad (1)$$

where c = signal propagation speed in a medium.

Because T_B is much larger than the flight time, the error can be simplified as

$$\Delta d = \frac{T_B \times (e_A - e_B)}{2} \times c \qquad (2)$$

TWR error is derived from clock synchronisation errors between two nodes.

2.2 SDS-TWR algorithm principle and error analysis

The IEEE 802.15.4a standard describes the location-based technique of ranging, which avoids the problem of time synchronisation between the transmitter and the receiver. The IEEE 802.15.4a standard has been described in Liu & Wan (2015).

The main steps of the SDS-TWR algorithm are as follows:

1. The whole process is divided into two transceiver cycles.
2. The first time, node A sends the message to node B.
3. Node A records the timestamp of the first transmission and reception, and calculates the time from transmission to the reception, recorded as T_{RA}.
4. Node B records the timestamp of the first reception and transmission, and calculates the time to reception of the transmission, recorded as T_{DB}.
5. The second time, node B sends the message to node A.
6. Node B records the timestamp of the second transmission and reception, and calculates the time from transmission to the reception, recorded as T_{RB}.
7. Node A records the timestamp of the second reception and transmission, and calculates the time from reception to transmission, recorded as T_{DA}.
8. A ranging loop is completed, as shown in Figure 1.

The distance between node A and node B can be calculated by

$$d = \frac{(T_{RA} + T_{RB} - T_{DA} - T_{DB})}{4} \times c \qquad (3)$$

Because T_{DB} or T_{DA} is much larger than the flight time, the error can be simplified as

$$\Delta d = \frac{(T_{DB} - T_{DA}) \times (e_A - e_B)}{4} \times c \qquad (4)$$

where e_A = clock offset error of node A; e_B = clock offset error of node B.

It can be seen from the comparison of the results of Formula (2) and Formula (4) that the

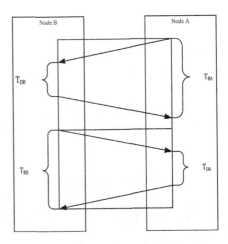

Figure 1. SDS-TWR method of ranging principle.

SDS-TWR algorithm can greatly reduce the ranging error, which can reduce node A and node B's requirements on the time synchronisation and equipment costs and improve the accuracy of ranging, as discussed in Chang (2015).

3 THE EXPERIMENT

3.1 Construction of experimental platform

Node antenna mounting height, UWB pulse repetition rate, signal centre frequency and bandwidth are the main factors influencing the precision of the SDS-TWR method based on UWB. Therefore, the theoretical analysis and experimental verification are carried out on the above factors in different environments.

Laboratory equipment: this article built the SDS-TWR method based on a UWB ranging system shown in Figure 2.

The system consists of a control module, UWB signal transceiver module, USB module, power module and JTAG module. The equipment frame is shown in Figure 3. The system selects a Cortex-M3 based STM32 F103 core as the core processor. It has a rich set of peripherals, low power and low cost advantages. It is mainly used for motor control, automation, and electronic measurement applications. The UWB signal transceiver module consists of the antenna and DW1000 RF chips. The most important UWB signal RF chip DW1000 complies with IEEE 802.15.4-2011 UWB standard. The USB module and the power module are used for power management. The JTAG module is used for debugging programs.

The system can modify UWB pulse repetition rate, UWB centre frequency and bandwidth by the

Figure 2. Experimental equipment.

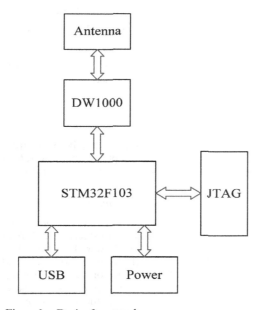

Figure 3. Device framework.

Table 1. Channels and their corresponding frequencies.

Channel	Centre frequency MHz	Bandwidth MHz
1	3494.4	499.2
2	3993.6	499.2
3	4492.8	499.2
4	3993.6	1331.2
5	6489.6	499.2
7	6489.6	1081.6

Figure 4. Complex experimental environment.

8.6 m * 3.5 m. The complex experimental environment is shown in Figure 4.

3.2 *Antenna mounting height experiment*

In Shukla & Volos (2008) it is shown that, in a UWB channel, the results of the analysis of different antenna heights by means of the path loss, the average receiving energy of the projected distance, the small scale attenuation show that the positive correlation between the antenna height and the attenuation.

In the complex environment, the experiment kept the other parameters invariant and changed the three antenna mounting heights, which were 1.20 m, 2.00 m and 2.88 m. More than 30 data points of every 2 m were collected in the experiment. The experimental data obtained from 2 m to 20 m are shown in Figure 5 and Figure 6.

Experimental data results showed that, according to the above experimental data, the antenna height at 1.20 m has higher accuracy. The antenna height at 1.20 m and 2.88 m produces significant distance error. When the antenna mounting height is 1.20 m, the UWB signal is affected by surface wave and nearby obstacles. When the antenna mounting height is 2.88 m, the UWB signal is affected by the roof and the roof hanging.

code. Its pulse repetition rate can be configured as 16 MHz or 64 MHz. UWB signal centre frequency and bandwidth can be configured as Channel 1, Channel 2, Channel 3, Channel 4, Channel 5 and Channel 7. Its corresponding centre frequency and bandwidth are shown in Table 1.

Complex experimental environment: the classroom with some people walking around, a lot of desks and chairs and other obstacles, all kinds of signal interference. The classroom size is 20 m *

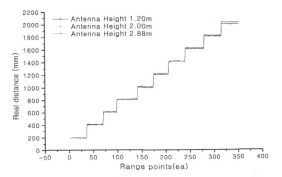

Figure 5.   Total data.

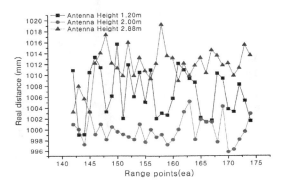

Figure 6.   Amplification at 10 m.

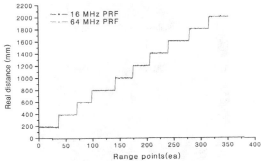

Figure 7.   Total data.

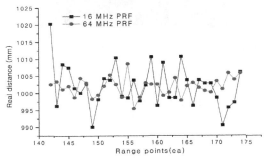

Figure 8.   Amplification at 10 m.

3.3   Pulse Repetition Frequency (PRF)

The system provides preamble extension. The number of preamble symbols is allowed to use a non-standard number; thus it extends the length of the preamble and optimises the system performance. The preamble sequence has a perfect periodic autocorrelation property, so that a receiver can receive the channel's accurate impulse response. This has two advantages. Firstly, the receiver can use the received multipath energy and turn multipath interference into an advantage. Secondly, it lets the receiver resolve the channel in detail and determine the arrival time of the first path, which brings precision advantages for RTLS applications.

In the complex environment, the experiment kept the other parameters and environment invariant and only changed the pulse repetition rate twice, to 16 PRF MHz and 64 PRF MHz. More than 30 data points of every 2 m were collected in the experiment. The experimental data obtained from 2 m to 20 m are shown in Figure 7 and Figure 8.

The experimental results can be seen from the above data. When the pulse repetition rate is 64 MHz, the range is more accurate and stable.

3.4   Centre frequency experiment

By ranging estimation of ideal channel model and the actual channel model considering multipath based on the Cramer-Rao Lower Bound (CRLB) and consideration of the receiver structure and the two protocols in the use of the synchronisation sequence of the accuracy of the measurement, the conclusion is that the two factors of ranging performance are the centre frequency of UWB signals and the bandwidth of UWB signals. Direct-sequence spread-spectrum UWB has better ranging performance at higher centre frequency and bandwidth; see Gong & Kong (2011).

The experiment kept the other parameters and environment invariant and changed the six channels (i.e. UWB signal, centre frequency and bandwidth), which are Channel 1, Channel 2, Channel 3, Channel 4, Channel 5 and Channel 7. Their corresponding centre frequencies and bandwidths are shown in Table 1.

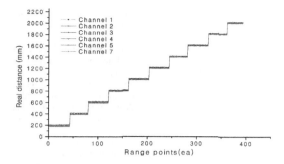

Figure 9.   Total data.

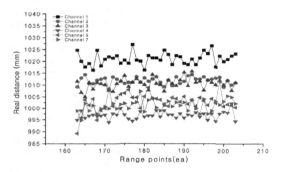

Figure 10.   Amplification at 10 m.

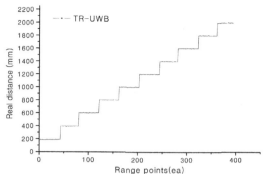

Figure 11.   Total data.

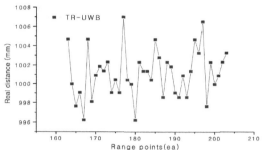

Figure 12.   Amplification at 10 m.

More than 30 data points of every 2 m were collected in the experiment. The experimental data obtained from 2 m to 20 m are shown in Figure 9 and Figure 10.

The experimental results can be seen from the above data. With the same bandwidth, the UWB signal's ranging accuracy is: Channel 5 > Channel 3 > Channel 2 > Channel 1. With the same centre frequency, the UWB signal's ranging precision and stability of Channel 7 are higher than that of Channel 5.

3.5   Ranging performance experiment of UWB ranging in complex environment

Based on the optimal conditions obtained from the previous three sets of experiments, the ranging performance of the SDS-TWR method based on UWB under optimal conditions was studied. In the complex laboratory, antenna height is set to 2.00 m, PRF is set to 64 MHz, centre frequency is set to 6489.6 MHz and bandwidth is set to 1081.6 MHz. The experiment collected a group of more than 30 points of data every 2 m. The experimental data obtained from 2 m to 20 m are shown in Figure 11 and Figure 12.

The experimental results can be seen from the above data. The errors of all points are within

±10 cm. This shows that the SDS-TWR method based on UWB can achieve high accuracy in a complex environment.

4   CONCLUSION

1. In the complex environment, with UWB pulse repetition rate, UWB signal centre frequency, UWB bandwidth and other factors constant, selection of antenna mounting height should be taken into consideration according to the corresponding environment. The antenna near the ground is easily influenced by ground wave and near-earth objects. The antenna near the roof is also easily influenced by the roof pendant. The installation location should avoid the influence of indoor walls and obstructions.
2. With other parameters and environment constant, high UWB PRF can receive more echo signal and get a more accurate mean echo intensity in a short period, which contributes to the UWB positioning node distance measurement. The experimental results show that higher PRF is more accurate and stable.
3. With other parameters and environment constant, the ranging accuracy of higher-frequency

channels for UWB signals is better than that of low centre frequency channels. With the same centre frequency, the ranging accuracy of wide bandwidth for UWB signals is higher than that of narrow bandwidth. The experimental results show that the higher the centre frequency and bandwidth of the UWB, the better the accuracy of the measurement.

4. The experimental data and the correlation theories show that the SDS-TWR method based on UWB is better than a current range system such as ZigBee. In appropriate configuration of the parameters, the error can be achieved within the range of ±10 cm in an indoor complex environment. The current RSSI ZigBee ranging method is controlled within ±0.5 m, as proposed by Zhang & Song (2011). The RSSI of the ZigBee ranging method has finite precision.

The SDS-TWR method based on UWB, under the optimal ranging conditions obtained in the experimental results, shows its excellent ranging performance in an indoor complex environment.

ACKNOWLEDGEMENTS

Supported by the National Natural Science Foundation of China (Grant No. 61540027).

REFERENCES

Chang, L. (2015). Application of SDS-TWR technology in coal mine personnel positioning system. *Industry and Mine Automation, 41*(10), 71–73.

Deng, Y. (2014). *UWB-based indoor accurate navigation research and implementation*. East China University of Technology.

Gong, S., & Kong, L. (2011). UWB signal ranging technology based on delay line model. *Journal of Chongqing University of Posts & Telecommunications*.

Hu, L., & Zhu, Z. (2005). Simulation and analysis of UWB signals is clustering property of indoor multipath channels. *Journal of Nanjing University of Posts and Telecommunications (Natural Science), 25*(6), 17–21.

Kristem, V., & Molisch, A.F. (2014). Coherent UWB ranging in the presence of multiuser interference. *IEEE Transactions on Wireless Communications, 13*(8), 4424–4439.

Liu, J., & Wan, J. (2015). A novel energy-saving one-sided synchronous two-way ranging algorithm for vehicular positioning. *Mobile Networks & Applications, 20*(5), 661–672.

Shukla, U.K., & Volos, H. (2008). On the effect of antenna height on the characterization of the indoor UWB channel. Global Telecommunications Conference, 1–5.

Ware, R.H., & Fulker, D.W. (2014). Realtime national GPS networks for atmospheric sensing. *Journal of Atmospheric and Solar-Terrestrial Physics, 63*(12), 1315–1330.

Zhang, C., & Song, X. (2011). Research on the ZigBee-based RSSI ranging accuracy. *ZigBee Journal of Hunan University of Technology, 25*(5), 37–41.

Zhang, L., & Yang, G. (2013). Ultra-wide-band based indoor positioning technologies. *Journal of Data Acquisition & Processing, 28*(6), 706–713.

Automotive, Mechanical and Electrical Engineering – Liu (Ed.)
© 2017 Taylor & Francis Group, London, ISBN 978-1-138-62951-6

Analysis of user influence in Sina Microblog

Jun Wang, Zewen Cao, Peiteng Shi & Wensen Liu
Science and Technology on Information Systems Engineering Laboratory, National University of Defense Technology, Changsha, P.R. China

ABSTRACT: This paper analyses the influence of users in Sina Microblog based on interaction of users in microblogging service. With the data from Open Weiboscope Data Access website, we focus on "active users" in original data and construct a probability transition matrix of information diffusion. Then we utilise the measurement of effective distance to define the users' influence, finding the top influence users in Sina Microblog. We find that the top users' effective distance to others are shorter than others. Usually, they also make more interactions (retweets) in microblogging service. Furthermore, we construct a net graph of users by selecting the minimum effective distance of users as their connection in graph, finding that top users are usually located at the centre of network. All the results demonstrate the measurement of effective distance as a novel perspective and effective tool, which can be well applied to the investigation of the social network of Sina Microblog.

Keywords: influence of user; Sina Microblog; effective distance; information diffusion

1 INTRODUCTION

Microblogging service is an online social platform where users share their information (texts, pictures, videos etc.) with others. One can follow another to create relationships with others. You will be someone's follower and receive all his/her update statuses if you follow him/her, but the one you followed does not have to follow you back as the relationship in microblogging service requires no reciprocation. Users can forward the status they are interested in, which is called retweets. This retweet mechanism enables information diffusing more quickly over the online social networks.

While microblogging service Twitter is becoming more popular all over the world, the information propagation on Twitter also attracts more attention from researchers. Researchers hope to figure out how the sretweet mechanism make information diffuse so fast and who the influential users are, and so on. Empirical studies have been done to solve these problem in recent years. Wu S. et al. (2011) classified "elite" users on Twitter into four categories and investigated the flow of information respectively, finding that different categories of users emphasise different types of content. Achananuparp et al. (2012) concentrated on connection in tweeting and retweeting activities, and the originating and promoting behaviours. Cha et al. (2010) investigated the influence of users in Twitter by users' in-degree, retweets and mentions,

and compared the three measurements' differences for computing the influence of users. Bakshy et al. (2011) studied the relative influence of Twitter users by tracking diffusion events that took place on the Twitter.

The Sina Microblog is a Twitter-like microblogging service which reached more and more attention of users in China in recent years since it was created in 2009. The majority of researchers in China also focus on the information diffusion in Sina Microblog. Some research (Wang R. et al., 2010; Chen Z. et al., 2012; Wu X. et al., 2014) have found influential users by focusing on the number of users' followers, so called in-degree. Due to the relations of users, all of users in Sina Microblog consist of a huge network. The topological characteristics of the social network have also have been investigated (Fan et al., 2011; Guo et al., 2011). Some other insights are offered to study the information propagation in Sina Microblog, for instance, Guan et al. (2014) selected 21 hot events which were widely discussed in Sina Microblog in 2011 and did some statistical analysis about the users and information. Yan et al. (2013) proposed an extended Susceptible-Infected (SI) propagation model to investigate the impact of human dynamics on the information propagation.

In our paper, we focus on the retweets of users in Sina Microblog, utilising the interactions of users to calculate the probability of message from one to another. This probability will be used to compute the "effective distance" from one to another. The "effective distance" represents the user's influence to others. Shorter "effective distance" means more influence of users. This method is supposed to help us to find the opinion leader in the social network of Sina Microblog.

The rest of this paper is organised as follows. Firstly, we explain the method to get the "effective distance" between users. Next, we introduce the data set of Sina Microblog and the data processing, followed by presenting the result we get from the data. We finally conclude our finding from the work.

2 METHODOLOGY

Effective distance is proposed to describe the connection between two places in a Global Mobility Network (GMN) by Dirk Brockmann and Dirk Helbing (Brockmann D et al., 2013). They applied the measurement of effective distance to the study for spread of epidemic (2009 H1 N1 influenza pandemic and 2003 SARS epidemic) and approved the method effectively. In their research, they constructed the GMN with the data of worldwide air traffic which contains 4,069 airports with 25,453 direct connections. In the GMN, countries are treated as nodes of network while the flow between nodes represents number of passengers between countries. Then they built up a probability transition matrix P between nodes in the network, which is denoted as follows:

$$P = \{p_{ij}\}_{N \times N}; i, j \in \{1, 2, ..., N\} \tag{1}$$

where i and j both represent the nodes (countries) in the GMN., and p_{ij} means the probability for a passenger travelling from i to j. According to the measurement for calculating the effective distance in the research of Dirk Brockman and Dirk Helbing, the effective distance between nodes is defined as:

$$d_{ij} = \min_{L=1}^{N}(L - \log(P^L)_{ij}) \tag{2}$$

Now, the d_{ij} denote the effective distance from i to j, but the d_{ij} is not equal to d_{ji} because of $p_{ij} \neq p_{ji}$. The detailed description for computing the effective distance can be found in the supplementary text of literature (Brockmann D et al., 2013).

Similar to the global mobility network, in social network of Sina Microblog, a probability transition matrix P can also be constructed with the data of users' tweets and retweets in the microblogging service. We identify the transition probability of information as follows. Firstly, considering the

number of statuses that user i posts, that is tweets of user i, as NT_i and the number of statuses that user j reposts from user i as his/her own tweets, that is retweets of user i from user j, as NRT_{ij}. Then we can conclude the probability of information diffuse from i to j as:

$$p_{ij} = \begin{cases} \dfrac{NRT_{ij}}{NT_i} & (i \neq j) \\ 0 & (i = j) \end{cases} \tag{3}$$

Note that we define the $p_{ij} = 0$ when $i = j$ which means one will not repost the tweets that were posted by himself/herself.

Finally, we utilise the method for calculating the effective distance, which is noted in formula (6), to the social network of Sina Microblog. Through the calculating of the effective distance between users in Sina Microblog, we can get an effective distance matrix D as follows:

$$D = \left\{d_{ij}\right\}_{N \times N}; i, j \in \{0, 1, 2, ..., N\} \tag{4}$$

where N denotes the number of users in Sina Microblog. Similar to the global mobility network, the d_{ij} is also not equal to d_{ji} here. And in Sina Microblog, actually the effective distance d_{ij} means the influence of user i to user j, where the shorter distance means the more influence.

3 DATA DESCRIPTION

3.1 Data source

The users' data we adopted in this paper is all from the Open Weiboscope Data Access website (http://147.8.142.179/datazip/). The Weiboscope is a data collection and visualisation project developed by the research team at the Journalism and Media Studies Centre (JMSC), The University of Hong Kong, which had investigated censorship on Sina Microblog with this data in 2013 (Fu et al., 2013). The dataset, which was collected during the whole of 2012, contains 226,841,122 tweets from 14,387,628 unique users. The structure of the original user data is as shown in Figure 1.

In the original user data, only the data in column "uid" and "retweeted_uid", which is marked with the red rectangle in Figure 1, is useful for our research. For ethical reasons, real user ID in Sina Microblog is replaced by pseudo ID, but there is no significant effect on the calculating for effective distance. We select the data which contains retweets to study, and then the data structure can be simplified as shown in Figure 2.

mid	retweeted_status_mid	uid	retweeted_uid	source	image	text	geo	created_at
mvRR3Gwo91	mkGsqA2Ne7	uFAJGSZL	uFAJGSZL	新浪微博	0	尼玛！你比我还强！！[怒] //@ukn：梁博是谁	(Null)	2012-10-01 00:01:50
mqeWAMdKtg	meByf955S3	uKB5I1OVK	uPVDO5B44	360安全浏览器	0	@uCBTJWYX4：//@uY044XKRU：没车的人	(Null)	2012-10-01 00:03:26
mu9pOmYnEk	(Null)	uSIAWSHUJ	(Null)	Weico.iPhone	0	菁这哪乾，莫愁没花了	(Null)	2012-10-01 00:02:12
m6Eeĺka2XX	mgGbs1yJZ	uRUWMWQR	uSIACIBCZ	皮皮时光机	0	12星座都脑两般船几面掛行。[微笑]	(Null)	2012-10-01 00:05:06
mEjVTmk6UI	mgY4Fvv87f	uRUWMWQR	uP2ZKANWN	皮皮时光机	0	内涵就在您的身边，开动脑筋获得的快乐会更成	(Null)	2012-10-01 00:00:10
mBc5vEpmoA	(Null)	uBIBS50FY	(Null)	皮皮时光机	1	【气场女人做人智慧——之一】行走如同，有些	(Null)	2012-10-01 00:00:18
mtqHmFot55	mdZ4IF1BAa	uDSCTY1V	uOQZQTZVB	Android客户端	0	转发微博	(Null)	2012-10-04 20:40:14
mxO1a8mGvY	mdOdQUPyk9	u3FTXZ0GL	uDGUIJQ4M	皮皮时光机	0	对女人来讲，爱情应系滋养品，而不是牢笼	(Null)	2012-10-01 00:05:03
mtq8E8CBRa	m7dGLWiaAp	u3FTXZ0GL	uGL4JPLFH	皮皮时光机	0	【学点客套话】	(Null)	2012-10-01 00:01:08
m4hQSByi4G	mGxIOdF172	uJ2Y4O3O0	uJ2Y4O3O0	iPhone客户端	0	回复@uJWI4OYBW：我是纳纳纳的声像插坤~	(Null)	2012-10-01 00:00:26
m604flBXW3	mnVWlxn0yD	uHRVMTK1G	uPA3SQI1G	WeicoPro.HD	0	转发微博 //@u5KYVFZHN：转发微博	(Null)	2012-10-01 00:05:57
m9RYK8IO7P	mu9pM3BEko	uHRVMTK1G	uB4GX5F4K	WeicoPro.HD	0	映，和「中国好声音」相比，bo的哪儿还真是厉	(Null)	2012-10-01 00:01:56

Figure 1.   The structure of the original user data.

uid	retweeted_uid
uKB5I1OVK	uPVDO5B44
uRUWMWQR	uSIACIBCZ
uRUWMWQR	uP2ZKANWN
uDSCTY1V	uOQZQTZVB
u3FTXZ0GL	uDGUIJQ4M
u3FTXZ0GL	uGL4JPLFH
uHRVMTK1G	uPA3SQI1G
uHRVMTK1G	uB4GX5F4K
u0ABXJG1G	uPKFOZJPZ
u0ABXJG1G	uBK14DAFY

Figure 2.   The structure of simplified user data.

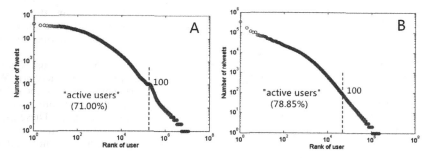

Figure 3.   The distribution of the number of users' tweets and retweets.

3.2   *Data processing*

The computing for effective distance in Sina Microblog is based on the interactions between users. Thus it is hard to determine influence of these users who post few tweets or whose tweets have hardly been reposted during the whole year. Then we have to focus on those "active users" which is a concept in the traditional media research (Levy et al., 1985). These "active users" are more likely to make contributions to information diffusion in Sina Microblog. So we ignored users who post fewer than 100 tweets and the users whose retweets

are fewer than 100. Both procedures are shown in Figure 3(A) and Figure 3(B) respectively.

In the Figure 3, the horizontal axis is the rank of users while the axis is the number of tweets or retweets. In both Fig. 3(A) and Figure 3(B), the left part of the dotted line represents the "active users". In the Figure 3(A), the tweets of all "active users" account for 71.00% of total tweets; similarly, the retweets of all "active users" account for 78.85% of total retweets in Figure 3(B). From the result, we can learn that the majority interactions (tweets and retweets) in the social network of Sina Microblog are created by minority "active users". That means, even ignoring these users, we will make no significant effect on investigating the influence of users in Sina Microblog and we also can display the main structure of the network graph of Sina Microblog more clearly by focussing on minority "active users".

4   RESULTS

After the filtering step in section 3, finally there are 4,977 "active users" left for our research. With the data of these users and the method for calculating effective distance, we get the effective distance matrix D of 4,677 users. In order to determine the influence of users in the whole network, we define the average effective distance d_i as the influence of user i.

$$d_i = \frac{1}{N} \sum_{j=1}^{N} 1/d_{ij} \tag{5}$$

By comparing the influence of all users, we sort the d_i of users and then get the rank of all users. Because the real user ID is replaced by pseudo ID, we only give the top 10 users' real ID in Sina Microblog, which is shown in Table 1, by looking for the text of users in the microblogging service.

Table 1.   Top 10 influential users.

Rank	Pseudo ID	Real ID
1	uP2ZBBHYQ	让你读懂男人心
2	uNEGDMAAR	做饭很简单
3	uPA3 LIZDC	种草小淘宝
4	uRLOV35GT	粤语·微情书
5	uK3RR2Y53	Cherry奈儿
6	uYC3IT2 JD	Unknown
7	uS5 WUDXRU	Unknown
8	uM1U0 NZOG	吃喝玩乐IN广州
9	uP2ZBBHYQ	让你读懂男人心
10	uNEGDMAAR	做饭很简单

In the Table 1, the No. 6 and No. 7 users' real ID are unknown due to their account now having been deleted. We also find that the top 10 users are almost all public accounts which are used for advertising and marketing, because these influential users can deliver information to more audiences than others. The capability of information diffusion has already attracted the attention of businessmen.

The description for influence of users is based on the interaction between users in microblogging service. Then we explore the relationship between number of users' retweets and influence of users, and the result is displayed in Figure 4.

In Figure 4(A), the axis represents the number of users' retweets NRT_i/NT_i while the horizontal axrepresents d_i (influence of user i). The distribution of the users' influence d_i is shown in Figure 4(B). In Figure 4(B), we find that the value of d_i is divided into two parts; the part below the dotted lines consist of influential users. Meanwhile, the left part of the dotted line in Figure 4(A) denotes that the value of NRT_i/NT_i, which is negative correlated with the users' influence d_i.

In order to show the clear structure of the network, we figure the net graph of the Sina Microblog. Based on the influence between the users in Figure 5, we select the minimum value of $d_{ij}(j = 1, 2, ..., N)$ as the only connection of user i. The steps for selection are as follows: firstly, if $d_{ik} = \min\{d_{ij}, (j = 1, 2, ..., N)\}$, then there is a connection from i to k. It means that each user only has one connection to others in the net graph, but that user can have more than one connection from others.

In Figure 5, we show several users' real ID which occupy the centre of a small network. We can find that these users mainly contain the celebrities and some interesting public accounts which have more audiences than normal users. These users are also "influential users" because they can deliver information to more normal users more quickly.

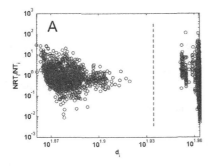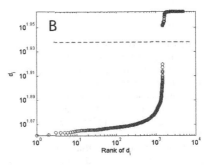

Figure 4.   The distribution of users' influence d_i and its relationship with NRT_i/NT_i.

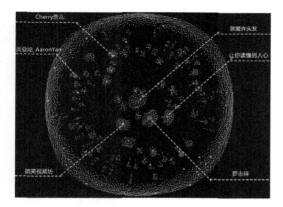

Figure 5. The structure of the social network of Sina Microblog.

5 CONCLUSION

This paper analysed the influence of users in the social network of Sina Microblog. To describe the information diffusion in microblogging service, we constructed a probability transition matrix based on the interaction of users in Sina Microblog. Then we defined the influence of users by computing the effective distance (Brockmann D et al., 2013) between them. We found that the effective distance can reflect the influence from one user to another.

Firstly, we filtered the original data by focussing on the user who makes more interactions in the microblogging service, so called "active users", finding that "active users" generate the majority interaction in the Sina Microblog. Then, we analysed the relationship between the influence and users' value of NRT_i/NT_i, finding that the influential users are exactly more likely to be very influential, but the normal users' influences are much smaller than influential users. Finally, we selected the minimum effective distance of user as one's connection in the net graph of Sina Microblog, which is used to display the clear structure of the social network. Through the net graph we can see the relationship between users more clearly. We found the influential user usually occupies the centre location of a small network of whole graph, meaning that the top user can deliver the information to others easier and faster.

We applied the measurement of effective distance to the analysis of influence of users in Sina Microblog and approved that the effective distance can be used to describe the influence of users. Because we did not get the more complete data of microblogging service, the analysis did not cover all users in Sina Microblog. Otherwise, the result could have been more complete in the paper.

REFERENCES

Achananuparp, P., Lim, E.P., Jiang, J. et al. (2012). Who is retweeting the tweeters? Modeling, originating, and promoting behaviors in the Twitter network. *ACM Transactions on Management Information Systems*, 3(3), 1–30.

Bakshy, E., Hofman, J.M., Mason, W.A. et al. (2011). Everyone's an influencer: quantifying influence on twitter. *ACM International Conference on Web Search and Data Mining* (pp. 65–74).

Brockmann, D. & Helbing, D. (2013). The hidden geometry of complex, network-driven contagion phenomena. *Science*, 342(6164), 1337–1342.

Cha, M., Haddadi, H., Benevenuto, F. et al. (2010). Measuring user influence in Twitter: The million follower fallacy. *International Conference on Weblogs and Social Media*. Washington, DC.

Chen, Z., Liu, P., Wang, X. et al. (2012). Follow whom? Chinese users have different choice. *Computer Science*.

Fan, P., Li, P., Jiang, Z. et al. (2011). Measurement and analysis of topology and information propagation on Sina-Microblog. *IEEE International Conference on Intelligence and Security Informatics*, 396–401.

Fu, K.W., Chan, C.H. & Chau, M. (2013). Assessing censorship on microblogs in China: Discriminatory keyword analysis and impact evaluation of the 'real name registration' policy. *IEEE Internet Computing*, 17(17), 42–50.

Guan, W., Gao, H., Yang, M., et al. (2014). Analyzing user behavior of the micro-blogging website Sina Weibo during hot social events.. *Physica A: Statistical Mechanics & Its Applications*, 395(4), 340–351.

Guo, Z., Li, Z. & Tu, H. (2011). Sina microblog: an information-driven online social network. *International Conference on Cyberworlds*, 160–167.

Levy, M., & Windhal, S. (1985). Media gratifications research. In *The Concept of Audience Activity*. Sage.

Wang, R &. Jin, Y. (2010). An empirical study on the relationship between the followers' number and influence of microblogging. *International Conference on E-Business and E-Government*. IEEE, *2010*, 2014–2017.

Wu, S., Hofman, J.M., Mason, W.A. et al. (2011). Who says what to whom on twitter *International Conference on World Wide Web*, 705–714.

Wu, X. & Wang, J. (2014). Micro-blog in China: Identify influential users and automatically classify posts on Sina micro-blog. *Journal of Ambient Intelligence & Humanized Computing*, 5(1), 51–63.

Yan, Q., Wu, L., Liu, C. et al. (2013). Information propagation in online social network based on human dynamics. *Abstract & Applied Analysis*, 2013(2), 173–186.

Cloud evolved packet core network architecture based on Software-Defined Networking (SDN)

Yaoyun Zhang & Zhan Xu
Beijing Information Science and Technology University, Beijing, China

Zhigang Tian
Tsinghua University, Beijing, China

ABSTRACT: This paper proposes a Cloud Evolved Packet Core (CEPC) architecture based on Software-Defined Networking (SDN)/OpenFlow technology, which separates the control plane and data plane of a Serving GateWay (SGW) and a PDN gateway (PGW) completely, and runs the control functions of a Mobility Management Entity (MME), SGW-C and PGW-C of the LTE-Evolved Packet Core (LTE/EPC) architecture in the cloud controller. By simulating the initial attach process of the CEPC architecture and the traditional LTE/EPC architecture, it is concluded that the CEPC architecture has a lower paging signalling load, and the network flexibility and programmability are higher.

Keywords: SDN; Openflow; CEPC; Signalling load

1 INTRODUCTION

With the expansion of modern mobile network services and the expansion of the scale of Internet applications, the operation and management of network complexity increases and the cost becomes high. This presents a huge challenge to the current mobile network architecture. The 3GPP LTE/EPC and LTE-A standards were released to cope with the increased demand for high-speed mobile networks. However, LTE/EPC architecture still faces some problems, such as the data plane at the SGW and the PGW is still tightly coupled with the control plane.

In this paper, based on the LTE/EPC architecture, the use of SDN/OpenFlow technology (Michael, 2011) to redesign the architecture, that is, the CEPC system architecture. The architecture is flexible and programmable by completely separating the control plane and the data plane. By comparing the initial attach process of a traditional LTE/EPC architecture and a CEPC architecture, the signalling load of the process (Nowoswiat, 2013) as an evaluation metric, the CEPC architecture is found to be superior to the traditional architecture in the case of a different number of users.

2 BACKGROUND AND RELATED WORK

2.1 *3GPP LTE/EPC architecture*

A 3GPP LTE/EPC (Krishna, 2015) architecture is mainly responsible for the overall control of the user terminal and the establishment of the relevant bearer. The EPC architecture is mainly composed of an MME, SGW, PGW and the Home Subscriber Server (HSS), as shown in Figure 1. The user sends data packets through the eNB and SGW to the PGW. The MME is a control node that handles signalling interactions between the UE and the core network. SGW and PGW are responsible for data forwarding, IP address assignment, and Quality of Service (QoS) assurance in the data plane.

2.2 *Initial attachment*

In this paper, we mainly study the signalling load in the initial attachment process of data plane management.

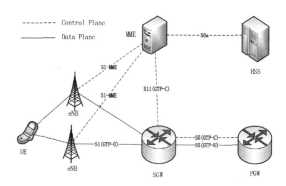

Figure 1.   Traditional LTE / EPC architecture.

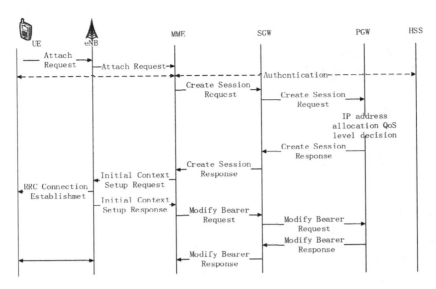

Figure 2.   Initial attachment procedure.

During the initial attachment procedure, the UE initiates an attachment request to register with the network. Figure 2 details the process including UE authentication, UE registration, and EPS bearer establishment procedures. After authentication, the UE registers with the network. The network assigns an IP address and establishes a PDN connection. When the UE needs to establish a new bearer, it sends an NAS Service Request message to the MME that sets up a new connection.

3   EPC NETWORK STRUCTURE BASED ON SDN

SDN (Kobayashi, 2014) is a new network innovation architecture, it is an indispensable way to realise network virtualisation. It is the core of the separation of the data plane and control plane in a mobile core network. The control plane is detached from the hardware and centrally controlled through the network, making programming easier and thus optimising the network management. SDN (Zheng Yihua, 2013) can provide a standard interface to control network resource access and network traffic. OpenFlow interface protocol is one of them. OpenFlow (OF) protocol is the core technology of the SDN, which defines the control layer and the forwarding device of the communication protocol.

The basic components of the SDN architecture are shown in Figure 3. The network device through control and data plane interfaces (such as Open-Flow) reports the capability of the device. The SDN service application through the API interface to tell the controller specific business needs, and the

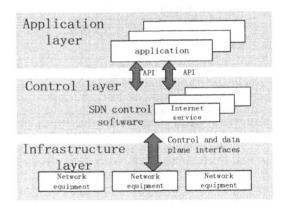

Figure 3.   SDN architecture.

controller through the business application needs to realise the control of the equipment. Through the optimal control strategy in the limited network resources to provide more competitive services.

In order to make control platform of LTE/EPC network architecture more flexibility and programmability. In this paper, we propose a mobile core network architecture based on SDN, which is called CEPC architecture.

4   CEPC ARCHITECTURE

4.1   *CEPC architecture overview*

The CEPC architecture is shown in Figure 4. The Cloud Controller (CC) is the main component in the architecture. CC runs all MME on the LTE/ EPC, SGW-C, and PGW-C control functions. CC runs all control functions of the MME, SGW-C, and PGW-C in the LTE/EPC. The main task of the CC is to manage the forwarding plane of eNB and SGW-D, which is responsible for the establishment of user sessions and the load monitoring of the data plane. According to the OF protocol, the main entities of the framework function as follows:

CC: It is responsible for setting up user sessions, installing the forwarding table for PGW-D and SGW-D and monitoring the network. MME is an application on the CC and is responsible for mobility management and UE authentication. It connects to the CC through the API interface.

SGW-C, PGW-C: They are the control functions of SGW and PGW. They are responsible for assigning the Tunnel Endpoint Identifiers (TEIDs) when the GTP tunnel is established, the UE IP is

allocated and session is established; these functions on the CC and the MME are virtualised and package them as applications.

SGW-D, PGW-D: GW-Ds (SGW-D and PGW-D collectively) are located between the CC and the PDN network or the Internet, and correspond to an OF switch that handles GTP packets. The purpose is to remove the GTP header of the packet to the network and add a GTP header to the packet arriving at the CC.

eNBs: The wireless function in 3GPP is maintained and the data plane is programmed according to the instructions received from the CC.

4.2 Initial attachment

During the initial attachment process of the PC architecture, the UE registers to the network which authorises the UE and obtains an IP address which is

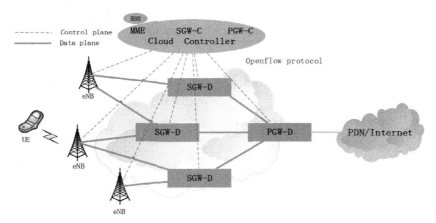

Figure 4.   CEPC architecture.

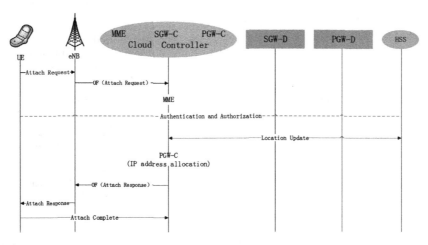

Figure 5.   CEPC structure of the initial attachment procedure.

used for data transmission. Compared to the 3GPP LTE/EPC initialisation attachment procedure, no session is established between the SGW and the PGW, thus reducing the exchange of information between the CC and the PGW for the establishment of this session. In this process, the CC assigns an IP address to the UE through the PGW-C application, used to initialise the attachment process, as shown in Figure 5. First, the UE sends an initial attach request to the eNB, this message is embedded in an OpenFlow Initial UE message and is sent to the CC. The CC sends the message to the MME and triggers the process of authenticating and authorising the UE. After the authentication and authorisation process, the PGW-C application allocates an IP address to the UE.

5 PERFORMANCE ANALYSIS

In this part, this paper analyses the signalling load of the control entity of the CEPC architecture and the 3GPP LTE / EPC architecture, and compares them. The control entity of the 3GPP LTE/EPC is an MME, the control entity of CEPC framework is a CC. In the initial attachment process, the signalling load of these control entities will be used as an evaluation metric.

In the signalling load analysis, the signalling analysis model described in Widjaja (2009) and Kumar (2012) is used in this paper to analyse the signalling load generated during the initial attachment process. Assume that each UE is a smart phone, which supports K application types, such as WeChat, e-mail, voice and so on. λk is the average arrival rate of the type-k session, P is the probability of the UE initiating the session, and a region is selected as the experimental region. C, A and ρ represent the number of eNBs in the experimental area, the area of the cell and the density of the UE in the test area. Sm and Sc represent the signalling load of MME, CC respectively.

5.1 *Traditional LTE/EPC architecture*

In a network architecture, the signalling load of the controlling entity is proportional to the number of messages entering and leaving these entities and the average arrival rate of the UE sessions. Because this paper assumes that each UE is a smart phone and capable of supporting a variety of application types, the signalling load is affected by the average arrival rate of sessions generated by an application in the UE. However, during the initial attachment procedure, the signalling load is not affected by the sessions arrival rate, since the initialisation attachment process does not depend on what type of application the user is using. So in

the traditional LTE/EPC architecture of the initial attachment process. In addition to the authentication and authorisation steps, the total number of messages entering and leaving the MME entity is ten messages. Thus, the total signalling load of the MME entity caused by the initial attach process is:

$$S_m(k) = 10 P \rho A C \qquad (1)$$

$$\rho = \frac{Nue}{S} \quad (\text{UEs/km}^2) \qquad (2)$$

where P is the probability that the UE initiates the attachment to the network, *Nue* is the total number of users, and S is the area of the experimental area.

5.2 *CEPC architecture*

In the same way, the total signalling load at the CC caused by the CEPC architecture during the initial attachment process is easily obtained. The signalling load caused by the initial attachment process is:

$$S_c(k) = 6 P \rho A C \qquad (3)$$

6 RESULTS ANALYSIS

This section compares the total signalling load of the traditional LTE/EPC architecture and the CEPC architecture and presents the numerical results of the equations described above. The values of the parameters in the equation are derived from (Krishna, 2015), as shown in Table 1. In this paper, the MME and CC of the total signalling load is evaluated.

The relationship between the number of users and the total signalling load of the two architectures is simulated through Equations (1) - (3) and the values given in Table 1. In Figure 6, the solid line and the dotted line respectively represent changes in the total signalling load of the control entities of the traditional LTE/EPC architecture

Table 1. Parameter for evaluation.

Parameter	Description	Value
P	Probability that a UE initiates an attachment procedure	0.2
A	The area of a cell	200
C	Total number of eNBs in a considered region	500
Nue	Total number of UEs	
S	Area of a considered region (km²)	500
ρ	UE density (UEs/km²)	

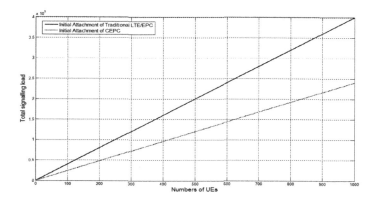

Figure 6.   Comparison of the results.

and the CEPC architecture as the number of users increases during the initial attachment process. The simulation results show that the total signalling load increases linearly with the number of users, and the growth of the total signalling load of the CEPC architecture is slower than that of the traditional LTE/EPC architecture. The CEPC architecture has a lower paging signalling load and is more flexible and programmable compared to the traditional LTE/EPC architecture.

7   SUMMARY

In this paper, a architecture based on SDN is proposed, and the control functions of the LTE/EPC with MME, SGW-C and PGW-C are run on the cloud controller. with the traditional LTE/EPC network architecture, the signalling load increases linearly with the number of users through simulation, but the growth of the total signalling load of the CEPC architecture is slower than that of the traditional LTE/EPC architecture. So the CEPC architecture that makes the network more optimised has a lower paging signalling load and is more flexible and programmable compared to the traditional LTE/EPC architecture.

ACKNOWLEDGEMENTS

This work was supported in part by the National Natural Science Foundation of China (No. 61402044), by China's 863 Plan Program (No. 2015 AA01 A706), by the Science Foundation of Beijing Education Commission (No. KM201511232011) and Bejing Nova Program (Z161100004916086).

REFERENCES

3GPP (2014). GPRS Enhancements for E-UTRAN Access [EB/OL]. http://www.tech-invite.com
3GPP (2015). General packet radio service (GPRS) enhancements for evolved universal terrestrial radio access network (EUTRAN) access [EB/OL]. http://www.3 gpp.org/
Cooney, M. (2011). Gartner: the Top 10 Strategic Technology Trends for 2012[EB/OL]. www.networkworld.com
Kobayashi, M., S. Seetharaman, G. Parulkar, G, Appenzeller, J, Little, J, van, Reijendam, P, Weissmann, and, N, McKeown. Maturing of OpenFlow and software-defined networking through deployments. Elsevier Computer Networks, (61), 151–175.
Krishna, M, Sivalingam, Sakshi, Chourasia. (2015) SDN based evolved packet core architecture for efficient user mobility support. IEEE Network Softwarization (NetSoft).
Kumar, C.S., R.V., Rajakumar, Pankaj, Kumar, Gupta. (2012). Analysis of impact of network activity on energy efficiency of 3GPP-LTE[C]. IEEE India Conference (INDICON), 665–669.
Nowoswiat, D., G. Milliken. (2013). Alcatel-Lucent: Managing the signaling traffic in packet core [EB/OL]. http://www.alcatel-lucent.com/techzine/.
Widjaja, I., P. Bosch, H. La, Roche. (2009). Comparison of MME signaling loads for long-term-evolution architectures. Vehicular Technology Conference Fall, (1), 1–5.
Zheng, Yi, Hua, Yiqiang, He, Xiaofeng. (2013). Characteristics, Development and Future of SDN. Telecommunications Science, (9), 102–107.

Automotive, Mechanical and Electrical Engineering – Liu (Ed.)
© 2017 Taylor & Francis Group, London, ISBN 978-1-138-62951-6

Construction of a military requirement system based on object-relational analytical approach

Xiaolei Zheng
Logistics College, Beijing, P.R. China

Ji Ren
Equipment Demonstration Center, Beijing, P.R. China

ABSTRACT: Given the characteristics of military information requirement systems, a military requirement system based on the reconstruction conjecture in graph theory was proposed to analyze the hierarchy of needs and resolve the relationships between objects. The steps of the military requirement system construction were determined. By combining quality and quantitative methods, the developed information system can fully meet the customers' requirements.

Keywords: Military requirement system; Hierarchical induction; Object-relation analytical

1 INTRODUCTION

When dealing with a demand management problem, it is important to logicalization and methodize the large number of chaotic military requirements in order to capture the basic profile of the system, grasp the relationship between the main demands and other demands, and build the military requirement system scientifically. There are two effective techniques to build the military requirement system. One is carrying out hierarchy of needs induction analysis for the military requirement set, and the other is resolving the relationships among each of the military requirement.

2 HIERARCHICAL INDUCTION ANALYSIS

The hierarchical induction analysis is a qualitative method for analysing the hierarchical structures of military requirement systems. The demand hierarchical structure can be defined as follows:

Definition 4.1 Set A is the nonempty set for military requirements. If the relationship r among the demands of A meets the conditions of transitivity and anti-reflexivity, the requirement system structure corresponding to <A, r> is called the hierarchical structure.

The induction analysis of the demand set is divided into the requirement system with a hierarchical structure, which can be obtained by dividing and classifying in detail the hierarchy of the demand set. According to the characteristics of a military requirement system, the concept of hierarchical induction and fine division of the military requirement system are presented as follows:

Definition 4.2 Set A is the nonempty set for the military requirement, π is a division of A, and $\pi(A) = \{A_1, A_2, ..., A_n\}$. If $A_{i+1} \subseteq A_i$ is the demand connotation, then π is called a hierarchical induction of A. If A_i is higher than A_{i+1} in the abstraction level, then π is called a hierarchical induction of A from top to bottom.

Definition 4.3 Set A is the nonempty set for the military requirement system, π is a division of A, $\pi(A) = \{A_1, A_2, ..., A_n\}$, and π_i is a division of A_i. $\pi_i(A_i) = \{A_{i1}, A_{i2}, ..., A_{in}\}$ if the demands in A_{ij} have close relationships; however, if the demands in $A_i - A_{ij}$ have no relationship or the relationships are not close, then π_i is called the fine division of π_i.

Here, a simple demand set of a data chain is presented to illustrate the application of demand hierarchical induction technology.

Data chain TC-4 supports the communication demand set $A = \{a_1, a_2, ..., a_{16}\}$. Herein, a_1 is the multi-service integrated communication ability, a_2 is the high quality voice communication function, a_3 is the connecting built-up time, a_4 is the communication function of high definition static image, a_5 is the communication function of the continuous dynamic image in real time, a_6 is the accurate data communication function, a_7 is the connecting release time, a_8 is the call blocking probability, a_9 is the code rate, a_{10} is the error rate, a_{11} is the dynamic allocation of bandwidth, a_{12} is the acceptable

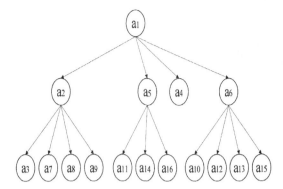

Figure 1.   Demand hierarchical structure of set A.

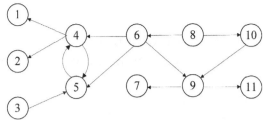

Figure 2.   The schematic diagram of the demand relationship.

packet error rate, a_{13} is the acceptable bit error rate, a_{14} is the message capacity, a_{15} is the performance ratio, and a_{16} is the communication bandwidth.

In practice, the hierarchical division $\pi(A) = \{A_1, A_2, A_3\}$ is first carried out, where $A_1 = \{a_1\}$ is the ability hierarchical demand, $A_2 = \{a_2, a_4, a_5, a_6\}$ is the function demand supporting a_1 multi-service integrated communication ability, and $A_3 = \{a_3, a_7, a_8, ..., a_{16}\}$ is the performance requirement.

Then, A_3 was refined and divided, and $\pi_i(A_3) = \{A_{31}, A_{32}, A_{33}\}$, where $A_{31} = \{a_3, a_7, a_8, a_9\}$ is the performance requirement supporting the high quality voice communication function, $A_{32} = \{a_{11}, a_{14}, a_{16}\}$ is the performance requirement supporting the communication function of a continuous dynamic image in real time, and $A_{33} = \{a_{10}, a_{12}, a_{13}, a_{15}\}$ is the performance requirement supporting the accurate data communication function.

Then, the hierarchical structure of this demand set is shown in Fig. 1.

3   OBJECT-RELATIONAL ANALYTICAL TECHNIQUE

The object-relational analytic technique is based on the reconstruction conjecture in graph theory. By analysing the relationship of the military requirements and performing operations on related graphs and matrices, the requirement system accessibility matrix can be obtained. Then, it can be resolved into multilevel hierarchical structures.

3.1   *The relational graph, relationship matrix and accessibility matrix of the military requirement system*

The military requirement system describes the relationship between military requirements, and the demands can be better described by digraph. The nodes are used to represent demands, and

the arrows to indicate the relationship between demands. Fig. 2 shows the diagram of a military requirement system.

In addition, the relationship matrix is used to present the direct relationships between every demand. Supposing the requirement system has n demands in total.

$$S = \{r_1, r_2, ..., r_n\} \qquad (1)$$

Then, the demand relationship matrix is

$$A = \begin{Bmatrix} a_{11} & a_{12} & \cdots & a_{1n} \\ a_{21} & a_{22} & \cdots & a_{2n} \\ \vdots & \vdots & & \vdots \\ a_{n1} & a_{n2} & \cdots & a_{nn} \end{Bmatrix}$$

Herein

$$a_{ij} = \begin{cases} 1, & \text{if } r_j \text{ and } r_i \text{ show a relationship;} \\ 0, & \text{if } r_j \text{ and } r_i \text{ show no relationship} \end{cases}$$

The corresponding relationship matrix of Fig. 1 is:

$$A = \begin{Bmatrix} 0 & 0 & 0 & 1 & 0 & 0 & 0 & 0 & 0 & 0 & 0 \\ 0 & 0 & 0 & 1 & 0 & 0 & 0 & 0 & 0 & 0 & 0 \\ 0 & 0 & 0 & 0 & 0 & 0 & 0 & 0 & 0 & 0 & 0 \\ 0 & 0 & 0 & 0 & 1 & 1 & 0 & 0 & 0 & 0 & 0 \\ 0 & 0 & 0 & 1 & 0 & 1 & 0 & 0 & 0 & 0 & 0 \\ 0 & 0 & 0 & 0 & 0 & 0 & 0 & 1 & 0 & 0 & 0 \\ 0 & 0 & 0 & 0 & 0 & 0 & 0 & 0 & 1 & 0 & 0 \\ 0 & 0 & 0 & 0 & 0 & 0 & 0 & 0 & 0 & 0 & 0 \\ 0 & 0 & 0 & 0 & 0 & 1 & 0 & 0 & 0 & 1 & 0 \\ 0 & 0 & 0 & 0 & 0 & 0 & 0 & 1 & 0 & 0 & 0 \\ 0 & 0 & 0 & 0 & 0 & 0 & 0 & 0 & 1 & 0 & 0 \end{Bmatrix}$$

This is an 11×11 square matrix. Each row or each column corresponds to one of the demands in Fig. 1 and reflects the relationship between the demands.

The concept of requirement system accessibility matrix is as follows:

Suppose D is the relational graph of the military requirement system $S = \{r_1, r_2, ..., r_n\}$ composed of n demands, then the element

$$m_{ij} = \begin{cases} 1, & \textit{if starting from } r_j \textit{ to } r_i \textit{ by passing} \\ & \textit{several branches}: \\ 0, & \textit{otherwise} \end{cases}$$

The n × n matrix M is called the accessibility matrix of the relational graph D. If it goes from rj to riby passing k branches, then it is accessible from rj to ri, and its length is k.

The computing method of the military requirement system accessibility matrix is

$$A \cup A^2 \cup \cdots \cup A^n \quad (2)$$

Supposing any demand to itself is accessible, then

$$M = I \cup A \cup A^2 \cup \cdots \cup A^n \quad (3)$$

The accessibility matrix can be calculated by the above formula. And

$$(I \cup A)^n = I \cup A \cup A^2 \cup \cdots \cup A^n \quad (4)$$

Thus, the even powers of $(I \cup A)$ can be calculated, namely $(I \cup A)^2 (I \cup A)^4, (I \cup A)^8, ...,$ if

$$(I \cup A)^{2^{i-1}} \neq (I \cup A)^{2^i} = (I \cup A)^{2^{i+1}} \quad (5)$$

Then

$$M = (I \cup A)^{2^i} \quad (6)$$

The flow diagram of the calculation procedure for the accessibility matrix and the detailed computing methods are described in literature.

The accessibility matrix of Fig. 1 was calculated by the above method, obtaining

$$M = \begin{Bmatrix} 1 & 0 & 1 & 1 & 1 & 1 & 0 & 1 & 0 & 0 & 0 \\ 0 & 1 & 1 & 1 & 1 & 1 & 0 & 1 & 0 & 0 & 0 \\ 0 & 0 & 1 & 0 & 0 & 0 & 0 & 0 & 0 & 0 & 0 \\ 0 & 0 & 1 & 1 & 1 & 1 & 0 & 1 & 0 & 0 & 0 \\ 0 & 0 & 1 & 1 & 1 & 1 & 0 & 1 & 0 & 0 & 0 \\ 0 & 0 & 0 & 0 & 0 & 1 & 0 & 1 & 0 & 0 & 0 \\ 0 & 0 & 0 & 0 & 0 & 1 & 1 & 1 & 1 & 1 & 0 \\ 0 & 0 & 0 & 0 & 0 & 0 & 0 & 1 & 0 & 0 & 0 \\ 0 & 0 & 0 & 0 & 0 & 1 & 0 & 1 & 1 & 1 & 0 \\ 0 & 0 & 0 & 0 & 0 & 0 & 0 & 1 & 0 & 1 & 0 \\ 0 & 0 & 0 & 0 & 0 & 1 & 0 & 1 & 1 & 1 & 1 \end{Bmatrix}$$

3.2 *The division of military requirement system accessibility matrix*

In order to analyse the hierarchical structure of the requirement system through the demand accessibility matrix, the relationships between demands given by the accessibility matrix need to be divided.

3.2.1 *Hierarchical division, $\pi_i(P)$*

First, the definitions of an accessible set and advanced set are given. For every demand r_i, the set composed of accessible demands is defined as the accessibleset $R(r_i)$ of r_i, namely $R(r_i) = \{r_j | r_j \in S, m_{ij} = 1\}$. The set composed of the demands r_i is defined as the advanced set $A(r_i)$ of r_i, namely $A(r_i) = \{r_j | r_j \in S, m_{ji} = 1\}$.

The intersection of $A(r_i)$ and $R(r_i)$ is the same as $R(r_i)$. Therefore, the condition of the uppermost required r_i is

$$R(r_i) = R(r_i) \cap A(r_i) \quad (7)$$

The requirement system of all hierarchies can be divided with the same method. Supposing $L_1, L_2, ..., L_l$ are hierarchies from top to bottom, then the hierarchical division of requirement system S can be expressed as:

$$\pi_l(P) = \{L_1, L_2, ..., L_l\} \quad (8)$$

The specific iteration steps are

1. Where, $L_0 = \emptyset$, and L_j means the j level, $j \geq 1$;

$$L_j = \{r_i \in P - L_0 - L_1 - \cdots - L_{j-1} | \\ R_{j-1}(r_i) \cap A_{j-1}(r_i) = R_{j-1}(r_i)\} \quad (9)$$

$$R_{j-1}(r_i) = \{r_i \in P - L_0 - L_1 - \cdots - L_{j-1} | m_{ij} = 1\} (10)$$

$$A_{j-1}(r_i) = \{r_i \in P - L_0 - L_1 - \cdots - L_{j-1} | m_{ji} = 1\} (11)$$

2. When $\{P - L_0 - L_1 - \cdots - L_j\} = \emptyset$, the division is completed.

If $\{P - L_0 - L_1 - \cdots - L_j\} \neq \emptyset$, then regard $j + 1$ as j and return to step (a).

Taking Fig. 1 as the example, the following are the hierarchical divisions of the accessibility matrix:

For the first hierarchical division, As shown in Table 1, from $R(r_i)$, $A(r_i)$ and $R(r_i) \cap A(r_i)$ obtained by $P - L_0$, the demands 3 and 8 meet the requirement of $R(r_i) = R(r_i) \cap A(r_i)$, and are the uppermost demands.

For the second hierarchical division, As shown in Table 2, from $R(r_i)$, $A(r_i)$ and $R(r_i) \cap A(r_i)$ obtained by $P - L_0 - L_1$, the demands 6 and 10 meet the requirement of $R(r_i) = R(r_i) \cap A(r_i)$, and are the uppermost demands.

Table 1. The hierarchical division of military requirement system.

Requirement i	$R(r_i)$	$A(r_i)$	$R(r_i) \cap A(r_i)$
1	1, 3, 4, 5, 6, 8	1	1
2	2, 3, 4, 5, 6, 8	2	2
3	3	1, 2, 3, 4, 5	3
4	3, 4, 5, 6, 8	1, 2, 3, 4, 5	3, 4, 5
5	3, 4, 5, 6, 8	1, 2, 4, 5	4, 5
6	6, 8	1, 2, 4, 5, 6, 7, 9, 11	6
7	6, 7, 8, 9, 10	7	7
8	8	1, 2, 4, 5, 6, 7, 8, 9, 10, 11	8
9	6, 8, 9, 10	7, 9, 11	9
10	8, 10	7, 9, 10, 11	10
11	6, 8, 9, 10, 11	11	11

Table 2. The second hierarchical division of military requirement system.

Requirement i	$R(r_i)$	$A(r_i)$	$R(r_i) \cap A(r_i)$
1	1, 4, 5, 6	1	1
2	2, 4, 5, 6	2	2
4	4, 5, 6	1, 2, 4, 5	4, 5
5	4, 5, 6	1, 2, 4, 5	4, 5
6	6	1, 2, 4, 5, 6, 7, 9, 11	6
7	6, 7, 9, 10	7	7
9	6, 9, 10	7, 9, 11	9
10	10	7, 9, 10, 11	10
11	6, 9, 10, 11	11	11

Similarly, the demands of the third and fourth hierarchies are 4, 5, 9, 1, 2, 7 and 11, respectively.

Thus, $\pi_l(P) = \{3, 8; 6, 10; 4, 5, 9; 1, 2, 7, 11\}$.

3.2.2 Regional division, $\pi_d(L_k)$

The regional division divides the demands at the same level into the disconnected subset and the strongly connected subset. If a demand does not belong to the strongly connected subset, then the accessible set of its hierarchy is composed of the demand itself, namely:

$$r_i = R_{L_k}(r_i) \qquad (12)$$

where, $R_{L_k}(r_i)$ exhibits an accessibility that approaches the L_k hierarchy. Thus, the accessibility matrix can divide the L_k demand in every hierarchy into two parts: $\pi_d(L_k) = [I; N]$, where I is included if the $r_i = R_{L_k}(r_i)$ condition is met;

otherwise, N is included. I includes the demands of disconnected parts, whereas N includes the demands of strongly connected parts. On special occasions, I and N can be empty sets, but not at the same time.

3.2.3 Circuit set division, $\pi_G(N)$

The goal of circuit set division is to find the circuit set from the strongly connected subset. Every demand in the circuit set can be reached from other demands and is able to reach other demands, namely $\pi_G(N)$ can be divided into:

$$\pi_G(N) = \{C_1, C_2, \ldots, C_y\} \qquad (13)$$

where, $C_i (i = 1, 2, \ldots, y)$ is the largest circuit set, and y is the number of the largest circuit set.

4 CONSTRUCTION STEPS OF THE MILITARY REQUIREMENT SYSTEM

To sum up, the steps for constructing the military requirement system's hierarchical structure are:

- Obtain the relational graph of the military requirement system according to the military requirements and their relationships.
- Translate the relational graph of military requirement system into a relationship matrix.
- Obtain the accessibility matrix of the requirement system calculated from the military requirement system relationship matrix.
- Divide all the demands according to hierarchical division $\pi_l(P)$, from the accessibility matrix of the military requirement system; then, the accessibility matrix is arranged according to the hierarchy as:

$$R = \begin{array}{c} \\ L_1 \\ L_2 \\ \vdots \\ L_l \end{array} \begin{array}{cccc} L_1 & L_2 & \cdots & L_l \\ \begin{bmatrix} N_{11} & 0 & \cdots & 0 \\ N_{21} & N_{22} & \cdots & 0 \\ \vdots & \vdots & & \vdots \\ N_{l1} & N_{l2} & \cdots & N_{ll} \end{bmatrix} \end{array}$$

- Carry out $\pi_d(L_k)$ and $\pi_G(N)$ for every bidiagonal submatrix $N_{ii} (i = 1, 2, \ldots, l)$ to find the largest circuit set. Choose one typical demand in every circuit set and remove other demands, and then the abbreviated accessibility matrix R' is obtained.
- Treat $N'_{i+1,i}, N'_{i+2,i}, \ldots, N'_{i+l-1,i}$ (herein, $i = 1, 2, \ldots, l$) from matrix R' individually. $N'_{i+1,i}$ exhibits the accessible situation from the later hierarchical demand. In $N'_{i+1,i}$, directional arrow lines are required to draw from the former to the later hierarchical demand and from demand 1.

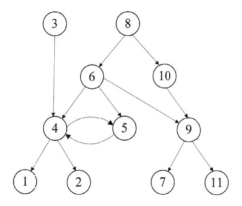

Figure 3. Hierarchical structure of the military requirement system.

– The element of 1 means the relationship in the cross layers for $N'_{i+2,i}, ..., N'_{i+l-1,i}$. If the existing arrow lines in adjacent layers can sufficiently describe the relationship, then the arrow lines are not required.

As an example, the hierarchical structure of the requirement system in Fig. 2 was obtained by the analysis described above, and is shown in Fig. 3.

5 CONCLUSION

This research present a new approach to structure military requirement system. It was constructed by combining the demand hierarchical induction technology and object-relation analytic technique. When the hierarchical relationships between demands were easily recognized, the demand hierarchical induction technology was adopted; when they were more complicated, the object-relation analytic technique was employed.

REFERENCES

Alan M. Ladwig, Gary A. Steinberg. Strategic Planning and Strategic Management within NASA. http://www.hq.nasa.gov/office/nsp/NSPTOC.html.

Alexander Egyed, Paul Grünbacher. Automating Requirements Traceability: Beyond the Record & Replay Paradigm. Proceedings of the 17th IEEE International Conference on Automated Software Engineering (ASE'02) 2012 IEEE.

Gotel O., Finkelstein A. An Analysis of the Requirements Traceability Problem Proc. of First International Conference on Requirements Engineering, 2004, pages 94–101.

Hamilton V.L. and M.L. Beeby, Issues of Traceability in Integrating Tools, Proc. IEE Colloquium Tools and Techniques for Maintaining Traceability during Design, Dec. 2011.

Jane Huffman Hayes. Improving Requirements Tracing via Information Retrieval. Proceedings of the 11th IEEE International Requirements Engineering Conference, 2013.

Medvidovic N. Bridging Models across the Software Lifecycle, to appear: Journal of Systems and Software, 2012.

Palmer, J.D. Traceability, Software Requirements Eng., R.H. Thayer and M. Dorfman, eds., pp. 364–374, 2007.

Thomas A. Finholt. NEESgrid Requirements Traceability Matrix. Whitepaper Version: 1.0 June 24, 2013.

Automotive, Mechanical and Electrical Engineering – Liu (Ed.)
© *2017 Taylor & Francis Group, London, ISBN 978-1-138-62951-6*

Data link layer design and verification based on PCIe 2.0

Zhiyong Lang, Tiejun Lu & Yu Zong
Beijing Microelectronics Technology Institute, Beijing, China

ABSTRACT: This article highlights the Function of Data Link Layer based on the PCIE 2.0 protocol. After the in-depth study of the protocol, Verilog is adopted for the forward design. Simulation was performed using NC which is the Cadence's simulation tool. The transmission rate is 5.0 G/second/lane/direction. In the end, we use SMIC 65 ns to synthesis. According to the simulation results, it is confirmed that the design works properly and meets the protocol requirements.

Keywords: PCI Express; Data Link Layer; ACK/NAK; TLP replay mechanism

1 INTRODUCTION

PCI Express is the third generation of high-performance I/O buses used to interconnect peripherals such as computing and communications platforms. It is primarily used for interconnection of peripheral I/O devices such as mobile devices, desktops, workstations, servers, embedded computing and communication platforms (Budruk R, 2004). PCIE bus has inherited powerful functions of the second generation, using the new features of the computer architecture. Compared with the original bus, PCI Express uses point-to-point serial to communicate, and the differential signal is used to transmit the data. The data is sent in the format of packet during transmission. PCIe bus uses a layered architecture for data transmission, and the hierarchical structure is shown in Fig. 1.

PCIe architecture is divided into Transaction Layer, Data Link Layer and Physical Layer from top to down (PCI-SIG P C I, 2006). Data packets are first generated in the device core, and then transmitted to the link through the Transaction Layer, data link layer and Physical Layer. The receiving side needs to receive data through the Physical Layer, Data Link Layer and Transaction Layer to reach the core layer. The data link layer is between the Transaction Layer and the Physical Layer, and its main function is to ensure that the TLP is transmitted correctly in the link. The Data Link Layer uses a series of mechanisms to ensure that the TLP can be transmitted correctly. The following sections focus on the Data Link Layer.

2 DATA LINK LAYER

Data Link Layer is responsible for the correct transmission of data packets, and provides a sound error correction mechanism. The functionality of the Data Link Layer is defined by the protocol as follows:

Figure 1. Layer of PCIe device.

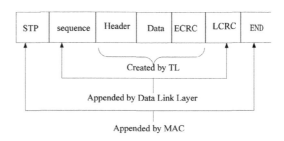

Figure 2. Transaction layer packet format.

1. Swapping the packets
2. Sending DLLP
3. Error detection and data retransmission

Data Link Layer achieves the main functions of processing the TLP and DLLP two types of data packets. The PCIe hierarchy layers of TLP processing is shown in Fig. 2 (PCI-SIG P C I, 2006). It can be seen that adding TLP sequence number and CRC in the Data Link Layer. The Data Link Layer uses all the fields of the received TLP to calculate the CRC value. The purpose of the CRC check is to ensure the integrity of the packet transmission. Adding serial number can target the wrong packets during transmission, so that ACK/NAK can realize the error replay mechanisms to ensure reliable transmission of data packets.

In the Data Link Layer transmission process will appear another kind of packet-DLLP, which includes ACK/NAK DLLP, InitFC1, InitFC2 and UpdataFC and so on. The ACK/NAN DLLP is used in the ACK/NAK protocol to inform the peer whether the received TLP is correct. InitFC1 and InitFC2 DLLP are the flow control initialization packets just after the link are established, used to inform the credit size to the peer. UpdataFC DLLP is an update packet, the Data Link Layer uses this packet to periodically update and inform the credit size to the peer. DLLP format is shown in Fig. 3 (Wilen A, 2003):

2.1 *DLCMSM*

The Data Link Layer monitors the status of the MAC through the Data Link Control and Management State Machine (DLCMSM), then informs the transaction layer. DLCMSM includes three states:

1. DL_Inactive: After reset, this state is initial state. In this state, the Data Link Layer is completely reset. It clears the TLP that exists in the replay buffer, and submits DL_Down status information to the transaction layer.
2. DL_Init: Flow control initialization phase. This state is divided into two parts, FC_INIT1 and FC_INIT2. In this state, DLCMSM will send

InitFC1 DLLP and InitFC2 DLLP for flow control initialization.
3. DL_Active: The link works normally. In this state, the link will normally process TLP.

DLCMSM model is shown in Fig. 4 (Wang Qi, 2011):

2.2 *ACK/NAK*

The Data Link Layer needs to use ACK/NAK to guarantee the correctness of the data packet transmission when handling TLP. In this paper, we use device A and device B as an example to illustrate. The transmitter is the sending part of the device A and the receiver is the receiving part of the peer device B. From the sending TLP to the receiving TLP, ACK/NAKDLLP is processed by order. This article describes the order. The elements of ACK/NAK are shown in Fig. 5 (Budruk R, 2004):

After receiving the TLP, the transmitter uses the counter to add a 12-bit sequence number to the received TLP. Once a TLP is sent, the sequence number will be incremented. When the value of the counter reaches 4095, the counter will return to 0 if an additional operation is performed. 32-bit CRC will calculate the redundant code according to the TLP and the newly added sequence number. The transmitter transmits the received TLP adding sequence number and CRC to the MAC, and stores the copies in the replay buffer. After Receiving TLP, the receiver will calculate the CRC to determine whether the calculated CRC value equals the CRC area's value. If they are equal, the next step is to determine whether the sequence number

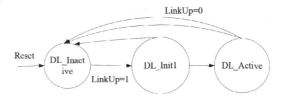

Figure 4. DLCMSM.

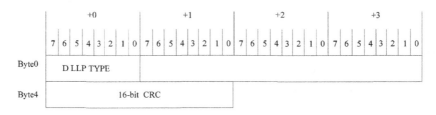

Figure 3. Data link layer packet format.

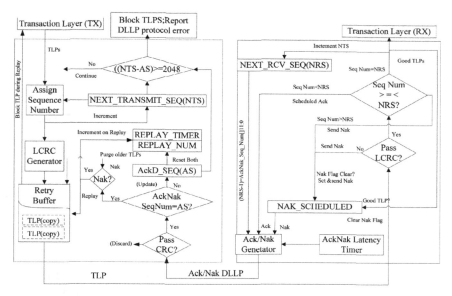

Figure 5.   Elements of the ACK/NAK protocol.

of TLP is the received sequence number. If there is no CRC error and the sequence number of the received TLP is the desired sequence number, the receiver will decide whether to send the ACK DLLP immediately or wait for more correct TLPs to be received before sending. The reason for this is that sending an ACK DLLP every time a correct TLP is received affects the efficiency of the link. In order to determine when to send ACK DLLP, the receiver uses a counter and a timer. If any one of them exceeds its threshold, the receiver sends an ACK DLLP. If the CRC values do not match or TLP sequence number is not the desired value, it indicates that the received TLP is wrong, then you need to generate NAK DLLP and transmit to the transmitter to inform its data packets are wrong. When the receiver detects that an error has occurred in the received TLP, it discards the TLP first, sets a NAK_SCHEDULED flag while discarding the packet, and generates a NAK DLLP. The receive flag is cleared only when the correct TLP is received again. When the flag bit is set high, all received TLPs are discarded.

When the transmitter receives the ACK DLLP, since the ACK DLLP contains the sequence number, the transmitter can determine whether the TLPs before the sequence number are correctly received and then clear the copies in the replay buffer. The size of the replay buffer should be large enough to ensure that the transaction layer will not be blocked due to the buffer overflow. When the transmitter receives the NAK DLLP, the transmitter will prevent the transaction layer to continue to

transmit TLP, according to the received sequence number to find the corresponding TLP in the replay buffer. The transmitter will also use the REPLAY TIMER and a 2-bit REPLAY_NUM counter to determine whether the link is faulty. The REPLAY TIMER is used to measure the time from the TLP is sent to the corresponding ACK or NAK DLLP is received. REPLY_NUM stores the number of attempts to retransmit due to an ACK DLLP or REPLAY TIMER timeout. When REPLY_NUM jumps from 2'b11 to 2'b00, it indicates there is an error in the link connection. The Data Link Layer notifies the MAC to retrain the link. The Data Link Layer will send multiple types of data packets. There is a certain priority relationship in sending packets. According to the recommendations of the agreement, we use the following transmission priority:

1. The DLL or TLP that is currently being sent
2. NAK DLLP
3. ACK DLLP
4. FC_Update DLLP
5. Replay TLP
6. Normal TLP

3   DATA LINK LAYER DESIGN

This paper is based on the PCIe 2.0 protocol to design endpoint mode Data Link Layer. Verilog language (Thomas D, 2008) is used for the design of the Data Link Layer. The overall block diagram of the design structure is shown in Fig. 6.

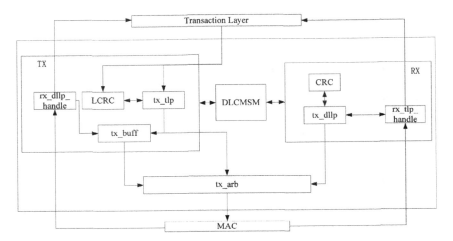

Figure 6.   Design structure.

DLCMSM: It is responsible for monitoring the status of the link. It receives the MAC signal indication and jumps to the appropriate state according to the indication signal. It generates the corresponding instruction signal in different states to control the other parts of Data Link Layer, and generates indicating signal to notify transaction layer whether it can send packets or not.

tx_tlp: When the transaction layer indicates that the beginning of transmission of TLP, it will add the sequence number to the TLP. At the same time, the TLP will also be sent to the LCRC. After the verification, CRC module will add32 bits CRC redundancy code to the end of the packet. As the serial CRC efficiency is too low to affect the performance of the design, in this design, we use a parallel mode. The CRC polynomial used has coefficients express as 04C11DB7H.

tx_buff: When a packet with a sequence number and a CRC is sent, a copy is stored in the replay buffer for use when an error occur during sending. In the design, the buffer will also store the start and end address of each packet. When the data packet needs to be replayed according to the sequence number, the buffer can find the corresponding data packet's start address, then lock the data packet. This mechanism can ensure the correctness of the replay mechanism.

rx_dllp_handle: The module will separate DLLP from the received packets. It will check the received CRC, and then determine the type of DLLP. If it is the correct data packet, this module will generate corresponding control signal to send the corresponding module. Error detection is performed for each step when parsing a DLLP packet, including CRC errors, sequence number errors and bad packet and so on. When an error occurs, the packet is discarded and an error signal is generated and reported to the error handling module for processing.

tx_arb: Only TLP and DLLP are transmitted on the link, and TLP is divided into replay TLP and normal TLP. In order to ensure the correctness of data transmission and improve the utilization rate of links, it is necessary to select the corresponding data packets according to the priority of the data packets. In this design, the module is responsible for the TLP and DLLP data arbitration. This module will also process the flag signal, which is transmitted to the TLP package module (tl_tlp module) and DLLP package module (tx_tllp module) and generates the corresponding signal which is sent to the MAC.

tx_dllp: The main function of the module is generating all kinds of DLLP based on different DLLP request. In this design, DLLP requests include flow control DLLP and ACK/NAK DLLP. The DLLP generated by this module also need to be added 16 bits CRC redundancy code and then sent to the arbitration module.

rx_tlp_handle: This module extracts the entire TLP packet from the received data according to the indication signal. It will calculate the CRC value, parsing 32 bits LCRC carried by the TLP, and then it will compare the calculated CRC with the carried CRC. The module will also detect the sequence number to determine whether it is the expected sequence number. After CRC and sequence number are checked, if there is no error in TLP, ACK DLLP transmission request is generated. If the TLP is repeated, a DUP ACK DLLP transmission request is generated; if the TLP is incorrect, an NAK DLLP transmission request is generated. If there is an error, the TLP errors on the Data Link Layer are reported to the upper layer, and bad packets are discarded.

4 SIMULATION WAVEFORM

After completing the code, we use Cadence's simulation tools (Cadence Design Systems, 2008) for function testing to check whether it meets the design requirements. The test results are as follows in Fig .7:

When linkup is high, DLCMSM jumps to state Init1.It starts receiving InitFC1p, InitFC1np, InitFC1cpl DLLP, after InitFC1cpl DLLP is received, then initFC1 initialization is completed. When receiving InitFC1cpl, dlrx_rcvd_fc1cpl is set to high, and the state jumps to InitFC2 state. In this state, DLCMSM starts receiving InitFC2, InitFC2p, and InitFC2np DLLP after InitFC2cpl DLLP is received, and hence InitFC2 initialization is completed similarly. Then dlrx_rcvd_fc2cpl is set to high, indicating that flow control initialization is completed. DLCMSM jumps to the dl_Active

state. In this state, dl_up is set to high, informing the transaction layer that it can send packets.

Fig. 8 shows the transmitting of ACK DLLP. When the condition of sending ACK DLLP is satisfied, dlrx_req_ack is set high, and tx_dllp will send ACK DLLP. It can be seenin Fig.8, dlrx_req_acknak_seq represents the sequence number to be sent, which is 2. dltx_mac_dllp_sot indicates the start signal of DLLP. dltx_mac_dllp_data is the packet to be transmitted, whose size is 16 bits per period. 16'h0000 indicates that the packet is ACK DLLP, while the next 16'h200 indicates that the received TLP with sequence numbers 0, 1, 2 is correct. Besides, replay buffer can clean up the copies of these TLPs. The signal dltx_mac_dllp_eot indicates the end of the packet.

When the packet error occurs, NAK DLLP will be sent. The simulation results is shown in Fig. 9.

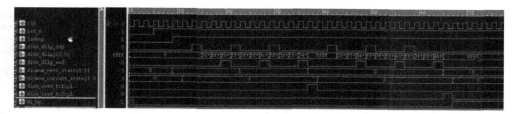

Figure 7. Link flow control DLLP.

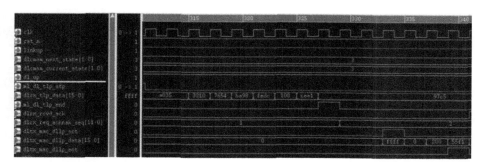

Figure 8. The process of ACK DLLP.

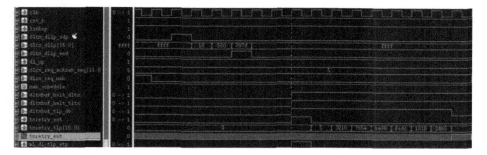

Figure 9. The process of NAK DLLP.

It can be seen from the Fig. 9 that when the error packet is received, the NAK SCHEDULED flag is set while the NAK DLLP is sent. dltx_mac_dllp_sot indicates the start signal of DLLP. dltx_mac_dllp_data is the packet to be transmitted, whose size is 16 bits per period.16'h0010 indicates that the packet is NAK DLLP, and the next 16'h500 indicates that the received TLP with sequence numbers 5 is wrong and needs to be resent. Moreover, the signal dltx_mac_dllp_eot indicates the end of the packet. When receiving ACK DLLP, dltx buf_halt_dltx is set high to prevent the transmitter sending new data packets. Besides, the dltx_buf_halt_tltx signal is also set high to prevent the transaction layer transmitting new packets. After the current TLP is completed, the retransmission mechanism is initiated. The TLP whose sequence number is 5 is extracted from the replay buffer and sent to the link. In Fig. 9, txretry_sot indicates the beginning of replay packet, while txretry_tlp indicates the data of replay packets, and txretry_eot indicates the end of the replay packet.

5   CONCLUSION

This paper introduces the design and implementation of Data Link Layer in PCIe2.0 endpoint mode. Based on the design, the function simulation is carried out. The transmission rate is 5.0G/second/lane/direction, and the data width of single channel is 16 bits. After transmitted to MAC, all packets will be stripped into two channels, and the data width of single channel is 8 bits. Therefore, the clock frequency is 500 MHz in our design and the data width is 16 bits. We also use SMIC 65 ns technology library to synthesis. The design can work normally under the 500 MHz clock, and achieves the goal that is expected.

REFERENCES

Budruk R, Anderson D, Shanley T. PCI express system architecture [M]. Addison-Wesley Professional, 2004.

Cadence Design Systems, Inc., "NC-Verilog Simulation Help," Product Version 8.2, Nov. 2008.

PCI-SIG P C I. Express Base Specification Revision 2.0[J]. 2006.

Thomas D, Moorby P. The Verilog ® Hardware Description Language [M]. Springer Science & Business Media, 2008.

Wang Qi. PCI Express architecture guide [M], 1st. Beijing: China machine press, 2011.

Wilen A, Thornburg R, Schade J P. Introduction to PCI Express [M]. Intel, 2003.

Automotive, Mechanical and Electrical Engineering – Liu (Ed.)
© *2017 Taylor & Francis Group, London, ISBN 978-1-138-62951-6*

Design of a new carrier tracking loop in a positioning receiver

Zhongliang Deng, Shu Jiang, Jun Mo & Shengchang Yu
Beijing University of Posts and Telecommunications, Beijing, China

ABSTRACT: In recent years, indoor positioning technology has been widely used in many fields. It changes people's lives deeply, and has become a hot topic in the field of LBS research. This paper discusses the problem of signal tracking in the TC-OFDM (Time & Code Division-Orthogonal Frequency Division Multiplexing) indoor positioning receiver, which is based on ground mobile radio network. In the TC-OFDM indoor positioning system, due to the complex structure of the indoor environment, and serious signal attenuation, a great challenge is presented to the performance of the carrier tracking loop, and also the positioning accuracy is greatly affected. This paper refers to the high dynamic carrier tracking loop, designs a dynamic threshold carrier tracking loop based on FLL and PLL, which is equipment with multiple status to adapt to the requirements of carrier tracking in different situations. This carrier tracking loop is designed with a joint judged threshold based on the residual frequency offset of FLL and the residual phase offset of PLL, with the convergence of the loop. The carrier tracking loop will correspondingly change the work status to track the signal stably when the discriminator value of the FLL and PLL and the threshold meet a certain condition. In addition, we emulate the algorithm with the TC-OFDM positioning receiver, and it can significantly improve the stability of the signal tracking on the complex indoor environment.

Keywords: Indoor positioning; Tracking; Phase-locked loop; Frequency-locked loop

1 INTRODUCTION

The carrier tracking loop is designed to accurately track the signal and is used to generate accurate local carriers, and finally to strip the carrier from the signal. The tracking loop is typically implemented by a Frequency-Locked Loop (FLL) and Phase-Locked Loop (PLL). The performance of the frequency-locked loop is equivalent to the performance of a high-order phase-locked loop, but the tracking precision is not as good as the phase-locked loop. The PLL has good noise immunity and high tracking accuracy under low dynamic conditions. However, if the Doppler shift is serious, and if the carrier is to be traced stably, the loop bandwidth needs to be amplified, which means more noise is entered, thus reducing the tracking accuracy.

At present, the popular design method is the frequency-locked loop auxiliary phase-locked loop, such as the classic second-order frequency-locked auxiliary third-order phase-locked, which has the advantages of PLL and FLL. However, in an indoor environment with severe signal attenuation, the working state of the carrier tracking loop frequently switches between the PLL and the FLL, which will cause the filter value to be discontinuous, and lead to serious tracking errors and even

signal loss. Considering that the switching process of this method is too stiff, a dynamic threshold carrier tracking loop based on PLL and FLL is proposed in this paper. This loop has multiple states, and in each state the weights of the FLL and the PLL are continuously adjusted according to the discriminator value of FLL and PLL. This is so that the indoor positioning signal under the dynamic environment can be continuously and stably tracked.

2 DESIGN OF THE NEW CARRIER TRACKING LOOP

2.1 *Architecture design of the carrier tracking loop*

The carrier tracking loop architecture designed in this paper is shown in Figure 1. The main structure of the tracking loop is composed of a phase-locked loop, a frequency-locked loop, a threshold calculating unit and a weight calculating unit. The main function of the threshold judgement unit is to calculate the appropriate threshold according to the value of the frequency discriminator and the value of the phase discriminator, and then give the different weights of the phase discriminator and the frequency discriminator, finally merging the

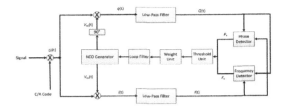

Figure 1. Loop filter schematics.

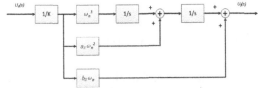

Figure 2. Third-order loop filter's architecture.

final filter output in the same filter. The output adjusts the carrier NCO so that the signal can be stably tracked (Deng et al., 2013).

$U_i(t)$ is the continuous time signal inputted by the system, $V_{os}(t)$ and $V_{oc}(t)$ are sinusoidal carrier and cosine carrier replica signals, respectively:

$$U_i(t) = \sqrt{2}aD(t) \sin(\omega_i t + \theta_i) + n \quad (1)$$
$$V_{os}(t) = \sqrt{2}a\sin(\omega_0 t + \theta_0) \quad (2)$$
$$V_{oc}(t) = \sqrt{2}a\cos(\omega_0 t + \theta_0) \quad (3)$$

When the input signal $U_i(t)$ is mixed with the sine carrier replica signal $V_{os}(t)$ and cosine carrier replica signal $V_{oc}(t)$ on the I-branch, the resulting products $i_p(t)$, $q_p(t)$

$$i_p(t) = U_i(t) V_{os}(t) = (\sqrt{2}aD(t)$$
$$\sin(\omega_i t + \theta_i) + n) \sqrt{2}\sin(\omega_i t + \theta_i) = -aD(t)$$
$$[\cos((\omega_i + \omega_0) t + (\theta_i + \theta_0)) - \cos(\omega_e t + \theta_e)] \quad (4)$$

$$q_p(t) = U_i(t) V_{oc}(t) = (\sqrt{2}aD(t)$$
$$\sin(\omega_i t + \theta_i) + n) \sqrt{2}\cos(\omega_i t + \theta_i) = -aD(t)$$
$$[\sin((\omega_i + \omega_0) t + (\theta_i + \theta_0)) - \sin(\omega_e t + \theta_e)] \quad (5)$$

$D(t)$ is the data code modulated on the carrier.

After mixing the results of $i_p(t)$ and $q_p(t)$ through the low-pass filter to filter out its high-frequency components, the filter results are as follows:

$$I_p(t) = aD(t) \cos(\omega_e t + \theta_e) \quad (6)$$
$$Q_p(t) = aD(t) \sin(\omega_e t + \theta_e) \quad (7)$$

The filtered $I_p(t)$ and $Q_p(t)$ are fed into the frequency discriminator and the phase discriminator to calculate the frequency value ω_e and the phase value Φ_e into the threshold calculation module to calculate the dynamic threshold. It will then be used to calculate the weight in the next module. Finally, the weighted value is added to the loop filter to perform the fusion filtering. The result of the filtering is used to adjust the carrier NCO generator, so as to realise the stable tracking of the signal (Deng et al., 2015).

2.2 Design of the PLL loop

In the frequency step excitation, the first-order phase-locked loop can track the signal stably, but the output signal has a constant phase tracking error

when compared with the input signal. However, the second-order or higher order phase-locked loop can accurately track the frequency step signal. In order to track the dynamic signal stably, the phase-locked loop is designed as a third-order loop. Because of the existence of message hopping in the navigation data signal, the phase-locked loop uses Costas loop, which is insensitive to the message jump.

The transfer function is:

$$F(s) = \frac{1}{k}\left(b_3\omega_n + \frac{a_3\omega_n^2}{s} + \frac{\omega_n^3}{s^2}\right) \quad (8)$$

The function of the phase discriminator is:

$$\Phi_e = Q_p \, \text{sign}(I_p) \quad (9)$$

Sign (I_p) is the symbolic function, its return value is the sign of I_p, that is, when I_p is greater than 0 to return +1, otherwise −1. Therefore, sign (I_p) is the level value of the data, and is then multiplied by Q_p to offset the data level jump on the impact of Q_p. The phase-detecting method requires the least amount of computation, and the phase-detecting result is proportional to sin (Φ_e) and is independent of the amplitude of the signal.

2.3 Design of the FLL loop

Because the phase-locked loop uses a third-order loop (Liu et al., 2014), so the second-order frequency-locked loop Jeffrey-Rutin filter, the architecture is shown below

The transfer function is:

$$F(s) = \frac{1}{k}\left(a_2\omega_n + \frac{\omega_n^2}{s}\right) \quad (10)$$

The symbol function sign () can detect the 180 degree phase transition caused by the data bit transition. This method is insensitive to the data bit transition and the computation is relatively small:

$$\omega_e(n) = \frac{P_{cross} \, sign(P_{dot})}{t(n) - t(n-1)} \quad (11)$$

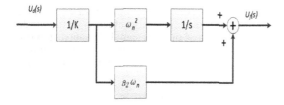

Figure 3. Second-order loop filter's architecture.

The dot product P_{dot} and the cross product P_{cross} are:

$$P_{dot} = I_P(n-1) I_P(n) + Q_P(n-1) Q_P(n)$$
$$= A_P(n-1) A_P(n) \cos(\varPhi e(n) - \varPhi e(n-1)) \quad (12)$$

$$P_{cross} = I_P(n-1) Q_P(n) + Q_P(n-1) I_P(n)$$
$$= A_P(n-1) A_P(n) \sin(\varPhi e(n) - \varPhi e(n-1)) \quad (13)$$

2.4 Threshold calculation

The design of the carrier tracking loop is to estimate the state of the loop operation based on the real-time variation of the value of the frequency discriminator and the value of the phase discriminator. If the value of the frequency discriminator ω_e is greater than the threshold value F_m, the loop is assumed to be in a coarse lock state, thereby strengthening the frequency-locked loop while weakening the phase-locked loop. If the value of the frequency discriminator ω_e is sufficiently small and less than the threshold value F_n, and the value of the phase discriminator $\varPhi_e$ is also smaller than the threshold value Pm, we estimate the loop has entered a stable locked state. The thresholds F_m, F_n and P_m are as follows:

$$F_m = a \frac{1}{n} \sum_{i=0}^{n} \phi_e(i) = a \frac{1}{n} \sum_{i=0}^{n} Q_p(i) sign(I_p(i)) \quad (14)$$

$$F_n = b \frac{1}{n} \sum_{i=0}^{n} \phi_e(i) = b \frac{1}{n} \sum_{i=0}^{n} Q_p(i) sign(I_p(i)) \quad (15)$$

$$P_m = c \frac{1}{n} \sum_{i=0}^{n} \omega_e(i) = c \frac{1}{n} \sum_{i=0}^{n} \frac{P_{cross}(i) sign(P_{dot}(i))}{t(i) - t(i-1)} \quad (16)$$

In order to reduce the computational complexity, the number of scoring n is generally less than one hundred, coefficient a, b, c, respectively, determined by the value of the phase discriminator and the value of the frequency discriminator.

2.5 Weight calculation

When ω_e is greater than the threshold F_m, the loop is in a high dynamic environment, so we need to strengthen the frequency-locked loop. Therefore, the

outputs of the phase discriminator and the frequency discriminator are respectively weighted as follows:

$$\Delta\omega_e = k \frac{\omega_e}{F_m} * \frac{P_{cross} sign(P_{dot})}{t(n) - t(n-1)} \quad (1 < k) \quad (17)$$

$$\Delta\varPhi_e = j \frac{\varPhi_e}{P_m} * Q_p sign(I_p) \quad (0 < j < 1) \quad (18)$$

If the frequency discriminating value ω_e is sufficiently small and smaller than the threshold value F_n, and the value of the phase discriminator $\varPhi_e$ is also smaller than the threshold value P_m, the frequency offset is low at this time and it is necessary to intensify the phase-locked loop accordingly. The outputs of the phase discriminator and the frequency discriminator are weighted respectively as follows:

$$\Delta\omega_e = j \frac{\omega_e}{F_n} * \frac{P_{cross} sign(P_{dot})}{t(n) - t(n-1)} \quad (0 < j < 1) \quad (19)$$

$$\Delta\varPhi_e = k \frac{\varPhi_e}{P_m} * Q_p sign(I_p) \quad (1 < k) \quad (20)$$

The weight changes in real time with the value of the phase detector $\varPhi_e$ and the value of the frequency discriminator ω_e, and is merged in the same filter. The output finally adjusts the NCO frequency according to the filter output, thus can keep the tracking to the signal stable under different situations.

3 SIMULATION AND ANALYSIS

The designed carrier tracking loop is simulated in MATLAB. When the input signal frequency is 10 MHz, the sampling frequency is 50 MHz, frequency step of 2 KHz occurs at 40 ms, frequency step of 1.5 KHz occurs at 80 ms. The simulated results are shown in Figure 4 as follows:

At the same time, we compare it with the traditional PLL carrier tracking loop. The test conditions are the same as above, in which the blue represents the new carrier tracking loop and the red represents the FPLL carrier tracking loop:

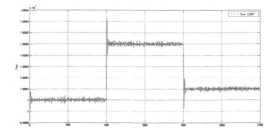

Figure 4. Performance of carrier tracking loop.

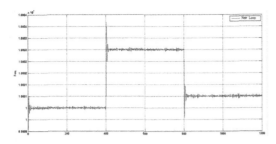

Figure 5. Performance of carrier tracking loop.

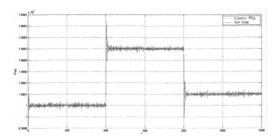

Figure 6. Comparison of new carrier tracking loop and FPLL carrier tracking loop.

Figure 4 and Figure 5 show that the new carrier tracking loop can stably track the 2 KHz step signal. Figure 6 shows that the convergence rate and tracking accuracy of the new carrier tracking loop are better than the traditional FPLL loop under the same conditions.

4 CONCLUSION

In this paper, a new carrier tracking loop is designed and simulated in MATLAB. Compared with the traditional PLL, FLL and FPLL carrier loops, the tracking performance is better under the excitation of the step signal. The next step is to simulate the Doppler shift of the ramp excitation to further optimise the performance of the carrier tracking loop.

ACKNOWLEDGEMENTS

Foundation project: The National High Technology Research and Development Program ("863" Program) of China (No.2015AA124101). The National Natural Science Foundation of China (No.61401040). The National Science and Technology Support Program of China (No.2014BAK12B00).

REFERENCES

Deng, Z., Li, X., Ma, W., Fang, Y., & Zeng, H. (2015). Design and Implementation of Indoor & Outdoor Positioning Service Platform Based on Beidou and Base Station System[C], The 5th China Satellite Navigation Conference, 2015.

Deng, Z., Yu, Y., Yuan, X., Wan, N., & Yang, L. (2013). Situation and development tendency of indoor positioning [J]. *China Communications*, 10(3), 42–55.

Liu, X., Chen, D., He, X., & Liu, J. (2014). FPLL carrier tracking loop simulation and high dynamic GPS signal test[C]. 11, 1329–1333.

Tian an Jian-ping Zhang Ruo-bing. Tracking error analysis and simulation of FLL-assisted PLL.

Xie, G. (2009). *Principles of GPS and Receiver Design [M]*. Beijing: Publishing House of Electronics Industry, 376–384.

Xu, Z., Cui, C., & Yu, J. (2012). Based on a combination of frequency-locked loop and phase locked loop carrier tracking technology. 52(4), 558–561.

Automotive, Mechanical and Electrical Engineering – Liu (Ed.)
© 2017 Taylor & Francis Group, London, ISBN 978-1-138-62951-6

Idealised Six Sigma software design and development

Hongbo Yan

National Technological Innovation Method and Tool Engineering Research Center, Hebei University of Technology, Tianjin, China
College of Mechanical Engineering, Inner Mongolia University of Science and Technology, Baotou, China

Runhua Tan

National Technological Innovation Method and Tool Engineering Research Center, Hebei University of Technology, Tianjin, China

Hang Song & Wuxiang Sun

College of Mechanical Engineering, Inner Mongolia University of Science and Technology, Baotou, China

ABSTRACT: Aimed at the shortcomings of the existing project management software Six Sigma and the training process, this paper describes the design and development of the Visual Basic (VB) based on the idealised Six Sigma software. First, by using the VB programming language on the basis of the Six Sigma DMAIC model, the corresponding tools are added to each stage to form an ideal Six Sigma theory. Then the VB programming language is used to combine the idealised Six Sigma software and Goldfire software, so that the Goldfire software simplifies the search step and across the language barriers. Finally, an intuitive simple window mode required to introduce the training content is presented. The use of the process shows that the software is easy to use, so that the ideal of Six Sigma is easier for enterprises and training institutions to implement, the enterprise staff to the idealised Six Sigma has a deeper understanding, in order to facilitate the use of the future.

Keywords: Idealised Six Sigma; Programing; Exploitation; DMAIC model; Goldfire

1   INTRODUCTION

Six sigma management is a kind of quality improvement method invented by American Motorola in the middle of 1980s, after the successful use by GE. It has become a widely used software by many world famous companies. Professor Chen Zishun and Professor Tan Runhua improved the "Six Sigma Theory" to "Idealized Six Sigma Theory" in China. It improved the core competence of Chinese companies and scientific research institutions.

Due to the shortage of the Six Sigma management software used in its current status, we had developed the Idealised Six Sigma software in the VB language environment. Idealized Six Sigma software added some innovative tools such as TRIZ, KT, TOC, etc. These innovative tools integrate into DMAIC in different ways, so they exhibit in different ways too. The software was then combined with Goldfire, to simplify the search steps and break down language barriers. The software demonstrated the Idealised Six Sigma theory by simple and intuitive windows, so that the Idealised Six Sigma theory can be easily accepted by company and training institutions.

2   A BRIEF INTRODUCTION TO THE THEORY AND TOOLS OF IDEALISED SIX SIGMA

As a management strategy, the meaning of the Six Sigma theory goes beyond the scope of statistics. It includes three aspects: statistical measurement, business strategy and quality culture. The Six Sigma theory shows different aspects for different purposes. There are many disadvantages of the traditional models of DMAIC. For example, there are delays in delivery and false choices in the definition phase, and during the analysis phase there may lack methods for searching for roots causes or core problems. All these disadvantages will reduce the Six Sigma'ability and efficiency to solve the problems. According to these problems, the Idealised Six Sigma theory has made some improvements, and it really has improved the ability and efficiency to solve the problems.

At the base of DMAIC, the Six Sigma has added and innovated some tools to improve the ability of solving problems. The core idea of the Six Sigma is to maximise the interest of a company in meeting customer requirements. The tools combined in

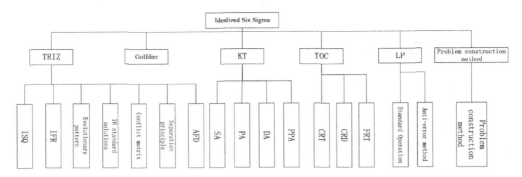

Figure 1. Fusion tool.

DMAIC are shown in Figure 1. Combined with the Goldfire, it makes the software more complete. Every tool plays different roles in the software.

1. The function of TRIZ

    ISQ and IFR output the report of the survey to the problems. Secondly, it ensures that the process characteristics and the current levels using 76 standard solutions and a conflict matrix. Thirdly, it makes innovative solutions by conflict matrix, evolution pattern, IFR and separation principle. Finally, it keeps the modifying effect using conflict matrix and failure prediction.

2. The function of KT is to find questions and sort them; it finds the change caused by Reasons of Problem; it eliminates the reason by the safest method; it formulates protection plans.

3. The function of TOC is to find the root reason of the questions, and to find the conflict inside the questions.

4. The function of LP is to improve the production efficiency by the best combination of person, machine and object, eliminate the mistakes and keep the disadvantages at the lowest level.

5. The function of Resolve is to provide a foundation for the rank of the solutions.

6. Goldfire is software to achieve CAI completely. It includes sophisticated concepts, problem solving, opportunity analysis, and knowledge acquisition of internal and external.

3   SOFTWARE DESIGN

The paper mainly uses VB programming language to design the software and integrate theories into a whole from different angles. ISQ and IFR in TRIZ theory are SA in KT theory, and are added in the definition period of the DMAIC model. 76 standard solutions and conflict matrices in TRIZ theory are added in the measurement period. PA in KT theory and CRT in TOC theory are added in the analysis period. IFR, evolution model, conflict model and separation principle in TRIZ theory, DA and PPA in KT theory, CRD and FRT in TOC theory and problem building method are added in the improvement period. Conflict matrix and AFD in TRIZ theory and standard work and anti-bugging method in LP theory are added in the control period. The added tools in different periods of DMAIC of Six Sigma are shown in Table 1.

The flows using the software to solve problems:

1. Definition (D) period

ISQ questionnaire is used to investigate in depth the information or materials relating to problems to provide the team with a full understanding at the initial period of project. IFR is used to get the final purpose and the goal. SA is used to sort out massive information and materials, and rank the goals of the project.

2. Measurement (M) period

Conflict matrix and 76 standard solutions are used to find the conflicts in the project and develop new methods when the measurement methods and tools are insufficient.

3. Analysis (A) period

PA in KT theory and CRT in TOC are used to improve problem analysis ability and determine the root causes or problems, and the core problems

4. Improvement (I) period

IFR, evolution model, conflict matrix and separation principle in TRIZ, DA and PPA in TRIZ, CRD and FRT in TOC theory, and problem building method, are used to increase the opportunities of getting innovative plans in the period, assess and select the plans and make relevant protective measures to improve the safety of plan implementation. TRIZ is mainly used to a produce innovative solution. TOC is used to identify variable conflicts. Problem building method is used to provide the improvement order of variables.

162

Table 1. Tools for each phase of DMAIC.

Methods and tools		definition(D)	measuring (M)	analysis (A)	Improve (I)	control (C)
TRIZ	ISQ	○				
	IFR	○			○	
	Evolutionary pattern				○	
	Standard solution		○			
	Conflict matrix		○		○	○
	Separation principle				○	
	AFD					○
KT	SA	○				
	PA			○		
	DA				○	
	PPA				○	
TOC	CRT			○		
	CRD				○	
	FRT				○	
LP	Standard Operation					○
	Anti-error method					○
Problem construction method					○	

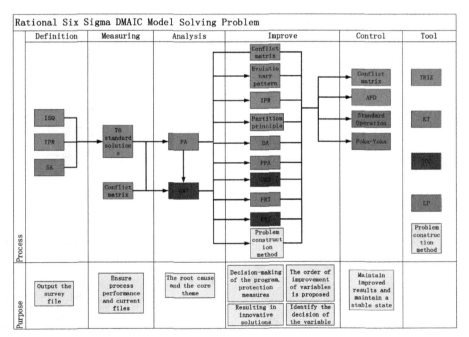

Figure 2. DMAIC model to solve the problem flow chart.

5. Control (C) period

Conflict matrix and AFD in TRIZ theory and standard works and anti-bugging method in LP are used. The addition of those tools aims to keep improvement effects and maintain the stable state. The flow chart describing the DMAIX model of idealised Six Sigma solving problems is shown in Figure 2.

The software finds ways to conveniently and rapidly solve problems by solving problems systematically and simplifying the complex process of problem solving. The software can export the information obtained in the previous investigation in the definition period. The nature of process and current level are determined in measurement period. The root causes and core problems are found in the improvement period through multiple methods to provide convenience for problem solving. Improvement effects and the stable state are maintained in the control period. The problems are solved well through the five steps.

4 SOFTWARE ENGINEERING

The main interface menu bar, quick entry, and access entry of mouse are designed. In the design of the menu bar, DMAIC model of Six Sigma, the tools integrated by Six Sigma and different classification ways are used to find the required modules more conveniently and rapidly. The design drawing of the main interface is shown in Figure 3 and the design drawing of menu bar is shown in Figure 4.

The previous design of the main window mentions that the design of the menu bar is divided into the DMAIC model of Six Sigma and the tools integrated by Six Sigma. The DMAIC model is divided into five parts: defined, measures, analyse, improve and control. The tools integrated by Six Sigma are divided into five parts: TRIZ, KT,

TOC, LP and problem building method. The convenience of software is improved through multiple classification and various interfaces. The structure chart of the menu bar of design of DMAIC model of idealised Six Sigma are shown in Figure 5.

All of those tools are integrated in the DMAIC model of Six Sigma, forming a new DMAIC model with innovation ability to eliminate the weaknesses of traditional the DMAIC model. Each module in the software is introduced in detail. The basic knowledge and use of methods of corresponding theories can be elaborated clearly. For example, in the evolution model, the software clearly expresses the content of each module of the evolution model; the software adds an example to each module to explain the content and the use of the module methods, making them easy to understand. In the menu of evolution model, modules can be added with the increase of theoretical knowledge to make software more per-

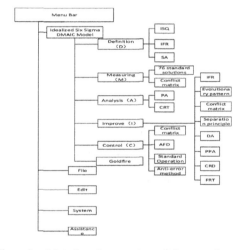

Figure 5. Main function structure of the menu bar.

Figure 3. The design drawing of the main interface.

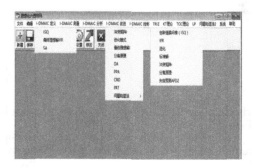

Figure 4. The design drawing of the menu bar.

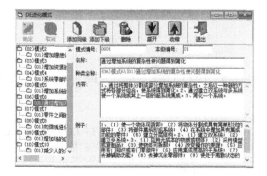

Figure 6. The sketch of evolution model software.

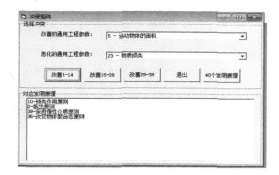

Figure 7. The sketch of conflict matrix software.

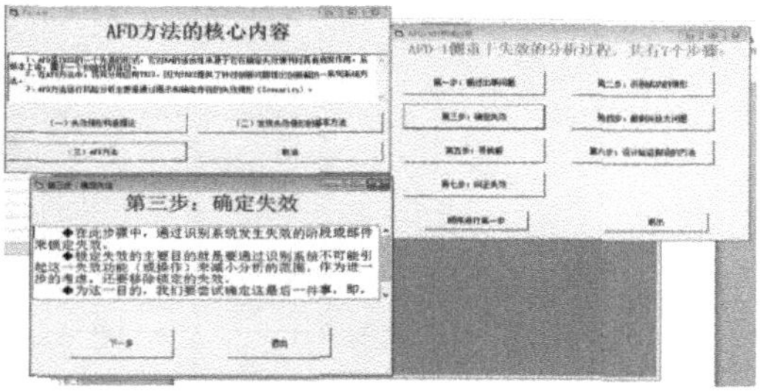

Figure 8. The sketch of the disabled predicting AFD software.

fect. The sketch of the evolution model software is shown in Figure 6. The checking of relevant knowledge can be done. For example, a corresponding invention principle can be obtained after choosing an improved universal engineering parameter and a worsened universal engineering parameter in the conflict matrix; this saves the time in table checking and improves the efficiency. The sketch of conflict matrix software is shown in Figure 7.

If you need to learn the knowledge of one module, the software can elaborate the theories of corresponding modules. For example, in the disabled predicting AFD, the module elaborates the knowledge content of the disabled predicting AFD step by step. Also, each window elaborates the knowledge of relevant modules correspondingly, and systematically introduces the content of the disabled predicting AFD. The sketch of the disabled predicting AFD software is shown in Figure 8.

5 CONCLUSION

The paper mainly introduces the functions, structures and especially the problem solving flows of idealised Six Sigma software. It improves the low degree of automation, problem solving ability, low efficiency and other defects of traditional DMAIC models of Six Sigma through the analysis of TRIZ theory, KT theory, TOC theory, problem building methods and lean production. Those theories are not only the tools solving problems but they have their own unique ideological connotation. Therefore, they can be used in the learning, reinforcement and checking of basic knowledge of idealised Six Sigma personnel in the company. They are also the tools helping idealised Six Sigma personnel find the root cause of problems and to more conveniently and rapidly solve problems.

ACKNOWLEDGEMENTS

Fund Project: Supported by the National Natural Science Foundation of China (51275153).

Author: Yan Hongbo, born in 1981, male, Luoyang, Henan, doctoral students. The main research directions for innovative design methods and implementation tools.

Tan Runhua, male, born in 1958, male, Hebei Renqiu, PhD., professor, doctoral tutor. The main research direction for innovative design.

REFERENCES

Chen, Z. & Tan R. (2013). *Principle and application of idealizing six sigma: The innovation method of product manufacturing process.* BeiJing: Higher Education Press.

Feng, C. (2007). Six Sigma theory and implementation. *Ship Science and Technology.*

Feng, F. (2009). Samsung group innovation method is applied to analysis of the experience. Science & Technology Progress and Policy.

Lin, M.. (2004). The six sigma management [M]. Beijing: Renmin University of China Publishing House.

Tan, R. Innovative design. (2002) Beijing: Mechanical Industry Press.

Tan, R. The invention problem solving theory [M]. Beijing: Science press. 2004.

Tan, R. (2010). TRIZ and applications—The process and methods of technological innovation. Higher Education Press.

Tan, R, Yang B. & Zhang J. (2008). Study on patterns of idea generation for fuzzy front end using TRIZ. China Mechanical Engineering.

Wei, M. (2011). Motorola 6 Sigma control software. *Printed Circuit Information.*

Yang, J. (2005). Toyota's business school: 68 details of the meticulous management. Beijing: Enterprise Management Publishing House.

Zhen, H.E., Zhou Y. & Gao X. (2006). Lean six sigma and its application. *Journal of Xidian University.*

Zhu, N. & Zhu, Y. (2009). Use TRIZ to avoid six sigma bottlenecks. Science and Management.

Zuo, T.Y. (2004). Resolve [M] Wang Haiyan, Makin, Beijing: Central Compilation and Translation Press, 2004.

Automotive, Mechanical and Electrical Engineering – Liu (Ed.)
© *2017 Taylor & Francis Group, London, ISBN 978-1-138-62951-6*

Message scheduling and network parameter optimisation for FlexRay

Y.F. Wang, X.H. Sun, J.B. Xia & J.L. Yin
Hefei University of Technology, Hefei, Anhui, China

ABSTRACT: FlexRay is recognised as the next generation of high speed vehicle communication standards. The real-time and reliability of the dynamic message transmission has an important effect on vehicular communication safety. For the FTDMA, the media access control of the dynamic segment, a two-layer message scheduling mechanism based on the Periodic Servers (PSS) is proposed. The worst case response times of messages under PSS are analysed, and also a sufficient schedule ability condition is deduced. The parameter configuration for the dynamic segment and PSS is modelled as the binary non-linear integer programming to maximise the utilisation under the constraint of schedulability of the dynamic segment. The experimental results show the effectiveness of the PSS and parameter configuration, which can effectively improve the utilisation and the real-time of the dynamic segment.

Keywords: FlexRay; Dynamic segment; Periodic server; Parameter configuration

1 INTRODUCTION

With the increasing demands of real-time communication between vehicular Electronic Control Units (ECUs), high speed busses have been widely used in vehicles. Compared with the CAN bus, the FlexRay bus contains both the time-triggered mechanism and the event-triggered mechanism. It has become the de facto standard of the next generation automotive bus due to its great advantages on bandwidth, fault tolerance and flexibility (Cheng et al., 2013; Wang & Zhang, 2016).

The Dynamic Segment (DS) is one of the important parts for the FlexRay communication cycle. It is suitable for the transmission of sporadic messages, such as the braking message and the fault diagnostic message (Ouedraogo & Kumar, 2013; Schmidt & Schidt, 2009). Hence, the design for the dynamic segment has attracted immense attention from researchers and engineers. An arbitration mechanism is proposed to schedule aperiodic message streams on dynamic that avoids the indefinite postponement of low-priority streams by Lange (Lange & Oliveira, 2014). Milbredty (Milbredty & Glab, 2012) analysed the dependence of internal parameters and message scheduling and proposed a method to select scheduling parameters. In Park & Sunwoo (2010), an indirect method of parameter configuration is presented by Park to make the static message latency communication cycle the smallest. Wang (Wang et al., 2012) formulates the parameter optimisation as the extremum problem of continuous function and an approximate analytical method is used to determine the length of the dynamic segment. Schmidt (Schmidt & Schidt, 2009) proposed

a reservation-based scheduling approach for the dynamic segment of FlexRay and formulated a nonlinear integer programming problem to compute the network parameters. Ouedraogo (Ouedraogo & Kumar 2014) presented a new ILP formulation to compute a precise value of the worst case response time of messages in the dynamic segment. All the above-mentioned methods were validated in the design of FlexRay. However, most of them separate message scheduling and parameter configuration into two independent design problems. Although the research difficulty is reduced, it is hard to directly apply it to real engineering design due to a lack of comprehensive consideration.

To address the problems in the previous works, this paper presents a two-layer message scheduling mechanism, and proposes a method to configure the parameters of the dynamic segment by considering the message scheduling. The rest of this paper is organised as follows: in Section 2, the FlexRay protocol is briefly described. The message scheduling model and its time analysis in the DS is presented in Section 3. In Section 4, a method to optimise parameters is proposed. The analysis of the experiments is shown in Section 5, and the conclusions are presented in Section 6.

2 BASIC PRINCIPLE OF FLEXRAY PROTOCOL

2.1 *Media access control*

The FlexRay bus transmits messages by communication cycle loop. A communication cycle contains the Static Segment (SS), the Dynamic Segment (DS), the Symbol Window (WS) and Network Idle Time (NIT). The static segment adopts the Time Division Multiple Access (TDMA) media access and consists of a number of isometric static slots; the dynamic segment adopts the Flexible Time Division Multiple Access (FTDMA) media access. Within the dynamic segment a set of consecutive dynamic slots that contain one or multiple mini-slots are superimposed on the mini-slots. The duration of a dynamic slot depends on whether or not frame transmission takes place. The dynamic slot only consists of one mini-slot if no message is transmitted; otherwise, it has a length equal to the number of mini-slots needed for the transmitted message. The message can be allowed to send on the dynamic segment if the following conditions are satisfied: 1) the slot counter of the dynamic slot is equal to the *frame ID* of the message; 2) the rest of the dynamic segment is enough to accommodate the transmission of the message.

2.2 Frame and coding

The FlexRay frame header contains 5 bits of indicator, 11 bits of frame ID, 7 bits of payload length, 11 bits of header CRC and 6 bits of cycle count. The payload segment is composed of 0–254 bytes and the frame tail consists of 3 bytes of frame CRC. The Transmission Start Sequence (TSS), the Frame Start Sequence (FSS), the Byte Start Sequence (BSS) and the Frame End Sequence (FES) are used in the frame encoding. TSS is used to mark the beginning of a transmission. FSS is used to compensate for a possible quantisation error. BSS is used to provide bit stream timing information for each byte and FES is used to mark the end of the last byte sequence of a frame. For a dynamic frame, the coding length l_b can be calculated by (1) where s is the payload of a frame and L_{TSS}, L_{FSS}, L_{BSS}, L_{FES} represents the length of TSS, FSS, BSS and FES, respectively.

$$l_b = l_{TSS} + l_{FSS} + (8 + 2s) \times (8 + l_{BSS}) + l_{FSS} + 1 \quad (1)$$

3 MESSAGE SCHEDULE FOR THE DS

3.1 System model

As specified in Wang et al. (2012), in this paper a system model for the FlexRay network system consists of FlexRay nodes and a FlexRay bus. The components of each node are a host and a Communication Controller (CC) and they are connected by a Controller–Host Interface (CHI). The host takes charge of the scheduling of the sporadic

messages in the DS, while the communication controller independently implements the FlexRay protocol services.

In terms of each FID allocated to a node, the host correspondingly maintains a message Queue (Q) and a Periodic Scheduler (PS). When the sporadic message in the DS is produced for the functionalities by the tasks in the host, it is not directly sent in the FlexRay bus but is put into the respective message queue in accordance with its predefined priority and waits for the scheduling of the periodic scheduler, which is periodically activated for the message transmission. When the periodic scheduler is in the activated state, the message with the highest priority is selected from the message queue and is put into the respective buffer of CHI to transmit in a given DYS. In the paper, the deadline-monotonic-priority is adopted, i.e. the message with the smallest deadline is assigned the highest priority and is put at the head of the message queue, in which the messages are sorted by decreasing priority (Schmidt & Schidt, 2009).

A single example is provided to demonstrate the message scheduling of the above model, as shown in Figure 1. The sporadic messages m_1, m_2, m_3, m_4, m_5 and their parameters are given in Figure 1(a). Let c and d be the length and the deadline of m_1-m_5. Furthermore, it is given that the DS is limited to 70 ms and a FlexRay cycle is 5 ms. The periodic schedulers PS_1, PS_2, PS_3, PS_4, PS_5 and their parameters are given in Figure 1(b). Let P, θ and m be the period, the phase and the corresponding message of PS.

Assume that all messages arrive at the beginning of the DS. Consider the case without PS in Figure 1(c). Although m_5 arrives in FC 0, it cannot be transmitted until FC 5 is coming and, hence, has to miss its deadline. This results from the transmission of m_2-m_4 with the smaller FID, which makes the remaining DS in FC0-FC4 shorter than

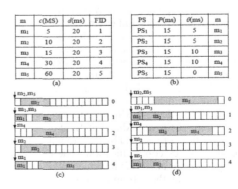

Figure 1. (a) Message set (b) Periodic scheduler set (c) Scheduling without PS (d) Scheduling with PS.

the length of m_5. Consider the case with PS in Figure 1(d). Although m_0, m_5 arrive at the same time in FC 0, due to the fact that the PS_1 is not activated, m_1 cannot be transmitted while m_5 is sent in FC 0. Actually, for this case it is easy to deduce that all messages can be transmitted before their deadlines.

From the above example, it can be seen that the message scheduling with PS can adjust the message transmission sequence and provide a guaranteed opportunity to each message to be transmitted before its deadline.

3.2 Periodic scheduler

In order to reduce the burden of the host, the host can implement each periodic scheduler by the clock-driven method (Liu, J. W. S., 2003). If the parameters of each PS are known, a periodic Scheduling Table (ST), for the hyperperiod H of each PS in a node, could be easily constructed. An item of ST consists of the decision time t_k and the activated PS at t_k, i.e. $PS(t_k)$. The decision time t_k can be got from a timer. The implementation for PS scheduling is shown in Figure 2.

3.3 System parameters

To support the mechanism of message scheduling in the above system model, three kinds of parameters, i.e. message, periodic scheduler and FlexRay network, should be taken into account. The parameters of the message are given as a priori knowledge and the optimal parameters of the FlexRay network and PS should be carefully designed according to the message scheduling to guarantee the requirements of the message deadline. For a sporadic message, it may be represented by the triples (T, d, c). T is denoted as a period which characterises the minimum time interval between two consecutive message releases. Like Figure 1(a), d is the deadline and used to describe the largest tolerable time interval between the release and the transmission of the message, while c is the length of the message that can be computed as in (2), including the payload s in the form of two-byte words, the DYS idle phase σ.

$$c = [l_b \times \tau_{bit} / t_{MS}] \qquad (2)$$

A periodic scheduler PS for a node n is a 5-tuple (n, P, θ, l, FID) with the period $P \in N$, the phase $\theta \in \{0, 1, \cdots, P-1\}$, the executing time l and the corresponding frame ID of the PS. The actual period of the PS is $P \cdot t_C$ and θ is used to denote the first activated time of the PS. At all FCs $(\theta + k \cdot P), k \in N_0$, the PS is activated to transmit the message and the executing time is l MS, while in the remaining FCs the PS is sleeping and no transmission of messages with the FID takes place.

Assume that some parameters, such as the duration of MT, mini-slot and dynamic slot idle phase, can be picked up from commercial FlexRay network design software (Wang & Zhang, 2016). In this paper, only two parameters are considered. The FlexRay cycle duration t_C is one important parameter. Considering the transmission of messages in the FlexRay static segment, t_C can be chosen as the *great common divisor* (gcd) of all messages, i.e. $t_C = \gcd(d_j)$. The other important parameter is the length of the dynamic segment t_{DS}. For the given sporadic message set and t_C, if all messages do not miss their deadlines, the smaller t_{DS} means that the utilisation of the dynamic segment is better.

If the Worst Case Response Time (WCRT) refers to the largest time interval between the release and the transmission of the message, two definitions are listed as follows:

Definition 1 Given a set of sporadic messages, the DS of FlexRay can be called being *schedulable* if the WCRT of each message is less than its deadline.

Definition 2 Given a set of sporadic messages, the sum of the WCRT of all sporadic messages is called *message total WCRT*, i.e. $\Sigma WCRT_l$.

According to the above definitions, the design of previous parameters can be easily viewed as an optimisation problem. That is, for a set of sporadic messages, how to assign each message to a PS and how to design P, θ, l, FID of each PS so that t_{DS} and message total WCRT is minimised and the DS is guaranteed to be schedulable.

Input: ST $(t_k, PS(t_k)), k \; 1, 2, 3, \ldots, N$

```
1: begin
2:    set i=1;
3:    set the timer to t_k;
4:    while(true)
5:       if(the timer is out) then
6:          select current PS as PS (t_k);
7:          i=i+1;
8:          compute the next item of ST k=i mod (N);
9:          set the timer to ⌊i/N⌋H+ t_k;
10:         get the message of Q corresponding to current PS;
11:         put the message to the buffer of CHI for transmission;
12:      end if
13:   end while
14: end
```

Figure 2.   Operation process of periodic scheduler.

4 TIMING ANALYSIS OF THE SCHEDULE

Let the FlexRay nodes in $N = \{1, 2, \cdots, n\}$ and the set of all sporadic messages in node $k \in N$ be $M_k = \{m_k^1, m_k^2, m_k^3, \cdots\}$. The entire set of sporadic messages is denoted as $M = \cup_{k=1}^{n} M_k$. the number of messages in $\mathbf{M}$ is denoted as $|\mathbf{M}|$ and $m_k^j \in M_k$ can be represented by triples (T_k^j, d_k^j, c_k^j). Assume that PSi is an arbitrary periodic scheduler in node k and its corresponding messages set is $G_i \subseteq M_k$. In the paper, the period P_i and the executing time l_i of PSi can be chosen as:

$$P_i = \min_{m_k^j \in G_i} \{ [d_k^j / t_C] - 1 \} \tag{3}$$

$$l_i = \max_{m_k^j \in G_i} \{ c_k^j \} \tag{4}$$

Define the binary variables $x_{ij} \in \{0, 1\}$ and $g_{ij} \in \{0, 1\}$. x_{ij} takes the value 1 if $M_k^j \in G_i$, i.e. m_k^j is assigned to Gi and is 0 otherwise. g_{ij} takes the value l if the PSi phase θi is in the j th FC and is 0 otherwise (Traian et al., 2006).

Theorem 1. For a set of sporadic messages $M = \bigcup_{k=1}^{n} M_k$, assume there are the periodic schedulers $PS_1, PS_2, \cdots, PS_i, \cdots$, then the DS is schedulable if the following constraints are satisfied: 1) The DS is long enough to accommodate the transmitting messages at any FC; 2) For $\forall m_k^j \in M, m_k^j \in G_i$, there exists:

$$\left(\left(\sum_{\substack{m_k^j \in M_k \\ d_k^j < d_k^j}} \left\lceil \frac{d_k^j}{T_k^s} \right\rceil \cdot x_{is} \cdot P \right) + 1 \right)_i \cdot t_C + \left(\left(\max_{\substack{\alpha = j \bmod P_u \\ j = 0, 1, \ldots, \beta-1}} \left(\sum_{u=1}^{FID_i - 1} (l_{\underline{u}} - 1) \cdot g_{u\alpha} \right) \right) + c_k^j \right) \cdot t_{MS} \leq d_k^j \tag{5}$$

where β is the *least common multiple* (lcm) of the period of all periodic schedulers whose FIDs are less than FIDi.

Proof. According to the message scheduling with the periodic scheduler in Section 3, a critical instant for any message in M is supposed to occur at the instant that the message is generated by its sender task and put into the head of the message queue immediately after its periodic scheduler has started. As a consequence, for $\forall m_k^j \in M, m_k^j \in G_i$, its WCRT can be given by the following equation:

$$WCRT(d_k^j) = \sigma_k + \omega_k(d_k^j) + \varphi_k \tag{6}$$

where σ_k is the longest delay suffered from the FC at which m_k^j is generated by the sender task, $\omega_k(d_k^j)$ is the worst case delay caused by higher

priority messages and φ_k denotes the time that passes in the last FC until m_k^j is sent. They are shown in Figure 4.

Clearly, from Figure 4, the delay σ_k is the longest when all periodic schedulers whose *FIDs* are less than *FID_i* do not send any messages at the FC. So the value of σ_k is:

$$\sigma_k = t_c - t_{SS} - (FID_i - 1) \cdot t_{MS} \tag{7}$$

In terms of $\omega_k(d_k^j)$, it can be produced because of the following causes: 1) Periodic schedulers with lower *FID* than *FID_i* send messages, the remainder of DS is not enough to transmit m_k^j, and m_k^j has to be delayed for the next FCs. 2) The higher priority messages in G_i, i.e. their deadlines are less than d_k^j, delay the transmission of m_k^j until all of them are transmitted. According to constraint: 1) in theorem 1, the DS is enough for transmitting messages at any FCs; 2) for $\omega_k(d_k^j)$ cannot exist. So $\omega_k(d_k^j)$ is only generated by the transmission of higher priority messages in G_i, and the value of $\omega(d_k^j)$ can be computed by:

$$\omega(d_k^j) = \left(\left(\sum_{\substack{m_k^j \in M_k \\ d_k^j < d_k^j}} \left\lceil \frac{d_k^j}{T_k^s} \right\rceil \cdot x_{is} \cdot P_i \right) + 1 \right) \cdot t_c \tag{8}$$

φ_k takes place in the FC at which m_k^j is sent and its value can be obtained by:

$$\varphi_k = t_{SS} + t_{ps} + (c_k^j \cdot t_{MS}) \tag{9}$$

where t_{ps} is the longest time that is taken to transmit the messages with lower *FID* less than *FID_i* before m_k^j is sent at the FC. If we assume that $\beta = \text{lcm}_{FID_u < FID_i}(P_u)$, then t_{ps} can be given as:

$$t_{ps} = \max_{\substack{\alpha = j \bmod P_u \\ j = 0, 1, \ldots, \beta-1}} \left(\sum_{u=1}^{FID_i - 1} (l_u \cdot g_{u\alpha} + 1 - g_{u\alpha}) \cdot t_{MS} \right) \tag{10}$$

Substitute equations (7), (8) and (9) for $\sigma k, \omega k, (d_k^j), \varphi k$ in equation (6), reformulate it, and it can be derived as follows:

$$WCRT(d_k^j) = \text{the left part of formula (5)} \tag{11}$$

Obviously, if $WCRT(d_k^j) < d_k^j, m_k^j$ must be transmitted in DS before its deadline. Since m_k^j is an arbitrary message in $\mathbf{M}$, according to definition 1, the DS should be schedulable. In the next section, equation (5) is employed as one constraint of the parameter optimisation.

5 PARAMETERS OPTIMISATION

For the set of sporadic messages **M**, we propose a two-step optimisation method to assign each message to a PS and design the parameters of each PS: P, θ, l, FID, G.

5.1 Minimise DS length t_{DS}

According to the previous description in 3, for a given message set, the smaller the DS length, the higher is the utilisation of DS. So the selection of all parameters should help minimise t_{DS}. Suppose PS_s is a periodic scheduler that corresponds to G_s, define the binary variables $\gamma_s \in \{0,1\}$ that take the value 1 if $G_s \neq \varnothing$ and is 0 otherwise. Then the *j*th cycle load L_j can be given as follows:

$$L_j = \sum_{s=1}^{|M|} \gamma_s \left[(l_s - 1) \cdot g_{sa} + 1 \right] \cdot t_{MS} \qquad (12)$$

where $a = j \bmod P_s$, $j = 1, 2, \cdots lcm(P_s)$. Clearly, there can exist the maximum cycle load for any FC $j = 1, 2, \cdots lcm(P_s)$.

Consider that t_{DS} should be enough to accommodate all message transmissions at any FC, the parameters optimisation can be formulated as follows:

$$t_{DS} = \min_Y \left(\max_{\substack{j=1,2,\cdots lcm(P_s) \\ G_s \neq \varnothing}} \sum_{s=1}^{|M|} \gamma_s \left[(l_s - 1) \cdot g_{sa} + 1 \right] \cdot t_{MS} \right) \qquad (13)$$

Subject to:

$$\text{For } q = 1, \ldots, |\mathbf{M}|, \quad \sum_{s=1}^{|M|} x_{s,q} = 1 \qquad (14)$$

$$\text{For } s = 1, \ldots, |\mathbf{M}|, \quad \sum_{k=0}^{P_s-1} g_{s,k} = \gamma_s \qquad (15)$$

$$\text{If } \forall \, m_k^q \in G_s, s = 1, \ldots, |\mathbf{M}|, \text{ then } G_s \subseteq M_k \qquad (16)$$

where **Y** is a vector with all variables $x_{s,j}$ and $g_{s,k}$. Only one periodic scheduler is selected for each m_k^q and exactly one FC can be determined as its phase for each PS_s. They are formulated by the constraints in (14) and (15), respectively. The constraint (16) denotes that all messages in G_s must be from the same node. The parameters t_{DS}, P, θ, l, G can be easily deduced from the solution of the above optimal problem with the objective function in (13) and the constraints in (14)-(16).

5.2 Minimise message total WCRT

If the t_{DS}, P, θ, l, G of each PS are determined, the message total WCRT of **M** has the close relation

```
Allocate_FID(FID[ng],G[ng], V_min, k, ng)
1:begin
2:  if(k>ng)then
3:    for  i∈{1,2,...,ng}   do
4:      compute WCRT(D) for each message in G[i];
5:    end for
6:    if(WCRT(d)≤d for each message)then
7:      calculate the sum of all WCRT(d) as Σ WCRT(d);
8:      if( Σ WCRT(d) < V_min )then
9:        V_min = Σ WCRT(d) ;
10:     end if
11:   end if
12:  else
13:    for (i = k; i <= ng; i++)
14:    swap the value of   FID [k] and FID[i];
15:    Allocate_FID(FID[ng],G[ng], V_min, k+1,ng);
16:    end for
17:  end if
18:end
```

Figure 3.   Algorithm for allocating FID of message.

with the *FID* of each *PS*. The appropriate allocation of *FID* to each *PS* can effectively cut down the message total WCRT. So an algorithm is developed to look for the optimal allocation that can minimise the message total WCRT. Denote the total number of $G \neq \phi$ as ng, number each *PS* in consequence, employ the array *FID[ng]* to hold the *FIDs* of PS_1, PS_2,..., PS_{ng} and employ the array *G[ng]* to keep G_1, G_2,..., G_{ng}. The following pseudo code in Figure 3 can describe the details of the algorithm. Note that $WCRT(d) \leq (d)$ in the algorithm is from theorem 1, which can guarantee that the dynamic segment is schedulable.

6 EXPERIMENTS AND ANALYSIS

All experiments are made in the CANoe, which is an automotive electronic network tool developed by the Vector Corporation in Germany. The transmission speed C is 10 Mbit/s, t_C is 5000 s, t_{SS} is more than 3000 s. The parameters of all the dynamic messages are shown in Table 1.

The optimised parameters of P, θ, l, FID, G can be obtained by solving the model of (13)-(16) and using the algorithm in Figure 4. The results are shown in Table 2. The messages $m_1, m_2, \ldots, m_{13}$ are allocated to 7 different periodic servers and the optimal t_{DS} is 546 μs.

To verify the method proposed in this paper, the message scheduling based on PSS and non-PSS is implemented in CANoe according to Table 1 and Table 2, respectively. It takes the same time to carry out the two experiments and the results are shown in Figure 4. The Mean Response Time (MERT) of a message from the non-PSS is less than that from the PSS. But the Maximum Response Time (MART) of the messages m_1, m_2, m_6, m_{10} is longer than their deadlines. Obviously all of them miss their deadlines and the dynamic segment is not scheduled under non-PSS scheduling. For the PSS scheduling, the under non-PSS scheduling. For the PSS scheduling, the MART of all messages is less than their deadlines and the dynamic segment can be schedulable. These indicate that the PSS can not only improve the utilisation of the dynamic segment, but also guarantee the schedulability of the dynamic segment.

To further verify the proposed method in this paper, the methods NIP and BIP are also used to

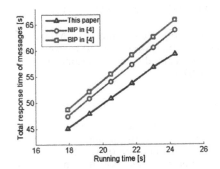

Figure 5. Total response time for different methods.

Table 1. The dynamic messages.

Node	Message	Length of payload (word)	$D[t_c]$	$T[t_c]$
N1	m_1	50	3	4
	m_2	125	5	6
	m_3	95	9	9
	m_4	97	17	20
	m_5	50	25	40
N2	m_6	80	5	6
	m_7	85	9	12
	m_8	50	25	26
	m_9	65	33	35
N3	m_{10}	65	3	5
	m_{11}	80	5	11
	m_{12}	110	5	20
	m_{13}	115	9	22

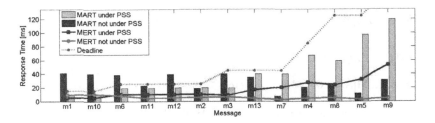

Figure 4. Response time of messages.

Table 2. The optimal configuration parameters.

N	Server	FID	l[MS]	$P[t_c]$	$\theta[t_c]$	G
N1	PS1	5	45	4	2	m_2
	PS2	1	36	2	0	m_1,m_3,m_4,m_5
N2	PS3	2	30	4	3	m_6,m_8
	PS4	3	32	8	7	m_7,m_9
N3	PS5	7	25	2	1	m_{10}
	PS6	4	42	4	1	m_{11},m_{13}
	PS7	6	40	4	0	m_{12}

design the same parameters according to Table 1. The related parameters from the design are configured to the CANoe, and the above experiments are also done. The total response time of message (TMRT) is shown in Figure 5. With the increase in time, the TMRT from three methods gradually increases. However, the TMRT from the proposed method in this paper is the shortest. This indicates that the proposed method can improve the response time of the system due to the optimisation on the message of FID.

7 CONCLUSION

The design of FlexRay is one of the most important works for vehicle design. Pointing to the dynamic segment of FlexRay, a two-layer message scheduling mechanism called a PSS is presented in this paper and the time analysis is made for the PSS. Considering both the network utilisation and the message schedulability, a method to design the parameter configuration of the dynamic segment and PSS is proposed. The experimental results validate the method. In the future, the optimisation of the PSS should be focused so that the proposed methods could be extended to the design of a larger and more complicated FlexRay network.

ACKNOWLEDGEMENTS

This work was supported by NSFC (61202096) and FRFCU (2013HGCH0014). This paper was also supported by STSP (15czz02039), NSF-HEI (KJ2012A226) and NSF (1708085MF157) of Anhui, Project of Jianghuai Auto (W2014 JSKF0086, W2016 JSKF0394).

REFERENCES

Cheng, C. et al. (2013). Analysis and forming construction of an on-vehicle communication protocol FlexRay network. *Journal of Automotive Safety and Energy*, 04(1), 75–81.

Lange, R., & De Oliveira, R.S. (2014). Probabilistic scheduling of the FlexRay dynamic segment. *IEEE Intern-ational Conference on Industrial Informatics IEEE, 2012-06.*

Liu, J.W.S. (2003). *Real-time systems*. USA: Prentice Hall.

Milbredty, P., & Glab, M. (2012). Designing FlexRay-based automotive architectures: A holistic OEM approach. *Proceedings of Design, Automation & Test in Europe Conference & Exhibition. Dresden, Germany*, 276–279.

Ouedraogo, L., & Kumar, R. (2014). Computation of the precise worst-case response time of FlexRay dynamic messages. *IEEE Transactions on Automation Science & Engineering*, 11(2), 537–548.

Park, I., & Sunwoo, M. (2010). FlexRay network parameter optimization method for automotive applications. *IEEE Transactions on Industrial Electronics*, 58(4), 1449–1459.

Schmidt, E.G., & Schidt, K. (2009). Message scheduling for the FlexRay protocol: The dynamic segment. *IEEE Transactions on Vehicular Technology*, 58(5), 2160–2169.

Traian, P. et al. (2006). Timing analysis of the FlexRay co-mmunication protocol. *Proceedings of Euro-micro Conf-erence on Real-Time Systems*, 203–216.

Wang, G. et al. (2012). Analytical method for dynamic segment time optimization in FlexRay network. *Computer Engineering*, 38(10), 241–243.

Wang, Y., & Zhang, Y. (2016). Communication fault detection and management methods of car FlexRay network under AUTOSAR. *Transactions of the Chinese Society of Agricultural Engineering*, 32(4), 105–111.

Automotive, Mechanical and Electrical Engineering – Liu (Ed.)
© 2017 Taylor & Francis Group, London, ISBN 978-1-138-62951-6

Micro-satellite attitude determination based on federated Kalman filter using multiple sensors

Feng Ni, Haiyang Quan & Wenjie Li
Beijing Microelectronics Technology Institute, Beijing, China

ABSTRACT: Considering of power, size and cost of micro-satellite, MEMS gyroscope, sun sensor and magnetometer are chosen as attitude sensors. In this paper, a Federated Kalman Filter, consisting of two sub-filters and one master filter, is designed to estimate micro-satellite's attitude. Simulation shows that it performs well and has a better accuracy when compared to traditional method of Runge-Kutta. Besides, it has less calculation load and good fault tolerance which is significant for micro-satellite application.

Keywords: Federated Kalman Filter; Attitude determination; Gyroscope; Magnetometer; Sun sensor

1 INTRODUCTION

The application of MEMS sensors in micro-satellite attitude determination has been one of the hottest issues recently (T Meng, et al., 2009). However, considering of MEMS gyroscope's large drift and noise, it's necessary to improve the accuracy of attitude determination using MEMS gyroscope and other sensors in micro-satellite application. A Federated Kalman Filter, consisting of two sub-filters and one master filter, is designed to estimate the attitude of micro-satellite in this paper. It's based on a set of measurement models of sensors and can deal with data from magnetometer and sun sensor in parallel. There are three main reasons for chosing Federated Kalman Filter: 1) higher accuracy. Sub-filters estimate micro-satellite's attitude between magnetometer and MEMS gyroscope and between sun sensor and MEMS gyroscope respectively. The master filter of Federated Kalman Filter realizes data fusion and feeds back the information to sub-filters, which can give a better attitude estimation at last. 2) Good fault tolerance. Two sub-filters are executed separately in Federated Kalman Filter. Therefore, the final outputs of attitude determination algorithm are still acceptable without one sub-filter working. 3) Less calculation load. Compared to traditional centralized Kalman Filter, such as Extend Kalman Filter (EKF) widely used in practical application, Federated Kalman Filter has smaller state dimension reducing calculation load. Simulation has conducted in this paper and results show that the Federated Kalman Filter presented performs better in accuracy than traditional methods such as Runge-Kunta, meeting the accuracy requirement 3°.

A set of measurement models of sensors are given in section 2. The main process and simulation results of attitude determination based on Federated Kalman Filter are presented in section 3 and 4. Conclusion is proposed in Section 5.

2 MEASUREMENT MODELS OF SENSORS

MEMS gyroscope's accuracy is much lower than traditional gyroscope such as fiber-optics gyroscope because of its large drift and noise (H Stearns, 2011). Therefore, magnetometer and sun sensor are chosen as assistant attitude sensors to improve the accuracy in this paper. It's important to choose a set of accurate measurement models of sensors in order to establish measurement equations of Federated Kalman Filter later in section 3.

2.1 *Measurement model of MEMS gyroscope*

Since MEMS gyroscope's large drift and noise, its measurement model can be given by:

$$\tilde{\boldsymbol{\omega}}_{bi} = \boldsymbol{\omega}_{bi} + \mathbf{b} + \boldsymbol{\eta}_g \quad E(\boldsymbol{\eta}_g \boldsymbol{\eta}_g^T) = \boldsymbol{\sigma}_v^2 \tag{1}$$

$$\dot{\mathbf{b}} = \boldsymbol{\eta}_b \quad E(\boldsymbol{\eta}_b \boldsymbol{\eta}_b^T) = \boldsymbol{\sigma}_u^2 \tag{2}$$

MEMS gyroscope's output is $\tilde{\boldsymbol{\omega}}_{bi}$, the angular velocity of micro-satellite's body relative to inertial coordinate system when fixed along the inertial axis of micro-satellite. $\boldsymbol{\omega}_{bi}$ is the real angular velocity and $\mathbf{b}$ is the drift of MEMS gyroscope. $\boldsymbol{\eta}_g$ and $\boldsymbol{\eta}_b$ are both assumed as zero-mean Gauss white noise. Besides, $\boldsymbol{\sigma}_v^2$ and $\boldsymbol{\sigma}_u^2$ can be referred to Angle Random Walk (ARW) and Rate Random Walk (RRW) according to articles (J.L. Crassidis, 2006).

In order to estimate the micro-satellite's attitude, $\boldsymbol{\omega}_{bi}$ should be transformed to $\boldsymbol{\omega}_{bo}$, the angular velocity of micro-satellite's body relative to orbit coordinate system:

$$\boldsymbol{\omega}_{bo} = \boldsymbol{\omega}_{bi} - \mathbf{A}_{bo}\boldsymbol{\omega}_{oi} \tag{3}$$

And $\boldsymbol{\omega}_{oi}$ is the orbit rate, $\mathbf{A}_{bo}$ is the rotation matrix from orbit coordinate system to satellite's body.

2.2 Measurement model of sun sensor

The measured sun vector $\tilde{\mathbf{S}}_b$ is given by sun sensor. If the aiming axis of sun sensor is fixed along pitch axis and other two measurement axis are fixed along micro-satellite's roll and yaw axis, its measurement model can be given by:

$$m_x = \frac{S_z}{S_y} + \eta_s, \quad m_y = -\frac{S_x}{S_y} + \eta_s \tag{4}$$

$$\tilde{\mathbf{S}}_b = \frac{1}{\sqrt{m_x^2 + m_y^2 + 1}}\begin{bmatrix} m_x & m_y & 1 \end{bmatrix}^T \tag{5}$$

where m_x and m_Y are the outputs of sun sensor and the noise of sun sensor $\boldsymbol{\eta}_s$ is assumed as a zero-mean Guass white noise.

The reference sun vector $\mathbf{S}_b$ in body frame can be also transformed from $\mathbf{S}_i$, the sun vector in inertial coordinate system which is calculated from solar ephemeris:

$$\mathbf{S}_b = \mathbf{A}_{bo}\mathbf{A}_{oi}\mathbf{S}_i = \mathbf{A}_{bo}\mathbf{S}_o \tag{6}$$

where $\mathbf{A}_{oi}$ the rotation matrix from inertial coordinate system to satellite are's orbit and $\mathbf{S}_o$ is the sun vector in orbit coordinate system.

The measurement model of sun sensor can be transformed to:

$$\tilde{\mathbf{S}}_b = \mathbf{S}_b + \eta_s \tag{7}$$

2.3 Measurement model of magnetometer

The measured geomagnetic vector $\tilde{\mathbf{B}}_b$ is the output of magnetometer when fixed along inertial axis of micro-satellite. On the other hand, the reference geomagnetic vector $\mathbf{B}_E$ can be calculated from International Geomagnetic Reference Field model (IGRF) given in the northward, eastward and radially inward coordinate system (NED). If the noise of magnetometer $\boldsymbol{\eta}_B$ is assumed as a zero-mean Guass white noise, the measurement model of magnetometer can be given by:

$$\tilde{\mathbf{B}}_b = \mathbf{A}_{bo}\mathbf{A}_{oe}\mathbf{B}_E + \boldsymbol{\eta}_B \tag{8}$$

where $\mathbf{A}_{oe}$ is the rotation matrix from NED to satellite's orbit coordinate system?

3 FEDERATED KALMAN FILTER IN ATTITUDE DETERMINATION

A Federated Kalman Filter is designed in this paper to estimate micro-satellite attitude determination based on data of MEMS gyroscope, magnetometer and sun sensor. It mainly consists of two sub-filters S1, S2 and one master filter M1, as shown in Fig. 1:

S1 is designed to estimate micro-satellite's attitude based on sun sensor and MEMS gyroscope while S2 is to output attitude based on magnetometer and MEMS gyroscope. Both of them are in process separately. Then, the master filter M1 outputs the final attitude information by fusing data of S1 and S2 and feeds back to them at the same time. Therefore, the Federated Kalman Filter in this paper has good fault tolerance. When micro-satellite is in invisible region where sun sensor is invalid, attitude determination algorithm can still work well.

3.1 The state equation of Federated Kalman Filter in attitude determination

The error quaternion $\Delta \mathbf{q}$ is defined by (9), where $\hat{\mathbf{q}}$ is the estimated quaternion and $\mathbf{q}$ is real quaternion:

$$\mathbf{q} = \hat{\mathbf{q}} \otimes \Delta \mathbf{q} \tag{9}$$

$$\Delta \mathbf{q} \approx \begin{bmatrix} 1 & \Delta q_1 & \Delta q_2 & \Delta q_3 \end{bmatrix}^T \tag{10}$$

Therefore, $\Delta \mathbf{q}$ can be replaced by a three dimension vector consisting of $\Delta q_1, \Delta q_2, \Delta q_3$: $\Delta \mathbf{q} = \begin{bmatrix} \Delta q_1 & \Delta q_2 & \Delta q_3 \end{bmatrix}^T$;

According to micro-satellite's quaternion kinematics equation (11) (LJ Zhang, 2013; SB Ni, 2011; YM Ge, et al., 2013), we substitute $\Delta \mathbf{q}$ to it:

$$\dot{\mathbf{q}}_{bo} = \frac{1}{2}\mathbf{q}_{bo} \otimes \boldsymbol{\omega}_{bo} \tag{11}$$

$$\Delta \dot{\mathbf{q}} = -\frac{1}{2}\hat{\boldsymbol{\omega}}_{bo} \otimes \Delta \mathbf{q} + \frac{1}{2}\Delta \mathbf{q} \otimes \Delta \boldsymbol{\omega}_{bo} + \frac{1}{2}\Delta \mathbf{q} \otimes \hat{\boldsymbol{\omega}}_{bo}$$
$$\approx -\hat{\boldsymbol{\omega}}_{bo} \otimes \Delta \mathbf{q} + \frac{1}{2}\Delta \boldsymbol{\omega}_{bo} \tag{12}$$

According of the measurement model of MEMS gyroscope in section 2.1, $\Delta \boldsymbol{\omega}_{bo}, \Delta \boldsymbol{\omega}_{bi}$ can by defined by:

$$\Delta \boldsymbol{\omega}_{bo} = \boldsymbol{\omega}_{bo} - \hat{\boldsymbol{\omega}}_{bo} = \Delta \boldsymbol{\omega}_{bi} - (\mathbf{A}(\mathbf{q}) - \mathbf{A}(\hat{\mathbf{q}}))\boldsymbol{\omega}_{oi}$$
$$\approx \Delta \boldsymbol{\omega}_{bi} + 2\Delta \mathbf{q} \otimes \mathbf{A}(\hat{\mathbf{q}})\boldsymbol{\omega}_{oi} \tag{13}$$

$$\Delta \boldsymbol{\omega}_{bi} = \boldsymbol{\omega}_{bi} - \hat{\boldsymbol{\omega}}_{bi} = \tilde{\boldsymbol{\omega}}_{bi} - \mathbf{b} - \boldsymbol{\eta}_g - (\tilde{\boldsymbol{\omega}}_{bi} - \hat{\mathbf{b}})$$
$$= -\Delta \mathbf{b} - \boldsymbol{\eta}_g \tag{14}$$

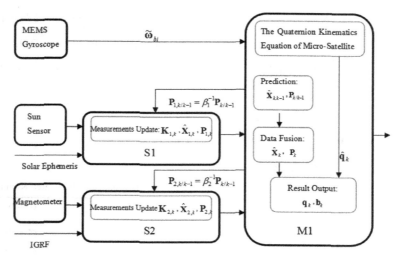

Figure 1. The main process of Federated Kalman Filter in attitude determination.

$$\Delta \dot{\mathbf{b}} = \boldsymbol{\eta}_b \qquad (15)$$

when the state vector is $\mathbf{X} = [\Delta \mathbf{q} \quad \Delta \mathbf{b}]^T$, the state equation of Federated Kalman Filter in micro-satellite attitude determination is:

$$\dot{\mathbf{X}} = \mathbf{F}(t)\mathbf{X} + \mathbf{G}(t)\mathbf{W} \qquad (16)$$

where

$$\mathbf{F}(t) = \begin{bmatrix} -[\hat{\boldsymbol{\omega}}_{bi} \times] & -0.5\mathbf{I}_{3\times3} \\ \mathbf{0}_{3\times3} & \mathbf{0}_{3\times3} \end{bmatrix}, \mathbf{G}(t) = \begin{bmatrix} -0.5\mathbf{I}_{3\times3} & 0 \\ 0 & \mathbf{I}_{3\times3} \end{bmatrix}$$

and $[\hat{\boldsymbol{\omega}}_{bi} \times] = \begin{bmatrix} 0 & -\hat{\omega}_{biz} & \hat{\omega}_{biy} \\ \hat{\omega}_{biz} & 0 & -\hat{\omega}_{bi} \\ -\hat{\omega}_{biy} & \hat{\omega}_{bix} & 0 \end{bmatrix}$;

Notice that the state equation (16) should be discretized by time.

3.2 The measurement equation of sub-filter S1 and S2

According to the measurement model of sun sensor in section 2.2, the error $\Delta \mathbf{S}_b$ between measured sun vector $\tilde{\mathbf{S}}_b$ and estimated one $\mathbf{S}_b$ can be defined by:

$$\Delta \mathbf{S}_b = \tilde{\mathbf{S}}_b - \hat{\mathbf{S}}_b \qquad (17)$$

$$\Delta \mathbf{S}_b = \mathbf{A}(\mathbf{q})\mathbf{S}_o + \boldsymbol{\eta}_s - \mathbf{A}(\hat{\mathbf{q}})\mathbf{S}_o = (\mathbf{A}(\hat{\mathbf{q}} \otimes \Delta \mathbf{q}) - \mathbf{A}(\hat{\mathbf{q}}))\mathbf{S}_o$$
$$+ \boldsymbol{\eta}_s \approx -2\Delta \mathbf{q} \otimes \mathbf{A}(\hat{\mathbf{q}})\mathbf{S}_o + \boldsymbol{\eta}_s = 2\hat{\mathbf{S}}_b \otimes \Delta \mathbf{q} + \boldsymbol{\eta}_s$$
$$(18)$$

Since the state vector of Federated Kalman Filter is $\mathbf{X} = [\Delta \mathbf{q} \quad \Delta \mathbf{b}]^T$, (18) should be transformed to (19), which is the measurement equation of sub-filter S1:

$$\Delta \mathbf{S}_b = \mathbf{H}_s \mathbf{X} + \boldsymbol{\eta}_s \qquad (19)$$

And $\mathbf{H}_s = [2[\hat{\mathbf{S}}_s \times] \quad \mathbf{0}_{3\times3}]$;

Similarly, according to measurement model of magnetometer mentioned in section 2.3, it's easy to derive the measurement equation of sub-filter S2:

$$\Delta \mathbf{B}_b = \tilde{\mathbf{B}}_b - \hat{\mathbf{B}}_b \qquad (20)$$

$$\Delta \mathbf{B}_b = \mathbf{A}(\mathbf{q})\mathbf{B}_o + \boldsymbol{\eta}_B - \mathbf{A}(\hat{\mathbf{q}})\mathbf{B}_o = (\mathbf{A}(\hat{\mathbf{q}} \otimes \Delta \mathbf{q}) - \mathbf{A}(\hat{\mathbf{q}}))\mathbf{B}_o$$
$$+ \boldsymbol{\eta}_B \approx -2\Delta \mathbf{q} \otimes \mathbf{A}(\hat{\mathbf{q}})\mathbf{B}_o + \boldsymbol{\eta}_B = 2\hat{\mathbf{B}}_b \otimes \Delta \mathbf{q} + \boldsymbol{\eta}_B$$
$$(21)$$

Substitute the state vector $\mathbf{X}$ into (21):

$$\Delta \mathbf{B}_b = \mathbf{H}_B \mathbf{X} + \boldsymbol{\eta}_B \qquad (22)$$

And $\mathbf{H}_B = [2[\hat{\mathbf{B}}_b \times] \quad \mathbf{0}_{3\times3}]$;

3.3 Process of Federated Kalman Filter in attitude determination

Because both of the state vector of two sub-filters is $\mathbf{X} = [\Delta \mathbf{q} \quad \Delta \mathbf{b}]^T$, the state vector $\mathbf{X}_{k/k-1}$ and error covariance $\mathbf{P}_{k/k-1}$ can be calculated in the master filter M1. Then, it will feed back $\mathbf{P}_{k/k-1}$ to two sub-filters according to information allocation factors β_1, β_2. Two sub-filters S1 and S2 complete the measurement update separately and give the estimated results to master filter M1. The master filter M1 finishes data fusion of two sub-filters and outputs the final attitude determination information. The main process of Federated Kalman Filter can be described as follows:

1. *Prediction in Master Filter:* One-step prediction of state $\hat{\mathbf{X}}_{k/k-1}$ can be calculated from $\boldsymbol{\omega}_{bi}$ and the error covariance $\mathbf{P}_{k/k-1}$ is:

$$\mathbf{P}_{k/k-1} = \boldsymbol{\Phi}_{k/k-1}\mathbf{P}_{k-1}\boldsymbol{\Phi}_{k/k-1}^T + \boldsymbol{\Gamma}_{k-1}\mathbf{Q}_{k-1}\boldsymbol{\Gamma}_{k-1}^T \quad (23)$$

2. *Information Allocation:* The master filter feeds back $\mathbf{P}_{k/k-1}$ to sub-filters according to factor β_1, β_2;

$$\mathbf{P}_{i,k/k-1} = \beta_i^{-1}\mathbf{P}_{k/k-1} \quad (24)$$

3. *Measurements Updating*: Sub-filters calculate the Kalman gain $\mathbf{K}_{i,k}$, the estimated state $\hat{\mathbf{X}}_{i,k}$ and the estimated error covariance $\mathbf{P}_{i,k}$:

$$\mathbf{K}_{i,k} = \mathbf{P}_{i,k/k-1}\mathbf{H}_{i,k}^T(\mathbf{H}_{i,k}\mathbf{P}_{i,k/k-1}\mathbf{H}_k^T + \mathbf{R}_k)^{-1} \quad (25)$$

$$\hat{\mathbf{X}}_{i,k} = \mathbf{K}_{i,k}(\mathbf{Z}_{i,k} - \mathbf{H}_{i,k}\hat{\mathbf{X}}_{i,k/k-1}) \quad (26)$$

$$\mathbf{P}_{i,k} = (\mathbf{I} - \mathbf{K}_{i,k}\mathbf{H}_{i,k})\mathbf{P}_{i,k/k-1} \quad (27)$$

4. *Data Fusion*: Master Filter completes the data fusion:

$$\mathbf{P}_k^{-1} = \mathbf{P}_{1,k}^{-1} + \mathbf{P}_{2,k}^{-1} \quad (28)$$

$$\hat{\mathbf{X}}_k = \mathbf{P}_k(\mathbf{P}_1^{-1}\hat{\mathbf{X}}_{1,k} + \mathbf{P}_2^{-1}\hat{\mathbf{X}}_{2,k}) \quad (29)$$

5. *Results Output*: correct the estimated quaternion and drift by $\hat{\mathbf{X}}_k$:

$$\mathbf{q}_k = \hat{\mathbf{q}}_k \otimes \begin{bmatrix} \sqrt{1-|\Delta\hat{\mathbf{q}}_k|^2} \\ \Delta\hat{\mathbf{q}}_k \end{bmatrix} \quad (30)$$

$$\mathbf{b}_k = \hat{\mathbf{b}}_k + \Delta\hat{\mathbf{b}}_k \quad (31)$$

4 SIMULATION

Simulation has been conducted in this paper and the orbit parameters of micro-satellite are presented in Table 1:

The parameters of Federated Kalman Filter are selected in Table 2:

The simulation time step of Federated Kalman Filter is 0.2 s and the initial state vector is $\mathbf{X} = [\Delta\mathbf{q} \quad \Delta\mathbf{b}]^T = [\mathbf{0}_{1\times6}]^T$. Fig. 2 shows the attitude

Table 1. Orbit parameters of micro-satellite.

Altitude	600 km
Inclination	97°
RAAN	220°
Orbit Period	5801.232 s
Orbit Rate ω_{oi}	0.00108 m/s

Table 2. Parameters of Federated Kalman Filter.

MEMS Gyroscope	$\sigma_v = 10^{-4}$, $\sigma_u = 10^{-6}$
Sun Sensor	$\sigma_s = 10^{-4}$
Magnetometer	$\sigma_s = 10^{-7}$
Real Velocity ω_{bi}	$\omega_{bi} = [0 \quad 0 \quad 0]^T$
Real Quaternion $\mathbf{q}_{bo}$	$\mathbf{q}_{bo} = [1 \quad 0 \quad 0 \quad 0]^T$
Simulation Time	2000 s

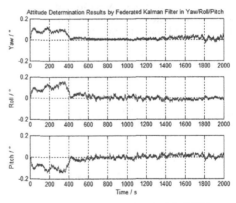

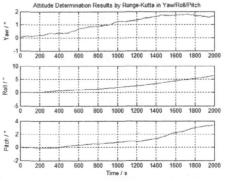

Figure 2. Attitude determination results given by Federated Kalman Filter and 4-order Runge-Kutta.

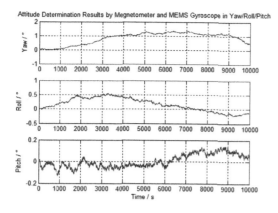

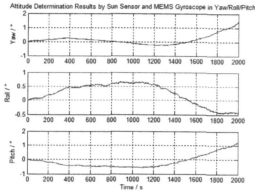

Figure 3. Attitude determination results only using magnetometer and MEMS gyroscope, sun sensor and MEMS gyroscope.

determination results given by Federated Kalman Filter designed in this paper and by 4-order Runge-Kutta, the traditional method. It's obvious that the Federated Kalman Filter has a much better accuracy in attitude determination than Runge-Kutta, meeting the accuracy requirement within 3°.

Fig. 3 shows the attitude determination results only generated by magnetometer and MEMS gyroscope and by sun sensor and MEMS gyroscope. When compared to Fig. 2, the accuracy

of Federated Kalman Filter is better than both of them. Therefore, the Federated Kalman Filter proposed in this paper has a better performance in micro-satellite attitude determination.

5 CONCLUSION

In this paper, a Federated Kalman Filter has been designed to estimate micro-satellite's attitude by using MEMS gyroscope, magnetometer and sun sensor. Simulation results show that it has a good performance in attitude determination and achieves a higher accuracy when compared to Runge-Kutta method. This method also has good fault tolerance and less calculation load because of its theory and structure. Since the algorithm has been simulated by software, its feasibility will be verified in physical platform of micro-satellite in following work.

REFERENCES

Cao, L, XQA Chen. An Algorithm for High Precision Attitude Determination When Using Low Precision Sensors, Sciece China Information Sciences, 2012, 55(3):626–637.

Cemenska, J, Sensor Modelling and Kalman Filtering Applied to Satellite Attitude Determination, University of California at Berkeley, 2004.

Crassidis, J.L, Sigma-point Kalman filtering for integrated GPS and inertial navigation, IEEE Transactions on Aerospace and Electronic Systems, 2006, 42(2): 750–756.

Ge, YM, YY Jiao, JQ Wang, et al. Satellite Attitude Determination Based On Federated Filter, Shanghai Aerospace, 2013, 30 (6): 23–27.

Meng, T, H Wang, ZH Jin, et al, Attitude Fusion Algorithm for Pico-Satellite Based On Low Cost MEMS Sensors, Journal of Astronautics, 2009, 30 (4): 1569–1573.

Ni, SB, C Zhang. Attitude Determination of Nano Satellite Based on Gyroscope, Sun Sensor and Magnetometer, Procedia Engineering, 2011, 15:959–963.

Stearns, H, M Tomizuka, Multiple model adaptive estimation of satellite attitude using MEMS gyros, Proceedings of the 2011 American Control Conference, IEEE, 2011: 3490–3495.

Zhang, LJ, SF Zhang, HB Yang, S Qian, Multiplicative filtering for spacecraft attitude determination, Journal of National University of Defense Technology, 2013.

Automotive, Mechanical and Electrical Engineering – Liu (Ed.)
© 2017 Taylor & Francis Group, London, ISBN 978-1-138-62951-6

Modeling and simulation of meteor burst communication network based on the OPNET

Hang Gao, Xiongmei Zhang & Xiaodong Mu
Xi'an Research Institute of Hi-Tech Hongqing Town, Xi'an, Shanxi, China

Jing Wan
The Second Artillery Armament Academy, Beijing, China

Yue Ren
Unit 96819, Beijing, China

ABSTRACT: In this paper, according to the characteristics of meteor burst communication network, a meteor burst network is designed based on OPNET simulation environment. Then the simulation results of the data amount of the corresponding nodes and the total amount of communication are obtained. Compared with the data of point-to-point node communication, the simulation results show that the simulation network can meet the characteristics of meteoric meteorological communication, and the design of the networking method improves the meteoric traces communication efficiency, which provides a reference for meteoric burst network construction.

Keywords: Meteor Burst Communication; OPNET Simulation; Communication Network Modeling

1 INTRODUCTION

Meteor burst communication technology plays an important role in the field of emergency communications. Its network structure has gone through the development process of data acquisition network, data communication network, fixed communication network, advanced communication network to interoperable network (Yavuz 1990). At present, the channel is the center of the simulation research of the meteor communication system, so in the demonstration of meteor burst communication system, according to the channel model established by the application, the meteorological data of the meteoric meteor is analyzed, the data sample is smoothed by the quintuple smoothing method and the channel distribution and parameter information are extracted, MFC VC++ 6.0 is used to develop a software on channel data processing and generate channel models (Feng 2014). In the literature (Hao 2009), the mathematical model of meteor satellite channel is studied deeply and the method on the combination of part of the realization and integration of the overall is adopted. Based on the SPW simulation platform, the sparse secret meteor channel, sparse dense meteor channel, multi-path meteor burst channel, short-time short-term secret class and long-term under currency channel are simulated, and the channel

model of meteor satellite communication system is established. Lacking of the simulation research on meteor burst network relatively and aiming at the stochastic problem of the meteor burst network at present, a design method of meteor burst network based on OPNET is proposed in this paper. The communication efficiency of the meteor burst is higher than the point-to-point communication.

2 THE STRUCTURE OF METEOR BURST COMMUNICATION NETWORK

2.1 *Single base station star topology*

The basic network structure of the meteor burst communication system is the star topological structure, which consists of a central node and a number of sub-nodes. The central node is the core of the whole network, and the child node can only communicate with the central node (Zhang & Zhong 2009).

2.2 *Multi-base station ring topology*

Ring topology consists of several nodes in the network through the point to point link to form a closed loop, the data in the loop in a direction between the transmissions of the nodes, the information is transported from one node to another node (Sheldon & Schilling 1992).

2.3 Tree topology

Tree topology is a structure whose nodes are linked hierarchically and information exchange is mainly between the upper and lower nodes, data exchange between adjacent nodes or peers is generally not carried out.

3 THE MODELING OF METEOR BURST COMMUNICATION NETWORK BASED ON OPNET

OPNET provides a comprehensive development environment that supports communication networks and discrete system modeling, it is a powerful network simulation tool (Chen 2004).

3.1 Meteor burst communication network model established

In the OPNET simulation of the network structure shown in Figure 1, there are four main stations, namely node 10, node 20, node 30, node 40, each master corresponds to two from the station, and such as the master node 10 subordinate two from the nodes are node 11 and node 12 respectively. The corresponding slave nodes of node 20 are node 21 and node 22 respectively. The corresponding nodes of node 30 are node 31 and node 32 respectively. The corresponding nodes of node 40 are node 41 and node 42 respectively. Node 0 as an interface to use the main role is to read the channel data.

3.2 Node model design and implementation

Node model of simulation experiment must have transceiver module and processing module. Because network emulation is required, a routing module must be included. However, the different modes of operation between the master and slave determine the node model is slightly different. The master station uses full-duplex mode, slave from half-duplex mode, so the main station needs a transmitter, while the need for six receivers and master and slave communication, and slave only connected with the main station half-duplex Therefore, only a pair of transceivers are required (Shen & Cai 2014).

3.2.1 Master node model

App module of the master node in the top-level represents the application layer and generates application packets at regular intervals. IP module which is below the app module represents the network layer and selects the corresponding route. IP module which is below the app module represents the network layer and selects the corresponding route. Cut module is mainly applied in dividing the application packets which is in the application layer into the packets transmitted on the physical channel in sender. What's more, at the receiving end, the received packet combination is restored to an application layer service packet to realize the conversion of the packet and the application packet. It is shown as Figure 2.

3.2.2 Slave node model

The difference between the master node model and the slave node model is the MAC layer which controls channel access. The slave station only sends its own information to the master or receive information from the same master, so the paper set a transmitter and a receiver. It is shown as Figure 3. In the design of antenna, the beam inclination, ground conductivity, antenna height and other factors are mainly considered. To facilitate the observation of the simulation results of the data, the transmission delay and receive packet rate attributes are added as the observation.

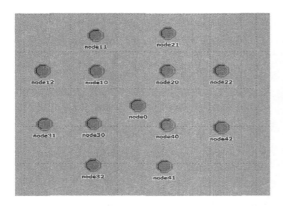

Figure 1.   Meteor burst network model.

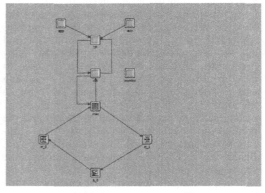

Figure 2.   Master node structure diagram.

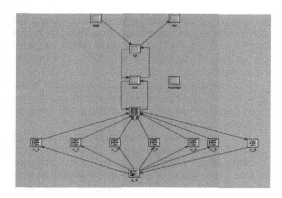

Figure 3. Slave node structure diagram.

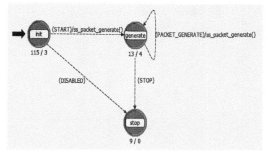

Figure 4. Slave node structure diagram.

4 PROCESS MODEL DESIGN AND IMPLEMENTATION

First, the processor is initialized, and the state machine receives packets. When the site receives the data packet, it will be kept as a copy of the time to repeat the reduction package, while the use of ARQ technology to address the address, which is added to the packet transmission queue. When the packet queue value is 0, the system will check the channel which is selected by the route. If the channel is available, the packet will be sent to the next hop address corresponding to the site, and when the signal of ACK is successfully received, the copy will be destructed. On the other hand, a copy of the call is required to transfer the data again.

App node, it is shown as Figure 4, which mainly used to generate data packets, is a data source and the simulation of the process model includes the init process, the generate process, the stop process. In the initialization process, the time of arrival of the package, the size of the package, the type, and the time of the package to produce and stop should be defined. And it goes into the process of generating the package, many other unpredictable circumstances will occur, so a closed line to return the generate package should be linked, to ensure the smooth formation process, and then enter the package to stop the process.

Mac layer is for channel access, which need to initialize the init process, send tra process, make idle process free and receive the rec process. It is shown as Figure 5, at the outset, the init process of the state machine should be initialized. The idle process is mainly used for the implementation of the package after the end of the process, back to the initial state, so this is a non-mandatory state.

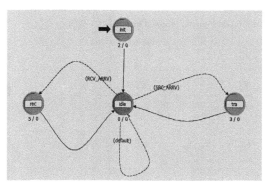

Figure 5. MAC layer state machine process diagram.

5 ANALYSIS OF SIMULATION RESULTS

It is assumed that the node 22 generates the information of the destination node 41, and the station 22 has to pass through the master station 20 to the master station 40 of the destination node 41. From the master station 20 to the master station 40 need via a hop relay. According to the path selection algorithm, the node 10 is preferentially selected as a relay node.

At the end of the simulation, the packet rate of node 20 and node 22 is obversed, as shown in the Figure 6.

Comparing the above two nodes, node 22 sends approximately 25 kbit of data, while node 20 sends 36 kbit of data, The difference value between them is that the amount of data transmitted by the node 21 is about 11 kbit, the graphs are stepped up, and the slopes are generally similar in the ascending phase, it is shown that the poisson distribution used in the channel is a good approximation to the uncertainty of the meteoric burst in the ascending phase. Analyzing the node 22, the node 22 begins to transmit packets in the channels of 0.5 to 1 s, 7.5 s to 11 s, in the simulation time of about 26 min, there are two significant

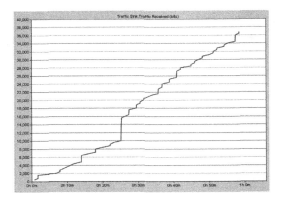

(a)The information received by node 20.

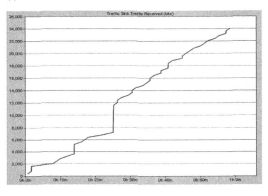

(b)The information received by node 22.

Figure 6.   Total amount of information received.

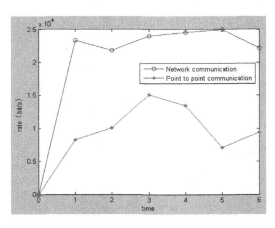

Figure 7.   Communication total simulation comparison chart.

amount of information jump stage, reflecting the meteor burst characteristics. And that the nodes 20, 0.5 s to 1 s, 1 s to 2 s are also consistent with the design of the link off to achieve the simulation result.

To compare the performance of meteor burst cluster and point-to-point communication, we need to collect the data of point-to-point communication. The results are shown in Figure 7.

The above figure is the simulation experiment under the same conditions for 18 s. The communication data of the networking model designed in this paper are selected every 3 s. It can be seen through the network experiment, effectively improve the information transfer rate.

6 SUMMARY

In this paper, a network communication method based on the OPNET environment is designed to model the master/slave nodes, processes and routing protocols. And a new method for the communication of meteor burst communication network is proposed, which is an important emergency communication method. The simulation results show that the proposed network improves the communication efficiency and enhances the connectivity and reliability of the network, and provides the basis for the construction of meteor burst communication network.

REFERENCES

Chen, M. 2004. OPNET Network Simulation. Beijing: Tsinghua University.
Feng, X.D. 2014. Research of MBC Modeling and Simulation Technology. Xi'an: Xidian University.
Hao, B.J. 2009. Meteor Burst Channel Model Building and System Simulation for Meteor Burst Communication. Xi'an: Xidian University.
Li, Z., Shen, J. & Cai, J. 2014. Semi-blind Joint Data Equalization and Channel Estimation for Meteor Burst Communication. IEEE 20th International Conference on Advanced Information Networking and Applications, 18–20 April 2006, 617–622.
Sheldon, S.L & Schilling, D.L. 1992. Efficient Communications Using the Meteor-Burst Channel. IEEE Transactions on Communications, Januray 40(1):119–128.
Yavuz, Davras. 1990. Meteor Burst Communications. IEEE Communications Magazine 28(9):40–48.
Zhang, W.X. & Zhong J.L. 2009. Wireless Tactical Communications Network Simulation Modeling based on OPNET. Computer Development and Applications 22(9):57–58.

Automotive, Mechanical and Electrical Engineering – Liu (Ed.)
© 2017 Taylor & Francis Group, London, ISBN 978-1-138-62951-6

Performance of wireless communication analysis at high speed

Xutao Li
School of Electronics and Information Engineering, Beihang University, Beijing, China

Yuxiang Liu
Department of Materials Science, Fudan University, Shanghai, China

ABSTRACT: With the developments in technique, people are paying more attention to high speed communication, which has excellent prospects for the future. As is known, the next type of communication system rate can be up to 10 Gb/s. In this paper, communication quality measured by bit error rate is discussed. We will analyse the results of simulations. The probability density function is discussed and the divergence angle and transmission diameter are also analysed. We will further analyse the relationship between parameters can bit error rate. This can help us understand the character of high-speed communication performance.

Keywords: Space wireless communication; BER; Atmospheric turbulence effect; MSK scheme

1 INTRODUCTION

With the progress in space wireless communication, space systems have attracted increased attention due to their advantages, such as excellent modulation rate, less transmission consumption and so on (Wei et al., 2009; Andrews et al., 2001). With the developments in technique, people are paying more attention to high-speed communication, which has excellent prospects for the future. As we know, the next type of communication system rate can be up to 10 Gb/s. Until now, many studies have been carried out on increasing the quality of communication (Le et al., 2014; Yang et al., 2014; Ma et al., 2008). To be more specific, the speed of communication systems is usually less than 1 Gb/s (Kikuchi, 2012). However, the next type of communication system rate can be enhanced to about 10 Gb/s (Aamer et al., 2012).

In a wireless communication system, as the transmission laser propagates through the aerosphere in the process, it is affected by atmospheric turbulence, which not only causes deterioration in performance of the laser beam propagation but also jeopardises the physical field (Djordjevic & Stefanovic, 1999). The signal can be affected by intensity scintillation and beam wander. Based on this, we consider the intensity scintillation and beam wander effect and use the Minimum Shift Keying (MSK) scheme to analyse bit error rate performance in a communication system. In this paper, the communication quality, measured by bit error rate, is discussed. We will analyse the results

from simulations. The probability density function is discussed and the divergence angle and transmission diameter are also analysed. We will further analyse the relationship between parameters can bit error rate. This will help us to understand the character of high-speed communication performance.

2 THEORY

When it comes to a wireless system, the communication signal is modulated by the MSK scheme. Hence, the bit error rate performance can be shown as (Masa and Akira, 2014):

$$BER_m = 1/2 \, erfc(m/2\sigma) \tag{1}$$

where m *is* average value, σ is variance.

In order to enhance signal amplification due to the long transmission distance, we usually use an Avalanche Photodiode (APD) to enlarge the received slight signal. Thus, the parameters are (Campbell et al., 2014).

$$m = G \cdot e \cdot \left(K_s(I) + K_b\right) + I_{dc}T_s \tag{2}$$

$$\sigma^2 = (G \cdot e)^2 \cdot F \cdot \left(K_s(I) + K_b\right) + \sigma_T^2 \tag{3}$$

where G is the photomultiplier gain factor, F is the noise factor, $K_b = \eta \, I_b T_s/hv$ is the photon number, v is frequency, η is quantum efficiency, $\sigma_T^2 = 2\kappa_c TT_s/R_L$ is the thermal noise, T_s is bit time, $1/T_s$ is the speed, and the other parameters can be seen in Campbell et al. (2014).

In a ground-to-satellite downlink wireless communication, the BER of the system is affected by the scintillation and beam wander. The PDF can be expressed as (Shu & John, 2015):

$$P_w(I) = \int_0^\infty \frac{1}{\sqrt{2\pi\sigma_1^2(r,L)}} \frac{r}{\sigma_r^2} \exp(-r^2/(2\sigma_r^2))$$

$$\times \frac{1}{I}\exp\left(-\left(\ln\frac{I}{\langle I(0,L)\rangle} + \frac{2r^2}{W^2} + \frac{\sigma_1^2(r,L)}{2}\right)^2 / 2\sigma_1^2(r,L)\right)$$

(4)

where $\langle I(0,L)\rangle = \alpha P_T D_r^2/2W^2$ is the average intensify, W is the radius, D_r is the received diameter, P_T is the transmission power, and the other indexes can be found in Shu & John (2015).

In conclusion, the overall BER is (Masa, 2014):

$$BER = \int_0^\infty BER_m \cdot P_w(I)dI$$

(5)

3 SIMULATIONS

The evolution of PDF versus received optical power is shown in Figure 1. In the figure, all four curves gradually increase to a point and then decline when the received optical power ranges from 0 nw to 600 nw. In the region of divergence 30 μrad, the PDF reaches its maximum of 8.8*10e6 when received optical power is approximately 220 nw. By increasing the divergence angle from 30 μrad to 45 μrad, the highest values of PDF of the curves increase while the received optical power gradually declines. PDF reaches its peak at 14*10e6 when the divergence angle is 45 μrad and received optical power is around 100 nw. It indicates that PDF is able to increase when the divergence angle increases and received optical power declines.

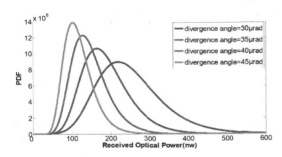

Figure 1. PDF versus power.

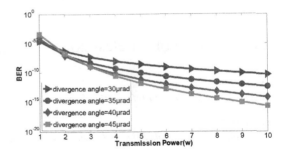

Figure 2. BER versus transmission power.

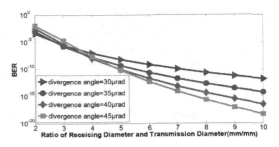

Figure 3. BER versus received ratio of receiving diameter and transmission diameter.

The evolution of BER versus transmission power is shown in Figure 2. As is shown in the figure, all four curves continually decline. In the region of divergence angle 30 μrad, the BER constantly reduces when transmission power ranges from 1 w to 10 w. Furthermore, in Figure 2, the highest BER is around 10e-5 when transmission power is 1 w. After increasing the divergence angle from 30 μrad to 45 μrad, the highest BER at the transmission power of 1 w does not change. However, the BER would decline when the divergence angle increases if the transmission power is fixed. This indicates that BER is more sensitive to transmission power when the divergence angle is higher.

In terms of specific communication system parameters, receiving diameter/transmission diameter versus BER is shown in Figure 3 with a different divergence angle. When we increase the ratio, the BER will correspondingly decrease in a different divergence angle. Thus, in order to increase the communication performance, the ratio should be controlled to a certain level. We can also figure out that the different divergence angles work when the ratio is larger than four.

4 CONCLUSION

In conclusion, when we take speed as 10 Gb/s, the bit error rate performance with consideration of atmospheric turbulence at various divergence angles is discussed. From the simulation results, we can see that the PDF is affected by the divergence angle. In the region of divergence 30 μrad, the PDF reaches its maximum of 8.8*10e6 when received optical power is approximately 220 nw. By increasing the divergence angle from 30 μrad to 45 μrad, the highest PDF values of the curves increase while the received optical power gradually declines. In the region of divergence angle 30 μrad, the BER constantly reduces when transmission power ranges from 1 w to 10 w. Furthermore, as seen in Figure 2, the highest BER is around 10e-5 when transmission power is 1 w. After increasing the divergence angle from 30 μrad to 45 μrad, the highest BER at the transmission power of 1 w does not change. When we increase the ratio, the BER will correspondingly decrease in a different divergence angle.

ACKNOWLEDGEMENTS

This work was financially supported by Beihang University fund.

REFERENCES

Aamer, M., Griol, A., Brimont, A., Gutierrez, A.M., Sanchis, P., & Hakansson, A. (2012). Increased sensitivity through maximizing extinction ratio of SOI delay-interferometer receiver for 10 G DPSK. *Opt. Express, 20*, 14698–14704.

Andrews, L., Phillips, R., & Hopen, C. (2001). *Laser beam scintillation with applications.* New York: SPIE Press.

Campbell, J.F., Lin, B., Nehrir, A.R., Harrison, F.W., & Obland, M.D. (2014). Binary phase shift keying on orthogonal carriers for multichannel CO2 absorption measurements in presence of thin clouds. *Opt. Express, 22*, A1634–A1640.

Djordjevic, I.B., & Stefanovic, M.C. (1999). Performance of optical heterodyne PSK systems with Costas loop in multichannel environment for nonlinear second-order PLL model. *J. Lightw. Technol., 17*, 2470–2480.

Kikuchi, (2012). Characterization of semiconductor-laser phase noise and estimation of bit-error rate performance with low speed offline digital coherent receivers. *Opt. Express, 20*, 5291–5302.

Le, S.T., Blow, K.J., Mezentsev, V.K., & Turitsyn, S.K. (2014). Bit error rate estimation methods for QPSK co-OFDM transmission. *J. Lightw. Technol., 32*, 2951–2959.

Ma, J., Jiang, Y.J., Tan, L.Y., Yu, S.T., & Du, W.H. (2008). Influence of beam wander on bit-error rate in a ground-to-satellite laser uplink communication system. *Opt. Lett., 33*, 2611–2613.

Masa, K. & Akira, M. (2014). Decision-directed Costas loop stable homodyne detection for 10 Gb/s BPSK signal transmission. *IEEE Photon. Technol. Lett., 26*, 319–323.

Shu, Z. & John, C.C. (2015). A SSP-based control method for a nonlinear Mach-Zehnder interferometer DPSK regenerator. *J. Lightw. Technol., 33*, 3788–3795.

Wei, C.C, Astar, W., Chen, J., Chen, Y.J., & Carter, G.M. (2009). Theoretical investigation of polarization insensitive data format conversion of RZ-OOK to RZ-BPSK in a nonlinear birefringent fiber. *Opt. Express, 17*, 4306–4316.

Yang, L., Gao, X.Q., & Alouini, M.S. (2014). Performance analysis of relay-assisted all-optical FSO networks over the strong atmospheric turbulence channels with pointing errors. *J. Lightw. Technol., 32*, 4011–4018.

Research on the key technologies for developing the internet of things

Caijuan Huang & Zhuohua Liu
Guangdong Jidian Polytechinc, School of Electrical Engineering, Guangzhou, Guangdong, China

ABSTRACT: Solving difficulties in kernel technologies is the critical issue in developing the Internet Of Things (IOT). The technologies related to IOT, such as RFID (Radio Frequency Identification Devices), wireless sensor network and semantic grid as well as by margining multiple key technologies to solve difficult applicatioin problems are researched in focus. The semantic grid integrates the machine readable of the semantic Web and the powerful computing capability of grid technology; the semantic P2P technology solves the communication bottleneck due to information overload. These make the deployment of sensor nodes easier and the sensor network can be expandable. Information semantic interaction and sharing in IOT are implemented by adopting ontology technology; this allows proper and prompt understanding of the information in IOT.

Keywords: Wireless sensor network; Semantic grid; Cloud computation; Internet of Things (IOT)

1 INTRODUCTION

The Internet Of Things (IOT) refers to various information sensing devices such as Radio Frequency Identification (Devices (RFID), infrared sensors, GPS, laser scanners, and various devices and the Internet combined with a huge network form. Its purpose is to let all the items be connected to the network together, to facilitate the identification and management. On all kinds of objects by means of electronic tags (RFID), sensor, through the two-dimensional code is connected with the wireless network interface, and to achieve intelligent objects, between people and objects of communication and dialogue, and between the object and the object of communication and dialogue. The Internet of things are widely used throughout intelligent transportation, environmental protection, government work, public safety, safe home, smart fire, industrial monitoring, elderly care and personal health and other fields.

Generally speaking, the technical architecture ofInternet of things can be divided into three layers: the perception layer, network layer and application layer. The sensing layer is composed of a variety of sensors and sensor gateways, including carbon dioxide concentration sensor, temperature sensor, humidity sensor, two-dimensional code label and RFID tag and reader, such as camera and GPS perception terminal. The perception layer is equivalent to the role of the human eye and skin nerve endings; it is the sources of the information networking, object recognition and its main function is to identify objects and collect information.

The network layer is composed of various private networks, Internet, wired and wireless communication networks, network management system and cloud computing platform, central nervous system; it is equivalent to the human brain, responsible for the transmission and processing of the information perception layer. The application layer is the interface between Internet of things and users (including people, organisations and other systems), which is combined with industry needs to realise the intelligent application of Internet of things.

As a kind of important network and Internet, industry experts predict that with the growing popularity of technology, mobile networks will develop into a trillion scale high-tech market for the Internet of things. It started late, before the two wave of information in the slow China, of great significance. This paper studies the key technologies of the Internet of things from the aspects of RFID, WSN and semantic grid, and adopts the method of fusing these key technologies to solve some practical problems.

2 KEY TECHNOLOGY OF THE INTERNET OF THINGS

2.1 Sensing technology

In the process of constructing the Internet of things, it is very important to realise the sense of perception.

Sensing technology can also be referred to as information acquisition technology; it is the basis for the realisation of the Internet of things.

At present, the information collection mainly uses electronic tags and sensors, etc. In the sensing technology, electronic tags are used for standardised identification to collect information, data acquisition and control equipment through the RFID reader, a two-dimensional code reader implementation.

RFID technology is a kind of communication technology, using radio signals to identify specific targets and to read and write data, established between the mechanical or optical contact without the recognition system and the specific goal; it is a kind of automatic recognition technology of non-contact. At present, the RFID technology in the field of manufacturing and assembly, aviation baggage handling, mail and express parcel, library management, document tracking and automatic recognition of moving vehicle logistics management has been, or is being, put into practice.

The concept of Internet of things, RFID tags are stored in the specification and interoperability of information through the wireless data communication network, collecting them automatically to a central information system, and achieving identification of goods. Then it realises the information exchange and sharing through the open computing network, as well as the transparent management of the goods. RFID has become a key technology in the application of Internet of things, and its application is still expanding. The most basic RFID system consists of three parts: the electronic tag (tag), the antenna (antenna) and the reader (Reader). Each tag has a globally unique Identification Number (ID). The ID cannot be modified and copied, so as to ensure its safety. The electronic tag is attached to the object to identify the target object; antenna for RF signal transmission between the reader and tag the ID, namely the label data information; the reader for reading (or writing) electronic label information equipment, can be designed to be portable or fixed.

At work, the RFID reader transmits a signal of a certain frequency through the antenna. When the RFID label into the field, by sending the obtained current energy product information stored in the chip, or by taking the initiative to send the signal of a certain frequency, then the reader reads and decodes the information and transmits the data to the central information system for the data processing.

The machine sensor is the awareness of the physical world. It is the "sense organ" to environmental parameters of the sensing information collection points. It can sense heat, light, electricity, sound, force and displacement signals, system processing, transmission, analysis and feedback to provide the original information for the Internet of things. With the development of electronic technology, the traditional sensor is miniaturised, intelligent, information and network gradually. At the same time, we are seeing a move from a traditional sensor to the intelligent sensor to the embedded Web sensor development process.

2.2 *Wireless sensor networks*

A number of communication, computing power of the sensor through the wireless connection, mutual cooperation, with the physical world to interact with each other to complete the specific application of the task, the composition of the sensor network (sensor network). Sensor network integrated sensor technology, embedded computing technology, distributed information processing technology and wireless communication technology, and other multi-disciplinary cross technology. In the sensor network, wireless communication independently constitutes a multi-hop self-organising system, providing data acquisition and processing network coverage of the sensing information, flexible monitoring. It deals with this information, and at the same time, with the transmission of information to the user, the possession of goods through the RFID networking equipment and Internet connection, and realises intelligent identification and management.

The main function of the sensor network is to collect the data of some kind of environmental things. The related research of sensor network includes three aspects: sensor, communication and Computing (including hardware, software and algorithm). On the one hand, a sensor network can be regarded as the Internet information, sharing function expansion to include integrated network information collection, information processing and information utilisation; the information collection function enhances the information processing process. On the other hand, a sensor network can be regarded as sensor nodes for the development of a new information processing network that has the interconnect structure.

In sensor networks, a large number of sensor nodes (node sensor) are scattered in a particular area, and are divided into different clusters (cluster) according to the geographical coordinates. Each cluster has a sensor node as the master node (master node), which is responsible for the cluster routing and other management. There is a host node to communicate with the Internet for such as the backbone network in all nodes (sink node). Compared with the traditional sensor, the sensor network is easy to deploy; that is, the sensor node position does not need to be determined in advance or carefully designed to allow arbitrary placement. At the same time, sensor network deployment and maintenance of low-cost, high-flexibility sensor networks consists of a large number of cheap nodes that can be placed in the physical phenomenon in the scope to obtain high accuracy observation; but this has a high price. A sensor network has a large number of redundant nodes. Even if some nodes fail, it will not affect the function of the whole system, so it has good robustness. In addition, the sensor network node also has the computing power, and can cooperate with each other to complete tasks which the traditional sensor cannot.

2.3 *Semantic grid and cloud computing*

Typically, the grid is considered to be the next generation of Internet technology, while the semantic Web is considered to be the next generation Web. Grid and semantic Web have many similarities, but their emphasis is different. The traditional meaning of the grid focuses on the computation, but the semantic Web is more inclined to reasoning and proof.

At present, the lack of machine readable and understandable data semantics, means lack of good cooperation between people and machine support. It is difficult for machines to process heterogeneous resources, realise resource sharing, and automatically generate knowledge according to the needs of users. While the semantic Web realises the sharing of resources (including hardware and software resources), but it is difficult to meet the growing computational requirements. While the semantic network has realised the sharing of resources (including hardware and software resources), it is difficult to meet the needs of KT benefits growth. The semantic grid describes all the resources, including the service using machine-understandable semantic information, to realise the interaction with no ambiguity. The development of the semantic grid cannot be separated from the development of the Web, semantic Web and grid. The grid is the improvement of Web in computing power, and the semantic grid is the extension of the semantic ability of the grid. From the other point of view, the semantic Web enhances the semantic function on the existing Web, and the semantic grid is the extension of the computing function of the semantic Web.

The emerging semantic grid, using the Internet and geographically widely distributed rules of resources (including computing resources, storage resources, bandwidth resources, software resources, data resources, information resources, knowledge resources, equipment and even people, generally represented by ontology (ontology is a kind of expression of resource specification)) together to make the network become a global information resource database. The information processing platform is used to provide integrated information services for users and Applications (computing, storage access, etc.). It can realise the comprehensive sharing and collaboration of network resources, and solve the problem of resource heterogeneity, distribution and dynamic change.

The semantic grid can be widely used in e-commerce and Internet of things. As a new interactive platform in the future, it has a strong effect, personalized, operational and interactive, the traditional plane and the Internet media have a strong impact. Grid technology has greatly improved the ability of mass information collection and classification processing. The development of the grid promotes the cloud computing (Computing Cloud) technology. Cloud computing will provide powerful computing capabilities for the Internet of things. Cloud computing is a distributed computing technology. The most basic concept is through the huge computing network which will automatically split into numerous smaller subroutines. Then the search and analysis in large systems composed of many servers will return the results back to the user. Through cloud computing technology, network service providers can be reached within a few seconds to deal with tens of millions or even billions of information, to achieve and "super computer" equally powerful network services.

The Peer To Peer (P2P) technology mainly refers to the information control technology which is formed by the hardware, and its representative form is the software. In a simple way, P2P makes it easier to communicate and share and interact more directly, eliminating the middlemen. Another important feature of P2P is to change the status of the Internet now as the centre of the Ethernet, to return to the non-centre, and the right to return to the user.

2.4 *Technology integration and intelligent technology*

The combination of RFID technology and sensor network technology is a development trend of the Internet of things in the future. Due to the poor anti-interference performance of RFID and the effective distance of generally less than 10 m, its application is limited. If the wireless sensor is combined with RFID, the effective radius of the former is as high as 100 m, the wireless sensor network is formed, and the application prospect of the wireless sensor network is immeasurable. Sensor networks generally do not care about the location of the node, so the nodes are generally not the overall use of the logo, and RFID technology for the node's logo has a unique advantage. Combined together, the network can mutually compensate for each other's shortcomings. The main focus will be concentrated on the data network, when there is a need for specific consideration to a specific node information, which can also use RFID functions to easily find the location of the node.

The Internet of things is made up of a large number of sensor network nodes. In the process of information sensing, it is not feasible to use individual nodes to transmit data to the sink node separately. Because there is a large amount of redundant information in the network, it will waste a lot of communication bandwidth and valuable energy resources. In addition, it will reduce the efficiency of the collection of information, the timeliness of the impact of information collection, and it is necessary to use the data fusion and intelligent technology to deal with it.

The so-called data fusion is the process of processing a variety of data or information, combined efficiently in-line with the needs of users of the data. Intelligent analysis and control information mass is relying on the advanced software engineering technology, mass storage and fast processing of all kinds of information on the Internet of things, and the results of various real-time feedback to the networking control unit.

Intelligent technology is to achieve a certain goal, by using knowledge analysis of the various methods and means. By implanting an intelligent system in the object, it can make the object have a certain intelligence, active or passive, to achieve communication with the user, which is also one of the key technologies of the Internet of things. According to the connotation of the Internet of things, we can know that the real things need to be aware, transmission, control and intelligent and many other technologies. The study of Internet of things will drive the whole industry chain or promote the common development of the industrial chain. Research and application of information technology, network communication technology, data fusion and intelligent technology, and cloud computing technology, will directly affect the development and application of the Internet of things. Only comprehensive research can solve these key problems; things can get rapid promotion, to benefit the human society, and provide good wishes to realise the wisdom of the earth.

In the P2P architecture, the information resources are oriented to the human rather than the machine. The information on the node is a chaotic data rather than knowledge, the ontology into the P2P, the formation of the semantic P2P. The proposed semantic P2P makes the information structured, easy to query and understand. Information is stored in the form of the body. Ontology is defined as the basic terms and relations that constitute the vocabulary of the relevant domain, and the definition of the rules of the extension of these terms by using these terms and relationships.

In addition, semantic P2P technology also solves the problem of the centralised architecture brought about by the difficulty to maintain. In a distributed environment, different nodes perform tasks such as processing and managing knowledge in the information exchange. Participants in the semantic P2P architecture, the management burden reduction, knowledge sharing and information retrieval are also simplified. Semantic P2P application of Internet of things can improve the speed of the network, to achieve resource sharing.

3 CONCLUSION

Internet of things is an important means of support from the manufacturing country to manufacturing powerhouse, and vigorously promotion of the Internet of things has important practical significance. In the advent of the information revolution, the Chinese government and enterprises should pay special attention to this opportunity, and strive to third wave as an opportunity to realise the rise of China's information industry, driven by the rapid development of China's economy. The focus of China's development of the Internet of things is to address the key technologies involved in the Internet, and the integration of these technologies to various industries. In this paper, a variety of technology characteristics are applied to the Internet of things, to lower the cost of the operation of the Internet of things. It has a strong use value.

REFERENCES

Atzoria, L. et al. (2014). The internet of things: a survey, *Computer Networks*, 54(15), 2787–2805.

Kevin, A. (2015). That 'Internet of Things' thing. *RFID Journal*, 22 July 2015.

Kyildiz, I. Su W. Sankara E. (2013). Wireless sensor networks: a survey. *Computer Networks*, 38(3), 393–422.

Automotive, Mechanical and Electrical Engineering – Liu (Ed.)

Research on analysis systems of gas sensors

Lixin Hou, Ji Li & Hongjun Gu
College of Information and Technology, Jilin Agricultural University, Changchun, China

ABSTRACT: It's very important to research gas-sensing analysis systems for the development of gas sensors. The early ceramic devices mainly based on semiconductor oxides have been widely used for commercialisation, while the gas-sensing analysis systems designed for the analysis of the ceramic devices cannot be appropriate for all of the novel devices. The new gas sensors based on the planar technology can make up for the deficiency in the fabrication process of ceramic sensors. The novel analysis systems exploited for planar devices will play an important role in the development of gas sensors. This paper mainly introduces several novel gas-sensing analysis systems.

Keywords: Gas sensors; Planar devices; Gas-sensing analysis systems

1 INTRODUCTION

Gas sensors are broadly employed in many applications including monitoring emissions from vehicles, detecting toxic gases, analysis for medical diagnosis, and quality control in the chemicals, food and cosmetics industries (Riegel et al., 2002; Temofonte et al., 1989). Early gas sensors were mainly sintering type devices based on semiconductor oxides. The ceramic tubes were coated with the materials by a grinding and mixing process (Li et al., 2005). The adhesives need be added in the poor adhesion of materials before coating, which limits the application of other novel materials. C making processes are also needed. Fortunately, all kinds of planar devices have been made by semiconductor processing technologies with the development of gas sensors (Chung W-Y. et al., 2000). The novel gas sensors are fabricated, such as field effect gas sensors, UV-activated gas sensors, micro- and nano-structured gas sensors (Lloyd et al., 2001; Ferrara, 2011). The early gas-sensing analysis systems designed for the ceramic devices have some limitations. So, the novel gas-sensing analysis systems based on planar devices have been emerging constantly. This paper mainly introduces several kinds of novel gas-sensing analysis systems.

2 PLANAR DEVICES

The traditional sensors based on ceramic tube can be directly converted to a commercial product with suitable film thickness and high consistency. However, such devices are not suitable to all sensing materials because of complex processes and grinding. The grinding processes may change the morphology and structure of materials and influence their sensitivity.

The gas sensors using semiconductor technology have many obvious advantages through daubing and growth on the substrate. The planar device is fabricated according to the following steps: (1) making planar electrodes; (2) directly growing or daubing on the substrate; and (3) connecting the probes. Au-doped In_2O_3 nanofibers are synthesised by electrospinning and calcination techniques. Micro-sensors are fabricated by spinning the nanofibers on Si-based substrates with Ti/Pt electrodes. A high sensitivity of 33 is observed to 1 ppm NH_3 at 160°C (Hou et al., 2015). Screen printing is a common method with several micrometres in minimum electrode spacing. Lithography electrode spacing can be reduced to a few microns, but lithography technology is complex, expensive and time-consuming. The Focused Ion/Electron Beam (FIB/EB) is used for micro-nano processing, which makes the micro-electrode. The processing method is convenient and quick with high precision, which is suitable for the development of micro-nano devices. The heating electrodes can be implanted during the production of the substrate using the new processing technology, however built-in heating electrodes are not practical. The main reason is that heating electrode has strict requirements on the thermal conductivity coefficient and the inflation rate at high temperature, which greatly increases the cost of the entire device. Therefore, the use of gas sensitive analysis systems with external temperature control, is a more practical and more economical way.

3 SEVERAL GAS-SENSING ANALYSIS SYSTEMS

3.1 *Chemical Gas Sensor-8 (CGS-8 series)*

The CGS-8 series were manufactured for ceramic devices by Beijing Elite Tech Co., Ltd, China. The series can meet the needs of a different range with low current, which avoids the damage of materials caused by self-heating. This system could change range automatically, which greatly facilitates the screening and testing of the device. The CGS-8 system is suitable for a resistance type gas sensor. The characteristics of the system are: resistance as the output; a range from 1 Ω to 500 MΩ; and a heating temperature from room temperature to about 400°C. The eight channels can be independently controlled, and the temperature of each channel is set by a heating current. Multiple devices can be measured simultaneously under the same temperature, they can also be tested independently under different temperatures. Test data can be transferred to an external computer via USB ports in real time. Sensitive properties of materials can be observed visually.

The CGS-8U system is suitable for a voltage type gas sensor. The properties of the system are: a voltage range from 100 μV to 1 V; a preamp output impedance from 1 Ω–1000 MΩ. In a nutshell, the CGS-8 series are the preferred systems for the analysis of ceramic devices.

3.2 *CGS-1TP series*

To meet the demands for the development of new gas-sensing devices, intelligent gas-sensing systems were fabricated including CGS-1TP (resistance type) and CGS-1TPU (voltage type).

The CGS-1TP system includes several modules such as temperature control, probe control, vacuum, distribution system, data acquisition and test software, as shown in Figure 1. The system cannot destroy the morphology and structure of the materials without grinding and roasting process, which measure the materials grown, assembled or coated on the substrate. The material and probe are observed by microscope in order to connect the probe and electrode accurately. The minimum adjustable distance of the probe is 5 microns. The gas-sensing properties of a planar device placed on the heating area are measured by adjusting the temperature. The CGS-1 TP system is able to meet testing requirements of low resistance, high resistance and wide resistance of materials. The range of resistance is from 1 Ω to 2000 MΩ, and the range of temperature is from room temperature to 500°C. Continuous measurements are carried out in the whole range. The system provides convenient conditions for the development of new types of gas sensors.

The output of CGS-1 TPU is a voltage signal in addition to measuring the voltage. The design is in good agreement with CGS-1TP.

Compared with the CGS-8 series, the CGS-1TP series are especially suitable for the analysis of sensitive gas-sensing materials, and the measurement of a variety of materials including powder, block, porous, thin film and thick film. Even micro-nano devices combined with the FIB techniques can be tested. In addition, the CGS-4 TPS series were designed with four groups of probes, except for keeping excellent performances of the CGS-1 TP series, as shown in Figure 2. The test speed is greatly accelerated because of the four devices tested simultaneously.

The ways of gas distribution for the CGS series are provided by Beijing Elite Tech Co., Ltd, China with the static distribution system including air chamber, injection drilling, heater, fan etc. The vacuum system can be connected to a dynamic distribution system by reserved air inlet and air outlet. Users can also customise the dynamic distribution system according to their needs. The static distribution system can be used to roughly analyse the property of quick response and screen a device. The dynamic distribution system is suitable for

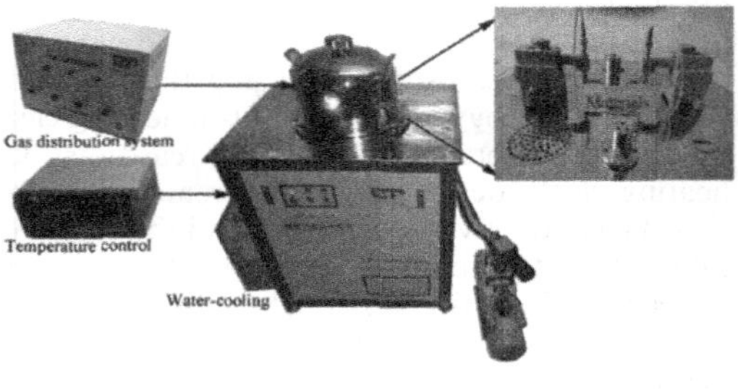

Figure 1. CGS-1TP system.

Figure 2. CGS-4TPS system.

measuring various gases with higher response value and less fluctuation, but the gas configuration process takes buffer time. Therefore, users can choose a distribution pattern according to their actual needs.

The gas-sensing analysis systems can realise diversified measurements (voltage, current, resistance), but the testing accuracy is not high, and it is difficult to achieve a wider range of measurements. The main direction of efforts will be more accurate, modes and integrated analysis in the future.

4 CONCLUSION

Over half of a century, ceramic devices based on semiconductor oxides have been commercialised. However, new planar devices based on semiconductor processing are more suitable to the demands of a variety of materials, such as powder, thin film and thick film. The novel analysis systems exploited for planar devices will play an important role in the development of gas sensors.

ACKNOWLEDGEMENTS

This work was supported by the 13th Five-Year Research Projects for Education Department of Jilin Province (No.2016168), the Scientific Research Staring Foundation for Jilin Agricultural University (No. 2015021). Beijing Elite Tech Co., Ltd, China is kindly acknowledged for introductions of gas-sensing systems.

REFERENCES

Chung, W.-Y., Lim, J.-W., Lee, D.-D., Miura, N., Yamazoe, N. (2000). Thermal and gas-sensing properties of planar-type micro gas sensor. *Sensors and Actuators B: Chemical, 64*, 118–123.

Ferrara V.L. (2011). Nanopatterned platinum electrodes by focused ion beam in single palladium nanowire based devices. *Microelectron. Eng., 88*, 3261–3266.

Hou, L., Gu, H., Li, S., Fang, D. (2015). Wireless NH3 sensors with high sensitivities fabricated from Au-doped In_2O_3 nanofibers. *Sensor Letters, 13*, 848–851.

Li, W.Y., L.N. Xu, Chen J. (2005). Co3O4 nanomaterials in Lithium-Ion batteries and gas sensors. *Advanced Functional Materials, 15*, 851–857.

Lloyd, A., Spetz, et al. (2001). SiC based field effect gas sensors for industrial applications. *Physica Status Solidi (a), 185*, 15–25.

Riegel, J., Neumann, H., Wiedenmann, H.M. (2002). Exhaust gas sensors for automotive emission control. *Solid State Ionics*, 152–153, 783–800.

Temofonte, T.A., Schoch, K.F. (1989). Phthalocyanine semiconductor sensors for room-temperature ppb level detection of toxic gases. *Journal of Applied Physics, 65*, pp. 1350–1355.

Automotive, Mechanical and Electrical Engineering – Liu (Ed.)
© 2017 Taylor & Francis Group, London, ISBN 978-1-138-62951-6

Research on clustering algorithms for wireless sensor networks

Huajun Chen & Lina Yuan
Institute of Big Data, Tongren University, Guizhou, China

ABSTRACT: With the emergence of cooperative communications for cellular networking, attention has been paid to clustering techniques, especially clustering algorithms. In this paper, we firstly simply introduce the state-of-the-art in Wireless Sensor Networks (WSNs) and clustering algorithms, and expound classical clustering protocols and the principle of the conventional K-means algorithm. Next, we propose a new clustering algorithm for WSNs based on the conventional K-means algorithm. Lastly, simulation studies show that our proposed clustering algorithm can adaptively get the number of clustering K, furthermore, the clustering result is more uniform, and the improved K-means algorithm is more suitable for high density network scenarios than the conventional one.

Keywords: Cooperative communications; Clustering techniques; K-means algorithm

1   INTRODUCTION

The purpose of Wireless Sensor Networks (WSNs) is real-time monitoring and control for the monitoring area, and makes a lot of sensor nodes with cheap, low power consumption and awareness, wireless communication and data processing functions deploy in monitoring area. WSNs are the networks where, according to the wireless routing protocol ad hoc network, the sensory data is sent to a gathering node after data fusion in a specified node through the ad hoc network between nodes, and is finally sent to the base station through the wired or wireless network, the users implement to real-time monitor for monitoring area by processing base station of data. Compared with the existing network, because the energy of the nodes is limited in WSNs, to reduce energy consumption and prolong the network life is a key issue in the routing protocol of WSNs, where at present it is mainly through clustering algorithms that the above problem is solved. Under the management mechanism of clustering topology, the remaining nodes of the network can be divided into two types: CHs and Cluster Members (CMs). Within each cluster, according to certain mechanism algorithms, select a node as the CH to manage or control the CMs within the whole cluster, coordinate among the CMs, and be responsible for information collection and data fusion processing in the intra-cluster and forwarding the inter-cluster.

Therefore, the recent advent of clustering has caused an upsurge in research activities for solving infrastructure deprived networks in various wireless communication systems, such as ad hoc networking and 5G networks. The strategy of clustering involves working nodes in virtual teams with one leader per team, in order that routing techniques take these teams as sole entities, thus decreasing a mass of network sources, destinations and possible routes, and hence increasing its stability.

There is an example of a clustering application in a multi-standard wireless network, as shown in Figure 1, where smart phones, wearable equipment, vehicular technology and cellular networks in general apply clustering to coordinate the nodes into groups to support efficient cooperative communication strategies.

A clustering algorithm is an effective management of network energy consumption, as well as one of the methods for improving the overall network performance. A clustering algorithm includes the election of Cluster Heads (CHs) and

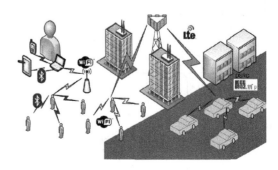

Figure 1.   Clustering for data aggregation in a multi-standard wireless network.

communication between the CHs and the base station. The existing election mechanism of CHs for a specific target to control is difficult to make a better trade-off among effectiveness in energy consumption, load balance and the number of CHs. According to the hop-count from the CHs to the base station, the cluster structure in general can be divided into single hop networks and multiple hop networks. A single hop communication mode intra-cluster needs all the nodes and the CHs to be in direct communication. Long distance communication can quickly consume the energy of the CHs, and lead CHs far away from the base station node to soon die, which reduces the network connectivity and decreases the network life cycle. While the multi-hop communication mode does not need CHs, it has the ability to directly communicate with the base station, which avoids the direct communication with the base station that takes a lot of energy. Of course, this increases the energy consumption of the CHs close to the base station. But compared with the single hop method of long distance communication, the multi-hop method can relatively reduce the network energy consumption.

The rest of this paper is organised as follows. In Section II we introduce the classical clustering protocols, and the K-means algorithm is described in Section III. Section IV proposes the improved K-means algorithm, and gives the numerical results compared with the conventional K-means algorithm. Finally, the paper is terminated with conclusions in Section V.

2 CLASSICAL CLUSTERING PROTOCOLS

The Low-Energy Adaptive Clustering Hierarchy (LEACH) algorithm is first suggested in the clustering routing protocol of WSNs. The basic idea is, through equal probability, to choose the CHs with random cyclic, equally distribute the energy load of the whole network to each node, so as to achieve the target of prolonging the network lifetime. When the energy appears inconsistent, it uses the counts of nodes becoming the CH as the selection standard, as shown in the formulation (1).

$$T(n) = \begin{cases} \dfrac{P}{1 - P * \left(r \bmod \dfrac{1}{p} \right)}, n \in G \\ 0, n \notin G \end{cases} \quad (1)$$

where P denotes the ratio of the number of CHs and the total number of network nodes, r denotes the beginning of the current rounds, and G denotes a node set without being a CH. Because the task of the CHs is more than that of the Cluster Members (CMs), and their energy consumption is bigger,

the selected CHs criteria of LEACH adapts the typical energy standards. But, in the LEACH algorithm, the threshold of the selected CHs does not take into account the energy factors and needs to assume that the initial energy of each node is equal, which is inconsistent with the actual situation.

The Deterministic Cluster-Head Selection (DCHS) algorithm improves the disadvantages of the LEACH algorithm by considering the energy factor with the threshold function of the selected CHs. The experiments show that the DCHS algorithm effectively increases the network lifetime by 20%~30% compared to the LEACH algorithm.

Heinzelman et al., based on the LEACH algorithm, present a LEACH-Centralised (LEACH-C) algorithm to resolve the problems that nodes, according to a random number, decide whether to be elected as the CH and per round the generated CH has no certain quantity, position and other issues, and greatly improves the generated quality of the CH. Because each node must periodically report its own energy, position and other information to the base station, the overhead of clusters is larger, and the network traffic, time delay and the probability of the LEACH-Fixed (LEACH-F) algorithm also centralises the CHs to produce an algorithm by the base station, being responsible for selecting CHs. Relative to the LEACH-C algorithm, the LEACH-F algorithm does not need to loop-construct new clusters per round, and reduces the overhead of the constructing clusters. However, because the LEACH-F algorithm does not dynamically process the joining, failing and mobility of the nodes, and increases the signal interference inter-cluster, it is not suitable for the real network application.

A Hybrid Energy-Efficient Distributed clustering (HEED) protocol was proposed by Ossama et al. in 2004. The selection of the CHs mainly draws on two parameters: primary and secondary. The primary parameter relies on the residual energy to randomly select the initial CHs set. The nodes with more residual energy will have the larger probability of temporarily becoming the CH, while whether this node becomes the CH is dependent on the residual energy and whether this is much larger than that of the surrounding nodes. Considering the intra-cluster communication overhead after clustering, the HEED takes the achievable average energy intra-cluster as measuring the standard of the communication cost intra-cluster.

3 K-MEANS ALGORITHM

The K-means is a typical clustering algorithm according to the distance to evaluate the similarity that, the higher the similarity, the nearer the distance of the object. The K-means divides the objects close to the location into the same cluster, so that the final result is to achieve a compact and independent cluster.

The K-means algorithm has two input parameters: the data object set and the number of clustering. The essence of the K-means algorithm is two process with recycle: First select the initial Cluster Head (CH), and allocate the remaining objects to the nearest CH. Then calculate the mean square error. If it does not achieve the minimum, it continues the calculation, the new CH is the average value of all objects in the intra-cluster, until the result of the clustering does not change. The concrete implementation process is as shown in Figure 2. Lastly, the achieved clustering result includes K clustering, the object layout of the single intra-cluster is closely, and different clusters are far apart and independent of each other. The K-means algorithm is suitable for a wide difference of intensive data sets between clusters.

The K-means algorithm has the following three disadvantages.

Firstly, the number of clustering K is an input parameter to be predetermined. In the K-means algorithm, the input parameter K needs to be predetermined. Under most conditions, we cannot evaluate the data set on the number of categories. This is the main deficiency in the K-means algorithm.

Secondly, the selection of the initial CH has an impact on the clustering results. The K-means algorithm firstly randomly selects K objects from the data set as the initial CH, and then loop-calculates the two-stage. The subsequent selecting CH is related to the initial CH, and a different initial CH may acquire very different clustering results. Therefore, the K-means algorithm has great dependence on the initial CH, and the selected initial CH should be a uniform distribution in the data set.

Lastly, the complexity of the repeated iteration time during two stages is higher. As can be seen from Figure 2, the K-means needs a continuous cycle object for classification and adjustment, and recalculating the CH. When the input data set is fairly big, the execution time of the algorithm is longer. Thus, the K-means algorithm requires an improvement according to the division of the big data set in order to reduce the complexity of time.

4 THE IMPROVED K-MEANS ALGORITHM

In the K-means algorithm, the number of clustering K needs to be predetermined, and then the selection of the initial CH has a severe effect on the clustering results. With regards to the above disadvantages, this paper proposes an improved K-means algorithm for the high density family base station network. The specific process flow is as shown in Figure 3.

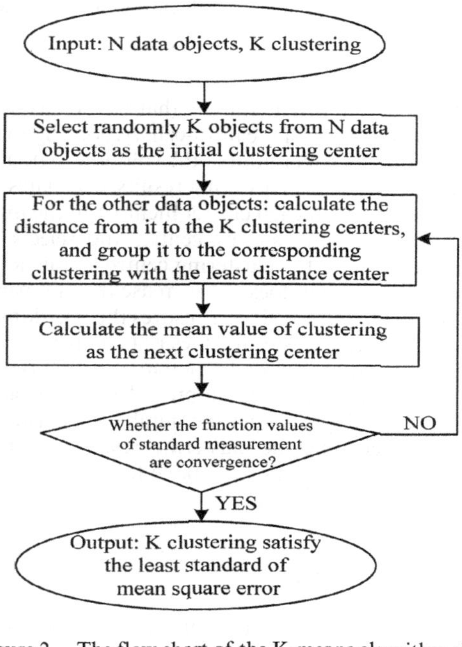

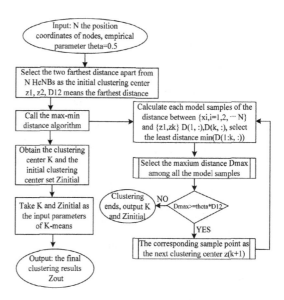

Figure 2. The flow chart of the K-means algorithm signal interference increase.

Figure 3. The flow chart of the improved K-means algorithm.

Based on the above algorithm flow in Table 1, we get the results of the two clustering algorithms through the simulation of MATLAB. In Figure 4, the black solid circle denotes the CHs, the black hollow circle denotes the Cluster Members (CMs), and the black solid line connects the CMs and their corresponding CHs. The parameters of the simulation use N = 40 and K = 4, i.e. when deploying 40 nodes within the scope of a small area with a radius of 200 metres, according to the improved K-means algorithm, the achieved number of clusters is 4. To make the results comparable, K = 4 is taken as an input parameter of the standard K-means algorithm, and at the same time the two schemes use the same position coordinates of the nodes. Note that the K-means takes the average value of all objects intra-cluster as the clustering centre of the new cluster, and the acquired clustering centre often conforms with the position coor-

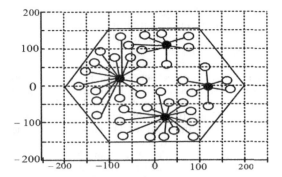

(a) K-means algorithm

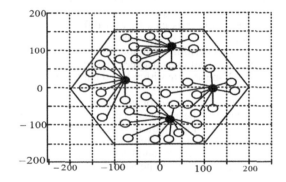

(b) The improved K-means algorithm

Figure 4. The comparison of clustering results with two algorithms.

dinates of the original node, that is, increases the number of nodes in the intra-cell. According to the problem, this paper adopts the solution that, after achieving the final clustering results, calculates the distance between each cluster member in the intra-cluster and the clustering centre, and selects the nearest node to the clustering centre as the actual cluster head, so it does not increase the number of nodes in the intra-cell, and ensures the positions of the nodes after clustering does not change.

As shown in Figure 4, compared to the typical K-means algorithm, the improved K-means algorithm can adaptively get the number of clustering K. Furthermore, the clustering results are more uniform, and it is more suitable for the high density network scenarios.

5 CONCLUSION

In a word, this paper has improved the traditional K-means algorithm. The numerical results show

Table 1. The improved K-means algorithm flow.

Input: N HeNBs's position coordinates set;
The empirical parameter $theta = 0.5$.

Implementation:

1) Select two with the farthest distance from N nodes as the initial centre z_1 z_2, D_{12} denotes the distance between them;

2) Call the max-min distance algorithm

   (1) Calculate each model samples the distance ($D(1,:)$ and ($D(2,:)$)) between $\{z_1, z_2\}$ and $\{x_i, i = 1, 2, \cdots, N\}$ one by one, and select the minimum distance $\min(D(1:2,:))$, the result is $1 * N$ matrix;

   (2) Select the maximum distance $D_{max} = \max(\min(D(1:2,:)))$ among the least distance of all model samples, if the maximum distance $D_{max} \geq theta \cdot D_{12}$, the corresponding sample point is taken as the third clustering centre z_3;

   (3) Repeatedly perform procedures (1) and (2), the new acquired clustering centre is parted to the clustering centre set, and the other calculated process does not change;

   (4) If it does not find the samples that satisfy the above demanded distance as the new clustering centre, the finding process of the clustering centre is over;

   (5) Output the CH set $Z_{initial}\{z_1, z_2, \cdots, z_K\}$, therein K is the number of clustering;

3) Execute K-means algorithm, $Z_{initial}\{z_1, z_2, \cdots, z_K\}$ as the input parameter;

4) Obtain the final clustering result $C = \{C_1, C_2, \cdots, C_K\}$, the CH set $Z = \{Z_1, Z_2, \cdots, Z_K\}$.

Output: K clustering centre;
The CH set $Z = \{Z_1, Z_2, \cdots, Z_K\}$;
The clustering result $C = \{C_1, C_2, \cdots, C_K\}$.

the superiority of our enhanced K-means algorithm to the reference algorithm. In future work, we will focus on applying it to different scenarios.

ACKNOWLEDGEMENTS

This work was supported by the Collaborative Fund Project of Science and Technology Agency in Guizhou Province Marked by the word LH on 7487 [2014], the National Natural Science Foundation of China (NO.61562703), and the project of education and cooperation for talent team word in Guizhou in 2015 (NO:[2015]67).

REFERENCES

Amis, A., et al. (2000). Max-min d-cluster formation in wireless ad hoc networks. *Nineteenth Annual Joint Conference of the IEEE Computer and Communications Societies Proceedings,* 1(20), 32–41, doi: 10.1109/INFCOM.2000.832171.

Jia, D., et al. (2016). Dynamic cluster head selection method for wireless sensor network. *IEEE Sens,* 16(8), 2746–2754, doi: 10.1109/JSEN.2015.2512322.

Klein, T.E. (2014). Energy Efficient Wireless Networks Beyond 2020. *//Proc of the Workshop on Research Views on IMT Beyond 2020,* 2(11). Geneva: ITU-R.

Lee, J.S., & Kao, T.Y. (2016). An improved three-layer low-energy adaptive clustering hierarchy for wireless sensor networks. *IEEE Internet Things J,* 1(9), 99–104, doi: 10.1109/JIOT.2016.2530682.

Lin, S., & Tian, H. Clustering based interference management for QoS guarantees in OFDMA femtocell. *//Proc of the WCNC,* 5(16), 649–654. Piscataway, N.J.

Parker, T., & McEachen, J. (2016). Cluster head selection in wireless mobile ad hoc networks using spectral graph theory techniques. *HICSS,* 7(24)2, 5851–5857, doi: 10.1109/HICSS.2016.724.

Sohn, I., et al. (2016). Low-energy adaptive clustering hierarchy using affinity propagation for wireless sensor networks. *IEEE Commun. Lett,* 20(3), 558–561, doi: 10.1109/LCOMM.2016.2517017.

Sucass, V., et al. (2016). A survey on clustering techniques for cooperative wireless networks. *Ad Hoc Network,* http://dx. doi. org/10.1016/j.adhoc.2016.04.008.

J. W et al. (2011). A cooperative clustering protocol for energy saving of mobile devices with Wlan and Bluetooth interfaces. *IEEE Trans. Mobile Comput,* 10(4), 491–504, doi: 10.1109/TMC.2010.161.

Automotive, Mechanical and Electrical Engineering – Liu (Ed.)
© 2017 Taylor & Francis Group, London, ISBN 978-1-138-62951-6

Research on real-time data centre reconstruction technology based on big data technology

Xianhui Li
NARI Group Corporation, Nanjing, China
China Realtime Database Co. Ltd., Nanjing, China

Sheng Zhou
State Grid Zhejiang Electric Power Research Institute, Hangzhou, China

Shengpeng Ji
NARI Group Corporation, Nanjing, China
China Realtime Database Co. Ltd., Nanjing, China

Gang Zhen
State Grid Zhejiang Electric Power Research Institute, Hangzhou, China

Yang He
NARI Group Corporation, Nanjing, China
China Realtime Database Co. Ltd., Nanjing, China

Wei Li
State Grid Zhejiang Electric Power Research Institute, Hangzhou, China

Yongsheng Wang & Weidong Tang
NARI Group Corporation, Nanjing, China
China Realtime Database Co. Ltd., Nanjing, China

Jun Chen & Ping Lou
State Grid Huzhou Power Supply Company, Huzhou, China

ABSTRACT: This paper presents a new method for the reconstruction of the real-time data centre of power grid enterprises based on the big data technology. Through the transformation of the real-time data centre data access interface, we can realise the real-time data access component based on Hbase data. Based on the Hbase table structure design and parameter tuning, we realise the standardisation of storage grid enterprises in the real-time data access interface. On account of the UAPI design of large data components with real-time data centre standard based on migration and smoothing the application of real-time data transition upper business centre. This method has been verified by the relevant power enterprises in the national Power Grid Corp, which confirms the feasibility of the method.

Keywords: Real-time data centre; Big data technology; Hbase; Real-time data management; Data access

1 INTRODUCTION

During the 12th Five-Year Plan period, power grid enterprises (such as the State Grid Corporation and the China Southern Power Grid Corporation) built a real-time data centre to realise the on-demand storage, integration, sharing and calcula- tion of real-time data generated by various business applications during the power production opera- tion, access to electricity information collection, SCADA, electric energy, power transmission line monitoring and other business systems real-time data, and to support a large number of business applications and real-time data access services.

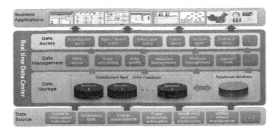

Figure 1. State Grid Corporation of real-time data centre architecture diagram.

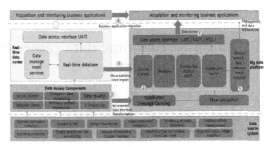

Figure 2. Overall structure diagram.

With the development of information technology, big data has become a hot topic in the field of data processing (Manyika et al., 2011; Mayer-Schnberger and Cukier, 2014; Tao et al., 2013). Grid enterprises, as information technology leaders, are also actively exploring big data based grid data processing, analysis, mining and application (Wang, J., 2015; Zhang et al., 2015; Zhu, Z., et al., 2015; Zhu, Y., et al., 2015; He et al., 2016; Pan et al., 2015). Time data and big data related technologies will have a greater value, as grid companies rely on big data, cloud computing and other new technologies to build a big data platform to support data storage, integration, calculation, analysis and mining strategic planning. Therefore, the existing real-time data centre needs to be reconstructed based on the large data technology. By integrating large data technology into the real-time data centre, it can optimise data integration, data storage, data calculation, data analysis and data service capability, and support business application construction.

2 REAL-TIME DATA CENTRE STATUS

Taking the State Grid Corporation real-time data centre as an example, it covers nine functional modules, such as data access, data processing, data quality and metadata management, and more than 180 sub-functional modules. Also the integration of 21 core business applications, such as grid operation monitoring and so on. The overall structure diagram is shown in Figure 1.

3 RECONSTRUCTION OF TECHNICAL ARCHITECTURE

According to the Ministry of Information's overall plan for the collection of monitoring data, based on big data platform technology architecture, real-time data centre component migration and transformation work includes six major works, as shown in Figure 2.

1. Data access transformation
Real-time data centre access to the original data components to transform, in support of real-time data centre access to data at the same time access to big data platform to complete the data access interface transformation to ensure data quality and effectiveness.

2. Data storage
Real-time data access requires high-time access and cross-section access. It requires comprehensive analysis of various real-time data usage scenarios to design a reasonable storage model to support efficient querying and fast storage of real-time data.

3. Query and sharing
In order to make real-time data, based on UAPI access to more than 40 business applications, smoothly transition to a big data platform, and to simplify the real-time data access complexity, the UAPI interface, based on the big data platform interface, is implemented to smooth the transition from the old system.

4. Historical data migration
Design and development of historical data migration tools, in order to complete the business system of real-time data storage of the migration of data from the real-time data centre migration to the big data platform, can also be migrated from the original business system database to the big data platform.

5. Real-time data management
The real-time data centre of the original real-time data management module to reconstruct and migrate to complete the migration and reconstruction for visualisation of real-time data management tools, data quality, access services, computing services and other modules.

6. Business application migration
Based on the real-time data centre unified promotion, self-built, personalised redevelopment and other business applications to migrate and transform, to achieve migration and reconstruction based on the large data platform.

4 TECHNOLOGY ARCHITECTURE

4.1 Data access transformation

For the analysis of the domain acquisition monitoring data that needs to be accessed in the existing collection and monitoring business system, the difficulties of data access are: the large number of systems, the large number of developers, the non-uniform data format and interfaces in various forms. Therefore, the workload for data access is huge. In a comprehensive analysis of various types of data sources, the data access programme mainly has the following two considerations:

1. Rebuild the access components of the original real-time data centre, upgrade it to the access module of the analysis domain, carry out the "once-send-twice-receive" operation through the reconstructed module, and provide data for the unified service data analysis domain.
2. Research and development of the new data access module, docking with the business system and sending the data to the analysis of the domain data centre.

The data acquisition access is shown in Figure 3.

At present, in the information communication department, the data access mode has been built mainly through the real-time data centre access to the corresponding monitoring data collection. The real-time data centre access components are shown in Figure 4.

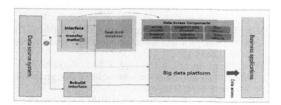

Figure 3. Data acquisition access in two ways.

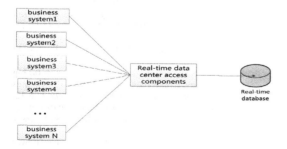

Figure 4. Real-time data centre access components.

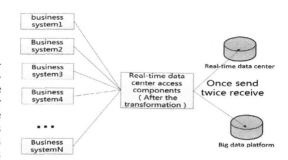

Figure 5. Transforming the real-time data centre access components.

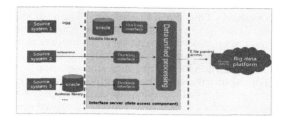

Figure 6. Data access components after refactoring.

Figure 5 shows the topology of transforming the real-time data centre access components and rebuilding the data access components.

The overall structure of the reconfigured data acquisition component is shown in Figure 6.

4.2 Data storage

The data storage part implements the distributed storage of the analysis of the domain acquisition monitoring data. In principle, it is suggested that the monitoring data be stored in the column database, and the recent data (currently half a day or one day) should be cached in the data cache to facilitate the application of real-time requirements. The real-time data storage diagram is shown in Figure 7.

Monitoring data acquisition has a large amount of data and the data has its fixed format. The query mode is based on batch query and cross-section query. In terms of data reading and writing, high data throughput is required for writing and low latency in data reading. Data storage design needs to have its cache mechanism to improve access efficiency, followed by the system having good scalability of the data in order to continue to grow. However, relational databases are also required to store some statistical information calculated by the flow computation or off-line calculation programme.

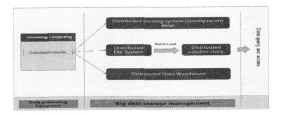

Figure 7. Real-time data storage diagram.

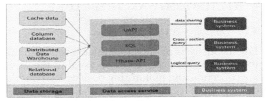

Figure 8. Application and query schematics.

1. Data storage policy

In order to improve the efficiency of data access, recent data should be stored in the data cache and long-term data should be stored in the "Column Database + Distributed File System". The relational database system mainly stores the computation result set that needs to be saved in the big data platform flow calculation or off-line calculation process.

2. Acquisition monitoring data storage model

For the distributed cache system, it is necessary to design its key-value model. For general acquisition of monitoring data, the recent data is written into the cache with the measured number + time stamp as RowKey and the measured value as Value.

The acquisition monitoring data goes into the system, and is finally stored in the "Distributed Column Database + Distributed File System". Taking into account that using the distributed column database to write the original interface throughput is not very satisfactory, in order to further improve the data loading efficiency, the data is distributed into the distributed column database through the distributed file system Batch Load. For the data storage model design of the distributed column database, we need to design it with the application scenario, such as the batch query business scene, so you can choose the point number + timestamp prefix for RowKey, column family for t, timestamp suffix for the column name, and the measured value for the corresponding column value.

4.3 Application query and sharing

For data application and sharing links, firstly achieve the built monitoring data acquisition system migration to the big data platform. Secondly, achieve data collection of the external unified shared services. Figure 8 shows the application to query real-time data.

For the part of the necessary monitoring data collection operation or the necessary subset of data sharing, to provide real-time data access in line with the national grid standard UAPI interface, big data platform common Hbase-API interface, similar to SQL JDBC data access interface, support for cross-section queries, batch queries and specific logic queries. Among them, through the API service, to achieve unified access to the column database and data cache system for the external business system, while achieving external systems and column database/data cache system decoupling.

5 CONCLUSION

This paper presents an overall solution to grid enterprise real-time data centre reconstruction, based on big data technology, through the real-time data centre access interface transformation, real-time data access Hbase, Hbase table structure through reasonable design and parameter tuning, and supporting real-time data fast access through the reconstruction, based on big data technology, for the real-time data centre UAPI to support the smooth application of the future transition.

REFERENCES

He, C., Wang, Y., & Wang, Y. (2016) Research on temperature monitoring system based on big data of power transmission network. *Zhejiang Electric Power*, 7.

Manyika, J., Chui, M., Brown, B., etc. (2011). Big data: The next frontier for innovation, competition, and productivity. *Analytics*.

Mayer-Schnberger, V., & Cukier, K. (2014). Big data: A revolution that will transform how we live, work, and think. *Information*, 17(1), 181–183.

Pan, J., Zhao, H., & Shi, Q. (2015). Application of multi —system monitoring and large data analysis in power supply repair. *Zhejiang Electric Power*, 8, 59–61.

Tao, X., Hu, X., & Liu, Y. (2013). A review of big data. *Journal of System Simulation*, (s1), 142–146.

Wang, J. (2015). Big data: The driving force of innovation development of power grid enterprises. *State Grid*, 12, 58–61.

Wang, Y., Tao, Y., Cai, Y., etc, Real—time Data Processing System for Smart Grid.

Zhang, D., Miao, X., Liu, L., etc, (2015). Research on big data technology development of smart grid. *Proceedings of the Chinese Society for Electrical Engineering*, 1, 2–12.

Zhu, Y., Huang, S., Cai, Y., etc, (2015). Research on coping strategies of regulatory management in times of power grid big data. *Zhejiang Electric Power*, 7, 30–32.

Zhu, Z., Wang, J., & Deng, C. (2015). Electric power research and design on big data platform. *Electric Power Information and Communication Technology*, 13(6), 1–7.

Automotive, Mechanical and Electrical Engineering – Liu (Ed.)
© 2017 Taylor & Francis Group, London, ISBN 978-1-138-62951-6

Research on a small wireless power transfer device via magnetic coupling resonant

Xin Liu, Ruqiang Dou, Yuben Yang, Ping Wu & Xuanyu Xiao
School of Mathematics and Physics, University of Science and Technology Beijing, Beijing, China

Sen Chen
Basic Experimental Center for Natural Sciences, University of Science and Technology Beijing, Beijing China

Jianing Sun
School of Automation and Electrical Engineering, University of Science and Technology Beijing, Beijing, China

ABSTRACT: The technology of wireless powers transfer is a potential method of electrical energy transmission in the future. We made a Wireless Power Transmission (WPT) device via strongly magnetic coupling resonant, and explained the multiple basic physical phenomena such as frequency splitting and skin effect etc. Furthermore, we used mathematical software to calculate our WPT model, which can offer us a theoretical support. We have set experiments courses for both undergraduates and junior students in Beijing.

Keywords: Wireless power transmission; Strongly magnetic coupling; Frequency splitting; Skin effect

1 INTRODUCTION

The technology of wireless power transmission has been humanity's dream for hundreds of years. In 1914, Nikola Tesla achieved Wireless Power Transfer (WPT) by using the electric charges in the earth's atmosphere to form an electric field with the ground (Tesla, 1914). However, he failed because of his funding. In 2006 an assistant professor, Marin Soljacic from MIT, came up with a way for wireless power transfer by magnetic coupling resonant (Kurs, 2007). Later they lighted a 60 W light-bulb from 2 m away with an efficiency of 40%. Research into wireless power transmission has become a frontier research topic again via strongly magnetic coupling resonant. In 2010, Japan Nagano Radio Company announced they managed 1 kW power transfer at a distance of 60 cm with a 90% transmission efficiency. In 2011, the research group of Fan Shanhui (Xiaofang Yu, 2011) from Stanford University made a theoretical calculation that 10 kW power transfer from 2 m away can keep its transmission efficiency up to 97%.

But in recent years, most research groups have focused their attention on high-power WPT. Our research group concentrates on several main factors, such as calculation of capacitance, inductance, quality factor and resonant frequency, easy design and making a small power wireless power transfer device via strongly magnetic coupling resonant, which is suitable for undergraduate-experiment teaching. At the same time, we conduct a number of experiments to explore frequency splitting and skin effect. Using mathematical software, we make a theoretical calculation about quality factor and resonant frequency, offering a simple way of making a wireless power transfer device for the intermediate and long distance.

2 INTRODUCTION OF WIRELESS POWER TRANSMISSION TECHNOLOGY

Compared with the different types of working principles, WPT can be divided into three main types (Xuezhe Wei, 2014) (Table 1).

Table 1. Comparison of different WPT technologies.

Principle	Distance	Maximum
Electromagnetic Induction	1 mm–20 cm	> 95%
Electromagnetic Coupling Resonant	10 cm–5 m	> 95%
Microwave or Laser	>5 m	

The first type is through electromagnetic induction. Its physical principle is similar to that of the transformer. It produces induced energy by Lenz's law, which has a high efficiency in transfer. However, the transmission distance is 1 mm–20 cm.

The second type is through electromagnetic coupling resonant. It can be mainly divided into two types: electric coupling and magnetic coupling. Generally, it transfers energy due to the same resonant frequency of resonators mutual coupling with each other in the form of a electromagnetic field in the space. Compared to the electric field, strong magnetic coupling is safer to explore as the magnetic field is safe for people inside. According to theoretical calculation, its transmission distance of strongly magnetic coupling is in the range of 10 cm to 5 m. Also, in terms of the characteristics of the magnetic field, it can penetrate the non-magnetic substance, which does not change the characteristic of the distribution of electromagnetic field. We think it is the theory model with the most potential.

The third type is electromagnetic radiation mode. Energy is transmitted through microwave and laser by using transmit and receive aerials, which can achieve a long-distance transmission. However, it has directivity and a very low efficiency.

3 KEY FACTORS OF SMALL WIRELESS POWER TRANSFER VIA STRONG MAGNETIC COUPLING RESONANT

Currently, the model of wireless power transfer device via magnetic coupling resonant is divided into two categories. The first is the high-power transfer device model with the basis of three-coil structure. Aanother is the four-coil (Kiani, 2011) structure and is suitable for small WPT. For small power transfer, the structure of four coils is more stable. As Figure 1 shows:

The driving coil carries a high-frequency signal transferred between the driving coil and the transmitting coil via mutual inductance. The transmitting coil and the receiving coil are coupled by the magnetic field. The coupling coefficient is denoted by k_s. Finally, mutual inductance happens between the receiving coil and the load coil to transmit energy to load. In terms of the strong coupling theory, transfer efficiency between two coupling resonant coils is determined by the degree of coupling of these two coils. We defined a ratio relation of strong coupling and dissipation capacity as follows:

$$\frac{|\kappa|}{\tau} = \frac{\omega/2L}{R/2L} = \frac{\omega M}{R} = \sqrt{\mu_0 \sigma \pi f} \cdot \frac{\pi N r^3}{(r^2 + d^2)^{3/2}} \cdot \frac{D}{2} \quad (1)$$

where μ_0 = Vacuum Permeability: $4\pi \times 10^{-7}$

σ = Electrical Conductivity of Copper Conductors: 5.998×10^7 S/m.

The strong coupling coefficient is κ. The dissipation capacity is τ. r is the radius of the coil. d is the distance between the centres of the device coils. D is the device coil diameter, and N is the number of coil turns. If the ratio between strong coupling and dissipation capacity is greater than 1, the coupling between those is very obvious, and the transmission efficiency will be greatly improved.

According to the strong coupling model, the higher the quality factor, the higher transmission efficiency will be, but the quality factor will also affect the waveform distribution of its frequency. In other words, it is possible to make the bimodal vicinity of frequency division become a single peak, making it difficult to control the resonant point.

According to the coupling model, the quality factor is calculated as follows:

$$\Gamma = \frac{R_0 + R_r}{2L} \qquad Q = \frac{\omega}{2\Gamma} \quad (2)$$

On the other hand, in the high frequency case, the coils will not only have ohm capacitances R_o, but also have a strong radiation resistance called R_r.

$$R_o = \sqrt{\frac{\mu_0 \omega}{2\delta}} \times \frac{l}{4\pi a} \quad (3)$$

$$R_r = \sqrt{\frac{\mu_o}{\varepsilon_o}} \left[\frac{\pi}{12} n^2 \left(\frac{\omega r}{c} \right)^4 + \frac{2}{3\pi^3} \left(\frac{\omega h}{c} \right)^2 \right] \quad (4)$$

We derived the relationship between quality factor Q and the resonant frequency f via Matlab, as is shown in Figure 2.

The quality factor Q and the resonant frequency f are mutually exclusive, so we can only choose an appropriate resonant frequency distribution, ensuring a high-quality factor. At the same time, it provides the designer with a good theoretical basis.

According to the electromagnetism, the capacitors and inductors of the coils are calculated as follows:

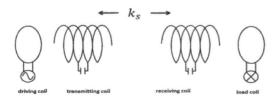

k_s

driving coil   transmitting coil   receiving coil   load coil

Figure 1. Four-coil model.

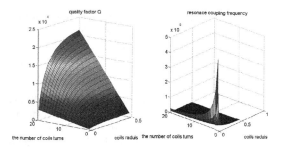

Figure 2. The relation between Q and f.

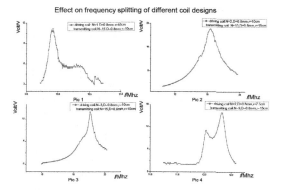

Figure 3. Data of frequency splitting of different designs.

$$L = \frac{\mu_0}{4\pi|I_0|^2}\iint drdr' \iint \frac{J(r)\cdot J(r')}{|r-r'|} \quad (5)$$

$$\frac{1}{C} = \frac{1}{4\pi\varepsilon_o|q_o|^2}\iint drdr'\frac{\rho(r)\rho(r')}{|r-r'|} \quad (6)$$

where $J(r)$ = spatial current $\rho(r)$ = charge density

For the phenomenon of frequency division in the transfer process, we conduct an experimental exploration. In the experimental range, we found an appropriate resonant frequency where the device has a certain bandwidth at a specific resonant frequency. As a result, it is easier to wirelessly transmit energy in an effective range.

According to (Pic 1), (Pic 2) and (Pic 3), we find that the number of turns of the driving coil has a certain effect on the resonant frequency of the system. And the more the turns, the less obvious the phenomenon of frequency division is. According to (Pic 2) and (Pic 4), it happens obvious frequency division when the radius of the driving coil is half of the transmitting coils. In our view, it is because most of the magnetic field lines of small coils are absorbed by the larger one and the conversion efficiency is increased. A large number of experiments show that the ratio of 1:2 provides the best

transmission effect. We observed a large transmission voltage between the two peaks and on both sides when it exhibits a certain frequency division characteristic, and that it provides a good bandwidth for device making. The bandwidth generated by frequency division is large due to high frequency, so we can find a resonant frequency with a reasonable value in the instrument error range to WPT.

4 THEORETICAL ANALYSIS AND ACHIEVEMENT

After theoretical analysis, we use copper conductors to spiral two same coils as our device coils with radius 10 cm, height 10.5 cm and wire diameter 0.1 mm. We calculate that the theoretical resonant frequency is 10.5 MHz. The ratio model of strong coupling to dissipation capacity is showed in Figure 4.

If the dissipation part is less than the coupling part, we can receive energy from the receive coil. Under our theoretical resonant frequency 10.5 MHz, we can predict that the distance of our model with the most effective transmission is about 100 cm.

When resonant frequency is an independent variable, as Figure 5 shows, we can enhance the transmission distance by improving resonant frequency, which provides us with an effective design method of our model.

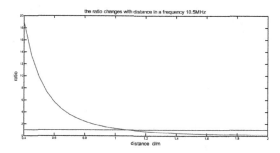

Figure 4. The ratio changes with distance.

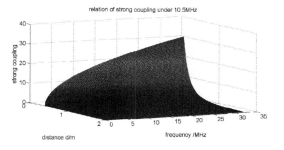

Figure 5. 3-Dimension ratio changes with distance.

In terms of our calculation above, we make a wireless power transmission device via magnetic coupling resonant. We can achieve wireless charging in a cell phone 70 cm away and illuminate a light-bulb at 98 cm away, as shown in Figure 6.

Two certain resistances are connected to separately driving and load coils. A diode is used to as a signal reflecting the energy transmission. We use an oscilloscope to measure the volts of the driving coil and load coil. According to the calculating formula:

$$P_1 = \frac{U_1^2}{R} \quad P_2 = \frac{U_2^2}{R} \quad \eta = \frac{P_1}{P_2} \times 100\%, \quad (7)$$

we can get a transmission efficiency distribution as shown in Figure 7. We can discern when the transmission distance is less than 70 cm that the transmission efficiency is above 80%, which can meet the basic requirement of the experiment.

When the receive coil is 100 cm away, the transmission efficiency is about 40%, so it causes a lot of energy waste. For the sake of keeping our power amplifier safe, we do not improve the output power. We therefore cannot ignite a light-bulb at that distance or charge a cell phone. However, we still can detect a voltage signal on the oscilloscope.

As a result of high-frequency alternating current, the whole device has a huge impact due to skin effect. And the formula of skin depth can be deduced due to Maxwell's equations.

$$\delta = \sqrt{\frac{2}{\omega\mu\sigma}} \quad (8)$$

In our experiment, we use two wires with different lengths connected to the two same light-bulbs. The longer one is darker than the shorter one. It can be explained by the skin effect, which plays a huge impact on the transmission efficiency.

5 CONCLUSION

We focus on several important aspects of the magnetic coupling resonant wireless power transmission model such as calculation of capacitance, inductance, quality factor and resonant frequency. We use the ratio of strong magnetic coupling to dissipation capacity to predict how far our device can transfer energy. Furthermore, we conduct experiments to explore the effect of resonant frequency between the driving coils' turns, size and transmitting coils' size, and we can observe skin effect using two different lengths of wires. Eventually, we make a small power wireless transmission device. We have opened experiments for undergraduates in University of Science and Technology Beijing (USTB) supported by Physics Experiment Center USTB. In addition, this device has become a sub-item course of Opening course held by Beijing Municipal Commission of Education. We have now set experiment courses for more than 1,000 junior students in Beijing.

ACKNOWLEDGEMENTS

We acknowledge help from Basic Experimental Center for Natural Science of USTB, and financial support from the Undergraduate Student Training Project of USTB.

Figure 6.   The real device.

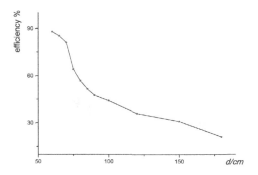

Figure 7.   Efficiency changes with distance.

REFERENCES

André, K. (2007). Wireless power transfer via strongly coupled magnetic resonances. *Science, 317*(5834), 83–86.

http://www.hqew.com/tech/news/181634.html

Kiani, M. (2011). Design and optimization of a 3-coil inductive link for efficient wireless power transmission. *IEEE Transactions on Biomedical Circuits and Systems, 5*(6), 579–591.

Tesla, N. (1914). Apparatus for transmitting electrical energy. U.S. patent number 1119732.

Xiaofang, Yu (2011). Wireless energy transfer with the presence of metallic planes. *Applied Physics Letters, 99*(214102).

Xuezhe, Wei (2014). A critical review of wireless power transfer via strongly coupled magnetic resonances. *Energies, 7*(4317).

Automotive, Mechanical and Electrical Engineering – Liu (Ed.)
© 2017 Taylor & Francis Group, London, ISBN 978-1-138-62951-6

Simulation and verification technology of shortwave communication based on OPNET

Xiongmei Zhang, Zhaoxiang Yi & Heng He
Xi'an Research Institute of Hi-Tech, Hongqing Town, Xi'an, China

Jing Wan
Equipment Academy of the Second Artillery Force, Beijing, China

Yue Ren
Unit 96819, Beijing, China

ABSTRACT: Aiming at the influence of jammers on shortwave communication, a simulation and verification method of shortwave communication based on OPNET is proposed. Analysing the working principle of shortwave communication, the node model of shortwave communication based on OPNET is established, and the module nodes are interconnected by modular programming. Accordingly, the shortwave communication scenario, based on the node model, is constructed to perform the commutation process as well as to verify the influence of the interference on shortwave communication, and validate that the average bit error changes under different conditions. The experimental results show that the proposed model can effectively validate the influence of interference on shortwave communication, and provide an effective basis for shortwave communication performance evaluation.

Keywords: Communication simulation; Shortwave communication; OPNET modelling

1 INTRODUCTION

Shortwave communication is a main means of long-range communication, as a result of the use of the sky wave propagation and ground wave propagation of communication, with a short communication distance, survivability, network flexibility and so on. Although one of the earliest means of communication used, because of strong anti-interference and easy to use, in the information warfare has been widely used (Yang & Wang, 2015).

The existing research mainly focuses on the application of shortwave communication and system design. Zuo has realised the simulation of an electromagnetic environment by introducing measured radio waves and local background noise, combined with the simulation of the signal-to-noise ratio of the modern to realise the systematic error code emulation, the two combining to form the system physical layer simulation basic platform (Zuo, 2014). Zhang detailed and analysed the principle of the shortwave OFDM communication system, proposed the method of realising OFDM modulation by DSP, established the basic frame model and studied the key technology (Zhang, 2006). Zhang put forward the proposal that shortwave communication can be implemented on the range of 0–200 km in a very wide frequency band without any interruption, when

graded lines are used in the antenna for an impedance match (Zhang, 2010). Yin and Jia put forward the proposal that fountain codes can be used in shortwave communication and also designed the shortwave communication system based on Raptor codes. This system can improve the transmission efficiency and reliability of shortwave communication due to the rateless property of fountain codes (Yin & Jia, 2011). On the other hand, in shortwave communication anti-jamming work is relatively scarce.

To this end, this paper focuses on the shortwave communication interference problem, the use of the OPNET simulation platform to study the performance of shortwave communication parameters and to achieve the communication network. The simulation results show that the communication is affected by interference. The experiment results provide a new modelling method and reference for shortwave communication and networks.

2 FUNDAMENTALS OF SHORTWAVE COMMUNICATION

The shortwave propagation mode is divided into ground wave and sky wave. Radio along the earth's surface is called the wave propagation, while waves directed to the sky through the ionosphere then

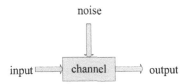

Figure 1. Shortwave channel model.

back to the ground are known as the sky wave. The shortwave channel model is shown in Figure 1.

The output of the model channel is shown as follows:

$$y(t) = k(t)x(t) + N(t) \tag{1}$$

In this model, the influence of the channel on the channel can be divided into two parts, one part is multiplied by $k(t)$ on the input signal, called multiplicative interference, another part is a certain noise interference $n(t)$ added to the input signal. For shortwave communication, in general, multiplicative interference can be divided into two quantities, the slow fading and the fast fading, respectively used to describe the long-term and short-term changes that can be described as a product of the two terms.

$$k(t) = k_L(t) \times k_S(t) \tag{2}$$

In the above equation, $k_L(t)$ represents the change of the signal amplitude in the slow fading, $k_S(t)$ represents the change of the signal amplitude in the fast fading.

3 MODELLING SHORTWAVE COMMUNICATION BASED ON OPNET

3.1 Modelling shortwave communication node

OPNET provides a comprehensive development environment that supports communication networks and discrete system modelling. Through the simulation of discrete events, the behaviour and performance of the model are analysed. Based on the OPNET network simulation, modelling is divided into a three-tier modelling mechanism, namely, network modelling, node modelling and process modelling. Network modelling mainly describes the topology of communication networks, node modelling implements the internal construction of the communication node itself, and process modelling describes the detailed process of network protocols, mechanisms and information processing (Li, 2004).

The antenna transceiver consists of 14 pipeline stages. The radio link is a broadcast medium, and each transmission may affect multiple receivers in the entire network model. The radio links

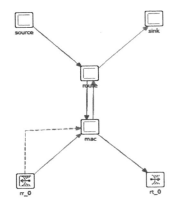

Figure 2. Normal node model diagram.

of each receiver also exhibit different behavioural characteristics. Therefore, most pipe stages must be performed according to the receiver whenever transmission is performed.

According to the OPNET three-layer modelling mechanism, we simulate 14 normal nodes and one interference node, and run in three different interference scenarios. Through the final statistical data, we get the influence of different interferences on shortwave communication. The network uses the form of self-organisation, all nodes are peer-to-peer, with the same network model, as shown in Figure 2. Node protocol is divided into four layers, namely the application layer, routing layer, data link layer and physical layer. Source module and sink module correspond to the application layer, route module corresponds to the routing layer, and mac module corresponds to the data link layer. The function of each module is as follows:

3.1.1 Source module

The source node generates services according to the Poisson distribution, as shown in Figure 3. The module generates traffic according to the Poisson distribution. The arrival interval of the packets follows an exponential distribution with mean X, where the value of X in the simulation is $1s$.

3.1.2 Sink module

As shown in Figure 4, the sink module is used to destroy and release the memory after the destination node receives the service. The sink module only contains INIT and DISCARD, two state machines. Only the discard state of the non-mandatory state for the destruction of the package. The sink module is responsible for destroying packets received from the input stream and returning a series of statistics about the packets.

3.1.3 Route module

The route module is used to realise packet routing, as shown in Figure 5.

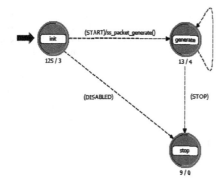

Figure 3.   Source module.

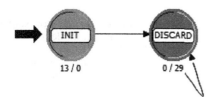

Figure 4.   Sink module.

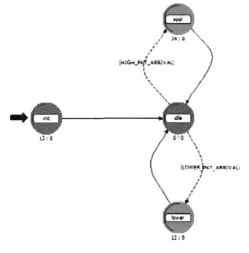

Figure 5.   Route module.

3.1.4 *Mac module*

Implementation of the slotted Aloha protocol, mac module, which is shown in Figure 6, is used to achieve multiple access and channel sharing. The slot size is equal to the frame length divided by the physical layer data transfer rate.

3.1.5 *Rt-0/Rr-0 module*

This module is a wireless transceiver, which calculates the transmission time, propagation time and SNR and frequency matching.

3.1.6 *Interference node*

This module is the interference packet generation module, including the transmitter and antenna.

3.2 *Modelling shortwave communication scenario*

The simulation scenario is shown in Figure 7. The network adopts the hierarchical structure and divides it into three layers, among which nodes 1 and 2 are the first layer, nodes 3~6 are the second layer, and nodes 7~14 are the third layer. Each node can send or receive data packets. The first layer node and the third layer communication must be relayed through the node of the second layer.

Shortwave communication will be subject to different types of interference, which may be long distance or may be close. Considering these situations, this chapter mainly simulates and analyses shortwave communication in the cases of interference and summarises the simulation structure. In Figure 7, the jammer is the interference node.

In the selection of simulation scenarios, before running the simulation, the statistical variables were selected and statistically analysed by the pipeline node model, including receiving power, background noise, interference noise, SNR, error bit and error correction ability.

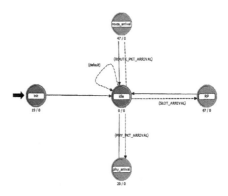

Figure 6.   Mac module.

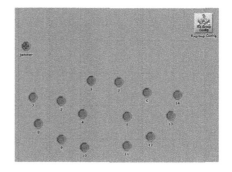

Figure 7.   Interference simulation scenarios.

213

4 SIMULATION AND ANALYSIS

The data obtained through the operation of simulation from first level of nodes 1 and 2 is shown in Figure 9.

As shown in Figure 9a, without interference, the frame error rates from nodes 1 and 2 rise rapidly in the first minute, respectively, because the network simulation starts running. A minute later the frame error rate begin to slowly rise, and tends to be gentle between 0.24 and 0.26. However, Figure 9b shows that under interference, the frame error rates not only begin to be gentle between 0.26 and 0.30, but also their slope is greater, so this phenomenon indicates that they change largely in the unit time.

Through the simulation of the shortwave communication, it proves that the interference which leads the error rate, the number of retransmissions to have a greater change on shortwave communication is more obvious.

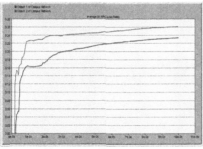

Figure 8.   Select the observation variable.

5 SUMMARY

To verify the influence of jammers on shortwave communication, a simulation model of shortwave communication based on OPNET is proposed, in which the module nodes are interconnected by modular programming. The shortwave communication network is constructed to perform the shortwave communication and validate the influence of the interference on the shortwave communication. When the communication network is affected by interference, the entire network will be seriously affected, and frame error rate, packet loss, and network delay have a serious fluctuation. The above simulation provides a theoretical basis for effectively guaranteeing the connectivity and reliability of shortwave communication.

REFERENCES

Li, M. (2004). *OPNET Network Simulation*. Beijing: Tsinghua University.
Yang, Z.Z. & Wang, L.W. (2015). The development trend of shortware communication and application in foreign studies. *Computer Engineering and Applications*, 51(S1), 179–184.
Yin, R.N. & Jia, K.B. (2011). *A shortwave communication system based on fountain codes*. The Second International Conference on Theoretical and Mathematical Foundations of Computer Science. Singapore, Singapore, 5–6 May 2011. Berlin: Springer.
Zhang, D.B. (2010). *Plannar Spiral Antennas Applied on Shortwave Communication*. Beijing: Beijing University of Posts and Telecommunication.
Zhang, W.G. (2006). *Simulation and Application of Military Communication Network based on the OPNET*. Xi'an: Xi'an University of Electronic Science and Technology.
Zuo, W. (2014). Review on pivotal technology and development of HF communication system. *Communication Technology*, 47(8), 847–853.

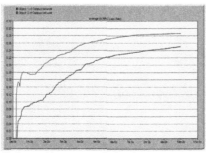

(a) Average bit error rate without interference.

(b) Average bit error rate under interference.

Figure 9.   Comparison of simulation results under different conditions.

Automotive, Mechanical and Electrical Engineering – Liu (Ed.)
© 2017 Taylor & Francis Group, London, ISBN 978-1-138-62951-6

The application of time-frequency analysis method in frequency hopping signal analysis

Qi Wu, Zhiliang Tan & He Shang
Ordnance Engineering College, Shijiazhuang, China

ABSTRACT: To solve the problem of frequency hopping signal detection and investigation, first the Short-Time Fourier Transform (STFT), Gabor transform, Wigner Distribution (WVD), Pseudo-Wigner Distribution (PWVD), Smoothed Pseudo-Wigner (SPWVD), Wavelet Transform (WT) and other common non-stationary random signal analysis methods are introduced. Then, the frequency hopping signal detection by these methods is simulated on computer, obtaining the map of each method. The advantages and disadvantages of every method when applied to the analysis of frequency hopping signal are obtained, providing the foundation for choosing the frequency hopping signal analysis method.

Keywords: Frequency hopping signal; Time-frequency analysis; MATLAB simulation; Electronic Warfare; Signal Detection

1 INTRODUCTION

The frequency hopping communication system has anti-interference ability, strong anti-interception capability, etc. It is therefore is widely used in the military communication field. With the development of electronic warfare technology, more information about frequency hopping signals needs to be acquired to come up with a better method for interference and jamming.

Time-frequency analysis is a commonly used method in time-varying non-stationary random signal description. The time domain information and frequency domain information of signals will be mapped to a two-dimensional plane; thisovercomes the disadvantage that the "positioning" of single time and frequency domain description method is not strong. The frequency hopping signal is a typical signal which is discontinuous and non-stationary in the time domain. Therefore, the time-frequency analysis method will be applied to analyse the frequency hopping signal to get more time-frequency information.

2 COMMONLY USED TIME-FREQUENCY ANALYSIS METHODS

2.1 Short-Time Fourier Transform

2.1.1 STFT of the continuous signals

Short-time Fourier (STFT) is that using a sliding window function to cut the signal. Then we can get the Fourier transform of the signal at different times:

$$STFT_S(t, f) = \int_{-\infty}^{+\infty} s(\tau)h(\tau - t)e^{-j2\pi ft}dt \qquad (1)$$

2.1.2 STFT of the discrete signals

The STFT of discrete signals via short-time Fourier is:

$$STFT_S(n, f) = \prod_{-\infty}^{+\infty} s(m)h(m - n)e^{-j2\pi fm} \qquad (2)$$

STFTS (t, f) is for the STFT of the signal; h(t) is for the window function.

2.2 The Wigner distribution

2.2.1 The Wigner distribution of continuous-time signal

The Wigner distribution of continuous-time signal is described:

$$W_x(t, \omega) = \int_{-\infty}^{+\infty} x(t + \tau/2)x^*(t - \tau/2)e^{-j\omega\tau}d\tau \qquad (3)$$

2.2.2 The Wigner distribution of discrete-time signal

The Wigner distribution of discrete-time signal is described:

$$W_x(n, \omega) = 2\sum_{k'=-\infty}^{+\infty} x(n + k'/2)x^*(n - k'/2)e^{-jk'\omega} \qquad (4)$$

W$_x$(t, w) is for the WVD of the signal.

2.3 Gabor transform

Gabor transform is a method which, similar to Fourier transform analysis, constructs a group locating easily in time and frequency domains.

2.3.1 The Gabor expand of the continuous signal

The Gabor expand of continuous signals is expressed as:

$$x(t) = \sum_n \sum_k G_x(n, k)g_{n,k}(t) \qquad (5)$$

2.3.2 The Gabor expand of the discrete signal

$$x(k) = \sum_{m=0}^{M-1} \sum_{n=0}^{N-1} G_x(n, n)g_{m,n}(k) \qquad (6)$$

$G_x(n, k)$ is the coefficient of the Gabor expand of the signal; $g_{m,n}(k)$ is the Gabor group.

2.4 Wavelet transform

The kernel function of the Wavelet transform is wavelet function, which is localised in time domain and frequency domain, and is known as "mathematical microscope". Haar wavelet, Mexican hat wavelet, Morlet wavelet, Daubechies wavelets and Symlets wavelet are common wavelet functions.

2.4.1 The wavelet transform of continuous signals

$$Wf(a,b) = \frac{1}{\sqrt{a}} \int_{-\infty}^{+\infty} f(t)\psi^*\left(\frac{t-b}{a}\right)dt \qquad (7)$$

2.4.2 The wavelet transform of discontinuous signals

$$c_{j,n} = <f(t), \psi_{j,n}(t)> = \int_{-\infty}^{+\infty} f(t)\psi_{j,n}^*(t)dt \qquad (8)$$

$Wf(a, b)$, $c_{j,n}$ is for the wavelet transform of signals; $\psi(t)$ is for the kernel function of wavelet transform.

3 SIMULATION RESULTS AND ANALYSIS

First of all, the frequency hopping signal is simulated on Matlab. The specific parameters are as follows:

The sampling rate: The sampling frequency is $f_s = 500$ KHz, the frequency hopping collection is {25, 100, 125, 75, 175, 150, 200, 500} kHz, the frequency hopping rate is 2000 H/s, the frequency hopping period is 0.5 ms, the modulation scheme is

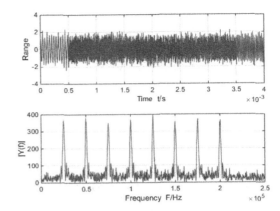

Figure 1. The waveforms of the frequency hopping signal in time and frequency domain.

2FSK, and the channel is 10 dB Gaussian white noise channel. The waveform of frequency hopping signal in time and frequency domain is shown Figure 1.

3.1 The STFT of frequency hopping signal

The frequency hopping signal is analysed by STFT transform in three conditions whose hamming window length is 1/4, 1/16 and 1/64 of the data:

The conclusion through simulation is that STFT can be used to describe the time-frequency information of the frequency hopping signal. However, the effect of analysis is related to the choice of the window length. The Figure 2(a), (b), (c) is the result when the window length is 1/4, 1/16 and 1/64 of the data in turn. The conclusion is as the window length is bigger, the frequency domain resolution is lower, but the time domain resolution is higher, and vice versa. So when STFT is applied to analyse the frequency hopping signals, we should select the appropriate window based on the actual situation to achieve the desired purpose.

3.2 The Winger-Ville distribution of frequency hopping signal

The WVD and its common improved methods including Pseudo-Wigner (PWVD) and smoothed Pseudo-Wigner (SPWVD) is respectively used to analyse the frequency hopping signal. The simulation results are as follows:

Figure 3(a) is the time-frequency map of the signal via WVD. The conclusion is that the frequency concentration is better but there is serious cross-term interference which seriously affects the analysis of the signal. When the plus smoothing window method is used, Figure 3(b) is obtained. The PWVD can reduce the interference of the

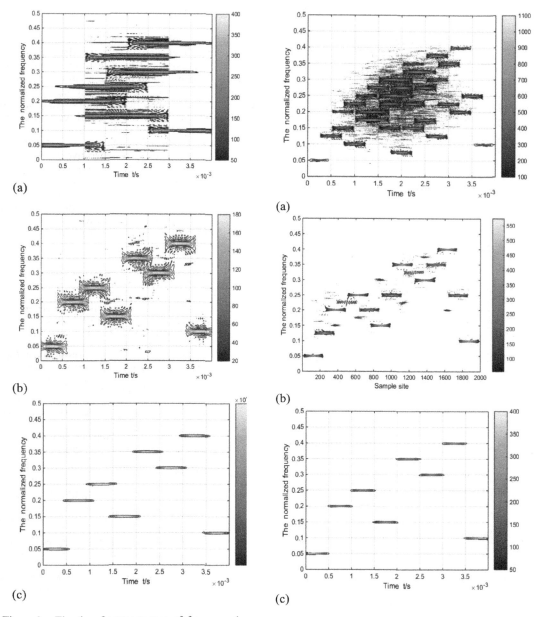

(a)

(b)

(c)

Figure 2. The time-frequency map of frequency hopping signal via STFT.

(a)

(b)

(c)

Figure 3. The time-frequency map of frequency hopping signal via WVD and its improved methods.

cross term, but cannot eliminate the cross term. The SPWVD is the method introduced in the base of the PWVD. The result is Figure 3(c). It can eliminate cross-interference completely. The characteristic of the frequency hopping signals are described better, but the time-frequency resolution is seriously reduced.

3.3 The Gabor transform of frequency hopping signals

From the Figure it can be learned that compared with the short-time Fourier transform, the higher time and frequency resolution can be derived when Gabor transform is used, but the algorithm

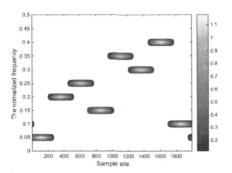

Figure 4. The time-frequency map of frequency hopping signal via Gabor transform.

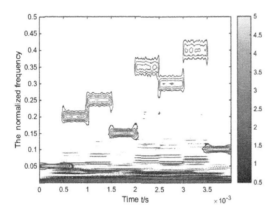

Figure 5. The time-frequency map of frequency hopping signal via WT.

is relatively complex, involving a large amount of calculation.

3.4 *The wavelet transform of frequency hopping signal*

The Morlet wavelet is selected to analyse the frequency hopping signals. The results are as follows:

From the simulation results, the conclusion we get is that by using wavelet transform analysis the frequency hopping signals can obtain a better analysis effect, and a better time and frequency resolution. However, the low frequency component of wavelet transform is produced, and the analysis effect is related to the choice of the wavelet kernel function. If an inappropriate function is chosen, it is difficult to achieve the desired analysis effect.

4 CONCLUSION

Through the theoretical analysis and simulation above, the conclusion is that the time-frequency analysis method could be used to analyse the frequency hopping signal but the analysis effect and speed are different when the methods are different. Therefore, in practical application, we should choose the reasonable method and set the parameters according to our own needs and the methods' characteristic to achieve the desired effect. For example, in the non-cooperative communication, we need to enemy the communications signals in time. Only its signal hopping pattern does not require a high time-frequency resolution, so the STFT method is a priority. If the time-frequency information is obtained for knowledge of interference, the high time-frequency resolution is needed, so the WT is the best choice.

REFERENCES

Chung, C.D. & Polydoros A. (1991) Detection and hop-rate estimation of random FH signals via autocorrelation technique [A]. *Military Communications Conference*, 345–349.

Chen, C. & Hao, Y. (2010) A high constitutional army estimated hopping signal WVD improved algorithm based on parameters. *Jilin University (Information Science Edition)*, 28(2), 124–130.

Chen, L. (2006). *Hopping signal reconnaissance technology research*, The National Defense University dissertation, Changsha.

Dan, T. & Zhao, W. (2013). Hopping communication interference and anti-jamming technology. *National Defense Industry Press*, Beijing.

Li, Z. (2008). Digital signal processing. *Tsinghua University Press*, Beijing.

Polydoros, A. & Woo, K.T. (1985). LPI detection of frequency hopping signal using autocorrelation techniques. IEEE *Journal on Selected Areas in Communication*, 3(5), 714–726.

Wang, H. (1999). Non-stationary random signal analysis and processing. *National Defense Industry Press*, Beijing.

Wu, Y. (2009). Hopping signals based on wavelet analysis of sorting identification. *Journal of Astronautics*, 30(2), 1500–1501.

Zhang, X. & Du, X. (2007). Frequency hopping signal analysis method based on Gabor spectrum [J]. *Data Acquisition and Processing*, 22(6), 150–155.

Zhang X. (2002). Modern signal processing [M]. *Tsinghua University Press*, Beijing.

Automotive, Mechanical and Electrical Engineering – Liu (Ed.)
© 2017 Taylor & Francis Group, London, ISBN 978-1-138-62951-6

Two strategies to improve the performance of two-way relay networks: Relay selection and power allocation

Tong Liu
Department of Information Engineering, Chongqing Vocational Institute of Engineering, Chongqing, China

Can Dong
Department of Electrical Engineering, Chongqing City Management College, Chongqing, China

Xi Jiang
Department of Information Engineering, Chongqing Vocational Institute of Engineering, Chongqing, China

ABSTRACT: Through relay selection and power allocation can achieve the purpose of improving the performance of Two-way relay system. This paper considered the probability of outage, proposed a Relay Selection Strategy based on Probability of Outage (RSSPO) in Two-way AF relay system. In order to enhance the sum-rate of Two-way relay system, a new Power Allocation Scheme based on Convex Optimization (PASCO) is proposed when total power is limited. Simulation result shows that both of the strategies can enhance the performance of the networks.

Keywords: Probability of outage; Power Allocation; Two-way relay

1 INTRODUCTION

Previous works has shown that a cooperative diversity gain is available in distributed wireless networks where nodes help each other by relaying transmissions. Shannon analyzed channel status of relay system in (Shannon C E, 1961). (Rankov B, 2007; Knopp R, 2007) analyzed the capacity of Two-way relay system and got the value boundary.

(Huang W, 2007) chose the best coordinated node by comparing the outage probability of every relay node, and regard the relay with minimum outage probability as the best transmit node. Author in (Hui Hui, 2009) introduced a two-step method to get the best relay node. We proposed a new way called RSSPO to choose the best cooperative node based on the probability of outage in each link. (Wonjae Shin, 2009) presented a power allocation method to improve the capacity of Two-way DF relay system. We proposed a power allocation strategy based on convex optimization in Two-way AF relay system.

2 STRATEGIES TO IMPROVE THE PERFORMANCE

We consider multiple cooperative relays networks with one source node N_1 and one destination node N_2 in the system, and the number of relay nodes is n. $h_{n1,r}$ is the channel coefficient of N_1 to the relay and from

N_2 to the relay this coefficients is $h_{n2,r}$. It's easily get $h_{n1,r} = h_{r,n1}$, $h_{n2,r} = h_{r,n2}$ because the channels are reciprocal.

In the first timeslot, N_1 and N_2 transport signals to the relay. Relay node broadcasts the received signals that is amplified in the second timeslot.

2.1 *RSSPO (Relay Selection Strategy based on Probability of Outage)*

The number of relay nodes in our research is n. The SNR threshold value is γ_{th}. Outage happened when the SNR value is lower than γ_{th}. We can get the expression of outage probability in the i link as

$$P_{out}^i = P(\bar{\gamma}_i < \gamma_{th}) \quad i = 1, 2, 3, 4. \tag{1}$$

$\bar{\gamma}_i$ is the SNR of the i link, $\bar{\gamma}_i = E_i |f_i|^2 / \sigma^2$. E_i means the power allocated to the transmitter of the link i. f_i is the channel coefficient and σ^2 means the noise and interference.

According to the math analysis, we can get the expression of outage probability of each link

$$P_{out}^i = 1 - \int_{\gamma_{th}}^{\infty} \frac{1}{\bar{\gamma}_i} e^{-(\gamma/\bar{\gamma}_i)} d\gamma \quad i = 1, 2. \tag{2}$$

The system get channel state information by sending the training sequence in all links between N_1 to N_{3i} and get the value of outage probability of each link by (2), then save the values in Γ:

$$\Gamma(M) = \left\{ M \mid M(i) = P_{out}^i \right\} \quad i = 1, 2, 3, \ldots, n \tag{3}$$

We can get set Φ of outage probability of the link N_2 to N_{3i} by the same way

$$\Phi(N) = \left\{ N \mid N(i) = P_{out}^j \right\} \quad j = 1, 2, 3, \ldots, n \tag{4}$$

In the next step, we define the function $S(t_k)$ (argument value from t_1, t_2, t_3, ..., t_n) which can rank the argument value from bottom to the top. The value of $S(t_k)$ is equal to the number of positions that t_n is ordered in all arguments when $k = n$.

Finally, we add the value of P_{out}^i and P_{out}^j, the relay with the lowest value will be chosen as the best partner. The last step of RSSPO can be expressed by

$$\min\left\{ S[\Gamma(P_{out}^i)] + S[\Phi(P_{out}^j)] \right\} \quad i = j \tag{5}$$

2.2 PASCO (Power Allocation Scheme based on Convex Optimization)

According to the conclusion in (Rankov B, 2007; Knopp R, 2007), the sum-rate of Two-way AF system is given by

$$R_{sum} = R_{12} + R_{21} \tag{6}$$

where

$$R_{12} = \frac{1}{2} \log_2 \left(1 + \frac{\dfrac{E_1 \left| h_{n1,r} \right|^2}{\sigma_1^2} \dfrac{E_3 \left| h_{r,n2} \right|^2}{\sigma_3^2}}{1 + \dfrac{E_1 \left| h_{n1,r} \right|^2}{\sigma_1^2} + \dfrac{E_2 \left| h_{n2,r} \right|^2}{\sigma_2^2} + \dfrac{E_3 \left| h_{r,n2} \right|^2}{\sigma_3^2}} \right) \tag{7}$$

$$R_{21} = \frac{1}{2} \log_2 \left(1 + \frac{\dfrac{E_2 \left| h_{n2,r} \right|^2}{\sigma_2^2} \dfrac{E_3 \left| h_{r,n1} \right|^2}{\sigma_3^2}}{1 + \dfrac{E_1 \left| h_{n1,r} \right|^2}{\sigma_1^2} + \dfrac{E_2 \left| h_{n2,r} \right|^2}{\sigma_2^2} + \dfrac{E_3 \left| h_{r,n1} \right|^2}{\sigma_3^2}} \right) \tag{8}$$

R_{12} means the sum-rage of $N_1 \rightarrow N_2$ and in $N_2 \rightarrow N_1$ it is R_{21}.

The variance of zero-mean Additive White Gaussian Noise (AWGN) is denoted by σ_1^2, σ_2^2 and σ_3^2, which is assumed to be 1 for simplicity. E_1, E_2, E_3 is the power allocated at N_1, N_2 and N_3 respectively.

The total power of the system is E_t, and the following mathematical expressions are established:

$$E_1 + E_2 + E_3 = E_t \tag{9}$$

The purpose of our strategy is maximum system capacity through reasonable power allocation. So we can get the optimization model

$$\begin{aligned} \max \quad & R_{sum} \\ s.t. \quad & E_1 + E_2 + E_3 \le E_t \end{aligned} \tag{10}$$

We regard the Two-way system as two one-way systems in our research which means the two-way relay system is consisted by $N_1 \rightarrow N_2$ and $N_2 \rightarrow N_1$. We can maximum the capacity of Two-way system by maximum the rate of two One-way. We decompose (10) into two sub-model.

Sub-model 1:

$$\begin{aligned} \max \quad & R'_{12} \\ s.t. \quad & E_1 + E_3 \le E_{t_1} \end{aligned} \tag{11}$$

E_{t1} is the power allocated in the link $N_1 \rightarrow N_2$. R'_{12} means the sum-rate of the link $N_1 \rightarrow N_2$ which can be expressed by

$$R'_{12} = \frac{1}{2} \log_2 \left(1 + \frac{\dfrac{E_1 \left| h_{n1,r} \right|^2}{\sigma_1^2} \dfrac{E_3 \left| h_{n2,r} \right|^2}{\sigma_3^2}}{1 + \dfrac{E_1 \left\| h_{n1,r} \right\|^2}{\sigma_1^2} + \dfrac{E_3 \left| h_{n2,r} \right|^2}{\sigma_3^2}} \right) \tag{12}$$

By introducing Lagrange multiplier, we can get the optimal solution is

$$\begin{cases} \dfrac{E_1}{E_3} = \dfrac{\sqrt{h_{r1,n}^2 h_{r2,n}^2} - h_{r2,n}^2}{h_{r1,n}^2 - \sqrt{h_{r1,n}^2 h_{r2,n}^2}} & h_{r1,n}^2 \ne h_{r2,n}^2 \\ \dfrac{E_1}{E_3} = 1 & h_{r1,n}^2 \ne h_{r2,n}^2 \end{cases} \tag{13}$$

Then we can get the sub-model 2 with the same analysis which is expressed by

$$\begin{aligned} \max \quad & R'_{21} \\ s.t. \quad & E_2 + E_3 \le E_{t_2} \end{aligned} \tag{14}$$

Lagrange multiplier is also introduced to obtain the optimal solution of the sub optimal model 2

$$\begin{cases} \dfrac{E_2}{E_3} = \dfrac{\sqrt{h_{r1,n}^2 h_{r2,n}^2} - h_{r1,n}^2}{h_{r2,n}^2 - \sqrt{h_{r1,n}^2 h_{r2,n}^2}} & h_{r2,n}^2 \ne h_{r1,n}^2 \\ \dfrac{E_2}{E_3} = 1 & h_{r2,n}^2 \ne h_{r1,n}^2 \end{cases} \tag{15}$$

Now we set $\alpha = E_1 / E_3$ and $\beta = E_2 / E_3$, we can further obtain that

$$\alpha\beta = 1 \tag{16}$$

We can get the relationship between channel gain and the ranges of value α.

$$\begin{cases} \alpha > 1 & \left| h_{r2,n} \right|^2 > \left| h_{r1,n} \right|^2 \\ \alpha = 1 & \left| h_{r2,n} \right|^2 = \left| h_{r1,n} \right|^2 \\ \alpha < 1 & \left| h_{r2,n} \right|^2 < \left| h_{r1,n} \right|^2 \end{cases} \qquad (17)$$

By introducing substitute (13), (15) and (16) into (9), we can get the optimal solution

$$E_1' = \frac{\alpha^2}{\alpha^2 + \alpha + 1} p_t$$

$$E_2' = \frac{1}{\alpha^2 + \alpha + 1} p_t \qquad (18)$$

$$E_3' = \frac{\alpha}{\alpha^2 + \alpha + 1} p_t$$

It seems that we allocate power to the relay node for twice in our analysis. In fact, in $N_1 \to N_2$ link we allocate the power in the relay node just for getting the proportional relationship between N_1 and N_3 but not the true power value of relay node. This is also true in the link of $N_2 \to N_1$.

2.3 *Simulation result*

To evaluate the true performance of the proposed scheme, computer simulation is conducted. Fading channel model satisfy $h_{r1,n} = v_1/d^{\alpha/2}$, $h_{r2,n} = v_2/(1-d)^{\alpha/2}$. $v_i \sim CN(0,1) i = 1,2$ denotes the AWGN of the channel. The channel coefficients $h_{r1,n}$ and $h_{r2,n}$ are generated by independent circularly symmetric complex Gaussian random variables with zero means and unit variances. Path loss factor α was set to 3 in our simulation.

We first compared the performance differences between random relay selection method and RSSPO. There are 10 candidate relays in our simulation. AS shown it Fig. 1, when the total power is lower than 10 W, there are little difference

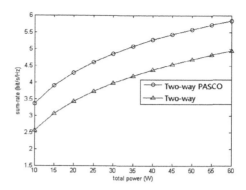

Figure 2. Comparison of PASCO and power equal allocation.

between two strategies. With the increase of the total power, the outage probability of the RSSPO strategy trends to smaller. The comparison shows that RSSPO is a useful tool in enhancing the stability of the system.

The following simulation focuses on the achievable sum-rate of the system. Fig. 2 proves that RSSPO increasing the capacity bound under the condition that total power is limited. PASCO have a channel gain about 1 bit/s/Hz compared with power is allocated to each node equally.

3 CONCLUSION

In this paper, we have investigated the relay selection and power allocation in Two-way AF relay network. We have proposed a three-step Relay Selection Strategy based on the Probability of Outage (RSSPO). We have further presented a Power Allocation Scheme based on Convex Optimization (PASCO), simulation analysis shows that the both of the two strategies can improve the performance of the system.

ACKNOWLEDGMENTS

This work was financially supported by the project of Chongqing City Management College.

REFERENCES

Huang W, Hong Y, Kuo Discrete C J, Power Allocation for lifetime maximization in cooperative networks. IEEE Vehicular Technology Conference, MD, 2007: 581–585.

Hui Hui, Zhu Shihua, Distributed Power Allocation Schemes for Amplify-and-Forward Networks. IEEE Wireless Communications and Networking Conference, 2009: 1–6.

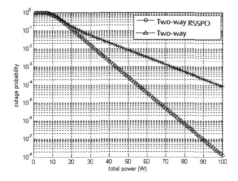

Figure 1. Comparison of RSSPO and random relay selection.

Knopp R, Two-way wireless communication via a relay station. GDRISIS meeting, 2007.

Rankov B, Wittneben A, Spectral efficient protocols for halfduplex fading relay channels. IEEE Journal on Selected areas in communications, 2007, 25(2): 379–389.

Shannon C E, Two-way communication channels. Proc. 4th Berkeley Symp. Math. Stat. Prob. 1961: 611–644

Wonjae Shin, Namyoon Lee, Jong Bu Lim, An Optimal Transmit Power Allocation for the Two-way Relay Channel Using Physical-Layer Network Coding. IEEE International Conference on Communications Workshops, 2009: 1–6.

Automotive, Mechanical and Electrical Engineering – Liu (Ed.)
© 2017 Taylor & Francis Group, London, ISBN 978-1-138-62951-6

Updated dynamic channel reservation scheme based on priorities in LEO satellite systems

Jiani Guo, Jinchun Gao & Jing Ran
School of Electronic Engineering, Beijing University of Posts and Telecommunications, Beijing, China

ABSTRACT: An updated dynamic channel reservation scheme based on priorities, which works for different classes of service in the Low Earth Orbit (LEO) satellite system, is proposed in this paper. With the deterministic and predictable LEO satellite motion, user location is not necessary. To obtain the overall system's optimal quality of service, the proposed scheme redefines the thresholds distribution for reserved channels and the method for solving the dynamic thresholds. Then, the analytical approach is described including the LEO satellite mobility model, the channel allocation model based on Markov chain and the solution of the thresholds with the method of genetic-particle swarm optimisation hybrid algorithm. It has been proved by simulation results that the proposed scheme has the capability of offering more accuracy and flexibility for the dynamic channel reservation of different classes of service as well as effectively reducing handover failures. Moreover, the quality of service for the overall system has been improved.

Keywords: LEO satellite systems; handover call; dynamic channel reservation

1 INTRODUCTION

With the advantages of lower transmission power and shorter transmission delay, Low Earth Orbit (LEO) satellite systems are capable of providing high-quality real time mobile communication services (Karaliopoulos et al., 2004; Albertazzi et al., 2003). The Iridium LEO-system, which works as the well-known commercial satellite system, could provide global voice services as well as data services (Fossa et al., 1998).

The LEO satellite systems generally use multispot beam for frequency reuse. As shown in Figure 1, the footprint of each satellite consists of contiguous circular cells. Despite mentioned advantages above, due to high-speed rotation of LEO satellites around the earth, users have to switch to adjacent cells (Fossa et al., 1998).

During the handover process, the lack of channels available in the adjacent cell will make handover drop. From a user's perspective, it feels more undesirable than the new call blocking. Thus, strategies prioritising handover calls are of great importance. In addition, there are users with different requirements of service, such as data service and voice service, which we take as class-1 service and class-2 service. The users of class-2 service require better quality of service while the users of class-1 service raise concerns about fairness. Therefore, in LEO satellite systems, both call types and the service class should be considered in the design of the channel reservation strategies.

The Guaranteed Handover (GH) scheme (Maral et al., 1998) is the earliest channel reservation scheme. A series of improved schemes based on the GH scheme have been proposed, such as the Dynamic Doppler-Based Handover Prioritisation (DDBHP) scheme (Papapetrou et al., 2005). However, those schemes own reserved channels with a fixed number, which causes unsatisfactory quality of service. (Boukhatem et al., 2003; Karapantazis et al., 2007; Wang, Zhipeng et al., 2001; Cho & Sungrae, 2000) have discussed various dynamic channel reservation schemes, but these schemes depend on positioning devices, e.g. GPS, which increases the signalling cost. Without user location, the Dynamic Channel Reservation scheme based on Priorities (DCRP) proposed by J. Zhou et al. (2015), can dynamically calculate the thresholds for reserved channels based on the traffic condition by using Genetic Algorithms (GA).

An Updated Dynamic Channel Reservation Scheme based on Priorities (UDCRP), which works for different classes of service in LEO satellite systems, is proposed in this paper. Compared with the DCRP scheme, the thresholds distribution for reserved channels and the method for solving the dynamic thresholds have been redefined. The proposed scheme prioritises handover calls of lower class service over new calls of higher class service. The Genetic-Particle Swarm Optimisation hybrid algorithm (GA-PSO), which outperforms the GA on the quality of solution, stability of convergence and time consumption (Salman, Ayed

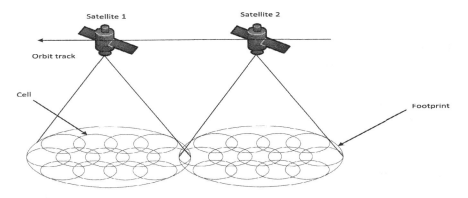

Figure 1.   LEO satellite systems coverage geometry.

et al., 2002; Gao et al., 2005), have been adopted. Similar to the DCRP scheme, with the deterministic and predictable LEO satellite motion, user location is not required in the updated scheme.

The remainder of this paper has been structured as follows. Firstly, the novel UDCRP scheme is introduced. Then, the analytical approach for the proposed scheme is described including the LEO satellite mobility model, the channel allocation model based on Markov chain and the solution of the thresholds. Then the simulation results are discussed. Finally, the summary of this paper is given.

2   PROPOSED CHANNEL RESERVATION SCHEME

In this system, Fixed Channel Allocation (FCA) is considered, and C channels are available for one cell. As shown in Figure 2, a set of thresholds $K = \{k_1, k_2, k_1', k_2'\}$ is used to reserve channels, where $0 \leq k_1 \leq k_2 \leq k_1' \leq k_2' \leq C$.

where, k_1 denotes the access threshold for class-1 service, k_2 denotes the access threshold for class-2 service, k_1' denotes the handover threshold for class-1 service and k_2' denotes the handover threshold for class-2 service. The process of UDCRP scheme is shown in Figure 3.

For different types of calls, the more reserved channels the handover call owns, the lower the handover failure probability for the system, while the higher is the new call blocking probability. For different classes of service, the more reserved channels the class-2 service obtains, the higher the quality of service for the class-2, while the lower is the quality of class-1 service. The mentioned problems obviously cause poor quality of service of the overall system.

Overall, the key of channel reservation is dynamically updating the thresholds K. The following paper is to find the solution.

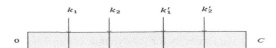

Figure 2.   Updated dynamic thresholds distribution for reservation channels.

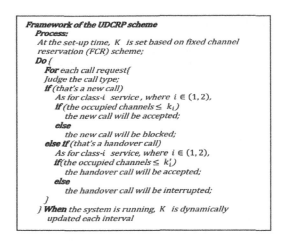

Figure 3.   The process of UDCRP scheme.

3   ANALYTICAL APPROACH

3.1   The mobility model

The ground-track speed of Iridium LEO-system V is 26,600 km/h; compared with that, user motion and earth rotation can be neglected (Dosiere et al., 1993). Therefore, the user speed is taken as V in the opposite direction of satellite orbit. As shown in Figure 4, the model of each cell is a rectangle whose length L is 425 km bounded by parallel sides of a strip.

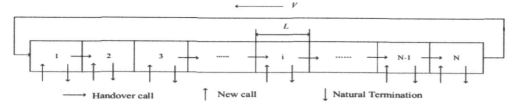

Figure 4. Rectangle cell model for considered LEO satellite systems.

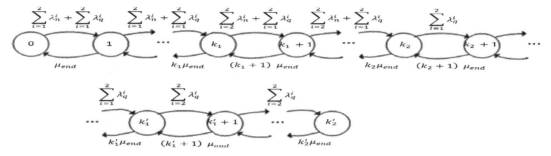

Figure 5. Markov chain of the proposed scheme.

This paper takes the following classic and well-accepted assumptions (Del Re, et al., 1995; Wang, Zhipeng et al., 2009). The new call arrival process meets the Poisson process with mean rate λ_n, and the handover call arrival process obeys the Poisson process independently to the former. Besides, the uniform traffic is considered and the call duration time follows exponential distribution with a parameter μ. Therefore, the parameters for the LEO mobility model (Cho & Sungrae, 2000; Zhou et al., 2015) are adopted as shown below.

3.2 Channel allocation model

Based on the mobility model above, one cell could be modelled as an M/M/C/C Markov process with state space, as shown in Figure 5. Among them, C is the total number of channels in each cell and the state denotes the number of occupied channels.

The global balance equations for the Markov process are shown below, where the steady state probability of the state j, which means j channels occupied, is denoted as π_j. Among them, $m \in (1,2)$, $k_2' = C$ and $\sum_{j=0}^{C} \pi_j = 1$.

$$\left(\sum_{i=1}^{2} \lambda_n^i + \sum_{i=1}^{2} \lambda_q^i \right) \pi_0 = \mu_{end} \pi_1 \tag{1}$$

$$\left[\left(\sum_{i=m+1}^{2} \lambda_n^i + \sum_{i=1}^{2} \lambda_q^i \right) + j\mu_{end} \right] \pi_j = \left(\sum_{i=m}^{2} \lambda_n^i + \sum_{i=1}^{2} \lambda_q^i \right) \pi_{j-1}$$
$$+ (j+1)\mu_{end} \pi_{j+1}, j = k_m \tag{2}$$

$$\left[\left(\sum_{i=m+1}^{2} \lambda_n^i + \sum_{i=1}^{2} \lambda_q^i \right) + j\mu_{end} \right] \pi_j = \left(\sum_{i=m+1}^{2} \lambda_n^i + \sum_{i=1}^{2} \lambda_q^i \right) \pi_{j-1}$$
$$+ (j+1)\mu_{end} \pi_{j+1}, k_m < j < k_{m+1} \tag{3}$$

$$\left(\sum_{i=1}^{2} \lambda_q^i + j\mu_{end} \right) \pi_j = \left(\sum_{i=1}^{2} \lambda_q^i \right) \pi_{j-1}$$
$$+ (j+1)\mu_{end} \pi_{j+1}, k_{m+1} < j < k_m' \tag{4}$$

$$\left(\sum_{i=m+1}^{2} \lambda_q^i + j\mu_{end} \right) \pi_j = \left(\sum_{i=m}^{2} \lambda_q^i \right) \pi_{j-1}$$
$$+ (j+1)\mu_{end} \pi_{j+1}, j = k_m' \tag{5}$$

$$\left(\sum_{i=m+1}^{2} \lambda_q^i + j\mu_{end} \right) \pi_j = \left(\sum_{i=m+1}^{2} \lambda_q^i \right) \pi_{j-1}$$
$$+ (j+1)\mu_{end} \pi_{j+1}, k_m' < j < k_{m+1}' \tag{6}$$

To discuss the performance of channel reservation schemes, the well-known parameters (Maral et al., 1998; Papapetrou et al., 2005; Boukhatem et al., 2003; Karapantazis et al., 2007; Wang, Zhipeng et al., 2001; Cho & Sungrae, 2000; Zhou et al., 2015) shown in Table 2 are generally referred to.

The quality of service of the overall system can be defined as

$$QoS = \sum_{i=1}^{2} \beta_i \left(\alpha_0 \left(1 - P_b^i \right) + \alpha_1 \left(1 - P_d^i \right) \right) \tag{7}$$

Table 1. Parameters for LEO mobility model.

Parameter	Description
$\lambda_n^i = l_i \lambda_n, \sum_{i=1}^{2} l_i = 1$	λ_n^i denotes the new call arrival rate of the i class service.
$t_{\max} = L/V$	$t_{\max}$ denotes the maximum time going through a cell.
$P_n^i = \left(1 - e^{\mu t_{\max}}\right)/\left(\mu t_{\max}\right), P_q^i = e^{-\mu t_{\max}}$	P_n^i and P_q^i respectively denote the new call and handover call handover probability of the i class service.
$\lambda_q^i = \left(N_n^i P_n^i + N_q^i P_q^i\right)/t_{\max}$	N_n^i and N_q^i respectively denote the new call and the handover call number of the i class service in the previous cell. λ_q^i denotes the handover call rate of the i class service.
$P_{NW} = \lambda_n/\left(\lambda_n + \lambda_q\right), P_{QW} = \lambda_q/\left(\lambda_n + \lambda_q\right)$	P_{NW} and P_{QW} respectively denote the probability of occurrence of new calls and handover calls.
$\mu_{end} = 1/\left(P_{NW} E\left(T_{NW}\right) + P_{QW} E(T_{QW})\right)$	μ_{end} denotes average departure rate of the calls.

Table 2. Parameters for quality of service.

Parameter	Description
$P_b^i = \sum_{j=k_i}^{C} \pi_j$	P_b^i denotes the new call blocking probability of the i class service.
$P_d^i = \sum_{j=k_i'}^{C} \pi_j$	P_d^i denotes the handover call failure probability of the i class service.
α_0, α_1	The scaling factors between new call and handover call.
β_i	The impact factor of the i class service.

According to the parameters in Table 2,

$$QoS = \sum_{i=1}^{2} \beta_i \left(\alpha_0 \left(1 - \sum_{j=k_i}^{C} \pi_j\right) + \alpha_1 \left(1 - \sum_{j=k_i'}^{C} \pi_j\right)\right), \text{ where}$$

π_j is a function of K. The calculation of λ_n^i and μ_{end} is shown in the previous section. Given the above, the following channel allocation model is established.

$$Max\ QoS = f\left(K\right), \quad s.t.\ 0 \le k_1 \le k_2 \le k_1'$$
$$\le k_2' \le C, k_i, k_i' \in Z, i \in \left(1, 2\right) \quad (8)$$

In fact, the computation requirements of the thresholds K quickly become excessive with the system capacity C increasing. With the exhaustive search algorithm, a very long solution time and great storage space are required. The proposed solution of thresholds based on the GA-PSO algorithm can effectively solve the above-mentioned problem.

3.3 *Solution of thresholds*

The GA-PSO algorithm introduces a genetic operator to the particle swarm optimisation algorithm (Salman, Ayed et al., 2002). The particles are

```
Framework of Solution Steps based on GA-PSO algorithm
Process:
  Use the real number encoding method. An ordered
  K = {k₁, k₂, k₃, k₄}  is a particle;
  Initialize randomly the particle swarm populations;
  For each particle, sort K in an ascending order to get a
  valid value;
  Evaluate each particle with fitness function
  QoS = f(K);
  Do {
    Set the individual best solution so far as pbest;
    Set the global best individual as gbest;
    For each particle K  in the population {
      if (rand() ≤ p)
        Make the single-point mutation;
      else if (rand() ≤ pc)
        Crossover with the individual best solution pbest;
      else
        Crossover with the global best individual gbest;
      Sort the new particle  K' in the ascending order;
      if (f(K') > f(K))
        K = K';
    }
    Evaluate the individuals in the updated population
    and update the  pbest  and  gbest;
  } While maximum iterations are not attained
```

Figure 6. Solution steps based on GA-PSO algorithm.

updated in order to search in the problem space and evaluate the candidate solution with the fitness value. By restricting evolution generations, the GA-PSO algorithm can find the suitable K in limited time. The key steps are shown in Figure 6, where p denotes the scaling factor between crossover and mutation operation, which generally decreases linearly from 0.04 to 0.01, while p_c denotes the scale factor between crossover operation with the best individual in history and the best individual of all the particles, which generally is a constant 0.3 (Gao et al., 2005).

4 SIMULATION RESULTS

The simulation results using the UDCRP scheme, compared with the FCR and DCRP schemes are discussed in this section. Referred to (Zhou, J et al., 2015), the following assumptions are made during the process. Class-1 service and class-2 service are considered. For different classes of service, the mean rate of new calls is $\lambda_n^1 = \lambda_n^1 = \lambda_n/2$. The fixed channel number of one cell is $C = 80$. The call duration $T_C = 1/180$ s, and the maximum time going through one cell is $t_{max} = 57.52$ s. In addition, for the DCRP scheme, at the set-up time, K is set to $\{k_1 = 65, k_1' = 70, k_2 = 75, k_2' = 80\}$ according to the FCR scheme. For the UDCRP scheme, K is set to $\{k_1 = 65, k_2 = 70, k_1' = 75, k_2' = 80\}$. Evolution generation is set to 100, and the population is set to 20. Generally, the QoS parameters are set as $\alpha_0 = 1, \alpha_1 = 10, \beta_1 = 0.4, \beta_2 = 0.6$.

The horizontal axis represents the traffic intensity λ_n, and the curves are described as a function of that.

Figure 7 and Figure 8 respectively compare the handover failure probability of class-2 and class-1 service between UDCRP, DCRP and FCR. For class-2 service, with the increasing traffic intensity, UDCRP performs better in terms of handover failure probability due to the more efficient and accurate algorithm for solution of thresholds. For class-1 service, UDCRP achieves obviously lower handover failure probability mainly because of higher priority of handover call.

As shown in Figure 9 and Figure 10, the new call blocking probability of UDCRP for class-2 and class-1 service is higher than FCR and DCRP because of lower priority of new call, which makes sacrifice for the quality of service of the overall system. Among them, DCRP has a lower new call blocking probability of class-2 service because of higher priority of class-2 service.

Figure 11 shows that the UDCRP scheme has a better performance of average quality of service for the system than FCR and DCRP, especially when the traffic intensity is heavy. Dynamic channel reservation helps reduce the cost of channel resources

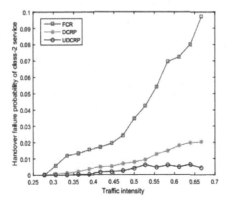

Figure 7. Handover failure probability for class-2.

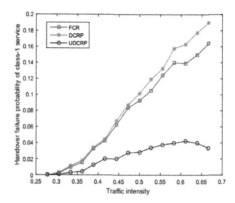

Figure 8. Handover failure probability for class-1.

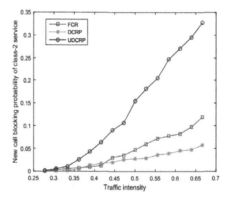

Figure 9. New call blocking probability for class-2.

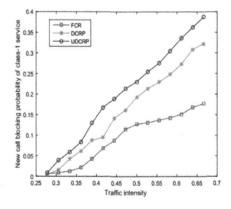

Figure 10. New call blocking probability for class-1.

which happens in FCR, and the more reasonable thresholds distribution for reserved channels and the more efficient method for solving the dynamic thresholds obviously help the UDCRP scheme obtain better QoS than DCRP.

227

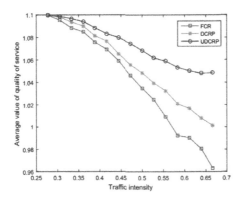

Figure 11. Average quality of service for the overall system.

5 CONCLUSION

A UDCRP which works for different classes of service in LEO satellite system, is proposed in this paper. Compared with DCRP scheme, the thresholds distribution for reserved channels and the method for solving the dynamic thresholds have been redefined. The proposed scheme prioritises handover calls of lower class service over new calls of higher class service. The GA-PSO algorithm has been adopted to calculate the reserved channel numbers according to traffic conditions. In addition, the channel allocation model has been established based on Markov process. Similar to DCRP scheme, with the deterministic and predictable LEO satellite motion, user location is not required in the updated scheme.

The simulation results have shown that, compared with the FCR and DCRP schemes, the proposed scheme can offer more accuracy and flexibility for the dynamic channel reservation of different classes of service as well as effectively reducing handover failures. Moreover, the quality of service for the overall system has been improved.

ACKNOWLEDGMENTS

The National Natural Science Foundation of China (No. 61272518) and the National 863 project (xxx) have supported this work, and the corresponding author is Jinchun Gao.

REFERENCES

Albertazzi, G., Corazza, G.E., Neri, M. & Vanelli-Coralli, A. (2003). Performance of turbo coding for satellite UMTS multimedia broadcast multicast services. *International Conference on Communication Technology Proceedings* (pp. 1078–1081). Beijing, China.

Boukhatem, L., Beylot, A.L., Gaïti, D. & Pujolle, G. (2003). TCRA: a timebased channel reservation scheme for handover requests in LEO satellite systems. *International Journal of Satellite Communications and Networking, 21*(3), 227–240.

Cho & Sungrae (2000). Adaptive dynamic channel allocation scheme for spotbeam handover in LEO satellite networks. *Vehicular Technology Conference* (pp. 1925–1929). Boston, MA.

Del Re, E., Fantacci, R. & Giambene, G. (1995). Efficient dynamic channel allocation techniques with handover queuing for mobile satellite networks. *IEEE Journal on Selected Areas in Communications, 13*(2), 397–405.

Dosiere, F., Zein, T., Maral, G. & Boutes, J.P. (1993). A model for the handover traffic in Low Earth-Orbiting (LEO) satellite networks for personal communications. *International Journal of Satellite Communications, 11*(3), 145–149.

Fossa, C.E., Raines, R.A., Gunsch, G.H. & Temple, M.A. (1998). An overview of the IRIDIUM (R) Low Earth Orbit (LEO) satellite system. *Aerospace and Electronics Conference IEEE* (pp. 152–159). Dayton, OH.

Gao, L., Gao, H. & Zhou C. (2005). General particle swarm optimization model. *Chinese Journal of Computers, 28*(12), 1980–1987.

Karaliopoulos, M., et al. (2004). Satellite radio interface and radio resource management strategy for the delivery of multicast/broadcast services via an integrated satellite-terrestrial system. *IEEE Communications Magazine, 42*(9), 108–117.

Karapantazis, S. & Pavlidou, F.N. (2007). Dynamic timebased handover management in LEO satellite systems. *Electronics Letters, 43*(5), 57–58.

Maral, G., Restrepo, J., Del Re, E., Fantacci, R. & Giambene, G. (1998). Performance analysis for a guaranteed handover service in an LEO constellation with a 'satellite-fixed cell' system. *IEEE Transactions on Vehicular Technology, 47*(4), 1200–1214.

Papapetrou, E. & Pavlidou, F.N. (2005). Analytic study of Doppler-based handover management in LEO satellite systems. *IEEE Transactions on Aerospace and Electronic Systems, 41*(3), 830–839.

Salman, A., Ahmad, I. & Al-Madani, S. (2002). Particle swarm optimization for task assignment problem". *Microprocessors and Microsystems, 26*(8), 363–371.

Wang, Z. & Mathiopoulos, P.T. (2001). Analysis and performance evaluation of dynamic channel reservation techniques for LEO mobile satellite systems. *Vehicular Technology Conference,* (pp. 2985–2989). Rhodes.

Wang, Z., Mathiopoulos, P.T. & Schober, R. (2009). Channeling partitioning policies for multi-class traffic in LEO-MSS. *IEEE Transactions on Aerospace and Electronic Systems, 45*(4), 1320–1334.

Zhou, J., Ye, X., Pan, Y., Xiao, F. & Sun, L. (2015) "Dynamic channel reservation scheme based on priorities in LEO satellite systems. *Journal of Systems Engineering and Electronics, 26*(1), 1–9.

Technologies in energy and power, cells,
engines, generators, electric vehicles

Automotive, Mechanical and Electrical Engineering – Liu (Ed.)
© *2017 Taylor & Francis Group, London, ISBN 978-1-138-62951-6*

A study on coordinated charging strategies for electric vehicles at the workplace under a uniform electricity price

Lixing Chen, Xueliang Huang & Feng Wen
School of Electrical Engineering, Southeast University, Nanjing, China

ABSTRACT: In order to coordinate relations between charging facility configuration and charging of Electric Vehicles (EVs), two coordinated charging strategies based on uniform electricity price or the conventional constant power charging ways were presented here for EVs according to conventional load types at workplace, parking characteristics of EVs and matching of parking space and charging infrastructure. In detail, they were parallel charging strategy and grouping serial charging strategy. Workplace power distribution systems were subsequently constructed for such two strategies. Besides, the average annual operating cost of charging station was used as an objective function to establish an EV charging optimization model. Finally, the corresponding simulation results indicated that, comparing with the parallel charging strategy, the grouping serial charging strategy was able to not only bring down operating costs of charging stations and charging costs of EVs, but alleviate impacts on a power grid. In this manner, an effective solution could be provided for charging facility configuration and EV charging.

Keywords: Workplace; EV; Charging Strategy

1 INTRODUCTION

Global warming and severe environmental pollution are major problems challenging the entire society at present. EV technology that aims at low carbon emissions and less environmental pollution has been rapidly developed and promoted. Currently, governments and automobile manufacturers of all countries all pay special attentions to this technology (Hu, et al. 2012, Sun, et al. 2013, Wang, et al. 2013). With large-scale popularization of EVs, oil-fueled automobiles can be replaced with quantities of EVs in urban areas. However, there still exists no effective approach that can be employed to properly deal with relations between EV charging strategy and charging facility configuration. Consequently, the problem of EV charging fails to be solved validly. In other words, charging facility configuration is implemented without taking the corresponding EV charging strategy into consideration, which leads to a waste of charging resources and increase in charging costs.

To solve charging issues related to EVs, charging of EVs should be controllable in the first place. An urban area is usually divided into workplaces, residential areas and commercial zones, etc. As for workplaces, standard labor time is 8-hour work per day. Here, parking duration in a parking lot can be uniformly set at 8:00–18:00 for EVs. Except EVs in particular cases (primarily referring to EVs leaving

the parking lot because of some unexpected reasons), most stay in the parking lot within work period. Therefore, charging demands of users can be satisfied by conventional charging in the case that EV charging is controllable.

Generally, a certain relation exists between EV charging strategies and charging facility configuration. According to the matching of charging facilities and parking space, two charging facility configuration schemes are proposed in this chapter. One is that the matching ratio between charging facilities and parking space is 1:1; the other is that such a ratio is 1: n. for both schemes, power outlets adopted by charging facilities are all single. Every charging facility involved in the former configuration scheme only corresponds to one EV and they cannot be shared by EVs at all. In this case, parallel charging control (or, referred to as parallel charging strategy) over EVs is carried out. With regard to the latter configuration scheme, each charging facility corresponds to multiple EVs. Therefore, there may exist a case that only some charging facilities are occupied. Under such a circumstance, grouping serial charging control can be performed for EVs (known as grouping serial charging strategy). Furthermore, the latter strategy requires that every EV must be charged during the time exclusively arranged for it. For exceptional cases, the charging duration can be negotiated and adjusted in advance. Normally, charging demands of users cannot be satisfied once they missed the specified charging duration.

Charging facility redundancies corresponding to the above two configuration schemes are different. As a result, diverse charging costs can be incurred to users. The reasons are as follows. (1) To increase the universality of charging facilities, manufacturers give considerations to charging demands of a variety of EVs at the time of charging facility design and manufacturing (mainly referring to power output range here) rather than being confined to charging demands of some particular EVs. In other words, it is less likely for manufacturers to customize a charging facility for an EV alone and the problem of charging facility redundancy exists to some extent. (2) The employment of diverse charging strategies can also give rise to charging facility redundancy. Charging facilities with different redundancies need different costs of configuration. When the redundancy is excessively large, operating cost of the charging facility can be dramatically increased. As a result, charging costs are brought up for users. Otherwise, charging costs of users may drop.

At present, charging facility planning and EV charging are studied separately in a majority of literatures (Sun, et al. 2013, Zhao, et al. 2015, Lu, et al. 2014, Xu, et al. 2014, Zhang, et al. 2014). Especially for studies on EV charging, it is always assumed that charging facilities have been configured completely, together with presenting a simple scheme. That is, a certain number of charging piles of the same type are allocated without considering relevant redundancy.

To sum up, workplace is selected as charging station running area targeted at EV charging problems that are still in presence at present. Then, impacts of EV charging on charging stations, users and power grids based on diverse charging strategies are comparatively analyzed. Accordingly, an effective solution is provided for charging facility configuration and coordinated charging of EVs. It is of extensive theoretical and application values.

2 WORKPLACE POWER DISTRIBUTION SYSTEM STRUCTURE

A workplace power distribution system is composed by power supply units and electric equipment. As for the former, the power is supplied by power distribution network in most cases. Under a circumstance that no charging stations have been constructed and EVs are promoted, electric equipment falls into the category of traditional electric equipment and load generated by it is referred to as conventional power load. After construction of charging stations in parking lots is completed

and there are parking EVs there, such parking lots can be adopted to park cars and satisfy charging demands of EV simultaneously. In this way, new elements are added into traditional electric equipment and its load. To be specific, while both charging facilities and EVs are integrated into electric equipment, conventional load is thus provided with charging loads of EVs.

According to related statistics, conventional loads required by different workplaces have different characteristics (Wang, et al. 2007). On one hand, some conventional loads are featured with significant peak valley fluctuations. On the other hand, others can be very smooth. Therefore, influences of conventional loads are not considered in this paper temporarily. In line with a constant power charging mode, parallel or grouping serial charging strategy is employed to charge EVs. The corresponding power distribution system structures are shown in Figure. 1.

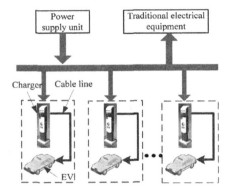

(a) Parallel charging strategy

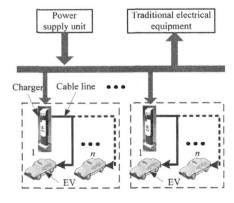

(b) Packet successive charging strategy

Figure 1. Power distribution system structures of two charging strategies.

Different charging strategies need different electricity price guidance mechanisms. EV users who make use of parallel charging strategy are not involved with a problem of charging order. Therefore, electricity price similar to peak-valley price or the uniform electricity price of charging can be used as electricity price of EV charging for these users. Comparatively, EV users of grouping serial charging strategy are associated with this problem and they have special requirements for electricity price forms of charging. In the case that the uniform electricity price of charging is adopted, such EV users can be insensitive to charging order that otherwise fails to affect their willingness of grouping charging. However, if an electricity price of charging sharing some resemblances with the peak-valley price, such users become sensitive to charging order so as to exert an impact on effective implementation of grouping charging. Considering this, uniform electricity price is employed in this paper with an aim to remove impacts of various electricity forms on EV charging based on diverse charging strategies.

3 EV CHARGING OPTIMIZATION MODEL

3.1 Optimization variable

According to statistics, a majority of EVs used for working have short mileages inside cities (Chen, et al. 2015) while only a very small number of them travelled for a long distance. In addition, they stay in workplaces for a very long time and diversified charging demands of EVs can be satisfied by conventional charging piles in most cases. In conformity with charging rate of these piles, they are classified into m levels, such as Level 1, Level 2, ..., and Level m. On this basis, charging piles of diverse levels are adopted by EV charging to meet the corresponding charging demands on the premise of taking diversity of EV charging demands and constraints over stopping time into account. Before optimization, the level of charging piles that every EV is equipped with is still unclear. Resultantly, optimization variable is set as the type of charging piles configured for each EV and it is also independent of charging strategies (that is, grouping serial/parallel charging). Thus, optimization variables for such two charging strategies are expressed into the following equation.

$$X = [x_1, x_2, ..., x_k, ..., x_n] \qquad (1)$$

where n is the serial No. of an EV; and, x_k is the type of a single charging pile matched with an EV k and its value can be denoted as below.

$$x_k \in \{1, 2, ..., u\}, \quad k = 1, ..., n \qquad (2)$$

where 1, 2 and u are all serial numbers of levels; among them, u depends on the highest level of a charging piles selected for study.

3.2 Objective function

Normally, operating cost of a charging station consists of two parts. One is static cost that covers purchase and installation costs; the other is dynamic cost that only contains maintenance cost. As far as charging piles are concerned, the lower their operating costs are, the more beneficial it will be to improve the related economic benefits. Provided that charging demands of EVs are satisfied, operating costs of charging stations can be minimized by means of EV charging control (the average annual operating cost of charging stations is adopted here). Objective functions corresponding to the parallel charging strategy can be written into a formula below.

$$F_1^b(X) = \min\left[\sum_{i=1}^{u} \rho(i) \sum_{k=1}^{n} H(x_k)\right] \qquad (3)$$

where $\rho(i)$ stands for the annual operating cost of a single charging piles of type I; and, min [·] for a function with a minimum value. H (x_k) can be figured out according to the below equation.

$$H(x_k) = \begin{cases} 1, & \text{if } x_k = i \\ 0, & \text{else} \end{cases} \qquad (4)$$

where x_k is the type of a single charging pile corresponding to an EV k.

Objective functions corresponding to the grouping serial charging strategy can be written into a formula below.

$$F_1^c(X) = \\ \min\left\{\sum_{i=1}^{u} \rho(i) \operatorname{Int}\left[\sum_{k=1}^{n} \frac{f(x_k)}{T} \Delta(x_k) + \Omega(x_k)\right]\right\} \qquad (5)$$

where Int [·] is an integer function that refers to the number of type i charging piles configured; T is operating cycle of a charging station (unit: hour); $f(x_k)$ is the time taken by an EV k to be charged by a charging pile x_k; $\Delta(x_k)$ and $\Omega(x_k)$ are both modified functions that can be worked out as follows.

$$\Delta(x_k) = \begin{cases} 1, & \text{if } x_k = i \\ 0, & \text{else} \end{cases} \qquad (6)$$

$$\Omega(x_k) = \begin{cases} 1, & \text{if Int } (x_k) < x_k \\ 0, & \text{else} \end{cases} \qquad (7)$$

Another optimization objective of charging is to reduce impacts of EV charging loads on power grid. As influences of conventional loads are not considered, reduction in volatility of EV charging loads is able to bring down the relevant impacts on power grid. Normally, power batteries of an electric vehicle are charged based on a two-stage charging method of constant current and constant voltage (Zhao, et al. 2015). Regarding electric vehicles of conventional charging approaches, most charging time are concentrated at the constant current charging stage (the constant voltage charging stage can be neglected) and charging power at this stage changes a little. Considering this, the constant power charging method is adopted for power batteries of EVs described in this paper. For EVs of parallel charging strategy, parking durations can be uniformly used as the charging time without optimizing the control over charging power. Charging demands of time of EVs are adopted to obtain the associated charging power so as to further define charging facility configuration. Evidently, the total charging load of EVs is at the minimal level, so is the volatility. But, when it comes to EVs of grouping serial charging strategy, they are charged by charging piles of the maximum output to determine the corresponding charging facility configuration. Then, charging powers related are adjusted to achieve the minimum total charging loads of electric vehicles. In this way, the volatility is also at the minimal level.

Based on the above analysis, the electric vehicle charging optimization model can be used to carry out optimizations according to an individual objective functions (to reduce operating costs of charging stations) in the first place. Subsequently, another objective function is adjusted (to reduce charging load volatility of EVs) to acquire the least value.

3.3 Constraints conditions

To meet charging demands of EVs, the charging time is not permitted to exceed the maximum charging time when charging piles are used by charging stations for EV charging. That is, the operating cycle of charging stations is,

$$f(x_k) \leq T, \quad k = 1, \cdots, n \qquad (8)$$

where $f(x_k)$ is the charging time taken by EV k to be charged by a charging pile x_k and it can be calculated in line with a formula below.

$$f(x_k) = \frac{\varepsilon L_k}{P(x_k)}, \quad k = 1, \cdots, n \qquad (9)$$

In this equation, ε is power consumption per mile of EV k; L_k is the average mileage of EV k going to work in every weekday (unit: mile); $P(x_k)$

is charging power of EV k. Moreover, in the case that EVs are charged by the corresponding charging piles, the charging power is not permitted to go beyond the upper limit of charging pile output that can be denoted as follows.

$$0 \leq P(x_k) \leq P_{max}(x_k), \quad k = 1, \ldots, n \qquad (10)$$

In addition, types of charging piles optimized must be within a type collection chosen for this study. In other words, value of charging pile type x_k should satisfy,

$$1 \leq x_k \leq u, \quad k = 1, \ldots, n \qquad (11)$$

Above constraint conditions (8)–(11) are all used for parallel/grouping serial charging strategy.

To sum up, EV charging optimization model can be solved by COMPLEX and unnecessary details will not be further given here.

4 SIMULATION

To be specific, a workplace parking lot is used as the operating area of charging station in this paper to construct a power distribution system correspondingly. Based on relevant statistical data, performance parameters involved with EVs and settings of this charging station, charging situations of EVs are simulated. The cycle lies between 8:00 and 18:00. It is divided into 10 durations and each duration lasts 1 hour. The simulation step size is 1 minute.

4.1 Basic parameter settings

In this chapter, the individual mileage survey statistics of vehicles derived from the National Household Travel Survey 2009 (NHTS2009) are utilized as the mileage of EVs going to work in the workplace under study (Chen, et al. 2015). Therefore, it is assumed based on mileage statistics of EVs that workplace EV mileages are distributed as shown in Figure 2. Size of workplace EVs is set to be 200.

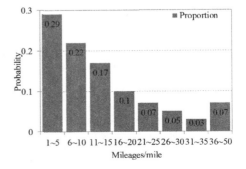

Figure 2. Mileages distribution of EVs at workplace.

Table 1. Workplace EV Mileage (unit: mile).

Group no.	Serial no. of EVs in each Group									
	1	2	3	4	5	6	7	8	9	10
1	2.0	2.4	2.6	3.3	4.0	4.9	9.9	13.2	19.4	33.0
2	1.5	2.5	3.7	3.9	4.0	4.4	8.5	17.5	18.6	32.0
3	1.8	2.0	2.5	2.8	2.8	3.3	8.3	14.6	24.5	34.7
4	1.6	2.0	2.3	2.4	3.3	3.3	9.5	16.0	24.2	33.4
5	1.1	2.8	3.5	4.1	4.6	4.8	8.9	12.6	23.4	32.5
6	1.8	2.7	2.9	3.0	4.3	6.1	7.9	13.9	23.6	32.3
7	1.5	1.6	2.6	4.4	4.4	6.3	8.6	14.7	22.9	31.9
8	1.5	2.1	3.5	3.7	4.4	4.8	8.0	13.6	28.7	28.9
9	1.9	2.0	2.4	3.0	3.1	6.8	8.5	13.7	24.3	34.1
10	1.5	2.7	3.5	4.1	4.4	4.7	8.6	14.9	23.0	32.5
11	1.0	1.7	2.6	5.0	7.6	8.4	9.3	13.2	19.6	32.2
12	2.4	2.5	2.8	3.5	3.9	8.1	8.5	16.2	18.9	33.9
13	2.0	2.9	3.2	4.1	4.6	4.8	11.7	13.2	19.9	35.0
14	1.1	3.1	3.3	3.3	4.4	6.7	8.9	17.8	23.0	30.0
15	1.9	2.1	2.4	2.6	2.8	4.7	6.2	17.4	28.4	33.7
16	1.1	4.0	4.0	4.1	4.1	4.9	11.4	13.2	21.5	34.0
17	1.3	1.8	2.1	2.7	3.6	3.8	11.6	17.7	24.8	33.1
18	3.1	4.1	4.7	5.0	6.9	7.2	7.7	9.1	21.2	34.6
19	1.4	2.6	3.3	3.4	4.4	7.6	12.5	17.9	18.1	32.5
20	3.5	3.6	4.2	4.8	7.8	7.8	9.2	12.6	17.2	34.7

Then, according to distribution of EV mileage, the mileage of 200 EVs can be acquired by Monte Carlo simulation, as given in Table 1. For the convenience of EV grouping in consistency with a grouping serial charging strategy, 200 EVs are divided into 20 groups and each group is equipped with 10 EVs on the premise of considering the diversity of charging demands. In addition, it is also required that proportions occupied by EVs of different charging demands should be the same as each other to the greatest extent. That is, minor differences are allowed. If the charging demand of an EV is electric quantity needed by a single travel, the charging demand of this EV can be acquired according to its mileage and power consumption per mile.

It is less likely to directly acquire the power consumption per mile of EVs and its value usually ranges between 0.24 kWh and 0.48 kWh (Chen, L.D. & Nie, Y.Q. 2015). In detail, it is under influences of multiple factors including battery life, road traffic, outdoor temperature and battery management system efficiency, etc. To simplify analysis, it is supposed here that battery lives, road traffics, outdoor temperatures and battery management system efficiencies of all EVs are identical. For the sake of being conservative, power consumptions per mile of EVs are all set at 0.48 kWh. Besides, charge characteristics of EV power batteries adopting conventional charging method depend on power output characteristics of the charging facility used. This is different from quick charge. Hence, restrictions of

Table 2. Cost of charger.

Level	Purchase cost/$	Installation cost/$	Maintenance cost/$
L1	600	300	60
L2	700	800	80
L3	1000	800	100

the permissible battery power characteristic curves over EV charging power are not taken into account. At present, there are a variety of charging piles on the market. Likewise, to simplify analysis, only charging piles of three typical power levels denoted as L1, L2 and L3 are selected. Correspondingly, the maximum outputs are 1.8 kW, 7.2 kW and 9.6 kW respectively. Charging pile configuration cost is constituted by purchase cost, installation cost and maintenance cost. For specifics, please refer to Table 2. Among them, the annual maintenance cost of a charger is generally 10% of its purchase cost.

As the total electric quantities charged by EVs studied are the same, the total charging costs are identical regardless of the charging strategies adopted. Meanwhile, the workplace charging stations are constructed on the original parking lots, extra civil engineering costs are not required. Therefore, charging station operating cost mainly depends on charging facility configuration cost in addition to costs of charging and civil engineering. Considering that the service life of a charging pile is 5–8 years, the average annual operating costs of charging piles in three levels can be figured out by supposing that the service life is 6 years and they are 210, 330 and 400 dollars separately. Additionally, the uniform electricity price of charging employed in this chapter is based on charging cost and charging facility operating cost. It plays a role in guiding coordinated charging of EVs. Nevertheless, specific setting of relevant electricity price of charging is not studied here.

4.2 *Analysis*

According to basic parameters given in Part 4.1, under a circumstance that coordinated charging of EVs in two charging scenarios is carried out based on different charging strategies (EV charging controllable), concrete simulation results and comparative analysis of impacts of charging stations, EV users themselves and power grid are as follows.

Regarding a charging station, 200 L1 charging piles are equipped in a condition of parallel charging strategy, while 3 L2 and 8 L3 charging piles are provided for the grouping serial charging strategy. Charging pile configuration costs and the average annual operating costs incurred by both charging strategies are given in Table 3 below.

Table 3 shows that, comparing with the parallel charging strategy, purchase cost, installation cost, maintenance cost and the average annual operating cost of charging piles in a condition of grouping serial charging strategy are all lower. Among them, the average annual operating cost reduces by 90% (that is, 37,810 dollars). Therefore, the grouping serial charging strategy has a better capability than the parallel charging strategy to enormously bring down the operating cost of charging stations. In addition, charging facility utilization factors of both charging strategies are 26.74% and 97.81% respectively (here, operating duration of a charging facility is 10 hours, not 24 hours). Resultantly, charging facilities based on a grouping serial charging strategy rather than the parallel charging strategy has lower redundancy.

As far as EV users are concerned, decrease in electricity price of charging can lead to reduction of charging cost under a circumstance that charging demands are invariant. However, charging station operators mainly consider operating cost, power purchase cost and yield of charging facilities at the time of formulating the electricity price of charging. Hence, in the case that there exist definite yield and power purchase cost of charging stations, grouping serial charging strategy is able to substantially bring down operating costs of charging facilities if compared with the parallel charging strategy. Furthermore, the electricity price of charging formulated for charging stations based on a grouping serial charging strategy is also lower. In other words, the grouping serial charging strategy can reduce charging costs for EV users. According to this strategy, the charging pile optimal distribution scheme is given in Fig. 3 where ellipses in orange represent EVs charged by L2 charging piles and blue ellipses stand for those charged by L3 charging piles.

Figure 3 indicates that 46 vehicles in a parking lot are charged by L2 charging stations according to a grouping serial charging strategy, while 154 vehicles are charged by L3. Considering EV parking positions and charging facility layout, EVs of two kinds are grouped as follows. EVs charged by L2 are divided into 3 groups and the number of vehicles in each group is 14, 16 and 16 respectively. With regard to those charged by L3, they are classified into 8 groups and the quantities of vehicle in each group are 18, 18, 18, 20, 20, 20, 20 and 20. There are varieties of EV grouping schemes in concrete. Here, only one optimal grouping scheme of EV charging is presented, as shown in Figure 4.

In consistency with Figure 4, EVs in every two lines are on behalf of a group from top to down. In total, there are 11 groups. Every group of EVs

Table 3. Cost of charger in two charging strategies.

Charging strategy	Purchase cost/$	Installation cost/$	Maintenance cost/$	Annual operating cost/$
Parallel	120000	60000	12000	42000
Serial	10100	8800	1040	4190

1.1	1.2	1.3	1.4	1.5	1.6	1.7	1.8	1.9	1.10
2.1	2.2	2.3	2.4	2.5	2.6	2.7	2.8	2.9	2.10
3.1	3.2	3.3	3.4	3.5	3.6	3.7	3.8	3.9	3.10
4.1	4.2	4.3	4.4	4.5	4.6	4.7	4.8	4.9	4.10
5.1	5.2	5.3	5.4	5.5	5.6	5.7	5.8	5.9	5.10
6.1	6.2	6.3	6.4	6.5	6.6	6.7	6.8	6.9	6.10
7.1	7.2	7.3	7.4	7.5	7.6	7.7	7.8	7.9	7.10
8.1	8.2	8.3	8.4	8.5	8.6	8.7	8.8	8.9	8.10
9.1	9.2	9.3	9.4	9.5	9.6	9.7	9.8	9.9	9.10
10.1	10.2	10.3	10.4	10.5	10.6	10.7	10.8	10.9	10.10
11.1	11.2	11.3	11.4	11.5	11.6	11.7	11.8	11.9	11.10
12.1	12.2	12.3	12.4	12.5	12.6	12.7	12.8	12.9	12.10
13.1	13.2	13.3	13.4	13.5	13.6	13.7	13.8	13.9	13.10
14.1	14.2	14.3	14.4	14.5	14.6	14.7	14.8	14.9	14.10
15.1	15.2	15.3	15.4	15.5	15.6	15.7	15.8	15.9	15.10
16.1	16.2	16.3	16.4	16.5	16.6	16.7	16.8	16.9	16.10
17.1	17.2	17.3	17.4	17.5	17.6	17.7	17.8	17.9	17.10
18.1	18.2	18.3	18.4	18.5	18.6	18.7	18.8	18.9	18.10
19.1	19.2	19.3	19.4	19.5	19.6	19.7	19.8	19.9	19.10
20.1	20.2	20.3	20.4	20.5	20.6	20.7	20.8	20.9	20.10

| charging EV by charger L2 | charging EV by charger L3 |

Figure 3. Optimized scheme of charger in packet successive charging strategy.

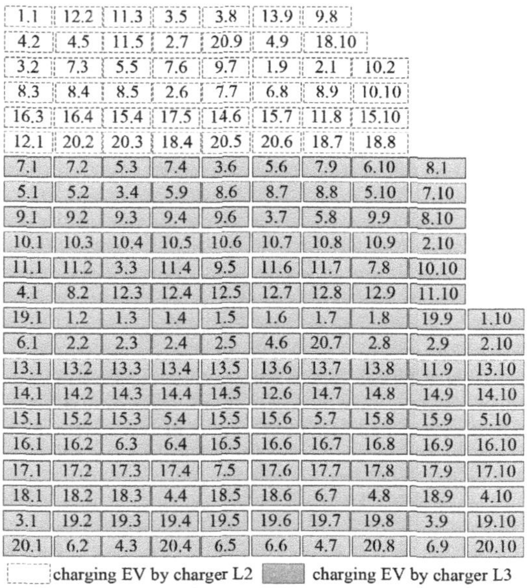

1.1	12.2	11.3	3.5	3.8	13.9	9.8			
4.2	4.5	11.5	2.7	20.9	4.9	18.10			
3.2	7.3	5.5	7.6	9.7	1.9	2.1	10.2		
8.3	8.4	8.5	2.6	7.7	6.8	8.9	10.10		
16.3	16.4	15.4	17.5	14.6	15.7	11.8	15.10		
12.1	20.2	20.3	18.4	20.5	20.6	18.7	18.8		
7.1	7.2	5.3	7.4	3.6	5.6	7.9	6.10	8.1	
5.1	5.2	3.4	5.9	8.6	8.7	8.8	5.10	7.10	
9.1	9.2	9.3	9.4	9.6	3.7	5.8	9.9	8.10	
10.1	10.3	10.4	10.5	10.6	10.7	10.8	10.9	2.10	
11.1	11.2	3.3	11.4	9.5	11.6	11.7	7.8	10.10	
4.1	8.2	12.3	12.4	12.5	12.7	12.8	12.9	11.10	
19.1	1.2	1.3	1.4	1.5	1.6	1.7	1.8	19.9	1.10
6.1	2.2	2.3	2.4	2.5	4.6	20.7	2.8	2.9	2.10
13.1	13.2	13.3	13.4	13.5	13.6	13.7	13.8	11.9	13.10
14.1	14.2	14.3	14.4	14.5	12.6	14.7	14.8	14.9	14.10
15.1	15.2	15.3	5.4	15.5	15.6	5.7	15.8	15.9	5.10
16.1	16.2	6.3	6.4	16.5	16.6	16.7	16.8	16.9	16.10
17.1	17.2	17.3	17.4	7.5	17.6	17.7	17.8	17.9	17.10
18.1	18.2	18.3	4.4	18.5	18.6	6.7	4.8	18.9	4.10
3.1	19.2	19.3	19.4	19.5	19.6	19.7	19.8	3.9	19.10
20.1	6.2	4.3	20.4	6.5	6.6	4.7	20.8	6.9	20.10

⬚ charging EV by charger L2　■ charging EV by charger L3

Figure 4. An optimal grouping scheme for charging EVs.

can be evenly distributed on both sides of a charging pile.

For power grid, the total charging demands of EVs based on both charging strategies are 962.49 kWh and the adjusted optimal constant charging load of EVs is 96.25 kW. Meanwhile, standard deviations of both loads are 0. Therefore, EV charging dependent on either charging strategy has the same influence on power grid.

Accordingly, charging effects of workplace charging stations that adopt a power supply mode of power grid and the grouping serial charging strategy are better than those obtained by the parallel charging strategy, provided that conventional load is far less than EV charging load (impacts of conventional loads are ignored). In detail, not only can the operating cost of charging station be dramatically reduced, but EV charging costs also drop. At the same time, such two strategies have the same influence on power grid and both are reduced to the lowest limits.

5　CONCLUSION

The charging strategy for EVs has been studied. In order to coordinate relations between charging facility configuration and charging of Electric Vehicles (EVs), two coordinated charging strategies based on uniform electricity price or the conventional constant power charging ways were presented here for EVs according to conventional load types at workplace, parking characteristics of EVs and matching of parking space and charging infrastructure. In detail, they were parallel charging strategy and grouping serial charging strategy. Workplace power distribution systems were subsequently constructed for such two strategies. Besides, the average annual operating cost of charging station was used as an objective function to establish an EV charging optimization model.

Finally, the simulation results indicated that, comparing with the parallel charging strategy, the grouping serial charging strategy was able to not only bring down operating costs of charging stations and charging costs of EVs, but alleviate impacts on a power grid. In this manner, an effective solution could be provided for charging facility configuration and EV charging.

ACKNOWLEDGMENT

This work was supported in part by Science and Technology Planning Project of Jiangsu Province (Grant no. BE2015004-4), and Major State Research Development Program of China (2016YFB0101800).

REFERENCES

Chen, L.D. & Nie, Y.Q. 2015. A model for electric vehicle charging load forecasting based on trip chains. Transactions of China Electrotechnical Society 30(04):216–225.

Hu, Z.C. & Song, Y.H. 2012. Impacts and utilization of electric vehicles integration into power systems. Proceedings of the CSEE 32(4): 1–10.

Lu, X.Y. & Liu, N. 2014. Multi-objective optimal scheduling for PV-assisted charging station of electric vehicles. Transactions of China Electrotechnical Society 29(8): 46–56.

Sun, X.M. & Wang, W. 2013. Coordinated charging strategy for electric vehicles based on time-of-use price. Automation of Electric Power System 37(1): 191–195.

Wang, X.F. & Shao, C.C. 2013. Survey of electric vehicle charging load and dispatch control strategies. Proceedings of the CSEE 33(1):1–10.

Wang, Z.Y. & Cao. Y.J. 2007. Electric power system load profiles analysis. Proceedings of the CSU-EPSA 19(3): 62–65.

Xu, Z.W. & Hu, Z.C. 2014. Coordinated charging strategy for PEV charging stations based on dynamic time-of-use tariffs. Proceedings of the CSEE 34(22): 1–10.

Zhang, L. & Yan, Z. 2014. Two-stage optimization model based coordinated charging for EV charging station. Power System Technology 38(4): 967–973.

Zhao, G. & Huang, X.L. 2015. Coordinated control of PV-generation and EVs charging based on improved DECell algorithm. International Journal of Photoenergy.

Automotive, Mechanical and Electrical Engineering – Liu (Ed.)
© 2017 Taylor & Francis Group, London, ISBN 978-1-138-62951-6

Analysis and study of an energy recycling segment erector hydraulic system

Jingang Liu & Xiaoqun Zhou
School of Mechanical Engineering, Xiangtan University, Hunan, Xiangtan, China

Kai Wang
China Railway Construction Heavy Industry Co. Ltd., Changsha, Hunan, China

ABSTRACT: The function of the segment erector's horizontal movement hydraulic system is to feed the segment for erection; however, due to the large overflow loss of the existing system, the hydraulic system produced a lot of heat, which led to low efficiency. To solve this problem, a new type of segment erector's horizontal movement hydraulic system was designed based on energy saving demands. The new system can realise energy recycling, and can ensure that the hydraulic cylinder is fast forwarded on a no-load condition and slowed down on a load condition, which can improve work efficiency. The simulation analysis and experimental verification of the new hydraulic control system were undertaken, and the results show that the new system is reasonable, and that it has a significant effect on energy recycling and the utilisation of the system.

Keywords: Segment erector; Hydraulic system; Overflow loss; Energy recycling

1   INTRODUCTION

An erector is a piece of mechanical equipment that operates the segment lining of the tunnel surface after a complete ring in the tunnelling shield machine. As an important supporting system of the shield machine, it has an important influence on the efficiency of the whole tunnel excavation construction. The hydraulic system is the power and control centre of the segment erector, and the horizontal movement hydraulic system achieves the level of feed plate segment function, which has a great influence on its efficiency. However, the overflow loss of the existing system is large, and it cannot achieve energy recovery and utilisation, which would result in serious heating system and low efficiency (Cui, G., 2009; Ding, S., 2005; Guan, C., 2012). Designing a hydraulic system with high power density and high efficiency is urgently needed.

To solve this problem, a new type of segment erector's horizontal movement hydraulic system was designed based on energy saving demands and the improvement of work efficiency. The system can realise energy recycling, achieve the ideal effect of working fast on no-load and slow on load, and enhance the working efficiency of the system. By using the numerical simulation software AMESim, the new hydraulic system is simulated and analysed.

Finally, test analysis of the new hydraulic system of the shield erector machine is carried out.

2   PRINCIPLE OF SEGMENT ERECTOR'S HORIZONTAL MOVEMENT HYDRAULIC SYSTEM

In this paper, the research object is a type of earth pressure balance shield machine. The segment erector's horizontal movement hydraulic system consists of four main parts: the variable pump, multi-way valve with load sensing, balance circuit and hydraulic cylinder (Hu, G., 2006). The diagram is as shown in Figure 1. Because of the optimisation of the balance circuit, the pump source and multi-way valve with load sensitive are simplified.

As is shown in Figure 1, the hydraulic cylinder works on the no-load condition when the multi-way valve works on the left, ignoring the friction that the hydraulic cylinder produces in the action process. At this time, the oil enters the rodless cavity of the hydraulic cylinder through the one-way valve of the balance valve, and the rod cavity oil returns to the tank through the balance valve and multi-way valve. It can be seen that the energy consumption of this system is huge, and it cannot meet the requirement of working fast on no-load. According to the field test, the temperature of the

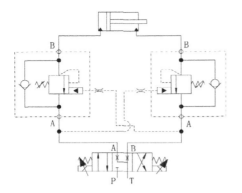

Figure 1. Principle of horizontal movement hydraulic system.

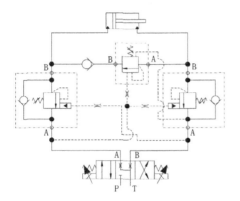

Figure 2. The principle of the improved horizontal movement hydraulic system.

balance valve returning oil and the oil itself are very high.

In order to achieve the requirements of working fast on no-load and slow on load (Ren, D., 2006; Shi, H., 2012), the horizontal movement hydraulic system is optimised based on the purpose of the energy recovery. The principle of the improved system is as shown in Figure 2.

As is shown in Figure 2, an external control pressure sequence valve is added to the return oil circuit of this system. The control oil for this pressure sequence valve and return oil balance valve is taken from the inlet of the A port of the balance valve. When the multi-way valve works on the left, part of the oil enters the rodless cavity of the hydraulic cylinder through the one-way valve of the balance valve, and a small portion of it controls the balance valve returning oil and pressure sequence valve as the control oil. Because the pilot ratio of the pressure sequence valve is bigger than that of the balance valve, the pressure sequence valve opens first. When the oil returns, the oil of the rod cavity of

the hydraulic cylinder communicates with the rod-less cavity through the pressure sequence valve in order to form a double-acting cylinder, so that the overflow loss of the system is greatly reduced and the speed of working on no-load is faster. When the system is disturbed, the load increases, and the balance valve returning oil opens in order to form the B type hydraulic half bridge, which makes the system keep a stable pressure and reduce the disturbance caused by the jitter of the system. When the multi-way valve works on the right, the pressure sequence valve closes under the inlet pressure and the pressure regulating spring. Then the oil enters the rod cavity of the hydraulic cylinder through the one-way valve of the balance valve, before turning back to the tank overflow through the balance valve inlet oil. Due to the throttling of the balance valve inlet oil, the system has back pressure, so that the hydraulic cylinder achieves the goal of working slow on load.

Through the analysis, we know that the improved hydraulic system can not only realise the energy recovery and utilisation, but also achieve the function of working fast on no-load and working smooth return on load. In addition, it can significantly improve the efficiency of the system (Wang, L., 2012; Zhang, S., 2010; Zhang et al., 2007).

3 ANALYSIS OF THE OPTIMISED SYSTEM BASED ON AMESIM

3.1 Simulation model

According to the principle of the segment erector's horizontal movement hydraulic system, the simulation model is built in AMESim. The segment erector's horizontal movement hydraulic system is used in a constant power variable pump with the pressure cutting off, according to the characteristic curve of the variable pump, the model of pump has also been built in AMESim (Zhu, J., 2011). By considering the external interference, the simulation model of the new type of segment erector's horizontal movement hydraulic system is built in AMESim, as shown in Figure 3.

3.2 Setting parameters

The main parameters of the AMESim model of the optimised segment erector's horizontal movement hydraulic system are shown in Table 1.

3.3 The analysis of simulation results

1. The comparison of the piston speed of the hydraulic cylinder before and after optimisation. By adding the pressure sequence valve to the back oil circuit of the segment erector's horizontal movement hydraulic system, the hydraulic cylinder becomes a double-acting cylinder

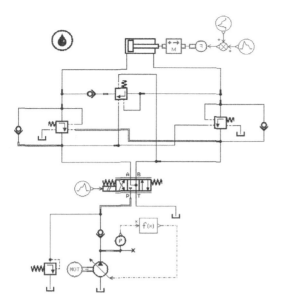

Figure 3. The simulation of the segment erector's horizontal movement hydraulic system.

Table 1. The main parameters of the horizontal movement hydraulic system.

Parameters	Value
Setting pressure of pressure relief valve (MPa)	25
Density of oil liquid (kg/m³)	850
Pilot ratio of balance valve	1:3
Pilot ratio of pressure sequence valve	1:1
Piston diameter of hydraulic cylinder (mm)	50
Piston rod diameter of hydraulic cylinder (mm)	80
Oil elastic modulus (MPa)	700
Maximum load force (KN)	80
Pump delivery (mL/rev)	140
Motor speed (rev/min)	1470

when it works on no-load, thus improving both the speed of the hydraulic cylinder piston and the efficiency. The simulation time is 6 s in total, of which 0~2.5 s is working no-load, and 2.5~6 s is returning on load. In addition, the simulation step size is 0.001 s. The contrasts between the speed of the hydraulic cylinder piston before and after optimisation are shown in Figure 4.

According to Figure 4, between 0~2.5 s the hydraulic cylinder works on no-load. The speed of the hydraulic cylinder piston is 0.06 m/s before the optimisation, and reaches 0.1 m/s after optimisation. Between 2.5~6 s, the hydraulic cylinder is back on load, and the speed is 0.06 m/s. Through the above analysis, the optimised segment erector's horizontal movement hydraulic system could improve the speed when

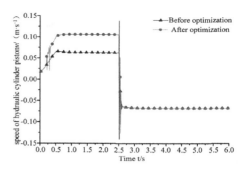

Figure 4. The speed of the hydraulic cylinder piston before and after optimisation.

working on no-load and reduce the speed when back on load. In this way, the work efficiency of this system can be improved significantly.

2. The comparison of the overflow loss of the system before and after optimisation.

Before the segment erector's horizontal movement hydraulic system is optimised, when the hydraulic cylinder works on no-load, the oil in the rod cavity of the hydraulic cylinder enters the tank through the balance valve returning oil and multi-way valve directly. In this way, it would cause a lot of overflow loss, increase the temperature of the system, and reduce the efficiency. After the optimisation, the system can recover and utilise energy, reduce the overflow loss when working on no-load, and improve the utilisation rate of energy. The simulation time is 6 s in total, of which 0~2.5 s is working no-load, and 2.5~6 s is returning on load. In addition, the simulation step size is 0.001 s. The comparison of the overflow loss of the system before and after optimisation is shown in Figure 5.

According to Figure 5, between 0~2.5 s the hydraulic cylinder works on no-load. After optimisation, the overflow loss can reduce by 1/3; between 2.5~6 s the hydraulic cylinder works back on load, the oil of both before and after optimised system is flowed into rod cavity of hydraulic cylinder through the one-way valve of balance valve returning oil. In this way, the overflow quantity is 0. Through the above analysis, we can draw the conclusion that the overflow loss can be reduced during the hydraulic cylinder working on no-load by optimising the segment erector's horizontal movement hydraulic system, and the energy can be recycled. Thus, improving the energy utilisation ratio and realising the targets of energy saving and emission-reduction.

3. The influence of external disturbance on the optimised system

The B type hydraulic half bridge is created in the return circuit of the system by adding the

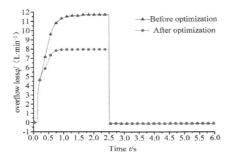

Figure 5. The comparison of the overflow loss of the system before and after optimisation.

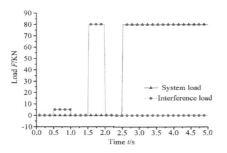

Figure 6. External load.

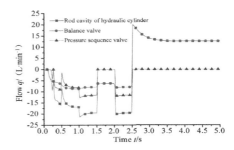

Figure 7. Flow rate of the system.

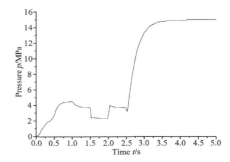

Figure 8. Pressure of rod cavity of the hydraulic cylinder.

pressure sequence valve in the segment erector's horizontal movement hydraulic system, which could reduce the systems jitters caused by external disturbance significantly. The load on the hydraulic cylinder is as shown in Figure 6. Between 0~2.5 s, it works on no-load. Between 2.5~6 s, when working on load, applying 8 KN interfering load in 0.5~1 s, and 80 KN interfering load in 1.5~2 s, the external load of the system is 80 KN.

The simulation time of the system is 5 s, and the simulation step size is 0.001 s, the varying curve of flow rate and the varying curve of pressure of the rod cavity of the hydraulic cylinder are shown in Figure 7 and Figure 8, respectively.

As shown in Figure 7, between 0.5~1 s an 8 KN interfering load is applied to the system. The oil flow through the balance valve is more than that of the pressure sequence valve. Because the balance valve is opened to realise the function of overflow due to the control pressure of the system increasing with the load going up, it could reduce the vibration of the system caused by external disturbance significantly. Between 1.5~2 s, the system is suffering an 80 KN interfering load, and there is no oil flowing through the pressure sequence valve, because the balance valve opened entirely, realising the over load protection of the system. Between

2.5~5 s, there is no oil flowing through the pressure sequence valve, because the pressure sequence valve closes due to the control pressure and spring force, and the oil is forced to enter the rod cavity of the hydraulic cylinder through the one-way valve of the balance valve returning oil.

It can be seen from Figure 8 that between 0.5~1 s, as the system is disturbed by 8 KN, the pressure of the rod cavity of the hydraulic cylinder has any obvious fluctuation. Between 1.5~2 s, the system is subjected to an 80 KN interference load, and due to the large interference, the system overflows rapidly and the pressure of the rod cavity decreases. Between 2.5~5 s, the pressure of the rod cavity is relatively stable. Through the above analysis, we can draw the conclusion that the horizontal movement hydraulic system after optimisation can obviously retard the phenomenon of dithering of the system by external disturbance, and could also be got system overload protection.

4 TEST ANALYSIS

In order to verify the validity of the simulation results, and provide the basis for the optimisation of the system, the tests for the segment erector's horizontal movement hydraulic system were carried out. The flow meter was fitted in the inlet of the rod cavity of the hydraulic cylinder and the

242

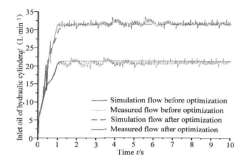

Figure 9.   Inlet oil of the hydraulic cylinder.

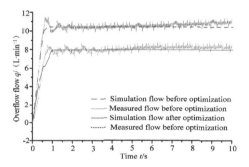

Figure 10.   Flow of overflow.

outlet of the balance valve returning oil, and the measured flow and the simulation flow are shown in Figure 9 and Figure 10.

From Figure 9 and Figure 10, the trend of the flow measured and simulated is basically consistent, which can verify that the simulation results are authentic and reliable.

As is shown in Figure 9, the inlet oil of the rodless cavity is about 20 L/min before optimisation, while the inlet oil of the rodless cavity is about 32 L/min after optimisation. A differential circuit is formed in this system on the no-load condition after optimisation, which could achieve the energy recovery and utilisation, and reach the work requirements of idler and fast forward.

As is shown in Figure 10, the overflow flow rate of the balance valve returning oil is about 11.5 L/min before optimisation, and is about 8 L/min after optimisation. The overflow loss of the system has decreased obviously, and the horizontal movement hydraulic system after optimisation can significantly increase the system's energy utilisation.

5   CONCLUSION

In this study a new type of segment erector's horizontal movement hydraulic system is optimised. It can not only achieve the goal of energy recovery and utilisation, but also ensure that the hydraulic cylinder is fast forward on no-load condition and slowed down on load condition, which can significantly improve the work efficiency of the system. The simulation analysis and test verification of the new type of hydraulic system are carried out, which shows that the design of the system is reasonable and it has a significant effect on energy utilisation and the improvement of efficiency.

ACKNOWLEDGEMENTS

This work is supported by the Major Project of Science and Technology of Hunan Province under Grant No. 2014FJ1002, the National Natural Science Foundation of China under Grant No. 51475402, and the Educational Commission of Hunan Province under Grant No. 15A185.

REFERENCES

Cui, G. (2009). *Research on Design and Dynamic Performance for Segment Assembly Erector of Shield Tunnel Machine*. Jilin University, Changchun (in Chinese).

Ding, S. (2005). *The Research of the Electro-Hydraulic Control System of the Shield's Erector*. Zhejiang University, Hangzhou (in Chinese).

Guan, C. (2012). Recovering system of swing braking energy in hydraulic excavator. *Journal of Zhejiang University (Engineering Science)*, 46(1), 142–149.

Hu, G. (2006). *Research into Electro-Hydraulic Control System for a Simulator Test Rig of Shield Tunneling Machine*. Zhejiang University, Hangzhou (in Chinese).

Ren, D. (2006). *The Research on Control System of the Shield Erector Test Rig*. Zhejiang University, Hangzhou (in Chinese).

Shi, H. (2012). *Investigation into Electrohydraulic Control Systems for Shield Tunneling Machine and Simulated Experiment Method*. Zhejiang University, Hangzhou (in Chinese).

Wang, L. (2012). Positioning Precision and Impact Force Control of Segment Erector for Shield Tunneling Machine. *Digital Manufacturing and Automation (ICDMA), 2012 Third International Conference on. IEEE*, 612–617.

Zhang, S. (2010). Energy flow of a hydraulic excavator with load independent flow distribution. *Mechanical Science and Technology for Aerospace Engineering*, 29(1), 94–99.

Zhang, Y., Wang, Q. F., & Xiao, Q. (2007). Hybrid hydraulic excavator hydraulic motor energy recovery of simulation and experiment. *Journal of mechanical engineering*, 43(8), 218–223+228.

Zhu, J. (2011). Analysis and experimental study on hydraulic system work efficiency for recovering the potential energy of electric forklift. *Machine Design and Research*, 27(6), 101–104.

Automotive, Mechanical and Electrical Engineering – Liu (Ed.)
© 2017 Taylor & Francis Group, London, ISBN 978-1-138-62951-6

Comprehensive evaluation of microgrid planning schemes based on the Analytic Hierarchy Process (AHP) method

Jun He
State Grid Hubei Electric Power Research Institute, Wuhan, China

Dahu Li
State Grid Hubei Electric Power Company, Wuhan, China

Kunpeng Zhou, Kun Chen & Can Cao
State Grid Hubei Electric Power Research Institute, Wuhan, China

Long Cheng
State Grid Shiyan Power Supply Company, Shiyan, China

ABSTRACT: A comprehensive evaluation method is proposed for optimising a microgrid power planning scheme considering dynamic time and space characteristics. First, a stochastic process model is built to describe the dynamic characteristics of the microgrid power planning project. Then, distributed energy generation and energy price risk index are proposed to evaluate the potential risk. Finally, the analytic hierarchy process method is used to determine the weight of each index, and to make a comprehensive evaluation of the different planning schemes in an independent microgrid. The simulation of an island microgrid shows that the proposed method can optimise a microgrid planning scheme from different time and space features, and provide reference for formulating the energy planning scheme and promoting the development of low carbon power.

Keywords: Microgrid power planning; Fuel price and technological uncertainty; Assessment metrics; AHP method

1 INTRODUCTION

How to establish the evaluation index system for microgrid planning, giving optimal decision basis, has become an urgent problem. In the current study on microgrid planning evaluation research, economic, environmental and reliability evaluation indexes were proposed. For example, metrics for assessing the reliability and economics of microgrids in a distribution system were proposed in paper. In the research on the reliability index, the output characteristics of intermittent power, and influence of power supply access capacity or position on reliability were analysed. In the research on the economic index, the supply capacity optimisation is based on the least investment cost. In the research on the environmental index, the literature shows the effect of emission price changes on low carbon power supply planning.

However, none of these evaluation index systems consider the investment characteristics of dynamic time during the microgrid planning.

If the starting year of a microgrid planning project does not start from the current year, but is delayed to a future year, due to the uncertainty of unit construction costs and fuel costs in the future, it is hard to calculate the economic index of the microgrid planning. Hence, it is hard to make the decision of "when to start a microgrid investment project".

In paper, investment factors considering the price of fuel and technology uncertainty were discussed, using the Nash-Cournot equilibrium model to assume that these two factors are constant or fluctuate by actual cases, which indicates that units and fuel price fluctuation have great influence on microgrid power planning. The research believes that the new energy power generation project investment income will be affected by the speed of technological progress, government investment subsidy policy, carbon taxes and other factors. In paper, an investment cost economics model changing with time is proposed considering technical progress and tax uncertainty. In paper a unit and fuel price uncertainty model was established based on the theory of real option, and qualitatively analysed the influence of power planning.

In addition, in the current study, efficiency indicators, and complementary benefit evaluation of different intermittent power supply. Furthermore, there is a lack of comprehensive evaluation of microgrid planning schemes.

In this paper, based on the traditional microgrid planning and evaluation index, a comprehensive evaluation method for optimising a microgrid power planning scheme considering dynamic time and space characteristics is proposed, mainly from the following aspects:

1. Based on the real option theory, a stochastic process model is proposed to describe the dynamic spatial and temporal characteristics in microgrid planning, and the energy price risk index is proposed.
2. A complementary efficiency indicator is proposed in power supply planning, in order to comprehensively evaluate the effectiveness of a variety of complementary energy supply.
3. The Analytic Hierarchy Process (AHP) method is introduced to define the weight of each index, and make a comprehensive evaluation of the microgrid planning scheme, considering different space-time characteristics.

2 STOCHASTIC PROCESS MODEL OF FUEL PRICE AND TECHNOLOGICAL UNCERTAINTY

2.1 Stochastic process model for describing technological progress uncertainty

To predict the future construction costs and the operation costs of a microgrid plan, a stochastic process model is proposed to assess the price and technological uncertainty based on the real option theory.

The initial investment cost of power units is the most important investment during power planning, but the progress on production technology directly affects the construction costs of the unit. Suppose a power planning project is started in year t: t is the first year when technology advances occur, and then the project will benefit from the decrease in the construction costs, this part of the cost causes the whole investment costs to reduce. Technological innovation is an external factor in cost reduction, assuming that technology changes over the course of time are random, and t > 0, the construction costs of the unit are:

$$I_t = I_0 \phi^{N_t} \tag{1}$$

where I_0 stands for the unit cost of construction when t = 1, N_t is the random Poisson variable,

assume λt is the amount of recording technology innovation, $\phi \in [0,1)$ is a constant, standing for the degree of technology innovation. Parameter λ is the arrival rate of technological innovation. Obviously, the unit cost I_t has an exponential relationship with time t:

$$E[I_t] = I_0 e^{-\lambda t(1-\phi)} = I_0 e^{-\gamma t} \tag{2}$$

In Formula (2), $\gamma = \lambda t(1 - \phi)$. Therefore, if you adjust the parameters λ or ϕ, such as increasing the arrival rate λ, while reducing the level of technology innovation ϕ, the variable $\lambda(1 - \phi)$ may not change. In other words, technical progress uncertainty can be adjusted by adjusting λ or ϕ.

2.2 Wiener process model for describing the fuel price uncertainty

In a microgrid, traditional high-carbon units such as gas turbine engines need to rely on fossil fuels to generate electricity, so the fuel price directly affects the traditional operation cost of the unit. In this paper, the fuel cost in year t is assumed that the process of change in the future to meet the geometric Brownian motion (Wiener):

$$\frac{dP_t}{P_t} = \mu dt + \sigma dz \tag{3}$$

where μ and σ represent this Brownian motion process drift rate and volatility parameters, dz is the amount change of each step in a standard Brownian motion process. Assume $0 \le \mu < r$, in which r is the discount rate. According to the mathematical properties of geometric Brownian motion, the future fuel price's expectation value is:

$$E[P_t] = P_0 e^{\mu t} \tag{4}$$

The fuel price uncertainty parameters can be adjusted by σ. From the above equation, adjusting the parameter does not change the motion paths of the Brownian process.

3 EVALUATION INDEX OF MICROGRID PLANNING CONSIDERING DYNAMIC TEMPORAL AND SPATIAL CHARACTERISTICS

3.1 Reliability metrics

The evaluation index system of microgrid planning proposed in this paper is shown in Figure 1.

In this paper, the reliability metrics mainly include loss of load probability, loss of load expectation, and loss of load frequency.

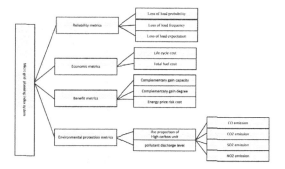

Figure 1.   Flow chart for the metrics calculation.

3.1.1 Loss of Load Probability, LOLP
LOLP stands for the probability of load demand not satisfied in a given time interval.

$$LOLP = \sum_{i \in s} P_i \qquad (5)$$

where P_i is the probability that the microgrid stays in state i; s stands for all stats that load demand is not satisfied in a given time interval.

3.1.2 Loss of Load Expectation, LOLE
LOLE stands for the expected load demand not supplied in a given time interval.

$$LOLE = \sum_{i \in s} P_i T \qquad (6)$$

where T stands for the number of hours or years, and the measure of LOLE is usually h/a or d/a.

3.1.3 Loss of Load Frequency, LOLF
LOLF stands for the frequency of the load demand not being supplied in a given time interval.

$$LOLF = \sum_{i \in s} F_i \qquad (7)$$

where F_i stands for the frequency of microgrids staying in state i.

3.2 Economic metrics

In this paper, life cycle cost and total fuel cost are introduced.

3.2.1 Life cycle cost
The economic indices include low-carbon forms of energy use (wind turbine, photovoltaic array, energy storage device) and other high-carbon energy use (mainly for thermal power units, gas turbine), while reliability meet the requirements of

Table 1.   Objective function symbol descriptions.

Symbol	Description
N	The number of the type of power generation
x_i	The number of power generation for i
J	Life cycle from j year
I_{ti}	Initial investment from t year
C_{pti}	Carbon punishment for power resource i
T	Life cycle years
O_{ti}	Operation costs for power resource i and the year i

the user, the mathematical model can be described as (8):

$$C_{CF_j}(x) = \sum_{t=j}^{T+j} \sum_{i=1}^{N} \frac{x_i C_{pt_i} + x_i I_{t_i} + x_i O_{t_i} + x_i M_{t_i} + x_i F_{t_i}}{(1+r)^t} \qquad (8)$$

The symbol descriptions are as Table 1:
where the construction cost of power units in year t is the expectation value I_t equal to $E[I_t]$ as discussed in Chapter II.

3.2.2 Fuel cost
For high-carbon energy, its operation cost and fuel costs in each year are in proportion to the international crude oil prices of that year.

$$O_{ti} = K_{FC_i} E_{ti} P_t \qquad (9)$$

where K_{FC_i} is the scale factor for each fuel cost of high-carbon energy, E_{ti} stands for the electric quantity by the number i type of power units in year i. The international crude oil price P_t is equal to $E[P_t]$ as discussed in the stochastic process model.

For low carbon energy generations, its annual operation costs are associated with the unit capacity.

3.3 Benefit metrics

In this paper, the benefit metrics include complementary gain capacity, complementary gain degree, the proportion of new energy power generation, the proportion of disposable resources, ratio of energy storage utilisation and energy price risk cost.

3.3.1 Complementary gain capacity
Due to the fact that wind and light resources are complementary in time and region, the scenario combining power generation rather than using wind power or solar power alone can make up for losses due to intermittent energy. In this paper, the

definition of complementary gain capacity and complementary degree to measure the benefits.

The scenario combined power generation system is arranged on the complementary capacity gain for

$$C_m = C_{un} - C_{wind} - C_{PV} \qquad (10)$$

where C_m is complementary gain capacity.

3.3.2 Complementary gain degree

Complementary gain degree λ_{un} is calculated as follows:

$$\lambda_{un} = \frac{C_m}{C_{un}} \qquad (11)$$

3.3.3 Energy price risk cost

Energy prices change in the future obey the Wiener process model, the change part is a random variable. If, in the early planning of the microgrid, there is no account of this change, the result will turn out to have deviation. In this paper we define energy price risk indices to characterise this deviation:

$$C_{risk} = \frac{O_{var} - O_{con}}{C_{CF_j}(x)}$$

$$= \frac{\sum_{t=t_0}^{t} K_{FC_i} E_{ti} P_t - \sum_{t=t_0}^{t} K_{FC_i} E_{ti} P_{con}}{C_{CF_j}(x)} \qquad (12)$$

where O_{var} stands for the account of future operation cost fluctuations in each year, O_{con} is the operation cost during the life cycle planning project, assuming the trend of energy prices in future years is constant. P_t is the expectation value of fuel price in year t, P_{con} stands for the data forecast by the U.S. Energy Department in a balanced scenario.

3.4 Environmental protection metrics

In this paper, the reliability metrics include loss of load probability, expected energy not supplied, loss of load frequency and loss of load expectation.

3.4.1 The proportion of high-carbon unit

The proportion of high-carbon unit is the ratio of power generation by high carbon to load:

$$K_{carbon} = \frac{\sum E_{carbon}}{\sum E_{load}} \qquad (13)$$

where $\sum E_{carbon}$ is power generation during the whole year; $\sum E_{load}$ is the load of microgrid during the whole year.

3.4.2 Pollutant discharge level

Pollutant emissions per year during the planning period include CO emissions, CO_2 emissions, SO_2 emissions, NO_2 emissions.

$$E_p = E_{carbon} * K_{var} \qquad (14)$$

where K_{var} is discharge of pollutants per kilowatt hour.

4 CALCULATION OF MICROGRID POWER PLANNING SCHEME BASED ON THE AHP METHOD

The weight vector of each index $W_k = (w_1 w_2 \cdots w_n)^T$ is proposed to describe different index values, and the comprehensive evaluation is calculated using the linear weighted sum method:

$$T = \sum_{k=1}^{n} W_k \mu(x_w) \qquad (15)$$

where $\mu(x_w)$ stands for the actual index normalised value; W_k is the weight of each index value.

The comprehensive evaluation process of the microgrid power planning index is shown in Figure 2.

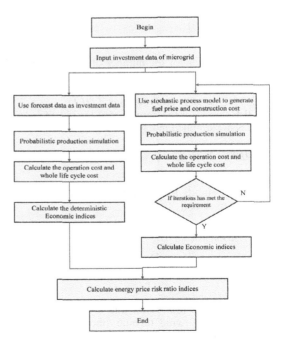

Figure 2. Flow chart for comprehensive evaluation of microgrid power planning.

248

The steps for calculating the comprehensive evaluation data value are as follows:

1. First, according to the power planning, we need to calculate the value of each index, and then compare the indicators of different plans to do the normalisation, so we can get an index coefficient matrix.
2. After the analysis and comparison of the index set, the decision-maker can construct the appropriate index weight matrix and the judgement matrix, then calculate the maximum eigenvalue and the characteristic vector, and determine the reasonable weight matrix.
3. Finally, the comprehensive evaluation value of the microgrid power supply planning scheme is calculated by using the linear weighted sum method.

5 SIMULATION EXAMPLE

Taking a real island grid as an example, the distributed generations are: wind turbines, photovoltaic cells, gas turbines, and energy storage batteries.

5.1 Microgrid planning cases

In order to facilitate the research, four configuration cases of distributed generation (Table 2) are chosen from reference, which can ensure the reliability of the microgrid in the required range:

In Case 1, there is least renewable generation, a high-carbon unit generates the most power, and no battery; Case 2, on the other hand, has the highest proportion of renewable generation; Cases 3 and 4 are compromise solutions.

Assuming that the microgrid planning starts from year 2014, the attribute decisions of the cases above are shown in Table 3:

W is the corresponding weight vector, which can be obtained by AHP method:

$$W = (w_1, w_2, ..., w_{13}) = (0.0822, 0.0822, 0.1643,$$
$$0.1643, 0.0822, 0.0235, 0.0205, 0.0411,$$
$$0.0235, 0.0041, 0.0088, 0.0044, 0.0044)$$

According to the sum weight, the comprehensive assessment values of the four cases are:

Table 2. Four configuration cases of distributed generation.

DGs/MW	Case 1	Case 2	Case 3	Case 4
Wind turbines	0.6	3.2	1.5	2
PV	0.4	1.8	0.5	1.2
Gas turbines	7.8	6.5	7.2	7
Batteries	0	1.2	0.6	0.6

Table 3. Attribute decisions of each case.

	Attribute decision	Case 1	Case 2	Case 3	Case 4
Reliability	LOLP	0.000	0.727	1.000	0.455
	LOLP	0.000	0.727	1.000	0.455
	LOLF	0.889	0.333	1.000	0.000
Economic	Life cycle cost	0.890	0.000	1.000	0.584
	Fuel cost	0.000	1.000	0.465	0.643
Benefit	Complementary gain capacity	0.000	1.000	0.259	0.630
	Complementary gain degree	0.000	0.058	0.038	1.000
	Energy price risk cost	0.000	1.000	0.586	0.731
Environmental protection	The proportion of high-carbon unit	0.000	1.000	0.338	0.559
	CO emissions	0.000	1.000	0.281	0.521
	CO_2 emissions	0.000	1.000	0.281	0.521
	SO_2 emissions	0.000	1.000	0.281	0.521
	NO_2 emissions	0.000	1.000	0.281	0.521

$$T_1 = 0.451, T_2 = 0.520, T_3 = 0.789, T_4 = 0.467$$

which can be sorted as:

$$T_3 > T_2 > T_4 > T_1$$

The results show that, in the case of the island microgrid planning, considering the dynamic space-time characteristics of the evaluation index, and considering the reliability, economic and environmental indicators, configuration Case 3 of distributed generation is the best choice for this microgrid.

5.2 Microgrid planning scheme

According to the future technology progress rate and international energy prices forecast provided by the U.S. energy information administration and the company Mott MacDonald, temporal and spatial characteristics can be divided into low carbon scenario, equilibrium scenario and high-carbon scenario.[21–22] In this paper, the technical progress of uncertainty degree can be adjusted by the parameter λ, and fuel price uncertainty can be adjusted by the parameter ϕ. The parameters are adopted to simulate balanced and high-carbon scenarios.

Based on these two scenarios, the simulation results can be seen in Table 2. The three planning programmes start respectively from year 2011, 2012, to year 2039.

The simulation of the balanced scenario is obtained for the future, and the comprehensive index change trend of the three kinds of planning scheme are shown in Figure 3.

The comprehensive evaluation change trend of the three kinds of planning scheme during high-carbon scenarios are shown in Figure 4.

As can be seen from the above simulation, considering multi-temporal characteristics, comprehensive evaluation of Case 2 will be better than Case 3 in the future. During the balanced scenario, Case 2 will be better than Case 3 in 2021, and in the high-carbon scenario, the singular point year may be delayed until 2032. This is because the uncertainty of the unit investment and construction cost is less in the high-carbon scenario, and new energy power generation accounted for a relatively high level in the balanced scenario, which will benefit to this uncertainty. Science and technology progress and policy support can accelerate the conversion

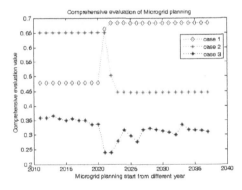

Figure 3. Comprehensive evaluation of microgrid in different years (balanced scenario).

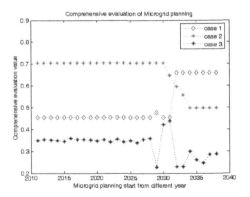

Figure 4. Comprehensive evaluation of microgrid in different years (high-carbon scenario).

of a high-carbon scenario to a balanced scenario, and make renewable energy power sources more cost-effective.

6 CONCLUSION

A comprehensive evaluation method is proposed for optimising a microgrid power planning scheme considering dynamic time and space characteristics. At first, the real options theory and a stochastic process model is introduced to describe the dynamic characteristics of microgrid power planning. Finally, the analytic hierarchy process method is used to determine the weight of each index, and a comprehensive evaluation of the different planning schemes in an independent microgrid is made. The simulation of an island microgrid shows that:

1. There are dynamic space and time characteristics in microgrid power planning, and equivalent capacity of power planning which start from different years can get different comprehensive evaluations.
2. The compromise solution is not necessarily optimal, the future comprehensive evaluations of Case 2 or Case 3 are better than Case 4.
3. Through policy support, when scientific and technological progress and tax incentives are encouraged, the dynamic characteristics of space and time can be adjusted, so that low carbon planning can be more cost-effective at an earlier time.

In summary, the evaluation method proposed can reflect the microgrid power optimisation considering the trend of the temporal dynamic characteristics, and answer the power planning question of "when to start power investment projects".

ACKNOWLEDGEMENTS

This work was supported by the Key Technology Research and Development Programme of the Ministry of Science and Technology of Hubei Province (2015BA A110), and Project of State Grid Hubei Electric Power Company (52153214001 J, 5215M0130467).

REFERENCES

Bøckman, T. & Fleten, S-E. (2008). Investment timing and optimal capacity choice for small hydropower projects. *European Journal of Operational Research*, 190, 255–267.

Chen Q., Kang C., & Xia Q. (2009). Key low-carbon factors in the evolution of power decarbonisation and their impacts on generation expansion planning.

Automation of Electric Power Systems, 33(15), 18–23 (in Chinese).

Dixit, A.K. & Pindyck, R.S. (1994). *Investment under uncertainty*. Princeton University Press.

Fuss, S. & Szolgayová, J. (2010). Fuel price and technological uncertainty in a real options model for electricity planning. *Applied Energy, 87*, 2938–2944.

Ge, S., Wang, H., & Xu, L. (2012). Reliability evaluation of distributed generating system including wind energy solar energy and battery storage using Monte Carlo simulation. *Power System Technology, 36*(4), 39–44 (in Chinese).

He, J., Deng C. et al. (2013). Assessment on capacity credit and complementary benefit of power generation system integrated with wind farm, energy storage system and photovoltaic system. *Power System Technology, 4*(4), *37*(11), 3030–3036 (in Chinese).

He, J., Deng, C. et al. (2013). Optimal configuration of distributed generation system containing wind PV battery power sources based on equivalent credible capacity theory. *Power System Technology, 37*(12), 3317–3324.

Jiang, Z. & Chen, J. (2013). Optimal distributed generator allocation method considering uncertainties and requirements of different investment entities. *Proceedings of the CSEE, 33*(31), 34–42 (in Chinese).

Lei, Z., Wei, G., Cai, Y., & Zhang, X. (2011). Model and reliability calculation of distribution network with zone-nodes including distributed generation. *Automation of Electric Power Systems, 35*(1), 39–43, 76 (in Chinese).

Li, B., Shen, H., & Tang, Y. (2011). Impacts of energy storage capacity configuration of HPWS to active power characteristics and its relevant indices. *Power System Technology, 35*(4), 123–128 (in Chinese).

Pauli, M. (2007). Timing of investment under technological and revenue-related uncertainties. *Journal of Economic Dynamics & Control, 31*, 1473–1497.

Pavee, S. & Jorge, V. (2011). Cournot equilibrium considering unit outages and fuel cost uncertainty. *IEEE Transactions on Power Systems, 26*(2), 747–754.

Reuter, W.H., Szolgayova, J., Fuss, S., & Obersteiner, M. (2012). Renewable energy investment: Policy and market impacts. *Applied Energy*, 249–254.

Shou xiang Wang, Zhi xin Li Lei Wu, et al. (2013). New metrics for assessing the reliability and economics of microgrids in distribution system. *IEEE Transactions on Power Systems, 28*(3), 747–754.

Tao, W., Zhang, L., & Huang, X. (2007). Dynamics model of generation investment planning in electricity market. *Proceeding of the CSEE, 27*(16), 114–118 (in Chinese).

U.S. Energy Information Administration. (2011). *International Energy Outlook*.

Wang, H., Bai, X., & Xu, J. (2012). Reliability assessment considering the coordination of wind power, solar energy and energy storage. *Proceeding of the CSEE, 32*(13), 13–20 (in Chinese).

Wang, X. & Lin, J. (2010). Reliability evaluation based on network simplification for the distribution system with distributed generation. *Automation of Electric Power Systems, 34*(4), 38–43 (in Chinese).

Wang, Z. & Su, H. (2014). A decision model for carbon-capture systems best investment opportunity based on real option theory. *Automation of Electric Power Systems, 38*(17), *37*(11), 137–142 (in Chinese).

Yuan, C., Li, F. & Kuri, B. (2011). Optimal power generation mix towards an emission target. *Proceedings of the IEEE Power and Energy Society General Meeting*. Detroit, USA: IEEE: 1–7.

Zhang, L., Tang, W., & Wang, S. (2011). Distributed generators planning considering benefits for distribution power company and independent power suppliers. *Automation of Electric Power Systems, 35*(4), 23–28.

Automotive, Mechanical and Electrical Engineering – Liu (Ed.)
© *2017 Taylor & Francis Group, London, ISBN 978-1-138-62951-6*

Finite element analysis of battery holder based on ANSYS

Guiyu Yang & Bin Zhao
Tianjin University of Technology, Tianjin, China

ABSTRACT: In this paper, the structure of a battery holder of electric vehicle is simplified, according to the operating conditions of electric vehicles. The battery holder is given three conditions of the situation and through ANSYS, the stress distribution under one condition and the location of maximum stress are analyzed and calculated.

Keywords: Battery holder; ANSYS; Mechanical analysis; Structure

1 INTRODUCTION

In order to cope with the energy and environment sustainable development, the improvement of Electric Vehicles (EV) seems to be a way out for the automotive industry in China. It also seems to be an important choice for China to cultivate strategic emerging industries. Currently, battery becomes one of the most important core components on EVs as well as the biggest challenge for EVs. It is the biggest obstacle for the marketization of EV. Hence, the research of battery becomes a hot issue among Original Equipment Manufacturers (OEM) and suppliers. The complex working condition puts forward high requirements on the installation of the battery pack. What's more, the quality and volume of battery will affect the vehicle layout and the reliability. This paper analyses the battery holder based on ANSYS. And the research mainly launches from two aspects: modal analysis and static analysis.

2 BATTERY HOLDER MODELING

Design of Batter Holder. The battery holder is designed based on envelope of the front floor and the installation of the battery pack. The battery holder includes two carlings and five beams which are made of rectangular steel. The section sizes of the carling and beam are 40 × 50 mm and 20 × 40 mm respectively. The 3D model of carling is shown in Figure 1 and the 3D model of beam is shown in Figure 2. On the welding surface of carling and beam, there are positioning threaded holes which are used to determine the relative position of carling and beam. At the same time, there are three installing threaded holes and three technical holes on the carling which are used

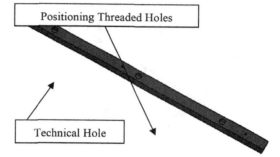

Figure 1.   Carling of the battery holder.

Figure 2.   Beams of the battery holder.

to install the battery pack. After determining the position of carlings and beams, the battery holder model is built and shown in Figure 3.

The Stress Analysis of Battery Holder. The battery holder connects the battery pack and the frame. And the load mainly results from the battery pack. Through analysing the setting relation, the load on the battery holder can be calculated. And the force situation is shown in Figure 4.

Figure 3.   3D model of the battery holder.

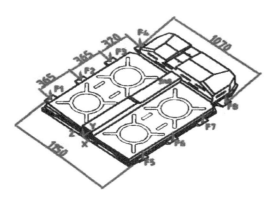

Figure 4.   Force situation of battery holder.

Table 1.   Acting forces from the mounting bolts. Unit: N.

Force	F1	F2	F3	F4	F5	F6	F7	F8
Value	−194.5	957.6	1138.8	1034.9	−194.1	957.0	1154.5	1022.7

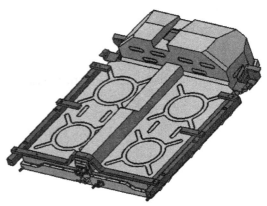

Figure 5.   Setting relations between pack and holder.

$$\{F'\}^e =$$
$$\left\{ X_i\ Y_i\ Z_i\ M_{xi}\ M_{yi}\ M_{zj}\ X_j\ Y_j\ Z_j\ M_{xj}\ M_{yj}\ M_{zj} \right\}^T$$
$$(2)$$

Every element has 6 degrees of freedom and a total of 12 degrees of freedom. Using node i, j and k, coordinate system can be determined. When the pre-processor is finished, the results will be calculated.

3   FINITE ELEMENT ANALYSIS OF THE BATTERY HOLDER

In the paper, two working conditions are analysed, namely the emergency braking condition and abrupt-turning condition on the bump road. With the help of these conditions, the stress distribution and the maximum stress can be figured out to test the reliability of the battery holder.

Emergency Braking Condition. When taking an emergency brake on a bump road, the deceleration is about 0.8 g and the dynamic load coefficient is about 2.0. So the vertical load now is doubled. Meanwhile, the vehicle bears inertial force as well. The results are shown in Figure 6 and Figure 7. Telling from the figures, the maximum stress which locates at the border of rear beam and carling is about 12.8 MPa. The maximum displacement which locates at the border of front beam and carling is about 3.88×10^{-3} m.

The battery pack bears dynamic load in Z direction, gravity and the acting forces from the mounting bolts. According to the gravity and working points, the acting force Fi can be obtained. It is a secondary statically indeterminate problem which can be solved by statics balance equations and geometric compatibility equations. The centre of gravity locates at (0,900.5, 107). The results of acting forces are shown in Table 1.

Simplified Model of Battery Holder. There are eight positioning holes on the battery holder. The loads from the battery pack transfer to the battery holder through the bolts. Figure 5 shows the setting relation between the battery pack and holder. It is easy to find that the stress state of the battery holder belongs to the space steel structure stress problem. On an issue like this, the element is made up of nodes i and j assisted by node k. The displacement and stress of the nodes are shown in Equation (1) and (2).

$$\{\vartheta\} = \left\{ u_i\ v_i\ w_i\ \theta_{xi}\ \theta_{yi}\ \theta_{zi}\ u_j\ v_j\ w_j\ \theta_{xj}\ \theta_{yj}\ \theta_{zj} \right\}^T \quad (1)$$

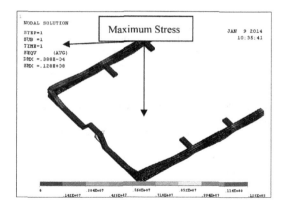

Figure 6. Stress reprogram.

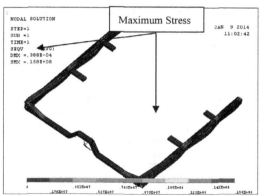

Figure 8. Stress reprogram.

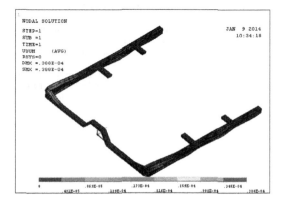

Figure 7. Displacement reprogram.

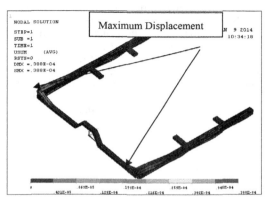

Figure 9. Displacement reprogram.

Abrupt-turning Condition. When taking an abrupt-turning on a bump road, the deceleration is about 0.6 g and the dynamic load coefficient is about 2.0. So the vertical load now is doubled. At the same time, the vehicle also bears centrifugal force as well. The results are shown in Figure 8 and Figure 9. Telling from the figures, the maximum stress which locates at the border of rear beam and carling is about 15.8 MPa. The maximum displacement which locates at the border of front beam and carling is about 3.88×10^{-3} m.

Analysis. According to the analysis results of the working conditions, the maximum stress locates at the border of rear beam and carling. Table 2 shows the results of the stress analysis. Regarding the yield strength and tensile strength of Q235, the battery holder in these working conditions can well satisfy the requirements. However, during the design process, the influence brought by the dynamic working condition cannot be ignored.

4 MODAL ANALYSIS

Modal analysis is mainly used to determine the natural frequencies of the battery holder. With the help of modal analysis, the natural frequencies can be avoided so that the noise and vibration can be improved. After defining the material property and constraints, the calculation can go on. The calculation shows the first six natural frequencies in Table 2. During the design process, these natural frequencies merit attention. To be more precise, these frequencies will result in resonance and the components should be installed far from the location where the amplitude is large. In this way, the mechanical damage caused by resonance can be avoided.

Economic Benefit. With the help of Finite Element Analysis (FEM), the manufacturers and institutions can gain a lot from it. Through FEM, the cost of the real vehicle test can be reduced to the lowest. Besides the cost of test, to be more precise,

255

Table 2. The first six natural frequencies of the battery holder.

Order	First Order	Second Order	Third Order	Fourth Order	Fifth Order	Sixth Order
Value (Hz)	237.9	251.7	406.3	406.3	476.2	700.5

the full product development life cycle can be shortened, that means companies can reduce time-to-market for new and improved products, solve business problems more quickly, and address client concerns right away. FEM has already become a common method among manufacturers and institutions which brings great benefits to them.

5 SUMMARY

This paper analyses the battery holder based on ANSYS. Through the CAE analysis process, it can make forecasts for the structure reliability and performance, so that the cost of development and risk can be greatly reduced. On the other hand, it can improve the quality of the product and shorten product development cycle. Currently, new energy vehicles are developing rapidly and domestic and international related manufacturers and

institutions are actively developing the new energy vehicles and components. However, CAE analysis carried out on the battery pack and holder is just on the start stage in China, hence, the work can be quite meaningful in the process of design and manufacture.

REFERENCES

Ma X., Y.S. Shen: The Finite Element Analysis and Optimization of the Frame Rigidity and Modal, Bus Technology and Research, Vol. 4 (2004).

Tao Y.P.: Application of CAE Technology in Electric Vehicle Battery Pack Design, Henan Automotive Engineering Technical Seminar (Henan, China, September, 2012).

Ye Q., Y.D. Deng, Y. Wang, W. Tan: Finite Element Analysis and Optimization of a Light Vehicle Frame, Journal of Wuhan University of Technology, Vol. 2 (2008).

Automotive, Mechanical and Electrical Engineering – Liu (Ed.)
© 2017 Taylor & Francis Group, London, ISBN 978-1-138-62951-6

Integrated control strategy of two operating modes switching based on energy storage

Tonggengri Yue & Xiangyang Yu
Xi'an University of Technology, Xi'an, China

ABSTRACT: Microgrid technology is one of the most effective ways to solve the grid-connected problem of a distributed generation system. A micro-network system based on optical storage has the characteristics of full utilisation of clean energy and reliable operation. Smooth switching between grid-connected and isolated island operation modes is an important guarantee of safe and stable operation. To solve this problem, a new master-slave and peer-to-peer control strategy is adopted to control the transition of the grid-connected/isolated operation mode, and the optical microgrid model is built on the PSCAD platform. The simulation results verify that Control strategy feasibility.

Keywords: Microgrid; Distributed Generation; PQ control; V/f control; Droop control

1 INTRODUCTION

Microgrid technology is one of the most effective ways to solve the grid-connected problem of a distributed generation system. A micro-network system based on complementary optical storage has the advantages of making full use of clean energy and reliable operation. Smooth switching between grid-connected and isolated island operation modes is an important guarantee of safe and stable operation.

At present, many studies have been done on the smooth switching control of microgrids at home and abroad. In Xiao (2009), a pre-synchronisation control method and control of the micro-network system with master-slave structure are given. In Lee (2013), a frequency-voltage recovery control and pre-synchronisation control based on decentralised control are presented. This method of adjustment of many inverters may cause uneven power distribution during the adjustment process and the system low-frequency oscillation and other issues. In Majumder (2010) and Vandoorn (2013), the back-to-back bi-directional transformer is used as the parallel interface between the microgrid and the large power grid, but this excuse is not conducive to the expansion of the system and the capacity of the grid-connected inverter is higher. In Chen (2014), the phase of the microgrid phase is synchronised with the large grid by using the Phase-Locked Loop (PLL) technique of the inverter in the microgrid, but this method is only applicable

to the small microgrid. At present, many studies on the microgrid are based on the smooth switching of the operating mode of the optical storage micro-network system.

To this end, this paper uses a new master-slave and peer-to-peer control of the integrated control strategy, the microgrid grid/island mode of operation to control the transition. A complementary microgrid model of optical storage is built on the PSCAD platform. The simulation results verify the feasibility of the control strategy.

2 MICROGRID STRUCTURE

In this paper, the structure of optical storage micro-grid is shown in Figure 1. The distributed power supply is connected to the 0.38 kV bus

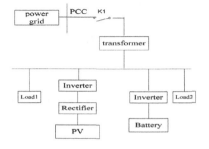

Figure 1. Basic structure of PV and battery micro grid.

through the inverter and then connected to the 10.5 kV distribution network through the transformer. The power supply of the load is isolated by two micro-source. Micro-grid by the public switch K1 action to achieve the grid and island switching. When K1 closed and run, when K1 open into the island running state, micro-grid and large power disconnected.

3 GRID-CONNECTED/ISOLATED ISLAND OPERATION CONTROL STRATEGY

3.1 *PQ control strategy*

Figure 2 is the design of the PQ controller structure. The active power and reactive power are decoupled to control the current, and the PI controller can make the steady-state error zero. The use of Phase-Locked Loop (PLL) technology can be used to control the DG PQ can get frequency support.

Given that, in the dqo coordinate system, the distributed power system injects the exchange network, the power is

$$\begin{cases} P_{grid} = u_d i_d + u_q i_q \\ Q_{grid} = u_q i_d - u_d i_q \end{cases} \quad (1)$$

If the Park transformation is selected the d-axis in the same direction as the voltage vector, the q-axis voltage component can be made zero. At this point, the power output expression can be simplified to active power only with the d-axis active current, and reactive power only with the q-axis reactive current, which can be through the power reference and AC network side voltage value calculated current reference Value, as shown in the following equation, this control is a simplified constant power control mode.

$$\begin{cases} P_{grid} = u_d i_d \\ Q_{grid} = -u_d i_q \end{cases} \Rightarrow \begin{cases} i_{dref} = \dfrac{P_{ref}}{u_d} \\ i_{qref} = -\dfrac{Q_{ref}}{u_d} \end{cases} \quad (2)$$

The inverter PQ control circuit topology is shown in Figure 2. If the d-axis and the voltage vector are chosen in the same direction, the q-axis voltage component can be made zero. At this time, the power output expression can be decoupled to obtain the reference current idrf and iqref of the voltage flow to the feeder. Corresponding to Figure 2, the relationship between the voltage and current across the line directly connected to the distributed power supply can be expressed as the following general form:

$$\begin{cases} u_{Fd} = u_d + R i_d + L' \dfrac{di_d}{dt} - wL' i_q \\ u_{Fq} = R i_q + L' \dfrac{di_q}{dt} - wL' i_d \end{cases} \quad (3)$$

Figure 2 is based on 1 and 2 design PQ controller, the controller current control and Figure 2 gives the same idea of the inner loop controller, but in the specific form of some adjustments, while omitting some non-critical Sexual links.

3.2 *Droop control strategy*

Droop control is mainly used in the peer control of microgrid to achieve coordination of multi-DG processing control, but droop control is a differential control and cannot make micro-network frequency or voltage to restore the original level of the grid.

In Figure 3, since the frequency signal facilitates the measurement, the frequency control is used instead of the phase angle control. The input power in the control loop is the output power of the distributed power supply. Among them, the distributed power output active power P and reactive power Q must be full of $0 \leq P \leq Pmax$ and $-Qmax \leq Q \leq Qmax$ these two conditions, the power controller output will be used as inner ring dq axis reference voltage.

In Figure 4, the controller parameters to the inverter output impedance was emotional, external voltage loop using PI controller to improve the steady-state accuracy; inner current loop using the

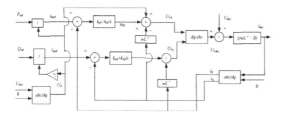

Figure 2.   Block diagram of PQ control.

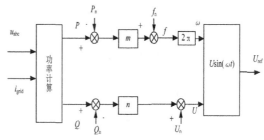

Figure 3.   Diagram of the droop power controller.

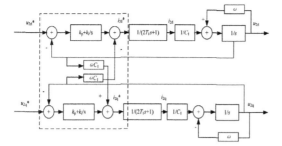

Figure 4. Link voltage control block.

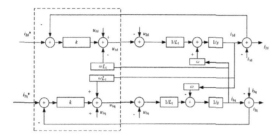

Figure 5. Link current control block.

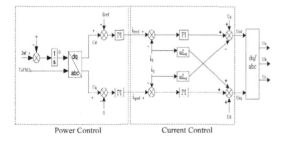

Power Control   Current Control

Figure 6. Structure block diagram of V/f controller.

proportional controller to improve the dynamic response. At the same time, since the load current and the current flowing to the feeder are in the forward channel in the current loop, it can be regarded as a disturbance effectively suppressed, so this current change has no effect on the steady state.

3.3 V/f control strategy

V/f control developed by the droop control is generally used in the controllable power supply, to ensure that the micro-network output voltage amplitude and frequency are constant. The frequency difference and the voltage deviation are used to calculate the power difference of the microgrid after the islanding. The output of the DG is determined by the frequency deviation and

voltage deviation. The PI controller is used to realise the frequency and voltage unequal control. The structure block diagram of the V/f controller is shown in Figure 6.

4 INTEGRATED CONTROL STRATEGY

For the microgrid system with multiple DGs, the existing research results mainly use master-slave and peer-to-peer control strategies. Peer-to-peer control is simple, reliable, and easy to implement, but at the expense of the frequency and voltage stability; master-slave control can support micronetwork voltage and frequency, but the main control unit has a strong dependence. In this paper, master-slave and peer-to-peer control of the integrated control strategy, as shown below.

To achieve the maximum utilisation of solar energy, the photovoltaic power generation system has always adopted the PQ control strategy, by the energy storage device to maintain the micronetwork mediation process stable operation. In grid-connected operation mode, the energy storage device adopts droop control. The advantage is that, when the energy storage device is connected or disconnected, there is no need to change other DG settings in the microgrid. When the voltage deviation and the frequency deviation are within the allowable range, the energy storage device will maintain droop control and adjust the output according to the frequency deviation and voltage deviation of the microgrid; if the voltage deviation or frequency deviation is out of the allowable range, the energy storage device inverter fast switch to V/f control, in order to maintain the microgrid voltage and frequency stability.

In this paper, the integrated control strategy can reduce the number of switching energy storage devices, reducing the possibility of failure to switch, and thereby improving the reliability of the microgrid operation.

When the controller is switched, in order to avoid large transient oscillations, a smooth switching control method based on controller state following can be adopted. As shown in Figure 7, the output of the V/f controller and droop controller is designed as A negative feedback as the V/f controller input, making the switch before the V/f controller at any time follow the droop controller output to ensure that the output state of the two controllers at the same time. K1 and K4 are closed, K2 and K3 are open, K1 and K4 are open, and K2 and K3 are closed when the control strategy is switched.

This control strategy not only exploits the advantages of DG autonomy in the peer-to-peer control strategy, reduces the dependency between

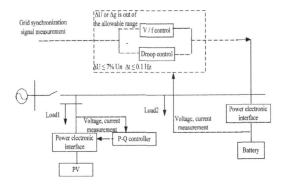

Figure 7. Diagram of integrated microgrid control strategy.

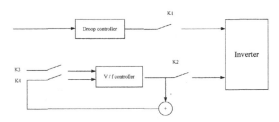

Figure 8. Smooth switching control method based on controller status followed.

them, and realises the automatic allocation of load power variation, but also plays the advantages of master-slave control strategy to maintain system voltage and frequency stability, leading to an increase in system reliability.

5 MODEL SIMULATION AND EXPERIMENTAL VERIFICATION

A microgrid model was built on PSCAD, and the PV output remained unchanged during mode switching. The microgrid runs for 5 s, when 0–2 s is connected to the grid, 2–3 s is the island and 3–5 s is the grid. The simulation under single droop control and integrated control is shown in Figure 9.

Figure 10 shows a single droop control microgrid battery output power curve, with a blue line for the active power and green line for the reactive power. It can be seen from the figure that a single sag control, the output power fluctuation is relatively large, but the integrated control of the energy storage device, the battery output power changes significantly smaller, indicating that the method can effectively reduce the power fluctua-

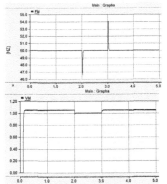

(a) The frequency and voltage at the PCC of the microgrid under a single droop control.

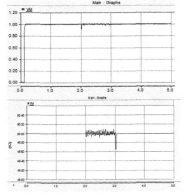

(b) The frequency and voltage at the PCC of the microgrid under integrated control.

Figure 9. Operation results of storage based on droop control strategy and integrated control strategy.

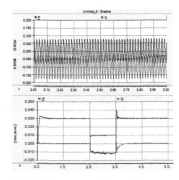

(a) Single droop control microgrid battery output active and reactive power.

(b) Integrated control of microgrid battery output of active and reactive power.

Figure 10. Waveform of active and reactive power of microgrid battery before and after improvement.

tions of the impact caused by the microgrid, and enhance the stability of the entire system.

6 CONCLUSION

Smooth switching between grid-connected and isolated island operation modes is an important guarantee of safe and stable operation. In view of the shortcomings of the traditional master-slave and peer-to-peer control, this paper uses an integrated control strategy for microgrid of the grid/island mode of operation to control the transition.

A complementary microgrid model is built on the PSCAD platform to realise the control strategy of this paper. The simulation results show that the proposed control strategy can ensure that the power balance, voltage and frequency of the microgrid remain within the allowable range before and after the operation mode switching, and can achieve the smooth switching of the microgrid operation mode.

REFERENCES

Chen, X., Ji, Q., & Liu, F. (2014). Smooth switching control strategy based on microgrid master slave. *Structure Transactions of China Electro technical Society, 60*(29), 163–170.

Lee, C.T., Jiang, R.P., & Cheng, P.T. (2013). A grid synchronization method for droop-controlled distributed energy resource converters. *IEEE Transactions on Industry Applications, 49*(2), 954–962.

Majumder, R., Ghosh, A., Ledwich, G. et al. (2010). Power management and power flow control with back-to-back converters in a utility connected microgrid. *IEEE Transactions on Power Systems, 25*(2), 821–834.

Vandoorn, T.L., De Kooning, J.D.M., Meersman, B. et al. (2013). Voltage-based control of transformer in a microgrid. *IEEE Transactions on Industrial Electronics, 60*(4), 1291–1305.

Xiao, Z., Wang, C., & Wang, S. (2009). Small-signal stability analysis of microgrid containing multiple micro source. *Automation of Electric Power Systems, 33*(6), 61–66.

Automotive, Mechanical and Electrical Engineering – Liu (Ed.)
© *2017 Taylor & Francis Group, London, ISBN 978-1-138-62951-6*

Load forecasting of expressway EV charging based on travel characteristics

Dapeng Yin, Yi Liu, Zhong Chen, Tao Zhou & Zhiqiang Shi
School of Electrical Engineering, Southeast University, Nanjing, China

ABSTRACT: The rapid development of Electric Vehicles (EVs) has had a prominent influence on the traffic system and distribution network. The charging load of electric vehicles is affected by quite a few factors and it has an important significance on modelling and forecasting of large-scale load. This paper firstly studies the impacts of each factor on charging power properties, including battery characteristics, SOC (State of Charge) and charging regimes, so as to guarantee the rationality of the input parameters of the power computation model and thus to establish the charging power computation model of a single electric vehicle. Then analysis is conducted on travel time, charging start and end time and vehicle flow properties, and the daily vehicle flow and traffic laws are studied for work days and non-work days based on the actual data of a practical expressway to establish the expressway EV charging load model. Finally, it is applied to a practical example using the Monte Carlo simulation method and the results verify the correctness and feasibility of the model.

Keywords: EVs; Expressway; Charging load forecasting; Traffic law; Monte Carlo simulation

1 INTRODUCTION

For the obvious advantages in environmental protection, new energy digestion and energy utilisation ratio and national strong support of government, relying on relative enterprises and research institutes, the software and hardware techniques of EV are maturing. At the same time, with the large-scale development of EV, the construction of expressway charging stations is a fundamental condition to realising EV long distance travel. The development of fast charging stations has made great progress since State Grid accelerated the construction of expressway charging stations. The construction of fast charging stations for electric vehicles is planned in the service areas of Beijing Shanghai Expressway (G2), Qingdao Yinchuan Expressway (G20), and Beijing to Hong Kong Macao Expressway (G4) in August 2014. The power supply network that electric vehicles need in long distance driving is about to be covered in North China, East China (Yangtze River Delta) and Southern China (Pearl River Delta) (Ruan, 2008). In mid-January 2015, an electric vehicle fast-charging network on the Beijing Shanghai high-speed was set up (Anonymous, 2015). The increase in electric vehicle charging load on the expressway has a great influence on the charging facilities and charging station construction, the operation of the charging station, and the planning and operation

of the distribution network (Shaoyun et al., 2013; Long et al., 2015). Effective charging power forecasting is the premise of load dispatch, which is related to the security, stability and economic operation of the power system.

The establishment of a model for charging load forecasting is of great significance based on the comprehensive analysis of the related factors affecting the charging load of electric vehicles. The domestic and foreign scholars in the research on load modelling of electric vehicle charging mainly concentrate on the Monte Carlo simulation method based on travel demand (Dai et al., 2014; Luo et al., 2011; Zhang et al., 2014), using the queuing theory to analyse the charging power of electric vehicle charging stations (Alizadeh et al., 2014; Chen et al., 2015; Zhang, 2014) and physical analysis (Peng et al., 2012). Zhang et al. (2014) proposed a new method of forecasting the Electric Vehicle (EV) charging load considering the spatial and temporal distribution based on driving and parking characteristics of private cars. Dai et al. (2014) presented a novel model based on Monte Carlo simulation to estimate uncontrolled energy consumption of the BSS by taking hourly number of EVs for battery swapping, the charging start time, the travel distance and the charging duration into account. Luo et al. (2011) calculated the charging time according to the charging mode (slow charging, fast charging, conventional charging), SOC, and different models, and then applied the Monte Carlo simulation method to calculate the charging load. Reference proposed a charging power demand model for electric bus and electric taxi based on the operating raw. Chen et al. (2015) investigated the interrelationship of multiple trips in one day based on Markov chain to give a detailed roadmap of daily routes, taking the influence of external conditions on energy consumption into account, and determined the charging criterion. The overall daily charging load forecast at different places is then obtained by Monte Carlo simulation. Alizadeh et al. (2014) proposed a stochastic model, based on queuing theory, for EV and Plug-in Hybrid Electric Vehicle (PHEV) charging demand and can provide more accurate forecasts of the load using real-time sub-metering data, along with the level of uncertainty that accompanies these forecasts and a mathematical description of load, as well as the level of demand flexibility that accompanies this load at the wholesale level. Peng et al. (2012) mathematically analysed the movement of two infinitesimal energy elements and a diffusion charging load model was proposed. The above research provides a foundation for the daily load modelling of an electric vehicle charging station. However, there are few studies on the modelling of the charging load on the expressway.

This paper analyses expressway EV travel time, start time and end time of charging, and traffic flow properties, and studies the daily vehicle flow and traffic laws of work days and non-work days based on actual data of a practical expressway to establish the expressway EV charging load model. Finally, considering EV battery characteristics, SOC and charging regimes, the correctness and feasibility of the model in this paper are verified by Monte Carlo simulation in the example of a practical expressway in Arizona, USA.

2 CHARGING CHARACTERISTICS OF A SINGLE ELECTRIC VEHICLE

2.1 Characteristics of the battery

So far, a large amount of research has been done to study the characteristics of the battery (Kintner et al., 2007; Wynne, 2009). According to a charge current id, the battery status can be described by the state of charge of the battery and can be calculated as follows (DeForest et al., 2009):

$$S(t_2) = S(t_1) + \int_{t_1}^{t_2} \frac{1}{C_a(i_d)} i_d dt \qquad (1)$$

where i_d is the charging current; the unit of i_d is A; $S(t_1)$ and $S(t_2)$ expressed in the form of a percentage stand for SOC at t_1 and t_2 moments

respectively; $C_a(i_d)$ is the effective capacity of the battery; the unit of $C_a(i_d)$ is Ah.

2.2 Initial SOC of the battery

In a charging cycle, the power demand of electric vehicle charging changes with time. To determine the load generated by the electric vehicle charging, the initial SOC of the battery must be obtained at the beginning of the charging. Initial SOC is a random function on the distance that the vehicle has travelled since the last charge. It can be indicated by probability density function $P(SOCi)$, which ranges from 0 to 100%.

Statistics show that the travel mileage of the car obeys lognormal distribution. So its probability density function is given as follows [10]:

$$g(d; \mu, \sigma) = \frac{1}{d\sqrt{2\pi\sigma^2}} \exp\left\{-\frac{(\ln d - \mu)^2}{2\sigma^2}\right\} \qquad (2)$$

where d is the mileage; the unit of d is km; μ and σ stand for the mean value and standard deviation of the probability density function respectively.

It is assumed that the SOC of the electric vehicle decreases linearly with the increase in mileage. Thus, the initial SOC can be estimated by the mileage of the vehicle.

2.3 Charging mode

When electric vehicles begin to charge, the charging power of the battery relates to the charging mode. Currently, there are three main charging modes, as follows: (1) Constant current charging mode. (2) Constant voltage charging mode. (3) Multi-stage charging mode.

Currently, two-stage charging mode is generally used in the electric vehicle charging station. Namely, the battery will be charged with constant current in the first stage and then constant voltage in the second stage. In the process of constant current charging, the battery terminal voltage is constantly changing. In the process of constant voltage charging, the charging current will be dynamic. Therefore, the charging power during the charging process is in a state of change. But, in order to be convenient, it is assumed that the power of the electric vehicle battery is constant when being charged according to the charging process of the lithium-ion battery. Accordingly, the electric vehicle charging load can be described by a constant value in the period of charging.

2.4 Charging power

The following will discuss the phosphoric acid battery 240100. The charging current is 0.30 C

(30 A) and the charging voltage may vary with the capacity. Thus, the power generated by the electric vehicle charging changes. In order to forecast the power of electric vehicle charging in a better way, the charging time can be divided into several periods. For the phosphoric acid battery that could be full after 3.3-hour charging, the charging time can be divided into 10 periods if 20 minutes is taken as a time period. At the beginning of each time period, the SOC is obtained as follows [10]:

$$\begin{cases} S(t_{k-}) = SOC_i \\ S(t_{k+}) = S(t_{k-}) + \left(\int_{t_{k-} \to t_{k+}} i_d dt \right) / C_a(i_d), k \geq 2 \\ S(t_{(j+1)-}) = S(t_{j+}), j \geq 1 \end{cases} \quad (3)$$

where t_{k-} and t_{k+} indicate the beginning and end of the kth period.

The charging voltage of $S(t_{k-})$ and $S(t_{k+})$ are $V(t_{k-})$ and $V(t_{k+})$ respectively. Obviously, in the charging curve of the 240100 phosphoric acid battery, the $V(t_{k+})$ is larger than $V(t_{k-})$. In order to forecast electric vehicle charging load more accurately, the charging voltage in the kth period can be indicated by the following equation (Haijuan, 2015):

$$V(t_k) = \frac{V(t_{k-}) + V(t_{k+})}{2} \quad (4)$$

Thus, the power generated in the kth period can be calculated as (Haijuan, 2015):

$$P_k = \sum_{j=1}^{N} (V_j(t_k) * i_{d,j} / \eta) \quad (5)$$

where N is the number of electric vehicles charging during the kth period; $V_j(t_k)$ and $i_{d,j}$ stand for charging voltage and charging current of the jth electric vehicle during the kth period respectively; η is the transmission efficiency of the electric vehicle charging, which is 0.85 in this paper.

3 MODEL OF ELECTRIC VEHICLE CHARGING LOAD ON EXPRESSWAY

3.1 Travel time and charging beginning time of Evs

Due to different economic levels, seasonal climatic changes and other factors, the traffic flow of an expressway demonstrates the law of monthly variation. The Monthly Variable Coefficient (MVC) is proposed to express the law of monthly variation.

According to relevant traffic statistics, for the monthly change in traffic, there are the following rules. Firstly, the MVC of a rural road is greater than that of an urban road, as it is affected by the busy farming season and the seasonal effects of crops. Secondly, the MVC of a tourist city is larger than that of an ordinary city due to the increase in holiday tourists. The main factor influencing the MVC is holidays. In the months of Ching Ming, Dragon Boat Festival, Mid-Autumn and Spring Festival, there is a sudden increase in traffic flow.

The traffic flow of an expressway is different between working days and non-working days in a week, so the traffic flow of the expressway demonstrates the law of weekly variation. In general, for tourist cities, the traffic flow is relatively large at weekends and relatively small on weekdays. On the contrary, there is the opposite situation for non-tourist cities. However, no matter what kind of city, both the working day and non-working day load demonstrate periodic variation characteristics. The traffic flow distribution is uneven in different time periods of the day. The traffic flow is larger in the rush hour and smaller at night. Different kinds of vehicles such as private cars, trucks, official vehicles and buses correspond to a different time-varying curve due to different travel characteristics. For most cases, the daily time-varying curve has two peaks which appear in the morning and evening rush hour separately, and the peak time period will not last for too long, usually for about 30 minutes to one hour.

3.2 Analysis of daily traffic flow on expressway

According to the above analysis of the traffic flow characteristic, the traffic flow of working days and non-working days show different probability characteristics, so the data of the two types should be analysed respectively.

The work-day traffic data are analysed through the MATLAB curve fitting toolbox (Curve Fitting Tool). The fitting result of the probability density function is obtained by segmenting at the lowest point of the traffic flow. The time scale is converted to obtain the probability density function of traffic flow on work days (Ma Rui, et al. 2015):

$$f_{workday}(t) = \begin{cases} a_1 \exp\left[-\left(\dfrac{4t - b_1 + 96}{c_1}\right)^2\right] \\ \quad + a_2 \exp\left[-\left(\dfrac{4t - b_2 + 96}{c_2}\right)^2\right], \\ \quad 0 \leq t < 3 \\ a_1 \exp\left[-\left(\dfrac{4t - b_1}{c_1}\right)^2\right] \\ \quad + a_2 \exp\left[-\left(\dfrac{4t - b_2}{c_2}\right)^2\right], \\ \quad 3 \leq t \leq 24 \end{cases} \quad (6)$$

where a_1, b_1, c_1, a_2, b_2, c_2 equal to the undetermined parameters of the probability density function of traffic flow on weekdays and t equals to the time scale.

The non-working day traffic data is analised through the MATLAB curve fitting toolbox (Curve Fitting Tool). The fitting result of the probability density function is obtained by segmenting at the lowest point of the traffic flow. The time scale is converted to obtain the probability density function of traffic flow at Non-workdays (Ma et al., 2015):

$$f_{nonworkday}(t) = \begin{cases} a\exp\left[-\left(\dfrac{4t-b+96}{c}\right)^2\right], 0 \le t < 3 \\ a\exp\left[-\left(\dfrac{4t-b}{c}\right)^2\right], 3 \le t \le 24 \end{cases} \quad (7)$$

where a, b, c are equal to the undetermined parameters of the probability density function of traffic flow on non-working days and t equals the time scale.

3.3 Modelling of expressway charging load

In the current transportation system, the main types of vehicles are bus, taxi, official vehicle and private car. The proportion of each type of vehicle is shown in Table 1.

It is assumed that taxis, private cars, official vehicles and lease cars will charge on the expressway and buses and other types of vehicles are negligible. With technological breakthroughs in the field of batteries, the battery capacity of electric vehicles has been significantly increased. Currently, the minimum capacity of a battery pack is about 16 KWh. Several typical types of batteries and their parameters are displayed in Table 2.

From Table 2, it can be derived that battery packs whose capacity ranges from 16 to 20 KWh

Table 1. Proportion of different types of electric vehicles.

Vehicle type	Proportion (%)	Vehicle type	Proportion (%)
Bus	65.52	Official vehicle	4.67
Taxi	15.52	Public vehicle	1.95
Private Car	11.24	Lease Car	1.11

Table 2. Capacity of several typical types of batteries.

Type	Capacity (KWh)	Type	Capacity (KWh)
Chevrolet Volt	16	Lifan 620	30
Nissan Leaf	24	Honda Fit EV	20
Ford Focus EV	23	iMiEV	16
Chery Riich M1	16–20	BYD E6	20

are in extensive use. Therefore, it is assumed that the capacity of all battery packs is 16 KWh. Generally, the number of battery packs can be increased or decreased by carmakers based on the owner's functional requirements and the vehicles' conditions. The capacities of four types of electric vehicles are as follows by reference to the mileage of the ordinary fuel vehicles: the capacity of private cars and lease cars is 32 KWh; the capacity of taxis and official vehicles is 64 KWh with reference to BYD E6.

According to the characteristics of the charging modes and the driving rules, charging beginning time, initial SOC and charging power of the electric vehicles can be obtained by the probability simulation. Then the total charging load can be calculated on the basis of vehicle flow. The structure of the charging load calculation is shown in Figure 1.

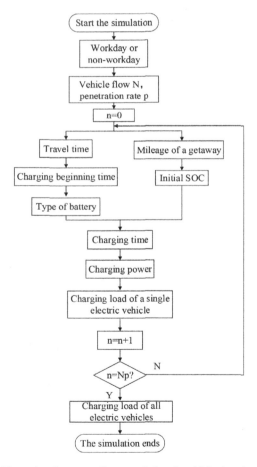

Figure 1. Structure diagram of electric vehicle charging load calculation model of expressway.

266

4 CASE STUDY

The case study is carried out on actual traffic data of the US No. 10 Expressway of Arizona 100115 segment sampled with interval of 15 min. Traffic data of working days and non-working days are analysed so that the parameters of (7) and (9) can be obtained and given as follows: a1 = 5,230.83, b1 = 61.271, c1 = 28.68, a2 = 4,147.46, b2 = 29.41, c2 = 11.801, a = 4,009.235, b = 59.017, c = 36.77.

Considering the characteristics of the user's driving behaviour, it is assumed that the start charging time of the vehicle follows uniform distribution in 24 periods. There are enough charging stations on expressways that can meet the demand for fast charging of electric vehicles. Considering the driving behaviour of the user and the maintenance and protection of the battery, the minimum value of SOC when charging begins is 0.2.

According to the above parameter settings, the simulation is conducted based on (5) and Figure 1. Viewing 15 minutes as a period of time, there will be 96 periods a day. The daily load demand of the EV charging station is analysed in the three cases that EV penetration rate is 20%, 50% and 80% respectively. The EV charging load demand under two time types of working days and non-working days is shown in Figure 2 and Figure 3.

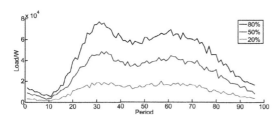

Figure 2. EV charging load on working days under different penetration rates.

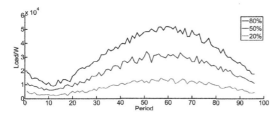

Figure 3. EV charging load on non-working days under different penetration rates.

5 CONCLUSION

This paper analyses expressway EV travel time, start time and end time of charging, traffic flow properties considering some factors such as EV battery characteristics, SOC and charging regimes and then establishes the expressway EV charging load model based on the daily vehicle flow and traffic laws of work days and non-work days. Simulation results reveal that, with the permeability rising, peak and valley phenomena occur in the expressway charging load and become more prominent. If the EV permeability is relatively high, the charging load will exert great pressure on the distribution network of the traffic system and additional peak phenomena may appear. As a result, the model proposed by this paper is proved to be feasible and practical and appropriate to be applied to analyse the influence of EV charging load on power grids.

ACKNOWLEDGEMENTS

This research was supported by the Project of State Grid Corporation of China (SGTYHT/14-JS-188).

REFERENCES

ADOT multimodal planning: transportation data management system. (2015, April 30). Retrieved from http:// adot.ms2 soft.com /tcds/tsearch.asp?loc = Adot&mod.

Alizadeh et al. (2014). A scalable stochastic model for the electricity demand of electric and plug-in hybrid vehicles. *IEEE Transactions on Smart Grid*, 5(2), 848–860.

Anonymous. (2015). Beijing-Shanghai high-speed fast charging network on January 15 from across the board. *FRP/Composite Materials*, 5(2), 97–98.

Chen, L. et al. (2015). Electrical vehicle charging load forecasting model based on travel chain. *Transactions of China Electrotechnical Society*, 30(4), 216–225.

Dai, Q. et al. (2014). Stochastic modeling and forecasting of load demand for electric bus battery-swap station. *IEEE Transactions on Power Delivery*, 29(4), 1909–1917.

DeForest, N. et al. (2009, August 31). Impact of widespread electric vehicle adoption on the electrical utility business: Threats and opportunities. Retrieved from http: // cet.berkeley.edu/dl/ Utilities _ Final_ 8–31–09.pdf.

Haijuan, L. (2015). Study on short term load forecasting of electric vehicle intelligent distribution network (dissertation). Southeast University.

Kintner, M et al. (2007). *Impacts assessment of plug-in hybrid vehicles on electric utilities and regional US power grids part 1: Technical analysis*. Electric Utilities Environmental Conference, Tucson, USA, 1–23.

Long, J. et al. (2015). Optimization of electric vehicle charging station on electric network. *Automation of Electric Power Systems*, (15).

Luo, Z. et al. (2011). *Forecasting charging load of plug-in electric vehicles in China*. IEEE Pes General Meeting, 1–8.

Ma, R. et al. (2015). Daily load random fuzzy modeling of highway electric vehicle charging stations. *Journal of Changsha University of Science and Technology: Natural Science Edition, the electric vehicle charging*, 12(4), 81–88.

Peng, L. et al. (2012). Charging load model based on diffusion theory for electric vehicles. *Electric Power Automation Equipment, 32*(9), 30–34.

Ruan, J. (2008). Beijing expressway network planning under the regional coordinated development strategy. *Beijing Planning & Construction, 2*, 34.

Shaoyun, G.E. et al. (2013). Design of charge station on expressway considering electricity distribution and driving range. *Electric Power Automation Equipment, 33*(7), 111–116.

Wynne, J. (2009). Impact of plug-in hybrid electric vehicles on California's electricity grid (dissertation). North Carolina: Nicholas School of the Environment of Duke University.

Zhang, H. et al. (2014). Prediction method of electric vehicle charging load considering space-time distribution. *Automation of Electric Power Systems, 38*(1), 13–20.

Zhang, X. (2014). Electric vehicle charging station power demand modeling (dissertation). North China Electric Power University.

Automotive, Mechanical and Electrical Engineering – Liu (Ed.)
© *2017 Taylor & Francis Group, London, ISBN 978-1-138-62951-6*

Machine-team quota measurement of a power transmission project based on GM and BPNN

Hongjian Li, Libin Zhang, Ke Lv & Wei Xiao
State Grid Jibei Electric Economic Research Institute, Beijing, China

ABSTRACT: Because the existing methods for machine-team quota measurement are time-consuming and laborious, it is very important to propose a new method. Considering the characteristics of power transmission project and quota measurement, this paper presents a hybrid method combined with Gray Model (GM) and Back Propagation Neural Network (BPNN) to measure the machine-team quota. Example results show that this method can not only reduce the work of quota measurement, but also make up for the lack of sample size.

Keywords: Power transmission project; Machine-team; GM; BPNN

1 INTRODUCTION

Since the construction process of a power transmission project is complicated, the quota measurement of machine-team is time-consuming and laborious. With the development of science and technology, the construction programmes of the same project are not exactly in common. In addition, the measurement of machine-team consumption is based on a large number of historical data. Thus, the advanced average level of quota is lagging behind the engineering practice, which makes the level of machine-team quota of power transmission projects not suitable in some cases. How to measure the consumption of machine-team quickly and efficiently is becoming an urgent problem.

At present, there are many methods commonly used in quota measurement of machine-team, such as statistical analysis, technical measurement and empirical estimation. However, these methods are always lacking in theoretical basis, or with overcomplicated calculation process, time-consuming and laborious. Statistical analysis is simple and easy to operate. Its drawback is that the calculated level of quota is unavoidably affected by abnormal factors. Technical measurement records the consumption of machine-team in detail, which. Its drawbacks are that it is time-consuming and laborious. Experience estimation is simple. Its shortcoming is poor accuracy. Therefore, a hybrid model combined with Gray Model (GM) and back propagation neural network is used to measure machine-team quota of a power transmission project, so as to provide a reliable referential method for the machine-team quota.

2 HYBRID MODEL

2.1 *GM*

Because of the limitation of manpower, material and financial resources, it is impossible to collect a large number of sample data to determine machine-team quota of a power transmission project. As a result, directed at the characteristic of "small sample, poor information", GM is proposed to solve this problem. GM is a newly emerging cross discipline, founded by Professor Deng Julong in 1982, which is based on the "small sample, poor information" uncertainty system where "some information is known and some information is unknown", and mainly through the generation, development and extracting the information of the "partial" known information to understand the real world.

Assume $x^{(0)}$ as a sequence with "n" elements:

$$x^{(0)} = [x^{(0)}(1), x^{(1)}(2), ..., x^{(0)}(n)] \tag{1}$$

Accumulation to a sequence:

$$x^{(1)} = \left[x_1^{(0)}, \sum_{t=1}^{1} x_t^0, \sum_{t=1}^{2} x_t^0, ..., \sum_{t=1}^{n} x_t^0 \right] \tag{2}$$

According to the new sequence $x^{(1)}$, set up whitening equation:

$$\frac{dx^{(1)}}{dt} + \alpha x^{(1)} = \beta \tag{3}$$

Solve the differential equation, and get GM (1, 1) sequence value. Through the GM we can explore the inherent law of things using a small part of information.

2.2 BPNN

As the GM does not have strong parallel computing capability, the external slight changes will lead to recalculation, which is not beneficial to the application of the GM. For this reason, the neural network model needs to be introduced. The neural network model has strong nonlinear capability, generalisation capability and fault-tolerance capability, and the parallel computing capability is powerful. In this paper, Back Propagation Neural Network (BPNN) is used to measure quota, which is one of the most widely used and successful neural networks at present.

BPNN is composed of input layer, hidden layer and output layer, as shown in Figure 1. In the network training phase, the processed samples are processed through the input layer. If the error requirements or the training times are not reached, the input layer, the hidden layer and the output layer are adjusted, making the network become a model of adaptive capacity. The BPNN mainly includes output model, action function model and error calculation model.

The specific steps of BPNN are presented as follows:

1. Initialise the network weights and thresholds;
2. Input training samples and target output;
3. Calculate the input for each layer;
4. Calculate network training error;
5. Correct network weights and thresholds;
6. Calculate the error index;
7. If the error meets the precision requirement, the training is finished, otherwise go to step (2).

2.3 GM-BPNN

The hybrid model is the combination of gray system and neural network, and learning from each other to improve the system's parallel computing capability and system modelling efficiency. The neural network needs enough sample data to train. It is obviously difficult for the actual power transmission project. However, the GM can deal with the small sample problem very well. The hybrid model calculation steps are given below:

1. After the original sequence $x^{(0)}$ is accumulated, the sequence $x^{(1)}$ is generated, then the gray differential equation of the model is established, and the whitening equation is obtained.
2. Solve the algebraic equation to obtain the grayscale discrete-time response function, and then transform the discrete-time response function;
3. The time response function is mapped to the BPNN to get the gray neural network structure diagram;
4. Network weights and thresholds are initialised;
5. For each input, sample for the forward calculation, reverse calculation, weight correction and threshold correction;
6. Do step (5), and the network training is performed until convergence;
7. Enter a new sample, do step (5) and (6) until the training of all the samples in the network ends, to the step (8);
8. The network forward calculation, get measured value.

3 EXAMPLE ANALYSIS

3.1 Influencing factors analysis

Take the process of ground groove excavation and backfill which is in an earthwork project below a 220 kV power transmission line, as an example to verify the rationality of the above model. Ground groove excavation and backfill work content are as follows: the excavation, soil replacement, soil filling, compaction, site cleaning, equipment transport etc. The mechanical factors are as follows: site mechanical utilisation rate, mechanical operator's driving experience, mechanical residue ratio, mechanical operator's technical level, measuring temperature and wind speed, terrain, work area, organisation level of construction, actual distance of soil transport.

3.2 Model validation

According to opinions of experts, remove the factors which have little effect and cannot be quantified. The final input sample contains the mechanical operator's driving year, mechanical residue ratio, mechanical operator's technical level, the actual distance of transport, work area and wind speed. Combining with 15 actual project

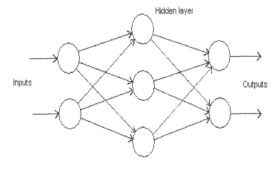

Figure 1.   The architecture of BPNN.

Table 1. The engineering sample data.

	Driving year	Residue ratio	Technical level	Transport distance	Work area	Wind speed
1	15	0.5	1	500	60	23
2	16	0.4	2	200	100	26
3	15	0.6	2	1000	150	32
4	13	0.7	1	800	80	21
5	19	0.5	2	700	200	30
6	14	0.6	2	400	130	28
7	17	0.7	2	1500	90	27
8	18	0.4	1	1200	160	29
9	15	0.6	2	600	100	24
10	14	0.4	3	800	120	25
11	18	0.7	1	1000	110	22
12	13	0.5	2	1200	140	27
13	15	0.6	3	1000	100	24
14	17	0.5	1	900	120	22
15	13	0.5	1	1200	150	26

Table 2. The results of machine-team quota measurement.

	Actual value	Measured value	Error
Sample 1	0.0034	0.0032	5.88%
Sample 2	0.0036	0.0038	5.55%

samples, 13 samples are used as learning samples and the other 2 samples are used to calculate samples. Project sample data is shown in Table 1.

Firstly, the GM is used to study the training samples, and the input value of BPNN is obtained. In this case, the node number of the input layer is 6 and the output layer is 1. Set the maximum number of learning to 10,000 times, the learning precision to 0.001, and the error sum of square of the learning goal to 0.1. When the number of training steps is 2,802, the training error is 0.001, which meets the requirement of the learning precision. Therefore, the structure of neural network is $6 \times 10 \times 1$.

The field measured data of "ground groove excavation and backfill" collected by "real record method" are compared with the measurement method proposed in this paper. The results are shown in Table 2.

From Table 2, it can be seen that the error between the measured value and the real value is

within 6%, which means that the machine-team quota of the power transmission project can be measured by the limited finite samples, effectively reducing the work. The hybrid method proposed in this paper can greatly improve the efficiency of the determination, and provide an effective reference for the quantitative measurement of machine-team quota for power transmission projects.

4 CONCLUSION

1. Using the combination model of GM and BPNN to measure machine-team quota can overcome the small sample defects, and improve the efficiency of the determination.
2. The combined model of GM and BPNN can be used not only to measure machine-team quota, but also to measure the material consumption quota.
3. The sample data used in this paper must be the same type or under the same conditions, as well as need to deal with the original data.

REFERENCES

Dai, Q.M. and Han, W.D. (2003). Inquisition into quotation of terms of works for transmission projects. *Electric Power Construction*, *24*(3), 64–65.

Ge, S.Y., Jia, O.S. and Liu, H. (2012). A gray neural network model improved by genetic algorithm for short-term load forecasting in price-sensitive environment. *Power System Technology*, *36*(1), 224–229.

Guo, Q. and He, X.J. (2016). Hydropower projects quota preparing model based on BP neural network. *Yangtze River*, *47*(5), 69–73.

Liang, J., Guo, Y.P., Ren, M. and Ma, X.C. (2014). Discussion on norm adjustment coefficient of crossing construction in detailed budget estimate norm of transmission line. *Guangxi Electric Power*, *37*(5), 35–38.

Ouyang, H.B. (2014). Generation of STEP-NC oriented working steps sequence based on BP neural networks. *Shanghai Dianji University*, *11*(7): 42–48.

Wang, S.X. and Zhang, N. (2012). Short-term output power forecasting of photovoltaic based on a grey and neural network hybrid model. *Automation of Electric Power Systems*, *36*(19), 37–41.

Xu, Q.S., Chen, N., Hou, W. and Wang, M.L. (2007). Application of transmission line dynamic heat-rating technology. *Electric Power Construction*, *28*(7), 28–31.

Automotive, Mechanical and Electrical Engineering – Liu (Ed.)
© 2017 Taylor & Francis Group, London, ISBN 978-1-138-62951-6

Man-hour quota measurement of power transmission project based on LSSVM and PSO model

Ke Lv, Fang Xu, Yin Xu & Jinwei Zhang
State Grid Jibei Electric Economic Research Institute, Beijing, China

ABSTRACT: Because the existing methods for man-hour quota measurement are time-consuming and laborious, it is very important to propose a new method. Considering the characteristics of power transmission project and quota measurement, this paper presents a hybrid method combined with Least Squares Support Vector Machine (LSSVM) and Particle Swarm Optimization (PSO) to measure the man-hour quota. Example results show that, this method can not only reduce the work of quota measurement, but also make up for the lack of sample size. It can provide some reference for quota measurement.

Keywords: Power transmission project; Man-hour quota; LSSVM; PSO

1 INTRODUCTION

Man-hour quota management is an important basic work of enterprise production management. It is an important basis for enterprise planning management, economic accounting, production schedule control, production cost control and reasonable quotation. The rationality of the measurement of man-hour quota not only affects the working hours, the utilization rate of the equipment and the labor remuneration of the employees, but also is one of the important means to improve the labor productivity of enterprises.

At present, the common methods of man-hour quota are look-up table method, experience estimation method, analogy method. In addition, artificial intelligence algorithms are introduced. A large number of scholars use the neural network model to measure man-hour quota and have achieved better results. However, neural network model have many shortcomings, such as poor convergence and easily trapped in local minimum. Although the neural network structure is reasonable, there is no evidence that it can quickly reach the optimal value. To overcome the defects of neural network model, this paper uses the LSSVM model to measure the man-hour quota. However, the LSSVM mode also has some defects. The main defect is the selection of model parameters. In the past, the selection of model parameter usually relies on the expert experience. It will inevitably affect the accuracy of the model. Thus, this paper uses PSO to optimize the model parameter. The man-hour quota is measured by the hybrid model combined with LSSVM and PSO.

2 THE HYBRID MODEL

2.1 LSSVM model

For a given set of samples, $(x_i, y_i), i = 1, 2, \cdots, l$, $x_i \in R^n, y_i \in R$, using nonlinear mapping $\phi(\cdot)$, which mapping the sample set from the input space to the feature space. Then, the linear regression is performed in the high dimensional feature space.

$$y(x) = \omega^T \phi(x) + b \tag{1}$$

According to the principle of structural risk minimization, regression problem can be presented as:

$$\min \frac{1}{2}\|\omega\|^2 + \frac{c}{2}\sum_{i=1}^{l}\xi_i^2 \tag{2}$$

$$y_i = \omega^T \phi(x_i) + b + \xi_i, i = 1, 2, \cdots, l \tag{3}$$

where ω the weight coefficient of LSSVM is, b is the constant deviation, c is the punishment factor, and ξ is the relaxation factor. In order to solve the constrained optimization problem, the Lagrange function is introduced. Then the final LSSVM regression function model can be defined as:

$$y(x) = \sum_{i=1}^{l} \delta_i K(x, x_i) + b \tag{4}$$

As discussed above, the LSSVM model has some drawbacks, such as the selection of model parameters. The main parameters affecting the LSSVM model are given below:

1. Penalty coefficient c. According to the properties of the sample data, the complexity of the model can be decided.
2. Insensitivity coefficient ε. It shows the expectation of the error of the estimation function on the sample data.
3. Kernel function parameter σ. It defines the structure of the high-dimensional characteristic space $\varphi(x)$.

Therefore, PSO is used to optimize the parameters (c, ε, σ) of the LSSVM model in this paper.

2.2 PSO

PSO is a kind of evolutionary computation method proposed by Dr. Eberhart and Dr. Kennedy in 1995. Its basic idea is to find the optimal solution by information transfer and information sharing among the individuals in the population.

Suppose the population size is M and each particle is flying in the D-space, the initial velocity is $V_i = [v_{i1}, v_{i2}, \cdots, v_{id}]$ and the initial position is a random variable $U_i = [u_{i1}, u_{i2}, \cdots, u_{id}]$, $i = 1, 2, \cdots, M$, $d = 1, 2, \cdots, D$. Then each particle will find the optimal solution through two extremes, one is the optimal solution of the particle itself p_{besti}, expressed by $P_i = [p_{i1}, p_{i2}, \cdots, p_{id}]$, the other is the entire population of the optimal solution g_{best}, expressed by $P_g = [p_{g1}, p_{g2}, \cdots, p_{gd}]$.

According to the particle fitness value, the following particle update speed and position can be obtained until the termination condition is satisfied:

$$v_{id}^{k+1} = \omega v_{id}^{k} + c_1 r_1 (p_{id} - u_{id}^{k}) + c_2 r_2 (p_{gd} - u_{id}^{k}) \quad (5)$$

$$u_{id}^{k+1} = u_{id}^{k} + v_{id} \quad (6)$$

where k represents the number of iterations, c_1, c_2 is the acceleration factor, which makes each particle close to the position of p_{besti} and g_{best}. r_1, r_2 is the random number between 0 and 1 w is inertia weight coefficient.

2.3 LSSVM-PSO

In this paper, the flow chat of the hybrid model is given in Figure 1, and the specific steps are presented as follows:

1. Initialize the size M of the particle swarm, the maximum allowable iteration number L, the inertia weight W, the learning factor D, and the velocity of each particle.
2. Initialize the particle position. Randomly generate a 3-dimensional vector, and generate N initial speeds.
3. If the particle fitness is superior to the individual extremum, then the fitness of the particle swarm is set to the new position.

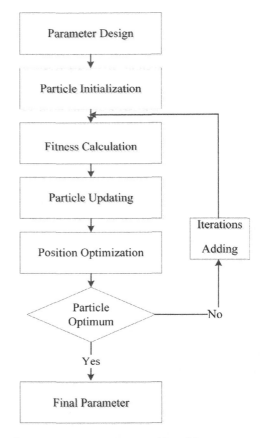

Figure 1. Flow chat of the hybrid model.

4. If the particle fitness is better than the global extremum, then the global extreme is set to the new position.
5. Update the velocity and position of the particle.
6. Optimize the optimal position.
7. If the number of iterations meet the maximum value, then the global optimal position is the parameter vector (L, W, D); otherwise, return to step (3).
8. According to the parameters of C, ε and σ, the sample variance is taken as the fitness function of the least squares support vector machine.
9. If the sample variance of the LSSVM model reaches the minimum value, then the corresponding parameters of C, ε and σ is the optimum parameters.

3 EXAMPLE ANALYSIS

This paper takes the automobile transportation as an example, and establishes the measurement model of man-hour quota. Since the historical sample size is 19, the population size of the particle swarm is set to 19, the maximum number of iterations L is 1000, the search range of W is set to [0, 100], and the search range for D is set to [0.1, 100]. The initial velocity of the particles is 2. The commonly used kernel function are radial basis function (RBF), polynomial function and linear function. Research shows that the radial basis function has stronger generalization ability. Thus, this paper uses radial basis function as kernel function in order to get the best evaluation model. The corresponding evaluation results are shown in Table 1.

Where, $e_{RMSE} = \left[\frac{1}{N} \sum_{i=1}^{19} (\hat{y}_i - y_i)^2 \right]^{1/2}$, $\hat{y}_i$ and y_i are the measured value and the actual value. As can be seen from Table 1, when C = 25, ε = 500, σ = 1.4, the error of the measured man-hour quota and the actual man-hour quota is the least. Therefore, we select the parameters of C = 25, ε = 500, σ = 1.4 as the final parameters.

The man-hour quota of the concrete pole transportation below the motor vehicle is measured as follows. Based on the 19 practical engineering samples, the hybrid model is built. 17 samples are used as learning samples and 2 samples are used as the testing samples. The measured results of the man-hour quota are as follows:

Table 1. Comparing results of different parameters.

	C	ε	Σ	e_{RMSE}
1	25	500	1.4	5.21
2	30	550	2.2	6.49
3	40	600	2.8	7.03
4	50	520	3.2	7.89

Table 2. Man-hour quota measurement results of concrete pole transportation.

Less than 500 kg	Actual value (work day)	Measured value (work day)	Error
Sample 1 common laborer	0.136	0.129	5.14%
Sample 1 Silled labour	0.015	0.014	6.66%
Sample 2 common laborer	0.136	0.127	6.61%
Sample 2 Silled labour	0.015	0.014	6.66%

From Table 2, it can be seen that the error between the proposed method and the real value is about 6%, which means that it can save a large amount of financial and material resources by using the hybrid model for measuring the man-hour quota. This paper provides an effective method for the man-hour quota measurement.

4 CONCLUSION

1. Using the LSSVM model combined with PSO to measure man-hour quota can avoid the disadvantage of local optimum. It can improve the efficiency of man-hour quota measurement.
2. The hybrid model can be used not only for the man-hour quota measurement, but also for machinery classes measurement.
3. The sample data used to measure the man-hour quota must be the same type or under the same conditions.

REFERENCES

Guo, C. & Zhou, D.C. 2009. The application of genetic neural network in calculating man-hour quota. Mechanical 36(2):15–17.
Li, S.J. & Li, Y. & Hong, W. 2000. A Neural Network Based Method for Determining Time Quota. Mechanical Science and Technology 29 (2): 266–268.
Liu, S.H. & Chen, J. 2007. Study on Man-hour Ration Calculation with Artificial Neural Networks. Machine tool & Hydraulics 35(1): 81–86.
Wang, G.Y. & Li, H. & Ding, J.X. 2008. Based Neural Network method of time quota compute research. Microcomputer Information 7(24):123–125.
Yi, D. & Chen, X.A. & Liu, L.L. 2007. Modeling of Man-hour Quote Information Management System Based on UML. Microcomputer Information 23(1):21–23.
Zhong, H.C. & Liu, J.F. 2003. Rapid Interim Product Man-hour Ration Estimation with Artificial Neural Network Method. Journal of East China Shipbuilding Institute (Natural Science Edition) 17(2):23–28.
Zhu, L.X & Zhou, J.T & Gao, J.J & Zhang, S.S. 2004. A Study on the Development of Man-hour Ration Based on Artificial Neural Networks. Mechanical Science And Technology 23(6):702–705.
Zhu, Q.Q & Xu, Y.Q. & Zhou, Y.J. 2008. Research and Development of Computer Man–hour Quota System. Mechanical manufacture and Automation Major 37(2):25–28.

Automotive, Mechanical and Electrical Engineering – Liu (Ed.)
© 2017 Taylor & Francis Group, London, ISBN 978-1-138-62951-6

Non-dissipative equalisation circuit research based on adjacent lithium-ion cell energy transfer

Renzhuo Wan, Ming Chen, Jun Wang & Fan Yang
Battery and Battery Management System Research Centre, School of Electronic and Electrical Engineering, Wuhan Textile University, Wuhan, Hubei, China

ABSTRACT: Battery equilibrium management is of significance to optimise the overall charging/discharging performance and prolong the battery lifespan. In this paper, a non-dissipative equalisation circuit incorporating MOSFETs and inductors is studied by using switching power supply technology. The proposed circuit is derived from the buck-boost converter. The energy transfer and the flow direction are controlled by adjusting the duty cycle of the corresponding transistors, which are triggered by PWM signal. The results based on MATLAB/Simulink simulation were performed to verify the feasibility and validity of the proposed circuit. The results indicate high balancing current and high balancing speed of the proposed circuit, which is potentially suitable for cell/pack balancing in a battery management system.

Keywords: Battery equalisation; MATLAB/Simulink simulation; Buck-boost converter; Battery management system

1 INTRODUCTION

Electric Vehicles (EVs), Hybrid Electric Vehicles (HEVs) and Plug-in Hybrid Electric Vehicles (PHEVs) are increasing in importance for economic development or financial investment, which may relieve the stress from the energy crisis and environmental pollution worldwide. Lithium-ion batteries as a relatively green energy source have been used widely, due to their high energy density, non-memory effect, low self-discharge rate, lightweight property, long lifetime, etc. (Tarascon, 2001). In practice, a majority of batteries in series and parallel are needed to meet the requirements of high voltage and large capacity (Lu, 2013). Based on the considerations of safety, practicability and stability, as well as to achieve superior power performance, a high uniformity of batteries is demanded. However, due to the nature of electrochemistry characteristics, manufacturing variance and actual complex working conditions, the capacities and internal resistances of batteries are variable, which may potentially cause a capacity imbalance of the cell/pack during the process of charging/discharging (Park, 2009; Anthony, 2014). At present, battery balancing technology is becoming necessary and a core index in the mainstream of a battery management system (Stuart, 2011; Dai, 2013; Manenti, 2011).

To compensate the inborn deficiency of the battery and obtain a uniform performance, equilibrium technologies have been developed in recent years (Lee, 2005). In general, equilibrium technologies can be classified into dissipative techniques and non-dissipative techniques according to the energy consumption (Yu). Dissipative techniques taking advantage of simple structure and easy control have been used widely (Gallardo-Lozano, 2014). However, the technique has extremely low efficiency due to the fact that all excess energy is converted to thermo through resistances. Non-dissipative techniques in accordance with different storage elements can be divided into capacitive-type, inductive-type, and transformer-type architecture. In these kinds of techniques, transfer energy could be well controlled with a relatively complex circuit and control stratagem from cells to cells, or pack to cell, or cell to cell, so as to achieve better equalisation behaviour with higher efficiency compared with the dissipative technique (Baronti, 2013).

In this paper, a proposed equalisation topology based on State Of Charge (SOC) is used to balance the lithium-ion cells or packs. A non-dissipative equalisation circuit incorporating MOSFETs and inductors is designed. Three equilibrium modes at standing, charging and discharging conditions are studied. The relationship between the circuit parameters of duty cycle and inductance with the battery balancing performance, such as balancing speed, balancing current and balancing efficiency, are well built and evaluated. In this paper, the relationship between SOC and OCV (Open-Circuit Voltage), as well as the battery parameters, are briefly introduced firstly in Section 2. Then the equilibrium working principle based on inductors is described in Section 3. The parameters used in the simulation and results are given in detail in Section 4.

2 BATTERY PARAMETERS

In the mainstream of battery management system, battery equilibrium is triggered by cell/pack voltage or SOC. Considering the voltage characteristic in practice, balancing based on SOC is employed in this study. There are several typical methods of estimating SOC, such as the ampere-hour counting method, open circuit method, Kalman filter method, particle filter method, etc. Each method has its merits and drawbacks. In addition, it is believed that there is a monotonous relationship between the SOC and the battery Electromotive Force (EMF) or OCV. In this article, the open circuit method to estimate the SOC is used in the simulation.

To make the simulation much closer to actual experiment conditions, the lithium-ion phosphate (LiFePO$_4$) battery (F8068260, from Xinxin Energy of Xiangyang, Hubei, China) with nominal capacity 10.8 Ah and a voltage cutoff of 3.65 for charging and 2.0 for discharging is used. The battery simulation performance is similar to the results verified by the battery test system (Neware CT-4001, Shenzhen, China). The platform voltage is 3.2 V. The internal resistance is set to 0.01 Ω, which is calculated in Equation (1) when the current drops to zero suddenly, and the voltage rises instantaneously from $V1 = 3.203$ V and $V2 = 3.147$ V.

$$R = (V2 - V1)/I \tag{1}$$

3 PWM CONTROLLED CONVERTER BASED ON INDUCTORS

The basic topology based on inductors is shown in Figure 1. Every module for equalisation is connected across each two adjacent cells to allow energy transfer from the cell with the higher SOC to the cell with the lower SOC. The energy is first transferred from the cell with the higher SOC to the common inductor, then the inductor releases the energy to the cell with the lower SOC. The proposed circuit is derived from the buck-boost converter. The flow direction of energy is determined by adjusting the duty cycle of the corresponding transistors, which are triggered by PWM signal.

For example, if the SOC of cell B1 is higher than the SOC of cell B2, two steps will be taken to transfer the electric charge from B1 to cell B2 in.

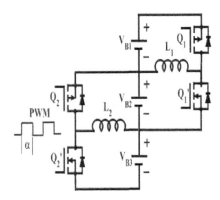

Figure 1. A typical equalisation circuit structure based on inductors.

Firstly, the energy is transferred from cell B1 to the common inductor when switch Q1 turns on. Secondly, switch Q1 is turned off while switch Q1' is turned on, then the inductor releases the energy to cell B2, the energy of cell B1 is transferred to cell B2 through inductor. In this way, the energy in cell B1 can be transferred to cell B3. The relevant circuit parameters are described by the following equations:

$$V_L = L\frac{di_L}{dt} \tag{2}$$

$$R_{on} = R + R_{sw} \tag{3}$$

$$V_B = i_L R_{on} + L\frac{di_L}{dt} \tag{4}$$

$$i_L = \frac{V_B}{R_{on}}\left(1 - e^{-\frac{R_{on}}{L}t}\right) \tag{5}$$

$$t_0 = L/R_{on} \tag{6}$$

$$\eta = (SOC1' + SOC2' + SOC3')\,/ \\ (SOC1 + SOC2 + SOC3) \tag{7}$$

where V_L is the voltage across the inductor, L is the inductance, i_L is the inductor current, V_B is the cell voltage, R_{on} is the total resistance of the loop, R is the internal resistance of the cell, R_{sw} is the switch resistance, t_0 is the time constant, and η is the balancing efficiency. SOC1, SOC2 and SOC3 stand for the initial state of charge respectively, and SOC1', SOC2' and SOC3' stand for the final state respectively.

The equations of relevant inductor current in the first switching cycle are as follows:

$$i_{DT} = \frac{V_{B1}}{R_{on}}\left(1 - e^{-\frac{R_{on}}{L}DT}\right) \tag{8}$$

278

$$i_T = \frac{V_{B1}}{R_{on}}\left(1 - e^{-\frac{R_{on}}{L}DT}\right) - \frac{V_{B2}}{R_{on}}\left(1 - e^{-\frac{R_{on}}{L}(1-D)T}\right) \qquad (9)$$

i_{DT} is the inductor current when t = DT, i_T is the inductor current when t = T. The average balancing current between two adjacent cells named I_b and the average inductor current named I_L are evaluated as the measurement index of battery balancing performance.

When $t/t_0 \approx 0$, then $i_L \approx V_B/Lt$, then

$$I_b = \frac{1}{2}\left(\frac{V_{B1}D^2 - V_{B2}(1-D)^2}{L}\right)T \qquad (10)$$

$$I_L = \frac{1}{2}(i_{DT} + i_T) * \frac{1}{1 - e^{-\frac{R_{on}}{L}T}} \qquad (11)$$

The balancing mechanism will work when the SOC difference of two adjacent cells reaches 0.2%, otherwise the balancing mechanism stops.

4 RESULTS AND DISCUSSION

In order to study the relationship between circuit parameters and balance performance, the simulation is performed by MATLAB/Simulink. In addition, according to the battery status, three modes at off working, charging and discharging state are simulated to evaluate the balancing ability simultaneously. Here, the balancing mechanism will be triggered when the SOC difference of two adjacent cells reaches 0.2%, otherwise the balancing process stops. In the simulation, the initial SOC are set to 90%, 89% and 88% orderly. The switch resistance, total resistance of the loop and switching cycle time are set to 0.09 Ω, 0.1 Ω and 100 us respectively.

To build the relation between duty cycle of PWM signal and average inductor current and average current between two adjacent cells, the circuit is simulated under off working mode. Here, L is set to 100 uH. The corresponding relationships working at standing condition are shown in Figure 2. It indicates that the average current between two adjacent cells keeps almost constant at 60 mA, while the average inductor current increases almost linearly when the duty cycle of PWM signal changes from 0.5 to 0.7. For safety considerations, it is suggested that the duty cycle should be less than 0.65. In the following study, it is set to 0.60 at charging/discharging mode.

The choice of inductance is also a key parameter to determine the inductor current and balance current. When the duty cycle is set to 0.60, the phenomenon that average inductor current decreases

with increasing inductance is shown in Figure 3. Considering the appropriate charging/discharging current for the batteries and balancing speed, the value of inductance is set to 50 uH in the following study. The corresponding average balancing current between two adjacent cells has a relationship with inductance as shown in Figure 4, which shows that the balancing current decreases almost exponentially with the increase in inductance. In the practical balance strategy, conditions the balancing mechanism is usually working at charging or discharging mode, the balance current should be constant with limited fluctuations. Here, when the duty cycle is set to 0.60 and the inductance is set to 50 uH, the average current between two adjacent cells is about 0.65 A.

Three equilibrium modes at off working, charging and discharging are simulated in similar configuration at D = 0.60 and L = 50 uH. A typical simulation circuit structure at off working mode is shown in Figure 5(a). The balancing result is shown in Figure 5(b). It can be seen that the cell condition

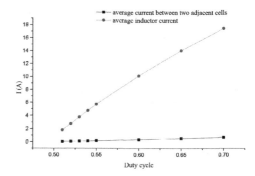

Figure 2. The relationship between duty cycle and relevant current.

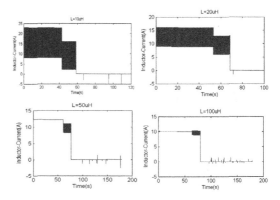

Figure 3. The results of inductor current working at different levels of inductance.

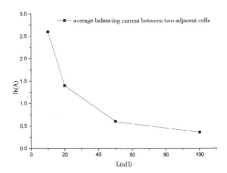

Figure 4. Relationship between inductance and the average current between two adjacent cells.

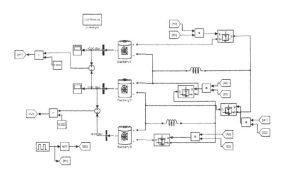

Figure 5(a). Equilibrium simulation circuit structure at off working condition.

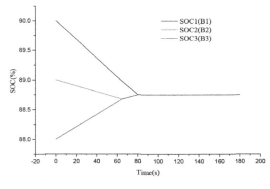

Figure 5(b). Equilibrium simulation result at off working condition.

is changed from $\Delta SOC_{max} = 2\%$ to 0.2% at about 80 seconds, and the equalisation efficiency goes up to a reasonable and high quality of 99.71%.

The additional two modes are simulated when the charging and discharging rates are set to about 1 C. The results are shown in Figure 6(a) for charging equilibrium and Figure 6(b) for discharging

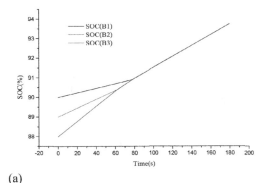

(a)

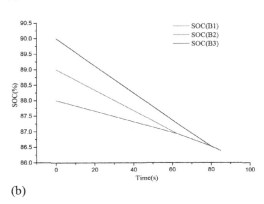

(b)

Figure 6. (a). Equilibrium simulation result at charging condition. (b). Equilibrium simulation result at off working condition.

equilibrium. By analysing the set parameter, $t/t_0 \approx 0$, Equation (10) can be applied to conclude that the external current supply has almost no effect on the average balancing current between two adjacent cells. It takes about 80 seconds to finish the balancing process whatever the work condition is.

5 CONCLUSION

MATLAB/Simulink simulation with actual battery parameters are used to verify the feasibility and validity of the proposed circuit. The simulation results indicate that the battery performance such as the average current between two adjacent cells and the balancing speed are determined by the switching duty cycle and the inductance. For practical consideration, the duty cycle is set to 60% for sustaining the average inductor current at about 1C rate and the average current between two adjacent cells at about 0.65 A. Moreover, the equilibrium mechanism can work in accompaniment with the charging/discharging process simultaneously, and the balancing operation has almost no relation with the outer current. The equalisation efficiency can achieve above 90%, the equilibrium circuit parameters could be well controlled to obtain high balancing current and the circuit is easy to be modularised for electric vehicles.

ACKNOWLEDGEMENTS

This work was supported by Natural Science Foundation of Hubei Provincial Education Department (Q20131603).

REFERENCES

Anthony, B. (2014). Statistical analysis for understanding and predicting battery degradations in real-life electric vehicle use. *Journal of Power Sources, 245*, 846–856.

Baronti, F. (2013). High-efficiency digitally controlled charge equaliser for series-connected cells based on switching converter and super-capacitor. *IEEE Transactions on Industrial Informatics, 9*, 1139–1147.

Dai, H. (2013). Cell-BMS validation with a hardware-in-the-loop simulation of lithium-ion battery cells for electric vehicles. *International Journal of Electrical Power & Energy Systems, 52*, 174–184.

Gallardo-Lozano, J. (2014). Battery equalization active methods. *Journal of Power Sources, 246*, 934–949.

Park, H-S. (2009). Design of a charge equalizer based on battery modularization. *IEEE Transactions on Vehicular Technology, 58*, 3216–3223.

Lee, Y-S. (2005). Intelligent control battery equalization for series connected lithium-ion battery strings. *IEEE Transactions on Industrial Electronics, 52*, 1297–1307.

Lu, L. (2013). A review on the key issues for lithium-ion battery management in electric vehicles. *Journal of Power Sources, 226*, 227–288.

Manenti, A. (2011). A new BMS architecture based on cell redundancy. *IEEE Industrial Electronics Society, 58*, 4314–4322.

Stuart, T.A. (2011). Modularized battery management for large lithium ion cells. *Journal of Power Sources, 196*, 458–464.

Tarascon, J.M. (2001). Issues and challenges facing rechargeable lithium batteries. *Nature, 414*, 359–367.

Automotive, Mechanical and Electrical Engineering – Liu (Ed.)
© 2017 Taylor & Francis Group, London, ISBN 978-1-138-62951-6

On-line electric vehicle charging system under complex environment of transmission lines and pedestrians with metallic implants

Feng Wen, Xueliang Huang & Lixing Chen
School of Electrical Engineering, Southeast University, Nanjing, China

ABSTRACT: On-Line Electric Vehicle (OLEV) is a dramatic scheme with infinite cruising ability by charging along its way using Wireless Power Transfer (WPT). There will be the possibility to lay the electric cables of the supply rails and ultra-high voltage transmission lines closely parallel to each other in a confined space. Just because of this sharing of the passageway, the transmission lines aerially may induce volts on the electric cables and supply rails setting in the same gallery and they can also result in more acute disadvantageous reactions. This paper studies reasonable impacts of the transmission line conductors, Extremely Low Frequency (ELF) magnetic field, induced volts, and Electro Magnetic Field (EMF) exposure on the WPT charging system for OLEV by the ultra-high voltage transmission lines. A pedestrian with metallic implants is considered in this paper. The EMF is computed using Finite Element Analysis (FEA) method, and by comparing with the values given by International Committee on Non-Ionizing Radiation Protection (ICNIRP) guidelines, the degree of influence can be judged.

Keywords: Wireless Power Transfer (WPT); On-Line Electric Vehicle (OLEV); Metallic implants; Electro magnetic Field (EMF) exposure

1 INTRODUCTION

The progress of the electrical vehicle is substantially restricted by many factors such as the finitude of storage battery capacity, heavy battery pack weight, long period of charging time and so forth. On-Line Electric Vehicle (OLEV) charging along the roadway by Wireless Power Transfer (WPT) has been arranged as a feasible solution to this problem. It is commodious, safe, light-weight, and has seen forward development in recent years. A demonstration engineering project (Choi et al., 2015) has been performed since 2009 by KAIST, Korea. Due to space restriction, as the project is implemented on the highway, there will be the possibility to lay the electric cables of the supply rails and ultra-high voltage transmission lines closely parallel to each other in a confined space. It is common sense to concern the electromagnetic coupling of the ultra-high voltage transmission lines with power cables or rails to assure that there exist no risky or damaging voltages, while other adverseness also needs to be considered.

In this paper, a WPT charging system for OLEV is designed which is also exposed closely under ultra-high voltage transmission lines. The influences of the transmission line conductors, Extremely Low Frequency (ELF) magnetic field, induced voltages, and Electromagnetic Field (EMF) exposure on the WPT charging system by

the transmission lines are studied. Finite Element Analysis (FEA) method is utilised in calculating the EMF. A 3D precise numeric human anatomic model with common tissues and metallic implants is considered. Human health hazard rating can be judged by comparing to the limits given by International Committee on Non-Ionizing Radiation Protection (ICNIRP) guidelines.

2 FUNDAMENTALS OF THE WPT CHARGING SYSTEM

The WPT charging system for OLEV constitutes of two sub-systems (Choi et al., 2015): one is the on-road system for furnishing electric power, and the other is the on-board system for picking electric power. In recent years, a few OLEV demonstration engineering projects have been performed by KAIST, Korea. In order to shorten the costs of construction, improve the property and safety, they use different types of cores for the electric power supply rail; for example, the E-type, I-type, W-type and S-type (Lee et al., 2011; Choi et al., 2015). The electric power supply rail can be sectionalised into several sub-rails, which can be activated independently on its own by supplying it with high frequency current when the electric vehicle to be charged is above, and be released not to squander electric power as the vehicle passes away. For simplicity, we achieve excellent performance of the WPT charging system by using an even higher frequency and refuse to utilise any type of magnetic core in the rail coils. A simplified WPT charging system model is shown in Figure 1a; the equivalent circuit is shown in Figure 1b; the pick-up coils and the sub-rail coils of the WPT charging system are shown in Figure 1c. The two coils have six turns with a space of 15 cm. The conductor diameter is 1 cm. At 1 MHz, the pick-up coil $R_2 + j\omega L_2 = 0.121 + j54.072$ Ω, the sub-rail coil $R + j\omega L_1 = 2.862 + j850.650$ Ω, the mutual inductance $\omega M = 17.258$ Ω, the matching capacitor $C_2 = 2.943$ nF, $C_1 = 0.187$ nF, so each coil is matched at 1 MHz.

While the power source frequency equals 1 MHz as well, the system is expected to have an excellent WPT efficiency using Equation 1.

$$\eta = \omega^2 M^2 R_L / [R_1 (R_2 + R_L)^2 + \omega^2 M^2 (R_2 + R_L)] \quad (1)$$

As the source voltage equals 1,200 V and resistance (r_s) is 50 Ω, the input and output power is shown in Figure 2 as the load resistance (R_L) is

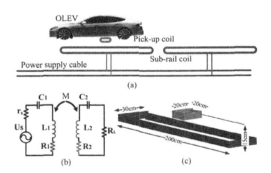

(a)

(b)

(c)

Figure 1. (a) A simplified WPT charging system model for OLEV. (b) Equivalent circuit of WPT system. (c) The pick-up coils and sub-rail coils of the WPT charging system.

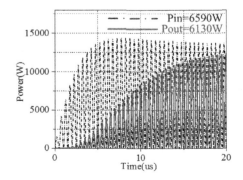

Figure 2. The input power and output power of the WPT charging system.

10 Ω. The pick-up coil gets 6,130 W while the sub-rail coil provides 6,590 W, and the efficiency reaches 93%.

3 TRANSMISSION LINES INFLUENCE

In this study, the 500 kV ultra-high voltage transmission lines as shown in Figure 3 are considered in extreme proximity to the WPT charging system. Equivalent radius of the phase conductors equals 0.211 m. The sub-rail coils are parallel to the ultra-high voltage transmission lines 20 m away, and they are 20 cm above the ground plane beyond the shield of any pipelines or concrete. The scenario can be much worse while the WPT charging system for OLEV is built on elevated highway. In this study, we adopt this scenario to figure out possible impacts on the WPT system.

As the EMF produced by the 500 kV ultra-high voltage transmission lines and the WPT charging system is time-harmonic, the Maxwell's equation can be written as

$$\nabla \times H = J + j\omega D$$
$$\nabla \times E = -j\omega B$$
$$\nabla \cdot B = 0 \quad (2)$$
$$\nabla \cdot D = \rho$$

where H is the magnetic field intensities, E is the electric field intensities, D is the electric flux densities, B is the magnetic flux densities, ρ is the volume charge density, J is electric current density of any external charges, ω is the solution frequency.

The electric field can be expressed by

$$E = -\nabla \phi - j\omega A \quad (3)$$

While

$$B = \nabla \times A \quad (4)$$

where A is the magnetic vector potential, and ϕ is an electric scalar potential.

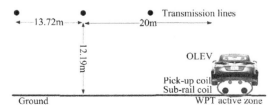

Figure 3. 500 kV ultra-high voltage transmission lines and WPT charging system for OLEV.

The first line in Equation 2 can be expressed by

$$\nabla \times (1/\mu)(\nabla \times A) = (\sigma + j\omega\varepsilon)(-\nabla\phi - j\omega A) \quad (5)$$

where ε, μ and σ are the electrical permittivity, the permeability, and conductivity of any material.

Total current in the conductor is

$$
\begin{aligned}
I_T &= \int_\Omega J^* d\Omega = \int_\Omega \sigma^* E d\Omega \\
&= \int_\Omega (\sigma + j\omega\varepsilon)(-\nabla\phi - j\omega A)d\Omega
\end{aligned}
\quad (6)
$$

where Ω is the solution region, and I_T includes the source current, induced current, and displacement current.

In this paper, the calculation is implemented by numerical method with FEA tool—ANSYS Electromagnetics.

3.1 Transmission line conductors

Research has been widely performed on conductive materials appearing close to the WPT active zone (Onar et al., 2013). The WPT system resonant frequency shifts and its transfer efficiency is also involved. As a stainless steel plate is set in the active zone on the axis of the coil centre, the plate is 200 cm by 10 cm, and its distance to the sub-rail coil changes as a: 0.1 m, b: 0.25 m, c: 0.4 m, d: 1 m. The peak transfer efficiency meets 68% at 1.09 MHz for a 80% for b, 85% for c, and 90% at 1 MHz for d, as Figure 4 shows. The efficiency is almost unaffected by the stainless steel plate as the distance of the plate and coil is larger than 1 m. Thus the transmission line conductors are far enough away to ensure that the WPT charging system will not be affected.

3.2 Extremely-low-frequency magnetic field

In the WPT charging system, electric power is transferred from the sub-rail coil to the pick-up coil through high frequency magnetic coupling while the transmission lines also a produce ELF magnetic field. The two magnetic fields are shown in Figure 5. The WPT charging system produces a much stronger magnetic field than the transmission lines, and the operating frequency of the two systems varies tremendously. Hence, the ELF magnetic field produced by the transmission lines will contribute little to the electric power collection by the pick-up coil.

3.3 Induced voltage

Influences arising from inductive, capacitive and resistive coupling occur as the electric power supply cables lie parallel to the 500 kV ultra-high voltage transmission lines (Kopsidas et al., 2008). Resistive coupling comes up in period of ground faults. During inductive coupling, the volts induced vary due to many factors such as the size and location of the cable. Even though the coil and cable are buried underground, the value can be very high as the length is long (Bortels et al., 2006). In the scenario shown in Figure 3, without considering the shielding effect of the earth, the voltage induced by capacitive coupling of the sub-rail coil which is set above the ground can be calculated using Equation 7. Maxwell's potential coefficient matrix '[P]' can be calculated according to the position of the ultra-high voltage transmission lines and the WPT charging system coils. The induced voltage can reach up to a few 100 volts, and it increases with its approach to the transmission lines and reduces as the coils are buried underground. The induced voltage on the pick-up coil is neglected as the coil is small in size and protected by the electric vehicle in charging.

$$[V] = [P][Q/(2\pi e_0)] \quad (7)$$

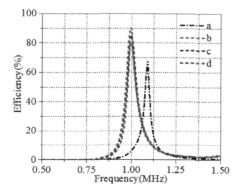

Figure 4.   The WPT efficiency changes with the stainless plate position.

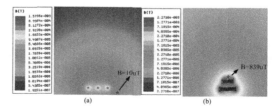

Figure 5.   (a) Magnetic field produced by the 500 kV ultra-high voltage transmission lines. Current in the lines is 1000 A. (b) Magnetic field produced by the WPT charging system. Input power is 6,590 W, and the current in the sub-rail coil is 15 A.

The input current I_{in} is shown in Figure 6, as we add a 2 kV power frequency voltage source, as total volts induced on the sub-rail coil. As a result of the resonant circuit characteristic, the induced volts can scarcely enhance the input current. Thus, the WPT charging system for OLEV will not be affected.

Nevertheless, the induced voltage by the 500 kV ultra-high voltage transmission lines on the coil can still lead to more possibility of the risk of suffering unpleasant electric shock. Relevant experiments are also carried out under the 500 kV transmission lines, where the unperturbed electric field is 5.8 kV/m, the ambient temperature is 35.6°C and the humidity is 29%. The induced voltage of the sub-rail coil U_{coil}, steady-state shock current I and transient-state shock current i(t) as the experimenter touches the coil, are shown in Figure 7. A need for concern is that large shock current may lead to seriously biological hazard. In the meantime, the induced voltage by the transmission lines will enhance the voltage of the capacitor banks, the WPT system coils and the series connected facilities. Therefore we should reconsider the withstand voltage, the electrical insulation, and erosion resistance in the construction of the WPT charging system.

3.4 Electromagnetic field exposure

The WPT system should completely conform to the limits such as ICNIRP guidelines (ICNIRP Guidelines 1998, 2010) in the application to charge OLEV.

With the purpose of avoiding adverse biological impacts, those regulations are fixed to protect humans from the excessive exposure to time-varying EMF. The 500 kV ultra-high voltage transmission lines overhead can worsen the electromagnetic environment, where a pedestrian would be inescapably close. As Figure 8 shows, we create a 3D precise finite element anatomic human model with organs such as brain, heart, liver, spleen, kidney,

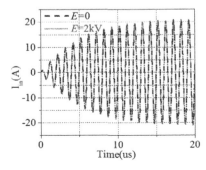

Figure 6. Input current of the WPT charging system in the sub-rail coil.

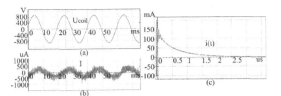

Figure 7. (a) Induced voltage of the sub-rail coil by the transmission lines. (b) Steady-state shock current as the experimenter keeps in touch with the coil. (c) Transient-state shock current at the moment the experimenter touches the coil.

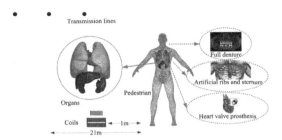

Figure 8. 3D precise finite element anatomic human model with common tissues and implantations.

lungs and so forth. The electromagnetic parameters of the tissues can be estimated under specific frequency (Gabriel et al., 1996). Titanium material in an artificial tooth, artificial arthrosis, and an artificial heart valve are also considered for implantation into human body. The EMF exposure from the WPT charging system and the ultra-high voltage transmission lines (Scorretti et al., 2004; Barchanski et al., 2006,; Scorretti et al., 2005; Motrescu et al., 2006) can be computed respectively as the pedestrian stands at 1 m from the coil and 21 m from the transmission line corridor centre.

According to the first two in Equation 2, we can get:

$$\nabla \times [(1/\mu)\nabla \times E] - \omega^2 \varepsilon_c E = -j\omega J_i$$
$$\nabla \times [(1/\varepsilon_c)\nabla \times H] - \omega^2 \mu H = \nabla \times [(1/\varepsilon_c)J_i] \qquad (8)$$

where $D = \varepsilon E$, $B = \mu H$, $J = \sigma E$, $\varepsilon_c = \varepsilon - j\sigma/\omega$, J_i is source current.

The current and power density can be obtained using Equation 9.

$$J = \sigma E$$
$$S = 0.5\text{Re}(E \times H^*) \qquad (9)$$

Current density of usual human tissues produced by WPT system is shown in Figure 9. Persons with metallic implants will not be more affected

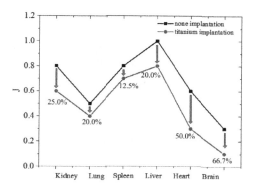

Figure 9. The current density of usual human tissues produced by WPT charging system.

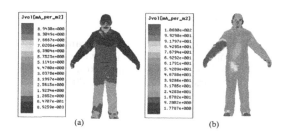

Figure 10. (a) The current density in pedestrian produced by the ultra-high voltage transmission lines and (b) by the WPT.charging system. ICNIRP limits for general public exposure: 2 mA/m² at 50 Hz and 2000 mA/m² at 1 MHz.

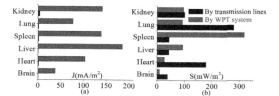

Figure 11. (a) The current density and (b) The power density of usual human tissues produced by the ultra-high voltage transmission lines and the WPT charging system.

due to the shielding effect of metals. Besides, the eddy currents in titanium are also weak considering its low permeability, and thus they will not cause the tissues to overheat. The influence by the transmission lines can be ignored as the implants are small in size.

The unperturbed EMF of 1.5 m above the ground plane at the human site caused by the ultra-high voltage transmission lines is 4.7 kV/m, 8 uT, and by the WPT charging system is 16 V/m, 0.5 uT. They are all in keeping with the restriction by ICNIRP for general public exposure: 5 kV/m, 100 uT for 50 Hz and 87 V/m, 0.92 uT for 1 MHz. It is important to note that the electric field produced by the ultra-high voltage transmission lines and the magnetic field produced by the WPT charging system are relatively close to the limits, which needs more attention. The current densities produced by the ultra-high voltage transmission lines and the WPT charging system are shown in Figure 10. Despite high value in local parts of the human body, such as the feet under the ultra-high voltage transmission lines and the body parts near to the WPT charging system, in most parts of the pedestrian the current density can meet the restriction. The current density and power density of human tissues are shown in Figure 11. As Figure 11a and b show, the current density produced by the ultra-high voltage transmission lines is comparatively small by contrasting with the one produced by the WPT charging system, while the power density produced by the ultra-high voltage transmission lines is so big that it needs to be considered.

According to the above research, although the EMF exposure of the pedestrian produced by the ultra-high voltage transmission lines and the WPT charging system on the whole meet the ICNIRP guidelines, the synthetic action of the EMF exposure might cause biological hazard due to the WPT

charging system providing higher electric power or the ultra-high voltage transmission lines being erected lower.

4 CONCLUSION

In conclusion, this paper studies the overhead 500 kV ultra-high voltage transmission lines' influence on the WPT charging system for OLEV lying closely parallel to each other in a confined space. The transmission lines conductors, extremely-low-frequency magnetic field, and induced voltage will produce little impact on the proper functioning of the WPT charging system. However, the induced voltage on the WPT coils by the ultra-high voltage transmission lines may lead to seriously biological hazard through electric shock and reconsideration of the withstand voltage, the electrical insulation, erosion resistance in the construction of the WPT charging system. In addition, human exposure to time-varying electromagnetic field also needs second thoughts, considering the synthetic action of the ultra-high voltage transmission lines as well as the WPT charging system. Persons with titanium implants will not be more affected.

ACKNOWLEDGMENTS

This work was supported by Science and Technology Planning Project of Jiangsu Province (Grant no. BE2015004-4), and Major State Research Development Program of China (2016YFB0101800).

REFERENCES

Barchanski, A. (2006). Local grid refinement for low-frequency current computations in 3-D human anatomy models. *IEEE Transactions on Magnetics, 42*(4), 1371–1374.

Bortels, L. (2006). A general applicable model for AC predictive and mitigation techniques for pipeline networks influenced by HV power lines. *IEEE Transactions on Power Delivery, 21*(1), 210–217.

Choi, S.Y. (2015). Advances in wireless power transfer systems for roadway-powered electric vehicles. *Emerging and Selected Topics in Power Electronics, 3*(1), 18–36.

Choi, S.Y. (2015). Ultra-slim S-type inductive power transfer system for roadway powered electric vehicles. *IEEE Transactions on Power Electronics, 30*(11), 6456–6468.

Gabriel, C. (1996). The dielectric properties of biological tissues: I. Literature survey. *Physics in Medicine and Biology, 41*(11), 2231.

ICNIRP Guidelines. (1998). Guidelines for limiting exposure to time-varying electric and magnetic fields (Up to 300 GHz).

ICNIRP Guidelines. (2010). Guidelines for limiting exposure to time-varying electric and magnetic fields (Up to 100 kHz).

Kopsidas, K. (2008). Induced voltages on long aerial and buried pipelines due to transmission line transients. *IEEE Transactions on Power Delivery, 23*(3), 1535–1543.

Lee, S. (2011). Active EMF cancellation method for I-type pickup of on-line electric vehicles, *Applied Power Electronics Conference and Exposition.*

Motrescu, V.C. (2006). Simulation of slowly varying electromagnetic fields in the human body considering the anisotropy of muscle tissues. *IEEE Transactions on Magnetics, 42*(4), 747–750.

Onar, O.C. (2013). A novel wireless power transfer for in-motion EV/PHEV charging. *Applied Power Electronics Conference and Exposition,* 3073–3080.

Scorretti, R. (2004). Computation of the induced current density into the human body due to relative LF magnetic field generated by realistic devices. *IEEE Transactions on Magnetics, 40*(2), 643–646.

Scorretti, R. (2005). Modeling of induced current into the human body by low-frequency magnetic field from experimental data. *IEEE Transactions on Magnetics, 41*(5): 1992–1995.

Automotive, Mechanical and Electrical Engineering – Liu (Ed.)
© 2017 Taylor & Francis Group, London, ISBN 978-1-138-62951-6

Optimisation of oxygen-enriched air intake system of an engine based on pressure swing adsorption

Dongmin Li

Department of Mechanical and Electronic Engineering, Shandong University of Science and Technology, Tai'an, Shandong Province, China

Zhao Gao & Yushan Li

College of Transportation, Shandong University of Science and Technology, Qingdao, Shandong Province, China

ABSTRACT: In order to improve the power performance of an engine, oxygen is produced with pressure swing adsorption and mixed with the air from the original intake pipe to realise oxygen-enriched air intake of the engine. A pressure swing adsorption device is designed based on the volume range of air intake at different oxygen/air ratio. The air intake system of the four-cylinder gasoline engine is modeled with GT-Power. Optimisation of air-fuel ratio and valve timing is carried out. During the process of optimisation, oxygen/air ratio and the maximum torque at different working points are taken as the optimisation parameter and the optimisation target respectively. Finally the optimisation results show that torque of the engine at middle and high speed is increased.

Keywords: Oxygen-enriched air intake; Pressure swing adsorption; Power performance; Optimisation

1 INTRODUCTION

Generally the oxygen needed with naturally aspirated engine state is only 20.93% of air, while the other gases accounting for up to 78.03% of the air, such as nitrogen, not only take a lot of heat, but also produce NOx, CO and other emissions, even causing serious air pollution and ecological deterioration. Oxygen-enriched engine technology is an effective way to solve this problem. In the 1960s, Wartinhee (1971) began to study the engine's oxygen-enriched combustion performance. It was found that the oxygen-enriched combustion in the engine can effectively reduce the production of HC and CO. Maxwell et al. (1993) found that oxygen combustion can significantly improve the gasoline engine combustion and reduce CO and HC emissions via analysis of the combustion mechanism of nitrogen and oxygen components.

Li et al. (2007) from Shanghai Jiao Tong University studied the oxygen-enriched air intake system of the LPG (Liquefied Petroleum Gas) engine and found that: the emissions of HC had not been significantly reduced, but the CO emissions were greatly reduced. Xiao et al. (2007) finished a study on emissions from the engine using membrane oxygen-enriched air under the condition of cold start, and found that the optimal concentration of oxygen-enriched air was about 23%. Baskar and

Senthilkumar (2016) increased the oxygen concentration in intake from 21% to 27%, and found that the engine's thermal efficiency was improved by 4–8%, and the brake fuel consumption was reduced by 5–12%; unburned HC, CO and density levels of flue gas were reduced by 40%, 55% and 60%. The oxygen-enriched combustion performance indicators of the gasoline engine studied by Liu et al. (2006) shows that: as soon as the oxygen concentration increases by 1%, the indicated power increases by 4.8%. Zhou (1995) studied the method of obtaining oxygen-enriched air using high-gradient magnetic field. Oxygen concentration was related to the flow speed of air through the magnetic field, so it can be done to keep low speed to obtain higher concentration of oxygen-enriched air. Based on the studies above, there are a variety of methods to achieve oxygen-enriched air intake, and their own characteristics are remarkable in improving the engine power performance and reducing emissions, but it has not been found that the pressure swing adsorption method applies to the engine oxygen-enriched air in the current literature.

Cui et al. (2004) and Liu et al. (2011) studied the oxygen content and recovery rate of pressure swing adsorption equipment, and the results showed that the type of molecular sieve and the ratio of height to diameter of the adsorption tower have a great influence on the production efficiency of oxygen. Rama et al. (2015) studied the performance of a new system for rapid pressure swing adsorption for the continuous production of approximately 90% oxygen concentration from compressed air. Santos et al. (2007) greatly improved the oxygen purity for the mixed gas of oxygen with less than 95% purity and argon, through selecting silver-exchanged zeolite with oxygen/argon adsorption selectivity as the adsorbent material. The size of the pressure swing adsorption device is determined by calculating the flow of intake air of the engine under rated condition, and the proportion of components fully mixed can be controlled.

In this paper, we provide a mixture of pure oxygen with the Pressure Swing Adsorption (PSA) and natural air intake. The intake system is modelled with GT-Power. It is done to achieve the engine oxygen-enriched air by changing the oxygen composition of the intake air. Then the air-fuel ratio and the valve timing in typical working conditions are optimised with Optimiser in GT-Power, and the actual air-fuel ratio and valve timing are determined. Finally, we compare the power performance and emissions with the original model of engine, and improve the overall performance of the engine.

2 OXYGEN PRODUCTION WITH PSA

Oxygen is produced by pressure swing adsorption, and mixed with air from the original engine intake pipe, in order to obtain the oxygen-enriched air intake system shown in Figure 1. This system consists of an oxygen generator with pressure swing adsorption and an original engine air intake pipe. Pure oxygen and air from the original intake are mixed in the stabilisation chamber indexed as component 6 in Figure 1, and the mixing ratio of the two kinds of gases is controlled by ECU indexed as component 8 in Figure 1.

The key structure of the adsorption tower is designed as follows:

The minimum fluidisation speed of the molecular sieve particles is calculated (Yuan Yi, 2001) according to Formula 1:

$$u_{min} = \frac{d_p^2 (\rho_s - \rho) g}{1650 \mu} \tag{1}$$

where:

ρ – air density, kg/m³;

G – acceleration of gravity;

μ – dynamic viscosity of the gas and the minimum fluidisation speed is 0.23 m/s via the corresponding calculation;

and the air flow rate in the empty tower is:

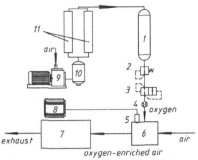

1-Oxygen storage tank   2-Pressure reducing valve   3-Check valve   4-Flow meter   5-Pressure sensor for intake air   6-Stabilisation chamber   7-Engine   8-ECU   9-Compressor   10-Air purifier   11-Adsorption tower

Figure 1.   Schematic diagram of the oxygen-enriched air intake system.

$$u = 70\% * u_{min} = 0.16 \, m/s$$

The minimum diameter of the adsorption tower is:

$$D_{min} = \sqrt{\frac{4 * Q_a}{\pi \mu}} = 0.087 m$$

Meanwhile, the height/diameter ratio of the absorption tower plays a crucial role in ensuring the adsorption device provides a stable oxygen supply efficiency and oxygen concentration. In the same height/diameter ratio, the higher the compressor power, the higher the oxygen purity. If the height/diameter ratio is too large, the adsorption process will take too long, the number of adsorption cycles per unit time will be reduced, the utilisation ratio of adsorbent will be reduced, and the bed resistance and compressor energy consumption will be increased. The height and diameter of the absorption tower should meet the following relationship:

$$2H_a \pi \rho_n \left(\frac{D}{2}\right)^2 = G$$

where:

G – lithium molecular sieve filling quality;

ρ_n – molecular sieve bulk density, 680 kg/m³;

D – adsorption tower diameter, cm

The height of the adsorption tower is 54 cm via calculation.

According to the overall structure of the pressure swing adsorption device and the empirical value, the height/diameter ratio is calculated as $H_a/D = 6.5$.

3 PERFORMANCE OPTIMISATION OF ENGINE WITH OXYGEN-ENRICHED INTAKE

3.1 Method of optimisation

Under different boundary conditions and operating conditions of the engine, the maximum torque is set as the optimisation target, then the corresponding air-fuel ratio and valve timing are solved. The stoichiometric air-fuel ratio and the range of original engine valve timing are set as 10 and 40 respectively, and the resolution is set as 0.1 and 1. The air-fuel ratio and the valve timing are calculated iteratively within the range, and the goal is to make the maximum torque reach a stable value; thus the corresponding air-fuel ratio and valve timing is the optimum value under this condition.

In order to obtain the best dynamic performance of the engine with oxygen-enriched intake, it is done to set the value for air pointer of "fluid-mixture" under keeping the structure of the engine and injector changeless. The ratio of oxygen/air is set as 21%, 22% and 23% respectively. The effects of these three typical working conditions on power performance of the engine are studied.

Assuming the air only consists of oxygen and nitrogen, then combustion of gasoline is expressed as Equation 2:

$$C_mH_nO_s+\left(m+\frac{n}{4}-\frac{s}{2}\right)O_2 \rightarrow mCO_2+\frac{n}{2}H_2O \qquad (2)$$

Assuming the volumes of oxygen and nitrogen in the oxygen-enriched air are set as V_{O2}, V_{N2}, and their substances are n_{O2}, n_{N2}, while the ratio of the molar weight is equal to that of volume under the standard state, that is:

$$\frac{n_{O2}}{n_{N2}}=\frac{V_{O2}}{V_{N2}}=\frac{V_{O2}}{1-V_{O2}}$$

Therefore, the air-fuel ratio of gasoline with oxygen-enriched air under complete combustion is:

$$L_0=\frac{n_{air} \cdot M_{air}}{n_{gasoline} \cdot M_{gasoline}}$$
$$=\frac{\left(1+\frac{1-V_{O2}}{V_{O2}}\right)\left(m+\frac{n}{4}-\frac{s}{2}\right)\left[32V_{O2}+28(1-V_{O2})\right]}{1\cdot(12m+n+16s)}$$

Gasoline is a mixture containing a variety of substances. In order to simplify the calculation, it is done to take 5–12 carbon atoms in gasoline, and the mass ratio of carbon, hydrogen, and oxygen is set as 0.855: 0.145: 0; thus the approximate

values m = 8, N = 16, and s = 0 are obtained. Then the above equation can be simplified as Formula 3:

$$L_0=\frac{3\left(7+V_{O_2}\right)}{7V_{O_2}} \qquad (3)$$

3.2 Optimisation process

According to the optimisation objectives of the performance for the engine oxygen-enriched intake, the following optimisation process is designed: (1) according to the original engine parameters, the engine model is built; (2) model check: if it is unqualified, return to the first step and remodel until the model meet the requirements; (3) setting of the parameters for the boundary environment of the model; (4) setting of RLT variables, independent variables and case variables respectively; (4) iterative calculation; (5) the final optimal RTL

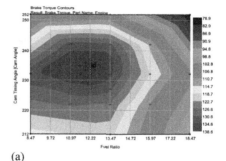

(a)

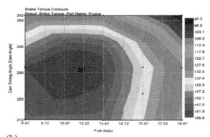

(b)

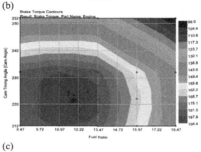

(c)

(Continued)

291

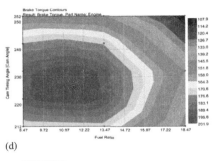

(d)

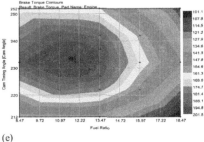

(e)

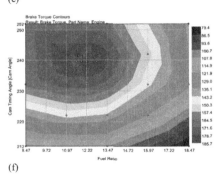

(f)

Figure 2.   Optimal air-fuel ratio and injection timing.

variable corresponding to the best air-fuel ratio and valve timing is obtained.

3.3 *Results of optimisation*

Based on the parameter setting, the system is simulated with GT-Power. The results show that, when the engine speeds are 1,000 r/min, 2,000 r/min, 3,000 r/min, 4,000 r/min, 5,000 r/min, and 6,000 r/min respectively, the optimal air-fuel ratio and injection timing are shown in Figure 2(a)–(f), and the numerical values are (12.46, 235), (11.99, 230) 221), (11.66, 226), (11.64, 233), (11.74, 241) respectively.

4 CONCLUSION

In order to improve the power of an engine, the oxygen-enriched air intake method is chosen to improve the intake conditions. Especially at full load, high-speed conditions, oxygen-enriched air intake can improve the speed of combustion and promote the formation of high-temperature environment, and the fuel combustion, cylinder pressure, and power performance are improved significantly. Oxygen-enriched air promotes the full burning of the fuel. Although oxygen-enriched combustion contributes to the power performance of the engine, a large amount of heat is simultaneously released to raise the temperature in the cylinder.

ACKNOWLEDGEMENTS

This work was supported by Project of Shandong Province Higher Educational Science and Technology Programme (Grant No. J14 LB11).

REFERENCES

Baskar, P. & Senthilkumar, A. (2016). Effects of oxygen enriched combustion on pollution and performance characteristics of a diesel engine. *Engineering Science and Technology, 19*(1), 438–443.

Cui, H., Liu, Y., Yue, K., & Zhao, Z. (2004). Mathematical simulation of small PSA oxygen process. *Cryogenic Technology*, (1), 20–23.

Li, G., Qiao, X., Li, L. et al. (2007). Experimental study of oxygen enriched combustion in the first cycle of cold start in a LPG SI engine. *Transactions of CSICE, 25*(1), 56.

Liu, Y., Feng, J., Le, K. et al. (2006). Investigation on the characteristics of gasoline engine with oxygen enrichment combustion. *Industrial Heating*, (6), 14–16.

Liu, Y., Zheng, X., Li, Y. et al. (2011). The concentration distribution in the axial-flow adsorber. *Low Temperature and Specialty Gases, 29*(1), 8–14.

Maxwell, T. T., Jones, J.C., Setty, V. et al. (1993). The effect of oxygen enriched air on the performance and emissions of an internal combustion engine. *International Fuels & Lubricants Meeting & Exposition*, 285–292.

Rama, R., Wu, K., & Sircar. (2015). Comparative performances of two commercial samples of LiLSX zeolite for production of 90% oxygen from air by a novel rapid pressure swing adsorption system. *Separation Science and Technology, 50*(10), 1447–1452.

Santos, J.C., Cruz, P., Regala, T., Magalhães, F.D., & Mendes, A. (2007). High-purity oxygen production by pressure swing adsorption. *Industrial and Engineering Chemistry Research, 46*(2), 591–599.

Wartinbee, W.J. (1971). *Emissions study of oxygen-enriched air*. SAE Paper 710606.

Xiao, G., Qiao, X., Sun, K. et al. (2007). Improvement on starting performance of a DI diesel engine by using membrane-based oxygen-enriched intake air. *Journal of Shanghai Jiaotong University, 41*(10), 1629–1632.

Yuan, Y. (2001). *Chemical engineers handbook*. Beijing: China Machine Press.

Zhou, B. (1995). Research on diesel engine for getting oxygen-enriched air. *Journal of Southwest Jiaotong University, 30*(3), 312.

Automotive, Mechanical and Electrical Engineering – Liu (Ed.)
© 2017 Taylor & Francis Group, London, ISBN 978-1-138-62951-6

Optimised state of charge estimation in lithium-ion batteries by the modified particle filter method

Renzhuo Wan, Binbiao Pan, Fan Yang, Jun Wang & Quan Chen
School of Electronic and Electrical Engineering, Battery and Battery Management System Research Centre, Wuhan Textile University, Wuhan, Hubei, China

ABSTRACT: Combining non-linear factors such as current, temperature, coulombic efficiency etc., a battery capacity correction equation for State of Charge (SOC) is established. The battery state space model is built by using the ampere-hour counting method and composite electrochemical principles. Based on the adaptability of a non-Gaussian and non-linear system, a Particle Filter (PF) algorithm is used to estimate the SOC of the battery. The simulation is performed under random dynamic charge/discharge condition. The result shows a good agreement between the experimental data processed by the PF algorithm and the raw value. Experiments also show that the estimation accuracy adopted with the PF method is 0.5% compared with that of the ampere-hour method, which may apply in a battery management system for runtime SOC estimation.

Keywords: SOC; Battery model; Particle filter

1 INTRODUCTION

The environmental pollution and energy crisis has forced us to search for new alternative energy sources. The secondary Lithium-Ion Battery (LIB), due to its high energy density, long lifetime, fast charging ability, memoryless and environmental-friendly features, is now successfully and widely used in consumer electronics, green energy vehicles and storage devices. In particular, with the extensive application of Hybrid Electric Vehicles (HEVs), Plug-in Hybrid Electric Vehicles (PHEVs) and Electric Vehicles (EVs), the Battery Management System (BMS) plays an increasingly key role in runtime monitoring the battery performance and controlling its behaviour [Lu, Megahed].

Since the highly emphasised safety and output power requirements, the State of Balance (SOB) to describe and adjust the battery uniformity in charging and discharging, State of Health (SOH), State of Power (SOP) and State of Function (SOF) are multi-functionally embedded into the BMS (Plett). All of these functions are strongly dependent on the SOC estimation of LIB; thus the precise estimation of SOC becomes crucial. However, the SOC is strongly dependent on ambient temperature, charge/discharge rate, battery aging, coulombic efficiency etc., and becomes a challenge in runtime working conditions. Most of the SOC estimation methods in use, such as the Ampere-Hour counting (AH) method (Ng, 2009), Open-Circuit Voltage

(OCV) method (Snihir, 2006), battery-model based method (Gu & Wang, 2000), PF method (Wang, 2015) and neural network model method (Chan) etc., as well as their hybridisation, have advantages and also drawbacks. The main research focuses on the stability and robustness of the SOC algorithms, as well as the resource consumption in MCU.

In this paper, the modified particle filter method is verified for SOC estimation of $LiMnO_2$. In Section 2, the particle filter is illustrated. The battery temperature and rate performance are shown in Section 3, to extract the battery parameters and model the battery. In Section 4, the results are discussed, which indicates the validation and stability of the algorithm.

2 PARTICLE FILTER METHOD

The traditional PF is based on the classical sequential Monte Carlo method. It can effectively solve the issues of non-linear correlation with a random tracking estimation algorithm, and can maintain good convergence property compared with other algorithms. Here, the PF is used to estimate terminal voltage and reduce the non-Gaussian noise produced by simplifying the battery model. According to the Bayesian rules, the random group of samples x_k (also known as SOC) related with the weights w_k^i, which is decided by measurement terminated voltage of battery y_k. Using the set of

Table 1. Particle filter algorithm for SOC estimation.

a. Initialisation: $k=0$. Randomly generate N initial particles x_0^i $(i=1....N_s)$ by the priori probability $P(x_0)$, and all the weights are set to $1/N$.

b. For k=1,2,3...

1) Update the weights at the time k by the following formulation: $\omega_k^i = \omega_{k-1}^i P\left(y_k|x_k^i\right) = \omega_{k-1}^i P\left(y_k - h\left(x_k^i\right)\right)$,

$i=1, 2, 3.., N_s$. Normal the importance weight:
$\omega_k^i = \omega_k^i / \sum_{i=1}^{N_s} \omega_k^i$.

Then the minimal square error estimation can be obtained by:

$x_k = \sum_{i=1}^{N_s} \omega_k^i x_k^i$.

2) Resampling according to importance: evaluate the effective sample s:

$N_{eff} = 1/\sum_{i=1}^{n} \left(w_k^i\right)^2$

3) Estimate x_{k+1} through the state equation.
4) $k=k+1$, return to Step 1.

random samples with weight, the posterior distribution of the initial particles, or $P(x_k|y_k^i)$, can be approximately evaluated. Especially, at the time of k. w_k^i can be modified by means of the weighted posterior probability density function. The sequential importance sampling principle is used here, and the phenomenon of weight concentration on individual particles can be avoided by resampling method, and this method can also improve the problem of non-convergence of iterative data. The simplified programme flow of the particle filter for SOC estimation used in this study is shown in Table 1.

3 EXPERIMENT

In general, SOC is defined as the percentage of the remaining capacity of the battery, as described in Equation 1.

$$SOC = \frac{Q_{remain}}{Q_{rate}} * 100\% \qquad (1)$$

where Q_{rate} is the rated capacity of the battery, which usually means the Factory Default Value (FDV). Q_{remain} is the remaining capacity of the battery. In fact, the Maximum Release Capacity (MRC) is not always equal to the FDV for two main reasons. First, the actual capacity of the battery is not exactly equal to the FDV. In particular, MRC will decrease with the aging of the battery. In the real working conditions of the battery, it varies strongly depending on temperature, current

rate, and coulombic efficiency, as well as its electrochemistry performance. In more practical terms, it is described by the AH method as Equation 2:

$$SOC(t) = SOC(t_0) - \frac{1}{Q_{true}} \int_{t_0}^{t} \eta_{i,T,c} I(t)dt \qquad (2)$$

where $SOC(t)$ is current SOC, $SOC(t_0)$ is initial SOC, Q_{true} is the actual maximum capacity of the battery, and $\eta_{i,T,c}$ are the correction factors related to current, temperature and coulombic efficiency, respectively.

3.1 Battery test bench

In order to obtain the battery performance data such as current, temperature, and voltage, the test bench is established and composed of a battery test system (Neware CT-4001, Shenzhen, China), programmable incubator (HDS-3120, Wuhan, China) and a host computer for online data acquisition. LiMnO2 (M7568132, from Xinxin Energy Company of Xiangyang, Hubei, China) with cut-off voltage of 4.2 V for charging and 3.0 V for discharging. Nominal capacity of 10 Ah per cell is used for the testing and SOC estimation. The battery Electromotive Force (EMF), which is also called OCV, is a monotonous function of SOC.

3.2 Current rates and temperature

In order to investigate the non-linear effect of temperature, current rate, and coulombic efficiency, the correction factor for the SOC estimation is introduced and defined as Equation 3:

$$\eta_{i,T,C} = Q_{true}/C_T * Q_{true}/C_i * Q_{true}/C_c \qquad (3)$$

Firstly, the state of charge/discharge experiment is conducted with a constant current of 5 A (about 0.5°C) and the average discharge capacity is calculated at intervals of 5°C, as shown in Figure 1(a). At each temperature, the process of charge/discharge is executed for five cycles, and then the battery is set aside for 24 h after the test is completed. As can be seen from the figure, the total discharge and charge capacity increase with the rise in temperature from −10 to 40°C. It is worth noting that the total discharge capacity keeps almost constant when the ambient temperature exceeds 25°C. A polynomial fitting is used to quantificate the relationship between the temperature (T) and capacity C_T at 0.5°C. The equation can be expressed as:

$$C_T(T) = a_0 + a_1 T + a_2 T^2 + a_3 T^3 \qquad (4)$$

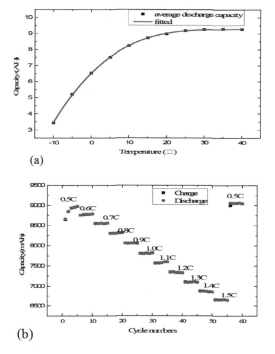

(a)

(b)

Figure 1. (a) The relationship between current rates and discharge/charge capacity at 25°C; (b) The relationship between temperature and discharge capacity at current 0.5 C.

Figure 1(b) shows the relationship between the current rate and discharge/discharge capacity. At a constant temperature of 25°C, current rates are set from 0.5 C to 1.5 C with an interval of 0.1 C and then return to 0.5 C charge and discharge. It shows that charge capacity and discharge capacity decrease linearly with the increase in the current rates, and it indicates that the capacity of the battery is seriously dependent on charge/discharge current. The correct current factor should be built. Therefore, the following linear equation is used to describe the relationship between the current rate and capacity C_i:

$$C_i(i) = b_0 + b_1 * i \qquad (5)$$

3.3 Coulombic efficiency and OCV-SOC

Coulombic efficiency is defined as a ratio of the total discharge capacity and charge capacity at a specific condition. Here, the coulombic efficiency is calculated at different temperatures using the above temperature-capacity data. Figure 2(a) shows the relationship between coulombic efficiency and temperature. We find that the coulombic efficiency

$\boxtimes_c$ keeps at 1 at intermediate temperature ranging from 5 to 30°C, and about 1.4% maximum fluctuations at very low temperature and above 35°C. It is described by five-order polynomial function as Equation 6:

$$C_c(T) = m_0 + m_1 T + m_2 T^2 + m_3 T^3 + m_4 T^4 \qquad (6)$$

EMF has no relationship with the temperature, current rates, and aging (PattiPati, 2014), the true value of the battery SOC is estimated through the relationship between OCV and SOC. Here the OCV-SOC curve is obtained in a generally used method, as illustrated in Wang (2015) and Zheng (2016). Figure 2(b) shows the relationship between SOC and OCV.

3.4 Battery model

In more practical terms, the internal SOC is associated with the external voltage, current, temperature and other quantities. Considering the intricate work conditions and complicated electrochemistry system of the battery, an accurate model of external characteristics should be established. Finding the numerical relationship between SOC and various directly measurable physical quantities is a high priority. The general battery model (Plett) in use, such

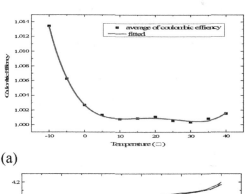

(a)

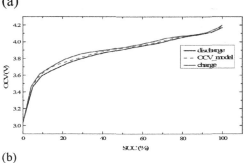

(b)

Figure 2. (a) Relationship between the coulombic efficiency and temperature; (b) Measured voltage and OCV modelled.

as the Shepherd model, Thevenin model, Unnewehr universal model, Rint model, hysteresis model and Nernst model, have their advantages and shortcomings under different circumstances. In this paper, the combined model embraced by the Shepherd model (Equation 7), Unnewehr model (Equation 8) and Nernst model (Equation 9) are used.

$$y_k = E_0 - Ri_k - K_i/x_k \tag{7}$$

$$y_k = E_0 - Ri_k - K_i x_k \tag{8}$$

$$y_k = E_0 - Ri_k + K_3 \ln x_k + K_4 \ln(1 - x_k) \tag{9}$$

Here, y_k is battery terminal voltage, R is battery internal resistance varying with SOC, K_i is polarisation resistance, x_k is the SOC. K_3 and K_4 are the coefficients of the model. Combined with Equations 7, 8, 9 and Equation 2, the battery state space model is obtained as below of the discrete observation equation and state equation respectively.

$$y_k = E_0 - Ri_k - K_1/x_k - K_2 x_k + K_3 \ln x_k \\ + K_4 \ln(1 - x_k) \tag{10}$$

$$x_k = x_{k-1} + \eta_{i,T,c} I(t) \Delta t / Q_{true} \tag{11}$$

The parameters of the battery model are identified and verified by the least squares method form the experimental data of the LiMnO$_2$ battery. The result is listed in Table 2 and the parameters' fitting progress is shown in Figure 3.

Table 2. Parameters of the battery model.

Parameters	LiMnO$_2$
E_0	4.1697 V
R	0.0062 Ω
K1	−0.0089 V
K_2	0.2381 V
K_3	0.3028 V
K_4	0.0517 V

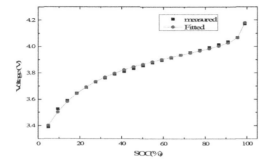

Figure 3. Relationship of terminal voltage and SOC; red curve is fitted by least squares method.

4 RESULTS AND DISCUSSION

In order to verify the proposed method, the experiments are designed and performed at a constant value of 25°C. A profile of the dynamic current condition is shown in Figure 4(a), and the voltage tracing data are shown in Figure 4(b). The battery

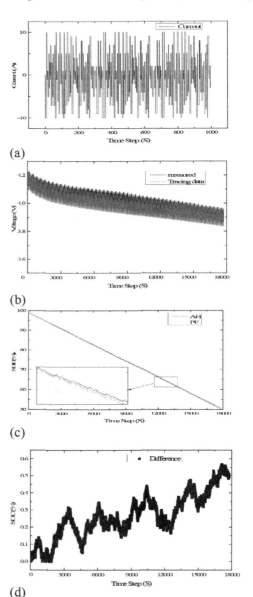

Figure 4. (a) The random dynamic current condition; (b) The tracing voltage data and measured battery terminal voltage; (c) The SOC estimation results of AH and PF; (d) The difference between AH and PF method.

was set aside for 6 h after finishing the test, and the SOC reference data are calculated using the relationship between OCV-SOC, which was 48.3% at the end of the experiment. Figure 4(c) shows the comparison of the SOC estimation results for the AH method and PF method. The AH result is 50.73%, and the PF result is 50.23%. Figure 4(d) shows the difference between the AH method and the PF method. As shown in the figure, the difference in the two methods shows a tendency to become larger, which may be caused by the accumulating error of the AH method.

5 CONCLUSION

SOC estimation is one of the core issues for BMS. Considering the significant temperature change, current rates and coulombic efficiency, the combined battery model is established based on the experimental data, and the PF method is proposed for SOC estimation.

The study shows that the PF algorithm could achieve a relative error of 0.6% compared with that of the AH method. Furthermore, runtime data acquisition and algorithm validation should be studied.

ACKNOWLEDGEMENTS

This work was supported by Natural Science Foundation of Hubei Provincial Education Department (Q20131603).

Corresponding author: Quan Chen.

REFERENCES

Chan, C.C. The available capacity computation model based on artificial neural network for lead-acid batteries in electric vehicles. *Journal of Power Sources, 87,* 201–204.

Gu, W.B., & Wang, C.Y. (2000). Thermal-electrochemical modeling of battery system. *Journal of the Electrochemical Society, 147*(8), 2910–2922.

Lu, L.G. (2013). A review on the key issues for lithium-ion battery management in electric vehicles. *Journal of Power Sources, 226,* 272–288.

Megahed, S., & Ebner, W. (1995). Lithium-ion battery for electronic. *Journal of Power Sources, 54,* 155–162.

Ng, K.S. (2009). Enhanced coulomb counting method for estimating state-of charge and state-of health of lithium-ion batteries. *Applied Energy, 86,* 1506–1511.

PattiPati, B. (2014). Open circuit voltage characterization of lithium ion batteries. *Journal of Power Sources, 269,* 317–333.

Plett, G.L. (2004). Extended Kalman filtering for battery management systems of LiPB-based HEV battery packs part 1– background. *Journal of Power Sources, 134,* 252–261.

Plett, G.L. (2004). Extended Kalman filtering for battery management systems of LiPB-based HEV battery packs part 2– modeling and identification. *Journal of Power Sources, 134,* 262–276.w3

Snihir, I. (2006). Battery open-circuit voltage estimation by a method of statistical analysis. *Journal of Power Sources, 159,* 1484–1487.

Wang, Y.J. (2015). A method for state-of charge estimation of LiFePO4 batteries at dynamic currents and temperatures using particle filter. *Journal of Power Sources, 279,* 306–311.

Zheng, F.D. (2016). Influence of different open circuit voltage tests on state of charge online estimation for lithium-ion batteries. *Applied Energy, 183,* 513–525.

Automotive, Mechanical and Electrical Engineering – Liu (Ed.)
© 2017 Taylor & Francis Group, London, ISBN 978-1-138-62951-6

Research on large power transformer electric field calculation analysis software

Huan Wang, Yan Li, Yongteng Jing & Bo Zhang
Research Institution of Special Electrical Machines, Shenyang University of Technology, Shenyang, China

ABSTRACT: The transformer electric field calculation is very important in transformer design. The paper establishes a winding equivalent circuit model, and deduces the winding equivalent circuit differential equations in matrix form by using the cut-set analysis and derives the calculation formula of the initial potential distribution of the equivalent circuit model. On the basis of equivalent circuit equations, the paper calculates the inductance and capacitance parameters, develops the design calculation software for the wave process engineer, calculates the initial potential distribution of transformer winding, shock response potential distribution, maximum potential distribution and oil potential gradient distribution, and outputs the results to the VB interface.

Keywords: Power transformer; Electric field; Winding

1 INTRODUCTION

The transformer is one of the most important pieces of equipment in the power system. Power plants, substations, transmission and distribution network, and the masses of users, all use the transformer. But in the actual operation of the UHV transformer, accidents from lightning impulse damage to the transformer insulation often happen. The main design method of transformer longitudinal insulation structure is the wave process calculation which is the basis of the transformer electromagnetic optimisation design. The so-called wave process usually expresses the lightning impulse along the transmission line inverses into the substation, making the transformer winding suffer shock effect, and produces a complex electromagnetic transient process in the transformer windings, resulting in the overvoltage between the turns, layers, lines, windings, and windings to ground. Therefore research into the wave process of the transformer windings under the effect of impulse wave is very necessary. Under the effect of impulse wave, for the different winding structure, arrangement and connection modes, research is required into the potential and gradient distribution to find out the winding insulation weak points, so that the designer can produce a reasonable design.

Finite element simulation technology is widely used in the engineering design, and to avoid spending a lot of time learning all kinds of simulation software, this paper uses the lumped parameter equivalent circuit to calculate and analyse the winding wave process of a large power transformer. This not only reduce the calculation time but also improve the efficiency of calculation. A general computing application is also written which is convenient for calculating various types of large transformer. Without training, designers can carry out the numerical simulation of the electric field within a short time, and get the distribution results of the intuitive electric field.

2 THE MAIN CALCULATION METHOD

2.1 *The principle of longitudinal insulation calculation*

Under the impulse condition, the transformer winding has the distributed parameter network property. It is a complex network which contains the distribution parameter such as resistance, capacitance and inductance. For convenient analysis, the lump parameters are used instead of the distribution parameters, and the chain network of lumped parameter is used as the equivalent circuit of transformer winding.

Each winding of the transformer is divided into several units, each line of breads is expressed by a self-inductance L, a longitudinal capacitor Cs, a ground capacitance Cg and the adjacent winding capacitance Cw. Based on a single winding, the magnetic coupling between line breads is expressed by mutual inductance, constitutes the conventional network, adds the inductance and capacitance of the windings in equivalent network, and forms a multi-winding equivalent network.

2.2 The derivation of longitudinal insulation calculation formula

The equivalent circuit of the single winding wave process calculation is shown in Figure 1.

L_1, L_2, ..., L_{N-1} is the self-inductance and mutual inductance of each calculation cell, C_1 and C_2, ..., C_{N-1} is the series capacitor of each calculation cell, C_{g1}, C_{g2}, ..., C_{gN-1} is the ground capacitance of each calculation cell.

Assume:

$$\Phi_{1J} = \int i_J dt, \Phi_J = \int i'_J dt \tag{1}$$

$$M\frac{d^2\Phi_{1J}}{dt^2} + C(\Phi_{1J} - \Phi_J) = 0 \tag{2}$$

In which:

$$\Phi_{1J} = \begin{bmatrix} \Phi_{1J}(1) \\ \Phi_{1J}(2) \\ \vdots \\ \Phi_{1J}(N) \end{bmatrix}, \Phi_{1J} - \Phi_J = \begin{bmatrix} \Phi_{1J}(1) - \Phi_J(1) \\ \Phi_{1J}(2) - \Phi_J(2) \\ \vdots \\ \Phi_{1J}(N) - \Phi_J(N) \end{bmatrix}$$

$$M = \begin{bmatrix} L_1 & M_{12} & \cdots & M_{1N} \\ M_{21} & L_2 & \cdots & M_{2N} \\ \vdots & \vdots & & \vdots \\ M_{N1} & M_{N2} & \cdots & L_N \end{bmatrix}, C = \begin{bmatrix} \frac{1}{C_1} & & & 0 \\ & \frac{1}{C_2} & & \\ \vdots & & \ddots & \vdots \\ 0 & & & \frac{1}{C_N} \end{bmatrix} \tag{3}$$

For the lower part of the mesh has the following equation:

$$Q\Phi_J - C\Phi_{1J} = \delta_J V(t) \tag{4}$$

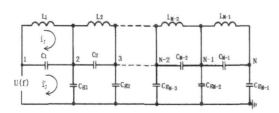

Figure 1. Single winding equivalent circuit.

In which:

$$Q = \begin{bmatrix} \frac{1}{C_1} + \frac{1}{C_{g1}} & -\frac{1}{C_{g1}} & \cdots & & 0 \\ -\frac{1}{C_{g1}} & \frac{1}{C_1} + \frac{1}{C_{g1}} + \frac{1}{C_{g2}} & \cdots & \ddots & \vdots \\ \vdots & -\frac{1}{C_{g1}} & \cdots & & \\ & & \ddots & \cdots & -\frac{1}{C_{gN-1}} \\ 0 & & \cdots & & \frac{1}{C_{gN-1}} + \frac{1}{C_{gN}} + \frac{1}{C_g} \end{bmatrix} \tag{5}$$

Take the (4) into (2), gives the second order differential equations:

$$M\frac{d^2\Phi_{1J}}{dt^2} + (C - CQ_1C)\Phi_{1J} = CQ_1\delta_J V(t) \tag{6}$$

$$\frac{d^2\Phi_{1J}}{dt^2} + M_1 W_1 \Phi_{1J} = M_1 W \tag{7}$$

In which:

$$W_1 = C - CQ_1C, \frac{d^2\Phi_{1J}}{dt^2} + M_1 W_1 \Phi_{1J} = M_1 W \tag{8}$$

Assume:

$$\Phi_J = \begin{bmatrix} \Phi(1) \\ \Phi(2) \\ \vdots \\ \Phi(N) \end{bmatrix} = \begin{bmatrix} y(1) \\ y(2) \\ \vdots \\ y(N) \end{bmatrix}, \frac{d}{dt}\begin{bmatrix} \Phi(1) \\ \Phi(2) \\ \vdots \\ \Phi(N) \end{bmatrix} = \begin{bmatrix} y(N+1) \\ y(N+2) \\ \vdots \\ y(2N) \end{bmatrix} \tag{9}$$

Substitute into (7), obtains:

$$\frac{d}{dt}\begin{bmatrix} y(1) \\ \vdots \\ y(N) \\ \vdots \\ y(2N) \end{bmatrix} = \begin{bmatrix} 0(N \times N) & E(N \times N) \\ -M_1 W_1 & 0(N \times N) \end{bmatrix}\begin{bmatrix} y(1) \\ \vdots \\ y(N) \\ \vdots \\ y(2N) \end{bmatrix} + \begin{bmatrix} 0(N \times N) \\ M_1 W_2 \end{bmatrix} \tag{10}$$

Write a calculation program of numerical methods and solve. The numerical matrix of Φ_J and Φ_{1J} can be obtained.

Potential difference is shown as:

$$V = C(\Phi_J - \Phi_{1J}) \qquad (11)$$

Equation (2) into Equation (7), obtains:

$$V = CQ_1 \delta_J V(t) - (C - CQ_1 C)\Phi_1 \qquad (12)$$

Node potential is shown as:

$$U = C(\Phi_J - \Phi_{J+1}) \qquad (13)$$

2.3 Calculate the initial distribution

Initial distribution is mainly determined by the K value:

$$\frac{d^2 u}{dx^2} - \frac{C}{K}u = 0, \quad \frac{d^2 u}{dx^2} - \alpha^2 u = 0 \qquad (14)$$

where $\alpha = \sqrt{\dfrac{C}{K}}$ is the spatial factor.

Equation (14) is an ordinary differential equation. Its general solution is:

$$u(x) = Ae^{\alpha x} + Be^{-\alpha x} \qquad (15)$$

A and B can be calculated from boundary conditions.

Neutral grounding should satisfy the following boundary conditions $x = 0, u = u_0; x = 1, u = 0$ the boundary conditions. Take it into Equation (15), can obtain:

$$U(x) = U_0 \frac{e^{\alpha(l-x)} - e^{-\alpha(l-x)}}{e^{\alpha l} - e^{-\alpha l}} = U_0 \frac{sha(l-x)}{shal} \qquad (16)$$

Neutral insulation should satisfy the following boundary conditions: $x = 0, u = u_0, x = 1, K\frac{dU}{dx} = 0$. According to the boundary conditions can obtain:

$$U(x) = U_0 \frac{e^{\alpha(l-x)} + e^{-\alpha(l-x)}}{e^{\alpha l} + e^{-\alpha l}} = U_0 \frac{cha(l-x)}{chal} \qquad (17)$$

3 THE MAIN FEATURES OF THE SOFTWARE AND THE BLOCK DIAGRAM

The developing interactive interface of the Visual Basic (VB) 6.0 programming language enables the users to easily input various model parameters and the material properties of the main insulation of the transformer in the interface. The core pro-

gram which is used to calculate electric field finite element is written in the FORTRAN language, through which the program can automatically divide the circuit elements, establish the equivalent circuit model, calculate the ground capacitance parameters and calculate the voltage distribution.

Then using the post-processing which is prepared by VB, the results are displayed, giving the dielectric strength criterion. The software supports the operating system which is much better than the Windows XP version. The softwarecan be used for all kinds of single-phase and three-phase power transformer insulation longitudinal electric field calculation and analysis. The software flow chart is shown in Figure 2.

The software is much more convenient than the currently popular commercial finite element soft-

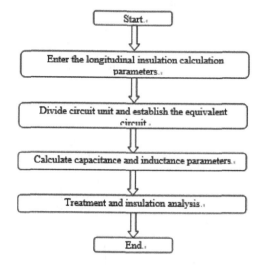

Figure 2. The software flow chart.

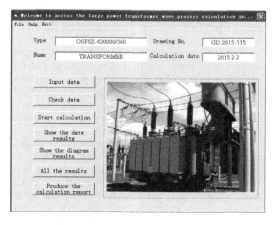

Figure 3. The main function interface.

301

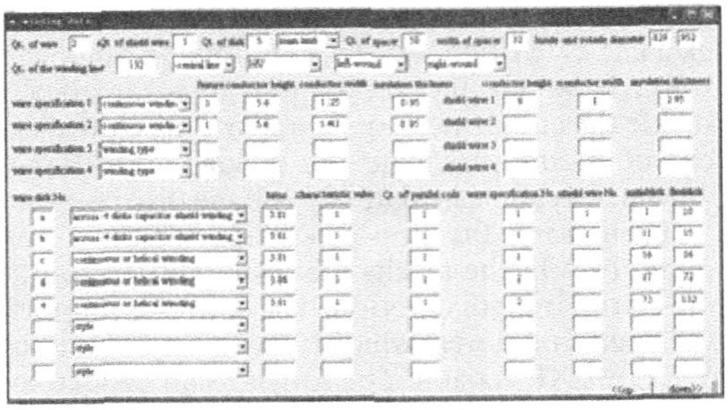

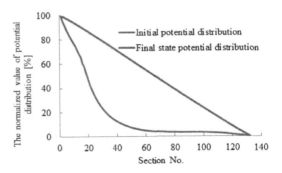

Figure 4.   The data input interface.

Figure 5.   HV winding initial potential distribution.

ware. Users only need input some necessary transformer insulation structure parameters, and the necessary material properties. Loading drive and boundary conditions step are automatically added within the software process code. The software has a powerful post-processing and insulation analysis function. The main function interface and the data input interface of the software is shown in Figure 3 and Figure 4.

4   CALCULATION EXAMPLE

To verify the accuracy of the calculation software, a power transformer model wave process of OSFSZ—420 MVA/360 kV is calculated and analysed. The transformer is end line, the HV is shielding continuous winding, with a total of 132 sections.

The HV winding initial potential distribution and the final steady state potential distribution are shown in Figure 5. The Figure shows that the HV winding initial potential distribution is not uniform. The head end part of the shield continuous winding potential drops fast, near the 16th section, and potential has dropped to 60%. The continuous part is from the 17th section, and potential drops from fast to slow. The maximum difference between the initial potential distribution and the final steady state potential can be seen near the 38th section. The drop speed is slower. Near the 68th section, potential has dropped to 4%. This shows that the capacitor shield winding can effectively improve the initial potential distribution and make it close to the final state potential. Continuous winding cannot achieve this effect.

As shown in Figure 6, under the effect of full-wave lightning impulse withstand voltage, the software computing results are compiled with enterprise computing results. The head end 16 sections are less than the maximum potential which is calculated by enterprise and the error is around 3.3%. From the 18th to the 60th section, the software value is much bigger than the calculated value, the

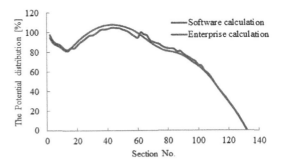

Figure 6.   HV full-wave maximum distribution.

error being between 2.5% to 5%. From the 62th to the 132th section, the error value is from 0% to 4.8%, which is less than the enterprise software calculated value. Figure 7 shows that under the effect of chopped wave lightning impulse withstand voltage, the calculated error value is from 0% to 6.5% less or more than the enterprise software calculate value. We can see the software computing results in comparison to the enterprise computing results.

The oil duct gradient distributions of HV winding, under the full wave and chopped voltage, are shown in Figure 8 and Figure 9 respectively.

Comparing the curves in Figure 8, at the full-wave, the software calculation and the enterprise calculation are mostly the same. The software calculates the maximum potential gradient in the 9th oil duct is 8.17, and comparing the value with the enterprise value, the relative error is 14.3%. Minimum potential gradient in the 87th oil duct is 3.17. The enterprise computing value maximum potential gradient appears in the 11th oil duct, where gradient value is 8.16. The minimum potential gradient in the 84th is 3.34. In Figure 9, at the chopped wave, the software calculates the maximum potential gradient in the 8th oil duct as 8.06. Minimum potential gradient in the 103th oil duct is 1.54. The enterprise computing value maximum potential gradient appears in the 10th oil duct, gradient

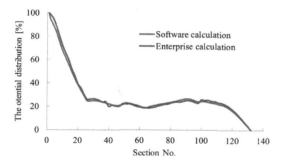

Figure 7. HV chopper-wave maximum distribution.

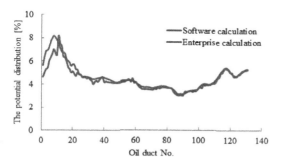

Figure 8. HV full-wave oil duct gradient.

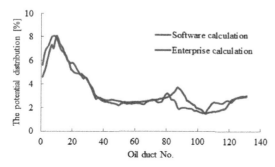

Figure 9. HV chopper -wave oil duct gradient.

value is 8.08, and the minimum potential gradient in the 106th is 1.55.

Using the development software, calculate and analyse the full-wave and impulse potential wave of the transformer, get the initial potential distribution, impact potential distribution, maximum potential distribution and the oil potential gradient distribution. The calculation results in contrast to the existing calculation software, show that the wave process of calculation software can be applied to engineering calculations.

5 CONCLUSION

a. The paper establishes a winding equivalent circuit model, and deduces the winding equivalent circuit differential equations in matrix form by using the cut-set analysis, and derives the calculation formula of the initial potential distribution of the equivalent circuit model.
b. Based on **VB** and **FORTRAN** programming ideas, wave process calculation software is developed. The software is easy to input and output, its calculation is accurate, and is widely applicable etc. It can be used to analyse the initial potential distribution, impact potential distribution, the biggest potential distribution and oil potential gradient of large transformer windings of different voltage grades and various forms.

REFERENCES

Jianping, D. et al. (2010). Research for the power transformer distributed capacitance parameter based on the finite element method. *The Sixth National Symposium on Transformer Technology Independent Innovation*, 106–109.
Popov, M. et al. (2003). Computation of very fast transient overvoltages in transformer windings. *IEEE Transactions on Power Deliver, 18*(4), 1268–1274.
Rao, M.M. et al. (2005). Frequency characteristics of very fast transient currents in a 245 kV GIS. *IEEE Transactions on Power Delivery, 20*(4), 2450–2457.
Xiqiang, X. et al. (2010). Calculation software of power transformer winding wave process design. *The Sixth National Symposium on Transformer Technology Independent Innovation*, 81–84.
Yan, L. et al. (2011). Design of calculation software for wave process power transformer winding. *Power and Energy Engineering Conference*, 1–4.
Yuanfang, W. et al. (2001). A principle and method of the protection of transformer longitudinal insulation. *Electrical Insulating Materials*, 821–824.
Yusheng, Q. et al. (2011). A novel diagnosis algorithm of maximum frequency for longitudinal insulation fault of transformer winding. *Power Engineering and Automation Conference*, 157–160.
Zhongyuan, Z. et al. (2008). The very fast transient simulation of transformer windings in parallel with MOV based on multi-conductor transmission lines theory. *Transmission and Distribution Conference and Exposition*, 1–5.

Automotive, Mechanical and Electrical Engineering – Liu (Ed.)
© 2017 Taylor & Francis Group, London, ISBN 978-1-138-62951-6

Research on the output-input relationship of Solid Oxide Fuel Cell (SOFC) stack with indirect internal reforming operation

Chao Yuan & Hai Liu

School of Mechanical, Electrical and Information Engineering, Shandong University, Weihai, Shandong Province, China

ABSTRACT: The Micro-CHP (Combined Heat and Power) cogeneration by Solid Oxide Fuel Cell (SOFC) integrated with Indirect Internal Reforming operation (IIR-SOFC) fuelled by methane is the focal point in this study. First, the paper introduces the framework of SOFC and the methane reformer respectively, and proposes the design of integrating SOFC with Indirect Internal Reforming operation (IIR-SOFC). Then, the numerical analysis of the system input-output relationship is performed by MAT-LAB software. The result shows that the output power and the stack temperature increase with the rising load current. Both the increase of the air input and methane input could raise the output voltage and power, but decrease the stack temperature. What's more, the output of the IIR-SOFC system is more sensitive to a change in the methane input than to a change in the air input, so it regulates the inlet flow rates to keep the temperature and the output at desired levels.

Keywords: SOFC; Indirect internal reforming; Modelling and simulation; Relationship analysis

1 INTRODUCTION

Since the second half of the 20th century, with the harsh reality of the problems of energy shortage and environmental issues becoming increasingly severe, countries around the world have increased their efforts in the research and exploitation of new clean energy. Therefore, the use of methane as an applicable energy carrier that is environmentally benign, efficient, cheap and readily available has been increasingly noticed and emphasised. In this context, as a new type of power generation technology, the fuel cell is regarded as one of the good candidates (Yu Li, 2016).

A Solid Oxide Fuel Cell (SOFC) is an electrochemical device which enables the conversion of chemical energy to electrical energy directly without combustion, which leaves SOFC free from the Kano cycle allowing it to achieve higher efficiency close to 80–90% in theory (Shulin Wang, 2016). Typically, SOFC is operated at a high temperature (between 700–1,100°C) under atmospheric or elevated pressures. What's more, as the reactants of SOFC are hydrogen and oxygen, the electrical generation process has almost no pollutant exhaust (Dokmaingam, et al., 2010). To simplify the model, the SOFC system model used in this paper is fuelled by steam-methane gas mixture. In the meantime, it should be ensured that the actual water-carbon ratio is

closed to 2.0~3.0, so as to realise and efficient process of reaction and avoid carbonisation (Qing Zhao, 2015).

In practical application, the SOFC stack can be used in combination with a fuel reformer. According to the reforming operation, there are three main approaches i.e. External Reforming (ER-SOFC), Direct Internal Reforming (DIR-SOFC) and Indirect Internal Reforming (IIR-SOFC) (Dokmaingam, 2010; Xiongwen Zhang, 2008), as shown in Figure 1. For the ER-SOFC approach, extra energy is needed to support the fuel reforming outside the SOFC stack. Although the external reforming has little effect on the original structure of the cell, the external reformer will vastly push up the operating costs. DIR-SOFC

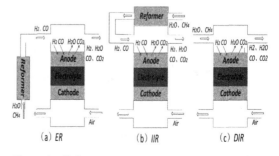

Figure 1. Reform structure diagram of SOFC stack.

simplifies the structure for the reforming reaction, but the anode material for DIR-SOFC must be optimised for all reactions as it could be easily poisoned by carbon deposition from the reforming of hydrocarbons (Dokmaingam, 2010). Furthermore, direct internal steam reforming in SOFC leads to inhomogeneous temperature distributions which can result in mechanical failure of the cermet anode (J. M. Klein, 2010). The IIR-SOFC makes full use of the advantages and offsets the limitation of the ER-SOFC and DIR-SOFC. Its specific structure not only makes an effective application of heat and reactant, but ensures every reaction goes smoothly, therefore, IIR-SOFC was chosen for the purposes of this paper.

2 PRINCIPLE OF IIR-SOFC

The schematic diagram of IIR-SOFC is presented in Figure 2. The parts that comprise a SOFC single cell are the modules of anode, cathode, electrolyte and reformer. The specific catalyst for reforming is centralised at the entrance to the reformer (Wenshu Zhang, 2012). According to this configuration, the steam-methane gas mixture is injected into the internal reformer, reacting to reform hydrogen and carbon monoxide. Then, the reformate gas is continuously fed back to the fuel channel of the SOFC. Meanwhile, the air is fed to the opposite flow direction through the air channel (Dokmaingam, 2010). An electrochemical reaction occurs in this process, releasing electrons as well as waste heat. In the process of the electrochemical reaction, the reduction reaction occurring at the cathode transforms the oxygen into oxygen ions which then reach the anode though the electrolyte. In the anode, hydrogen reacts with oxygen ions to form water, and releases electrons which move to the cathode though the external circuit and generate an electric current.

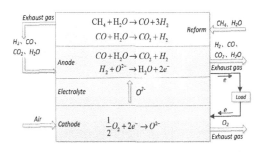

Figure 2. Schematic diagram of IIR-SOFC.

3 MATHEMATICAL MODELS OF IIR-SOFC

To better achieve the operation control of the SOFC stack, it is essential to establish a complete dynamic model, which has a better reflection of the input-output relationships of the IIR-SOFC system. The output voltage of a single SOFC is only about 1 V, so a certain number of single cells are usually piled up to form the IIR-SOFC stack to meet the demand of higher voltage and power.

The stack model will be based on the following assumptions (Dokmaingam, 2010; Xiongwen Zhang, 2008; Wenshu Zhang, 2012):

a. All the gases are ideal;
b. Gas leakage and heat transfer are ignored;
c. The temperature of the gas output is the same as inside the stack;
d. The temperature, mole fraction and partial pressure of every gas are uniformly distributed;
e. The difference of the cells is negligible, so all the cells of the stack make no difference.

3.1 Reforming model

The reforming reactions mainly include methane steam reforming and water-gas shift reaction. They are both reversible and reach equilibrium soon under suitable conditions (Shulin Wang, 2016; Van Nhu Nguyen, 2016):

$$CH_4 + H_2O \leftrightarrow CO + 3H_2 \quad \Delta H_1 = 206.1 KJ \cdot mol^{-1} \quad (1)$$

$$CO + H_2O \leftrightarrow CO_2 + H_2 \quad \Delta H_2 = -41.1 KJ \cdot mol^{-1} \quad (2)$$

Flow conservation equation:

$$w_{out,re} = w_{in,re} + 2r_{re,1} \quad (3)$$

Molar flow state equation of reforming:

$$\frac{dn_{i,re}}{dt} = w_{i,re,in} - w_{i,re,out} + \bar{R}_{i,re}$$
$$i \in \{CH_4, CO, CO_2, H_2, H_2O\} \quad (4)$$

$$\bar{R}_{re} = \left[-r_{re,1} \ r_{re,1} - r_{re,2} \ r_{re,2} \ 3r_{re,1} + r_{re,2} - r_{re,1} - r_{re,2} \right] \quad (5)$$

Due to the high temperature in the stack, it is assumed that all the methane entered into it will be reformed, thus the equilibrium constant of water-gas shift reaction can be calculated as:

$$K_{an,s} = \exp\left(A_s T_s^4 + B_s T_s^3 + C_s T_s^2 + D_s T_s + E_s \right)$$
$$= \frac{\left(w_{H2,re} + 3r_{re,1} + r_{re,2} \right) \cdot \left(w_{CO_2,re} + r_{re,2} \right)}{\left(w_{CO,re} + r_{re,1} - r_{re,2} \right) \cdot \left(w_{H_2O,re} - r_{re,1} - r_{re,2} \right)} \quad (6)$$

where i is the reformate gas; $\bar{R}_{re}\left(mol \cdot s^{-1}\right)$ is the flow generated in the reforming reaction; $r_{re,1}$ and $r_{re,2}\left(mol \cdot s^{-1}\right)$ are the rate of methane steam reforming and water-gas shift reaction in the reformer respectively. $w_{i,re,in}\left(mol \cdot s^{-1}\right)$ is the input flow while $w_{i,re,out}\left(mol \cdot s^{-1}\right)$ is the output flow; $n_{i,re}\left(mol\right)$ is the mole of the gas i in the reformer.

The coefficient is shown by the data in Table 1:

Based on the above formulae, the reforming model can be established using MATLAB software, as shown in Figure 3.

3.2 The model of anode flow channel

The reaction occurring at the anode is based on the combination of two main reactions including water-gas shift reaction (2) and hydrogen oxidation reaction (7).

$$H_2 + O^{2-} \rightarrow H_2O + 2e^- \quad \Delta H_3 = -241.8 KJ \cdot mol^{-1} \quad (7)$$

The relationship between entrance flow $w_{in,an}$ and exit flow $w_{out,an}$ is as follows:

$$w_{out,an} = w_{in,an} \quad (8)$$

Molar flow state equations of anode:

$$\frac{dn_{j,an}}{dt} = w_{j,an,in} - w_{j,an,out} + \bar{R}_{j,an} \quad j\epsilon\left\{CO, CO_2, H_2, H_2O\right\} \quad (9)$$

$$\bar{R}_{an} = \left[-r_{an,1} \; r_{an,1} \; r_{an,1} - r_{an,2} \; -r_{an,1} + r_{an,2}\right] \quad (10)$$

Table 1. The coefficient of equilibrium constant.

Coefficient	Reform	Shifting
A	-2.63121×10^{-11}	5.47301×10^{-12}
B	1.24065×10^{-7}	-2.57479×10^{-8}
C	-2.25232×10^{-4}	4.63742×10^{-5}
D	1.95028×10^{-1}	-3.91500×10^{-2}
E	-6.61395×10	1.32097×10

Faraday's law could be applied to calculate the rate $r_{an,2}$:

$$r_{an,2} = \frac{I \cdot N}{2F} \quad (11)$$

where I (A) represents the stack current; $F(96486C/mol)$ is Faraday's constant and N is the number of cells in the stack; $r_{an,1}$ and $r_{an,2}\left(mol \cdot s^{-1}\right)$ are the rate of water-gas shift reaction and hydrogen oxidation reaction in the anode respectively.

The simulation model of anode flow channel is shown in Figure 4.

3.3 The model of cathode flow channel

In the cathode (Hongliang Cao, 2012), the reaction process is that oxygen in the air has a reaction with the electron:

$$\frac{1}{2}O_2 + 2e^- \rightarrow O^{2-} \quad (12)$$

The relation between entrance flow $w_{in,an}$ and exit flow $w_{out,an}$ of the cathode is as follows:

$$w_{out,ca} = w_{in,ca} - \frac{1}{2}r_{an,2} \quad (13)$$

Molar flow state equations of cathode:

$$\frac{dn_{k,ca}}{dt} = w_{k,ca,in} - w_{k,ca,out} + \bar{R}_{k,ca} \quad k\epsilon\left\{N_2, O_2\right\} \quad (14)$$

$$\bar{R}_{re} = \left[0 \; -\frac{1}{2}r_{an,2}\right] \quad (15)$$

The simulation model of the cathode is developed, as shown in Figure 5.

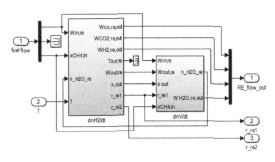

Figure 3. The reforming model.

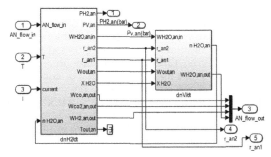

Figure 4. The model of anode flow channel.

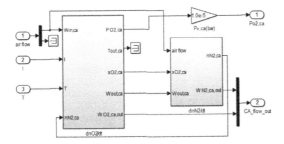

Figure 5. The model of cathode flow channel.

3.4 Electrochemical model

The output voltage of a single cell can be evaluated from Gelen (2013), and Jafarian (2010):

$$V_{cell} = E_N - \eta_{ohmic} - \eta_{act} - \eta_{con} \tag{16}$$

Where E_N is the reversible cell voltage that can be calculated from the Nernst equation:

$$E_N = E_0 + \frac{RT}{2F} \ln\left(\frac{P_{H_2,an} \times P_{O_2,ca}^{1/2}}{P_{H_2O,an}}\right) \tag{17}$$

$$E_0 = 1.2723 - 2.7645 \times 10^{-4} T_s \tag{18}$$

Where E_0 is the standard electromotive force; E_{act}, E_{ohmic} and E_{conc} are the main parts of the cell polarisation loss, and these losses represent activation loss, ohmic loss and concentration depletion, which are calculated from the following expressions, respectively:

$$\eta_{ohmic} = r(T)I = \left\{r_0 \exp\left[\alpha\left(\frac{1}{T_0} - \frac{1}{T}\right)\right]\right\}I \tag{19}$$

$$\eta_{con} = -\frac{RT}{nF} \ln\left(1 - \frac{I}{I_L}\right) \tag{20}$$

$$\eta_{act} = a + b\log I \tag{21}$$

Where T is the temperature of cell working process; $r_0 = 0.126\ \Omega$, $\alpha = -2870\ K$ and $T_0 = 973\ K$ are all the constant of the cell. $a = 0.05$ and $b = 0.11$ is respectively Tafel constant and Tafel slope; $I_L = 350\ A$ is the limit current value of concentration polarisation loss, $n = 2$ is the number of electronic transferred per-mole in the electrochemical reaction (Shulin Wang, 2016).

The output voltage can be obtained by multiplying the voltage of N cells:

$$V_{stack} = V_{cell} \times N \tag{22}$$

The output power of the SOFC stack is:

$$P_{stack} = V_{cell} \times I \tag{23}$$

Then, the electric characteristic model of the stack is established, as shown in Figure 6.

3.5 Temperature model

In the temperature model, the heat of inflow gas and outflow gas are considered, as well as the power and specific heat produced by the electrochemical reaction.

Energy conservation equation:

$$AC_p \frac{dT}{dt} = H_{in} - H_{out} - P_{stack} + \sum_m Q_m \tag{24}$$

Where Q_m represents the heat generated in the reaction m:

$$Q_m = r \cdot \Delta H_m \{r \in \{r_{re,1}, r_{re,2}, r_{an,1}, r_{an,2}\} \tag{25}$$

$$\Delta H_m = [\Delta H_1\ \Delta H_2\ \Delta H_2\ \Delta H_3] \tag{26}$$

H_{in} and H_{out} stand respectively for the enthalpy of the inflow gas and the outflow gas:

$$H_{in} = w_{air,in} h_{air,in} + w_{CH_4,in} h_{CH_4,in} + w_{H_2O,in} h_{H_2O,in} \tag{27}$$

$$H_{out} = w_{O_2,out} h_{O_2,out} + w_{N_2,out} h_{N_2,out} + w_{j,out} h_{j,out}$$
$$j \in \{CO, CO_2, H_2, H_2O\} \tag{28}$$

The temperature model mentioned above can be established as shown in Figure 7:

3.6 The model of the IIR-SOFC system

We have so far accomplished all parts models of the IIR-SOFC stack. Then when the models are connected according to the principle shown in Figure 2, a complete dynamic model of the IIR-SOFC system as shown in Figure 8 is realised.

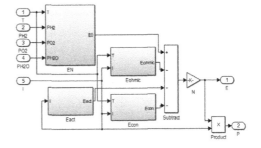

Figure 6. The electric model.

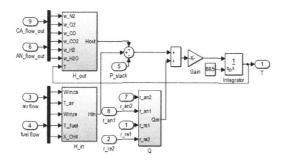

Figure 7. The temperature model.

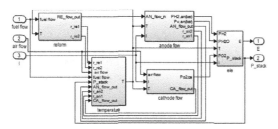

Figure 8. The model of IIR-SOFC system.

4 RESULTS AND ANALYSIS

The output-input relationship of the IIR-SOFC system can be investigated based on the models above, and the associated parameters are shown in Table 2:

To simplify the model, methane is chosen as the fuel and air is chosen as the oxidant of the IIR-SOFC system. Since the output voltage and stack temperature are two essential parameters, the system is identified with output voltage, output power and temperature as outputs, and inlet methane flow rate, airflow rate and current as inputs. The research of the input-output relation of the IIR-SOFC system is to observe the effect of input changes on the output of the system.

4.1 *Effect of load current*

Keeping the conditions above constant but changing the load current over time, the output response curves of the defined reference cases i.e. the temperature, output voltage and output power of the stack, are shown in Figure 9 respectively.

Near zero load current, since triggering a reaction has to smash the energy barrier that generates activation polarisation voltage (Chunyan Li, 2011), the output voltage dropped quickly and the stack temperature rose for the heat released from the electrochemical reaction. With the current increasing, the decreasing amplitude of the voltage went down. The voltage depended almost linearly on the current. The ohm polarisation voltage loss was the main part of the stack loss in this stage. The range of the ohm polarisation decreased with the increase in current, which made the output power rise and the increasing range smaller. Once the current reached a certain level, too much water in the cathode prevented the diffusion rate of oxygen from keeping up with the speed of electrochemical reaction, which prejudiced the reaction working smoothly, so the voltage dropped rapidly, and the power reduced accordingly. But as the heat emission was increasing, the temperature remained growth.

4.2 *Effect of air input*

According to the simulation results above, the load current in the following research is set to 200 A and the inlet methane flow rate is set to 2.3 $mol \cdot s^{-1}$. The simulation results are shown in Figure 10.

The output response curves along with the changing air inlet are shown in Figure 10. When the inlet airflow rate increased from 5.0 $mol \cdot s^{-1}$ to 10.0 $mol \cdot s^{-1}$, the output voltage of IIR-SOFC rose from 629 V to 650 V, the output power went from $1.258 \times 10^5 W$ to $1.3 \times 10^5 W$ and the stack temperature decreased gradually. The amount of hydrogen generated by methane reforming is about 2.314 $mol \cdot s^{-1}$ and the amount of reactant consumed in the electrochemical reaction is certainty for the stable current. With the air inlet increasing, the partial oxygen pressure rose, which increased the output voltage. And the growth trend paralleled the trend of the output power based on the equation (23). For the higher stack temperature, more heat was required to warm the gases entered into the stack. Therefore the tack temperature gradually decreased, but it was within the normal range.

4.3 *Effect of methane input*

With the same load current as above, the air input flow rate was set to 6.8 $mol \cdot s^{-1}$, and the output response curves with the methane input increasing was observed.

Table 2. Main parameters of IIR-SOFC stack model.

Parameter	Value
Air inlet flow rate	2.3 $mol \cdot s^{-1}$
Fuel inlet flow rate	6.8 $mol \cdot s^{-1}$
The cell number	712
The inlet gas temperature	861.35 K
Methane humidity	70.1%

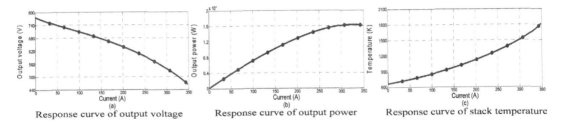

Figure 9. Response curve of load current input changes.

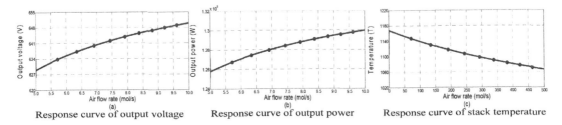

Figure 10. Response curve of air input changes.

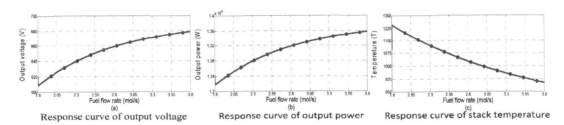

Figure 11. Response curve of methane input changes.

Observe the simulation curves shown in Figure 11. When the methane input flow rate increased from 1.8 $mol \cdot s^{-1}$ to 3.8 $mol \cdot s^{-1}$, the output voltage of IIR-SOFC increased from 610 V to 679 V, the output power went from $1.219 \times 10^5 W$ to $1.358 \times 10^5 W$, and the stack temperature gradually lowered. As described in previous sections, the methane entered into the stack would be heated, which would lead to a reduction in stack temperature. The gas entered into the anode was a mixture of hydrogen, carbon monoxide, carbon dioxide and water vapour. The hydrogen content was more than the water content, so the growth of the partial pressure of hydrogen was faster than water vapour, which finally made the output voltage increase. The output power had the same growth trend as the output voltage, and the growth rate declined gradually.

Comparing the output characteristic curves shown in Figures 10 and 11, both the increase of the air input and methane input could raise the output voltage and power, and decrease the stack temperature. When the increasing range of the methane input is 2 $mol \cdot s^{-1}$, the IIR-SOFC output voltage rises by about 69 V and the output power rises by about $0.319 \times 10^5 W$. When the growth margin of the air input is 5 $mol \cdot s^{-1}$, the IIR-SOFC output voltage rises by about 21 V and the output power rises by about $0.042 \times 10^5 W$. It can then be projected that the output of the IIR-SOFC system is more sensitive to a change in the methane input rather than a change of the air input.

5 CONCLUSION

The IIR-SOFC fuelled by methane has been intensively applied in practice. Therefore it is especially meaningful to research the input-output relationships of IIR-SOFC. This paper first introduced how the indirect internal reforming operation and each component of SOFC works, proposing the SOFC based on indirect internal reforming unit.

Then the mathematical model of IIR-SOFC was developed on the platform of MATLAB/Simulink, and the simulation results were presented. Finally, the IIR-SOFC input-output relationships were analysed though the simulation results.

A set of output response curves obtained respectively by changing methane input, air input and load current are presented. It can be indicated from the analysis results that, in the allowable range, the output voltage and power increase gradually with the rising methane input flow rate and air input flow rate, but the temperature experiences a significant reduction. The output and the working performance of the IIR-SOFC system are more sensitive to the change in methane input than the air input. For a rising load current, the output power and temperature of the IIR-SOFC system increased but the output voltage decreased gradually. So it can regulate the inlet methane flow rate to keep the temperature and the output at desired levels. The above-mentioned results provide an insight and some basic data for further understanding of the IIR-SOFC system. Furthermore, according to the research results, the input and operating parameters of the IIR-SOFC system can be adjusted to meet the users' requirements better and faster.

ACKNOWLEDGEMENTS

This work is supported by the National Natural Science Foundation of China (No. 61473174, 61105100, 51376110), the Specialized Research Fund for the Doctoral of Higher Education (No: 2013013 1130006) and China Postdoctoral Science Foundation (No.2014M551907.)

REFERENCES

Chunyan, Li. (2011). Research on the Modelling and Control Strategies of Solid Oxide Fuel Cell Generation System in Microgrid. Chongqing: Chongqing University.

Dokmaingam, P. et al. (2010) Modelling of IT-SOFC with indirect internal reforming operation fuelled by methane: Effect of oxygen adding as autothermal reforming. *International Journal of Hydrogen Energy.* 35, 13271–13279.

Gelen, A. T. and Yalcinoz, A. (2013). dynamic model for solid oxide fuel cell system and analyzing of its performance for direct current and alternating current operation conditions. *International Journal of Energy Research.* 37, 1232–1241.

Hongliang, Cao. (2012). Dynamic modeling and control of solid oxide fuel cell systems. Wuhan: Huazhong University of Science and Technology.

Klein, J. M., Georges, S. and Bultel, Y. (2010). SOFC fuelled by methane without coking: optimization of electrochemical performance. *Journal of Applied Electrochemistry.* 40, 943–954.

Qing Zhao, lv Xiaojing, Wang Weiguo, and Weng Yiwu. (2015). Effect of Reforming Conditions Performance of the SOFC Stack. *Journal of Chinese Society Power Engineering.* 35, 929–933.

Shulin Wang. (2016). Research on the Modelling and Performance Simulation for DCCHP System based on *SOFC/MGT Technology.* Shandong: Shandong University.

Jafarian, S.M. and Haseli, P. (2010). Performance analysis of a solid oxide fuel cell with reformed natural gas fuel, J. G Karimi: *International Journal of Energy Research.* 34, 946–961.

Van Nhu Nguyen, Robert Deja and Roland Peters. (2016). Methane/steam global reforming kinetics over the Ni/YSZ of planar pre-reformers for SOFC systems. Ludger Blum: *Chemical Engineering Journal.* 292, 113–122.

Wenshu Zhang. (2012). The Simulation of the Novel Hybrid System of Solid Oxide Fuel Cell and Gas Turbine. Shanghai: Shanghai Jiao Tong University.

Xiongwen Zhang. (2008). Multi-level Modelling and Simulation for the Power Generation of Solid Oxide Fuel Cell and Hybrid System Xi'an: Xi'an Jiao Tong University.

Yu Li, Shuang Ye and Weiguo Wang, (2016). Performance analysis of SOFC system based on natural gas autothermal reforming. *CIESC Journal.* 67, 1557–1564.

Automotive, Mechanical and Electrical Engineering – Liu (Ed.)
© *2017 Taylor & Francis Group, London, ISBN 978-1-138-62951-6*

Simulation of a heat pipe receiver with high-temperature latent heat thermal energy storage during the charging process

Hongjie Song, Wei Zhang, Yaqi Li, Zhengwei Yang & Anbo Ming
Xi'an Hi-Tech Research Institute, Xi'an, Shaanxi, China

ABSTRACT: The charging process of Phase Change Material (PCM) packaged in a cylindrical container that melts by absorbing the heat with the Heat Pipe (HP) is simulated in the paper. A two-dimensional physical model is built to couple with the problem that the heat transmission of HP with PCM. An enthalpy model and VOF method is separately introduced to couple with the melting of PCM, the evaporation and the condensation in HP. The numerical simulation reveals that the temperature of the evaporation section rises faster than the condensation section at the beginning and the gap decreases as the continuum flow is established in the entire HP. The PCM temperature distribution is closely related to the flow in the HP. The melting time of the PCM meets the demands of the Solar Dynamic Space Power System (SDPSS). The work in this paper lays a good foundation for further research on the character of the heat transfer of the HP receiver integrated with High-Temperature Latent Thermal Energy Storage (HTLHTES).

Keywords: PCM; HP; High-temperature latent heat thermal energy storage; VOF (Volume of Fluid)

Nomenclature			
u	velocity (m/s)	Φ	Viscous dissipation (J/kg m^2)
ρ	Density (kg.m^{-3})	μ	Dynamic viscosity (pa.s)
p	pressure (pa)	v	kinematic viscosity (m^2 / s)
x	axial coordinate	Subscripts	
r	radial coordinate	r	radial direction
t	time (s)	x	axial direction
c	heat capacity (j.kg^{-1}.k^{-1})	p	PCM
k	Thermal Conductivity (w.m^{-1}.k^{-1})	m	melt temperature
H	latent heat (J.kg^{-1})	l	liquid
f	liquid fraction	s	solid
T	temperature (K)	v	vapour
a	discrete equation coefficients	eff	effective
Greek		w	wall
ε	porosity of wick structure	W, E, N, S	west, east, north, and south faces of control volumes

1 INTRODUCTION

With the development of human activities in space, the electricity demand of the space station and other large space vehicles is also growing quickly. But the main power source of the space equipment is the power system is at present supplied by solar power. The space solar power system contains Photovoltaic System and SDPSS. Because of its high power-mass ratio, small size, long service life and other advantages, SDPSS is taking up more and more promising worldwide technology for future power generations. As a prospective space energy system, SDPSS was proposed owing to the large power demands of the space station. The receiver with integrated HTLHTES is one of the key components during the heat transfer process of energy transfer to energy and storage to energy conversion, which accounts for about 60% mass of the total system (Wang & Hou, 2014). The new proposed HP receiver with HTLHTES has high heat transfer efficiency, is not affected by gravity and has been used to cool electronic components of the space station and other large space vehicles. Also, the HTLHTES utilises latent heat when PCM phase transition occurs at a constant temperature with high energy density, which can solve the problem of energy supply during eclipse periods and provides the possibility of developing SDPSS (Makki et al., 2016).

The phase change problem is covered in the HTLHTES and HP, which contains evaporation and condensation. The character of the phase change is that the solid-liquid interface in the solution area changes over time, which is a strongly non-linear problem and can only be treated by numerical analysis. It is necessary to study the effect of the PCM property and the HP boundary condition on the charging process of the HP receiver with HTLHTES, which can determine the PCM mass and the melting time of the phase change. The phase transition problem is crucial in the optimisation and effective utilisation of PCM.

Adine & Qarnia (2009) built a mathematical model and deduced the conservation energy equations by introducing the control volume approach based on the finite difference method to discrete energy equations during phase change. Wang et al. (2015) studied the phase change and energy efficiency problem that happened in the shell-and-tube unit. The governing equations were discretised by the finite volume method. Similarly, the two-dimensional mathematical model of the shell-and-tube latent heat thermal energy storage unit with three phase change materials were simulated and optimised by Li et al. (2013).

At present, the key step is to find a suitable high-temperature PCM. The more suitable PCM is pure molten salt and mixed molten salt, and most of the current research is mainly concentrated in low temperature PCM paraffin wax. The HP-heat exchanger problem was analysed by adopting an experimental method and stearic acid was chosen as the PCM (Tm = 52.1°C) in Liu et al. (2006 a, b). Respectively, the charging problem and discharging problem were considered. It should be noticed that the SDPSS will obtain high efficiency if the melting temperature of PCM is high enough. Also it found that the efficiency of the system increases as the operating temperature increases (Tiari et al. 2015). The heat conductivity of some salts and the mixture were investigated by Tufen et al. (1985). Kenisarin (2010) studied high-temperature PCM that is perspective for heat thermal storage and solar power, and the temperature of which ranges from 393 K to 1,273 K.

But now the low thermal conductivity limits the application, one of the way is to add the high thermal conductivity material. Fukai et al. (2002, 2003) studied that carbon brushes were packaged in the PCM to improve the thermal conductivity of the PCM. Measured by instruments, the transient responses of the carbon brushes packaged in the PCM increased as the growth of the brush

diameter. It found that natural convection was prevented by the brushes during the charging process. However, compared with no fibre, the rate of the charging process is 10% –20% higher.

Another way is using HP to enhance heat transfer (Jegadheeswaran & Poheka, 2003; Zhang et al., 2010; Lopez et al., 2010; Bayon et al., 2010). The effective thermal conductivity of HP is up from 5,000 W/m.K to 200,000 W/m.K. Also HP is a highly reliable and efficient energy transport device, which is widely applied in many nuclear and space applications because of its light weight, low cost and variety of shape and size options (Kim et al., 2013; Rao & More, 2015; Pooyoo et al., 2014). To simulate the behaviour of HP, many different complexity models have been proposed (Faghri, 1995).

The algorithm of either the VOF or level-set method is thought to be a useful method for solving two-phase flow in the HP by Computational Fluid Dynamic (CFD) (Al-abidi et al. 2013; Riffat & Gan 1998 a, b). The VOF algorithm was adopted by Yang et al. (2008) to simulate R-141B flow boiling along a coiled tube. Compared with the experiment data, the results of the simulation showed good agreement. Alizadehdakhel et al. (2010) studied the two-phase flow of evaporation and condensation in thermosyphon using CFD. The VOF model was adopted by Kuang et al. (2015) to simulate the boiling flow of a separate type of HP in low heat flux. The improved VOSET Method was proposed by Guo et al. (2011) to solve the boiling bubbles arising from a vapor film.

There are few numerical simulations of the complex physical phenomena present in HP-PCM systems. Previous studies of HP-PCM systems were mainly based on lumped capacitance analyses and network modelling (Shabgard et al., 2010; Nithyanandam & Pitchumani, 2011). In the last decades, computer hardware and numerical algorithms have been developed rapidly. Following that, the complex two-phase flow phenomenon were solved mainly with the help of numerical simulations. In Sharifi et al. (2015), the different structures of the HP and PCM were studied, and the influence of the gap between the HP during temperature change was discussed in three modes of operation. Shabgard et al. (2014) adopted the FVM method to simulate the transient response of the HP-assisted latent heat thermal energy storage unit in Dish-Stirling solar power generation systems. Taking the natural convection into account, Sharifi et al. (2015) solved the two-dimensional transient response problem of a conjugate HP-PCM system.

From the above it can be seen that most of the investigations are concerned about the low temperature PCM. Also, few HP-PCM studies take the effects of microgravity into account. In order to simulate PCM heat exchange with HP and two-phase flow in HP, the VOF model is introduced in the paper. The enthalpy method is introduced to solve the phase change problem. By coupling with the HP-PCM energy equation, the heat transfer between HP and PCM are studied to provide a reference for the analysis and optimisation of the SDPSS.

2 GOVERNING EQUATIONS

Figure 1 presents the two-dimensional schematic of the physical model. The HP contains two sections: the evaporation section and the condensation section. During the sun period, the sunlight is collected and reflected by the concentrator to the evaporation section of HP, which heats sodium from the evaporation section into vapour. Part of the energy carried by sodium vapour is used to melt PCM, the rest of which drives the heat engine. With the total HP length of 1.2 m, $L_e = 0.36$ m, $L = 0.41$ m, $L_1 = 0.43$ m corresponding to each length. The thickness of PCM is 0.015 m and HP is in the vacuum inside. Also $r_v = 20$ mm, $r_w = 36.7$ mm. Sodium (Na) is chosen as the working fluid of the high-temperature HP. Compared with other alkali metals, Sodium has higher boiling points (1156 K) and higher thermal conductivity (Wang et al., 2013).

Due to the complexity of the physical model, some assumptions were proposed to simplify the model:

1. It assumes that the properties of PCM are constant and unequal when PCM is in solid and liquid state. The sodium and PCM's thermal and physical properties do not change as temperature changes;
2. In microgravity, regardless of the effect of natural convection to the PCM melting during phase change happens;
3. Because the specific heat of vapour sodium is much smaller than liquid sodium, the steady vapour flow occurred quickly.

2.1 For PCM

The following continuity, momentum and energy equations of PCM are described in the following, (Faghri et al., 2010, Cao & Faghri, 1991).
The continuity equation is:

$$\frac{\partial u_r}{\partial r} + \frac{u_r}{r} + \frac{\partial u_x}{\partial x} = 0 \tag{1}$$

The momentum equations are:

$$\frac{\partial u_r}{\partial t} + u_r \frac{\partial u_r}{\partial r} + u_x \frac{\partial u_r}{\partial x} = -\frac{1}{\rho} \frac{\partial p}{\partial r} + v\left(\frac{\partial^2 u_r}{\partial r^2} + \frac{1}{r}\frac{\partial u_r}{\partial r} - \frac{u_r}{r^2} + \frac{\partial^2 u_r}{\partial x^2}\right) \tag{2}$$

$$\frac{\partial u_x}{\partial t} + u_r \frac{\partial u_x}{\partial r} + u_x \frac{\partial u_x}{\partial x} = -\frac{1}{\rho} \frac{\partial p}{\partial x} + v\left(\frac{\partial^2 u_x}{\partial r^2} + \frac{1}{r}\frac{\partial u_z}{\partial r} + \frac{\partial^2 u_x}{\partial x^2}\right) \tag{3}$$

The energy equation (Tao & He, 2011; Stririh, 2003) is:

$$(\rho_p c_p)_p \frac{\partial \theta}{\partial t} = \frac{\partial}{\partial x}\left(k_p \frac{\partial \theta}{\partial x}\right) + \frac{1}{r}\frac{\partial}{\partial r}\left(k_p r \frac{\partial \theta}{\partial r}\right) - \rho_p \Delta H \frac{\partial f}{\partial t} \tag{4}$$

Where $\theta = T - T_m$, the f is defined as:

$$\begin{cases} f = 0, & \theta < 0 \\ 0 < f < 1, & \theta = 0 \\ f = 1, & \theta > 0 \end{cases} \tag{5}$$

And the k and c are:

$$k = \begin{cases} k_s, & \theta < 0 \\ k_s(1-f) + k_l, & \theta = 0 \\ k_l, & \theta > 0 \end{cases} \tag{6}$$

$$c = \begin{cases} c_s, & \theta < 0 \\ c_s(1-f) + c_l, & \theta = 0 \\ c_l, & \theta > 0 \end{cases} \tag{7}$$

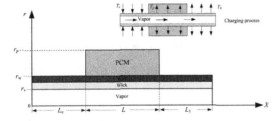

Figure 1. Physical model.

Table 1. The thermophysical properties of PCM (Hall, 1998).

	Solid	Liquid
ρ(kg / m³)	2590	2190
c(J / (kg.k))	1770	1770
k(W/(kg.k))	3.82	1.7
$T_m(K)$	1040	
ΔH(J/kg)	816000	
μ(m²/s)	1.05×10^{-6}	

2.2 For HP

The continuity equation of HP vapour flow is described as:

$$\frac{\partial \rho}{\partial t} + \frac{1}{r}\frac{\partial}{\partial r}(\rho r u_r) + \frac{\partial}{\partial x}(\rho u_x) = 0 \tag{8}$$

The r and x momentum equations are:

$$\frac{\partial}{\partial t}(\rho u_r) + \frac{1}{r}\frac{\partial}{\partial r}(\rho r u_r^2) + \frac{\partial}{\partial x}(\rho u_r u_x)$$
$$= -\frac{\partial p}{\partial r} + \frac{1}{r}\frac{\partial}{\partial r}\left(\mu r \frac{\partial u_r}{\partial r}\right) + \frac{\partial}{\partial x}\left(\mu \frac{\partial u_r}{\partial x}\right)$$
$$-\frac{\mu u_r}{r^2} + \frac{1}{3}\mu \frac{\partial}{\partial r}\left(\frac{1}{r}\frac{\partial}{\partial r}(r u_r) + \frac{\partial u_x}{\partial x}\right)$$
$$+\frac{\partial \mu}{\partial r}\left(\frac{\partial u_r}{\partial r} - \frac{2}{3}\left(\frac{1}{r}\frac{\partial}{\partial r}(r u_r) + \frac{\partial u_x}{\partial x}\right)\right) + \frac{\partial \mu}{\partial x}\frac{\partial u_x}{\partial x} \tag{9}$$

$$\frac{\partial}{\partial t}(\rho u_x) + \frac{\partial}{\partial x}(\rho u_x^2) + \frac{1}{r}\frac{\partial}{\partial r}(\rho r u_r u_x)$$
$$= -\frac{\partial p}{\partial x} + \frac{1}{r}\frac{\partial}{\partial r}\left(\mu r \frac{\partial u_x}{\partial r}\right) + \frac{\partial}{\partial x}\left(\mu \frac{\partial u_x}{\partial x}\right)$$
$$+\frac{1}{3}\mu \frac{\partial}{\partial x}\left(\frac{1}{r}\frac{\partial}{\partial r}(r u_r) + \frac{\partial u_x}{\partial x}\right)$$
$$+\frac{\partial \mu}{\partial x}\left(\frac{\partial u_x}{\partial x} - \frac{2}{3}\left(\frac{1}{r}\frac{\partial}{\partial r}(r u_r) + \frac{\partial u_x}{\partial x}\right)\right) + \frac{\partial \mu}{\partial r}\frac{\partial u_r}{\partial x} \tag{10}$$

Introducing the ideal gas equation and the enthalpy-internal energy equation, so the modified energy equation is shown as:

$$\frac{\partial}{\partial t}(\rho c_v T) + \frac{1}{r}\frac{\partial}{\partial r}(\rho r u_r c_v T) + \frac{\partial}{\partial x}(\rho u_x c_v T)$$
$$= \frac{1}{r}\frac{\partial}{\partial r}\left(k r \frac{\partial T}{\partial r}\right) + \frac{\partial}{\partial x}\left(k \frac{\partial T}{\partial x}\right)$$
$$- p\left(\frac{1}{r}\frac{\partial}{\partial r}(r u_r) + \frac{\partial u_z}{\partial x}\right) + \mu \Phi \tag{11}$$

The term of viscous dissipation is:

$$\Phi = 2\left[\left(\frac{\partial u_r}{\partial r}\right)^2 + \left(\frac{u_r}{r}\right)^2 + \left(\frac{\partial u_x}{\partial x}\right)^2\right] + \left(\frac{\partial u_r}{\partial x} + \frac{\partial u_x}{\partial r}\right)^2$$
$$- \frac{2}{3}\left[\frac{1}{r}\frac{\partial}{\partial r}(r u_r) + \frac{\partial u_x}{\partial x}\right]^2 \tag{12}$$

It assumes that the HP wall is filled with fluid and the velocity is negligible (Cao & Faghri, 1991). Therefore, the heat conduction equation for the wick is:

$$(\rho c)_{eff}\frac{\partial T}{\partial t} = k_{eff}\left[\frac{1}{r}\frac{\partial}{\partial r}\left(r \frac{\partial T}{\partial r}\right) + \frac{\partial^2 T}{\partial x^2}\right] \tag{13}$$

The temperature distribution governing the equation in the HP wall is shown as (Sharifi et al., 2015):

$$(\rho c)_w\frac{\partial T}{\partial t} = k_w\left[\frac{1}{r}\frac{\partial}{\partial r}\left(r \frac{\partial T}{\partial r}\right) + \frac{\partial^2 T}{\partial x^2}\right] \tag{14}$$

For the porous wick, it is difficult to calculate the thermal conductivity. So the effective thermal conductivity of the wick can be deduced by (Wang et al., 2013):

$$k_{eff} = \frac{k_l\left[(k_l + k_s) - (1-\varepsilon)(k_l - k_s)\right]}{\left[(k_l + k_s) + (1-\varepsilon)(k_l - k_s)\right]} \tag{15}$$

3 EQUATIONS DISCRETISATION

In the paper, governing equations are discretised by adopting the finite volume method. In order to better understand the process, Figure 2 shows the discrete schematic diagram.

For PCM (Adine & Qarnia. 2009):

$$a_p^{i+1}\theta_p = a_p^i\theta_p + a_E^{i+1}\theta_E + a_W^{i+1}\theta_W + a_N^{i+1}\theta_N + a_S^{i+1}\theta_S \tag{16}$$

$$a_p^i\theta_p = (\rho c_P)_P r \frac{\Delta x \Delta r}{\Delta t}\theta_P^{-1} +$$
$$\rho_P \Delta H r \frac{\Delta x \Delta r}{\Delta t}(f_P^{i-1} - f_P) \tag{17}$$

$$r = \frac{r_N + r_S}{2} \tag{18}$$

$$a_E = \frac{k_{P,E}x\Delta r}{x_E - x_P}, \quad a_w = \frac{k_{P,W}x\Delta r}{x_P - x_W}$$
$$a_N = \frac{k_{P,N}r_N\Delta x}{r_N - r_P}, \quad a_S = \frac{k_{P,W}r_S\Delta x}{r_P - r_S} \tag{19}$$

$$a_P = a_E + a_W + a_N + a_S + (\rho c_P)_P r \frac{\Delta x \Delta r}{\Delta t} \tag{20}$$

It is found that the independent solutions can achieve sufficient accuracy when the time step is

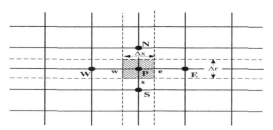

Figure 2. The grid schematic diagram in the HP-PCM latent thermal energy storage unit.

$t = 0.0001$ s. Coupling the HP-PCM energy equation with the boundary conditions, an iteration procedure is introduced in every time step to achieve the convergence. The convergence residual value is set to less than 10^{-4} for continuity, momentum and energy equations.

4 SIMULATION RESULTS AND DISCUSSION

The initial HP-PCM temperature is set to 500 K. At this time, the PCM is in solid phase and sodium is in liquid phase.

4.1 Flow characteristic in HP

The flow characteristics of the vapour in the HP have an important effect on the HP performance. Because of the viscous limit, sonic limit and entrainment limit caused by the vapour flow, it is significant to study the flow characteristics of the vapour in HP. Figure 3 shows the streamline of the outlet section of the evaporation section and the condensing section, respectively.

It can be seen that the streamline diagram of the two sections is quite different. Due to the expansion of the vapour pressure drop, the streamline of velocity spreads and disturbed flow happens, which is different from the laminar flow of the evaporating part. Here, the velocity reached the maximum in the outlet. The vapour velocity is a continuous reduction process after the vapour enters the condensation section. With the decrease of the kinetic energy, the pressure increases along the direction of flow. The vapour velocity appears along the radial direction, which is mainly because of the vapour condensation in the condensation section, and the vapour velocity distribution is close to constant.

The velocity vector is shown in Figure 4. The fluctuation of the vapour velocity can be seen in the velocity vector of the HP vapour flow. The axial distribution of the velocity in 0.5 s has been achieved in the condensing section. The velocity distribution only achieves to the PCM section in 1 s. The results of 1.5 s and 2 s are similar to the above. The velocity vector diagram also verifies the conclusion from Figure 3.

4.2 Temperature distribution

Figure 5 shows the temperature contour of HP-PCM, and it can be concluded that the temperature distribution inside the HP shows good agreement with the distribution of the vapour velocity in the HP. As shown in Figure 5, the temperature contour of the vapour flow shows great difference along

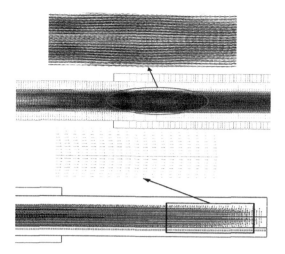

Figure 3. Streamline diagram.

Figure 4. Velocity vector.

the radial direction. But the temperature contour of the wick increases along the axial direction. The pressure drop of the axial flow in the tube also makes the corresponding vapour saturation temperature difference, which leads to a certain difference of the HP outer surface temperature along the axial direction. The temperature of the wall of the evaporating section is the highest and drops along the radial direction. Also the PCM temperature distribution closed to evaporation section outlet is much higher. But it is uniform along the radial direction. This is because the HP has the isothermal advantage to heat the PCM evenly, which can make PCM melt evenly along the radial direction.

It is clear that Figure 6a shows the temperature contour of the evaporation outlet and the PCM outlet. Also it clearly shows the heat transfer process of the HP and PCM. The temperature of PCM changes little along the axial direction but there is a large radial temperature difference. Because there is no convection, the temperature of PCM is uniform by heat conduction. The temperature of the

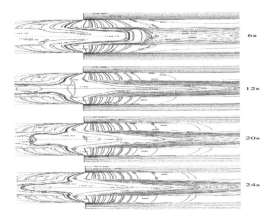

Figure 5. Temperature contour of HP-PCM.

Temperature: 550 625 700 775 850 925 1000 1075 1150

(a) Temperature contour of evaporation outlet

Temperature: 550 625 700 775 850 925 1000 1075 1150

(b) Temperature contour of PCM outlet

Figure 6. Temperature contour of HP.

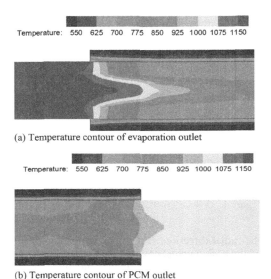

symmetric axis of the evaporation section is the highest. The temperature drops along the radial and decreases a little along the axial direction.

Because the vapour enters the condenser section with the changing of the flow pattern, this leads to the axial velocity and the radial velocity decreasing with the increase of the axial direction. The HP continuum flow establishes quickly and reaches to steady flow in Figure 6b.

4.3 PCM liquid fraction varied as time

Liquid fraction of PCM varied as time is shown in Figure 7. The PCM begins to melt at the time of 305 s and it all melts to liquid in 3,379 s. The whole

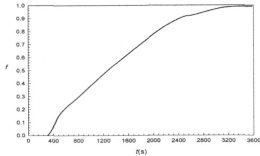

Figure 7. Liquid fraction of PCM varies as time.

process takes about 3,074 s and the sun period of the space vehicle in low Earth orbit is about 54 min (3,240 s). The HP-PCM can meet the demands of the SDPSS to PCM, which melts to liquid completely during the period of sunshine.

5 CONCLUSION

To simulate the heat transfer problem of HP-PCM, a numerical model has been developed to simulate the heat transfer problem of HP-PCM. The VOF model and finite volume method are introduced to simulate the vapor flow in the HP and research the heat transfer process between the HP and PCM. The following conclusions are shown:

The fluctuation of the vapour velocity can be seen in the velocity vector of the HP vapour flow. The distribution of PCM is closely related to the vapour flow in the HP. The temperature contour of the vapour flow shows a great difference in radial direction. But the temperature contour of wick along the axial direction. Also, the whole PCM melting time takes about 3,074 s and the sun period of the space vehicle in low Earth orbit is about 54 min (3,240 s), which can meet the SDPSS demand to the PCM melting time. In future work, many optimal and design works should be done to make the SDPSS achieve its best service.

REFERENCES

Adinberg, R., Zvegilsky, D., & Epstein, M. (2010). Heat transfer efficient thermal energy storage for steam generation. *Energy Conversion and Management*, 51(1), 9–15.
Adine, H.A., & Qarnia, H. El. (2009). Numerical analysis of the thermal behaviour of a shell-and-tube heat storage unit using phase change materials. *Applied Mathematical Modelling*, 33(4), 2132–2144.
Al-abidi, A.A., Mat, S.B., Sopian, K., Sulaiman, M.Y., & Mohammed, A.T. (2013). CFD applications for latent

heat thermal energy storage: A review. *Renewable and Sustainable Energy Reviews*, 20, 353–363.

Alizadehdakhel, A., Rahimi, M., & Alsairafi, A.A. (2010). CFD modeling of flow and heat transfer in a thermosyphon. *International Communications in Heat and Mass Transfer*, 37(3), 312–318.

Bayon, R., Rojas, E., Valenzuela, L., & et al. (2010). Analysis of the experimental behaviour of a 100 kWh latent heat storage system for direct steam generation in solar thermal power plants. *Applied Thermal Engineering*, 30(2), 2643–2651.

Cao, Y., & Faghri, A. (1990). A numerical analysis of phase change problem including natural convection. *Journal of Heat Transfer*, 112(3), 812–816.

Cao, Y., & Faghri, A. (1991). Transient two-dimensional compressible analysis for high-temperature heat pipes with pulsed heat input. *Numerical Heat Transfer. Part A: Applications*, 18(4), 483–502.

Faghri, A., Zhang, Y., & Howell, J. (2010). *Advanced Heat and Mass Transfer*. Global Digital Press.

Faghri, A. (1995). *Heat Pipe Science and Technology*. Washington, DC: Taylor and Francis.

Fukai, J., Hamada, Y.C., Morozumi, Y.S., & Miyatake, O. (2002). Effect of carbon-fiber brushes on conductive heat transfer in phase change materials. *International Journal of Heat and Mass Transfer*, 45(24), 4781–4792.

Fukai, J., Hamada, Y.C., Morozumi, Y.S., & Miyatake, O. (2003). Improvement of thermal characteristics of latent heat thermal energy storage units using carbon-fiber brushes: Experiments and modeling. *International Journal of Heat and Mass Transfer*, 46(23), 4513–4525.

Gan, G., & Riffat, S.B. (1998a). A numerical study of solar chimney for natural ventilation of buildings with heat recovery. *Applied Thermal Engineering*, 18(12), 1171–1187.

Guo, D.Z., Sun, D.L., Li, Z.Y., & Tao, W.Q. (2011). Phase change heat transfer simulation for boiling bubbles arising from a vapor film by the VOSET method. 59(11), 857–881.

Hall, C.A. (1998). *Thermal state-of-charge of solar heat receivers for space solar dynamic power*, Ph.D. Thesis, Howard University, USA.

High Energy Advanced Thermal Storage (HEATS) Funding Opportunity Announcement. (2011). *Advanced Research Projects Agency (ARPA-E), Department of Energy, DE-FOA-0000471 CFDA Number 81135*.

Jegadheeswaran, S., & Poheka, S.D. (2003). Performance enhancement in latent heat thermal storage system: A review. *Renewable and Sustainable Energy Reviews*, 13(9), 2225–2244.

Kenisarin, M.M. (2010). High-temperature phase change materials for thermal energy storage. *Renewable and Sustainable Energy Reviews*, 14(3), 955–970.

Kim, T.Y., Hyun, B.S., Lee, J.J., & Rhee, J.H. (2013). Numerical study of the spacecraft thermal control hardware combining solid–liquid phase change material and a heat pipe. *Aerospace Science and Technology*, 27(1), 10–16.

Kuang, Y.W., Wang, W., Zhuan, R., & Yi, C.C. (2015). Simulation of boiling flow in evaporator of separate type heat pipe with low heat flux. *Annals of Nuclear Energy*, 75, 158–167.

Li, Y.Q., He, Y.L., Song, H.J., Xu, C., & Wang, W.W. (2013). Numerical analysis and parameters optimization of shell-and-tube heat storage unit using three phase change materials. *Renewable Energy*, 59, 92–99.

Liu, Z., Wang, Z., & Ma, C. (2006a). An experimental study on heat transfer characteristics of heat pipe heat exchanger with latent heat storage. Part I: Charging only and discharging only modes. *Energy Conversion and Management*, 47(7–8), 944–966.

Liu, Z., Wang, Z., & Ma, C. (2006b). An experimental study on the heat transfer characteristics of a heat pipe heat exchanger with latent heat storage. Part II: Simultaneous charging/discharging modes. *Energy Conversion and Management*, 47(7–8), 967–991.

Lopez, J., Caceres, G., Palomo Del Barrio, E., & et al. (2010). Confined melting in deformable porous media: A first attempt to explain the graphite/salt composites behaviors. *International Journal of Heat and Mass Transfer*, 53(2), 1195–1207.

Makki, A., Omer, S., Su, Y.H., & Sabir, H. (2016). Numerical investigation of heat pipe-based photovoltaic–thermoelectric generator (HP-PV/TEG) hybrid system. *Energy Conversion and Management*, 112, 274–287.

Nithyanandam, K., & Pitchumani, R. (2011). Analysis and optimization of a latent thermal energy storage system with embedded heat pipes. *International Journal of Heat and Mass Transfer*, 54(21–22), 4596–4610.

Pooyoo, N., Kumar, S., Charoensuk, J., & Suksangpanomrung, A. (2014). Numerical simulation of cylindrical heat pipe considering non-Darcian transport for liquid flow inside wick and mass flow rate at liquid–vapor interface. *International Journal of Heat and Mass Transfer*, 70, 965–978.

Rao, R.V., & More, K.C. (2015). Optimal design of the heat pipe using TLBO (teaching learning-based optimization) algorithm. *Energy*, 80, 535–544.

Riffat, S.B., & Gan, G. (1998b). Determination of effectiveness of heat-pipe heat recovery for naturally-ventilated buildings. *Applied Thermal Engineering*, 18(3), 121–130.

Shabgard, H., Bergman, T.L., Sharifi, N. & et al. (2010). High temperature latent heat thermal energy storage using heat pipes. *International Journal of Heat and Mass Transfer*, 53(15–16), 2979–2988.

Shabgard, H., Faghri, A., Bergman, T.L., & Andraka, C.E. (2014). Numerical simulation of heat pipe-assisted latent heat thermal energy storage unit for Dish-Stirling systems. *Journal of Solar Energy Engineering*, 136(2).

Sharifi, N., Faghri, A., Bergman, T.L., & Andraka, C.E. (2015). Simulation of heat pipe-assisted latent heat thermal energy storage with simultaneous charging and discharging. *International Journal of Heat and Mass Transfer*, 80, 170–179.

Stririh, U. (2003). Heat transfer enhancement in latent heat thermal storage system for buildings. *Energy Build*, 35(11), 1097–1104.

Tao, Y.B., & He, Y.L. (2011). Numerical study on thermal energy storage performance of phase change material under non-steady-state inlet boundary. *Applied Energy*, 88(11), 4172–4179.

Tiari, S., Qiu, S.G., & Mahdavi, M. (2015). Numerical study of finned heat pipe-assisted thermal energy storage system with high temperature phase change material. *Energy Conversion and Management*, 89, 833–842.

Tufen, R., Petitet, J.P., Denielou, I., & Le Neindre, B. (1985). Experimental determination of the thermal conductivity of molten pure salts and salt mixtures. *International Journal of Thermophysics*, 6(4), 315–330.

Wang, C.L., Zhang, D.L., Qiu, S.Z., Tian, W.X., Wu, Y.W., & Su, G. H. (2013). Study on the characteristics of the sodium heat pipe in passive residual heat removal system of molten salt reactor. *Nuclear Engineering and Design*, 265, 691–700.

Wang, L., & Hou, X.B. (2014). Key technologies and some suggestions for the development of space solar power station. *Spacecraft Environment Engineering*, 30(4), 343–350.

Wang, W.W., Wang, L.B., & He, Y.L. (2015). The energy efficiency ratio of heat storage in one shell-and-one tube phase change thermal energy storage unit. *Applied Energy*, 138, 169–182.

Yang, Z., Peng, X.F., & Ye, P. (2008). Numerical and experimental investigation of two-phase flow during boiling in a coiled tube. *International Journal of Heat and Mass Transfer*, 51(5–6), 1003–1016.

Zhang, P., Song, L., Lu, H. et al. (2010). The influence of expanded graphite on thermal properties for paraffin/high density polyethylene/chlorinated paraffin/antimony trioxide as a flame retardant phase change material. *Energy Conversion and Management*, 51(12), 2733–2737.

Automotive, Mechanical and Electrical Engineering – Liu (Ed.)
© *2017 Taylor & Francis Group, London, ISBN 978-1-138-62951-6*

Temperature distribution of wire crimp tubes under different loads

Hao Zhang, Bin Wang, Miao Wang, Yubing Duan, Xiaoli Hu & Hui Liu
State Grid Shandong Electric Power Research Institute, Jinan, China

ABSTRACT: Wire hydraulic tube is a vital part of the transmission lines, at the same time; it is an element that causes defects and faults easily. Once broken, it will lead directly to the falling of the conductor; hence a series of serious secondary disasters may occur. The operating condition of electric transmission line is very complex. Meteorological conditions and running conditions of the lines could cause the rise and fall of the hydraulic tubes' temperature. In this paper, we use the heat balance equation to study the factors that affect the heat dissipation of the crimp tube. We put the focus on the heating of the different locations rather than the overall of the strain clamp. The study also analyzes the influence of load on temperature distribution of crimp joint.

Keywords: Transmission Line; Hydraulic tube; Temperature Distribution; Infrared temperature measurement; Internal defect

1 INTRODUCTION

As the conductor crimping is a concealed project, it is difficult to detect whether there is any defect inside the crimping after the whole work is done. The infrared temperature measurement is a more intuitive detection method (Guang Dong Electric Power, 2012; Liu Minghui, et al., 2014; Ding Xiliang, et al., 2008). If there is a defect in the crimp tube, it will generate heat during operation. Due to the different thermal conductivity, infrared camera can be used to find the presence of injury.

Overhead conductor is twised by aluminum-cable and steel-reinforced conductor. Small air gaps exist among bundled conductors, meanwhile; the contact resistance lies on the contact surface. Contact thermal resistance and air gap will affect the thermal conductivity and thermal conductivity, which can cause the distribution of temperature uneven (Fujii O, 2007).

Abnormal heating of the crimping tube is current-heating-type defect (Tao Wenquan, 2006). Due to the presence of resistance, there must be power loss when the current flows through the conductor, which cause the conductor temperature to rise (Chen Lan, et al., 2014). The formula of heating power is shown in equation (1).

$$P = K_f I^2 R \qquad (1)$$

In the formula (1), P is the heating power (W); I is the current intensity (A); R is the DC resistance of the conductor (Ω); K_f is the additional loss factor, which indicates the coefficient of increasing resistance because of the skin resistance effect and the proximity effect in the AC circuit.

In this paper, the focus of the test is on the heating of different locations of strain clamp, instead of the overall clamp. This will be helpful for analyzing the internal defects of the clamp.

2 CURRENT CARRYING AND TEMPERATURE RISE TEST

Current test select constant current source to output different high currents to analyze the influences of load on the crimp tube's temperature.

The samples used in the test are the crimping tubes which have been operated for at least 10 years. Crimp should be banded on both sides of the wire ends, and a 3-meter-long wire will be left from the end of the aluminum tube, as is shown in Figure 1.

Eight clamps were selected to connect into four groups, which were recorded as A/B/C/D. After connected the experimental circuit and accessed current, we tested the temperature of the reference

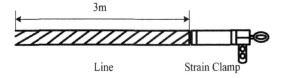

Figure 1. The samples' schematic diagram of strain clamp.

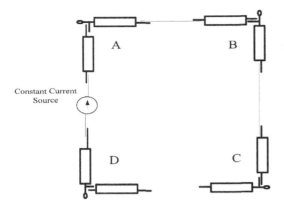

Figure 2. The schematic diagram of current carrying and temperature rise test.

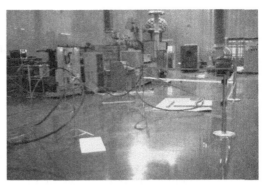

Figure 3. The testing field of current carrying and temperature rise test.

wire and samples when the temperature was stable for 15 minutes. The experiment was repeated 3–5 times, and the average value of the test results was taken as the test result.

The current of the test was 600A and 1200A, and every 5minutes we used infraredthermometer to measure the data. The test results are shown below.

Test results show that the temperature rise of the normal strain clamp is less than the conductor's, and the closer to the wire, the higher the temperature we get. The highest temperature of pressure pipe is at the end of the tension clamp which connects with the wire. The temperature of current plate is slightly lower than the wire. The reason is that the surface area of the current plate is larger than wire and the heat dissipation is faster.

Dr. Ying of Nanjing University of Science and Technology proposed an axial thermal circuit model in the form of state equation, which is based on thermal equilibrium principle (Ying Zhanfeng, et al., 2015). As shown in equation (2).

$$\begin{cases} C = mc \\ R = \dfrac{l}{S\lambda} \\ R_E = \dfrac{1}{Ah_c} \end{cases} \quad (2)$$

In the formula, C is the thermal capacity of the wire's micro-element; R is the axial thermal resistance of the micro-element; R_E is the convective thermal resistance of the micro element; m is the mass of the wire's micro-element; c is the specific heat capacity; l is the length of the micro-element; S is the radial sectional area; λ is thermal conductivity; A is the area of convective heat transfer; h_c is the coefficient of convective heat transfer.

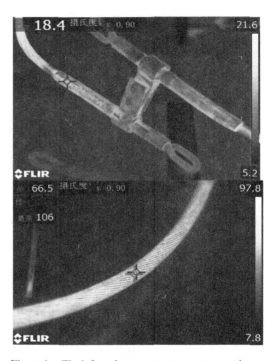

Figure 4. The infrared temperature measurement image of clamp and wire.

3 ANALYSIS OF THE DATA AND CONCLUSION

The temperature test points were selected as shown in the following table.

The temperature rise of the four groups at 600A and 1200A is shown in the following figure.

From the data of Figure 6 and Figure 7, we can draw the conclusions as following:

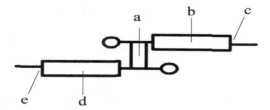

Figure 5. The temperature measurement points of strain clamp.

Table 1. The temperature measurement points.

Symbol	Position
a	The Center of the Current Plate
b	the middle of Clip's aluminum tube (small side)
c	The end of connection of Clamp and wire (small side)
d	the middle of Clamp aluminum tube (large side)
e	The end of connection of Clamp and wire(large side)
f	The maximum temperature of wire

1. The crimping technology of crimp has a great influence on heat loss. It directly shows as the distance of the maximum temperature curve of the wire (orange) and the curve of pressure pipe. The shorter the distance between the curves is, the better the crimping technology processes.
2. The greater the current passes, the shorter time it takes for the balance of wire and pressure pipe temperature to rise, which is related to its thermal balance. The large current is helpful to stimulate the defects of the pressure pipe.
3. Besides the abnormal data, the higher the temperature rises, the smaller the data dispersion is. It is because that the surface of the wire is seriously oxidized and the surface roughness of the wire is severe, especially when the line has been operated for a long time. The part of temperature rise is more likely to accumulate contamination, which makes an effective insulation effect. Good insulation effect makes the data of temperature rise stable.
4. Precise temperature measurement should pay more attention to the reflectivity of the object temperature. The oxidation of metal surface has a great impact on the measured data. It is generally only concerned about the maximum temperature when we measure on field, so we should pay attention to measure from different angles.

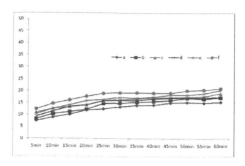

(1)Group A

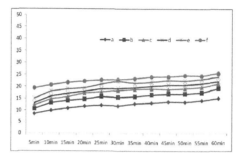

(2) Group B

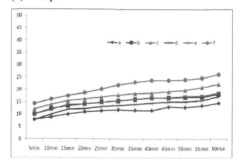

(3) Group C

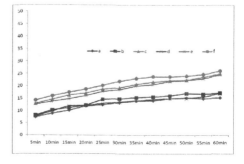

(4) Group D

Figure 6. The data of temperature rise test (I = 600A).

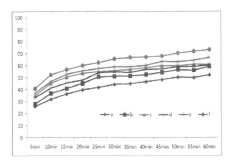

(1) Group A

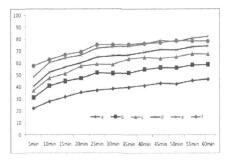

(2) Group B

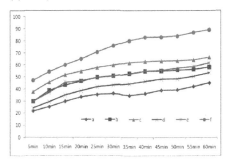

(3) Group C

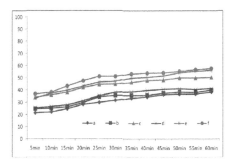

(4) Group D

Figure 7. The data of temperature rise test (I = 1200A).

4 THE ANALYSIS OF WIRE BREAKING

Conductor heating is a current-heating-type effect (Tao Wenquan, 2006). Due to the presence of resistance, a part of the electrical energy will convert into heat (Xu Qingsong, et al., 2006). And the heat is the reason of the wire and pressure pipe's temperature rise. In normal circumstances, the resistance of link fitting in the transmission line is lower than the wire. So the temperature of the link fitting is lower than that of the wire. Only when there is a defect in the pressure pipe, the contact resistance is abnormal. The temperature of the abnormal part will be higher than the conductor part, and the resistance will increase when the temperature increases. When the temperature exceeds a certain value, the oxidation rate of the metal will increase sharply (Peng Xiangyang, et al., 2007; Guo Xiaofei, 2010). The oxidation products increase the resistance more rapidly conversely. This is a vicious cycle, which leads to the wire breakage.

5 CONCLUSION

In this paper, we proposed a method to judge the situation of pressure pipe based on the infrared temperature measurement results. We focus on the heat of different locations of the tension clamp, instead of the overall clip. By studying the temperature distribution of the crimp tube, it is used to analyze the crimp condition and internal defects. The conclusions are as following:

1. The temperature rise of the normal pressure joint is less than the temperature rise of the conductor, and the closer to the wire, the higher the temperature. The connection of the cable clamp's end and wire is the highest temperature of crimp tube. The temperature of drainage plate is slightly lower than the wire. When there is a defect in the crimping tube, the temperature of abnormal part will be higher than the conductor part.
2. The large current is helpful to find the defects of the pressure pipe. The crimping technology of crimp has a great influence on heat loss. The shorter the distance between the curves is, the better the crimping technology processes.

REFERENCES

Carlos Alberto Cimini Jr, Beline Quintino Araujo Fonseca. Temperature profile of progressive damaged overhead electrical conductors [J]. Electrical Power and Energy Systems, 2013(49): 280–286.

Chen Lan, Bian Xingming, Wan Shuwei, et al. Influence of Temperature Character of AC Aged Conductor on Current Carrying Capacity [J]. High Voltage Engineering, 2014, 40(5) 1499–1506.

Ding Xiliang, Han Xueshan, and Zhang Hui, et al. Analysis on electro thermal coordination power flow and transmission line temperature variation process [J]. Proceedings of the CSEE, 2008, 28(19): 138–143.

Experimental Investigation on Impact of Ambient Wind Velocity on Ampacity of Overhead Transmission Lines and Temperature Rise [J]. Guang Dong Electric Power, 2012, 25(3): 20–25.

Fujii O, Mizuno Y, Naito K. Temperature of insulators as heated by conductor [J]. IEEE Transactions on Power Delivery, 2007, 22(1): 523–526.

Guo Xiaofei, Li Yongchun, Zhao Yuanlin. The Analysis and Process about High Temperature of Elementary Streams [J]. Electrical technology, 2010, 8(1):136–139.

Liu Minghui, LI Chenming, WANG Changchun et al. The Compression Analysis of LGJ-300/40 Wire Splicing Sleeve. Northeast Electric Power Technology, 2014, 13 (1) 26–29.

Peng Xiangyang, Zheng Xiaoguang, Zhou Huamin, et al. Experimental Study of Current Carrying and Temperature rise of Overhead Conductors [J]. Proceedings of the CSEE, 2007, 27 (S1):23–27.

Tao Wenquan. Heat transfer [M]. Xi'an: Northwestern Polytechnical University Press, 2006: 4–15.

Xu Qingsong, Ji Hongxian, Hou Wei, et al. The Novel Technique of Transmission Line's Capacity Increase by Means of Monitoring Conductor's Temperature [J]. Power System Technology, 2006(S1):171–176.

Ying Zhanfeng, Feng Kai, Du Zhijia et al. Thermal Circuit Modeling of the Relationship between Current and Axial Temperature for High Voltage Overhead Conductor [J], Proceedings of the CSEE, 2015, 11 (35) 2887–2895.

Automotive, Mechanical and Electrical Engineering – Liu (Ed.)
© 2017 Taylor & Francis Group, London, ISBN 978-1-138-62951-6

Optical property research on $Mg_{0.15}Zn_{0.85}O$ film by sol-gel technology

Jide Zhang
College of Physics and Electronic Information, Baicheng Normal University, Baicheng, China

ABSTRACT: By preparing $Mg_{0.15}Zn_{0.85}O$ film on quartz substrates using sol-gel technology, this paper focuses on the effect of annealing treatment on the optical property of film. The result shows that the film is an amorphous structure without annealing treatment, and it is a polycrystalline structure under four kinds of annealing temperature: 600°C, 700°C, 800°C and 900°C. The crystal quality of film under 800°C and 900°C is worse than under 600°C and 700°C. All films at UV range exist strong edge absorptions, and the absorption edges exhibit red shift with annealing temperature rising; all films have strong UV emission peaks, and the emission peaks exhibit red shift with annealing temperature rising. Moreover, there are green emission peaks around 500 nm.

Keywords: Sol-gel technology; $Mg_{0.15}Zn_{0.85}O$ film; annealing treatment; optical property

1 INTRODUCTION

In recent years, the wide band semiconductor ZnO (3.3 eV at room temperature), which has a large exciton binding energy (60 meV (Ohtomo et al., 1999; Lee et al., 2004; Zhang et al., 2002)), has been widely discussed because of the needs of short wavelength devices (Zhao et al., 2001; Lin Wei et al., 2007). In order to achieve the high output efficiency of short wavelength devices, scientists are committed to improving the UV transmittance and adjusting the band gap width of ZnO. The forbidden band width of MgO is 7.8 eV, Mg^{2+} ionic radius (0.057 nm) is similar to Zn^{2+} ionic radius (0.074 nm). Mg and its oxides are nonpoisonous and harmless. The crystal lattice mismatch of MgZnO is small, and the property of MgZnO alloy, which is formed by ZnO and MgO, is similar to ZnO (Muthukumar et al., 2004). Therefore, many research groups adjust the forbidden band width, improve the UV transmittance and research the property of ZnO by Mg ion doping method, but few research reports about the effect of annealing treatment on optical property can be found.

This paper focuses on the effect of annealing treatment on structural property, absorptive character and photoluminescence property of $Mg_{0.15}Zn_{0.85}O$ film prepared by the sol-gel spin-coating process.

2 EXPERIMENT

In this experiment, the film is prepared by sol-gel spin-coating. Its sol system uses zinc acetate dihydrate $(Zn (CH_3COO)_2 \cdot 2H_2O)$ and magnesium acetate tetrahydrate $(Mg (CH_3COO)_2 \cdot 4H_2O)$ as precursor, anhydrous ethanol (C_2H_5OH) as solvent, monoethanolamine $(C_2H_7 NO)$ as stabiliser, and glacial acetic acid (CH_3COOH) as catalyst. To add $Zn (CH_3COO)_2 \cdot 2H_2O$ and $Mg (CH_3COO)_2 \cdot 4H_2O$ powder to agitated reactor in proper proportions, meanwhile, respectively add cholamine and glacial acetic acid whose mole number is same with the sum of Zn^{2+} and Mg^{2+}; next, add a certain amount of anhydrous ethanol, and then mix for 2.5 hours at 70°C, 1,000 *r/min* to make the powder dissolved sufficiently, which forms the uniform and transparent sol (the ionic concentration in the sol is about 0.45 *mol/L*), then leave it in a clean environment for use.

Quartz substrate is dipped into cleaning mixture, toluene solution, acetone, anhydrous ethanol and deionised water respectively, cleaned with ultrasonic washer, and finally dewatered by anhydrous ethanol. Put the clean substrate on the spin coater, under the 350 *r/min* condition, add ageing sol to the substrate, and speed up to spin the sol for 20 seconds to make sure the sol covers the surface of the substrate evenly. Then, put it into a 120°C drying baker for ten minutes, and repeat the coating five times. Finally, put the gel film after drying treatment into a box-type resistance furnace, under temperatures of 0°C, 600°C, 700°C, 800°C and 900°C, with a heating rate of 10°C/min, under constant temperature for 150 minutes, then remove the $Mg_{0.15} Zn_{0.85}O$ film.

To analyse the film structure by D/max 2500 X-Ray Diffraction (XRD), Cu *Ka* radiation ($\lambda = 0.1542$ nm), pipe pressure is 40 kV and pipe flow is 100 mA, use a Cary-50 UV-Vis spectrophotometer to measure the absorption spectrum of film with 600 nm/min scanning speed and 1 nm scanning step. Analyse the photoluminescence of the film using a Traix 320 fluorescence spectrophotometer, and use a He-Cd laser with 325 nm excitation wavelength as excitation source.

3 RESULT ANALYSIS

3.1 Effect of annealing temperature on $Mg_{0.15}Zn_{0.85}O$ film structure

Figure 1 illustrates the XRD diffraction spectrum of $Mg_{0.15}Zn_{0.85}O$ film after annealing treatment under temperatures of 0°C, 600°C, 700°C, 800°C and 900°C. As the figure shows, no peak position exists on the films without annealing treatment, which means the films are amorphous, while the films express as polycrystalline structure under four kinds of annealing temperature: 600°C, 700°C, 800°C and 900°C, and three diffraction peaks appear: (100), (002), and (101). Cubic phase and other impurity peaks associated with MgO do not exist in the diffraction spectrum and, compared with un-doped ZnO (Tang et al., 1997), all diffraction angles of the films shift to the large angle direction, which means Mg^{2+} has replaced Zn^{2+} and entered the ZnO lattice. The shift angles of diffraction peak (002) under annealing temperatures 800°C and 900°C are larger than under 600°C and 700°C, and strength decreasing, which illustrates that the crystal quality of the film under 800°C and 900°C is worse than under 600°C and 700°C. The annealing treatment (under temperatures of 800°C and 900°C) changes the ratio of Zn and Mg in film lead to (002) diffraction shift; therefore, the optimised growth temperature of this film should not exceed 700°C.

3.2 Effect of annealing temperature on $Mg_{0.15}Zn_{0.85}O$ film absorption spectrum

Figure 2 displays the absorption spectrum of $Mg_{0.15}Zn_{0.85}O$ film at room temperature. (The curves

correspond to the following annealing temperatures: a: 0°C, b: 600°C, c: 700°C, d: 800°C, e: 900°C). As Figure 2 shows, the absorption of all the films is very small in the visible region, while at UV range there exists strong edge absorption. The absorptivity rate of film increases with annealing temperature, but not by much. The absorption edge of the film shifts towards the longer wavelength side with increasing annealing temperature, namely red shift. This shows that the particle size of film increases as the annealing temperature rises.

3.3 Effect of annealing temperature on $Mg_{0.15}Zn_{0.85}O$ film photoluminescence

In order to study further the effect of annealing temperature on $Mg_{0.15}Zn_{0.85}O$ film optical quality, we have measured the photoluminescence of film under five different annealing temperatures, as displayed in Figure 3. As the figure shows, all films exhibit a UV emission peak at UV range; the corresponding peak positions are: a: 368.1 nm, b: 369.5 nm, c: 370.9 nm, d: 371.7 nm, e: 373.5 nm, and these emission peaks are mainly derived from the excitonic transition. Therefore, with the increasing annealing temperature, the UV emission of the sample shows a red shift; this is due to an increase in the sample's particle size during annealing, which is consistent with

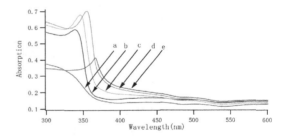

Figure 2. Absorption spectrum of $Mg_{0.15}Zn_{0.85}O$ films at different temperatures.

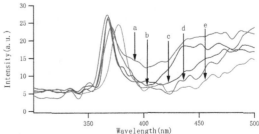

Figure 3. Photoluminescence spectrum of $Mg_{0.15}Zn_{0.85}O$ films at different temperatures.

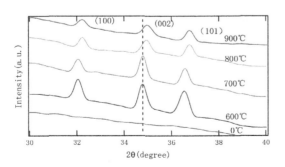

Figure 1. XRD spectra of $Mg_{0.15}Zn_{0.85}O$ films at different temperatures.

the analysis result of the absorption spectrum. In addition, we found all films have green emission peaks around 500 nm. There are different explanations as to the source of the green emission peaks (Deng et al., 2003) and, in our view, the green emission is caused by the composite between shallow donor level formed by oxygen vacancy (V_0) and shallow acceptor level formed by zinc vacancy (V_{Zn}).

4 CONCLUSION

By preparing $Mg_{0.15}Zn_{0.85}O$ film on quartz substrates using sol-gel technology, this paper focuses on the effect of annealing treatment on the structure and optical property of film. The result shows: (1) No peak position exists on the films without annealing treatment, which means the films are amorphous, while the films express as a polycrystalline structure under four kinds of annealing temperature: 600°C, 700°C, 800°C and 900°C, and three diffraction peaks appear: (100), (002), and (101). The crystal quality of film under 800°C and 900°C is worse than under 600°C and 700°C; (2) All films at UV range exhibit strong edge absorptions, and the absorption edges exhibit red shift as the annealing temperature rises; (3) All films have strong UV emission peaks, and the emission peaks exhibit red shift as the annealing temperature rises. Moreover, there are green emission peaks around 500 nm.

REFERENCES

Deng, H., Gong, B., Petrella, A.J. et al. (2003). Characterization of the ZnO thin film prepared by single source chemical vapor deposition under low vacuum condition. *Science in China Series (E), 463,* 255–360.

Lee, J-H., Yeo, B-W., & Park, B-O. (2004). Effects of the annealing treatment on electrical and optical properties of ZnO transparent conduction films by ultrasonic spraying pyrolysis. *Thin Solid Films, 457,* 333–337.

Lin, W., Ma, R., Shao, W. et al. (2007). Structural, electrical and optical properties of Gd doped and undoped ZnO: Al (ZAO) thin films prepared by RF magnetron sputtering. *Applied Surface Science, 253,* 5179–5183.

Muthukumar, S., Chen, Y., Zhong, J. et al. (2004). Metalorganic chemical vapor deposition and characterizations of epitaxial MgxZn1-xO (0≤x≤0.33) films on r-sapphire substrates. *Journal of Crystal Growth, 261,* 316–323.

Ohtomo, A., Kawasaki, M., Ohkubo, I. et al. (1999). Structure and optical properties of ZnO/Mg0.2Zn0.8O Superlattiees. *Applied Physics Letters, 75*(7), 980–982.

Tang, Z.K., Yu, P., Wong, G.L. et al. (1997). Ultraviolet spontaneous and stimulated emissions from ZnO microcrystallite thin films at room-temperature. *Solid State Communications, 103,* 459–462.

Zhang, X.T., Liu, Y.C., Zhang, L.G. et al. (2002). Structure and optically pumped lasing from nanocrystalline ZnO films prepared by thermal oxidation of ZnS Films. *Journal of Applied Physics, 92*(6), 3293–3298.

Zhao, D.X., Liu, Y.C., Fan, X.W. et al. (2001). Photoluminescence properties of MgxZn1- xO alloy thin films fabricated by the sol-gel deposition method. *Journal of Applied Physics, 90*(11), 5561–5563.

System test and diagnosis, monitoring and identification,
video and image processing

Automotive, Mechanical and Electrical Engineering – Liu (Ed.)
© 2017 Taylor & Francis Group, London, ISBN 978-1-138-62951-6

A novel remote sensing image segmentation algorithm based on the graph theory

Yirong Zhou
School of Management, University of Science and Technology of China, Hefei, China

Yansen Han
Department of Mathematics, Hefei University of Technology, Hefei, China

Ji Luo
Department of Mathematics Sciences, Zhejiang University, Hangzhou, China

ABSTRACT: Image segmentation is to divide the image into a number of specific, unique character-istics of the region and to put forward the technology and process of the target of interest. It is the key step of preprocessing in the regime of image recognition and computer vision. Without correct image segmentation, there wouldn't be correct identification. However, the only basis for segmentation is the brightness and color of pixels in the image. Therefore, when the image is automatically processed by the computer segmentation, a variety of difficulties will be encountered. Because remote sensing image has many characteristics such as high resolution and unequal noise distribution, the traditional segmentation algorithm cannot achieve satisfactory accuracy when analyzing these images. This paper proposes a new image segmentation framework based on the graph theory. During the recent years, the use of graph theory in many mature theories and mathematical tools for image segmentation has become a hot topic in the field of image segmentation research. The graph theory is a branch of applied mathematics, and have a good relationship with the mapping between the images. This paper also includes experimental result on the high resolution that is spatial resolution 0.1~0.3 m for remote sensing image.

Keywords: Region Connection Calculus; Image function; Convex association; Concave association

1 INTRODUCTION

On account of the assorted qualities of entities along with the expansion of their spatial deter-mination and drearily, the programmed investiga-tion of pictures stands as an urgent resource in the field of remote sensing. It is expected to outline and actualize new image investigation procedures which can perform complex image preparing in a productive way (Drăguţ, L., 2014; Wang, Haoxi-ang, and Jingbin Wang, 2014). These components come frequently from handling at a low level. This sort of items can't be productively portrayed by surfaces, edges, and so on (Forestier, 2012). This is because they are not prepared to hook up with lavishness and multifaceted nature of great deter-mination images.

Late productions show fascinating advances in the acknowledgment of specific protests as struc-tures or urban regions, yet this sort of methodolo-gies are not sufficiently nonexclusive to manage diverse sorts of articles. It is hence valuable to utilize procedures which can manage more elevated amounts of reflection for the representation and the control of the data contained in the images. In the work, the authors will utilize spatial thinking procedures with a specific goal to depict complex items (Inglada, J. 2007; Katartzis, A., & Sahli, H. 2008).

The scope of the paper includes the outstand-ing Region Connection. For any qualitative three-dimensional demonstration and intellectual, the Region Connection Calculus (RCC) is projected. Region Connection Calculus (RCC) abstractly describes regions in Euclidean space, or in a topological space by their possible relations to each other. RCC8 consists of 8 simple associa-tions. They are disconnected that is represented by DC, externally connected that is represented by EC, equal that is represented by EQ, partially overlapping that is represented by PO, tangential proper part that is represented by TPP, tangential

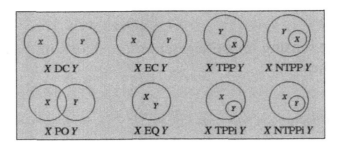

Figure 1. Instance for proper part.

o	DC	EC	PO	TPP	NTPP
DC	*	DC,EC,PO,TPP,NTPP	DC,EC,PO,TPP,NTPP	DC,EC,PO,TPP,NTPP	DC,EC,PO,TPP,NTPP
EC	DC,EC,PO,TPPi,NTPPi	DC,EC,PO,TPP,TPPi,EQ	DC,EC,PO,TPP,NTPP	EC,PO,TPP,NTPP	PO,TPP,NTPP
PO	DC,EC,PO,TPPi,NTPPi	DC,EC,PO,TPPi,NTPPi	*	PO,TPP,NTPP	PO,TPP,NTPP
TPP	DC	DC,EC	DC,EC,PO,TPP,NTPP	TPP,NTPP	NTPP
NTPP	DC	DC	DC,EC,PO,TPP,NTPP	NTPP	NTPP
TPPi	DC,EC,PO,TPPi,NTPPi	EC,PO,TPPi,NTPPi	PO,TPPi,NTPPi	PO,TPP,TPPi,EQ	PO,TPP,NTPP
NTPPi	DC,EC,PO,TPPi,NTPPi	PO,TPPi,NTPPi	PO,TPPi,NTPPi	PO,TPPi,NTPPi	PO,TPP,NTPP,TPPi,NTPPi,EQ
EQ	DC	EC	PO	TPP	NTPP

Figure 2. Consistency-based composition table.

proper part inverse that is represented by TPPi, non-tangential proper part that is represented by NTPP, non-tangential proper part inverse that is represented by NTPPi, From these basic relations, combinations can be built. For instance, proper part that is represented by PP is the union of Tangential Proper Part and non-Tangential Proper Part. Figure 1 shows the Instance for proper part. Figure 2 shows the Consistency-based composition table.

The numerical approach is chosen as light of the fact that, as opposed to direct multiscale approaches, it permits to choose protests as far as their size. The meter and sub-meter determination images are more and more popularly used for identifying vehicles, structures, and others (Zhong, P., & Wang, R. 2007). Novel calculation can deliver locales superimposing over the scales. This permits misusing the full degree of the Region Connection Calculus (RCC) with 8 fundamental connection set. The locales and their connections are spoken to by hubs speaking to the districts and the circular segments speak to the Region Connection Calculus (RCC) with 8 essential connection set (Yi, L., Zhang, G., &

Wu, Z. 2012). The Attributed Relational Graph can be made more entire by adding area ascribes to the hubs, with respect to occasion shape and radiometric highlights. At last, question discovery and acknowledgment can be viewed as a graph coordinating issue for which effective calculations can be executed. In these unique situations, a voracious pursuit joined with a graph metric which can utilize all the data contained in the ARG has been executed (Syed, A. H., Saber, E., & Messinger, D. 2012).

2 MULTISCALE IMAGE SEGMENTATION

Region Connection Calculus (RCC) with 8 fundamental connection set manages locales of space and how they associate and cover each other. Along these lines, the initial move toward the utilization of this framework is to acquire a reasonable arrangement of areas (Tison, C., Pourthié, N., & Souyris, J. C. 2007). The prerequisite for covering avoids a solitary layer segmentation technique, where every one of the areas would be separated.

The defining point is focused on the segmentation that is to take place at various levels.

The multiscale segmentation process is progressive. The details are incorporated at a level present in the low level, inappropriate for Region Connection Calculus (RCC) with 8 essential connections set (Tuia, D., Muñoz-Marí, J., & Camps-Valls, G. 2012; Yuan, J., Wang, D., & Li, R. 2014).

3 RECOMMENDED TECHNIQUE FOR SEGMENTATION

The loss along with the solitary layer technique is focused upon. Favorable circumstances of iterative, pyramidal edifice that is present are assumed. Extra powerful curve along with arched participation elements of second consideration is used for a better segmentation. The arched enrollment capacity for any given pixel is viewed as probability of pixel that is a piece of a curve that is better than encompassing foundation structure. Comparable description is connected with the curved that is darker when compared to encompassing foundation participation work. The leveling capacity L(i) is a disentanglement of image 'i'. The convex association utility $Convex_{au}(i)$ have the function of an image and the structuring element that is present in the opening of reconstructed operator is defined in equation 1.

$$Convex_{au}(i) = i - OR_n(i) \qquad (1)$$

The concave association utility $Concave_{au}(i)$ have the basic information of an image. The structuring element that is present in the closing of reconstructed operator is defined in equation 2.

$$Concave_{af}(i) = CR_n(i) - (i) \qquad (2)$$

The leveling function of an image as represented by L(i) is defined by having convex association utility $Convex_{au}(i)$ and concave association utility $Concave_{au}(i)$ in equation 3.

$$L(i) = \begin{cases} ORn(i): Convexmf(i) > Convexmf(i) \\ ORn(i): Convexmf(i) < Convexmf(i) \\ (i): Convexmf(i) = Concavemf(i) \end{cases} \qquad (3)$$

The segmentation scheme has been projected for which the already mentioned image function is used.

The segmentation scale is ranged from 1 to x. where the 'x' is total analysis number that is used in segmentation. The structuring element size is also considered. The image is segmented using the convex association utility Convexau(i) having the

function of an image and the structuring element that is present in the opening of reconstructed operator. The image is segmented using the concave association utility Concaveau(i).

The scope of scales can be picked by keeping in mind the goal to choose objects of a given size. Despite the fact that a choice control is proposed for the segmentation, in the means the authors utilize a fundamental thresholding system.

4 GRAPH MATCHING

Graph coordination has been broadly utilized for question acknowledgment and image investigation. For example, one can refer to the works of an analyst where graph coordination on a multi-determination order of nearby Gabor segments is utilized. A comparable approach is proposed. Similarly, the authors amplify this work where the various levelled graph coordinating takes into consideration extent along with invariant protest acknowledgment. Every hub in the graph speaks to an element and circular segments depict the separations between the elements. A Hopfield twofold system is utilized to perform sub-graph coordinating (Renz, J., & Ligozat, G. 2005). Concerning case, it constitutes of a decent prologue to protest acknowledgment by plot coordination. The plots are arranged in a novel manner by the authors. A diagram showing the representation of the graph-based methods is presented. This is done for preparing the image along with the examination that is given.

Be that as it may, the approach is constrained to the acknowledgment of the exceptionally same question and for extremely basic shapes (Sorlin, S., & Solnon, C. 2005). An examination of graph coordination and common data amplification for protest location limited was anticipated to the instance of multidimensional wavelet developed by Dennis Gabor (Wang, Jingbin, et al., 2015).

Inventiveness of approach exhibited takes a shot at graphs which depict subjective spatial connections, which will permit us to speak to classes of items, as opposed to numerous perspectives of a specific question. Three diverse methodologies, for example, the graph, arranged graph and non-situated graph to graph examination can be recognized.

Auxiliary method misuses finest graph topology deprived of issue subordinate, yet is somewhat hard towards actualizing. Figure 3 shows the effect of applying the levels of segmentation.

Original Image

Convex image

Concave image

Dark image

Bright Image

Figure 3.   Effect of applying the levels of segmentation.

5   EXAMINATION OF REGIONS

Fundamental calculations are at first considered in the investigation for an arrangement of areas. The utilization of the calculation of the relationship between couples of districts to every one of the sets of locales in the image might be extremely tedious. So as to lessen the quantities of processed connections, the authors will consider the symmetry properties of the connections. This permits

diminishing by a component of two the quantity of figured connections. Authors additionally utilize such a way the each scale is considered in the segment separately.

Technique is then investigated. Utilizing the figure advance strategies can be executed. Keeping in mind the goal to completely misuse this data, before registering the relationship between a locale and a district, the authors search for a middle of the road area for which the connections which interface it to what have as of now been figured. In the event that such an area exists, two cases may show up.

The learning of the as of now processed connections permits bounces the different level that are present in the choice tree. The investigation is done on the data of the middle of the road areas found goes on. The learning as of now figured connections permits explicitly deciding novel association with the calculation done.

In the event that towards the completion, the examination of transitional locales was found to be difficult in deciding what is required for the relationship. A calculation utilizing the techniques for the past segment is finished by utilizing all the data put away amid this pre-calculation step.

6 ASSOCIATED AND BI-ASSOCIATED COMPONENTS

With a specific goal to do that, the authors use various ideas such as associated graphs and bi-associated graphs. In the Associated Graph, the authors call a graph associated "AG" if for any couple of hubs, for example, n1 and n2 of 'AG', one can discover a way from n1 to n2 through an arrangement of circular segments and middle of the road hubs.

In Bi-associated Graph, the authors call a graph bi-associated "Pack" in the event that it is related and if the evacuation of a solitary hub makes it get to be not related. Authors are in this manner intrigued by dividing a graph in an arrangement of associated graphs of greatest size that is to which one can't include any hub deprived of losing the affiliation. The bi-association is stronger than association. In reality, associated graph acts naturally part of an arrangement of bi-associated graphs. The hubs which permit the association between the bi-associated components of graph are known as enunciation focuses. Figure 4 shows the Associated and Bi Associated components.

Since the items in the neighboring range might be stirred up with it, it is in this manner intriguing to break down the got graph into associated segments keeping in mind the end goal to segregate the graph of the question of intrigue.

In expansive image of the quick bird image, associated parts are not sufficient segregation of objects of intrigue. By utilizing bi-associated segments, a better outcome is obtained...

Additionally, the current connection is not same as DC. At long last, marks are given to the hubs and bends of the meta-graph with a specific end goal to point which hub and which curve is as of now part of the arrangement. Utilizing this structure, authors can exploit the properties of graphs on the off chance that one considers arrangements

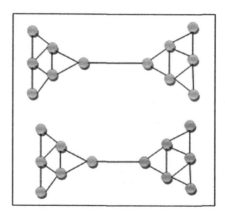

Figure 4. Associated and bi-associated components.

of size x, the neighboring arrangements of size x added to 1 are specifically available. To be sure, one just needs to consider the arrangement of meta-hubs connected by a circular segment to the hubs which as of now have a place with the arrangement. One can likewise ensure that an associated graph will be coordinated with another associated graph.

7 CONCLUSION

In the manuscript, the utilization spatial thinking methods with a specific goal to portray complex items is exhibited. The authors utilize the notable Region Connection Calculus (RCC) with 8 essential associations aiming at constructing a graph-based depiction connections among locales of an image. This approach empowers the execution of question acknowledgment calculations utilizing graph coordinating systems. The utilization of locale shape demonstrates the enthusiasm of presenting this sort of data in the graph coordinating methodology keeping in mind the end goal to create more powerful recognitions.

The contribution to the calculation is the consequence of a multi-scale segmentation in view of morphological profiles. It is demonstrated through test cases that the exhibitions of the question discovery are firmly subject to the nature of the segmentation. The calculation displayed here has a calculation many-sided quality which is good with genuine applications.

REFERENCES

Drăguț, L., O. Csillik, C. Eisank, and D. Tiede. "Automated parameterisation for multi-scale image segmentation on multiple layers." ISPRS Journal of Photogrammetry and Remote Sensing 88 (2014): 119–127.

Forestier, Germain, Anne Puissant, Cédric Wemmert, and Pierre Gançarski. "Knowledge-based region labeling for remote sensing image interpretation." Computers, Environment and Urban Systems 36, no. 5 (2012): 470–480.

Inglada, J. (2007). Automatic recognition of man-made objects in high resolution optical remote sensing images by SVM classification of geometric image features. ISPRS journal of photogrammetry and remote sensing, 62(3), 236–248.

Katartzis, A., & Sahli, H. (2008). A stochastic framework for the identification of building rooftops using a single remote sensing image. IEEE Transactions on Geoscience and Remote Sensing, 46(1), 259–271.

Region connection calculus. (2016, January 24). In Wikipedia, the Free Encyclopedia. Retrieved 00:27, January 24, 2016, from https://en.wikipedia.org/w/index.php?title=Region_connection_calculus&oldid=701345827.

Renz, J., & Ligozat, G. (2005, October). Weak composition for qualitative spatial and temporal reasoning. In International Conference on Principles and Practice of Constraint Programming (pp. 534–548). Springer Berlin Heidelberg.

Sorlin, S., & Solnon, C. (2005, April). Reactive tabu search for measuring graph similarity. In International Workshop on Graph-Based Representations in Pattern Recognition (pp. 172–182). Springer Berlin Heidelberg.

Syed, A. H., Saber, E., & Messinger, D. (2012, May). Encoding of topological information in multi-scale remotely sensed data: applications to segmentation and object-based image analysis. In Proceedings of International Conference on Geographic Object-Based Image Analysis (GEOBIA), Rio de Janeiro, Brazil (Vol. 79, p. 102107).

Tison, C., Pourthié, N., & Souyris, J. C. (2007, July). Target recognition in SAR images with Support Vector Machines (SVM). In 2007 IEEE International Geoscience and Remote Sensing Symposium (pp. 456–459). IEEE.

Tuia, D., Muñoz-Marí, J., & Camps-Valls, G. (2012). Remote sensing image segmentation by active queries. Pattern Recognition, 45(6), 2180–2192.

Wang, Haoxiang, and Jingbin Wang. "An effective image representation method using kernel classification." 2014 IEEE 26th International Conference on Tools with Artificial Intelligence. IEEE, 2014.

Wang, Jingbin, et al. "Multiple kernel multivariate performance learning using cutting plane algorithm." Systems, Man, and Cybernetics (SMC), 2015 IEEE International Conference on. IEEE, 2015.

Yi, L., Zhang, G., & Wu, Z. (2012). A scale-synthesis method for high spatial resolution remote sensing image segmentation. IEEE Transactions on Geoscience and Remote Sensing, 50(10), 4062–4070.

Yuan, J., Wang, D., & Li, R. (2014). Remote sensing image segmentation by combining spectral and texture features. IEEE transactions on geoscience and remote sensing, 52(1), 16–24.

Zhong, P., & Wang, R. (2007). A multiple conditional random field's ensemble model for urban area detection in remote sensing optical images. IEEE Transactions on Geoscience and Remote Sensing, 45(12), 3978–3988.

Automotive, Mechanical and Electrical Engineering — Liu (Ed.)
© *2017 Taylor & Francis Group, London, ISBN 978-1-138-62951-6*

A subpixel edge detection algorithm for the vision-based localisation of Light Emitting Diode (LED) chips

Yuanhong Qiu & Bin Li
School of Mechanical Science and Engineering, Wuhan, China

ABSTRACT: LED chip localisation is an important process during the testing and sorting of LED chips. Many template matching have been widely used for LED chip localisation. However, these methods require that the template image is artificially made, which may lead to misjudgement due to human fatigue. In order to overcome these shortcomings, we proposed an edge detection-based algorithm to detect the pose (position and orientation) of an LED chip automatically. The algorithm consists of three steps. Firstly, smooth the image histogram by Gaussian filter to segment the LED chip from background. Meanwhile, apply the image pyramid method to reduce computation. Secondly, according to morphology operator on the LED chip region to obtain the Region of Interest (ROI) of edge detect and detect the subpixel edge precisely. Finally, take the RANdom SAmple Consensus (RANSAC) algorithm as weighting factors to fit the subpixel edge and obtain the pose of LED chip. The results shows that the proposed approach successfully locates the pose of the LED chip. Personnel costs and misjudgement due to human fatigue can be reduced using the proposed approach.

Keywords: Blob analysis; Edge detection; Curve fitting; Position

1 INTRODUCTION

With the rising importance of global environmental protection in recent years, the Light Emitting Diode (LED) has been replacing conventional lamps owing to its excellent characteristic of environmental friendliness. Regarded as the fourth generation of green lighting source (Zheludev, N, 2007). Vision-based location is an important machine-vision technology and it is widely applied in LED manufacturing to detect the accuracy of LED chip sorting.

In recent years, a plethora of template matching algorithms have been applied for the object localisation (L. Di Stefano et al., 2003), such as the sum of squared differences (K. Nickels et al., 2002), the sum of absolute difference (V.-A. Nguyen et al., 2006), and the Normalised Cross Correlation (NCC) (K. Briechle et al., 2001). The NCC based template matching algorithm is robust to illumination variation and locates the object with subpixel precision. It has therefore been widely used in LED chip localisation. Tsai et al. (2003) used the NCC method to position the LED chip and reduce the runtime complexity by applying the image pyramid method. Buyang Zhang et al. (2015) proposed a region-based normalised cross correlation algorithm to position the elongated chip, that traditional NCC method are not effective (Zhong, F. et al., 2015). Xinyu Xu et al. (2014) propose a new object matching method that improves efficiency over existing approaches by

decomposing, orientation and position estimation into two cascade steps. It achieves a good performance. All these algorithms compare a template image with the target image to find the best match. However, the template image requires producing the image, which is cropped from a reference image by an experienced human. But human error may affect the accuracy of localisation.

Therefore, this paper presents an edge detection-based algorithm for LED chip localisation. The remainder of this paper is organised as follows. The second section describes the edge detection-based algorithm. The third section of the paper presents the experimental results. Conclusions are drawn in the last section.

2 PROPOSED ALGORITHM

In this section, an edge detection-based algorithm for the LED chip localisation is described in detail. The goal of the algorithm is to detect the pose (position and orientation) of the LED chip with high speed and high pose accuracy.

2.1 *Image segment*

The acquired LED chip image contained multiple LED chips as shown in Figure 1 (a). An LED chip consists of two main components: the light-emitting area and the electrode area. In our attempt to get the pose of the LED chip, we segment the LED chip from the background at first. In this paper, we propose an automatic segmentation method based on image histogram smoothing, first, calculating the image histogram and then convoling it with a Gaussian filter, the function is defined in Eq. (1).

$$H_{smoothed} = H * G \qquad (1)$$

$$G = \frac{1}{\sigma\sqrt{2\pi}} e^{\frac{-s^2}{2\sigma^2}}$$

where H is the original grey histogram, $H_{smoothed}$ is the smoothed grey histogram convoled with a Gaussian filter G, σ is the smoothing parameter. The larger the chosen value of sigma, the fewer regions will be extracted, In order to find the suitable sigma automatically, a good strategy is to increase the value of sigma until two unique local minimums are obtained. Figure 1(b) shows the grey histogram smoothed result (N. Orlov et al., 2007).

After image segmentation, the holes and indentations are at all inevitable that will affect the position accuracy. It is therefore necessary to use a closing operation to smooth the boundary of the region as well as blob analyses to select the LED chip region from the background. The feature of each blob contained area and rectangularity feature can be calculated as the selection criteria (L. Zhao

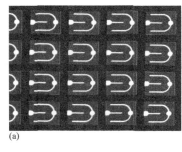

(a)

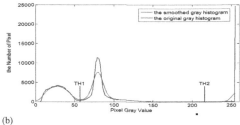

(b)

Figure 1. LED image and grey histogram: (a) the acquired LED image, (b) the original and smoothed grey histograms.

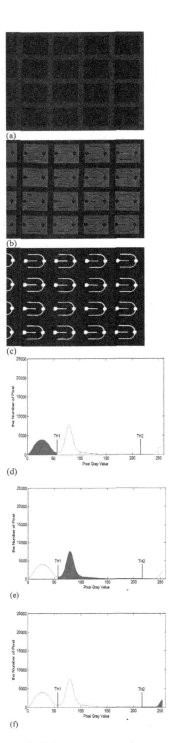

Figure 2. LED image segmentation result and corresponding histogram: (a) Background; (b) light-emitting area; (c) electrode area.

et al., 2005). The area feature is simply the number of points in the region after image segmentation, if the value is set between 2000 and 99999 according to the specification of the LED chip. The rectangularity feature calculates the similarity of the rectangle of the input regions, in order to exclude the noise. The threshold in this paper is set from 0.5 to 1. Figure 2 shows the image segmentation.

In the end, for the sake of reducing the computation time of the image segmentation, the image pyramid method was used. The smaller the segment region is, the less time the algorithm consumes.

2.2 *Subpixel edge extraction*

Edge detection is a fundamental task in many image processing applications such as motion analysis and pattern recognition, The widely used algorithms of edge extraction are Sobel, Robel, Prewitt and Canny operators. Especially the Canny operators, almost the most occasion for edge extraction precisely are applied to it. In this paper, the Canny operators are used for the edge detection of the LED chip (Canny 1986). Because the edge extraction is relatively costly, it is necessary to reduce the search space. The morphology method is therefore applied on the blobs after the image segmentation, a circle of diameter 3 is used as the basic structuring element to dilate and erode the boundary, we then calculate the difference between the two regions to obtain the ROI for the edge extraction as shown in Figure 3(a). The region between the two green contours is the ROI we need.

Extracting the edges with pixel accuracy is sufficiently accurate for LED chip position detection in the industrial manufacturing. Subpixel precision is required. To extract edges with subpixel accuracy, the paper presents the subpixel edge detection based on a polynomial three times fitting (Feipeng et al., 2010). The algorithm first fetches some points near the edge, then gets the grey level of these points, and fits the grey curve through the polynomial curve three times by Eq. (2), the edge detection results are shown in Figure 3 (b) and (c).

$$f(x) = ax^3 + bx^2 + cx + d \qquad (2)$$

$$S = \sum_{i=1}^{n} (y_i - \hat{y})^2 =$$
$$\sum_{i=1}^{n} [y_i - (ax^3 + bx^2 + cx + d)]^2 (i = 1, 2 \ldots, n)$$

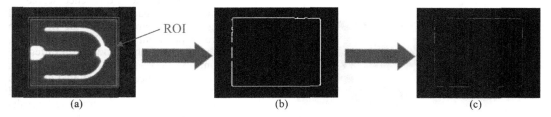

Figure 3. Subpixel edge detection process: (a) obtain the ROI of edge detection. (b) edge detection result. (c) subpixel edge detection result.

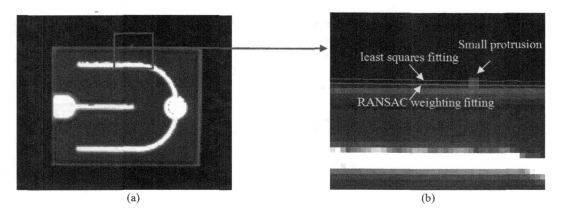

Figure 4. The fitting image result: (a) the LED chip image with small protrusion. (b) the fitting result of the least squares fitting and RANSAC weighting fitting.

2.3 Fitting the subpixel edge

The subpixel edge extraction will create an enormous data owing to the noise around the edge. Therefore, the paper takes the line fitting method to detect the boundary of the LED chip accuracy. The formula representing the boundary is given as follows:

$$\alpha r + \beta c + \gamma = 0 \tag{3}$$

To fit the boundary through a set of points (r_i, c_i), $i = 1, ..., n$ and minimise the sum of the squared distances of the points to the boundary. However, in practice, when the LED chip boundary contains a small protrusion, the fitting boundary will deviate from the actual boundary and severely affect the accuracy of the LED chip localisation. To reduce the influence of the small protrusion edge point, we introduce a weight w_i for each point. What is more, the weight should be smaller than one for the protrusion edge point. The paper applies the RANdom SAmple Consensus (RANSAC) algorithm to define the weights w_i (Douxchamps et al., 2009), a method which constructs a solution from the minimum number of points and checks how many points are consistent with the solution, continues the process until the solution with the largest number of consistent points is found.

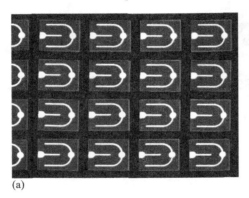

(a)

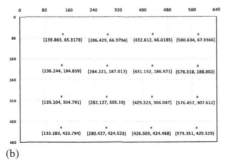

(b)

Figure 5. The image localisation result: (a) the subpixel edge detection result. (b) the pose of the LED chip.

After fitting the boundary in the subpixel accuracy, then calculating the pose of the LED chip according to Eq. (4), the result of the LED chip localisation result is shown in Figure 5.

$$\bar{x} = \frac{\int_A x dA}{A} \quad \bar{y} = \frac{\int_A y dA}{A} \tag{4}$$

3 RESULTS AND DISCUSSION

To evaluate the effectiveness of the proposed algorithm, various experiments which test the accuracy and the time consumption were conducted. In the paper, we used a camera (resolution: 640*480, pixel size 4.7 um*4.7 um) and industrial lens to acquire the LED chip image. The proposed algorithm was programmed in the Visual Studio 2010 environment. We describe some preliminary experiments to evaluate the proposed algorithm.

3.1 Position accuracy tests for proposed algorithms

In the actual manufacture, the LED chips were attached to a blue film with difference degree, so the test images were made to simulate the actual LED image. Firstly, some parts of the LED chip image were cropped out as the LED chip model, then rotated the LED chip model with difference degree from −3° to 3°, the rotation step is 0.5°, then locate the difference degree rotated model to the appointed position in advance. Figure 6 and Table 1 show the position accuracy test results of the the proposed method. The detect rate can reach 100% with the accuracy below 0.25 pixel.

3.2 Speed tests for the proposed algorithms

We obtain the LED chip image for five times in repeat; calculate the speed for our proposed algorithms. As Table 2 shows, the speed for positioning the LED chip applied the proposed algorithm was 43 ms in average.

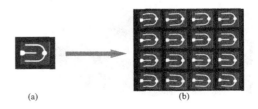

(a)   (b)

Figure 6. Position accuracy test image. (a) LED chip model, (b) rotate and translate the chip model to the position in advance.

Table 1. Position accuracy test result.

No.	x	$\bar{x}$	Δx	y	$\bar{y}$	Δy
1	60.1537	60	0.1537	95.2006	95	0.2006
2	60.1598	60	0.1598	245.202	245	0.202
3	60.1731	60	0.1731	395.193	395	0.193
4	60.1712	60	0.1712	545.177	545	0.177
5	180.167	180	0.167	95.1859	95	0.1859
6	180.174	180	0.174	245.183	245	0.183
7	180.191	180	0.191	395.189	395	0.189
8	180.178	180	0.178	545.152	545	0.152
9	300.194	300	0.194	95.1864	95	0.1864
10	300.213	300	0.213	245.17	245	0.17
11	300.215	300	0.215	395.155	395	0.155
12	300.216	300	0.216	545.148	545	0.148
13	420.217	420	0.217	95.1364	95	0.1364
14	420.223	420	0.223	245.131	245	0.131
15	420.226	420	0.226	395.116	395	0.116
16	420.238	420	0.238	545.121	545	0.121

Table 2. Speed test result.

	1	2	3	4	5	6	7	8	Average
Speed (ms)	45.2022	44.4244	43.8599	42.441	52.2087	41.5881	39.9601	39.5495	43

4 CONCLUSION

In the paper, we proposed a subpixel edge detect algorithm for the vision-based positioning of LED chips. The proposed algorithm consists of three phases. Firstly, a smoothed grey histogram was used to segment the LED chip image with the image pyramid method to reduce the computation. Secondly, the subpixel edge detection method was applied to get the boundary edge of the LED chip, a morphology operator could obtain the ROI to reduce the computation time. Thirdly the RANSAC algorithm was applied to fit the edge to get the pose of the LED chip. The experimental analysis demonstrates that our proposed algorithm results can locate the LED chip image effectively with a position accuracy below 0.25 pixel, and the average time consumed is 43 ms.

REFERENCES

Briechle K., and U.D. Hanebeck, (2001). Template matching using fast normalized cross correlation. *Proc. Aerosp. Defense Sens. Simulat. Controls*, Orlando, FL, USA, 95–102.

Canny J.A. (1986). Computational approach to edge detection. *IEEE Trans. Pattern Anal. Mach. Intell. 1986 8(6)*, 679–698.

Di Stefano L., and S. Mattoccia, (2003). Fast template matching using bounded partial correlation. *Mach. Vis. Appl., 13(4)*, 213–221.

Douxchamps, D., and K. Chihara, (2009). High-accuracy and robust localization of large control markers for geometric camera calibration. *IEEE Trans. Pattern Anal. Mach. Intell. 31 (2)*, 376–383.

Feipeng, D., and H. Zhang, (2010). Sub-pixel edge detection based on an improved moment. *Image Vis. Comput. 28 (12)*, 1645–1658.

Nguyen V.-A., and Y.-P. Tan, (2006). Efficient block-matching motion estimation based on integral frame attributes. *IEEE Trans. Circuits Syst. Video Technol., 16(3)*, Mar., 375–385.

Nickels K., and S. Hutchinson, (2002). Estimating uncertainty in SSD-based feature tracking. *Image Vis. Comput., 20(1)*, 47–58.

Orlov, N., J. Johnston, T. Macura, L. Shamir, I. Goldberg, (2007). Computer vision for microscopy applications, In: Vision Systems – Segmentation and Pattern Recognition. ARS Press. 221–242.

Tsai, D.M., and Lin, C.T. (2003). Fast normalized cross correlation for defect detection. *Pattern Recognition Letters, 24*, 2625–2631.

Xu, X., P. van Beek and X. Feng (2014). High-speed object matching and localization using gradient orientation features. *Proc. SPIE 9025, Intelligent Robots and Computer Vision XXXI: Algorithms and Techniques, 902507.*

Zhao, L., and L.S. Davis (2005). Closely coupled object detection and segmentation. *Tenth IEEE International Conference on Computer Vision (ICCV'05), Volume 1*, 454–461.

Zheludev, N. (2007). The life and times of the LED — A 100-year history. *Nature Photonics, 1(4)*, 189–192.

Zhong, F., S. He and B. Li, (2015). Blob analyzation-based template matching algorithm for LED chip localization. *The International Journal of Advanced Manufacturing Technology.*

Automotive, Mechanical and Electrical Engineering – Liu (Ed.)

Active learning framework for android unknown malware detection

Hua Zhu

College of Engineering, University of Michigan, Ann Arbor, MI, USA

ABSTRACT: There are a lot of unknown labels Android examples in real-world applications, and they will cost much to mark manually. In this paper, we propose an active learning framework to solve this problem, so that Android malware are detected. In the active learning framework, Naive Bayes (NB), Decision Tree (DT), Logistic Regression (LR) and Support Vector Machines (SVM) are used to mark the labels for Android examples. The results indicate that this approach is effective to detect Android malware.

Keywords: Android; Malware static detection; Feature selection; Particle Swarm Optimization; Information Gain

1 INTRODUCTION

In order to mitigate the threat on mobile device, various efforts have been made to detect and analyze malicious applications. There are three main approaches: static analysis, dynamic analysis and hybrid analysis. Static analysis approach is implemented through the source code without the execution of Android malware, which is used by Enck et al. (Enck W, 2009) and Felt et al. (Felt A P, et al., 2011). Dynamic detection works through monitoring the execution of Android malware activity at runtime, which is used in Crowdroid (Burguera I, 2011) and TaintDroid (Enck W, et al., 2014). Hybrid analysis methods combine static and dynamic techniques.

Because Android malicious applications grow rapidly and emerge into various kinds, traditional detection methods cannot identify new unknown malicious applications. It is necessary to apply some mechanism based on active learning framework to detect new unknown malicious applications and ensure high efficiency and accuracy (Felt A P, et al., 2011).

Active learning model mainly imitates people to learn. It is through a certain method to extract the most informative data for a sample, and then the labels are manual marked on these selected sample data, which can train classifier with more abundant information. Next, the active learning classifier marks other the unlabeled data samples, so it can reduce the cost of artificial label. Most important of all, the active learning framework can effectively solve the problem of the lack of data (Freeman S, et al., 2014; Settles B. 2010).

According to the unlabeled samples in different ways, active learning algorithms are divided into two types: based on stream and based on pool. In the stream-based model, the unlabeled samples sequentially are submitted to the selection engine, which determines whether the current annotation submitted the samples. If the samples are not marked, they are discarded. In the pool-based model, selection engine maintains a set of unlabeled examples, and selects the unlabeled samples (Millis B J. 2012). Popular machine learning selection engine methods include k-Nearest Neighbor (KNN), Naive Bayes (NB), Decision Tree (DT), Logistic Regression (LR), Support Vector Machines (SVM), Adaboost, k-Means, and more. They can identify potentially malicious applications with high probability (Wang Y, et al., 2013; Aafer Y, 2013).

This paper mainly focuses on malware detection based on active learning model. When new unknown applications are coming, they are processed according to the extracted features. Then several machine learning methods are used to predict the class labels for the new applications. If some ones have the same predicted results from different machine learning methods, they are added to the original training dataset to generate a new training dataset. If not, the ones are testing letter with the new training model, which is from the new training dataset.

2 METHODOLOGY

The proposed model based on active learning framework, as shown in Fig. 1, consists of three major parts. The first one is detection models trained using the original training sample dataset, which containing two steps: extracting features from decompiled samples, and training detection models using machine learning methods. The second one is predicting the new coming samples without class labels based on the

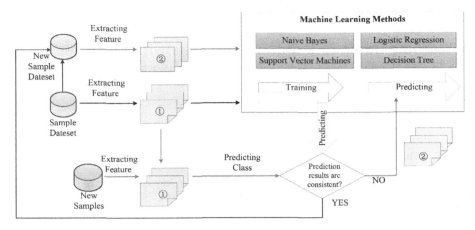

Figure 1.   Structure of android malware online detection model.

trained detection models. If the prediction results of the new samples from all the trained detection models are consistent, the new samples are viewed as high quality samples and added to the original training dataset to create a new training dataset. Otherwise, the controversial samples are taken as the test dataset. The third part is training new model with the new training dataset and testing the test dataset. In this part, the procedure of processing new training dataset contains extracting features. In addition, the used machine learning methods are NB, LR, SVM and DT. They all have good performance for malware detection (Tong S, 2001; Freeman S, et al., 2014).

3   EXPERIMENT AND DISCUSSION

3.1   *Sample dataset for experiments*

The datasets we use are described in Table 1. The dataset contains benign applications and malicious applications. Benign applications are downloaded from Google Play Store in March 2015. Malicious applications are collected from Tencent safety lab and http://sanddroid.xjtu.edu.cn:8080/#overview in August 2015, covering many kinds of malware families. At last, the size after clearing shown in Table 1.

Make sure a thorough comparison, we create two datasets. Dataset S1 is the training dataset, containing 700 benign and 680 malicious applications. Dataset S2 is the testing dataset, containing 144 benign and 120 malicious applications. At first, Dataset S1 is used to train models and Dataset S2 is used as testing dataset. According to the testing result, Dataset S2 is separated into two parts according to the consistency of the predicted results. One is added to Dataset S1 and becomes Dataset S3 for training, and the other one turns into Dataset S4 for testing.

Features are extracted not only from Android permissions but also from SDK API. In the experi-

ments, 106 Android permissions are extracted. After statistics, we find malicious applications clearly tend to request more SMS-related permissions, such as READ_SMS, WRITE_SMS, RECEIVE_SMS, and SEND_SMS. Moreover, we scan all. smali files to collect 9342 Android SDK functions. Totally, the features are 9448.

3.2   *Statistics for the coincident predicted result with training dataset*

In this part, Dataset S1 is taken as the training dataset and Dataset S2 is taken as the testing dataset. The class labels of samples in Dataset S2 are unknown. Firstly, we use the Dataset S1 to train classification model. Then the trained model is applied to predict the class labels of Dataset S2. In addition, SVM, LR, DT, and NB are used to be classification models. The predicting procedures are running 10 times, and we record the results every time, containing the number of predicted benign and malware applications.

Firstly, we record the predicted result every time. Then, we count the coincident results of each classifier running 10 times, containing the number of the samples predicted as benign samples, and the number of the samples predicted as malicious samples. Finally, we get the coincident results of all classifiers, containing the 85 samples are predicted as benign samples, 80 the samples are predicted as malicious samples, and 99 samples are not predicted to coincident results, as shown in Table 2. As a result, we add the samples with coincident predicted result to Dataset S1 to generate Dataset S3, and let the samples without coincident predicted results to generate Dataset S4.

Comparing the true class labels of Dataset S2, we find the predicted coincident results keep the consistency with the truth. It means that there is no samples predicted wrong in the coincident results.

3.3 *Performance of the new training dataset*

In this part, we will further detect the samples based on the samples without coincident predicted results in last part. Dataset S3 is taken as the training dataset and Dataset S4 is taken as the testing dataset. Dataset S3 is processed like Dataset S1, including extracting feature, training classification models, and predicting the class labels of the samples in Dataset S4. The predicting procedures are running 10 times, and we record the observed results every time, containing the number of predicted benign and malware applications.

In Table 3, we record the predicted coincident results of each classifier and all four classifiers. At last, there are 69 samples predicted as benign ones, no one predicted as malicious ones, and 30 samples are predicted without coincident results.

Comparing with the true class labels of Dataset S3, we can get the correct predicted results of Dataset S3, as shown in Table 4. We find that the number of correct predicted benign samples

decreases. In the predicted coincident results, there are 39 malicious samples incorrectly predicted as benign ones. And no malicious samples correctly predicted as malicious ones. On other hand, SVM, LR and DT are better than NB.

3.4 *Performance of the new training dataset*

According to the correct predicted results of in Table 2 and Table 4, we can get the whole correct predicted results of Dataset S2, as shown in Table 5. Meanwhile, we can get the predicted accuracy based on the size of the correct predicted results divided by the size of Dataset S2.

In Table 5, we count the number of samples in Dataset S2, which are predicted correctly by any SVM, LR, DT and NB classifiers, and the number of the coincident results, which are predicted correctly by SVM, LR, DT and NB classifiers. The correct predicted applications contain 124 benign and 80 malicious applications. And the incorrect predicted applications contain 20 benign and 40 malicious applications.

Table 1. Descriptions of dataset and experimental settings.

Dataset	Set name	Content	For training/testing	Size after cleaning	Description
S1	B1	Benign	Training	644	Collected from Google Play Store in March 2015
	M1	Malware	Training	600	Collected from Tencent Safety Lab in August 2015
S2	B2	Benign	Testing	200	Collected from Google Play Store in March 2015
	M2	Malware	Testing	200	Collected from http://sanddroid.xjtu.edu.cn:8080/ #overview in October 2015 in August 2015
S3	S1 + S2(part)	Benign + Malware	Training	–	One part of S2 is the coincident predicted results from S1 training model
S4	S2 (the other part)	Benign + Malware	Testing	–	The other part of S2 is the inconsistent predicted results from S1 training model

Table 2. The observed coincident number of the predicted benign and malware applications after predicting procedure running 10 times.

The predicted class	SVM	LR	DT	NB	Coincident number	Inconsistentnumber
Benign	122	130	112	99	**85**	–
Malware	104	107	91	104	**80**	–
total	226	237	203	203	**165**	**99**

Table 3. The observed coincident number of the predicted benign and malware applications.

The predicted class	SVM	LR	DT	NB	Coincident number	Inconsistent number
Benign	97	95	99	69	**69**	–
Malware	0	0	0	22	**0**	–
total	97	95	99	91	**69**	**30**

Table 4.　The correct predicted results of Dataset S3.

The correct predicted class	SVM	LR	DT	NB	Coincident number	Inconsistent number
Benign	58	57	59	40	**39**	**20**
Malware	0	0	0	8	**0**	**40**
total	58	57	59	48	**39**	**60**

Table 5.　The correct predicted results of Dataset S2.

The correct predicted class	SVM	LR	DT	NB	Coincident number	Inconsistent number
Benign	143	142	144	125	**124**	**20**
Malware	80	80	80	88	**80**	**40**
total	223	222	224	213	**204**	**60**

Table 6.　The correct predicted accuracy of Dataset S2.

The correct predicted accuracy	SVM	LR	DT	NB	Coincident accuracy	Error rate
Accuracy of benign	0.9931	0.9861	1.0	0.8681	**0.8611**	**0.1389**
Accuracy of Malware	0.6667	0.6667	0.6667	0.7333	**0.6667**	**0.3333**
Accuracy	0.8447	0.8409	0.8485	0.8069	**0.7727**	**0.2273**

In Table 6, we list the correct predicted accuracy of Dataset S2. The coincident accuracy from SVM, LR, DT and NB classifiers is taken as the final accuracy of malware detection model, which is a conservative estimate. As shown in Table 6, we get that the accuracy of the Dataset S2 is 77.27%, the accuracy of the benign applications in Dataset S2 is 86.11%, and the accuracy of the malicious applications in Dataset S2 is 66.67%. In addition, among SVM, LR, DT and NB classifiers, we find DT can correctly predict benign applications with the highest accuracy, and DT gets the best performance of the whole Dataset S2. However, NB prefers to predict benign applications correctly.

4　CONCLUSION

With the development of mobile devices, Android security attracts much attention, and Android malware detection is regarded as an important research task. In this paper, we propose a model for Android malware static detection with feature selection algorithms, which uses active learning containing several machine learning methods. Then we carry out series of experiments to indicate that the active learning framework can effectively detect Android malware.

REFERENCES

Aafer Y, Du W, and Yin H. DroidAPIMiner: Mining API-level features for robust malware detection in Android [M]//Security and Privacy in Communication Networks. Springer International Publishing, 2013: 86–103.

Burguera I, Zurutuza U, Nadjm-Tehrani S. Crowdroid: behavior-based malware detection system for Android [C]//Proceedings of the 1st ACM workshop on Security and privacy in smartphones and mobile devices. ACM, 2011: 15–26.

Enck W, Gilbert P, Han S, et al. TaintDroid: an information-flow tracking system for realtime privacy monitoring on smartphones [J]. ACM Transactions on Computer Systems (TOCS), 2014, 32(2): 5.

Enck W, Ongtang M, McDaniel P. On lightweight mobile phone application certification [C]//Proceedings of the 16th ACM conference on Computer and communications security. ACM, 2009: 235–245.

Felt A P, Chin E, Hanna S, et al. Android permissions demystified [C]//Proceedings of the 18th ACM conference on Computer and communications security. ACM, 2011: 627–638.

Freeman S, Eddy S L, McDonough M, et al. Active learning increases student performance in science, engineering, and mathematics [J]. Proceedings of the National Academy of Sciences, 2014, 111(23): 8410–8415.

Freeman S, Eddy S L, McDonough M, et al. Active learning increases student performance in science, engineering, and mathematics [J]. Proceedings of the National Academy of Sciences, 2014, 111(23): 8410–8415.

Millis B J. Why Faculty Should Adopt Cooperative Learning Approaches [J]. Cooperative Learning in Higher Education: Across the Disciplines, Across the Academy, 2012: 1.

Settles B. Active learning literature survey [J]. University of Wisconsin, Madison, 2010, 52(55–66): 11.

Tong S, Koller D. Support vector machine active learning with applications to text classification [J]. Journal of machine learning research, 2001, 2(Nov): 45–66.

Wang Y, Zheng J, Sun C, et al. Quantitative security risk assessment of Android permissions and applications[M]//Data and Applications Security and Privacy XXVII. Springer Berlin Heidelberg, 2013: 226–241.

Automotive, Mechanical and Electrical Engineering – Liu (Ed.)
© 2017 Taylor & Francis Group, London, ISBN 978-1-138-62951-6

An accelerated testing method based on similarity theory

Shumin Li
Key Laboratory of High Performance and Complex Manufacturing, Central South University, Changsha, China
College of Mechanical and Electrical Engineering, Central South University, Changsha, China

Yingying Zhang
Key Laboratory of High Performance and Complex Manufacturing, Central South University, Changsha, China
Jinan Kunrui Optical Science and Technology Co. Ltd., Jinan, China

Ailun Wang
Key Laboratory of High Performance and Complex Manufacturing, Central South University, Changsha, China
College of Mechanical and Electrical Engineering, Central South University, Changsha, China

ABSTRACT: Aiming at the problem that accelerated testing can often not be executed in the prototype and is always at high cost, a new accelerated testing method based on similar theory is proposed. By ensuring the similarity of geometry, mechanics, kinematics and service environment between prototype and model, a small-scale model can be established, and an accelerated testing of this small-scale model can be conducted in the laboratory. A similarity model with single parameter distortion or multiple parameter distortion can be built to shorten life by means of increasing certain stress to deteriorate the model service environment. Finally, through the rod bolt accelerated life example under alternating load, the correctness of the proposed method is verified.

Keywords: Accelerated life testing; Model testing; Similarity theory; Distortion model

1 INTRODUCTION

In the general case, there are two kinds of failure modes, respectively sudden failure mode and degradation failure mode. So accelerated testing could be identified as accelerated life testing and accelerated degradation testing (Chen X. & Zhang C. H. 2013; Deng A. M. & Chen X., 2007; Yurkowsky W. S. et al., 1967; Zhang C. H. & Wen X. S., 2004). For sudden failure mode, product failure time can be derived from accelerated life testing. For degradation mode, product degradation data can be derived from accelerated degradation testing (Chen X. & Zhang C. H., 2013).

Essentially, life testing is time statistics of life actual consumption, which makes little significance of the use and maintenance of equipment, because of the long test cycle and high cost (Chen X. & Zhang C. H., 2013). As accelerated testing can derive life data before the end of real life, judgment of product life can be acquired in advance. In other words, accelerated testing has the ability to predict real life (Chen X. & Zhang C. H., 2013; Deng A. M. & Chen X., 2007; Yurkowsky W. S. et al., 1967; Zhang C. H. & Wen X. S., 2004). The statistical analysis of accelerated testing began in the 1960s, and constant stress testing was developed first, as

it was simple. With the establishment of statistical analysis for constant stress testing, the application of constant stress testing also saw large development (Zhang C. H. & Wen X. S., 2004). Accelerated life testing and accelerated degradation testing have been defined and researched systematically by Chen Xun (Chen X. & Zhang C. H., 2013). The inference method of dynamic linear model has been proposed by Mazzuchi T. A. to improve the analysis precision of constant stress testing (Mazzuchi T. A. & Soyer R., 1990). Threshold stress was introduced by non-linear accelerated testing to improve the analysis precision of constant stress testing (Hirose H., 1993). The solving model of Weibull constant stress testing analysis has been simplified by Watkins A. J. (Bugaighis M. M., 1995). The analysis precision of statistics has been improved by introducing constraint relation of life distribution which is of a different magnitude (McLinn J. A., 1999). The WK-MLE numerical solution method has been established by Wang W. in order to solve the initial value sensitivity problems of Weibull accelerated model maximum likelihood estimation numerical iterative method (Wang W & Kececioglu D. B., 2000).

The above scholars mainly researched the accelerated testing stress loading method of product or equipment and how to transform life data from high stress condition to normal stress condition. However, there still exists a problem in accelerated testing in that much of it cannot be carried out on work site. On the one hand, the service environment is too complex and severe to collect life testing data. On the other hand, if life testing data is collected on equipment directly, the equipment will scrap too early and costs will be incalculable. Aiming at the above disadvantages, a new accelerated testing method will be proposed in this paper to overcome the problems which exist in current accelerated testing methods.

2 ACCELERATED TESTING BASED ON SIMILARITY THEORY

2.1 *Physical quantities which affect life*

Product failure is always determined by some physical quantities, such as temperature, pressure, frequency, speed, current and voltage etc. which are called stress variables in this paper. Stress variables can reflect the service degree of environment where the equipment operates. Furthermore, life can also be influenced by geometric parameters and some materials properties, such as elastic modulus, strength limit and yield limit etc.

2.2 *Proposing π equation of accelerated testing based on similarity theory*

In classical similarity theory, similarity should be the same in content and equation description (Yurkowsky W. S. et al., 1967). For instance, a product degradation data can be represented by η, and η is influenced by stress variables $A_1, A_2, \cdots, A_{n1}$, geometric parameters $B_1, B_2, ..., B_{n2}$, materials properties $C_1, C_2, ..., C_{n3}$ and service time t. If the total number of the physical quantity is n and there are k independent physical quantities, then similarity criteria $\pi_1, \pi_2, ..., \pi_{n-k}$ will be derived by using classical similarity theory. Let π_1 be π term which includes degradation data η. Let $\pi_2, \pi_3, ..., \pi_{n-k}$ be π term which includes above the stress variable. Finally, π equation of life can be written as:

$$\pi_1 = f\left(\pi_2, \pi_3, ..., \pi_{n-k}\right) \tag{1}$$

As long as the prototype and model have the same life π equation, they are similar in life.

3 RESEARCH ON A COMPLETELY SIMILAR LIFE TESTING METHOD

Complete life similarity refers to the model and prototype with the same π term and π equation. This is:

$$\begin{cases} \pi_{2m} = \pi_{2p} \\ \vdots \\ \pi_{n\text{-}km} = \pi_{n\text{-}kp} \\ \pi_{1m} = f\left(\pi_{2m}, \pi_{3m}, ..., \pi_{n\text{-}km}\right) \\ \pi_{1p} = f\left(\pi_{2p}, \pi_{3p}, ..., \pi_{n\text{-}kp}\right) \end{cases} \tag{2}$$

Annotation: in this π equation, life variables are in π_1 and other physical quantities are in $\pi_2, \pi_3, ..., \pi_{n-k}$. π_1 is called the dependent π term, $\pi_2, \pi_3, ..., \pi_{n-k}$ are called the independent π term.

For example, η is the degradation quantity of a product, its π term can be written as $\pi_1 = \frac{\eta}{M_1}$, and M_1 is an expression which has the same dimension with composed with η. The independent π term with time variable t is $\pi_i = \frac{t^\alpha}{M_i}$, and M_i has the same dimension with t^α. In order to ensure that the model and prototype are completely similar, these π terms should be:

$$\begin{cases} \pi_{2p} = \pi_{2m} \\ \pi_{3p} = \pi_{3m} \\ \vdots \\ \pi_{ip} = \pi_{im} \\ \vdots \\ \pi_{(n-k)p} = \pi_{(n-k)m} \\ \pi_{1p} = f\left(\pi_{2p}, \pi_{3p}, ... \pi_{ip}, ..., \pi_{n\text{-}kp}\right) \\ \pi_{1m} = f\left(\pi_{2m}, \pi_{3m}, ... \pi_{im}, ..., \pi_{n\text{-}km}\right) \end{cases} \tag{3}$$

$\pi_{1p} = \pi_{1m}$ can be ensured.

Above it, $\pi_{1p} = \frac{\eta_p}{M_{1p}}$, $\pi_{1m} = \frac{\eta_m}{M_{1m}}$, $\pi_{ip} = \frac{t_p^\alpha}{M_{ip}}$, $\pi_{im} = \frac{t_m^\alpha}{M_{im}}$.

When establishing a geometric small-scale model, ensure $M_{1P} = M_{1m}$ and $M_{im} = \frac{1}{\lambda} M_{ip} (\lambda \geq 1)$ using dimensions analysis.

According to the above deduction, when model and prototype are completely similar, $\eta_p = \eta_m$ (for $\pi_{1P} = \pi_{1m}$, $M_{1P} = M_{1m}$), $\pi_{ip} = \frac{t_p^\alpha}{M_{ip}} = \frac{t_m^\alpha}{M_{im}} = \pi_{im}$, and $t_m = \sqrt[\alpha]{\frac{1}{\lambda}} \cdot t_p (\lambda \geq 1)$ can be derived, so the similar model can predict prototype life time.

4 RESEARCH ON ACCELERATED LIFE TESTING WITH DISTORTED SIMILARITY

The life of the time can be shortened to $\sqrt[\gamma]{\frac{1}{\lambda}}$ by complete similarity model testing. However, it cannot meet the requirement of engineering practice. In order to shorten testing time significantly, some stress levels can be increased to deteriorate the service environment. Therefore some π terms of the model are no longer the same with prototype; this model is called distortion model, and the distortion degree can be defined as $\varphi_i = \frac{\pi_{im}}{\pi_{ip}}$, $\varphi_i \neq 1$.

4.1 Research on accelerated life testing with single parameter distortion

A similarity model with single parameter distortion can be built to shorten life by improving some stress levels to deteriorate service environment. For example, σ_j is a stress variable which can be temperature, pressure, force etc. that affect the speed of degradation. Its π terms are $\pi_j = \frac{\sigma_j^{\beta}}{M_j}$, and M_j has the same dimension with σ_j^{β}. When $\pi_2, \pi_3, \cdots \pi_{i-1}, \pi_{i+1} \cdots, \pi_j, \cdots, \pi_{n-k}$ remains constant, $\pi_1 = \frac{\eta}{M_1}$ is changed with $\pi_i = \frac{t^{\alpha}}{M_i}$, and the relation between π_1 and π_i can be expressed as this equation:

$$\pi_1 = f\left(\bar{\pi}_2 ... \bar{\pi}_{i-1}, \pi_i, \bar{\pi}_{i+1} ..., \bar{\pi}_j ... \bar{\pi}_{n-k}\right) \quad (4)$$

Generally, degradation path can be fitted by linear function, single-logarithmic linear function, double-logarithmic linear function, exponential function, composite exponential function etc. When research π_1 term changes with π_i term, model degradation path can be derived in a short time by improving the stress level of σ_j to distort π_j term, and π_j distortion degree can be expressed as φ_j. For example, $\pi_1 = \exp\left[-b \cdot \pi_i^a\right]$ can be used to fit one product degradation curve, and the function between φ_j and b or a can be built by analysing the data of a and b with different stress levels $\left(\varphi_{j1}, \varphi_{j2}, \varphi_{j3}\right)$. After establishing the function between a and φ_j, and the function between b and φ_j, degradation data with normal stress can be calculated from the function built, such as in Figure 1.

4.2 Research on accelerated life testing with multiple parameter distortion

One stress variable cannot be increased infinitely in many cases, so single parameter distortion cannot meet the demand of engineering practice. For

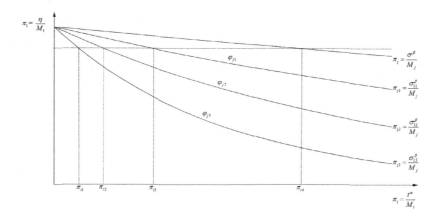

Figure 1. Degradation curve of one parameter distortion in different distortion data level.

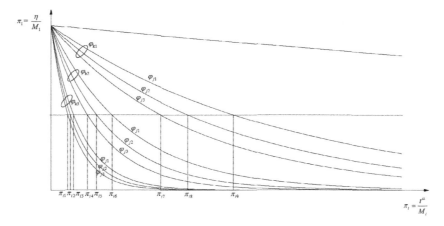

Figure 2. Degradation curve of double parameter distortion in different distortion data level.

example, rotating speed can affect rotor degradation (the higher the rotating speed, the faster the degradation), but it is very difficult to increase rotating speed infinitely. In order to quicken the rate of life testing, we can improve another stress variable, such as the service temperature, to carry out the double or multiple parameter distortion and to meet the demand of accelerated degradation.

For example, σ_k is another stress variable which affects the rate of degradation. Its π term is $\pi_k = \frac{\sigma_k{}^\gamma}{M_k}$, and M_k has the same dimension with $\sigma_k{}^\gamma$. The stress level of σ_k and σ_j can be improved together to distort π_j and π_k, and their distortion degree can be expressed as φ_j and φ_k respectively. Also, $\pi_1 = \exp\left[-b \cdot \pi_i^a\right]$ can be used to fit one product degradation curve, the function between φ_j, φ_k and a, b by analysing the data of a and b with different distortion $\left(\varphi_{k1}, \varphi_{k2}, \varphi_{k3}, \varphi_{j1}, \varphi_{j2}, \varphi_{j3}\right)$. After establishing the function between a, b and φ_j, φ_k, degradation data with normal stress can be calculated from the function built, such as in Figure 2.

5 CONCLUSION

Aiming at the problem that often accelerated testing cannot be executed in the prototype and are high cost, a new accelerated testing method based on similar theory is proposed from the aspect of similarity theory. The corresponding relation between the model life and the prototype life is found by establishing a geometry small-scale model. A model with single parameter distortion or multiple parameter distortion can shorten the life significantly, and the method will simplify the accelerated test and reduce test costs. As a result, the method has important implications for the estimating lifetime of the product with high reliability.

ACKNOWLEDGMENTS

This research is supported by the 973 program (2013CB 035706), National Natural Science Fund (51175517) and Central South University graduate Students Innovation Fund (2014zzts186).

REFERENCES

Bugaighis, M.M. (1995). Exchange of censorship types and its impact on the estimation of parameter of a Weibull regression model. *IEEE Transactions on Reliability*, 44(3), 496–499.
Chen, X. and Zhang, C.H. (2013). *Accelerated Life Testing Technology and Application*. Beijing: National Defense Industry Press: 1–12.
Deng, A.M. and Chen, X. (2007). A comprehensive review of accelerated degradation testing. *Acta Armamentarii*, 28(8), 1002–1007.
Hirose, H. (1993). Estimation of threshold stress in accelerated life testing. *IEEE Transactions on Reliability*, 42(4), 650–657.
Mazzuchi, T.A. and Soyer, R. (1990). Dynamic models for statistical inference from accelerated life tests. *IEEE Proceedings of Annual Reliability and Maintainability Symposium* (pp. 67–70).
McLinn, J.A. (1999). New analysis methods of multilevel accelerated life tests. *IEEE Proceedings of Annual Reliability and Maintainability Symposium* (pp. 38–42).
Wang, W. and Kececioglu, D.B. 2000. Fitting the Weibull log-linear model to accelerated life test data. *IEEE Transactions on Reliability*, 49(2), 217–223.
Yurkowsky, W.S. et al. (1967). *Accelerated testing technology*. (Report No. RADC-TR-67–420). New York: Rome Air Development Center.
Zhang C.H. & Wen X.S. (2004). A comprehensive review of accelerated life testing. *Acta Armamentarii*, 25(4), 485–490.

Automotive, Mechanical and Electrical Engineering – Liu (Ed.)
© *2017 Taylor & Francis Group, London, ISBN 978-1-138-62951-6*

Analysis of factors influencing laser ranging accuracy

Y.G. Ji, C.Y. Tian, X.G. Ge, C.D. Ning & Z.H. Lan
Air Force Aviation University, Changchun, China

ABSTRACT: According to the research on the principle of pulse laser ranging, some factors and reasons arousing the measured errors can be found. At the same time, aiming at the primary factors, several effective methods to solve the precision of pulse laser ranging were brought forward.

Keywords: Laser; Pulse ranging; Ranging precision

1   INTRODUCTION

With the development of society, distance measurement plays an important role in aviation, military, construction and other fields. Laser has the characteristics of good monochromaticity, good coherence and strong brightness and direction, which makes the laser range finder come into our sight, and it has developed rapidly. It gets widely used because the laser range finder has such characteristics as being of small size, light weight and easy to carry. So the study of laser ranging technology is very meaningful.

Laser ranging is one of the most mature technologies among military usages. Compared with generic optic ranging, laser ranging has these advantages: easy manipulation, simple structure and usable during both day and night. Compared with radar ranging, it is advanced in anti-jamming and precision.

Laser ranging contains two types: pulse and continuous. Continuous ranging can reach a high level of precision but has a limited working distance. It is used mainly in high precision measure so it has limitations in its application.

2   PRINCIPLE OF PULSE LASER RANGE FINDER

In general, a metering range consists of three units. They are the laser transmitting unit, receiving unit and distance calculation unit, as shown in Figure 1.

The laser transmitting unit is the place where it transmits the laser. It usually consists of two parts: the frequency synthesis part and the laser emission part. The receiving unit generally includes: a photoelectric receiving part, a D/A conversion part, some filter parts and so on. Finally, the key is the distance calculation unit. Laser ranges mainly include the pulse laser ranging method and the laser phase method.

The principle of pulse laser ranging is very simple. Mainly through the calculation of pulse sequence of laser pulses launch unit with photoelectric detection receiving unit receives the pulse sequence for time interval between the distance measurements. The measuring principle is shown in Figure 2.

Pulse laser ranging is mainly composed of a laser pulse transmission unit, a photoelectric detection receiver unit of timing and distance calculation units. First laser pulse transmission unit on the measured target after a laser pulse sequence, after that reflection object to be tested by the photoelectric detection receiving unit for receiving, following the timing unit start the clock, the final target distance is measured by formula.

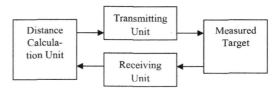

Figure 1.   Laser range finder block diagram.

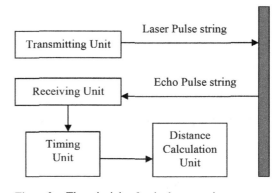

Figure 2.   The principle of pulse laser ranging.

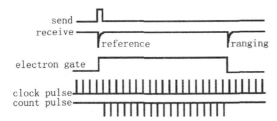

Figure 3. Time measurement schematic diagram.

The pulse laser range finder sends a set of short optic pulses (width<50ns) to the target (50ns), measures the time that the pulse takes to return from the target, then computes the distance of the target. If target distance is R, pulse return time is t, velocity of light is c, then:

$$R = ct/2 \qquad (1)$$

In a pulse laser range finder, t is measured by a counter that counts the number of clock pulses between the sending and receiving return pulse. As shown in Figure 3, the laser range finder generates the clock pulse inside, the electron gate rises when sending the laser, at the same time the counter begins to count the clock pulse; the electron gate declines when receiving the returned laser, and at the same time the counter stops counting. If between sending and receiving there are n clock pulses, the interval is τ, and the frequency of the clock pulse is $f = 1/\tau$, then:

$$R = \frac{1}{2}cn\tau = \frac{cn}{2f} = \ln \qquad (2)$$

In Formula (1), $c/2f$ expresses the distance increment of every clock pulse, count n clock pulses can get distance R.

3 ANALYSIS OF RANGING ERROR

According to the analysis of the basic ranging equation (1), laser ranging is essentially the measure of time t, so the precision of the measure of time is also the precision of the ranging.

$$\Delta R_1 = R\Delta t/t_0 \qquad (3)$$

Because there are great differences between the intensities of reference waves when sending laser and return signals, time delays when they act on the range finder and response times when they trigger the gate circuit, the time error caused can be 10^{-8} s maximally, the limited bandwidth always causes a pulse rising edge delay error Δt_2.

In addition, the mix of various noises may cause a time error Δt_3. All of these will cause inaccuracies in the counter, the relevant error being $\Delta R_t = \Delta t_1 + \Delta t_2 + \Delta t_3$.

However, according to the principle and structure of the pulse laser range finder, and the analysis of Formula (2), the pulse laser ranging error is:

$$\Delta R_2 = R\Delta c/c_0 + R\Delta n/n_0 + R\Delta f/f_0 \qquad (4)$$

In the formula, $\Delta c = c - c_0$: c_0 is the velocity of light in the vacuum, approximately 2.9979×10^8 m/s, c is the laser transmit velocity in the air. In practice this value would have a difference with the velocity of light because of some influence factors;

$\Delta f = f - f_0$: f_0 is the design frequency value of the distance counter, f is the instantaneous real frequency value of the oscillator. This is caused by the in-stability of the counter.

Δn: Δn is the error caused by the inaccuracy of the count pulse. For example, taking an oscillator of 30 MHz, the ranging error in one clock pulse will be ΔR, $c \approx 3 \times 10^8$ m/s, $f = 30$ MHz, $\Delta R = \pm 5$ m.

According to Formulas (3) and (4), the pulse laser ranging error mainly contains: timing error ΔR_t, velocity of light error ΔR_c, counter error ΔR_n, counter frequency error ΔR_f, etc. All of the error factors are random, according to the error theory. The system compositive error is:

$$\Delta R = \Delta R_c + \Delta R_n + \Delta R_f + \Delta R_t \qquad (5)$$

It is proved that generically ΔR_c, ΔR_n, ΔR_f are very small, while ΔR_t is always primary, and should be strictly controlled.

4 INCREASE OF SYSTEM PRECISION

According to the analysis of the ranging error, ranging precision is relative to the factors of atmosphere refraction, oscillation frequency, non-vacuum velocity of light, stability of oscillation frequency and return wave rising edge vibration.

4.1 Increasing counter frequency

The most ordinary laser ranging information disposing device is called the counter, it uses main waves as the open gate signal and return waves as the close gate signal. The counter counts the number of clock pulses when the gate is open. According to Formula (2), the measured distance can be written as follows:

$$R = \frac{c(n \pm 1)}{2f}$$

Commonly, $c/2f$ is designed as intuitionistic long measure, as 1 m, 10 m, etc. However ±1 in the formula is the counter error caused by the synchronism between the frequencies of the main and return waves and the clock.

So if we want a resolution of ±1 m, there should be a clock of 150 MHz. If we want it to be ±1 cm, the clock should be 15 GHz. In practice, it is hard to get an oscillation frequency of more than 150 MH, so to get a ranging precision of 1 m is too hard at present.

4.2 Increasing resolution of counter timing

Signal pulse trigger counter gate circuit, timing precision is influenced by mantissa time of switch and clock non-synchronisation time which is less than one clock quantity.

Time inner-insert method's principle is shown in Figure 4. And pending time interval T_x is

$$T_x = T_3 + T_1 - T_2$$

T_3 is the integer multiple of the clock pulse, and has the same phase with it. T_1, $T_2 < T_0$. T_1, T_2 is just which direct counter can't count accurately. The main function of inter-insert is to get the time interval and T_3 from the clock pulse, sample pulse and return pulse. Apparently T_3 can be measured by the counter accurately without any error. The end of T_1 and T_2 has a fixed phase relationship with the clock pulse, so $T_1 - T_2$ is a random time interval. If we multiply T_1 and T_2 by K, then:

$$T_1' = KT_1;\ T_2' = KT_2$$

Then use T_1' and T_2' to control two electron gates to make clock pulse with same frequency go through them. If the clock pulse that goes through during T_1' and T_2' is N_1 and N_2, the total measure error of $T_1 - T_2$ is ±1 clock pulse. Because T_1' and T_2' have already multiplied by K, so the total measure ΔT is:

$$\Delta T = T_0/K$$

Apparently, the ranging precision of this method is K times more than the direct counter method with the same clock pulse frequency.

4.3 Reducing the error of differences of trigger point by constant-fraction timing

Because of the width of the laser pulse, there will be a ranging error caused by the position changes of the counter open and close trigger point in the rising edge. Measure error changes when the trigger mode is varied.

When adopting a constant-fraction timing trigger mode, constant-fraction timer input signals will be divided into two routes; one route enters the comparator after dividing by K, another route enters the comparator after a t_d delay, and the point of intersection is the trigger point time.

The constant-fraction timing trigger does not change with the signal range; it is only relative to the pulse width. After deduction and validation, the trigger point variable caused by the changes of pulse width is:

$$\Delta t_s = \frac{\tau \ \ln K}{2.7225 \ t_d} \Delta \tau$$

If the signal pulse width is 10 ns, $K = 2$, $t_d = 4$ ns, $\Delta \tau = 0.5$ ns, then $\Delta t_s = 0.31$ ns, and the corresponding ranging error is 4.7 cm.

So we can see that the precision is very high when adopting constant-fraction timing, which can effectively reduce the error caused by position changes of the counter open and close trigger point in the rising edge.

In addition, velocity of light error can be corrected by level and tilted distance measure. In level measure, the corrected formula can be got from ground temperature, pressure and humidity. In tilted distance measure, the formula should be corrected by weather data with the atmosphere delaminating method. This error will be 5 cm~10 cm when $E \geq 10°$; 10 cm~20 cm when $E \geq 0°~10°$ after correcting.

5 CONCLUSION

According to the above-mentioned analysis, there are many factors that can influence the precision of laser ranging, mainly: measure time error ΔR_t, velocity of light error ΔR_c, counter error ΔR_n, counter frequency error ΔR_f, etc. In order to reduce these errors, respective methods such as increasing counter frequency, increasing resolution

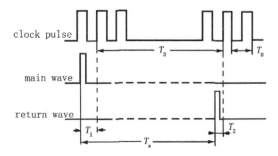

Figure 4.   Time inner-insert method's principle.

and adopting constant-fraction are used, however, ΔR_c can be corrected by formula according to different instances, and all these methods can obviously improve ranging precision.

Additionally, many other system errors can be corrected when used in a high precision ranging system.

REFERENCES

An Weining, Zhang Fumin, Wu Hanzhong, et al (2014). Absolute distance measurement method based on frequency-domain interference using a femtosecond optical frequency comb [J]. *Chinese Journal of Scientific Instrument*, 35(11).

Marius A. Albota Richard M. Heinrichs, David G. Kocher, et al. (2002). Three-dimensional imaging laser radar with a photon-counting avalanche photodiode array and microchip laser. *Applied Optics,* 41(36).

Ou-Yang, M., Huang, C. Y., & Chen, J (2006). High dynamic range laser range finders based on a novel multimodulated frequency method [J]. *Optical Engineering*, 45(12).

Wang, Y. Z. (2003). Modern military optic technology [M]. BeiJing: science publisher, 2003.01

Wu Hanzhong, Cao Shiying, Zhang Fumin, et al (2015). Spectral interferometry based absolute distance measurement using frequency comb [J]. *Acta Physica Sinica*, 64(2).

Wu, H., Zhang, F., Li, J., et al. (2015). Intensity evaluation using a femtosecond pulse laser for absolute distance measurement [J]. *Applied Optics*, 54(17).

Automotive, Mechanical and Electrical Engineering – Liu (Ed.)
© 2017 Taylor & Francis Group, London, ISBN 978-1-138-62951-6

Application of graph databases in the communication and information asset management in power grid

Xuming Lv, Shanqi Zheng & Zhao Li
State Grid Liaoning Electric Power Supply Co. Ltd., Shenyang, China

Siyan Liu & Yue Wang
Global Energy Interconnection Research Institute, Beijing, China

ABSTRACT: An increasing amount of communication and information equipment is added to the power grid. The new situation brings great challenges to the communication and information asset management in the power grid. Traditional relational database could not manage the complex relations in a reasonable efficiency. An exploration is made to use a popular graph database, Neo4j, to store, manage, and query those data in the paper. The requirement of the management and several graph products are surveyed. Besides, a graph model of communication and information asset in the power system and its visualization are presented in detail.

Keywords: Communication and information, database

1 INTRODUCTION

The equipment of communication and information is deployed to realize high performance control and automated management in the power system. With the penetration of smart grid technologies, a large amount of power devices is widely applied, including the distributed energy resources, smart transportation stations, and E.V charging stations (EPRI, 2014).

However, such new trends of power system developments introduce great challenges for communication and information asset management. Firstly, it is necessary to manage such power devices in the application systems of power system. Much more servers, interchangers and other communication devices are required with the increasing number of power devices. Traditional relational databases cannot handle the large scale dataset very well anymore. Secondly, the connections among the communication and information devices are complex. It is impossible for users to dig out the impact area of a specific device. To realize reliable control and flexible management, an innovative communication and information asset management method is greatly required.

Graph databases emerge as a sophisticated tool to storage, manage, visualize and analyze highly connected data (EPRI, 2014). A variety of data applications can be modelled as graph structure, e.g., social networks, web pages,

biological networks, chemical compounds, and road networks (Miller, 2013). Graph databases are inspired by graph theory. Three basic elements of graph databases are node (or vertex), edge, and property. Nodes are utilized to define entities for keeping record, e.g., people, city, and order records. Nodes in graph databases are roughly equivalent to one column in relational databases. Edges are utilized to define the relationship between two nodes. Edges can be directed or undirected, and can represent different meanings. Edges are the key concept in graph databases, since the connections between items are not directly implemented in other types of databases. The relationships allow data in the store to be linked together directly, and in most cases the collated data can be retrieved with a single operation. Properties are associated information that relates to nodes. For instance, if a person is a node in graph databases, one may define the properties as the gender, age, or education background, depending on which aspects of people are associated in the particular database.

The graph structure provides a natural way to present connections. In relational database, links between data are stored in the rational data itself, and queries search for this data within the store and use the JOIN concept to collect the related data. Graph database allows simple and rapid retrieval of complex hierarchical structures that are difficult to model in relational systems.

To the best of knowledge, graph databases have not been widely applied in power system yet. However, communication and information in power system can be modelled as property graph. Devices, e.g., server, storage device, and network switch can be modelled as node. The transmission line and the logistical connection between devices can be modelled as edge. The configuration of the device, the power grid management system it belongs to, the actual address can be modelled as properties. To the best of our knowledge, we firstly utilized graph databases to study the communication and information asset management in power system. Firstly, we investigate the actual requirements of communication and information asset management in power system and conduct modelling. Secondly, we make a brief survey of the latest popular graph databases, i.e., Neo4j, GraphFrame, GraphSQL. Thirdly, we realize the modelling of the topological graph of the communication and information asset of power system. Fourthly, we achieve the visualization of communication and information asset managements

2 SECTION I: A BRIEF SURVEY OF GRAPH PROCESSING PLATFORMS IN POWER SYSTEM

Graph processing platforms have played more and more important roles in large online system with deep links between records. Pregel is the system built by Google to power PageRank, which is a fundamental algorithm for web searching engine. It is also the inspiration for Apache Giraph, which Facebook uses to analyze their social graph. However, Pregel is owned by Google and unavailable to other developers. Apache Giraph only releases the first version in 2015 and releases no update.

With the rapid development of graph processing platforms, a lot of Graph systems have emerged to meet the market requirements of storing and managing a big graph, and most of the systems claim that they are superior to other graph processing platforms with somewhat unfair compassion tests. For example, some comparisons are not conducted with the latest version. Some comparisons are conducted with the community version rather than business versions.

Although a lot of work compares the most substantial graph database functionality and obtains the benchmark, e.g., load graph data, page-rank algorithm, the shortest path algorithm, a forward step to utilize of graph databases in power systems has not been realized yet.

Among a number of graph database system, we choose Neo4j, GraphFrame and GraphSQL to design communication and information asset management system in power grid. Before implement those graph databases, we conduct a brief study of all the mentioned graph databases (Singh, et al., 2015) (Pabon, 2014) (Jouili, et al., 2013) (Jain, et al., 2013) (McColl, et al., 2014) (Kolomičenko, et al., 2013) (Vicknair, et al., 2010) (Beis, et al., 2015).

2.1 Neo4j

Neo4j, a graph database management system developed by Neo Technology Inc, is one of the most popular graph databases according to db-engines.com. The first version is released in 2010, and the latest version, i.e., the third version is released in 2016. Neo4j supports graph model called "Property graph", which includes nodes, edged and attribute (i.e., properties). Neo4j is written in Java, and provides APIs which are exposed through a whole range of various languages, e.g. Java, Python, Ruby, JavaScript, PHP, .NET, etc. It is noteworthy that C and C++ are not in the above list. Cypher is a declarative graph query language for the graph database Neo4j, and is roughly equivalent to SQL querying language in relational databases. The database system also implements the Blue-prints interface and a native REST interface to further expand the ways to communicate with the database. Neo4j provides three editions: Community, Enterprise, and Government. Community version is provided for individuals to learn graph databases and conduct smaller projects that do not require high levels of scaling. However, it excludes professional services and support. The Neo4j Enterprise edition offers incredible power and flexibility, with enterprise-grade availability, management and scale-up & scale-out capabilities.

Neo4j provides sustainable competitive advantage in internal applications for enterprises, e.g., master data management, network and IT operations and fraud detection. Neo4j has provided sophisticated supports for customer, e.g., Walmart, ebay, and LinkedIn.

2.2 GraphFrame

GraphFrame is a database that could manage data with a graph structure. It is a module on Spark SQL in Spark. The data structure is based on Spark Resilient Distributed Dataset (RDD). The data in GraphFrame could be transformed to Spark GraphX. GraphX is a distributed graph processing framework based on Apache Spark. GraphX unifies ETL, exploratory analysis, and iterative graph computation within a single system. The data can be presented as both graphs and collections. RDDs are efficiently utilized to transform and join graphs. Developers can use Pregel API to write custom iterative graph algorithms. GraphX emerges with a number of graph algorithms, e.g., page rank, connected components, SVD++, strongly connected components, and triangle count. Since GraphX is built as a library on top of Spark, no modification to Spark is required. To achieve superior performance parity with specialized graph processing systems, GraphX introduces a range of optimizations techniques: distributed graph representation, implementing the triplets view, optimization to mrTriplets. GraphX has been successfully utilized to analyze QQ social data by Tencent. It is noticed that graph theory has been utilized to design power flow computer algorithm. GraphX is a potential solution to large scale power flow computing problems since GraphX achieves great performance in graph computing.

2.3 *GraphSQL*

GraphSQL databases and associated analytical engine are claimed as a complete, distributed, parallel graph computing platform for web-scale data analytics in real-time. Map reduced and parallel graph processing are unified to in GraphSQL to accelerate scalable parallel graph algorithms. User defined functions are available with the vertex-centric map-reduce style API. In addition, a SQL-like graph query language, i.e., GSQL is provided for ad-hoc exploration and interactive analysis of Big Data. Powerful parallelization mechanism is provided for scaling massive graph processing. Based on the above techniques, GraphSQL achieves fast data loading speed to build graph, and fast execution of parallel graph algorithm. Large scale offline data processing and real-time analytics are unified in GraphSQL. The unique feature of GraphSQL is that it has the real-time capability for streaming updates and inserts. In general, there exist three key functional components in Graph-SQL: graph-powered real-time analytics platform, enterprise data engine, and customer intelligence solutions. The graph-powered real-time analytics platforms collects, stores, and processes massive data to form a property graph. The enterprise data engine supports robust analytics by graph theory based analytical algorithms. The customer intelligence solutions focus on decision support system and data visualization with tools, e.g., Tableau.

3 SECTION II: DATA MODELLING OF APPLICATION SYSTEMS

In this chapter, the composition of the application systems is described. Then, the modelling of the facilities and their relationship are discussed. The fault analysis function and the visualization of the facilities are based on the graph model.

3.1 *Physical elements*

The devices in the Application System of the power system and the connection among devices should be shown precisely in the graph model. The vertices and edges corresponding to the physical devices and their connections are discussed in the section.

a. Vertices Modelling

Three types of physical vertices are considered in the graph model of application systems. The vertices include Server vertices, Storage Device vertices, and Network Device vertices. Various kinds of servers are considered in the model. Therefore, server vertices contain application servers (App), data servers (Data), port servers (Port), map servers (Map), data agent servers (Agent), and F5 load balancing devices (F5). Besides, a network device refers to interchanger in the graph model.

Different types of vertices have different properties. The table lists the properties of the vertices:

Vertex Type	Device ID	Subcategory	Model No.	Location	IP address
Server	√	√	√	√	√
Storage Device	√	√	√	√	√
Network Device	√	√		√	√

Vertex Type	Port	Operation System	Configuration	Type of Cluster
Server		√	√	√
Storage Device				
Network Device	√			

Device ID, subcategory, model No., and location are universal properties. In other words, whatever the vertex type is, a vertex has the four properties to identify the device.

As for IP address and port, an IP address of a server or storage device corresponds to a port of network device (interchange). The detailed physical connections are not shown by the properties of vertices. Denote that a vertex could have more than one IP addresses or more than one ports.

The operation system, configuration and type of cluster are unique properties of server vertices. The operation system could be either Windows or Linux. The configuration refers to the hardware configuration of the server, for example, the number of CPU cores. The type of cluster

Table 1. Shows the Comparison of Neo4j, GraphX, GraphSQL in detail.

	Neo4j	GraphX	GraphSQL
	The most well-known graph database implemented in Java. Storing graph data in its own native graph format and provides graph traversal APIs.	A graph computing framework built in Spark. Utilizing all the resource and operations in Spark.	A complete parallel graph database built from scratch optimized for performance, providing both real time and large-scale graph analytics.
Pros	• Comprehensive graph database functionality • SQL-like query language • Detailed technical documentation • Large developer community	• Powerful capability of parallel computing in memory • Including implementation of basic graph algorithms • With mature development modules and APIs, easy to build up new function	• Fast graph loading speed and graph traversal speed • Rapid real-time incremental graph update capability • Node-centric MapReduce style API with built-in parallelism support • Graph data computing functions, with Arithmetic, logical, union operations, and logical control • Support User-Defined Functions • SQL-like graph query language
Cons	• Lack of built-in parallelism for large scale graph analytics • Slow incremental graph update	• Support Java or Python, complicated in programming • Poor in real-time data update • Lack of graph data storage and output	• C/C++ application development only, lack of Java support • Insufficient technical documentation • Poor maturity and stability

indicates the topological structure of a server cluster, e.g., F5 or RAC. All servers in a cluster share the same type of cluster properties.

b. Edges Modelling

Bidirectional edges are used to model the connections among vertices. The physical correspondence of the edges is cable. The vertex types connected by the different edges are listed in the table below:

Type of Edge	Vertex 1	Vertex 2
1	Server	Network device
2	Network device	Network device
3	Storage device	Network device
4	Storage device	Storage device

c. Virtual Elements

In order to describe the function of the devices, two groups of virtual elements are defined in the graph model. The one are system vertices and belonging edges. Another one are business logic edges.

1. System Vertices and belonging edges

A system vertex is a super node that represents an application system. The super node consists of the devices that belongs to the application system. In the graph model, a vertex and a set of belonging edges are used to represent the system-level super node.

2. Business logic edges

Another type of virtual edges is business logic edges. The edges indicate the business logical relationship between two devices. With the business logic edges, the impact area of a fault node can be determined in a query. Denote that business logic edge is unidirectional.

4 SECTION III: COMMUNICATION DEVICES VISUAL REPRESENTATION

4.1 Introduction to visualization

The goal is to present the relations and connections between communication devices, which is beneficial for device management, communication system analysis and network planning. This allows people to have a good view of all the operating devices and their working conditions. In this section we load data and achieve data visualization with the model already built before. We use the dataset which is synthesized based on real device connection.

4.2 Neo4j vs. GraphSQL

Neo4j Community Edition and GraphSQL both have visualization tools. In Neo4j, since Cypher is essentially a declarative query language, it allows the developers to write queries and update graph store efficiently without paying much attention on data structure. Users can write queries in Cypher language and run query command in Neo4j's editor. Neo4j generates visualization automatically in the result frame to display the query result. In GraphSQL platform, user can submit query in web browser after creating the query job, and the query visualization comes out. Considering the flexibility of Cypher language and highly customized visualization, we choose Neo4j as our tool.

4.3 Loading data and visualization

We use Neo4j database to create graph and visualize graph data. Before loading data, we have got our data in CSV format based on the real connections between different devices. As discussed above, nodes represent communication devices and edges represent physical connections between them. The main steps are as follow:

4.4 Create graph

Load device data → Create device node → Load connection data → Create edges

We have four types of nodes and all edges are undirected. The four types of nodes represent F5, Server, Switch and Storage devices. Each node has four properties, name, type, subcategory and system, except for Switch node, which only has name, type and system.

We load data and create device node for only one type of devices at a time and then move to the next type. Once all nodes are created, edges can be built. It is necessary to notice that all edges are undirected, because the edge here only means the two devices at the end are connected physically. However, in our graph, we created directed edges, and we can just ignore the direction. At the moment we do not use system node and logic edges.

a. Example of creating nodes
   load csv with headers from
   "file:/Users/yw/Desktop/device_Server.csv" as vertex1
   create(n:Server{name:vertex1.Device_ID, type:vertex1.Device_type, subcategory:vertex1. Subcategory, system:vertex1.Dev_system})
b. Example of creating edges
   load csv with headers from
   "file:/Users/yw/Desktop/Work/Liaoning/ LiaoNing/com_edge1.csv" as edge1
   match(a:Server{name:edge1.Device_ID1}),(b: Storage{name:edge1.Device_ID2}) create (a)-[r:S erver2Storage{type:"Physical"}] –> (b)

The execution times of the above two examples are 50 ms and 49 ms respectively.

4.5 Visualization

Run query → Customize Graph style

We can write different queries for different visualization application scenarios. It is very convenient for to get a thorough display of all devices by the visualization of the whole graph. Querying all the node whose property is server can visualize all server devices. Querying all the node whose system

property is PMS can visualize all the devices in PMS system. Users can simply write these kind of queries to make their own visualization.

It is convenient to change graph style using Graph Style Sheet provided by Neo4j. In Graph Style Sheet we can set color, caption, text color, font size for nodes and edges. Users can import already modified Graph Style Sheet to customize their graph visualization. The nodes layout is not clear when they first come out, so users must drag nodes to suitable places to get desired visualization.

a) Visualization of all switch devices

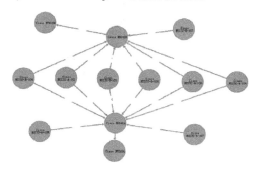

MATCH (n:Switch) RETURN n

b) Visualization of all IMS devices

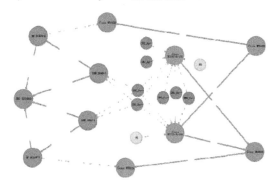

```
MATCH (n: F5)
WHERE n. system = 'IMS'
RETURN n
UNION MATCH (n: Server)
WHERE n. system = 'IMS'
RETURN n
UNION MATCH (n: Switch)
WHERE        n.system="IMS"      OR
n.system="IMS,PMS"               OR
n.system='IMS,PMS,GIS'
RETURN n
UNION MATCH (n: Storage)
RETURN n
```

4.6 *Remark*

By using Neo4j database, we can make visualization of graph data which contains information about devices status and connections between devices via the web browser. Visualization will supports the fault affected area analysis. Node can be made in striking color to remind people of the fault affected area.

5 CONCLUSION

We have surveyed GraphSQL, GraphFrame, and Neo4j. A graph model is put forward in order to model communication and information devices in the power system. We have made an exploration to use Neo4j to store and manage the graph model. A visualization of the physical graph topology is also put forward.

REFERENCES

Beis S, Papadopoulos S, Kompatsiaris Y. Benchmarking graph databases on the problem of community detection [M]//New Trends in Database and Information Systems II. Springer International Publishing, 2015: 3–14.

EPRI Smart Grid Demonstration Initiative. Two year update [J]. Electric Power Research Institute (EPRI).– USA, California, 2014.

Jain R, Iyengar S, Arora A. Overview of popular graph databases[C]//Computing, Communications and Networking Technologies (ICCCNT), 2013 Fourth International Conference on. IEEE, 2013: 1–6.

Jouili S, Vansteenberghe V. An empirical comparison of graph databases[C]//Social Computing (Social-Com), 2013 International Conference on. IEEE, 2013: 708–715.

Kolomičenko V, Svoboda M, Mlýnková I H. Experimental comparison of graph databases [C]//Proceedings of International Conference on Information Integration and Web-based Applications & Services. ACM, 2013: 115.

McColl R C, Ediger D, Poovey J, et al. A performance evaluation of open source graph databases[C]//Proceedings of the first workshop on Parallel programming for analytics applications. ACM, 2014: 11–18.

Miller J J. Graph database applications and concepts with Neo4j[C]//Proceedings of the Southern Association for Information Systems Conference, Atlanta, GA, USA. 2013, 2324.

Pabon O S. A comparison of NoSQL graph databases[C]//2014 9th Computing Colombian Conference (9CCC). 2014.

Robinson I, Webber J, Eifrem E. Graph Databases: New Opportunities for Connected Data [M]. "O'Reilly Media, Inc.," 2015.

Singh M, Kaur K. SQL2 Neo: Moving health-care data from relational to graph databases[C]//Advance Computing Conference (IACC), 2015 IEEE International. IEEE, 2015: 721–725.

Vicknair C, Macias M, Zhao Z, et al. A comparison of a graph database and a relational database: a data provenance perspective[C]//Proceedings of the 48th annual Southeast regional conference. ACM, 2010: 42.

Automotive, Mechanical and Electrical Engineering – Liu (Ed.)
© 2017 Taylor & Francis Group, London, ISBN 978-1-138-62951-6

Application of support vector regression in surge arrester fault diagnosis

Yu He
State Grid Yancheng Power Supply Company, Yancheng, Jiangsu, China

Yifan Sun & Hao Jiang
School of Electrical Engineering, Southeast University, Nanjing, Jiangsu, China

ABSTRACT: A surge arrester fault diagnosis method based on Support Vector Regression (SVR) is put forward. In the method, the influence of environmental factor and power network factor on leakage current value is taken into account and as the input feature vector of support vector machine, and the fundamental and third harmonic waves of resistive current are taken as the diagnosis characteristic of surge arrester dampness and aging respectively. The regression model of unaged surge arrester is determined via multivariable experiment, and the predictive value of the diagnosis characteristic is obtained in combination with online monitoring data at the time of diagnosis. With the predictive value and the measured value, the deviation between the current state and the unaged state can be calculated, and then the occurrence and severity of fault can be judged on this basis. At last, the method is verified with an example for accuracy and effectiveness.

Keywords: Surge arrester; Support vector regression; Fault diagnosis

1 INTRODUCTION

Surge arrester is widely used in power system, which can effectively protect lightning overvoltage and switching overvoltage. It is one of the most important electric equipment. The performance and reliability of surge arrester are of great significance to safe and stable operation of power network, and the running state of surge arrester must be grasped accurately from time to time.

Traditional surge arrester fault diagnosis methods mainly include off-line test, infrared temperature measurement, and total current and resistive current measurement, etc. Off-line test is complicated and time-consuming, with which fault cannot be identified timely, but fault degree may progress and fault coverage may expand. The precision of infrared temperature measurement is susceptible to the influence of various factors, and there exists measuring blind zone. The limitation of total current and resistive current measurement is that since surge arrester is installed in rugged physical environment and electromagnetic environment, many on-the-spot factors will interfere current measurement, which leads to deviation, and then affects the accuracy of fault diagnosis.

In recent years, both domestic and foreign scholars started trying surge arrester fault diagnosis in combination with intelligence algorithm, and have made some achievements. Zhang Pei et al. put forward a surge arrester fault diagnosis method based on BP neural network. In this method, the influence of multiple physical environment parameters on leakage current surge arrester is taken into account, and the output result is a fault probability ranging from 0 to 1, but the fault type is not determined. Lira, G. R. S. et al. simulated multiple types of fault, created a fault database, and judged surge arrester fault occurrence and fault type with self-organizing map network with the measured total current as input. The accuracy of this method, however, will lower if the power grid has harmonic wave. Khodsuz, M. & Mirzaie, M. took the each current component and the ratios of multiple current components as input, and applied multi-class support vector machine to realize judgment of fault type, but didn't take the influence of physical environment on input into account.

On this basis, an SVR-based surge arrester fault diagnosis method is put forward in this paper. In this method, the influence of power harmonic and physical environment on leakage current of surge arrester is taken into account, different eigenvectors are selected as input based on fault type, and the deviation between the current state and unaged state of surge arrester is calculated in combination with monitoring data, to determine the occurrence and severity of fault.

2 SVR-BASED SURGE ARRESTER FAULT DIAGNOSIS

2.1 Selection of input and output eigenvectors

For modeling surge arrester with SVR algorithm, the input and output eigenvectors should be determined. Output eigenvector refers to a set of characteristic quantities that are able to reflect fault state of equipment, while input eigenvector to a set of variables influencing the value of the fault characteristic quantities. The selected fault characteristic quantities must be able to reflect the type of a certain fault, and have a large change relative to the normal value in the case of fault. Since many surge arrester faults occur due to dampness or aging, this paper focuses on the diagnosis method of this two types of fault. Experimental study shows that fundamental wave of resistive current is sensitive to dampness change, but doesn't change significantly in the case of valve block aging, while the case of third harmonic resistive current is the opposite. Hence, the fundamental wave and third harmonic of resistive current are selected as the fault characteristic quantities of dampness and aging of surge arrester respectively in this paper.

The research of Ding Guocheng et al. has verified the influence of surface filth, ambient temperature and humidity on leakage current of surge arrester. Concerning power network factor, among the voltage components of working voltage, the fundamental component and the third harmonic component have significant influence on the value of resistive current. Besides, interphase interference will also cause certain error in measurement of leakage current. Since this factor is linked with the installation of surge arrester, however, it can be regarded as a known quantity after the surge arrester goes into operation. To sum up, surface filth, ambient temperature, relative humidity, fundamental and third harmonic wave of working voltage are selected as input characteristic quantities in this paper.

2.2 Steps of fault diagnosis

The SVR-based surge arrester fault diagnosis method mainly includes the following steps:

Step 1: Take training samples required for SVR algorithm via multivariable experiment. Specifically, measure and record the fundamental resistive current I_{R1} and third harmonic resistive current I_{R3} of a surge arrester in initial healthy state under the condition that the ambient temperature, relative humidity, surface filth, fundamental voltage and third harmonic voltage have small-step change one by one. The variable factors are taken as input characteristic quantities of training samples, and I_{R1} and I_{R3} as output characteristic quantities.

Step 2: Perform data preprocessing over the training samples obtained in Step 1. The dimensions of different influencing factors are different, and the values widely differ from each other. If the values are directly adopted without processing, it is highly impossible to figure out the real weight of influence of each factor on resistive current of surge arrester. Therefore, the training samples must be subject to normalization processing, so as to enable comparison and analysis at the same order of magnitude. For purpose of this paper, linear independence of information normalization is adopted to normalize to the interval of [−1, 1], of which the computational formula is:

$$x_{iN} = \frac{x_i - \min(x)}{\max(x) - \min(x)} \times 2 - 1 \tag{1}$$

where x_i and x_{iN} are values of a data before and after normalization respectively, and $\max(x)$ and $\min(x)$ are the maximum and the minimum of the same type of data.

Step 3: Solve the regression model with SVR algorithm based on the normalized training samples.

Step 4: Obtain test samples via online monitoring of surge arrester. Specifically, sample the target surge arrester regularly in a continuous period, covering the values of the variable factors as set forth in Step 1, I_{R1} and I_{R3}. The monitoring data of variable factors are taken as input characteristic quantities of the test samples, and I_{R1} and I_{R3} as output characteristic quantities. Perform data preprocessing over all test samples according to the normalization method as set forth in Step 2.

Step 5: Diagnose the state of the target surge arrester based on the regression model. Specifically, put the input characteristic quantities of the test samples into the regression model to calculate out the predictive values of the fundamental component and the third harmonic component of resistive current at each sampling point of the surge arrester, and then calculate the MSE and squared correlation coefficient R^2 to realize state diagnosis. The larger the MSE is or the closer R^2 is to 0, the larger the degree of the target surge arrester deviating from the initial healthy state is. Specifically, the MSE or R^2 of I_{R1} is taken to judge the degree of dampness of the surge arrester; the MSE or R^2 of I_{R3} to judge the aging degree of the surge arrester. The computational formula of the MSE and squared correlation coefficient R^2 is:

$$\text{MSE} = \frac{1}{n} \sum_{i=1}^{n} \left(y_p - y_m \right)^2 \tag{2}$$

$$R^2 = \frac{(n\sum_{i=1}^{n} y_p y_m - \sum_{i=1}^{n} y_p \sum_{i=1}^{n} y_m)^2}{\left[n\sum_{i=1}^{n} y_p^2 - (\sum_{i=1}^{n} y_p)^2\right]\left[n\sum_{i=1}^{n} y_m^2 - (\sum_{i=1}^{n} y_m)^2\right]} \quad (3)$$

where n refers to the number of sampling points, and y_m and y_p to the measured value and predictive value of the ith sampling point respectively.

3 EXAMPLE

To check the accuracy and effectiveness of the method put forward herein, the experimental data of surge arrester in the research of Khodsuz, M. & Mirzaie, M. were adopted, which has clear conclusions. The experimental data were divided into training samples and test samples for fault diagnosis. In the experiment process, constant working voltage and temperature were kept, and the influencing factors were surface filth and relative humidity. For the training samples, the experiment data of unaged surge arrester were taken, as shown in Table 1. The test samples were divided into two groups, namely experimental data of unaged and aged surge arresters respectively, as shown in Table 2.

The training samples were normalized first, and the optimal values of parameters C and γ of SVR were searched by means of grid search and cross validation. Grid range: C, $\gamma \in [2^{-8}, 2^{8}]$; search range: 0.1; optimal parameter values: C = 42.2243, γ = 0.0059. The parameter optimization process is shown in Figure 1. After the values of C and γ were determined, a regression model was built with the training samples.

The third harmonic resistive current I_{R3} of the first group of test samples was predicted with the regression model, and the results are as shown in Figure 2. Since the first group of test samples are experimental data of an unaged surge arrester, the difference between the measured value and the predictive value based on the regression model is not significant. According to calculations, the MSE equals to 0.01134, and the squared correlation coefficient R^2 equals to 0.978, consistent with the conclusion in the figure.

Table 1. The training sample data.

No.	Pollution level	RH%	State	$I_{R3}(\mu A)$
1	1	65	Normal	1.226
2	1	75	Normal	1.495
3	1	85	Normal	2.869
4	1	92	Normal	3.17
5	2	65	Normal	1.113
6	2	75	Normal	2.641
7	2	85	Normal	3.132
8	2	92	Normal	3.168

Table 2. The test sample data.

No.		Pollution level	RH%	State	$I_{R3}(\mu A)$
1	1	4	65	Normal	1.215
	2	4	75	Normal	1.748
	3	4	85	Normal	2.861
	4	4	92	Normal	3.508
2	1	4	65	Aged	1.632
	2	4	75	Aged	1.912
	3	4	85	Aged	3.054
	4	4	92	Aged	5.267

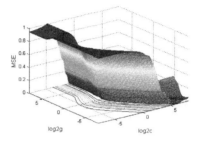

Figure 1. 3D view of parameter optimization for SVR.

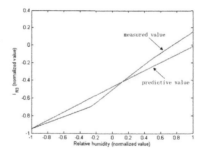

Figure 2. The measured value and the predictive value of I_{R3} (Un-aged).

The prediction results of the second group of test samples are as shown in Figure 3. According to Figure 3, the difference between the measured value and the predictive value of I_{R3} is large. This is because the I_{R3} of the test sample was taken from the experimental data of an aged surge arrester. Thus, the difference between the I_{R3} value and the regression model predictive value representing the initial healthy state is large. According to calculations, the MSE equals to 0.27012, and the squared correlation coefficient equals to 0.78039. Comparing with the unage state, the MSE largely rises, while the squared correlation coefficient significantly declines, which verifies the effectiveness of the method put forward.

To compare the performance of the SVR-based regression model put forward herein, BP neural network was adopted for modeling the same train-

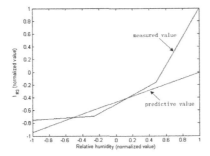

Figure 3. The measured value and the predictive value of I_{R3} (Aged).

Table 3. Comparison of the predictive values for different models.

$I_{R3}(\mu A)$

No.	Measured	SVR	Error%	BPNN	Error%
1	2.869	2.848	−0.73	2.872	0.10
2	3.17	3.149	−0.66	3.151	−0.60
3	1.113	1.133	1.80	1.165	4.67
4	2.641	2.620	−0.80	2.616	−0.95
5	3.132	3.112	−0.64	3.159	0.86
6	3.168	3.148	−0.63	3.184	0.51
7	1.215	1.236	1.73	1.21	−0.41
8	1.748	1.769	1.20	1.671	−4.41

Table 4. Comparison of modeling performance for different models.

	SVR	BPNN
MSE	0.000426	0.00133
R^2	0.999944	0.99808

ing set for comparison. The BP network is a single-hidden layer network, of which the node number of the input layer, hidden layer and output layer are 2, 3, 1 respectively, and for which tansig function is adopted as the transfer function of the hidden layer, and Levenberg-Marquardt algorithm as the training function, the learning rate is 0.1, and the target error is 10^{-4}. The comparison between the measured value and the predictive value of I_{R3} of the two models are as shown in Table 3, and the comparison of MSE and squared correlation coefficient is as shown in Table 4. According to the comparisons, SVR algorithm is more accurate when used for modeling the leakage current characteristics of surge arrester, and has a stronger extensive ability comparatively.

4 CONCLUSION

In the SVR-based surge arrester fault diagnosis method put forward herein, the influence of physical environment and power harmonic on leakage current of surge arrester is taken into account, and the fundamental wave component and third harmonic component of resistive current are taken as characteristic quantities of fault diagnosis. A regression model was built on the basis of the experimental data of a surge arrester in healthy state. Online monitoring data was put into the regression model for diagnosis. The predictive values of corresponding diagnosis quantities were obtained and compared with the measured values. The severity of aging and dampness of surge arrester were judged by calculating the MSE and squared correlation coefficient. Comparing with BP neural network, the regression model built with this method are closer to the actual electrical characteristics of surge arrester.

REFERENCES

Ding, G.C., Tian, Y. & Chen, Q.T. et al. 2015. Study on effectiveness test of MOA charged detection technique. Insulators and Surge Arresters.04:105~108.

Kang, Y.L., Wang, X. & Ding, X.Y. 2015. Effects of interphase interference on MOA charged test. Ningxia Engineering Technology. 03:249~251.

Khodsuz, M. & Mirzaie, M. 2015. Condition Assessment of Metal Oxide Surge Arrester Based on Multi-Layer SVM Classifier. Iranian Journal of Electrical & Electronic Engineering. 11(4):354~362.

Khodsuz, M. & Mirzaie, M. 2015. Monitoring and identification of metal oxide surge arrester conditions using multi-layer support vector machine. Iet Generation Transmission & Distribution. 9(16):2501~2508.

Khodsuz, M. & Mirzaie, M.2015. Evaluation of ultraviolet ageing, pollution and varistor degradation effects on harmonic contents of surge arrester leakage current. Iet Science Measurement Technology. 9(8):979~986.

Lira, G.R.S., Costa, E.G. & Almeida, C.W.D. 2010. Self-organizing maps applied to monitoring and diagnosis of ZnO surge arresters. Transmission and Distribution Conference and Exposition: Latin America. 659–664.

Lira, G.R.S., Costa, E.G. & Ferreira, T.V. 2014. Metal-oxide surge arrester monitoring and diagnosis by self-organizing maps. Electric Power Systems Research, 108(3), 315~321.

Mokhtari, K., Mirzaie, M. & Shahabi, M. 2014. Evaluation of polymer housed metal oxide surge arrester's condition in humid ambient conditions. Synthetic Communications an International Journal for Rapid Communication of Synthetic Organic Chemistry. 21(1):31~42.

Shao, T. Zhou, W.J., Yan, H.G. et al. 2004. Influence of voltage harmonics on leakage and its resistive component of MOA. Power System Technology, 28(8): 55~59.

Wang, X.B. 2015. Fault diagnosis and example analysis of MOA based on charged test technique. Insulators and Surge Arresters. 03:69~73.

Wooi, C.L., Abdul-Malek, Z. & Mashak, S.V. 2013. Effect of Ambient Temperature on Leakage Current of Gapless MOA. Jurnal Teknologi. 64(4):157~161.

Zhang, P., Wang, L.M. & Zhao, X.Y. 2013. Metal oxide arrester fault diagnosis based on back-propagation neural network. Shanxi Electric Power.04:22~25.

Automotive, Mechanical and Electrical Engineering – Liu (Ed.)
© *2017 Taylor & Francis Group, London, ISBN 978-1-138-62951-6*

Effect of heave plate on wave piercing buoy

Dong Jiang, Jianfeng Zhang, Laihao Ma & Haiquan Chen
Dalian Maritime University, Dalian, China

ABSTRACT: This paper proposed a new type of wave piercing buoy with a heave plate on the base of a traditional columnar buoy. The mathematical model of the buoy is established by using the frequency domain analysis method and the numerical simulation is carried out in order to analyse the effect of the heave plate on the added mass and the RAO response on the heave direction of the buoy, thus providing more choices for improving the buoy's performance.

Keywords: Wave piercing buoy; Heave plate; Heave; Added mass; RAO

1 INTRODUCTION

The buoy is mainly used to mark channels or obstacles to ensure the safety of navigation. It is an important facility for navigation security. The buoy can also be equipped with sensors and a data transmission system to form a marine monitoring network (Wang et al., 2014). The United States and other countries began buoy research work in the 40 s of the last century, and have developed a variety of buoys such as the NOMAD ship buoy, which can meet the requirements of different water depths (Wang et al., 2016). The development work of buoys in China started in the 60 s of the last century, but the performance and survivability of the existing buoys is not sufficient to meet the needs of navigation and ocean monitoring compared with foreign countries. Therefore, it is necessary to further improve the structure and performance of the buoy.

This paper proposed the design of a wave piercing buoy on the base of traditional column buoy. The heave plate is added to the design based on the structure of the Spar platform, to analyse the influence of the heave plate on the performance of the buoy.

2 THE STRUCTURAL DESIGN OF WAVE PIERCING BUOY

In order to reduce the heave amplitude of the buoy, it is necessary to avoid resonance between the buoy and the wave. Mostly, the heaving inherent frequency should be kept away from the main energy frequency of the wave, so as to improve the movement stability of the buoy.

2.1 Design of the main structure

The amplitude of the heaving motion of the columnar buoy is smaller compared to other buoys, because the waterline area of it is much smaller. This makes the columnar buoy become the common high stability buoy. The Newman had studied on the heave inherent period of columnar buoy and given the calculation equation:

$$w_0 = \sqrt{\frac{\rho g A_0}{M}} \tag{1}$$

where W_0 = natural heaving frequency of buoy, ρ = density of seawater, A_0 = waterline areas of buoy, M = weight of the buoy. The equation shows that reducing the quality of the buoy can improve the heave inherent period of it, therefore reducing the heave amplitude and improving performance of the buoy significantly. This paper puts forward the design of the wave piercing buoy on the basis of a traditional columnar buoy. The new type of buoy is processed by UHMWPE and uses a multi-pontoon design in order to reduce the quality of the buoy, which can improve the natural heave frequency of the buoy and avoid resonance of it with the wave. The use of a multi-pontoon design also can get a larger wet surface area, which can increase the resistance and resilience of the buoy in the water to reduce the swing angle.

2.2 Heave plate

The heave inherent period of the buoy is related to the draft and the added mass of it.

$$T = 2\pi \sqrt{\frac{T_d (1 + C_a)}{g}} \tag{2}$$

where T_d = draft of buoy, C_a = added mass factor, g = acceleration of gravity (Wei et al., 2010). There are two ways to increase the heaving inherent period of the buoy according to the equation above, increasing the draft or the additional mass of it. However, increasing the draft will increase the quality and size of buoy, leading to an increase in construction costs and difficulties in transport and installation. Reference design of Spar platform, the heave plate which installed at the bottom of the buoy is proposed (Sang et al., 2015). The heave plate is a circular plate coaxial with the buoy, whose diameter is greater than the buoy's diameter. The installation of the heave plate is more economical and practical than that of adding the buoy draft. The structure of the wave piercing buoy with heave plate is shown in Figure 1.

According to the RANS Equation, the influence of the heave plate on the hydrodynamic performance of the buoy mainly depends on the dimensionless parameter Kc (Keulegan-Carpenter Number) and β:

$$K_c = \frac{2\pi a}{D} \tag{3}$$

$$\beta = \frac{D^2 f}{v} \tag{4}$$

where D = diameter of circular heave plate, a = amplitude, f = frequency, v = Viscosity coefficient of fluid motion (Zhou et al., 2015). In this paper, the influence of the heave plate on the hydrodynamic performance of the buoy is studied through changes to the diameter of the heave plate. Four groups of heave plates with the same thickness and a larger diameter were selected for comparison and analysis. The parameters of the heave plates are shown in Table 1.

3 HYDRODYNAMIC ANALYSIS

The added mass and RAO response curves of the buoy in the heave direction are obtained by mathematical modelling and numerical simulation using the frequency domain analysis method. The influence of the heave plate on the buoy's added mass and heave direction motion performance was studied by comparative analysis.

3.1 Analysis of the added mass

The results of the hydrodynamic analysis of the buoy with heave plates of different diameters are shown in Figure 3 and Figure 6, and the results are compared with the buoy without the heave plate, shown in Figure 2. It is found that the heave plate can effectively improve the added mass of the buoy in the heave direction, and the value of the

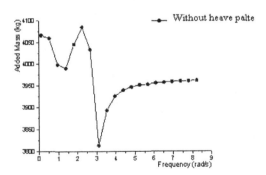

Figure 2. Added mass of buoy without heave plate.

Figure 1. Wave piercing buoy with heave plate.

Table 1. Parameters of heave plate.

Diameter (m)	Thickness (mm)	Quality (Kg)
3.0	20	128.3
4.0	20	232.2
5.0	20	367.8
6.0	20	529.1

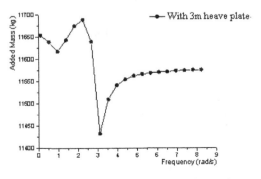

Figure 3. Added mass of buoy with 3 m heave plate.

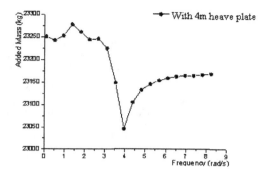

Figure 4. Added mass of buoy with 4 m heave plate.

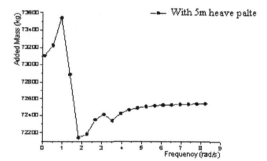

Figure 5. Added mass of buoy with 5 m heave plate.

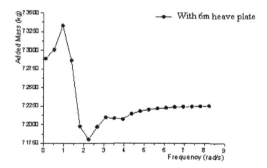

Figure 6. Added mass of buoy with 6 m heave plate.

added mass increases as the heave plate diameter increases. The reason for this is because when the buoy moves in the heave direction, the existence of the heave plate will produce a lot of vortices, which will increase the added mass of the heave motion. The bigger the diameter of the heave plate, the more vortices will be generated, thus the added mass will be larger. The amplitude of the added mass is also affected by the wave frequency, with the increase of frequency present the change trend of increasing first and then decreasing and then increasing. In the main energy frequency of waves

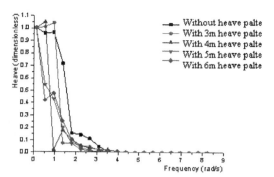

Figure 7. Heave RAO of buoy with different diameters of heave plate.

within 2–4 rad/s, the fluctuation of the added mass of the buoy is relatively obvious and the minimum value of it appears. The minimum value of the added mass of the design scheme with the heave plate is obviously larger than that without the heave plate, and the magnitude of the heave damping value, both at the high frequency and the low frequency, is also much larger than that without the heave plate.

3.2 Analysis of the heave RAO

The hydrodynamic analysis of the design scheme with different diameter heave plates was carried out and the RAO curve was obtained, as shown in Figure 7.

The heave plates of different diameters have different effects on the heave response of the buoy in the whole wave frequency, but in the main energy frequency range of the wave, the design scheme of the heave plates with different diameters can both remarkably improve the heave response of the buoy and reduce the severity of heave motion, with those using the 5 m heave plate having the best results. The heaving RAO of it has a decreasing tendency in the whole frequency, which shows that the buoy does not resonate with the wave, and its RAO is only one third of that of the design without the heave plate at the same frequency.

4 CONCLUSION

In this paper, a new type of wave piercing buoy with heave plate is proposed.

The heave plate can produce a lot of vortices when the buoy moves in the heave direction, therefore effectively improving the added mass of the buoy in the heave direction and reducing the amplitude of heave motion. The bigger the diameter of the heave plate, the larger the added mass that will be generated. In the main energy frequency of the wave the fluctuation of added mass is relatively obvious and the minimum value appears. This minimum value of the buoy with the heave plate is obviously larger than that without the heave plate. Therefore, the scheme with different diameters of heave plates can improve the heave response of the buoy, reduce the amplitude of heave RAO, improve the performance and survival ability of the buoy, and meet the needs of navigation and marine monitoring.

ACKNOWLEDGEMENTS

This work was supported by the Fundamental Research Funds for the Central Universities and Natural Science Foundation of Liaoning Province.

REFERENCES

Sang Song, Yi Shuyu, Shi Xiao, Tian Yi & Liu Wei, (2015). Study on motion performance effect of semi-submersible platform derives from heave plates structures. *China Offshore Platform*, 30(5).

Wang, B., Li, M., Liu, S.X., Chen, S.Z., Zhu, Q., & Wang, H.G. (2014). Current status and trend of ocean data buoy observation technology applications. *Chinese Journal of Scientific Instrument*.

Wang, J., Wang, Z., Wang, Y., Liu, S., & Li, Y. (2016). Current situation and trend of marine data buoy and monitoring network technology of China. *Acta Oceanologica Sinica*.

Wei, Y., Yang, J., & Chen, X. (2010). A review of the hydrodynamic performance of heave damping plates on Spar platform. China Offshore Platform, 25(6).

Zhou Guolong, Ye Zhou, Cheng & XinLi Chun, (2015). Influence of heave plate on hydrodynamic characteristics of traditional spar platform. *Journal of Water Resources & Water Engineering*, 26(4).

Automotive, Mechanical and Electrical Engineering – Liu (Ed.)
© 2017 Taylor & Francis Group, London, ISBN 978-1-138-62951-6

Height detection in smart baby weighing system using machine vision

Qiang Fang & Huanrong Tang
Xiangtan University, Xiangtan, China

ABSTRACT: At present, the height of a baby is detected manually by a nurse. In order to realise rapid non-destructive testing for the detection of a baby's height and weight, a module was established to detect the height of baby based on machine vision in a smart baby weighing system. This paper proposed a fast detection method for finding a baby's height using the GrabCut segmentation algorithm, based on the image captured by a monocular camera in a smart baby weighing system. First, a baby was marked by preprocessing and a body detector. Second, by accurately extracting the feature points of the top of the head and the feet, the real height of the baby is calculated accurately using the theory of the GrabCut segmentation algorithm. The test results proved that the method has obtained the precision of the manual method at present and been adaptive to the angle of camera and baby moving posture. The simple and efficient structure of the detecting system ensures its application.

Keywords: Height Detection; Smart Baby Weighing System; Monocular Machine Vision; Image Processing

1 INTRODUCTION

Image processing has had a rapid development in our society. Plenty of productions applied to the electrical engineering field use image processing technology. A newborn is the hope of the whole family, so there are special and high concerns to enforce the special requirements of the quality of babies. The height of babies, the weight of them, etc. must to some extent meet the requirements of maternity hospitals. Therefore, it is an important step to detect the quality of babies during the newborn growth period. At present, they are detected manually, one by one. This outdated operation mode makes the speed of detection slow and the measurement unstable. It is urgent for nurses to adopt an automatic detecting technology instead of a manual one due to the ever-increasing demands of users.

Nowadays, it is rare to find automatic detecting methods and devices around the world. Most of the existing weight detecting devices are electronic, but the baby's height detecting devices are mechanical. Some experts have developed a video-based real-time auto body height measurement system. A method using a CCD sensor (Mingxin et al., 2015) was proposed by Qiulei Dong when he was undertaking a Ph.D. at the Chinese Academy of Science. It can measure the babies with different camera angles. Another method, named the multi-target tracking body height measure-

ment system (Qiulei et al., 2009), was developed by Mingxin Jiang, a researcher from Dalian University of Technology. There are both creatively make the height measurement function come true. Both are smart devices, however, there is still little research about height measurement using machine vision technology.

Machine vision inspection technique is a detecting technology, which is a simulation of human visual function. It obtains the required information by means of analysis and calculation of the scene image. This technique is excellent for extracting information of two-dimensional shapes; the contour, size, features and other aspects of an object. The technique has the merits of high speed, high precision, abundant information, non-contact detection, etc., which has a lot of successful applications for a variety of objects and a number of industries (Rother et al., 2008). Image segmentation is one of the basic problems for integration with machine vision, which is the key step in image analysis. The application of image segmentation technologies for detecting height has principally focused on utilising hyperspectral imaging techniques to test the baby. The nature of image segmentation is in accordance with certain criteria, which is divided into different areas. These disjointed regions have the same or similar characteristics, while the adjacent and different characteristics are separated by the border regions between the zones. Now, the common image segmentation methods are an edge-based approach and a region-based approach (Hui & Xiujie, 2005), and the best performing one is the region-based approach.

This paper established a module using a region-based algorithm named GrabCut (Rother et al., 2004) to detect the height of a baby in a smart baby weighing system, which is based on monocular machine vision technology. The system takes a picture of a baby and obtains the precise size of the brick by using an image processing technique. It can improve the efficiency and stability of the detection greatly, which can bring the manufacturers profit and improve the competitiveness of their products.

The principles and structure of the height detecting system are introduced in the second section, and the two key algorithms of body detecting and GrabCut segmentation are presented in Sections III and IV separately. The experimental results of the detecting system are given in Section V. At the end, the conclusion is given in Section VI.

2 STRUCTURE AND PRINCIPLE OF HEIGHT DETECTING SYSTEM

In this paper, the height detecting system proposed is a monocular machine vision system, whose structure belongs to the simplest of the machine vision systems. Since the length and weight of babies to be detected are scalar parameters, the monocular machine vision system is the most appropriate to be adopted. Not only is its structure simple but also its mathematical model, calculation method and calibration process are simple and easily implemented. Therefore, it is easy to ensure the reliability of the system.

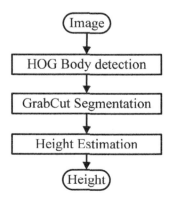

Figure 1. The flowchart of baby height detection by machine vision.

The basic structure of the monocular machine vision system, as shown in Figure 1, is composed of a camera and a computer. The camera faces to one of the planar surfaces of a detected object and takes a picture, then the image is transferred to the computer where it is analysed and calculated to get the desired information. Because a point in a plane of the detected object and its projection point in image plane is one to one mapping illustrated in Figure 1, the principles it is based on are the body detecting and segmentation algorithm in the picture. Therefore, we apply the body detecting and image segmentation technologies, take the smart baby weighing system, and integrate with the Internet network technology, for the purpose of getting a higher measurement efficiency.

3 HOG BODY DETECTION ALGORITHM

As mentioned above, Histograms of Oriented Gradients (HOG) is one of the key methods for detecting the body of a baby in the system. During video processing, HOG has been widely applied in detecting pedestrian. HOG features proposed by Dalal and Triggs (Dalal & Triggs, 2005) are adopted for our application. To detect the area of the baby's body, we firstly calculate the gradient orientations of the pixels in the cells of the image. Then in each cell, we calculate a 9-dimensional histogram of gradient orientation of the features. Each block is represented by a 36-dimensional feature vector, which is normalised by dividing each feature bin with the vector module, named HOG feature vector.

After HOG feature representation, baby detection is formulated as a linear classification problem in a high-dimensional feature space. We detect the body of the baby using a linear SVM with Gaussian kernel SVM to increase performance. This approach ensures a lower empirical risk than other classifiers. The samples of detecting results are shown in Figure 2.

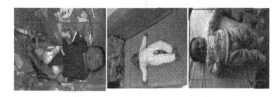

Figure 2. The samples of body detecting results (red rectangles).

4 GRABCUT SEGMENTATION ALGORITHM

The GrabCut algorithm is an image segmentation algorithm that gives an effective extraction of foreground objects from a complex background in interactive for dividing a high accuracy and efficiency. The GrabCut has made two enhancements to the graph cuts mechanism: iterative estimation and incomplete labelling, which together allow a considerably reduced degree of user interaction for a given quality of result. This allows GrabCut to simply detect a rectangle around the desired object. Then a new mechanism for alpha computation has been developed for border matting, whereby alpha values are regularised to reduce visible artefacts.

As it is impractical to construct adequate colour space histograms, we follow a practice that is already used for soft segmentation and use GMMs. Each GMM, one for the background and one for the foreground, is taken to be a full-covariance Gaussian mixture with K components. In order to deal with the GMM, the additional vector $k_n \in \{1,...,K\}$ is introduced in the optimisation framework, assigning to each pixel a unique GMM component, one component from either the background or the foreground model, according as $\alpha_n \in \{0,1\}$. An energy function $E(\underline{\alpha},k,\underline{\theta},z)$ is defined so that its minimum should correspond to a good segmentation depending on k as follows:

$$\hat{\underline{\alpha}} = \arg\min_{\underline{\alpha}}(E(\underline{\alpha},k,\underline{\theta},z)) = \arg\min_{\underline{\alpha}}(U(\underline{\alpha},k,\underline{\theta},z)+V(\underline{\alpha},z)) \quad (1)$$

The data term U is defined for segmentation, where $\pi(\cdot)$ are the mixture weighting coefficients, the parameters of the model are $\underline{\theta} = \left\{\pi(\alpha,k),\mu(\alpha,k),\sum(\alpha,k)\right\}$ that consists of weights, π, means μ and covariances Σ.

The structure of the algorithm in Rother (2004) guarantees proper convergence properties. This is because each of the steps of iterative minimisation can be shown to be a minimisation of the total energy E with respect to the three sets of variables k, $\underline{\theta}$, , and $\underline{\alpha}$ in turn. Hence E decreases monotonically, and the algorithm is guaranteed to converge at least to a local minimum of E and to terminate iteration automatically. Then, an alpha-map for the strip without generating artefacts is estimated and recovered from the foreground colour. After the border and foreground estimation, illustrated in Figure 3, the heights of the babies are the maximum value of white pixels in the black and white image.

5 EXPERIMENTAL RESULTS

In order to verify the accuracy of this automatic measurement algorithm, the comparison experiments are made in pictures from maternity hospitals. They are carried out on the client of the developed system prototype. We ran the experiments on a Pentium(R) E700@3.2GHz CPU unit, dual CPU core, 2GB memory drive with 64-bit windows file system in Microsoft Windows. Various results obtained by the automatic algorithm (shown in Figure 4) are compared with manual and automatic measurement results, respectively. The manual measurement results are obtained by two different nurses, and each person repeats three times.

With regards to the automatic method, the coordinates of the rectangles are detected automatically in the images for GrabCut segmentation. This is because both the manual and automatic measurement results of the height H are highly dependent on the results of the ventral straight length. The percentage difference between the manual method and automatic method is calculated as follows:

$$diff = \frac{|X - X_m|}{X} \times 100\% \quad (2)$$

where $diff$ is the percentage difference, and X and X_m are the automatic and manual measured value of the baby, respectively. Sixteen baby images are

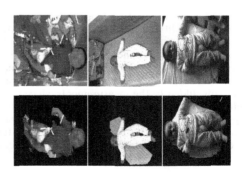

Figure 3.   The border estimation results.

Figure 4.   The foreground segmentation results.

373

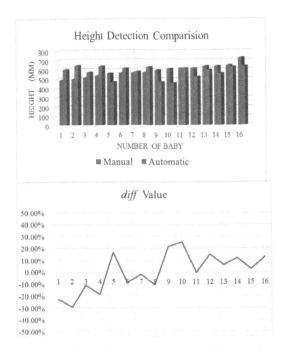

Figure 5. The performance of the two different methods for height detection.

utilised for the validation experiments to evaluate the accuracy of the automatic method.

The performance of the two different methods for the height determination of the baby are presented in Figure 5. The *diff* was used to evaluate the precision of the automatic method. The mean and standard deviation of *diff* in Figure 5 are 0.5% and 0.16, respectively. It could also be found that the mean and standard deviation of the difference among manual results and automatic ones are 9.99 mm and 44.39, respectively. It indicated that the precision of automatic methods to detect baby height is less than 1 cm. For the automatic method, it could be observed that half of the automatic results are a little higher than the manual ones and half are lower. The reason might be that some subjectivity existed in the method used to find the coordinates of the rectangles in images automatically. Besides, the existence of surface curvature of the baby added to the measurements, since it was difficult to eliminate the measurement error caused by curved surfaces. Recently, some high-end cameras could be used to eliminate the curvature effect within a certain object distance by cameras.

But these cameras have not been used widely due to their limited application fields.

6 CONCLUSION

In this paper, the HOG body detecting and Grab-Cut segmentation algorithms are developed in the smart baby weighing system to determine the height of the baby. First, a smart baby weighing system is taken as an example to introduce the height detection problem. Second, the HOG and Grab-Cut algorithm are applied to estimate the height. Finally, the conclusion is given by comparing this with the results of manual detection. The proposed height measurement algorithm is proved to be precise in application. In terms of the accuracy of the automatic algorithm, the mean value of the height difference is less than 1 cm. Consequently, the GrabCut height measurement algorithm is acceptable for baby height determination and the implementation of the algorithm would promote height measurement technology and improve standards formulations of social application.

ACKNOWLEDGEMENTS

This research has been supported by the construct programme of the key discipline in Hunan province (computer science and technology).

REFERENCES

Dalal N., & Triggs B. (2005). *Histograms of Oriented Gradients for Human Detection [C]*. International Conference of Computer Vision Pattern Recognition, pp. 886–893.

Hu M., Dong Q., Malakar P K, et al. (2014). Determining banana size based on computer vision [J]. *International Journal of Food Properties*, 18(3), 508–520.

Hui Y., & Xiujie Q. (2005). Survey of image segmentation method [J]. *Computer Development & Applications*, 18(3), 21–23.

Mingxin J., Peichang W., & Hongyu W. (2015). Height estimation algorithm based on visual muti-object tracking [J]. *Acta Electronica Sinica*, 43(3), 591–596.

Qiulei D., Yihong W., & Zhanyi H. (2009). Video-based real-time automatic measurement for the height of human body [J]. *Acta Automatica Sinica*, 35(2), 137–144.

Rother C C E., Kolmogorov V., & Blake A. (2008). Border matting by dynamic programming: EP, US 7430339 B2 [P].

Rother C., Kolmogorov V. &, Blake A. (2004). "Grab-Cut": Interactive foreground extraction using iterated graph cuts [J]. *ACM Transactions on Graphics*, 23(3), 307–312.

Automotive, Mechanical and Electrical Engineering – Liu (Ed.)
© 2017 Taylor & Francis Group, London, ISBN 978-1-138-62951-6

Lamb wave metal plate defect detection based on COMSOL and finite element analysis

Peng Qin
Institute of Signal Capturing and Processing Technology, North University of China, Taiyuan, China
School of Computer Science and Control Engineering, North University of China, Taiyuan, China

ABSTRACT: With the rapid development of China's economy, and with metal plates being widely used in aerospace, the national defence industry, shipbuilding industry, high-speed railway and so on, the domestic demand for metal plates is increasing. However, in sheet metal processing and production processes, due to the influence of the environment, there will inevitably be a variety of different types of defects. As a result, it is of great significance to choose a reliable method to carry out a large area detection of the metal plate. The electromagnetic ultrasonic guided wave transducer, because of possessing the advantages of both the Electromagnetic Ultrasonic Transducer (EMAT) and the ultrasonic guided wave, which is namely non-contact, large detection area, high sensitivity, small attenuation, and long distance transmission, is very suitable for non-destructive testing of large areas of sheet metal. However, its energy-change efficiency is not high. In order to improve the non-destructive testing of the electromagnetic ultrasonic Lamb wave transducer, this paper studies the optimisation method of the Lamb wave transducer in the detection of metal plate, carries out simulation calculations of the Lamb wave transducer after optimisation, and applies the optimised Lamb wave transducer in the detection of sheet metal defects. Simulation and experimental results show that for the Lamb wave, in the face of defect mode conversion, as the defect depth increases, the received echo signal amplitude increases, which characterises the defect echo size.

Keywords: Lamb wave; Defect detection; Finite element analysis

1 INTRODUCTION

The metal plates are widely used in aerospace, the national defence industry, shipbuilding industry, high-speed railway and so on, and the domestic demand for the metal plate is increasing. From 2011 to 2013, the production of metal plate in our country increased day by day; China's steel production increased from 6.13 tons to 11 tons; China's aluminium production increased from 17.5537 million tons to 2.41 million tons (Imano and Endo, 2013). However, in sheet metal processing and production processes, due to the influence of the environment, there will be a variety of different types of defects, and thus choosing a reliable method to detect metal plates is important. Ultrasonic guided wave detection technology is a kind of technology that has advantages such as time-saving, labour-saving, simple operation, high sensitivity, low detection device and low production cost, thus it can be applied in high temperatures and in high speed online detection, metal pipe defect detection, pipeline thickness detection and so on, under harsh environments.

Therefore, ultrasonic guided wave has advantages with which many ultrasonic guided waves cannot compare: small attenuation and long transmission distance, which can reach 10–100 metres. With ultrasonic guided wave, in metal plate defects detection, the particle vibration can spread across the board body, the particle can vibrate when the wave is in the internal plate body, and it can detect defects in the internal metal plate. The electromagnetic ultrasonic guided wave detection technology combines the respective advantages of electromagnetic ultrasonic detection technology and guided wave, which has non-contact, high sensitivity, small attenuation, and long distance transmission, and thus is widely used in non-destructive testing of large areas of sheet metal. Therefore, the electromagnetic ultrasonic guided wave technology is of great economic and academic significance in engineering applications. It is one of the new directions for the future development of testing technology.

2 ANALYSIS OF DEFECT DETECTION OF ULTRASONIC LAMB WAVE EMAT BASED ON COMSOL

There are two kinds of methods of ultrasonic Lamb wave testing, namely the projection method and the pulse echo method. The projection method is highly sensitive to defects, which can avoid the detection blind area. Based on the pulse echo method, it can accurately locate the position of the defect by using COMSOL finite element simulation software to simulate the electromagnetic ultrasonic Lamb wave transducer applied in flaw detection of aluminium plate. Through the particle displacement observation of the Lamb wave in the defect position, it can analyse the Lamb wave defect propagation form.

2.1 Finite element simulation of EMAT based on COMSOL

The basic starting point of the finite element analysis method is to simplify the complex problem, and then solve the problem. The solution domain is divided into several small regions, each of which is called a finite element. Solve the approximation solution of each finite element, and then according to the boundary conditions that the whole region needs to meet, solve the problem, which is also an approximate solution. In practical engineering, there are many problems only having an approximate solution, so it is possible to use the finite element analysis method to solve the problem (Huang et al., 2016). Its calculation accuracy is very high, and it can solve all kinds of problems in a complex condition. Therefore, finite element analysis has been widely used in the fields of electromagnetic field, fluid mechanics, heat conduction and so on. COMSOL Multiphasic is a multiple-physical field coupling analysis software, as shown in Figure 1.

The main features of COMSOL Multiphasic software are: 1) it can choose a variety of languages on the interface and automatically establish coupling physical field; 2) users can automatically construct the model that they themselves use; 3) it can be seamlessly connected with MATLAB, strong two times development function; 4) it can interactively model and simulate the external real environment, and there are a large number of preset physical application modes and PDE application modes (Nagy et al., 2014); 5) it can carry out

different dimensions modelling; 6) preprocessors have powerful functions, able to simply lead in the CAD format of geometric modelling form and they have powerful grid partition function; 7) it has a powerful solver and post processor.

The steps of finite element analysis are as follows: 1) to determine the problem and its solution domain; 2) to discrete solution domain; 3) to determine the state variables and control method; 4) derived units; 5) assembly solution; 6) to solve the equations and explain the results.

2.2 Detection and analysis of Lamb wave EMAT defect based on COMSOL

2.2.1 Establishment of finite element model

After completing the global definition parameter, the excitation current is substituted into COMSOL Multipgysics. Choose the space dimensions, which are divided into three-dimensional, two-dimensional axisymmetric and one dimension, two dimension, and zero dimension. The simulation in this paper selects the two-dimensional modelling, and uses the physical field needed. The permanent magnet chooses the magnetic field (mf) in the low frequency electromagnetic field, and the preset solution chooses the stable one (Stationary); the alternating magnetic field also chooses the magnetic field (mf) in the low frequency electromagnetic field, and the preset solution chooses the transient (Time Dependent); the transient pipe chooses solid mechanics (solid) in structural mechanics, and the preset solution chooses the transient to carry out the modelling calculation (Zhang et al., 2015). Carry out the two-dimensional model simulation of the electromagnetic ultrasonic Lamb wave, the basic model is composed of a permanent magnet, coil winding, aluminium plate and air layer, as shown in Figure 2.

The establishment model process is to establish bodies in different shapes by direct definition of size. It refers to make use of COMSOL Modelling—Variables (defined parameters)—Geometry—Surface to define the size of different shapes and structures. In addition, adopt Boolean operations, such as difference sets and set operations to complete the establishment of the pipe model containing defects. In the Figure 2 model, the permanent magnet is perpendicular to the surface of the aluminium plate, the material for the Nd-Fe-B permanent mag-

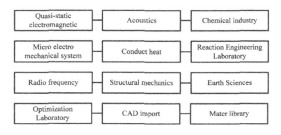

Figure 1. Multi-physics coupling of COMSOL Multiphasic.

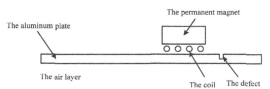

Figure 2. The electromagnetic ultrasonic Lamb wave model.

net alloy; the excitation coil with zigzag coil. In order to ensure the simulation results are more accurate, add an air field in the external, used to describe the dissipation of the electromagnetic field in a dissipative far-field region. The basic structure of the electromagnetic ultrasonic Lamb wave transducer uses a uni-polar permanent magnet, return type coil and the aluminium plate. The material properties are as follows: the relative permeability of the coil is $\mu r = 1$, the resistivity is $\rho = 1 \times 10^{-7}\Omega$ m; the aluminium resistivity is $\rho = 4.032 \times 10^{-6}\Omega$ m. The Poisson's ratio is 0.33, Young's modulus E = 6.9×1010 Pa, density $\rho = 2750$ kg/m³; relative permeability of air is $\mu r = 1$; permanent magnet: coercive force Hc = 895000.

2.2.2 *Grid division*

Before the solution of the finite element, the material parameters of the model are assigned to each model to define the material properties and carry out the mesh generation. The static bias magnetic field is applied in the direction of Y, and Br = −B0; in the 1, 3, 5 and 7 coils, the current I is applied, and in the 2, 4, 6 and 8 coils, the current −I is applied; do not apply a low reflection boundary condition on both ends of the aluminium plate; the mesh partition is divided into free mesh partition (free triangle mesh subdivision and free quadrilateral) and mapping mesh partition. Mesh partition can improve the simulation accuracy and calculation accuracy. In particular, the irregular shape model should be carried out with refined mesh partition. The free mesh partition makes use of the COMSOL smart size to automatically control the scale and mesh density of the mesh partition; the mapping mesh partition is the regular division of the regular model. There are both advantages and disadvantages for the two methods, the free mesh partition and the mapping mesh partition (Huang et al., 2014). The method for the mesh partition is supposed to be selected according to the shape and boundary conditions of the model established. The aluminium plate selects the manual mesh partition, and uses the customised unit size (both the largest and the smallest element size select 1 mm) to divide the aluminium plate containing defects into five layers, making the results for analysing specific areas (such as defects) more accurate after the solution. The permanent magnet, the coil and the air field are divided by using a smart grid, and the shape of the unit is a free split triangle. The mesh of the crack defect is shown in Figure 3.

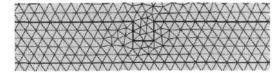

Figure 3. Crack defects meshing.

2.2.3 *Simulation results of EMAT analysis of Lamb wave*

After solution calculation, it can obtain the Lamb wave particle displacement nephogram and analyse the case and communication form of the Lamb wave in the defect. The Lorenz force is acted to the aluminium plate to carry out structure analysis, and then obtain the aluminium plate X direction particle displacement nephogram, as shown in Figure 4.

Figure 4 and Figure 5 show the maximum displacement of the particle in the area that the EMAT coil is located in. By comparing the particle displacement in the X and Y direction at the same moment, it can be seen that the motion amplitude of the particle on the X axis is far greater than the displacement amplitude on the Y axis. The particle is mainly propagated along the X axis, which indicates that the whole wave is propagated along the X axis (Masserey et al., 2014). The wave is propagated along the X axis, and the closer it gets to the positive and negative direction of the X axis, the smaller the displacement. The overall particle shape of displacement is also the propagation shape of the wave in the internal of the plate. The total displacement nephogram of the aluminium plate particle is shown in Figure 6.

In the figure, with the increase of time, the Lamb wave gradually spreads to the defect position. When the time is 12 µs, the particle displacement of the left wave of the defect is larger than that in the other positions. When the time is 10 µs, the particle displacement of the left wave of the defect is smaller than that in the other positions. It can be obtained that, when the wave meets the defect, the waveform is converted, and the new mode is generated, thus the particle displacement of the wave increases (Zhang et al., 2016). The particle displacement in the left of the defect is larger than that in the right side. In consequence, it can be obtained that, when the wave encounters defects, most of the waves will be reflected, and only a few waves will continue to move forward.

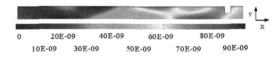

0	20E-09	40E-09	60E-09	80E-09
	10E-09	30E-09	50E-09	70E-09	90E-09

Figure 4. Displacement contours of different particle.

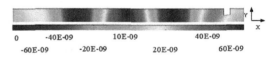

0	-40E-09	10E-09	40E-09
	-60E-09	-20E-09	20E-09	60E-09

Figure 5. Displacement contours of different particle.

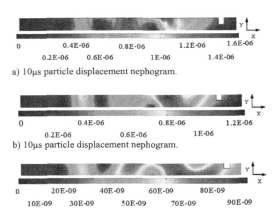

a) 10μs particle displacement nephogram.

b) 10μs particle displacement nephogram.

c) 12μs particle displacement nephogram.

Figure 6. Displacement contours of different particles.

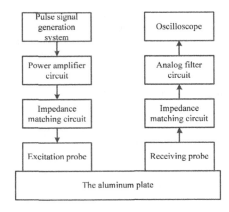

Figure 7. Electromagnetic ultrasonic Lamb wave system block diagram.

3 EXPERIMENTAL VERIFICATION AND RESULT ANALYSIS OF METAL PLATE TEST

On the basis of simulation, the metal plates under the simulation parameters are studied, and the detection of the Lamb wave on the aluminium plate and the steel plate are respectively studied. The results of the experiment coincide with the calculated numerical results, which can be used to guide the parameter selection for the Lamb wave transducer.

3.1 *Aluminium plate detection experiment*

The electromagnetic ultrasonic detection system is composed of a RAM-5000 pulse generator, receiver, impedance matching and other parts. The receiving circuit is composed of the preamplifier circuit and the analogue filter circuit. The structure block diagram is shown in Figure 7.

The pulse generator and receiver, as the excitation sources of the EMAT probe, generate RF tone burst signals, and the amplitude and frequency of the signal can be adjusted arbitrarily. In the role of a horseshoe shaped permanent magnet, a transmitting coil with high frequency current excites the Lamb wave in the internal of the aluminium plate, and through the impedance matching circuit and amplifying filtering circuit, the results are displayed by the oscilloscope. The real-time signal of the oscilloscope is read out by the ultrasonic wave detector, and through the analysis of the echo signal, obtained to explore the propagation characteristics of the Lamb wave in the metal plate.

The purpose of the matching of the coil is to enable the coil to obtain the maximum energy from the excitation source, that is, the maximum coil current under the given excitation. In order to improve the energy conversion efficiency and SNR of the Electromagnetic Acoustic Transducer (EMAT),

the distributed capacitance of the coil will have an impact on the matching of the coil. In allusion to the equivalent circuit of the EMAT coil, the matching capacitance of the coil is calculated and capacitance C is the matching capacitance required. When the high frequency excitation signal is connected, the excitation coil produces inductance, and at the same time, there is distributed capacitance existing between the coils. The calculation formula of capacitance C is shown as follows:

$$C = \frac{L_{eq} + L_t}{R^2_{eq} = \left[\omega_0\left(L_{eq} + L_t\right)\right]^2} - C_d \qquad (1)$$

In Formula (1), Req is the excitation coil equivalent resistance, Cd is the excitation coil distributed capacitor, Leq is the coil equivalent inductance, Lt is the secondary equivalent inductance of the transformer, and ω0 is the angular frequency of the excitation signal. The excitation coil and the capacitor meet the resonant conditions, the output power reaches the maximum value, and the signal-to-noise ratio is the highest. After calculation, the impedance matching capacitance is 3.9*104p F.

The bias magnetic field in the experiment is generated by the horseshoe shaped permanent magnet, the volume is 50 mm * 50 mm * 30 mm, and remanence is 1.26T. The tested plate selects the aluminium alloy plate with a length of 2000 mm, width of 1000 mm, and thickness of 4 mm. The defect distribution in the experiment is shown in Figure 8.

The crack parameters are shown in Table 1. In the experiment, the system adopts manual winded sending and receiving integrated loop back coil, and the coil spacing is 3 mm. The probe is sending and receiving integrated, and the distance to the probe is 18 cm.

In the actual analysis of the echo signal, excessive noise may be masking the signal features effectively,

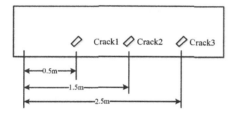

Figure 8.   The distribution of defects in aluminium.

Table 1.   The defect parameters and distribution in aluminium.

No.	Defect length	Crack depth
1	30 mm	2 mm
2	30 mm	1.5 mm
3	30 mm	1 mm

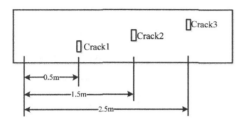

Figure 9.   The distribution of defects in the steel plate.

and the more serious is that the signal is submerged in noise. In order to improve the reliability of the signal, it is necessary to carry out noise-elimination processing of the output signal [8]. When the pulse signal is in the frequency domain, frequency dispersion is inevitable. The general pulse signal is frequency-scattered and becomes more serious with the increase of distance. The frequency dispersion phenomenon can be weakened by the matching mode of the uni-polar permanent magnet and the winding coil.

3.2   *Steel plate detection experiment*

The ferromagnetic material is jointly acted on by the Lorenz force mechanism and the magnetostrictive force mechanism. The detection of ferromagnetic materials and the detection of non-ferromagnetic materials are basically the same, using the sending and receiving integrated probe for the detection of defects in the steel plate.

Carry out defect detection of different depths on the 6 mm steel plate, the distance between the ultrasonic transducer and the defect is kept constant, and the defect distribution in the experiment is shown in Figure 9.

Table 2.   The defect parameters and distribution in steel plate.

No.	Defect length	Crack depth
1	30 mm	3 mm
2	30 mm	2.5 mm
3	30 mm	2 mm

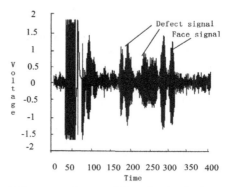

a) The measured wave of the defect depth of 3 mm.

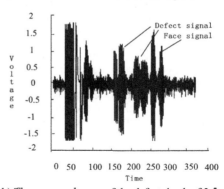

b) The measured wave of the defect depth of 2.5 mm.

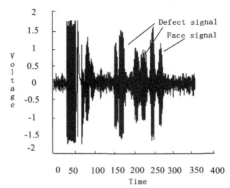

c) The measured wave of the defect depth of 2 mm.

Figure 10.   The measured waveform of different depth defects.

In the experiment, the system adopts manual winded sending and receiving integrated loop back coil, and the coil spacing is 3 mm. The probe is integrated, and the distance to the probe is 18 cm. The crack parameters are shown in Table 2.

At this time, the frequency of the excitation coil is 0.5 MHz, and the echo signal is obtained as shown in Figure 10.

In Figure 10, when the Lamb wave encounters defects, waveform conversion will occur, and generate four kinds of modes of the Lamb waves. As the defect depth increases, the amplitude of the mode increases. The experimental results verify the correctness of the Lamb wave detection of defects.

4 CONCLUSION

In this paper, the application of electromagnetic ultrasonic guided wave in metal plate is studied by numerical calculation, simulation and experimental verification. Through the COMSOL finite element simulation software and experiment, the propagation characteristics and detection ability of the Lamb wave are analysed. The research results of this paper are as follows: (1) the combination mode of the unipolar permanent magnet and return type coil whose spacing is 3 mm can excite the Lamb wave in sheet metal. By adjusting the matching circuit capacitance and inductance value, the output power reaches the maximum value, and the signal-to-noise ratio is the highest, realising the optimisation of the echo signal. (2) Simulation and experiment results show that the Lamb wave mode will be converted when defects occur, and a new Lamb wave mode will be generated. The depth of defects increase, the amplitude

of the received echo signal will also increase, and the amplitude of the echo signal characterises the scale of the defects.

REFERENCES

Burrows, S.E., Dutton, B., & Dixon, S. (2012). Laser generation of Lamb waves for defect detection: Experimental methods and finite element modeling. *IEEE transactions on ultrasonic, ferroelectrics, and frequency control*, 59(1), 82–89.

Huang, S., Wei, Z., Zhao, W., & Wang, S. (2014). A new Omni-directional EMAT for ultrasonic Lamb wave tomography imaging of metallic plate defects, *Sensors*, 14(2), 3458–3476.

Huang, S., Zhang, Y., Wang, S., & Zhao, W. (2016). Multi-mode electromagnetic ultrasonic Lamb wave tomography imaging for variable-depth defects in metal plates. *Sensors*, 16(5), 628.

Imano, K., & Endo, (2013). T. Experimental study on the mode conversion of Lamb wave using a metal plate having a notch type defect. *International Journal of the Society of Materials Engineering for Resources*, 19(1_2), 20–23.

Masserey, B., Raemy, C., & Fromme, P. (2014). High-frequency guided ultrasonic waves for hidden defect detection in multi-layered aircraft structures. *Ultrasonic*, 54(7), 1720–1728.

Nagy, P.B., Simonetti, F., & Instanes, G. (2014). Corrosion and erosion monitoring in plates and pipes using constant group velocity Lamb wave inspection, *Ultrasonic*, 54(7), 1832–1841.

Zhang, J., Ma, H., Yan, W., & Li, Z. (2016). Defect detection and location in switch rails by acoustic emission and Lamb wave analysis: A feasibility study. *Applied Acoustics*, 105, 67–74.

Zhang, Y., Wang, S., Huang, S., & Zhao, W. (2015). Mode recognition of Lamb wave detecting signals in metal plate using the Hilbert-Huang transform method. *Journal of Sensor Technology*, 5(1), 7–14.

Automotive, Mechanical and Electrical Engineering – Liu (Ed.)
© *2017 Taylor & Francis Group, London, ISBN 978-1-138-62951-6*

Lane detection and fitting using the Artificial Fish Swarm Algorithm (AFSA) based on a parabolic model

Xiaojin Wang & Zengcai Wang
School of Mechanical Engineering, Shandong University, Jinan, China
Key Laboratory of High Efficiency and Clean Mechanical Manufacture, Shandong University, Jinan, China

Lei Zhao
School of Mechanical Engineering, Shandong University, Jinan, China

ABSTRACT: This paper presents a novel optimisation algorithm for lane detection and fitting. We use the Artificial Fish Swarm Algorithm (AFSA) which is based on a parabolic lane boundary model to solve our boundary detection problem. Initially, the RGB road image is transformed into intensity images. We then use the Finite Impulse Responses (FIR) filter to eliminate noise. Next, the Otsu's method is used to convert the intensity images into binary images. Thirdly, an objective function is constituted with gradient character and grey level of the binary images, coming along with the road boundaries are fitted by parabola model. The major point of the AFSA is to optimise the parameters of the quadratic parabola based on the objective function, and setting a frame's parameters' values as the initial value of the next frame, which is equivalent to the tracking method. Experimental results of the real-time image sequences show that the method presented in this paper is capable of robustly and accurately detecting the road boundaries on highways roads. The accuracy of the algorithm has reached a high level. The mean processing speed of each image is 37 ms.

Keywords: Image preprocessing; Parabolic model; FIR filter; AFSA

1 INTRODUCTION

Road safety is a major social issue. According to the global status report on road safety of the WHO in 2015, the total number of road traffic deaths worldwide has plateaued at 1.25 million per year, with over 3,400 people dying on the roads all around the world every day. Advanced Driver Assistance Systems (ADAS) may help reduce this huge number of human fatalities which has received considerable attention since the mid-1980s. A vision-based Lane Departure Warning System (LDWS) play a significant role in an ADAS. As a matter of course, lane detection and tracking are considered as an important basic module in LDWSs.

In the past twenty years, many countries have successfully developed plenty of vision-based lane detection and tracking systems. To-date, many scholars have also proposed a large number of vision-based road boundaries and lane mark detection methods. We grouped all these methods into three types as listed below.

The first type are the feature-based lane detection methods which mainly identifies some traits of the road, such as gradient and colour. High-level primitives (gradient) that refer to environmental features are used to detect the edge of the urban roads (R. Turchetto and R. Manduchi, 2003). The papers (S. G. Jeong, 2001; Q. Lin, 2010; J. Canny, 1986; Q.-B. Truong and B.-R. Lee, 2008; B. Yu and A. K. Jain, 1997) used different gradient operators (Sobel, Canny, Roberts and so on) to extract edge information of lane markings. The colour feature with local statistical characteristics is combined to achieve the road recognition (Cheng, 2003). The main advantage of the feature-based lane detection methods is the lack of sensitivity to the shape of the road. But it will fail to illumination changes, damaged and discontinuous road markings, shadows and water area.

The second type is vision-based and multiple sensor fusion lane detection methods which perceive road information by sensor to achieve road recognition depending on various image features. Aufrere (2001) who facilitated four different sensors for the understanding of vehicle surroundings successfully used three modules for automotive purposes. Model-based multi-sensor fusion is used for simultaneous detection of lane and pavement boundaries (Ma, 2000). Currently, this method is a research hotspot, but it has the demerits of high cost.

The third type are the model-based lane detection methods which start with the hypothesis of a road model. Models used the literature include: A global linear model which is proposed to identify road boundaries (Xu, 2004), B-Snake model which is used for lane detection and tracking (Wang, 2004), a hyperbola-pair model which is used for real-time lane detection (Q. Chen and H. Wang, 2006; O. O. Khalifa, 2010), a linear-parabolic model which is used for lane detection and lane departure (C. R. Jung and C. R. Kelber, 2005), and the Catmull-rom spline model which is proposed to detect lane boundaries (Wang, 1998). Compared with the above-mentioned two kinds of methods, the model-based lane detection methods have a huge number of merits. Uniquely, they are more comprehensive since only a few parameters are needed to model the lane edges. In addition, the model-based lane detection methods are more robust against outside environmental influences (lighting changes, shadows, water areas and so on).

Taking the many advantages into account, we adopt the model-based lane detection method. Note that model selection is vital. If the model selected is not appropriate, lane detection becomes invalid. A linear (straight) model which is fast is often fitted by the means of a Hough transform, however this method constitutes relatively simplistic structures which are weak at curve recognition. Splines (Wang, 2004; Wang, 1998; Zhao, 2012) can process more complex lane marking shapes, but they require large computational resources resulting in too much time being consumed. Hyperbolic models (Chen and Wang, 2006; Khalifa, 2010) have a strong anti-noise capability, however there is a curve fitting large fluctuation shortcoming when the road encounters more noise. Different models have been studied for the purpose of determining a trade-off between complexity and accuracy. The parabolic model (A. Linarth and E. Angelopoulou, 2011) is treated as being medium between complexity and accuracy. Combined with the subsequent AFSA novel optimisation approach, we chose the parabolic model to achieve lane detection and fitting. The processing of detecting lanes is then equal to the processing of calculating those model parameters. Most noteworthy is the fact that the technique is much more robust against noise, and

does not have the disadvantages that the existing methods have.

The whole process can be summarised as follows. In the first place, we pre-process the sequential image; the RGB road images are transformed into intensity images, coming along with Finite Impulse Response (FIR) filter to eliminate noise. Secondly, the Otsu's method is used to convert intensity images to binary images. Finally, based on the information in the binary images, an AFSA's objective function is constituted with a gradient character and grey level. After the constitution of the objective function, AFSA is used to optimise the parameters of the quadratic parabolic model based on the objective function, followed by a setting of frame parameters' values as the initial values of the next frame, which is the same as the tracking method. The contribution of this paper is that the FIR filter and the AFSA are creatively used, which are not found in the reference literatures. Figure 1 shows the complete process flowchart of a frame.

The rest of this paper is organised as follows. Section 2 gives a detailed description of the parabolic model. Section 3 presents the image pre-processing including filtering and thresholding. Section 4 shows the lane detection based on the AFSA algorithm. Section 5 presents the experiments and analysis. Finally, a conclusion is given in section 6.

2 ROAD MODEL

This paper uses a quadratic parabolic model which can integrate both linear and curved shapes to fit lane edges. Formula is expressed as follows:

$$x = ay^2 + by + c \qquad (1)$$

We defined the upper left point of the image as the origin of coordinates. A horizontal line through the origin point is defined as the x-axis, and the perpendicular direction to the x-axis through the origin point is defined as the y-axis. Factor 'a' represents a lane curvature, factor 'b' represents the slope of the lane boundary at the bottom of the image, factor 'c' represents the intersection point of the lane boundary and the x-axis. It is worth mentioning that all three factors 'a', 'b', and 'c'

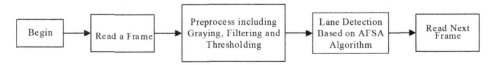

Figure 1.   The complete process flowchart.

can be calculated from the shape of the lane on the ground based on the AFSA algorithm.

It should be pointed out that the plus or minus of factor 'a' represents lane trends. If a > 0, roads turn right, in contrast, if a < 0, roads turn left. In the following sections, we will calculate the parameters of left and right lanes respectively using the AFSA algorithm.

3 IMAGE PROCESSING

The input vision sequences are RGB images in our processing procedure. The first step is to convert the RGB images to intensity images. During the process of converting the image from RGB to grey-scale, that is to say, specific weights to channels R, G, and B ought to be applied. These weights are: 0.2989, 0.5870, and 0.1140.

3.1 Filtering

Since it would affect subsequent processing if noise appeared in intensity images, noise removal is very important. As our select, the transformed intensity images are then filtered by a two dimensional Finite Impulse Response (FIR) filter. The FIR filter consumes much less time than the median filter which is commonly used in many studies. The most important thing is that the FIR filtering resulting image is much better than the median filter. The result of FIR filtering is shown in Figure 2.

3.2 Thresholding

Thresholding is critical, because it allows the images to be recognised more efficiently. We use the adaptive Otsu's method to alter intensity images into binary images. The adaptive Otsu's method determines the threshold by splitting the histogram of the input image to minimise the variance for each of the pixel groups. Equation (2) gives the main idea of the adaptive Otsu's method (Y. Zhang and L. Wu, 2011).

Figure 2. FIR filtering result.

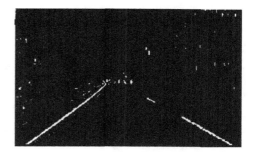

Figure 3. Thresholding result.

$$\sigma_\omega^2(t) = \omega_1(t)\sigma_1^2(t) + \omega_2(t)\sigma_2^2(t) \qquad (2)$$

Parameter t is the threshold separating the two classes, $\omega_i(t)$ denotes the probabilities of the two classes separated by the threshold t, $\sigma_i^2(t)$ denotes the variances of the two classes.

Otsu's method works better compared to other thresholding methods, as the results of thresholding a binary image are shown in Figure 3.

4 LANE DETECTION

4.1 Determining the objective function

The AFSA's objective function is used to evaluate the accuracy of lane detection fitted by a quadratic parabolic model. So it plays a key role throughout the lane detection. All sequences tested in this paper were captured with a resolution of 240×320 pixels. Since the lane boundaries are mainly concentrated on the lower half of the images, we only detect the 150 pixels below so that the real-time performance is improved, as shown in Figure 4.

According to the basic requirements of road construction, a lane marking line's width is generally 0.15 metres which is approximately 5 pixels in the image. In order to accurately reflect the statistical characteristics of the binary images, we set 5×5 pixel blocks as the minimum statistical units. When detecting the left or right lane parabolic equations, we sequentially take out N points downward as the centre of pixel blocks starting from the 90th binary image row in accordance with the principle of the same y coordinate intervals (5 pixels). The whole pixel blocks' centre points thus fulfil the parabola. If the x coordinates of any points were not within the scope of [3,318], it indicated that the blocks were not on the binary images and should be removed. Finally, the valid blocks number is N (N≤30). The left and right lane blocks are shown in Figure 5.

According to Figure 5, the valid blocks have a high gradient character and grey level, therefore, the objective function is determined by considering

Figure 4. Coordinate system and the lane detection area.

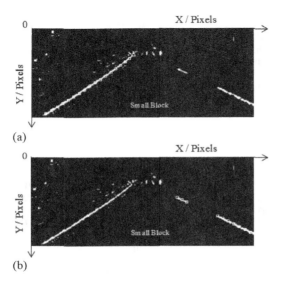

(a)

(b)

Figure 5. The left and right lane blocks.

the difference between the mean grey of the blocks and the mean grey of the image. The average grey value of the kth block is calculated by the following Equation (3).

$$\bar{H}_k = \frac{\sum\limits_{m,n=-2}^{2} h(x+m, y+n)}{4 \times 4 \times 255} \qquad (3)$$

where $\bar{H}_k$ is the average grey value of the kth block; $h(x+m, y+n)$ is the grey value at coordinates $(x+m, y+n)$.

The mean grey of an image is calculated by the following Equation (4).

$$\bar{H} = \frac{\sum\limits_{y=90}^{240}\sum\limits_{x=0}^{320} h(x, y)}{150 \times 320 \times 255} \qquad (4)$$

where $\bar{H}$ is the mean grey of an image, $h(x,y)$ is the grey value at coordinates (x,y).

Considering the fact that the lane boundary at the bottom of the images is much clearer and the lane boundary becomes blurred higher up, we set different weighting coefficients for each $\bar{H}_k$ depending on the position of the blocks. The weighting coefficients formula is as follows:

$$\rho_k = 1 - \frac{k}{2N} \qquad (5)$$

The ultimate objective function can be expressed as:

$$F(a, b, c) = \sum_{k=1}^{N} \left(\rho_k \bar{H}_k - \bar{H} \right) \qquad (6)$$

Factors a, b, c are the parameters of the parabolic model.

4.2 Parameter initialisation

The lane detection proposed herein is to optimise a, b, c to make the parabola as close as possible to the lane lines. In the image sequence, the lane boundaries of a frame change slowly compared with the next frame, so we set a frame parameters' value as the initial value of the next frame, which is equivalent to the tracking method. At the first few frames, lane recognition is not yet stable, so we have adopted the following approach to set parameters.

1. The first frame has a search blindness, however considering the high-frequency driveway of the vehicle and the signification of a, b, c we set the region of interest (ROI) of a, b, c as shown in Table 1.
2. Because of minor changes of lane edges between two adjacent frames, we set ROI of a, b, c except for the first frame as $a_{last} - 0.004 \leq a \leq a_{last} + 0.005$, $b_{last} - 0.4 \leq b \leq b_{last} + 0.5$, $c_{last} - 20 \leq c \leq c_{last} + 25$, where a, b, c jump respectively at 0.001, 0.1, 0.5.

4.3 Artificial Fish Swarm Algorithm (AFSA)

AFSA, which was first proposed in 2002 (X.-l. Li, 2002), is one of the optimisation methods among the Swarm Intelligence (SI) algorithms. This algo-

Table 1. Region of Interest (ROI) of a, b, c.

Parameters	Left	Right
a	(−0.005, 0.005)	(−0.005, 0.005)
b	(0.3, 2.3)	(−2.3, −0.3)
c	(−100, 150)	(200, 450)

rithm is inspired by the collective movement and various social behaviours of fish which result in an intelligent social behaviour. The basic idea of AFSA is to imitate the fish's behaviour including preying, swarming, following and moving with local search of fish individual, matter of course, reaching the global optimum (X.-L. Li, 2003). AFSA has many merits including high convergence speed, high accuracy, flexibility, autonomous and fault tolerance.

Artificial Fish (AF) is a fictitious entity of true fish, which is used to seek out the blocks. With the help of the object-oriented analytical method, we can regard the AF as an entity encapsulated with one's own data and a series of behaviours, which can accept stimulating information from the environment by sense organs. The environment where the AF lives is the solution space and the states of other AFs. Its next behaviour relies on the current state and environmental state itself; meanwhile, it influences the environment by virtue of its own activities and other companions' activities.

AF realises external environment by its vision shown in Figure 6. X s the current state of an AF, $Visual$ is the visual distance, $Step$ is the moving step length, X_v is the visual position at some moment, X_{next} is the next position to be reached, X_i is another AFs. If the state at the visual position is better than the current state, X goes forward a $Step$ in this direction, and arrives at X_{next}; otherwise, X continues an inspection tour within its vision. The processing can be expressed as follows:

$$X_v = X + Visual \cdot Rand() \qquad (7)$$

$$X_{next} = X + \frac{X_v - X}{\| X_v - X \|} \cdot Step \cdot Rand() \qquad (8)$$

where $Rand()$ produces random numbers between 0 and 1.

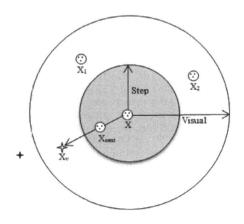

Figure 6.   Vision concept of the artificial fish.

Depending on the object-oriented analytical method, we define AFs encapsulated with two parts: variables and functions. The variables include: N is the number of AFs, X represents the current position of AF, $Visual$ is the visual distance, $Step$ represents the moving step length, Try_number represents the try number, $iterate \_ times$ is the maximum number of iterations. $\delta (0 < \delta < 1)$ is the crowding factor, $d_{ij} = | X_i - X_j |$ represents the distance between two AFs i and j. The functions include: the objective function $Y = F(a, b, c)$ which represents the food concentration in the current position; the behaviours of AF contain preying, swarming, following, moving and evaluating.

1. Preying behaviour: This is a basic biological behaviour of tending to the food; generally, a fish perceives the food concentration in water to determine the movement by vision or sense shown in Figure 8.

   Behaviour description: as displayed in Equations (9) and (10), let X_i be an AF current state and select a state X_j randomly in its visual distance, the objective function Y is the food concentration.

$$X_j = X_i + Visual \cdot Rand() \qquad (9)$$

   If $Y_j > Y_i$, it goes forward a step in this direction as displayed in Equation (9).

$$X_i^{t+1} = X_i^t + \frac{X_j - X_i^t}{\| X_j - X_i^t \|} \cdot Step \cdot Rand() \qquad (10)$$

   Otherwise, select a state X_j randomly again and judge whether it satisfies the forward condition, if it cannot satisfy the condition after Try_number times, it moves a step randomly as shown in Equation (11).

$$X_i^{t+1} = X_i^t + Visual \cdot Rand() \qquad (11)$$

2. Swarming behaviour: Fish naturally gather in groups during the moving process, which is a kind of living habit to guarantee the survival of the group and avoid dangers.

   Behaviour description: let X_i be an AF current state, X_c is the centre position, and n_f is the number of its companions in the current neighbourhood ($d_{ij} < Visual$).

   If $\frac{Y_c}{n_f} > \delta Y_i$, which means the companion centre has more food and is not very crowded, X_i goes forward a step as per Equation (12) to the companion centre.

$$X_i^{t+1} = X_i^t + \frac{X_c - X_i^t}{\| X_c - X_i^t \|} \cdot Step \cdot Rand() \qquad (12)$$

Otherwise, execute the preying behaviour.

3. Following behaviour: In the swimming process, the neighbourhood partners will trail and reach the food quickly, when a single fish or several ones find food.

Behaviour description: let X_i be an AF current state, and it explores the companion X_j in the neighbourhood $(d_{ij} < Visual)$, which has the greatest Y_j.

If $\frac{Y_c}{n_f} > \delta Y_i$, which means the companion state X_i has higher food concentration and the surrounding is not very crowded, X_i goes forward a step to the companion X_j.

$$X_i^{t+1} = X_i^t + \frac{X_j - X_i^t}{\| X_j - X_i^t \|} \cdot Step \cdot Rand() \qquad (13)$$

Otherwise, execute the preying behaviour.

4. Moving behaviour: Fish move randomly in water; in fact, they are seeking food or companions in larger ranges.

Behaviour description: Choose a state randomly in the vision distance, and then move towards this state as displayed in Equation (14). In fact, it is a default behaviour of preying behaviour.

$$X_i^{t+1} = X_i^t + Visual \cdot Rand() \qquad (14)$$

5. Evaluating function: We evaluate the behaviours of preying, swarming, following and moving using a heuristics method. We then select the behaviours that have the biggest objective function value to execute. After every action, the objective function value is compared with the bulletin board which is set to record the optimum value of selected behaviour that has been performed, and then retain the biggest one. After *iterate_times*, the value on the bulletin board is the optimisation value that is nearest to the true. And the best a, b, c is out.

4.4 Lane detection based on AFSA

We detect lane boundaries based on AFSA according to the steps below.

1. Initialise parameters, including *Total* which is the total number of AFs; *iterate_times* which is the maximum number of iterations; variables of AFs.
   Set *Total* = 20, *iterate times* = 30, *Visual* = 1.5, *Step* = 0.3, δ = 0.6, *Try_number* = 30.
2. Get the optimal a_0, b_0, c_0 according to the lane detection results of the previous frame.
3. Evaluate all the AFs depending on the objective function and select one of the behaviours of preying, swarming, following and moving that have the maximum objective function to execute.

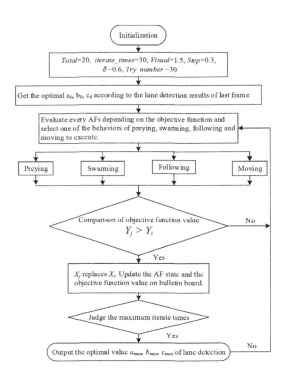

Figure 7. Flowchart of lane detection based on AFSA.

4. Compare objective function values with the bulletin board. If $Y_j > Y_i$, X_j replaces X_i. Update the AF state and the objective function value on the bulletin board, otherwise, step to step three.
5. Judge the maximum iterate times, if the program has achieved the maximum *iterate_times*, output the optimal value $a_{max}, b_{max}, c_{max}$ which is closest to the true value of lane detection, and the loop ends. Otherwise, step to step three.

The processing steps of each frame are the same as shown above. A flowchart of lane detection based on AFSA is shown as Figure 7.

5 EXPERIMENTS AND ANALYSIS

In order to verify the feasibility of the proposed algorithm, we have taken a large number of image sequences on urban roads and highways. During the experiment stage, our programming platform was a desktop computer with Inter (R) Pentium (R) CPU G3220 @3.00GHz and RAM was 4.00 GB. The programming was done in MAT-LAB programming software, and the algorithm was tested on the images with 320*240 pixels image resolution. As mentioned earlier, and illustrated in Figure 4, we set the ROI and then we only detected the lower 150 pixels of an image in order to improve real-time performance.

Figure 8.   Lane detection experiment results.

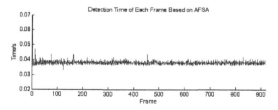

Figure 9. Detection time of each frame based on AFSA.

Figure 8 sshows some of our experimental results where the detected lane boundaries are superimposed onto the original images. The results are very good.

The results of our test show the robust performance of our algorithm. The real-time performance of our algorithm is shown in Figure 9. The figure presents the elapsed time of the first frame exceed more than 70ms to achieve 0.17s why the main reason is that it includes programming initialisation time; however subsequent thousands of frames' processing time was stabilised at a mean of 37 ms.

6 CONCLUSION

In this paper, a novel optimisation algorithm AFSA based on a parabolic lane boundary model was proposed to solve our lane boundary detection problem. Prior to the adoption of AFSA, we firstly preprocessed each frame of sequential videos in order to get ideal binary images, which involved setting the region of interest, removing noise using anFIR filer and binarisation using the Otsu's method. Through the theoretical analysis and experimental results, we can see the proposed algorithm showed excellent lane detection in the application to lane images on highway roads. The experiments and analysis indicate the robustness and real-time performance of our algorithm. As a highlight of this paper, we used the ROI method and ingeniously setting of objective function. And the contribution of this paper is that the FIR filter and AFSA are creatively used, which are not found in the reference literatures.

REFERENCES

Aufrere R., R. Chapuis, and F. Chausse (2001). A model-driven approach for real-time road recognition. *Machine Vision and Applications*, 13(2), 95–107.

Canny J. (1986). "A computational approach to edge detection". *IEEE Transactions on Pattern Analysis and Machine Intelligence*, 6, 679–698.

Chen Q. and H. Wang. (2006). A real-time lane detection algorithm based on a hyperbola-pair model. *IEEE Intelligent Vehicles Symposium*, 510–515.

Cheng H., N. Zheng, Q. Ling, and Z. Gao. (2003). Road recognition algorithm using principal component neural networks and k-means. *Third International Symposium on Multispectral Image Processing and Pattern Recognition, International Society for Optics and Photonics*, 77–80.

Jeong S. G., C. S. Kim, D. Y. Lee, S. K. Ha, D. H. Lee, M. H. Lee, and H. Hashimoto (2001). Real-time lane detection for autonomous vehicle. In, 2001. Proceedings of the *IEEE International Symposium on Industrial Electronics (ISIE)*, 1466–1471.

Jung C. R. and C. R. Kelber, (2005). Lane following and lane departure using a linear-parabolic model. *Image and Vision Computing*, 23(13), 1192–1202.

Khalifa O. O., I. M. Khan, A. A. Assidiq, A.-H. Abdulla, and S. Khan. (2010). A hyperbola-pair based lane detection system for vehicle guidance. *Proceedings of the World Congress on Engineering and Computer Science*, 978–988.

Li X.-L. (2003). A new intelligent optimization method-artificial fish school algorithm. Un-published Doctoral Thesis, Zhejiang University, 1–95.

Li X.-l., Z.-j. Shao, and J.-x. Qian (2002). An optimizing method based on autonomous animats: Fish-swarm algorithm. *System Engineering Theory and Practice, 22(11)*, 32–38.

Lin Q., Y. Han, and H. Hahn (2010). Real-time lane departure detection based on extended edge-linking algorithm. *IEEE Second International Conference on Computer Research and Development*, 725–730.

Linarth A. and E. Angelopoulou. (2011). On feature templates for particle filter based lane detection. 14th International IEEE Conference on Intelligent Transportation Systems (ITSC), 1721–1726.

Ma B., S. Lakshmanan, and A. O. Hero, (2000). Simultaneous detection of lane and pavement boundaries using model-based multisensor fusion. *IEEE Transactions on Intelligent Transportation Systems*, 1(3), 135–147.

Truong Q.-B. and B.-R. Lee (2008). New lane detection algorithm for autonomous vehicles using computer vision. *IEEE International Conference on Control, Automation and Systems (ICCAS)*, 1208–1213.

Turchetto R. and R. Manduchi (2003). Visual curb localization for autonomous navigation". *Proceedings of the IEEE/RSJ International Conference on Intelligent Robots and Systems (IROS)*, 1336–1342.

Wang Y., D. Shen, and E. K. Teoh. (1998). Lane detection using catmull-rom spline. *IEEE International Conference on Intelligent Vehicles*, 51–57.

Wang Y., E. K. Teoh, and D. Shen (2004). Lane detection and tracking using B-Snake. *Image and Vision Computing, 22(4)*, 269–280.

World Health Organization (2015). Global status report on road safety 2015. Available from: http://www.who.int/violence_injury_prevention/road_safety_status/2015/en/.

Xu Y., R. Wang, K. Li, and Y. Zhao (2004). A linear model based road identification algorithm. *Journal of Image and Graphics, 9(7)*, 859–864.

Yu B. and A. K. Jain (1997). Lane boundary detection using a multiresolution hough transform. Proceedings of the International Conference on Image Processing, 748–751.

Zhang Y. and L. Wu (2011). Fast document image binarization based on an improved adaptive Otsu's method and destination word accumulation. *Journal of Computational Information Systems, 7(6)*, 1886–1892.

Zhao K., M. Meuter, C. Nunn, D. Muller, S. Muller-Schneiders, and J. Pauli. (2012). A novel multi-lane detection and tracking system. IEEE Intelligent Vehicles Symposium (IV), 1084–1089.

Automotive, Mechanical and Electrical Engineering – Liu (Ed.)
© 2017 Taylor & Francis Group, London, ISBN 978-1-138-62951-6

Method for supplementing incomplete traffic flow data in hazy conditions based on guidance data

Yan Gong, Jie Zhang & Sujian Li
School of Mechanical Engineering, University of Science and Technology Beijing, Beijing, China

Jinhui Lan
School of Automation and Electrical Engineering, University of Science and Technology Beijing, Beijing, China

ABSTRACT: This study proposes a vehicle guidance data-based method for supplementing incomplete data of traffic flow in hazy conditions. An urban traffic monitoring system was used to determine the completeness of traffic flow data. The incomplete traffic flow data were classified according to road network location. A GPS was used to identify vehicle position and obtain guidance data. The guidance data were to be used in supplementing the incomplete traffic flow data. The useful vehicle guidance data were retained. A statistical vehicle guidance data method was established to generate traffic flow data under hazy conditions. The traffic flow data were obtained and entered into the urban traffic monitoring system and GPS. Then, we established an index for supplementing incomplete data of traffic flow under hazy conditions. According to the simulation result, the proposed method based on vehicle guidance data can supplement and enrich incomplete data of traffic flow in hazy conditions.

Keywords: Traffic Engineering; Vehicle Guidance; Haze; Supplementary Incomplete Data; GPS; Urban Traffic Monitoring System

1 INTRODUCTION

Haze frequently causes a significant inconvenience in transportation. Specifically, haze significantly affects urban traffic video monitoring systems and leads to incomplete traffic flow data collection. To solve this problem, scholars have extensively studied image processing under hazy conditions using various methods. Sun (2013) discussed three major factors influencing fog removal for a single image in detail based on the atmospheric scattering model. Li (2014) proposed a novel monochrome atmospheric scattering model in order to get a clear view of the image taken under bad weather. These methods improve image clarity, but they require heavy calculations. Thus, such methods are only suitable for small-scale road networks and are geared towards the intersection of data acquisition; moreover, they are not conducive to lane data acquisition.

At the same time, comprehensive research must be undertaken to obtain traffic information data based on GPS floating car data in non-hazy conditions. Jenelius (2013) presents a statistical model for urban road network travel time estimation using vehicle trajectories obtained from low-frequency GPS probes as observations. Velaga (2009) developed an enhanced weight-based topo-logical map-matching algorithm for intelligent transport systems. Ramaekers (2013) proposed the travel path estimation method to identify the relationship between activity patterns and route choice decisions. Tang (2014) showed that, without haze, 3–5% of floating car data can provide 95% basic road network data and lead to a precise traffic state estimation related to floating car proportions. The aforementioned methods are applied under non-hazy conditions. However, Gong (2015) found that the traffic detection data based on floating car data do not change, even though the literature states that drivers are affected in different ways in hazy situations. Therefore, we must consider traffic flow characteristics in hazy conditions to supplement traffic flow data.

In the present work, GPS vehicle guidance data statistics are used to supplement the incomplete traffic data of an urban traffic video monitoring system under hazy conditions. The method has ten steps, as follows: S1. Urban traffic monitoring system determines whether traffic flow data are complete; S2. Retain traffic flow data with high integrity; S3. Determine missing traffic data and location of the road network; S4. According to the position of the road network, classify the missing status of traffic flow data; S5. According to GPS, determine vehicle position and access to guidance data; S6. Determine vehicle guidance data to supplement missing road network data; S7. Retain useful vehicle guidance data; S8. Establish appropriate statistical vehicle guidance data method for supplementing traffic flow data; S9. Supplementary traffic flow data. S10. Traffic flow data output. In these steps, traffic flow data include road network traffic flow, speed, and length fleet. Guidance data include vehicle speed and traffic flow.

2 SPECIFIC STEPS

Step S1 comprises the following:

S1-1. Mark the location of the road network. Label the intersections and lanes. Suppose that three intersections i, j, and k exist. Sections ij represent the direction of the road from intersection i to intersection j. When a road connects intersection j to intersection k, sections ij include the lane through which the intersections i and j connect to intersection k; we mark the lane as $ij - k$. On sections ij, the intersection part that is connected to lane $ij - k$ is marked as intersection $ij - k'$. All lanes and intersections in accordance with per unit length are divided into several units and labelled accordingly. Each lane $ij - k$ in accordance with per unit length is divided into several units in accordance with the vehicle travelling direction order marked as $ij - k - p$, $p = 1, 2, ..., n$. Each intersection $ij - k'$ in accordance with per unit length is divided into several units in accordance with the vehicle travelling in a reverse direction order marked as $ij - k' - q$, $q = 1, 2, ..., n$. If the length of the last unit of the lane and the intersection is less than the per unit length, then the last unit is marked as a cell.

S1-2. Mark the related network traffic flow data of road network location: the traffic flow data of each lane unit $ij - k - p$ are referred to as $\rho(ij - k - p)$, and the vehicle speed of each lane unit $ij - k - p$ is referred to as $v(ij - k - p)$. The traffic flow data of each intersection unit $ij - k' - p$ are referred to as $\rho(ij - k' - q)$, and the vehicle speed of each lane unit $ij - k' - p$ is referred to as $v(ij - k' - q)$. The motorcade length of each intersection $ij - k'$ is referred to as $l(ij - k')$.

S1-3. The urban traffic monitoring system determines whether the traffic flow data on the road network location are complete. If complete, the data do not have to be supplemented. If incomplete, the following steps should be performed.

Step S4 comprises the following:

S4-1. The missing status of the traffic flow data is classified as follows according to the road network location of the missing road network traffic flow data. According to the missing characteristics of traffic flow data under hazy conditions and hazy status, the road network location of missing traffic flow data is determined if it is continuous. The lane unit and intersection unit have discontinuous missing traffic flow data called missing breakpoints. Within the same lane or the same intersection, the multiple lane unit and multiple intersection unit have continuous missing traffic flow data called continuous missing points. The lanes and intersection have continuous missing traffic flow data called locally missing points.

S4-2. According to the classification result of step S4-1, missing traffic data are combined into one collection: the missing traffic flow data collection with missing breakpoints is recorded as D. The collection of traffic flow data with continuous missing points is recorded as X. The collection of traffic flow data with locally missing data is recorded as J. Collections D, X, and J are stored separately.

Step S5 comprises the following:

S5-1. According to the GPS, determine vehicle location. According to the GPS, the respective lane unit or intersection unit owned by each vehicle is determined, and vehicles in the road network are marked as $c(ij - k - p - m)$, $m = 1, 2, ..., n$ or $c(ij - k' - p - h)$, $h = 1, 2, ..., n$ in accordance with the travelling direction order.

S5-2. According to the GPS, obtain guidance data: according to the GPS, the speed of each vehicle can be obtained with $vc(ij - k - p - m)$, $m = 1, 2, ..., n$ and $vc(ij - k' - p - h)$, $h = 1, 2, ..., n$, and vehicle density data can be obtained with $\rho c(ij - k - p)$ and $\rho c(ij - k' - q)$.

Step S7 comprises the following:

According to steps S4 and S6, the related vehicle guidance data of collections D, X, and J are respectively stored in the corresponding guidance data collections D', X', and J'.

Step S8 comprises the following:

S8-1. Collections D', X', and J' respectively correspond to collections D, X, and J.

S8-2. For collection D through collection D', a vehicle guidance data statistics method to supplement traffic flow data is established. A method is also established for missing lane unit data and missing intersection unit data.

A. For missing lane unit data:

a1. When the cycle time is long, collection D related to the guidance data of this cycle is used.

a1-1. When the road network position of the guidance data in this cycle is the same as the road network position of the missing traffic flow data in collection D, we directly supplement collection D with the guidance data. Suppose that the corresponding traffic flow data breakpoints of lane unit $ij - k - p$ are missing. We directly take the guidance data $vc(ij - k - p - m), m = 1, 2, ..., n$ and $\rho c(ij - k - p)$ as $v(ij - k - p)$ and $\rho(ij - k - p)$.

a1-2. When the road network position of guidance data in this cycle is different from the road network position of missing traffic flow data in collection D, we obtain the traffic flow data breakpoints of the unit according to the traffic flow data before and after the unit breakpoints in this cycle and haze visibility. The result is supplemented to collection D. The specific methods are as follows:

Suppose that the corresponding traffic flow data breakpoints of lane unit $ij - k - p$ are missing and that the cycle is the s cycle. Haze visibility exerts an effect on drivers, thereby affecting vehicle speed. We determine the relationship between driver lane visibility $d(ij - k - p)$ and the maximum driving speed limit $v(ij - k - p, sT)_{vis}$.

$$
\begin{aligned}
&if \quad d(ij - k - p, sT) > d_{ij0}^k \\
&v(ij - k - p, sT)_{vis} = vl_{max} \\
&else \quad if \quad d_{ij1}^k < d(ij - k - p, sT) \leq d_{ij0}^k \\
&d(ij - k - p, sT) = [al_{ij0} \cdot (vl(ij - k - p, sT))^2 \\
&\quad\quad + al_{ij1} \cdot vl(ij - k - p, sT) + al_{ij2}] \cdot lv \quad (1) \\
&else \quad if \quad d_{ij2}^k < d(ij - k - p, sT) \leq d_{ij1}^k \\
&v(ij - k - p, sT)_{vis} = al_{ij3} \cdot d(ij - k - p, sT) + al_{ij4} \\
&else \quad if \quad d(ij - k - p, sT) > d_{ij2}^k \\
&v(ij - k - p, sT)_{vis} = vl_{min}
\end{aligned}
$$

where vl_{max} = highest traffic speed limit of the lane; vl_{min} = the lowest traffic speed limit of the lane; lv = the average length of the vehicle; $al_{ij0}, al_{ij1}, al_{ij2}, al_{ij3}, al_{ij4}$ are coefficients; and d_{ij0}^k, d_{ij1}^k, and d_{ij2}^k are the criterion values of d_{ij}^k.

We can obtain the ratio range of driver lane visibility $d(ij - k - p)$ and lane visibility $d(ij - k - p)_{vis}$ under hazy conditions. According to the measured range, we assume that the ratio is $1/10$.

$$
d(ij - k - p) = d(ij - k - p)_{vis} / 10 \quad (2)
$$

By substituting Formula (2) into Formula (1), we can derive the maximum driving speed $v(ij - k - p, sT)_{vis}$ under hazy conditions $v(ij - k - p, sT)$.

The corresponding vehicle speed $v(ij - k - p, sT)$ of the lane unit $ij - k - p$ should be less than the maximum driving speed and is related to the vehicle speed $v(ij - k - (p-1), sT)$ of the breakpoints of the former unit $ij - k - (p-1)$ in this cycle and to the vehicle speed $v(ij - k - (p+1), sT)$ of the breakpoints of the latter unit $ij - k - (p+1)$ in this cycle.

$$
\begin{aligned}
v(ij - k - p, sT) = \min\{&[v(ij - k - (p-1), sT) \\
&+ v(ij - k - (p+1), sT)] / 2, v(ij - k - p, sT)_{vis}\} \quad (3)
\end{aligned}
$$

The minimum traffic flow values $\rho(ij - k - p)_{vis}$ under hazy conditions are related to driver lane visibility $d(ij - k - p)$.

$$
\rho(ij - k - p)_{vis} = \frac{1}{d(ij - k - p) + lv} \quad (4)
$$

The lane unit $ij - k - p$ corresponding to $\rho(ij - k - p)$ should be higher than the minimum vehicle flow value and is related to the vehicle flow $\rho(ij - k - (p-1), sT)$ of the breakpoints of the former unit $ij - k - (p-1)$ in this cycle and to the vehicle flow $\rho(ij - k - (p+1), sT)$ of the breakpoints of the latter unit $ij - k - (p+1)$ in this cycle.

$$
\begin{aligned}
\rho(ij - k - p, sT) = \max\{&[\rho(ij - k - (p-1), sT) \\
&+ \rho(ij - k - (p+1), sT)] / 2, \rho(ij - k - p, sT)_{vis}\}
\end{aligned}
$$

$$(5)$$

a2. When the cycle time is short, collection D is related to the guidance data of this cycle and the former cycle.

a2-1. When the road network position of the guidance data in this cycle is the same as the road network position of the missing traffic flow data in collection D, we directly supplement collection D with the guidance data. Suppose that the corresponding traffic flow data breakpoints of lane unit $ij - k - p$ are missing. We directly take the guidance data $vc(ij - k - p - h), h = 1, 2, ..., n$ and $\rho c(ij - k - q)$ as $v(ij - k - p)$ and $\rho(ij - k - p)$.

a2-2. When the road network position of the guidance data in the former cycle is the same as the road network position of the missing traffic flow data in collection D, we obtain the traffic flow data value of the breakpoint unit according to the guidance data of the breakpoint position in the former cycle, the traffic flow data before and after the breakpoint unit in this cycle and the former cycle, and haze visibility. The result is supplemented to collection D. The specific methods are described as follows:

Using Formulas (1) and (2), when this cycle is the s cycle and the traffic flow data of lane unit $ij-k-p$ represent the missing breakpoints under hazy conditions, we obtain visibility $d(ij-k-p)_{vis}$ corresponding to the maximum driving speed limit $v(ij-k-p,sT)_{vis}$.

The lane unit $ij-k-p$ corresponding to vehicle speed $v(ij-k-p,sT)$ should be less than the maximum driving speed, and its value is related to the vehicle speed $v(ij-k-(p-1),sT)$ of the breakpoints of the former unit $ij-k-(p-1)$ in this cycle, the vehicle speed $v(ij-k-(p+1),sT)$ of the breakpoints of the latter unit $ij-k-(p+1)$ in this cycle, the vehicle speed $v(ij-k-p,(s-1)T)$ of this unit $ij-k-p$ in the former cycle, the vehicle speed $v(ij-k-(p-1),(s-1)T)$ of the breakpoints of the former unit $ij-k-(p-1)$ in the former cycle, and the vehicle speed $v(ij-k-(p+1),(s-1)T)$ of the breakpoints of the latter unit $ij-k-(p+1)$ in the former cycle.

$$v(ij-k-p,sT) = \min\{v(ij-k-p,(s-1)T) \\ -[v(ij-k-(p-1),(s-1)T)+v(ij-k-(p+1), \\ (s-1)T)-v(ij-k-(p-1),sT) \\ -v(ij-k-(p+1),sT)]/2, v(ij-k-p,sT)_{vis}\}$$

(6)

As shown in Formula (4), the minimum traffic flow values $\rho(ij-k-p)_{vis}$ under hazy conditions are related to driver lane visibility $d(ij-k-p)$.

The lane unit $ij-k-p$ corresponding to $\rho(ij-k-p)$ should be higher than the minimum vehicle flow value, and it is related to vehicle flow $\rho(ij-k-(p-1),sT)$ of the breakpoints of the former unit $ij-k-(p-1)$ in this cycle, the vehicle flow $\rho(ij-k-(p+1),sT)$ of the breakpoints of the latter unit $ij-k-(p+1)$ in this cycle, the vehicle flow $\rho(ij-k-p,(s-1)T)$ of this unit $ij-k-p$ in the former cycle, the vehicle flow $\rho(ij-k-(p-1),(s-1)T)$ of the breakpoints of the former unit $ij-k-(p-1)$ in the former cycle, and the vehicle flow $\rho(ij-k-(p+1),(s-1)T)$ of the breakpoints of the latter unit $ij-k-(p+1)$ in the former cycle.

$$\rho(ij-k-p,sT) = \max\{\rho(ij-k-p,(s-1)T) \\ -[\rho(ij-k-(p-1),(s-1)T+\rho(ij-k-(p+1), \\ (s-1)T)-\rho(ij-k-(p-1),sT) \\ -\rho(ij-k-(p+1),sT)]/2, \rho(ij-k-p,sT)_{vis}\}$$

(7)

a2-3. When the road network position of the guidance data in this and the former cycle are not the same as the road network position of the missing traffic flow data in collection D,

we obtain the traffic flow data value of the breakpoint unit according to the traffic flow data before and after the breakpoint unit in this cycle and haze visibility. The result is supplemented to collection D. The specific methods are specified in (a1-2).

B. For missing intersection unit data:

In accordance with the marked order, when the speed of each former intersection unit is zero, the intersection unit is preset as the occupied lane of a waiting vehicle for supplementing data. For the missing intersection unit data, the approach to supplementing traffic flow and vehicle velocity data is similar to that used for the missing lane unit data in (A). The difference is that the velocity of the former unit with missing intersection data in this cycle is zero. The methods for supplementing the convoy length data of the intersection unit are described as follows: determine if the intersection speed of the original missing traffic flow data is zero according to the supplemented results of the traffic flow data. If the speed is zero, the unit belongs to the lane part of the intersection occupied by a waiting vehicle. We then determine if the vehicle speed of the latter unit is zero, and so on. When the value is not zero, the unit does not belong to the said lane. Thus, we determine the convoy length $l(ij-k')$ of each intersection.

S8-3. For collection X through collection X', we establish a statistical vehicle guidance data method to supplement traffic flow data. According to the statistical vehicle guidance data method for supplementing traffic flow data in step S8-2, collection X' collection X twice.

S8-4. For collection J through collection J', we establish a statistical vehicle guidance data method to supplement traffic flow data. According to the statistical vehicle guidance data method for supplementing traffic flow data in step S8-2, collection J' supplements collection J' three times.

Step S9 comprises the following: collections D', X', and J' supplement collections D, X, and J.

3 TRAFFIC FLOW DATA SUPPLEMENT RATE

The road network model is divided into the lane model and the intersection model. According to road network visibility and an urban traffic data monitoring system, Gong (2015) shows statistics on traffic flow missing rates for the two parts of the model, including the lane data missing rate lo_l and the intersection data missing rate lo_c.

For lanes, the lane data supplement rate is related to lane data missing rate lo_l; the rate is the difference in the lane data missing rates before and after supplementation.

$$LO_l = lo_l^b - lo_l^a \qquad (8)$$

where lo_l^b = lane data missing rate before supplementation; lo_l^a = the lane data missing rate after supplementation.

For the intersection, the intersection data supplement rate is related to the intersection data missing rate lo_c; the rate is the difference in the intersection data missing rate before and after supplementation.

$$LO_c = lo_c^b - lo_c^a \qquad (9)$$

where lo_c^b = intersection data missing rate before supplementation; lo_c^a = intersection data missing rate after supplementation.

4 SIMULATION

4.1 Simulation data

The proposed method is verified using MATLAB with the simulation lasting 1,800 s, during which the method is verified once every 10 s.

4.1.1 Road network data
In this study, the road network model data for simulation are the same as those in Gong's paper. The simulation area covers the road sections as part of a road network consisting of the Kehui Road, Kehui South Road, Tatun North Road, and Tatun Road near the Olympic Green in the city of Beijing. Figure 1 shows the position of the road sections and the structure of the network.

4.2 Haze data

The haze effect of missing traffic flow data is mainly reflected in two aspects: haze influence degree and haze affected area. Thus, the simulation is divided into two portions: simulation of haze influence degree of road network and simulation of haze affected area.

Simulation data of haze influence degree of road network: the visibility values of the lanes and intersections with mild, moderate, and severe impact haze are set to 1,000, 500, and 150 m, respectively.

Simulation data of haze influence area of road network: moderate haze and severe impact haze are called blind areas. According to the missing data, intersection 12 is set as the point-blind area. The areas surrounded by intersections 12, 13, and 17 are set as the local-blind areas. In road networks, the visibility values of lanes and intersections are set to 1,000 m in non-blind areas and to 400 m in blind areas.

4.2.1 Guidance data related to the method
Take 5% of all vehicles in the road network as floating cars. To closely approximate the actual situation, the initial positions of the floating cars in the road network are randomly distributed. After marking the position of the road network, the per unit length of the lane units is set to 20 m, and the per unit length of the intersection units is set to 10 m.

4.3 Simulation results and analysis

4.3.1 Simulation results and analysis of the haze effect degree on the road network
According to the simulation data, the simulation results of the haze effect degree of the road network are shown in Figure 2.

According to the simulation results, the proposed method supplements the missing data of traffic flow under mild impact, moderate impact, and severe impact haze. The supplementation rate of the intersection and lane data under severe impact haze is the highest, and the supplementation rate of the intersection and lane data under mild impact haze is the lowest. The supplementation rate of the intersection data under severe impact haze is about 40–60%. These results show that the method under severe impact haze is more prominent than that under the other two conditions.

4.3.2 Simulation results and analysis of the haze effect on the regional road network
According to the simulation data, the simulation results of the haze affecting the regional road network are shown in Figure 3.

According to the simulation results, the proposed method supplements the missing traffic flow data under point-blind, local-blind, and full-blind haze. The supplementation rate of the intersection and lane data under full-blind haze is the highest,

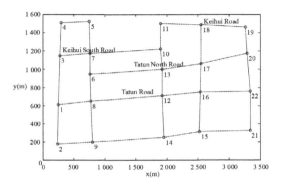

Figure 1. Simulation road network diagram.

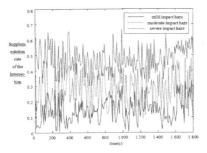

(a) Supplementation rate of incomplete intersection data

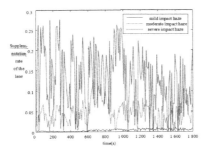

(b) Supplementation rate of incomplete lane data

Figure 2. Simulation results under different degrees of influence of haze on the road network.

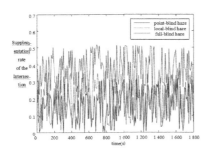

(a) Supplementation rate of incomplete intersection data

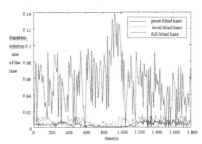

(b) Supplementation rate of incomplete lane data

Figure 3. Simulation results under different influence areas of haze on the road network.

and the supplementation rate of the intersection and lane data under point-blind haze is the lowest. The supplementation rate of the intersection data under full-blind haze is about 25–50%. The results show that the method in full-blind haze is more prominent than that in the other conditions.

5 CONCLUSION

We established indicators for the supplementation rate of traffic flow data under hazy conditions, including those of intersection data and lane data. These indicators are targeted and can reflect the effect of supplemental traffic flow data under hazy conditions.

According to the simulation results, the vehicle guidance data-based proposed method for supplementing incomplete traffic flow data under hazy conditions is effective, especially under severe haze and full-blind haze. The supplementation rate of the traffic flow data of the proposed method is relatively high and leads to remarkable results.

Some traffic flow data may still exist in cases with missing data. Determining the proportion and regions of floating cars according to the haze influence degree and haze-affected area is one of the main issues that should be explored in future research.

REFERENCES

Gong, Y. (2015). A road network model in haze and haze situation influence on traffic networks. *Journal of Transportation Systems Engineering and Information Technology, 15*(5), 114–122.

Jenelius, E., & Koutsopoulos, H.N. (2013). Travel time estimation for urban road networks using low frequency probe vehicle data. *Transportation Research Part B: Methodological, 53*, 64–81.

Li, Q.-H. (2014). Haze degraded image scene rendition. *Acta Automatica Sinica, 40*(4), 744–750.

Ramaekers, K. (2013). Modelling route choice decisions of car travellers using combined GPS and diary data. *Networks and Spatial Economics, 13*(3), 351–372.

Sun, W. (2013). Fast single image fog removal based on atmospheric scattering model. *Optics and Precision Engineering, 21*(4), 1040–1046.

Tang, K.-S. (2014). A simulation-based evaluation of traffic State estimation accuracy by using floating car data in complex road networks. *Journal of Tongji University (Natural Science), 42*(9), 1347–1351+407.

Velaga, N.R. (2009). Developing an enhanced weight-based topological map-matching algorithm for intelligent transport systems. *Transportation Research Part C-Emerging Technologies, 7*, 672–683.

Automotive, Mechanical and Electrical Engineering – Liu (Ed.)
© *2017 Taylor & Francis Group, London, ISBN 978-1-138-62951-6*

Micro-satellite attitude determination based on MEKF using MEMS gyroscope and magnetometer

Feng Ni, Haiyang Quan & Wenjie Li
Beijing Microelectronics Technology Institute, Beijing, China

ABSTRACT: There is a recent trend in applying MEMS sensors in micro-satellite attitude determination. Considering MEMS sensors' large bias drift and noise, a Multiplicative Extended Kalman Filter (MEKF) has been designed to estimate a micro-satellite's attitude, based on a series of measurement models of MEMS sensors in this paper. Simulation results show that this algorithm performs well in estimating micro-satellite's attitude information; even though there is an initial error it achieves better accuracy when compared to traditional methods. Moreover, it reduces computation load which is significant for micro-satellite practical missions.

Keywords: Gyroscope; Magnetometer; MEKF; Attitude determination; Micro-satellite

1 INTRODUCTION

Attitude determination systems play an important role in micro-satellite missions. Recently, MEMS sensors applied to micro-satellite attitude determination systems have been paid more attention by researchers all over the world. As a typical attitude sensor, the MEMS gyroscope has obvious advantages in energy, size and cost compared with conventional gyroscopes like the fibre-optics gyroscope (Stearns, 2011). However, since its larger bias drift and noise in accuracy, a high performance algorithm is necessary. In contrast with traditional methods, estimation algorithms, such as EKF, UKF, particle filter (Sekhavat, 2007; Crassidis, 2015; and Feng Yu, 2008), have a better performance in accuracy and solving uncertainty. But, on the other hand, they always have a greater computation load, which limits the application in micro-satellites. In this paper, a Multiplicative Extended Kalman Filter (MEKF) is proposed to accomplish attitude determination that can reduce the computation load due to the smaller dimension of the state vector. Simulation results show that the attitude determination algorithm performs well in estimating the micro-satellite's attitude and achieves better accuracy than traditional methods, even though there is an initial error regarding attitude information.

The organisation of this paper is as follows: measurement models of MEMS sensors, including the gyroscope and magnetometer, are presented in Section 2. The main process and equations of MEKF are presented in Section 3. Simulations are conducted after that, and the results are presented in Section 4, and finally, a conclusion is proposed in Section 5.

2 MEASUREMENT MODELS OF MEMS SENSORS

2.1 *Measurement model of magnetometer*

Considering the MEMS gyroscope's poor performance in long-term attitude determination, the magnetometer was chosen as an assistant sensor for attitude estimation. When fixed along the inertial axis of the micro-satellite's body frame, the magnetometer outputs the geomagnetic vector measurement $\tilde{\mathbf{B}}_b$.

In order to determine the micro-satellite's attitude, a reference geomagnetic vector should be generated from an accurate mathematical model of the geomagnetic field. In this paper, the International Geomagnetic Reference Field (IGRF) model, which is generally revised every five years by the International Association of Geomagnetism and Aeronomy (IAGA) was chosen. The IGRF model simulates the geomagnetic field of a spherical harmonic model with a series of Gauss coefficients g_n^m, h_n^m:

$$\mathbf{V} = R \sum_{n=1}^{N} \sum_{m=0}^{n} \left(\frac{R}{r}\right)^{n+1} \left(g_n^m \cos m\lambda + h_n^m \sin m\lambda\right)$$
$$\times P_n^m(\cos\theta) \tag{1}$$

R = 6371.2 km, r is the radial distance between the micro-satellite and the centre of the Earth, λ is east longitude and θ is co-latitude. $P_n^m(\cos\theta)$ is the Schmidt quasi-normalised associated Legendre function of degree n and order m. The generation 12th IGRF is a spherical harmonic degree 13 model and was published in 2014.

According to articles (Bak, 1999; Simpson, 1989; Castellanos, 2014; and Tohami, 2005), it is easy to calculate the geomagnetic vector by taking the gradient of magnetic scalar potential V:

$$B_x^E = \frac{1}{r}\frac{\partial V}{\partial \theta} \quad B_Y^E = -\frac{1}{r\sin\theta}\frac{\partial V}{\partial \lambda} \quad B_z^E = -\frac{\partial V}{\partial r} \quad (2)$$

$\mathbf{B}_E = \begin{bmatrix} B_x^E & B_y^E & B_z^E \end{bmatrix}^{\mathrm{T}}$ is the geomagnetic vector which is given in the Northward, Eastward and Radially Inward Frame (NED). It can be transformed to the micro-satellite body frame by a series of rotation matrices $\mathbf{C}_o^b, \mathbf{C}_e^o, \mathbf{C}_d^e$. Therefore, the measurement model of magnetometer can be established as follows:

$$\tilde{\mathbf{B}}_b = \mathbf{C}_o^b\mathbf{C}_e^o\mathbf{C}_d^e\mathbf{B}_E + \mathbf{V}_B \quad (3)$$

where $\mathbf{V}_B$ is the measurement error of magnetometer.

2.2 *Measurement model of gyroscope*

If the gyroscope is fixed along the inertial axis of the satellite body frame, its measurement model can be given by:

$$\tilde{\boldsymbol{\omega}}_{bi} = \boldsymbol{\omega}_{bi} + \boldsymbol{b} + \boldsymbol{\eta}_g \quad (4)$$

$$\dot{\mathbf{b}} = \boldsymbol{\eta}_b \quad E(\boldsymbol{\eta}_b\boldsymbol{\eta}_b^T) = \boldsymbol{\sigma}_u^2 \quad (5)$$

$\tilde{\boldsymbol{\omega}}_{bi}$ is the angular velocity vector measured by the gyroscope, and $\boldsymbol{\omega}_{bi}$ is the real angular velocity of the micro-satellite body frame relative to the inertial frame. $\mathbf{b}$ is the bias drift of the gyroscope. $\boldsymbol{\eta}_g$, $\boldsymbol{\eta}_b$ are assumed as a zero-mean Gauss white noise.

According to articles by Crassidis (2006) and Cemenska (2004), the measurement model of gyroscope proposed above can be discretised as:

$$\mathbf{b}_{k+1} = \mathbf{b}_k + \boldsymbol{\sigma}_u\Delta t^{1/2}\mathbf{N}_u \quad (6)$$

$$\tilde{\boldsymbol{\omega}}_{k+1} = \boldsymbol{\omega}_{k+1} + \frac{1}{2}(\mathbf{b}_{k+1} + \mathbf{b}_k)$$
$$+ \left(\frac{\boldsymbol{\sigma}_v^2}{\Delta t} + \frac{1}{12}\boldsymbol{\sigma}_u^2\Delta t\right)^{1/2}\mathbf{N}_v \quad (7)$$

$\mathbf{N}_u$ and $\mathbf{N}_v$ are both zero-mean and unit variance white noise and Δt is the gyroscope sample rate.

Generally, $\boldsymbol{\sigma}_v^2$ is regarded as Angle Random Walk (ARW) and $\boldsymbol{\sigma}_u^2$ is referred to as Rate Random Walk (RRW). Notice that the scale factor error and other time-correlation drifts are ignored in this model.

It is necessary to transform $\boldsymbol{\omega}_{bi}$ to the angular velocity of the micro-satellite body frame relative to the orbit frame $\boldsymbol{\omega}_{bo}$ during the process of attitude determination:

$$\boldsymbol{\omega}_{bo} = \boldsymbol{\omega}_{bi} - \mathbf{C}_{ob}\boldsymbol{\omega}_{oi} \quad \boldsymbol{\omega}_{oi} = [0 \quad -\omega_{oi} \quad 0]^T \quad (8)$$

where $\boldsymbol{\omega}_{oi}$ is the orbit rate (the mean angular velocity of the micro-satellite rotation around Earth (Sekhavat, 2007; and Bak, 1999).

3 MULTIPLICATIVE EXTENDED KALMAN FILTER IN ATTITUDE DETERMINATION

Multiplicative Extended Kalman Filter (MEKF), compared with traditional EKF, not only avoids singularities of the co-variance matrix, but also reduces the computation load of algorithm. The state equation and measurement equation of MEKF in attitude determination are derived in this section based on the quaternion kinematics equation of micro-satellites and the measurement models of MEMS sensors proposed in Section 2.

3.1 *The quaternion kinematics equation of micro-satellites*

Quaternion is chosen as the attitude parameter in this paper, which is defined as:

$$\mathbf{q} = q_0 + \mathbf{i}q_1 + \mathbf{j}q_2 + \mathbf{k}q_3 = \begin{bmatrix} q_0 & \mathbf{q}_v \end{bmatrix}^T \quad (9)$$

And it must satisfy the constraint $\|\mathbf{q}\| = 1$.

According to theoretical derivations (in Castellanos, 2014; Pham, 2015; Lijun Zhang, 2013; and Lam, 2004) and (8), the quaternion attitude kinematics equation can be given by (10):

$$\dot{q}_{bo} = \frac{1}{2}q_{bo} \otimes \boldsymbol{\omega}_{bo} \quad (10)$$

$$\boldsymbol{\omega}_{bo} = \boldsymbol{\omega}_{bi} - \mathbf{A}(\mathbf{q}_{ob})\boldsymbol{\omega}_{oi} \quad (11)$$

$\mathbf{q}_{bo}$ is the quaternion from the micro-satellite body frame to the orbit frame and $\mathbf{A}(\mathbf{q}_{ob})$ is the rotation matrix expressed in quaternion from the body frame to the orbit frame.

3.2 State equation of MEKF in attitude determination

A multiplicative error quaternion $\Delta \mathbf{q}$ is derived from the estimated quaternion $\hat{\mathbf{q}}$ and real quaternion $\mathbf{q}$ by:

$$\mathbf{q} = \hat{\mathbf{q}} \otimes \Delta \mathbf{q} \tag{12}$$

Since the error quaternion is small, it is reasonable to assume that $\Delta q_0 \approx 1$ and then, the error quaternion $\Delta \mathbf{q}$ can be defaulted as a vector consisting of three independent components Δq_1, Δq_2, Δq_3. The derivative of $\Delta \mathbf{q}$ is:

$$\Delta \dot{\mathbf{q}} = \dot{\hat{\mathbf{q}}}^{-1} \otimes \mathbf{q} + \hat{\mathbf{q}}^{-1} \otimes \dot{\mathbf{q}} \tag{13}$$

Substituting (10) into (13) gives:

$$\begin{aligned}
\Delta \dot{\mathbf{q}} &= \frac{1}{2} \hat{\mathbf{q}}^{-1} \otimes \hat{\boldsymbol{\omega}}_{bo} \otimes \mathbf{q} + \frac{1}{2} \hat{\mathbf{q}}^{-1} \otimes \mathbf{q} \otimes \boldsymbol{\omega}_{bo} \\
&= -\frac{1}{2} \hat{\boldsymbol{\omega}}_{bo} \otimes \hat{\mathbf{q}}^{-1} \otimes \mathbf{q} + \frac{1}{2} \hat{\mathbf{q}}^{-1} \otimes \mathbf{q} \otimes \boldsymbol{\omega}_{bo} \\
&= -\frac{1}{2} \hat{\boldsymbol{\omega}}_{bo} \otimes \Delta \mathbf{q} + \frac{1}{2} \Delta \mathbf{q} \otimes \boldsymbol{\omega}_{bo}
\end{aligned} \tag{14}$$

Default $\Delta \boldsymbol{\omega}_{bo} = \boldsymbol{\omega}_{bo} - \hat{\boldsymbol{\omega}}_{bo}$ and substitute it into (14):

$$\begin{aligned}
\Delta \dot{\mathbf{q}} &= -\frac{1}{2} \hat{\boldsymbol{\omega}}_{bo} \otimes \Delta \mathbf{q} + \frac{1}{2} \Delta \mathbf{q} \otimes \Delta \boldsymbol{\omega}_{bo} + \frac{1}{2} \Delta \mathbf{q} \otimes \hat{\boldsymbol{\omega}}_{bo} \\
&\approx -\hat{\boldsymbol{\omega}}_{bo} \otimes \Delta \mathbf{q} + \frac{1}{2} \Delta \boldsymbol{\omega}_{bo}
\end{aligned} \tag{15}$$

Considering (11), gives: $\Delta \boldsymbol{\omega}_{bo}$:

$$\begin{aligned}
\Delta \boldsymbol{\omega}_{bo} &= \boldsymbol{\omega}_{bo} - \hat{\boldsymbol{\omega}}_{bo} = \Delta \boldsymbol{\omega}_{bi} - (\mathbf{A}(\mathbf{q}) - \mathbf{A}(\hat{\mathbf{q}})) \boldsymbol{\omega}_{oi} \\
&= \Delta \boldsymbol{\omega}_{bi} - (\mathbf{A}(\hat{\mathbf{q}} \otimes \Delta \mathbf{q}) - \mathbf{A}(\hat{\mathbf{q}})) \boldsymbol{\omega}_{oi} \\
&\approx \Delta \boldsymbol{\omega}_{bi} + 2 \Delta \mathbf{q} \otimes \mathbf{A}(\hat{\mathbf{q}}) \boldsymbol{\omega}_{oi}
\end{aligned} \tag{16}$$

Substituting (16) into (15) gives:

$$\begin{aligned}
\Delta \dot{\mathbf{q}} &= -(\hat{\boldsymbol{\omega}}_{bi} - \mathbf{A}(\hat{\mathbf{q}}) \boldsymbol{\omega}_{oi}) \otimes \Delta \mathbf{q} + \frac{1}{2} \Delta \boldsymbol{\omega}_{bi} + \Delta \mathbf{q} \otimes \mathbf{A}(\hat{\mathbf{q}}) \hat{\boldsymbol{\omega}}_{bi} \\
&= -\hat{\boldsymbol{\omega}}_{bi} \otimes \Delta \mathbf{q} + \frac{1}{2} \Delta \boldsymbol{\omega}_{bi}
\end{aligned} \tag{17}$$

According to (4), it is easy to derive $\Delta \boldsymbol{\omega}_{bi}$:

$$\begin{aligned}
\Delta \boldsymbol{\omega}_{bi} &= \boldsymbol{\omega}_{bi} - \hat{\boldsymbol{\omega}}_{bi} = \tilde{\boldsymbol{\omega}}_{bi} - \mathbf{b} - \boldsymbol{\eta}_g - (\tilde{\boldsymbol{\omega}}_{bi} - \hat{\mathbf{b}}) \\
&= -\Delta \mathbf{b} - \boldsymbol{\eta}_g
\end{aligned} \tag{18}$$

$$\Delta \dot{\mathbf{b}} = \boldsymbol{\eta}_b \tag{19}$$

where $\Delta \mathbf{b} = \mathbf{b} - \hat{\mathbf{b}}$ and the estimated angular velocity $\hat{\boldsymbol{\omega}}_{bi} = \tilde{\boldsymbol{\omega}}_{bi} - \hat{\boldsymbol{b}}_{bi}$;

Finally, if the state vector is $\mathbf{X} = [\Delta \mathbf{q} \quad \Delta \mathbf{b}]^T$, the state equation is given by:

$$\dot{\mathbf{X}} = \mathbf{F}(t)\mathbf{X} + \mathbf{G}(t)\mathbf{W} \tag{20}$$

where

$$\mathbf{F}(t) = \begin{bmatrix} -[\hat{\boldsymbol{\omega}}_{bi} \times] & -\mathbf{0.5I}_{3\times3} \\ \mathbf{0}_{3\times3} & \mathbf{0}_{3\times3} \end{bmatrix}, \mathbf{G}(t) = \begin{bmatrix} -\mathbf{0.5I}_{3\times3} & \mathbf{0} \\ \mathbf{0} & \mathbf{I}_{3\times3} \end{bmatrix},$$

$$[\hat{\boldsymbol{\omega}}_{bi} \times] = \begin{bmatrix} 0 & -\hat{\omega}_{biz} & \hat{\omega}_{biy} \\ \hat{\omega}_{biz} & 0 & -\hat{\omega}_{bix} \\ -\hat{\omega}_{biy} & \hat{\omega}_{bix} & 0 \end{bmatrix};$$

Since the state equation (20) is continuous, it is necessary to discretise it:

$$\mathbf{X}_k = \boldsymbol{\Phi}_{k,k-1} \mathbf{X}_{k-1} + \boldsymbol{\Gamma}_{k-1} \mathbf{W}_{k-1} \tag{21}$$

$$\boldsymbol{\Phi}_{k,k-1} = \mathbf{I} + T\mathbf{F}_{k-1} + \frac{T^2}{2} \mathbf{F}_{k-1}^2 \tag{22}$$

$$\boldsymbol{\Gamma}_{k,k-1} = T\left(\mathbf{I} + \frac{1}{2} T\mathbf{F}_{k-1} + \frac{T^2}{3!} \mathbf{F}_{k-1}^2\right) \mathbf{G}_{k-1} \tag{23}$$

where T is the Kalman filter time step.

Apparently, the dimension of the state equation in MEKF is 6×1. There is a reduction on the computation load when compared to the traditional EKF whose dimension is 7×1.

3.3 Measurement equation of MEKF in attitude determination

The estimated magnetic vector $\hat{\mathbf{B}}_b$ is defined by:

$$\hat{\mathbf{B}}_b = \mathbf{A}(\hat{\mathbf{q}})\mathbf{B}_o \tag{24}$$

And $\mathbf{B}_o$ is the reference magnetic vector. The error $\Delta \mathbf{B}_b$ between the magnetometer measurement $\tilde{\mathbf{B}}_b$ and the estimated one $\hat{\mathbf{B}}_b$ can be given by:

$$\begin{aligned}
\Delta \mathbf{B}_b &= \tilde{\mathbf{B}}_b - \hat{\mathbf{B}}_b = \tilde{\mathbf{B}}_b - \mathbf{A}(\hat{\mathbf{q}})\mathbf{B}_o \\
&= \mathbf{A}(\mathbf{q})\mathbf{B}_o + \mathbf{V}_B - \mathbf{A}(\hat{\mathbf{q}})\mathbf{B}_o \\
&= (\mathbf{A}(\hat{\mathbf{q}} \otimes \Delta \mathbf{q}) - \mathbf{A}(\hat{\mathbf{q}}))\mathbf{B}_o + \mathbf{V}_B \\
&= \mathbf{A}(\hat{\mathbf{q}})(\mathbf{A}(\Delta \mathbf{q}) - \mathbf{I})\mathbf{B}_o + \mathbf{V}_B \\
&\approx -2\Delta \mathbf{q} \otimes \mathbf{A}(\hat{\mathbf{q}})\mathbf{B}_o + \mathbf{V}_B \\
&= -2\Delta \mathbf{q} \otimes \hat{\mathbf{B}}_b + \mathbf{V}_B \\
&= 2\hat{\mathbf{B}}_b \otimes \Delta \mathbf{q} + \mathbf{V}_B
\end{aligned} \tag{25}$$

Considering the state vector $\mathbf{X} = [\Delta \mathbf{q} \quad \Delta \mathbf{b}]^T$, the measurement equation can be derived from (25):

$$\Delta \mathbf{B}_b = \mathbf{H}_b \mathbf{X} + \mathbf{V}_B \tag{26}$$

where $\mathbf{H}_b = [2[\hat{\mathbf{B}}_b \times] \quad \mathbf{0}_{3\times3}]$. Actually, the error vector $\mathbf{V}_B$ includes errors from the model of IGRF,

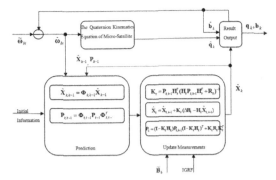

Figure 1. The process of MEKF in micro-satellite attitude determination.

measurement error, and so on. In this paper, it is simply assumed that it is a zero-mean white noise.

3.4 *Process of MEKF in attitude determination*

There are three parts regarding the process of MEKF in micro-satellite attitude determination: 1) Prediction: calculate the one-step prediction of state $\hat{\mathbf{X}}_{k,k-1}$ and its mean square error matrix $\mathbf{P}_{k,k-1}$; 2) Update measurements: calculate the estimated state $\hat{\mathbf{X}}_k$ by Kalman gain $\mathbf{K}_k$ and measurement information; and 3) Result output: correct the estimated quaternion $\hat{\mathbf{q}}_k$ and bias $\hat{\mathbf{b}}_k$ by the estimated state:

$$\mathbf{q}_k = \hat{\mathbf{q}}_k \otimes \begin{bmatrix} \sqrt{1 - |\Delta\hat{\mathbf{q}}_k|^2} \\ \Delta\hat{\mathbf{q}}_k \end{bmatrix} \qquad (27)$$

$$\mathbf{b}_k = \hat{\mathbf{b}}_k + \Delta\hat{\mathbf{b}}_k \qquad (28)$$

4 SIMULATION RESULTS

Simulation was conducted to evaluate the attitude determination algorithm's performance in this paper. By reference to existing MEMS gyroscopes and magnetometers, a set of parameters about them were chosen for simulation purposes:

1. Magnetometer: $\sigma_b = 10^{-7}$;
2. MEMS gyroscope: $\sigma_v = 10^{-4}$, $\sigma_u = 10^{-6}$;
3. Micro-satellite's orbit: Altitude is 600 km, inclination is 97°, RAAN is 220°, and the orbit period of micro-satellite is 5801.232s.

The real angular velocity $\boldsymbol{\omega}_{bi}$ is $[0 \quad 0 \quad 0]^T$ and the real attitude quaternion is $\mathbf{q}_{bo} = [1 \quad 0 \quad 0 \quad 0]^T$. The initial state vector of MEKF is $\mathbf{X} = [0 \quad 0 \quad 0 \quad 0 \quad 0 \quad 0]^T$ and the time step of MEKF is 0.2s. Total simulation time is 2000s.

Figure 2 shows the geomagnetic filed components in the NED frame calculated from the IGRF model according to the positions of the micro-satellite. Figure 3 shows the results of attitude determination by MEKF in attitude angles yaw, roll and pitch.

Simulation results show that the attitude determination algorithm of MEKF above, based on gyroscope and magnetometer, gives a good estimation of the satellite's attitude. And it can approach accuracy, especially for roll and pitch angle.

Figure 4 shows the result of attitude determination by the traditional method—Runge-Kutta. It shows that the determination errors of attitude angles by Runge-Kutta increase over time. However, MEKF in this paper avoids this problem well. In addition, there is obvious accuracy promotion for MEKF when compared to Runge-Kutta.

Figure 5 shows the results of attitude determination when there are big errors in initial angles: [10°, 10°, 10°] and [15°, 15°, 15°]. Results show that despite this, MEKF can still maintain output of acceptably accurate results, which is fundamentally important for practical satellite missions.

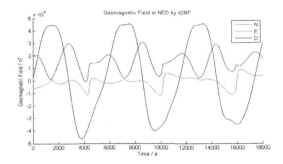

Figure 2. Geomagnetic filed components in NED frame calculated from the IGRF model.

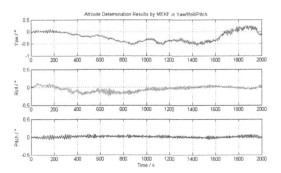

Figure 3. Attitude determination results by MEKF in attitude angles yaw/pitch/roll.

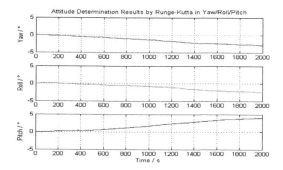

Figure 4. Attitude determination results by Runge-Kutta in yaw/roll/pitch.

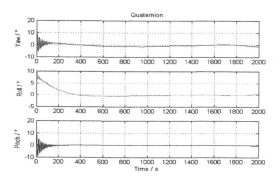

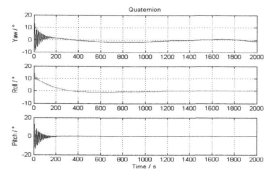

Figure 5. Attitude determination Results when initial angle errors are 10° and 15°.

5 CONCLUSION

This paper presents an attitude determination algorithm based on MEKF by making use of data from a gyroscope and magnetometer. According to the experiments, the algorithm can estimate the satellite attitude information well, meeting accept-able accuracy within 2°. Moreover, even if there is a big error in the initial attitude angles, the algorithm can still work well. It is also important that the algorithm reduces consumption of computation, which is meaningful for micro-satellite practical missions. In the following work, the feasibility of the algorithm will be verified in the physical platform of micro-satellite.

REFERENCES

Bak, T. (1999). *Spacecraft attitude determination - a Magnetometer Approach.*

Castellanos, C.A. and Aparicio, L.E. (2014). Pico, design and simulation of an attitude determination system based on the Extended Kalman Filter. for *Cube-Sat Colombia I, Revista Facultad de Ingeniería Universidad de Antioquia. (70)*, 146–154.

Cemenska, J. (2004). Sensor Modeling and Kalman Filtering Applied to Satellite Attitude Determination. Berkeley: University of California.

Crassidis, J.L. and Markley, F.L. (2015). Unscented filtering for spacecraft attitude estimation. *Journal of Guidance Control & Dynamics.* 26(4),536–542.

Crassidis, J.L. (2006). Sigma-point Kalman filtering for integrated GPS and inertial navigation. *IEEE Transactions on Aerospace and Electronic Systems.* 42(2), 750–756.

Feng Yu, Jianye Liu, Zhi Xiong. (2008). Application of predictive filtering algorithm in micro satellite attitude determination. *Journal of Astronautics.* 29 (1), 110–114.

Lam, Q.M., Lakso, J., Hunt, T. and Vanderham, P. (2004). Enhancing attitude estimation accuracy via system noise optimization. *Defense and Security, International Society for Optics and Photonics*, 553–564.

Lijun Zhang, Shifeng Zhang, Huabo Yang, S Qian. (2013). Multiplicative filtering for spacecraft attitude determination, *Journal of National University of Defense Technology.*

Pham, M.D., Low, K.S., Goh, S.T. and Chen, S. (2015). Gain-scheduled extended kalman filter for nanosatellite attitude determination system, *IEEE Transactions on Aerospace and Electronic Systems.* 51(2), 1017–1028.

Sekhavat, P., Gong, Q and Ross, I.M. (2007). Npsat1 parameter estimation using unscented kalman filtering. *Proceedings of the 2007 American Control Conference Marriott Marquis Hotel.* Times Square New York City, USA, July 11–13, 2007, IEEE.

Simpson, D.G. (1989). *Spacecraft attitude determination using the earth's magnetic field.*

Stearns, H. and Tomizuka, M. (2011). Multiple model adaptive estimation of satellite attitude using MEMS gyros, *Proceedings of the 2011 American Control Conference, IEEE*, 2011, 3490–3495.

Tohami, S. and Brembo, E.M. (2005). *Sensor modeling, attitude determination and control for micro-satellite.* M. Sc. Thesis, Norwegian University of Science and Technology.

Automotive, Mechanical and Electrical Engineering – Liu (Ed.)
© 2017 Taylor & Francis Group, London, ISBN 978-1-138-62951-6

Research and implementation of a vehicle top-view system based on i.MX6Q

Yang Xu, Xiyang Zuo, Anyu Cheng & Rongdi Yuan
College of Automation, Chongqing University of Posts and Telecommunications, Chongqing, China

ABSTRACT: With the rapid development of chip multiprocessors and the popularity of private cars in recent years, all kinds of solutions and products related to vehicle top-view systems have appeared. Because the systems widely use several microprocessors or heterogeneous structure CMP as core control units, there are some problems has widely exits on the systems, such as development difficulty and complex surrounding construction etc. This paper describes a vehicle top-view system that uses a monolithic four-channel TW6865 video decoder to replace the traditional single channel based on i.MX6Q. Compared with other solutions, this one has the advantage of simple design and development. Finally, we have applied lens distortion corrections, bird-view transformation and image fusion algorithms on the system. In the test, the system achieved the desired effect that a surrounding bird-view picture, which could be used to monitor the surrounding conditions, was displayed on the screen.

Keywords: Vehicle top-view; i.MX6Q; Lens distortion corrections; Bird-view transformation; Bird view

1 INTRODUCTION

With the increase in the number of cars and the complexity of the driving environment in cities, more and more people are focusing their energies on the research of driving assistant systems to improve safety. The driver assistance system uses advanced sensors, electronic control and a variety of mechanical systems, methods for the purpose of improving the convenience and safety of driving (Jeon et al., 2014).

A vehicle top-view system uses multiple cameras installed in the car body to collect the image data, while perspective transformation and image stitching algorithms make up the panoramic bird view of the vehicle. This can assist the driver by reducing accidents in parking, narrow lanes and other complex traffic environments, to guarantee their safety.

2 THE PRINCIPLE OF THE SYSTEM

A vehicle top-view system consists of a central processing unit, a display and four wide-angle cameras installed all around the vehicle (Yu et al., 2012). A diagram of the installation of the camera position and the angle of overlap region is shown in Figure 1.

The original image is collected by four wide-angle cameras which cover the automobile body, after which the data is sent to the central processing

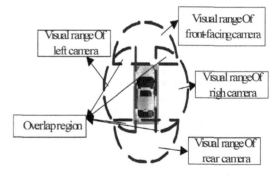

Figure 1. Camera placement and overlap region.

unit. Because of wide-angle lenses are good, these data has the characteristic of distortion. Firstly, we need to correct and perspective transformation of the image data to support the stitching algorithm. The four wide-angle cameras on the car have an overlap area; to take advantage of this characteristic, image fusion splicing into a panoramic bird's eye-view of the vehicle ensures that the driver in the complex traffic environment (such as parking, tunnel, etc.) is driving safely.

3 THE STRUCTURE OF THE SYSTEM

Figure 2 is the block diagram of a vehicle top-view system based on i.MX6Q. The system collects

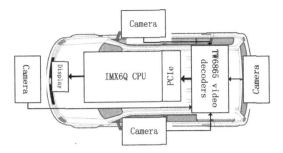

Figure 2. Block diagram of vehicle top-view system based on i.MX6Q.

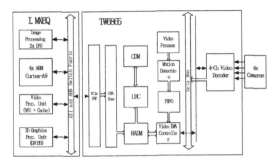

Figure 3. The signal flow graph of top-view system.

primitive data from four cameras through the TW6865 video decoding chip of image data transmission produced by the Intersil company. After decoding, these data would be sent to i.MX6Q through the PCIe interface. An image fusion splicing algorithm and image correcting of distortion would be transplanted in the i.MX6Q embedded platform. Finally, the system would present the driver with a tableau of a vehicle with 360-degree panoramic aerial view, which shows a panoramic scope of vision for safe driving.

3.1 *The flowchart of data processing*

Figure 3 is the signal flow graph of a top-view system. As shown in the figure, four-channel video decoders that internal integration in TW6865 decode four-way analogue signals into a digital YCbCr signal, and use high-performance adaptive 4H comb filters to separate luminance and chrominance. The TW6865 contains a high-performance dedicated DMA controller, optimises fully PCIex1 bandwidth utilisation, and enables it to maintain a high throughput. Based on V4L2 drivers use DMA can grab the camera data directly, because of the four road cameras at the same time, the amount of data researched to megabytes-per-second, PCIe interface can fully meet the requirements. Then

the application processor of Freescale i.MX6Q uses the Image Processing Unit (IPU) to catch the interlaced camera image solution mixed get to sawtooth smooth after the image processing, and run on processors of image processing algorithm. Four ARM architecture A9 kernel and three Graphics Processing Units (GPUs) are embedded in the i.MX6Q, with the ability of parallel processing programmable pipeline processing, which could guarantee the smooth implementation of the algorithm.

4 ALGORITHMS OF THE SYSTEM

The algorithm process of the vehicle top-view system is shown in Figure 4.

4.1 *Correcting the geometrical distortion of fisheye image*

Images taken by cameras with fisheye lenses tend to be seriously distorted. If we want to use these distorted images, we need to correct the distorted images to normal ones that people are used to. The process of correcting is related to the parameters of fisheye lenses (Kannala & Brandt, 2016). In order to obtain the intrinsic and extrinsic parameters of the camera, we placed a calibration board under the camera to get the images of objects. The image pixel of fisheye (p, q) and the pixel of the rectifying images (x_c, y_c) have the following mathematical expressions:

$$\begin{bmatrix} (p - c_x) \\ (q - c_y) \\ f(\rho) \end{bmatrix} = \begin{bmatrix} x_c - W_c/2 \\ y_c - H_c/2 \\ Z_c \end{bmatrix} \quad (1)$$

where W_c and H_c is the width and height of the corrected distortion image, c_x, c_y represents the fisheye deformation of the image centre, and x_c, y_c and $\mathbf{a} = [a_0, a_1, ..., a_n]$ are the intrinsic parameters for the camera. $\rho = \sqrt{(\mathbf{p} - c_x)^2 + (q - c_y)^2}$ is the length between the pixel points and the geometry distortion centre, and the curve surface equation camera of Taylor's expansion would be given as: $f(\rho) = a_0 + a_1\rho + a_2\rho^2 + \cdots + a_n\rho^n$.

4.2 *Image perspective transformation*

With the restriction installation position gene of cameras, every camera needs to be placed on the

Figure 4. Algorithm process of the system.

car body all around in the vehicle top-view system, a certain viewing angle would be existed between optical axis and ground. Hence the image data from the cameras have the effect of perspective. If we want an aerial view of the vehicle, we must transform the perspective of all the image data from the four cameras, to lay the foundation for the next step of the stitching algorithm.

In order to get the top view of the video image, there are some steps to establish a generated overlooking the pixel mapping relationship between the image and the original image (Rose et al., 2009). Above we have described the coordinate point under the camera system (x_c, y_c, z_c) and the bird's-eye view of the world (x_w, y_w, z_w).

$$\begin{bmatrix} x_c \\ y_c \\ z_c \end{bmatrix} = s[R;t] \begin{bmatrix} x_w \\ y_w \\ z_w \\ 1 \end{bmatrix} = sM \begin{bmatrix} x_w \\ y_w \\ z_w \\ 1 \end{bmatrix} \qquad (2)$$

4.3 Image mosaics

After correcting the images and perspective transformation, the primitive image data have been transformed to virtual vertical view without optical distortion. There are three steps we need to do for the purpose of getting four correction image stitching around the vehicle.

1. The coordinate system as shown in Figure 5 would be set up, according to the experimental vehicle conductor L1 and body width W1, and the wide panoramic aerial views W2 and L2 are set up;

2. Set the effective area, that is, the dimensions of the four trapezoidal, effective areas relative to their respective correction chart of displacement, intercepting the trapezoidal area effectively;

3. The trapezoidal intercept figure at panoramic aerial view coordinate system (X, Y, O) in configuration (Maik et al., 2014), according to, from top to bottom, from left to right scan, the coordinates of the four ladder diagram are:

$$(0,0), \quad (0,0), \quad (W2,0), \quad \left(\frac{W2-L1}{2}, \frac{L2+W1}{2} \right).$$

4.4 Experimental tests

In the experimental scenario, we simulated car size to build the platform; the experimental platform is 3.1 metres long and 1.5 metres wide. The height is 0.7 metres of four cameras, the visual field is 2.1 metres, and the original image from the four cameras was shown in Figure 6. Figures 7–8 showed the processed images of one direction of the camera after a fisheye image correction and perspective transformation algorithm of Rendering.

After correction and perspective transformation for the four directions, the image fusion algorithm

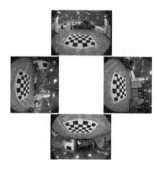

Figure 6. Four original images.

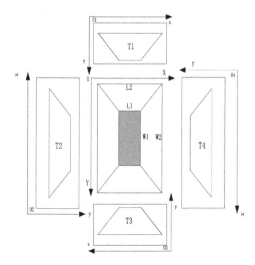

Figure 5. Coordinate system.

Figure 7. Corrected image.

Figure 8. Transformation image.

Figure 9. Mosaic image.

would be used to process the image data. The top-view of the simulation of the vehicle is shown in Figure 9.

5 CONCLUSION

We have put forward a vehicle top-view system based on i.MX6Q to improve driving safety. This system uses a single processor to replace heterogeneous CMP or more processors, which made its hardware frame simpler. Because of the quad-core architecture, we do not need to deal with a communication mechanism between multi-core processors. This paper also made some related analysis and research about the algorithms of a vehicle top-view system. After the whole design had been finished, the system was tested in the vehicle laboratory which proved that it is feasible. Therefore, this project has engineering value. The author will make more improvement to the algorithms to adapt it to more complex environments.

ACKNOWLEDGEMENTS

This project was supported by the Doctoral Scientific Research Foundation of Chongqing (A2016-27), The Science and Technology Project Affiliated to the Education Department of Chongqing Municipality (KJ120511) and Achievement Transfer Programme of Institutions of Higher Education in Chongqing (KJZH 14207).

REFERENCES

Jeon, B. et al. (2014). A memory-efficient architecture of full HD around view monitor systems. *IEEE Transactions on Intelligent Transportation Systems, 15*(6), 2683–2695.

Kannala, J. & Brandt, S.S. (2016). A generic camera model and calibration method for conventional, wide-angle, and fish-eye lenses. *IEEE Transactions on Pattern Analysis & Machine Intelligence,* 28.8(2006):1335–40 IEEE Computer Society, 2014:676–681.

Maik, V. et al. (2014). Automatic top-view transformation for vehicle backup rear-view camera. *The IEEE International Symposium on Consumer Electronics IEEE, 2014,* 1–2.

Rose, M.K. et al. (2009). Real-time 360 degrees imaging system for situational awareness. *Proceedings of SPIE—The International Society for Optical Engineering, May 2009.*

Yu, C. et al. (2012). An imaging method for 360-degree panoramic bird-eye view. *IEEE Intelligent Control and Automation,* 4902–4906.

Automotive, Mechanical and Electrical Engineering – Liu (Ed.)
© 2017 Taylor & Francis Group, London, ISBN 978-1-138-62951-6

Research of the fault detection and repair technology for submarine cable

Zhenxin Chen, Zhifei Lu, Xutao He & Yanjie Le
Zhoushan Power Supply Company, State Grid Zhejiang Electric Power Corporation, Zhoushan, China

ABSTRACT: The purpose of this paper is to study the fault detection and repair technology of submarine cable. The cause of the fault is analysed and the type of fault is put forward. The fault detection methods and the repair technology for submarine cable are introduced in detail, and also some protective measures for after the submarine cable is repaired are put forward.

Keywords: Submarine Cable; Fault Detection; Repair Technology

1 INTRODUCTION

The transmission engineering of submarine cable is an important part of cross-sea networking construction, which plays an important role in the realisation of internationalisation of the grid and the interconnection of regional grids. Because of the concealment and importance of submarine cable, once the submarine cable fails, it will not only have huge direct and indirect economic losses, but also cause a serious social impact. Therefore, how to detect and repair submarine cable accurately and in a timely manner is particularly important.

2 THE FAULT CAUSES AND TYPES OF SUBMARINE SELECTING A TEMPLATE

The causes of faults in submarine cable are various. From the submarine cable itself, its material or manufacturing process may not be in accordance with the specifications at the time of installation etc., which will eventually have a negative impact on the cable's operation. Thus, a series of problems caused by power and heat cause the cable to age fast and seriously affect the performance of the cable.

The faults caused to the submarine cable are mostly affected by external forces. The actions of external forces damage the cable. According to the location of the fault, the faults of submarine cable can be divided into the core disconnection fault, the main insulation fault and the sheath fault. According to the nature of the fault (impedance nature), submarine cable faults can be divided into low resistance and high resistance faults. The low resistance fault, also known as the short-circuit fault, refers to the insulation resistance down to the characteristic impedance of the cable, even the DC resistance is zero at the point of fault. The high resistance fault refers to the DC resistance of the fault point being greater than the characteristic impedance of the cable, and this can be divided into an open circuit fault, a high resistance leakage fault and a flashover fault (Bauch et al., 2004).

3 THE FAULT DETECTION OF SUBMARINE CABLE

We detect the faults of the submarine cable through three steps. The first step is fault diagnosis, the second step is preposition and the third step is precise positioning. In the event of a fault, the first thing you need to do is to make a diagnosis of the nature of the fault by way of measuring insulation resistance. Then, the corresponding ways to implement fault detection should be used, combined with the basic type of fault (Chen et al., 2006). The fault detection of submarine cable is shown in Figure 1.

After determining the fault cut-off point of the submarine cable, the empirical value and manual comparison method is usually used to calculate the length of the spare submarine cable required. In a shallow water area (water depth of less than 50 m), the most commonly used inlet angle α and approximate simulation is used to calculate the length of the required submarine cable. The minimum length of the spare submarine cable required and the minimum length of the cable to be removed is shown in the following formula (Couderc et al., 1996):

$$L_{min} = B + D + 2\delta + \frac{H}{\sin \alpha}$$

$$W_{min} = \left(\frac{2H}{\sin \alpha} + 3B + D \right) - 2\left(\frac{H}{\sin \alpha} + B + \delta \right) \qquad (1)$$
$$= B + D - 2\delta$$

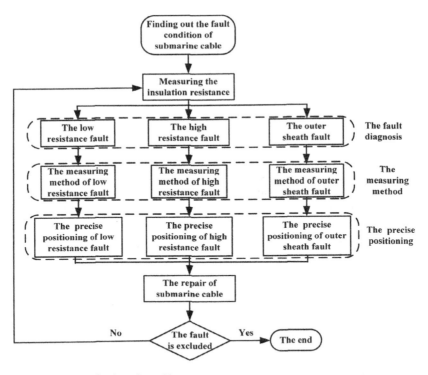

Figure 1. The fault detection of submarine cable.

The α is the inlet angle of the submarine cable during fault repair. The H is the vertical distance from the sea floor to the maintenance of the ship's deck. The B is half of the length of the submarine cable maintenance platform (Chandrabhan et al., 2007). The D is the broken section found by fault detection. The δ is the required redundancy length for submarine cable repair.

The diver determines the location of the submarine cable fault underwater. They then draw out the submarine cable along the bottom of the cable at the proposed fault point by using a high-pressure water gun or other blowing equipment. The length of the submarine cable needed to be drawn out should be determined by the depth of the water, the bending radius of the cable, the margin of the fault point and so on (Tian et al., 2004).

4 THE REPAIR TECHNOLOGY OF SUBMARINE CABLE

4.1 The repair technology

The repair technology of different submarine cables is different. This paper will focus on the repair technology of XLPE insulated lead sheathed thick steel wire armoured submarine cable in a shallow water area.

The typical repair technology steps of submarine cable are shown as follows:

a. After the location of the submarine cable fault is found, the ship is in place at the proposed fault point.
b. Calculate the length of the spare submarine cable required.
c. The diver cuts the cable underwater, removes the damaged submarine cable and then installs the waterproof component at both ends and marks (Mark No. 1 and No. 2).
d. Remove the faulty submarine cable from the water and fix it on the working platform by using a floating crane near the mark of No. 1.
e. After the damage point and the water inlet part is removed, eliminate other faults by conducting the DC resistance, insulation resistance and withstand voltage tests (S. et al., 2007).
f. Use the special connector of the submarine cable to connect the bottom of the submarine cable marked No. 1 and then put it back.
g. Move the floating crane to the submarine cable breakpoint marked No. 2. Remove the faulty submarine cable from the water and fix on the platform. After the damage point and the water inlet part are removed, eliminate other faults by conducting the DC resistance, insulation resistance and withstand voltage tests.

Use the special connector of the submarine cable to connect the bottom of the submarine cable marked No. 2 and then put it back (Tatsuo et al. 1996).

h. After the repair of conductor DC resistance, insulation resistance, lead sheath DC resistance uniformity test and pressure test to confirm whether the test results meet the latest APISpec17E (K. et al., 2006).

4.2 *The preventive measures*

Submarine cable is the indispensable power transmission facility between various platforms in the sea. Its importance is self-evident. However, due to the particularity of its environment (marine environment), the maintenance of it and any maintenance work needed is difficult. Once damaged, it will cause serious losses.

The strengthening of the submarine cable used in the process of maintenance work and reasonable submarine cable construction are particularly important in order to prevent submarine cable faults and measurement errors in the event of failure. Therefore, the following precautions are recommended to prevent submarine cable faults and measurement errors (Henningsen et al., 1996):

1. The ship should first obtain the permission anchor coordinates and break down in the designated area strictly according to the regulations, when the ship needs to break down in the oil area.
2. The submarine cable protection should be unified, inspected and the damage immediately repaired to prevent the fracture along the cable protection pipe falling in the sea.
3. The anti-wear measures should be taken at the intersection of submarine cable and the intersection of submarine cable and submarine pipeline.
4. Fix and support the submarine cable under the long-floating platform.
5. Collect the construction information of the submarine cable and the direction of each, the bottom of the amount of surplus, the actual length of the detailed to facilitate accurate measurement and fault location.
6. The length of the submarine cable should be regularly patrolled by the oil field ship on duty, to prevent passing ships, fishing operations and others from endangering the safety of the submarine cables and to strengthen the safety of the submarine cables.
7. Make emergency plans and contingency plans, so that the submarine cable fault can be repaired in time to minimise losses.

5 METHODS OF DETECTION AFTER REPAIR OF SUBMARINE CABLE

Measure the insulation of the three values by using megger until the cable connectors are all made and the measured value should be in the range of more than 1,000 megohm. If the leakage current is less than 75 mA to 80 mA during the 10 minute voltage regulation test under the 15 kV, it means that the cable is qualified (P. et al., 2007).

6 CONCLUSION

In this paper, the submarine cable fault detection and repair technology is discussed. Finally, the detection method of submarine cable repair is introduced to ensure the cable's qualification after repairing. The detection and repair technology proposed in this paper has been applied many times in the submarine cable dimension repair project.

REFERENCES

Bauch, A., Piester, D., Moudrak, A., et al. (2004). Time com-parisons between USNO and PTB: A model for the determination of the time offset between GPS time and the future galileo system time. *Frequency Control Symposium and Exposition.*

Chen, L., Zhu, X., & Li, T. (2006). Choice of Submarine Cable of Hainan Interconnection Project, 32(7), 39–42.

Couderc, D., Bourassa, P., & Muiras, J. M. (1996). Gas-in-oil criteria for the monitoring of self-contained oil-filled power cables. *IEEE 1996 Annual Report of the Conference on Electrical Insulation and Dielectric Phenomena.*

Czupryna, S., Everaere, J., & Delattre, M. (2007). Improvements in super absorbent water blocking materials for new power cable applications. in Jicable Conf.

Du, P., & Fu, S. Z. (2007). Brief analysis on several kinds of detection technologies of optical fiber submarine cable. *Guangxi Journal of Light Industry*, 1(12), 67–68.

Henningsen, C. G., et al. (1996). Experience with an on-line monitoring system for 400 kV XLPE cables. *Proceedings of 1996 IEEE Transmission and Distribution Conference.*

Sasaki, T., & Kurihara, M. (1996). Oil-filled cable surveillance system using newly developed optical fiber gas sensor. *IEEE Transactions on Power Delivery.*

Sharma, C., & Singh, P. (2007). Contribution of loads to low frequency oscillations in power system operation 2007 iREP symposium. *Revitalizing Operational Reliability*, 19(24), 1–8.

Tian, Y., Lewin, P. L., Wilkinson, J. S., et al. (2004). Continuous on-line monitoring of partial dischangees in high voltage cables. *Conference Record of the 2004 IEEE International Symposium on Electrical Insulation.*

Yung, K. M., Wong, K. C., & Ip, S. L. (2006). Quality plan for transmission plant and equipment. *CLP Power Hong Kong Limited.*

Automotive, Mechanical and Electrical Engineering – Liu (Ed.)
© 2017 Taylor & Francis Group, London, ISBN 978-1-138-62951-6

Research on blind identification method of error correcting code type

Xinhao Li, Min Zhang & Shunan Han
School of Electronic Engineering Institute, Hefei, China
Key Laboratory of Anhui Electronic Restricting Technique, Hefei, China

ABSTRACT: Based on the distribution of run feature and rank difference feature of coding sequence, this paper proposed a method to identify a type of common coding sequence. Using the concept of run distribution in combination with analysis of random nature within the coding sequence, it can be distinguished whether the encoding type belongs to block codes or others. Based on the structural features of turbo codes, rank difference analysis following the split of the coding sequence can distinguish the encoding type from convolutional and turbo codes. Herein the simulation experiment is described, the results of which demonstrate the validity and applicability of the proposed method.

Keywords: Block codes; Convolutional codes; Turbo codes; Run feature; Rank difference statistical feature

1 INTRODUCTION

The error correcting code is also known as anti-interference encoding and the main purpose of it is to reduce the bit error rate and increase the reliability of communication (Zhang Yongguang et al., 2010). Block code, convolutional code, and turbo code are common schemes for channel coding. The block code is a forward error correcting code and it has already been used in the fields of mobile communication, deep space communication, satellite communication and modern digital communication systems, etc. (Gallager, 1963; Li Yi, 2009; Vladislav Sorokine et al., 2000; Hua Li, 2007; Wang Kaiyao, 2013; Yang Xiaowei et al., 2012; and Zhang Chengchang et al., 2014).The convolutional code is a memory error control code and it has been used in Digital Video Broadcasting systems (DVB), Digital Audio Broadcasting systems (DAB) and Digital Multimedia Broadcasting systems (DMB) (Tang Qi, 2013; He Xianguo, 2013). The turbo code has an excellent decoding performance that is close to the Shannon limit. Due to the advantages of the turbo code, it has been widely used in deep space communication, satellite communication and third generation mobile communication (e.g. WCDMA, CDMA2000 and TD-SCDMA) (Xu Zhong et al., 2001; (Bringer et al., 2012; Moosaviet al., 2011; Sung-Joon et al., 2009; Sherif Welsen Shaker, 2014; and Debessu et al., 2012).With the development of cognitive radio and cognitive communication, channel coding recognition has been widely used in various fields, such as information interception, intelligent communication, network confrontation and communication reconnaissance (Xie H et al., 2013). The field has been flourishing in recent years as new identification algorithms of error correcting codes keep emerging ('endless stream' has a very strong negative hint as it is boring and tedious, which is obviously not the case here.) (Wang Xinmei et al., 2006; QI Lin et al., 2011; WEN Niancheng et al., 2011; ZHOU Pan et al., 2013; LIU Jiancheng et al., 2013; CHAI Xianming et al., 2010; WEI Yuejun, 2013; Wang D Kobasyashi, 2009; and RENG Defeng, 2013). They are mainly about the construction of check matrix, optimisation of encoding and decoding algorithms, and decoding performance analysis. Nonetheless, they do not take the identification of encoding type into consideration. However, the type of error correcting code is unknown in the context of non-cooperative communication. In the protocol of the Consultative Committee for Space Data Systems (CCSDS), not only block codes, but also convolutional and turbo codes have widespread application (LI Xiangying, 2011). In order to identify the protocol, the encoding type must be identified, so it is necessary to develop an effective method for the blind identification of error correcting codes.

To address the challenge, a method for blind identification among the types of block code, convolutional code and turbo code is described herein. Theoretical analysis is validated by experiments and simulation results. Effectiveness of the identification methods is also proved.

2 BRIEF INTRODUCTION OF ERROR CORRECTING CODES

There are many types of error correcting codes and the practical applications include block code, convolutional code and turbo code. The main focus here is the type identification of these codes (Wang Xinmei et al., 2006).

2.1 Block codes

The (n, k) block code is a kind of algebraic encoding, and each code word includes k information bits and $n - k$ check bits. There is an algebraic relationship between information bits and check bits. As shown in Figure 1, the input of encoder is M (swap the positions 'M' and 'is') called information bits, which includes k bits. The output of the encoder is C (swap the positions 'C' and 'is') called code word, which includes n bits.

2.2 Convolutional codes

The most important feature of convolutional codes is memory, which differentiates it from block codes. Its check bits not only relate to its own information bits, but also relate to the existing encoder. The structure of convolutional codes' encoder is shown in Figure 2.

2.3 Turbo codes

The most pre-eminent feature of turbo codes is random encoding, which is achieved by introducing an interleaving device. Because of the effective combination of short codes and long codes,

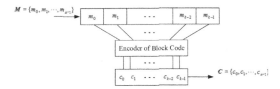

Figure1. Structure of block codes' encoder.

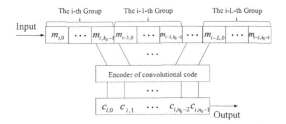

Figure 2. Structure of convolutional codes' encoder.

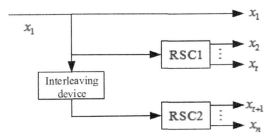

Figure 3. Structure of turbo codes' encoder.

a performance close to the Shannon theoretical limit can be achieved. At present, the most widely used arrangement of turbo codes is parallel-concatenated convolutional code, the structure of which is shown in Figure 3. The information bits that will be encoded are divided into three parts: the first part is x_1 which will not be encoded, the second part includes $x_2, x_3, ..., x_t$, which are generated by information bits through RSC1, and the third part includes $x_{t+1}, x_{t+2}, ..., x_n$, which are generated by information bits through the interleaving device and RSC2. Turbo codes can be yielded by multiplexing these three parts (Zhang Yongguang et al., 2010).

3 PROPOSED METHOD

3.1 Related definitions

Definition 1: Suppose that a is a periodic sequence of v in the field of GF(2). Loop arrange a's one period $a = (a_0, a_1, a_2, ..., a_{v-1})$ in cyclic manner. Hence, a_{v-1} and a_0 are adjacent. The shaped item, such as 100...001 or 011...110 is referred as zero or one run of a's one cycle. The number of 0s in zero run or the number of 1 in one run is the length of this run (Zhang Yongguang et al., 2010).

Definition 2: The number of great linearly independent vectors of matrix C is referred to as the rank of matrix C. $R(C) = r$ denotes that the rank of matrix C is r. Similarly, the rank of matrix zero is written as $R(0) = 0$ (Xie H et al., 2013).

Definition 3: The $p \times q$ matrix C whose rank is r and its rank difference defined as $N = p - r$.

For clarity, the rank distribution of matrix will be replaced by rank difference distribution in this paper.

3.2 Analysis of run feature

3.2.1 Run feature of random sequence

In terms of run and the number of 0, 1, the m sequence X_n whose cycle is $2^n - 1$ has the following properties (Zhang Yongguang et al., 2010):

1. If the number of one element was 2^{n-1} in one period of the m sequence whose period is $2^n - 1$, the other would be $2^{n-1} - 1$.
2. In one period of the m sequence whose period is $2^n - 1$, the total number of run is 2^{n-1}. Each number of zero run and one run composes 50% of the period. The number of the run whose length is $k(0 < k \leq n-2)$ is 2^{n-k-1}.

This shows that the number of zero runs and one runs in the m sequence is decreasing with the increase of the run length, in accordance with the 1/2 law. Otherwise, their numbers are basically the same when the lengths of zero and one run are equal.

3.2.2 *Run feature of block codes*

It is exhibited in Figure 4a that the (n, k) block code, where the length of information bit is k, has 2^k kinds of output code words. While in Figure 4b, the binary random sequence with length of n has 2^n kinds of output code words. Thus the (n, k) block code has $2^n - 2^k$ kinds of disable code words, which leads to invalidation of the random nature of the block code. Since the k-bits information symbols can be randomly generated, the run whose length is less than k can be obtained from the code word. In comparison, the run with a length greater than or equal to k can only be obtained from the position where two code words are spliced. Therefore, different from random sequence, the number of block code's run that is around the information bit length would have a certain amount of distortion, which is different from random sequence.

3.2.3 *Run feature of convolutional codes*

The most important feature of convolutional codes is memory, which makes it different from other block codes. The check bit of convolutional codes does not only relate to its information bits, but also relates to the information bits that were previously inputted into the encoder. The structure of convolutional codes is similar to linear feedback shift registers, which can generate an m sequence. Due to the above reasons, convolutional codes bear

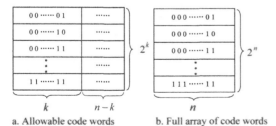

| a. Allowable code words | b. Full array of code words |

Figure 4. Distribution of block code.

similar features as random sequence. The feature performs on run is that the zero and one run number are decreasing in accordance with 1/2 law when length of the run is increasing.

3.2.4 *Run feature of turbo codes*

The biggest feature of turbo codes is random encoding, which is achieved by introducing an interleaving device. It therefore has an excellent decoding performance that is close to the Shannon limit (Zhang Yongguang et al., 2010). Because of the above-mentioned reason, the random nature of turbo codes is similar to random sequences. Performing on run is that the zero and one run number are decreasing in accordance with 1/2 law when the length of run is increasing. So the run feature cannot be used to distinguish the encoding type between convolutional codes and turbo codes.

By analysing these coding sequences' randomness, two conclusions can be drawn as follows:

1. The run feature of block codes is different from that of convolutional codes and turbo codes in terms of randomness. The randomness feature of convolutional codes and turbo codes is better than that of block codes and is similar to that of random sequence. The zero and one run number of convolutional codes and turbo codes decreases in accordance with 1/2 law.
2. With the increase of the run length, block codes' run number presents an irregular descending, which is around the information bit length would have a certain amount of distortion.

3.3 *Analysis of rank feature*

3.3.1 *Rank feature of convolutional codes*

Considering that the $p \times q(p > 2n, q < p)$ order matrix consists of (n_0, k_0, m) convolutional code, if q was an integer multiple of n, the upper left corner matrices would have the same dimension and the rank of the $p \times q$ matrix would not be equal to q (Zhang Yongguang et al., 2010).

From the definition of convolutional codes it is known that the output vector is the linear transformation of the input vector. The linear constraint relation expressed by any complete convolutional codes is exactly the same and is equal to the system form $[I_k \ P]$. Given the $p \times q(p = tn_0 + l, q < p, t > 2)$ order matrix C which consists of (n_0, k_0, m) convolutional code.

• The matrix C consists of a complete code word when $q = an_0 \ (a \geq 1)$ and the code word starting point is correct. After simplifying this matrix, the formula (1) is reached:

$$C = \begin{pmatrix} c_{11} & c_{12} & \cdots & c_{1q} \\ c_{21} & c_{22} & \cdots & c_{2q} \\ \vdots & \vdots & \vdots & \vdots \\ c_{p1} & c_{p2} & \cdots & c_{pq} \end{pmatrix} \rightarrow$$

$$C = \begin{pmatrix} I_{k_0} P_{n_0-k_0} & 0_{k_0} & 0_{k_0} & \cdots & 0_{k_0} \\ 0_{p-k_0} & 0_{p-k_0} & 0_{p-k_0} & \cdots & 0_{p-k_0} \\ 0_{k_0} & I_{k_0} P_{n_0-k_0} & 0_{k_0} & \cdots & 0_{k_0} \\ \vdots & \vdots & \vdots & \ddots & \vdots \\ 0_l & 0_l & 0_l & \cdots & 0_l \end{pmatrix} \tag{1}$$

Rank of matrix C is $nk(n \geq 1)$.

- The matrix C consists of a number of sub-matrices when $q = an_0 (a \geq 1)$ and the code word starting point is incorrect. At the left of matrix C, there is a sub-matrix that has p rows and $n_0 - i + 1$ columns. In the middle of matrix C, there are $a - 1$ sub-matrices that have p rows and n_0 columns. At the right of matrix C, there is a sub-matrix that has p rows and $i-1$ columns. After simplification, formula (2) is reached:

$$C = \begin{pmatrix} c_{1,i} & c_{1,i+1} & \cdots & c_{1,i+q} \\ c_{2,i+1} & c_{2,i+2} & \cdots & c_{2,i+q} \\ \vdots & \vdots & \vdots & \vdots \\ c_{p,i+1} & c_{p,i+2} & \cdots & c_{p,i+q} \end{pmatrix} \rightarrow$$

$$C = \begin{pmatrix} I_\alpha & 0_\alpha & 0_\alpha & \cdots & 0_\alpha \\ 0_{k_0} & I_{k_0} P_{n_0-k_0} & 0_{k_0} & \cdots & 0_{k_0} \\ 0_{k_0} & 0_{k_0} & I_{k_0} P_{n_0-k_0} & \cdots & 0_{k_0} \\ \vdots & \vdots & \vdots & \ddots & \vdots \\ 0_\beta & 0_\beta & 0_\beta & \cdots & I_\beta \end{pmatrix} \tag{2}$$

The left-side sub-matrix and right-side sub-matrix have a fixed rank value α and β. When the number of columns is an integer multiple of the code length, the rank of matrix C increases k_0 $(R = \alpha + \beta + nk_0)$.

- The linear correlation between each row is poor when $q \neq an_0 (a \geq 1)$. Simplifying matrix C will get a unit matrix I_q or $\begin{bmatrix} I_{q-1} \\ 0_l \end{bmatrix}$. At this time, rank of matrix C is q or $q - l$.

From the above analysis it can be concluded that if the number of columns is an integer multiple of code length, the rank of matrix C consists of convolutional codes would obtain the minimum value periodically.

3.3.2 Rank feature of turbo codes

The encoder of turbo codes consists of two sub-encoders and one interleaving device. The Recursive

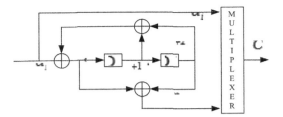

Figure 5. Structure of RSC.

Systematic Convolutional (RSC) code encoder is always used to take the place of sub-encoders and its structure is showed in Figure 5.

From the relationship between output and input of RSC in Figure 5, formulae (3) and (4) are derived:

$$s_k = d_{in} + \sum_{j=1}^{m_c} s_{k+j} g_{b,j} \quad \mod 2 \tag{3}$$

$$d_{out} = \sum_{j=0}^{m_c} s_{k+j} g_{f,j} \quad \mod 2 \tag{4}$$

In formulae (3) and (4), the variable $g_{f,j}$ and $g_{b,j}$ indicate coefficients of feed-forward polynomial and feedback polynomial. Length of the register is expressed as m_c. Transform the formula (3) and one will get formula (5):

$$d_{in} = s_k + \sum_{j=1}^{m_c} s_{k+j} g_{b,j} \quad \mod 2 \tag{5}$$

From the formulae (4) and (5) it is known that if s_k was the input sequence, d_{out} and d_{in} would be the output sequence of 1/2 convolutional codes.

As shown in Figure 3, turbo codes whose rate is 1/n can be divided into n outputs. Combining any two groups of these outputs, one can get three kinds of split modes as shown in Figure 6.

In Figure 6(a), one way of input and the other way of RSC check consist the first split mode. From formulae (3) and (4) it is known that the output of the first split mode has the same properties as convolutional codes. In Figure 6(b), because of the application of interleaving device, the linear relationship between the original information bits and the output of RSC is destroyed. So the output of the second split mode does not have properties as convolutional codes. While in the first road of Figure 6(b), the output x_1 and input x_1 have the mapping relationship $x_1 = x_1 \cdot G_1$. Similarly in the second road, the output x_t and input x_1 have the mapping relationship $x_t = x_1 \cdot G_2$. After cross multiplexing, the output B consisting of these two road sequences also has a mapping relationship $B = x_1 \cdot MUL(G_1 \ G_2)$ with the input x_1, which

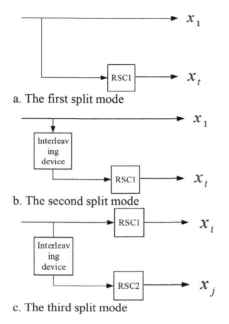

a. The first split mode

b. The second split mode

c. The third split mode

Figure 6.   Three kinds of split modes.

can be referred to as a block code. Likewise, the third split mode that is revealed in Figure 6(c) also shares the properties of block codes.

Through the analysis of these coding sequences' rank feature, two conclusions can be drawn:

1. Since the code length of convolutional codes is generally less than eight and the encoder of turbo codes is similar to convolutional codes, it is possible to construct a $p \times q$ $(p = mn + 1, q < p, m > 2)$ matrix the rank of which can be periodically minimised after simplification.
2. Because of the unique encoding method of turbo codes, the rank difference statistics of split turbo codes can detect the nature of block codes and convolutional codes.

3.4   *Identification step*

Through the analysis of run feature and rank feature of coding sequence, steps of identification are proposed as following:

1. Statistic run distribution $v_i (i = 1, 2, ..., t)$ of coding sequence, i is run length and v_i is the number of run;
2. Drawing run distribution of coding sequence and random sequence;
3. Observing run distribution: if run distribution of coding sequence is similar to random sequences, and zero and one run number decreased in accordance with 1/2 law, the coding

sequence could be either convolutional or turbo codes; on the contrary, it would be block codes;

4. If the coding sequence was not block codes, splitting this sequence and constructing a $p \times q$ $(p = mn + 1, q < p, m > 2)$ matrix;
5. Simplifying the $p \times q$ matrix and statistic rank $R_j (j = 1, 2, ..., k)$ of each matrix;
6. Statistic rank difference $C_j = q_j - R_j$ and drawing a map of it. Observing the period of rank difference on the map and estimating encoding type of each split sequence;
7. If one of the split sequences was convolutional code and other two split sequences were block codes, the coding sequence could be turbo codes. On the contrary, it could be convolutional codes.

4   EXPERIMENTS

Experiment 1: run statistics of coding sequence

Under the condition of no error bits, the following were generated: (15, 5) block code, (3, 2, [4 3]) convolutional code and 1/3 turbo code whose polynomial was (7, 5) and random interleaving device's length was 15 bits. Length of each test sequence was 33,600 bits. Extracting test sequence's run number when length of run was varied from 1–10. Run distribution of each coding sequence and random sequence are indicated in Figure 7:

It can be observed for convolutional code (Figure 7 (b)) and turbo code (Figure 7 (c)) that the number of run decrease smoothly with the increase of the length of run, while for block code (Figure 7(a)) there is significant distortion.

Thus it can be drawn from the experimental results that:

1. There is a big difference of run distribution between block codes and random sequence. The block codes' run number around the information bit length would have a certain amount of distortion.
2. The run distributions of convolutional codes and turbo codes coincide with random sequences, and the zero and one run number decreased in accordance with 1/2 law.
3. Since the better random performance of convolutional and turbo codes' run distributions are similar to each other, the characteristic of run distribution to distinguish encoding type between convolutional codes and turbo codes cannot be used.

Experiment 2: rank difference statistics of coding sequence

Under the condition of no error bits, the following were generated: (2,1,6) convolutional code and 1/3 turbo code whose polynomial was (7,5), and random interleaving device's length was 15 bits.

413

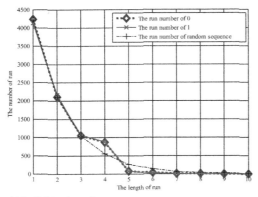

a. (15, 5) block code

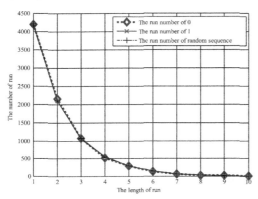

b. (3, 2, [4 3]) convolutional code

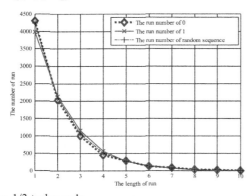

c. 1/3 turbo code

Figure 7. Run distribution of coding sequence.

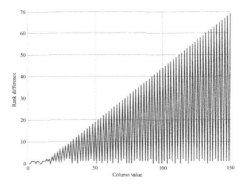

a. Convolutional codes

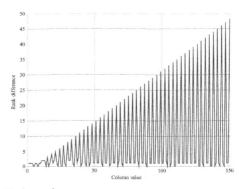

b. Turbo codes

Figure 8. Rank difference distribution of coding sequence.

The length of each test sequence was 33,600 bits. With no sequence split, the statistic rank difference distribution of these two sequences and simulation results are shown in Figure 8.

Under the condition of no sequence split, the rank difference distributions of convolutional codes and turbo codes have periodical peak values. The column value which corresponds to the periodical peak is an integer multiple of the code length. So the rank difference feature cannot be used to distinguish encoding type between convolutional codes and turbo codes without sequence split.

From the experimental results, the following can be concluded:

1. There is a linear relationship in convolutional codes and turbo codes under the condition of no sequence split. Their rank differences present periodical distributions with peak values.
2. Rank difference distribution cannot be used to distinguish encoding type between convolutional codes and turbo codes without sequence split.

Experiment 3: rank difference statistics after sequence split

Under the condition of no error bits, the following were generated: (2,1,6) convolutional code, split into three modes. The length of each split sequence was 33,600 bits and their rank difference distributions are shown in Figure 9.

414

As shown in Figure 9, rank difference distributions of these three split sequences have no periodical peak values and they all have the feature of random sequence.

Under the condition of no error bits, 1/3 turbo code with polynomial (7, 5) and 15 bits random interleaving device length was generated. Splitting it into three modes and each test sequence's length was 33,600 bits. Rank difference distributions of these sequences are shown in Figure 10.

As shown in Figure 10, rank difference distributions of these three split sequences have periodical

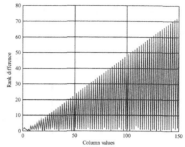

a. The first split mode

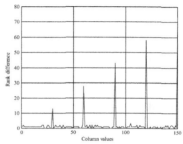

b. The second split mode

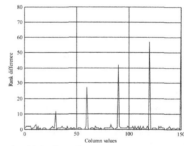

c. The third split mode

Figure 10. Rank difference distribution of split turbo codes.

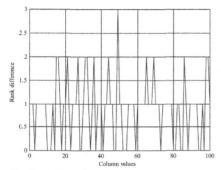

a. The first split mode

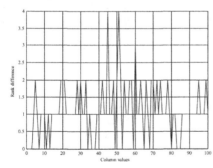

b. The second split mode

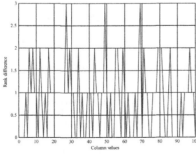

c. The third split mode

Figure 9. Rank difference distribution of split convolutional codes.

peak values. Figure 10(a) shows that under the condition of the first split mode, turbo codes have the same rank difference distribution as convolutional codes. Figures 10(b) and 10(c) demonstrate that turbo codes have the same rank difference distribution with block codes under the conditions of second and third split modes.

From the experimental results, it can be concluded as follows:

1. There is no linear relationship in the split convolutional codes, and rank difference of them does not present periodical distributions with peak values.
2. The rank difference of split turbo codes has the same distributions with convolutional codes and block codes. So we can split turbo codes

Table 1. Recognition rate under different BER.

Bit error rate	0.001	0.002	0.003	0.004	0.005	0.006	0.007	0.008	0.009	0.01
Recognition rate	1	1	1	0.99	0.98	0.96	0.90	0.77	0.52	0.21

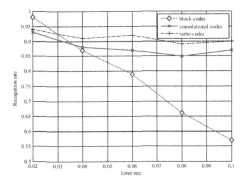

Figure 11. BER analysis of run.

and then statistic its rank difference distribution to distinguish the encoding type between convolutional codes and turbo codes.

Experiment 4: analysis of BER
The following were generated: (15, 5) block code, (2, 1, 6) convolutional code and 1/3 turbo code whose polynomial was (7, 5) and random interleaving device's length was 15 bits. The length of each test sequence was 33,600 bits. Under different bit error rate conditions, 100 times Monte Carlo simulations were carried out to verify run distribution method's capability of resisting BER and the result was showed in Figure 11:

As shown in Figure 11, the correct recognition rate of block codes stays above 80% and both convolutional and turbo codes stay above 85% when the bit error rate is less than 0.06. With the increase of the bit error rate, the correct recognition rate of block codes descends rapidly while convolutional codes and turbo codes still stay above 85%.

Generating 1/3 turbo code whose polynomial was (7, 5) and random interleaving device's length was 15 bit. The length of test sequence was 33,600 bits. Under different bit error rate conditions, 100 times Monte Carlo simulations of the sequences that had been split were carried out to verify rank difference distribution method's capability of resisting BER, and the result was shown in Table 1:

In Table 1, the correct recognition rate comes down when the bit error rate goes up. It still stays around 0.9 when the bit error rate is as high as 0.007. However, in higher bit error rate range it drops steeply to 0.21 at bit error rate 0.01.

From the experimental results, it can be concluded that:

1. The run feature has a good capability of resisting BER. It can distinguish the encoding type between block codes and convolutional codes, turbo codes effectively when the bit error rate is 0.05.
2. The rank difference's correct recognition rate is more than 0.75 when the bit error rate is less than 0.007.

However, it decreases significantly when the bit error rate is more than 0.08.

5 CONCLUSION

An unprecedented and efficient identification method of encoding type among block, convolutional and turbo codes is presented in this paper. The encoding type of these codes can be distinguished by studying their run and rank features. Simulation results showed that this method had significant advantages, such as high robustness and accurate recognition, indicating its promising potential in future engineering applications.

ACKNOWLEDGEMENTS

This work was financially supported by the National Natural Science Foundation of China (No.61171170) and Natural Science Foundation of Anhui Province (No.1408085QF115).

REFERENCES

Bringer, J. and Chabanne, (2012). H. Code reverse engineering problem for identification codes. *IEEE Transactions on Information Theory*. 58(4), 2406–2412.

Chai Xianming, Cai Kai, Shouye, L.V., et al. (2010). Research on blind identification of convolutional codes. *Journal of Circuits and Systems*. 15(4), 38–44.

Debessu, Y.G., Wu, H.C. and Jiang, H. (2012). Novel Blind Encoder Parameter Estimation for Turbo Codes. *IEEE Communications Letters*. 12(16), 1917–1920.

Gallager, R.G. (1963). Low Density Cheek Codes. Cambridge, Mass: MIT Press.

He Xianguo. (2013). Design and optimisation of pseudo random interleaver in Turbo code. Hangzhou: Hangzhou Dianzi University. 12.

Hua Li. (2007). Encoding and Implementation of LDPC Codes in the DVB-S2. Changsha: Information &

Communication Engineering Graduate School of National University of Defense Technology. 11.

Liu Jiancheng and YANG Xiaojing. (2013). Research on channel coding of convolutional codes.. Hefei: Electronic engineering institute. 6.

Li Xiangying. (2011). Key technologies of protocol identification for CCSDS data link layer. Beijing: Center for Space Science and Applied Research, Chinese Academy of Sciences.

Li Yi. (2009). Study on Encoding and Decoding Algorithm of LDPC Code Based on WiMAX. Harbin: Harbin Institute of Technology. 06.

Moosavi, R. and Larson, E.G. (2011). A fast scheme for blind identification of channel codes. *Global Telecommunications Conference*, Linkoping, Sweden. 1–5.

Qi Lin, Hao Shiqi and Li Jinshan. (2011). Recognition method of RS codes based on Euclidean algorithm in Galois field. *Journal of Detection & Control*. 33(2), 63–67.

Reng Defeng. (2013). High speed turbo decoding and cooperative spectrum sensing for next generation wireless communications. Xian: Xidian University.

Sherif Welsen Shaker. (2014). DVB-RCS: Efficiently Quantised Turbo Decoder. *ICACT Transaction on Advanced Communications Technology*. 2(3), 426–433.

Sung-Joon, Park, Jun-Ho and Jeon. (2009). Interleaver Optimisation of Convolutional Turbo Coder for 802.16 Systems. *IEEE Communications Letters*. 5(13), 339–341.

Tang Qi. (2013). Research on Convolutional Code Based on 3G. Wuhan: Central China Normal University. 05.

Vladislav, Sorokine, Frank, R. and Kschischang, Subbarayan Pasupathy. (2000). Gallager Codes for CDMA Applications [J]. *IEEE Transactions on Communications*. 48(10). OCTOBER.

Wang, D. and Kobasyashi, H. (2009). Low-complexity MAP Decoding for Turbo Codes. *Proceedings of IEEE Vehicular Technology*, 1035–1039.

Wang Kaiyao. (2013). The Design of High Performance LDPC Codes and its Application for Cognitive Radio Networks. Beijing: Beijing Jiaotong University. 04.

Wang Xinmei and Xiao Guozhen. (2006). Error Correcting Code—Principle and Method. Xian: Xidian University Press.

Wen Niancheng, Yang Xiaojing and Bai Yu. (2011). A new recognition method of RS codes. *Electronic Information Warfare Technology*. 26(2), 36–40.

Wei Yuejun. (2013). Research on the decoding algorithms of channel codes in 3GPP UMTS and LTE systems. Shanghai: Shanghai Jiaotong university.

Xie, H., Wang, F.H., Huang, Z.T. (2013). A Method for Blind Recognition of Convolutional Interleaver. *Journal of Electronics and Information Technology*. 35(8), 1952–1957.

Xu Zhong, Zhang Kaiyuan, Lu Quan and Leng Guowei. (2001). Matrix Theory Concise Guide. Beijing: Science Press. 01.

Yang Xiaowei, Gan Lu. (2012). Blind Estimation Algorithm of the Linear Block Codes Parameters. Based on WHT. *Journal of Electronics & Information Technology*. 34(7), 1642–1646.

Zhang Chengchang, Peng Wanquan and Wei Bo. (2014). Construction of a new class of (2k, k, 1) convolutional codes. *Journal on Communications*. 35(6), 200–206.

Zhang Yongguang and Lou Caiyi. (2010). Channel Coding and Recognition Analysis. Beijing: Publishing House of Electronics Industry.

Zhou Pan and Gan Lu. (2013). Blind recognition and parameter estimation of cyclic codes. Chengdu: University of Electronic Science and Technology of China. 5.

Automotive, Mechanical and Electrical Engineering – Liu (Ed.)
© *2017 Taylor & Francis Group, London, ISBN 978-1-138-62951-6*

Research on the detection system for dynamic ship draft on the basis of ultrasonic diffraction effect

Mudi Xiong & Lei Lu
School of Dalian Maritime University, Dalian, China

Weili Zheng & Ran Li
Three Gorges Navigation Authority, Yichang, China

ABSTRACT: A method to measure the dynamic ship draft by adopting ultrasonic diffraction effect is proposed in the view of the current state of detection technology for ship draft: the depth data of ship cross section is obtained using the diffraction effect produced by dynamic ship's occlusion of the ultrasonic transmission path under the water. Through the synchronous control of emitter array as well as receiving array and the data processing method of corresponding diffraction curve fitting, the real-time draft value of dynamic ship is measured. And according to the method mentioned above, the detection system for dynamic ship draft is developed on the basis of ultrasonic diffraction effect. The system is verified in a pool with the length of 25 meters, and the measurement precision of underwater standard occlusion is 0.02 m; during the measurement of dynamic ship in the gate of Gezhouba Dam ship lock, the contrast difference between the draft measured and the water gauge of ship reaches to 0.1 m, which proves that the system and data processing method are practical and effective.

Keywords: detection of draft; ultrasonic diffraction effect; synchronous control

1 INTRODUCTION

With the rapid development of inland shipping industry, the transportation demand also increased, part of the ship owners for the sake of more profit, usually lie the actual loading capacity of the shipping vessel or modify the plimsoll line of ship, in order to avoid the maritime law enforcement inspection. This will bring a great security risk, so the detection of the draft is very necessary. Existing draft detection methods have some disadvantages, such as inconvenient installation, low measurement accuracy, poor realtime ability and so on (ZHAO W et al., 2015). In order to overcome these disadvantages, this paper proposes a new detection method and develops a corresponding ship draft detection system. The system is provided with a corresponding ultrasonic emission sensor array and the ultrasonic receiving sensor array on both sides of the channel, with the aid of the diffraction effect caused by the occlusion of the ship to the ultrasonic transmission path, analyze and calculate the draft of the ship. Due to the above characteristics, the system is called the side scan type ship draft detection system. In addition, water quality, ultrasonic multipath effect, and the effect of crosstalk between the sensor arrays will affect

the measurement accuracy. The data processing method proposed in this paper can reduce the influence of the above factors to a certain extent, and improve the accuracy of the measurement.

2 THE PRINCIPLE AND COMPOSITION OF THE SIDE SCAN TYPE SHIP DRAFT DETECTION SYSTEM

When the ultrasound meets the occlusion, it will be scattered, and the scattered wave will contain the information of the shield and the medium. In this paper, the side scan type ship draft detection system is the use of the one side of the ultrasonic array sensor to transmit the ultrasonic wave, the other side of the ultrasonic receiving sensor array to receive ultrasound. Through the diffraction effect of the ship to the transmission path of the ultrasonic transmission, we can calculate the ship draft.

When the ultrasonic wave in the water encounter the occlusion and geometry size of occluded objects is much larger than the ultrasonic wavelength, a large number of sound waves are reflected by the shelter and form a shadow area behind the occlusion, while the unmasked area forms a bright area. In the shadow and light areas at the junction, a part of sound waves from the side shield to the diffraction effect of the diffraction phenomenon (SHI J W et al., 2009) (LU M H et al., 2015) (CHI D ZH) occurred. Through real time detection and analysis of diffraction patterns, it can be calculated to draw the depth of the occlusion.

Straight edge diffraction formula:

$$E(x, y) = \frac{E_\infty}{1+i}\left[F\left(x\sqrt{\frac{2}{\lambda z_1}} \right) - F(-\infty) \right] \quad (1)$$

$F(\omega)$ is fresnel equation,

$$F(\omega) = \int_0^\omega \exp(i\pi i^2/2)\,dt \quad (2)$$

The intensity of the sound $I = |E(x, y)|^2$. It can be concluded that the sound intensity distribution of Fresnel straight edge diffraction.

When the distance between the receiving screen horizontal distance 3 m occluder occlusion 500 K ultrasonic frequency, the relative intensity of the sound receiving sensor simulation diagram as shown in Figure 1, the horizontal axis shows occlusion depth, occlusion depth covering relative receiving screen center depth is 0 m, the vertical axis represents the relative acoustic normalized intensity.

When the received sound intensity is lower than 0.2 times of its own, the sound field is affected by the diffraction wave, which is recognized as a shelter area. When the sound intensity is 0.2 times stronger than the maximum sound intensity, the sound field is greatly affected by the diffraction wave. The intensity of the sound wave changes with the change of position. When the sound intensity is approximately equal to the maximum sound intensity, the sound field is mainly based on the incident wave, and is recognized as a no shelter area. Through the determination of the occlusion area, the depth of the occlusion can be roughly

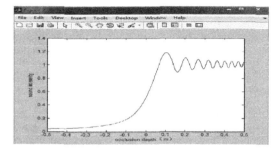

Figure 1. The relationship between sound intensity and occlusion depth of simulation.

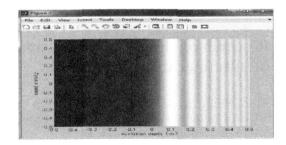

Figure 2. Simulation received light intensity pattern.

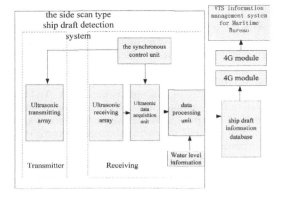

Figure 3. System composition.

calculated, and the analysis of the diffraction area can be more accurate to calculate the depth of the occlusion.

Based on the above principles, the side scan type ship draft detection system is shown in Figure 3, mainly including the following parts: the ultrasonic transmitting array composed of a plurality of ultrasonic transmitting sensor, the synchronous control unit, the ultrasonic receiving array composed of a plurality of ultrasonic receiving sensor, ultrasonic data acquisition unit, data processing unit and ship draft information database.

As shown in Figure 4, the ultrasonic transmitting sensor array is installed on one side of the navigable channel, and the corresponding ultrasonic receiving sensor array is installed on the other side of the navigable channel. The number and the horizontal position of the two sensor arrays are corresponding to one by one, and the distance between the sensors in the array is the same as the D.

The first synchronization control unit controls the ultrasonic transmitting sensor array to transmit the ultrasonic wave, and simultaneously control the ultrasonic data acquisition unit to capture the ultrasonic data received in the ultrasonic receiving sensor array. Then data processing unit to deal

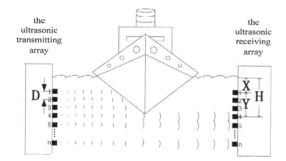

Figure 4. System schematic diagram.

with water level information and ultrasonic information, and calculate the depth of the ship draft. Finally, the ship draft information is transmitted to the ship draft information management database through long distance transmission.

The measurement principle is as follows:

$$H = X + Y \qquad (3)$$
$$Y = N * d + \Delta h \qquad (4)$$

In the formula: H is the ship's draft measured by the detection device; X is the ultrasonic transmitter and receiver array in the depth of the first sensor under the water (When the water level is changed, X can be obtained by measuring the water level meter installed at the first sensor position); Y is based on the ultrasonic diffraction effect of the ship to calculate the depth of the transmission of the ultrasonic transmission path; N is the number of acoustic wave intensity received by the ultrasonic wave receiving sensor array in the shelter area; D is the installation distance between the sensors in the sensor array, and the distance between the installation should be smaller than the distance from the diffraction effect caused by the diffraction effect; Δh is calculated by the depth of the acoustic intensity calculated by the first sensors in the diffraction area, which is relative to the depth of the sensor in the area.

3 SYNCHRONOUS CONTROL UNIT FOR TRANSMITTING AND RECEIVING SYNCHRONOUS CONTROL OF ULTRASONIC SENSOR

The ultrasonic wave sensor is provided with a certain angle of opening. In the receiving sensor array, a plurality of receiving sensors can receive the ultrasonic wave, which is called the crosstalk between the sensors (DONG H T et al., 2016) (ZH M et al., 2013). In addition, Ultrasonic encounter other object in water will also occur along different paths to reach the ultrasonic receiving sensor, but

from these different paths ultrasonic does not contain the ship's draft information, this phenomenon is called multipath effect. The crosstalk between the sensors and the multipath effect will affect the ultrasonic receiving sensor array to accurately determine the path of ultrasonic occlusion of the ship, in order to solve this problem, the synchronous control unit is very important for the synchronous control of ultrasonic transmitting and receiving arrays.

The control unit of the synchronous control unit adopts the method of time division control, and only one path ultrasonic emission sensor is used to arrive at the receiving sensor array at a certain time period. As shown in Figure 5 is the simulation of synchronous control unit for the control signal of the 6 ultrasonic sensors, a total cycle time is t, the low level is the corresponding way of the ultrasonic emission sensor is working.

The synchronous control unit controls the ultrasonic transmitting array at the same time, the ultrasonic data acquisition unit collects the voltage signal receiving ultrasonic detector output in the array. After the operation of the ultrasonic emission sensor, the time of the delay direct ultrasonic transmission is T2, the ultrasonic data acquisition card can start the data acquisition of the corresponding ultrasonic receiving sensor. The signal acquisition of an ultrasonic wave receiving sensor is completed by the data processing after the acquisition of the time t_1. t_2 is the direct wave transmission time, which can be measured by the identification of the first direct wave transmitted from the ultrasonic transmitting sensor array to the ultrasonic receiving sensor array.

As shown in Figure 6, the CH1 is a synchronous control unit control signal to an ultrasonic transmitting sensor, CH2 voltage signal corresponding to the ultrasonic receiving sensor to collect ultrasonic sensor output, 0V–5V is the ultrasonic intensity of different voltage amplitude, higher representative ultrasonic intensity. Observed from the figure, since the launch of post t_2 time for the first time the sensor receives the ultrasonic signal, the signal for the direct ultrasonic signal, and after t_3 time can also receive the ultrasonic signal, but is not the direct wave signal, using the method of

Figure 5. Control chart of synchronous control unit to ultrasonic transmitting sensor.

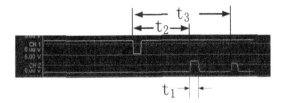

Figure 6. Control chart of data acquisition unit with synchronization control unit.

ultrasonic propagation delay can well eliminate these no direct wave signal.

4 DATA PROCESSING METHOD

4.1 Data processing in a single receive cycle

The output voltage signal detection in the ultrasonic receiver array will have certain fluctuations, in order to reduce the affect caused by the fluctuation of the need for appropriate treatment in the ultrasound data acquisition unit. In the ultrasonic data acquisition unit, the voltage of a receiving period ultrasonic receiving sensor detect the average voltage, which can replace the entire cycle of the ultrasonic receiving sensor output value. It can also eliminate the interference of voltage fluctuation in the output to reduce the amount of data at the same time. Use the following formula:

$$P(i) = (f(1) + f(2) + \ldots\ldots + f(n-1) + f(n))/n \quad (5)$$

In the formula: $P(i)$ is a period of ultrasonic data after the mean value of the ultrasonic data; $f(x)$ is the ultrasonic voltage value collected by the data acquisition unit; n is a total of n ultrasonic voltage values collected in this cycle.

4.2 Data processing between multiple receive periods

Channel of the algae, fish and marine debris, driving the bubbles produced, the output signal will make the ultrasonic receiving sensor abnormal data in a large number of doping these abnormal data points, we call the noise data points. In order to remove the noise data points, an improved median filtering algorithm is proposed based on the idea of median filter in image processing (ZHAO G CH et al., 2011) (HU Y L et al., 2008) (ZHOU D CH, 2010). The size of the filter window is N, and the threshold value is a, then the median filtering results of X points are as follows:

$$g(x) = (f(x - N/2) + f(x - N/2 + 1) + \ldots f(x) \\ + \ldots + f(x + N/2))/(N+1) \quad (6)$$

If a data satisfies the formula $|f(x) - g(x)| > a$, the point is considered as noise data points, the value of the use of $g(x)$ value; If a data satisfies the formula $|f(x) - g(x)| <= a$, the point is considered noise data points, the use of the original value f(X). As shown in Figure 7, for the 4 channel ultrasonic sensor data drawn under multiple cycles.

As shown in Figure 8, is to deal with the noise of the data points after the effect diagram. In comparison with Figure 7, it is found that the improved median filtering algorithm can effectively remove the noise data points, and retain the original data.

4.3 Occlusion depth subdivision method

When the ultrasonic receiving sensor receives the ultrasonic intensity in the diffraction region affected by ultrasonic diffraction, occlusion depth and ultrasonic intensity is not linear change. Based on the analysis of ultrasonic diffraction, combined with thousands of sets of calibration results in the Dalian Maritime University survival Museum, fitting subdivision depth and normalized occlusion ultrasonic receiving sensor receives the ultrasonic intensity between the regression fitting using least square method (FAN W B et al., 2013) (TIAN L et al., 2012). As shown in the following formula, Δh is a subdivision in the diffraction region relative a road in the depth of the ultrasonic sensor occlusion blocking region, receiving sound wave intensity sensor receives the ultrasonic in the diffraction region normalized in ultrasonic receiving array.

$$\Delta h = a_3 x^3 + a_2 x^2 + a_1 x + a_0 \quad (7)$$

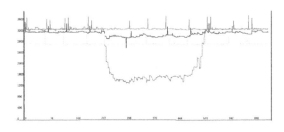

Figure 7. Data graph drawn from multiple cycles.

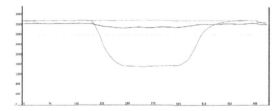

Figure 8. Based on the improved median filtering algorithm to deal with multiple cycles of data.

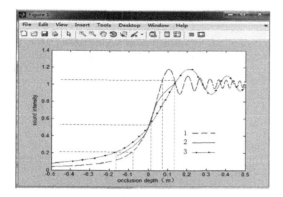

Figure 9. Fitting of diffraction region sound intensity and subdivision occlusion depth curve.

As shown in Figure 9, the relationship between the relative sound intensity and the depth of occlusion is simulated.

In the figure 9, the relationship between the relative sound intensity and the shielding depth of the block 3 M is simulated in the curve 1; the relationship between the relative sound intensity and the shielding depth of the block 22 M is simulated in the curve 3; the curve 2 shows a large number of experimental data fitting subdivision curve. It can be seen from the simulation, the distance between the receiver and the receiver is not the same, and the relationship between the sound intensity and the shielding depth is different after the diffraction effect. But the fitting of the experimental data is located between the sound intensity and the depth curve of the maximum distance and the minimum distance. This shows that the subdivision curve can represent the relationship between the relative sound intensity and the depth of the occlusion in the case of ignoring the error caused by the occlusion distance.

5 MEASUREMENT EXPERIMENT

By using this system, the measurement of the simulation of the ship draft and the measurement of the ship draft in the field environment are completed. As shown in Figure 10, the length of 25 meters, a width of 25 meters, 4 meters deep in the swimming pool. 650 sets of experiments were carried out to simulate the ship draft of 0.5 meters to 1 meters, which are different from the receiving end. In each experiment, the same amount of draft to simulate the ship repeated measurement three times.

As shown in Table 1, the measurement error of the different distances between the ship and the ultrasonic receiving array is simulated. A large number of experimental data show that the measurement error of the system under laboratory conditions is better than 0.02 M.

The field experiment is working with the Yangtze River Three Gorges Navigation Administration Department of maritime affairs in Gezhouba Dam

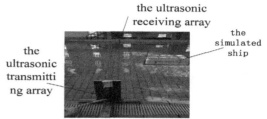

Figure 10. Testing system in laboratory environment.

Table 1. Experimental data error of different occlusion distance in laboratory.

The distance between the simulated ship and the ultrasonic receiving array/m	Root mean square error of the measured value and the measured value of the ship's draft/cm
3	1.298305
8	1.21468
13	1.337038
18	1.656673
22	1.264351

Figure 11. The site measured ship draft.

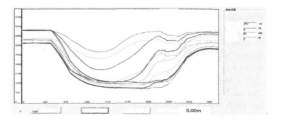

Figure 12. Ultrasonic sensor data collected after treatment.

No. 2 Ship Gate test 29 ships, and the measuring errors of the system. Due to the limitation of the field condition, the same ship can not be measured for many times, so the measured data are the result of single measurement.

Figure 11 is the draft image collected in the field of ship "hung chun No. 2", by the method of man-ual reading ship draft is 4.05 meters. As shown in Figure 12, is installed in the field of the data processing unit for ultrasonic receiving sensor array to receive the 12 channel ultrasonic signal processing results. Data can be seen from the figure, the use of the data processing methods in this paper no significant noise, more easy to ship draft calculation. Figure 13 is based on the ultrasonic wave receiving sensor array data and water level information to simulate the ship's underwater cross section map, to calculate the ship's draft value is 4.08 m. The results of the ship draft at the site are shown in Table 2.

From the field measured data, the contrast difference is –0.09 M. After analysis, the main error sources are the following: uncertainty of the true shape of the ship, read the draft measurement error and artificial water level system error. In general, the amount of water that the system measured can be used as a basis for judging the actual ship draft in the maritime sector.

Figure 13.   Drawing of ship dynamic waterline.

Table 2.   Field measurement ship draft results.

Measured ship name	Time	System measurement draft/m	Read draught value/m	Contrast deviation/m
guanhai009	4/2	3.56	3.45	0.09
hanwan	4/2	4.21	4.3	–0.09
yuanyang	4/2	3.62	3.56	0.06
changjiangjiyun	4/2	3.8	3.9	–0.1
Hunchun No. 2	4/21	4.08	4.05	0.03
huijin	4/21	3.7	3.8	–0.1
nanhaihuihuang	4/21	3.68	3.75	–0.07
shunzou	4/21	3.79	3.8	–0.01
xiangfu	4/21	3.8	3.85	–0.05
xintao	4/21	3.69	3.6	0.09
minjin	4/23	3.69	3.68	0.01
Bingang No. 888	4/23	3.54	3.6	–0.06
zhongyuan	4/23	3.32	3.3	0.02
yuanlong	4/24	3	3	0
hongtai	4/24	3.36	3.36	0
jiangjiyun	4/25	3.81	3.79	0.02
xinhuaxin	4/25	3.42	3.46	–0.04
Hengda No. 1198	4/25	3.61	3.7	–0.09
ezijiangyun	4/25	3.87	3.95	–0.08
jiangshunda	4/26	4.04	4.12	–0.08
changtai	4/26	4.19	4.27	0.08
shuguang	4/26	4.27	4.22	0.05
taigang No. 899	4/26	4.39	4.42	–0.03
hainiu No. 25	4/27	4.28	4.36	–0.08
huihuang No. 168	4/27	3.52	3.44	0.08
huahai No. 6	4/27	3.19	3.1	0.09
hainiu No. 669	4/28	3.73	3.82	0.09
yichangliming	4/28	3.48	3.4	0.08
haohang No. 1003	4/28	3.74	3.74	0

6   CONCLUSION

The field experiment shows that the method can be widely used in the measurement of the ship draft based on the ultrasonic diffraction effect. The side scan type ship draft detection system has the advantages of convenient installation, high measurement accuracy and high real-time performance.

ACKNOWLEDGMENTS

This work was financially supported by Liaoning province Talents Project fund.

REFERENCES

Chi DZH. Research on characterization of weld defect based on ultrasonic TOFD [D]. Harbin: Harbin Institute of Technology.

Dong HT, Ma YY. Design of the automatic detection system based on ultrasonic sensor array for quantity of crowd [J]. Process Automation Instrumentation, 2016, 37(1): 43–46.

Fan WB, et al. Methods for Least Squares Fitting of a Straight Line and Their Application in Geochronology [J]. Geological Review, 2013, 59(9): 801–815.

Hu YL, et al. Research on Image Filtering Algorithm and Its FPGA implementation [J]. Computer Measurement & Control, 2008, 16(11): 1672–1675.

Indukumar K, Reddy V. Broad-band DOA estimation and beamforming in multipath environment[C]// Radar Conference, 1990. Record of the IEEE 1990 International. IEEE, 1990: 532–537.

Lu MH, Pan WCH, Liu XF. The quantitative method of hole-type defects with diffracted echo by phased array ultrasonic technology [J]. Journal of Applied Acoustics, 2015, 34(5):385–390.

Sazontov AG, Matveyev AL, Vdovicheva NK. Acoustic coherence in shallow water: Theory and observation. Oceanic Engineering, IEEE Journal of, 27(3): 653–664.

Shi JW, Liu SP. Ultrasonic TOFD Technology for Detection [J]. Testing Technology, 2009, S1: 95–100.

Shi J, Yang DS, Shi SHG. Research on noise sound source localization method in shallow water based on the multi-path model match [J]. Acta Electronic Sinica, 2013, 3(3): 575–581.

Shu XL, et al. Effect of the multipath in the shallow water to the source bearing estimation [J]. Ship Science and Technology, 2009, 31(9): 121–124.

Tian L, Liu ZT. Least-squares method piecewise linear fitting [J]. Computer Science, 2012, 39(6): 482–484.

Wang SHJ, et al. FPGA implementation for solving linear least square problem [J]. Chinese Journal of Scientific Instrument, 2012, 33(3):701–707.

Zh M, Shao FQ, Zhang WB. Ultrasonic imaging sensor design and sound field measurement for settlement process [J]. Control and Instruments in Chemical Industry, 2013, 40(7): 845–859.

Zhang HL. Theoretical Acoustics. Higher Education Press, 2012: 325–327.

Zhao GCH, Zhang L, Wu FB. Application of improved median filtering algorithm to image de-noising [J]. Journal of Applied Optics, 2011, 32(4): 678–682.

Zhao W, Li YQ. Inland river ships draught problem and testing technology research[J]. China Water Transport: 2015, 15(12): 20–21.

Zhou DCH. Research on ship draft information collection and process system [D]. Dalian: Dalian Maritime University, 2010.

Automotive, Mechanical and Electrical Engineering – Liu (Ed.)
© 2017 Taylor & Francis Group, London, ISBN 978-1-138-62951-6

Research on key technologies of the unmanned-helicopter-born obstacle avoidance radar for power line inspection

Chaoying Li
State Grid Shandong Electric Power Research Institute, Jinan, China

Bo Yang
State Grid Shandong Electric Power Company, Jinan, China

Chaoying Li & Zongyu Li
Shandong Luneng Intelligence Technology Co. Ltd., Jinan, China

Ting Ge
Nanjing University of Science and Technology, Nanjing, China

ABSTRACT: In a power line inspection system, obstacle avoidance is an urgent problem which needs to be solved when an unmanned helicopter flying at low altitude which will threat the flying safety with complex background. Aiming at this problem, this paper presents a high-voltage power line detection scheme, using MMV broadband LFM technology. In the obstacle avoidance system, we focus on the signal accumulation technology of a small target, CFAR technique and polarisation detection. Based on the theoretical analysis A system simulation was done and the radar based on obstacle detection system was developed. The flight test results show that this method is effective.

Keywords: MMV radar; Power line inspection system; Small target detection

1 INTRODUCTION

Unmanned helicopter power line inspection systems have been used for 3D flying SPC technology, but it can't avoid the outside emergency obstacle. If the flight level of the unmanned helicopter is not high, it is also likely to collide with the cross of the power line in front. So it is necessary to develop an obstacle avoidance system for the unmanned helicopter patrol. Helicopter collision avoidance radar technology has certain technical accumulation at present. Since the 1960s, a variety of airborne systems have been developed for obstacle detection, obstacle avoidance and obstacle warning in other countries (Wang Jiaxiu, 2011). The millimetre wave obstacle avoidance radar has many advantages, such as large bandwidth, small volume, and light weight. It is more suitable for airborne and the millimetre wave wavelength is between 1 and 10 millimetre (Zheng Xinglin, 2007). There are several typical of millimetre wave collision avoidance radar for foreign as follows: MARCEL DASSALILT

produced the SAIGA. In the 80 s, Germany's AEG —TELEFUNKEN has developed U-band collision avoidance radar. The Canadian Eminem Teck Company developed the Ka-band Oasys collision avoidance radar in 2000 (Cao Peiyong, 2009). An obstacle avoidance millimetre wave radar was applied to the unmanned helicopter patrol system and it will improve the efficiency of the unmanned helicopter

Patrol, security and open more broad application prospects.

2 ANALYSIS OF MILLIMETRE WAVE RADAR OBSTACLE AVOIDANCE SYSTEM

2.1 *Structure of millimetre wave radar obstacle avoidance system*

The bandwidth of the signal which the linear frequency modulation continuous wave radar system

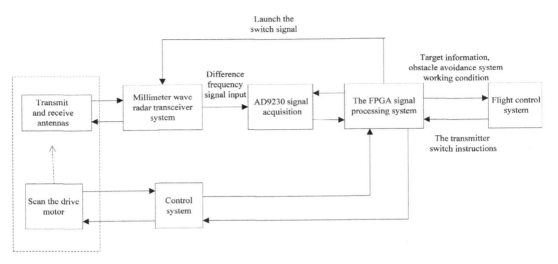

Figure 1.   Block diagram of the millimetre wave obstacle avoidance radar system.

uses, is 500 MHz. The received signal is converted to an intermediate frequency signal through the linear frequency modulation, that will be sent into the signal processing system based on FPGA to extract the distance information of obstacles and then the obstacles information is sent into the flight control system in order to achieve obstacle avoidance actions. The flight control system sends the radar control instructions at the same time. When the signal processing system receives the command, it will control the obstacle avoidance radar system state. The specific system structure is shown in Figure 1.

2.2   Technology of small target signal processing

In the work of the unmanned helicopter for power line inspection, the background is complex and changeable, the extent to which the radar beam strips may from open terrain to alpine vegetation zone, forest, and from water to land. Serious changes have taken place in the characters of the background clutter. Millimetre wave obstacle avoidance radar mainly detects on the power line, electrical wiring in front of the tower, the obstacles such as trees, mountains that could be encountered on the flight path and the front. The RCS of the power line is small, and due to the particularities of the power structure itself, when the beam incident Angle away from the Angle of Prague, it is smaller. So the power line is difficult to detect. In order to better detect the transmission line, polarisation detection technology can be used. But the technology system is very complex and costly. This paper mainly studies the signal accumula-

tion and constant false alarm algorithm to improve the detection abilities of targets (especially small targets).

a. Target echo signal which is gained by accumulation

In radar signal detection, signal detection performance is closely related to the signal-to-noise ratio, the higher the signal-to-noise ratio, the better the signal detection performance (Ulaby, 1990). Signals can effectively improve the target echo signal-to-noise ratio, and improve the ability of detecting targets (small targets), which generally can be divided into the coherent accumulation and the incoherent accumulation.

The coherent accumulation is completed before the envelope detecting, it can improve signal-to-noise ratio by M times than before by accumulating the amplitude and phase of M echo signal that is sampled from the same distance from the door. So coherent accumulation is adopted in this paper.

b. Constant false alarm detection

In the process of radar signal detection, an echo signal is often submerged in the receiver noise or clutter, such as: building, grass, trees, mountains and other background which scatters the echo. So it is necessary to detect the target signal and make an accurate judgement under the false alarm probability of expected. In the automatic test system of the radar, in order to enable the clutter and interference to have a minimal effect on the false alarm probability of the system, constant false alarm detection technology should provide a detection threshold (Wu Yi, 2009) to detection strategy.

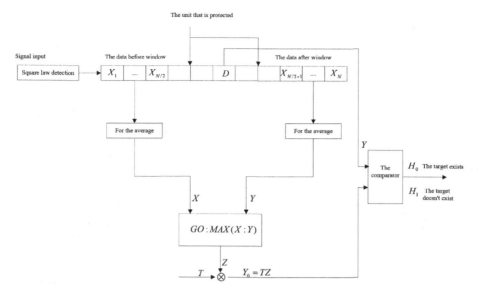

Figure 2. Block diagram of GO—CFAR detection.

A GO—CFAR detector is not sensitive to the background clutter wave, that is to say GO—CFAR detector of different mixed ratio of false alarm control basic consistent. In the airborne obstacle avoidance radar system, detection of target background, including earth, vegetation, and mountains, this kind of clutter wave will produce dramatic changes in space and time. The common CA—CFAR detection has limitations in the detection of clutter wave edge (Luo Chengquan, 2007; Rohling, 1983; Zhang Wei, 2009; Marcum, 1960; Mark A. Richards, 2010; P. P. GANDHI, 1988). The ideal block diagram of a GO—CFAR detection is shown in Figure 2.

3 OUTFIELD EXPERIMENT

In adopting a system prototype, we developed some experimental base in the test flight of unmanned helicopter patrol system. Millimetre wave radar system obstacle avoidance is a good way to detect millimetre power cables, ranging accuracy is 0.4 metres, distance to distance imaging and transmission line can be in horizontal direction distinguish two parallel 1 metre apart wires. By the obstacle avoidance system of unmanned helicopter, the radar target information through the UART module to flight control system, the flight control system of unmanned helicopter and route planning system is for implementation of control power.

Figure 3. The complex environment of detection.

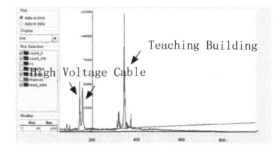

Figure 4. The transmission lines result that detected by pitch angle of the obstacle avoidance system.

Figure 3 shows the complex environment of detection. Figure 4 and Figure 5 show the detection results. Figure 6 shows the outfield test environment.

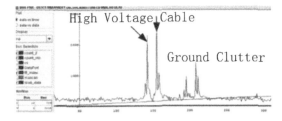

Figure 5.   The transmission line result that detected by pitch angle of the obstacle avoidance system under the radar detection.

Figure 6.   A field of test environment in Fuzhou.

4   CONCLUSION

Through the accumulation of small target signals of high tension line technology, analysis and simulation of constant false alarm detection technology, we could solve the obstacle avoidance problem when the power line detection system of an unmanned helicopter in a low-level flight by adopting the millimetre wave broadband linear frequency modulation technology. As the results of the outfield experiment show, this method is effective and reliable, and has the best reference value for subsequent research and equipment.

REFERENCES

Cao Peiyong (2009). Airborne millimeter wave collision avoidance radar antenna array [D]. Harbin Institute of Technology.

Gandhi, P.P., S.A. Kassam. (1988). Analysis of CFAR processors in nonhomogeneous background. *IEEE Transaction on Aerospace and Electronic Systems, 24(4)*, 427–445.

Luo Chengquan. (1960). Several kinds of constant false alarm detector applied research [D]. Nanjing Southeast University.

Marcum J. (1960). A Statistical theory of target detection by pulsed radar and mathematical. *IRE Transactions on Information Theory, 6(2)*, 59–267.

Mark A. Richards. (2010). Radar signal processing. Beijing: Electronic Industry Press.

Rohling H. (1983). Radar CFAR thresholding in clutter and multiple target situations. *IEEE Transaction on Aerospace and Electronic Systems, 1983, 19(4)*, 608–621.

Ulaby, F.T., and C. Elachi. (1990). Radar polarimetry for geoscience applications. Dedham, MA, USA: Artech House.

Wang Jiaxiu. (2011). The technology research of helicopter millimeter wave collision avoidance system. Nanjing: Nanjing University of Science and Technology.

Wu Yi. (2009). Radar signal processing research of constant false alarm detection. Nanjing: Nanjing University of Science and Technology.

Zhang, Wei. (2009). The technology research of constant false alarm detection under complex clutter background. Nanjing University of Aeronautics and Astronautics.

Zheng, Xinglin (2007). Key techniques of millimeter-wave automotive anti-collision radar signal processing [D]. Wuhan: national University of Defense Technology.

Automotive, Mechanical and Electrical Engineering – Liu (Ed.)
© *2017 Taylor & Francis Group, London, ISBN 978-1-138-62951-6*

Research on the key technology of automatic target detection for visual vehicles

Qiong Ren, Hui Cheng & Junming Chang
School of Mathematics and Computer Science, Jianghan University, Wuhan, China

ABSTRACT: In the face of the continuously deteriorating traffic environment, this research studies the key technology of automatic navigation target detection for visual vehicles. The image collected by binocular vision system is carried out with real-time processing, and MATLAB is used to simulate analysis of the related algorithms. It verifies that the method of combining threshold segmentation and morphological processing can effectively segment part of a road; quantitative symmetry obstacle calculation method can effectively detect the obstacles and achieve the automatic driving of intelligent vehicles for the purpose of effective obstacle avoidance.

Keywords: Visual vehicle; Automatic navigation; Target detection

1 INTRODUCTION

With the rapid development of China's economy, the process of urbanisation is accelerating. In today's society with rapid development, vehicles have become a very important means of transportation in people's lives and production. However, the ensuing problems are increasingly apparent. As a result, more new technologies will be added to the navigation system of intelligent vehicles (Gao, J. et al., 2013), so as to realise unmanned driving as soon as possible, which has become a hot issue in modern society.

The main road environment studied in this paper is the structural road. Binocular stereo vision system is used in this paper. First of all, the road area and non-road area are segmented, and then the road area is extracted. Then in the extracted road part, obstacle detection is carried out. This paper uses the method based on the obstacle symmetry to detect the obstacles on the road (Cabo, C. et al., 2014). Finally, binocular stereo vision distance measurement method is used to measure the distance between the obstacle and the car body. In the description of the algorithm, the actual image is used to illustrate, and MATLAB is used to simulate the experiment.

2 VISION VEHICLE AUTOMATIC NAVIGATION SYSTEM

2.1 *Machine vision theory*

Current methods based on vision are studied under the guidance of the framework of Marr's vision theory, as shown in Figure 1 (Wang, T., et al., 2013):

Marr completely introduces the basic methods of recovering the existing two-dimensional image information to the depth information of 3D object, to lay a solid foundation for the establishment of the discipline of machine vision theory, which is of epoch-making significance (Hancock, P. A. et al., 2013). Although this theory has a lot of controversies in many aspects, and needs to continue to be improved and perfected, our research in the field of computer vision theory is still completed under the guidance of this theory. As a result, the machine vision theory has a very important position.

2.2 *Overall design of automatic target detection system for visual vehicles*

In this paper, we focus on the detection of obstacles in structured road and binocular vision distance measurement using an algorithm for obtaining obstacle and body distance (Duan, H. et al., 2013).

The overall design of the automatic target detection system for visual vehicle is divided into two parts. First of all, the road part of the vehicle is detected. After finding the road part of the vehicle, it can narrow the search range of obstacles. We only need to detect obstacles on the road, and the part of the road outside is not required to be considered. The extraction of pavement and the detection of obstacles on the road are introduced respectively. In Figure 2, the overall design block diagram of the automatic target detection system for visual vehicle is shown (Chen, Y. L. et al., 2013):

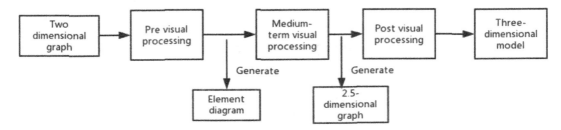

Figure 1. Marr's vision theory framework.

Figure 2. Overall design block diagram.

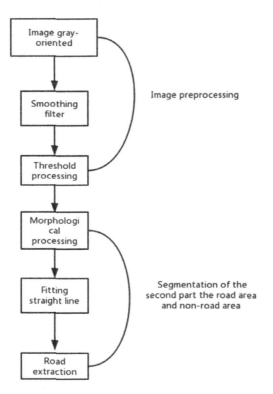

Figure 3. Structure block diagram of road extraction.

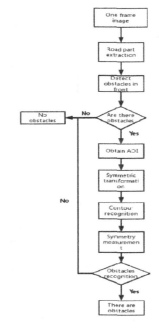

Figure 4. Obstacles detection block diagram.

The whole intelligent vehicle obstacle detection system is composed of a hardware part and software part. The hardware composition studied mainly includes a notebook computer, a camera two, a three-foot platform and the image acquisition card. These devices constitute the detection system of binocular stereo vision.

The first part of the software for this study is the study of the road extraction algorithm, as shown in Figure 3 (Zarco-Tejada, P. J. et al., 2014). It is mainly divided into two parts, which are the preprocessing of the collected images and the segmentation and extraction of road and non-road area. In the first part, the first thing to do is the gray-scale of the image. That is to say, the original image collected is converted to gray-scale image. Then the gray image is filtered, thus reducing the error of subsequent image processing. The last and the most essential step is to carry out threshold segmentation of the filtered image. Since the vehicle driving road part in the structured road image occupies a large proportion, and has a great difference in the gray with most of the sidewalk and the scenery on both sides, so it can use the threshold method to process the image and segment the borders on both sides of

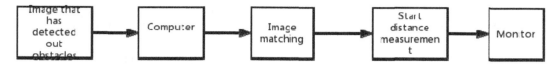

Figure 5.   Structure block diagram of the ranging system.

the running road. After the segmentation, the image is extracted. This is the segmentation of the second part the road area and non-road area shown in Figure 3 and the completely extracted road sections.

According to the detection of pavement part obstacles, this paper puts forward effective a detection method, and describes it in detail in the following chapters. This method is a method of obstacle detection based on symmetry, because obstacles in front in the structural road are mainly the vehicle, and this special vehicle has symmetric feature. The obstacle detection algorithm, as shown in Figure 4, is in allusion to the main obstacle—vehicles on the structured road to detect (Sharma B. et al., 2014). This is of a great help in the improvement of detection algorithm speed. It is only needed to analyse and judge based on the symmetry of vehicle.

Figure 5 (Chevalier, M. et al., 2016) shows the structure of the distance measurement system. In this paper, after detecting obstacles, it is necessary to detect some information on obstacles, and this paper mainly measures the distance between obstacles and vehicle body. Binocular stereo vision ranging method is used, which first of all needs to calibrate the binocular stereo vision system. In this paper, a variety of calibration methods are compared and analysed, and then the direct linear calibration method selected for binocular stereo vision system calibration. Finally, through the analysis of the principle of binocular stereo vision distance measurement and image matching, obtain the depth information of the obstacles.

3   RESEARCH ON OBSTACLE DETECTION ALGORITHM BASED ON SYMMETRY FEATURE

In this paper, we study the detection of obstacles in the structured road, then the main obstacles in the structured road is the vehicle running ahead.

3.1   *Establishment of the region target detection obstacle of interest*

In accordance with the image after threshold segmentation, determine the interested region. The black area is pavement section, the white line on the black area is the location of the obstacle vehicle (Zhang, D. et al., 2015). Then it is only needed to

set up AOI based on the white line, and further position obstacles in AOI.

3.2   *Identification and recognition of obstacles in target detection*

In the process of building, in order to include all possible obstacles, it is supposed to be greater than the actual obstacle area. To further determine the location of obstacles, a lot more work is required. The next thing to do is to identify obstacles. Then, this paper proposes a detection algorithm based on vehicle symmetry.

1. Symmetry analysis

If a plane image is reflected by a linear symmetric axis along its axis, and the image remains the same, the plane image is called the mirror symmetry.

Based on the following vehicle back model, give a simple introduction to the symmetry. First of all, the detection system is still running under the binocular parallel mode that we set up. As shown in Figure 6, the whole vehicle back is a symmetrical plane model. According to the symmetry property, we can start the follow-up work.

Next, the area that needs to be measured in symmetry is what has been found. This improves the effectiveness and real-time performance of the algorithm.

2. Symmetry measurement and symmetry axis

In this paper, a quantitative method is established to measure the symmetry by using the concept of continuous symmetry. The quantitative symmetry

Figure 6.   Diagram of region of interest.

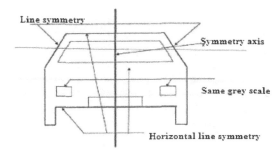

Figure 7. Symmetry characteristic of the vehicle.

description method is as follows. In the image, a line of gray data can be regarded as one-dimensional function g(x) of the horizontal pixel coordinates. Any function can be written as an even function $g_e(x)$ and an odd function $g_0(x)$ (Srinivas U, 2014).

$$g(u) = g_e(u) + g_0(u)$$
$$g_e(u) = \frac{g(u) + g(-u)}{2}$$
$$g_0(u) = \frac{g(u) - g(-u)}{2}, u \in \left[\frac{-d}{2}, \frac{d}{2}\right]$$

The importance of even function and odd function $g_e(x)$ and $g_0(x)$ can explain the symmetry of the image, which is the quantitative description of symmetry (Zhang R. et al., 2013). The more symmetric the image, the greater the proportion of even function.

To establish the new symmetric coordinate system u ($-w/2 \le u \le w/2$) that takes symmetry axis x_g as the origin, there are the following coordinates between u and x:

$$u = x - x_s$$

For certain x_s and w, the even function and the odd function of the function g(x) = g (xs + u) are respectively (Wang T., 2012):

$$e(u, u_s, w) = \frac{g(x_s + u) + g(x_s - u)}{2}$$
$$0(u, u_s, w) = \frac{g(x_s + u) + g(x_s - u)}{2}$$

By using the concept of energy function, the importance of the odd function and the even function is compared:

$$E(f(x)) = \int f^2(x) dx$$

The energy function of the odd function and the even function is shown below (Lee G. et al., 2015):

$$E_e(x_s, w) = \sum_{u=-w/2}^{w/2} e^2(u, x_s, w)$$

By using the energy function, the measured values of symmetry are as follows:

$$S(x_s, w) = \frac{E_n(x_s, w) - E_0(x_s, w)}{E_n(x_s, w) + E_0(x_s, w)}$$

Symmetric measurement S is an arbitrary number in the range of (−1, 1). S = 1 suggests the complete symmetry, S = 0 represents asymmetry (Ontañón S., 2015), S = −1 refers to anti-symmetric, and the measure x_s corresponds to the parameter w.

According to the above method, to calculate the symmetry of AOI there is a statistical Table 1 of the AOI symmetry measurement in Figure 6 (Shen W. et al., 2016). Taking into account the algorithm speed, S is the symmetry measurement that each of the four lines carries out.

The average values of symmetry measurement value calculated in Table 1 are compared with the threshold set. If S is smaller than the threshold set in advance, then it is believed that the vehicle barrier does not exist in the AOI, and so this area can be removed (Hsieh J. W., 2014); if S is greater than a given threshold, then determine that the vehicle obstacles exist in the AOI. After the comparison, the determination of specific location of the vehicle can be carried out.

3.3 Experimental results and analysis

The results of this study are analysed as follows:
Choose conditions in fine weather and gloomy weather for analysis. Fine weather means the situation that the sky is bright and full of light; in gloomy conditions, there is almost no sunlight (Akita T., 2014). The two cases are carried out with video capture analysis.

1. Condition that the weather is fine and the vehicle distance is moderate.
2. Condition that the weather is gloomy and the vehicle distance is moderate.

From the above experimental results, it can be found that the difference in weather conditions may have been resulted in that threshold segmentation being inaccurate, thus making the road extraction in the first stage in error; a direct consequence of which is that obstacle detection is not accurate. Just the difference of a weather condition factor reduces correct-detection rate by 4%, which shows that the effect of light on the system is very

Table 1. AOI symmetry measurement.

h	1	2	3	4	5	6	7	8	Average value
s	0.994	−0.114	0.913	0.938	0.987	0.931	0.938	0.931	0.7901

Table 2. Fine weather data statistics table.

Video	Number of frame	Times of missing detection	Times of wrong detection	Correct-detection rate	Wrong-detection rate
1	320	11	18	91%	9%
2	428	10	25	92%	8%
Average	374	10	21	92%	8%

Table 3. Gloomy weather data statistics table.

Video	Number of frame	Times of missing detection	Times of wrong detection	Correct-detection rate	Wrong-detection rate
1	260	15	15	88%	12%
2	340	18	17	89%	11%
Average	300	17	16	88%	11%

large. Under low light conditions, the gray-scale of obstacles and the road part may be relatively close, which leads to the occurrence of missed and false detection. The results show that the stability and reliability of the system are relatively high when the weather is sunny (Shiru Q., 2016); the system also has certain adaptability when the weather is gloomy, but it needs further improvement.

In this section, the key technologies of automatic target detection for a visual vehicle are studied and analysed, and the method of object detection based on symmetry is researched.

4 CONCLUSION

Research on the key technologies of vision vehicle automatic navigation target detection is a hot issue in the field of intelligent vehicle research in the world, and it is also a basic problem. In this paper, the hardware structure of the whole system is designed reasonably. Comprehensively considering the requirement of each performance index, the binocular parallel mode is designed for the whole vision system. The results show that the system in this study can extract the road part to obtain the position lines of targeted obstacles; then according to the line length and the position information generation, at last, through the quantitative analysis of each symmetry (Yao Y., 2013), remove vehicles with no obstacle, and further determine the

specific location of the obstacle, so as to achieve the purpose of automatic vehicle navigation.

ACKNOWLEDGMENTS

Hubei Provincial Department of Education Guidance Project of Scientific Research Program (No. B2016 281).

REFERENCES

Akita, T. & Yamada, Y. (2014). Image recognition of vehicle applying fusion of structured heuristic knowledge and machine learning. *International Journal of Automotive Engineering*, 5(3), 101–108.

Cabo, C., Ordoñez, C., García-Cortés, S. et al. (2014). An algorithm for automatic detection of pole-like street furniture objects from mobile laser scanner point clouds. *ISPRS Journal of Photogrammetry and Remote Sensing*, 87, 47–56.

Chen, Y.L., Chen, T.S., Huang, T.W. et al. (2013). Intelligent urban video surveillance system for automatic vehicle detection and tracking in clouds. *Advanced Information Networking and Applications (AINA), IEEE 27th International Conference on Advanced Information Networking and Applications (AINA)*. 814–821.

Chevalier, M., Thome, N., Cord, M. et al. Low resolution convolutional neural network for automatic target recognition. (2016). *7th International Symposium on Optronics in Defence and Security*.

Duan, H., Deng, Y., Wang, X. et al. (2013). Small and dim target detection via lateral inhibition filtering and artificial bee colony based selective visual attention. *PLOS ONE, 8*(8), e72035.

Gao, J., Blasch, E., Pham, K. et al. (2013). Automatic vehicle license plate recognition with color component texture detection and template matching. *SPIE Defense, Security, and Sensing. International Society for Optics and Photonics, 8739,* 0Z–87390Z–6.

Hancock, P.A., Mercad, J.E., Merlo, J. et al. (2013) Improving target detection in visual search through the augmenting multi-sensory cues. *Ergonomics, 56*(5), 729–738.

Hsieh, J.W., Chen, L.C. & Chen, D.Y. (2014). Symmetrical surf and its applications to vehicle detection and vehicle make and model recognition. *IEEE Transactions on Intelligent Transportation Systems, 15*(1): 6–20.

Lee, G., Yun, U., Ryang, H. et al. (2015). Multiple minimum support-based rare graph pattern mining considering symmetry feature-based growth technique and the differing importance of graph elements. *Symmetry, 7*(3): 1151–1163.

Ontañón, S. & Meseguer, P. (2015). Speeding up operations on feature terms using constraint programming and variable symmetry. *Artificial Intelligence, 220,* 104–120.

Sharma, B., Katiyar, V.K., Gupta, A.K. et al. (2014). The automated vehicle detection of highway traffic images by differential morphological profile. *Journal of Transportation Technologies, 4*(2).

Shen, W., Bai, X., Hu, Z. et al. (2016). Multiple instance subspace learning via partial random projection tree for local reflection symmetry in natural images. *Pattern Recognition, 52,* 306–316.

Shiru, Q. & Xu, L. (2016). Research on multi-feature front vehicle detection algorithm based on video image. *Control and Decision Conference (CCDC),* Chinese. (pp. 3831–3835).

Srinivas, U., Monga, V., Raj, R.G. (2014). SAR automatic target recognition using discriminative graphical models. *IEEE Transactions on Aerospace and Electronic Systems, 50*(1): 591–606.

Wang, T. & Zhu, Z. (2012). Real time moving vehicle detection and reconstruction for improving classification. *Applications of Computer Vision (WACV), IEEE Workshop on. IEEE* (pp. 497–502).

Wang, T., Zhu, Z. & Taylor, C.N. (2013). A multimodal temporal panorama approach for moving vehicle detection, reconstruction and classification. *Computer Vision and Image Understanding, 117*(12), 1724–1735.

Yao, Y, Xiong G, Wang, K, et al. (2013). Vehicle detection method based on active basis model and symmetry in ITS. *16th International IEEE Conference on Intelligent Transportation Systems (ITSC 2013).* (pp. 614–618).

Zarco-Tejada, P.J., Diaz-Varela, R., Angileri, V. et al. (2014). Tree height quantification using very high resolution imagery acquired from an unmanned aerial vehicle (UAV) and automatic 3D photo-reconstruction methods. *European Journal of Agronomy, 55,* 89–99.

Zhang, D., Han, J., Cheng, G. et al. Weakly supervised learning for target detection in remote sensing images. *IEEE Geoscience and Remote Sensing Letters, 12*(4), 701–705.

Zhang, R., Ge, P., Zhou, X. et al. (2013). A method for vehicle-flow detection and tracking in real-time based on Gaussian mixture distribution. *Advances in Mechanical Engineering, 5,* 861321.

Automotive, Mechanical and Electrical Engineering – Liu (Ed.)
© 2017 Taylor & Francis Group, London, ISBN 978-1-138-62951-6

Research on machine learning identification based on adaptive algorithm

Qiong Ren, Hui Cheng & Junming Chang
School of Mathematics and Computer Science, Jianghan University, Wuhan, China

ABSTRACT: This paper studies the machine learning sign recognition based on the adaptive algorithm. In allusion to the problem that the contrast degree of lane image obtained at night is low, a night-time lane mark identification algorithm based on adaptive threshold segmentation is put forward. First of all, the preprocessed image is divided into blocks to enhance the edge information. In combination with the Otsu threshold method and the neighbourhood value method, the adaptive threshold is obtained by weighted distribution, and the lane image is segmented. Finally, partition domain search and Hough transformation are used to extract lane marking line accurately. The field test shows that, as for lane line of structured road, the lane line extraction method is highly accurate and provides good in real-time performance.

Keywords: Adaptive; Machine learning; Lane mark identification

1 INTRODUCTION

With the rapid development of highway, the traffic accident rate is rising, which brings great loss to people's life and property. As a result, to design a system that can give the driver timely warning when the vehicle is not aware of its lane departure can effectively reduce the occurrence of lane departure accidents. In this system, the accurate and fast identification of lane marking is a key step (Juneja, P. et al., 2012; Juneja, P. et al., 2013). Up to now, there are a lot of lane detection algorithms based on vision. They usually use the road model with straight line or curve, as well as Hough transformation, curve fitting, neural network and other techniques to achieve results. Although the success rate of detection and recognition of algorithm is high, these algorithms lack the ability to adapt to the changes of light, and the effect is not good when the light is uneven and under the night environment. In the night lane image, the road gray pixel value is not uniform, and the contrast degree is relatively low (Hamdi, M. A., 2014). The opposite vehicle lights or roadside signs reflection increase the light spot on the image, which makes the lane marking line difficult identify. In this paper, through the image edge enhancement technology and local adaptive image segmentation technology, the lane marking line of lane image is segmented. And the lane marking line is identified by the improved Hough transformation.

2 IMAGE PREPROCESSING

The lane image that a vehicle mounted camera collects not only contains information on the remote road and the sky, but also random noise of the environment. The purpose of lane image preprocessing is to obtain the effective area of lane (lane area) (Jumb, V. et al., 2014; Iglesias, J. E., 2015), to eliminate the interference of random noise on the image, and to enhance the identification feature of the lane mark edge, providing good precondition for the subsequent recognition of lane marking line.

Because during the shoot, the camera is fixed in the latter mirror position in the cab (horizontal centre position of the vehicle), the lane image obtained is divided into two parts. The upper part belongs to the non-road area, containing a lot of useless information; the other section is the lane area needing to be identified, such as shown in Figure 1 (Das, A. & Sabut, S. K., 2016). In this paper, the processing of lane image is only aimed at the lane area of the image.

2.1 *Gray scale and denoising of lane image at night*

Gray image only uses one colour channel, occupying a small memory space (Wang, F. et al., 2014). Taking into account the real-time requirements of the algorithm, this paper uses gray that is matches

(a)The original image   (b) Effective image segmentation of lane

Figure 1.　Lane image region division.

-2	-1	0
-1	0	1
0	1	2

(a) 45° gradient direction

0	1	2
-1	0	1
-2	-1	0

(b) 135° gradient direction

Figure 2.　Improved sobel operator template.

best with the characteristics of human vision to carry out gray processing of road area, and then uses the improved median filtering method to reduce the disturbance of noise.

2.2 *Edge detection of lane image at night*

Sobel operator is a first-order differential operator, which can effectively eliminate most of the useless information in the road image and keep the edge information of lane marking. In order to highlight the edge of the left and right lane marks, the improved Sobel operator is used to process the lane image, as shown in Figure 2 (Manikandan, S. et al., 2014). For the left and right lanes, the template operator in Figure 2(a) and Figure 2(b) is used to enhance the edge of the lane.

3 LOCAL ADAPTIVE IMAGE SEGMENTATION

The global threshold can well distinguish the background and the target, but the resistance ability of noise in the image or the uneven illumination is poor; while the local threshold is determined by the median value of the gray value that the investigation point corresponds to, whose adapta-

bility is wider than the global threshold, if a neighbourhood is full of background then it is necessary that a part is sentenced as the goal, thus exaggerating the new noise emerging in image details.

In view of the above shortcomings, this paper proposes a method combining an adaptive global threshold and local threshold to deal with image after edge enhancement. The global threshold is determined by the maximum class variance method, and the local threshold value is determined by the neighbourhood median value method. The flow chart of the adaptive algorithm is shown in Figure 3 (Rahman, M. H. & Islam, M. R., 2013).

The neighbourhood median value method sorts several pixel values of the neighbourhood window W of a pixel f (i, j) on image in ascending order, from which the value of TM is taken as the segmentation threshold of the window, as shown in Table 1.

$$T_M = MedianPixel\{W\} \tag{1}$$

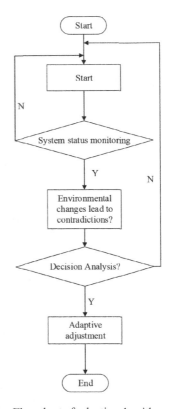

Figure 3.　Flow chart of adaptive algorithm.

Table 1.　Threshold of the window w.

$f(i+1, j-1)$	$f(i+1, j)$	$f(i+1, j+1)$
$f(i, j-1)$	$f(i, j)$	$f(i, j+1)$
$f(i-1, j-1)$	$f(i-1, j)$	$f(i-1, j)$

An image segmentation algorithm based on global threshold and local threshold is described as follows (Ju. Z. et al., 2013):

1. The preprocessed lane image is divided into the left and right lane area, and respectively divided into equal parts. The improved window gray stretching method is used to draw the area.
2. By using the method of maximum class variance, the segmentation threshold value TL and TR of the left and right lane regions are obtained respectively, and the following formula is used to obtain the global threshold T0.

$$T_M = \begin{cases} T_L & j < \text{center} \\ T_R & j \geq \text{center} \end{cases} \quad (2)$$

where j is the number of columns, and center is the boundary threshold of the left and right lanes.

3. A new threshold T is obtained by using the global threshold T0 and the local threshold TM calculated by formula (1) to weight, which is the adaptive segmentation threshold (Bhandari, A. K. et al., 2014).

$$T = T_0 \times (1-w) + T_M \times w \quad (3)$$

where w is the weight. This paper takes the empirical value $w = 0.5$.

This algorithm can effectively avoid the influence of environment light intensity change on image, and improve the ability of anti-noise. In Figure 4(a), we use the maximum inter class variance method, the neighbourhood median method and the algorithm of this paper to carry out segmentation processing. The processing results are shown in Figure 4(b) to Figure 4(d) (Rodtook, A. & Makhanov, S. S., 2013). Figure 4(c) exaggerates the details of the image, add increases the noise interference of lane image, and Figure 4(d) is the image segmentation result, from the details, is better than that of Figure 4(b), clearly retaining the lane mark edge information.

4 LANE MARKING IDENTIFICATION

The structured road meets a series of strict industry standards in design and construction. The lane mark has a smooth curve in the extending direction, and the curvature is small. In general, myopia field lane can be approximated as a linear, thus taking line as the lane model and using Hough transformation to detect lane. Hough transformation is not easily affected by the noise and the curve discontinuity, and it can easily realise the identification of lane marking line.

In the practical application, the parameters formula of the linear polar coordinates is often used to represent the straight line, such as in Formula (4) (Chen, T., et al., 2014).

$$\rho = x\cos\theta + y\sin\theta \quad (4)$$

where p represents the distance from origin to the straight line, θ represents the angle of the straight line with x-axis. Expressed in polar coordinates, in image coordinates space, points with collinear line are intersected at a point in parameter space after transformation, and at that time the polar coordinate parameters ρ and θ for linear are obtained. In fact, the Hough transformation is to complete the linear detection in image space by means of statistics and accumulation in the parameter space.

Considering that the time consumed of Hough transformation in the lane mark identification process is great, this paper proposes a regional search type Hough transformation (Deng, L. & Li, X., 2013), which not only has high robustness of random noise and is not sensitive to local information loss, but also improves the running time of the algorithm. The specific algorithm steps are as follows:

1. Take the lowest edge width centre of the lane image to determine the coordinates of x o y. The image is divided into two parts, the left and right parts.
2. The suitable angle range and the polar diameter of the lane marking line is selected on the lane image. The right lane mark angle range is selected as $\theta \in (25°, 75°)$, the left lane mark angle range is selected as $\theta \in (115°, 165°)$; and the lane mark pole size range is selected as $\rho \in$ (weightmin × Distance, weightmax × Distance) (Schuler, C. J., 2013).

(a)Preprocessed lane image

(b)Otsu method segmentation

(c)Domain median segmentation

(d)The segmentation algorithm

Figure 4. Three segmentation threshold processed images.

3. The angle θ is taken as circular reference, respectively statistics the points on a line of different polar radius p in lane image, and they are stored in the corresponding cumulative space Mem (ρ, θ).
4. Carry out the cumulative space search, and check out the two larger values of the left and right lane storage space, converted into lane marking parameters and recorded.

5 EXPERIMENTAL RESULTS

A night-time lane video image with 720×576 pixel is obtained from the camera, with the speed of 25 frames per second to select 2,000 frames to do the experiment (Anguita, D. et al., 2012). The experimental results as shown in Figure 5. Figure 5(a) is a normal night road mark recognition, Figure 5(b) is the night after tunnel mark identification, Figure 5(c) is the night after the bridge mark identification, Figure 5(d) is the night double lines mark identification, Figure 5(e) is the night road sign interference mark identification, and Figure 5(f) is the night partial bend mark identification (Tian, T. et al., 2014).

From the above detection results of the lane line in all kinds of circumstances, it is known that this algorithm has strong anti-interference ability, capable of overcoming the interference of road signs at night, capable of adapting to the complex road conditions, and has a strong recognition performance.

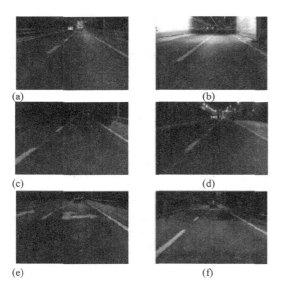

(a)   (b)

(c)   (d)

(e)   (f)

Figure 5. The recognition results of lane marking line.

Table 2. Average time for two Hough transforms.

Methods	Time (ms)
Basic Hough transform	58.48
Hough transform in this paper	29.37

Table 3. Average time for the algorithm with different steps.

Methods adopted	Average time (ms)
Image preprocessing	16.51
Local adaptive image segmentation	11.19
Hough transform in this paper	29.37

In engineering applications, real-time performance is an important indicator to measure the overall performance of an algorithm. In order to verify the real-time performance of the detection algorithm, this paper presents the average time of the basic Hough transform and the partial search Hough transform algorithm in this paper. Table 2 (Senior, A. et al., 2013) shows the running time of steps of the algorithm in this paper.

Experiments took an average of 57.07 ms on recognition algorithm on CPU of Intel (R) Core (TM) i3–21203.3 GHz (Hazar, M. A. et al., 2015), achieving the real-time processing speed of 15 frames per second.

6 CONCLUSION

In this paper, in allusion to the problems that the night average gray degree is low, contrast degree is low, night lamp light interference and other complex environment problems, the method of adaptive segmentation threshold is proposed to segment out lane marking line, and then use regional search Hough transformation to extract the lane mark parameters (Lara, Ó. D. et al., 2012). The algorithm can extract the parameters of lane marking line on the structured road at night, and it can effectively eliminate the interference between vertical line and parallel line in the lane, having better robustness and real-time performance.

ACKNOWLEDGEMENTS

Hubei provincial department of education guidance project of scientific research programme (no. b2016 281).

REFERENCES

Anguita, D., Ghio, A., Oneto, L. et al. (2012). Human activity recognition on smartphones using a multiclass hardware-friendly support vector machine. *International Workshop on Ambient Assisted Living.* Springer Berlin Heidelberg, *7657*, 216–223.

Bhandari, A.K., Singh, V.K., Kumar, A. et al. (2014) Cuckoo search algorithm and wind driven optimization based study of satellite image segmentation for multilevel thresholding using Kapur's entropy. *Expert Systems with Applications, 41*(7), 3538–3560.

Chen, T., Du, Z., and Sun, N. et al. (2014). Diannao: A small-footprint high-throughput accelerator for ubiquitous machine-learning. ACM *Sigplan Notices. ACM, 49*(4), 269–284.

Das, A. & Sabut, S.K. (2016). Kernelized fuzzy C-means clustering with adaptive thresholding for segmenting liver tumors. *Procedia Computer Science, 92,* 389–395.

Deng, L. & Li, X. (2013). Machine learning paradigms for speech recognition: An overview. *IEEE Transactions on Audio, Speech, and Language Processing, 21*(5), 1060–1089.

Hamdi, M.A. (2014). Modified algorithm marker-controlled watershed transform for image segmentation based on curvelet threshold. *Middle-East Journal of Scientific Research, 20*(3), 323–327.

Hazar, M.A., Odabaşioğlu, N., Ensari, T. et al. (2015). Evaluation of Machine Learning Algorithms for Automatic Modulation Recognition. *International Conference on Neural Information Processing,* Springer International Publishing, 208–215.

Iglesias, J.E., Augustinack, J.C., Nguyen, K. et al. (2015). A computational atlas of the hippocampal formation using ex vivo, ultra-high resolution MRI: application to adaptive segmentation of in vivo MRI. *Neuroimage, 115,* 117–137.

Ju, Z., Zhou, J., Wang, X. et al. (2013) Image segmentation based on adaptive threshold edge detection and mean shift. Software Engineering and Service Science (ICSESS), 2013 4th IEEE International Conference on Software Engineering and Service Science. (pp. 385–388).

Jumb, V., Sohani, M., & Shrivas, A. (2014). Color image segmentation using K-means clustering and Otsu's adaptive thresholding. *International Journal of Innovative Technology and Exploring Engineering, 3*(5).

Juneja, P., Harris, E.J., Kirby, A.M. et al. (2012). Adaptive breast radiation therapy using modeling of tissue mechanics: a breast tissue segmentation study. *International Journal of Radiation Oncology*Biology*Physics, 84*(3), e419-e425.

Korzynska, A., Roszkowiak, L., Lopez, C. et al. (2013). Validation of various adaptive threshold methods of segmentation applied to follicular lymphoma digital images stained with 3, 3'-Diaminobenzidine & Haematoxylin. *Diagnostic Pathology, 8*(1), 1.

Lara, Ó.D., Pérez, A.J., Labrador, M.A. et al. (2012). Centinela: A human activity recognition system based on acceleration and vital sign data. *Pervasive and Mobile Computing, 8*(5), 717–729.

Manikandan, S., Ramar, K., Iruthayarajan, M.W. et al. (2014). Multilevel thresholding for segmentation of medical brain images using real coded genetic algorithm. *Measurement, 47,* 558–568.

Rahman, M.H. & Islam, M.R. (2013). Segmentation of color image using adaptive thresholding and masking with watershed algorithm. *Informatics, Electronics & Vision (ICIEV), 2013 International Conference on. IEEE,* (pp. 1–6).

Rodtook, A. & Makhanov, S.S. (2013). Multi-feature gradient vector flow snakes for adaptive segmentation of the ultrasound images of breast cancer. *Journal of Visual Communication and Image Representation, 24*(8), 1414–1430.

Schuler, C.J., Burger, H.C., Harmeling, S. et al. (2013). A machine learning approach for non-blind image deconvolution. *Proceedings of the IEEE Conference on Computer Vision and Pattern Recognition.* (pp. 1067–1074).

Senior, A., Heigold, G. & Yang, K. (2013). An empirical study of learning rates in deep neural networks for speech recognition. *IEEE International Conference on Acoustics, Speech and Signal Processing,* 6724–6728.

Tian, T., Sethi, I. & Patel, N. (2014). Traffic sign recognition using a novel permutation-based local image feature. *International Joint Conference on Neural Networks,.* 947–954.

Wang, F., Li, J, Liu, S. et al. (2014). An improved adaptive genetic algorithm for image segmentation and vision alignment used in microelectronic bonding. *IEEE/ ASME Transactions on Mechatronics, 19*(3), 916–923.

Automotive, Mechanical and Electrical Engineering – Liu (Ed.)
© 2017 Taylor & Francis Group, London, ISBN 978-1-138-62951-6

Research on main insulation monitoring of submarine cable based on the low frequency signal of the system

Zhenxin Chen, Zhifei Lu, Xinlong Zheng & Weilong Peng
Zhoushan Power Supply Company, State Grid Zhejiang Electric Power Corporation, Zhoushan, China

ABSTRACT: In order to reduce the difficulty of measuring insulation by the dielectric loss method, this paper studies the method of measuring the main insulation tanδ under low frequency according to the structure characteristics of submarine cable. Firstly, the causes, selection and principle of the main insulation monitoring of submarine cable is introduced. Then a specific embodiment of main insulation monitoring of submarine cable based on the system low frequency signal is provided.

Keywords: Main insulation; Submarine cable; Low frequency

1 INTRODUCTION

The insulation performance will show a certain degree of deterioration after a certain number of years, which is called insulation ageing. The reasons for the ageing of insulating materials are diverse and complex. The most representative reasons include thermal ageing, mechanical ageing, electrical ageing, etc. The main results of the ageing performance of insulation material includes the decrease of insulation resistance and the increase of dielectric loss.

1. Thermal ageing. Thermal ageing refers to the chemical structure of the insulating medium under the action of heat changes, making the insulation performance of the insulating medium decline. The essence of thermal ageing is the chemical change of the insulating material under the influence of heat, so thermal ageing is also called chemical ageing. In general, the rate of the chemical reaction is accelerated with the increase of ambient temperature. The polymer organic material used for insulation will occur thermal degradation under the long-term effect of thermal, mainly oxidation reaction. This reaction is also known as the autoxidative free radical chain reaction. Such as the oxidation reaction of polyethylene is the H bond from the C-H (Chandrabhan et al. 2007).
2. Mechanical ageing. Mechanical ageing is the process of ageing of the solid insulation system by a variety of mechanical stresses in the production, installation and operation. This ageing is mainly due to the micro defects of insulating material caused by mechanical stress. These small defects slowly deteriorate with the passage of time and the continuing effect of mechanical stress. The small cracks will gradually expand until they cause partial discharge and other phenomenon of insulation damage. This phenomenon is also known as "electrical-mechanical breakdown".
3. Electrical ageing. Electrical ageing refers to the ageing of the electrical equipment insulation system when under the electric field for a long time. The mechanism of electrical ageing is complex, and involves a series of physical and chemical effects generated by insulation breakdown.

Insulation ageing is a very complicated process caused by the electric field, heat, mechanical force, the environment (moisture, sunlight, etc.) and many other combined factors. The above factors should be integrated when calculating the service life of insulation materials (Bauch et al., 2007).

2 SELECTION OF MAIN INSULATION MONITORING

From the monitoring point of view, there are dielectric loss, partial discharge and chemical analysis methods of main insulation monitoring for high voltage oil-filled cable. The partial discharge method is more applicable to the laboratory, because the partial discharge signal is weak, the interference signal is strong, the difficulty of implementation is relatively large and the effect is not very satisfactory. The chemical analysis method is similar to the chromatographic analysis method of dissolved gas in transformer oil. The dissolved gas diffusion rate of oil-filled cable is relatively slow and the distribution is not uniform. It is therefore difficult to accurately analyse the sampling. In comparison, the electrical parameters required by the dielectric loss method are relatively easy to obtain. Therefore, the dielectric loss method is mainly considered for the main insulation monitoring of the submarine cable (Phung et al., 2007).

It is feasible to monitor the tanδ value of the main insulation of the submarine cable by the dielectric loss method. However, the tanδ can only reflect the integrity defect of the insulation or the serious local defect of a small-volume test. It is difficult to use this method to find the local fault of a submarine cable for long-distance and large-capacity submarine cables. But it can reflect the overall ageing of the cable and help to determine the remaining life of the submarine cable. So we can take appropriate measures in advance. This paper puts forward an idea to reduce the frequency of the voltage signal, so that the capacitive current decreases and the resistive current is basically unchanged, so as to increase the ratio between the resistive current and the capacitive current to facilitate the purpose of detection (Oyegoke et al., 2007).

3 PRINCIPLE OF MAIN INSULATION MONITORING

The insulation medium under the AC electric field has energy loss in its interior due to dielectric conductance, partial discharge, dielectric polarisation and other reasons. Therefore, the current I through the medium is not ahead of the voltage $\pi/2$ in the application of a sinusoidal voltage V, but $\varphi = \pi/2 - \delta$. The equivalent circuit is shown in Figure 1. I is the full current, Ic is the capacitive current component, and I_R is the resistive current component. The full current vector I deviates at an angle δ from the capacitive current vector Ic (Yu et al. 2007).

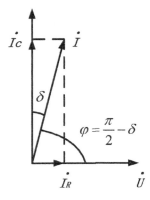

Figure 1.   Electrical parameters vector of the insulation test through the alternating current.

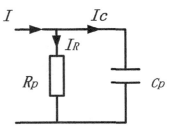

Figure 2.   The parallel equivalent circuit.

$\tan\delta = \frac{I_r}{I_c} = \frac{1}{\omega C_p R_p}$ can be seen from Figure 1 and Figure 2. The operating voltage is the main frequency in the operation of the power system in which the other frequency voltage content is small, so that the equivalent insulation resistance and capacitance value of the insulating material is known as the resistance, and the capacitance value under the power frequency is not affected by the non-frequency signal. Therefore, the tanδ value of the insulating material increases with the decrease of the current and voltage frequency. The measurement principle of ultra-low frequency dielectric loss of submarine cable. The difference is that when the signal frequency is lower, the tanδ value measured is greater. The increase of the tanδ value is helpful for detecting the resistive current and capacitive current flowing through the insulation, and to distinguish the phase difference between current and voltage, so as to reduce the difficulty of field measurement. The phase of the current and voltage is actually measured by the device, which reduces the difficulty of distinguishing the phase angle of the current and voltage. This paper presents a method to detect the main insulation (Ohki et al., 2007).

4   MONITORING OF TAN DELTA BASED ON THE SYSTEM LOW FREQUENCY SIGNAL

With the increasing scale of the interconnected power system, the problem of low frequency oscillation caused by system interconnection has become one of the most important factors that jeopardise the safe operation of the power grid and restrict its transmission capacity (Tatsuo et al., 2007).

It is reasonable to use the natural low frequency signal to avoid inputting the low frequency signal to the power system. After the cable laying is completed, the low frequency signal can be extracted by the low pass filter after the voltage signal and current signal are measured by the sensor. The measured low frequency current and voltage signal are

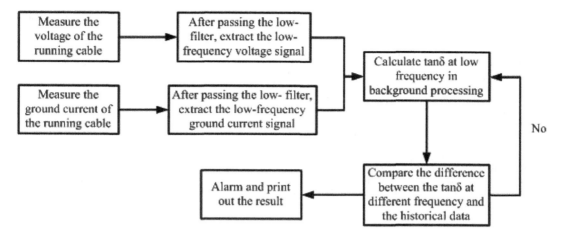

Figure 3.   The monitoring logic diagram of submarine cable main insulation.

converted by A/D and sent to the host computer to calculate tanδ. The monitoring logic diagram of the submarine cable main insulation is shown in Figure 3 (M. et al. 2007).

Because the value of tanδ measured at low frequency is larger than the value measured at the power frequency, the original tanδ measured at the power frequency is no longer applicable. A new ageing failure criterion needs to be established. Therefore, the tanδ measured at low frequency should be compared with the historical data. When the tanδ value changes over a certain range, the alarm signal is issued to remind the operating unit to take oil and other necessary measures for cable. The following units of the main submarine cable insulation monitoring system are included (Lei et al. 2007):

The current sampling module: Used to detect the low frequency current in the ground wire. The module requires high measuring precision and anti-disturbance ability and should be convenient for the spot installation.

The voltage sampling module: The low frequency voltage signal can be obtained from the output by the system voltage transformer. The voltage signal acts as the time synchronisation reference signal measured at both ends of the cable so that the phase of the voltage is calculated.

The synchronisation A/D module: In order to ensure the accuracy of the conversion calculation, the synchronous A/D module must be used to ensure that the phase of low frequency current and voltage is unchanged after the A/D conversion (Fothergill et al., 2007).

The microcomputer unit: After the phase of low frequency current and voltage is unified after A/D conversion, it will be sent to the main computer

to calculate and deal with. Due to the phase angle calculation of the current and voltage, the digital signal after conversion is required to have high precision.

The communication module: Used to transfer the data measured at one end to the other end for unified data processing. It adopts the wireless communication mode or the unused channel of the submarine cable.

5   CONCLUSION

The monitoring of tanδ based on the system low frequency signal has no precedent in China. The biggest feature of this scheme is the use of a large-scale system low frequency signal without the introduction of additional signal sources to avoid changing the system wiring, which ensures the safety and reliability of the primary system. This scheme can achieve the function of long-term online monitoring.

REFERENCES

Bauch, A., Piester, D., Moudrak, A., et al. (2004). Time comparisons between USNO and PTB: A model for the determination of the time offset between GPS time and the future galileo system time. *Frequency Control Symposium and Exposition.*

Fothergill, J.C., Montanari, G.C., Stevens, G.C., Laurent, C., Teyssedre, G., Dissado, L.A., Nilsson, U.H., Platbrood, G. (2003). Electrical, microstructural, physical and chemical characterization of HV XLPE cable peelings for an electrical ageing diagnostic data base. *IEEE Transactions on Dielectrics and Electrical Insulation.*

Given, M.J., Fouracre, R.A., Gregor, S.J., et al. (2001). Diagnostic dielectric spectroscopy methods applied to water-treed cable. *IEEE Transactions on Dielectrics and Electrical Insulation*, 8(6), 917–920.

Lei, M., Zeng, J., & Liu, G. (2009). Dynamic simulation of running test to long distance 500 kV oil-filled submarine cable. *Proceedings of the 9th International Conference on Properties and Applications of Dielectric Materials*.

Ohki, Y., Yamada, T., & Hirai, N. (2011). Diagnosis of cable ageing by broadband impedance spectroscopy, Electrical Insulation and Dielectric Phenomena (CEIDP) 2011 Annual Report Conference on.

Oyegoke, B., Hyvonen, P., Aro, et al. (2003). Application of dielectric response measurement on power cable systems. *Dielectrics and Electrical Insulation, IEEE Transactions on, IEEE Transactions on* 10(5), 862–873.

Phung, B.T., et al. (1999). On-line partial discharge measurement on high voltage power cables. *Eleventh International Symposium on High Voltage Engineering*.

Sasaki, T., & Kurihara, M. (1996). Oil-Filled cable surveillance system using newly developed optical fiber gas sensor. *IEEE Transactions on Power Delivery*, 11(2), 656–662.

Sharma, C., & Singh, P. (2007). Contribution of loads to low frequency oscillations in power system operation 2007 irep symposium. *Revitalizing Operational Reliability*, 19(24), 1–8.

Yu, H., & Wang, S.Y. (2008). Study and Application on Cable Duct Gas Online Monitoring System Based on GPRS, Distribution Equipments.

Automotive, Mechanical and Electrical Engineering – Liu (Ed.)
© *2017 Taylor & Francis Group, London, ISBN 978-1-138-62951-6*

Research on object re-identification with compressive sensing in multi-camera systems

Yongfeng Huang, Qiang Liu & Cairong Yan
School of Donghua University, Shanghai, China

ABSTRACT: Object tracking in multi-camera systems has been widely used in social life, however object re-identification between cameras is difficult and inefficient. It will improve the efficiency of object re-identification by applying fast and real-time compressive sensing feature into multi-camera. Due to the lighting conditions and projection size of an object in different cameras being different, while a compressive feature is sensitive to the greyscale and size, for the greyscale and size problems, this paper puts forward the idea of unification. It maps the compressive feature value of the sample frame to the value of the original frame by the ratio of the average greyscale value and the ratio of the area of two frames. Experiments show that the algorithm proposed in this paper behaves quickly and accurately in object re-identification when lighting and object size changed.

Keywords: Multi-camera; Object re-identification; Compressive sensing

1 INTRODUCTION

With the extensive application of multi-cameras, target tracking has gradually been developed from single-camera target tracking to multi-camera cooperative target tracking (LI Caihui, 2011). When the tracked object disappears from one camera, it needs to find the object in another camera in a non-overlapping multi-camera system. This is called object re-identification.

The traditional object re-identification algorithms are based on constructing and learning object models. The constructions are usually characterised in three aspects, colour, shape and texture (Wang 2013). The colour histogram model analyses the colour distribution of the whole image by counting the pixel value, so it is real-time with small calculations (Wang et al. 2007). However, it is sensitive to the variations of lighting conditions and discrimination of multiple objects. The shape model characterises the local outline by capturing edges and gradient structures, HOG and Sobel are based on it (Schwartz, 2009). It is robust against different rotation and lighting conditions, but sensitive to occlusion. The robustness of the shape model promoted by Wang (2007) is used to partition human bodies into constitutes for person-identification by learning a shape dictionary. The texture model is constructed by extracting special regions or points from images with filters. These filters are: Gabor filter, LBP, SIFT, SURF (Hamdoun et al., 2008). It is robust to lighting, size, occlusion, and

discrimination of multiple objects. But it needs hug calculation. Usually, they are combined to use.

It will improve the efficiency of object re-identification by applying fast and real-time compressive sensing (Zhang et al., 2012) feature into object re-identification. However, the compressive feature is sensitive to greyscale and size (Wu Yutong, 2013), since the com it mainly extracts the Haar-like feature of the partial region of the image. For the problems of greyscale and size, this paper puts forward the idea of unification. It maps the compressive feature value of the sample frame to the value of the original frame by the ratio of the average greyscale value and the ratio of the area of two frames. Experiments show that the algorithm proposed in this paper behaved quickly and accurately in object re-identification when lighting and object size changed.

2 OBJECT RE-IDENTIFICATION WITH COMPRESSIVE SENSING

Compressive sensing, which is based on sparse sensing theory in signal processing, extracts a discrete sample from the original signal by a random Gaussian matrix that satisfies the Restricted Isometry Property (RIP) then reconstructs the original signal perfectly by a non-linear reconstruction algorithm. It discards the redundant information in image features, extracts effective discrete compressive features directly from the original signal, and then combines the compressive features for object re-identification. It is sensitive to greyscale and size.

2.1 Unification of greyscale

When an object crossed from camera A to camera B, system will call camera B to find the missed object. Due to the differences in environment, appearances of the same object in different camera behave differently. But one thing is invariable, that is ratio of the greyscale value between different pixels of the same object. Thus, as long as we calculate the average greyscale value of two frames to get the greyscale ratio, we can map compressive feature value of the frame B to value of the frame B by the average greyscale ratio.

Figure 1 shows that before unification of greyscale, the compressive feature vector of the original frame expressed as $V_o = (v_1^o, v_2^o, v_3^o, ..., v_t^o)$, the compressive feature vector of the sampling frame expressed as $V_s = (v_1^s, v_2^s, v_3^s, ..., v_t^s)$. Then the average greyscale values l_o, l_s of the original frame and sampling frames respectively are calculated.

$$l_o = \frac{\sum_{i=1}^{w_o} \sum_{j=1}^{h_o} P_{ij}}{w_o * h_o}, \quad l_s = \frac{\sum_{i=1}^{w_s} \sum_{j=1}^{h_s} P_{ij}}{w_s * h_s} \quad (1)$$

$$r = \frac{l_o}{l_s} \quad (2)$$

After obtaining the ratio r, we can unify the compressive feature vector.

$$v_i' = r * v_i^s \quad (3)$$

Compressive feature vectors which are mapped are presented as $v_s' = (v_1', v_2', v_3', ..., v_t')$. After mapping operations, we realised the unification of greyscale in compressive feature between the original frame and sampling frame.

2.2 Unification of multiple sizes

The system reserved a size ratio table, size ratios of the same object in different cameras. It means that if size of object O projected in A camera is S_a, size of object O projected in camera B is S_b, we save the ratio $r = \frac{s_a}{s_b}$ in the size ratio table. Based on the size ratio table, we used 11 different size frames to sample with 0.04 step size, it can improve the fault tolerance and promote the robustness. If original size ratio is r, 11 different size ratio are showed.

The size of the random matrix changed with different sampling scales, each element in the compressive feature vector is the grey-scale convolution of the random matrix and fixed region of the frame. So the compressive feature vector changed with the sampling scales, we should unify the compressive feature value of the different sampling scale frames. Here we can map the compressive feature value of frame B to the value of frame A by the size ratio according to the size ratio table, then it unified the compressive feature of frame B and compressive feature of frame A.

As shown in Figure 2, the original frame size in camera A is $w_o * h_o$, the sampling frame size in camera B is $k^2 * w_o * h_o$ (k presents the ratio of two cameras according to the size ratio table). Before unification of multiple sizes, the compressive feature vector of the original frame expressed as $V_o = (v_1^o, v_2^o, v_3^o, ..., v_t^o)$, the compressive feature vector of sampling frame expressed as $V_s = (v_1^s, v_2^s, v_3^s, ..., v_t^s)$. We map the compressive feature value of frame B to the value of frame A by the size ratio.

$$K = \frac{w_d * h_d}{k^2 * w_d * h_d} = \frac{1}{k^2} \quad (4)$$

$$v_i' = K * v_i^s \quad (5)$$

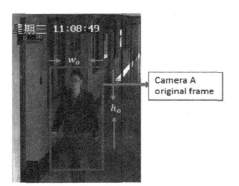
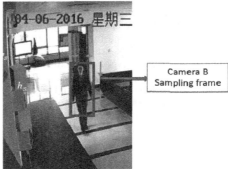

Figure 1. Unification of greyscale.

Table 1. Size ratio table.

1	2	3	4	5	6	7	8	9	10	11
0.80 * r	0.84 * r	0.88 * r	0.92 * r	0.96 * r	1.00 * r	1.04 * r	1.08 * r	1.12 * r	1.16 * r	1.20 * r

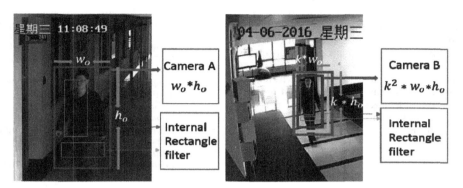

Figure 2. Unification of multiple sizes.

The compressive feature vectors which are mapped are presented as $v'_s = (v'_1, v'_2, v'_3, \ldots, v'_1)$. After the mapping operations, we achieved unification of multiple sizes in the compressive feature between the original frame and the sampling frame.

After the unification of greyscale and multiple sizes, we could classify the compressive feature vector mapped by the naive Bayes classifier to get the maximum classification response frame in the sampling frames. By comparing the maximum classification value with the given threshold, this checked the result of whether the object was detected or not. This far, we complete the object re-identification in multi-camera.

3 EXPERIMENTS

In order to verify the effectiveness of the proposed algorithm in this paper, the experimental scenario included four cameras, A, B, C, and D. The lighting conditions and projection size of the object in different cameras were different. We used 30 groups of objects crossing the four cameras.

It compares the compressive sensing based on unification of greyscale and multiple sizes (CS-u), original compressive sensing which was not unified (CS), and traditional three kinds of algorithms in object re-identification in multi-camera. The HSV colour histogram is used to represent the colour histogram model. The HOG (Histogram of Oriented Gradient) is used to represent the shape model, The SIFT (Scale-Invariant Feature Transform) is used to represent the texture model.

Table 2. Re-identification rate.

Scenes/ Rate	HSV	HOG	SIFT	CS	CS-u
A->B	0.40	0.73	0.90	0.43	0.87
B->C	0.47	0.77	0.93	0.50	0.90
C->D	0.40	0.67	0.90	0.53	0.83
Average	0.42	0.72	0.91	0.49	0.87

Table 3. Re-identification time.

Scenes/ Time (ms)	HSV	HOG	SIFT	CS	CS-u
A->B	150	486	1062	173	204
B->C	75	310	603	90	103
C->D	135	452	956	157	185
Average	120	416	874	140	164

In the experiment, we contrast the five algorithms in terms of re-identification time and re-identification accuracy rate.

Table 2 summarises the successful re-identification rate of the 5 algorithms by 30 groups of objects. According to the experimental result, SIFT, CS-u, HOG, CS and HSV dropped in sequence at the performance of accuracy. We can conclude that compressive sensing based on unification of greyscale and multiple sizes play high accurately in object re-identification.

According to the experimental results from Table 3, SIFT, HOG, CS-u, CS and HSV improved in sequence at the performance of real-time in all the scenes. SIFT and HOG need huge calculation work and process complexly, they performed worst real-time. HSV showed little better than CS-u and CS, because compressive sensing requires training of the positive and negative samples before the process. S-u is an improvement of CS by the unification in greyscale and multiple sizes, it needs more calculation than CS, so CS-u shows worse than CS but it is only little. Finally, we can conclude that compressive sensing based on unification of greyscale and multiple sizes has great real-time in object re-identification.

In summary, in a multi-camera environment, compressive sensing based on unification of greyscale and multiple sizes improved the real-time on the base of keeping high accuracy in object re-identification.

4 CONCLUSION

In this paper, we proposed an object re-identification algorithm with compressive sensing theory in multi-cameras. Due to the lighting conditions and projection size of an object in different cameras being different, while compressive feature is sensitive to the greyscale and size. For the greyscale and size problems, this paper puts forward the idea of unification. It maps the compressive feature value of the sample frame to the value of the original frame by the ratio of the average greyscale value and the ratio of the area of two frames. Experiments show that the algorithm proposed in this paper behaves quickly and accurately in object re-identification when lighting and object size changed.

REFERENCES

Hamdoun, O., Moutarde, F., Stanciulescu, B., et al. (2008). Person re-identification in multi-camera system by signature based on interest point descriptors collected on short video sequences. Proceedings of the Second ACM/IEEE International Conference on Distributed Smart Cameras (ICDSC), 1–6.

Caihui Li, Qun Zhang, Qiyong Lu (2011). Object tracking algorithm based on adaptive background information. *Journal of Terahertz Science and Electronic Information Technology*, 9(5), 596–599.

Schwartz, W.R., Davis, L.S. (2009). Learning discriminative appearance-based models using partial least squares, pp. 322–329.

Wang, X., Doretto, G., Sebastian, T., et al. (2007). Shape and appearance context modeling. *IEEE International Conference on Computer Vision*, 1–8.

Wang, X., Ma, X., Grimson, E. (2007). Unsupervised activity perception by hierarchical Bayesian models. *IEEE Conference on Computer Vision & Pattern Recognition*, pp. 1–8.

Wang, X. (2013). Intelligent multi-camera video surveillance: A review. *Pattern Recognition Letters*, 34(1), 3–19.

Wu Yutong (2013). Research and application of target tracking algorithm in multi camera environment. University of Electronic Science and technology of China.

Zhang, K., Zhang, L., Yang, M.H. (2012). Real-time compressive tracking. Computer Vision—ECCV. Berlin: Springer pp. 864–877.

Automotive, Mechanical and Electrical Engineering – Liu (Ed.)
© 2017 Taylor & Francis Group, London, ISBN 978-1-138-62951-6

Research on plugging technology in the Yingtai Area of Daqing oilfield

Yunjie Li & Zhonglin Dong
Gas Extraction Branch, Daqing Oil Field Co. Ltd., China

ABSTRACT: Serious lost circulation often occurs with relatively fast lost circulation velocity during drilling construction in Yingtai Area, Daqing oilfield. The maximum lost circulation velocity may reach 100–120 m³/h. The lost circulation volume is large, and the highest lost circulation volume may reach 1100 m³/well. The main cause of the lost circulation in this area lies in the relatively developed natural fracture in the Yingtai Area due to its geological structure. As to the characteristics of the lost circulation in the Yingtai Area, a test on plugging materials was carried out in the laboratory to evaluate six plugging materials of low-density rubber particles, dander plugging material, 801 plugging material, complex plugging material, liquid casing plugging material and shielding temporary plugging material. Through analysis of the effect of six plugging materials for fracture plugging of simulated formation, the plugging material suitable for the Yingtai Area, Daqing oilfield was found, based on which a field application was conducted, obtaining a good effect and effectively resolving the lost circulation problem during well exploration and drilling in this area.

Keywords: Yingtai Area; Lost circulation; Plugging and fracture

1 INTRODUCTION

Lost circulation is a common problem during well drilling, and complex cases caused by and various downhole vicious accidents induced by lost circulation because great harm on the drilling engineering. Lost circulation may delay the drilling operation and lengthen the drilling cycle (Xu Jianjun, et al. 2013). Mud or plugging materials and great lost circulation may lead to huge material damage during lost circulation. If lost circulation happens in the reservoir stratum, it will damage the reservoir stratum as well as the productivity (Xu Jian-Jun, 2013; Longchao, 2016). Lost circulation also disturbs the geological logging work and the normal maintenance and treatment of mud performance. Due to complex causes and constraints of lost circulation as well as strong pertinence of plugging technology, this problem has so far not been resolved (Yan Limei et al., 2014). Complex cases such as well collapse and lost circulation often happen during drilling construction in the Yingtai Area, Daqing oilfield, causing lots of inconvenience for the drilling construction. In order to solve the lost circulation in the Yingtai Area, a laboratory test analysis was made to evaluate plugging materials, and finally the plugging material suitable for Yingtai Area was found.

2 CHARACTERISTICS OF LOST CIRCULATION IN THE YINGTAI AREA

30 exploratory wells have been constructed in Yingtai Area since 2002, and complex cases happened in 6 wells during construction, mainly showing as well collapse and lost circulation. Lost circulation is naturally fractured lost circulation with fast lost circulation velocity. Its average velocity is 40–50 m³/h, and the maximum may reach 100–120 m³/h; with a great lost circulation volume. The average lost circulation volume of a single well is 500–600 m³/well, and the maximum may reach 1100 m³/well. It showed through drilling coring observation, logging imaging interpretation and drilling lost circulation condition research, the main reason of lost circulation in Yingtai Area lies in the relatively developed vertical natural fracture due to the geological structure in this area. There are relatively developed vertical fractures in the sandstone of Section II, III and I of the Yaojia Formation and Section II and III of Qingshankou Formation, and the width of fracture is 0.75–1.3 mm. Low regional fracture pressure is another cause of lost circulation in the Yingtai Area. Once annulus is slightly blocked during drilling, it is easy to appear induced lost circulation.

3 LABORATORY STUDY ON PLUGGING TECHNOLOGY

In view of the above characteristics of lost circulation in the Yingtai Area, an evaluation was made of six plugging materials of low-density rubber particles, dander plugging material, 801 plugging material, complex plugging material, liquid casing plugging material and shielding temporary plugging material.

3.1 *Test instrument*

The test instrument used is a QD type plugging material tester produced by Special Instrument Plant, Qingdao Haixin Optical Communication Co., Ltd. This instrument is mainly used to help evaluate materials used to re-build circulation, it may effectively simulate various different formations by applying a series of fracture plates and bed layers in different sizes to determine the plugging effect and the lost circulation volume before plugging formation. The simulated bed particles used in the instrument are stainless steel beads (Φ14.3 mm and Φ4.39 mm). The simulated formation fracture is a set of fracture plates of $1^{\#}$–$6^{\#}$ (fracture width: 0.5–5 mm).

3.2 *Test principle*

During the test, the prepared packed bed or selected fracture plate were put in the assigned position of the main unit, the prepared plugging fluid was poured into the plugging material tank as required. The plugging fluid would then be squeezed into the simulated lost circulation channel under pressure effect. If the plugging material is properly applied, it will form blocking. continue to apply pressure to the rated pressure of the unit, and it may form the bearing capacity of blocking in test. If the plugging material is improperly applied, the plugging fluid will be wholly squeezed out of the lost circulation channel. After the test is done, the simulated lost circulation channel is dismantled and the plugging condition of the plugging material observed.

3.3 *Test method*

Based on the operation principle of a QD type plugging metre, four test methods of static simulated bed, static simulated formation fracture, dynamic simulated bed and dynamic simulated formation fracture may be carried out.

Static test: pour the prepared plugging fluid into the plugging material tank as required. Screw the tank cap down, open the ball valve, record the volume V_0 (ml) of fluid discharged without pressure applied. Open the pressure source as specified after stopping fluid discharge, and apply pressure to 0.69 MPa with a velocity of 0.014 MPa/s, record the volume $V_{0.69}$ (ml) of fluid discharged and observe the minimum pressure that may lead to plugging at the same time. Then increase the pressure with a velocity of 0.069 MPa/s until it reaches 6.9 MPa or until the plugging is damaged and the drilling fluid in the instrument is emptied. Record the volume $V_{6.9}$ (ml) of the drilling fluid discharged and the maximum pressure $P_{rupture}$ (MPa). If plugging is successfully done, maintain this pressure for 10 min and record the final volume of the drilling fluid.

Dynamic test: stabilise the pressure at 0.69 MPa first, and then record the volume $V_{0.69}$ (ml) of drilling fluid discharged and the time $t_{plugging}$(s) to realise the plugging. Increase with velocity of 0.069 MPa/s to 6.9 MPa or until the plugging is damaged and the drilling fluid in the instrument is emptied. Record the volume $V_{6.9}$ (ml) of the drilling fluid discharged and the maximum pressure $P_{rupture}$ (MPa). If the plugging is successfully done, maintain this pressure for 10 min and record the final volume of drilling fluid.

3.4 *Test result and analysis*

The Zwitter-ion drilling fluid used on site was taken as the basic mud in the test. Six plugging materials were then added for test. The added amount of plugging material was respectively 8%, 10%, 12% and 15%. After the addition of the above plugging materials the drilling fluid was fully stirred and prepared into plugging slurry, and then an evaluation using the static and dynamic tests was carried out separately.

1. A stainless steel bead bed was adopted to simulate irregular fracture, and the fracture width was 0.5 mm–3 mm. Under a pressure of less than 6.9 MPa, a dynamic and static test was carried out for the six plugging materials. It was discovered in the test that for large fracture, a plugging slurry prepared with complex plugging material had the optimal plugging effect. The test data of the complex plugging material are shown in Figure 1. As for the small fracture, the plugging slurry prepared with a fluid casing plugging material had the optimal effect and the test data of the liquid casing plugging material are shown in Figure 2. In this test, the effects of the static test are better than those of the dynamic test. As for each plugging material, it had a better plugging effect with the increase of the added amount.

2. Fracture plates were applied to simulate regular fracture with a fracture width of 0.5 mm–5 mm, and a dynamic and static test were carried out for six plugging materials under a pressure of

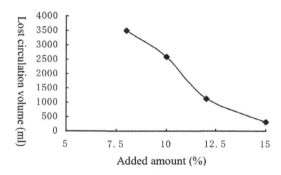

Figure 1. Added amount of static complex plugging material versus. Leakage volume.

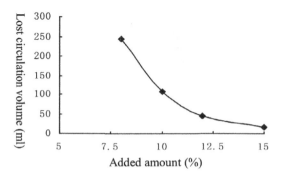

Figure 2. Added amount of static liquid casing plugging material versus. Leakage volume.

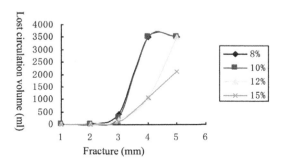

Figure 3. Fracture of static complex plugging material versus. Leakage volume.

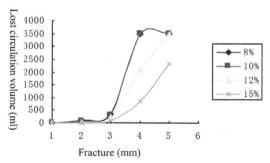

Figure 4. Fracture of static dander plugging material versus. Leakage volume.

less than 6.9 MPa. It was discovered that as for large fracture of 3 mm–5 mm, both the complex and dander plugging materials had a better plugging effect, and may effectively plug the fracture. The test data of the complex and dander plugging materials are shown separately in Figures 3 and 4. It can be seen from the figures that when the fracture width is larger than 3 mm, the lost circulation volume will increase sharply. As for small fracture, all six plugging materials may play the role of plugging. In this test, the effects of the static test are also better than those of the dynamic test. As for each plugging material, it has a better plugging effect with the increase of the added amount.

4 FIELD APPLICATION

Lost circulation happened when the Ying Well 901 in the Yingtai Area, Daqing oilfield was drilled to 1639 m. The initial lost circulation velocity was 0.2 m³/min and the lost circulation velocity increased to 0.5 m³/min when further drilled to 10 m. During this course the total lost circulation of drilling fluid reached 17.5 m³. Liquid casing plugging material, 801 plugging material while drilling and complex plugging material matched with ultrafine calcium were prepared into plugging slurry (6 m³) and injected into the well. Another 4 m³ leaked during the injection of the plugging slurry. After the plugging was done, a leakage test was done with a small displacement first, and no lost circulation was visible. Then the drilling displacement was reduced (reduced to 27l/s from the original 30l/s) and drilling was recovered. The downhole condition showed no abnormality, and plugging of the well was successful. The total lost circulation of the drilling liquid was only 21.5 m³ from measures adopted upon discovery of the lost circulation until the final successful plugging. The measures greatly reduced the lost circulation of the well, saving drilling cost and preventing the occurrence of complex accidents.

5 CONCLUSION

1. As for relatively large irregular fracture, shielding plugging material, 801 plugging material,

low-density rubber particle plugging material and liquid casing plugging material cannot be used for effective plugging, because the relatively small particle size of these four plugging materials is inappropriate for the fracture width. The particle size of complex and dander plugging material is appropriate for large fracture widths, so the two may play the role of effective plugging. Furthermore, the plugging effect of complex plugging material is better than that of dander plugging material.

2. As for small irregular fracture, except that shielding temporary plugging material and 801 plugging material cannot form effective plugging, all the remaining four plugging materials can be applied for effective plugging. The plugging effect becomes better with the increase of the added amount. Liquid casing plugging material and complex plugging material have a better plugging effect, and the former one has the optimal plugging effect.

3. As for regular fracture, under a pressure of less than 6.9 Mpa, shielding plugging material and 801 plugging material can only be used to effectively plug tiny fractures. Low-density active rubber particle plugging material and liquid casing plugging material can only effectively plug fractures below 3 mm. As for large fractures between 3 mm–5 mm, the plugging effect of complex and dander plugging materials is better.

4. It can be discovered through static fracture plate and dynamic fracture plate that the test results under the static condition are slightly better than that under dynamic condition.

5. Plugging material selected through tests exerted a successful plugging role during drilling construction in the Yingtai Area, obtaining a good plugging effect.

ACKNOWLEDGMENTS

This work is supported by Petro China Innovation Foundation (2016D-5007-0201).

REFERENCES

Longchao, Zhu Jianjun, Xu; Limei, Yan. (2016). Research on congestion elimination method of circuit overload and transmission congestion in the internet of things. *Multimedia Tools and Applications,* 27, June, 1–20.

Lv, Kaihe, Qiu, Zhengsong; Song, Yuansen, (2009). Study on auto-adapting lost circulation curing drilling fluid with reservoir protection characteristics in drilling process. Yingyong Jichu yu Gongcheng Kexue Xuebao/*Journal of Basic Science and Engineering,* 17(5), October, 683–689.

Xu J.J., Gai D., Yan L.M. (2016). A new fault identification and diagnosis on pump valves of medical reciprocating pumps. *Basic & Clinical Pharmacology & Toxicology,* 118(1), pp. 38–38.

Xu Jianjun, Xu Yan-chao, Yan, Li-me, et al. (2013). Research on the method of optimal PMU placement. *International Journal of Online Engineering,* 9(7), 24–29.

Xu Jian-Jun, Y. Y. Zi., (2013). Numerical modeling for enhancement of oil recovery via direct current. *International Journal of Applied Mathematics and Statistics,* 43(13), 318–326.

Yan Li-mei, Cui Jia, Xu Jian-jun, et al. (2014). Power system state estimation of quadrature Kalman filter based on PMU/SCADA measurements. *Electric Machines and Control,* 18(6), 78–84. (In Chinese).

Yan Limei, Zhu Yusong, Xu Jianjun, et al. (2014). Transmission lines modeling method based on fractional order calculus theory. *Transactions of China Electrotechnical Society,* 29(9), 260–268. (In Chinese).

Yan Limei, Xie Yibing, Xu Jianjun, et al. (2013). Improved forward and backward substitution in calculation of power distribution network with distributed generation. *Journal of Xi'an Jiaotong University,* 47(6), 117–123. (In Chinese).

Automotive, Mechanical and Electrical Engineering – Liu (Ed.)
© 2017 Taylor & Francis Group, London, ISBN 978-1-138-62951-6

Research on vision technology: Introducing an intelligent destacker based on an ARM-based laser scanner

Lujin Wang, Yechang Peng & Hao Li
School of Wuhan University of Technology, Wuhan, China

ABSTRACT: Destackers are essential equipment in an automated logistics system, however the structure of traditional destackers is simple. In this paper, we deliver an intelligent destacker with an ARM-based laser scanner, which improves machine vision technology and makes up for the deficiencies in current research. First, we introduce the main structure and design of the destacker, including hardware and software design. The controller section, the drive section and the scanning section are then elaborated. The key techniques of the technical system, including data acquisition of goods, data analysis, imaging acquisition, automatic unpacking, are illustrated. The properties and advantages of this technology are analysed using experiments. Finally, the conclusions drawn from the experimental results are summarised. New problems and challenges, which are essential in developing more intelligent and precise automatic equipment, are also discussed.

Keywords: Intelligent destackers; machine vision technology; ARM-based laser scanner; Automatic unpacking

1 INTRODUCTION

The application of ARM-based laser scanners is a very important method in intelligent destackers. It focuses on how to make destackers have vision in the automatic unpacking. It means that the destackers can accurately know the location and shape of goods, especially the spacing between the goods. Even if the goods are in an erroneous position because of wrong stacking, the destacker can adjust itself to grab them without manual adjustment. This meets the demand of efficiently unloading goods without damaging them. At the same time, it is common in factories that different goods have various stacking modes. In the past, one type of destacker woule be used in one stacking mode (Li XG et al., 2011). Now, based on this method, one destacker can grab goods in various stacking modes. Finally, we make progress in the adaptive capacity of destackers and achieve the goal of "double zero": zero waiting in vain and zero manual adjustment.

Recently, many papers have shown several methods and implementations for the vision technology of machines. In terms of the laser scanner, as an important exteroceptive sensor for environmental perception, it has widely been applied in many fields such as object tracking or environment mapping (Forsyth et al., 2004). In terms of the vision, many researchers are employing contact force measurements to predict, detect and model the goods (Kim et al., 2004; Liu et al., 2004; Vogler et al., 2003). In terms of applications, machine vision technology has been widely used in precision manufacturing lines (HU YM et al., 2003), industrial product quality online automated testing, and subtle operations (FANG ZJ et al., 2014). As above, many domestic enterprises have tried to transform the finished products automated logistics system, but still in the ascendant. The combination of visual technology and destackers is rare, so most destackers still only have a single function, poor adaptability and poor visual ability (Li XG et al., 2011).

Thus, this study proposes a new model for destackers, by applying visual technology to the machine based on a laser scanner. On the hardware side, we use an ARM Cortex-M3 as the core processor. An I-shaped two-dimensional structure is used for the mechanism. On the software side, using a PWM algorithm to control the motors and laser scanner is determined for data acquisition in two-dimensional space. T breadth of traversal algorithm is then used to analyse the data and send it to the host computer using Bluetooth. In the host computer, users are able to observe the locations by greyscale. Finally, a PID algorithm is used for the automatic control of motors, to arrive at the specified location and start unpacking accurately. Both the hardware and software have higher open and advanced to ensure a better development of the machines.

This paper is organised as follows: Section 2 describes the main structure and design, including hardware and software. The process of how the destacker works, especially data acquisition and analysis is presented in this section. In Section 3, the properties and advantages of the system are analysed by experiment and compared with traditional destackers. Finally, the conclusions drawn from the experimental results are summarised in Section 4.

2 THE MAIN STRUCTURE AND DESIGN

The main structure and design of the technology are shown in Figure 1. This section is divided into hardware and software to describe our system. The details are as follows.

2.1 Hardware design

The hardware platform for smart destacking system design is divided into three parts: the controller section, the drive section and the scanning section. They are as described below.

1. Controller hardware design. This is based on an ARM Cortex-M4 processor. This design not only meets the demand of the low cost and low power consumption, but it can also achieve the goal of real-time performance. Moreover, it is convenient for algorithm migration and comparative studies of various control methods (Luo et al., 2007; Chen, 2009). In order to improve the accuracy of the demolition chop, using high-precision encoders captures the number of pulses of the motor. Converted data for the ARM processor, the system can improve the accuracy of the control demolition chop. By implementing the PID algorithm for precise control of the motor, it can unpack precisely.

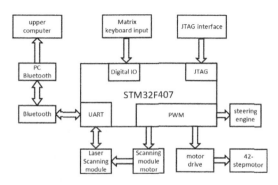

Figure 1. The main structure and design.

2. Drive hardware design. Figure 2(a) shows the drive isolation circuit: BTN7971 and MOS-FET are used to drive the motors in order that the ARM processor can control the speed of the motors more conveniently. Meanwhile, the drive circuit achieves isolated measures of the motor-driven by 74 LVC245 chip to prevent ARM processor burnout.

As is shown in Figure 2 (b), the voltage regulator circuit of the source realises 5 V voltage regulation by means of an LM2940 chip. 5 V is used to supply the SCM and other sensor peripherals. The LM2576 chip provides a 6 V

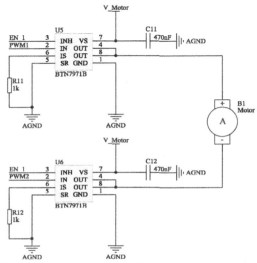

(a) The drive isolation circuit of the motor

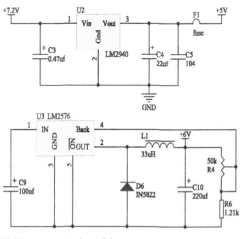

(b) The power section of the motor

Figure 2. The circuit of the motor.

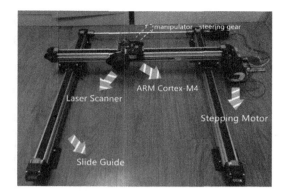

Figure 3. The structure of the destacker.

adjustable voltage, mainly to feed the motors. The reason why we choose 6 V is that the motors work more sensitively under this higher voltage.

3. Scanning hardware design. An I-shaped two-dimensional slide is used for the mechanism. The essence of the slide is to convert the rotational motion of the motors into the liner motion of the sliders. Without the need of moving very fast, using a DC geared motor with an encoder is determined for precise control of the movement of sliders.

To grab the goods accurately in the I-shaped range, we raise the slides by the support structure and secure the laser scanner and manipulator to the sliders. The manipulator is controlled by two steering gears. One is to control the manipulator joint, and another is focused on the claw. Thus, it is able to grab goods of any height and any size.

A whole precise system model for the destacker based on a laser scanner is established, as shown in Figure 3. Our manipulator, which is not shown in Figure 3, is tied to the section of the ARM-based laser scanner so that they move together.

2.2 Software design

As shown in Figure 4, the software of the unpacking control system consists of the following functional modules: laser scanning signal processing module, adjusting grab to the target position module, stepping motor speed regulator module, speed loop and energy loop of motors module, as well as other protection function modules, etc. Wherein, protection function modules include maximum PWM output of motors protection, security control protection and watchdog protection etc.

Software design includes the upper layer software of the host computer and the lower layer software of the ARM. The host computer gives the specific location and shape of the goods through

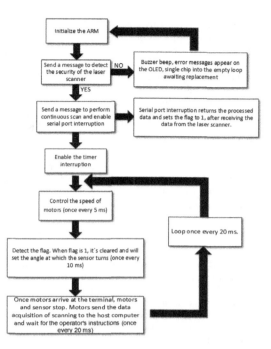

Figure 4. Soft flow chart of the system.

greyscale. The lower layer ARM controls the grab by processing signals from the laser scanner. The details are as follows.

First, the ARM process sends out a PWM plus with fixed frequency in order to make stepping motors move at a certain speed and enable the manipulator and sensor to move in any position of the whole I-shaped range. The speed of the stepping motors v is given by Equation (1):

$$v = \frac{p * \theta_o}{360 * m} \tag{1}$$

where P is the pulse frequency, θ_0 is the fixed step angle, m is the subdivision number.

In addition, the PWM generator cycle Period is forecast by calculations of Equation (2):

$$Period = (T/F)/PWM\_DIV \tag{2}$$

where T is the sys clock of the ARM, F is the set-up pulse frequency, PWM_DIV is the frequency division of the PWM.

The laser scanner then scans the goods in the I-shaped range. We adopt a low-cost 360-degree two-dimensional laser scanning ranging system (RPLIDAR) as our laser scanner (Li et al., 2014). The sensor adopts serial communication and can choose a suitable speed by entering an appropriate

PWM. We can use this sensor to scan a one-dimensional plane, but it is not enough. Thus, through the mechanism mentioned above, which can drive the sensor to move, we can finish the two-dimensional data acquisition.

However, RPLIDAR is a 360-degree full-angle scanner and we only need data acquisition within no more than 120 degrees. To improve this, when ARM detects the laser head in this range, ARM is able to control the sensor to slow down the speed of acquisition. Once beyond this range, it speeds up and discards the data. This can improve the efficiency of data acquisition. The data is the distance from point to surface, and by Equation (3), it is converted to the distance from plane to plane:

$$Direct\_dst = Distance\_linear * \cos(\theta * \pi/180) \quad (3)$$

where Direct_dst is the distance from plane to plane, Distance_linear is the linear distance, θ is the angle at which the sensor turns.

Next, it is data analysis. Set a distance error before, and the distance within the error range can be regarded as the same distance. Using a breadth traversal algorithm quickly traverses all the data in order to divide the two-dimensional data into several areas. Then, through the analysis of the different areas, ARM can quickly find the area where the goods are. Finally, ARM can get the precise position of the goods, such as the shape of the goods and the distance of the goods from the height of the manipulator. The advantages of the breadth traversal algorithm are that it processes data quickly and has obvious data division (Valtteri Heiskanen et al., 2008; Yang et al., 2014; Song et al., 2011; Jiang, 2011).

The processed data is then sent to the computer by Bluetooth. After the data is received by the serial port, it is passed to the host computer. By setting different greyscale for different distances, the controllers are able to observe the goods clearly.

To grab the goods automatically, we use a PID algorithm to set up the software of the manipulator claw. First we set a flag, and through time interruption to detect the flag in order to tell ARM that the data processing is completed. Then, the stepping motor is controlled to reach the specified location using the PID algorithm. The two steering gears mentioned above are ready to control the manipulator to unpack the goods. The formulas of the position PID algorithm are forecast using Equations (4)–(8):

$$\delta = Distance\_Set - Distance \quad (4)$$

$$K = \delta * Kp \quad (5)$$

$$D = (\delta - \delta_o) * Kd * 100 \quad (6)$$

$$I+ = \delta * Ki * 100 \quad (7)$$

$$Distance\_CtrOut = K + I + D \quad (8)$$

where δ is the error between the speed of the last acquisition and the speed of the acquisition now, Distance_set is the initial set-up distance of the goods, Distance is the distance moving now, k is the proportional value, Kp is the proportional constant, D is the differential value, δ_o is the error received before, Kd is the differential constant, I is the sum of the integrate, Ki is the integration constant. Distance_CtrOut is the output distance.

3 EXPERIMENTAL VERIFICATION

In order to verify the applicability and precision of the technique proposed in the paper, we accomplish the scanning experiment. Table 1 shows the results calibrated by the ARM-based laser scanner and manipulator. The measuring accuracy of the destacker is 0.01 mm.

The table above clearly demonstrates that the proposed technique can achieve the precision required by an automated logistics system. Meanwhile, the average time of measurement of the proposed technique is 90.5 s. Compared to traditional destackers, of which the average time is 58.4 s, our destacker is slower than the traditional destacker. This needs to be improved.

For the test, first, we put goods (several plugs) in the scanning range and turn on the power. Next, we connect the host computer with Bluetooth and send instructions. After that, stepping motors control laser scanner moving together for data acquisition and send analysis to the host computer. Our grab experiment has two modes.

Mode one: The pictures of goods are presented on the host computer. Meanwhile, ARM controls manipulator to grab target goods.

Table 1. Results of the scanning experiment.

Code	Goods spacing (mm)	Crawl time by the proposed technology (s)	Traditional crawl time (s)
1	135.75	90	58
2	167.25	87	57
3	213.50	92	56
4	203.75	83	59
5	189.25	94	62
6	182.50	93	61
7	120.00	95	56
8	148.50	88	57
9	155.25	94	61
10	232.75	89	57

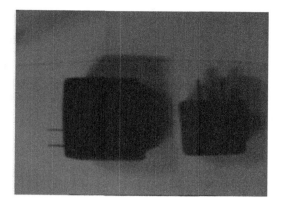

(a) Two objects

(b) The image presented on the host computer

Figure 5.   Sample of the image processing.

Mode two: The pictures are divided into multiple regions and numbered by the ARM. The host computer displays all the regions and sends the instructions of the number of the target region to grab.

Figure 5 shows a simple sample of the image after data acquisition and analysis.

As we can see, the technique based on ARM-based laser scanner and machine vision just needs a high-resolution camera. It can work without the projective light source, which means that our destacker can work in place even with black goods or reflective objects (Song et al., 2011). It is also suitable to achieve the goal of "double zero" and calibrate for use with goods of various sizes or configurations.

4 CONCLUSION

A vision technology about the application of ARM-based laser scanner in destackers is presented. The technology not only has solved the issue of damaging goods using traditional destackers, but also improves the efficiency of unpacking. According to the experiment, the destacker needs more time to grab the goods, compared to traditional destackers. The combination of the I-shaped slide,

ARM-based laser scanner and manipulator are simplified, and the algorithms of scanning and controlling are introduced in the software section. The PWM algorithm is used to control the stepping motors and laser scanner. To extract and analyse more accurate features of goods, the breadth traversal algorithm is used to automatically grab the goods and locate them, the PID algorithm is evaluated. Experimental results show that the proposed algorithms improve vision technology and are suitable for industrial applications. With the advantageous features like "double zero" and high accuracy, they can be widely utilised in the real world. However future research first needs to improve and solve problems occurring in the model, such as lower speed.

ACKNOWLEDGMENTS

This work was financially supported by WHUT National Undergraduate Training Program for Innovation and Entrepreneurship in 2016 (ID: 2016 1049 711002).

REFERENCES

Chen, B. (2009). Design of data acquisition system based on ARM Processor. Wuhan University of Science and Technology, 2009.

Fang, Z.J., Xia, L., Chen, G.D., et al. (2014). Vision-based alignment control for grating tiling in petawatt-class laser system. *IEEE Transactions on Instrumentation and Measurement, 63(6)*, 1628–1638.

Forsyth, D.A., Ponce, J., Lin, X.Y., et al. (2004). Computer vision: A modern approach. Beijing: Publishing House of Electronics Industry.

Hu, Y.M., Du, J., Wu, X.S. et al. (2003). Program realization of high-speed and high-precision surface mounting system based on vision. *Computer Integrated Manufacturing Systems, 9(9)*, 760–764.

Improved method for the feature extraction of laser scanner using genetic clustering. *Journal of Systems Engineering and Electronics, 2*, 280–285.

Jiang, Z.H. (2011). Research and design of bulk material volume measurement system based on laser scanning. Huazhong University of Science and Technology.

Kim, C.J., Mayor, J., and Ni, J. (2004). A static model of chip formation in microscale milling.*Transactions of the ASME, 126(11)*, 710–718.

Li, J., Xu, C., and Zhu, J. (2014). The application of CCD and two-dimensional laser scanner in the design of pipeline inspection system. *Mechanical Design and Manufacturing, 1*, 18–20+24.

Li, X.G. and Liu, J.H. (2011). Palletizing robot research and application actuality, *Problem and Countermeasure. Packaging Engineering, 2011, 3*, 96–102.

Liu, X., de Vor, R.E., Kapoor, S.G., et al. The mechanics of machining at the microscale: Assessment of the current state of the science.*Transactions of the ASME, 126(11)*, 666–678.

Luo, J., Sun, Y.Q. (2007). Comparison of 51 series micro-controllers and ARM processor [J]. Industrial control computer, 5, pp. 64–66.

Song, L.M., Zhang, C.B., Wei, Y.Y., Chen, H.W. (2011). Technique for calibration of chassis components based on encoding marks and machine vision metrology. *Optoelectronics Letters, 1*, 61–64.

Valtteri Heiskanen, Kalle Marjanen, and Pasi Kallio. (2008). Machine vision based measurement of dynamic contact angles in microchannel flows. *Journal of Bionic Engineering, 4*, 282–290.

Vogler, M.P., de Vor, R.E., Kapoor, S.G. (2003). Microstructure-level force prediction model for micro-milling of multi-phase materials. *Transactions of the ASME, 125*, 202–209.

Yang, Z.D., Wang, P., Li, X.H. and Sun, Z.K. (2014). Flexible calibration method for 3D laser scanner system. *Transactions of Tianjin University, 1*, 27–35.

Automotive, Mechanical and Electrical Engineering – Liu (Ed.)
© 2017 Taylor & Francis Group, London, ISBN 978-1-138-62951-6

Review of the image quality assessment methods and their applications to image defogging and stabilisation

Jiaqi Cai
Shanghai Jiao Tong University, Shanghai, China

Wenhan Yan
Zhengzhou University, Zhengzhou, China

Jiaming Liu
University of Electronic Science and Technology of China, Chengdu, China

ABSTRACT: The meaning of image quality mainly includes two aspects: the fidelity and the intelligibility of the image. Image quality directly depends on the optical performance of imaging equipment, image contrast, instrument noise and other factors. Image quality assessment research, which aims at reasonable assessments of various aspects of image processing, has become one of the basic technologies of image information engineering. The two typical models of image degradation are fuzzy and jitter, and the effectiveness of different algorithms is analysed through the analysis of the existing image defogging. A referenceless perceptual fog density prediction model has been proposed based on statistics obtained by Environmental Scenery (SES) and fog responsive arithmetic structures. The projected idea is known as Fog Responsive Thickness Evaluator (MRTE). The proposed display forecasts the perceivability of a foggy scene on or after a solitary image that has no reference for comparison with an image that is free of fog, without reliance on remarkable questions in a scene, deprived of side geological camera data, deprived of assessing a profundity subordinate broadcast outline, without preparing on human-evaluated judgments. In this paper, the authors test a few defogging methods and raise a practical design to use SVM to predict if a defogging method is suitable for a fogging image.

Keywords: Fog; Environmental scenery; Fog; Fog Responsive Thickness Evaluator; Mean Deducted; Difference Regularized; Multi-Variant Gaussian

1 INTRODUCTION

Estimation of visual quality is important for various image and video preparing applications, where the unbiased image quality assessment calculations should be in concurrence with excellent human judgments. Throughout the years, various strategies have been proposed to tackle the issue. Many have their unique advantages (He Kaiming, 2013). It is critical to assess their execution and examine their qualities and shortcomings. The objective of quality assessment research is to outline calculations for target assessment of quality in a way that is reliable with subjective human assessment (Nishino, 2012). The view of open air environmental scenes is vital for comprehension of the environmental environment and for effectively executing visual exercises. For example, question discovery, acknowledgment, and route. In awful climates, the retention or diffusing of light by barometrical particles (for example fog, cloudiness, or fog), can significantly diminish the perceivability of scenes. Subsequently, protests in images caught under terrible climate conditions suffer from low complexity, black out shading, and moved luminance (Kratz, 2009; Ancuti, 2013).

Since the diminishment of perceivability can be misleading for judgments in vehicles guided by camera images, and can initiate wrong detection in remote observation frameworks, there has been serious concentration on programmed techniques for perceivability forecast and upgrade of foggy images.

2 RELATED WORK

Realising the above mentioned challenge, Fattal R. et al. (2008) presented a method for estimating an optical transmission in hazy scenes in a single image. They formulated a refined image formation model that accounts for surface shading in addition to the transmission function. They proposed the dark channel prior to removing haze from a single input image (He Kaiming, 2011; He Kaiming, 2013). They estimated the thickness of the haze and used a high-quality depth map to improve the output. Kratz et al. introduced a method that leverages natural statistics of both the albedo and depth of the scene (Kratz, 2009; Nishino, 2012). The image is modelled with a factorial Markov random field where the two variables are statistically independent latent layers. Tarel et al. (2009) introduced a fast algorithm to handle both colour image and gray level images. Their algorithm has high processing speed which allows visibility restoration to be applied for real-time processing applications.

Current perceivability forecast prototypes that work on a foggy image require either a fog-free image taken in similar scene under various climate conditions to determine perceivability, or recognised striking articles in a foggy image, for example, path markings or movement signs to supply separate signals. However, achieving enough images is tedious, and it is hard to locate the most extreme and least level of polarisation amid quick scene changes (Mitchell, 2010). A programmed technique for fog identification and of estimation of perceivability separation utilising side topographical data acquired from a locally available camera have been shown. While this strategy eased the requirement for numerous images, it is still hard to apply due to the fact that making precise three-dimensional symmetrical prototypes that catch dynamic true structure is difficult. What is more, this approach works just under constrained suspicion (Gibson, 2013; Bovik, 2013).

As to upgrading images with fog, assorted defogging models are required. The most punctual methodologies use a dull protest subtraction strategy to handle environmental dispersing rectification of multispectral information or various images of similar scene under various climate conditions. A more proficient and alluring methodology is to utilise just a solitary foggy image; in any case, coordinate expectation of fog thickness from a solitary foggy image is troublesome. In this manner, most defogging calculations use an extra evaluated profundity delineate a profundity subordinate transmission guide to enhance perceivability utilizing presumptions from, for example the air scrambling model. Specialist anticipated scene by boosting neighbourhood differentiate while assuming a smooth layer of air light. However the outcomes had a tendency to be excessively immersed, making radiance impacts (Ancuti, 2010).

The researcher enhanced perceivability by accepting that transmission and surface shading are measurably uncorrelated. This strategy requires considerable shading and luminance vari-

ety to happen in the foggy scene. The imperative commitment was made of the dull channel earlier. Sending this requirement conveys more fruitful outcomes by refining the underlying transmission outline using a delicate tangling technique; however, delicate tangling is computationally costly, despite the fact that it can be accelerated utilising a guided channel (Mittal, 2012; Mittal, 2013). A quick arrangement was fabricated utilising an edge protecting middle of middle channel, yet the separated profundity delineate be smooth aside from along edges that are incidental with substantial profundity bounced.

A Bayesian defogging model was recommended by a specialist for foreseeing both scene of albedo alongside the profundities. These profundities depend on the factorial Markov irregular fields (Zhang Qieshi, 2012). Results are satisfying, yet this strategy creates some dim ancient rarities at locales drawing closer vast profundity. The execution of defogging calculations has just been assessed subjectively because of the nonattendance of any fitting perceivability assessment apparatus.

Humans are viewed as definitive mediators of the quality of visual signs. Hence, in order to get the best sign for the defogging, human judgments are given the highest importance (Groen, 2013). Human subjective assessments are difficult, tedious, and are not valuable for substantial, remote, portable information. Such bottlenecks have led the researcher to create target execution assessment techniques for defogging calculations. As of late, pick up parameters demonstrating recently obvious edges, and the rate of pixels that get to be dark or white subsequent to defogging (Choi, 2014). Target IQA calculations have additionally been utilised in order to assess the upgraded differentiate along with the auxiliary deviations that are present in the image that is defogged. However, these correlation strategies necessitate first an image that is foggy as a kind of perspective to assess image that is defogged. In addition, present image quality assessment measurements are for the most part wrong for the submission as it is intended in evaluating the bending levels instead of the perceivability of images that are foggy (Duda, 2012). Henceforth, no-reference and defogging-purposed nonspecific perceivability assessment apparatuses are alluring objectives.

3 CONTRIBUTION

The whole image database was derived from an arrangement of source images that reflects satisfactory assorted qualities in image content. Twenty-nine high determination and excellent shading images were gathered from the Internet.

Although it is generally not essential, frequently it is appealing towards having the capacity to evaluate naturally, and lessens fog in a perceptually pleasant way (Bovik, 2010). Towards accomplishing discernment driven exact perceivability expectation, the authors have built up another model: Fog Responsive Thickness Evaluator (MRTE), in view of models of statistic derived from by environmental landscape and haze-aware measurable elements. As compared to other techniques, the proposed method has clear favourable circumstances. In particular, MRTE can anticipate perceivability on a foggy scene that has no reference, without different images that are foggy, with no reliance on pre-recognised notable questions that are present in a scene with fog, deprived of lateral geological data acquired from a locally available camera, without assessing a profundity subordinate transmission outline, and without preparing on human-appraised judgments. MRTE just uses quantifiable deviations from measurable regularities seen on environmental haze free images. The thicknesses of the fog are related to the fog aware elements. Two sets of images were considered. The first set of images are said to be foggy images whereas the second set of images are said to be fog-free images. Components are obtained from dependable space area statistics from an environmental view model, and on watched qualities of foggy images including low difference,

Figure 1. Various source pictures considered in the study.

black out shading, and moved luminance. To assess the execution of MRTE, a study was performed with the students in the universities. By way of effective applications, MRTE can precisely assess the execution of defogging calculations by anticipating related fog thickness in the image that is defogged. These are utilied in developing models that support image defogging and that are intended towards improving the perceivability of Foggy Images (FI).

This paper also has contributions as follows.

First, the authors review four classic defogging solutions under common conditions and verify the effectiveness of each method.

Second, the authors raise a practical design, using SVM to predict whether a certain method is useable to defog a given image.

4 IMAGE QUALITY ASSESSMENT

Some of the quality assessment along with the remarks has been projected. Peak Signal to Noise Ratio (PSNR) is one of the image quality assessments, where the luminance component alone is considered. Quality assessment such as Sarnoff, Just Noticeable Differences (JND), and Metrix works on colour images. DCTune was originally designed for JPEG optimisation. It also works on colour images. NQM works with luminance only. Fuzzy S7 also works with luminance only. BSDM works on colour images as well as multi-scale. SSIM, IFC, and VIF work with luminance only (Achanta, 2009).

5 VISUAL MODEL SHOWING THE FOGGY IMAGE

Foggy Image with the Fog Formation is discussed in this area. Precise demonstration of optical dissipating is an unpredictable issue involvess the wide assortment of sorts, dimensions, and introductions, along with the circulations of elements establishing a media. In this way, the improved climatic disseminating model is broadly meant for visual foggy image development. At the point when sun oriented light penetrates in the environment that is foggy, the reflected rays from the article are straightforwardly weakened. Furthermore, they are captured in the camera that are diffusely scattered. Scientifically, FI might decay into segments such as coordinate weakening and air light.

Expectation Model of Perceptual Fog Density is considered. The non-references forecast model of perceptual fog thickness, MRTE extricates haze aware factual elements from a test FI, turns fog aware elements to a multi-variant Gaussian model, and after that registers the abnormalities that has been obtained from environmental fog images and fog-free images. Abnormalities are processed utilising separation extent amongst multi-variant Gaussian attack of the haze aware elements, acquired from assessment image against a multi-variant Gaussian model of haze aware elements removed from a set of images that are related to the fog, and another set of images that are fog-free images.

The portion of the level of the foggy is related to the level of fog-free. The proportion technique epitomises measurable components from FI and Fog-Free Images (FFI) together, and in this manner can foresee perceptual fog thickness over a more extensive territory than by utilising images at foggy level alone.

Fog Mindful Arithmetical Sorts is examined. Having the neighbourhood image patches, the three initial fog mindful measurable are obtained. Fundamental small request measurements of FI & FFI, that are relately applicable, separated on or after a 3-D area statistic acquired by environmental landscape model of neighbourhood Mean Deducted, Distinction Regularised (M2DR) coefficients.

In the presence of environmental FI, the authors have established that the change of M2DR coefficients diminishes as haze thickness increments. The nearby standard deviation is a critical descriptor of basic image data that evaluates neighbourhood sharpness. In any case, the perceptual effect of neighbourhood standard deviation differs with the nearby mean esteem. Henceforth, the coefficient of variety,

$$\xi(s, t) = \sigma(s, t)/\mu(s, t) \qquad (1)$$

which measures standardised scattering is processed. $\sigma(s, t)$ is a representation of the nearby standard deviation and the $\mu(s, t)$ is a representation of neighbourhood mean esteem.

Expectation of related density of the fog is studied. A test foggy image is parcelled into FXF lattice. Every one of the squares is then used to process the normal component values, in this way yielding an arrangement of fog aware measurable elements for every fix. The mean and covariance lattice are assessed utilising a standard most extreme probability estimation strategy.

6 NOVEL DEFOGGING CALCULATION AND COMPARATIVE STUDY OF EXISTING METHODS

The authors propose an intense and valuable direct utilisation of MRTE, related defogging of perceptual image defogging, named Novel Defogging calculation. Novel Defogging calculation uses factual regularities seen in FI & FFI to separate notice-

able data from pre-handled images. There are three pre-handled images that are considered. Out of the three, one image is white adjusted and two are complexity upgraded images.

Immersion, related fog thickness, fog mindful luminance, along with complexity weight maps are connected on the pre-prepared images. The weight maps are connected with the help of the Laplacian multi-scale modification. Figure 2 shows the Input FI and the result is the images with defogged, as itemised in the accompanying.

The principal pre-handled image is white adjusted to change the environmental version of the yield by taking out chromatic throws brought on by air shading. The shades-of-dim shading steadiness procedure is utilised on the grounds that it was quick and strong.

The Weight Maps (WM) specifically define the weight of the furthermost sharp districts present in the pre-prepared images. The three WM were characterised in light of estimations of immersion, along with saliency. The authors utilise the arrangement of target WM. Then they propos the utilisation of another arrangement of related

Figure 2. Input Foggy images (FI) and the resultant defogged images.

roused fog aware WM. Fog aware WM precisely catches the related perceivability of pre-processed images, in this way creating the edges that are clear on perceivability upgraded images.

The immersion weight outlines the immersion pick up between neighbourhood immersion and the greatest immersion is equivalent to solidarity in hue, saturation and value shading space. The fog thickness weight delineates the additional WM to precisely adjust FI and FFI. Related fog thickness outline one is anticipated utilising MRTE on covered patches, and after that a guided channel is connected to diminish clamour. The scope of the denoised haze thickness guide is scaled from 0 to 1.

The difference weights outline image points of interest by appointing advanced weights next to locales of great angle values. Gaussian weighting capacity is tested out to three standard deviations. Standardised weight NW of number list maps are acquired to guarantee that they aggregate to solidarity as takes after:

$$NWi = Wi/\Sigma i \; Wi \qquad (2)$$

where Wi incorporates chrominance weight, immersion weight, saliency weight, haze thickness weight, fog aware luminance weight, differentiate weight.

Fattal (2008) assumes that it is a constant albedo in a local area of an image, and the surface shading is not statistically related to the medium transmission. He uses independent component analysis to estimate the constant albedo. The algorithm is a solution to a nonlinear inverse problem, and its performance depends greatly on the statistical properties of input data. It will lead to an unreliable estimate that the change of independent component being not obvious or lack of colour information.

He et al. (2011; 2013) assume that in the local area of at least one colour channel, the scene albedo tends to 0, and they use the minimum value filtering to roughly estimate the propagation function. They also use the propagation function which is refined by image matting algorithm. This refinement method is essentially a large—scale sparse linear equation used to solve the problem, with a high time complexity and spatial complexity. However, the purpose of introducing α into image matting is to soften or anti-aliasing the transition regions of foreground and background, and the media propagation function is the exponential decay factor of scene radiance. Therefore, it is unreasonable to a use image matting algorithm for the refinement of a media propagation function. And in the function used, the data item is not important. However, if we increase the value of the regular parameter, it is easy to create overshoot distortion in the colour at the edge of the depth-of-field mutation.

Kratz et al. (Kratz, 2009; Nishino, 2012) assume that scene albedo and Depth-Of-field (DOF) are statistically independent, and can be modelled using a regular prior probability. The gradient of the scene albedo is modelled as a heavy-tailed prior, while the depth-of-field is determined by the specific scene, modelled basing on the natural scene feature as δ segment constant function or Gaussian smoothing function. By solving a MAP to estimate the problem, we can estimate the scene albedo and the DOF. This algorithm needs us to select the DOF prior model according to the specific image, and use experience to give the parameters of the prior model.

Tarel et al. (2009) propose a fast image dehazing algorithm, by assuming the atmospheric veil function approximates the maximum in the feasible region, and the local variation is flat. This algorithm uses the deformation of the median filter to estimate the atmospheric veil function. However, median filtering is not always a good edge-preserving filtering method. Therefore, inappropriate parameter settings are prone to halo effects.

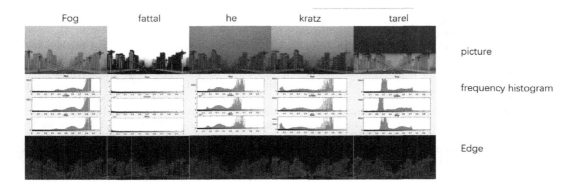

Figure 3. Some features of results of the methods.

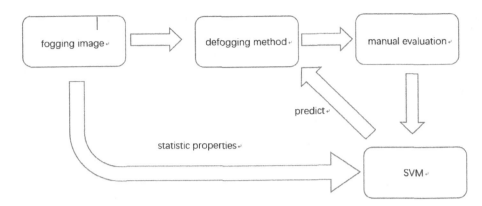

Figure 4. Solution predicting flowchart.

In addition, the parameters of this algorithm are difficult to adjust.

For all these dehazing solutions, it is difficult to predict which method is suitable in a future haze scene. So the authors also introduce SVM method to predict if a certain method is suitable. To build the SVM samples for predicting, there are ten different and separate experimenters, to classify the result of each solution, "1" for acceptable and "0" for unacceptable. If a result gets more than six in total, the method is considered to be practical for the input fogging image.

Also, few features of an image are required in building SVM: variance, mean, skewness, and kurtosis in all RGB channels. And the defogging images are separated into two groups according to the previous classification. After using these inputs to build the machine, it can be predicted whether a method is effective for a given image.

7 EXPERIMENT

Since past perceivability expectation prototypes are needed, different foggy, various divergence pictures, and side topographical data is acquired by utilising an on-board camera, it is unrealistic to straightforwardly look at the execution of MRTE with other forecast models. Rather, the authors assessed the execution of MRTE in contradiction of aftereffects related to the study related to human. Towards dispassionately assessing execution of Novel Defogging calculation, the authors utilise the complexity improvement assessment technique and the perceptual fog thickness of MRTE.

Test images considered here are 100 shading images that were chosen to catch sufficient differing qualities of image substance and fog thickness as of recently documented FI, surely understood FI, and by comparing defogged images. A few

images were caught by an observation camera. The sizes of the image changed from 450×750 pixels (that is 1.5″ to 2.5″) to 1200×1800 pixels (that is 4″ to 6″).

The test methodology incorporates subjects, gear and the data about the show. A sum of 20 understudies at the university went to the subjective study. No fiscal remuneration was given to the subjects.

Hardware and show setup wer produced by the authors. The real-time simulator Matlab was used. This simulator takes care of enhancing building and logical issues. The screen was set at a determination of high definition, that is 1920×1080 pixels. The test images that are considered are shown at the focal point of an LED screen for few moments at their local image determination, to keep any mutilations because of scaling operations by means of programming or equipment. No blunders, (for example, latencies) were experienced while showing the images. The rest of the zones of the show were dark.

8 RESULTS AND PERFORMANCE COMPARISON

The proposed show MRTE not just predicts perceptual fog thickness of a whole image, but additionally gives a neighbourhood perceptual haze thickness forecast on every fix. The fix magnitude could cover relying upon an application involving distinctive thickness estimations. Figure 5 exhibits the aftereffects of applying MRTE. Related fog density for various pixel sizes. 500×333 pixels, 300×199 pixels, 100×66 pixels and 50×33 pixels are considered.

Figure 6 exhibits the different calculation examinations for various foggy images. It can be seen that the defogged images created by Kopf (2008) look oversaturated and contain radiance impacts (Tan, Robby T, 2008). Images close to horizons of the scene were paritally defogged, while (R. Fattal, 2008) yields darker sky locales. The images defogged by He Kaiming (2011) and Novel Defogging calculation re-establish more environmental hues. Amongst these the defogged images conveyed by Novel Defogging calculation uncover all the sharper points of interest. Despite the fact that the technique for single image dehazing by multiscale combination accomplishes the best decrease of perceptual haze thickness after rebuilding, most defogged images delivered by that strategy lose noticeable edges yielding higher estimations of the measurements because of oversaturation. By and

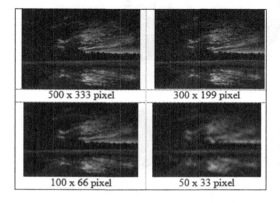

Figure 5. Related fog density for various pixel size.

large, the subjective and target correlation brings about.

In the next part, the authors use 20 images to test different dehazing methods and three more images to test whether the SVM works. The images are in two different groups: city images and country images.

As a defogging solution evaluation, ten independent experimenters are involved to classify each result. They have two choices on each defogging image: "acceptable" means 1 point, and "unacceptable" means 0 points. Then, the total score of a certain result is calculated. If it reaches 6, this solution to this specific image is considered useful.

Based on the classification part, the authors come to the conclusion that different methods have different usage. Fattal's solution works well in places with many subjects, as buildings, though it does not always work, performing badly in the thick fog. Kaiming He's method works in most conditions, but the fog is not entirely swept. Kartz's method requires choosing parameters appropriately and takes the longest time of the four, but can be the most accurate under some conditions. Tarel's method works fast and recognises the background and the foreground well, but the image may generate a fake edge.

The result of evaluation is taken advantage of to build an effective SVM. This SVM can predict whether a certain defogging method is suitable for a given image. The parameters of this SVM is variance, mean, skewness, and kurtosis of all channels. After building the machine, three more

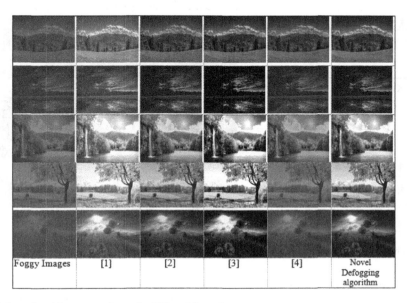

Figure 6. Various algorithm comparisons for different foggy images.

Figure 7.   Different defogging solutions.

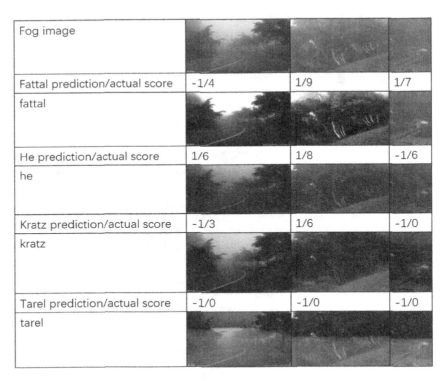

Fog image			
Fattal prediction/actual score	-1/4	1/9	1/7
fattal			
He prediction/actual score	1/6	1/8	-1/6
he			
Kratz prediction/actual score	-1/3	1/6	-1/0
kratz			
Tarel prediction/actual score	-1/0	-1/0	-1/0
tarel			

Figure 8.   Comparison between prediction and test.

testing images are used to verify the effectiveness of the machine. Comparing the prediction to the rank given by those ten independent experiment- ers, only one is incorrect amongst twelve predic- tions ("acceptable" in the prediction is "1" while in the actual score is over "6").

9 CONCLUSION AND FUTURE ENHANCEMENT

The authors have portrayed a forecast model of perceptual haze thickness called Fog Responsive Thickness Evaluator, and a perceptual image defogging calculation known as Novel Defogging calculation; both are in the light of image statistics acquired by environmental landscape and fog aware measurable elements. MRTE predicts the level of perceivability of a foggy scene from a solitary image, while Novel Defogging calculates upgrades the perceivability of a foggy image with no reference data. For example, numerous foggy images of similar scene, diverse degrees of polarisation, remarkable questions in the foggy scene, assistant land data, a profundity subordinate transmission delineate, situated presumptions, and even without preparing on human-appraised judgments. MRTE saw in environmental foggy and haze free images, just by using quantifiable deviations from measurable regularities. The authors detailed the model and the fog aware factual components, and exhibited how the haze thickness expectations delivered by MRTE associate well with human judgments of fog thickness taken in a subjective study on an extensive foggy image database.

As an application, the authors showed that MRTE can be a helpful, using NR apparatus for assessing the execution of defogging calculations. Future work could include creating equipment inviting variants of Novel Defogging calculation, appropriate for incorporated circuit usage and the advancement of portable image defogging applications.

REFERENCES

Achanta, R., Hemami, S., Estrada, F. & Susstrunk, S. (2009). Frequency-tuned salient region detection. *Computer Vision and Pattern Recognition, 2009. IEEE Conference on Computer Vision and Pattern Recognition*, 1597–1604.

Ancuti, C., O., Ancuti, C., Hermans, C. & Bekaert, P. (2010). A fast semi-inverse approach to detect and remove the haze from a single image. *Asian Conference on Computer Vision* (pp. 501–514). Heidelberg: Springer.

Ancuti, C. O. & Ancuti, C. (2013). Single image dehazing by multi-scale fusion. *IEEE Transactions on Image Processing*, 22(8), 3271–3282.

Bovik, A. C. (2010). Perceptual video processing: Seeing the future [Point of view]. *Proceedings of the IEEE*, 98(11), 1799–1803.

Bovik, A. C. (2013). Automatic prediction of perceptual image and video quality. *Proceedings of the IEEE*, 101(9), 2008–2024.

Choi, L. K., You, J. & Bovik, A. C. (2014). Referenceless perceptual fog density prediction model. *IS&T/SPIE Electronic Imaging* (pp. 90140H–90140H). International Society for Optics and Photonics.

Duda, R. O., Hart, P. E. & Stork, D. G. (2012). *Pattern classification*. New York: John Wiley & Sons.

Fattal, R. (2008). Single image dehazing *ACM Transactions on Graphics*, 27(3), Article ID 72.

Gibson, K. B. & Nguyen, T. Q. (2013). A no-reference perceptual based contrast enhancement metric for ocean scenes in fog. *IEEE Transactions on Image Processing* 22(10), 3982–3993.

Groen, I. I. A., Ghebreab, S., Prins, H., Lamme, V. A. F. & Scholte, H. S. (2013). From image statistics to scene gist: evoked neural activity reveals transition from low-level environmental image structure to scene category. *The Journal of Neuroscience*, 33(48), 18814–18824.

Guo, F., Tang, J. & Cai, Z. (2014). Objective measurement for image defogging algorithms. *Journal of Central South University*, 21(1), 272–286.

He, K., Sun, J. & Tang, X. (2011). Single image haze removal using dark channel prior. *IEEE Transactions on Pattern Analysis and Machine Intelligence*, 33(12), 2341–2353.

He, K., Sun, J. & Tang, X. (2013). Guided image filtering. *IEEE Transactions on Pattern Analysis and Machine Intelligence*, 35(6), 13971409.

Kopf, J., Neubert, B., Chen, B., Cohen, M., Cohen-Or, D., Deussen, O., Uyttendaele, M. & Lischinski, D. (2008). Deep photo: Model-based photograph enhancement and viewing. *ACM Transactions on Graphics* 27(5), 116.

Kratz, L. & Nishino, K. (2009). Factorizing scene albedo and depth from a single foggy image. *2009 IEEE 12th International Conference on Computer Vision*, 1701–1708.

Mitchell, H. B. (2010). *Image fusion: theories, techniques and applications*. Heidelberg: Springer Science & Business Media.

Mittal, A., Moorthy, A. K. & Bovik, A. C. (2012). No-reference image quality assessment in the spatial domain. *IEEE Transactions on Image Processing*, 21(12), 4695–4708.

Mittal, A., R., Soundararajan, R. & Bovik, A. C. (2013). Making a "completely blind" image quality analyzer. *IEEE Signal Processing Letters*, 20(3), 209–212.

Nishino, K., Kratz, L. & Lombardi, S. (2012). Bayesian defogging. *International Journal of Computer Vision*, 98(3), 263–278.

Saleem, A., Beghdadi, A. & Boashash, B. (2012). Image fusion-based contrast enhancement. *EURASIP Journal on Image and Video Processing*, 1(2012):1.

Tan, R. T. (2008) Visibility in bad weather from a single image. *2008 IEEE Conference on Computer Vision and Pattern Recognition*, (pp. 1–8).

Tarel, J. & Hautiere, N. (2009). Fast visibility restoration from a single color or gray level image. *2009 IEEE 12th International Conference on Computer Vision*.

Zhang, Q. & Kamata, S. (2012). Improved optical model based on region segmentation for single image haze removal. *International Journal of Information and Electronics Engineering*, 2(1), (2012): 62.

Automotive, Mechanical and Electrical Engineering – Liu (Ed.)
© 2017 Taylor & Francis Group, London, ISBN 978-1-138-62951-6

Study of a driver's face and eyes identification method

Yuhang Zhang, Hui Zhang, Zhipeng Ding & Jianguo Zhang
China FAW Corporation R&D Center, Changchun, China

ABSTRACT: This work-in-progress paper introduces a method for detecting the face and eyes. An Active Shape Model (ASM) is utilised to identify the driver's facial features; the AdaBoost cascade classifier is adopted to classify the Local Binary Pattern (LBP) features of the human eye corners and fit eyelids by the parabola method and the Sobel edge pupil detection. After a subjective evaluation of a large number of samples, it is proved that the method of the driver's face and eye detection and positioning is efficient and accurate.

Keywords: Corner detection; eyelid fitting; pupil detection

1 INTRODUCTION

The proportion of serious accidents caused by fatigue driving is more than 40%. Prevention of fatigue driving is an important direction in the research on human-computer interaction and intelligent security. Methods for the detection of fatigue driving include: detection of physiological parameters of the driver (EEG, ECG, EMG, etc.); testing driver's operating characteristics (such as steering wheel angle, steering wheel angle rate, accelerator pedal angle, etc.) (King et al., 1998); the driver state (the state of the driver's eyes (Wirewille et al., 1994)) and mouth etc. Detection of eye state is more direct than the operating characteristics. Compared with the detection of the normal physiological state, a detection of the eye state does not result in invasive activities on the drivers. The accuracy of the results of this type of testing is good.

Detection of a driver's eyes is based on a dynamic image capture and processing technology, the detection techniques include three categories: Study-based methods which include the AdaBoost, the SVM, the LDA and the PCA methods, extraction of eye training samples and non-eye training samples; training a human eye classifier then using it to detect the eye image samples. The second category is based on the method of characteristics, including geometric features of the human eye positioning method, EyeMap eye location method, and the horizontal and vertical projection method, etc. The third category is based on a template method, including the contour characteristic template eye localisation and the left and right eye synthesis template localisation, etc.

Face detection based on an Active Shape Model (ASM) algorithm, corner detection based on an AdaBoost cascade classifier to classify LBP (local two valued Pattern) features, Sobel edge pupil detection

and an eyelid fitting method based on parabola fitting is adopted in this paper, to gradually realise the accurate capture of the eyes and their position.

2 DRIVER FACE VIDEO IMAGE ACQUISITION DESIGN

The video capture device consists of two parts: the camera and the infrared light source.

Part 1: The camera is installed on the window of the instrument panel to avoid affecting the driver's view, as shown in Figure 1.

Part 2: 70 degrees infrared ring light is used to avoid localised strong light that would simulate the driver's eyes, as shown in Figure 2.

Figure 1. Camera installation.

Figure 2. Infrared light.

3 EYE DETECTION AND LOCATION RESULTS

3.1 *Face detection results*

An ASM algorithm based on the point distribution model is used for the face detection (Cootes et al., 2008). The shape of a human face can be expressed as $x = (x_1, y_1, x_2, y_2, ..., x_n, y_n)$, where (x_i, y_i) represents the coordinates of point i, the training set contains K samples, and the sample covariance matrix is:

$$W = (x - \bar{x})(x - \bar{x})^T \tag{1}$$

$\bar{x} = \frac{1}{K} \sum_{i=1}^{K} x_i$ is the average value of the sample training set. The eigenvalues and eigenvectors of W are:

$$Wp_k = \lambda_k p_k \tag{2}$$

where λ_k is the first k eigenvalue of W, p_k is the eigenvector.

The greater the λ_k, the more important it is for the p_k. The eigenvalues are sorted by large to small, a new principal axis system composed of eigenvectors corresponding to the extracted M eigenvalues is given by $P = [p_1, p_2, ..., p_M]$, $\lambda_1 > \lambda_2 > ... > \lambda_M$.

Any face shape in the subspace of the face set can be approximated by the average value coupled with the new coordinate system and weighted sum of a set of control parameters:

$$x = \bar{x} + P \cdot b \tag{3}$$

$b = (b_1\ b_2\ ...\ b_M)^T$ is the weight factor.

After training study, average value $\bar{x}$ and principal component matrix P are definite value, therefore the shape vector only depends on the model parameter b. Changing parameter b can generate a series of different shapes of the model of the human face.

The training samples for the calibration of the 77 feature points in the face region are shown in Figure 3.

Figure 3. ASM feature point calibration.

3.2 *Corner detection results*

After positioning the face, the eye corner detection uses the AdaBoost cascade classifier on the corner of the eye LBP (Local Binary Pattern, local two value model) feature classification (Tang Xumian et al., 2007). Each layer of the AdaBoost cascade classifier is a strong classifier. They form a multi-layer classification structure, constitute the classifier parameters (Wang Wei et al., 2008), can make each layer through almost all of eye corner samples, and refuse to a large number of non corner samples. Thus, in the presence of the corner of the eye area to spend more time, reduce the overall computational time.

During the training process on the inner and outer corners of human eyes, a total of 4000 positive samples and 10000 negative samples were collected, some of the positive samples as shown in Figure 4.

Figure 5 shows the results of using a cascade classifier to locate eye corners.

3.3 *Results of the recognition of pupils and eyelids*

Since the contour of the pupil is a circle, we fit the contour of the pupil by its centre and radius. According to the characteristics of the human eye,

Figure 4. Sample set of the inner and outer corners of left and right eyes.

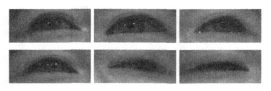

Figure 5. Results of the inner and outer corner of the eye point detection.

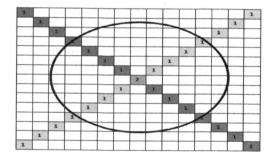

Figure 6. Voting schematic.

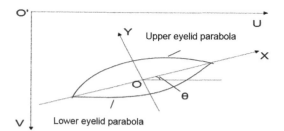

Figure 7. Upper and lower eyelid parabolic model.

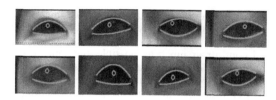

Figure 8. Eyelid fitting optimisation results.

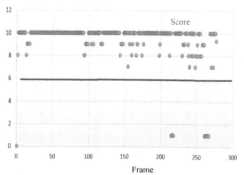

A. face recognition

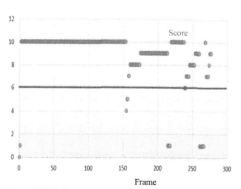

B. eyelid fitting

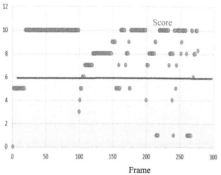

C. pupil detection

Figure 9. Subject evaluation results.

we select the pixels between the upper and lower eyelids in a Sobel edge image (Cermak et al., 2004) to determine the pupil's position, see Figure 6.

The normal extension of line for each pixel point passes through some points, the number of votes plus 1 on the points. We select the point with the most votes as the centre point (x_0, y_0) and calculate the distance of each point (x_i, y_i) between the upper and lower eyelids to the centre point of the pupil (x_0, y_0). Statistically we calculate all the range of the integer value r and get the maximum value r_0 as the pupil radius.

The parameter equation of the pupil is obtained by:

$$(x - x_0)^2 + (y - y_0)^2 = r_0^2 \qquad (4)$$

Using two parabolic fittings to the eyelids (Daugman, 2007). In order to simplify the model, the image coordinate system $O'UV$ is translated and rotated to the coordinate system OXY, make the upper and lower eyelids parabolic symmetry about the axis. θ Is the rotation angle (Figure 7).

The parabolic model of the upper and lower eyelids can be expressed as:

$$y = ax^2 + c \qquad (5)$$

In combination with the Sobel operator, we first select the edge image of the human eye then use statistical methods to calculate the parameters of the upper and lower eyelids parabolic equation a and c. In order to keep the points on the eyelid contour as much as possible and remove the redundant points, the gradient images obtained by the Sobel operator are improved.

The absolute value of the gradient in the X direction and the gradient in the Y direction are summed up:

$$sobelNew(i, j) = \left|sobelx(i, j)\right| + \left|sobely(i, j)\right| \quad (6)$$

where $sobelx(i, j)$ and $sobely(i, j)$ are the gradients in X and Y direction of point (i, j). Then, $sobelNew$ is normalised to 0~255, and the threshold to obtain the image's edge is calculated. Separating the upper and lower eyelids in the edge image respectively, to solve the parameter a and c of the parabolic equation. The fitting results are shown in Figure 8.

Using the method of subjective validation, 276 frames were randomly selected to evaluate the face recognition, eyelid fitting and pupil fitting results. According to the 10 point scoring, less than 6 recorded points failed. Statistical results show that the success rate of the face recognition is 95.65%, and the eyelid fitting is 93.48%; Taking the circumstances of failure into account, including closed eyes and dark light, the success rate of iris recognition is 81.88% (Figure 9).

4 CONCLUSION

In this paper, a driver fatigue detection method for safely driving by face recognition, corner detection, eyelid fitting and pupil detection is utilised after analysing a large number of samples which were randomly selected in video images, and getting the result of optimisation of the human eye. Subjective validation shows that the rate of face recognition and eyelid fitting are over 95% and 93%, respectively. In the case of eyes closed and dark light condition, the accuracy rate of pupil identification is more than 80%. According to current research results, this method has shown a good detection accuracy, and has obtained a certain amount reference values for further research.

REFERENCES

Cermak, M., and Skala, V. (2004). Adaptive edge spinning algorithm for polygonization of implicit surfaces. *Proceedings of the IEEE International Conference on Computer Graphics, Los Alamitos*. IEEE Computer Society Press, pp. 211–221.

Cootes, T. F., Twining, C. J., and Taylor, C. J. (2008). Diffeomorphic statistical shape models. *Image and Vision Computing, 26(3)*, 326-332.

Daugman, J. (2007). How iris recognition works. *IEEE Trans. on Systems, Man, and Cybernetics-Part B: Cybernetics, 37(5)*, 1167–1175.

King, D. J., Mumford, D. K., and Siegmund, G.P. (1998). An algorithm for detecting heavy-truck driver fatigue from steering wheel motion. *16th International Technical Conference on the Enhanced Safety of Vehicles*, Windsor, 1998, 873–882.

Tang Xumian, Ou Zongying, Su Tieming, Hua Shungang (2007). Fast face location algorithm based on AdaBoost and genetic algorithm. *Journal of South China University of Technology (Natural Science Edition), 35(1)*.

Wang Wei, Huang Feifei, Li Jianwei, Feng Hailiang (2008). Face description and recognition using multiscale LBP feature. *Optics and Precision Engineering, 2008, 16(4)*, 696–705.

Wirewille, W. W., and Ellsworth, L. A. (1994). Evaluation of driver drowsiness by trained observers. *Accident Analysis and Prevention, 26(5)*, pp. 571–581.

Automotive, Mechanical and Electrical Engineering – Liu (Ed.)
© *2017 Taylor & Francis Group, London, ISBN 978-1-138-62951-6*

Testability modeling for a remote monitoring system of marine diesel engine power plant

Xuedong Wen
Naval University of Engineering, Wuhan, China
Equipment Mending Office of the Navy in Qingdao Area, Qingdao, China

Guo He
Naval University of Engineering, Wuhan, China

Kun Bi & Peng Zhang
Equipment Mending Office of the Navy in Qingdao Area, Qingdao, China

ABSTRACT: In order to solve the problem of Design For Testability (DFT) for a Remote Monitoring and Alarming System of Marine Diesel Engine Power Plant (RMASMDEPP), a testability modeling method based on Multi-signal flow graph is studied. After establishing function signal set S, fault pattern sample set F, test point set TP and executable test set T of each subsystem for the power plant, the Multi-signal directed graph model of the system can be obtained as well as the "Fault-Test" dependency matrix. The research results of this paper provide basis for DFT and fault diagnosis of a RMASMDEPP and some other similar equipment.

Keywords: Ship; Diesel engine power plant; Testability modeling; Multi-signal flow graph model

1 INTRODUCTION

RMASMDEPP is one of the most severe working environment and the most widely used in the monitoring and alarming system of ship platform. Under these circumstances, the ability to detect and isolate faults and the testability of the system are the important factors which restrict the performance of the power plant. DFT technology can greatly improve the ability of test, diagnosis and maintenance for large complex equipment. That is, it is the key technology to solve the problem of testing and diagnosis for RMASMDEPP. For DFT, The establishment of testability model is the basic work, which is also a hot research area. At present, both of Information-flow model and Multi-signal flow graph model are more general and more extensive studied.

Information-flow model using a single signal dependent relationship is easily detached from the actual system, especially for the CAN bus system. Multi-signal flow graph model basing on system structure model describe dependency relation between functional signals, and complete modeling work by tracing the signal flow of each component and each test node of researched system. With respect to the former, in practical application, the latter has advantages of simpler modeling process,

more concise for structure description and faster reasoning speed.

Aircraft, missile, radar and other electronic systems have been studied for Multi-signal flow graph model of DFT in the existing literatures. But for RMASMDEPP, the related research about DFT has not formed systematic and mature achievements. In this paper, Multi-signal flow graph model of a RMASMDEPP is studied, which lays the foundation for the further DFT research of the system.

2 MULTI-SIGNAL FLOW GRAPH MODEL

2.1 *Composition of multi-signal flow graph model*

As mentioned above, Multi-signal flow graph model is based on analysis of system structure and function, which is in a hierarchical directed graph form to express the signal flow orientation, failure modes of system components and their mutual connection relations. By defining series of structure units (failure modes), function signals (unit fault characterizations), test points, test sets, signal sets and their correlations, the model can be used to realize the mathematical expression of elements and their correlation for a RMASMDEPP.

A typical Multi-signal flow graph model mainly consists of the following components:

1. A finite fault source set of system $\boldsymbol{F} = \{f_1, f_2, ..., f_l\}$ ($l \in \boldsymbol{Z}^+$), which represents the total number of failure modes;
2. A set of functional signal $\boldsymbol{S} = \{S_{i1}, S_{i2}, ..., S_{ik}\}$ for representing the change of system state;
3. Available test point set for n_p dimension $TP = \{TP_1, TP_2, ..., TP_{n_p}\}$, which means a test point can achieve a number of test functions;
4. The n-dimensional executable test set $\boldsymbol{T} = \{t_1, t_2, ..., t_o\}$ ($o \in \boldsymbol{Z}^+$), which represents the total number of available tests of the system assuming that each test are independent under given failure mode condition;
5. Test point TP_i contains a set of executable test sets $\boldsymbol{PT}(\boldsymbol{TP_i})$, $\boldsymbol{PT}(\boldsymbol{TP_i}) \subseteq \boldsymbol{T}$;
6. Fault source f_i affects a set of signal attribute variables $\boldsymbol{MS}(f_i)$: failure rate λ_i, Fault criticality sc_i, Failure position fl_i etc., and $\boldsymbol{MS}(f_i) \subseteq \boldsymbol{S}$;
7. Test t_j detect a set of signal attribute variables $\boldsymbol{TS}(t_j)$: test reliability γ_j, test cost v_j etc., and $\boldsymbol{TS}(t_j) \subseteq \boldsymbol{S}$;
8. Directed graph model: $\boldsymbol{DG} = \{\boldsymbol{F}, \boldsymbol{TP}, \boldsymbol{E}\}$. Among them, $\boldsymbol{E}$ is the edge of the directed graph, which indicates physical connection of the electronic circuit and the pipeline of system.

According to the influence range for the system function, Multi signal flow graph model has two kinds of fault type, that is, local fault (short for F) and global fault (short for G). When local fault occurs, the partial function of the system is abnormal or failure. Or else, global fault occurs.

2.2 Modeling steps of multi-signal flow graph model

Based on the model definition, the process of testability modeling can be divided into three steps:

Step 1. Familiar with the modeling object. That is, identifies and extracts the relevant information of the test model. Such as system structure, signal function, and test information, etc.

Step 2. Build model. First of all, set up model with system composition as well as node attributes, and add the functional attributes of the signal. And then, determine input, output and internal links of system units according to the function of signal flows. Finally, set test points and related test sets on the basis of test requirements.

Step 3. Adjust, modify and calibrate the constructed model, and then generate directed graph model and fault-test dependency matrix.

3 MULTI-SIGNAL FLOW GRAPH MODELING FOR A RMASMDEPP

The system of RMASMDEPP consists of 12 subsystems, and the diagram of structure and test points distribution as shown in Fig. 1. The system uses the field bus network of CAN, whose structural composition is a double layer network. The lower layer is used for data acquisition and processing, and the upper one is the data transmission and display network. According to the cabin structure, the lower net is also divided into several independent sub networks, which maintain data link with the upper network through each own router.

3.1 Modeling elements analysis for a RMASMDEPP

In the subsystems of a RMASMDEPP, except for the monitoring PC which is no need to set test points because of its created advanced DFT, the other 11 subsystems are all potential fault sources. Among them, the following parts are prone to failure: five types of sensors, power conversion module, A/D conversion module, photoelectric isolation unit, module calibration unit, clock/reset circuit, CAN control module, router and monitoring board etc. In the process of modeling, this paper chooses 22 kinds of test units, which are easy to produce faults, as the sources of the faults that to be tested. And the sources can produce 18 measured points, see Figure 1.

Modeling elements for a RMASMDEPP are as follows:

1. Finite fault source set of the system $\boldsymbol{F} = \{f_1, f_2, ..., f_{22}\}$.
   Where: f_1: thermal resistive sensor; f_2: thermocouple sensor; f_3: 4~20 mA current signal sensor; f_4: on-off quantity sensor; f_5: pulse quantity sensor; f_6: high temperature signal output unit; f_7: power supply module I; f_8: power supply module II; f_9: indicator light display module; f_{10}: buzzer alarm module; f_{11}/f_{12}: clock/reset circuit module III; f_{13}: CAN interface circuit III; f_{14}/f_{15}: clock/reset circuit module II; f_{16}: CAN interface circuit I; f_{17}/f_{18}: clock/reset circuit module I; f_{19}/f_{20}: calibration circuit for signal processing module; f_{21}: 24 V voltage conversion circuit module; f_{22}: data conversion circuit.
2. Signal feature set $\boldsymbol{S} = \{S_1, S_2, ..., S_{34}\}$ generated by the fault source set $\boldsymbol{F}$ which cause system function change.
   S_1: mV voltage input signal of thermal resistance sensor; S_2: drift signal of thermal resistance sensor; S_3: mV voltage input signal of thermocouple sensor; S_4: drift signal of thermocouple sensor; S_5: 4~20 mA current input signal;

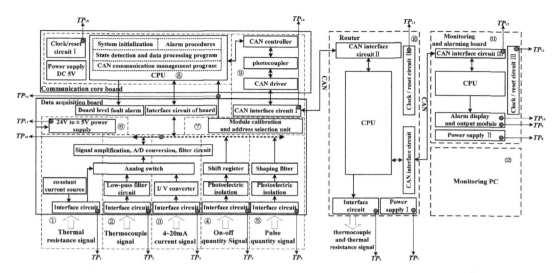

Figure 1. The diagram of structure and test points distribution for RMASMDEPP.

S_6: drift signal of 4~20 mA current input sensor; S_7: input signal of on-off quantity sensor; S_8: detection signal of the on-off quantity sensor; S_9: input signal pulse quantity sensor; S_{10}: drift signal of pulse sensor; S_{11}: output signal of mV—temperature; S_{12}: power supply voltage signal I; S_{13}: critical signal of insulation for power line I; S_{14}: reduced signal of insulation for power line I; S_{15}: power supply voltage signal II; S_{16}: critical signal of insulation for power line II; S_{17}: reduced signal of insulation for power line II; S_{18}: voltage signal of alarm lamp; S_{19}: current signal for buzzer; S_{20}: frequency signal of clock circuit III; S_{21}: pulse signal of reset circuit III; S_{22}: data signal of CAN communication III; S_{23}: frequency signal of clock circuit II; S_{24}: pulse signal of reset circuit II; S_{25}: data signal of CAN communication I; S_{26}: frequency signal of clock circuit I; S_{27}: pulse signal of reset circuit I; S_{28}: value of jumper signal for calibration circuit; S_{29}: CPU working signal; S_{30}: critical signal of insulation for circuit board; S_{31}: reduced signal of insulation for circuit board; S_{32}: voltage signal of 24V conversion power supply; S_{33}: output characteristics of A/D converter; S_{34}: output of on-off and pulse quantity signal.

3. Test set $T = \{t_1, t_2, \ldots, t_{24}\}$, which is capable of performing in 24 dimensional.

t_1: input voltage test of thermal resistance sensor; voltage signal to meet the rating of 0~250°C requirements, and then determine the test pass, otherwise as a fault. t_2: drift measurement of thermal resistance sensor; the drift signal needs to meet the test requirements, and then determine the test pass, otherwise as a fault. t_3: voltage test of thermocouple sensor; detected voltage signal should meet rated range value of 0~800°C, and then determine the test pass, otherwise as a fault. t_4: signal drift test of thermocouple sensor; signal drift values need to satisfy test requirements, and then determine the test pass, otherwise as a fault. t_5: input current test; the tested current signal should satisfy in the range of 4~20 mA, and then determine the test pass, otherwise as a fault. t_6: drift measurement of 4~20 mA current sensor; signal drift values need to satisfy test requirements, and then determine the test pass, otherwise as a fault. t_7: voltage test of on-off quantity sensor; tested voltage value should meet requirements, which is normally open display ±5 V and normally closed equal 0 V, otherwise determine it as a fault. t_8: test for on-off quantity sensor; tested voltage signal values need to satisfy test requirements, and then determine the test pass, otherwise as a fault. t_9: signal test of pulse sensor; the pulse signal (voltage VDD, pulse width, baud rate) meets with the required range of 0 Hz~5000 Hz, or convince it as a fault. t_{10}: drift measurement of pulse sensor; the drift signal needs to meet the test requirements, and then determine the test pass, otherwise as a fault. t_{11}: analog voltage signal test; the measurement analog value satisfy the requirement, and then determine the test pass, otherwise as a fault. t_{12}: power voltage test I; voltage values meet requirements of the rated input, and then determine the test pass, otherwise as a fault. t_{13}: circuit insulation test I; the insulation resistance is not less than 153 kΩ, which is recognized as the test pass. And less than 148 kΩ is determined to reduce the insulation. t_{14}: test of power supply voltage II; To meet the requirements of the rated input voltage value is passed, or convicted of failure. t_{15}: circuit insulation test II; The insulation resistance is not less than 153 kΩ, which is recognized as the test pass. And less than 148 kΩ is determined to reduce the insulation. t_{16}: alarming lamp voltage test; work voltage value is at the rated value, then the test is passed, otherwise the sentence is a failure. t_{17}: buzzer current test; 4~20 mA current signal meet the rating requirements, and then the test is passed, otherwise the sentence as a failure. t_{18}: crystal oscillator circuit characteristic test III; the oscillator frequency meets the rated value, then the test is passed, or otherwise the test failed. t_{19}: Reset circuit characteristic test III; when turn on or turn off the power, the reset pulse is an ideal square wave, or otherwise the reset circuit is faulty. t_{20}: communication data test III; the communication data signal is in conformity with the standard communication data, then the test is passed, or otherwise the test failed. t_{21}: crystal oscillator circuit characteristic test II; the oscillator frequency meets the rated value, then the test is passed, or otherwise the test failed. t_{22}: Reset circuit characteristic test II; when turn on or turn off the power, the reset pulse is an ideal square wave, or otherwise the reset circuit is faulty. t_{23}: communication data test I; the communication data signal is in conformity with the standard communication data, then the test is passed, or otherwise the test failed. t_{24}: crystal oscillator circuit characteristic test I; the oscillator frequency meets the rated value, then the test is passed, or otherwise the test failed. t_{25}: Reset circuit characteristic test I; when turn on or turn off the power, the reset pulse is an ideal square wave, or otherwise the reset circuit is faulty. t_{26}: test of characteristic value for calibration circuit; characteristic value of digital circuit meets the established requirements value, then the test is passed, or otherwise the test failed. t_{27}: test for CPU working status; If read permission signal can meet the jumper clock signal requirement, the test is passed, or otherwise identify it as fault. t_{28}: circuit insulation test; The insulation resistance is not less than 153 kΩ, which is recognized as the test pass. And less than 148 kΩ is determined to reduce the insulation. t_{29}: voltage test of voltage conversion module; DC/DC parameters meet the requirements of the rated output range, then the test is passed, or otherwise the test failed. t_{30}: test of input/output signal characteristic; If characteristics value of gating channel input, A/D conversion out-

put and on-off quantity output/pulse quantity output can meet the requirements, then determine the test pass, or otherwise recognize as a fault.

4. 18 dimensional available test point set $TP = \{TP_1, TP_2, ..., TP_{18}\}$;
5. The corresponding test set $PT(TP_i)$ of each test point TP_i is as follows: $PT(TP_1) = \{t_1, t_2\}$; $PT(TP_2) = \{t_3, t_4\}$; $PT(TP_3) = \{t_5, t_6\}$; $PT(TP_4) = \{t_7, t_8\}$; $PT(TP_5) = \{t_9, t_{10}\}$; $PT(TP_6) = \{t_{11}\}$; $PT(TP_7) = \{t_{12}, t_{13}\}$; $PT(TP_8) = \{t_{14}, t_{15}\}$; $PT(TP_9) = \{t_{16}\}$; $PT(TP_{10}) = \{t_{17}\}$; $PT(TP_{11}) = \{t_{18}, t_{19}\}$; $PT(TP_{12}) = \{t_{20}\}$; $PT(TP_{13}) = \{t_{21}, t_{22}\}$; $PT(TP_{14}) = \{t_{23}\}$; $PT(TP_{15}) = \{t_{24}, t_{25}\}$; $PT(TP_{16}) = \{t_{26}, t_{27}\}$; $PT(TP_{17}) = \{t_{28}, t_{29}\}$; $PT(TP_{18}) = \{t_{30}\}$.
6. The set of $FS(f_i)$ corresponding to the 17 module f_i signal is as follows: $FS(f_1) = \{S_1\}$; $FS(f_2) = \{S_2\}$; $FS(f_3) = \{S_3\}$; $FS(f_4) = \{S_4\}$; $FS(f_5) = \{S_5\}$; $FS(f_6) = \{S_6\}$; $FS(f_7) = \{S_7, S_8\}$; $FS(f_8) = \{S_9, S_{10}\}$; $FS(f_9) = \{S_{11}, S_{12}\}$; $FS(f_{10}) = \{S_{13}, S_{14}\}$; $FS(f_{11}) = \{S_{15}\}$; $FS(f_{12}) = \{S_{16}, S_{17}\}$; $FS(f_{13}) = \{S_{18}\}$; $FS(f_{14}) = \{S_{19}, S_{20}\}$; $FS(f_{15}) = \{S_{21}\}$; $FS(f_{16}) = \{S_{22}, S_{23}\}$; $FS(f_{17}) = \{S_{24}, S_{25}\}$; $FS(f_{18}) = \{S_{26}\}$.
7. The functional signal set $TS(t_j)$ that each test t_j capable of detecting is listed as follows: $TS(t_1) = \{S_1\}$; $TS(t_2) = \{S_2\}$; $TS(t_3) = \{S_3\}$; $TS(t_4) = \{S_4\}$; $TS(t_5) = \{S_5\}$; $TS(t_6) = \{S_6\}$; $TS(t_7) = \{S_7\}$; $TS(t_8) = \{S_8\}$; $TS(t_9) = \{S_9\}$; $TS(t_{10}) = \{S_{10}\}$; $TS(t_{11}) = \{S_{11}\}$; $TS(t_{12}) = \{S_{12}\}$; $TS(t_{13}) = \{S_{13}, S_{14}\}$; $TS(t_{14}) = \{S_{15}\}$; $TS(t_{15}) = \{S_{16}, S_{17}\}$; $TS(t_{16}) = \{S_{18}\}$; $TS(t_{17}) = \{S_{19}\}$; $TS(t_{18}) = \{S_{20}\}$; $TS(t_{19}) = \{S_{21}\}$; $TS(t_{20}) = \{S_{22}\}$; $TS(t_{21}) = \{S_{23}\}$; $TS(t_{22}) = \{S_{24}\}$; $TS(t_{23}) = \{S_{25}\}$; $TS(t_{24}) = \{S_{26}\}$; $TS(t_{25}) = \{S_{27}\}$; $TS(t_{26}) = \{S_{28}\}$; $TS(t_{27}) = \{S_{29}\}$; $TS(t_{28}) = \{S_{25}, S_{31}\}$; $TS(t_{29}) = \{S_{32}\}$; $TS(t_{30}) = \{S_{33}, S_{34}\}$.
8. According to the above mentioned parameters, combine the modeling elements and their corresponding relations can lead to the directed graph $DG = \{F, TP, E\}$ of Multi-signal flow graph model, see Figure 2.

Figure 2 shows the dependence between the test points and the fault sources. It can be seen from the flow direction of the function signal that each test point of the set can detect all of the fault sources corresponding to the point.

3.2 Fault-test dependency matrix based on multi-signal flow graph model

In this system, the fault source which may occur with G type fault can be listed as follows: $f_4, f_{11}, f_{12}, f_{13}, f_{14}, f_{15}, f_{16}, f_{17}, f_{18}, f_{19}, f_{20}, f_{22}$. Possible fault sources for the F type fault are listed as follows: f_6, f_9, f_{10}. The fault sources which are likely to occur in the G type fault as well as the F type fault can be listed as follows: $f_1, f_2, f_3, f_5, f_7, f_8, f_{21}$. After analyzing on

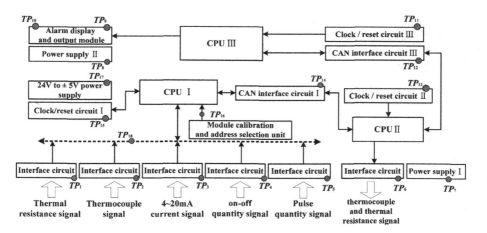

Figure 2.   The directed graph of multi-signal flow graph model for a RMASMDEPP.

Table 1.   Fault-test dependency matrix for a RMASMDEPP.

	t_1	t_2	t_3	t_4	t_5	t_6	t_7	t_8	t_9	t_{10}	t_{11}	t_{12}	t_{13}	t_{14}	t_{15}	t_{16}	t_{17}	t_{18}	t_{19}	t_{20}	t_{21}	t_{22}	t_{23}	t_{24}	t_{25}	t_{26}	t_{27}	t_{28}	t_{29}	t_{30}
$f_1(G)$	1	1	0	0	0	0	0	0	0	0	1	0	0	0	0	0	0	0	0	0	1	0	0	1	0	0	0	0	0	1
$f_1(F)$	0	1	0	0	0	0	0	0	0	0	0	0	0	0	0	0	0	0	0	0	0	0	0	0	0	0	0	0	0	0
$f_2(G)$	0	0	1	1	0	0	0	0	0	0	1	0	0	0	0	0	0	0	0	0	1	0	0	1	0	0	0	0	0	1
$f_2(F)$	0	0	0	1	0	0	0	0	0	0	0	0	0	0	0	0	0	0	0	0	0	0	0	0	0	0	0	0	0	0
$f_3(G)$	0	0	0	0	1	1	0	0	0	0	1	0	0	0	0	0	0	0	0	0	1	0	0	1	0	0	0	0	0	1
$f_3(F)$	0	0	0	0	0	1	0	0	0	0	0	0	0	0	0	0	0	0	0	0	0	0	0	0	0	0	0	0	0	0
$f_4(G)$	0	0	0	0	0	0	1	1	0	0	1	0	0	0	0	0	0	0	0	0	1	0	0	1	0	0	0	0	0	1
$f_5(G)$	0	0	0	0	0	0	0	0	1	1	1	0	0	0	0	0	0	0	0	0	1	0	0	1	0	0	0	0	0	1
$f_5(F)$	0	0	0	0	0	0	0	0	0	1	0	0	0	0	0	0	0	0	0	0	0	0	0	0	0	0	0	0	0	0
$f_6(F)$	0	0	0	0	0	0	0	0	0	0	1	0	0	0	0	0	0	0	0	0	0	0	0	0	0	0	0	0	0	0
$f_7(G)$	0	0	0	0	0	0	0	0	0	0	1	1	1	0	0	0	0	0	0	0	1	1	1	1	0	0	0	0	0	0
$f_7(F)$	0	0	0	0	0	0	0	0	0	0	1	0	1	0	0	0	0	0	0	0	0	0	0	0	0	0	0	0	0	0
$f_8(G)$	0	0	0	0	0	0	0	0	0	0	0	0	0	1	1	1	1	1	1	1	0	0	0	0	0	0	0	0	0	0
$f_8(F)$	0	0	0	0	0	0	0	0	0	0	0	0	0	0	0	1	0	0	0	0	0	0	0	0	0	0	0	0	0	0
$f_9(F)$	0	0	0	0	0	0	0	0	0	0	0	0	0	0	0	0	1	0	0	0	0	0	0	0	0	0	0	0	0	0
$f_{10}(F)$	0	0	0	0	0	0	0	0	0	0	0	0	0	0	0	0	0	1	0	0	0	0	0	0	0	0	0	0	0	0
$f_{11}(G)$	0	0	0	0	0	0	0	0	0	0	0	0	0	0	0	0	0	1	0	1	0	0	0	0	0	0	0	0	0	0
$f_{12}(G)$	0	0	0	0	0	0	0	0	0	0	0	0	0	0	0	0	0	0	1	1	0	0	0	0	0	0	0	0	0	0
$f_{13}(G)$	0	0	0	0	0	0	0	0	0	0	0	0	0	0	0	0	0	0	0	1	0	0	0	0	0	0	0	0	0	0
$f_{14}(G)$	0	0	0	0	0	0	0	0	0	0	1	0	0	0	0	0	0	0	0	1	1	0	0	0	0	0	0	0	0	0
$f_{15}(G)$	0	0	0	0	0	0	0	0	0	0	1	0	0	0	0	0	0	0	0	1	0	1	0	0	0	0	0	0	0	0
$f_{16}(G)$	0	0	0	0	0	0	0	0	0	0	1	0	0	0	0	0	0	0	0	1	0	0	1	0	0	0	0	0	0	0
$f_{17}(G)$	0	0	0	0	0	0	0	0	0	0	1	0	0	0	0	0	0	0	0	1	0	0	1	1	1	0	0	0	0	1
$f_{18}(G)$	0	0	0	0	0	0	0	0	0	0	1	0	0	0	0	0	0	0	0	1	0	0	1	0	1	1	0	0	0	1
$f_{19}(G)$	0	0	0	0	0	0	0	0	0	0	1	0	0	0	0	0	0	0	0	1	0	0	1	0	0	0	1	0	0	1
$f_{20}(G)$	0	0	0	0	0	0	0	0	0	0	1	0	0	0	0	0	0	0	0	1	0	0	1	0	0	0	0	1	0	1
$f_{21}(G)$	1	0	0	0	0	0	0	0	0	0	1	0	0	0	0	0	0	0	0	1	0	0	1	1	1	1	1	1	1	1
$f_{21}(F)$	0	0	0	0	0	0	0	0	0	0	0	0	0	0	0	0	0	0	0	0	0	0	0	0	0	0	0	1	0	0
$f_{22}(G)$	0	0	0	0	0	0	0	0	0	0	1	0	0	0	0	0	0	0	0	1	0	0	1	0	0	0	0	0	0	1

the signal flow and function of each fault source in Figure 2, it can get the "fault-test" dependence matrix of the system based on Multi-signal flow graph model, as shown in Table 1.

4   CONCLUSION

In this paper, based on the DFT optimization problem for a RMASMDEPP, the system testability modeling method using Multi-signal flow graph model has been studied. On the basis of the comprehensive study for the unit structure, modeling elements and test sets of the system, the structure-function model of the system is transformed into the testability directed graph model, and then a "fault test" dependency matrix is obtained which can be used to optimize the selection of test sets. The research results of this paper provide a technical reference for the test model of similar equipment, which is conducive to further in-depth analysis of the connotation of DFT and improvement measures for equipment's testability.

REFERENCES

Deb S. & Pattipatik R. 1995. Multi-signal flow graphs: a novel approach for system testability analysis and fault diagnosis. IEEE Aerospace and Electronic Magazine 10(5): 14–25.

Lei H.J. & Qin K.Y. 2013. Quantum-inspired evolutionary algorithm for analog test point selection. Analog Integrated Circuits and Signal Processing 75(3): 491–498.

Oborski P. 2014. Developments in Integration of Advanced Monitoring Systems. The international Journal of Advanced Manufacturing Technology 75(9): 1613–1632.

Su Y. D. et al. 2011. Missile testability requirement analysis and index determination oriented to mission. Journal of National University of Defense Technology 33(2): 125–129.

Wen X. d. et al. 2016. Research on method of modeling complex electro-mechanical system for testability based on contract index. Journal of naval university of engineering 24(4): 54–58.

Zhang S.G. et al. 2013. Optimal selection of imperfect tests for fault detection and isolation. IEEE Transactions on Systems Man and Cybernetics: Systems 13(6): 1370–1384.

Zhang X.S. et al. 2015. Hierarchical hybrid testability modeling and evaluation method based on information fusion. Journal of Systems Engineering and Electronics 26(3): 523–532.

Automotive, Mechanical and Electrical Engineering – Liu (Ed.)
© 2017 Taylor & Francis Group, London, ISBN 978-1-138-62951-6

The dynamic target of Yellow River ice tracking algorithm based on wireless video streaming

Lukai Xu & Shuxia Li
Yellow River Institute of Hydraulic Research, Yellow River Conservancy Commission, Zhengzhou, China
Research Center of Levee Safety and Disaster Prevention Engineering Technology, Ministry of Water Resources, Zhengzhou, China

ABSTRACT: In recent years, the melting ice flood situation of the Yellow River was grim due to the influence of extreme weather. Therefore, ice flood prevention work become very important. Flood control departments could obtain real-time video streams about the Yellow River's ice through remote video surveillance systems. On this basis, by using target tracking technology and subsequent video measurement technology, they could get the parameter of the ice velocity, which is needed by the ice flood hazard prediction model for early warning. However, this paper found that there would be a phenomenon of false targets and loss targets in the practical application by using the traditional methods of target detection and tracking. Also it would directly impact on the calculation precision of the ice velocity. The reasons for this problem are the characteristics of the Yellow River ice images as well as the problems of lost frames in the process of remote wireless video streaming transmission. This article considers that introducing the pyramid structure of the L-K optical flow method, which is based on a strong angular point feature point set, can solve this problem. This article further verifies the effectiveness and robustness of the algorithm through experiments.

Keywords: Ice velocity; The Yellow River; Image recognition; Target tracking

1 INTRODUCTION

A melting ice flood frequently occurs in the Yellow River, which is caused by its special geographical position, hydro-meteorological condition and river channel characteristics. So it is very necessary to get the ice information quickly and accurately. The Yellow River ice remote video surveillance system has been built and covers the important section of the Yellow River. The decision makers on flood control can directly monitor the key section of the Yellow River ice by relying on this system. Unfortunately, the system does not have the function to aid intelligent decision. This paper argues that by using image processing technology and monocular video measuring technology for data processing and analysis of the Yellow River ice images, the demands of automation extraction with ice velocity can be met.

In practice, we found that the traditional tracking algorithm, as well as the ordinary optical flow method, cannot be steady to much ice target tracking or large tracking error occur. The causes of these problems are due to the particularity of the Yellow River ice target, complicated harsh working environment, and the lost frame phenomenon in the process of streaming video remote transmission.

After full analysis, this paper thinks that, based on the strong corner feature point set of the L-K pyramid optical flow method, we can effectively avoid a large amount of false tracking, which was produced by the traditional tracking methods. Also the improved optical flow method can maintain relatively stable and continuous tracking in the face of the complex target background and rapid flow.

2 THE YELLOW RIVER ICE TARGET IMAGE FEATURE ANALYSIS

The image of river ice is more complex than the image of sea ice and higher request to the algorithm. The Yellow River ice image has the following characteristics:

1. The Yellow River ice image has a more complex and changeable target background. As everyone knows, the Yellow River is a sediment-laden river. The different sections of the river have different sediment concentrations and a different geographical environment, so part

of the river ice and the water on the image is indistinguishable.

2. Yellow River ice has characteristic of various type, shape with no regularity, and easy to keep out each other. When ice target collisions produce deformation, we cannot advance to shape modelling.

3. The Yellow River ice target shows, for the most part, a weak target form. A lot of ice is in the form of an ice ball, which is concealed in the water. People could not rely on their edge information or grey information to extract intact ice targets.

4. Due to all the surveillance cameras being arranged in the wild, and easily affected by local climate conditions, such as strong winds, therefore the ice images will appear to have an obvious shake, which is not conducive to sustained and stable tracking.

5. The video streams about the Yellow River ice image will inevitably be affected by lost frame phenomenon in the remote wireless transmission process. The above phenomenon will cause ice targets to appear inconsistent, a phenomenon of large-scale motion in the images.

3 MOVING TARGET DETECTION AND TRACKING ALGORITHM ANALYSED

In terms of moving target detection, the relatively mature motion target detection method includes the background difference method, the frame difference method and the optical flow method. The background difference method is applicable to a fixed or not easily changed background of the application. Due to the background being a changeable river, in this paper using this method is not suitable.

The continuous inter-frame difference method algorithm is simple, easy to implement, and has better adaptability with a dynamic environment. But in the inter-frame difference method obtained, the feature point set is not the most appropriate tracking feature point set for the application of this paper. Through comparative experiments, this paper argues that the strong corner feature point set with ice target tracking effect is significantly better than that of the interframe difference method.

The optical flow method does not need to predict any movement information in the scene. It may be applicable to the paper application.

In terms of the tracking algorithm, based on the expression of moving targets and similarity measure, the target tracking algorithm can be divided into four categories: active contour tracking, region-based tracking, model tracking and feature-based tracking.

Based on the analysis for this paper of the characteristics of the wireless video stream of the Yellow River ice image, the former three kinds of tracking algorithms are not applicable. This paper argues that algorithms that are based on the characteristics of strong angular point set tracking have good distinguishing features, invariance of translation and rotation and scale change. In addition, the feature point set can be obtained quickly. So the algorithm is an ideal ice tracking algorithm.

4 L-K PYRAMID OPTICAL FLOW METHOD BASED ON STRONG CORNER FEATURE POINT SET

4.1 *L-K optical flow computation*

The Lucas-Kanade algorithm (hereinafter referred to as L-K) is one of the main algorithms of the sparse optical flow method. Sparse optical flow method calculating just interested in points (feature points) around the small window are available on the local information of optical flow value, thereby reducing the time cost in the process of calculation.

The L-K algorithm, which is based on local smoothing, is more efficient. This algorithm avoids the fuzzy phenomenon of object borders, which is caused by optical flow smooth in the global. However, the L-K algorithm is based on the continuous time or movement that is "a little movement" on the premise of this assumption, namely time relative to the percentage of target motion in the image is small enough, and this goal in the interframe motion is small, so for this project to study the ice quickly without a coherent movement target happening, simply using the L-K optical flow method for target tracking of the actual effect is not good.

Based on the above reasons, this paper argues that the image pyramid algorithm could effectively solve the problems. If the traditional optical flow method cannot track large movement in the high resolution image layer, it will need the image pyramid to find a suitable low resolution image layer, and compute optical flow, then map the light flow into the original image to realise the optical flow computation for the larger movement.

4.2 *Obtain strong corner feature point set*

This paper chose strong angular points as feature points. A strong angular point is a point of brightness that can change the sharp or edge of the maximum curvature point on the curve.

The angular point is located in the place with the two maximum eigenvalues of the second derivative autocorrelation matrix in the image, proving that there are around this point as the centre of the edge, at least two different directions (or textures), so the problem of effective selection of the angular point is converted to the algebra problems about calculating and selecting larger eigenvalues at corresponding points in the image matrix. The paper algorithm steps to obtain feature points are as follows:

1. Calculate the second derivative autocorrelation matrix about the neighbourhood of the input image pixel, namely image I (x, y) of the two-dimensional Hessian matrix is:

$$H(p) = \begin{bmatrix} \dfrac{\partial^2 I}{\partial x^2} & \dfrac{\partial^2 I}{\partial x \partial y} \\ \dfrac{\partial^2 I}{\partial x \partial y} & \dfrac{\partial^2 I}{\partial y^2} \end{bmatrix} \qquad (1)$$

And then calculate the two corresponding characteristic values of the matrix.

2. Keep the local maximum of the neighbourhood. This pater puts some smaller eigenvalue to compare with the threshold for selecting feature points. If the characteristic value is greater than the threshold, it is considered a strong angular point, otherwise ruled it out.

3. In order to improve the stability and real-time performance of the algorithm, this paper can use distance measure to eliminate some angular points at close range to ensure the selected points for keeping sufficient distance.

4.3 The calculation of optical flow

After the extraction of the strong angular point, mark the angular point position in the image sequence. By calculating the displacement of the feature points between two adjacent frames, get the movement of the optical flow field. Assume that point M, whose grey value was I, in the image, at time t, and its grey value changes slowly with x, y, t, could get the basic equation of optical flow:

$$I_x u + I_y v + I_t = 0 \qquad (2)$$

$u = \frac{dx}{dt}$, $v = \frac{dy}{dt}$ on behalf of optical flow with direction of x, y, $I_x = \frac{\partial I}{\partial x}$, $I_y = \frac{\partial I}{\partial y}$, $I_t = \frac{\partial I}{\partial t}$ represent the partial derivatives about image grey level for x, y, t. Its vector form is:

$$\nabla I \cdot v_M + I_t = 0 \qquad (3)$$

$\nabla I = (Ix, Iy)$ for image gradient at point M, $VM = (u, v)$ is at point M's optical flow. This type is called the optical flow constraint equation, after the extraction of the strong angular point, based on the L-K method to calculate light flow.

4.4 The construction of image pyramid

The Yellow River ice images are based on the remote for 3 g wireless transmission, thus inevitably appear to have the lost frame phenomenon, which results in the phenomenon of the incoherent large-scale movement under the condition of high resolution target images. This phenomenon will not be able to meet the requirements of the L-K method of optical flow constraint equation.

In order to solve the problems, this article adds the image pyramid model to the L-K optical flow method, based on the multi-scale map layered pyramid structure and optical flow transformation for optical flow tracing.

Pyramid structure is a kind of computer vision multi-resolution representation, a different resolution pyramid layer can parse out the different scales of the target. At the same time, the low resolution of the high-level (coarse scale) information can also be used to guide the high resolution analysis of low-rise (fine scale). This article uses the image pyramid structure; the current layer is lower after smoothing down sampling results, and the original image layer number is zero. When the image layer gets to a certain level, the target's displacement between the two frame quantities is small enough and can satisfy the constraint conditions of the optical flow computation, which is used to estimate optical flow.

5 THE TEST RESULTS AND ANALYSIS

In this tracking test, using the inter-frame difference method and the proposed based on strong corner feature point set of L-K optical flow method to the same video data have been carried out. The video data were captured from Cao Jiawan monitoring stations in Inner Mongolia within reach of the Yellow River, the video data is affected by strong winds and produced obvious shaking. The background of the targets changed rapidly. For Matlab2010a test environment, the system hardware is i5 processor, 4 GB memory, the operating system is Windows 7. The test results are shown in Figure 1 and Figure 2:

From Figure 2, we found that using the inter-frame difference method tracking feature point sets successfully in the absolute number is more than the number of feature point sets by using the

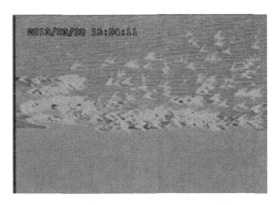

Figure 1. Using inter-frame difference method for tracking.

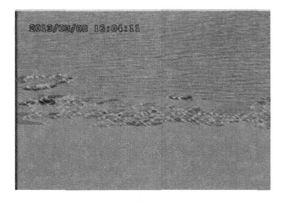

Figure 2. Using the proposed algorithm for tracking.

proposed algorithm. However, many of them were wrong. Because of the complex background, rapid flow of water, and the action of the wind shaking the cameras constantly, there are important interference factors for the inter-frame difference method. In comparison, the proposed algorithm is accurate and stable. The above factors would have no impact on the tracking result.

6 CONCLUSION

The result of the contrast test found the following conclusion. Based on the strong corner feature point set of the L-K optical flow method, this can avoid complex background interference effectively. The method is accurate and effective, therefore, it is suitable for the video monitoring of the Yellow River ice. On this basis, the precise measurements of the Yellow River ice velocity could be obtained.

REFERENCES

Cordelia, S., Roger, M., & Christian, B. (2000). Evaluation of interest point detectors. *International Journal of Computer Vision*, 37(2), 151–172.

Horn, B. K. P., & Schunck, B. G. (1981). Determining opticak flow. *Artificial Intelligence*, 17, 185–203.

Liao, S., & Liu, B. (2010). An edge-based approach to improve optical flow algorithm. *Proc of the 3rd International Conference on Advanced Computer Theory and Engineering*, 45–51.

Meng, W., Guo, Y., & Wang, L. (2010). Ice flood characteristics and prevention measures in Yellow River. *Journal of North China Institute of Water Conservancy and Hydroelectric Power*, 31(6), 27–42.

Muramoto, K., Matsuurak, Endoh, T. (1993). Measuring sea-ice concentration and floe-size distribution by image processing. *Annals of Glaciology*, 1(18), 33–38.

Richard, J. H., Nick, H., & Peter, W. (2002). A systematic method of obtaining ice concentration measurements from ship-based observations. *Gold Regions Science and Technology*, 34, 97–102.

Shi, J., & Tomasi, C. (1994). Good features to track. *Ieee Computer Society Conference on Computer Vision and Pattern Recognition*. IEEE: 593–600.

Wei, B., & Li, J. (2012). Fast optical flow algorithm for large-scale motion. *Application Research of Computers*, 29(9), 3551–3557.

Automotive, Mechanical and Electrical Engineering – Liu (Ed.)
© 2017 Taylor & Francis Group, London, ISBN 978-1-138-62951-6

The on-line fault diagnosis technique for the radar system based on one-class support vector machine and fuzzy expert system theory

M.S. Shao, X.Z. Zhang & G.H. Fan
College of Wuhan Mechanical Technology, Wuhan, China

ABSTRACT: A joint on-line fault diagnosis method based on the one-class Support Vector Machine (SVM) and the fuzzy expert system theory was proposed. To the measured data during the on-line faulty diagnosing, first a one-class SVM is applied to judge whether the device is faulty. Then the fuzzy expert system is applied to reason and accurately locate the fault. The result from the conventional fault diagnosing for certain radar indicates that the proposed method has a diagnosis accuracy higher than 90%. The new method has a high fault location accuracy and real-time processing capability, and improves on-line fault diagnosis capabilities.

Keywords: On-line fault diagnosis; SVM; Fuzzy expert system; Location accuracy

1 INTRODUCTION

The high degree of integration and the complex structure are the main characteristics of the modern radar. Inharsh environments the risk of radar fault increases due to the increase in precision electronic components used in radar systems. The frequent maintenance-support of the equipment is therefore significant. However, the fault diagnosis presently relies on the radar's self-test capability. This conventional fault detecting technique has low accuracy, and it reduces the technicians' repairing efficiency. In order to solve this problem, varieties of on-line fault diagnosis equipment have been developed by means of virtual instrument technology. Encountering the same problem, the equipment's functions are still relatively simple and cannot satisfy the requirements of the fault diagnosis for the radars (Wang Xujing. 2015; Sun Fuan, 2014).

A joint detecting method was successfully used by our work team to design new online radar fault diagnosis equipment. The method is based on the one-class Support Vector Machine (SVM) (Babenko B., 2009) and the fuzzy expert system theory (Shen C. & Lu C., 2007) and will be illustrated in detail in this paper. Whether a radar works with failure will be identified by the one-class SVM theory. Then the fault will be accurately located in a specific unit based on the method of the fuzzy expert system.

2 PRINCIPLE OF THE JOINT FAULT DIAGNOSIS TECHNIQUE

The principle of the detecting method is shown in Figure 1. Whether the radar system works properly is checked by a one-class SVM classification technique. If the system is identified with failure,

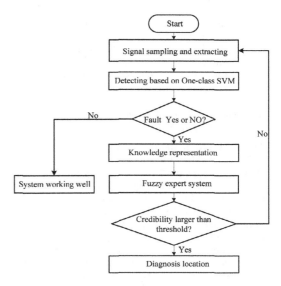

Figure 1. The flowchart of the fault diagnosis.

the fault will be accurately located by the method for a fuzzy expert system. Finally, the disabled work-unit will be repaired or replaced.

2.1 Working state identification based on one-class SVM classification method

As a matter of fact, a radar system can always be divided into several sub-systems according to its real function, the number of which being represented by k. We establish a mathematical model to describe the radar system working state based on one-class SVM theory. Testing data from the normally working radar can be used as the sample vector data to train the model that is training k groups of one-class SVM. Each corresponding group has an output value so as to indicate the normal working state of the radar system. Therefore, a real fault diagnosis problem is changed into a mathematical classification and identification problem by the one-class SVM method.

There are two steps to determine whether a radar on power is faulty, as follows:

Step 1: Using the known sample vector data in training the k groups of one-class SVM. In order to obtain the optimal classifier, the SVM parameters are trained and optimised using the artificial bee colony algorithm.

Step 2: For a new real test and unknown data of the radar, as the input data, it will be sent to the k groups of one-class SVM to determine whether the radar works normally. The radar will be identified working without failure only if all the outputs of the k groups of the SVM are all 1. The value 1 indicates the result is true in logic. Otherwise there exists a fault and the radar system cannot work well.

2.2 Fault location based on the fuzzy expert system

The structure of the fuzzy expert system is shown in Figure 2. The fault location based on the mathematical fuzzy expert system mainly includes three core issues: the knowledge represented, the fuzzy reasoning, and the result confidence calculated.

In order to help the artificial intelligence programs make proper reasoning and decisions, the knowledge representation is significant in designing the data structures in the fuzzy expert system mathematical model. The radar system diagnostics knowledge comes mainly from three aspects. First is the general technical knowledge, such as the design and the principles of the radar system, and the accumulated faulty case of the equipment. Second is the experience and knowledge of the experts. Third is the real measuring data by the

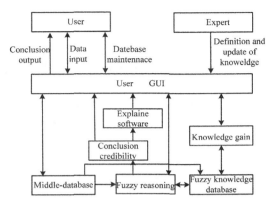

Figure 2.   Structure of the fuzzy expert system.

equipment. The various types of knowledge are mixed together by the method's hierarchical mesh comprehensive knowledge representation and form a frame structure. The frame structure is the actual composition of the radar system. In the overall hierarchical structure, the inner layer mesh nodes connect with each other and the node is the basic unit of knowledge.

The fuzzy reasoning, a method of uncertain reasoning, is used to diagnose the uncertain faulty information from the radar system. The procedures of the fuzzy inference need to follow the fuzzy rules. All the fault data measured from the radar systems are treated as the prerequisites of the rules, and will be rooted in the fault database in a parallel way. With regard to a group of new measured fault data, it may not be possible to directly give more details about the fault; for example, the kind or the location of the fault in the system. Generally, the fault diagnosis and location procedure can be explained as follows. When the calculated confidence value of the measured data is larger than the threshold of a certain fuzzy rule, this rule will be activated. All the activated rules will be superimposed and according to this phenomenon, the confidence value of the rule conclusion will be calculated. Finally, the classification and location of the faults will be succeeded.

The details of the definition of the fuzzy rule and the procedure of the confidence calculation are discussed next.

The general mathematical form of fuzzy rules can be explained as follows:

$$\text{IF} \quad (u_1, P_1) \text{ and } (u_2, P_2) \text{ and } \cdots \text{ and } (u_n, P_n)$$
$$\text{THEN} \quad Q(CF, \alpha) \tag{1}$$

P_i is the measured fault sign, also known as the fault premise feature; it is fuzzy and cannot be directly determined as belonging to which kind of faults. u_i stands for the weight coefficient of the fault sign P_i, and it shows the contribution of P_i to the whole system fault. u_i also obeys the relationship:

$$\sum_{i=1}^{n} u_i = 1 \quad \text{and} \quad u_i \geq 0 \tag{2}$$

Maybe kinds of faults share one fault sign, so the value of u_i is different from the faults; that is, the more contribution, the larger the weight coefficient. The fuzzy conclusion of the rule which always points to the fault, is indicated by Q. The confidence of the rule is CF and $CF \in [0,1]$. α stands for the threshold value. When the confidence of the fault sign, σ, is larger than α, then this corresponding rule will be activated.

As the key point during all the fuzzy reasoning process, there are many ways to calculate the confidence, such as the forward and the backward reasoning (Pan J. Y., 1998). In the following part, taking the generation of fuzzy rules in the fault diagnosis expert system of the radar system as the example, the calculation process of forward-reasoning is discussed (Li W., 2008).

The confidence calculation algorithm for the forward fuzzy reasoning involves the following two aspects:

1. When one rule is activated, there are three mathematical methods to estimate the confidence of the conclusions in fuzzy diagnosis. These methods include the maximum membership method, the weighted average method and the fuzzy distribution method. Fault diagnosis for a radar system solves the problems with a multi-factor comprehensive evaluation and decision-making process. Therefore, the weighted average method was preferentially adopted.

   Assuming β_i is the confidence of the i-th premise feature P_i, then the confidence, σ, of all the premise features in this rule, can be calculated by Equation 3. According to Equation 2, σ can be simplified as presented in Equation 4.

$$\sigma = \frac{\sum_{i=1}^{n} u_i * \beta_i}{\sum_{i=1}^{n} u_i} \tag{3}$$

$$\sigma = \sum_{i=1}^{n} u_i * \beta_i \tag{4}$$

If there is $\sigma \geq \alpha$, this rule will be applied and its conclusion Q has the confidence β calculated by Equation 5.

$$\beta = CF * \sigma = CF * \sum_{i=1}^{n} u_i * \beta_i \tag{5}$$

2. When m rules with the same conclusion are activated, the total confidence of the rules is calculated as follows. All the m rules have the same conclusion, that is:

$$u_{11} * P_{11} \wedge u_{12} * P_{12} \wedge \cdots \wedge u_{1n} * P_{1n} \longrightarrow Q(CF_1, \alpha_1)$$
$$\vdots$$
$$u_{m1} * P_{m1} \wedge u_{m1} * P_{m2} \wedge \cdots \wedge u_{mn} * P_{mn} \longrightarrow Q(CF_m, \alpha_m)$$
$$\tag{6}$$

The premise feature P_{ji} has the confidence β_{ji} and the weight coefficient u_{ji}. There are $i = 1, 2, ..., n; j = 1, 2, ..., m$. If the entire σ_j is larger than the applied threshold α_j indicated by Equation 7, we can obtain the total confidence β of all the rules' conclusions by Equations 8–10.

$$\left. \begin{aligned} \sigma_1 &= \sum_{i=1}^{n} u_{1i} * \beta_{1i} \geq \alpha_1 \\ &\vdots \\ \sigma_m &= \sum_{i=1}^{n} u_{mi} * \beta_{mi} \geq \alpha_m \end{aligned} \right\} \tag{7}$$

$$\beta = \sum_{j=1}^{m} CF_j' * \sigma_j \tag{8}$$

$$CF_j' = \frac{CF_i}{\sum_{i=1}^{m} CF_i} \tag{9}$$

Obviously there is:

$$\sum_{i=1}^{m} CF_i = 1 \tag{10}$$

3 EXPERIMENT RESULTS AND CONCLUSION

A set of fault diagnosis equipment was developed by our work team. The measurement equipment has been successfully applied in fault diagnosis for a certain type of radar. The average correct rate is more that 90% when the equipment was applied to diagnose the faults which originated from the frequency synthesiser and the signal process unit, and even the hydraulic system of the radar. It obviously decreases the time that the technicians spend on radar maintenance-support. The new method proposed in this paper improves the location accuracy and the efficiency of the radar on-line fault diagnosis.

REFERENCES

Babenko B. (2009). Visual tracking with on-line multiple instance learning. *Proceedings of IEEE Conference* on *Computer Vision and Pattern Recognition* (pp. 983–990).

Pan J.Y. (1998). Fuzzy shell: A large-scale expert system shell using fuzzy logic for uncertainty reasoning. *IEEE Transactions on Fuzzy Systems*, *6*(4), 563–581.

Sun F. (2014). Design of the fault diagnosis expert system for a certain type of radar system. *Modern Radar*, *36* (9), 74–78 (In Chinese).

Shen C. & Lu C. (2007). A voltage sag index considering compatibility between equipment and supply. *IEEE Transactions on Power Delivery*, *22* (2), 996–1002.

Wang, X. (2015). Improved LS-SVM based radar fault diagnosis technology. *Fire Control and Command Control*, I(2), 63–65 (In Chinese).

W, L. (2008). Fuzzy models of overhead power line weather-related outages. *IEEE Transactions on Power System*, *23*(3), 1529–1531.

Applied and computational mathematics, methods, algorithms and optimization

Automotive, Mechanical and Electrical Engineering – Liu (Ed.)
© 2017 Taylor & Francis Group, London, ISBN 978-1-138-62951-6

A highly robust power window anti-pinch algorithm based on approximate integral method

Heng Fu & Jianguo Liu
School of Automotive Engineering, Wuhan University of Technology, Wuhan, China
Hubei Key Laboratory of Advanced Technology for Automotive Components, Wuhan, China
Hubei Collaborative Innovation Center for Automotive Components Technology, Wuhan, China

ABSTRACT: Reliability is an important indicator for measuring the anti-pinch function of the power window. Under circumstances such as pit roads, pulse width of the Hall signal is disturbed and the pulse width curve can generate spike noise. Considering the noise, a high reliability anti-pinch algorithm based on approximate integral is proposed. Firstly, the position of the approximate integral of the Hall signal pulse width curve is determined and then the curve is approximately integrated. Finally the integral area and threshold are compared; most of the spike noise can be filtered out, improving the correctness of judgement as to whether to prevent the clip. Using MATLAB to establish the simulation model of the power window, the simulation results show that this algorithm has much more reliability.

Keywords: Electric glass elevator; Anti-pinch control system; Hall signal; Gauss filter; Integration method

1   INTRODUCTION

In 2015, global auto sales reached 82.9 million and sales of power windows in the cars exceeded 300 million. With the widespread adoption of power windows, the number of injuries to passengers (especially to children) caused by power windows is increasing year by year. 74/60/EEC regulations developed by Europe and MVSS18 regulations developed by the United States stipulate requirements of power windows on normal lifting and automatic anti-pinch.

Vehicle driving conditions are complex, because of diverse unpaved roads (such as asphalt road, pebble road and Belgium road), window strip ageing and battery voltage fluctuations, among others. Generally an anti-pinch algorithm based on the Kalman filter is adopted to estimate pinch conditions through torque change rates, which are converted from Hall signal pulse width. This algorithm is prone to make misjudgements of pinch under conditions of large vibration amplitude of the car body and spike noise in the Hall signal.

To solve the problems mentioned above, a highly robust power window anti-pinch algorithm based on approximate integral is proposed. When the Hall signal pulse width increases beyond a certain threshold, approximate integration of the Hall pulse width curve is started. When the integral area value exceeds a certain threshold, the current state is determined as the anti-pinch state. As a result of the approximate integral method, integral area and the work done by the obstruction force maintain a certain relationship; therefore, most spikes caused by vehicle vibration are eliminated. Accordingly the robustness of the anti-pinch algorithm is improved significantly.

2   MATHEMATICAL MODEL OF POWER WINDOW

The structure of the power window is shown in Figure 1. The permanent magnet is mounted on the DC motor output shaft, which is driven to rotate as the DC motor rotates. The Hall sensor and the DC motor output shaft are in the same plane. Each time the permanent magnet sweeps the

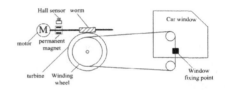

Figure 1.   Structure of the power window.

Hall sensor, the Hall sensor exports a Hall signal pulse. This Hall sensor exports two Hall signals.

In the process of rising, window glass encounters resistance and the rate of rise decreases. Through the mechanical transmission structure, the DC motor output shaft speed decreases, which causes the Hall signal pulse width to increase. Therefore, the encountered resistance of the window glass can be indirectly estimated by analysing the Hall signal pulse width.

2.1 State space model of motor torque

Window motors are generally two-phase brush DC motors. The DC motor voltage balance equation is as follows:

$$\begin{bmatrix} u_a \\ u_b \end{bmatrix} = \begin{bmatrix} R_a & 0 \\ 0 & R_b \end{bmatrix}\begin{bmatrix} i_a \\ i_b \end{bmatrix} + \begin{bmatrix} L_a & L_{ab} \\ L_{ba} & L_b \end{bmatrix}P\begin{bmatrix} i_a \\ i_b \end{bmatrix} + \begin{bmatrix} e_a \\ e_b \end{bmatrix} \quad (1)$$

where u_a, u_b is the phase voltage; R_a, R_b is the phase resistance; i_a, i_b is the phase current; L_a, L_b is the coil self-inductance; L_{ba}, L_{ab} is the coil mutual inductance; P is the differential operator; and e_a, e_b is the back electromotive force.

According to the symmetry of the motor structure:

$$\begin{cases} L_a = L_b = L; & L_{ab} = L_{ba} = M \\ R_a = R_b = R; & i_a + i_b = 0 \end{cases} \quad (2)$$

the DC motor voltage balance equation can be reduced to:

$$\begin{bmatrix} u_a \\ u_b \end{bmatrix} = \begin{bmatrix} R & 0 \\ 0 & R \end{bmatrix}\begin{bmatrix} i_a \\ i_b \end{bmatrix} + \begin{bmatrix} L-M & 0 \\ 0 & L-M \end{bmatrix}P\begin{bmatrix} i_a \\ i_b \end{bmatrix} + \begin{bmatrix} e_a \\ e_b \end{bmatrix} \quad (3)$$

The motor torque equation is:

$$T_e = \frac{P_e}{w}; \quad P_e = \begin{bmatrix} e_a & 0 \\ 0 & e_b \end{bmatrix}\begin{bmatrix} i_a \\ i_b \end{bmatrix} \quad (4)$$

where w is motor angular speed? The coil motion equation is:

$$T_e = T_L + Bw + J\frac{d_w}{d_t} \quad (5)$$

where B is the coil damping coefficient.

It is assumed that the motor speed changes very slowly so that the relationship between the electromagnetic torque of the DC motor, T_e, and the resistance torque of the motor, T_L, can be simplified as:

$$T_e = T_L + Bw \quad (6)$$

2.2 Power window mechanical transmission

The research object of this anti-pinch algorithm is a rope-wheel electric glass elevator. Through the turbo-worm mechanism, the DC motor output shaft will transfer torque to the reel. The motor output shaft and the worm shaft are coaxial. The reel and the turbine are coaxial. The reel is driven by the wire rope to raise and lower the window glass.

The force balance equation of the transmission system is:

$$\begin{cases} (mg + F_f + F)\dfrac{D}{2} = T_L K \\ F = K_N \Delta x \end{cases} \quad (7)$$

where m is the window glass quality, g is the acceleration of gravity, F_f is the friction of the window glass, F is the obstacle clamping force, K_N is the stiffness of the obstacle, Δx is the deformation of the obstacle, D is the diameter of the roll, and K is the worm gearing ratio. Motor angular velocity takes the following form:

$$w = \frac{2\pi}{Pw_d} \quad (8)$$

where P is the number of Holzer signals generated by each motor rotation, and w_d is the Hall signal pulse width.

Worm gear tooth number is Z_1, turbine tooth number is Z_2, and transmission ratio is $K = Z_2/Z_1$. The turbine and reel are coaxial so that the two angular velocity is consistent. Turbine diameter is D_1, and the diameter of the roll is D. The window lift distance corresponding to each Holzer signal takes the following form:

$$d = \frac{Z_1\pi D_2}{Z_2 P} = \frac{\pi D}{PK} \quad (9)$$

Accordingly, the window position can be determined by the number of statistics of the Holzer signal. d is about 0.43 mm, according to the calculation using practical parameters, as shown in Table 1. The calculated window position accuracy

Table 1. Actual parameters of a power window.

Motor parameters		Mechanical part parameters	
e_a	12 V	m	5.3 Kg
i_a	2.4 A	g	9.8 m/s²
e_b	12 V	F_f	5.2 N
i_b	2.4 A	K_N	10 N/mm
P	1	K	11
		D	0.054 mm

492

has met the requirements of this anti-pinch algorithm.

2.3 *Relationship between Hall pulse width and obstacle deformation*

According to the 74/60/EEC and MVSS18 regulations, barrier clamping force cannot exceeding 100 N and barrier stiffness is $(10 + 0.5)$ n/mm when measuring window anti-pinch function. Obstacle deformation is positively correlated with Holzer pulse width. Assuming that the car body does not vibrate ideally, the window encounters obstacles during rising. Taking the formulae from Equations (4), (6), (7), and (8), the relationship between Hall pulse width and obstacle deformation takes the following form:

$$Cw_d^2 - Aw_d - E = \Delta x \qquad (10)$$

where C, A, and E are the two term coefficients, first order coefficients and constant terms on the left side of the formula, respectively:

$$C = \frac{KP}{DK_N}\begin{bmatrix} e_a \\ e_b \end{bmatrix}^T \begin{bmatrix} i_a \\ i_b \end{bmatrix} A = \frac{1}{K_N}\begin{bmatrix} m \\ F_f \end{bmatrix}^T \begin{bmatrix} g \\ 1 \end{bmatrix} E = \frac{4K\pi}{DPK_N}$$

Without obstacles, the power window Hall signal pulse width is about 42 ms. Bringing the parameters in Table 1 into Equation (10), it is clear that the values of the two term and the constant term on the left side of Equation (10) are one order of magnitude higher than the value of the first term, so the linear term can be ignored in Equation (10).

The relationship between Hall pulse width and obstacle deformation can be simplified as followed:

$$Cw_d^2 - E = \Delta x \qquad (11)$$

$$w_d = \sqrt{(E + \Delta x)/C} \qquad (12)$$

3 APPROXIMATE INTEGRATION OF ANTI-PINCH ALGORITHM

Whether the window is moving up automatically is currently estimated by detecting the switch signal of the window. If the window is currently moving up automatically, calculation of the integral area is started when the window enters the anti-pinch zone. As long as $\Delta s > \Delta s_0$, the judgement of the clip is made and window is controlled to drop a certain distance automatically.

3.1 *Selection of the anti-pinch zone*

According to 74/60/EEC and MVSS18 regulations, the anti-pinch zone is the area below the top

of the electric vehicle window and within the area 4–200 mm to the top of the window. According to Equation (9), the electric car window area for the pinch is the area below the top of the electric vehicle window and within the area 9–465 Hall signal pulses to the top of the window.

3.2 *Selection of integral start position*

When obstacles are encountered during automatic ascent, motion of the window glass slows down and the Hall pulse width increases, which shows a slope rising edge on the Hall pulse width curve, as shown in Figure 2.

The position of integration is determined by the difference method. The sum of the first five Hall pulse widths and the next five are first found respectively, and then the difference between these calculated.

$$\Delta w_d = \sum_{n=6}^{10} w_{dn} - \sum_{n=1}^{5} w_{dn} \qquad (13)$$

When the sum of the Hall pulse width exceeds a certain threshold, Δw_0, the first Hall signal of the next five is set as the start position of integration.

$$\Delta w_d > \Delta w_0 \qquad (14)$$

3.3 *Calculation of integral area*

In the uniform rising stage of an electric vehicle window, the Hall signal pulse width curve basically remains unchanged.

In the obstacle holding stage of the power window, the Hall signal pulse width curve rises significantly. When the difference (Δw_d) of the Hall pulse width exceeds a certain threshold (Δw_0), approximate integration is started. If the contrary occurs, approximate integration is ended and Δs is reset. Approximate integration is started when clamping force is 0 N and ended when clamping force is 100 N. The region of integration is the area below the rising edge of the Hall pulse width curve, as shown in the black triangle in Figure 2.

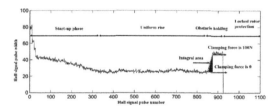

Figure 2.   Integral area of pulse width curve.

493

$$\Delta s = \int_{\Delta w_d > \Delta w_0}^{\Delta w_d \leq \Delta w_0} (w_{dn} - w_{d0}) dn \qquad (15)$$

where w_{d0} is the Hall signal pulse width before starting integration? w_{dn} is the Hall signal pulse width during integration. n is the number of pulses of Hall signal contained in the integral region.

When the integral area, Δs, exceeds a certain area threshold, Δs_0, the judgement of the clip is made; the window is controlled to drop a certain distance automatically.

$$\Delta s > \Delta s_0 \qquad (16)$$

4 ANTI-CLIP JUDGEMENT OF POWER WINDOW

According to 74/60/EEC and MVSS18 regulations, when measuring the window anti-pinch function, the barrier-clamping force shall not exceed 100 N. The calculation of the threshold, Δw_0, of the integral start position and the threshold, Δs_0, of the integral area on the standard condition is based on the barrier clamping force, $F \leq 100\ N$, and barrier stiffness of $(10 + 0.5)$ N/mm.

4.1 Threshold of integral start position

According to Equation (12), when $\Delta x = 0$, the Hall signal pulse width, w_d, is 42 ms; when $\Delta x = F_{\max} / K_N = 10^{-2}\ m$, the Hall signal pulse width, w_d, is 62 ms. Approximate integration is started when the clamping force is 0 N. Approximate integration is ended when the clamping force is 100 N. The number of pulses of the Hall signal contained in the integral region is as follows:

$$\Delta n = \frac{\Delta x}{d} \approx \frac{\Delta x K_1 K}{\pi D} = 24 \qquad (17)$$

For the parameters, refer to Table 1.

The five Hall signal pulse widths before integration started are $w_{d1} \sim w_{d5}$. The five Hall signal pulse widths after integration started are $w_{d6} \sim w_{d10}$, as shown in Table 2. According to Equation (13),

Table 2. Pulse width of the first five Hall signals.

Hall	Width (ms)	Hall	Width (ms)
w_{d1}	42	w_{d6}	43
w_{d2}	41	w_{d7}	44
w_{d3}	42	w_{d8}	45
w_{d4}	43	w_{d9}	45
w_{d5}	42	w_{d10}	46

Δw_d obtained by the difference method is 13. Therefore, the threshold Δw_0 of the integral start position is set as 13.

4.2 Threshold of integral area

The area of integration is:

$$\Delta s = \int_{F=0}^{F=100\,N} (w_{dn} - w_{d0})\, dn = \int_0^{\Delta n} (w_{dn} - w_{d0})\, dn \qquad (18)$$

As n is a discrete random variable, Equation (18) can be converted to:

$$\Delta s = \sum_{n=0}^{\Delta n} (w_{dn} - w_{d0}) \qquad (19)$$

Calculation of the pulse width of Hall signals in the integral region is done according to Equation (12). Δs is 240, according to Equation (19). Therefore, the threshold, Δs_0, of the integral area is 240.

5 SIMULATION AND ANALYSES

A power window simulation model was built to test the highly robust anti-clip algorithm. The simulation model includes a control algorithm, test conditions, Hall signal output, mechanical part of the window and model of the DC motor. The DC motor outputs two Hall signals with a phase difference, so that the control algorithm can detect the phase difference and pulse width of the Hall signal, which can be used to determine whether or not to perform the anti-pinch command.

During test conditions, obstructions are designed in accordance with the European anti-seize standards and the stiffness is 10 N/mm. In the proposed algorithm, the pulse width curve is approximately integrated so that the integral area can be calculated. The pulse width is related to the resistance value, while the integral area is related to the clamping force work.

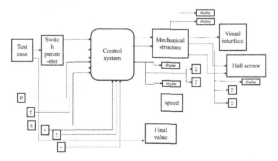

Figure 3. MATLAB model of the anti-pinch window.

5.1 Bumpy road

'Bumpy road' includes Belgium pavement and gravel pavement, among other similar road types. The car body vibration amplitude gets greater on these types of road, which results in bigger noise of the Hall signal pulse width. One of the hallmarks of the Hall signal pulse width curve is the rising edge.

As shown in Figure 4(c), as a result of the integration method, the anti-clamping force mutation caused by the transient tremor is effectively filtered out. Simulation results show that the algorithm proposed can accurately determine the anti-pinch when the car driving on a bumpy road; there is no false anti-pinch and it has good robustness.

5.2 Pit road

When driving in and going through the deceleration zone, the car body generates a larger amplitude of instantaneous vibration and a spike. As shown in Figure 5(c), even under the conditions of a larger instantaneous car body vibration amplitude, the proposed anti-pinch algorithm can still make an accurate judgement and no false anti-pinch. The proposed anti-clipping algorithm has good robustness on pit road.

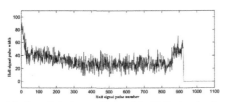

(a) The original waveform of the Hall signal pulse width.

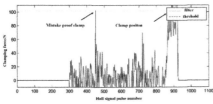

(b) Anti-pinch Judgement of the Kalman Filter Algorithm.

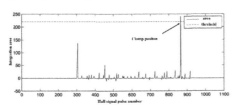

(c) The integral of the algorithm in this paper.

Figure 4. Results of the anti-clip algorithm simulation.

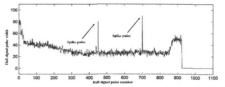

(a) The original waveform of the Hall signal pulse width.

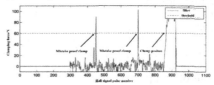

(b) Anti-pinch Judgement of the Kalman Filter Algorithm.

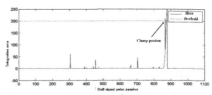

(c) Anti-pinch Judgement of the Approximate Integral Algorithm.

Figure 5. Results of the anti-clip algorithm simulation.

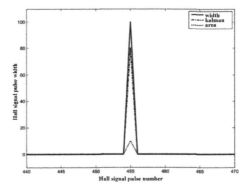

Figure 6. Filtering analysis of spike noise.

5.3 Mathematical analysis of robustness

Figure 6 shows the results of the Kalman filter algorithm and the integration algorithm for spike noise. The core of the Kalman filter algorithm is the best estimate of the next state obtained by the covariance of the measured value and the last estimate. The covariance is continually updated iteratively. The filtering effect of spike noise is not obvious. From Figure 6, the Kalman filter algorithm only slightly reduces the peak value, while the integral value of spike noise is small and the integral rule has a good filtering effect on spike noise.

6 CONCLUSION

An anti-clip algorithm based on an approximate integral is proposed. When the windows encounter obstacles in the automatic rise process, the Hall signal pulse width curve will immediately rise. The bit value at the beginning of the integration is judged by detecting the rising edge, whether or not the integral area has reached the threshold value is used as the judgement criterion as to whether or not to initiate anti-pinch control. The results show that the proposed anti-pinch algorithm based on approximate integral has much higher robustness under the conditions of bumpy road and deep pavement.

REFERENCES

Chlebek, C., Steinbring, J., & Hanebeck, U.W. (2016). Progressive Gaussian filter using importance sampling and particle flow. *IEEE International Conference on Information Fusion, V42,* 2043–2049.

EU Council Directive 74/60/EEC (1973). On the Approximation of the Laws of the Member States Relating to the Interior Fittings of Motor Vehicles [S].

Fu K., Liu, B., & Li E. (2008). Research and design of LIN-Bus based on anti-pinched power window in vehicle [J]. *Journal of Hangzhou Dianzi University, 28*(3), 39–42.

Gui, Z. & Han, F. (2004). Wavelet packet-maximum entropy spectrum estimation and its application in turbineps fault diagnosis [J]. *Automation of Electric Power Systems, 28*(2), 62–66.

Gui-yang F. (2014). *Research on window controlling system of automobile and related anti-pinch algorithm* [D]. Harbin: Harbin Institute of Technology.

Guojun, D., Xiang, Z., Huaixiang Z., Ertao, L., & Hong, Z. (2008). Modeling and Implementation of Power Window Anti-pinch System [J]. *Automotive Engineering, 30*(6), 539–542.

Ra, W.S., Lee, H.J., Park, J.B., & Yoon, T.S. (2008). Practical pinch detection algorithm for smart automotive power window control systems. *IEEE Trans. Industrial Electronics, 55*(3), 1376–1384.

Sollmann, M., Schurr, G., Duffy-Baumgartner, D., & Huck, C. (2004). Anti-pinch protection for power operated features. *SAE Technical Paper Series, 2004-01-1108.*

Wang, Z., Dou, Li., & Chen, J. (2007). An adaptive Gaussian filter with scale adjustable [J]. *Optical Technology, 32*(3), 395–397.

Weizel, M., Zhang, S., & Wang, H. (2008). The design of anti-pinch car windows using hall sensor [J]. *Automotive Engineering, 30*(12), 1122–1124.

Xu J. (2007). Study on approximation theory and application of Gaussian filter [D]. Harbin: Harbin Institute of Technology.

Automotive, Mechanical and Electrical Engineering – Liu (Ed.)
© 2017 Taylor & Francis Group, London, ISBN 978-1-138-62951-6

A probabilistic aircraft conflict resolution method using stochastic optimal control

Shiyu Jia & Xuejun Zhang
School of Electronic and Information Engineering, Beihang University, Beijing, China
Beijing Laboratory for General Aviation Technology, Beijing, China

Xiangmin Guan
Civil Aviation Management Institute of China, Beijing, China

ABSTRACT: With the high economic growth in China, the field of civil aviation transportation has witnessed a rapid development. But the efficiency of current centralized air-traffic management is limited, so billions of yuan was wasted by flight delay and fuel waste every year. It is a significant problem to ensure the aircraft keep safe separation as the solutions must meet the requirements of the feasibility and timeliness, but the existing algorithms can't balance them in actual scenarios. In this paper, an algorithm based on stochastic optimal control is proposed to improve the search capability. The experimental studies which using real UAVs shown that the proposed algorithm provides robustness against uncertainties in the system and is suitable for real applications.

Keywords: Conflict resolution; Stochastic optimal control; Intelligent computation

1 INTRODUCTION

Conflict is one of the main causes of flight delays and flight safety hazards. Aircraft conflict resolution cost the airline industry billions of dollars annually in delays and wasted fuel. In order to improve the airspace capacity and reduce the collision risk, the main task of planners is to plan airspace and air-routes. Hence, research on collision resolution is very significant in airspace planning and air-routes planning.

Probabilistic methods are earliest method which used in this field. Collision risk model for long range parallel air-routes over oceanic areas was provided by Reich in 1966 (Reich P G., 1966). The Reich model had a great influence on estimation of collision risk. The problems of safe separations for parallel air-routes were solved well by the Reich model.

To improve the safety and efficiency of flight, some optimization approaches have been proposed to conflict resolution with the lowest cost (changes of velocity, heading angle or fuel). The conflict resolution problem was formulated as a mixed integer linear program, so that the optimization approaches could be implemented in real time. In (L. Pallottino, 2002), two integer linear optimization models are proposed by allocating aircraft to change speeds or heading angles, but not the

both. Antonio et al. (A. A.-Ayuso, 2010) extend VC model in (L. Pallottino, 2002) and develops a model by allowing aircraft to perform speed and altitude changes. Not all kinds of conflict could be resolved by these approaches. For example, aircraft could not avoid head-to-head conflict by VC model. Meanwhile, all of the resolution maneuvering dimensions including speed change, lateral and vertical should be allowed.

JK Archibald et al. (Archibald J K, 2008) using the satisfaction of game theory to solve the problem of multi-machine conflict resolution, resolution models are changing the heading angle, speed remains constant. To establish a "social relationship" through the method of the conditional probability that the decisions made by the current aircraft impact on other aircraft. Each decision maker (aircraft) in decision-making, will be affected by the decision which made by higher priority aircraft.

Potential field method is Simple and convenient, it has been widely researched recent years. J. Kosecka (Tomlin C, 1998) use this method into conflict resolution field in 1997, basic idea is build motion planning based on potential and vortex field methods to generate conflict resolution maneuver strategy. Artificial potential field method can use relatively simple mechanical equations to obtain a continuous path of conflict resolution. But some adjustment of heading angle is too big in conflict resolution plan, so that real aircraft could not fly like this.

In this paper, a stochastic approximation algorithm based on the Jacobi iteration is proposes to solve the stochastic optimal control problem. This approach is suitable for decomposition of optimization, which is able to reduce the computational complexity. This paper studies the decomposition within each pair of conflicting aircraft, which is able to reduce the computational complexity further for centralized conflict resolution. In the last part of this paper, two real UAVs was used to analysis and verify the effectiveness of the algorithm.

2 PROBLEM DESCRIPTION

2.1 Conflict and conflict resolution

Aircraft conflict refers to the risk of collision with other aircraft during flight. Aircraft collision and aircraft conflict are two different concepts, aircraft collision means two aircraft have physical contact; while aircraft conflict imply that the distance between two aircraft is less than the security separation standard.

As we know the flight status of aircraft i and j, we can predict their flight track in subsequent time Tw according to it. As it shown in Figure 1, the distance between two aircraft at time t is d (t) (i, j):

$$d(t)(i,j) = \sqrt{(x_i(t) - x_j(t))^2 + (y_i(t) - y_j(t))^2} t \in [0, T_w] \quad (1)$$

If within the time Tw the minimum distance between two aircraft dmin (i, j) is less than the security separation, namely dmin (i, j) <dsafe, we can define the aircraft conflict happened.

Aircraft conflict resolution means when two or more aircraft are predicted to be conflict, we to have quickly finding an effective and ideal trajectory to avoid conflict, maintain safety distance between aircraft by changing their heading, speed and height.

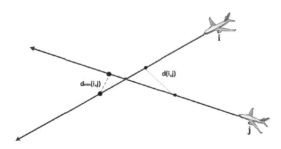

Figure 1.   Example of aircraft conflict.

That means for the any t [0, Tw], inequality d (t) (i, j) > dsafe holds.

2.2 Optimal control using the Jacobi iteration

We use the Jacobi iteration (H. J. Kushner and P. Dupuis, 2001) in this paper. The Jacobi iteration is a kind of gradient descent algorithm whose iterations converge to the optimal solution. For a given cost function:

$$\hat{V}_0(q) = \begin{cases} anyvalue, \forall q \in Q \\ \hat{r}(q), \forall q \in \partial Q \end{cases} \quad (2)$$

Which is the initial condition for the Jacobi iteration, define iterative function $\forall q \in Q$:

$$\begin{cases} \hat{V}_{n+1}(q) = \min_{u \in \mu} \left\{ \sum_{q' \in N(q)} \Pr\{Q_{k+1} = q \mid Q_k = q', \\ \hat{u}_n(q)\} \times \hat{V}_n(q') + \hat{c}(q) \right\} \\ \hat{u}_{n+1}(q) = \arg\min_{u \in \mu} \left\{ \sum_{q' \in N(q)} \Pr\{Q_{k+1} = q \mid Q_k = q', \\ \hat{u}_n(q)\} \times \hat{V}_n(q') + \hat{c}(q) \right\} \end{cases} \quad (3)$$

For a given small number $\delta > 0$, the following inequality can be used as a stopping condition for the formula (3):

$$\max_{q \in Q} \left\| \hat{V}_{n+1}(q) - \hat{V}_n(q) \right\| \leq \delta \quad (4)$$

The optimal feedback control $u^*(x)$ for X_t^μ can be approximated by:

$$u^*(x) \approx \hat{u}^*(q), \forall x \in G(q) \quad (5)$$

Figure 2 shown the control law presented in (5). The aircraft flies through the grids G (q), G (q'), and G (q") centered at x (q), x (q'), and x (q"). For each point, the optimal control input has been computed by the Jacobi iteration. Then, when the aircraft is located in G (q), the optimal control input $u^*(x)$ is constant $\hat{u}^*(q)$. When the aircraft enter G (q), the optimal control input is changed to $\hat{u}^*(q')$ and is constant until the aircraft leaves G (q').

2.3 Experiment environment

The experiments were made in two computers which using Bluetooth serial device for communication. The computer configuration is 4xIntel Core i5-2430M, 2.40 Ghz, 4Gb RAM and windows 7 spl 32 OS. As shown in Figure 3, two

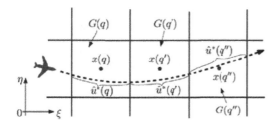

Figure 2. Optimal control at different sections of a trajectory.

Figure 3. UAV used in the experiment.

UAVs were used in this experiment. The UAV's hardware platform using Pixhawk platform, software platform adopts APM control, and the transmission protocol is MAVLINK protocol. Mission Planner 1.3.38 which shown in Figure 4 is used as Ground control station software.

2.4 Experimental field

The experiment was conducted in a school in Beijing, the size of the site is less than 200*200 m. Two drones take off at the same time, and use two computers to control. UAV reports its flight status to the computer in real time, including the location, speed and other information. The computer uses the Bluetooth serial port to obtain the information of the other UAV, and calculating the movement direction for next moment. Thus, the conflict resolution problem is realized by distributed algorithm.

Fig. 5 and Fig. 6 show the flight trajectory of the aircraft 1 and 2 respectively.

2.5 Coordinate transformation

In order to transform the coordinates of the WGS-84 standard into the form of XoY, the Gauss projection formula is needed. Gauss projection formula is shown as follows:

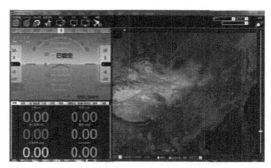

Figure 4. Mission Planner 1.3.38.

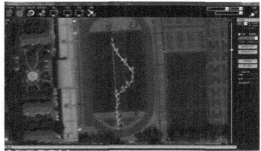

Figure 5. Flight trajectory of aircraft 1.

Figure 6. Flight trajectory of aircraft 2.

$$\begin{cases} x = X + \dfrac{N}{2\rho''^2}\sin B\cos Bl''^2 + \dfrac{N}{24\rho''^4}\sin B \times \\ \qquad \cos^3 B(5-t^2-9\eta^2)l''^4 \\ y = \dfrac{N}{\rho''}\cos Bl'' + \dfrac{N}{6\rho''^3}\cos^3 B(1-t^2+\eta^2)l''^3 + \\ \qquad \dfrac{N}{120\rho''^5}\cos^5 B(5-18t^2+t^4)l''^5 \\ t = \tan B, \eta^2 = e'^2\cos^2 B, \rho'' = 206265 \end{cases} \quad (6)$$

Assuming that the latitude and longitude coordinates of two points are (b, l) and (b', l'). The coordinates of two points in the XoY coordinate system are (x, y) and (x', y'). Since $\cos(B - B') \approx 1, \cos(L - L') \approx 1$, bring into formula (6), the obtained as followed:

$$\begin{cases} (x - x') \approx R\cos B \sin(L - L') \approx \dfrac{\pi R(l - l')\cos B}{180} \\ (y - y') \approx R\sin(B - B') \approx \dfrac{\pi R(b - b')}{180} \end{cases} \tag{7}$$

where R ≈ 6371393 m is the radius of the earth. The experiment was conducted in 40 degrees north latitude, so π take as 3.1416. So the formula (7) can be simplified as:

$$\begin{cases} (x - x') \approx 85181(l - l') \\ (y - y') \approx 111202(b - b') \end{cases} \tag{8}$$

Take southwest corner of the playground as the origin of the XoY coordinate system, the trajectory of the aircraft can be obtained by coordinate transformation as it shown in Figure 7.

The distance between two aircraft is visible in Figure 8. It can be seen from Figure 8 that the

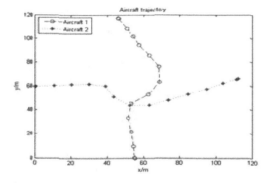

Figure 7.   Aircraft trajectory in XoY system.

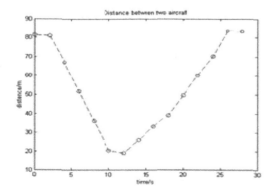

Figure 8.   Distance between the two aircrafts.

minimum distance between two aircraft is 20 m, no conflict risk. The experimental studies which using real UAVs shown that the proposed algorithm provides robustness against uncertainties in the system and is suitable for real applications.

3 CONCLUSION

This paper has considered the problem of aircraft conflict resolution, the stochastic optimal control was applied to solving the conflict resolution problem, which has the capability of incorporating various and uncertainties control strategy. The optimal control problem has been solved a stochastic approximation algorithm based on the Jacobi iteration. At last part of this paper, the real time and reliability of the distributed algorithm are verified by two real UAVs. So it can be concluded that, the algorithm can solve the problem of aircraft conflict resolution well in the real environment where wind and measurement uncertainties is existence.

ACKNOWLEDGEMENTS

This work was supported by the National Natural Science Foundation of China (Grant No. U1433203, Grant No. U1533119), and the Foundation for Innovative Research Groups of the National Natural Science Foundation of China (Grant No. 61221061).

REFERENCES

Archibald J K, Hill J C, Jepsen N A, et al. A satisficing approach to aircraft conflict resolution [J]. Systems, Man, and Cybernetics, Part C: Applications and Reviews, IEEE Transactions on, 2008, 38(4): 510–21.

Ayuso, A.-A., L.F. Escudero and F.J.M.-Campo, "Collision Avoidance in Air Traffic Management: A Mixed-Integer Linear optimization Approach," IEEE Trans. Intell. Transp. Syst., 2010.

Kushner H.J. and P. Dupuis, Numerical Methods for Stochastic Control Problems in Continuous Time, 2ed. New York, NY, USA: Springer-Verlag, 2001.

Pallottino, L., E. Feron, and A.Bicchi, "Conflict resolution problems for air traffic management systems solved with mixed integer programming," IEEE Trans. Intell. Transp. Syst., vol. 3, no. 1, pp. 3–11, Mar. 2002.

Reich P G. Analysis of long-range air traffic systems: separation standards—I [J]. Journal of Navigation, 1966, 19(01): 88–98.

Tomlin C, Pappas G J, Sastry S. Conflict resolution for air traffic management: A study in multiagent hybrid systems [J]. Automatic Control, IEEE Transactions on, 1998, 43(4): 509–21.

Automotive, Mechanical and Electrical Engineering – Liu (Ed.)
© 2017 Taylor & Francis Group, London, ISBN 978-1-138-62951-6

A study on wind speed probability distribution models used in wind energy analysis

Y.D. Li, Y.T. Sun, K. Li & M. Liu
State Grid Shandong Electric Power Research Institute, Jinan, China

ABSTRACT: The wind speed probability distribution is one of the most important wind characteristics for the performance of wind energy conversion systems and for the assessment of wind energy potential. Several kinds of wind speed probability distribution models widely used in wind energy analysis are studied in this paper. The structural form, the estimation method and the simulation approach of different models are introduced. The accuracy and simulation speed are compared using data from four wind sites. The applicability of different models on reliability evaluation of power system including wind energy is verified with the standard IEEE reliability test system (IEEE-RTS79).

Keywords: Wind speed model; Probability distribution; Probability density function; Wind farm

1   INTRODUCTION

Wind power has been the fastest growing renewable energy source in the last few years due to its non-exhaustive nature and environmental benefits. The global installed capacity of wind power has reached 432.4 GW up to the end of 2015. The wind speed model is a prerequisite for wind energy research and utilisation. However, the wind is a kind of physical phenomena affected by many factors, and it is difficult to realise the accurate prediction of wind speed. Most of the wind speed models can only mimic the major statistical properties of the recorded data, such as the probability distribution characteristics and the auto-correlation characteristics (Klockl B. & Papaefthymiou G. 2010; Papaefthymiou G. & Klockl B., 2008; L D. & Goic R. & Krstulovic J., 2010; Li Y. & Xie K. & Hu B., 2013.). Studies (Safari B. & Gasore J., 2010; José A. & Penélope R. & Sergio V., 2008; Chang T. P., 2011; Akpinar S. & Akpinar E., 2009) show that the probability distribution characteristics are the most important piece of information needed in the wind energy study, and the wind speed models based on probability distribution can meet the requirements of the wind energy research in most situations.

In the past two decades, researchers have been devoted to developing a large number of statistical models to describe wind speed probability distribution. Three kinds of models are widely used in the engineering, namely one-component distribution model, mixture distribution model and Kernel Density Estimation (KDE) model. The one-component distribution model is a kind of model composed of one probability density function. A mixture distribution function is a mixture of two or several component distributions. A KDE model is a non-parametric density estimation methods which does not make any assumptions on the underlying wind speed distributions, and has applications in load uncertainty analysis (Zhao Y. & Zhang X. & Zhou J., 2010) and renewable energy sources study (Yan W. et al., 2013; Qin Z. & Li W. & Xiong X., 2011).

In this paper, the research progress of probability distribution models for wind speed is introduced. The structure and principle of wind speed probability distribution models are summarised. The accuracy and applicability of each model is illustrated. The drawbacks of different probability distribution models are analysed.

2   WIND DISTRIBUTION MODELS

2.1   *One-component distribution model*

The one-component distribution model is a kind of model composed of one Probability Density Function (PDF). The Weibull function is a commonly used one-component distribution model for fitting the measured wind speed data.

Let $V = (v_1, v_2, \ldots, v_n)$ denote wind speed data samples and follows Weibull distribution. Its PDF can be expressed as:

$$f(v) = \left(\frac{k}{c}\right) \times \left(\frac{v}{c}\right)^{k-1} \exp\left[-\left(\frac{v}{c}\right)^k\right] \qquad (1)$$

where v stands for the wind speed random variable, c is the scale parameter and k is the shape parameter. When k = 2, the Weibull distribution becomes the Rayleigh distribution, which can be expressed as:

$$f(v) = \left(\frac{v}{c^2}\right)\exp\left(-\frac{v^2}{2c^2}\right) \qquad (2)$$

2.2 Mixture distribution model

A mixture distribution function is a mixture of several component distributions. The PDF of a k-component finite mixture distribution can be expressed as:

$$f(v\,|\,\boldsymbol{\Theta}) = \sum_{j=1}^{m} w_j f_j(v\,|\,\boldsymbol{\theta}_j) \qquad (3)$$

where the parameters $\boldsymbol{\Theta} = (w_1,\ldots,w_m, \boldsymbol{\theta}_1,\ldots,\boldsymbol{\theta}_m)$ are such that $w_j > 0$ and $\sum_j w_j = 1$. w_j is a mixing weight and $f_j(v|\boldsymbol{\theta}_j)$ is a component density function parameterised by θj. m is the number of components. When m = 2, the mixture distribution becomes a two-component mixture distribution, which can be expressed as:

$$f(v\,|\,\boldsymbol{\Theta}) = w_1 f_1(v\,|\,\boldsymbol{\theta}_1) + (1-w_1) f_2(v\,|\,\boldsymbol{\theta}_2) \qquad (4)$$

The Weibull-Weibull (W-W) mixture distribution is commonly used for fitting the measured wind speed data, which can be expressed as

$$f(v) = \omega \left(\frac{k_1}{c_1}\right)\left(\frac{v}{c_1}\right)^{k_1-1}\exp\left[-\left(\frac{v}{c_1}\right)^{k_1}\right] \\ + (1-\omega)\left(\frac{k_2}{c_2}\right)\left(\frac{v}{c_2}\right)^{k_2-1}\exp\left[-\left(\frac{v}{c_2}\right)^{k_2}\right] \qquad (5)$$

It can be seen that W-W model, which has five parameters, is more complex than the one-component distribution model. The W-W model can be estimated using the maximum likelihood method by:

$$LL = \sum_{i=1}^{n} \ln\left(wf(v_i;c_1,k_1) + (1-w)f(v_i;c_2,k_2)\right) \qquad (6)$$

2.3 Kernel density estimation model

A. Basic principle of kernel density estimation

Let $V = (v_1, v_2,.., v_n)$ represent a wind speed sample data with PDF $f(v)$. Its Kernel Density Function (KDF) $\hat{f}(v)$ can be expressed by (Prakasa B.L.S., 1983; Silverman W., 1986):

$$\hat{f}(v) = \frac{1}{nh}\sum_{i=1}^{n} K\left(\frac{v-v_i}{h}\right) \qquad (7)$$

where n is the size of wind speed sample, h is the smooth parameters and $K(\cdot)$ is a known kernel function.

When $n\to\infty$, $h\to 0$ and $nh\to\infty$, $\hat{f}(v)$ is converged to $f(v)$ with probability one. The performance of a KDE depends on the kernel function and smooth parameters. Reference indicates that the selection of kernel function is insensitive in KDE. When n is larger enough, the kernel function has a limited influence on the evaluated results. Therefore, it generally chooses a kernel function as a certain type for KDE.

B. Selection of smooth parameter

The selection of smooth parameter is critical to KDE. The $\hat{f}(v)$ is not only related to sample size, but also refers to the function of observation point v. Studies show that the optimal smooth parameter can be calculated by the following formula:

$$h_{AMISE} = \left[\frac{R(K)}{\mu_2(K)^2 R\left(f''(v)\right) n}\right]^{1/5} \qquad (8)$$

It can be seen from Equation 8 that the optimum smooth parameter h_{AMISE} contains a second derivative f″(v) of the unknown function f(v), so the calculation of the second derivative must be carried out before calculation of smooth parameter. Silverman (1986) presents a simple approach on calculating smooth parameter, which is to take a normal function as the reference distribution to the unknown function $f(v)$, and achieve $h_{AMISE} = 1.06\hat{\sigma}n^{-1/5}$.

C. Wind speed simulation

The PDF originated from KDF is extremely complicated, which means that direct sampling would face huge difficulties. This paper presents an accept-rejection sampling approach to generate wind speed data. The basic idea of the proposed technique is as following:

Construct a reference probability density function $g(v;\,\alpha)$ (α represents model parameter vector), and make it approaching to wind speed KDF in form and satisfying $f(v) \leq Cg(v;\,\alpha)$, where C is a constant factor with $C \geq 1$. The sampling process can be illustrated as following steps:

a. Generate a random sample u from U (0, 1) uniform distribution;
b. Generate a random sample v from the reference probability density function $g(v;\,\alpha)$;
c. Test whether the conditions $u < f(v)/(C \times g(v;\,\alpha))$ are satisfied:

If satisfied, accept v as a sample for f(v); otherwise, reject v, and repeat the simulation steps a)c).

The simulation of wind speed based on the KDF can be performed through the above procedures.

3 RESULTS AND DISCUSSION

3.1 *Information of wind speed data*

The actual hourly wind speeds at four wind sites (designated as A, B, C and D) for three years were used to illustrate the establishment, accuracy and applicability of the models. Weibull, Rayleigh, W-W model, and KDE model are used to fit the actual speed data. The calculation results of different model parameters of the four sites are shown in Table 1.

3.2 *Accuracy analysis*

The comparison between the four models including the Weibull, Rayleigh, W-W model, and KDE model, is made using the statistic R_a^2 and the results are shown in Table 2. The PDF plot of wind speed data for the four sites using the histogram estimation, Weibull, Rayleigh, W-W model, and KDE are illustrated in Figure 1.

It is shown that the PDF of station A and B have one hump and all the test probability functions match well with the observed histogram. According to the Ra² summarised in Table 2, the discrepancy of statistical errors among different models is basically not very significant for station A and B.

It can be seen that the wind regimes show two humps on the PDF plot for station C and D, and the W-W and the KDE match still very well with the observations and present relatively smaller statistical errors, while the Weibull and Rayleigh model seems to be unsuitable for describing the wind speed data of station C and D.

3.3 *Simulation speed analysis*

The inverse-transformation method is used to simulate the one-component model and two-component model. The approach proposed in section 2.3.3 is used to simulate KDE model. Five million wind speed samples are generated ten times using Matlab and the simulation time is achieved by the average of the ten times. The simulation times of different models are shown in Table 3.

It can be seen that the one-component model shows the least simulation time costing while the KDE model needs the longest simulation time. The time consumed by KDE is far more than the one-component model mixture component model, which limits the application of this method

3.4 *Application analysis*

The applicability of different models on reliability assessment of power system including wind power are investigated using the IEEE-RTS79 (Billiton R. & Kumar S., 1989) by the addition of a wind farm with 400 MW capacity. The cut-in, the rated, and the cut-off wind speed of wind turbine generators are 3.3 m/s, 10.6 m/s, and 22.2 m/s, respectively. The reliability study on the test system is conducted using KDE, Weibull model, Rayleigh, and W-W model. Table 4 and Table 5 show the reliability evaluation results.

Table 4 and Table 5 show that the reliabiliy indices including the Loss Of Load Expectation (LOLE) and Loss Of Energy Expectation (LOEE) exhibit great differences, which implies that the selection of an appropriate probability function would be of importance for wind power assessment. The differences between the KDE model and W-W model are not significant, which means that the two models have a similar performance.

Table 2. Calculation results of R_a^2 of different models.

Sites	KDE	Weibull	W-W	Rayleigh
A	0.9989	0.9119	0.9801	0.9106
B	0.9994	0.9753	0.9784	0.9285
C	0.9929	0.8062	0.9936	0.7867
D	0.9963	0.8353	0.9891	0.8333

Table 1 Calculation results of model parameters.

Sites	KDE	Weibull		W-W					Rayleigh
	h	c	k	c_1	k_1	c_2	k_2	w	c
A	0.32	9.24	1.99	10.38	2.23	5.10	3.35	0.78	6.54
B	0.33	8.25	1.84	9.11	2.05	3.97	2.58	0.83	5.96
C	0.30	8.68	1.93	10.75	3.15	3.66	2.22	0.70	6.18
D	0.33	11.10	2.01	12.91	4.14	10.23	1.72	0.28	7.84

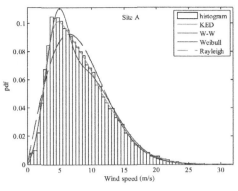

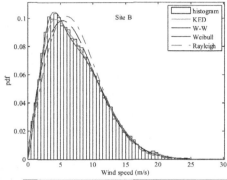

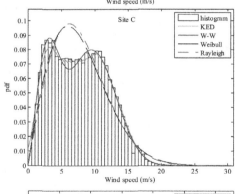

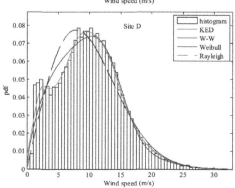

Figure 1.   PDF plots of wind speed for different sites.

Table 3.   Simulation time of different models.

Model	Rayleigh	Weibull	W-W	KDE
time (sec.)	0.8	1.4	2.4	119

Table 4.   LOLE results using different models (h/a).

Sites	KDE	Weibull	W-W	Rayleigh
A	5.76	5.60	5.72	5.57
B	6.23	6.18	6.19	6.05
C	5.63	5.97	5.61	5.89
D	4.55	4.81	4.56	4.81

Table 5.   LOEE results using different models (MWh/a).

Sites	KDE	Weibull	W-W	Rayleigh
A	702.11	677.51	699.81	670.50
B	756.21	749.34	752.11	730.61
C	682.45	717.12	689.32	715.34
D	546.32	576.54	545.33	576.60

4   CONCLUSION

In this paper, three kinds of probabilistic distribution of wind speed that have been proposed in the scientific literature related to renewable energies are discussed. The structural form, the estimation method and the simulation method are introduced. The accuracy, the simulation speed and the applicability are illustrated using actual wind data. The conclusions are summarised as follows:

a. One-component model is suitable to wind speed with unimodal mode, while mixture component model and KDE model are flexible to any wind regime.
b. The KDE model comsumes a large amount of simulation time which is far more than one-component model mixture component model.
c. The reliability indices calculated with KDE and mixture component model present relatively smaller difference, which means the performance differences between the two modes are not significant.

ACKNOWLEDGEMENTS

This research was financially supported by the National Science Foundation (NO.61503216).

REFERENCES

Akpinar, S. & Akpinar, E. (2009). Estimation of wind energy potential using finite mixture distribution. *Energy Conversion and Management. 50*, 877–884.

Billiton R., Kumar, S. et al. (1989). A reliability test system for educational purposes-basic data. IEEE Transactions on Power Systems, *4*(3), 1238–1244.

Chang, T. P. (2011). Estimation of wind energy potential using different probability density functions. *Applied Energy*, *88*(5), 1848–1856.

Carta, J. A., Ramírez, P., & Velázquez, S. (2008). Influence of the level of fit of a density probability function to wind-speed data on the WECS mean power output estimation. *Energy Conversion and Management. 49*(10), 2647–2655.

Jakus, D., Goic, R. & Krstulovic, J. (2011). The impact of wind power plants on slow voltage variations in distribution networks. *Electric Power Systems Research, 81*(2), 589–598.

Klöckl, B. & Papaefthymiou, G. (2010). Multivariate time series models for studies on stochastic generators in power systems. *Electric Power Systems Research, 80*(3), 265–276.

Li, Y., Xie, K. & Hu, B. (2013). Copula function-based dependent model for multivariate wind speed time series and its application in reliability assessment. *Power System Technology*, *37*(3), 840–846.

Papaefthymiou, G. & Klöckl, B. (2008). MCMC for wind power simulation. *IEEE Transactions on Energy Conversion*, *23*(1), 234–240.

Prakasa, B. L. S. (1983). *Nonparametric function estimation*. London: Academic Press.

Qin, Z., Li, W., & Xiong, X. (2011). Estimating wind speed probability distribution using kernel density method. *Electric Power Systems Research. 81*(12), 2139–2146.

Safari, B. & Gasore, J. (2010). Statistical investigation of wind characteristics and wind energy potential based on the Weibull and Rayleigh models in Rwanda. *Renewable Energy*, *35*(12), 2874–2880.

Silverman, B. W. (1986). Density estimation for statistics and data analysis. London: Chapman and Hall.

Yan, W., Ren, Z., Zhao, X. et al. (2013). Probabilistic photovoltaic power modeling based on nonparametric kernel density estimation. Automation of Electric PowerSsystems, *37*(10), 35–40.

Zhao, Y., Zhang, X., & Zhou, J. (2010). Load modeling utilizing nonparametric and multivariate kernel density estimation in bulk power system reliability evaluation. *Proceedings of the CSEE. 29*(31), 27–33.

Automotive, Mechanical and Electrical Engineering – Liu (Ed.)
© 2017 Taylor & Francis Group, London, ISBN 978-1-138-62951-6

A tunnel inspection robot localisation algorithm based on a wireless sensor network

Lin Zhang & Ronghui Huang
Shenzhen Power Supply Bureau Co. Ltd., Shenzhen, China

Lingfen Zhu & Yikun Tao
Zhejiang Guozi Robotics Co. Ltd., Hangzhou, China

Senjing Yao
Shenzhen Power Supply Bureau Co. Ltd., Shenzhen, China

Hongbo Zheng
Zhejiang Guozi Robotics Co. Ltd., Hangzhou, China

Bo Tan
Shenzhen Power Supply Bureau Co. Ltd., Shenzhen, China

ABSTRACT: The tunnel inspection robot has been studied widely. The localisation of the inspection robot is very important when an accident happens. Localisation technology based on a wireless sensor network can be used in this situation. In this work, the Fast KNN algorithm is proposed to obtain the real-time position of the inspection robot. The inspection robot is based primarily on a fingerprint localisation algorithm. The experimental results indicate that the proposed algorithm is superior to the traditional NN and KNN algorithms.

Keywords: Tunnel inspection robot; Localisation technology; Fast KNN algorithm; Fingerprint localisation

1 INTRODUCTION

With the development of modern cities, landscape space and public resources are increasingly important. Electric transmission lines and electric equipment occupies much of the living space of a city. In order to solve the problem of power demand and city living space, the power sector deploys the electric transmission lines under the ground. The tunnel cable of China has achieved great progress in the construction of urbanisation. However, the environment of tunnel cable is very complex, and causes aging and corrosion of the lines. This potential risk may cause a serious fire disaster that spreads quickly and burns violently. It is hard to fight fire while the accident happens.

In order to reduce the risk of fire disaster, the power sector should inspect the tunnel lines frequently. Nowadays, the major of the inspection style is based on manpower verbosely. It is not only inefficient, but also very dangerous. Hence tunnel inspection robots take on more and more importance in intelligent electric inspection (Victores et al., 2011; González et al., 2009). Firstly, inspecting the tunnel lines using an inspection robot can avoid injuries for electric workers. The robot can inspect the phenomena of water seepage, aging lines, corrosion and damage. Secondly, the study of the tunnel inspections root can guarantee the reliable transmission of underground cables.

When finding the risk in the tunnel using an inspection robot, we should locate the position of the inspection robot and then send the information to the control centre. Hence the localisation of the inspection robot is also a key issue. For outdoor localisation, GPS has been used widely, but it works extremely well in indoor space that cannot receive the signal of the GPS satellites. To solve this problem, localisation technology based on a wireless sensor network has been researched widely (Wu et al., 2015), such as infrared, Ultrawideband (UWB) (Zhang et al., 2006), ultrasonic, Bluetooth and Radio-Frequency Identification (RFID) (Yang et al., 2014), ZigBee and Wireless Fidelity (WiFi) (Sun et al., 2014) localisation technology.

Besides, there are many different measurement methods proposed in the indoor localisation system, such as Time Of Arrival (TOA) (Horiba, Manato, et al. 2014), Time Difference Of Arrival (TDOA) (Han et al., 2010) and Received Signal Strength (RSS) (Xiong, 2015). RSS is the first choice for indoor localisation due to its low calculated cost.

In our work, the inspection robot realises the localisation function based on a fingerprint localisation algorithm that includes two stages: offline measurement and online localisation. In the offline measurement stage, the fingerprint information at every reference point should be measured before the localisation stage. In the online localisation stage, the real-time position information is estimated by the proposed Fast k Nearest Neighbor (Fast KNN) algorithm.

The remainder of this work is organised as follows: Section 2 discusses the related works on the localisation algorithm. Section 3 describes the Fast KNN algorithm. The experimental results and analysis are listed in Section 4. Finally, Section 5 concludes the work.

2 RELATED WORKS

Although there is a plentiful of indoor localisation algorithm applied in the different field. According to the different localisation theory, the algorithm can be divided into two categories: triangle and fingerprint localisation algorithms.

2.1 Triangle localisation algorithm

The triangle localisation algorithm (Yang & Chen, 2009) is based on the signal attenuation model to estimate the distance between the sending device and the receiving device. The traditional attenuation model has been used widely, as follows:

$$P_r = P_0 - 10n \lg \frac{d}{d_0} + X$$

where P_0 indicates the power measured at d_0 (1 metre) distance; d is the distance between sending device and receiving device; the path attenuation exponent is expressed by n; X is a Gaussian random variable.

The RSS includes the distance information. In order to obtain the position information of the unknown point, three known points are used to calculate the distance. The principle can be described in Figure 1.

In Figure 1, the A, B and C is the known point, and the (x_1, y_1), (x_2, y_2), (x_3, y_3) is the coordination of known points. The distance from the unknown

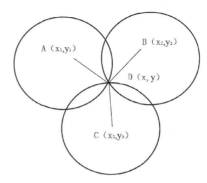

Figure 1. The principle of triangle localisation algorithm.

point D to known points is d_1, d_2 and d_3. So the coordination of unknown point D (x, y) can be calculated as follows:

$$\begin{aligned} (x - x_1)^2 + (y - y_1)^2 &= d_1^2 \\ (x - x_2)^2 + (y - y_2)^2 &= d_2^2 \\ (x - x_3)^2 + (y - y_3)^2 &= d_3^2 \end{aligned}$$

2.2 Fingerprint localisation algorithm

The fingerprint localisation algorithm (Rahman et al., 2012; Sorour et al., 2012) includes two stages: offline measurement and online localisation stage. In the offline measurement stage, the fingerprint map is built at every reference point. The RSS of reference points is received in the localisation region. The match algorithm is processed in the online localisation stage (Esposito & Ficco, 2011; Ismail et al., 2016). The traditional match algorithm is k Nearest Neighbor (KNN) (Mirzaei Azandaryani, 2013). The k points are used to calculate the coordination unknown point.

3 FINGERPRINT LOCALISATION BASED ON FAST KNN ALGORITHM

In our work, the Fast KNN is proposed as a match algorithm. After building the fingerprint map, the inspection robot can use the proposed algorithm to obtain the real-time position. The detailed algorithm is described as follows:

Algorithm: Fast KNN
Input: fingerprint; the real-time RSS x;
Output: optimal localisation result;

1: m is the level of fingerprint map; p is the number of node m; L is the final level of fingerprint map.

2: function obtain the optimal node (m,p,x)
3: select all the subsequent nodes of the p-st node, and add into list.
4: for i = 0 to list.size
5: if D(x,M_list)>B+r_list
6: move the point i out from the list
7: end
8: min = D(x,M_list) the smallest node
9: end
10: if list.size = = 0
11: m = m-1
12: if m = = 0
13: stop calculation
14: end
15: end
16: if m ! = L
17: obtain the optimal node (m+1,min,x)
18: else
19: obtain the coordination of min, return the average value
20: end

4 EXPERIMENTAL RESULTS AND ANALYSIS

We simulate the experiment in an indoor corridor that is similar to the tunnel environment. The total dimension of the corridor is 10.6*2.2. The corridor is equipped with WiFi. Thirty reference points are selected to evaluate the performance of localisation accuracy and speed.

In this work, we compare the Fast KNN against the Nearest Neighbor (NN) and KNN algorithms to evaluate the localisation accuracy. The experimental results are shown in Figures 2 and 3.

From Figures 2 and 3, we can see that Fast KNN achieves the optimal comprehensive performance. Not only the localisation accuracy, but also the localisation speed of the proposed algorithm achieves the best performance.

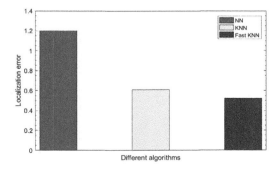

Figure 2. The accuracy performance comparison of different algorithms.

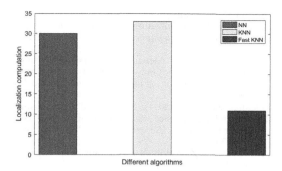

Figure 3. The localisation computation comparison of different algorithms.

5 CONCLUSION

In this paper, we propose the Fast KNN algorithm for tunnel inspection robot localisation. It can give the position information of the inspection robot to the control centre in real-time while the accident happens. The experimental results demonstrate that the Fast KNN achieves the superior localisation performance. In further work, we plan to use the proposed algorithm in a real tunnel environment, and improve the performance of the proposed algorithm further.

REFERENCES

Esposito, C., & Ficco, M. (2011). Deployment of RSS-based indoor positioning systems. *International Journal of Wireless Information Networks, 18*(4), 224–242.
González, J.C. et al. (2009). Robot-aided tunnel inspection and maintenance system. *International Symposium on Automation and Robotics in Construction,* 629–636.
Han, T., Lu, X., & Lan, Q. (2010). Pattern recognition based Kalman filter for indoor localization using TDOA algorithm. Applied Mathematical Modelling, *34*(10), 2893–2900.
Horiba, M. et al. (2014). A study on improvement of NLOS detection accuracy in TOA-based indoor localization. IEICE Technical Report. ASN Ambient Intelligence & Sensor Networks, 113.
Ismail, A.H. et al. (2016). WiFi RSS fingerprint database construction for mobile robot indoor positioning system. *IEEE International Conference of Systems, Man and Cybernetics.*
Mirzaei Azandaryani, S. (2013). *Indoor localization using Wi-Fi fingerprinting.* Dissertations & Theses—Gradworks.
Rahman, M.S., Park, Y., & Kim, K.D. (2012). RSS-based indoor localization algorithm for wireless sensor network using generalized regression neural network. *Arabian Journal for Science and Engineering, 37*(4), 1043–1053.

Sorour, S., Lostanlen, Y., & Valaee, S. (2012). RSS based indoor localization with limited deployment load. *Global Communications Conference IEEE*, 303–308.

Sun, Y., Liu, M., & Meng, Q.H. (2014). WiFi signal strength-based robot indoor localization. *IEEE International Conference on Information and Automation*, 250–256.

Victores, J.G. et al. (2011). Robot-aided tunnel inspection and maintenance system by vision and proximity sensor integration. *Automation in Construction, 20*(5), 629–636.

Wu, C., Yang, Z., & Liu, Y. (2015). Smartphones based crowdsourcing for indoor localization. *IEEE Transactions on Mobile Computing, 14*(2), 444–457.

Xiong, J., Sundaresan, K., & Jamieson, K. (2015). ToneTrack: leveraging frequency-agile radios for time-based indoor wireless localization. *International Conference on Mobile Computing and Networking ACM*, 537–549.

Yang, J., & Chen, Y. (2009). Indoor localization using improved RSS-based lateration methods. *IEEE Conference on Global Telecommunications*, IEEE Press, 1–6.

Yang, Z., & Lei, C. (2014). Adaptive fitting reference frame for 2-D indoor localization based on RFID. *6*(2), 114–118.

Zhang, C. et al. (2006). Accurate UWB indoor localization system utilizing time difference of arrival approach, 515–518.

Automotive, Mechanical and Electrical Engineering – Liu (Ed.)
© 2017 Taylor & Francis Group, London, ISBN 978-1-138-62951-6

Behavior-learning based semi-supervised kernel extreme learning machine for classification

Haibo Yin, Junan Yang & Yunxiao Jiang
Electronic Engineering Institute, Hefei, China

ABSTRACT: Semi-supervised Extreme Learning Machine is proved to be an efficient algorithm for multiple semi-supervised classification tasks. However, the weights connecting the input layer and the hidden layer, and the bias vector of the hidden layer are generated randomly, making the classifier unstable. Besides, an improper choice of heat kernel parameter will badly influence the performance of the classifier when constructing the Laplacian matrix. In this paper, we introduced kernel theory to eliminate the uncertainty of input weights and adopted human behavior learning strategy to make it easier to find proper heat-kernel parameters. The simulation experiments show that the behavior-learning based semi-supervised extreme learning machine achieves a better classification performance compared with Semi-supervised Extreme Learning Machine.

Keywords: extreme learning machine; kernel function; semi-supervised learning; behavior learning

1 INTRODUCTION

As we have entered a new era of big data, it's too costly to label all the data in practice. But it is a waste if we only take account of the limited labeled data while throwing away those unlabeled ones. Semi-supervised classification methods are proposed which utilizes both labeled and unlabeled data at the same time, such as the transductive SVM (TSVM) (Joachims T, 1999) and the Laplacian SVM (LapSVM) (Mikhail Belkin et al, 2004). With the manifold assumption, they manage to find out the structure relationship between labeled data and the unlabeled data and construct the optimal separate hyperplane using both of them so as to improve the classification performance. But they have to face the problem of high computation burden, especially for a multi-classification application. Semi-Supervised ELM (SSELM) (Huang, G et al, 2014) is proposed with the introduction of manifold learning in the framework of ELM (Huang, Guang Bin 2015), where ELM is a novel neural network without iterative training process. Compared with those SVM based semi-supervised algorithms such as TSVM and LapSVM, SSELM can achieve similar or even better classification performance with less time. However, the parameter selection for heat kernel function during Laplacian matrix construction is very important and SSELM directly chooses the parameter by experience. We introduced human behavior learning strategy (X.J Zhu et al, 2010) (Khan, Faisal et al, 2011) so as

to guide the parameter selection. In addition, since the weights connecting the input layer and the hidden layer and the biases of the hidden layer are generated randomly, the performance tends to be unstable. In this paper, we replaced the randomly generated weights and biases of SSELM by the different kernel functions, inspired by the concept of Kernel Extreme Learning Machine (Huang, Guang Bin, 2015), to make the classification more stable.

2 BEHAVIOR LEARNING BASED SEMI-SUPERVISED KERNEL ELM

2.1 *Semi-supervised kernel ELM*

As demonstrated in (Huang, G et al, 2014), the objective function of SSELM is shown in Equation (1).

$$\mathbf{f} = \begin{cases} \mathbf{G}_o(\mathbf{I}_{L \times L} + \mathbf{G}^T\mathbf{C}\mathbf{G} + \lambda\mathbf{G}^T\mathbf{L}\mathbf{G})^{-1}\mathbf{G}^T\mathbf{C}\hat{\mathbf{T}} & L_1 > N \\ \mathbf{G}_o\mathbf{G}^T(\mathbf{I}_{L \times L} + \mathbf{C}\mathbf{G}\mathbf{G}^T + \lambda\mathbf{L}\mathbf{G}\mathbf{G}^T)^{-1}\mathbf{C}\hat{\mathbf{T}} & L_1 < N \end{cases} \quad (1)$$

where L_1 and N denote the number of labeled training instances and all training instances, $\mathbf{G}_o$ and $\mathbf{G}$ are the output matrix of the hidden layer when instances of training set and testing set are taken as input, respectively. $\mathbf{L}$ is the Laplacian matrix and $\hat{\mathbf{T}}$ is the label matrix with respect to instances of the training set. λ, $\mathbf{C}$ are parameters specified by the user. More information can be found in the work of Huang et al. (Huang, G et al, 2014). Motivated by Kernel Extreme Learning Machine (KELM) which utilizes kernel function for data mapping to reduce the instability during network training, we rewrite the objective function in Equation (1), as shown in Equation (2).

$$\mathbf{f} = \begin{bmatrix} K(\mathbf{x},\mathbf{x}_1) \\ \vdots \\ K(\mathbf{x},\mathbf{x}_L) \end{bmatrix}^T \left(\mathbf{I}_{L\times L} + \mathbf{C}\Omega_{ELM} + \lambda \mathbf{L}\Omega_{ELM} \right)^{-1} \mathbf{C}\hat{\mathbf{T}} \tag{2}$$

where $K(\mathbf{x}_i,\mathbf{x}_j)$ is called the kernel function which maps $\mathbf{x}_i$ and $\mathbf{x}_j$ onto a higher dimensional space, making instances easier to be separated. As a result, the fluctuation caused by randomly generated input weights will not influence the training process.

2.2 Behavior-learning based parameter selection method for heat kernel function

It is quite clear that when we get to know the world, part of the knowledge comes from teachers and books, but more of it comes from the deduction and judgment based on the knowledge that has been learnt, using the old knowledge to explore the unknown so as to generate new knowledge. After doing that, when we meet new problems, the old knowledge as well as the new one can be used together to solve the new problems. That is to say, the process that we human understand the world is also semi-supervised. Thus, if we can use the approaches and patterns of human to understand the world in machine learning for classification, the performance may be improved.

When we calculate the Laplacian matrix $\mathbf{L} = \mathbf{D} - \mathbf{W}$, the similarity matrix $\mathbf{W}$ is needed to be calculated first. Usually, we take heat kernel function to calculate the elements in $\mathbf{W}$. But for heat kernel function, there is a parameter δ that should be carefully chosen. An improper δ will badly affect the performance of SSELM for classification. Naturally, when we human manage to solve a problem, the problem is not taken in isolation. We always try to find the connections between different items and then find out the rules in depth so as to solve the problem. Based on this fact, Zhang et al. (Zhang, C et al, 2014) proposed a local behavior learning based parameter searching strategy to find a proper parameter δ by calculating the local behavior learning parameter σ by redefining the heat kernel function between Instance $\mathbf{x}_i$ and Instance $\mathbf{x}_j$.

$$w_{ij}(k) = \exp\left(\frac{-\left\| \mathbf{x}_i - \mathbf{x}_j \right\|^2}{2\sigma_i(k)\sigma_j(k)} \right) \quad i,j = 1,2,...,L \tag{3}$$

In Equation (3), $\sigma_i(k) = d(\mathbf{x}_i, \mathbf{x}_k)$ is called the local behavior parameter, which measures the distance between the sample $\mathbf{x}_i$ and the sample that is the k-th nearest sample $\mathbf{x}_k$ in Euclidean Space, where k is user specified. The introduction of the local behavior parameter σ utilizes the probability distribution information around $\mathbf{x}_i$ and $\mathbf{x}_j$, so as to improve the classification performance.

2.3 Behavior learning based semi-supervised kernel ELM

Combine with the ideas above, we propose a Behavior Learning based Semi-supervised Kernel ELM (BLSSKELM).

The effectiveness of the algorithm will be tested by experiments in the next section.

3 EXPERIMENTS

In order to testify the effectiveness of the proposed algorithm, we have used some multi-labeled UCI datasets (iris, wine and air) to conduct some experiments so as to compare the classification efficiency of kernelELM (Huang, Guang Bin, 2015), SSELM (Huang, G et al, 2014) and BLSSKELM. All the simulation experiments were carried out in MATLAB R2010a environment running in a machine with an Xeon E7500 1.87 GHZ CPU and a 16 GB RAM.

3.1 Comparison for classification accuracy

For each dataset, we randomly choose 3/5 of the total instances for training, 1/5 for testing and the rest for parameter selection. Different labeled instance numbers are set. The Gaussian kernel function is chosen as the kernel function for kernelELM and BLSSKELM while the kernel function parameter is set to be 1 by experience. 1000 is set to be the number of hidden nodes of SSELM. Other parameters are set by the parameter selecting set where $\lambda, C \in [10^{-6}, 10^{-5}, ..., 10^5, 10^6]$ and $K = [5,10,20,30]$. Repeat the process for 20 times, compare the classification accuracy of kernelELM, SSELM and BLSSKELM for each testing set.

From Table 2, we can conclude that:

1. The more labeled training samples, the better classification accuracy.
2. Compared with the all-supervised kernelELM algorithm, SSELM and BLSSKELM can use unlabeled training samples efficiently and improve the classification performance.
3. BLSSKELM outperforms SSELM under the same condition.

512

Table 1. main steps of LBSSKELM.

Input: labeled training samples $((\mathbf{X}_l)_{L_1 \times M_1}, (Y)_{L_1 \times 1})$, unlabeled training samples $(\mathbf{X}_u)_{L_2 \times M_1}$, testing samples $(\mathbf{X}_t,)_{L_1 \times M_1}$, tradeoff parameter C and λ, local behavior parameter k;
Output: $(\mathbf{Y}_t)_{L_1 \times M_1}$, which are the output labels of $(\mathbf{X}_t,)_{L_1 \times M_1}$

Main steps:

1. Calculate the similarity matrix $\mathbf{W}$ of the training samples $\begin{bmatrix} (\mathbf{X}_l)_{L_1 \times M_1} \\ (\mathbf{X}_u)_{L_2 \times M_1} \end{bmatrix}$ with the local parameter k;

2. Calculate the Laplacian matrix $\mathbf{L} = \mathbf{D} - \mathbf{W}$, where $\mathbf{D}$ is a diagonal matrix whose diagonal elements are $D_{ii} = \sum_{j=1}^{L} w_{ij} \ (i = 1, 2, ..., L)$;

3. Choose a proper kernel function $K(\mathbf{x}, \mathbf{y})$ with proper parameters;

4. Calculate $\boldsymbol{\beta}^* = (\mathbf{I}_{L \times L} + C\boldsymbol{\Omega}_{ELM} + \lambda \mathbf{L}\boldsymbol{\Omega}_{ELM})^{-1} \mathbf{C}\hat{\mathbf{T}}$, where:

$$\boldsymbol{\Omega}_{ELM} = [K((\mathbf{X}_{l+u})_i, (\mathbf{X}_{l+u})_j)]_{L \times L}$$

$$i, j = 1, 2, ..., L, \mathbf{C} = \begin{bmatrix} C & & & \\ & C & & \\ & & \ddots & \\ & & & C \end{bmatrix};$$

5. Calculate the output when the testing samples are used as input of the network:

$$\mathbf{f} = \begin{bmatrix} K(\mathbf{x}, \mathbf{x}_1) \\ \vdots \\ K(\mathbf{x}, \mathbf{x}_L) \end{bmatrix}^T \boldsymbol{\beta}^* = \begin{bmatrix} K(\mathbf{x}, \mathbf{x}_1) \\ \vdots \\ K(\mathbf{x}, \mathbf{x}_L) \end{bmatrix}^T (\mathbf{I}_{L \times L} + C\boldsymbol{\Omega}_{ELM} + \lambda \mathbf{L}\boldsymbol{\Omega}_{ELM})^{-1} \mathbf{C}\hat{\mathbf{T}};$$

where $\mathbf{x} \in \mathbf{X}_t$, $\mathbf{x}_i \in \mathbf{X}_{l+u}, i = 1, ..., L \ (L = L_1 + L_2)$.

Table 2. Classification accuracy comparison with different numbers of labeled training samples.

	2 labeled instances			5 labeled instances			10 labeled instances		
	kernelELM	SSELM	BLSSKELM	kernelELM	SSELM	BLSSKELM	kernelELM	SSELM	BLSSKELM
Iris	79.3%	86.7%	91.7%	85.3%	87%	95%	84.7%	89%	98.3%
Wine	88.3%	85.3%	90.7%	91.3%	75.7%	93%	89%	83.3%	96.3%
Air	55.4%	58%	65.6%	73.8%	73.4%	74.4%	78.4%	82.3%	84%

3.2 Comparison with classification time

Next we compare the training time and the testing time between SSELM and BLSSKELM. We randomly choose 5 samples in the training set as the labeled training samples and the rest is considered as the unlabeled training samples, other parameters are set as Experiment 1 does. The experiment is repeated for 20 times, the average training time and testing time of BLSSKELM is shown in Table 3.

From Table 3, we can conclude that:

1. The training time and the testing time of both algorithms will increase when the number of samples or the number of feature dimension increases.
2. Generally, the training time and the testing time of BLSSKELM is less than that of SSELM under

Table 3. the comparison of training time and testing time of SSELM and BLSSKELM.

	SSELM		BLSSKELM	
	Training time (ms)	Testing time (ms)	Training time (ms)	Testing time (ms)
Iris	15.0	1.7	3.4	0.4
Wine	19.7	3.0	5.4	0.4
Air	116.3	5.6	31.0	2.5
X8D5K	794.0	10.2	248.4	5.3

the same condition. This is due to the fact that a lot of hidden neurons are added for SSELM so as to improve the classification performance.

4 CONCLUSION

In order to overcome the instability of SSELM caused by randomly generating weights between the input layer and the hidden layer, and the improper parameter selection for Laplacian matrix construction, this paper introduces kernel function and human behavior learning so as to bring up the BLSSKELM algorithm. The simulation experiments show that the classification accuracy of BLSSKELM is better than that of SSELM under the same condition, as well as less training and testing time. The next research is to analyze the effectiveness of the algorithm the theoretically.

ACKNOWLEDGEMENTS

This work is supported by National High-tech R&D Program (863 Program), and Anhui Provincial Natural Science Foundation (NO.1308085QF99, NO. 140 8085MKL46).

REFERENCES

Huang, G, Song, S, Gupta, J.N.D, et al. Semi-supervised and unsupervised extreme learning machines. IEEE Transactions on Cybernetics, 44(12) (2014), 2405–2417.

Huang, Guang Bin. "What are Extreme Learning Machines? Filling the Gap Between Frank Rosenblatt's Dream and John von Neumann's Puzzle." Cognitive Computation, vol. 49, no. 22, pp. 263–278, 2015.

Joachims T. Transductive inference for text classification using support vector machines. Proceedings of the 16th International Conference on Machine Learning. (Bled, Slovenia, 1999) 200–209.

Khan, Faisal, X. Zhu, and B. Mutlu. "How Do Humans Teach: On Curriculum Learning and Teaching Dimension." Advances in Neural Information Processing Systems (2011): 1449–1457.

Mikhail Belkin, Niyogi P. Semi-Supervised Learning on Riemannian Manifolds[J]. Machine Learning, 56(1–3) (2004) 209–239.

UCI Machine Learning Repository [http:// archive. ics. uci. edu/ml]. Irvine, CA: University of California, School of Information and Computer Science.

Zhang, C; Yang, JN; Zhang, JY; Li, DS; Yong, AX. Semi-Supervised Learning by Local Behavioral Searching Strategy. Applied Mathematics & Information Sciences, 8 (4)(2014) 1781.

Zhu, X.J., B.R. Gibson, K.S. Jun, Cognitive Models of TestItem Effects in Human Category Learning, ICML, (2010).

Automotive, Mechanical and Electrical Engineering – Liu (Ed.)
© 2017 Taylor & Francis Group, London, ISBN 978-1-138-62951-6

Big data storage and processing method on the coal mine emergency cloud platform

L. Ma, S.G. Li & S.C. Tang
Xi'an University of Science and Technology, Xi'an, Shaanxi, China

B.F. Yi
Computer Science Department, Salem State College, Salem, MA, USA

ABSTRACT: With the data explosion comes the need for big data storage and processing in digital coal mine. This paper proposed a frame of big data management based on coal mine emergency cloud platform and a method of big data storage and processing with the use of HBase and Hive. On this platform, an abnormal model of environment parameter in coal mine safe production was designed and implemented. As a test instance of this model, a big monitoring data from Huang Ling Coal Mine was used to store with HBase in distributed database and processed with Hive in distributed warehouse. By comparing the real-time monitoring value with the empirical value obtained from the model analysis, it was judged whether the value is abnormal. The result shows that the cloud computing technology in the construction of the digital coal mine is not only a useful practice, but also has important practical significance and research value.

Keywords: Coal mine; Big data; Cloud computing; Distributed storage; Distributed processing

1 INTRODUCTION

In recent years, cloud computing technology has made tremendous development under the joint promotion of industry and academy, which is changing the way people think using computers and creating unprecedented social computing power, leading the information service into a new mode (Chen Kang & Zheng Weimin, 2009) (Zhang Jianxun, et al., 2010). Domestic emergency management experts believe that computing sharing integration platform should be built through the usage of cloud computing and cloud storage technologies to meet the increasingly complex analysis requirements of the cross-disciplinary, cross-regional emergency, providing more intelligent and more efficient simulation analysis services (Zhang Hui, 2012) (Li Congdong, et al., 2011). The modern coal mine informatization has produced vast amounts of data resources. Therefore, mining companies need to create new idea and mode on emergency management according to the changes in demand. As a service-oriented computing model, cloud computing is the latest stage of the development of information technology, which will provide new ideas and strong support for the emergency management of coal mine (Guo Deyong, 2009).

2 BIG DATA MANAGEMENT FRAMEWORK OF COAL MINE CLOUD PLATFORM

Mine emergency cloud (Ma Li, et al., 2013) is the base platform and the environment for the future coal mine emergency management informatization.

This paper adopts hierarchical design, the system uses HDFS as the underlying file system, on the basis of which the Hadoop cloud computing platform for the storage and processing of the big data is built. The huge storage and processing power the coal mine emergency cloud platform needs to provide big data mining service is distributed to each node of the Hadoop cluster, making full use of the distributed framework features of Hadoop cluster. The computing power and the storage capacity at the bottom is called transparently by the appropriate interface at the top.

3 IMPLEMENTATION METHODS FOR THE STORAGE AND PROCESSING OF THE BIG DATA ON COAL MINE EMERGENCY CLOUD PLATFORM

From engineering or technical point of view, how to store, analyze and mining the big data to solve practical problems is the core to big data. SQL on Hadoop thus has become an important tool to query and analyze TB/PB level data in the era of big data. Currently, Hive is the most commonly used SQL on Hadoop solution in big data processing and warehouse building for the Internet companies. Hadoop cluster is deployed in many companies not to run the native MapReduce program but to execute the Hive SQL query.

The paper integrates the advantages of HBase technology and Hive technology. When the data set gets updated, HBase is used to ensure the data to be fast inserted. When the big data gets processed, it can be queried using HiveQL like SQL by importing the data from HBase to Hive.

4 THE IMPLEMENTATION OF THE VARIATION MODEL OF MINE SAFETY AND PRODUCTION OPERATING ENVIRONMENT PARAMETER

4.1 Analysis on the variation model of mine safety and production operating environment parameter on the coal mine emergency cloud platform

The change in coal mine safety production operating environment parameters is an obtained empirical value through the analysis of long-term

1. Calculate the average value of current hour

$$\bar{A}_{(n)hour} = \frac{\bar{A}_{(n-1)hour} * Count_{n-1} + \bar{A}_n}{Count_{n-1} + 1} \quad (1)$$

$\bar{A}_{(n)hour}$ is an average value of the n hour, $Count_{n-1}$ is the cumulative number of times of the (n−1)th hour. The current inspection cycle of the mine environment parameters is 1/0 s (that is, the number of cumulative sensor inspection is 120 times per hour). Therefore, $\bar{A}_{nhour}$ is the average value of the 120 times sampling in the current nth hour, and the cumulative number of times and the present value of $\bar{A}_{nhour}$ are cleared at the end of the hour after the statistics, starting to calculate the average value of the next hour.

2. Calculate the average value of the month

$$\bar{A}_{(n)month} = \frac{\bar{A}_{(n-1)month} * Count_{n-1} + \bar{A}_{(n)hour}}{Count_{n-1} + 1} \quad (2)$$

At the end of each hour, calculate $\bar{A}_{(n)month}$, which is the average value of the nth hour of the current month, $\bar{A}_{(n)hour}$ is the average value of the nth hour.

3. Calculate the empirical value

$$\bar{A}(n)empirical = \frac{\bar{A}_{(n-3)month} * Count_{n-3} * w_3 + \bar{A}_{(n-2)month} * count_{n-2} * w_3 + \bar{A}_{(n-1)month} * count_{n-1} * w_1 + \bar{A}_{(n)month} * count_n}{Count_{n-3} * w_3 + Count_{n-2} * w_3 + Count_{n-1} * w_1 + Count_n} \quad (3)$$

environmental parameters in the working place. The results can be obtained through the comparison with the empirical value. When the mine production environment parameters change, in the meanwhile, there's no alarm but the change rate is relatively large, the paper take this situation as the exception handling, which need for timely reminder of regulators to prevent accidents.

The algorithm model uses the average value of the historical data for a period of time as the variation reference value to compare with the current value and set a certain extent of changes and duration as the auxiliary reference, setting different weights to the calculation of the data from different time periods (He Yaoyi, 2007).

The cumulative time: Every day is divided into several stages, the average value of each stage is cumulative, take per hour for a tentative stage (i.e. 24 times a day).

The change model of Huang Ling coal mine production environment parameter is as follow:

$\bar{A}_{(n)empirical}$ is the empirical value of the nth hour, which is obtained by the weighted average of the first three months, the first two months, last month and the current month. W3 is the weight of the first three months, W2 is the weight of the first two months, and W1 is the weight of the last month. The average value of the first three months has less influence on the empirical value of the hour than that of the last month, thus the common situation is W3 <W2 <W1. $\bar{A}_{(n)empirical}$ is stored into the database per hour.

The determination of the parameter change

IF (*(1+S%)) and ($T_{lower\ limit} < T_{continue} < T_{upper\ limit}$)
Then $V_{present}$ = usual $\qquad (4)$

S is the floating range of the parameters that the mine managers set according to the actual situation. $T_{continue}$ represents the duration of the parameter, $V_{present}$ represents the real-time data collected by sensors every 30 seconds. When the value is greater than the hourly empirical floating range and duration is within the scope of an exception period set manually, it is determined that the data is abnormal.

"Scenario-Response-Contingency" handling mechanism.

$$Min(A_{manual}, \overline{A}_{(n)empirical}) \qquad (5)$$

When the empirical value $\overline{A}_{(n)empirical}$ that the variation model calculates is high, the "Scenario-Response-Contingency" handling mechanism can be applied. The empirical value is set by the expertise manually experience according to the actual situation of the monitored area. Take Min (A_{manual}) as the practical empirical value to determine whether the monitoring parameters change

4.2 The design and implementation of the variation model of the coal mine production parameter in the cloud environment

In this paper, the variation model of the coal mine production environment parameters is extended from a single one to the Hadoop cloud computing platform. The data that the coal mine safety production monitoring system has uploaded can be acquired and stored through HBase. The statistical data analysis of the model based on Hbase can be fast implemented by the HQL query language that Hive has offered.

1. HBase database design

All kinds of sensor sampling data from subordinate coal mines can be stored in the storage system. When new data is received, put command in HBase is used to insert data into the data table and transfer parameters like table name, RowKey, nomen: column name, the values to HBase.

2. The work process of the Hive data warehouse

Hive data warehouse creates the table and loads data by HQL language that is similar to SQL, the table name, the field and the metadata that have been generated are stored in the Derby or MySQL database. When a client needs to carry on the data analysis, Hive executes metadata query from Derby or MySQL database according to the content of the request, thus further querying the table and the its attributes in the corresponding data file directory to get the attribute values that meet the conditions of the query. For big data, hive contains query, analysis, statistics and report generation operation, n returning the generated data processing results back to the client.

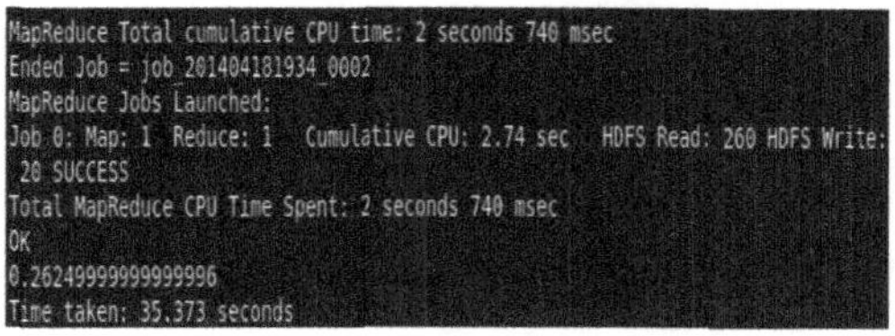

Figure 1. Data processing results of Hive.

Hive executes the query operation through the external connection to the coaldatahbase that has been established in Hbase after the integration of Hbase and Hive. The command is as the following:

Create external table coaldatahive (value float, time string) partitioned by (coalsensorid) STORED

BY 'org. apache. hadoop. hive. hbase. HBaseStorageHandler'

WITH SERDEPROPERTIES ("hbase. columns. mapping" = "sensor: value, sensor: status")

TBLPROPERTIES ("hbase.table. name""coaldatahbase");

3. Write UDAF class of Hive

There are only a few dozen embedded function in HiveQL. For the variation model of the coal mine production environment parameters, the function of the algorithm cannot be realized just through the existing functions. Therefore, UDAF written by Java is used to realize the algorithm of the model.

The calculation result shows = 0.2625. As shown in Figure 1.

When the value of the real-time data collected by every 30 seconds exceeds the value of to a certain range, and the duration is within some specified period of time, it is determined that the value is the variation value, appropriate warning against the variation value should be made.

5 DATA WAREHOUSE BI SYSTEM PERFORMANCE TEST ON THE COAL MINE EMERGENCY CLOUD PLATFORM

The paper continue to use 50 monitoring sites of different orders of magnitude data collected from Huang Ling coal mine within 1 year, like 110 thousand, 1 million, 3.5 million, 10.36 million, 51.84 million to test the effect that different numbers of nodes of the data work on the data processing performance of the coal mine emergency cloud platform. The experiment takes three nodes and six nodes as the cluster size of the test.

517

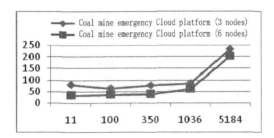

Figure 2. The comparison of the data processing performances of different Hive cluster sizes on the coal mine emergency cloud platform.

First, different order of magnitude of data is imported into the three nodes and six nodes coal mine emergency cloud platform in the hive distributed data warehouse. Then aggregation operation is executed upon the data to test the data processing performances of different cluster sizes. The comparison of the performances is shown in Figure 2. Among them, the abscissa is the amount of data (million), the ordinate is the processing time (s) that hive database executes the average calculation.

6 SUMMARY

The storage and processing of the big data on the coal mine emergency cloud platform has become a tough problem for the construction of the coal mine emergency management informatization. How to store and process the increasing structured, semi-structured and unstructured big data to provide scientific and rational decision-making at all stages of the coal mine emergency management determines the informatization level of coal mine emergency management.

ACKNOWLEDGMENTS

The paper is supported by the Science and Technology Special Project of the Education Department in Shaanxi Province (No. 14 JK1486), and Ph.D Research Startup Foundation of Xi'an University of Science and Technology (No. 6310115032).

REFERENCES

Chen Kang, Zheng Weimin, Cloud Computing: Examples of systems and research status. Journal of software, 2009. 20 (5): 1337–1348.

Guo Deyong, Mine Emergency Rescue Information Management System banse on Oracle. Journal of University of Science and Technology Beijing, 2009 (3): 281–284.

He Yaoyi, The design and implementation of Coal digital remote integrated monitoring system, 2007. Nanjing University of Science and Technology.

Li Congdong, Xie Tian and Liu Yi, Emergency cloud-New mode of Intelligent Emergency Managemen. China Emergency Management, 2011 (5): 27–32.

Ma Li, Li Shugang, Tang Shancheng, Research on the construction of coal mine accident emergency rescue resources. Coal cloud engineering, 2013. 45 (6): 122–125.

Wang Dewen, Xiao Kai, Xiao Lei, Power device status information data warehouse based on Hive. Power system protection and control, 2013 (9): 125–130.

Zhang Hui, Basic scientific issues and integration platform on national emergency platform system based on "scenario response". Systems engineering theory & practice, 2012 (5)947–953.

Zhang Jianxun, Gu Zhimin and Zheng Chao, The research progress of cloud computing. The research and application of computer, 2010 (02): 429–433.

Automotive, Mechanical and Electrical Engineering – Liu (Ed.)
© 2017 Taylor & Francis Group, London, ISBN 978-1-138-62951-6

Clustering based random over-sampling examples for learning from binary class imbalanced data sets

Shi Chen, Zhiping Huang & Xiaojun Guo
School of Mechatronics and Automation, National University of Defense Technology, Changsha, P.R. China

ABSTRACT: The data imbalance problem has become a challenge in many real-life classification applications. Although numerous synthetic over-sampling techniques have been put forward to alleviate this problem, most of them do not consider the distribution of the minority examples and may generate noisy synthetic minority examples which overlap the majority examples. In this regard, an improved synthetic over-sampling algorithm, named Clustering Based Random Over-Sampling Examples (CBROSE) algorithm, for balancing the binary class data sets is presented in this paper. CBROSE generates synthetic minority examples by combining Kmeans clustering algorithm with the basic mechanism of existing synthetic over-sampling methods. The synthetic minority examples created by CBROSE always be located in an elliptical area centered at the observed minority example. The experimental results based on 5-folder cross validation show the effective-ness of CBROSE on some real-life data sets in terms of AUC.

Keywords: Imbalanced classification; Machine learning; K-means clustering; Synthetic over-sampling

1 INTRODUCTION

Imbalanced data exists widely in practical application domains, such as financial fraud detection (Wei et al. 2013), network intrusion detection (Tesfahun & Bhaskari 2013), spam detection, text classification (Tan 2005) and in all circumstances in which the examples of one class (majority class) outnumber that of another class (minority class). The performance of most standard machine learning algorithms can be badly affected by the problem of data imbalance, because there are not enough minority samples for training the classifier. In most of these application domains, people are more interested in the recognition of the minority. For example, the number of legal transactions is far more than that of the fraudulent in credit card fraud detection. Because the numbers of two types of transactions are severely skewed, fraudulent transactions will be easily identified as legal by the classifiers. If a lawful transaction is misclassified as fraudulent, it can be corrected easily with additional manpower. On the contrary, if a fraudulent transaction is misclassified as legal, the economic losses would be many times of the former.

As a result, a number of researchers have focused attention on solving the data imbalance problems in recent years. The state-of-the-art research techniques to settle class imbalance can be categorized as: data level techniques (Chawla et al. 2002, Zheng et al. 2015, Menardi & Torelli 2014, Qazi & Raza 2012), cost sensitive learning (Ting 2002, Zhou & Liu 2006) and ensemble techniques (Chawla et al. 2003, Seiffert et al. 2010). In this paper, we only follow with interest data level techniques. Therefore, this paper makes a brief overview on the data-level techniques only, the details of the other two types of techniques can be found in the related literature (Peng et al. 2014).

Data-level techniques are independent of the classification algorithm, but to balance the class distribution based on resampling methods which can be split into two categories: under-sampling and over-sampling. The former alleviates class imbalance by deleting some majority examples or by applying parametric and nonparametric techniques based on statistical theory (Yen & Lee 2006), and the later balances the class distribution by resampling rare examples at random or generating synthetic minority examples (Chawla et al. 2002, Menardi & Torelli 2014). Over-sampling techniques usually outperform under-sampling techniques, because the latter easily cause the loss of some useful information of the majority class examples. However, over-sampling methods usually cause over-fitting problem and increase the training time of the classification model. In order to overcome over-fitting problem of random over-sampling, (Chawla et al. 2002) proposed a famous and effective over-sampling technique named Synthetic Minority Oversampling Technique (SMOTE), in which new examples are generated by random linear interpolation between original minority examples. Subsequently, A variety of SMOTE-based improved algorithms have been proposed, such as Borderline-SMOTE (Han et al. 2005), ADASYN (He et al.), and CBSO (Barua et al. 2011).

Considering the risk of new decision region generated by SMOTE algorithm for minority class in the feature space is not large enough, (Menardi & Torelli 2014) proposed a novel artificial examples generation method, named ROSE, which combined with under-sampling and over-sampling techniques. In ROSE, synthetic examples are generated based on a function of kernel density estimation rather than based on k-nearest neighbors. SMOTE creates synthetic examples for the minority class only, and the synthetic examples are added to the original data set. However, in ROSE, synthetic examples generation is performed for both classes and the synthetic examples completely replace the original examples to generate a new training set. Although ROSE is superior to SMOTE under certain conditions, the synthetic examples of one class generated by ROSE may be located in the area of the other class if the feature space do not completely follow Normal distribution, which will increase overlapping area between classes. As shown in Figure 1(a), circles and squares represent the majority and minority examples, respectively. Because the minority examples are distributed in two areas, some synthetic examples (stars) created by ROSE overlap the majority examples in dashed area. Although ROSE can be combined with different kernel functions to adapt to different examples distribution, it is difficult to recognize in advance the distribution of the examples for an unknown data set, and thus cannot have a base to select the appropriate kernel function to generate new examples, which will largely affect the adaptability and robustness of ROSE.

For the purpose of overcoming the drawback of ROSE, we propose an improved algorithm, named Clustering Based Random Over-Sampling Examples (CBROSE) algorithm, which combines the k-means clustering algorithm with the examples creation mechanism of ROSE. Firstly, in the training data set, k-means algorithm is adopted to cluster the minority examples, and then a pre-defined number of original minority examples are randomly selected to compose a temporary minority data set in each cluster. Finally, the ROSE method is performed in this temporary minority data set for creating synthetic examples. As a desired result (as show in Figure 1(b)), the generated examples should be only located around each cluster, so that it will reduces the overlap between new synthetic examples and the majority examples.

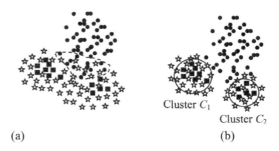

Cluster C_1

Cluster C_2

(a)

(b)

Figure 1. Distributions of the synthetic examples. (a) Based on ROSE. (b) Based on CBROSE.

This rest of this paper is organized as follows. Section 2 presents the proposed algorithm. Section 3 depicts the experiments including the characteristics of the data sets, performance metrics, the results and analysis. And then we summarize some conclusions in the study finally.

2 PROPOSED CBROSE ALGORITHM

CBROSE is an improved algorithm for ROSE [7], and it can effectively avoid that the generated minority examples overlapping the majority examples. [Algorithm CBROSE] presents the pseudo-code of CBROSE.

[Algorithm CBROSE]

Input:

Training set D_{tr} of examples $(x_1, y_1),...,(x_m, y_m)$, where x_i is an example that belonging to d dimensional feature space $\mathbf{X}$, and $y_i \in \mathbf{Y}$, $|\mathbf{Y}| = 2$;

The balanced level factor: p, where $p \in [0,1]$. $p = 1$ means that the number of generated synthetic minority examples is equal to M_{maj} (M_{maj} denotes the number of majority examples).

Step:

1. Count the quantity of new minority examples that should be created:

$$N_s = round(M_{maj} \times p)$$

2. Find the clusters of the minority class based on K-means algorithm.

3. For $j = 1, ..., K$ #For each cluster C_j

a. Count the quantity $n_s^{(j)}$ of new examples that should to be created in cluster C_j:

$$n_s^{(j)} = round(N_s \times \frac{m_{min}^{(j)}}{M_{min}})$$

where M_{min} and $m_{min}^{(j)}$ denote the quantity of minority instances in D_{tr} and cluster C_j, respectively.

b. Perform the synthetic examples generation mechanism of ROSE in cluster C_j.

If $m_{min}^{(j)} > 1$, then

1. For $q = 1, ..., d$
Calculate the sample standard deviation estimation of the q-th dimension of the minority examples.
2. Calculate the diagonal smoothing matrices:

$$\mathbf{H}^{(j)} = diag(h_1^{(j)}, ..., h_d^{(j)})$$

where,

$$h_q^{(j)} = \left(\frac{4}{(d+2)m_{min}^{(j)}}\right)^{1/(d+4)} \hat{\sigma}_q^{(j)}$$

3. Create a temporary set $D_t^{(j)}$ of $(\mathbf{X}^{(j)}, \mathbf{Y}^{(j)})$ that having $n_s^{(j)}$ examples based on random over-sampling with replacement.
4. Generate $n_s^{(j)}$ new synthetic examples for minority class in cluster C_j:

$$\mathbf{X}_s^{(j)} = \mathbf{X}^{(j)} + \mathbf{N}(\mu = 0, \sigma = 1) \cdot \mathbf{H}^{(j)}$$

where $\mathbf{N}(\mu = 0, \sigma = 1)$ is a matrix of random numbers following standard Normal distribution.
Else if $m_{min}^{(j)} = 1$, then
Duplicate the minority example of cluster C_j $n_s^{(j)}$ times for the minority class.
4. Return all synthetic examples for minority class. Like ROSE, the actual conditional densities of the examples are assumed to be correspondent to Gaussian distribution by CBROSE. For avoiding the loss of information of majority examples, CBROSE only creates synthetic examples for the minority class rather than both classes. Before the generation of synthetic examples, the original minority examples of the training set are divided into several non-overlapping clusters by K-means algorithm (Step 2 of [Algorithm CBROSE]). The quantity of new rare instances that should be created for each cluster is calculated in step 3a. On one hand, If the number of minority examples is more than one in a cluster, the examples generation mechanism of ROSE is performed in four steps (Step 3b (1)–(4)). Because Gaussian Kernels with diagonal smoothing matrices $\mathbf{H}$ is considered for each cluster, the generated examples are in fact random numbers following Gaussian distribution centered at the random resampling minority examples. Compared to SMOTE, CBROSE can produces a better decision space for classifier. On the other hand, if there is only one minority example in a cluster, we will consider the only member to be a noisy example. For avoiding overlapping between new minority examples and the majority examples, the generated examples will be duplications

of the original minority example. In Step 4, all artificial examples are outputted to replace the original minority examples, and combined with majority examples into a new training set.

Taking into account that the value of K is one of the factors that affect seriously the performance of K-means algorithm used in Step 2 of [Algorithm CBROSE]. So, for reducing the effect of different value of K, an objective evaluation method suggested by (Pham et al. 2005) is adopted to obtain a suitable value of K for K-means algorithm. In this method, an clustering result evaluation function $f(K)$ is applied to choose the most suitable value of K corresponding to the minimum clustering distortion.

The clustering distortion in K-means algorithm can be defined as:

$$D_j = \sum_{t=1}^{N_j} [d(x_{jt}, w_j)]^2 \tag{1}$$

where D_j denotes the distortion of cluster C_j, N_j is the number of examples of cluster C_j, x_{jt} is the t-th example of cluster C_j, w_j is the center of cluster C_j, and $d(x_{jt}, w_j)$ is the Euclidean distance between example x_{jt} and the center w_j of cluster C_j.

Then, the total distortions of all clusters can be calculated as follow:

$$S_K = \sum_{j=1}^{K} D_j \tag{2}$$

where, K is the pre-defined cluster number for K-means algorithm.

$$f(K) = \begin{cases} 1 & if\ K = 1 \\ \dfrac{S_K}{\alpha_K S_{K-1}} & if\ S_{K-1} \neq 0, \forall K > 1 \\ 1 & if\ S_{K-1} = 0, \forall K > 1 \end{cases} \tag{3}$$

$$\alpha_K = \begin{cases} 1 - \dfrac{3}{4N_d} & if\ K = 2\ and\ N_d > 1 \\ \alpha_{K-1} + \dfrac{1 - \alpha_{K-1}}{6} & if\ K > 2\ and\ N_d > 1 \end{cases} \tag{4}$$

where N_d is the feature numbers used for clustering, and α_K is a weight factor. When the distribution of examples is uniform, $f(K)$ is approximately equal to one. If some concentrated areas exist in the examples distribution, $f(K)$ will decrease. The more concentrated the example distribution is, the smaller is $f(K)$. Therefore, for K-means algorithm, the value of K which yields the smallest $f(K)$ can be selected as the pre-defined cluster number.

521

3 EXPERIMENTS

In our experiments, the effectiveness of CBROSE is evaluated and compared with SMOTE (Chawla et al. 2002) and ROSE (Menardi & Torelli 2014) algorithm. Classification tree (CART) and Support Vector Machine (SVM) are chose as two base classifiers. The experiments run on a Win7 operating system with 2.9 GHz of Intel core and 4GB of RAM. The codes of CBROSE, ROSE, and SMOTE are written in R programming language.

3.1 Imbalanced data sets and preprocessing

For a binary class data set, we use the Imbalance Ratio (*IR*) (Peng et al. 2014), which is the ratio of the number of majority examples to the number of minority examples (as shown in formula (5)), to describe its class imbalance degree.

$$IR = M_{maj} / M_{min} \tag{5}$$

In our experiments, ten binary class data sets (see Table 1) with different imbalance ratio are random selected from the KEEL-dataset repository (http:// www.keel.es.datasets.php).

For removing the redundant and irrelevant features in each data set, a feature selection method based on Information Gain (IG) is used. The IG based feature selection method runs on WEKA. In addition, data normalization of numerical features is a very effective step for eliminating influence of dimension. In this paper, we use formula (6) to normalize the numeric features.

$$\hat{x} = \frac{x - \bar{x}}{\sigma} \tag{6}$$

where σ and $\bar{x}$ are the standard deviation and mean of feature x, respectively.

3.2 Performance metrics

For reflecting the actual classifier's performance in imbalanced data sets, the area under the ROC curve (AUC), which has been widely selected as a performance metric in the researching on imbalanced data sets (Menardi & Torelli 2014, Peng et al. 2014), is used to assess our proposed method. AUC can be calculated by

$$AUC = \frac{1 + TPR - FPR}{2} \tag{7}$$

where, True Positive Rate (TPR) is the proportion of positive (minority) examples that are correctly classified and False Positive Rate (FPR) is the proportion of negative (majority) examples that are incorrectly classified as positive examples. Usually, AUC $\in$ [0.5, 1]. Note that an AUC of 0.5 implies that the result of classification equals to a sheer random conjecture, and an AUC of 1 implies that the learner perfectly identifies all class examples.

In our experiments, 5-folder cross validation is performed. Generally, the n value of the n-folder cross validation is selected to 10 in traditional classification tasks. The reason of choosing 5-folder cross validation rather than 10-folder cross validation is that there are not enough minority examples in several extremely imbalanced data sets. For instance, there are only 9 minority instances in the glass5 data set (see Table 1). A subset without any positive example will be produced based on 10-folder cross validation.

3.3 Result and analysis

In order to make a comparison between different algorithms under the same conditions, the model complexity of CART algorithm is set as 0 with pruning and the kernel function of SVM is set as Radial Basis (Gaussian) function, in which

Table 1. Experimental data sets with different imbalance ratio.

Datasets	Features	Examples	Majority	Minority	IR
WisconsinImb	9	683	444	239	1.86
Vehicle3	18	846	634	212	2.99
Glass-0-1-2-3_vs_4-5-6	9	214	163	51	3.20
New-thyroid1	5	215	180	35	5.14
Ecoli2	7	336	284	52	5.46
Glass6	9	214	185	29	6.38
Yeast-2_vs_4	8	514	463	51	9.08
Vowel0	13	988	898	90	9.98
Abalone 9-180.7	8	731	689	42	16.00
Glass5	9	214	205	9	22.78

Table 2. Comparison of average AUC on 10 binary class data sets.

	CART				SVM			
	None	SMOTE	ROSE	CBROSE	None	SMOTE	ROSE	CBROSE
WisconsinImb	0.9376	0.9503	0.9395	**0.9627**	0.9685	0.9694	0.9694	**0.9725**
Vehicle3	0.5998	**0.7182**	0.6818	0.6248	0.6925	**0.8236**	0.7425	0.7780
Glass-0-1-2-3_vs_4-5-6	0.8878	0.8811	0.8811	**0.8902**	0.8639	0.9301	0.9099	**0.9424**
New-thyroid1	0.8833	0.9262	**0.9437**	0.9087	0.9544	0.9944	0.9690	**0.9972**
Ecoli2	0.8076	0.8581	0.8536	**0.8967**	0.9011	0.9167	0.9158	**0.9177**
Glass6	**0.9375**	0.8749	0.8695	0.9003	0.9283	0.9229	0.9094	**0.9375**
Yeast-2_vs_4	0.8236	0.8838	0.9000	**0.9154**	0.8475	0.8957	0.8978	**0.9065**
Vowel0	0.8850	0.9194	0.9265	**0.9594**	0.9711	0.9911	0.9749	**0.9916**
Abalone9-18 0.7	0.5000	0.7551	**0.7776**	0.6940	0.5465	**0.7609**	0.7195	0.7297
Glass5	0.5000	0.9220	0.9488	**0.9878**	0.7729	0.9201	0.9030	**0.9951**
Winning Times	1	1	2	**6**	0	2	0	**8**

the smoothing parameter sigma is set to 0.05. For SMOTE, $k = 5$ (Chawla et al. 2002), $N = 300$. For CBROSE, the balance coefficient p is set to 1, so that the quantity of generated minority examples is equal to that of majority examples. The number of experiments on each data set is set to 100, and the average value of the AUC is chose as the final results for each algorithm. The experimental results of the algorithms are shown in Table 2 where the bests are shown in bold.

Form Table 2, in terms of AUC, we can see that CBROSE wins more times than the other methods, which implies that CBROSE is superior to the other methods in most of the data sets. There are still several data sets, such as vehicle and abalone 9-18, on which CBROSE cannot obtain satisfactory results. By observing the example distributions of these data sets, we find that they do not obey Normal distribution. In addition, in most of the data sets, the performance of CBROSE using SVM as the base classifier is higher than that of CBROSE using CART as the base classifier, indicating that the proposed data balanced algorithm may be more suitable for combining with SVM.

4 CONCLUSION

A novel synthetic over-sampling algorithm, named CBROSE, for balancing the binary class data sets is presented in this paper. CBROSE is an improved algorithm based on ROSE. Unlike ROSE, CBROSE creates synthetic examples combining with K-means algorithm which guarantees that the created minority examples always be located in an elliptical area centered at the observed minority example, thus, avoiding any created minority examples overlap the majority. In addition, for getting the optimal cluster-

ing results, an objective evaluation method is adopted to obtain the suitable value of K for K-means algorithm. By comparing with SMOTE and ROSE, the application of CBROSE to some real word data sets shows satisfactory performance in terms of AUC. Ensemble techniques, such as Adaboost, can be easily combined with CBROSE by relating the weights update approach to the probability of example creation, which will be an object of our further research.

ACKNOWLEDGMENTS

This work is supported by the National Natural Science Foundation of China (Grant No. 61374008).

REFERENCES

Barua, S., Islam, M. M. & Murase, K. 2011. A Novel Synthetic Minority Oversampling Technique for Imbalanced Data Set Learning. Berlin, Heidelberg: Springer Berlin Heidelberg.

Chawla, N.V., Bowyer, K.W., Hall, L.O. & Kegelmeyer, W.P. 2002. SMOTE: Synthetic minority over-sampling technique. Journal of Artificial Intelligence Research, 16: 321–357.

Chawla, N.V., Lazarevic, A., Hall, L.O. & Bowyer, K.W. 2003. SMOTEBoost: Improving prediction of the minority class in boosting. 7th European Conference on Principles and Practice of Knowledge Discovery in Databases, September 22, 2003 - September 26, 2003. Cavtat-Dubrovnik, Croatia: Springer Verlag.

Han, H., Wang, W.Y. & Mao, B.H. 2005. Borderline-SMOTE: A new over-sampling method in imbalanced data sets learning. Intelligent Computing, 2005. 878–887.

He, H., Bai, Y., Garcia, E.A. & Li, S. ADASYN: Adaptive synthetic sampling approach for imbalanced learning. 2008. IEEE, 1322–1328.

Mazurowski, M.A., Habas, P.A., Zurada, J.M., Lo, J.Y., Baker, J.A. & Tourassi, G.D. 2008. Training neural network classifiers for medical decision making: The effects of imbalanced datasets on classification performance. Neural Networks, 21(2): 427–436.

Menardi, G. & Torelli, N. 2014. Training and assessing classification rules with imbalanced data. Data Mining and Knowledge Discovery, 28(1): 92–122.

Peng, L.Z., Zhang, H.L., Yang, B. & Chen, Y.H. 2014. A new approach for imbalanced data classification based on data gravitation. Information Sciences, 288: 347–373.

Pham, D.T., Dimov, S.S. & Nguyen, C.D. 2005. Selection of K in K-means clustering. Proceedings of the Institution of Mechanical Engineers Part C-Journal of Mechanical Engineering Science, 219(1): 103–119.

Qazi, N. & Raza, K. 2012. Effect of Feature Selection, SMOTE and under Sampling on Class Imbalance Classification. 14th International Conference on Modelling and Simulation. IEEE.

Seiffert, C., Khoshgoftaar, T.M., Van Hulse, J. & Napolitano, A. 2010. RUSBoost: A Hybrid Approach to Alleviating Class Imbalance. IEEE Transactions on Systems, Man, and Cybernetics, 40(1): 185–197.

Tan, S. 2005. Neighbor-weighted K-nearest neighbor for unbalanced text corpus. Expert Systems with Applications, 28(4): 667–671.

Tesfahun, A. & Bhaskari, D.L. 2013. Intrusion Detection Using Random Forests Classifier with SMOTE and Feature Reduction. International Conference on Cloud & Ubiquitous Computing & Emerging Technologies. IEEE.

Ting, K.M. 2002. An instance-weighting method to induce cost-sensitive trees. IEEE Transactions on Knowledge and Data Engineering, 14(3): 659–665.

Wei, W., Li, J., Cao, L., Ou, Y. & Chen, J. 2013. Effective detection of sophisticated online banking fraud on extremely imbalanced data. World Wide Web, 16(4): 449–475.

Yen, S.-J. & Lee, Y.-S. 2006. Under-Sampling Approaches for Improving Prediction of the Minority Class in an Imbalanced Dataset. Berlin, Heidelberg: Springer Berlin Heidelberg.

Zheng, Z.Y., Cai, Y.P. & Li, Y. 2015. Oversampling Method for Imbalanced Classification. Computing and Informatics, 34(5): 1017–1037.

Zhou, Z.-H. & Liu, X.-Y. 2006. Training cost-sensitive neural networks with methods addressing the class imbalance problem. IEEE Transactions on Knowledge and Data Engineering, 18(1): 63–77.

Automotive, Mechanical and Electrical Engineering – Liu (Ed.)
© 2017 Taylor & Francis Group, London, ISBN 978-1-138-62951-6

Comparative study on directional and reverse directional expert systems

Zhiheng Lin, Haiping Huang & Pin Wang
Mathematics and Information Science Department, Guangxi College of Education, Nanning, Guangxi, China

ABSTRACT: The mathematical modelling method calculates and analyses four indexes of expert systems, namely, RSI, W%R, MA and MACD by taking annual net profit margin, rate of return and winning rate as the management goal. According to the results, when compared to the winning rate of one-time buy-in of all funds, the step-by-step buy-in with 30% funds increases by a mere 9.29%, 6.31%, 1% and 6.31% respectively. However, the annual rate of return and net profit margin drop significantly to 41.85%, 50.93%, 50.94% and 50.95%. Thus, step-by-step buy-in is meaningless for the above four systems. As can be seen from the results, the reverse directional expert system RSI is superior to W%R, and the directional expert system MA is superior to MACD. The calculation result of reverse directional expert systems is generally better than that of directional expert systems. For instance, the winning rate, annual rate of return and net profit margin of the reverse directional RSI expert system are 1.19, 1.40 and 1.401 times that of the directional MACD expert system.

Keywords: Directional; Reverse directional; Calculation and analysis; Comparative research

1 INTRODUCTION

Directional indexes in the technical analysis of financial trade refer to the index category that is used to determine the price tendency of transacted species. As a main index category for references in technical analysis of the financial market, it is also the most common and widely used category, the core of which is to take advantage of the tendency. A representative of directional indexes is the weighted Moving Average (MA). In this paper, only the most extensive and commonly used MA and Moving Average Convergence Divergence (MACD) are discussed.

The other category is reverse directional indexes, the major function of which is to issue the entrance or departure signals. Meanwhile, through the comparative result, it can also imply whether the price has been in a depressed state. Its core is the Eastern philosophy that supposes things will develop in the opposite direction when they become extreme, as well as the interchange of the strong and weak. Here, only the most extensive and commonly used RSI and W%R is discussed.

When carrying out a technical analysis of investment stocks, futures and precious metals, it can be understood from the basic investment principle that there's no index or method with 100% success rate. Thus, how can the yield rate be increased, investment risks avoided, and uncertainties prevented when analysing indexes? Funds management, also known as risk management, is a common method adopted by many investors.

By taking A-share data in the Shanghai stock market of China in the period 26 August 2015 to 25 November 2016 as the basis, this paper attempts to test relevant data with the method of mathematical modelling by taking the winning rate, annual rate of return and net profit margin as the management goal, thus conducting comparative analysis about MA, MACD, RSI and W%R systems. The advantages and disadvantages of the two systems as well as the optimal usage will be discovered, so as to provide an optimised investment scheme for investors.

The mathematic formula of relative strength index is:

$$RSI = \frac{100 \times RS}{1 + RS} \tag{1}$$

$$RS = \frac{\text{Average Rise Point in i days}}{\text{Average Dropped Point in i days}} (i = 1, 2, \cdots n) \tag{2}$$

The W%R mathematical formula of Williams' overbought/oversold (WMS%R) formula (LI YONG-chun, 2013; WANG Pin, 2013; WANG Pin, 2011) is:

$$\%R = \frac{high_{Ndays} - close_{today}}{high_{Ndays} - low_{Ndays}} \times (-100\%) \tag{3}$$

$$(close_{today} - low_{Ndays}) - (close_{today} - high_{Ndays}) = high_{Ndays} - low_{Ndays} \tag{4}$$

$$W\%R = \frac{C_n - H_n}{H_n - L_n} \times 100\% \qquad (5)$$

where, n is the transaction period set by the dealer, C_n is the closing price in n day(s), L_n is the lowest price in n day(s), and H_n is the highest price in n day(s).

The mathematical formula of the MACD is as follows:

$$MACD = 12\text{-daysEMA-26-daysEMA} \qquad (6)$$

The mathematical formula of the Exponential MA (EMA) index is as follows:

$$EMA_{today} = \frac{p_1 + (1-\alpha)p_2 + (1-\alpha)^2 p_3 + (1-\alpha)^3 p_4 + \cdots}{1 + (1-\alpha) + (1-\alpha)^2 + (1-\alpha)^3 + \cdots}$$
$$(7)$$

$$\alpha = \frac{2}{N+1} \text{ N represents the periodicity} \qquad (8)$$

Where, $p_i (i = 1, 2, \cdots n)$ refers to the closing price of day i, and n refers to the MA periodicity.

Mathematical formula of the MA is:

$$MA = \frac{c_1 + c_2 + \cdots\cdots + c_n}{n} \qquad (9)$$

where $c_i (i = 1, 2, \cdots n)$ refers to the closing price of day i, and n refers to the MA periodicity.

2 EMPIRICAL ANALYSIS OF RSI EXPERT SYSTEM, W%R EXPERT SYSTEM, MACD EXPERT SYSTEM AND MA EXPERT SYSTEM

2.1 *The experiment and results*

1. Experimental procedure:

Source code of the RSI expert system
N1 1 100 14
LL 0 40 20
LH 60 100 80
LC: = REF (CLOSE, 1);
RSI:SMA(MAX(CLOSE-LC,0),N1,1)/SMA(ABS(CLOSE-LC),N1,1)*100,colorwhite;
ENTERLONG: CROSS (RSI, LL);
EXITLONG: CROSS (LH, RSI)

Source code of the W%R expert system
N 2.00 100.00 14.00
LL 0.00 100.00 20.00
LH 0.00 100.00 80.00
WR:=100*(HHV (HIGH, N)-CLOSE)/(HHV (HIGH,N)-LLV(LOW,N));
ENTERLONG: CROSS (WR, LH);

EXITLONG: CROSS (WR, LL);

Source code of MA expert system
SHORT 1 30 5
LONG 5 100 30
CROSS (MA (CLOSE, SHORT), MA (CLOSE, LONG))
CROSS (MA (CLOSE, LONG), MA (CLOSE, SHORT))
ENTERLONG: CROSS (MA (CLOSE, SHORT), MA (CLOSE, LONG));
EXITLONG: CROSS (MA (CLOSE, LONG), MA(CLOSE, SHORT))

Source code of MACD expert system
LONG 10 200 26
SHORT 2 200 12
M   2 200 9
DIFF: = EMA (CLOSE, SHORT) – EMA (CLOSE, LONG);
DEA: = EMA (DIFF, M);
MACD: = 2*(DIFF-DEA);
ENTERLONG: CROSS (MACD, 0);
EXITLONG: CROSS (0, MACD);

2. Experiment platform: V5.00 version of DZH securities information platform
3. Experimental parameters: one-time position opening or full position closing for funds meeting the conditions, with a transaction cost of 0.5%
4. Experimental samples: all A-share daily line data in Shanghai stock market (26 August 2015–24 November, 2016)
5. Experimental process, time and results:

Table 1. The first test result of RSI expert system.

System test setting
Test method: Technical index – RSI (14)
Test time: 26/08/2015–24/11/2016   Calculating the forced liquidation
Test share: 1,111 shares in total   Initial input: 40,000.00yuan
Buy-in conditions:
One of the following groups is tenable:
1. The following conditions are tenable simultaneously
1.1 Technical index: WP (14) index line WP cross and dncross [daily line]
When the condition is satisfied according to the middle market value, the closing price is bought in with all funds
In case of continuous signal: no more buy-in
Sell-out condition: no sell-out condition
Conditions for closing a position: (according to the closing price)
Share selection by indexes: technical index: WP (14) index line WP dncross and cross [daily line]
System test report
Tested stock number: 1,111
Net profit: 9,543,976.00yuan   Net profit margin: 21.48%

(Continued)

Table 1. (*Continued*)

Total profits: 10,353,684.00yuan   Total losses: -809,816.69yuan

Transaction times: 1,122   Winning rate: 80.57%

Annual transaction times: 897.60   Profit/ loss transaction times: 904/218

Total transaction amount: 46,369,628.00yuan   Transaction fee: 33,692.84yuan

The highest one-time profit: 524,012.50yuan The highest one-time loss: -18,582.60yuan

Average profit: 9,227.88yuan   Average loss: -721.76yuan

Average profit: 8,506.22   Average profit/average loss: -1,278.52

Maximum continuous profit times: 41   Maximum continuous loss times: 4

Average transaction periodicity: 103.79

Average transaction periodicity with profits: 90.28   Average transaction periodicity with losses: 159.81

Profit coefficient: 0.85

Maximum floating profits: 53,951,048.00yuan Maximum floating losses: 0.00yuan

Maximum difference between floating profits and losses: 53,951,048.00yuan

Total input: 44,440,000.00yuan

-----------------------------Statistics of buy-in signals--

(Create statistics of the situation of all buy-in signal points, excluding signal deletion caused by funds and strategies in the transaction test)

Success rate: 78.74%

Signal number: 1223   Annual average signal number: 978.40

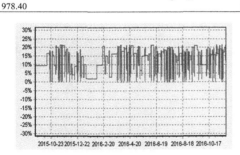

Figure 1. RSI earnings curve of one-time buy-in with all funds.

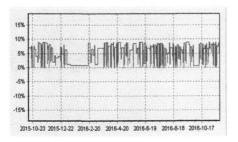

Figure 2. RSI earnings curve of step-by-step buy-in with 30% funds.

Table 2. The second test result of RSI expert system.

System test setting

Test method: Technical index – RSI (14)

Test time: 26/08/2015–24/11/2016   Calculating the forced liquidation

Test share: 1,111 shares in total   Initial input: 40,000.00yuan

Buy-in conditions:

One of the following groups is tenable:

1. The following conditions are tenable simultaneously

1.1 Technical index: WP (14) index line WP cross and dncross [daily line]

When the condition is satisfied, according to the middle market value, the closing price is bought in with all funds

In case of continuous signal: buy-in on an equal basis

Sell-out condition: no sell-out condition

Conditions for closing a position: (according to the closing price)

Share selection by indexes: technical index: WP (14) index line WP dncross and cross [daily line]

System test report

Tested stock number: 1,111

Net profit: 3,993,072.00yuan   Net profit margin: 8.99%

Total profits: 4,182,184.25yuan   Total losses: -189,127.28yuan

Transaction times: 1,017   Winning rate: 89.28%

Annual transaction times: 813.60   Profit/loss transaction times: 908/109

Total transaction amount: 18,416,068.00yuan   Transaction fee: 38,637.22yuan

The highest one-time profit: 254,867.52yuan The highest one-time loss: -10,933.96yuan

Average profit: 4,112.28yuan   Average loss: -185.97yuan

Average profit: 3,926.32   Average profit / average loss: -2,211.31

Maximum continuous profit times: 76   Maximum continuous loss times: 3

Average transaction periodicity: 102.25

Average transaction periodicity with profits: 94.61   Average transaction periodicity with losses: 165.96

Profit coefficient: 0.91

Maximum floating profits: 48,395,508.00yuan Maximum floating losses: 0.00yuan

Maximum difference between floating profits and losses: 48,395,508.00yuan

Total input: 44,440,000.00yuan

-----------------------------Statistics of buy-in signals--

(Create statistics of the situation of all buy-in signal points, excluding signal deletion caused by funds and strategies in the transaction test)

Success rate: 82.34%

Signal number: 2248   Annual average signal number: 1,798.40

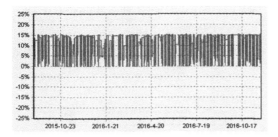

Figure 3. W%R earnings curve of one-time buy-in with all funds.

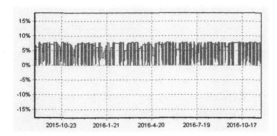

Figure 4. W%R earnings curve of step-by-step buy-in with 30% funds.

Table 3. The first test result of the MA expert system.

System test setting

Test method: Technical index – MA

Test time: 26/08/2015–24/11/2016   Calculating the forced liquidation

Test share: 1,111 shares in total   Initial input: 40,000.00yuan

Buy-in conditions:

One of the following groups is tenable:

1. The following conditions are tenable simultaneously

1.1 Technical index: MA (5, 10, 20, 30, 120, 250) index line MA2 cross MA3 [daily line]

When the condition is satisfied, according to the middle market value, the closing price is bought in with all funds

In case of continuous signal: no more buy-in

Sell-out condition: no sell-out condition

Conditions for closing a position: (according to the closing price)

Share selection by indexes: technical index: MA (5, 10, 20, 30, 120, 250) index line MA2 dncross MA3 [daily line]

System test report

Tested stock number: 1,111

Net profit: 9,186,380.00yuan   Net profit margin: 20.67%

Total profits: 18,482,934.00yuan   Total losses: -9,296,557.00yuan

Transaction times: 7947   Winning rate: 46.67%

(*Continued*)

Table 3. (*Continued*)

Annual transaction times: 6,357.60   Profit/loss transaction times: 3709/4238

Total transaction amount: 325,691,264.00yuan   Transaction fee: 242,536.97yuan

The highest one-time profit: 134,826.13yuan   The highest one-time loss: -24,304.87yuan

Average profit: 2,325.77yuan   Average loss: -1,169.82yuan

Average profit: 1,155.96yuan   Average profit / average loss: -198.81

Maximum continuous profit times: 9   Maximum continuous loss times: 11

Average transaction periodicity: 18.90

Average transaction periodicity with profits: 26.99   Average transaction periodicity with losses: 11.82

Profit coefficient: 0.33

Maximum floating profits: 53,597,300.00yuan   Maximum floating losses: 0.00yuan

Maximum difference between floating profits and losses: 53,597,300.00yuan

Total input: 44,440,000.00yuan

------------------------------Statistics of buy-in signals--

(Create statistics of the situation of all buy-in signal points, excluding signal deletion caused by funds and strategies in the transaction test)

Success rate: 46.67%

Signal number: 7948   Annual average signal number: 6,358.40

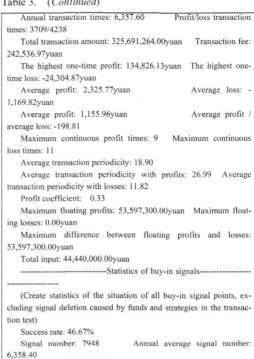

Figure 5. MA earnings curve of one-time buy-in with all funds.

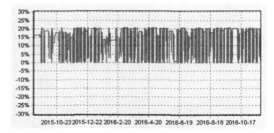

Figure 6. MA earnings curve of step-by-step buy-in with 30% funds.

Table 4. The second test result of the MA expert system.

System test setting
Test method: Technical index – MA
Test time: 26/08/2015–24/11/2016   Calculating the forced liquidation
Test share: 1,111 shares in total   Initial input: 40,000.00yuan
Buy-in conditions:
One of the following groups is tenable:
1. The following conditions are tenable simultaneously
1.1 Technical index: MA (5, 10, 20, 30, 120, 250) index line MA2 cross MA3 [daily line]
When the condition is satisfied according to the middle market value, the closing price is bought in with 30.00% funds
In case of continuous signal: buy-in on an equal basis
Sell-out condition: no sell-out condition
Conditions for closing a position: (according to the closing price)
Share selection by indexes: technical index: MA(5, 10, 20, 30, 120, 250) index line MA2 dncross MA3 [daily line]
System test report
Tested stock number: 1,111
Net profit: 2,420,980.00yuan   Net profit margin: 5.45%
Total profits: 4,405,408.00yuan   Total losses: -1,984,437.00yuan
Transaction times: 6994   Winning rate: 47.61%
Annual transaction times: 5,595.20   Profit/loss transaction times: 3330/3664
Total transaction amount: 74,004,224.00yuan   Transaction fee: 146,465.08yuan
The highest one-time profit: 28,088.78yuan   The highest one-time loss: -4,340.15yuan
Average profit: 629.88yuan   Average loss: -283.73yuan
Average profit: 346.15yuan   Average profit/average loss: -222.00
Maximum continuous profit times: 9   Maximum continuous loss times: 11
Average transaction periodicity: 19.14
Average transaction periodicity with profits: 27.20   Average transaction periodicity with losses: 11.82
Profit coefficient: 0.38
Maximum floating profits: 46,822,580.00yuan   Maximum floating losses: 0.00yuan
Maximum difference between floating profits and losses: 46,822,580.00yuan
Total input: 44,440,000.00yuan
----------------------------Statistics of buy-in signals----------------------------------
(Create statistics of the situation of all buy-in signal points, excluding signal deletion caused by funds and strategies in the transaction test)
Success rate: 47.71%
Signal number: 7067   Annual average signal number: 5,653.60

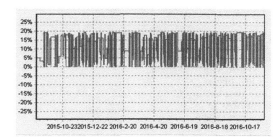

Figure 7. MACD earnings curve of one-time buy-in with all funds.

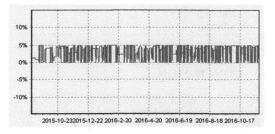

Figure 8. MACD earnings curve of step-by-step buy-in with 30% funds.

2.2 *Comparative analysis of results*

Table 5. Comparative analysis table.

	Winning rate	Annual rate of return	Net profit margin	Annual transaction times
One-time buy-in of RSI expert system with all funds	80.57	17.18	21.48	897.60
Step-by-step buy-in of RSI expert system with 30% funds	89.28	7.19	8.99	813.60
One-time buy-in of W%R expert system with all funds	67.52	12.27	15.33	7,605.60
Step-by-step buy-in of W%R expert system with 30% funds	73.83	6.25	7.81	6,847.20
One-time buy-in of MA expert system with all funds	46.67	16.54	20.67	6,357.60

(*Continued*)

Table 5. *(Continued)*

Step-by-step buy-in of MAexpert system with 30% funds	47.61	4.36	5.45	5,595.20
One-time buy-in of MACD expert system with all funds	67.52	12.27	15.33	7,605.60
Step-by-step buy-in of MACD expert system with 30% funds	73.83	6.25	7.81	6,847.20

3 CONCLUSION

By taking the winning rate, annual rate of return and net profit rate (that are of most concern to investors) as the management goal, this paper attempts to calculate and analyse the practicability of RSI, W%R, MA and MACD expert systems with a full sample size in the premise of funds management. Regarding RSI, W%R, MA and MACD expert systems in comparison to the winning rate of one-time buy-in with all funds, that of step-by-step buy-in with 30% funds merely increases by 9.29%, 6.31%, 1% and 6.31%. However, the annual rate of return and net profit margin drops sharply to 41.85%, 50.93%, 50.94% and 50.95% with the dramatic reduction of annual transaction times (see Figures 1–8). The market implication of this calculation result, is that step-by-step buy-in has proved meaningless for the above four systems. Regarding the winning rate, annual rate

of return and net profit margin, the reverse directional expert system RSI is superior to W%R, and the directional expert system MA is superior to MACD (regardless of the poor performance on the winning rate). Seen in Table 5, the calculation result of reverse directional expert systems is generally better than that of directional expert systems. For instance, the winning rate, annual rate of return and net profit margin of the reverse directional RSI expert system are 1.19, 1.40 and 1.401 times that of the directional MACD expert system. Thus, it can be seen that the RSI system should be the first choice for investors.

REFERENCES

Cao Yubo. (2010). Application of MACD index in securities investment. *Manager' Journal (in Chinese),* the 4th issue in, 166–167.

http://en.wikipedia.org/wiki/Moving_average

http://en.wikipedia.org/wiki/Moving_average.

http://en.wikipedia.org/wiki/Relative_strength_index

http://en.wikipedia.org/wiki/Williams_%25R

http://forex-indicators.net/macd

Liu, Y. Empirical Study on Senior Managers and Performances in Companies of High-Tech based on SPSS Software Regression Analysis. *Journal of Software,* 7(7), 1569–1576.

Li Yong-chun and Cheng Hao. (2013). Study on Effectiveness of William Overbought/Oversold in Technical Analysis. *Commercial Times,* 7, 70–71

Wang Pin. (2011). Empirical Analysis of RSI Transaction System Based on Historical Data. *Securities & Futures of China,* 9,

Wang Pin. (2012). RSI Investment Decision-Making Based on Consolidation Market. *Journal of Quantitative Economics,* 1,.

Wang Yumei. (1998). Application of Moving Average in the Stock Market. *Zhe Jiang Statistics (in Chinese),* the 2nd issue in 1998: 27–28.

Automotive, Mechanical and Electrical Engineering – Liu (Ed.)
© 2017 Taylor & Francis Group, London, ISBN 978-1-138-62951-6

Dynamic modelling and simulation of a centrifugal compressor for diesel engines

Lin Huang, Gang Cheng & Guoqing Zhu
Research Institute of Equipment Simulation Technology, Naval University of Engineering, Wuhan, China

Shilong Fan
Army of PLA, Sanya, China

ABSTRACT: In this paper a novel approach for modelling and simulation of a centrifugal compressor based on bond graph theory is proposed. The heat-work conversion, heat transfer, mass transfer and other energy interactions in the system were analysed by the bond graph method which offers a different perspective to pervious work. The main parts of the compressor system were mathematically modelled for developing compressor characteristics. The comparison of the modelled and measured data shows good agreement both in steady working condition and deep surge condition. Furthermore, the effects of different plenum volume on the amplitude and frequency of the outlet pressure when the compressor operates in deep surge were simulated and analysed.

Keywords: Centrifugal compressor; MS Word Bond graph; Surge; Turbocharger

1 INTRODUCTION

Centrifugal compressors have been widely used in turbocharged diesel engines, automobile engines, gas turbines etc. It is the main component of the turbocharger for marine diesel engines. Under some special working conditions, namely emergency stop, fault deceleration and sudden load drop, the compressor may deviate from the set point, resulting in surge or rotating stall which could affect the safety and stability of the diesel engine.

F. K. Greitzer and E. M. Moore established the famous compressor dynamic model in 1986 (Moore & Greitzer, 1986), which was later called the MG model. The lumped parameter method is used in the model to simplify the compression system into three parts, namely the compressor, the storage space for compressed air and the front and back pipe network system. The dimensionless parameter B was proposed to predict the compressor stability, which can reflect the characteristics of compressor rotating stall and surge. D. A. Fink built a non-fixed speed surge model of a centrifugal compressor, the calculated results of which were in good agreement with the measured results (Fink et al., 1992). Later, J. T. Gravdahl, F. Willems and others studied the compressor surge control theories and put forward a variety of surge control methods (Gravdahl, 1998; Galindo et al., 2006).

The mathematical model of a compressor system contains the knowledge of thermodynamics, fluid mechanics, dynamics, and the mechanical

properties of the impeller. Bond graph is a kind of multi-domain modelling method based on power energy flow. By summing up four state variables, which are effort, flow, generalised displacement and generalised momentum from a variety of physical parameters in different energy fields, bond graph theory can describe the transmission, conversion, storage and dissipation of system energy in a unified way; this is quite suitable for the modelling of complex systems. Therefore, the modelling of a compressor system based on power bond graph is proposed in this paper.

2 INTRODUCTION TO BOND GRAPH

2.1 *Overview*

Bond graph theory was originally proposed by Professor Paynter in the 1960s, and is also called power bond graph. After generalisation, the theory has gradually become a systematic dynamic modelling method for describing systems that are coupled in multi-energy domains. However, the application of the traditional power bond graph is limited in practical processes especially in thermal systems since the systems are quite nonlinear. For example, taking the working fluid temperature and entropy flow as a pair of effort variable and flow variable does not meet the requirements of the traditional bond graph theory, because the entropy variable is not conservative. Therefore, the concept of pseudo bond graph was proposed by Karnopp et al. In pseudo bond graph, the product of effort and the flow is no longer power but the generalised power (Galindo et al., 2006; Moore F. K. & Greitzer E. M., 1986). As long as the effort, flow, generalised momentum and generalised displacement are connected correctly in pseudo bond graph, the bond graph theory can be effectively applied to any systems. In this paper, a dual-channel bond graph model is used to describe the system and the pressure p and mass flow $\dot{m}$ temperature T and enthalpy flow $\dot{H}$, were chosen as the power variables, as shown in Figure 1.

2.2 *Basic elements*

Based on the same physical properties, bond graph abstracts nine basic elements from different energy fields to describe various ideal components. Figure 2 shows the schematic diagrams of the basic elements of bond graph. Power is a scalar with no direction, but people tend to use power flow to represent the energy flow (Afshari et al., 2010), so the direction of the energy flow is referred to as the direction of power flow as shown in Figure 2 (a). Figure 2 (b), (c) represent the 0 junction and 1 junction respectively in bond graph. The 0-junction is an effort equalising junction while the algebra sum of the flows is 0. The 1-junction is a flow equalising junction while the algebra sum of the efforts is 0. TransFormer (TF) is a kind of energy converter. It is mainly used to convert the same or different types of energy as shown in Figure 2 (d). Fig. 2 (e) is GYrator (GY). It is used to describe the transformation relationship between the effort variables and the flow variables in energy transfer process. Figure 2 (f) represents the resistive element, compliance element and inertia element, which are used respectively to simulate the components of energy consumption, storage and release.

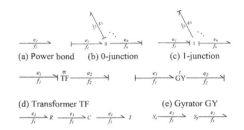

Figure 1.   Dual channel bond graph.

(a) Power bond   (b) 0-junction   (c) 1-junction

(d) Transformer TF   (e) Gyrator GY

(f) Resistive, compliance, inertia elements (g) Source of effort and flow

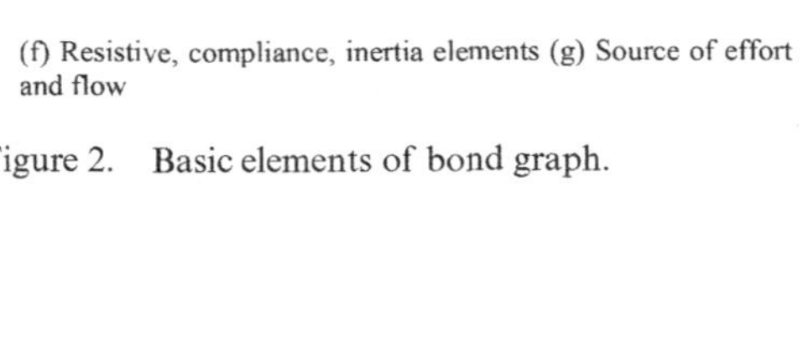

Figure 2.   Basic elements of bond graph.

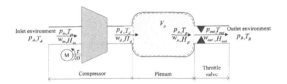

Figure 3.   Sketch of compressor system.

3   SYSTEM WORKING PRINCIPLE

The compressor system is composed of impeller, diffuser, volute, inlet and outlet pipe, plenum and throttle valve etc. For modelling and analysis, the system structure was simplified as shown in Figure 3, where $p, w, T, \dot{H}$ represent the pressure, flow, temperature and enthalpy flow of the air, subscript *in*, *d*, *p* and *out* correspond to the air parameters of compressor inlet, plenum inlet, plenum outlet and throttle valve outlet respectively, and subscript *A*, *B* denote environmental parameters.

Assume that: (1) The flow of air in the compressor is one-dimensional and incompressible. (2) Pressure in the plenum is uniformly distributed. (3) Internal thermodynamic process of plenum is isentropic. (4) Air flow through the compressor and the throttle valve is quasi steady state behaviour.

4   BOND GRAPH MODELLING

Figure 4 shows the overall word bond graph model of compressor system. Each block in the graph represents a component, and each component is connected by a power bond and marked with the corresponding effort variables and flow variables. The corresponding mathematical models of each component will be given as follows.

4.1   *Compressor*

For simplicity, the compressor impeller, diffuser, volute and others were simplified as the compressor module. The function of the module is to raise the pressure and temperature of the air. The module can be represented by a Multi-ports Modulated Transformer (MTF) as shown in Figure 5, where γ_1 and γ_2 respectively represent the transformation ratio of the pressure energy channel and the heat energy channel of the MTF; R_s, R_s correspond to the fan-shaped loss and the friction loss, variables with Rome number denote intermediate variables between each component and node.

4.1.1   *Pressure energy channel of MTF*
Γ_1 is the transformation ratio of the pressure energy channel of the MTF. It equals the pressure ratio ε

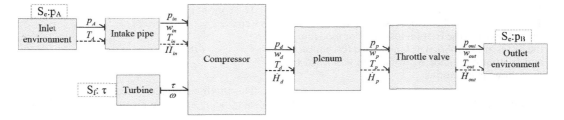

Figure 4.   Word bond graph model of the compressor system.

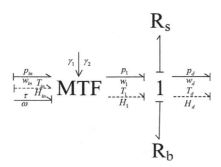

Figure 5.   Bond graph model of the compressor.

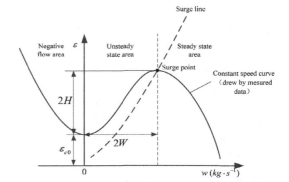

Figure 6.   Compressor operating characteristic.

of the compressor, namely $\gamma_1 = \varepsilon$. The characteristic curves of the compressor only provide data for the steady state region which is on the right side of the surge line. In order to calculate the pressure ratio from the surge line to the negative flow area, cubic interpolation and quadratic interpolation were used to fit the pressure ratio respectively:

$$\gamma_1 = \varepsilon_{c0} + H\left[1 + \frac{3}{2}\left(\frac{w_{in}}{W} - 1\right) - \frac{1}{2}\left(\frac{w_{in}}{W} - 1\right)^3\right] \quad (1)$$

$$\gamma_1 = \varepsilon_{c0} + H\frac{w_{in}^2}{2W^2} \quad (2)$$

With H and W the semi-height and semi-width of the cubic interpolation as shown in Figure 6. ε_{c0} the pressure ratio at zero flow rate:

$$\varepsilon_{c0} = \left[1 + \frac{\kappa - 1}{2\kappa R T_{in}}\omega^2(r_l^2 - r_v^2)\right]^{\frac{\kappa}{\kappa - 1}} \quad (3)$$

With κ the air specific heat ratio, R the gas constant, r_l and r_v the outlet radius and the average inlet radius of the impeller.

4.1.2   Heat energy channel of MTF

γ_2 is the transformation ratio of the pressure energy channel of the MTF. It equals the ratio of inlet and outlet temperature of the compressor:

$$\gamma_2 = \frac{T_1}{T_{in}} \quad (4)$$

The relationship between the inlet and outlet temperature is written:

$$T_1 = T_{in} + \frac{M_c\omega}{w_{in}\kappa R}(\kappa - 1) \quad (5)$$

With M_c the compressor torque and can be calculated as follow:

$$M_c = \frac{30\kappa R w_{in} T_{in}}{\pi n_c (\kappa - 1)\eta_c}\left(\varepsilon^{(\kappa-1)/\kappa} - 1\right) \quad (6)$$

With η_c the compressor efficiency, n_c the compressor speed.

The compressor is driven by the exhaust gas turbine. According to the conservation of angular momentum, the compressor speed can be obtained by the following equation:

$$\frac{dn_c}{d} = \frac{30}{\pi I_c}(M_t - M_c) \quad (7)$$

With M_t the driving torque, I_c the moment of inertia of the compressor.

4.1.3 Compressor resistance loss

The energy loss in the compressor includes the fan-shaped loss and the friction loss etc. For modelling, the losses are simplified as fan-shaped resistance R_s and friction resistance R_b, which are described by the following equations:

$$R_s = 0.7\left(L_{cl}/2r_v\right)^2 \tag{8}$$

$$R_b = 16\mu_c r_v^5 n_c^3 \rho \tag{9}$$

With L_{cl} the axial length of stator blade, μ_c the coefficient of friction loss, ρ the air density (Wang Wei-cai & Wang Yin-yan, 2007).

4.2 Plenum

The temperature and pressure of the air are changed when diffusing in the plenum. For describing the storage effect of the plenum, the compliance element C was utilised and the air leakage R_l loss and blast loss R_g were considered as shown in Figure 7.

The energy conservation equation and the ideal gas equation of the plenum are as follows:

$$\frac{\mathrm{d}\left(u_2 m_2\right)}{\mathrm{d}t} = h_d w_d - h_3 w_3 \tag{10}$$

$$p_2 = \frac{m_2 R T_2}{V_p} \tag{11}$$

With h_d and h_3 respectively the air enthalpy of the inlet and outlet of the plenum, u_2 the air specific internal energy, m_2, T_2 and V_p respectively the air mass, temperature and volume of the air in plenum.

The air is continuously flowing in the plenum, so its mass change rate is so small that can be ignored. T thus the air temperature in the plenum is calculated from Equation (10):

$$\frac{\mathrm{d}T_2}{\mathrm{d}t} = \frac{\kappa}{m_2}\left(T_d w_d - T_3 w_3\right) \tag{12}$$

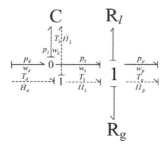

Figure 7.   Bond graph model of the plenum.

The continuity equation of the air in the plenum is written:

$$w_d - w_3 = V_p \frac{\mathrm{d}\rho}{\mathrm{d}t} \tag{13}$$

The thermodynamic process of the air in the plenum being adiabatic reversible, the relationship between the density and the pressure of the air is written:

$$\frac{\mathrm{d}\rho_p}{\mathrm{d}t} = \frac{\rho_p}{\kappa p_2}\frac{\mathrm{d}p_2}{\mathrm{d}t} \tag{14}$$

Substitute Eq. (14) into Eq. (13):

$$\frac{\mathrm{d}p_2}{\mathrm{d}t} = \frac{c^2}{V_p}\left(w_d - w_3\right) \tag{15}$$

With c the speed of sound.

The capacitance parameters of pressure channel C_1 and heat channel C_2 are calculated respectively from Eq. (15) and Eq. (12):

$$C_1 = \frac{V_p}{\rho c^2} \tag{16}$$

$$C_2 = \frac{m_2}{\kappa} \tag{17}$$

The air leakage R_l loss and blast loss R_g are evaluated as follows:

$$R_l = 3.1\frac{b_c}{h_c} \tag{18}$$

$$R_b = 2\mu_w\left(1 - \zeta_w\right)r_l L_c^{1.5} n_c^3 \rho \tag{19}$$

With b_c and h_c respectively the blade radial clearance and the average blade height, μ_w and ζ_w respectively the coefficient of blast loss and the coefficient of local air intake loss.

4.3 Throttle valve

The throttle valve of the compressor can be described by the resistive element R, as shown in Figure 8.

Under the adiabatic condition, the variables in Figure 8 meet the following formulas:

$$\dot{H}_p = \dot{H}_{out} \tag{20}$$

$$\frac{T_{out}}{T_p} = \left(\frac{p_{out}}{p_p}\right)^{\frac{\kappa-1}{\kappa}} \tag{21}$$

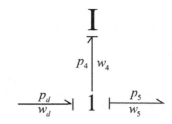

Figure 8.   Bond graph model of throttle valve.

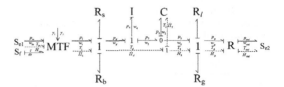

Figure 9.   Flow inertia of the air.

If $\frac{p_{out}}{p_p} \le \left(\frac{2}{\kappa+1}\right)^{\frac{\kappa}{\kappa-1}}$, which is the condition corresponding to a supersonic flow, the air mass flow rate is written:

$$w_{out} = \mu_{out} A_{out} \frac{p_p}{\sqrt{RT_p}} \left(\frac{2}{\kappa+1}\right)^{\frac{1}{\kappa-1}} \sqrt{\frac{2\kappa}{\kappa+1}} \qquad (22)$$

If $\frac{p_{out}}{p_p} \ge \left(\frac{2}{\kappa+1}\right)^{\frac{\kappa}{\kappa-1}}$, which is the condition corresponding to a subsonic flow, the air mass flow rate is written:

$$w_{out} = \mu_{out} A_{out} \frac{p_p}{\sqrt{RT_p}} \sqrt{\frac{2\kappa}{\kappa-1}\left[\left(\frac{p_{out}}{p_p}\right)^{\frac{2}{\kappa}} - \left(\frac{p_{out}}{p_p}\right)^{\frac{\kappa+1}{\kappa}}\right]} \qquad (23)$$

With μ_{out} the exhaust flow coefficient, A_{out} the exhaust effective cross-sectional area.

4.4   Flow inertia of air

The flow inertia of air in the compressor is described by the inertia component I as shown in Figure 9.

The flow inertia of air is related to several factors, which can be evaluated by the one-dimensional momentum conservation equation (SUN Jian-bo et al., 2008):

$$\frac{d(\rho A_c L_c C_c)}{d} = A_c(p_A - p_d) + A_c(p_5 - p_A) \qquad (24)$$

With L_c the equivalent length of the compressor, C_c the axial velocity of the compressor, A_c the compressor inlet area and is calculated as follow:

$$A_c \rho \frac{dC_c}{d} = \dot{w}_4 \qquad (25)$$

Substitute Eq. (25) into Eq. (24):

$$\dot{w}_4 = \frac{A_c}{L_c}(p_d - p_5) \qquad (26)$$

Thus, the inertia coefficient of air flow in the compressor is written:

Figure 10.   Overall bond graph model of compressor system.

$$I = \frac{\rho L_c}{A_c} \qquad (27)$$

4.5   Overall bond graph model of the compressor system

As the compressor inlet duct is quite short, we can assume that the air is sucked into the compressor directly from the environment. Figure 10 shows the overall bond graph model of the compressor system.

5   MODEL VERIFICATION AND SURGE CHARACTERISTICS ANALYSIS

The bond graph model of a compressor, which is a part of a 16PA6STC diesel engine for a certain type of marine ship, is built and simulation is carried out. For verifying the validity of the model, the simulation results are compared with the test data. Part of the structural parameters of the supercharger compressor is shown in Table 1.

5.1   Model verification

5.1.1   Steady state
Figure 11 shows the comparison of the simulation results and measured steady compressor map at speeds of 15,000, 17,500, 20,000, 22,500, 24,500, 26,500, 28,000 rpm. The measured map is extended to surge and negative flow zones which are obtained by Equation (1) and Equation (2), as

shown in Figure 11. The graph shows quite a good agreement, especially for low speed.

5.1.2 *Surge performance verification*

For the validation of the model in surge condition, the surge phenomenon is simulated by closing the valve from fully open to 0.15 gradually. Figure 12 shows the evolutions of the outlet pressure and mass flow rate; by looking at these it can be seen that the model predicts well the deep surge development. Figure 13 shows a zoom of the previous one and the calculated values are compared with the measured data. The results show good agreement in terms of amplitude and frequency of the pulses of pressure and mass flow rate.

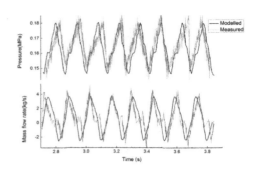

Figure 13. Comparison of calculated and measured data in surge state.

Table 1. Compressor structure specifications.

Parameter	Value
Number of impeller blades	20
Outer diameter of impeller inlet	238.9 mm
Inner diameter of impeller inlet	118.5 mm
Impeller outlet diameter	349.7 mm
Blade width	1.6 mm
Sweepback angle	3.17°

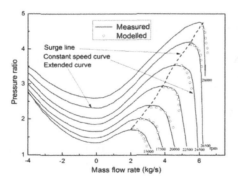

Figure 11. Comparison of simulation results with the measured data in steady state.

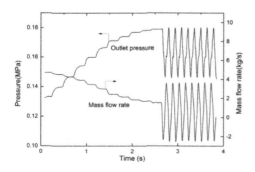

Figure 12. Evolution of compressor surge simulation.

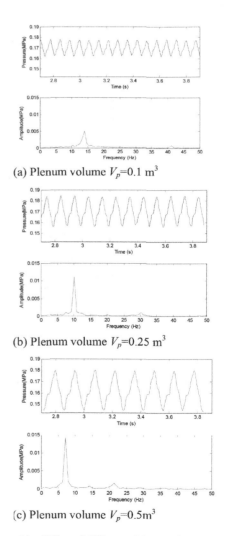

(a) Plenum volume V_p=0.1 m^3

(b) Plenum volume V_p=0.25 m^3

(c) Plenum volume V_p=0.5m^3

Figure 14. Effect of different plenum volume on surge behaviour.

5.2 Surge characteristics analysis

The 16PA6STC diesel engine is a sequentially turbocharged engine. The volume of plenum can be changed in a short time when one of the turbochargers is forced to blow against a closed valve. For this reason, an additional simulation was performed to analyse the effect of the plenum volume on surge behaviour. Figure 14 shows the evolution of outlet pressure and corresponding spectrum map for the plenum volume of 0.1 m³, 0.25 m³ and 0.5 m³ respectively (Galindo J. et al., 2006). It can be seen from the graph that the amplitude of outlet pressure increases and the surge frequency decreases with the increase of the volume. The fundamental wave of the frequency signal gradually moves to the left. Compared to Figure 14(a), the amplitude of the pressure in Figures 14(b) and (c) increased 291.5% and 359.2% respectively. The fundamental wave frequency is reduced from 14.0 Hz, to 10.1 Hz and 7.1 Hz respectively and the amplitude increased to 120.4% and 187.8%.

6 CONCLUSION

An experimental and modelling work of a compressor system based on bond graph has been presented in this paper aimed to analyse the surge phenomenon of the compressor. Firstly, the modelling of the compressor was introduced. Then the simulation and experiment of both the steady state and the surge state were conducted to verify the model. Finally, the effects of different plenum volume on surge behaviour were analysed. Simulation results show that:

1. The energy flow shown in the bond graph method has obvious physical meaning and gives a simple description of compressor working process, which is quite suitable for the modelling of complex physical systems that coupled among multi-energy domains.

2. The established compressor model has high precision and can be used to analyse the influence of several parameters on the performance of the compressor, so as to provide reference for the analysis and optimisation of the compressor system.

ACKNOWLEDGMENTS

Supported by the National Nature Science Fund of China (51579242).

REFERENCES

Afshari H.H., Zanj A., Novinzadeh A.B. (2010). Dynamic analysis of a nonlinear pressure regulator using bondgraph simulation technique. *Simulation Modelling Practice and Theory*, 18(2), 240–252.

Fink D.A., Cumpsty N.A., Greitzer E.M. (1992). Surge dynamics in a free-spool centrifugal compressor system. *Journal of Turbomachinery*, 114, 321–332.

Galindo J., Serrano J.R., Guardiola C. et al. (2006). Surge limit definition in a specific test bench for the characterization of automotive turbochargers. *Experimental Thermal and Fluid Science*, 30, 449–462.

Gravdahl J.T. (1998). *Modeling and control of surge and rotating stall in compressors*. Norwegian University of Science and Technology, Trondheim.

Moore F.K. and Greitzer E.M. (1986). A theory of post-stall transients in axial compression systems: Part I-development of equations. *Journal of Engineering for Gas Turbines and Power*, 108, 68–76.

Sun J., Guo C., Wei H. et al. (2008). Calculation and characteristic analysis on turbocharger surging of marine diesel engine. *Journal of Dalian Maritime University*, 34(1), 28–31.

Wang W. and Wang Y.. (2007) Establishment of a dynamic model for a compressor and analysis of the surge process. *Journal of Engineering for Thermal Energy*, 22(2), 124–128.

Automotive, Mechanical and Electrical Engineering – Liu (Ed.)
© 2017 Taylor & Francis Group, London, ISBN 978-1-138-62951-6

Exploration of the operational management of a pig farm based on mathematical modelling

Ni Ruan & Luling Duan
Department of Mathematics and Information Science, Guangxi College of Education, Nanning, Guangxi, China

ABSTRACT: Optimisation is a common issue facing people in the fields of engineering technology, scientific research and economic management. This paper, taking the operational management of a pig farm as an example, makes simplifying assumptions, changes the original breeding process diagram, and takes one month as the interval to obtain the recurrence relation. According to the required targets, optimisation models are respectively established to obtain the result that seven piglets per sow per year can reach or exceed the breakeven point and three years' total profit of 8,854,166 RMB, namely an average annual profit of 2,951,388 RMB. The whole analysis and solution process reasonably reflects the situations occurring in the operational management of this pig farm, providing theoretical support for determining its operating strategy.

Keywords: Pig farming; Breakeven; Optimisation

1 INTRODUCTION

The issue of optimisation is extremely common in the real world. For instance, to meet strength requirements and other conditions, a designer needs to select the size of the material in order to minimise the light of total weight. According to production cost and market demand, a factory manager needs to determine the product price in order to obtain the greatest profit. In the case of a forest fire, taking into consideration the minimum sum of forest damage and rescue cost, the fire station needs to determine the number of firefighters to be dispatched. When facing a solution optimisation issue, most people tend to rely on experience and combine too many personal subjective factors, but cannot confirm the optimality of the result; or they spend a lot of money and manpower and carry out repeated tests, but the result is still in expectation. At present, the country strongly advocates the elimination of backward production capacity and is seeking to develop the economy. Given today's increasing emphasis on scientific and quantitative decisions, solving optimisation issues using mathematical modelling methods undoubtedly meets the need of situation development.

The methods and steps for establishing an optimisation model are as follows: (1) Clarify the objective of optimisation; (2) Find the decision-making method for the solution and make a reasonable simplifying assumption for the issue; (3) Restrictions in decision-making; (4) Express by mathematical tools, namely establish the objective function and solve; (5) Perform qualitative and quantitative analysis of the results and check the rationality of the model.

Now, by combining teaching, explain the process of establishing an optimisation model in terms of the operational management issue of a pig farm in the modelling contest.

2 CASE

Background Knowledge. A pig farm can breed up to 10,000 pigs and use its boars for breeding. The general process of pig breeding is that, after a pregnancy of about 114 days, the fertilised sow will give birth to piglets, and the piglets will grow into porkets after lactation. Some of the porkets will be selected as boars, and will undertake the task of reproduction in the pig farm. Sometimes, a number of the porkets will be sold as piglets in order to control the breeding scale. After castration, most of the porkets will grow into pork pigs for slaughter. The reproduction period of a sow is generally three to five years. Sows and boars losing fertility will be harmlessly disposed of. Boars and pork pigs need feeding every day, but the feed cost of boars is higher. According to the market situation, farms optimise operating strategy by determining the number of boars reserved, fertilisation time, livestock scale and so on in order to improve profitability. Please collect relevant data, establish

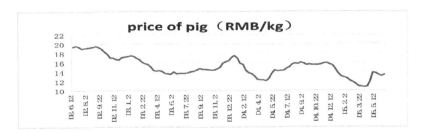

Figure 1. The prediction curve of the price.

a mathematical model and answer the following questions:

1. Assuming the cost of pig breeding and the price of a pig remain unchanged, piglets are not sold, and all porkets are turned into boars and pork pigs, to reach or exceed the breakeven point, how many piglets will a sow have on average per year?

2. Given that it takes nine months from fertilisation to slaughter, assuming that the prediction curve of price change of pig within three years after slaughter is show in the figure, please determine, according to the price prediction, the optimal operating strategy of this pig farm, calculate the average profit within three years, and provide the curve of number of livestock for sow and pork pig in this strategy.

Question Analysis. (1) The key to this question is to reasonably define the variables (pig breeding cost, pig price, breeding stock of boar and breeding stock of sow), define the relationship between annual slaughter and breeding stock in this farm as well as the annual elimination rate of boar, and establish the optimisation model with the minimum annual average farrowing amount as the objective function, and reaching the breakeven point as the restriction. (2) The key factor to consider with this question is that the breeding stock of sows and the breeding rate of sows at the breeding age directly affect the breeding stock of live pigs after nine months, thus affecting the profit ration. Therefore, first be clear that the income source of this pig farm is selling pork pigs and piglets. According to the prediction curve of the pig price provided, determine the reserve percentage to control the quantity of produced pork pigs as well as the quantity of piglets and give the corresponding breeding process diagram. Take one month as the interval to establish the recurrence model. Based on the recurrence results, establish the optimisation model with optimal profit as the objective, calculate the annual average profit within three years, determine the breeding stocks of sows and pork pigs in a certain period, and obtain the optimal operating strategy of this pig farm through prediction.

Model Assumption. (1) The farm uses the advanced artificial insemination technology (breeding stock of boar: breeding sock of sow = 1:100). (2) Each sow gives birth twice. Each time, nine piglets survive. (3) Pig farming costs remain unchanged; the feeding cost of a pork pig and spare boar is 100 RMB/head/month; the feeding cost of a boar in the growth period is 150 RMB/head/month; the price of a piglet is 300 RMB/head; the weight of a pork pig at slaughter is 100 kg.

3 MODELLING AND SOLUTION

Problem 1. Given the following notations: F: annual average number (heads) of piglets each sow has per year; Z: breeding stock (heads) of boar during growth period; Y: breeding stock (heads) of pork pig; C: total breeding stock (heads).

According to the assumption, the annual slaughter is two times the breeding stock and the pigs slaughtered are pork pigs; the elimination period is four years, then the annual elimination quantity of boars is $\frac{1}{4} \times Z$. Breeding stock of boar: breeding stock of sow = 1:100, then breeding stock of sow is $\frac{100}{101} \times Z$. Calculate the annual average farrowing amount of each sow by dividing the annual total farrowing amount (farrowing amount equals sum of annual breeding stock and number of eliminated boars) by breeding stock of sows, and establish the (1) 1 of annual minimum total farrowing amount:

$$\min F = \left(2C + \frac{1}{4} \times Z\right) \div \left(\frac{100}{101} \times Z\right)$$
$$s.t. 0 \le (2C \times 100 \times 12) - 2C \times 100 \times 9 - Z \times 150 \times 12 \quad (1)$$
$$0 \le Z \le Y \le Y + Z \le 10000$$

Use Lingo software to solve the above model and obtain $F = 7$ Under the given condition, it is obtained that seven piglets per sow per year on average can reach or exceed the breakeven point and will not affect the scale of this pig farm.

Problem 2. According to the result analysis of Problem 1, determine the reserve percentage to control the quantity of pork pigs as well as the quantity of piglets, and change the original breeding process Diagram 1 to the breeding process Diagram 2.

Assume that each sow gives birth to 18 piglets per year, then a sow can give birth to 1.5 piglets per month. According to the question, given the fertility period of a sow is generally three to five years, averaging three years, so assume the elimination rate of sow and boar is 1/48. According to the literature (Shouzhi, 2003), the lactation period of the piglet is 35 years, it takes 35 days for porkets to turn into piglets, pork pigs and spare sows, and it takes 110 days for a pork pig to grow until slaughter. According to the predicted data for pig prices within three years in the question, we take one month as the interval, and establish the following recurrence (2):

$$
\begin{aligned}
S_{n+1} &= S_n - \frac{1}{48}S_n + \frac{30}{110} \times \frac{25}{26} \times H_n \\
G_{n+1} &= G_n - \frac{1}{48}G_n + \frac{30}{110} \times \frac{1}{26} \times H_n \\
H_{n+1} &= C_n L_n \times \frac{30}{35} \\
R_{n+1} &= \frac{3}{2}S_n \\
L_{n+1} &= \frac{30}{35}R_n \\
Y_{n+1} &= C'_n L_n \times \frac{30}{35} \\
P_{n+1} &= \frac{30}{110} \times Y_n \\
Q_{n+1} &= (1 - C_n - C'_n) \times L_n \times \frac{30}{35}
\end{aligned}
\tag{2}
$$

Wherein: S_i: breeding stock of sow; G_i: breeding stock of boar; R_i: breeding stock of piglet; L_i: breeding stock of porket (nursing pig); Y_i: breeding stock of pork pig; H_i: breeding stock of spare boar; Q_i breeding stock of piglet; P_i slaughter quantity of pork pig; x_i: average pig price per month within three years after nine months; C_i: percentage of porkets selected as boars; C'_i: percentage of porkets selected as pork pigs, $(i, j = 1, 2, \ldots, 36)$.

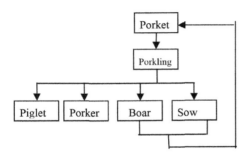

Figure 2. Original breeding process.

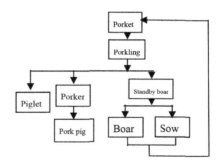

Figure 3. New breeding process.

From the literature (Shouzhi, 2003), according to the recurrence (2), by adjusting the percentage of reserve C_i, the percentage of pork pig C'_i and the number of piglets sold, establish (3) with the highest total profit within three years as the objective function:

$$
\begin{aligned}
\max = \sum_{i=1}^{36} \big(& P_i \times x_i \times 100 + Q_i \times 300 - (S_i + G_i) \times 150 \\
& - (L_i + H_i + Y_i \times 100) \big)
\end{aligned}
$$
$$
P_i + Q_i + S_i + G_i + L_i + H_i + Y_i \leq 10000, i = 1, 2, \cdots, 36
\tag{3}
$$

Use Lingo software for the solution and obtain the total profit 8,854,166 RMB, within three years, namely an annual average profit of 2,951,388 RMB. Meanwhile, give the breeding stock of sow and the breed stock of pork pig within three years according to this result, and draw the corresponding curve graph.

Model Evaluation: Problem 1, make simplifying assumptions, consider the relationship between various cost parameters and pig price and, according to the model, relatively ideally, calculate the farrowing amount of each sow per year. This result, to a certain degree, can determine the breeding stock of sow, thereby saving the breeding cost. Problem 2, clarify that the income source of this pig farm is selling pork pigs and piglets, according to the prediction curve of pig price given in the question, give the corresponding breeding process diagram, establish a clever recurrence mode, obtain the optimisation model with optimal profit as the objective, calculate the annual average profit within three years, and determine the breeding stock of sow and breeding stock of pork pig within three years. However, in the modelling process, we should note that there is a certain deviation between the predicted pig price given in the question and the actual price. Therefore, by considering the deviation, we can better determine the optimal operating strategy of this pig farm.

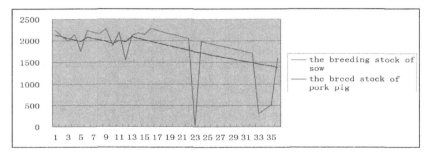

Figure 4. The curve graph of the breeding stock of sow and breed stock of pork pig.

4 CONCLUSION

From the abovementioned assumed conditions of the mathematical model: 1) boar breeding stock: sow breeding stock = 1:100; 2) each sow gives birth twice annually with nine piglets surviving each delivery; 3) the feeding costs of pigs remain unchanged with the feeding cost of pork pigs and breeding pigs being 100 RMB/head/month, and the feeding cost of pigs in the period of duration being 150 RMB/head/month; the selling price of piglets is 300 RMB each; the weight of pigs that are ready for slaughter is 100 kg. $F = 7$ is obtained by utilising Lingo software to solve the mathematical model, that is, as long as the annual average farrowing quantity of each sow reaches seven, the breakeven point can be achieved or surpassed, and there will be no impact on the scale of the livestock farm. Under the condition that each sow delivers 18 piglets, each sow can give birth to 1.5 piglets monthly. As the period of duration for sows generally lasts for three to five years, the monthly elimination rate of boars and sows is 1/48 by taking the average duration period of four years. Through making use of Lingo software again, it can be obtained by solution that the total gross profit of the three years is 8,854,166 RMB, that is to say, the annual average profit is 2,951,388.67 RMB. In the meantime, the sow breeding stock and the

pork pig breeding stock within the three years have also been given under this result. In addition, the corresponding curve has been drawn.

ACKNOWLEDGEMENTS

This paper is supported by the Guangxi Office for Education Sciences Planning, China (project approval number: 2013C108), and Guangxi Provincial Natural Science Research Project for Universities (project approval number: 2013YB378), and Characteristic Professional Project fund of the Education Department of Guangxi, China (project approval number: GXTSZY277).

REFERENCES

Lijie, Z., & Xibiao, W. (2009). Simulation study of dynamic change of pig herd in pig farm. *Northeast Agricultural University* (in Chinese), *40*(6).

Ni, R. (2013). Primary investigation of ordinary differential equation teaching by integrating mathematical modeling thought. *University Education* (in Chinese), (4).

Qiyuan, J., & Jinxing, X. (2003). *Mathematical model.* Higher Education Press, Beijing.

Shouzhi, Z. (2003). Pig herd structure in modern pig breeding and calculation of pigsty configuration. *Pig Breeding* (in Chinese) (2).

Automotive, Mechanical and Electrical Engineering – Liu (Ed.)
© 2017 Taylor & Francis Group, London, ISBN 978-1-138-62951-6

Global large behaviour on a 2-D equation in turbulent fluid

Hongen Li
Department of Basic Courses, Zhengzhou University of Industrial Technology, Zhengzhou, China

Shuxian Deng
Department of Basic Courses, Zhengzhou University of Industrial Technology, Zhengzhou, China
College of Science, Henan University of Engineering, Zhengzhou, China

ABSTRACT: We consider global large behaviour on a 2-D equation in turbulent fluid. We use a so-called energy perturbation method to establish results of numerical simulations of some selected unsteady flows via the analysis of the governing PDE. We show that boundary conditions necessary to solve the incompressible 2-D equation are conditions either for the normal or alternatively for the normal velocity with discretised algebraic method, by estimating the relationship between energy inequalities and attenuating property of weak solutions. Under various assumptions on u, we obtain estimates for the size of its large behaviour.

Keywords: 2-D; Behaviour; Global

1 INTRODUCTION

We consider the control problem of weak decay to the following so-called quasi-linear wave equation in a smooth bounded domain $\Omega \subset R^2$:

$$u_{tt} - \lambda \Delta u_t + u \cdot \nabla u = f(x, t) \qquad (1)$$

$$\operatorname{div} u = 0 \qquad (2)$$

$$u(x,0) = u_0, \quad u_t(x,0) = u_1 \qquad (3)$$

where u_0, u_1 and f are care given functions, Δu is a Laplacian with respect to the variable $x \in \Omega$, $u = u(t, x)$ is an unknown function, $\lambda > 0$ is a fixed positive number, $f \in L^2$ and (Ω) are given external forces, and satisfying the following conditions:

$$\frac{2}{p} + \frac{n}{q} \leq 1$$

For some positive $a, p \in [1/2, 2)$, $C > 0$, and $q > 0$, we have a weak solution, which fulfils additionally:

$$d + a \mid x \mid^p \leq f'(x,t) \leq d(1 + \mid x \mid^q), \forall s \in \mathbb{R}^2 \qquad (4)$$

And:

$$u_{tt} - \operatorname{div} \lambda \nabla u^2 - \Delta u_t = f(x, t) \qquad (5)$$

We have several techniques to prove the existence of weak decay solutions with respect to the phase space $X^t \times X^r$, $X^e := D(\Omega^{q/p})$; we have additional nice properties with energy inequality for almost all times or solutions with weak decay properties for $t \to \infty$. This has been studied recently by several people, for example Qin (2013), Qin, Liu and Deng (2015), Nakao (2015), Nakao (2016), Friedman and Necas, (2014), and Beirao da Veiga (2014), among others.

As a model of the quasi-linear wave equation, for $N = 1/3, 2$ and $f = 0, 1$, Equations (1)–(3) admit a global weak decay solution as large initial data, which was previously proved (Qin et al., 2015; Nakao, 2015). Wang, Zhang, and Wang simplified the above arguments to give the proof of control with exponential decay. From the perspective and background of physics, this represents an implementation of an axial movement of the viscoelastic material; this causes the form of the above equations, in the one-dimensional case, and their model of longitudinal vibration of a uniform rod with nonlinear stress function, *f*. In two, 1-D case, they describe the viscoelastic solid of anti-plane shear action. While $n = 1$ and $f = 0$, Li proved the existence of weak periodic decay strong solution on the periodicity condition; Su and Zhang (Nakao, 2015; Nakao, 2016; Friedman & Necas, 2014; Beirao da Veiga, 2014) proved the controllability of a smooth solution in the method of Cauchy problems, in the case of smooth and small data.

While $n > 1/3$ and $f = 0$, Wang and Yang gave the proof of global large behaviour with a smooth solution, in the case of small initial data. Making use of combining the L^p-theorem of Sobolev space and semi-group theorem of operators, Nakao (2016) and Kato (Friedman & Necas, 2014) devised a certain decay rate of the energy of global solutions with large data under a specific condition, which is certainly satisfied if the mean curvature of the boundary $\partial\Omega$ is non-positive. For $n > 1$ and $f = 1$, nonlinear elliptic equation with periodicity conditions was studied (Kato, 1984):

$$x'(t) + \tau_0 \left(1+t\right)^{(t+\frac{d}{2})} x(t)^{\frac{(2+\frac{\gamma}{p})}{p}} \leq \sigma y(t)^{\frac{\nu}{2}} + \tau_1 (1+t)^{\delta} \quad (6)$$

$$\left\| w_t \right\| - \Delta w + \left\| u(r) \right\|_q = \frac{n}{2} + 1 - pn + 2p\delta \quad (7)$$

With $w(t) = u(t) - u_0(t)$ c, and

$$y(t) = \left\| w(t) \right\|_p^m \quad (8)$$

$$w(x,\ t+\omega) = w(x,\ t) \quad (9)$$

$$d'(t) + k_1 \left\| w_1(t) - w_2(t) \right\|_p^\nu d(t) \leq k_2 d(t)^{p+q-1} + \kappa d(t)^\mu \left\| w_1(t) \right\|_2^{\nu+1}$$

In the case of $\left\| w(t) \right\| \leq k_0 (1+t)^{-\frac{\lambda}{2}}$, and:

$$\left\| w(t) \right\|_p \leq \tau (1+t)^{(\frac{\tau}{2}+\frac{n}{2}(\frac{1}{2}-\frac{1}{p}))} \quad (10)$$

Thus, we can conclude immediately the following results (Beirao da Veiga, 2014):

Lemma

Let $f(x,\ t) \in H^1\left(I,\ L^2(\Omega)\right)$ and $u_0,\ u_1 \in H^2(\Omega) \cap H_0^1(\Omega)$, then under assumption H^1, the problem admits a unique solution $u(t)$ in the class

$C^2(I,\ L^2) \cap C^1(I,\ H^2 \cap H_0^1) \cap H^2(I,\ H_0^1)$:

$$M_i^2 = \int_0^\omega \left\| \frac{\partial^i}{\partial t^i} f(s) \right\|^2 ds,\ i = 0,1 \quad (11)$$

And further, if $\alpha > 0, t \to \infty$ for some constant as $\left\| f(t) \right\|^2 + \left\| f_t(t) \right\|^2 = O(e^{-\alpha t})$, there holds the estimate:

$$\left\| u_{tt} \right\|^2 + \left\| \Delta u \right\|^2 + \left\| \nabla u_t(t) \right\|^2 + \left\| \Delta u_t \right\|^2 + \int_t^\infty \left\| \nabla u_{tt} \right\|^2 ds \leq Ce^{-\beta t}$$

$$\sup_{t\in[0,\omega]} \left\{ \left\| u(t) \right\|^2 + \left\| \nabla u_t(t) \right\|^2 \right\} + \int_0^\omega \left\| u_t(s) \right\|^2 ds \leq CM_0^2.$$

For some constant $\beta > 0$, where $\left\| u_0 \right\|_{H^2} + \left\| u_1 \right\|_{H^2}$ is a constant depending on:

$$\sup_{t\in[0,\omega]} \left\{ \left\| u_t(t) \right\|^2 + \left\| \nabla u_{tt}(t) \right\|^2 \right\} + \int_0^\omega \left\| u_{tt}(s) \right\|^2 ds \leq C\left(M_0^2 + M_1^2\right).$$

It turns out that this has an influence on the control of weak solution with structural damping. We can continue the following argument.

2 MAIN RESULTS

We note that Equation (6) implies the existence of u_0, $u_1 \in H^2$ as a distribution. Also Equation (4) is known as the energy inequality. This concept of a solution appears too weak to guarantee that $C^2(I,\ L^2)$ satisfies local energy type inequalities or that p is an integrable function. In order to describe the manoeuvrability, let us define global weak solution with decay for Equations (1)–(3).

In the above theorem, we can denote the Hausdorff function ω, one measure that uses parabolic cylinders for coverings (as defined above in Equation (7)); we say that Equation (5) is a suitable weak solution to Equations (1)–(3), if u and ω are Lebesgue measurable functions on R^2 and:

(a)

$$E(t) \equiv \frac{1}{2}\left\{ \left\| \omega_t(t) \right\|^2 + \int_\Omega \int_0^{|\nabla\omega|^2} \tau(\xi)\,d\xi dy \right\}$$

$$= \int_0^t \left\| \nabla \omega_t(t) \right\|^2 ds + \frac{1}{2}\left\{ \left\| \nabla \omega_1(t) \right\|^2 + \int_\Omega \int_0^{|\nabla\omega_0|^2} \tau(\xi)\,d\xi dy \right\}$$

(b)

$$\int_0^t \left\{ \left(\text{div}\left(\tau\nabla\omega\right), \mu\right) - \left(\Delta\omega_t, \mu\right) - \left(f, \mu\right) \right\} = 0$$

and (c)

$\omega \in \Theta$.

If the initial data satisfy u_0, $u_1 \in H^2 \cap H_0^4$, our particular interest has been the initial boundary value problem on a bounded smooth domain. We focus on the control of weak decay stability in Equations (1)–(3), to describe the manoeuvrability. Let us define global weak solution with decay for Equations (1)–(3), and construct a weak solution space with:

$$\Theta = \left\{ \Gamma : \Omega \times w_{1/2}^2 \to R \mid_\Gamma \in L^\infty,\ \Gamma' \in L^\infty(w_{1/2}^2,\ H_0^4(\Omega)) \right\}$$

Obviously, we will construct the controllability of the local weak solution with decay for the semigroup generated by weak energy solutions:

$$\int_0^t \left\{ (\omega_t, \mu_t) - (\mathrm{div}(t \nabla \omega), \mu) - (\Delta \omega_t, \mu) - (f, \mu) \right\} = 0$$

$$t^2 \|\partial_t v\|_H^2 \le \lambda t^2 \|\partial_t v\|_{L^\infty}^{1/2} \|\partial_t v\|_{L^2}^{3/2} \le C^3 \rho^2 \delta^{-2} \|\partial_t \omega\|_{L^\infty}^2$$

$$(\partial_t^2 u, \phi) = \rho(\partial_t v, \Delta_x \phi) + (v, \Delta_x \phi) - (f(u), \phi) + (g, \phi)$$

Theorem

Let ω be a weak solution with the decay of Equations (1)–(3). Denoted by $\omega_0(t) = e^{-t\Delta}$, the solution of the wave equation and suppose $\|\omega_0(t)\| \le \lambda_0 (1+t)^{-\tau}$, $\omega_0 \in H^2(\Omega) \cap H_0^4(\Omega)$, $v(t)$, $f'(x,t)$, $\omega_t \in H_0^4(\Omega)$ $f(x,t) \in H_0^2(\Theta, L^2(\Omega))$ and $\in W_0^2(\Theta, L^2(\Omega))$, then Equations (1)–(4) hold a unique local weak solution, $\omega(t)$, with the following estimate:

$$y'(t) + y(t)^{\left(1 + \frac{\varepsilon}{2}\kappa\right)} \le \lambda_1 y(t)^{\frac{\gamma}{2}\mu} + \lambda_2 (1+t)^{-q} \cdot \nabla x \omega(t) \tag{12}$$

$$\omega \in L^\infty([0,T], H^2(\Omega) \cap H_0^4(\Omega)) \tag{13}$$

$$\omega' \in L^\infty(\Theta, H_0^4(\Omega)) \tag{14}$$

$$\omega'' \in W_0^2(\Theta, L^2(\Omega)) \tag{15}$$

3 CONCLUSION

To verify the correctness of the above theorem, we will give its proof. In fact, for $n=3$ $\omega \in L_1 \cap L_p$, and $\|v(t) - v_0(t)\| \le \sigma(1+t)^{\frac{z}{3}\left(1 - \frac{1}{q}\right)}$ note that the initial data are dense, hence the module $\|v_0(t)\|$ attenuates exponentially fast. Indeed, using $|\omega|^{2p/n(p-2)}$ and $\partial_t \omega$, multiply Equation (7) and integrate by parts over $x \in \Omega$. One gets:

$$\frac{\partial}{\partial t} \left(\|\partial_t(\omega)\| + \lambda(\theta(\nabla u), 1) + \sigma \|\nabla \omega\|_{L^\infty}^2 \right.$$
$$\left. + \frac{1}{2}(G(v), 1) + (\varphi, \varepsilon) \right) + \eta \|\nabla \partial_t h\|_{H_0^2}^2 = 0 \tag{16}$$

$$\delta'(t) + \lambda_1 (1+t)^{\kappa \frac{d}{2}} y(t)^{\lambda + \frac{\kappa}{p}}$$
$$\le \lambda_2 \delta(t)^{\frac{n}{2}\left(\frac{1}{r} - \frac{1}{2}\right)} + \varepsilon(1+t)^{-(p-r)} \delta(t)^{\frac{1}{q}} \tag{17}$$

Where ε is a small positive number, which will be fixed. Then we arrive at:

$$\frac{\partial}{\partial t} \left(\frac{\rho}{2} \|\partial_t(\omega)\|^2 + \lambda(\partial_t(\omega), \omega) \right) - \|\partial_t \omega\|_{L^\infty}^2 + \theta \|\nabla u\|^2$$
$$+ \tau(\varphi'(\nabla \omega), \nabla \omega) + \frac{1}{2} \upsilon(G(v), v) = \beta(g, \omega)$$
$$\tag{18}$$

And notice that:

$$E(\xi_u(t)) = \frac{p-2}{p} \|\partial_t \omega\|_{L^2}^p + (\varphi(\nabla \omega), \kappa)$$
$$- \frac{pn+4}{p+4-n} \|\nabla \omega\|_{L^2}^2 + \lambda(\partial_t \omega, \omega) \tag{19}$$

Let λ, α be small enough, one gets:

$$\int g(|v|^2 + |\omega_0|^4)|u|^{\frac{pm}{n-2}} dx \le \sigma \left(\|v\|_{L^2}^{\frac{p}{2}} + \|\omega_0\|_p^4 \right) \|\omega\|_{H_0^2}^{\frac{p}{2}-1}$$

$$\le C \left(1 + \|g\|_{L^2}^n \right) + \|\partial_t v\|_{L^2}^{n-2} + \|\nabla x^u\|_{L^p}^{n+1} + \|v\|_{L^q}^{\frac{p}{2}} \right)^{\frac{n+1}{p}}$$

$$\le C \|v\|_{L^p}^{n+1-p} \left(\beta(u)^{\frac{p}{n}} + \|v_0\|_{L^p}^{p+1} \right) \beta(u)^{2 + \frac{n}{q}} \tag{20}$$

where the constants C_ε and λ_ε depend only on ε.

Hence by the standard Galerkin method, inverse domination principle, interpolation inequality, and Young and Sobolev inequality, this section can be estimated by:

$$(\varphi'(\xi_2) - \varphi'(\xi_1))(\xi_2 - \xi_1) \le \theta(|\xi_2| - |\xi_1|)^p |\xi_2 - \xi_1|^2,$$
$$\forall \xi_1, \xi_2 \in \mathbb{R}^2 \tag{21}$$

Thus:

$$\alpha \|u\|_{L^q}^{n+1} \left(\|v_0\|_{L^{2p}}^4 + \|u\|_{L^2}^{\frac{1}{2}\frac{1}{p}} \right)$$
$$\le \alpha \|u_0\|_{L^{2p}}^{p+2} \|v_0\|_{L^p}^{p-1} + \lambda \varepsilon + C_\varepsilon \|u\|_{L^p}^{\lambda p} \tag{22}$$

By the properties of heat kernel, we deduce that:

$$\partial_t \left[\frac{pn - 2n + 2p\alpha}{p+4-n} \|\nabla v\|_{L^2}^{p+2} + (\partial_t v, v) \right]$$
$$+ \tau_0 (|\nabla_x \omega_1| + |\nabla_x \omega_2|)^{\frac{pn}{n-2}} |\nabla_x u|^2 + (|v_1| + |v_2|)^p$$
$$\le C_1 \varphi(t)^{\frac{1}{p}} + C_2 (1+t)^{-\mu \tau} \varphi(t)^{\frac{1}{\gamma}} \tag{23}$$

$$\partial_t \left(\|\partial_t v\|_{L^2}^4 + \|u\|_{L^2}^2 \right) + \rho \|\partial_t u\|_{L^{2p}}^{p+\mu} + C \|u\|_{L^2}^p + \|\partial_t u\|_{L^2}^\mu$$
$$\le \kappa(|g(v_1)|, -|\partial_t v|)^{1+\tau} + \frac{1}{2}(|\varphi'(\nabla v_1)|, |\nabla_x \partial_t v|)^{-\frac{n}{2}}$$
$$\le C(1+t)^{-p\upsilon} + C_\varepsilon (1+t)^{-\frac{n}{2}\left(\frac{1}{2} - \frac{1}{q}\right)} + C_0 (1+t)^{-\gamma} \tag{24}$$

Noting the energy solution of Equation (1) is unique and the Lipschitz continuity in a weak space, there holds Equations (20)–(24).

Analogously, for some positive ε, υ and ρ, together with the Hölder inequality and the interpolation:

$$\|u\|^2 + \|g\|^2 + \frac{C\varepsilon^2}{4}\|\nabla u\|^2 + \|g\|^2 + \kappa_1\|\partial_t u\|_H^2 + \|u\|_H^2$$
$$\leq (g,u_t) - \|\nabla u_t(t)\|^2 + \kappa_1\|\partial_t u\|_H^2 + \|u\|_H^2 - \varepsilon\|\nabla u\|^2 \quad (25)$$
$$\leq \varepsilon\|u_t\|^2 - a\|\nabla u_t\|_{L^{2-a}}^{p+\nu} + \kappa_2\|\partial_t u\|_H^2$$

Similar to above, we derive the differential inequality as long as the local weak solution $u(t)$ exists, inserting the above estimates, and using the energy estimate for estimating the energy norms. One can get:

$$u_{tt} - \operatorname{div}\{(a\,|\nabla u\,|\,2r + 1)\nabla u\} - \Delta u_t + \Delta_x u = g(x,t)$$
$$(|g(u_1) - g(u_2)|, |\Delta_x \partial_t v|) + \Delta\omega_n + m_0\,|\nabla\omega_n|^q \quad (26)$$
$$\leq \varepsilon\|g(u_1) - g(u_2)\|_L^2 + C_\varepsilon\|\partial_t v\|_H^{2-s}$$

Alternatively, by Young's inequality, choosing $\varepsilon > 0$ small enough and $E(t)$ is bounded, we finally deduce the following:

$$g'(t) + E(t) + a(1+t)^\tau g(t)^{2+\sigma}$$
$$\leq E(0) + \beta(1+t)^{1-\tau} + C_6\int_0^\infty \|g(t)\|^2 dt \quad (27)$$

$$G_0'(t) + C_3 G(t) \leq C_4 M_1(1+t)^{-A} \quad (28)$$

$$g(t) \leq C_i(1+t)^{-pk} + KG(0)(1+t)^{-\lambda_1} \quad (29)$$

$$g'(t) < +C_4(1+t)^{\frac{\lambda\rho}{2}} + C_5 g(t)^\beta \quad (30)$$

Using now Equations (20)–(22) together with the interpolation inequality, we infer from Equations (25)–(30) that:

$$g'(t) + C_3 g(t) + G(t) \leq C_4 M_1(1+t)^{-\lambda_1}$$
$$\qquad + C_3\big(g(u_1) - g(u_2), |u|\big) \quad (31)$$

$$g'(t) + \|\xi_u(y)\|_{L^2}^{p+2} + C_0 g(t)^{p+2-\nu}$$
$$\leq C_6 g(t)^{\frac{n}{p(p+2)}} + C_7 e^{K(L-s)}\|\xi_u(x)\|_{L^2}^{p+2-n} \quad (32)$$

Now fix C and M. Then, taking advantage of the smoothing property together with the obviously bounded $E(t)$:

$$E(t) + G_0(t) + C_7\Phi(0) \leq 2\Phi(t)$$
$$\leq 2g(t) + C_8 M_3 e^{-\lambda t} + 2KG(0)(1+t)^{-K} \quad (33)$$

$$\|u\|_{L^{p+2}}^{r+\frac{1}{\nu}} + |g''(t)| + |g'(t)|$$
$$\leq C_8 G(u) + C_9\|u\|_{L^p}^{\frac{2\rho}{p+2}} + C_{10}\big(1 + \|g(t)\|\big)^{\frac{1}{2}} \quad (34)$$

Multiplying both sides of Equation (31) by e^{Ct} and integrating from 0 to t, we derive:

$$\int_0^t e^{C_0 s}\big(G'(s) + C_6\Phi(s)\big)ds + \int_{B(x_0,r)} \tau^2|\nabla u_k|^q dx$$
$$\leq \int_0^t C_7 K e^{(C_7-\lambda)s}ds + \int_{B(x_0,r)} \delta^2\Delta u_k dx \quad (35)$$

Obviously, from Equations (27)–(32), we can easily establish the control of stability for polynomial decay to Equations (13)–(15). In comparison to Equation (12), we give the strong stability estimates. Add the limit $t \to \infty$ to the dissipative Equation (12), for the approximations $\omega(t)$, and together with Sobolev embedding theorem, we can immediately conclude that the limit weak solution $\omega(t)$ also satisfies:

$$\|\nabla\omega\|_2 + \|\partial_t u\|_{L^2}^2 + (\partial_t v, v)\|\nabla\omega\|_H^2$$
$$\leq \|\partial_t u_1\|_{L^\infty}^2 + \int_0^t \|\nabla h(x)\|_{L^2} dx + \lambda(1+t) \quad (36)$$

Let the assumptions in Equations (2) and (4) be satisfied with $\theta = 2$, and $f(t) \leq \kappa e^{-2\rho t}$. Then, the local weak solution with attenuation $\omega(t)$ of Equations (1)–(3), which satisfies the additional regularity, and exists constants $M > 0$, $v > 0$, such that:

$$E(t) \leq Me^{-3vt}$$

Choosing θ appropriately, one deduces from Equations (12) the Caccioppoli inequality:

$$\int_{Q_r(x,t)} |\nabla u|^2 dx + \sup_{|s+t|\leq r^2}\int_{Br(x)} |v|^2\,|s-t|(v\cdot t)(u,s)dx$$
$$\leq c\theta^{-1}\int_{Q_{2r}(x,t)}(\cdot,s)^2 dxdt \quad (37)$$

$$cr^{-2}\int_{Q_8(x,t)} |s|^3(s,t)dtds$$
$$\leq \int_{Q_8(x,t)} |s|^3|\mu|]dxdt + \lambda\int_{Q_8(x,t)} |\rho|^3|f|dxdt \quad (38)$$

Let the assumptions of f being satisfied, and if $f(t) \leq \sigma_0(1+t)^{-\tau}$. Then, the following dissipative estimate holds for the unique local weak solution $\omega(t)$ of Equation (2):

$$E(t) \leq \lambda(1+t)^{-\tau} \quad (39)$$

Under the above hypothesis and supposing that $\omega(x,t)$ is a sufficiently regular weak solution of Equation (3), the following estimates hold:

$$t^2\|\partial_t v\|_H^2 + \int_n^t \|\partial_t^2\omega(\xi)\|_{L^2}^2 d\xi + \|\omega(t)\|^2 + \|\partial_t^2\omega(\xi)\|_H^2$$
$$\leq S\|\partial_t v\|_{L^\infty}^{1/2}\|\partial_t v\|_{L^2}^{3/2} + \delta t^3\|\partial_t v\|_{L^2}^3 + C^3\rho^2\delta^{-2}\|\partial_t\omega\|_{L^\infty}^2$$
$$+ e^{\epsilon(t-a)}\|\omega(t)\|_0^4 + \lambda v(t)u\big(\|f\|_{L^2}\big) \quad (40)$$

Here, repeated indices denote summation.
Similar to above, we derive the differential inequality as long as the local weak solution $u(t)$ exists, inserting the above estimates, and using the energy estimate for estimating the energy norms.

$$2C_\varepsilon e^{-C_\varepsilon t} \leq E(u(t)) \leq C_\sigma E\big(e^{-\lambda_\sigma t}\big)$$

The integral involving Equation (35) is understood as a principal value. Noting the energy solution of Equations (1)–(3) is unique and the Lipschitz continuity is in a weak space, there holds Equations (12)–(15). Hence, the theorem is completed, and we come to the conclusions on the global large behaviour.

ACKNOWLEDGEMENTS

This work was supported by the Ph.D. Foundation of Henan University of Engineering (No. D2010012).

REFERENCES

Beirao da Veiga, H. (2014). Existence and asymptotic behavior for strong solutions of the N-S equations. *Indiana Univ. Math. J., 36*, 149–166.

Cheng, K.-S. & Ni, W.-M. (2012). On the structure of the conformal scalar curvature equation on RN. *Indiana Univ. Math. J. 41*(1), 261–278.

Friedman, A. & Necas, J. (2014). Systems of nonlinear wave equations with nonlinear viscosity. *Pacific J. Math.,* 135, 29–55.

Friedman, A. & Necas, J. (2013). Systems of nonlinear wave equations with nonlinear viscosity. *Pacific J. Math.,* 135, 29–55.

Kalantarov, V. & Zelik, S. (2013). Finite-dimensional attractors for the quasi-linear strongly-damped wave equation. *J. Differential Equations* 1–36.

Kato, T. (1984). Strong LP-solutions of the Navier-Stokes equations in Rn, with applications to weak solutions. *Math. Z., 187*, 471–480.

Nakao, M. (2015). Energy decay for the quasilinear wave equation with viscosity. *Math. Z., 219*, 289–299.

Nakao, M. (2016). On the strong solution of some quasilinear wave equations with viscosity. *Advances in Mathematical Sci. Appl., 6*, 267–278.

Qin, Y., Liu, X., & Deng. S. (2015). Decay rate of quasilinear wave equation with viscosity. *Acta Math App Sinica, 17*, 147–152.

Qin. Y. (2013). Global existence of a classical solution to a nonlinear wave equation [J]. *Acta Math. Sci., 17*, 121–128.

Automotive, Mechanical and Electrical Engineering – Liu (Ed.)
© *2017 Taylor & Francis Group, London, ISBN 978-1-138-62951-6*

Glonass almanac parameters algorithm model

Xiaogang Xie & Mingquan Lu
Department of Electronic Engineering, Tsinghua University, Beijing, China

ABSTRACT: Aimed at the complexity of the GLONASS (Global Navigation Satellite System) almanac algorithm, this paper presents a simple and fast almanac parameters user algorithm model. This algorithm is reconstructed based on the GPS almanac parameters user algorithm and it considers the effects of harmonic perturbation, orbital period and the change rate of satellite orbital period. Meanwhile, the related calculation formula is derived and the calculation steps are given in detail. Finally, the algorithm is proved to be simple, effective and highly accurate using the precise orbit data of IGS (International GNSS Service, IGS). The results show that the algorithm can fully meet the user's requirement for position prediction precision.

Keywords: Satellite Navigation; GLONASS; Almanac Parameters User Algorithm

1 INTRODUCTION

Almanac parameters are an important part of the satellite navigation system, and plays a very important role during navigation signal acquisition process (Wang & Huang, 2013). In the absence of auxiliary information, the satellite receiver can estimate its compendium of position and velocity information according to almanac parameters. Thus the visible satellite can be reappeared and search directly. At the same time, according to the satellite velocity, the outline Doppler frequency can be estimated, which can assist the navigation receiver to search for the navigation signal quickly at the signal acquisition phase. It also greatly shortens the time to capture the satellite signals, so as to shorten the first positioning time (Li & Zeng, 2016; Wang & Zhang, 2015). Therefore, the simplicity and high efficiency of the almanac parameters user algorithm can directly affect satellite navigation signal tracking performance of receivers (Li & Wang, 2014; Anyaegbu & Townsend, 2014). Aimed at the complexity of the GLONASS (Global Navigation Satellite System) almanac parameter user algorithm (GLONASS-ICD, 2008), this paper presents a simple and fast GLONASS almanac parameters user algorithm. The calculation steps and specific formula are given in detail. The availability and accuracy of the almanac parameters user algorithm are proved to be high based on IGS (International GNSS Service, IGS) precise orbit data.

2 GLONASS ALMANAC PARAMETERS

The GLONASS satellite almanac parameters include the GLONASS time parameters, satellite

Table 1. GLONASS almanac parameters.

$t_{\lambda n}^A$	An instant of a first ascending node within a day
λ_n^A	Greenwich longitude of ascending node at instant
Δi_n^A	Correction to the mean value of inclination at instant
ΔT_n^A	Correction to the mean value of draconian period at instant
$\Delta \dot{T}_n^A$	Rate of change of orbital period
ε_n^A	Eccentricity of satellite orbit at instant
ω_n^A	Argument of perigee of satellite orbit at instant

orbit elements, satellite health information and so on. The satellite almanac parameters can predict the satellite position and velocity to enhance rapid acquisition of the navigation signal for the navigation receiver. The GLONASS satellite almanac parameters are shown in Table 1 (GLONASS—ICD, 2008).

3 GLONASS ALMANAC PARAMETERS USER ALGORITHM

According to the GLONASS interface control document, the almanac parameters user algorithm is very complicated. Based on the motion rule of navigation satellite, a fast prediction algorithm of satellite motion state is proposed. The specific steps are summarised as follows (Tang & Chen, 2002; Wen, 2009; Jing & Wu, 2015; Xie & Song, 2015):

1. Calculate the normalisation time:

$$t_k = t - t_{\lambda n}^A \tag{1}$$

2. Calculate the satellite orbit period:

$$T_k = \Delta T_n^A + 43200 + \frac{\Delta \dot{T}_n^A \cdot t_k}{\Delta T_n^A + 43200} \tag{2}$$

3. Calculate the average angular velocity:

$$n_{0k} = 2\pi/T_k \tag{3}$$

4. Calculate the semi-major axis:

$$A_k = \sqrt[3]{\mu/n_{0k}^2} \tag{4}$$

5. Calculate the eccentric anomaly and mean anomaly at the moment of satellite crossing the ascending node:

$$E_{0k} = 2\arctan\left(\frac{\sqrt{1-\varepsilon_n^A}}{\sqrt{1+\varepsilon_n^A}}\tan\left(-\frac{\omega_n^A}{2}\right)\right) \tag{5}$$

$$M_{0k} = E_{0k} - \varepsilon_n^A \cdot \sin E_{0k} \tag{6}$$

6. Calculate the change rate of the longitude of ascending node:

$$\dot{\Omega}_{0k} = -1.5n_{0k}\frac{J_2 R_e^2 \cos(i_k)}{\left(A_k\left(1-\varepsilon_n^A \cdot \varepsilon_n^A\right)\right)^2} \tag{7}$$

7. Calculate the change rate of average angular velocity:

$$\dot{n}_k = -2\pi \cdot \frac{\Delta \dot{T}_n^A}{T_k^3} \tag{8}$$

8. Calculate the eccentric anomaly and mean anomaly at normalisation time:

$$M_k = M_{0k} + n_{0k} \cdot t_k + \frac{\dot{n}_k \cdot t_k^2}{2} \tag{9}$$

$$E_k = M_k + \varepsilon_n^A \cdot \sin E_k \tag{10}$$

The mean anomaly should be calculated by iterative method.

9. Calculate the argument of latitude, the radius vector and orbit inclination respectively:

$$\phi_k = 2\arctan\left(\sqrt{\frac{1+\varepsilon_n^A}{1-\varepsilon_n^A}} \cdot \tan\frac{E_k}{2}\right) + \omega_n^A \tag{11}$$

$$r_k = A_k\left(1 - \varepsilon_n^A \cos E_k\right) \tag{12}$$

$$i_k = \Delta i_n^A + 0.35\pi \tag{13}$$

10. Calculate the change rate of the mean anomaly and the change rate of the eccentric anomaly:

$$\dot{M}_k = n_{0k} + 2\dot{n}_k \cdot t_k \tag{14}$$

$$\dot{E}_k = \frac{\dot{M}_k}{1-\varepsilon_n^A \cos E_k} \tag{15}$$

11. Calculate the change rate of argument of latitude, the change rate of radius vector and the change rate of right ascension of ascending node:

$$\dot{\phi}_k = \dot{E}_k\frac{\sqrt{1-\varepsilon_n^A \cdot \varepsilon_n^A}}{1-\varepsilon_n^A \cos E_k} \tag{16}$$

$$\dot{r}_k = \dot{A}_k\left(1 - \varepsilon_n^A \cos E_k\right) + A_k \varepsilon_n^A \sin E_k \dot{E}_k \tag{17}$$

$$\dot{\Omega}_k = \dot{\Omega}_{0k} - \omega_e \tag{18}$$

$$\dot{A}_k = -\frac{2}{3} \cdot GM^{1/3} \cdot n_{0k}^{-5/3} \cdot \dot{n}_k \tag{19}$$

12. Calculate the satellite position and velocity in orbit coordinate system:

$$x_k' = r_k \cos\phi_k \tag{20}$$

$$y_k' = r_k \sin\phi_k \tag{21}$$

$$z_k' = 0 \tag{22}$$

$$v_x' = \dot{r}_k \cos\phi_k - r_k\dot{\phi}_k \sin\phi_k \tag{23}$$

$$v_y' = \dot{r}_k \sin\phi_k + r_k\dot{\phi}_k \cos\phi_k \tag{24}$$

$$v_z' = 0 \tag{25}$$

13. Calculate the longitude of ascending node:

$$\Omega_{0k} = \lambda_n^A + \dot{\Omega}_k \cdot t_k - \omega_e \cdot t_k \tag{26}$$

14. Calculate the satellite position and velocity in PZ-90:

$$x_k = x_k' \cos\Omega_k - y_k' \cos i_k \sin\Omega_k \tag{27}$$

$$y_k = x_k' \sin\Omega_k + y_k' \cos i_k \cos\Omega_k \tag{28}$$

$$z_k = y_k' \sin i_k \tag{29}$$

$$v_{xk} = v_{xk}' \cos\Omega_k - v_{yk}' \cos i_k \sin\Omega_k - y_k\dot{\Omega}_k \tag{30}$$

$$v_{yk} = v_{xk}' \sin\Omega_k - v_{yk}' \cos i_k \cos\Omega_k - x_k\dot{\Omega}_k \tag{31}$$

$$v_{zk} = v_{yk}' \sin i_k \tag{32}$$

where, μ is the gravitational constant in PZ-90, J_2 is the harmonic coefficients of gravity field, and R_e is the earth's radius.

550

4 EXAMPLE AND ANALYSIS

The availability and accuracy of the GLONASS almanac parameters user algorithm mentioned above in this paper is verified using the precise orbit data of IGS (International GNSS Service, IGS) and STK (Satellite Tool Kit, STK) based on the corresponding GLONASS almanac parameters of IGS respectively. The interval of precise orbit data generated by STK software is one minute.

The satellite almanac parameters and corresponding ephemeris parameters of GLONASS satellite number 17 in IGS are shown in Table 2 and Table 3. The almanac parameters are used to predict the satellite position, to compare with the precise orbit data of IGS and the precise orbit data generated based on ephemeris parameters of IGS by STK.

The verification method is to contrast the satellite orbit data predicted using the almanac parameters of IGS above based on the GLONASS almanac parameters user algorithm proposed in this paper, with the satellite precise orbit data of IGS. The satellite orbit prediction interval is 15 min and its prediction time length is one day which is the GLONASS satellite orbit period. Since the reference time UTC (SU) of almanac parameters is not the same as the reference time UTC (USNO) of precise satellite orbit data, it needs to consider the conversion relationship between the two kinds of time:

Table 2. GLONASS almanac parameters of IGS.

Almanac parameters	Data
$t_{\lambda n}^A$	11388.0937
λ_n^A	0.2944613
Δi_n^A	0.01045895
ΔT_n^A	−2655.971
$\Delta \dot{T}_n^A$	−0.0003051758
ε_n^A	0.001801491
ω_n^A	0.9805298

Table 3. GLONASS ephemeris data of IGS.

Position parameters	Data
Reference time	2010,01,02 00h45 min
X (m)	1.493771777344E+04
Y (m)	1.936091552734E+04
Z (m)	7.411822265625E+03
Vx (m/s)	−6.023426055908E−01
Vy (m/s)	−8.344507217407E−01
Vz (m/s)	−8.344507217407E−01

$$UTC(SU) + 3h = UTC(USNO) - leap\,seconds \quad (33)$$

Figure 1 shows the variation curve of satellite position accuracy which is predicted using the almanac parameters user algorithm. From the Figure, within the validity period of almanac parameters, three-axis position prediction error is less than 3 km. By analysing the data, the mean values of three-axis position prediction errors are 1,291.2 m, 1,482.5 m, 463.5, m and the absolute error is 739.8, m, 756.4, m, 270.1, m respectively. The simulation results show that the prediction accuracy of the almanac parameters user algorithm mentioned in this paper is high, and it can fully meet the need of the satellite fast acquisition and forecasting satellite position (Li & Wang, 2014; Ma & Zhou, 2014).

The satellite position and velocity are predicted based on the almanac parameters of IGS, which are used to compare with the STK precise orbit data. Figure 2 and Figure 3 are the satellite position

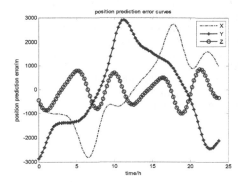

Figure 1. Error curves for the position prediction data using the almanac parameters of IGS based on the almanac parameters user algorithm is compared to IGS precise orbit data.

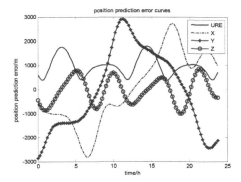

Figure 2. Error curves for the position prediction data using the almanac parameters of IGS based on the almanac parameters user algorithm is compared to STK precise orbit data.

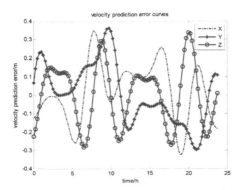

Figure 3. Error curves for the velocity prediction data using the almanac parameters of IGS based on the almanac parameters user algorithm is compared to STK precise orbit data.

and velocity accuracy curves respectively. They are shown as follows:

Figure 2 and Figure 3 show that within validity period of almanac parameters, the three-axis position prediction error is less than 3 km, the three-axis velocity prediction error is less than 0.4 m/s and the URE (User Range Error, URE) (Ala-Luhtala & Seppanen, 2012; Montenbruck & Steigenberger, 2015; Pukkila & Ala-Luhtala, 2015) caused is less than 2 km. The mean of URE is 941.7 m and the mean square deviation of URE is 402.4 m.

5 CONCLUSION

This paper presents a simple and fast GLONASS almanac parameters user algorithm to solve the problem that the existing GLONASS almanac parameters user algorithm is complicated and difficult. The simulation results show that the proposed algorithm can predict the satellite position and velocity with high accuracy, and it can fully meet the requirements of practical application of the navigation receiver.

The GLONASS almanac parameters user algorithm can be applied to predict satellite position and velocity for the navigation receiver. The algorithm in this paper can provide the reference for the design of navigation system almanac parameters.

REFERENCES

Ala-Luhtala, J., Seppanen, M., & Piche, R. (2012). An empirical solar radiation pressure model for autonomous GNSS orbit prediction. *Proceedings of the IEEE/ION Position, Location and Navigation Symposium* (pp. 568–75).

Anyaegbu, E., Townsend, B. R., & Zou, R. (2014). Optimal search strategy in a multi-constellation environment. *Proceedings of 27th International Technical Meeting of the Satellite Division of the Institute of Navigation (*(pp. 321–329). Florida.

Jing, S. R., Wu, P., Liu, W. X., et al. (2015). Fitting method of improved almanac parameters with semi-major axis compensation. *Journal of Central South University (Science and Technology)*, 46(7), 2504–2509.

Li, J., Zeng, G., Qiang, W., et al. (2016). Research on almanac ephemeris parameters fitting arithmetic for navigation satellites in engineering. *China Satellite Navigation Conference*. China Satellite Navigation Office, Changsha.

Li, X. B., Wang, Y. K., & Chen, J. Y. (2014). Research on the acquisition search strategy of navigation constellation inter-satellite link. *Journal of Astronautics*, 35(8), 946–952.

Ma, L. H. & Zhou, S. L. (2014). Positional accuracy of GPS satellite almanac. *Artificial Satellites*, 49(4), 225–231.

Montenbruck, O., Steigenberger, P., & Hauschild, A. (2015). Broadcast versus precise ephemerides: a multi-GNSS perspective. *GPS Solutions*, 19(2):321–333.

Pukkila, A., Ala-Luhtala, J., Piche, R., et al. (2015). GNSS orbit prediction with enhanced force model. *2015 International Conference on Localization and GNSS (ICL-GNSS)*. Gothenburg.

Russian Institute of Space Device Engineering (2008). Interface control document (Edition 5.1).

Tang, X. S., Chen, Y. Y., et al. (2002). Orbit determination and reentry control for manned spacecraft. National Defence Industry Press, 309–367. Beijing.

Wang, E. S., Zhang, S. F., Cai, M., et al. (2015). GPS receiver autonomous integrity monitoring algorithm using the genetic particle filter. *Journal of Xidian University*, 42(1), 136–141.

Wang, L. X., Huang, Z. G., & Zhao, Y. (2013). Two Sets of GPS almanac on time-to-first-fix influence. *Geomatics and Information Science of Wuhan University*, 38(2), 140–143.

Wen, Y. L. (2009). Satellite navigation system analysis and simulation technology, *China Aerospace Press*, 242–260. Beijing.

Xie, X. G., Song, Y. Y., Long, T., et al. (2015). Prediction algorithm of GLONASS almanac based on a model conversion. *Transactions of Beijing Institute of Technology*, 35(7), 755–761.

Automotive, Mechanical and Electrical Engineering – Liu (Ed.)
© *2017 Taylor & Francis Group, London, ISBN 978-1-138-62951-6*

Local stability of solutions for a class of viscous compressible equations

Shuxian Deng
Department of Basic Courses, Zhengzhou University of Industrial Technology, Zhengzhou, China
College of Science, Henan University of Engineering, Zhengzhou, China

Hongen Li
Department of Basic Courses, Zhengzhou University of Industrial Technology, Zhengzhou, China

ABSTRACT: This work presents some results regarding the stability of local solutions for the system of equations. The method is based on damping in multi-dimensions and introduces a mixed Fourier pseudospectral and Hermite finite difference scheme in space, and an efficient projection method to solve stress velocity coupling. The local existence and pointwise estimates of the solutions are obtained. Furthermore, the optimal stability and convergence rate of the solution when it is a perturbation of a constant state is obtained. It is shown that for some compressible equations, the corresponding local stability result holds and the properties of the analytic semi-group are employed to show the compactness for the semi-process generated by Euler solutions. Provided that the compressible equations are strictly stabilised.

Keywords: Stability; Local

1 INTRODUCTION

In this paper, we prove the regularity, local stability of a class of compressible equations generated by the local solutions for the following non-linear viscous equations and by use of the existence of a two-parameter family of operators.

Consider a strictly compressible system:

$$y_{tt} = \sigma(y_x)y_{xx} + f(x,t), \; (x,t) \in (0,1) \times (\tau, +\infty) \quad (1)$$

$$y(0,t) = y(1,t) = 0, \; t \geq \tau \quad (2)$$

$$u_t + f(u)_x = 0 \quad (3)$$

where $y = y_\tau(x,t)(\tau \in R^+)$ an unknown function, σ is a real function defined on R, $f = f(x,t)$ is an external forcing term. It is assumed that all characteristic fields are genuinely non-linear. Call λ_1 (u) $< \cdots < \lambda_n$ (u) the eigenvalues of the Jacobian matrix $A(u) = Df(u)$. Where the unknowns v > 0, u, θ > 0, p, e, and s represent the specific volume, the velocity, absolute temperature, pressure, internal energy, and the entropy of the gas, respectively. The coefficients of viscosity and heat conductivity, μ and κ are positive constants.

Here the u_0, u_1, f are given functions, Δu is a Laplacian with respect to the variable $x \in \Omega$, $u = u(t, x)$ is an unknown function, $\lambda > 0$ is a fixed positive number, f are given external forces. From which, if (v, θ), (v, s), or (v, e) are chosen as independent variables and respectively, then it

can be deduced that they satisfying the following conditions:

$$f \in L_2, \; u \in L_q\left(\mathbb{R}^+, L_p\right) \quad (4)$$

$$\theta ds = de + pdv \quad (5)$$

$$\sigma_v(v, \; \theta) = p_\theta(v, \theta) \quad (6)$$

$$p_x(v, \; s) = e\theta(v,\theta) - \theta p_\theta(v,\theta) \quad (7)$$

$$\frac{2}{p} + \frac{n}{q} \leq 1 \quad (8)$$

Using uniform compact attractors of semi-processed, the weak expansion waves for (1)–(3) will be shown. Therefore, it is easy to use the equation Entropy and absolute temperature, such as,

$$u_t + p(v)_v = -\delta u, \; \gamma > 0. \quad (9)$$

Here $u_t, p(v)_v, \delta, u$ and $\gamma > 0$ represent the specific volume and velocity respectively; the pressure p(v) is assumed to be a smooth function of v with $p(v)_{tt} \geq \frac{1}{\tau}$.the local stability of the solution is captured by the following equation:

$$\nabla \zeta_i(t)\sigma_i(\tau) \equiv 1 \quad (10)$$

Observing (4)–(8), a vector $\vartheta_x(t)$ will be found such that

$$y_{tt} = \sigma(y_x) + \lambda y_{xt} \tag{11}$$

With different boundary conditions from (2). Qin et al. proved the local existence of solutions to initial-boundary value problems, or periodic boundary or Cauchy problems of the equation (1)–(3).

For some positive $a, p \in [1/2, 2)$ $C > 0$, and $q > 0$ there is a weak solution, which fulfils additionally

$$d + a|x|^p \le f'(x,t) \le d(1+|x|^q), \forall s \in \mathbb{R}^3 \tag{12}$$

$$u_{tt} - div\,\overline{\Lambda}\nabla u^2 - \Delta u_t = f(x,\ t) \tag{13}$$

Several techniques to prove the existence of weak decay solutions with respect to the phase space are available; and in addition have suitable properties with energy inequality for almost all times or solutions with weak decay properties for $t \to \infty$ (this has been studied recently by several researchers, e.g. Qin, Ebihara, Xin Liu, etc.).

This represents an implementation of an axial movement of the viscous material, and cause the form of above equations, in two, three-dimensional case, they describe the viscoelastic solid of anti-plane shear action. While $n = 1$ and $f = 0$, ST Li prove the existence of weak periodic decay strong solution on the periodicity condition, X.K Su and J.L. Zhang proved the controllability of a smooth solution.

For some constants ρ_0, $M > 0$, $\tau \in H^4[0, +\infty)$ satisfies:

$$x'(t) + \tau_0(1+t)^{(\ell + \frac{d}{2})} \le \sigma y(t)^{\frac{v}{2}} + \tau_1(1+t)^\delta \tag{14}$$

$$\tau(v^2) + 2\sigma_0(v^2)v + \tau''(v^2) \le M < \infty \tag{15}$$

Thus, a so-called energy perturbation method can be used to establish weak controllability of solutions in terms of energy norm for a class of non-linear functions.

The controllability in a steady state will be shown with the help of differential inequalities, by estimating the relationship between energy inequalities and the attenuating property of weak solutions. Furthermore, a small positive number will be determined and differential inequalities derived by using a perturbation of energy, from which the following results are concluded.

2 MAIN RESULTS

Exploit the property of an analytic semi-group and delicate estimates to establish the existence of weak uniform compact attractors of a two-parameter family of operators. It is well-known that continuous dependence of solutions on initial data is very important, especially when the compressible infinite-dimensional dynamics system is studied.

First, the local stability of weak solutions is focused upon in order to describe the manoeuvrability; and local weak solution with decay for (1)–(3) is defined:

If the initial data satisfies $u_0, u_1 \in H^2 \cap H_0^4$, and the function ω is said to be a weak solution of problem (1)–(3), if it satisfies the following conditions:

$$\|\delta_i(x) - \delta_i(x+\lambda)\| = o(1) \tag{16}$$

$$\sup_{meas(\Omega)} \kappa(\Omega) \le \mu(A) \le \sup_{meas(E)} \zeta(E) \tag{17}$$

The measure μ_i can thus be defined equivalently as

$$T_s = \kappa\theta_x + \delta\theta_v + v\theta_x^2 \tag{18}$$

In fact, for smooth solutions, (1)–(3) are equivalent to (16)–(18). In what follows, (16)–(18) will be considered with the initial data

$$\gamma_t + div(\gamma_u) = 0 \tag{19}$$

$$(\theta h^j)t + div(\theta h h^j) + \rho(\delta)_x = \alpha\Delta u^j \tag{20}$$

$$\|f\|_{W^{m,p}} = \sum_{n=0}^{k} \|\partial_m^k f\|_{L^p} \tag{21}$$

where

$$\theta = \theta^\tau(x,t) = \theta_x^\tau(x,t)$$

$$\delta = \delta^\tau(x,t) = \delta_t^\tau(x,t)$$

The following is the main result in this paper:
Theorem. Let

$$\mu_i(\{x\}) = \sigma_i, \theta(s) \in C^1(R),\ \theta(s) > 0 \tag{22}$$

And set

$$p_i = m_i(x) \bullet (v(x+) - v(x-)) \tag{23}$$

Then for any fixed $f \in \Re^2$ and for any

$$Y(t) = Q(u(t)) + C_0 G(u(t))$$

The problem (1)–(3) admits a unique local (regular) weak solution $(u(t), v(t)) \in W_2^{m,p}$, which generates a unique semi-process on H_2 of a family of operators such that for any $t \ge \tau \ge 0$,

$$U_g^{(1)}(t,\tau)(u_0^\lambda, v_0^\tau) = (u^\lambda(t), v^\lambda(t)) \in W_2^{m,p} \tag{24}$$

554

$$w_t + (\frac{w^2}{\lambda})_x = -\eta sgn(x) \cdot \frac{d}{dt} G(u(t)) \tag{25}$$

$$p_\tau(u,x) = p_\tau(u,s) - \frac{\lambda(p_\lambda(u,t))^2}{e_\theta(x,t)} \tag{26}$$

$$\theta_x(u,t) = \frac{\theta}{e_t(x,\tau)} \tag{27}$$

Here and hereafter

$$p_v(v,s) = p_v(v,x) - p(v,\theta)p_v(v,\tau) \tag{28}$$

Is a generic constant. If it is further assumed that

$$(v,u,s)(t,x)|_{t=0} = (v_0,u_0,s_0)(x)$$

For some positive constants ζ_1 and κ_1, then there exists a constant $\rho_1 = \rho_1(\delta_1(\tau)) > 0$ such that
For any fixed $\beta \in (0,\beta_1]$, and any $t \ge \tau \ge 0$, there is

$$u_t + \xi(v,s)_x = 0 \tag{29}$$

There are two families of expansion solutions for (29) which are the compressible Euler equations with Riemann data. Here and hereafter, use

$$W^{m,2} = H^m, \ \|\cdot\| = \|\cdot\|_{L^2} \ and \ \|\cdot\|_m = \|\cdot\|_{H^m}$$

3 CONCLUSION

Without loss of generality, the initial-boundary value problem for the following non-linear compressible equation is considered:

$$u_{tt} - div\{(a|\nabla u|2r + 1)\} - \Delta u_t = g(x,t) \tag{30}$$

$$\Delta \omega_n + m_0 |\nabla \omega_n|^q \le C_\varepsilon \|\partial_t v\|_{H^2}^2 + m_0 |\nabla v_n| \tag{31}$$

$$(g(u),\partial_t v) + \Delta \omega_n \le \varepsilon |v|^{2q} + m_0 |\nabla v_n| \tag{32}$$

Consider a closed linear operator $A = \frac{\partial^2}{\partial x^2}$:

$$H^2(0,1) \to H^2(0,1)$$

With the domain

$$D(A) = w^{2,p}(0,1) \cap w_0^{2,p}(0,1)$$

Analogously, for some positive $\varepsilon, \upsilon, \rho$, together with the Hölder and Young inequality and the Euler interpolation

$$Cg(t)^{1+\frac{\upsilon}{p}} \le C(1+t)^{-\rho} g(t)^\upsilon \le \varepsilon \|\partial_t v\|_{L^2}^2 \tag{33}$$

$$g'(t) \le (1+t)^{\frac{\upsilon d}{2n}} g(t)^\upsilon \le C_\varepsilon \|\partial_t v\|_{H^2}^{p+1} \tag{34}$$

Similar to the above, the differential inequality is derived as long as the local weak solution $u(t)$ exists, inserting the above estimates, and using the energy estimate for estimating the energy norms, one can get

$$\|\partial_t u\|_{L^2}^{np-\gamma} \le \frac{C_4}{2}\Phi \tag{35}$$

$$\frac{\gamma}{2}\|\nabla_x u\|_{L^2}^{\frac{1}{v}+\frac{d}{2}} \le \frac{\varepsilon}{4}M\|u\|^2 \tag{36}$$

$$a\|\nabla u\|_{L^2}^{2r} \le C_2 E(t) \tag{37}$$

On the other hand, by Young's inequality, choosing $\varepsilon > 0$ small enough and $E(t)$ is bounded, and finally the following deduced

$$G_0'(t) + C_3 G(t) \le C_4 M_1(1+t)^{-\rho_1} \tag{38}$$

$$g'(t) + C_4(1+t)^{\frac{\lambda \rho}{2}} < C_5(1+t)^{-\frac{1}{1+\gamma}} g(t)^\beta \tag{39}$$

Using now estimate (30)–(35) together with the interpolation inequality, from (34)–(37) it is inferred that

$$g'(t) \le C_6 g(t)^{\frac{n}{p(p+2)}} \tag{40}$$

$$C_0 g(t)^{p+2-\upsilon} \le C_7 e^{K(L-s)} \|\xi_u(x)\|_{L^2}^{p+2-n} \tag{41}$$

$$\|\xi_u(y)\|_{L^2}^{p+2} \le C_6 e^{(L-s)} \|\xi_u(x)\|_{L^2}^{p+2-n} + C_7 g(t)^{\frac{n}{p(p+2)}} \tag{42}$$

Now fix k, m, c. Then, taking advantage of the smoothing property together with the obviously bounded $E(t)$

$$E(t) \le 2\Phi(t) \le 2g(t) \tag{43}$$

$$C_7 \Phi(0) \le G_0(t) \le 2KG(0)(1+t)^{-K} \tag{44}$$

$$G_0(t) \le G(0)(1+t)^{-K} \le C_8 M_3 e^{-\lambda t} \tag{45}$$

Multiplying both sides of (44)–(45) by e^{Ct} and integrating from 0 to t, will derive

$$C_6 \Phi(s) \le \int_{B(x_0,r)} \delta^2 f(u_k)p(x)\Delta u_k dx \tag{46}$$

$$\int_{B(x_0,r)} \tau^2 g(u_k)p(x)|\nabla u_k|^q dx \le \int_{B(x_0,r)} p(x)\Delta u_k dx \tag{47}$$

$$\int_{B(x_0,r)} \tau^2 |\nabla u_k|^q dx \le \int_{B(x_0,r)} \delta^2 f(u_k)dx \tag{48}$$

And finally obtain

$$(\|u(t)-\tau\|_{H^2}^2 + \|v(t)-\delta\|_{H^2}^2 + \int_{u_0}^{u_1}(\|v(t)-\delta\|_{H^2}^2) + \|v(t)-\delta\|_{H^2}^2 d\tau \le c_7 \|u(0)\|_{H^2}^2 \tag{49}$$

555

$$(\|u(t) - \tau\|_{H^2}^2 + \|v(t) - \delta\|_{H^2}^2 + \int_{u_0}^{u_1} (\|v(t) - \delta\|_{H^2}^2 \tag{50}$$
$$+ \|v(t) - \delta\|_{H^2}^2) d\tau \leq c_8 \|u(0)\|_{H^2}^2$$

$$(\|\rho_t(t)\|_{H^2}^2 + \int_{u_0}^{u_1} (\|v(t) - \delta\|_{H^2}^2 + \|\rho_t(t)\|_{H^2}^2) d\xi \leq c_9 (\|\rho_t(0)\|_{H^2}^2) \tag{51}$$

$$\|v(t) - \delta\|_{H^2}^2 + \int_{u_0}^{u_1} (\|v(t) - \delta\|_{H^2}^2 + \|v(t) - \delta\|_{H^2}^2) dt \leq c_{10} \|\tau(0)\|_{H^2}^2 \tag{52}$$

Moreover, if for $\delta \leq 3/2$, there exists a sufficiently small constant α such that

$$\frac{\theta p_v(v, \theta)}{q_\theta(v, \theta)} > 1$$

Here we have used the fact that in (43)–(45). Thus the approximate compressible equation is constructed as above.

Given a suitably small but fixed constant $\beta > 0$, let u(t, x) be the unique local smooth solution to the Cauchy problem.

$$s_t + \rho s_x = 0 \tag{53}$$

$$S(t, x) + \lambda_1(u(t, x), s) = s(t, x) + (\rho(t, x), s) \tag{54}$$

For $n = 2$ $\omega \in L_2 \cap L_p$, and $\|v(t)\| \leq \sigma(1 + t)^{\frac{\gamma_1}{3q}}$, note that the initial data is dense, hence $\|v_0(t)\|$ attenuates exponentially fast indeed, using $\omega |\omega|^{n(p-2)}$ and $\partial_t \omega$ multiply the equation (7) and integrate by parts over $x \in \Omega$. One gets

$$\delta'(t) + \lambda_1(1 + t)^{\kappa \frac{d}{2}} \delta(t)^{\frac{1}{q}} \tag{55}$$
$$\leq \varepsilon(1 + t)^{-(p-r)} y(t)^{\lambda + \frac{\kappa}{p}} + \lambda_2 \delta(t)^{\frac{n}{2}(\frac{1}{r} - \frac{1}{2})}$$

where ε is a small positive number which will be fixed, then one arrives at

$$\frac{1}{2}(p - n)\delta'(t) + \int_\Omega |\nabla \omega^2|^p dx \tag{56}$$
$$\leq \sigma_\varepsilon \int_\Omega (|v_0|^4 + |\nabla \omega|^{p-2}) dx + \frac{1}{2} v(G(v), v)$$

And notices that

$$u_{tt} - \lambda^2 \Delta u + \gamma u_t = R(v, u) \tag{57}$$

$$\Phi(x, y) = div\left(\frac{1}{1+x} \nabla \partial(1 + y) - v^\beta\right) \tag{58}$$
$$-div(\delta xy + (xy)_t)) + (\sum_{\mu,k} u^k v^\lambda + \sum_{\mu,k} u^k div v v^\delta)$$

Under *a priori* assumption, the following will be estimated,

$$E(t) = sup\{\|u(x,t)\|_m^2 + \|v(x,t)\|_m^2 +\}\|\delta(x,t)\|_m^2 \leq \sigma_0 e^{\lambda t} \tag{59}$$

By using Sobolev and Hölder inequality, it is known that (41)–(44) implies

$$\|u_t\| \leq C(\|u\| + \|\nabla v\| + \|\nabla w\|) \tag{60}$$

$$v_0(x) + \theta_0(x) \leq V(t, x) + \Theta(t, x) \tag{61}$$

For all $(t, x) \in R^+ \times R$ and some positive constants, the Cauchy problem admits a unique local smooth solution $(u, v, \tau)(t, x)$ satisfying

$$\lim_{t \to +\infty} sup_{x \in R}\{\|(u(x, t), v(t, x), \tau(t, x))\|\} = 0 \tag{62}$$

$$\frac{1}{2}(p - n)\delta'(t) \leq \sigma_\varepsilon \int_\Omega (|v_0|^4 + |\nabla \omega|^{p-2}) dx \tag{63}$$

$$\int_\Omega |\nabla \omega^2|^p dx \leq \frac{1}{2} v(G(v), v) \tag{64}$$

And notice that

$$E(u(t)) = \frac{1}{2}\alpha\left\{\|\nabla \omega_1(t)\|^q + \int_\Omega \int_0^{|\nabla u_1|^2} \sigma(\xi) \omega(t) d\xi dx\right\}$$

$$E(\xi_u(t)) = \frac{p-2}{p}\|\partial_t \omega\|_{L^2}^p + \left(G(\omega), \frac{1}{q-1}\right) - \frac{pn+4}{p+4-n}\|\nabla \omega\|_{L^2}^2$$

Let λ, α be small enough, one gets

$$\frac{d}{dt} E(\xi_u) + \rho\|\nabla_x u\|_{H_0^2}^p - \alpha\|\partial_t u\|_{H_0^2}^2$$
$$= \mu(\phi(\nabla_x u), \nabla_x u) + \beta(g(u), \omega)$$

$$\int |u|^{\frac{pn}{n-2}} dx \leq \|\omega\|_{H_0^2}^{\frac{p}{2}-1} + \sigma\|\nabla_x u\|_{L^p}^2$$
$$\leq \frac{C_\varepsilon}{\beta} \zeta(\omega) + \lambda_\varepsilon \|\omega\|_{L^p}^\gamma + \delta E\|\xi_u(t)\|$$

where the constant C_ε, λ_ε depend only on the ε.

Hence by the standard Galerkin method, interpolation inequality, Young and Sobolev's inequality, this section can be estimated by

$$\alpha\|u\|_{L^q}^{n+1}\left(\phi(\xi_2) + \|u\|_{L^2}^{\frac{1}{2}-\frac{1}{p}}\right) \tag{65}$$
$$\leq \alpha\|u_0\|_{L^2p}^{p+2} \|v_0\|_{L^p}^{p-1} - \phi(\xi_1) + C_\varepsilon\|u\|_{L^p}^{\lambda p}$$

By the properties of heat kernel, it can be deduced that

$$\partial_t\left[\frac{pn - 2n + 2p\alpha}{p+4-n}(\partial_t v, v)\right] + \tau_0(|\nabla_x \omega_1| + |\nabla_x \omega_2|)^{\frac{pn}{n-2}}$$
$$\leq C_1\phi(t)^{\frac{1}{p}} + C_2(1 + t)^{-\mu\tau}|v|^{p+2} \tag{66}$$

556

$$\rho \|\partial_t u\|_{L^{2p}}^{p+\mu} + C\|u\|_{L^2}^p \le \frac{1}{2}\kappa\varphi'(\nabla v_1)$$
$$\le C(1+t)^{-p\upsilon} + C_\varepsilon(1+t)^{-\frac{n}{2}\left(\frac{1}{q}-\gamma\right)} \qquad (67)$$

It is well-known that (64)–(67) generates an analytic semi-group of bounded linear operators $\hat{E}(t)$ on $D = H^2(0, 1)$ verifying

$$(E^{-1}u)(x) = \int_0^\tau (x-\xi)u(\xi)d\xi + \int_\tau^\delta (\xi-1)v(\xi)d\xi \qquad (68)$$

$$\|E(t-\tau)W_0^\tau\|_{H^2} \le Ke^{-n\alpha t/3}\|W_0^\tau\|_{H^2} < +\infty \qquad (69)$$

Which together with the proposition (30)–(39) implies that (65) (69) is the translation compact in $\hat{E}(t)$. Furthermore, the conclusions (57)–(64) also follow from propositions (44)–(50) respectively. Noting that the uniqueness of exponent solution and the Cauchy conditions in a strong solution space, there holds (68), (69). Thus, the conclusion of local stability of the solutions can be drawn.

ACKNOWLEDGEMENTS

This work was supported in part by grants from the Ph.D. Foundation of Henan University of Engineering (No. D2010012).

REFERENCES

Beirao da Veiga, H. (2014). Existence and asymptotic behavior of strong solutions of the N-S equations. *Indiana Univ. Math. J.*, 36, 149–166.

Friedman, A. and Necas, J. (2015). Systems of non-linear wave equations with non-linear viscosity. *Pacific J. Math.* 135, 29–55.

Greenberg J.M., MacCamy R.C., and Misel V.J. (1968). On the existence, uniqueness and stability of solutions of the equation σ (ux) uxx + λuxtx = ρ0utt. *Math. Mech.* 17, 707–728.

Kuttler, K. and Hicks, D. (1988). Initial-boundary value problems for the equation utt = (σ(ux))x+(α(ux)uxt) x+f. *Quart. Appl. Math.* 46(3), 393–407.

Liu, Y. and Liu, D. (1988). Initial boundary value problem, periodic boundary problem and initial value problem of equation utt = uxxt+σ(ux)x. *Chin. Ann. Math.* 9 A, 459–470.

Nakao, M. (2005). Energy decay for the quasilinear wave equation with viscosity. *Math. Z.* 219, 289–299.

Automotive, Mechanical and Electrical Engineering – Liu (Ed.)
© 2017 Taylor & Francis Group, London, ISBN 978-1-138-62951-6

Maximum entropy-based sentiment analysis of online product reviews in Chinese

Hanqian Wu & Jie Li
Institute of Computer Science and Engineering, Southeast University, Nanjing, China

Jue Xie
Southeast University-Monash University Joint Graduate School, Suzhou, China

ABSTRACT: Sentiment analysis has become an important research area with the rapid growth of Chinese online reviews. This paper proposes a supervised machine learning method based on Maximum Entropy for the classification of Chinese reviews. Maximum Entropy is a probability distribution estimation algorithm which has been applied to a variety of Natural Language Processing tasks. Our experimental results demonstrate that the classifier based on Maximum Entropy is ideal for the classification of Chinese reviews, which can both raise accuracy and recall rate to considerable levels. Pre-processing of the reviews for the purpose of training and testing is also introduced in this paper.

Keywords: Chinese reviews; Sentiment Analysis; Maximum Entropy

1 INTRODUCTION

Nowadays, there are abundant Chinese reviews available in online documents. The majority of the online reviews are regarded as disorganised and unstructured. Merely using the artificial methods to understand their sentiment orientation is almost impossible. As a result, sentiment analysis hase an important role to play in opinion mining and product recommendation.

Sentiment classification classifies a review as giving a positive or negative position through excavating and analysing the subjective information such as viewpoint, opinion, attitude and emotion etc. Businesses can make better use of this information resources to improve the quality of the products or services, and this information can also encourage prospective clients to make purchase decisions. Thus, getting an accurate understanding of sentiment expressed in Chinese reviews is mutually beneficial.

Extensive research efforts have been made to analyse sentiment in languages such as English and other European languages. In this paper however, we analyse the sentiment of Chinese reviews by applying a supervised classification technique of Maximum Entropy. Maximum Entropy as a supervised machine learning approach is a probability distribution estimation technique and has been extensively applied to various Natural Language Processing tasks. The challenge of this work is that training a Maximum Entropy model is usually a computationally intensive task.

The organisation of the remainder of this paper is as follows. Section 2 presents the related work and methods about sentiment analysis. A Maximum Entropy-based model is proposed in Section 3. Meanwhile Section 4 discusses the data pre-processing for sentiment analysis, which includes segmentation, conjunction rules, negation handling and feature selection. Section 5 presents experimental results and discussion. Finally, our work is concluded in Section 6.

2 BACKGROUND OF THE SENTIMENT ANALYSIS METHODOLOGY

In general, sentiment analysis methods can be divided into two groups, depending on which techniques they are based on; either lexicons or machine learning techniques.

Lexicon-based sentiment analysis methods, which use a dictionary to measure the sentiment orientation of reviews, have attracted wide publicity of many researchers. For these methods, a lexicon of sentiment words with their labelled orientation is required. Examples of the commonly used lexicons include: General Inquirer Lexicon, Opinion Lexicon, SentiWordNet and etc. Turney et al. (2002) calculate the sentiment orientation of sentences using Point-wise Mutual Information (PMI). It computes the sentiment orientation of the words in a subjective sentence based on pre-defined seed words. A sentence is grouped into positive or negative category on the basis of the average semantic orientation. Xiong et al. (2008) propose the method which is semantic distance-based to calculate sentence tendentiousness based on the semantic similarity calculation of HowNet. The semantic distance reflects the semantic relations between words in a sentence, which can obtain the orientation of sentences. Wen et al. (2010) raise another sentiment analysis approach based on semantics where adverbs of degree and negation words are investigated. Wan (2011) uses a bilingual co-training approach for sentiment classification of Chinese product reviews utilising both the English view and the Chinese view.

On the other hand, some studies have been concentrating on training sentiment classifiers using machine learning techniques. Nakagawa, Inui and Kurohashi (2010) introduce a dependency tree-based classification method using Conditional Random Fields with hidden variables. Zhou, Chen and Wang (2010) present a novel semi-supervised learning algorithm called Active Deep Networks (ADN) and then exploit it to tackle the semi-supervised sentiment classification problem. Xu Qunling (2011) proposes a new sentiment orientation calculation model, which is based on the analysis of the characteristics of Chinese text. The model uses an improved point-by-point analysis method SO-PMI to classify category sentiment orientation using words. Li et al. (Li 2011) investigate a semi-supervised learning method for imbalanced sentiment classification. In particular, it expresses the imbalanced class distribution problem via various dynamically generated random subspaces.

As a machine learning algorithm, Maximum Entropy is widely used for Natural Language Processing tasks. Maximum Entropy combines contextual features in a principled way and allows unrestricted use of them. Moreover, Wang and Acero (2007) propose that the Maximum Entropy model can obtain global optimisation due to the properties of the convex objective function. Based on the above-mentioned factors, we train a Maximum Entropy classifier to analyse the sentiment of Chinese reviews. Our experimental results show that Maximum Entropy as a text classification algorithm warrants further investigation.

3 MAXIMUM ENTROPY CLASSIFICATION

Maximum Entropy is a probability distribution estimation algorithm that is extensively used for Natural Language Processing tasks. The motivation of Maximum Entropy classification is that we should calculate the frequencies of individual joint-features without making any unfounded assumptions. That means the probability is uniformly distributed when there is no pre-knowledge.

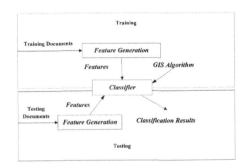

Figure 1. Model of maximum entropy classifier.

In this paper, we estimate the sentiment orientation of Chinese reviews utilising the Maximum Entropy classification algorithm.

There is no difference from other learning technique; the outputs of machine learning technique rely on the given training data set of input. The first step of Maximum Entropy classification is to get the constraints from a labelled training data set for the model distribution.

Figure 1 is the model of Maximum Entropy classifier for Chinese reviews classification, which is made of a training and testing process. In a training process, we utilise a labelled training data set to derive a set of constrains for building the model that characterises the class-specific expectations for the distribution. Finally, we use the General Iterative Scaling (GIS) algorithm to find the probability distribution that is in keeping with the given constraints. In the process of classification, the testing data set is denoted in features, and then the review classification is obtained by using the classifier.

4 DATA PRE-PROCESSING FOR SENTIMENT ANALYSIS

4.1 Segmentation

The section headings are in boldface capital and lowercase letters. Unlike European languages, Chinese sentences are made up of strings of Chinese characters. Each Chinese character has a specific meaning, but two or more characters can be combined into a word that has a different meaning. For this reason, segmentation plays a key role in Chinese sentiment analysis, which belongs to the category which utilises Natural Language Processing techniques.

In this paper, Institute of Computing Technology, Chinese Lexical Analysis System (ICTCLAS) is adopted for segmentation and Part-Of-Speech tagging (POS). The purpose of segmentation and POS tagging is to prepare for identifying what characters should be combined to label the POS. After that, we select the words which have the specific part-of-speech, using a Chinese part-of-speech tagging set.

It is obvious that Chinese sentiment analysis cannot continue without a proper segmentation algorithm. The Chinese sentence must be decomposed into words before any tasks can take place. This is the first phase of pre-processing of Chinese sentiment analysis.

4.2 *Conjunction rules*

Ordinarily a sentence only expresses one sentiment orientation if it does not contain some conjunction words such as BUT, ONLY, ALTHOUGH, HOWEVER, WHILE etc. These conjunction words will change the orientation of sentences. In this case, we obtain the accurate meaning from a given sentence using conjunction rules.

For an example:

I have always trusted Jingdong because of its well-service, but the computer is too slow, it is a very bad shopping experience, really too disappointed.

Under these circumstances, we will cut off the phrase before 'BUT', and the rest of the sentence will be retained to represent the sentiment orientation of the whole sentence. Although some of the sentence information is lost, the accuracy for the sentence sentiment analysis will be improved.

Conjunction words will affect the sentiment orientation of the sentence. Using conjunction rules can makes the sentence more explicit and comprehensible.

4.3 *Negations handling*

Negation is a very common linguistic construction that widely appears in all languages. Therefore, negation should be taken into consideration in Chinese sentiment analysis. When negation words such as 'not' and 'hardly' occur in the sentence, the sentiment orientation will be changed. Negation detection is used for distinguishing the factual and non-factual information that is extracted from sentences.

A negation word such as 'hardly' will invert the orientation of the sentiment word. For example, 'hardly know' means 'unknown'. Das and Chen (2001) raise a technique, which uses the tag 'NOT_' to replace the negation word and the first blank space following the negation word. After that, a new corpus was achieved. In Chinese, we substitute '不_' for 'NO_', and the list of the negative words contains '无', '不', '没有', '非', '没', '未必'.

4.4 *Feature selection*

Feature selection is an important stage in sentiment analysis. Feature selection aims to select features which have the most representative and excellent classification performance from original feature information. Therefore, the appropriate feature selection method will largely determine the quality of the final classification results.

There is a huge quantity of words in data processing. The dimension of the feature space will be too large if all words are selected as the feature items. It not only increases the amount of computation, but also affects the accuracy of classification. Therefore, it is necessary to choose the appropriate feature selection method to reduce the dimension.

In feature selection, the features that are irrelevant to the emotions and the weak category correlation need to be removed to eliminate unnecessary interference. In this paper, the method of sentiment lexicon is adopted. The Bilingual Knowledge Dictionary provided by HowNet and National Taiwan University School of Dentistry are applied to distinguish the sentiment words from candidate feature words.

5 EXPERIMENT

5.1 *The analysis data*

Our corpus is originated from the Jingdong shopping site (www.jd.com), a leading online B to C trading platform in China. The corpus contains 450,417 reviews: 80% of reviews are exploited as the training data, while 20% of reviews are used for testing purposes.

All reviews have been labelled by the starValue attribute that is rated by customers. The reviews are then manually set to positive when starValue >= 3 and negative when starValue < 3.

5.2 *Model evaluation*

It is very important to evaluate the experiment results. The balance of various factors is considered in the selection of evaluation indicators to make an objective and impartial evaluation. The evaluation indicators in the information retrieval field can be utilised to evaluate the effectiveness of classification. In this paper, five evaluation indicators are used to evaluate the experiment results. They are precision, recall, F-score, accuracy and specificity respectively.

5.3 *Experiment 1*

One word may have opposite meaning in different contexts, such as "pride".

For example:

They can look back on their endeavours with pride.

Pride is the parent of all evils.

In the above example, "pride" has two opposite meanings: In the first sentence "pride" has commendatory meaning while "pride" has derogatory meaning in the second sentence.

Thus, two scenarios are taken into consideration in the phase of training. One situation is that some

Table 1. Result from Experiment 1.

Indicator	Precision	Recall	F-score	Accuracy	Specificity
Result	0.9806535	0.9777033	0.97917616	0.95956135	0.31958762

Table 2. Result from Experiment 2.

Indicator	Precision	Recall	F-score	Accuracy	Specificity
Result	0.9890121	0.81236035	0.8920245	0.8087564	0.68162525

of the extracted features are labelled as both positive and negative. For instance, feature "pride" has both positive and negative sentiment polarities.

In this scenario, a high recall can be guaranteed. The recall means that recognition of positive reviews accounts for the proportion of the total positive reviews. However, the capacity of the classifier to recognise the negative reviews is lower, namely specificity. The result is shown in Table 1 below.

5.4 Experiment 2

Another scenario is that the labelled sentiment words are either positive or negative. For instance, feature "pride" is either of the positive or negative category.

In this situation, the capacity of the classifier to recognize the negative reviews is improved, which has increased by approximately 40%. However, the indicators of recall and accuracy are reduced by 15 percentage points. The result is shown in Table 2 below.

From the experimental results, we can see that the performance of the classifier is better in the second scenario, where the capacity of the classifier to recognise positive reviews and negative reviews is well matched. In summary, the indicators of precision, accuracy, and F-score are considerable in both scenarios. We can conclude that the classifier based on the Maximum Entropy can achive satisfying classification performance on the Chinese reviews.

6 CONCLUSION

Maximum Entropy is an algorithm that is widelyapplied in Natural Language Processing. Its overriding principle is that probability distributions should be estimated from the training data set.

In this paper, we investigate the approach for sentiment analysis of online product reviews in Chinese based on Maximum Entropy. Our approach is mainly developed in three aspects of pre-processing, feature extraction and classification arithmetic based on the Maximum Entropy model.

The experimental results demonstrate that the classifier is ideal for the classification of Chinese reviews, which can yield a high accuracy and recall.

ACKNOWLEDGMENTS

This work is supported by National High-tech R&D Program of China (863 Program) (Grant No. 2015 AA015904) and the Open Project Program of the State Key Lab of CAD & CG (Grant No. A1429), Zhejiang University.

REFERENCES

Das, S., & Chen, M. (2001). Yahoo! for Amazon: Extracting market sentiment from stock message boards. *Proceedings of the Asia Pacific Finance Association Annual Conference, 35*, 43.

Li, S., Wang, Z., Zhou, G., Lee, S.Y.M. (2011). Semi-Supervised learning for imbalanced sentiment classification. *Proceedings of International Joint Conference on Artificial Intelligence (IJCAI-2011)*.

Nakagawa, T., Inui, K. & Kurohashi, S. (2010). Dependency tree-based sentiment classification using CRFs with hidden variables. *Proceedings of Human Language Technologies: The 2010 Annual Conference of the North American Chapter of the ACL (HAACL-2010)* (pp. 786–794). Los Angeles, CA.

Turney, P.D. (2002). Thumbs up or thumbs down? semantic orientation applied to unsupervised classification of reviews. *Proceedings of Annual Meeting of the Association for Computational Linguistics (ACL-2002)* (pp. 417–424). Philadelphia, PA.

Wan, X. (2011). Bilingual co-training for sentiment classification of Chinese product reviews. *Computational Linguistics, 37(3), 587–616.*

Wang, Y.Y., & Acero, A. 2007. Maximum entropy model parameterization with TF*IDF weighted ector space model. *IEEE Automatic Speech Recognition and Understanding Workshop*, 213–218. Kyoto, Japan.

Wen, B. et al. (2010). Text sentiment classification research based on semantic comprehension. *Computer Science, 37*(6), 261–264.

Xiong, D., Cheng, J. & Tian, S. (2008). The research of sentence sentiment tendency based on HowNet. *Computer Engineering and Applications, 44*(22), 143–145.

Xu, Q. (2011). A new model of Chinese text sentiment computing. *Computer Application Technology and Software*, 6.

Zhou, Shusen, Chen, Q. & Wang, W. (2010). Active deep networks for semi-supervised sentiment classification. *Proceedings of Coling,* Poster Volume (pp. 1515–1523).

Automotive, Mechanical and Electrical Engineering – Liu (Ed.)
© 2017 Taylor & Francis Group, London, ISBN 978-1-138-62951-6

Node spatial distribution and layout optimization based on the improved particle swarm algorithm

Hongyang Yu
College of Electronic Information and Optical Engineering, Nankai University, Tianjin, China

Fangwei Cui
College of Science, Purdue University-West Lafayette, West Lafayette, IN, USA

Yan Wang
College of Mathematics and Computer Science, Hunan Normal University, Changsha, China

ABSTRACT: Particle Swarm Optimization algorithm (PSO) is a kind of adaptive random optimization algorithm based on search strategies. Because the algorithm is simple yet effective, it catches the interest of many scholars. It has showed wide application in function optimization, neural network training, etc. The PSO algorithm development was firstly proposed a decade ago, and is far from being mature neither theoretically nor practically up to now. This paper focuses on improving the traditional particle swarm algorithm and applying it in the wireless sensor network node spatial distribution and layout optimization. Wireless Sensor Network (WSN) is a distributed sensor network which includes sheer numbers of sensors that detects and monitors the outside world. In WSNs, the network settings and the equipment locations can change at any time, can also vary with the Internet connection cable or wireless link. This manuscript adapts the PSO algorithm into WSN design and provides a novel node spatial distribution and layout optimization strategy for optimal network coverage.

Keywords: Node spatial distribution; Particle swarm algorithm; Wireless sensor network; Distance Vector-Hop; Novel Improved PSO & DVH (NIPD)

1   INTRODUCTION

Wireless Sensor Network (WSN) incorporates sheer numbers of sensors which captures information from the environment, and has caught global interest during the recent years. The special distribution of the remote sensor has been a key question in the study of WSNs (X. Li, Y. Xu and F. Ren, 2007). Regarding this issue, researchers have proposed considerable examinations on hub restriction issues of remote sensor systems, and it basically incorporates range-based and go free limitation calculation (J. A. Costa, 2006), without range restriction calculation incorporates centroid calculation (N. Bulusu, 2008), DV-Hop calculation (D. Niculescu and B. Nath, 2003), Approximate Point-In-Triangulation test calculation (APIT) (R. Nagpal, 2003) and Amorphous calculation (Z. Ma, 2008). Because of the points of interest of the equipment necessities, system arrangement cost, vitality utilization and different viewpoints, sans range limitation calculation is more reasonable for Wireless Sensor Networks (W. Yang and W. Pan, 2013). The Vector-Hop Distance (DV-Hop) calculation, which

changes separation estimation between hubs to the result of bounce tally and normal jump separation, is a standout amongst the most generally concentrated on calculations (J. Wen, 2014).

DV-Hop calculation can be separated into three phases. In the first and second phase of DV-Hop calculation, the separation between the obscure hub and the grapple hub is acquired by the jump check and the normal bounce separation between hubs. Then in the third stage, the area of the obscure hub is assessed by the separation between hubs. There are two approaches to enhance the accuracy of the DV-Hop calculation: the first is to make more precise estimation of the separation between hubs in the chief and another stage; the other one is to reduce hub position estimation error in the third stage. Most papers focus on the second one, concentrating on the extensive mistake issue of the slightest square technique in the third stage. To take care of this issue, molecule swarm calculation (Particle Swarm Optimization, PSO) have been proposed to supplant the slightest squares strategy to appraise the hub area. The estimation mistake of the slightest square strategy in conventional Distance Vector-Hop (DV-Hop) calculation is too expansive and the Particle Swarm Optimization (PSO) calculation easy to be trapped in a neighborhood ideal. In order to solve this issue, a combination calculation of enhanced molecule swarm calculation and DV-Hop calculation is proposed in this paper and referred to as Novel Improved PSO and DVH (NIPD). PSO calculation was enhanced from molecule speed, latency weight, learning methodology and variety, which improved the capability to hop out of neighborhood ideal of the calculation and expanded the hunt rate of the calculation in later iterative stage. The hub limitation result was then advanced by utilizing the enhanced PSO calculation as a part of the third phase of the DV-Hop calculation.

2 RELATED WORK

Liu et al proposed a hub self-situating with enhanced molecule swarm calculation in light of turmoil aggravation, which utilize the Tent guide tumultuous to hunt (Z. Liu, Z. Liu and X. Tang, 2012). It diminished the likelihood of falling into neighborhood ideal and enhanced pursuit calculation exactness, yet at the same time indicated low situating precision. Li et al (L. Li and Y. Du, 2014) proposed a self-restriction strategy in view of enhanced molecule swarm advancement, in which the idleness weight and learning element is set to direct decrease, and utilized the technique for segment variety to escape from nearby optima when experiencing seek stagnation. Likewise this strategy enhanced Positioning execution, yet not being perfect and included a specific computational many-sided quality. Wang et al (Y. Wang and J. Yang, 2014) proposed hub confinement calculation in view of an enhanced molecule swarm streamlining. The chronicled global best position was transformed part by part when experiencing looking stagnation, and obscure hubs that has been situated was taken as the stay hub to do iterative inquiry. Although the situating exactness was enhanced, there is still opportunity for further improvement by including a specific correspondence overhead and computational unpredictability.

Different from the abovementioned methods, the approach proposed in this paper, which is an enhanced molecule swarm calculation and DV-Hop combination calculation, provides a viable answer for the enormous minimum squares estimation blunder issue in the third phase of DV-Hop calculation. This calculation exhibits higher situating precision and better solidness; moreover, this calculation just enhances number-crunching operations in the third phase of DV-Hop calculation, and the first and second stage are the same with the traditional DV-Hop calculation,

consequently no correspondence overhead and equipment expenses are included.

3 ENHANCED PARTICLES SWARM OPTIMIZATION

3.1 *Standard particle swarm optimization*

PSO calculation is a global improvement calculation. Taking into account common populace look procedure (L. Li and Y. Du, 2014), it has favorable circumstances of less parameters, quick hunt speed, and so forth and is broadly utilized as a part of logical exploration and viable designing. Molecule Swarm can be particularly portrayed as taken after: Suppose the quantity of the molecule populace is P, the measurement of inquiry space is S, the i-th molecule position is $yi = (yi1, yi2, ..., yiS)$, speed is u, $ui = (ui1, ui2, ..., uiS)$, best recorded position of molecule i is $bhpi = (Pi1, Pi2, ..., PiS)$, global best verifiable position is $gbhp = (Pg1, Pg2, ..., PgS)$. Speed and position redesign recipes of molecule i are as per the following:

$$ui(t+1) = \omega ui(t) + t1f1(bhpi-yi(t)) + t2f2(gbhp-yi(t)) \tag{1}$$

$$yi(y+1) = yi(t) + ui(t+1) \tag{2}$$

where $i = 1, 2, ..., N$, Inertia ω is the latency weight; t is the present cycle time; t1, t2 are learning variables; f1, f2 are irregular numbers consistently dispersed in [0,1]. In conventional PSO, the particles tend to total to the global best position and the individual's best position, then the populace will have union impact and untimely merging, then pursuit stagnates in later iterative stage. It is hard to escape from nearby ideal, causing low hunt precision of PSO calculation. To upgrade the execution of molecule swarm calculation, this paper separately enhance 4 aspects: speed, idleness weight, learning methodology and variability, to enhance the accuracy and security of the calculation.

3.2 *Enhancement of speed*

In prior iterative stage, the particles can be effectively guided by the global best position to untimely accumulation; Search speed turns out to be moderate in later iterative stage and the calculation can't hop out of nearby ideal. To take care of this issue, this paper utilizes a great arrangement, which is gainful to global inquiry of the underlying emphasis and nearby union of the later cycle, to supplant the global best position, it can be depicted as the accompanying: haphazardly select k particles from the molecule populace and think about the best positions of k elements, then select the its individual best position rather than the global best position to manage the molecule movement. The estimation of k increments directly with emphasis: littler estimation of k exacerbates fantastic arrangement, which wells to global pursuit and expansions the differences of the populace; the more prominent estimation of k improves high caliber, which wells to neighborhood merging and enhance the joining rate. Enhanced pace overhaul equation is as per the following:

$$ui(t+1) = \omega ui(t) + t1f1(bhpi-yi(t))+t2f2 (gbhpn-yi(t)) \quad (3)$$

$$Z = Zmin+(Zmax-Zmin)t/T \quad (4)$$

where: pbestm is the best of the Z particles' own best history positions; t is the present emphasis time; T is the most extreme time of cycle; Zmax and Zmin are the greatest and least estimations of Z, Zmax esteem takes populace of particles, Zmin esteem takes 1.

3.3 *Inertia weight advances*

In order to adjust global hunt and nearby advancement capacities of Particle Swarm Optimization, idleness weight takes a bigger quality in beginning emphasis to advantage global pursuit, and it takes a bigger worth in later iterative stage to advantage neighborhood improvement, so Shi et al. proposed a strategy for straightly diminishing dormancy weight, The equations are computed as takes after:

$$InertiaW = Wmax- (Wmax–Wmin) t/T \quad (5)$$

where: t is the present emphasis time; T is the most extreme time of emphases; Wmax and Wmin are the greatest and least of the latency weight w. Bigger idleness weight in the underlying cycle benefits global inquiry, yet creates stun, builds correspondence overhead and reductions seek proficiency; In later iterative stage, smaller dormancy weight benefits neighborhood advancement, yet increases the difficulty to escape from nearby ideal calculations and reduces the pursuit exactness. To take care of this issue, a directly diminishing stochastic unpredictability system, whose instability diminishes with cycles, is proposed in NIPD: In the early emphasis, exceptionally vary idleness weight enhances the seek effectiveness of the calculation; in later iterative stage, low change dormancy weight upgrades the capacity to escape from neighborhood ideal.

3.4 *Mutation*

To solve the PSO calculation's problems such as easily trapped into nearby ideals, low limits of neighborhood advancement and different inadequacies in the advanced stage, unsettling influences to existing situation can both speed up the meeting and

enhance the precision of the ideal arrangement. Every measurement of the present position is not the best, so any measurement can be been transformed. Lévy appropriation has unlimited change which comprises of regular little esteem and aleatory substantial quality, visit little esteem is helpful for the last some portion of neighborhood improvement, and aleatory expansive worth can escape from nearby goals, so a Lévy circulation step is utilized to transform the present best esteem, the transformation equation is as per the following equation:

$$gbhpn = gbhpn +a.s \quad (6)$$

where: n is an irregular number in 1~D; D is the measurement of the pursuit space; an is the progression size element, whose worth is 0.1; s is step that is liable to Lévy dissemination.

4 ALGORITHMIC PROCESS

The steps in the algorithm process are:

Step 1: Aimlessly send various sensor hubs in the objective territory, then get the separation between obscure hub to the grapple hub by the first and second phase of DV-Hop calculation.

Step 2: Fix the limits. Counting the most extreme number of times of cycles Imax, learning element t1, t2, greatest Zmax and least Zmin of latency weight, the most extreme Wmax and least Wmin number of particles.

Step 3: Initialize populace. The underlying position of molecule 'I' is yi = (yi1,yi2…,yiS), velocity is ui = (ui1,ui2,…,uiS), its own best position is gbhpni = yi, Evaluate swarm wellness estimation of molecule as indicated by condition (3), the global best chronicled situation gbest is best starting position of molecule populace, so that t = 0.

Step 4: Fix t = t+1, redesign molecule position x and velocity v as per recipe (5) to (10), and do cross-outskirt treatment on y, u.

Step 5: Implement learning techniques. Actualize learning techniques for particles as per pseudo-code in segment 2.4 and condition (11), (12).

Step 6: Apprise particles' own particular best position and the global finest situation.

Step 7: Mutation. Change the ideal area gbhp as per condition (13), redesign gbhp if the transformed position is better.

Step 8: Place the global Antiquity ideal arrangement if the emphasis end state is met, then skip to step 9; generally, come back to step 4 and proceed with the iterative streamlining.

Step 9: Repeat the steps from step 3 to step 8 until the evaluated area of all the obscure hubs are put out.

5 SIMULATION AND INVESTIGATION

Matlab simulation and investigation is performed to exhibit the situating execution of the enhanced molecule swarm calculation. Fix the limits of the calculation: Learning component t1 = t2 = 1, the quantity of molecule populace was 50, the most extreme number of cycles is 70 and the greatest pace of the molecule is 10 m/s, then conform the proportion of stay hubs, all out number of hubs, correspondence span separately to analyze the execution of four area calculations. Arbitrarily circulate 150 sensor hubs in the 200 m × 200 m square region, with correspondence sweep 25 m.

By conforming the proportion of grapple hubs, we compared the proposed calculation Novel Improved PSO and DVH (NIPD) with the customary DV-Hop number, (Reformed PSO) RPSO1-DV-Hop calculation and RPSO2-DV-Hop calculation. As appeared in Figure 1, with the expansion in the extent of stay hubs, four calculations situating blunder Deviations are abatement, and the proposed calculation indicates greatest drop in situating mistakes. Contrasted and DV-Hop calculation, RPSO1-DV-Hop calculation and RPSO2-DV-Hop calculation, the situating precision of the proposed Novel Improved PSO and DVH (NIPD) calculation averagely moves forward.

The effect of the Nodes in The Positioning Accuracy is appeared in Figure 2. Haphazardly convey 200 sensor hubs in the 150 m × 150 m region, with correspondence range 30 m and the stay hub proportion of 30%, by changing the quantity of hubs, think about the situating execution of four calculations. The situating blunder of the proposed Novel Improved PSO and DVH (NIPD) calculation is constantly not exactly the conventional DV-Hop calculation, RPSO1-DV-Hop calculation and RPSO2-DV-Hop calculation.

In correlation, the normal situating precision of the proposed strategy Novel Improved PSO and DVH (NIPD) increments.

Correspondence Radius Effect on the Positioning Accuracy Randomly disperse 2s00 sensor hubs in the 150 m × 150 m region, wherein the quantity of stay hubs is 30, with correspondence sweep of 15 m~40 m, analyze the Location execution of four calculations by modifying the correspondence span, as appeared in Figure 3. With the expansion of correspondence sweep, system availability increments and going mistake decreases, in this way the situating exactness of the calculation enhanced; when the correspondence range is more prominent than 45 m, with the increment of correspondence span, the normal hop separation blunder and the separation mistake build, so the bend demonstrates an upward pattern.

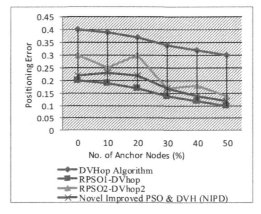

Figure 2. The effect of the nodes in the positioning accuracy.

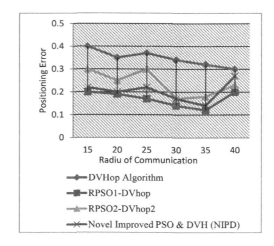

Figure 3. Effect on the positioning accuracy by the radius of communication.

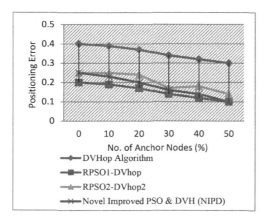

Figure 1. Proving the positioning accuracy.

Table 1. Stability in analysis.

Algorithm	Avereage	Std Deviation
DVHop Algorithm	0.38	0.038
RPSO1-DVhop	0.28	0.028
RPSO2-DVhop2	0.2	0.02
Novel Improved PSO & DVH (NIPD)	0.16	0.016

Haphazardly appropriate 200 sensor hubs in the 150 m*150 m region, wherein the quantity of grapple hubs is 25, with correspondence range of 40 m. Separately execute 30 reproduction tests on conventional DV-Hop calculation, RPSO1-Hop calculation, RPSO2-DV-Hop calculation and the proposed calculation Novel Improved PSO and DVH (NIPD), to get the mean and standard deviation of the normal situating contrast of 30 investigations, and the normal running time of finding every obscure hub, as appeared in Table 1. The mean and standard deviation of the uniform confinement blunder of 30 examinations are appeared in Table 1.

As can be seen from Table 1, contrasted with the customary DV-Hop calculation, RPSO1-DVHop calculation and RPSO2-DV-Hop calculation, this calculation dependably keeps up high-accuracy situating and better soundness. global pursuit and neighborhood improvement of molecule swarm calculation hnve diverse proficiency to locate a great answer for every emphasis, and area productivity and populace related particles and emphasess, The effectiveness is identified with molecule area and the quantity of cycles, yet it is not an entirely direct relationship, learning methodologies choice in light of likelihood is to choose a global inquiry or nearby advancement by the input data before the ebb and flow emphasis, so the calculation has more grounded versatile capacity, therefore has higher hunt productivity and better solidness.

6 CONCLUSION

The manuscript proposes a combination calculation of enhanced molecule swarm calculation and DV-Hop. When considering equipment expenses and consistent correspondence overhead, it has better situating precision and strength, to some degree, the issue of unacceptable execution of PSO's DV-Hop calculation Novel Improved PSO and DVH (NIPD) is tackled. The enhanced calculation has enhanced pace, inactivity weight, learning methodologies and figure variety, despite the fact that the computational burden builds a bit, the process did not expand the quantity of series,

the time multifaceted nature is in the same request of extent with the first calculation, in this manner the enhanced calculation has the calculation time with customary DV-Hop calculation and does not influence its uses in situation of wireless sensor network. The future studies will concentrate on diminishing the measure of count and vitality utilization.

REFERENCES

Bulusu, N., J. Heidemann and D. Estrin, "GPS-less low cost outdoor localization for very small devices", IEEE Personal Communications Magazine, vol. 7, no. 5, (2008), pp. 28–34.

Costa, J.A., N. Patwari and A.O. Hero, III, "Distributed weighted multidimensional scaling for node localization in sensor networks", ACM Transactions on Sensor Networks, vol. 2, no. 1, (2006), pp. 39–64.

Li, L. and Y. Du, "Application of Modified Particle Swarm Optimization in Node Locating of Wireless Sensors Networks", Computer Applications and Software, vol. 31, no. 4, (2014), pp. 69–72.

Li, X., Y. Xu and F. Ren, "Techniques for wireless sensor network", Beijing: Beijing Institute of Technology Press, (2007), pp. 191–192.

Liu, Y., X. LV and X. Wang, "Application of hybrid genetic algorithm in WSNs localization", Transducer and Microsystem Technologies, vol. 33, no. 2, pp. 150–153.

Liu, Z., Z. Liu and X. Tang, "Node self-localization algorithm based on modified particle swarm optimization", Journal of Central South University (Science and Technology), vol. 43, no. 4, (2012), pp. 1371–1275.

Ma, Z., Y. Liu and B. Shen, "Distributed locating algorithm for wireless sensor networks- MDS-MAP(D)", Journal on Communications, vol. 29, no. 6, (2008), pp. 58–62.

Nagpal, R., H. Shrobe and J. Bachrach, "Organizing a global coordinate system from local information on an ad hoc sensor network", The 2nd International Workshop on Information Processing in Sensor Networks (IPSN '03)[C]. Palo Alto., (2003), pp. 1–16.

Niculescu, D. and B. Nath, "DV based positioning in ad hoc networks", Journal of Telecommunication Systems, vol. 22, no. 1–4, (2003), pp. 267–280.

Wang, S.-S., K.-P. Shih and C.-Y. Chang, "Distributed direction-based localization in wireless sensor networks", Computer Communications, vol. 30, no. 6, (2007), pp. 1424–1439.

Wang, Y., and J. Yang, "Localization in wireless sensor network based on improved particle swarm optimization algorithm", Computer Engineering and Applications, vol. 50, no. 18, (2014), pp. 99–102.

Wen, J., X. Fan and X. Wu, "DV-Hop localization algorithm based on the RSSI hop correction", Chinese Journal of Sensors and Actuators, vol. 27, no. 1, (2014), pp. 113–117.

Yang, W., and W. Pan, "DV-Hop localization algorithm based on RSSI ratio correction in wireless sensor network", Transducer and Microsystem Technologies, vol. 32, no. 7, (2013), pp. 26–135.

Automotive, Mechanical and Electrical Engineering – Liu (Ed.)
© 2017 Taylor & Francis Group, London, ISBN 978-1-138-62951-6

Optimal load control in direct current distribution networks

Shan Liu, Suli Zou & Zhongjing Ma
School of Automation, Beijing Institute of Technology, Beijing, China

Yunfeng Shao & Shiqiang Feng
The State Grid Lvliang Power Supply Company, Lvliang, China

ABSTRACT: With the promising development of Direct Current (DC) distribution networks, load control problems in DC networks need great attention. In this paper, we focus on the optimisation of energy use in DC networks. We first model the Power Electronic Load (PEL) units as the controllable impedance, and establish the voltage model of the distribution network. By applying this model, we formulate the underlying load control problem as an optimisation problem that coordinates the bus voltage by minimising the system costs, consisting of the line losses costs and the deviation costs of PELs. Compared with the traditional optimal power flow problems, our model involves one control variable only, i.e. voltage, which makes it easier to solve. Nevertheless, as the resulting optimisation problem is non-convex, we use a Particle Swarm Optimisation (PSO) algorithm to solve it, and the developed results are demonstrated via a numerical example.

Keywords: DC distribution networks; Load control; Non-convex optimisation; The PSO algorithm

1 INTRODUCTION

Recently, DC distribution networks have attracted more and more attention due to their advantages in accommodating DC devices and distributed generation units, energy storage units and plug-in electric vehicles, see (Justo et al., 2013). G distributed energy resources, such as DC distribution networks, have lower line costs and losses, and higher power supply reliability (Jiang & Zheng, 2012; Song et al., 2013; Wang et al., 2008). When loads are regarded as adjustable units, distribution networks can meet the power demand flexibly by implementing the load control in response to system operations. Hence, the system can potentially reduce the peak load demand, mitigate the negative effects of fluctuations from distributed generations and assure fair power delivery among individual units, while ensuring security and quality, e.g. (Mohsenian-Rad & Davoudi, 2014; Ramanathan & Vittal, 2008; Shi et al., 2014; Zhu et al., 2015).

Conventionally, a load in the DC distribution network behaves as a constant impedance or power load (Weaver & Krein, 2009). With the increasing penetration of Power Electronic Loads (PELs) that are supplied with power electronic converters, they can be controlled to reflect a desirable impedance by adjusting their internal load characteristics (Mohsenian-Rad & Davoudi, 2014). In this paper, following the modelling issue introduced in

Mohsenian-Rad and Davoudi (2014) and Weaver and Krein (2009), we consider each PEL as a variable resistor and each Fixed Load (FL) as a constant resistor, and then develop the relationship between the bus voltage and the resistances of all PEL units. Based on this relationship, we formulate an optimisation load control problem, such that the system costs are minimised by controlling the voltages of all the PEL buses. Specifically, the system costs include the line costs and power deviation costs of PELs. Compared with the traditional Optimal Power Flow (OPF) problems, it is easier to solve the underlying optimisation problem.

Furthermore, the resulting underlying optimisation problem is a non-convex optimisation problem involving a convex set and a multivariate non-convex cost function. The heuristic search algorithms, which are enlightened by the nature laws and human experiences, can solve different kinds of NP-hard optimisation problems. The heuristic algorithms include genetic algorithms (Goldberg, 1989), simulated annealings (Kirkpatrick et al., 1983), ant colony algorithms (Duan et al., 2004), Particle Swarm Optimisation (PSO) algorithms (Trelea, 2003), and so on. Among them, the PSO algorithms are widely applied in various fields, such as system identification (Dub & Stefek, 2014) and motor control (Dong & Jin, 2005), because of their simple concepts, little parameter adjustment, high calculation speed and strong global search capability (Wu & Wang, 2016; Yang & Cheng, 2013).

In this paper, we apply a PSO algorithm to solve the underlying non-convex optimisation problem. Generally, the standard PSO algorithm is used to calculate the unconstrained optimisation problems (Sortomme & El- Sharkawi, 2009). In this paper, an improved PSO algorithm is designed to solve this optimisation problem with constraints. Here we introduce a check function into the algorithm to ensure that the variables can satisfy their constraints. To demonstrate our model, we apply an IEEE test distribution system to verify our results. The simulation results demonstrate the effectiveness of our proposed optimal model.

The rest of the paper is organised as follows. In Section 2, we develop a voltage model of PEL units in the DC distribution networks. Considering physical and operational constraints, we formulated a constrained optimisation problem in Section 3. In Section 4, we design an improved PSO method to solve the underlying load control problem and give some simulation results. Conclusions are given in Section 5.

2 FORMULATION OF DC DISTRIBUTION NETWORKS

In this paper, we consider a DC distribution network that consists of a group of DC buses, denoted by $N = \{1,...,|N|\}$. In Figure 1, we give an example of a DC distribution network modified from (http:// www.ee. washington.edu/research/pstca/). The DC distribution network consists of different types of load units, including Fixed Loads (FLs), which are modelled as fixed resistors, and Power Electronic Loads (PELs) (Mohsenian-Rad & Davoudi, 2014; Weaver & Krein, 2009). In particular, a PEL unit with a switch-mode converter can be considered as a variable resistor.

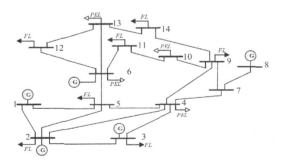

Figure 1. A DC distribution network with 14 buses. In this figure, we use hollow and solid triangles to represent the PEL and FL respectively.

As is widely applied in the literature, e.g. (Mohsenian-Rad & Davoudi, 2014; Weaver & Krein, 2009), we can affect the bus voltage level, line losses and power flow in the network by controlling the variable resistors of PEL units. In the rest of this section, we would like to establish the voltage model, which is a description of the distribution network by using the voltages of PEL units.

Denoted by L, F and P are the set of load buses, FL buses and PEL buses respectively, and we have $L = F \cup P$. B denotes the set of all the non-load buses; then $N = L \cup B$. For each single bus $i \in N$, let N_i with $N_i \subset N$ denote the set of all its neighbouring buses. In addition, some buses are equipped with power sources and the set of such buses is denoted by S. A DC power source at bus i is modelled as a voltage source with a fixed voltage level V_i^s, and a fixed internal resistor R_i^s. If bus i does not have a power source, we assume $V_i^s = 0$ and $R_i^s = \infty$.

Based on the above formulations, we can obtain the relationship between the bus voltage and load resistor by Kirchhoff's current law, such that:

$$\frac{V_i}{R_i} + \frac{V_i - V_i^s}{R_i^s} + \sum_{k \in N_i} \frac{V_i - V_k}{R_{ik}} = 0, \forall i \in L \quad (1)$$

$$\frac{V_j - V_j^s}{R_j^s} + \sum_{k \in N_j} \frac{V_j - V_k}{R_{jk}} = 0, \forall j \in B \quad (2)$$

where V_i represent the voltage of bus i, and R_{ij} is the resistance of the distribution line between bus i and j. Using (2), for each $j \in B$, it gives:

$$\frac{V_j - V_j^s}{R_j^s} + \sum_{k \in N_j} \frac{V_j}{R_{jk}} - \sum_{k \in N_j \cap B} \frac{V_k}{R_{jk}} = \sum_{k \in N_j \cap L} \frac{V_k}{R_{jk}} \quad (3)$$

Then rewrite it as:

$$V_j = \sum_{k \in L} \alpha_{jk} V_k + \beta_j, \forall j \in B \quad (4)$$

By (1), we have:

$$R_i = V_i / \left(\frac{V_i^s - V_i}{R_i^s} + \sum_{k \in N_i} \frac{V_k - V_i}{R_{ik}} \right) \quad (5)$$

By which together with (4), we have:

$$\frac{V_i}{R_i} = \frac{V_i^s - V_i}{R_i^s} - \sum_{k \in N_i} \frac{V_i}{R_{ik}} + \sum_{k \in N_i \cap L} \frac{V_k}{R_{ik}}$$
$$+ \sum_{k \in N_i \cap B} \frac{1}{R_{ik}} \left(\sum_{l \in L} \alpha_{kl} V_l + \beta_k \right)$$
$$= \sum_{k \in L} \mu_{ik} V_k + \nu_i, \forall i \in L \quad (6)$$

As (6) holds for all $i \in$ F, since the resistors of FLs are constant, there are |F| equations. Collating the |F| equations, we can derive that:

$$V_f = \sum_{k \in P} \lambda_{fk} V_k + \eta_f, \forall f \in F \qquad (7)$$

Furthermore, the following expressions are obtained by (4), (6) and (7):

$$V_j = \sum_{k \in L} a_{jk} V_k + b_j, \forall j \in B \qquad (8)$$

$$\frac{V_p}{R_p} = \sum_{k \in P} c_{pk} V_k + d_p, \forall p \in P \qquad (9)$$

In summary, (7)–(9) specify how to adjust the bus voltages by controlling the resistances of PELs.

By applying (9), we can obtain the relationship between the power and voltage of PEL p:

$$P_p = \frac{V_p^2}{R_p} = V_p \left(\sum_{k \in P} c_{pk} V_k + d_p \right), \forall p \in P \qquad (10)$$

3 OPTIMAL LOAD CONTROL PROBLEMS

In this section, we will implement the optimal solution of the load control problems. By (10), the power of PEL units can be adjusted by controlling the voltage of PEL buses. Hence, the objective is to specify the voltages of PEL buses by minimising the total system costs under the operational and physical constraints of power sources and PELs.

3.1 Power source constraints

For each bus $k \in$ S, suppose that there is a maximum power limit $P_k^{\max}$; then:

$$V_k^s I_k^s = V_k^s (V_k^s - V_k) / R_k^s \leq P_k^{\max} \qquad (11)$$

Hence, we have V_k shall satisfy the following inequality constraint (12). Also it is easy to verify that (12) also holds for those buses $i \notin$ S, due to $V_i^s = 0$ and $R_i^s = \infty$.

$$V_k \geq V_k^s - P_k^{\max} R_k^s / V_k^s \qquad (12)$$

For the bus $p \in$ P, it also satisfies $V_p \leq \max_{k \in S} \{V_k^s\}$, by which together with (12), we have:

$$V_p^s - P_p^{\max} R_p^s / V_p^s \leq V_p \leq \max_{k \in S} \{V_k^s\}, \forall p \in P \qquad (13)$$

Also, by (7), (8) and (12), we can obtain that:

$$\sum_{k \in P} \lambda_{fk} V_k \geq V_f^s - P_f^{\max} R_f^s / V_f^s - \eta_f, \forall f \in F \qquad (14a)$$

$$\sum_{k \in P} a_{jk} V_k \geq V_j^s - P_j^{\max} R_j^s / V_j^s - b_j, \forall j \in B \qquad (14b)$$

3.2 Constraints and preferences of PEL units

Suppose that $R_p^{\min} \leq R_p \leq R_p^{\max}$ is satisfied for each bus $p \in$ P. Then by (9), the following inequalities hold for all $p \in$ P:

$$\sum_{k \in P} m_{pk} V_k \geq R_p^{\min} d_p \qquad (15a)$$

$$\sum_{k \in P} n_{pk} V_k \leq R_p^{\max} d_p \qquad (15b)$$

where $m_{pk}(n_{pk}) = \begin{cases} 1 - c_{pk} R_p^{\min}(R_p^{\max}), & \text{if } k = p \\ -c_{pk} R_p^{\min}(R_p^{\max}), & \text{otherwise} \end{cases}$.

In this paper we apply a form of deviation costs considered in Li et al. (2012) to describe the preference of each PEL unit:

$$W_p(P_p) = \theta_p (P_p - \bar{P}_p)^2 \qquad (16)$$

where $\bar{P}_p$ and θ_p denote a desired power of PEL p and a weighting parameter respectively. We consider $\mathbf{V} \equiv (V_p, p \in P)$; then by (10) and (16), we have:

$$W_p(\mathbf{V}) = \theta_p \left(V_p \left(\sum_{k \in P} c_{pk} V_k + d_p \right) - \bar{P}_p \right)^2 \qquad (17)$$

3.3 Optimal problem

The distribution line losses are denoted by $H(V)$, and

$$H(\mathbf{V}) = \sum_{k \in N} \sum_{l \in N_k} (V_k - V_l)^2 / R_{kl} \qquad (18)$$

In (18), the control variables are V_p, for all $p \in$ P, since, by (7) and (8), each V_i with $i \in F \cup B$, can be expressed by $V_p, p \in P$.

Let $J(V)$ denote the system cost, such that:

$$J(\mathbf{V}) \triangleq \sum_{p \in P} W_p(\mathbf{V}) + \rho H(\mathbf{V}) \qquad (19)$$

where ρ is used to deal with the trade-off between the distribution line losses and the total deviation costs of PELs. Then we can establish the optimisation model for the underlying load control problems.

Problem 1:

min $J(V)$

s.t. (13), (14) and (15)   (20)

Problem 1 is a non-convex optimisation problem under a convex set. By (13), (14) and (15), the constrained set is convex as all the constraints are linear. However, $J(V)$ is a (fourth-order) non-convex function due to the convexity of $H(V)$ and the non-convexity of $W_p(V)$, which can be verified by using their Hessian matrixes. Therefore, we will use a PSO algorithm to implement the global optimal solution.

4 THE PARTICLE SWARM OPTIMISATION ALGORITHM AND CASE STUDY

In the following section, we apply the Particle Swarm Optimisation algorithm and demonstrate the results through a numerical simulation.

4.1 *The Particle Swarm Optimisation (PSO) algorithm*

PSO is a population-based, self-adaptive search method introduced by Kennedy and Eberhart (Dub & Stefek, 2014).

Algorithm 1 The PSO Algorithm.

Require:
    The maximum iteration z;
    The number of particles s;
    Set the constraints factor *cheak* = 0;

Ensure: The optimal or approximate optimal position $Pgx(t)$.

1:   **while** *check* = 0 **do**
2:       Randomly select the position $x_i(0)$ and the velocity $v_i(0)$, $i = 1, 2, \dots, s$ between the limit (13).
3:       Compute the fitness (objective function) of each particle in the swarm by (19).
4:       Initialise the local best position $Plx_i(0)$ and the global optimal position $Pgx(0) = \arg\min Plx_i(1)$, $i = 1, 2, \dots, s$.
5:       **for** $t = 1$ to z **do**
6:         **for** $i = 1$ to s **do**
7:           Update $v_i(t)$ and $x_i(t)$;
8:           Compute the fitness;
9:           Update the local best position $Plx_i(t)$;
10:         **end for**
11:         Update $Pgx(t) = \arg\min Plx_i(t)$, $i = 1, 2, \dots, s$.
12:       **end for**
13:       **if** $Pgx(t)$ satisfy the constraints then
14:         *check* = 1
15:       **end if**
16:   **end while**
17:   **return** $Pgx(t)$.

In the PSO algorithm, the solution can be regarded as the position of a point, which is called a particle in the search space. We summarise the standard PSO process as below:

- Select the appropriate iterations.
- The PSO begins with a set of random points, which is called a swarm in the search space.

- In each iteration, each particle will calculate its position value according to its objective function and adjust its position and velocity through its best location and the swarms global best location.

For the underlying optimal problem, the voltages of the PEL units as the particles, the power flow is implemented to compute the line flows and the system power transfers. However, since the standard PSO algorithm is used to solve the unconstrained optimisation problem, a check function is established to ensure the best strategy to be feasible. Specifically, the improved PSO algorithm is displayed in Algorithm 1.

4.2 *Case study*

Consider an IEEE 14-bus network with four PEL units, which is modified by ignoring the line reactance and reactive power flows, as shown in Figure 1. The parameters are given in Table 1. For each PEL p, suppose that $0.025 \leq R_p \leq 2.025$, $\theta_p = 2$, and $\rho = 1$.

By applying the PSO algorithm, we initialise 50 particles with 1000 iterations; weights for the local and global minima are both assumed to be 1.4962, and the inertial weight of each particle is initially set to be 0.7298. As displayed in Figure 2, the system converges to the optimal solution after about 30 iteration steps. Compared with the traditional OPF problem, which involves at least 14 voltage

Table 1. The parameters of the IEEE 14-bus network.

Line data			Source data			
From Bus.	To Bus.	R	Bus.	V^s	P_{max}	R^s
1	2	0.01938	1	1	3.5	0.02
1	5	0.01403				
2	3	0.04699	2	1	2	0.04
2	4	0.01811	3	1	3.5	0.01
2	5	0.05695				
3	4	0.02701	6	1	10	0.03
4	5	0.01335	8	1	3	0.02
4	7	0.05228				
4	9	0.01034	Load data			
5	6	0.0230	FL data		PEL data	
6	11	0.01498	Bus.	R	Bus.	P_i
6	12	0.02291	2	1.00	4	1.55
6	13	0.01615	3	1.00	6	0.95
7	8	0.04404	5	1.00	10	0.57
7	9	0.03667	9	1.00	13	1.25
9	10	0.03181	11	1.00		
9	14	0.02711	12	1.00		
10	11	0.01205	14	1.00		
12	13	0.02092				
13	14	0.01938				

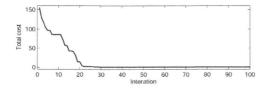

Figure 2. The first 100 steps of the PSO iterative process.

Table 2. The implemented optimal results for the IEEE 14-bus network.

Bus.	Bus type	V_i	R_i	P_i
1	B&S	0.9532	–	–
2	F&S	0.9442		0.8915
3	F&S	0.9685		0.9380
4	P	0.9237	0.5610	1.5209
5	F	0.9269		0.8592
6	P&S	0.9038	0.8916	0.9162
7	B	0.9355	–	–
8	B&S	0.9799	–	–
9	F	0.9070		0.8226
10	P	0.8880	1.2914	0.6106
11	F	0.8891		0.7905
12	F	0.8830		0.7798
13	P	0.8826	0.6557	1.1880
14	F	0.8828		0.7793

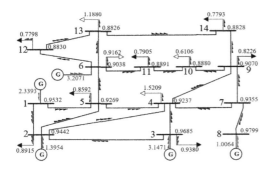

Figure 3. The power flow of the IEEE 14-bus network.

variables, the underlying optimal problem with four variables can be solved in less time by using the PSO algorithm.

The corresponding solution is listed in Table 2. The optimal power flow of the distribution network is displayed in Figure 3. Especially, at bus 2, the bus voltage is 0.9442, and the power from the power sources are connected to bus 1 and 2 to the FLs at bus 2, 4 and 5. Moreover, the actual power of the PEL units is close to their desired power and the load (FL and PEL) power consumption accounts for about 91% of the total generation capacity, as shown in Figure 4.

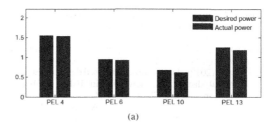

(a)

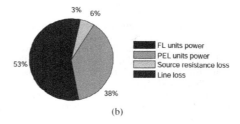

(b)

Figure 4. The results. (a) The actual power of PEL units near to the desired power; (b) The line losses and the source resistance losses little.

5 CONCLUSION

In this paper, we model the PELs as variable resistors and formulate the relationship between the bus voltage and PEL resistance in DC distribution networks. Based on these, we propose the optimisation problems that determine an optimal voltage strategy by regulating the PELs, to minimise losses of lines and deviation costs of PELs. Compared with the traditional OPF problems, the underlying optimisation problems have fewer variables, and can more easily be solved. To complement our theoretical model, we apply the proposed optimal model to an IEEE 14-bus distribution network system, and obtain an optimal voltage strategy or a resistance strategy of PELs by using a PSO algorithm.

REFERENCES

Dong, H.K., & Jin, I.P. (2005). *Loss Minimization Control of Induction Motor Using GA-PSO*. Berlin Heidelberg: Springer.

Duan, H. et al. (2004). Development on ant colony algorithm theory and its application. *Control and Decision*, 19(12), 1321–1320.

Dub, M., & Stefek, A. (2014). *Using PSO Method for System Identification*. Springer International Publishing: 143–150.

Goldberg, D.E. (1989). *Genetic Algorithms in Search, Optimization, and Machine Learning*. Addison-Wesley Pub. Co.

Jiang, D., & Zheng, H. (2012). Research status and developing prospect of DC distribution network. *Automation of Electric Power Systems*, 36(8), 98–104.

Justo, J.J. et al. (2013). AC-microgrids versus dc-microgrids with distributed energy resources: A review. *Renewable and Sustainable Energy Reviews,* 24(10), 387–405.

Kirkpatrick, S. et al. (1983). Optimization by simulated annealing. *Science,* 220(4598), 671–680.

Li, N. et al. (2012). An optimization-based demand response in radial distribution networks, in IEEE Globecom Workshops: 1474–1479, Anaheim, CA, 3–7 Dec. 2012.

Mohsenian-Rad, H., & Davoudi, A. (2014). Towards building an optimal demand response framework for DC distribution networks. *IEEE Transactions on Smart Grid,* 5(5), 2626–2634.

Power systems test case archive, Available: http:// www. ee.washington.edu/research/pstca/.

Ramanathan, B., & Vittal, V. (2008). A framework for evaluation of advanced direct load control with minimum disruption. *IEEE Transactions on Power Systems,* 23(4), 1681–1688.

Shi, W. et al. (2014). Optimal residential demand response in distribution networks. *IEEE Journal on Selected Areas in Communication,* 32(7), 1441–1450.

Song, Q. et al. (2013). An overview of research on smart DC distribution power network. *Proceedings of the CSEE,* 33(25), 9–19.

Sortomme, E., & El-Sharkawi, M.A. (2009). Optimal power flow for a system of microgrids with controllable loads and battery storage, in 2009 Power Systems Conference and Exposition: 1–5.

Trelea, I.C. (2003). The particle swarm optimization algorithm: convergence analysis and parameter selection. *Information Processing Letters,* 85(6), 317–325.

Wang, F. et al. (2008). AC vs. DC distribution for offshore power delivery, in IEEE Industrial Electronics Society: 2113–2118.

Weaver, W.W., & Krein, P.T. (2009). Game-theoretic control of small-scale power systems. *IEEE Transactions on Power Delivery,* 24(3), 1560–1567.

Wu, B., & Wang, D. (2016). Traffic signal networks control optimize with pso algorithm, in 12th International Conference on Natural Computation: 230–234.

Yang, B., & Cheng, L. (2013). Study of a new global optimization algorithm based on the standard pso. *Journal of Optimization Theory and Applications,* 158(3), 935–944.

Zhu, L. et al. (2015). Direct load control in microgrids to enhance the performance of integrated resources planning. *IEEE Transactions on Industry Applications,* 51(5), 3553–3560.

Automotive, Mechanical and Electrical Engineering – Liu (Ed.)
© 2017 Taylor & Francis Group, London, ISBN 978-1-138-62951-6

Research on attribute-based encryption scheme for constant-length ciphertexts

Lu Han, Zhongqin Wang & Yan Wang
University of Science and Technology Beijing, Beijing, China

ABSTRACT: Ciphertext-Policy Attribute-Based Encryption (CP-ABE) is especially suitable for access control system in cloud computing environment. In most of the existing CP-ABE schemes, the length of the ciphertexts increases linearly with the complexity of the access policy, so that the expenses are very expensive. To tackle the challenge above, we propose an attribute-based encryption scheme for constant-length ciphertexts in this paper. The algorithm limits the length of the ciphertext and the computation of the bilinear pairing in fixed value, improving the efficiency of the system. Apart from this, it introduces hierarchical authorizations to reduce the burden and risk of the single authorization. As a result, it achieves a high efficiency, fine and scalable access control. The scheme supports the situation which has a plurality of attribute values and wild-card *AND* strategy. Furthermore, the scheme is proved secure under the assumption of $(t, \varepsilon, l) - BDHE$.

Keywords: Cloud computing; constant-length ciphertexts; hierarchical authorization

1 INTRODUCTION

Cloud computing is an important part in the development of information technology. With the application of the cloud computing, the security problem has raised lots of concerns. Cloud computing service providers claim that it can improve the security of the data, while the users' concerns about the problem have never stopped. In a sense, how to solve the security problem is the key to the development of the cloud computing industry. As an important part of the security of cloud computing, access control restricts users to access or use certain resources or functions in accordance with the users' identity and the sets of groups which are predefined by the administrator. Because of the unique characteristics of the cloud computing, it is necessary to make researches on the security measures adapted to the cloud environment. In view of the threats faced by the current cloud computing, it is meaningful to strengthen the security of the data by effective access control in the field of cloud.

1.1 *Our contribution*

Firstly, we propose an algorithm which is suitable for the cloud computing environment. The main idea is that AES algorithm is used to encrypt the data, and CP-ABE algorithm is used to encrypt the secret key, which is used in the *AES*. The users whose attributes match the access policy can decrypt the secret key and data in turn. Cloud computing providers have no secret key, so it can only store the data, and can't decrypt it, thus ensuring the data stored in the cloud security. Secondly, we prove that the proposed scheme is secure under the assumption of $(t, \varepsilon, l) - BDHE$ by the complexity and security analysis.

1.2 *Related work*

Since the concept of "attribute" was firstly put forwards in 2005 (Sahai A et al., 2005), the entities that used to identify the identity in the traditional encryption system can be expressed through it. Two kinds of Attribute-Based Encryption were proposed in 2006 (Goyal V et al., 2006), one is Key-Policy Attribute-Based Encryption; the other is Ciphertext-Policy Attribute-Based Encryption. There are a lot of access control schemes applying the ABE algorithm in recent years. The first official CP-ABE scheme was proposed in (Bethencourt J et al., 2007). The scheme is easy to implement and can only be applied to the condition that the access control rules are not strict. In order to overcome this shortcoming, Cheung et al. (Cheung L et al, 2007) proposed a new CP-ABE scheme that supported $AND^*_{+,-}$ policy, and proved the security of the scheme in the standard model. However, the length of the ciphertext in most of the CP-ABE schemes existed is very large, which increases linearly with the increase of the complexity of the access policy. Considering the practical application scenarios with limited bandwidth resources, many scholars have studied the ABE scheme with constant-length ciphertext. Emura et al. (Emura K et al., 2009). proposed the first CP-ABE scheme with constant-length ciphertext. The access strategy $AND_{+,-}$ is that it only supports positive and negative of the attributes, and it does not support wildcards. Only if his attributes set is the same as the access strategy, he can decrypt the file encrypted. As a result, the scheme can't be used in one-to-many systems. Similarly, the schemes in (Rao Y S et al., 2013) supports AND_m and in (Han J et al., 2012) supports AND_+ also have the same problem. Compared with $AND_{+,-}^*$ strategy, AND_m^* has higher efficiency. As a result, the CP-ABE scheme supports the AND_m^*, which ensures a constant-length ciphertext is extremely meaningful.

2 PRELIMINARIES

2.1 Bilinear pairings

On the assumption that G_1, G_2, G_T are cyclic multiplicative groups of a large prime order p, bilinear pairing $\hat{e}: G_1 \times G_2 \to G_T$ is a deterministic function. Input the elements of G_1 and G_2 respectively, and the output is an element in G_T. The rules of bilinear pairing are as follows:

1. Bilinear: $\hat{e}(x^a, y^b) = \hat{e}(x, y)^{ab}$ for all $x \in G_1$, $y \in G_2$, $a, b \in Z_p$.
2. Nondegenerate: $\hat{e}(g_1, g_2) \neq 1$ where g_1 and g_2 are generators of G_1 and G_2 respectively.
3. Computable: It is easy to compute $\hat{e}(g_1, g_2)$.

We call $\hat{e}$ is an effective bilinear map if it is a map with the above attributes.

2.2 Complexity assumptions

The security of the CP-ABE algorithm is based on a complexity assumption called $(t, \varepsilon, l) - BDHE$.

On the assumption that G is a cyclic multiplicative group of a large prime order p, g and h are two independent generators of G, and there is a bilinear pairing $\hat{e}: G \times G \to G_T$. For some unknown $a \in Z_p$, denote $\vec{y}_{g,\alpha,l} = (g_1, g_2, \cdots, g_l, g_{l+2}, \cdots, g_{2l}) \in G^{2l-1}$, where $g_i = g^{(\alpha^i)}$. B is an algorithm that can solve the problem of $(t, \varepsilon, l) - BDHE$ and outputs $\mu \in \{0, 1\}$. We call B has advantage ε if formula (1) is workable.

$$\left| \Pr\left[B\left(g, h, \vec{y}_{g,\alpha,l}, \hat{e}(g_{l+1}, h)\right) = 1 \right] - \Pr\left[B\left(g, h, \vec{y}_{g,\alpha,l}, Z\right) = 1 \right] \right| \geq \varepsilon \quad (1)$$

If there is no algorithm that has an advantage at least ε in solving the problem of $(t, \varepsilon, l) - BDHE$ within t, we claim that $(t, \varepsilon, l) - BDHE$ assumption holds in G.

3 ACCESS POLICY

For a list of attributes L and an access policy W, $L \models W$ denotes that L matches W, and $L \not\models W$ denotes that L doesn't match W. In this paper, the access policy is AND_m^*, where $L = [L_1, L_2, \cdots, L_n]$ and $W = [W_1, W_2, \cdots, W_n] = \bigcap_{i \in I_W} W_i$. If $L_i = W_i$ or $W_i = *$, we say $L \models W$, else $L \not\models W$ for $1 \leq i \leq n$ in which $I_W = \{i | 1 \leq i \leq n, W_i \neq *\}$. An example of AND_m^* policy is shown in Table 1. On the assumption that there are n attributes in the system, attribute set is $U = \{\omega_1, \omega_2, \cdots, \omega_n\}$, where each attribute has more than one value. The th attribute ω_i has n_i values, whose value set is $S_i = \{v_{i,1}, v_{i,2}, \cdots, v_{i,n_i}\}$. In Table 1, AND_m^* policy is $CP = v_{1,4} \cap v_{2,3} \cap * \cap v_{4,1}$.

4 SYSTEM SCHEME

4.1 System model

As is shown in Fig. 1, there are five components in system model. The root authority has the highest authority and is fully trusted. It is responsible for generating the system parameters, including the public key and the master key, and author-

Table 1. An example of AND_m^* policy.

Attribute	ω_1	ω_2	ω_3	ω_4
Name	A	B	C	D
Value	A_1	B_1	C_1	D_1
	A_2	B_2	C_2	D_2
	A_3	B_3	C_3	
	A_4			
CP	A_4	B_3	*	D_1

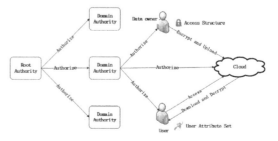

Figure 1. System model.

576

izing the top-level domain authority the secret key. The domain authorities authorize the secret keys to the subordinate authorities, data owners and users in the jurisdiction of the domain. Data owner is a provider of privacy data, and he will select a specific access structure to encrypt the privacy data. At last, the encrypted data will be uploaded to the cloud server for storage. The ciphertext contains the access structure, therefore, only if the attributes set contained by the secret key that authorized by domain authority matches the access structure, the user can decrypt the data. Cloud service provider is responsible for the storage of the data that uploaded by the owner, and provides download service for users who put forward the request to access the file. The system model is a multi-level structure with root authority and domain authority. Thus the calculation of the authorized key can be effectively dispersed and the burden of the root authority will be reduced. In addition, it makes the scheme more scalable to cope with the huge growth of users in cloud environment.

4.2 Detailed scheme

The scheme mainly includes five algorithms: Setup, KeyGen, Delegate, Encrypt and Decrypt.

1. Establish system: $Setup(\lambda, U) \rightarrow (PK, MK)$
Suppose that G and G_T are cyclic multiplicative groups of a large prime order p, and g is a generator of G. Function $\hat{e}: G_1 \times G_2 \rightarrow G_T$ is a bilinear pairing. There are n attributes in the system, attribute set is $U = \{\omega_1, \omega_2, \cdots, \omega_n\}$, where each attribute has more than one value. The ith attribute ω_i has n_i values, whose value set is $S_i = \{v_{i,1}, v_{i,2}, \cdots, v_{i,n_i}\}$. Define two hash functions: $H_0: Z_p \times \{0,1\}^{\log_2 n} \times \{0,1\}^{\log_2 n_m} \rightarrow Z_p$ and $H_1: Z_p \rightarrow G$ where $n_m = \max_{i=1}^{n} n_i$. The root authority chooses $x, y \in Z_p$. For $1 \leq i \leq n$ and $1 \leq k_i \leq n_i$, then calculate $X_{i,k_i} = g^{-H_0(x\|i\|k_i)}$ and $Y_{i,k_i} = \hat{e}(g,g)^{H_0(y\|i\|k_i)}$. The public key is $PK = (g, \{X_{i,k_i}, Y_{i,k_i}\}_{1 \leq i \leq n, 1 \leq k_i \leq n_i})$, and the master key is $MK = (x, y)$.

2. Authorize for top-level domain authority: $KeyGen(PK, MK, L) \rightarrow SK_{DA}$
When a top-level domain authority whose attribute set is $L = [L_1, L_2, \cdots, L_n]$ where $L_i = v_{i,k_i}$ requests to join the system, the root authority verify its legitimacy at first. If it is legal, the root authority applies KeyGen algorithm to generate the private key. It chooses $r \in Z_p$. For $1 \leq i \leq n$ and $1 \leq k_i \leq n_i$, then calculate $h^r = H_1(r)$, $d_0 = g^{H_0(y\|i\|k_i)}, d_1^r = (h^r)^{H_0(x\|i\|k_i)}$ and $\bar{\sigma}_i^r = d_0 \cdot d_1^r$. The private key of the top-level domain authority is $SK_{DA} = (r, h^r, \{\bar{\sigma}_i^r\}_{1 \leq i \leq n, 1 \leq k_i \leq n_i})$.

3. Authorize for new domain authority/user: $Delegate(SK_{DA}, L') \rightarrow SK_{DA}/SK_u$
When a new domain authority/user whose list of attributes is $L' = [L_1', L_2', \cdots, L_n']$ requests to join the system, where $L_i' = v_{i,k_i}$, the top-level authority verify its legitimacy at first. If it is legal, the top-level authority applies Delegate algorithm to generate the private key. It chooses $r' \in Z_p$. For $1 \leq i \, n$ and $1 \leq k_i \, n_i$, then calculate $h_1^r = H_1(r'), h^{r'} = h^r \cdot h_1^r, d_1^{r'} = (h_1^r)^{H_0(x\|i\|k_i)}$ and $\bar{\sigma}_i^{r'} = d_0 \cdot d_1^r \cdot d_1^{r'}$. The private key of the new member is $SK_{DA}/SK_u = (r', h^{r'}, \{\bar{\sigma}_i^{r'}\}_{1 \leq i \leq n, 1 \leq k_i \leq n_i})$.

4. Create a new file: $Encrypt(PK, M, W) \rightarrow CT_W$
The access policy is $W = [W_1, W_2, \cdots, W_n]$, where $W_i = v_{i,k_i}$, and the message to be encrypted is M. For $1 \leq i \leq n$ and $1 \leq k_i \leq n_i$, the owner calculates $\langle X_W, Y_W \rangle = \left\langle \prod_{i \in I_W} X_{i,k_i}, \prod_{i \in I_W} Y_{i,k_i} \right\rangle$. It chooses $s \in Z_p$, and calculates $C_0 = MY_W^s$, $C_1 = g^s$ and $C_2 = X_W^s$. The ciphertext is $CT_W = (W, C_0, C_1, C_2)$. The length of CT_W is $2L_0 + L_1$, where L_0 is the bit-length of an element in G and L_1 is the bit-length of an element in G_T.

Because of the complexity of CP-ABE algorithm, it isn't suitable to encrypt large files. To solve the problem, we use Symmetric Encryption Key (DEK) to encrypt data files to get cipher data, and then use CP-ABE algorithm to encrypt the DEK to get cipher key. Users can get the data files by decrypting the cipher key and cipher data in proper order. Before the data file is uploaded to the cloud server, the owner should manipulate the data file as follows:

a. Select a unique ID for the data file.
b. Select a DEK from keyspace randomly, the DEK is used to encrypt the file.
c. Encrypt the DEK using Encrypt algorithm, output is CT_W. Regard cipher data and cipher key as a whole file, and upload it to the cloud service.

Data files' format in the cloud storage is as shown in Fig. 2.

5. Access files: $Decrypt(PK, SK_u, CT_W) \rightarrow M$
When cloud service provider receives a request to access the encrypted file that is stored in the

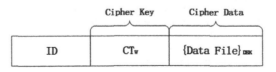

Figure 2. Data files' format in the cloud storage.

cloud server from a user, it sends the corresponding ciphertext file to the user. At first, the user detects L and W. If $L \neq W$, the algorithm returns $\perp$. Otherwise, $L| = W$. When $L| = W$, user calculates $\sigma_W = \prod_{i \in I_W} \bar{\sigma}_i^{r'}$ and $M = \frac{C_0}{\hat{e}(\sigma_W, C_1) \cdot \hat{e}(h', C_2)}$, where M is the DEK. At last, the user uses the DEK to decrypt the encrypted file to access the data.

5 SECURITY ANALYSIS

Theorem 1: Assume that the adversary A requests q_{H_1} random oracle and q_K key generation at most. If it is true that $(t, \varepsilon, l) - BDHE$ holds in G, the proposed scheme is (T', ε, m) secure, where m is an upper bound of the total number of users in the system, $N = \sum_{i=1}^{n} n_i$, $T' = T + o(q_{H_1} + m \cdot q_k + N)T_1 + o(N) T_2$ and T_1, T_2 represent the time complexity of modular exponent operation in group G, G_T respectively.

Proof: Suppose that there is a t-time adversary A so that $Adv_{CP-ABE}^{IND-sCP-CPA}(A) \geq \varepsilon$. We build a simulator S that has advantage ε in solving the decision $m - BDHE$ problem in G. S inputs a random decision $m - BDHE$ challenge $(g, h, \bar{y}_{g,\alpha,m}, Z)$, where $\bar{y}_{g,\alpha,m} = (g_1, g_2, \cdots, g_m, g_{m+2}, \cdots, g_{2m})$ and Z is either $\hat{e}(g_{m+1}, h)$ or a random element in G_T. S plays the role of challenger in the $IND-sCP-CPA$ game, and it interacts with A as follows:

Init. A sends a challenge policy $W^* = \bigcap_{i \in I_{W^*}} W_i$ to S, where $I_{W^*} = \{i_1, i_2, \cdots, i_w\}(w \leq n)$ represents all index of the attributes that appear in W^*.

Setup. S generates the public key PK. At first, S chooses $j^* \in \{1, 2, \cdots, w\}$ and $x, x', y, y' \in Z_p$. Then it calculates as follows.

If $i_j \in I_{W^*} - \{i_{j^*}\}$ where $W_{i_j} = v_{i_j, k_{i_j}}$. Calculate $(X_{i_j, k_{i_j}}, Y_{i_j, k_{i_j}}) = (g^{-H_0(x\|i_j\|k_{i_j})} g_{m+1-i_j}^{-1}, \hat{e}(g,g)^{H_0(y\|i_j\|k_{i_j})})$.

For i_{j^*} where $W_{i_{j^*}} = v_{i_{j^*}, k_{i_{j^*}}}$. Calculate $(X_{i_{j^*}, k_{i_{j^*}}}, Y_{i_{j^*}, k_{i_{j^*}}}) = (g^{-H_0(x\|i_{j^*}\|k_{i_{j^*}})} \prod_{t \in I_{W^*} - \{i_{j^*}\}} g_{m+1-t}, \hat{e}(g,g)^{H_0(y\|i_{j^*}\|k_{i_{j^*}})} \hat{e}(g,g)^{\alpha^{m+1}})$.

If $i_j \notin I_{W^*}$, for $1 \leq k_{i_j} \leq n_{i_j}$, S calculates $(X_{i_j, k_{i_j}}, Y_{i_j, k_{i_j}}) = (g^{-H_0(x\|i_j\|k_{i_j})}, \hat{e}(g,g)^{H_0(y\|i_j\|k_{i_j})})$.

The public key is $PK = (g, \{X_{i,k_i}, Y_{t,k_i}\}_{1 \leq i \leq n, 1 \leq k_i \leq n_i})$, and is sent to A.

Phase1. A initiates the following questions:

Hash Oracle $O_{H_0}(\cdot)$: when A requests "." for H_0, S checks whether there is "." in L_0 at first. If it is true, then return the previous value. If not, choose $a \in Z_p$, add $\langle \cdot, a \rangle$ to L_0, and return a.

Hash Oracle $O_{H_1}(r)$: when A requests r for H_1, S checks whether there is "r" in L_1 at first. If it is true, then return the previous value. If "r" isn't in L_1 and r corresponds to a property list L in the key generation query, S adds $\langle r, g_{i_j} g^z \rangle$ to L_1 and

returns $g_{i_j} g^z$ where $z \in Z_p$ and $L_{i_j} \notin W_{i_j}$. Or else S chooses $i_j \in \{1, 2, \cdots, n\}$ and adds $\langle r, g_{i_j} g^z \rangle$ to L_1 and returns $g_{i_j} g^z$ where $z \in Z_p$.

KeyGen Oracle $O_{KeyGen}(L)$: on assumption that A requests L to generate the private key, if $L| \neq W^*$, there must be $i_j \in I_{W^*}$ that makes $L_{i_j} \notin W_{i_j}$ existed. Supposed $L_{i_j} = v_{i_j, \hat{k}_{i_j}}$ and $W_{i_j} = v_{i_j, k_{i_j}}$, S chooses $r \in Z_p$ and calculates $\bar{\sigma}_{i_j} = \sigma_{i_j, \hat{k}_{i_j}} = g^{H_0(y'\|i_j\|\hat{k}_{i_j})} (g_{i_j} g^z)^{H_0(x'\|i_j\|\hat{k}_{i_j})}$. For $t \neq i_j$, choose $z \in Z_p$ and calculate as follows.

Case 1. If $t \in I_{W^*} - \{i_{j^*}\}$, on assumption that $L_t = v_{t,k_t}$, S calculates (2).

$$\bar{\sigma}_t = \sigma_{t,k_t} = g^{H_0(y\|t\|k_t)} (g_{i_j})^{H_0(x\|t\|k_t)} g_{m+1-t+i_j} (\bar{X}_t)^{-z} \quad (2)$$

Case 2. If $t = i_{j^*}$, on assumption that $L_{i_{j^*}} = v_{i_{j^*}, k_{i_{j^*}}}$, S calculates (3).

$$\bar{\sigma}_{i_{j^*}} = \sigma_{i_{j^*}, k_{i_{j^*}}} = g^{H_0(y\|i_{j^*}\|k_{i_{j^*}})} (g_{i_j})^{H_0(x\|i_{j^*}\|k_{i_{j^*}})} \times \left(\prod_{k \in I_{W^*} - \{i_{j^*}\}} g_{m+1-k+i_j}^{-1} \right) (\bar{X}_{i_{j^*}})^{-z} \quad (3)$$

Case 3. If $t \notin I_{W^*}$, on assumption that $L_t = v_{t,k_t}$, S calculates (4).

$$\bar{\sigma}_t = \sigma_{t,k_t} = g^{H_0(y\|t\|k_t)} (g_{i_j} g^z)^{H_0(x\|t\|k_t)} \quad (4)$$

Challenge. Suppose the formulas (5) (6) (7) (8).

$$x_{W^*} = \sum_{t \in I_{W^*}} H_0(x\|t\|k_t) = \sum_{j=1}^{w} H_0(x\|i_j\|k_{i_j}) \quad (5)$$

$$y_{W^*} = \sum_{t \in I_{W^*}} H_0(y\|t\|k_t) = \sum_{j=1}^{w} H_0(y\|i_j\|k_{i_j}) \quad (6)$$

$$X_{W^*} = \bar{X}_{i_{j^*}} \prod_{t \in I_{W^*} - \{i_{j^*}\}} \bar{X}_t$$
$$= \left(g^{-H_0(x\|i_{j^*}\|k_{i_{j^*}})} \prod_{t \in I_{W^*} - \{i_{j^*}\}} g_{m+1-t} \right) \prod_{t \in I_{W^*} - \{i_{j^*}\}} g^{-H_0(x\|t\|k_t)} g_{m+1-t}^{-1} = g^{-x_{W^*}} \quad (7)$$

$$Y_{W^*} = \bar{Y}_{i_{j^*}} \prod_{t \in I_{W^*} - \{i_{j^*}\}} \bar{Y}_t = \hat{e}(g,g)^{H_0(y\|i_{j^*}\|k_{i_{j^*}})} \hat{e}(g,g)^{\alpha^{m+1}} \prod_{t \in I_{W^*} - \{i_{j^*}\}} \hat{e}(g,g)^{H_0(y\|t\|k_t)} = \hat{e}(g,g)^{y_{W^*} + \alpha^{m+1}} \quad (8)$$

A submits two messages of equal length M_0 and M_1 to S. S chooses $b \in \{0, 1\}$ and calculates $C_0^* = M_b Y_{W^*}^s = M_b \cdot Z \cdot \hat{e}(g,h)^{y_{W^*}}, C_1^* = h$ and $C_2^* = h^{-x_{W^*}}$. The challenging ciphertext is $CT_{W^*} = (W^*, C_0^*, C_1^*, C_2^*)$. If $Z = \hat{e}(g_{m+1}, h)$, CT_{W^*} is an effective ciphertext. If Z is a random element in G_T, CT_W is independent of b.

Phase 2. Phase 2 is the same as Phase 1.

Guess: The output of A is b' that is the speculation about b. If $b = b'$, it represents that $Z = \hat{e}(g_{m+1}, h)$ in the $m - BDHE$ game. Or else it output 0, which represents that Z is a random element in G_T. As a result, if $Z = \hat{e}(g_{m+1}, h), CT_{W^*}$ is an effective ciphertext, where (9) is workable.

$$\Pr\left[S\left(g, h, \bar{y}_{g,\alpha,m}, \hat{e}(g_{m+1}, h)\right) = 1 \right]$$
$$= \frac{1}{2} + Adv_{CP-ABE}^{IND-sCP-CPA}(A) \geq \frac{1}{2} + \varepsilon \qquad (9)$$

If Z is a random element in G_T, M_b is completely hidden from A, and $\Pr[S(g, h, \bar{y}_{g,\alpha,m}, Z) = 1] = \frac{1}{2}$. In the process of solving the problem of determining $m - BDHE$, the advantage of A is at least ε, where $T' = T + o(q_{H_1} + m \cdot q_K + N)T_1 + o(N)T_2$.

6 SUMMARY

In this paper, we propose an access control scheme for encrypted files based on cloud storage system. It is a hierarchical based on the attribute and the length of the ciphertext is constant. And it is proved that the new scheme is secure in the random oracle model. The scheme supports the strategy that has a plurality of attributes and wildcard AND_m^* strategy. The scheme limits the length of the ciphertext and the computation of bilinear pairing to the constant value, and introduces hierarchical authorization structure, reducing the burden and risk of a single authority.

REFERENCES

Bethencourt J, Sahai A, Waters B. Ciphertext-policy attribute-based encryption[C]//2007 IEEE symposium on security and privacy (SP'07). IEEE, 2007: 321–334.

Cheung L, Newport C. Provably secure ciphertext policy ABE [C]//Proceedings of the 14th ACM conference on Computer and communications security. ACM, 2007: 456–465.

Emura K, Miyaji A, Nomura A, et al. A ciphertext-policy attribute-based encryption scheme with constant ciphertext length[C]//International Conference on Information Security Practice and Experience. Springer Berlin Heidelberg, 2009: 13–23.

Goyal V, Pandey O, Sahai A, et al. Attribute-based encryption for fine-grained access control of encrypted data[C] // Proceedings of the 13th ACM conference on Computer and communications security. Acm, 2006: 89–98.

Han J, Susilo W, Mu Y, et al. Attribute-based oblivious access control [J]. The Computer Journal, 2012, 55(10): 1202–1215.

Rao Y.S, Dutta R. Recipient anonymous ciphertext-policy attribute based encryption[C] //International Conference on Information Systems Security. Springer Berlin Heidelberg, 2013: 329–344.

Sahai A, Waters B. Fuzzy identity-based encryption[C]// Annual International Conference on the Theory and Applications of Cryptographic Techniques. Springer Berlin Heidelberg, 2005: 457–473.

Automotive, Mechanical and Electrical Engineering – Liu (Ed.)
© 2017 Taylor & Francis Group, London, ISBN 978-1-138-62951-6

Research on attribute-based encryption scheme for constant-length decryption key in Hadoop cloud environment

Lu Han, Yan Wang & Yuxia Sun
University of Science and Technology Beijing, Beijing, China

ABSTRACT: Cloud computing is a new application model. It can significantly reduce the operating costs and improve operational efficiency. With the continuous development of cloud computing technology, security and privacy have become the biggest obstacle to cloud computing applications. Access control technology is an important measure and means to protect the access and data security of cloud services in cloud computing environment. In the cloud computing environment, massive data grows with each passing day, which makes the data of Internet more and more abundant. At the same time, it also leads that the storage space of the third party cloud service provider has enormous challenges. Based on the Hadoop cloud platform, a hierarchical Ciphertext-Policy Attribute-Based Encryption (CP-ABE) access control model with constant-length decryption key is proposed and implemented in Hadoop cloud environment. The model not only has the characteristics of constant-length decryption key, hierarchical authorization structure, reduction of the computation quantity, but also is proved that the model can realize the efficient access control of encrypted data in the cloud computing environment, and solve the problem of limited cloud storage space.

Keywords: Constant-length decryption key; hierarchical access control; cloud computing; Hadoop

1 INTRODUCTION

Cloud computing is a new network computing model based on Internet (Zhang Q et al., 2010). Cloud computing provides super large scale of cloud resource pool such as computing, storage and software (Armbrust M et al., 2010), including Infrastructure-as-a-Service, Platform-as-a-Service and Software-as-a-Service. As the cloud computing technology is in line with the world government to advocate and promote the development of low carbon economy and green computing, has become a new economic growth point in the Internet field (Ali M, 2009). However, cloud computing itself is also facing many problems in the process of development, in which the security issues bear the brunt and show a gradual upward trend which has become an important factor restricting its development. The application of cloud computing is bound to contain a large number of personal or business sensitive information transferred to the cloud. Once the user stores the sensitive information in the cloud server, the data and other sensitive information will be out of control. Data security issues will cause great concern. Cloud servers are typically operated by commercial cloud service providers, and once the sensitive data leakage may cause significant economic losses or catastrophic consequences. In the cloud computing environment, users need to upload the data file encrypted to the cloud server for storage and management. Many scholars have done research on cloud computing data encryption (Odelu V et al., 2016) (Liu Z et al., 2015) (Chung H H et al., 2015). In addition, the procurement of the server, the establishment and maintenance of the data center requires a lot of manpower and financial resources, thus reducing the size of the data file that uploaded by the owner is an urgent problem to solve to reduce the cost of the third party cloud service provider.

2 BACKGROUND

2.1 *Summary of Hadoop*

Hadoop is an open source distributed computing platform (Zhao J, 2009), in which MapReduce and Hadoop Distributed File System (HDFS) provide users distributed basic framework that the details at the bottom of the system are transparent. High scalability and high fault tolerance of HDFS allow users to deploy Hadoop on inexpensive machines to form a distributed file system. MapReduce distributed programming model allows users to develop parallel applications without understanding the underlying details of the distributed system. So users can easily use Hadoop to organize computer resources, so as to build their own distributed computing platform.

HDFS cluster adopts the Master-Slave structure, which is composed of a NameNode and a number of DataNodes. NameNode is the main server, which manages the file system namespace and the users to access to the file. DataNodes manages the data stored in the cluster. The architecture of HDFS is shown in Fig. 1.

Hadoop mainly has the following advantages: (1) High reliability. Hadoop computing power is worthy of trust. (2) High extensibility. Hadoop clusters can be easily extended to thousands of

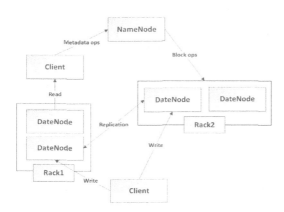

Figure 1. The architecture of HDFS.

nodes. (3) High efficiency. Hadoop can guarantee the dynamic balance among the nodes. (4) Strong fault tolerance. Hadoop automatically saves multiple copies of the data, and can automatically redistribute of failed tasks.

2.2 Summary of CP-ABE algorithm

CP-ABE algorithm with constant-length decryption key consists of five polynomial time algorithms: Setup, KeyGen, Delegate, Encrypt and Decrypt, achieving to establish the system, authorize for the root authority, authorize for the domain authority/users, and create new files and access files respectively.

$Setup(1^{\lambda}, U) \rightarrow (PK, MK)$: The inputs are the security parameter λ in the system and all the real attributes set U, and output the system public key PK and master key MK.

$KeyGen(PK, MK, L) \rightarrow SK_{DA}$: The inputs are the public key PK, the master key MK and the attributes set of the domain authority L, and output the private key of the domain authority SK_{DA}.

$Delegate(SK_{DA}, L_1) \rightarrow SK_{DA'}/SK_u$: The inputs are the private key of the domain authority SK_{DA} and the attributes set of the domain authority/users L_1, and output the private key of the domain authority $SK_{DA'}$ or the decryption key of users SK_u.

$Encrypt(PK, M, W) \rightarrow CT_W$: The inputs are the public key PK, the data to be encrypted M and the access structure W, and output the ciphertext CT_W.

$Decrypt(PK, SK_u, CT_W) \rightarrow M$: The inputs are the public key PK, the decryption key SK_u and the ciphertext CT_W, and output the data to be encrypted M.

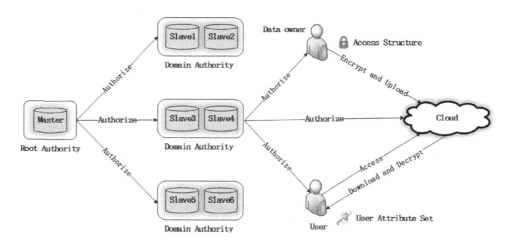

Figure 2. System model.

3 SYSTEM SCHEME

3.1 *System model*

The system model is based on the Hadoop cloud platform, which is shown in Fig. 2.

The system mainly consists of five types of entities: the root authority, the domain authority, data owner, user and cloud service providers. The root authority has the highest authority, which is responsible for the generation of system parameters, authentication domain authority and its authorization. The domain authority manages owners in the domain, and authorizes for the general users. The hierarchical attribute authority structure distributes the computation. At the same time, combined with the access authentication mechanism between Master node and Slave nodes in Hadoop, the burden and risk of the root authority are reduced. The owner obtains cipher keys and other related parameters, encrypts the file to be shared, and then uploads it to the cloud service provider. Because the cipher text contains the file access structure, only the users with the corresponding attributes can decrypt the file.

3.2 *Detailed scheme*

Master node, which is the root authority of the authority structure in Hadoop, establishes the system using Setup algorithm.

Algorithm 1: $Setup(1^\lambda, U) \rightarrow (PK, MK)$

G and G_T are cyclic multiplicative groups of a large prime order p, and g is a generator of G. $\hat{e}: G_1 \times G_2 \rightarrow G_T$ is a bilinear pairing. The system attributes set is $U = \{\omega_1, \omega_2, \cdots, \omega_n\}$, where the ith attribute ω_i has n_i values, whose value set is $S_i = \{v_{i,1}, v_{i,2}, \cdots, v_{i,n_i}\}$.

Step 1: Define two hash functions: $H_0: Z_p \times \{0,1\}^{\log_2 n} \times \{0,1\}^{\log_2(\max_{i=1}^n n_i)} \rightarrow Z_p$ and $H_1: Z_p \rightarrow G$.

Step 2: The root authority chooses $x, y \in Z_p$.

Step 3: The root authority calculates $X_{i,k_i} = g^{-H_0(x\|i\|k_i)}$ and $Y_{i,k_i} = \hat{e}(g,g)^{H_0(y\|i\|k_i)}$ ($1 \le i \le n$, $1 \le k_i \le n_i$). Output: $PK = (g, \{X_{i,k_i}, Y_{i,k_i}\})$

$MK = (x, y)$.

Algorithm 2: $KeyGen(PK, MK, L) \rightarrow SK_{DA}$

A domain authority requests to join the system, whose attribute set is $L = [v_{1,k_1}, v_{2,k_2}, \cdots, v_{n,n_i}]$.

Step 1: The root authority verifies legitimacy of the domain authority.

Step 2: The root authority chooses $r \in Z_p$.

Step 3: The root authority calculates $h^r = H_1(r)$, $d_0 = g^{H_0(y\|i\|k_i)}$, $d_1^r = (h^r)^{H_0(x\|i\|k_i)}$ and $\bar{\sigma}_i^r = d_0 \cdot d_1^r$ ($1 \le i \le n, 1 \le k_i \le n_i$).

Output: $SK_{DA} = (r, h^r, \{\bar{\sigma}_i^r\})$.

Algorithm 3: $Delegate(SK_{DA}, L_1) \rightarrow SK_{DA}/SK_u$

A new domain authority/user requests to join the system, whose attribute set is $L' = [v_{1,k_1}', \cdots, v_{n,n_i}']$.

Step 1: The top-level domain authority verifies legitimacy of the new domain authority/user.

Step 2: The top-level domain authority chooses $r', r^* \in Z_p$.

Step 3: The top-level domain authority calculates $d_1^{r'} = (h_1^r)^{H_0(x\|i\|k_i)}$, $\bar{\sigma}_i^{r'} = d_0 \cdot d_1^r \cdot d_1^{r'}$, $D = \prod_{i \in I_{L_1}} [\bar{\sigma}_i^{r'} \cdot H_1(r^*)^{H_0(x\|i\|k_i)}]$ and $E = H_1(r) \cdot H_1(r') \cdot H_1(r^*)$ ($1 \le i \le n, 1 \le k_i \le n_i$).

Output: $SK_{DA}/SK_u = (D, E)$.

Algorithm 4: $Encrypt(PK, M, W) \rightarrow CT_W$

The access policy is $W = [v_{1,k_1}, v_{2,k_2}, \cdots, v_{n,n_i}]$ and the data file to be encrypted is M.

Step 1: The owner calculates $\langle X_W, Y_W \rangle = \langle \prod_{i \in I_W} X_{i,k_i}, \prod_{i \in I_W} Y_{i,k_i} \rangle$ ($1 \le i \le n, 1 \le k_i \le n_i$).

Step 2: The owner chooses $s \in Z_p$.

Step 3: The owner calculates $C_0 = MY_W^s$, $C_1 = g^s$ and $C_2 = X_W^s$.

Output: $CT_W = (W, C_0, C_1, C_2)$.

Algorithm 5: $Decrypt(PK, SK_u, CT_W) \rightarrow M$

A user requests to access the encrypted file that is stored in the cloud server.

Step 1: The user obtains the corresponding ciphertext file from the cloud server.

Step 2: The user detects L and W.

Step 3: If $L = W$, the user calculates $M = \frac{C_0}{\hat{e}(D, C_1) \cdot \hat{e}(E, C_2)}$.

Output: The data file M.

4 EXPERIMENT AND ANALYSIS

4.1 *Experimental environment*

CPU: Inter(R) Core(TM) i7–4712MQ 2.3GHz
RAM: 16GB
Platform: VMware Workstation 12 Pro
Hadoop version: 2.5.1

In the experiment, there are three machines of the above configuration. Each machine invents two Ubuntu systems. The Hadoop cluster is the cloud service provider. Table 1 shows the configuration of the experimental machines.

Table 1. The configuration of the experimental machines.

Name	OS	IP	Indentity
1	CentOS	192.168.211.128	Master
2	CentOS	192.168.211.129	Slave1
3	CentOS	192.168.211.130	Slave2
4	CentOS	192.168.211.131	Slave3
5	CentOS	192.168.211.132	Slave4
6	Ubuntu	192.168.211.133	Owner
7	Ubuntu	192.168.211.134	User

4.2 *Data analysis*

As is shown in Fig. 3, the time that the system establishes is proportional to the number of the attributes in the access structure, and the time that the generation of the domain authority's private key is proportional to the number of the attributes in the access structure, too.

As is shown in Fig. 4, the time that the generation of the user's decryption key is proportional to the number of the attributes in the access structure, and the execution time of the encryption algorithm is nearly a constant value, which has nothing to do with the number of attributes in the access structure.

As is shown in Fig. 5, the execution time of the decryption algorithm is nearly a constant value, which has nothing to do with the number of attributes in the access structure. Besides, the length of the decryption key is constant, which is not affected by the number of attributes in the access structure and other factors.

The experimental results show that the scheme limits the length of the decryption key, the computation of the encryption and the decryption to

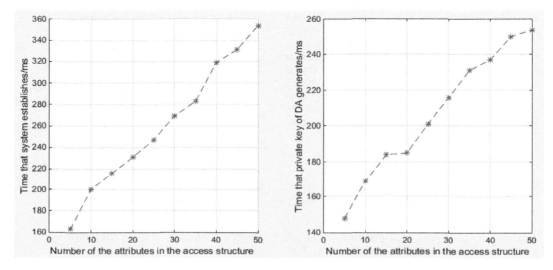

Figure 3.   Time of the *Setup* and the *KenGen* algorithms.

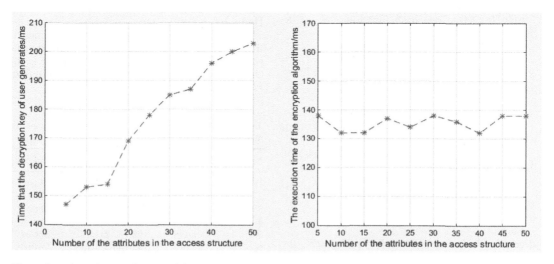

Figure 4.   Time of the *Delegate* and the *Encrypt* algorithms.

584

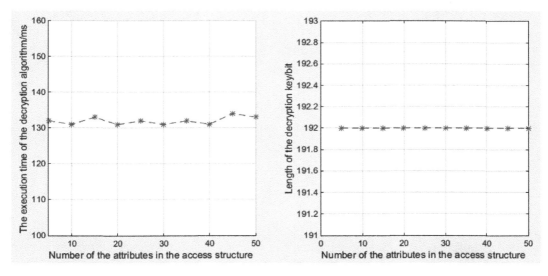

Figure 5. Time of the Decrypt algorithm and the length of the decryption key.

a constant value. At the same time, it is combined with the Hierarchical Authorization Model Based on Hadoop. When the number of users and attributes in the system increases, it will perform obvious advantages.

5 CONCLUSION

The problem of data security in the cloud computing environment has become more and more serious with the popularity of cloud computing. In this paper, the encryption algorithm based on the constant-length decryption key is used to maintain the length of the decryption key and the calculation of encryption and decryption with a fixed value. It not only improves the efficiency of the system, but also reduces the computational overhead of owners and users. At the same time, a hierarchical authorization model based on Hadoop is established, which not only reduces the burden and risk of a single root authority, but also solves the problem of limited storage space in the actual cloud storage environment.

REFERENCES

Ali M. Green cloud on the horizon[C]//IEEE International Conference on Cloud Computing. Springer Berlin Heidelberg, 2009: 451–459.

Armbrust M, Fox A, Griffith R, et al. A view of cloud computing [J]. Communications of the ACM, 2010, 53(4): 50–58.

Chung H H, Wang P S, Ho T W, et al. A secure authorization system in PHR based on CP-ABE[C]//E-Health and Bioengineering Conference (EHB), 2015. IEEE, 2015: 1–4.

Liu Z, Cao Z, Wong D S. Traceable CP-ABE: how to trace decryption devices found in the wild [J]. IEEE Transactions on Information Forensics and Security, 2015, 10(1): 55–68.

Odelu V, Das A K, Rao Y S, et al. Pairing-based CP-ABE with constant-size ciphertexts and secret keys for cloud environment [J]. Computer Standards & Interfaces, 2016.

The GNU Multiple Precision arithmetic library (GNU MP) [EB/OL]. [2013–05–09]. http://gmplib.org/.

Zhang Q, Cheng L, Boutaba R. Cloud computing: state-of-the-art and research challenges [J]. Journal of internet services and applications, 2010, 1(1): 7–18.

Zhao J, Pjesivac-Grbovic J. MapReduce: The programming model and practice [J]. 2009.

Automotive, Mechanical and Electrical Engineering – Liu (Ed.)
© 2017 Taylor & Francis Group, London, ISBN 978-1-138-62951-6

Research on Bayesian network structure learning method based on hybrid mountain-climbing algorithm and genetic algorithm

Wei Xu, Gang Cheng & Lin Huang
Power and Engineering Academy, Naval University of Engineering, Hubei, China

ABSTRACT: In order to solve the problem of Bayesian network structure learning, this paper proposes a Bayesian network structure learning problem to solve the problem of adjacency matrix. The adjacency matrix and 1-step dependency coefficient are given, and the adjacency matrix is constructed. The advantage of this method is that the local optimisation ability of the climbing algorithm is applied to the genetic algorithm to fully release the global optimisation ability of the genetic algorithm to achieve the fast convergence of the algorithm. The simulation results show that the algorithm has the effectiveness of the optimisation algorithm.

Keywords: Bayesian Network Structure; Mountain-climbing Algorithm; Genetic Algorithm

1   INTRODUCTION

Genetic algorithm is a classical method to deal with complex optimisation problems. It simulates the law of natural evolution for the survival of the fittest, through selection, crossover, and mutation to find the optimal solution. It has a strong adaptability and robustness. In recent years, many scholars have applied it to the Bayesian network structure learning and has made a wealth of research results. Campos et al (2002) established a Bayesian belief network by induction learning, and optimised the adaptive Bayesian network learning model by genetic algorithm. Wang et al. (2013) proposed a restricted unconstrained Bayesian network structure, and so on, for the Bayesian optimisation algorithm deficiencies proposed a new hybrid Bayesian optimisation method to solve the optimal undirected graph in the use of genetic algorithm optimisation to obtain the optimal Bayesian network structure. Non-stationary DBN structure learning model of non-stationary stochastic system is constructed by the Lees optimization method (Gao, 2007). The non-stationary DBN structure frame is given by genetic algorithm optimisation. In this paper, we propose a structural learning method based on hill-climbing method and genetic algorithm. Firstly, we use the local structure search of climbing method to find out the parent node and child node as candidate set points, and transform the Bayesian network structural learning problem into adjacency matrix. The optimal parent-child nodes are obtained by selecting, crossing and mutation of the genetic algorithm, and then the optimal

adjacency matrix is obtained. Finally, the optimal structure of the Bayesian network is obtained.

2   BN ADJACENCY MATRIX AND 1-STEP DEPENDENCY COEFFICIENT

2.1   *Adjacency matrix representation*

The Bayesian structure is represented by the corresponding directed acyclic graph DAG, whose network relation can be expressed by the adjacency matrix 0–1. It is assumed that any point X_i in the set of random variables $X = \{X_1, X_2 \cdots X_n\}$, corresponds to the other $i - 1$ nodes in the DAG. Figure 1 shows the BN structure and its corresponding adjacency matrix.

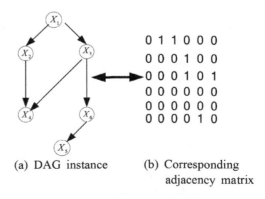

(a) DAG instance

(b) Corresponding adjacency matrix

Figure 1.   BN and its adjacency matrix.

Figure 1(b) is the corresponding adjacency matrix A of (a), a_{ij} is the row i and column j of the adjacency matrix, that is:

$$a_{ij} = \begin{cases} 1 & (\textbf{Node a is the parent of node b}) \\ 0 & (\textbf{Other cases}) \end{cases} \quad (1)$$

2.2 1-step dependence coefficient

Let $\chi^2_{ij|k}$ denote the χ^2 distribution with degrees of freedom $df = (n_i - 1)(n_j - 1)n_k$, n_j is the number of state n of the variable set X; $\chi^2_{df,\alpha}$ means degrees of freedom $df = (n_i - 1)(n_j - 1)n_k$; the significance level is α distribution of χ^2 values. For variables X_i and X_j, $c_{ij\alpha}^{(1)} = \min_{k \neq i,j}\{\chi^2_{ij|k} - \chi^2_{df,\alpha}\}$ is defined as a 1-step dependency coefficient between X_i and X_j. In the actual calculation, according to the theorem in Wang et al. (2013), it is approximated as:

$$c_{ij\alpha}^{(1)} = \min_{k \neq i,j}\left\{2NMI_d\left(X_i, X_j | X_k\right) - \chi^2_{df,\alpha}\right\} \quad (2)$$

where: N is the number of sample data, $MI_d(X_i, X_j | X_k)$ is the conditional probability mutual information between variables X_i and X_j, X_k.

To calculate the interdependencies between nodes X_2 and X_4 in Figure 1, assuming that the distribution significance level of χ^2 is $\alpha = 0.005$, the degree of freedom is $df = (2-1) \times (2-1) \times 2 = 2$, through the look-up table to get the distribution value of χ^2 is 10.6. Substituting it into equation (2) to compute $c_{ij\alpha}^{(1)}$. If $c_{ij\alpha}^{(1)} > 0$, the probability of an edge between nodes X_2 and X_4 is 0.995.

2.3 1-step dependence coefficient matrix

The 1-step dependence coefficient matrix can be derived from the 1-step dependency coefficient:

$$c_{ij} = \begin{cases} c_{ij\alpha}^{(1)}, & i \neq j; \\ 0, & i = j. \end{cases} \quad (3)$$

From the Bayesian adjacency matrix and the 1-step dependency coefficient matrix between the nodes, we can calculate the mutual influence between the nodes:

$$H(X,\alpha) = \sum_{i=1}^{n} \sum_{j=1, j \neq i}^{n} c_{ij}a_{ij} \quad (4)$$

The influence relation obtained by the formula (4) is used as the candidate edge of the Bayesian network structure to construct the unconstrained optimisation problem:

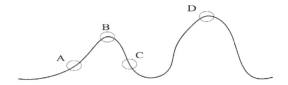

Figure 2.   Climbing algorithm search process.

$$\max H(X,\alpha) = \sum_{i=1}^{n} \sum_{j=1, j \neq i}^{n} c_{ij}a_{ij} \quad (5)$$

3   CLIMBING ALGORITHM

The hill-climbing algorithm is a heuristic greedy algorithm. First, randomly select any point in the search space as the initial point, and search for the new position instead of the previous position by performing the function calculation with other points in the field. Generally, the height difference The maximum direction as the search direction, until the search cannot be better than the current value of the value of the stop search, through the roundabout forward to achieve the ultimate solution (Wang et al., 2013).

The advantage of this method is that it can get the relative optimal value quickly, but the disadvantage is that the multimodal function is easy to fall into the local optimal solution and cannot search for the global optimal solution. Figure 2 is the search process of hill-climbing algorithm, in which point A is the current search point. After searching for the local optimal solution, it cannot get a better solution than the current one. It is difficult to get the global optimal solution D point. Although the global search ability of mountain-climbing algorithm is not enough, it has a strong non-differentiable single peak function search level, and does not need to use gradient information, so climbing algorithm can be applied in the local optimisation of genetic algorithm.

4   GENETIC ALGORITHM

GA is a computational model of the evolutionary process of natural selection and genetic mechanism of Darwinian biological evolution. It is a method of searching for the optimal solution by simulating the natural evolutionary process. The genetic algorithm starts with a population that represents the potential solution set of the problem, while a population consists of a number of individuals encoded by the gene (Wang, 1999). Each individual is actually a chromosome with the characteristics of the entity. Because of the complexity of the work of genetically coding, it is often reduced to a binary-coded form in practice. After the initial population is generated, according to the survival of the fittest and survival of the fittest principle, generation by generation produces better and better approximation. In each generation, the individuals are selected according to the fitness of the individuals in the problem domain, and the crosses and variants are generated by means of the genetic operators of natural genetics to produce the population representing the new set. This process will lead to the evolution of the population as the natural evolution of the post-generation population is more adaptable to the environment than the previous generation. The best individuals in the final population can be decoded as the problem approximate optimal solution.

4.1 Code

In the algorithm, the individual population is represented by an adjacency matrix. The Bayesian network structure shown in Figure 1(a) can be expressed as a matrix in Figure 1(b). When the individuals in the network are subject to mutation and crossover, the matrix form of the network structure coding is first transformed into the row vector form. If the adjacency matrix of the Bayesian network structure is $X = (x_{ij})$, the corresponding row vector form of the rectangular structure is $x_{11}x_{12}\cdots x_{1n}x_{21}x_{22}\cdots x_{2n}\cdots x_{n1}x_{n2}\cdots x_{nn}$: Figure 2.8 (b) matrix form corresponding to the row vector is 01100000 01000001010000000000000 00 10.

4.2 Initial population selection

There are three ways to choose the initial population: randomly generated, by the experts in the field based on experience, or proposed by the Madigan variant population selection techniques. The variational population selection technique assigns *a priori* distribution to the network structure hypothesis. A computer program is used to generate a complete data hypothesis set. The initial population is selected by selecting the partial data according to the pattern structure of the domain expert in the case of complete data. In this paper, an undirected graph is obtained by solving an unbounded optimisation problem, and an initial population is randomly generated from an undirected graph. Each individual satisfies the designed chromosome structure.

4.3 Fitness function

The BIC measure function is used as the fitness function. The network structure is encoded as a vector, and the genes on each chromosome are measured separately. At each evolution, the structural measure of the offspring is estimated according to the expected network statistic.

4.4 Genetic operators

1. Select the mode
For the genetic model to be selected to be arranged, according to the fitness value of each monomer as a parent to determine the size of the possibility of a larger fitness value will become the next generation of population, Let $I_r(j)$ denote the j th monomer of the population at time t, $rank\left[I_t(j)\right]$ denote its fitness function value arrangement, and λ denote the size of the population. The probability $p_{j,t}$ that the monomer can be selected as a parent during the genetic process is:

$$p_{j,t} = \frac{rank\left(I_t(j)\right)}{\lambda(\lambda+1)/2} \tag{6}$$

2. Crossover mode
The crossover operation is to select the operator from the random pairing of the population, and the average quality of the population can be increased by the crossover operation, which will facilitate the global optimisation of the next generation genetic as fast as possible. The crossover probability p_a is crossed by the uniform parameterisation during the generation of the parent to the offspring and the gene having the high fitness value can be exchanged with higher probability in the parent.

3. Mutation operation
The mutation operation can randomly generate new traits of the individual, and avoid the algorithm from getting into the local optimum by increasing the diversity to achieve the global optimisation. In the process of mutation, the probability of changing the expression value of the individual gene makes the adjacency relation between nodes change between $0\rightarrow1$ and $1\rightarrow0$ to find the optimal dependency between the child nodes. The operation of the mutation operator in the algorithm is to calculate the 1-step dependency coefficient between the reversal arc nodes, and its value can be used to judge the edge deletion or inversion.

5 A GENETIC ALGORITHM BASED ON EMBEDDED HILL-CLIMBING ALGORITHM

This genetic algorithm is a good global optimisation algorithm through the fitness value of the next generation "best" structure as a new candidate to cross and a certain probability of variation to improve individual environmental adaptability to promote the continuous evolution of the population. In the iterative process of genetic algorithm, the method of local climbing is introduced to reduce the iteration step to improve the efficiency of searching.

1. Initialise each node in the network and its fitness function;
2. According to the network structure, the initial undirected graph is determined and its optimisation objective function is constructed;
3. Randomly generate N chromosomes as the initial population C_0;
4. The vector structure of the network coding;
5. The second-generation population C_1 is obtained from the initial population with crossover probability p_a;
6. A new population C_2 is obtained from C_1 with mutation probability p_b;
7. Calculate the fitness value and selection probability of each generation to eliminate the parent;
8. The population C_t of the generation t is obtained by local climbing algorithm;
9. After the evolution of the n generation or after the iterative generation of m-generation network structure is not changed after the termination of iteration, the resulting network structure will be the optimal algorithm to optimise the network.

6 SIMULATION EXPERIMENT AND ANALYSIS

In order to verify the effectiveness of the algorithm Figure 1(a) is an example of structural learning. First, the number and position of the nodes are determined according to the observed training samples and the relational data are generated. The 1-step dependency coefficient between nodes was calculated. The results are shown in Table 1.

In this paper, we use Matlab to solve the initial solution of the relationship between nodes. In the calculation, the parameters include the initial population size, significance level, crossover probability, mutation probability, and iteration algebra. The population size $N = 100$, significance level $\alpha = 0.005$, crossover probability $p_a = 0.98$, and mutation probability $p_b = 0.01$. The use of BIC scoring function, when the iteration accuracy is less than

Table 1. 1-step dependence coefficient calculation result.

	X_1	X_2	X_3	X_4	X_5	X_6
X_1	0	80	127	−18	32	−2.5
X_2	80	0	−1.0	3.5	−12	−20
X_3	127	−1	0	70	−30	140
X_4	−18	3	70	0	8	−2
X_5	32	−12	−30	8	0	80
X_6	−2	−20	140	−2	80.	0

10–5, then stops the operation. Finally, the network structure shown in Figure 1 (a) is obtained by the method proposed in this paper.

7 CONCLUSION

In this paper, Bayesian network structure is learned through improved hill-climbing algorithm and genetic algorithm, and the optimal results are obtained by solving the interdependent coefficient of 0–1 matrix. The experimental results show that the proposed method is reasonable and effective, which provides a reliable foundation and broad application prospect for Bayesian network structure learning in large-scale and complex systems in the future.

REFERENCES

Campos, L.M., Gamez, J.A. & Moral S. (2002). Partial abductive inference in Bayesian belief networks—an evolutionary computation approach by using problem-specific genetic operators. *IEEE Transactions on Evolutionary Computation, 6*(2), 105–131.

Chai, Y. & Zhou, Y.Z. (2014). Improvement of climbing-hill method by genetic. *Journal of Liaoning Technical University (Natural Science) 33*(7), 997–999.

Gao, X.G. & Xiao, Q.K. (2007). Non-stationary stochastic system dynamic Bayesian structure learning relation network. *Acta Aeronautica et al., Astronautica Sinica, 28*(6), 1408–1418.

Wang, C.F.A. (2013). Bayesian network learning algorithm based on unconstrained optimization and genetic algorithm. *Control and Decision, 28*(4), 618–622.

Wang, X.Y. & Wang, X.L. (1999). Improvement and implementation of heuristic search strategy. (1999). *Journal of Shan Xi Normal University (Natural Science Edition), 27*(1), 58–60.

Automotive, Mechanical and Electrical Engineering – Liu (Ed.)
© 2017 Taylor & Francis Group, London, ISBN 978-1-138-62951-6

Research of the evaluation of electric power construction projects based on fuzzy comprehensive evaluation

Xingguang Feng & Yuheng Sha
State Grid Corporation of China, Beijing, China

Xi Wang & Mingwei Li
State Grid Sichuan Electric Power Company, Chengdu, Sichuan, China

Kaijiang Cao & Hao Xu
Research Institute of Economics and Technology, State Grid Sichuan Electric Power Company, Chengdu, Sichuan, China

Ze Sun & Yan Wang
College of Economics and Management, North China Electric Power University, Beijing, China

ABSTRACT: In recent years, people's living standards have steadily improved. Thus, the demand for electricity is growing. Besides, electric power system reform is constantly advancing, so the evaluation of power construction projects is particularly important. In this paper, we establish an understanding of the current situation and basic theory of power construction project evaluation. We then set up the comprehensive evaluation index system of a power construction project, establish a three-level index system, and use the analytic hierarchy process to calculate the weight. The fuzzy comprehensive evaluation method is used to evaluate the electric power construction project, and the evaluation results are obtained by using first-level fuzzy evaluation and two grade fuzzy comprehensive evaluation.

Keywords: Electric power construction project; Analytic hierarchy process; Fuzzy comprehensive evaluation

1 INTRODUCTION

The electric power industry is an important basic industry, which greatly affects the development of China's economy. As China's economic development and residents' living standards continue to improve, it puts forward higher requirements for the reliability, security, stability and quality of China's power grid. In 2015, a new round of China's electric power system reform started. The reform of the electric power system is a significant part of deepening the Chinese reform, and is conducive to stabilising growth, adjusting structure, promoting safety and scientific development of Chinese electric power enterprises. With the continuous reform of the power industry, the comprehensive evaluation of the electric power construction projects is of great significance to avoid improper project investments and prevent the waste of resources.

The comprehensive evaluation of electric power projects is to establish a comprehensive evaluation index system for the economic, technical, policy and environmental impact. It carries the objective of fair and reasonable comprehensive evaluation

from many aspects. This paper establishes the comprehensive evaluation index system of electric power construction projects, using fuzzy comprehensive evaluation and Analytic Hierarchy Process (AHP) evaluation methods, in order to make a comparative analysis and guidance on the plan and audit of the electric power enterprises' construction projects.

2 THE METHOD OF COMPREHENSIVE EVALUATION

2.1 Analytic Hierarchy Process

AHP is a system analysis method, proposed by the professor A. L. Saaty in the 1970s, which combines qualitative and quantitative analysis. It simulates the human decision-making process, and it will make the subjective judgement systematic, quantitative and model. This method has the advantages of clear thinking, simplicity and convenience, wide application, strong system and so on. It is a powerful tool for the analysis of a multi-objective, multi-factor and multi-criteria complex large system.

The basic steps of AHP are as follows:

1. To establish the hierarchical structure model
According to the characteristics of the research content, a multi-level structure model is needed to be organised, and then set up. These levels can be divided into three categories: the top layer, the middle layer and the bottom layer. There is only one element in the top level, which is usually the intended target of the analysis. The middle layer contains the intermediate links to achieve the objectives, including the criteria and sub-criteria. The bottom layer includes a variety of measures to achieve the goal of choice, such as decision-making.
2. Structure comparison judgment matrix
We should compare the middle layer and the bottom layer of each model. Note A_i the relative importance of A_j to a_{ij}. It usually takes digital 1~9 and its reciprocal as the scale, and ultimately we need to construct a pairwise comparisons judgement matrix.
3. Solving the judgement matrix obtains the characteristic root and the characteristic vector, and carries on the consistency check.

a. Solving the judgement matrix A characteristic root:

$$A\omega = \lambda_{max}\omega \qquad (1)$$

In this formula, λ_{max} is the maximum eigenvalue of the matrix A.
b. Check the consistency of the judgement matrix and calculate the consistency index CI:

$$CI = \frac{\lambda_{max} - n}{n - 1} \qquad (2)$$

Find the corresponding average random consistency index RI. On n = 1, 2,..., 9, its RI value is shown in the table:

n	1	2	3	4	5	6	7	8	9
RI	0	0	0.58	0.9	1.12	1.24	1.32	1.41	1.45

c. Calculate the consistency ratio CR:

$$CR = \frac{CI}{RI} \qquad (3)$$

When CR<0.1, the consistency of the judgement matrix is acceptable, otherwise the corresponding judgement matrix is properly modified.

2.2 *Fuzzy comprehensive evaluation method*

Fuzzy comprehensive evaluation method is an application of fuzzy mathematics. By means of fuzzy mathematics, it can quantify the factors that are not easy to quantify. Then it carries on the comprehensive appraisal of the evaluated object.

The basic principle of the fuzzy comprehensive evaluation method is based on the theory of fuzzy set, and then values single factor, combined with the weight of each factor to obtain the comprehensive evaluation results. The four essential factors are the set of factors, the judgement set, the single factor evaluation and the weight set, and the concrete steps are as follows:

1. Establish factor set
Factor set is the index system, a classic combination, usually expressed by U, factor set U = {u1, u2,..., un}.
2. Set up evaluation set
V = {v1, v2,...,vm}. Each level represents a variety of possible evaluation results. V represents a series of fuzzy relations. The description of the state of the evaluation system and the state of various factors has fuzziness, which not only expresses the degree of the system state, but also conforms to people's acceptance habits and understanding methods.
3. Single factor evaluation
In general, experts discuss the evaluation of individual factors, then constitute the evaluation matrix R according to the corresponding membership relation.

$$R = \left(r_{ij}\right) = \begin{pmatrix} r_{11} & r_{12} & \cdots & r_{1m} \\ r_{21} & r_{22} & \cdots & r_{2m} \\ \cdots & \cdots & & \cdots \\ r_{n1} & r_{n2} & \cdots & r_{nm} \end{pmatrix} r_{ij} \in [0,1]$$

r_{ij} is the views of all experts to be integrated, analysed and obtained, It means evaluation object in the Ui for the Vj membership degree.
4. Establish weight collection
In general, the weight set is established by the analytic hierarchy process, and the normalisation is carried out before the synthesis. A = (a1, a2,..., an).

$$\sum_{i=1}^{n} a_i = 1, a_i > 0$$

5. Comprehensive evaluation
After finding out the weight set and evaluation matrix, the following is the comprehensive evaluation. Evaluation model:

$$B = \left(b_1, b_2, \cdots, b_m\right) = AoR = \left(a_1, a_2, \cdots, a_n\right) o \begin{pmatrix} r_{11} & r_{12} & \cdots & r_{1m} \\ r_{21} & r_{22} & \cdots & r_{2m} \\ \cdots & \cdots & & \cdots \\ r_{n1} & r_{n2} & \cdots & r_{nm} \end{pmatrix}$$

3 CASE STUDY

3.1 Establish the evaluation indexes

Due to the establishment of a comprehensive evaluation index system of a power enterprise project involving the consideration of a lot of factors, the uncertainty is serious and some factors are difficult to describe quantitatively, making the problem more difficult to solve. Therefore, according to the principle of index selection, and the establishment of a power enterprise project evaluation index system from domestic and foreign research results, we use an expert investigation method, through the investigation and discussion of the decomposition, all possible factors, and eliminate those weak effect, small factors, summary of expert opinion survey, concludes that comprehensive opinion, again to seek expert advice, after repeatedly, can form the same or close to the group of expert opinion consistent.

In this paper, we set up the index system for the power supply reliability, power quality, safety and stability, development coordination, service social development and economy. Specifics are shown in Table 1.

3.2 Determination of index weight based on Analytic Hierarchy Process

AHP needs to build judgement matrix of determining, the construction of judgment matrix is by invite relevant experts to rate the importance of the various indicators, the indicators of the relative importance of evaluation. Finally, experts get consistent results through negotiations.

3.2.1 Determination of the weight of the first class index

According to experts' opinions and scores, we should structure a comparison judgement matrix, and put the data into the AHP method in determining weight formula. Then we calculate the eigenvalues and the corresponding eigenvectors and carry on the consistency check. Finally, we get the weights of acceptable values. Detailed results are shown in Table 2.

According to Formulas (2) and (3), we can calculate CI = 0.01612 and CR = 0.01. Because CR<0.1, the consistency of the judgement matrix can be accepted through consistency checking.

3.2.2 Determination of the weight of the secondary indexes

By using a similar method, the weight of each level two indexes is determined, and the comparison judgement matrix is constructed. The results are as follows:

$W_1 = (0.6, 0.2, 0.2)$,
$W_2 = (0.4, 0.4, 0.2)$,
$W_3 = (0.26, 0.1, 0.64)$,
$W_4 = (0.64, 0.26, 0.1)$,

Table 2. First level index weight.

O	A_1	A_2	A_3	A_4	A_5	A_6	Weight
A_1	1	2	1	3	3	1	0.24
A_2	1/2	1	1/2	3	3	1	0.17
A_3	1	2	1	3	3	1	0.24
A_4	1/3	1/3	1/3	1	1	1/3	0.07
A_5	1/3	1/3	1/3	1	1	1/3	0.07
A_6	1	1	1	3	3	1	0.21

Table 1. Evaluation index system of electric power construction project.

Objective layer	Criterion layer		Scheme layer	
Evaluation of electric power construction project	Reliability index of power supply	A_1	Reliability rate of power supply	A_{11}
			Average user outage time	A_{12}
			Average interruption times of customer	A_{13}
	Power quality index	A_2	Voltage deviation	A_{21}
			Supply voltage qualification rate	A_{22}
			Frequency deviation	A_{23}
	Safety stability index	A_3	N-1 calibration qualified rate	A_{31}
			Transformer load rate	A_{32}
			Transformer capacity ratio	A_{33}
	Development coordination index	A_4	One-Line and One-Transformer	A_{41}
			Ratio of linear variable capacity	A_{42}
			The ratio of the upper and lower variable electric capacity	A_{43}
	Service social development index	A_5	Provide direct employment	A_{51}
			Environmental improvement	A_{52}
	Economic index	A_6	Net Profit Margin	A_{61}
			Internal rate of return	A_{62}
			Line loss rate	A_{63}
			Unit power supply cost	A_{64}

$W_5 = (0.75, 0.25),$
$W_6 = (0.07, 0.15, 0.39, 0.39),$
$W = (0.24, 0.17, 0.24, 0.07, 0.07, 0.21).$

After calculation, the above CR are less than 0.1. So the consistency of the judgement matrix can be accepted and can pass the consistency test.

3.3 *First level fuzzy comprehensive evaluation*

According to the characteristics of electric power construction projects and expert opinion, establish remark set $V = \{v1, v2, v3, v4\} = \{$excellent, good, medium, poor$\}$, correspondence to $V = \{1, 0.8, 0.6, 0.4\}$. Then select certain quantity experts who have appropriate qualifications to score each indicator. Establish fuzzy judgement matrixes of R1, R2, R3, R4, R5, and R6. The relation matrix of power supply reliability is as follows:

$$R_1 = \begin{pmatrix} 0.6 & 0.3 & 0.1 & 0 \\ 0.5 & 0.5 & 0 & 0 \\ 0.6 & 0.4 & 0 & 0 \end{pmatrix}$$

Similarly, the relation matrix of power quality, security stability, development harmony, development of service society, and economic can be concluded:

$$R_2 = \begin{pmatrix} 0.4 & 0.3 & 0.2 & 0.1 \\ 0.5 & 0.4 & 0 & 0.1 \\ 0.3 & 0.5 & 0.2 & 0 \end{pmatrix},$$

$$R_3 = \begin{pmatrix} 0.7 & 0.3 & 0 & 0 \\ 0.8 & 0.1 & 0.1 & 0 \\ 0.6 & 0.2 & 0.2 & 0 \end{pmatrix},$$

$$R_4 = \begin{pmatrix} 0.4 & 0.4 & 0.1 & 0.1 \\ 0.5 & 0.4 & 0 & 0.1 \\ 0.5 & 0.3 & 0.2 & 0 \end{pmatrix},$$

$$R_5 = \begin{pmatrix} 0.4 & 0.4 & 0.1 & 0.1 \\ 0.2 & 0.3 & 0.3 & 0.2 \end{pmatrix},$$

$$R_6 = \begin{pmatrix} 0.4 & 0.3 & 0 & 0.3 \\ 0.5 & 0.5 & 0 & 0 \\ 0.6 & 0.3 & 0.1 & 0 \\ 0.5 & 0.2 & 0.2 & 0.1 \end{pmatrix}.$$

According to the formula "B = W×R", we can get the final evaluation set of the project.

$B_1 = (0.58, 0.36, 0.06, 0),$
$B_2 = (0.42, 0.38, 0.12, 0.08),$
$B_3 = (0.646, 0.216, 0.138, 0),$

$B_4 = (0.436, 0.39, 0.148, 0.026),$
$B_5 = (0.35, 0.375, 0.15, 0.125),$
$B_6 = (0.532, 0.291, 0.117, 0.06).$

3.4 *Second fuzzy comprehensive appraisement*

According to the results of the first-level fuzzy evaluation, the level two fuzzy evaluation will be carried out. The first level index weight is:

$$W = (0.24, 0.17, 0.24, 0.07, 0.07, 0.21)$$

$$R = \begin{pmatrix} 0.58 & 0.36 & 0.06 & 0 \\ 0.42 & 0.38 & 0.12 & 0.08 \\ 0.646 & 0.216 & 0.138 & 0 \\ 0.436 & 0.39 & 0.148 & 0.026 \\ 0.35 & 0.375 & 0.15 & 0.125 \\ 0.532 & 0.291 & 0.117 & 0.06 \end{pmatrix}$$

The final evaluation set can be obtained according to the formula "B = W×R".
$B = (0.5324, 0.3175, 0.1134, 0.0368).$

3.5 *Result analysis*

In accordance with the principle of maximum membership degree, the evaluation results of the power construction project can be seen in the membership of the four grades, the maximum value of 0.5324. Therefore, the comprehensive evaluation of the power construction project is 0.5324, and the evaluation value is "excellent", indicating that the power construction project feasibility is higher. But there is still 15.02% (11.34%+3.68%) showing that the project feasibility is not very high, so it should be studied and analysed to further improve the quality of all aspects, in order to achieve higher requirements.

4 CONCLUSION

In this paper, the current evaluation situation has carried out a simple introduction, focusing on the analytic hierarchy process, and fuzzy comprehensive evaluation and index system construction content were studied. According to the characteristics of the electric power construction project, we adopt the fuzzy comprehensive evaluation method for comprehensive evaluation. We set up the index system for the power supply reliability, power quality, safety, stability, balanced development, service and social development and the economy from six aspects, using AHP to calculate index weight, and finally some conclusions. The evaluation results can provide reference for the operation, planning, and construction of electric power construction projects, so as to ensure the scientific reliability of the decision.

ACKNOWLEDGEMENTS

The authors acknowledge the support of the "Key Technology and System Research of Company Project Plan Audit Analysis" project.

REFERENCES

Du, J. (2015). 220 kV intelligent substation construction project comprehensive evaluation and empirical research. *Beijing: North China Electric Power University.*

He, Y. (2011). Electric power comprehensive evaluation method and application. *Beijing: China Electric Power Press.*

Liu, F. (2014). The comprehensive evaluation of Qinhuangdao Changli 500 K substation construction project. *Beijing: North China Electric Power University.*

Wang, X. (2012). The comprehensive evaluation of Inner Mongolia Qixiaying power transmission project. *Beijing: North China Electric Power University.*

Wang, X. (2013). Comprehensive evaluation study on the expansion project of Beijing urban substation expansion project. *North China Electric Power University.*

Zhong, Y. (2011). Electric power safety risk assessment based on fuzzy analytic hierarchy process. *Journal of Chongqing Electric Power College.*

Automotive, Mechanical and Electrical Engineering – Liu (Ed.)
© 2017 Taylor & Francis Group, London, ISBN 978-1-138-62951-6

Research on vehicle routing optimisation in the emergency period of natural disaster rescue

Junchi Ma, Xifu Wang & Lieying Zhao
School of Traffic and Transportation, Beijing Jiaotong University, Beijing, China

ABSTRACT: Based on summarising the current situation of related theories at home and abroad, we put forward the research content of this paper, that is, research on vehicle routing optimisation in the period of natural disaster rescue emergency. The paper determines the objective function of the emergency period and establishes a model based on the function. During the modelling process, we use dimensionless method and multi-objective optimisation to transfer the multi-objective problem into a single-objective problem. Finally, we get the total weights and put them into the emergency network, then we are able to transfer the emergency vehicle routing problem into shortest road problem. And we can use Lingo to solve this problem. In this paper, we also regard the rescue of Wenchuan earthquake is an example used to illustrate and apply the method and make some recommendations.

Keywords: Emergency Logistics; Vehicle Routing Optimisation; Multi-objective Optimisation

1 INTRODUCTION

For thousands of years, mankind and nature have existed side by side. Even though part of nature, human beings still seem so small when encountering natural disasters. Whenever there is a mutation in the natural environment, humans are often caught off guard. After the occurrence of natural disasters, the primary task is to calmly carry out emergency management effectively so as to control the effects of the disaster, by starting the appropriate emergency rescue system immediately. At this time, a large number of emergency supplies and personnel for emergency rescue of various types is needed, so there is a wide range of emergency resource mobilisation activities to assist in emergency rescue.

The purpose of this paper is to improve the effectiveness of emergency vehicle routing optimisation. Based on the objective of optimisation during a disaster emergency period, and by taking into account the external factors of the impact of vehicle traffic, the value of emergency rescue vehicles can be assessed then used as a reference for emergency vehicle routing decision analysis and emergency rescue decision-making, as well as emergency management.

2 IDEAS ON VEHICLE ROUTING OPTIMISATION IN THE EMERGENCY PERIOD OF NATURAL DISASTER RESCUE

During the emergency period of rescue, the restraints around emergency vehicle routing mainly concern the reliability and time urgency of the vehicles. In building the model, attention should be paid to the factors which influence the emergency road network to the natural disaster. The reliability and actual driving time of the emergency vehicle needs to be taken into account.

2.1 Evaluation of vehicle driving reliability in an emergency road network

Due to the occurrence of uncertain events after natural disasters, the safety and reliability of emergency vehicles differ, so decision-makers need to consider all potential situations when choosing the emergency vehicle route. The choice of the emergency vehicle route in natural disasters should focus on the path which satisfies the most requirements, but not necessarily the shortest path. In this paper, the fuzzy comprehensive evaluation method combined with the analytic hierarchy process, is used to evaluate the emergency vehicles' reliability in the road network, and then to assist the emergency vehicle routing optimisation.

2.2 Consider the travel time when the road network is congested

In the initial stage of a rescue emergency period after a natural disaster, the emergency road network has not been repaired, so the road traffic conditions are relatively poor, prone to serious congestion. At the same time, there might be heavy traffic or congestion, as a result of heavy casualties and the subsequent transfer of affected persons and spontaneous people, which will have a greater impact on the entire emergency road network. Therefore, the actual situation in the study of an emergency vehicle routing optimisation problem needs to consider road congestion as a constraint on the impact of vehicle travel time. In this paper, it is assumed that the roads are all connected, but there remains a phenomenon of traffic jam.

3 OPTIMISATION MODEL OF VEHICLE ROUTING IN THE EMERGENCY PERIOD OF NATURAL DISASTER RESCUE

3.1 Model assumptions

The model assumptions are as follows:

1. Emergency logistics distribution centre has m emergency vehicles, which is sufficient. That is, all emergency vehicles are the same and load capacity is known, regardless of the cost of vehicle distribution;
2. Emergency vehicles start from the emergency logistics distribution centre, and after a series of nodes (transit points) to complete the distribution task, return to the emergency logistics centre (each time only responsible for one route);
3. One emergency distribution of emergency vehicles involves an emergency logistics distribution centre, an emergency material demand point, and n transit points (nodes), whose geographical location is known;
4. Emergency network is a complete network, that is, the sections are connected;
5. In order to simplify the problem, only the travel time that the emergency vehicles use through the various nodes to the demand point is considered, regardless of service time and the time that vehicles use to return to the emergency logistics distribution centre;
6. Emergency logistics distribution centre meets the stock requirements of the disaster site and will not run out of stock;
7. Emergency road network is complex, there is a certain pressure on traffic and the phenomenon of road congestion exists.

3.2 Symbol description

z: Objective function;
S: The definition variable of the total weight value;
S_k: The total weight value of each link before the node with the permanent number is obtained;
C: Emergency Logistics Distribution Centre;
D: Emergency supplies demand point;
N: Collection of all nodes, $C \in N, D \in N, i \in N, j \in N$;

$\bar{v}$: The average travel speed of the vehicle from node i to node j;
∂_{ij}: The speed constraint coefficient of emergency vehicles on the road between node i and node j;
$Q_{ijADAPT}$: The adapt traffic volume between node i and node j;
$Q_{ijVOLUME}$: Real-time traffic volume on the road between node i and node j;
d_{ij}: Distance between node i and node j;
t_j: The time that emergency vehicles use from the emergency logistics distribution centre to node j;
t_{ij}: Travel time for rescue vehicles at different stages of emergency from node i to node j;
t'_{ij}: Dimensionless value of t_{ij};
r_{ij}: Ranking Value of Vehicle Driving Reliability in Road Network;
r'_{ij}: Dimensionless value of r_{ij};
l_{ij}: Section between node i and node j;
g: The corresponding demand for emergency supplies for a disaster-stricken area;
q: Load capacity of single emergency vehicle;
Y_{ij}: 0–1 variable;
$y_{ij} = 1$ means emergency vehicle selects the path from node i to node j;
$y_{ij} = 0$ means emergency vehicle doesn't select the path from node i to node j.

3.3 Model construction

The model is as follows:

$$\min z = S \tag{1}$$

$$S = S_k + \left(t'_{ij} + r'_{ij}\right)y_{ij}, \forall i \subset N, \forall j \subset N \tag{2}$$

$$t_{ij} = \frac{d_{ij}}{\partial_{ij}v}, \forall i \subset N, \forall j \subset N \tag{3}$$

$$\partial_{ij} = \frac{Q_{ijADAPT}}{Q_{ijVOLUME}} \tag{4}$$

$$t'_{ij} = \frac{t_{ij} - \min\{t_{ij}\}}{\max\{t_{ij}\} - \min\{t_{ij}\}} \tag{5}$$

$$r'_{ij} = \frac{r_{ij} - \min\{r_{ij}\}}{\max\{r_{ij}\} - \min\{r_{ij}\}} \tag{6}$$

$$s.t. q \leq g \leq mq \tag{7}$$

$$\sum_i y_{ij} = \sum_i y_{ji}, \forall i \subset N, \forall j \subset N \tag{8}$$

$$\sum_i y_{ij} = 1, \forall j \subset N \tag{9}$$

$$\sum_j y_{sj} > 0, \forall j \subset N \tag{10}$$

$$y_{ij} = \{1,0\}, \forall i \subset N, \forall j \subset N \tag{11}$$

$$t_k, t_{ij} \geq 0, \forall d_{ij} \geq 0 \tag{12}$$

In the above model, the objective function (1) indicates that the total weight value is minimum, that is, the total weight of the total travel time of the emergency rescue vehicle on the emergency road network and the emergency vehicle driving reliability is the least.

Equation (2) represents the total weight value on the emergency road network, and the total weight value takes into account the reliability factor and the time factor of the emergency vehicle driving. Equation (3) shows the travel time of the emergency rescue vehicle on each section of the emergency network. Equation (4) represents the constraint factor of the speed of the emergency rescue vehicle on a congested road network. Equations (5) and (6) are dimensionless processing of the reliability and the time of the emergency vehicle. Constraints (7) indicate that more than one emergency rescue vehicle needs to be allocated for emergency resource demand at the disaster demand point, and the emergency logistics distribution centre's inventory is able to meet the demand. Constraints (8) indicate that the emergency rescue vehicles are moving in and out of the same disaster-affected area to ensure that the sections of the emergency rescue vehicle are continuous with each other. Constraints (9) indicate that the emergency rescue vehicle only passes once at the same disaster site. Constraints (10) indicate that the emergency rescue vehicle must initially start from the emergency logistics distribution centre. (11) is a 0–1 integer variable constraint. (12) is a constraint on time and distance in the real world.

3.4 The solution of the model

The solution of the model is as follows:

1. Count the ranking value r_{ij} of vehicle driving reliability in road network;
2. Calculate the average travel time t_{ij} of the vehicle on each section when the road is congested;
3. Non-dimensionalise r_{ij} as well as t_{ij}, gets r'_{ij} and t'_{ij};
4. Calculate the total weight value $S_k = t'_{ij} + r'_{ij}$;
5. S_k is expressed as the weight value in the network diagram, and the whole problem is transformed into the problem of establishing the shortest path in the network diagram;
6. Lingo programming solution.

4 VEHICLE ROUTING OPTIMISATION IN THE EMERGENCY PERIOD OF WENCHUAN EARTHQUAKE RESCUE

In this paper the Wenchuan earthquake was chosen as the research object, to analyse and solve the problem of the disaster situation and emergency rescue vehicle routing in a selected area of Shifang City in the disaster area. This paper takes into account the basic situation of the disaster, the division of the affected area, the grade of the road and highway between the affected points, the road damage degree between the affected points, the traffic volume between different regions, and the distance between the affected points. Then the fuzzy comprehensive evaluation method, based on Analytic Hierarchy Process (AHP), is used to evaluate the reliability of emergency rescue vehicles in an emergency road network. Twelve road sections can be selected in the road network, and the fuzzy comprehensive evaluation value of each line is obtained.

4.1 Count the ranking value r_{ij} of vehicle driving reliability in road network

Based on the fuzzy comprehensive evaluation value of each road segment, the driving reliability of each road section is sorted; the higher the rank of the road, the higher the reliability of the vehicle and the more inclined to select the road. The ranking value of each road is counted and the extreme value method (13) is adopted to non-dimensionalise r_{ij}, by which we can get r'_{ij}. As shown in Table 1.

$$r'_{ij} = \frac{r_{ij} - \min\{r_{ij}\}}{\max\{r_{ij}\} - \min\{r_{ij}\}} \qquad (13)$$

4.2 Calculate the average travel time t_{ij} of the vehicles in the emergency period

According to the traffic flow at this period, we can determine ∂_{ij}, combined with the formula $t_{ij} = d_{ij}/\partial_{ij}v$, we are able to get the vehicle driving time and the length of each section and average running speed and average travel time of vehicles in different road sections. Non-dimensionalise t_{ij} gets t'_{ij}. As shown in Table 2.

Table 1. Road vehicle reliability ranking.

Section	B	r_{ij}	r'_{ij}
l_{12}	0.885	2	0.125
l_{13}	0.894	1	0
l_{14}	0.887	2	0.125
l_{23}	0.852	3	0.25
l_{34}	0.85	3	0.25
l_{25}	0.845	4	0.375
l_{26}	0.828	5	0.5
l_{35}	0.793	7	0.75
l_{36}	0.778	8	0.875
l_{45}	0.818	6	0.625
l_{46}	0.773	8	0.875
l_{56}	0.768	9	1

Table 2. Road traffic conditions.

Section	d_{ij} (km)	$\bar{v}$ (km/h)	∂_{ij}	t_{ij} (h)	t'_{ij}
l_{12}	5.94	60	1/3	0.297	0
l_{13}	7.14	60	1/3	0.357	0.088
l_{14}	8.47	60	1/3	0.424	0.186
l_{23}	10.21	40	1/2	0.511	0.314
l_{34}	8.24	40	1/2	0.412	0.169
l_{25}	12.14	60	1/3	0.607	0.455
l_{26}	13.61	60	1/3	0.681	0.563
l_{35}	16.98	40	1/2	0.849	0.811
l_{36}	15.08	40	1/2	0.754	0.671
l_{45}	19.56	60	1/3	0.978	1
l_{46}	17.59	40	1/2	0.88	0.855
l_{56}	9.1	60	2/5	0.379	0.12

Table 3. Total weight value.

Section	S_{ij}
l_{12}	0.125
l_{13}	0.088
l_{14}	0.311
l_{23}	0.564
l_{34}	0.419
l_{25}	0.830
l_{26}	1.063
l_{35}	1.561
l_{36}	1.546
l_{45}	1.625
l_{46}	1.730
l_{56}	1.120

4.3 Calculate the total weights $S_k = t'_{ij} + r'_{ij}$

The total weights $S_k = t'_{ij} + r'_{ij}$ can be derived from the results of the calculations in the above sections, as shown in Table 3.

The total weight value is marked in the emergency road network to solve the shortest road network. As shown in Figure 1. Because Jiandi, Luoshui and Yinghua are hit the hardest, and the emergency rescue centre is placed in a suburb in Yuanshi Town, the paper focuses on the shortest road from Yuanshi to these three towns.

Using Lingo programming to solve the problem, output the optimal solution is A-B1-C2, which is the shortest road between Yuanshi and Jiandi. Similarly, we can get the shortest road between Yuanshi and Luoshui is A-B1-C1. Because only by Luoshui can we get to Yinghua, so the shortest road to Yinghua is the same as to Luoshui.

5 SUMMARY AND PROSPECT

During the emergency period of the rescue, time is the most significant factor, and is crucial to the

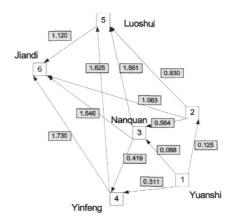

Figure 1. Road network of the emergency period.

success of rescue operations. We should take into account the reliability of the emergency vehicles under the complex situation to avoid the occurrence of secondary accidents. Research in this paper has the reference value for the relevant research in emergency logistics.

Whether the vehicle path is reasonable or not directly affects the efficiency, cost and benefit of emergency logistics. Efficient solutions to the emergency logistics vehicle routing problem to reduce logistics costs and improve service standards are of great significance. Natural disasters are inevitable, but through the corresponding study in this paper, the loss caused by natural disasters can be reduced, and in the meantime, help mankind to better adapt to nature.

REFERENCES

Alumur, S. and Kara, B.Y. (2007). A new model for the hazardous waste location-routing problem. *Computers & Operations Research*. 34(5), 1406–1423.

Mete, H.O. and Zabinsky, Z.B. (2010). Stochastic optimisation of medical supply location and distribution in disaster management. *International Journal of Production Economics*. 76–84. Netherlands: Elsevier science bv.

Murali, P., Ordóñez, F. and Dessouky, M.M. (2012). Facility location under demand uncertainty: Response to a large-scale bio-terror attack. *Socio-Economic Planning Sciences*. 46(1), 78–87.

Nahleh, Y.A., Kumar, A. and Daver, F. (2013). Facility Location Problem in Emergency Logistic. *Waset Org* 7(10), 1045–1050.

Thorleuchter, D., Schulze, J. and Poel, D.V.D. (2012). Improved emergency management by a loosely coupled logistic system. 5–8. Berlin, Heidelberg: Springer.

Automotive, Mechanical and Electrical Engineering – Liu (Ed.)
© 2017 Taylor & Francis Group, London, ISBN 978-1-138-62951-6

Research on the Zigbee routing algorithm based on link availability prediction

Zhiyong He
College of Electrical Engineering, Nanjing Institute of Industry Technology, Nanjing, China
Jiangsu Wind Power Engineering Technology Center, Nanjing, China

ABSTRACT: The fault local route repair mechanism of AODVjr is after the link is broken in Zigbee network, the lagging repair process causes the cache data packet loss and the data end-to-end delay. In this paper, a Zigbee link availability prediction algorithm (APR-AODVjr) is proposed to predict the link availability of the AODVjr routing algorithm, which starts the route discovery process before the link breaks. Simulation results show that the routing algorithm based on link availability prediction can effectively reduce the network fault recovery response time, prolong the network lifetime and reduce the network congestion and delay.

Keywords: Zigbee network; Fault repair; AODVjr routing algorithm

1 INTRODUCTION

Zigbee is a low-power, low-cost and good reliability wireless network technology featuring with little delay and large network capacity (Baront P & Pillai P & Chook V W C, 2007; Ren Xiuli & Yu Haibin, 2007). Presently it is mainly used in medical, mineral, environmental testing and military areas. IEEE 802.15.4 protocol standard is adopted in the physical layer and MAC layer of wireless sensor network. There are three types of network topology: star, tree and mesh (Ma Haichao, 2014; Chakeres I D, 2002). The network layer mainly adopts Cluster-Tree routing Algorithm and AODVjr routing algorithm, and cluster-Tree routing algorithm is based on the parent-child relationship between nodes in the tree structure. Data packets need to be forwarded according to the parent-child relationship between nodes, and the routing method need not to store the routing table, which is the static route. Through AODVjr routing algorithm (Zhang Fenghui & Zhou Huiling, 2009) the optimal path between the source node and the destination node could be found. The node route discovery process sends the RREQ message to set up the route through the broadcast packet. The node energy depends on the battery. Most of the communication overheads for packet forwarding between nodes, due to the limited overall energy of the network, especially at the upper nodes near the coordinator, the energy consumption of the nodes increases with the running time of the network, which leads to interruption of the

network link. AODVjr routing algorithm using local repair mechanism, the fault node to restart a route discovery process by broadcasting RREQ data packets to find a new route (Ren Weil & Yeung D.Y, 2006). The local route repair process initiates from the downstream node automatically, and the link disconnection leads to the loss of node data. Therefore, the link availability prediction is very important to guarantee the normal link communication.

Researchers on Zigbee network mainly focus on the optimization of routing algorithms and node energy protection (Xie Chuan. S, 2011), propose measures to improve AODVjr routing algorithm, carry on the selection of data and packet forwarding by avoiding low energy node, and limit the broadcast RREQ packets. However, the algorithm only protects the energy of the node during the route discovery process. It only delays the network link disconnection and does not analyze the link availability. In article (Bai Le-Qiang & Zhang Shi-Hong, 2014), nodes are divided into different types according to the strength of node signals. The nodes transmit the broadcast packets selectively and reduce the energy consumption of the network. However, this algorithm does not take into account the link disruption caused by node mobility. In article (Bai Le-Qiang & Sun Jing-Jing, 2015), we propose an algorithm to improve the Cluster-Tree routing based on neighbor table, and select the next-hop node address according to link quality. This algorithm can reduce the number of forwarded nodes, but this algorithm does not consider link break. The probability of data packet loss is increased.

In this paper, we propose a Zigbee Link Availability Prediction Routing Algorithm (APR-AODVjr) based on the node energy consumption threshold model and the remaining energy of the next-hop node stored in the neighbor table of the routing node, which selects the energy-rich nodes as the next-hop route, by calculating the energy consumption of the entire link *Cenergy* value to improve the local link repair time and reduce the loss of data packets. This paper is organized as follows: Section 2 analyzes AODVjr routing mechanism; Section 3 presents link availability prediction algorithm APR-AODVjr; Section 4 simulates and analyzes the results; At last, Section 5 concludes the paper.

2 AODVjr ROUTING MECHANISM ANALYSIS

AODVjr routing algorithm establishes the path of the process as shown in Figure 1. The source node S need send data packets to the destination node D, S looks for their own routing table to find the path to the destination node firstly. If the node routing table cannot find the destination node, then the source node S initiates the route discovery process to establish the route through the broadcast RREQ packets. When the destination node receives the RREQ broadcast message, it establishes the route in reverse direction {3, 4, 6}. AODVjr routing algorithm selects next-hop routing principle which node returns the RREP message quickly, the routing algorithm does not consider the node's remaining energy, the link is interrupted because of the node energy is exhausted along with the network running time increasing. Especially in high-level nodes in the network, data forwards frequently and node energy loss is very fast, the

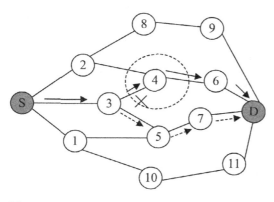

Figure 1. AODVjr local repair algorithms.

network will appear split phenomenon in advance. As shown in Figure 1, the node 4 is in the middle node of link {2, 4, 6} and {3, 4, 6}, the node energy is exhausted and the link is interrupted. In AODVjr local repair algorithm, the upstream node 3 of the faulty node restarts the route initiation process and broadcasts the RREQ packets to perform the route re-establishment process (Zhou Shao-qiong, et al., 2011). The new path {3, 5, 7} of the node 3 to the destination node D is established. The AODVjr routing algorithm restarts local repair only after node 4 energy is exhausted, which leads to the data packets lost in the network, having an impact on network transmission efficiency and increasing network latency.

3 LINK AVAILABILITY PREDICTION ALGORITHM APR-AODVjr PROPOSED

The main idea of the APR-AODVjr link availability prediction algorithm proposed in this paper is that when the node selects the next-hop route, the node can find out the remaining energy of the next-hop node stored in the neighbor table. If the link energy consumption is lower than 2% of the total link energy, the link between the node and its upstream node is considered to be in danger, and the upstream node initiates the rerouting discovery process.

3.1 *Energy consumption model*

Ariticle (LI Gang, 2016) proposed a node energy consumption calculation model, the node sends and receives k bit data packet information. When the transmission distance is d, the energy required for data packet transmission is:

$$T_t(k,d) = E_{T-elec}(k) + E_{T-amp}(k,d)$$
$$= \begin{cases} E_{elec} \times k + E_{fs} \times k \times d^2, d < d_0 \\ E_{elec} \times k + E_{mp} \times k \times d^4, d > d_0 \end{cases} \quad (1)$$

where E_{elec} the energy is consumed to receive and transmit a unit data packet, E_{fs} and E_{mp} are the energy of the network nodes that amplify the gain, d_0 is the node distance threshold value. The energy required to receive the data packet is $E_R(k)$:

$$E_R(k) = E_{R-elec}(k) = E_{elec} \times k \quad (2)$$

In the process of sending and receiving n data packets between node S and node D, the energy consumption of the whole chain is:

$$C_{energy} = \sum_{i=0}^{n} E_R(k) + T_t(k,d) \quad (3)$$

Table 1. Node neighbor table format.

Name	Size/bit	Function
NodeList	16	Record node number
NeighborPanID	16	Neighbor node Pan identifier
Extend Address	64	64-bit IEEE address
Node Type	16	Node type
RemainPower	8	Remaining battery
PowerType	8	Node power type

3.2 APR-AODVjr algorithm

In order to solve the problem of route discovery and route maintenance in AODVjr routing algorithm, APR-AODVjr algorithm saves the node energy value in the process of route discovery through the node neighbor table, avoiding to select the node which is about to run out or low energy as the next hop The improved algorithm adds a neighbor table to save the information of neighbor nodes when a node has routing capability. And update them dynamically through routing maintenance messages. The neighbor table structure is shown in Table 1.

After the link between the source node and the destination node is established, the route enters the maintenance phase. The traditional AODVjr does not repair the local route until the node energy is completely dead with energy is exhausted. In order to reduce the data packet loss in the network packet, AODVjr route repair mechanism to predict the energy on the link and predict the routing outage in advance. Through the energy consumption model to calculate the energy consumption of the entire link *Cenergy* value by KEEP_ALIVE packets during link maintenance process. In this paper, when the energy of the link is less than 2% of the total energy of the link, the probability that the route is interrupted is very large, the node restarts the routing process to establish a new link.

4 SIMULATION ANALYSIS

The APR-AODVjr algorithm proposed in this paper is validated by the NS2 platform. The platform implements the PHY layer and MAC layer of the IEEE 802.15.4 protocol. In order to facilitate the statistics of the number of death nodes in the network, the initial energy of the node is set to 5J, the simulation time is 400s, the entire network range is set to 150 m × 150 m, uses 160 protocol nodes, single-hop transmission distance is 5m. The simulation results show that this simulation in the experiment, the node energy is less than 2% of the initial energy, the node is regarded as the death node. In this paper, a comparison is made between the network node mortality rate and the

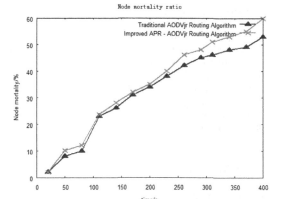

Figure 2. Node mortality comparison.

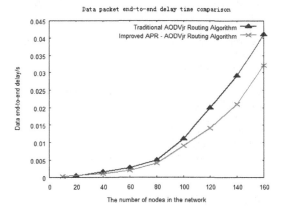

Figure 3. Data packet end-to-end time delay comparison.

data packet end-to-end delay with the traditional AODVjr algorithm.

4.1 Node mortality comparison

From the simulation results in Figure 2, we can see that with the network running time increasing, the energy consumption of node appears, the death nodes increases with the running time of network. APR-AODVjr routing algorithm, the nodes with sufficient energy are selected as the next hop of the route through the remaining energy value stored in the neighbor table when the next hop route is adopted, thus APR-AODVjr protects the nodes with low energy in the network and extends the network lifetime.

4.2 Data packet end-to-end delay

From the simulation results in Figure 3, it could be seen that with the increasing of the node data in the network and data end-to-end delay is increasing, the APR-AODVjr routing algorithm proposed in this paper calculates the energy of the link through the energy consumption model, the routing discovery process is started in advance for the nodes to be death and the network load is balanced, the number of active nodes in the network is increased and broken links is reduced. This avoids the loss of broken data packets and effectively shortening the data packet end-to-end time.

5 SUMMARY

In Zigbee networks, the limited energy of a node directly affects the performance and stability of the network. In this paper, we propose a Zigbee link availability prediction algorithm APR-AOD-Vjr, by improving the traditional AODVjr in the route discovery process which does not consider the remaining energy of the next hop, adding neighbor table to store energy information for each node.The link energy consumption is calculated and the link availability is predicted by the total energy of the link. The route discovery process is started before the link breaks to reduce the loss of data packets. Simulation results show that the routing algorithm based on link availability prediction effectively reduces the number of dead nodes in the network and prolongs the network lifetime.

ACKNOWLEDGMENTS

This work is sponsored by Jiangsu Wind Power Engineering Technology Center Open Fund Project (No: ZK16-03-13) and Nanjing Institute of Industrial Technology Research Fund Project (No: ZK16-02-01). The author gratefully acknowledge their supports.

REFERENCES

AODV routing protocol with high energy and high signal strength node priority [J]. Computer Engineering and Applications, 2014, 50 (17): 86–89.

Bai Le-Qiang & Sun Jing-Jing. Improved routing algorithm for ZigBee network tree based on neighbor table [J]. Computer Engineering and Design, 2015, 36 (5): 1156–1160.

Bai Le-Qiang & Zhang Shi-Hong. Optimized Broadcasting Algorithm for ZigBee Network Node Selection [J]. Control Engineering, 2014, 21 (3): 403–408.

Baront P & Pillai P & Chook V W C. Wireless sensor networks: A survey on the state of the art and the 802.15.4 and ZigBee stand-stards [J]. Computer Communications, 2007. 30(7): 1655–1695.

Chakeres I D, Klein Berndt. AODVjr AODV Simplified [J]. Mobile Computing and Communication Review, 2002, 6 (3):100–101.

LI Gang. Improved AODV protocol based on energy in Manet [J]. Journal Of Huaqiao University (Natural Science), 2016, 37 (4): 503–506.

Ma Haichao. ZigBee network scalability and clustering routing protocol research [D]. Xi'an: Xidian University, 2014.

Ren Weil & Yeung D.Y. TCP performance evaluation over AODV and DSDV in RW and SN mobility models [J]. Journal of Zhejiang University, 2006, 7(10):1683–1689.

Ren Xiuli & Yu Haibin. Study on the Implementation of ZigBee Wireless Communication Protocol [J]. Computer Engineering and Applications, 2007, 43 (6): 143–145.

Xie Chuan. Study on AODVjr algorithm based on Zig-Bee [J]. Computer Engineering, 2011, 37 (10): 87–89.

Zhang Fenghui & Zhou Huiling. A Routing Algorithm for ZigBee Network Based on Dynamic Energy Consumption Decisive Path: Proceedings of the 2009 International Conference on Computational Intelligence and Natural Computing, 2009 [C]. Washington, IEEE Computer Society, 2009: 429–432.

Zhou Shao-qiong & Xu Yi & Jiang Li, et al. Application of ant colony optimization algorithm in Ad Hoc network routing [J]. Journal of Computer Applications, 2011, 31 (2): 332–334.

Short-term load forecasting based on fuzzy neural network using ant colony optimization algorithm

Zhichao Ren, Qian Chen, Lipin Chen, Kaijiang Cao & Haiyan Wang
Research Institute of Economics and Technology of State Grid Sichuan Electric Power Company, Chengdu, Sichuan, China

Rundong Chen & Yun Tian
Beijing China-power Information Technology Co. Ltd., Beijing, China

Yan Wang & Haichao Wang
College of Economics and Management, North China Electric Power University, Beijing, China

ABSTRACT: On the basis of fuzzy neural network load forecasting, the weights and thresholds of fuzzy neural network and the number of neurons in hidden layer are optimized by ant colony algorithm. A short-term load forecasting model of fuzzy neural network based on ant colony optimization is proposed. Through case analysis, it shows that this algorithm can obtain high accuracy, and it is an effective method for short-term load forecasting.

Keywords: Ant colony optimization algorithm; Fuzzy neural network; Short-term load forecasting; Parameter optimization

1 INTRODUCTION

Short-term load forecasting of electric power system is an important basis for the comprehensive plan audit, and it is also an important part of the key technology of the project audit analysis, which has a very important influence on the intelligent audit and analysis of the project.

The methods of the traditional load forecasting include time series method, regression analysis method and the state space method and so on. These methods are simple, fast speed, and widely used, but because of the power load variation affected by weather conditions (such as the seasons, a sudden change of weather factors) and people's social activities (such as major recreational activities, holidays etc.) and other factors, there are a lot of nonlinear relationship that make these methods difficult to accurately describe the actual load forecasting. The other method is the traditional artificial neural network, which has strong memory ability, nonlinear mapping ability and strong self-learning ability, so it can quickly fit the curve of the load change. However, the network training is usually a complex large-scale optimization problem. Because of the change of the training set and the initial weights of network, the training results of the network are random, which

has some disadvantages such as slow convergence and being easy to fall into local extremum. Fuzzy control is a kind of controller based on fuzzy rules, whose core problem is the membership functions of the fuzzy sets, fuzzy knowledge planning and fuzzy set. But it has no self-study ability, and the selection of parameter is too subjective.

Ant colony optimization is a robust evolutionary algorithm based on groups. It combines the positive feedback structure, distributed computing and some heuristic factors, and can find a better solution. This paper uses the ACO algorithm to train the parameters of the fuzzy neural network, so as to realize the automatic optimization of fuzzy neural network. A short term load forecasting model based on ant colony algorithm for fuzzy neural network is constructed, and it is applied to a city.

2 BASIC THEORY

2.1 *Ant colony optimization algorithm*

Ant Colony Optimization algorithm (ACO) is affected by the ant colony search heuristic process of food and produce. Through the research on the behavior of ants, it is found that the individual behavior of ants is very simple, but the groups composed of simple individuals show a very complex behavior. The ants are transmitted by means of the substance of the hormone, and ants are able to guide their movement through the perception of the substance. As a result, the more ants walk on a path, the higher the probability that an ant will choose the path. The purpose of searching food is realized through constant information exchange among the ants. Its mechanism is shown in Figure 1.

A is the nest, E stands for food sources, and there are only two paths between A and E. The length of road ABDE and ABCDE are 6 and 8 respectively. It's supposed that each time unit there are 15 ants from A to B and there are 15 ants from E to D. The amount of hormones (commonly known as Element Information, IE) left behind after each ant is set to L.

At the initial time of t0, because there is no pheromone on the three ways where ants can randomly choose the path. From the point of view of statistics, the three paths can be chosen with the same probability. After a time unit, the BCD is two times the length of the BD, therefore in the case that individual sow is equivalent to IE, BD pheromone quantity is two times of BCD. As a result, in the T1 time, there will be 20 ants to select the path ABDE (EDBA), while there will be 10 ants to select the path ABCDE (EDCBA). With the passage of time, there will be more and more probability that ants choose the path ABDE (EDBA), and ultimately choose the path ABDE (EDBA), in order to find the shortest path from their nest to the food source.

ACO algorithm is a general purpose internal heuristic algorithm, which can be used to solve large-scale combinatorial optimization problems. In the ACO algorithm, a proxy agent is set up in each set. The members of the agent are like mutual cooperation artificial ants, which exchange information through the IE of the "search path" and finally realize the solution of the problem. When the artificial ants move, the solution of the problem is set up, and the description of the problem is modified by the addition of the new IE. It must be pointed out that there are some differences between the artificial ants and the real ants in the ACO algorithm, and the artificial ants are used as an optimal tool to operate, rather than the simulation of the actual ants in the natural world.

Thus, the main characteristics of the ACO algorithm are: positive feedback helps to quickly find a better solution; distributed computing can be avoided in the iterative process of premature convergence; inspired makes the search process earlier found acceptable solution possible; heuristic convergence makes it possible to find an acceptable solution earlier in the search process.

2.2 Fuzzy neural network

It is supposed that a fuzzy subset of the domain X is A, and for $\forall x \in X$ to specify a number $\mu_A(x) \in [0,1]$ corresponding to the x, it is called the degree of membership of the A on the x, mapping: $X \to [0, 1]$ is the membership function of A. $\mu_A(x)$ reflects the degree of x to the fuzzy subset A, and it is more close to 1 which indicates that the x is more dependent on the A. On the set of real numbers, the distribution of membership functions, such as triangle, rectangle, trapezoidal and parabolic, can be used in the practical application, which can be selected according to the characteristics of the research object.

Fuzzy neural network gives the fuzzy input signal and fuzzy weight is for the conventional neural network (such as the forward feedback neural network, Hopfield neural network, etc.), which usually includes the following types: the fuzzy neural network with real input signal and fuzzy weight; the fuzzy neural network with fuzzy input signal and real weight; the fuzzy neural network with fuzzy input signal and fuzzy weight. Learning algorithm of the fuzzy neural network is usually the same as the learning algorithm of conventional neural network or it is its promotion algorithm, mainly including: BP learning algorithm; fuzzy BP learning algorithm; the learning algorithm of alpha set based on BP; the random search algorithm; genetic algorithm. Based on the analysis of actual data, this paper adopts the third kind of fuzzy neural network and BP algorithm, which establishes the forecasting model with the fuzzy input data and the fuzzy weight, in order to get the accurate and stable prediction results.

3 SHORT-TERM LOAD FORECASTING

3.1 Modeling

The number of parameters you need to train is n, denoted as p1, p2, ..., pn. For any of these

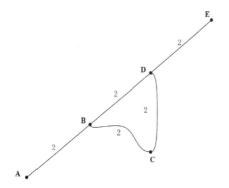

Figure 1. Schematic diagram of ants path searching.

parameters which are represented by pi (1≤i≤N), and it is set to L random non-zero so as to form a set of Ipi. S ants set out to look for food from the ant nest according to each element in the set of pheromone path state and the following formula selection rules:

$$P(\tau_j^k(I_{p_i})) = \frac{\tau_j(I_{p_i})}{\sum\limits_{j=1}^{N} \tau_j(I_{p_i})} \tag{1}$$

One element is selected randomly and independently from each of the set Ipi. When the ant completes the selection of the element in all the collection, it reaches the food source, and thereafter, the pheromone of each element in the set is adjusted according to the following formula:

$$\tau_j(I_{p_i})(t+l) = \rho\tau_j(I_{p_i})(t) + \Delta\tau_j(I_{p_i}) \tag{2}$$

$$\Delta\tau_j(I_{p_i}) = \sum_{m=1}^{m} \Delta\tau_j^m(I_{p_i}) \tag{3}$$

$$\Delta\tau_j^m(I_{p_i}) = \begin{cases} Q/e^m \\ 0 \end{cases} \tag{4}$$

where $0 \le \rho < 1$ expresses the persistence of pheromone, $\Delta\tau_j^m(I_{p_i})$ expresses that the information ele-

ments of the mth ant in this cycle left on the jth elements of the sey Ipi. Q is a constant which is used to adjust the pheromone adjustment speed. The maximum output error of each training sample is defined as the weight of the neural network, which is used as the weight of the neural network, which can be defined as the M. e^m indicates the maximum output error of each training sample when the elements that ant m chooses is taken as the weight of the neural network, whose formula is as follows:

$$e^m = \max_{l=1}^{h} |O_l - O_q| \tag{5}$$

where O_l is the actual output of FNN and O_q is the desired output.

This process is repeated until the optimal solution is found. Optimal solution can be found when the evolutionary trend is not obvious or the maximum number of iterations is reached.

3.2 Case study

In order to verify the feasibility and effectiveness of the fuzzy neural network model based on Ant Colony Optimization in this paper, the short-term load

Table 1. Prediction results and errors.

Time	Actual load/MW	Predictive load of ACO-FNN model/MW	Prediction error of ACO-FNN model/%	Predictive load of FNN model/MW	Prediction error of FNN model/%
0	965.68	973.12	0.77	980.59	1.54
1	921.95	924.52	0.28	934.26	1.34
2	796.52	790.21	−0.79	809.66	1.65
3	768.18	765.1	−0.40	785.65	2.27
4	789.68	776.12	−1.72	800.21	1.33
5	816.21	818.63	0.30	825.32	1.12
6	766.23	771.21	0.65	750.22	−2.09
7	756.12	750.13	−0.79	738.85	−2.28
8	819.36	820.51	0.14	809.25	−1.23
9	858.61	858.63	0.00	839.98	−2.17
10	913.12	911.56	−0.17	900.05	−1.43
11	869.36	866.12	−0.37	858.52	−1.25
12	839.68	835.15	−0.54	850.52	1.29
13	856.15	850	−0.72	860.1	0.46
14	838.56	839.99	0.17	850.12	1.38
15	851.92	856.21	0.50	832.65	−2.26
16	858.68	852.12	−0.76	835.69	−2.68
17	888.69	890.18	0.17	873.66	−1.69
18	981.26	981.86	0.06	996.56	1.56
19	921.36	921.09	−0.03	926.96	0.61
20	966.88	963.51	−0.35	950.21	−1.72
21	989.62	986.25	−0.34	980.65	−0.91
22	856.58	859.61	0.35	850.01	−0.77
23	951.12	956.26	0.54	938.97	−1.28
Average relative error/%	0.46			1.51	

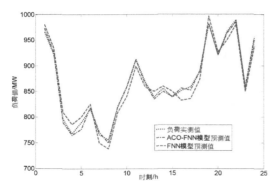

Figure 2. The load forecast and actual values.

forecasting of a substation in Sichuan province was calculated and analyzed. Load, weather data, and holiday information of June to May 2016 are used to form a forecast model so that the value of 24 points in June 30th can be predicted. For the fuzzy neural network, a supervised learning method is adopted. Seven input samples correspond to one output sample, a total of 24 groups, the number of neurons in the hidden layer is determined by the ant colony optimization algorithm, the convergence factor is 0.02, the hidden layer activation function and the output layer as a nonlinear function of Sigmoid. The number of neurons in the hidden layer is determined by the ant colony optimization algorithm. The error convergence factor is 0.02, and the implicit and output layer is the nonlinear Sigmoid function. The number of connection weights of the network is 58. Ant colony algorithm parameters are set as follows, $S = 30$, $Q = 30$, fuzzy neural network weights and thresholds are the random number between $[-1, 1]$ and L takes 30.

The predicted results are shown in Table 1 and Figure 2.

It can be seen from Table 1 that the average error of the proposed ACO-FNN prediction model is 0.46% and the maximum error is −1.72%. The average relative error of the traditional BP algorithm which uses the gradient descent training weight is 1.51%, and the maximum error is −2.68%. Therefore, the proposed method can get higher accuracy of load forecasting.

4 CONCLUSION

Application of ant colony algorithm to optimize the weight of fuzzy neural network, threshold and the number of hidden layer neurons overcome the defect that the parameter selection of fuzzy neural network is too subjective, which significantly improves the accuracy of load forecasting. Practical example shows that the improved ant colony neural network algorithm is an effective method for short term load forecasting of power system, and it has a certain application value in engineering.

ACKNOWLEDGMENTS

The authors acknowledge the support of the "Key Technology and System Research of Company Project Plan Audit Analysis" project.

REFERENCES

Chen, Ke. 2015. Short term load forecasting of power system based on Fuzzy Neural Network. Southeast University.

Cui, Yanyan & Cheng, Yawei. 2014. Short term load forecasting of power system based on fuzzy control for RBF neural network. Journal of Jingchu University of Technology, 02:32–35.

Le, Huan. 2009. Study on the vacant area inference based method of spatial load forecasting for city distribution network planning. Chongqing University.

Li, Long. 2015. Load model prediction based on artificial neural network. Transactions of China Electrotechnical Society, 08:225–230.

Wang, Fangfang. 2011. The establishment of the load forecasting model and the load forecasting based on the regression analysis. China High-Tech Enterprises, 34:56–58.

Yin, Xin. 2011. Research on swarm intelligence algorithms and power load forecasting. Hunan University.

Zhang, Chao. 2015. Research on the improved method of ultra-short term load forecasting based on time series method. Liaoning University of Technology.

Zhang, Jian. 2014. A study of ant colony optimization for community detection in complex networks. Xidian University.

Automotive, Mechanical and Electrical Engineering – Liu (Ed.)
© 2017 Taylor & Francis Group, London, ISBN 978-1-138-62951-6

Soft measurement of emulsion matrix viscosity based on GA-BP neural network theory

Yuesheng Wang & Jian Zhang
School of Hang Zhou Dianzi University, Hangzhou, China

ABSTRACT: The emulsion matrix viscosity determines the quality of an emulsion explosive. However, in the production process of the emulsion explosive, the emulsion matrix viscosity cannot be measured directly. In order to solve this problem, a soft measuring method based on a BP neural network is proposed in this paper. However, the simple BP neural network cannot accurately measure the viscosity of the emulsion matrix. So, a Genetic Algorithm (GA) is used to optimise the network and then train the network prediction model to get the optimal solution. The predicted results with the predictions of the conventional BP neural network prediction model are compared. The BP neural network optimised by the GA can obtain more accurate results in a shorter period of time.

Keywords: Soft measurement; GA-BP neural network; Emulsion matrix viscosity

1 INTRODUCTION

As the basis of an emulsion explosive, the emulsion matrix viscosity determines the quality of the emulsion explosive. As an important parameter characterising matrix rheology, the absolute viscosity of the matrix (or apparent viscosity) and the performance of the emulsion explosive are closely related. The viscosity emulsion matrix not only affects the shear strength, pumping, sensitisers, charge and other processes, but it also relates to the performance of the explosives explosive and storage stability. (Xvguang Wang et al., 2008). The emulsion matrix viscosity must be controlled within a suitable range. If the viscosity is too small, it is not conducive to maintaining the tiny sensitised bubbles. Sensitised bubbles have a tendency to coalesce and spill. It reduces the emulsion explosives detonation sensitivity and performance; if the viscosity is too large, the emulsion explosives become "ageing" easily. It is not conducive to storage the production. In the production process of emulsion explosives, the emulsion matrix viscosity is measured by using a viscometer. The sampling period is generally 4 to 8 hours. So it is difficult to realise the real-time detection of the matrix viscosity.

The basic idea of soft measurement is to use some kind of mathematical method and computer software to estimate the measured variables on the basis of the acquired information, using the mathematical relationship between measurable information and measured variables (Dominant variable). This paper presents the factors which affect the viscosity of an emulsion matrix and uses a BP neural network to build a soft measurement model of emulsion matrix viscosity. However, the BP neural network falls into a local minimum through the experiment. So the GA is used to optimise the network. The GA-BP neural network model is applied to the prediction of the matrix emulsion viscosity. The prediction results were compared with the traditional prediction model of a BP neural network. More accurate results were obtained in a shorter period of time by the GA optimised BP neural network.

2 FACTORS AFFECTING THE EMULSION MATRIX VISCOSITY

1. The ratio of oil phase and water phase: Improper ratio causes breaking. The ratio of oil and water phases is 1:8, which can form a good water in oil emulsion system.
2. Emulsifying temperature: If the emulsifying temperature is too low, it increases the emulsion viscosity and influences the fluidity of the matrix. It also causes emulsification uneven, inadequate, and leads to reduced quality or even emulsion failure, affecting the safety of production. If the temperature is too high, the emulsion matrix viscosity is reduced, making the emulsion stability become poor, resulting in a waste of energy consumption. At the same time it is not conducive to safe production.
3. Emulsifier speed: Under higher rotating speeds, material in the emulsion chamber is subjected to more shear number at per unit time. Homogenisation is more intense, the internal phase is

more dispersed, resulting in the matrix viscosity increasing. In a certain range, under the higher emulsion speed, a greater viscosity of the emulsion matrix is achieved.

4. Emulsifying time: Sufficient emulsifying time to ensure the material sufficiently emulsified. It obviously affects the emulsion matrix viscosity. The longer the emulsifying time, the more number of times the material being cut, the greater emulsion matrix viscosity is made.

Through the above analysis, the water phase flow, the oil phase flow, emulsifier speed, emulsifying temperature and emulsifying time were used as auxiliary variables and a soft sensor model was used to measure emulsion matrix viscosity.

3 BP NEURAL NETWORK

There are many advantages of a BP neural network, such as simple structure, more tuneable, more training algorithm, and good operability. So there are 80% to 90% of neural networks use BP neural network or its variants. It is one of an artificial neural network of the most important network.

3.1 BP network topology

According to the theory of the BP neural network, the satisfactory results can be obtained by using the single hidden layer in the general case (Qiyuan Ning et al., 2012). So a single three-layer neural network hidden layer is used. According to a previous analysis, five input layer nodes, respectively, water phase flow and oil phase flow, emulsifying temperature, emulsifying time and emulsifier speed. According to the following empirical formula to select the BP neural network's hidden layer neurons number: $s = \sqrt{m+n} + a$. s represents the hidden nodes number; m represents the input layer neurons number, in this case m = 5; n represents the output layer neurons number, n = 1; a represents a constant between 1 and 10. We can get a range of values hidden layer neurons numbers based on an empirical formula, and then through the simulation test obtain the number of neurons as the last best network performance prediction of the number of neurons network. In this paper, through the simulation test, the value of a is 2. The hidden layer nodes number is 5. In this paper, the structure of the BP neural network is as follows:

3.2 Training data processing

In actual measurement, due to sudden fluctuations in instrumentation, it can cause abnormal results. For the abnormal processing of the data, we use the 3σ guidelines. This paper selects 100 sets of processed data to model, each data set contain five auxiliary variables and a sample test value of emulsion matrix viscosity. Taking into account the training of BP neural network may occur in the case of over fitting, the sample data is cut into two parts, respectively, to train and test the network. In this paper, 90% of the data was used for training, and 10% for testing the network.

Before training the neural network, the data should be normalised. The effect of normalisation is to sum up the statistical distribution of the sample. In the normalisation process all the input and output data are normalised to the range of [–1, 1]. MATLAB comes with a normalisation function premnmx:

$$[P, \min p, \max p, T, \min t, \max t] = premnmx(P,T);$$

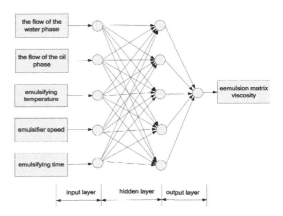

Figure 1. Structure of BP.

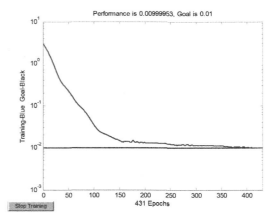

Figure 2. BP neural network training error performance diagram.

where P is the input variable, T is the target variable.

The data after normalisation is used to train the network, and greatly improves the accuracy of the model.

3.3 *The neural network training of emulsion matrix viscosity*

In this example, the BP network is trained using the momentum gradient method, using the traingdx function in the MATLAB neural network toolbox, the hidden layer transfer function using the tansig function, and the output layer using the linear transfer function purelin. The target error is set as 0.01; and the number of maximum iterations is 500.

The training error variation is shown in Figure 2. After 431 iterations, the network training error reaches the desired level.

4 GA-BP SOFT MEASUREMENT MODEL

GA is a stochastic global search and optimisation method that simulates the evolution mechanism of natural biological evolution. Its essence is an efficient, parallel, global search optimisation method.

4.1 *Structure of GA-BP model*

A genetic algorithm is a random search algorithm for global optimisation. Genetic operators simulate the phenomenon of reproduction, crossover

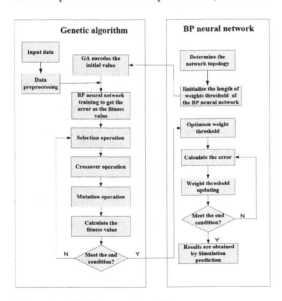

Figure 3. GA-BP network model structure.

and mutation in the genetic process to obtain the optimal individual (Zhang yin et al., 2013). Its structure is shown in Figure 3.

First of all, the initial values of the network parameters are generated randomly. The genetic algorithm is used to calculate the initial value of these network parameters. After several generations of evolution to find the most suitable network parameters of the BP network. Finally the network parameters are used to train the BP network to achieve the required precision output.

4.2 *The realisation of the GA-BP neural network*

The input layer nodes number is 5; the output layer nodes number is 1; the hidden layer nodes number is 5. The genetic algorithm optimisation neural network weights threshold are as shown in Table 1.

The weights and thresholds are used by 10-bit binary number. The binary code lengths of the individual are 360. Among them, the former 250 are connected to the input layer and the hidden layer weights coding, 251 to 300 is the threshold of the hidden layer coding, 301 to 350 is a hidden layer and output layer connection weights coding, 351 to 360 are the output layer threshold coding (Shi Feng et al., 2013). The population size of the genetic algorithm is set to 30, and the genetic algebra is set to 100. The fitness function is shown in the Equation (1).

$$fitness = \frac{1}{SSE} \tag{1}$$

where SSE (sum-squared errors) is the square of the error, and its expression is as follows:

$$SSE = \sum (Y_d - Y)^2 \tag{2}$$

For individuals with a smaller error square, the fitness function value is greater, and the individual is better.

As shown in Figure 4 and Figure 5, the evolution of the first group 10 generation convergence rapidly, and then tend to be stable close to the best. The change of population mean is shown by the red line; the blue one indicates the optimal solution.

Table 1. Weights and threshold number.

Connection weights of input layer and hidden layer	Hidden layer threshold	Connection weights of hidden layer and output layer	Output layer threshold	Total
25	5	5	1	36

611

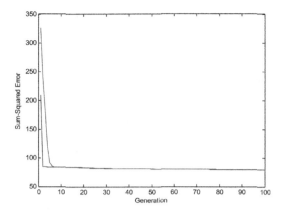

Figure 4.　Sum-squared error evolution curves.

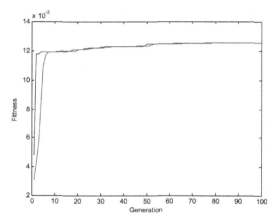

Figure 5.　Fitness function curves.

The initial weight and threshold which has been optimised by GA are applied to the BP neural network training. The training error performance variation diagram is obtained as shown in Figure 6.

From Figure 6, we can see that the network trains 178 steps to reach the desired level of error. And the network training stops.

The convergence rate of the GA-BP neural network is faster than the simple BP neural network. Reducing the number of training steps is beneficial to improve the generalisation ability of the network. Over-training may achieve a higher matching effect on the training sample set, but for a new input sample and the target may have a greater difference in output. A BP neural network optimised by GA can obtain more prediction accuracy in a shorter period of time.

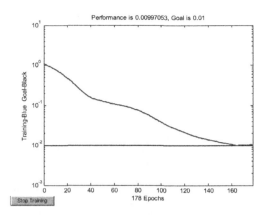

Figure 6.　GA-BP training error performance.

5　CONCLUSION

In this paper, a soft measurement method is proposed for the problem that the viscosity of an emulsion matrix cannot be measured online. A BP neural network is used to build a soft sensing model to measure the viscosity of the emulsion matrix by the five parameters which are easy to measure in the production of emulsion explosives, and are closely related to the viscosity of the emulsion matrix. A GA is used to optimise the BP neural network. It can be seen by MATLAB simulation analysis that after the optimisation of the GA, the BP network can improve the accuracy of the latex matrix viscosity prediction.

REFERENCES

Qiyuan Ning, Zude Liu. (2012). Prediction model of coal and gas outburst intensity based on BP neural network. *Coal Mining.*

Shi Feng, Wang Hui, Yu Lei. (2013). MATLAB intelligent algorithm. Beijing: Beihang University Press.

User's Guide: Genetic algorithm and direct search toolbox for use with MATLAB. The Math Works, 49–56.

Xvguang Wang. (2008). Emulsion matrix (Version 2). Beijing: Metallurgical Industry Press.

Yang Yang, Yuesheng Wang. (2010). Study on control optimization strategy of powdery emulsion explosive production process. Hang Zhou: Hangzhou Dianzi University.

Yingjie Lei. (2005). MATLAB genetic algorithm toolbox and its application. Xi'an: Xi'an Electronic and Science University Press.

Zhang Yin, Zhan Ran. (2013). Soft sensing method for algae propagation condition based on genetic algorithm optimization neural network. Shanghai: Shanghai Maritime University.

Automotive, Mechanical and Electrical Engineering – Liu (Ed.)
© 2017 Taylor & Francis Group, London, ISBN 978-1-138-62951-6

Stability control algorithm for unmanned aerial vehicle based on dynamic feedback search

Xinghua Lu & Jianbo Ke
Huali College Guangdong University of Technology, Guangzhou, Zengcheng, P.R. China

ABSTRACT: The optimal control design of Unmanned Aerial Vehicles (UAV) is an important technology to ensure flight stability. UAV has sharp attitude variation and the airflow, causing flight stability is not good. The dynamic feedback search on flight longitudinal plane can improve the stability. A stability control algorithm for UAV is proposed based on dynamic feedback search. Firstly, a longitudinal motion mathematical model is constructed of UAV. Then, the control objective function and control constraint parameters are analysed, and an adaptive inversion pitch angle tracking method is used for integral stability functional analysis. The angular velocity is taken as the virtual control input. Adaptive dynamic feedback error tracking and search is realized in greatly attitude change. A PID adaptive control law is designed, to eliminate the effect of swinging inertia force and moment on the stability of the system. The control algorithm optimisation design is created. Finally, the simulation test is performed. The simulation results show that the stability of control is optimised, and that pose tracking and error compensation are obtained effectively. It can improve the stability during the flight, and the attitude angle tracking error can converge to zero rapidly.

Keywords: dynamic feedback; search; Unmanned Aerial Vehicle (UAV); control; self adaptation

1 INTRODUCTION

Unmanned aircraft is referred to as Unmanned Aerial Vehicle (UAV). It is an aircraft which uses radio remote control equipment and a self-contained program control device for remote command control. UAV is a new type of aircraft and has higher application value in the field of military and civilian areas. In military application areas, it is divided into man-machine reconnaissance and target. It can carry out air strikes and intelligence gathering tools. In civilian areas, it can be used as a tool for disaster relief, forest fire detection, search field, etc. Attitude control system design is one of the key technologies of unmanned aerial vehicles, and the design of attitude control system for unmanned aerial vehicle is essentially the stability control design problem of aircrafts (Li, T. et al., 2010; Mahmoud, E. E., 2014; Zhao, S. et al., 2015).

At present, the UAV attitude control design is widely used in fixed controller and gain-scheduling (gain-scheduling) control schemes. The design process is divided into three steps: the first step is to change the object model, select different feature points in the whole flight segment and get linear of corresponding points. The second step is to design specific points of the controller. The last step is to fix the controller structure interpolation processing

parameters of the controller, to achieve control of the whole flight (Ming, P. et al., 2016; Palomares, I. et al., 2014; Dong, B. et al., 2016). The difficulty is the design of the fixed point controller. In recent years, along with the rapid development of UAV design and control technology, the UAV intelligent control and limit control requirements are higher. Although the system design of UAV attitude control problem is a fairly mature research field, in fact there are still many problems. Mainly in that the design method of fixed controller is relatively old, using the frequency method of classical control theory, which is required to have a rich experience in the design and debug off the controller parameters repeatedly (Lu, X. et al., 2015; Lizhi, G. E., 2015). Optimal design of unmanned aerial vehicles is an important technology to ensure flight stability, unmanned aircraft by sharp attitude variation and the airflow, causing flight stability is not good, through dynamic feedback search on flight longitudinal plane, and it can improve the stability of UAV. Aiming at the drawbacks existing in the design of the flight control, this paper proposes an unmanned vehicle stability control algorithm based on dynamic feedback search. The longitudinal motion mathematical model is constructed first, the constraint parameters analysis is taken, the PID neural network model is constructed to design the controller, the angular velocity is used as the virtual control input, and the PID control adaptive law is designed to eliminate the influence of swing inertia force and moment on the stability of the system. The simulation test is carried out, and the effective conclusion is obtained.

2 CONTROL THEORY AND MATHEMATICAL MODEL

2.1 *Analysis of stability control principle of UAV*

The control system of unmanned aerial vehicle is built on the basis of the mathematical model of the motion of the vehicle. In order to facilitate the treatment of the problem, some simplifications are made in the modelling of UAV. (a) The shape of UAV is symmetry on x_1Oy_1 plane. (b) The damping force and hydrodynamic force are linear change of position. (c) The error caused by the machining and installation is neglected. In the control system design of this paper, we mainly use the following coordinate system: speed coordinate system $Ox_3y_3z_3$, body coordinate system $Ox_1y_1z_1$, ballistic coordinate system $Ox_2y_2z_2$, and ground coordinate system $Axyz$.

1. Speed coordinate system $Ox_3y_3z_3$, centroid of unmanned aerial vehicle is the origin of the coordinate system, velocity vector V is selected as Ox_3 axis; Oy_3 axis and Ox_3 axis are vertical, and located in the UAV longitudinal symmetry plane, upward is positive; Oz_3 axis perpendicular to the plane Ox_3y_3, the direction is determined by the right-rule.
2. Body coordinate system $Ox_1y_1z_1$: the body coordinate system is connected to the unmanned aerial vehicle. The centroid UAV is the origin of coordinates, Ox_1 points to the head, and unmanned aircraft longitudinal overlap; Oy_1 upward is positive, in the longitudinal symmetry plane, Oz_1 is determined by the right-hand rule.
3. Ballistic coordinates $Ox_2y_2z_2$: centroid UAV is the coordinate origin; velocity vector V is Ox_2 axis, Oy_2 axis located in velocity vector. The plane contains the velocity vector, Oz_2 axis is determined by right-hand rule.

The standard trajectory of unmanned aerial vehicles is taken as a benchmark, on the basis of goal orientation method of beam split. Calculate the relative standard deviation of unmanned aircraft trajectory. The adaptive backstepping control method is used to get a set of integral description of flight stability of linear differential equations, namely small disturbance equation of aircraft, and unmanned spacecraft attitude dynamics nonlinear equations:

$$\begin{cases} m\dot{V} = -mg\sin\theta - c_x qS_M + P \\ mV\dot{\theta} = -mg\cos\theta + c_y^\alpha qS_M\alpha + P(\alpha + \delta_\varphi) + m_R l_R\ddot{\delta}_\varphi \\ J_{z1}\ddot{\varphi} = -c_{y1}^\alpha qS_M(x_g - x_T)\alpha - qS_M m_{dz} l_k^2\dot{\varphi}/V \\ \quad -P(x_R - x_T)\delta_\varphi - m_R\dot{W}_{x1}l_R\delta_\varphi - m_R l_R\ddot{\delta}_\varphi(x_R - x_T) - J_R\ddot{\delta}_\varphi \end{cases}$$
(1)

Then, in the space of array manifold, the equations are linearised. Select the input state variables for flight control $x = [\varphi, \dot{\varphi}, \theta]^T$. Constructing a fitting relationship, the attitude dynamics equations were obtained $\dot{x} = f(x, u)$. The standard trajectory state x_0 ($x_0 = [\varphi_0, \dot{\varphi}_0, \theta_0]^T$) is the balance state of the equation. According to the subspace fitting, the balance conditions of the flight stability is obtained as $f(x_0, u_0) = 0$. The longitudinal motion state vector is $x = [\varphi_0 + \Delta\varphi, \dot{\varphi}_0 + \Delta\dot{\varphi}, \theta_0 + \Delta\theta]^T$, $\delta_\varphi = \Delta\delta_\varphi$. The linearised perturbation equations for the stability control of unmanned aerial vehicles are:

$$\begin{cases} mV\Delta\dot{\theta} = (c_y^\alpha qS_M + P)\Delta\alpha + mg\sin\theta\Delta\theta + P\Delta\delta_\varphi \\ \quad + m_R l_R\Delta\ddot{\delta}_\varphi + F_{gr} \\ J_{z1}\Delta\ddot{\varphi} = -c_{y1}^\alpha qS_M(x_g - x_T)\Delta\alpha - qS_M m_{dz}l_k^2\Delta\dot{\varphi}/V \\ \quad - P(x_R - x_T)\Delta\delta_\varphi - m_R\dot{W}_{x1}l_R\Delta\delta_\varphi \\ \quad - m_R l_R\Delta\ddot{\delta}_\varphi(x_R - x_T) - J_R\Delta\ddot{\delta}_\varphi + M_{gr} \end{cases}$$
(2)

Finally, the weighted subspace fitting linear equation is obtained on the flight's longitudinal plane:

$$\begin{cases} \Delta\dot{\theta} = c_1\Delta\alpha + c_2\Delta\theta + c_3\Delta\delta_\varphi + c_3''\Delta\ddot{\delta}_\varphi + \bar{F}_{gr} \\ \Delta\ddot{\varphi} + b_1\Delta\dot{\varphi} + b_2\Delta\alpha + b_3\Delta\delta_\varphi + b_3''\Delta\ddot{\delta}_\varphi = \bar{M}_{gr} \\ \Delta\varphi = \Delta\theta + \Delta\alpha \end{cases}$$
(3)

Where:

$$c_1 = \frac{1}{mV}(57.3c_y^\alpha qS_M + P) \quad [\text{s}^{-1}], \ c_y^\alpha \ 1/\text{Degree}$$

$$c_2 = \frac{1}{V}g\sin\theta \quad c_3 = \frac{P}{mV}$$

$$c_3'' = \frac{m_R l_R}{mV} \ b_1 = \frac{57.3}{J_{z1}V}m_{dz}qS_M l_k^2$$

$$b_2 = \frac{57.3}{J_{z1}}c_{y1}^\alpha qS_M(x_g - x_T)$$

$$b_3 = \frac{1}{J_{z1}}(P(x_R - x_T) + m_R\dot{W}_{x1}l_R)$$

$$b_3'' = \frac{1}{J_{z1}}(m_R l_R(x_R - x_T) + J_R)$$

According to the above control principle, the control objective function and control parameters are analysed, and then the optimal control algorithm is designed.

2.2 *Mathematical model of longitudinal motion*

Ballistic unmanned aircraft is basically flying in the plane of fire. The airframe rolling angle and angular velocity are smaller. There is larger coupling between the pitch and yaw motion. The attitude motion equation of UAV can be decomposed into longitudinal motion, lateral motion and roll motion of three independent equations. Unmanned aircraft in pitch plane motion is called longitudinal motion; describing the longitudinal motion variables are $\varphi, \dot{\varphi}, \alpha, \theta, \delta_\varphi$, etc., which focus on the longitudinal motion in the plane of the flight vehicle stability control, under small perturbations. The flight pitch motion equation is constructed in the longitudinal motion. The motion equations of an unmanned aircraft for three separate channels are described as:

Pitch:

$$\begin{cases} mV\dot{\theta}\cos(\sigma) = F_y \\ J_z\dot{\omega}_{z1} + (J_y - J_x)\omega_{x1}\omega_{y1} = M_{z1} \\ \varphi = \theta + \alpha \end{cases} \quad (4)$$

Yaw:

$$\begin{cases} -mV\dot{\sigma} = F_z \\ J_y\omega_{y1} + (J_x - J_z)\omega_{z1}\omega_{x1} = M_{y1} \\ \phi = \sigma + \beta \end{cases} \quad (5)$$

Roll:

$$J_x\dot{\omega}_{x1} + (J_z - J_y)\omega_{y1}\omega_{z1} = M_{x1} \quad (6)$$

On the basis of this, perturbation linearisation processing is performed, and the three channel model of unmanned aerial vehicle can be expressed as:

$$\begin{cases} \ddot{\varphi}_a = -(b_1 + \Delta b_1)\dot{\varphi}_a - (b_2 + \Delta b_2)\varphi_a - (b_3 + \Delta b_3)\delta_\varphi + fd_1 \\ \ddot{\psi}_a = -(b_1 + \Delta b_1)\dot{\psi}_a - (b_2 + \Delta b_2)\psi_a - (b_3 + \Delta b_3)\delta_\psi + fd_2 \\ \ddot{\gamma} = -(d_3 + \Delta d_3)\delta_\gamma + fd_3 \end{cases} \quad (7)$$

where, $\varphi_a, \psi_a, \gamma$ are attitude angle of unmanned aerial vehicle, $\dot{\varphi}_a, \dot{\psi}_a, \dot{\gamma}$ are attitude angular velocity of unmanned aerial vehicle, $\ddot{\varphi}_a, \ddot{\psi}_a, \ddot{\gamma}$ are acceleration, b_1, b_2, b_2, d_3 are known coefficients of the model, $\Delta b_1, \Delta b_3, \Delta d_3$, are undetermined coefficients, fd_1, fd_2, fd_3 are interference signals, and $\delta_\varphi, \delta_\psi, \delta_\gamma$ are control inputs of pitch, yaw and roll channels. Furthermore, it is simplified as:

$$\begin{cases} \ddot{\varphi}_a = -b_1\dot{\varphi}_a - b_2\varphi_a - b_3\delta_\varphi + \rho_1 \\ \ddot{\psi}_a = -b_1\dot{\psi}_a - b_2\psi_a - b_3\delta_\psi + \rho_2 \\ \ddot{\gamma} = -d_3\delta_\gamma + \rho_3 \end{cases} \quad (8)$$

where, $\rho_1 = -\Delta b_1\dot{\varphi}_a - \Delta b_2\varphi_a - \Delta b_3\delta_\varphi + fd_1, \rho_2 = -\Delta b_1\dot{\psi}_a -\Delta b_2\psi_a - \Delta b_3\delta_\psi + fd_2, \rho_3 = -\Delta d_3\delta_\gamma + fd_3$ are uncertain items. According to the above-mentioned model, the mathematical model of the aircraft's longitudinal motion is built, and the stability control design of the aircraft is carried out.

3 IMPROVED DESIGN AND IMPLEMENTATION OF CONTROL ALGORITHM

3.1 *PID neural network controller*

On the basis of construction of longitudinal motion control mathematical model and parameter constraints analysis, the control algorithm is designed. This paper proposes a dynamic feedback unmanned vehicle stability control algorithm based on a PID neural network. A PID neural network is used as the controller prototype. The basic form of PID structure is shown in Figure 1.

The control model of UAV is a multi-variable controlled model of an m input n output, using an m PID output neuron sub-network cross parallel constitute as a new neural network control model. The input layer control model has 2n same input neurons. The neurons inputs are:

$$\begin{cases} net_{s1}(k) = r_s(k) \\ net_{s2}(k) = y_s(k) \end{cases} \quad (9)$$

The state of neurons is:

$$u_{si}(k) = net_{si}(k) \quad (10)$$

The output of input layer neurons is:

$$x_{si}(k) = \begin{cases} 1, & u_{si}(k) > 1 \\ u_{si}(k), & -1 \le u_{si}(k) \le 1 \\ -1, & u_{si}(k) < -1 \end{cases} \quad (11)$$

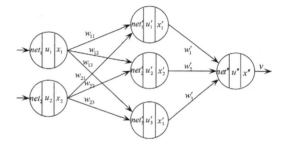

Figure 1. Basic form of PID structure.

Where, $r_s(k)$ is a given value of the longitudinal motion and the lateral movement, $y_s(k)$ is system controlled quantity, s is pitch channel controlled volume serial number ($s = 1,2,...n$); i is input layer number of UAV body posture dynamic ($i = 1,2$).

The input value of the hidden layer of the neural network is:

$$net'_{sj}(k) = \sum_{i=1}^{2} w_{sij} x_{si}(k) \qquad (12)$$

The state of the PID neural network is proportional to:

$$u'_{s1}(k) = net'_{s1}(k) \qquad (13)$$

The state of the PID neural network integral element is:

$$u'_{s2}(k) = u'_{s2}(k-1) + net'_{s2}(k) \qquad (14)$$

The state of the PID neural network is:

$$u'_{s3}(k) = net'_{s3}(k) - net'_{s3}(k-1) \qquad (15)$$

The output of each neuron in the hidden layer is obtained by calculating the longitudinal motion parameters and the transverse rudder angle:

$$x'_{sj}(k) = \begin{cases} 1, & u'_{sj}(k) > 1 \\ u'_{sj}(k), & -1 \le u'_{sj}(k) \le 1 \\ -1, & u'_{sj}(k) < -1 \end{cases} \qquad (16)$$

The output layer of the controller is obtained by using the dynamic feedback search and the inversion integral control:

$$net''_h(k) = \sum_{s=1}^{n} \sum_{j=1}^{3} w'_{sjh} x'_{sj}(k) \qquad (17)$$

Output quantity x_1, x_2,..., x_n respectively by weight value w_{1j}, w_{2j},..., w_{nj}, and the state of the output neuron of the controller is obtained:

$$u'''_h(k) = net''_h(k) \qquad (18)$$

Under disturbance, the output of the PID output layer neuron is:

$$x''_h(k) = \begin{cases} 1, & u''_h(k) > 1 \\ u''_h(k), & -1 \le u''_h(k) \le 1 \\ -1, & u''_h(k) < -1 \end{cases} \qquad (19)$$

where,

$$v_h(k) = x''_h(k) \qquad (20)$$

By using PID neural network control method, the adaptive dynamic feedback error tracking and searching are taken.

3.2 Implementation of adaptive control law for PID control

The integral stability of adaptive backstepping tracking of pitch angle stability is used for functional analysis, angular velocity is taken as the virtual control input, network weights are adjusted online, and learning step size is set as η. After n step training and learning, the iterative equation of the pitch angle tracking of adaptive inversion integral control is:

$$W(n+1) = W(n) - \eta \frac{\partial E}{\partial W} + \partial \Delta W(n) \qquad (21)$$

Choosing appropriate step size and initial value, the control system with error back propagation constraint is obtained. The weight of the hidden layer to the output layer is changed into:

$$\begin{aligned} \frac{\partial E}{\partial w'_j} &= \frac{\partial E}{\partial v} \cdot \frac{\partial v}{\partial w'_j} = -\frac{2}{m} \sum_{k=1}^{m} [r(k) - y(k)] \frac{\partial y}{\partial v} x'_j(k) \\ &= -\frac{2}{m} \sum_{k=1}^{m} [r(k) - y(k)] \frac{\partial y_E}{\partial v} x'_j(k) \end{aligned} \qquad (22)$$

Taking the pitch channel as an example, the weight of the input layer to the hidden layer is changed into the value of the input layer to the hidden layer in the weight value learning step of each layer:

$$\begin{aligned} \frac{\partial E}{\partial w_{ij}} &= \frac{\partial E}{\partial v} \frac{\partial v}{\partial x''} \frac{\partial x''}{\partial u''} \frac{\partial u''}{\partial I''} \frac{\partial I''}{\partial x'_j} \frac{\partial x'_j}{\partial u'_j} \frac{\partial u'_j}{\partial I'_j} \frac{\partial I'_j}{\partial w_{ij}} \\ &= -\frac{1}{m} \sum_{k=1}^{m} \delta''(k) w'_j \, \mathrm{sgn} \frac{u'_j(k) - u'_j(k-1)}{I'_j(k) - I'_j(k-1)} x_i(k) \\ &= -\frac{1}{m} \sum_{k=1}^{m} \delta_j(k) x_i(k) \end{aligned} \qquad (23)$$

The stability conditions of the PID control are: initial stability and appropriate learning step size. According to the principle of Lyapunov stability, as long as the step stability is in the control range, by adjusting the dynamic weights, the input sequence of neural network control system is $r(k) \in R^n$ in long-term incentive R^n, and meet the learning step η:

$$0 < \eta < \frac{1}{\varepsilon^2} \qquad (24)$$

where, $\varepsilon = \frac{1}{2\sqrt{E}} \frac{\partial E}{\partial w}$, the UAV is axisymmetric, pitching motion equation and small disturbance equation of structure is exactly the same, small disturbance yaw motion can be used to adjust the weights of the adaptive gradient algorithm for training. To get the weight controller hidden layer to output layer value adjustment algorithm:

$$w'_{sjh}(n_0+1) = w'(n_0) - \eta'_{sjh} \frac{\partial J}{\partial w'_{sjh}} \qquad (25)$$

Variable structure neural network control of learning step size η_{sjh} must be satisfied:

$$0 < \eta_{sjh} < \frac{1}{\varepsilon_{sjh}^2} \qquad (26)$$

Where,

$$\varepsilon_{sjh} = -\frac{\sum\limits_{p=1}^{n}\sum\limits_{k=1}^{l} \delta'_{hp}(k)x'_{sj}(k)}{2\sqrt{l\sum\limits_{p=1}^{n}\sum\limits_{k=1}^{l}[r_p(k)-y_p(k)]^2}} \qquad (27)$$

In the rolling motion, according to the Lyapunov function, the parameter self-tuning control method is used to train the appropriate weight, and the weight of the input layer to the hidden layer is adjusted as:

$$w_{sij}(n_0+1) = w_{sij}(n_0) - \eta_{sij} \frac{\partial J}{\partial w_{sij}} \qquad (28)$$

At this point, the control system of UAV includes PID Neural Network Controller (PID-NNC) and PID neural network identifier, and η_{sij} of the weight of PID neural network identifier must be satisfied:

$$0 < \eta_{sij} < \frac{1}{\varepsilon_{sij}^2} \qquad (29)$$

where,

$$\varepsilon_{sij} = -\frac{\sum\limits_{p=1}^{n}\sum\limits_{h=1}^{m}\sum\limits_{k=1}^{l} \delta_{sjh}(k)x_{si}(k)}{2\sqrt{l\sum\limits_{p=1}^{n}\sum\limits_{k=1}^{l}[r_p(k)-y_p(k)]^2}} \qquad (30)$$

Through the above processing, the PID control adaptive law is designed, the influence of the swing inertia force and torque on the stability of the system is eliminated, and the control algorithm is optimized.

4 SIMULATION ANALYSIS

In order to test the performance of the control algorithm in improving the stability control of an unmanned aerial vehicle, the simulation experiment is carried out. The simulation platform is based on MATLAB Simulink, which is shown in Figure 2.

The simulation platform consists of seven modules. 1) unmanned aircraft weight and centre of gravity and moment of inertia weight parameter by inertia moment input module; 2) aerodynamic parameter module to establish the aerodynamic derivatives database; 3) the equation of six degrees of freedom module is used to solve the equation of kinematics and dynamics; 4) control commands through the navigation control module based on the comparison of guidance the law and the actual

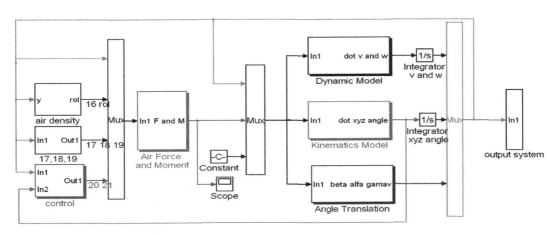

Figure 2.   Unmanned aerial vehicle control simulation platform based on MATLAB Simulink.

trajectory of energy management is calculated; 5) flight height, distance closer to the earth and the latitude and longitude information through the geophysical model calculation module; 6) calculation of different height under atmospheric density, pressure, temperature and speed through the gas module; 7) the simulation system also includes the interference of wind.

The pitch channel is taken as an example. The initial pitch angle φ is assumed 0°, the value of the angle tracking is 2.8° step signal; the flight height, distance closer to the earth and the latitude and longitude information are calculated. The pitch angle simulation results with different control methods is shown in Figure 3.

It can be seen from the diagram, using this method for UAV control, pitch angle can fast track the given reference value, and pitch angle tracking error converges to zero. The control performance is superior to traditional methods. Further, compari-

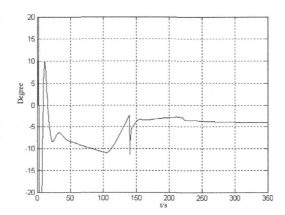

Figure 5.   Rudder angle δ_φ with traditional method.

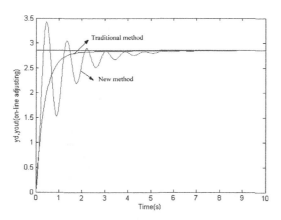

Figure 3.   Tracking performance of pitch angle parameters using different control methods.

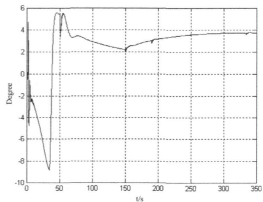

Figure 6.   Attack angle α with new method.

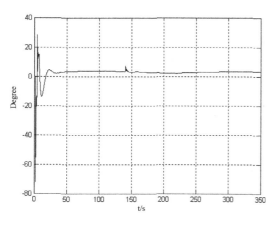

Figure 4.   Attack angle α with traditional method.

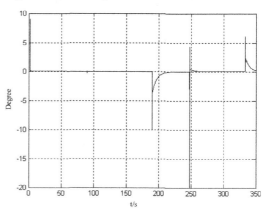

Figure 7.   Rudder angle δ_φ with new method.

son of other control parameters results are shown in Figures 4–7.

The simulation results show that adjusting time of unmanned aircraft flight attitude parameters is shorter, it has smaller oscillation, and the distortion can be effectively suppressed, it has small overshoot, with fast dynamic feedback searching of the flight parameters, and the fast and stable control of UAV is achieved.

5 CONCLUSION

In order to improve the stability of the unmanned aerial vehicle, a stability control algorithm for UAV is proposed based on dynamic feedback search. Firstly, a longitudinal motion mathematical model is constructed of UAV. The control objective function and control constraint parameters are analysed, adaptive inversion pitch angle tracking method is used for integral stability functional analysis, the angular velocity is taken as the virtual control input, adaptive dynamic feedback error tracking and search is realised in greater attitude change, PID adaptive control law is designed, eliminating the effect of swinging inertia force and moment on the stability of the system, and the control algorithm optimisation design is realised. Finally, the simulation test is performed The simulation results show that the stability of control is optimised, and pose tracking and error compensation are obtained effectively. It can improve the stability during the flight, the attitude angle tracking error can converge to zero rapidly, and it has good control performance and control quality.

ACKNOWLEDGMENTS

This project is supported by Foundation for Distinguished Young Talents (Natural Science, 2015K QNCX218) of 2015 Key Platforms and Research Projects of Department of Education of Guangdong Province and 2016 Undergraduate Scientific and Technological Innovation Project Fund of Guangdong Province (pdjh2016b0940).

REFERENCES

Dong, B., Liu, K. & Li, Y. (2016). Decentralized integral sliding mode control for time varying constrained modular and reconfigurable robot based on harmonic drive transmission. *Control and Decision, 31*(03), 441–447.

Ge, L. (2015). Visual simulation of UUV attack model based on whole trajectory control analysis. *Ship Electronic Engineering, 35*(3), 137–141.

Li, T. & Zhang, J. (2010) Consensus conditions of multi-agent systems with time-varying topologies and stochastic communication noises. *IEEE Transactions on Automatic Control, 55*(9), 2043–2057.

Lu, X. & Chen P. (2015). Traffic prediction algorithm in buffer based on recurrence quantification union entropy feature reconstruction. *Computer Science, 42*(4), 68–71.

Mahmoud, E. E. (2016). Complex complete synchronization of two nonidentical hyperchaotic complex nonlinear systems. *Mathematical Methods in the Applied Sciences, 37*(3), 321–328.

Ming, P. & Liu, J. (2016). Consensus stability analysis of stochastic multi-agent systems. *Control and Decision, 31*(03), 385–393.

Palomares, I., Martinez, L. & Herrera, F. (2014). A consensus model to detect and manage non-cooperative behaviors in large scale group decision making. *IEEE Transactions on Fuzzy Systems, 22*(3), 516–530.

Zhao, S., Guo, H. & Liu, Y. (2015). Fault tolerant control for linear time-delay system based on trajectory tracking. *Information and Control, 44*(4), 469–473.

Automotive, Mechanical and Electrical Engineering – Liu (Ed.)
© 2017 Taylor & Francis Group, London, ISBN 978-1-138-62951-6

Study of MA and MACD directional expert systems based on funds management

Fang Yu, Haiping Huang & Pin Wang
Mathematics and Information Science Department, Guangxi College of Education, Nanning, Guangxi, China

ABSTRACT: Using the method of mathematical modelling, this study takes authentic public securities data as the basis to test the MA expert system and MACD expert system. By taking annual net profit margin, rate of return and winning rate as the management goal, the two directional indexes are calculated and analysed, thus obtaining the MA expert system and MACD expert system. Compared with the winning rate of one-time buy-in of all funds, step-by-step buy-in with 30% funds can merely increase 1% and 6.31% respectively. However, the annual rate of return and net profit margin drop significantly to 26.36% and 50.93%, and the number of annual transactions is reduced by 762.4 and 758.4 times. According to the market implication of this calculation result, the method of funds management is meaningless for MA expert system and MACD expert system.

Keywords: Funds management; MA; MACD; Expert system

1 INTRODUCTION

When carrying out technical analysis about investment in stocks, futures and precious metals, directional indexes such as Moving Average (MA) and Moving Average Convergence Divergence (MACD) are commonly used. The basic investment principle states that there is no index or method whose success rate is 100%. Thus, how can we increase the yield rate, avoid investment risks, and prevent uncertainties when analysing indexes? Funds management, also known as risk management, is a common method adopted by many investors.

For investors who expect to survive in the investment market for the long run, especially those who intend to make a living by transactions, funds management is also a critical part in addition to excellent transaction skills. To a certain extent, funds management is even more important than transaction skills and technical analysis. Positive funds management policies may not help you gain profits immediately, but it enables your funds to increase steadily by accumulating long-term profits.

By taking A-share data from the Shanghai stock market of China from 26 August 2015 to 25 November 2016 as the basis, this paper attempts to test relevant data with the method of mathematical modelling by taking the winning rate, annual rate of return and net profit margin as the management goal, thus conducting comparative analysis of the MA system and the MACD system. The advantages and disadvantages of the two systems, as well as the optimal usage, will be examined in order to provide an optimised investment scheme for investors.

A review of the relevant literature on the expert systems MA and MACD has not been carried out due to the limitations of the paper length.

The MACD system was proposed by Geral Appel in 1979, and is calculated by the difference between the fast and slow exponential moving average (EMA). "Fast" refers to the EMA in a short period, and "slow" refers to the EMA in a long-term period. The EMA of the 12th and the 26th is applied for most.

The mathematical formula of MACD is as follows:

$$\text{MACD} = 12\text{-daysEMA} - 26\text{-daysEMA} \tag{1}$$

The mathematical formula of the EMA index is as follows:

$$\text{EMA}_{today} = \frac{p_1 + (1-\alpha)p_2 + (1-\alpha)^2 p_3 + (1-\alpha)^3 p_4 + \cdots}{1 + (1-\alpha) + (1-\alpha)^2 + (1-\alpha)^3 + \cdots} \tag{2}$$

$$\alpha = \frac{2}{N+1} \quad \text{N represents the periodicity} \tag{3}$$

where $p_i(i = 1,2,\ldots,n)$ refers to the closing price of day i, and n refers to the moving average periodicity.

MA is based on the theory of "average cost concept" by Dow • Jones with the adoption of the "moving average" principle in statistics to connect the average value of the stock price in a period to the curve for display of the historical volatility of the share price, so as to indicate the technical analysis method for future development trend of the share price index. Hence it can be seen as an appropriate representation of the Dow Theory.

The mathematical formula of MA is as follows:

$$MA = \frac{c_1 + c_2 + \cdots\cdots + c_n}{n} \qquad (4)$$

Where $c_i = (1,2,\ldots,n)$ refers to the closing price of day i, and n refers to the moving average periodicity.

2 EMPIRICAL ANALYSIS FOR MACD AND MA EXPERT SYSTEMS

2.1 *Experiment and result*

1. Experimental procedure:
The development of the MA expert system is based on the eight laws of Granville (Wang Yumei, 1998), whose transaction law shows the above connection of the MA (5) line and MA (10) line to form the golden cross for buying, and the below breaking of the MA (10) line and MA (30) line to form the death cross for selling (see Figure 2).

Source code of MA expert system:
SHORT 1 30 5
LONG 5 100 30
CROSS (MA (CLOSE, SHORT), MA(CLOSE, LONG))
CROSS (MA (CLOSE, LONG), MA (CLOSE, SHORT))
ENTERLONG: CROSS (MA (CLOSE, SHORT), MA (CLOSE, LONG));
EXITLONG: CROSS (MA (CLOSE, LONG), MA (CLOSE, SHORT))

The development of the MACD expert system is based on the law of Geral Appel (Cao Yubo, 2010), whose transaction law shows the above connection of the EMA lower line on the 12th and the buying of it on the 26th, and the below breaking of the EMA higher line on the 12th and the selling of it on the 26th (see Figure 1).

Source code of MACD expert system:
LONG 10 200 26
SHORT 2 200 12
M 2 200 9
DIFF: = EMA (CLOSE, SHORT)- EMA (CLOSE, LONG);
DEA: = EMA (DIFF, M);
MACD: = 2*(DIFF-DEA);
ENTERLONG: CROSS(MACD,0);
EXITLONG: CROSS(0, MACD);

2. Experimental platform: the great wisdom securities information platform of V5.99 version.
3. Experimental parameters: one-time position opening or entire position closing to meet the

capital conditions, 0.5% of the transaction costs shall be taken.
4. Experimental samples: all A-share daily-line data for the Shanghai stock market (26 August 2015 – 24 November 2016).
5. Experimental process, time and results:

Figure 1.   MACD expert system.

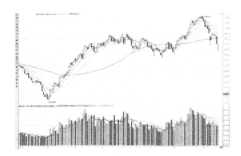

Figure 2.   MA expert system.

Table 1.   The first test result of MA expert system.

System test setting
Test method: Technical index – MA
Test time: 26/08/2015 – 24/11/2016   Calculating the forced liquidation
Test share: 1,111 shares in total   Initial input: 40,000.00 yuan
Buy-in conditions:
One of the following groups is tenable:
1. The following conditions are tenable simultaneously
1.1 Technical index: MA (5, 10, 20, 30, 120, 250) index line MA2 cross MA3 [daily line]
When the condition is satisfied, according to the middle market value, the closing price is bought in with all funds
In case of continuous signal: no more buy-in
Sell-out condition: no sell-out condition
Conditions for closing a position: (according to the closing price)
Share selection by indexes: technical index: MA (5, 10, 20, 30, 120, 250) index line MA2 dncross MA3 [daily line]
System test report
Tested stock number: 1,111
Net profit: 9,186,380.00 yuan   Net profit margin: 20.67%

(*Continued*)

Table 1. (*Continued*)
Total profits: 18,482,934.00 yuan   Total losses: -9,296,557.00 yuan
Transaction times: 7,947   Winning rate: 46.67%
Annual transaction times: 6,357.60   Profit/ loss transaction times: 3,709/4,238
Total transaction amount: 325,691,264.00 yuan   Transaction fee: 242,536.97 yuan
Highest one-time profit: 134,826.13 yuan   Highest one-time loss: -24,304.87 yuan
Average profit: 2,325.77 yuan   Average loss: -1,169.82 yuan
Average profit: 1,155.96 yuan   Average profit / average loss: -198.81
Maximum continuous profit times: 9   Maximum continuous loss times: 11
Average transaction periodicity: 18.90
Average transaction periodicity with profits: 26.99   Average transaction periodicity with losses: 11.82
Profit coefficient: 0.33
Maximum floating profits: 53,597,300.00 yuan   Maximum floating losses: 0.00 yuan
Maximum difference between floating profits and losses: 53,597,300.00 yuan
Total input: 44,440,000.00 yuan
-----------------------------Statistics of buy-in signals------------------ -------------------
(Make statistics of the situation of all buy-in signal points, excluding signal deletion caused by funds and strategies in the transaction test)
Success rate: 46.67%
Signal number: 7,948   Annual average signal number: 6,358.40

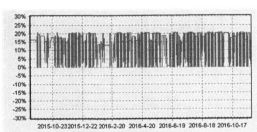

Figure 3.   MA earnings curve of one-time buy-in with all funds.

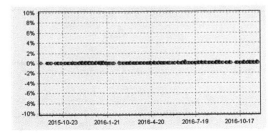

Figure 4.   MA buy-in signal distribution of one-time buy-in with all funds.

Table 2.   The second test result of MA expert system.

System test setting
Test method: Technical index – MA
Test time: 26/08/2015 – 24/11/2016   Calculating the forced liquidation
Test share: 1,111 shares in total   Initial input: 40,000.00 yuan
Buy-in conditions:
One of the following groups is tenable:
1. The following conditions are tenable simultaneously
1.1 Technical index: MA (5, 10, 20, 30, 120, 250) index line MA2 cross MA3 [daily line]
When the condition is satisfied, according to the middle market value, the closing price is bought in with 30.00% funds
In case of continuous signal: buy-in on an equal basis
Sell-out condition: no sell-out condition
Conditions for closing a position: (according to the closing price)
Share selection by indexes: technical index: MA (5, 10, 20, 30, 120, 250) index line MA2 dncross MA3 [daily line]
System test report
Tested stock number: 1,111
Net profit: 2,420,980.00 yuan   Net profit margin: 5.45%
Total profits: 4,405,408.00 yuan   Total losses: -1,984,437.00 yuan
Transaction times: 6,994   Winning rate: 47.61%
Annual transaction times: 5,595.20   Profit/ loss transaction times: 3,330/3,664
Total transaction amount: 74,004,224.00 yuan   Transaction fee: 146,465.08 yuan
Highest one-time profit: 28,088.78 yuan   Highest one-time loss: -4,340.15 yuan
Average profit: 629.88 yuan   Average loss: -283.73 yuan
Average profit: 346.15 yuan   Average profit / average loss: -222.00
Maximum continuous profit times: 9   Maximum continuous loss times: 11
Average transaction periodicity: 19.14
Average transaction periodicity with profits: 27.20   Average transaction periodicity with losses: 11.82
Profit coefficient: 0.38
Maximum floating profits: 46,822,580.00 yuan   Maximum floating losses: 0.00 yuan
Maximum difference between floating profits and losses: 46,822,580.00 yuan
Total input: 44,440,000.00 yuan
-----------------------------Statistics of buy-in signals------------------ -------------------
(Make statistics of the situation of all buy-in signal points, excluding signal deletion caused by funds and strategies in the transaction test)
Success rate: 47.71%
Signal number: 7,067   Annual average signal number: 5,653.60

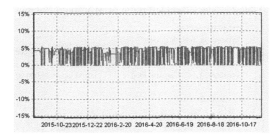

Figure 5. MA earnings curve of step-by-step buy-in with 30% funds.

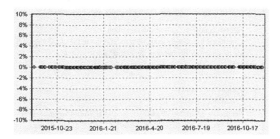

Figure 6. MA buy-in signal distribution of step-by-step buy-in with 30% funds.

Table 3. The first test result of MACD expert system.

System test setting
Test method: Share selection formula – MACD buy-in
Test time: 26/08/2015 – 24/11/2016   Calculating the forced liquidation
Test share: 1,111 shares in total   Initial input: 40,000.00 yuan
Buy-in conditions:
One of the following groups is tenable:
1. The following conditions are tenable simultaneously
1.1 Share selection formula: MACD buy-in (26, 12, 9) [daily line]
When the condition is satisfied, according to the middle market value, the closing price is bought in with all funds
In case of continuous signal: no more buy-in
Sell-out condition: no sell-out condition
Conditions for closing a position (according to the closing price):
Share selection by indexes: share selection formula: MACD sell-out (12, 26, 9) [daily line]
System test report
Tested stock number: 1,111
Net profit: 8,555,344.00 yuan   Net profit margin: 19.25%
Total profits: 17,263,908.00 yuan   Total losses: -8,708,693.00 yuan
Transaction times: 5,303   Winning rate: 55.31%

(Continued)

Table 3. (*Continued*)

Annual transaction times: 4,242.40   Profit/ loss transaction times: 2,933/2,370
Total transaction amount: 217,884,224.00 yuan   Transaction fee: 160,373.05 yuan
Highest one-time profit: 869,525.69 yuan Highest one-time loss: -40,317.07 yuan
Average profit: 3,255.50 yuan   Average loss: -1,642.22 yuan
Average profit: 1,613.30 yuan   Average profit / average loss: -198.24
Maximum continuous profit times: 12   Maximum continuous loss times: 11
Average transaction periodicity: 26.76
Average transaction periodicity with profits: 30.71   Average transaction periodicity with losses: 21.86
Profit coefficient: 0.33
Maximum floating profits: 52,969,072.00 yuan   Maximum floating losses: 0.00 yuan
Maximum difference between floating profits and losses: 52,969,072.00 yuan
Total input: 44,440,000.00 yuan
------------------------------Statistics of buy-in signals-----------------------------------
(Make statistics of the situation of all buy-in signal points, excluding signal deletion caused by funds and strategies in the transaction test
Success rate: 55.31%
Signal number: 5,303   Annual average signal number: 4,242.40

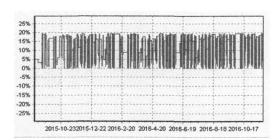

Figure 7. MACD earnings curve of one-time buy-in with all funds.

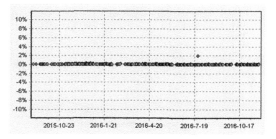

Figure 8. MACD buy-in signal distribution of one-time buy-in with all funds.

Table 4. The second test result of MACD expert system.

System test setting

Test method: Share selection formula – MACD buy-in

Test time: 26/08/2015 – 24/11/2016   Calculating the forced liquidation

Test share: 1,111 shares in total   Initial input: 40,000.00 yuan

Buy-in conditions:

One of the following groups is tenable:

1. The following conditions are tenable simultaneously

1.1 Share selection formula – MACD buy-in (26, 12, 9) [daily line]

When the condition is satisfied, according to the middle market value, the closing price is bought in with 30.00% funds

In case of continuous signal: buy-in on equal basis

Sell-out condition: no sell-out condition

Conditions for closing a position (according to the closing price):

Share selection by indexes: share selection formula: MACD sell-out (12, 26, 9) [daily line]

System test report

Tested stock number: 1,111

Net profit: 2,210,740.00 yuan   Net profit margin: 4.97%

Total profits: 4,069,719.25 yuan   Total losses: -1,859,097.63 yuan

Transaction times: 4,716   Winning rate: 55.94%

Annual transaction times: 3,772.80   Profit/ loss transaction times: 2,638/2,078

Total transaction amount: 49,384,82.00 yuan   Transaction fee: 98,921.71 yuan

Highest one-time profit: 114,978.61 yuan   Highest one-time loss: -5,760.25 yuan

Average profit: 862.96 yuan   Average loss: -394.21 yuan

Average profit: 468.77 yuan   Average profit / average loss: -218.91

Maximum continuous profit times: 12   Maximum continuous loss times: 11

Average transaction periodicity: 26.86

Average transaction periodicity with profits: 30.92   Average transaction periodicity with losses: 21.70

Profit coefficient: 0.37

Maximum floating profits: 46,613,628.00 yuan   Maximum floating losses: 0.00 yuan

Maximum difference between floating profits and losses: 46,613,628.00 yuan

Total input: 44,440,000.00 yuan

-----------------------------Statistics of buy-in signals----------

----

(Make statistics of the situation of all buy-in signal points, excluding signal deletion caused by funds and strategies in the transaction test)

Success rate: 55.94%

Signal number: 4,716   Annual average signal number: 3,772.800

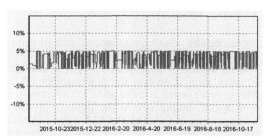

Figure 9. MACD earnings curve of step-by-step buy-in with 30% funds.

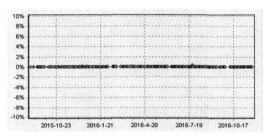

Figure 10. MACD buy-in signal distribution of step-by-step buy-in with 30% funds.

2.2 Comparative analysis of results

Table 5. Comparative analysis table.

	Winning rate	Annual rate of return	Net profit margin	Annual transaction times
One-time buy-in of MA expert system with all funds	46.67	16.54	20.67	6,357.60
Step-by-step buy-in of MA expert system with 30% funds	47.61	4.36	5.45	5,595.20
One-time buy-in of MACD expert system with all funds	67.52	12.27	15.33	7,605.60
Step-by-step buy-in of MACD expert system with 30% funds	73.83	6.25	7.81	6,847.20

3 CONCLUSION

By taking the winning rate, annual rate of return and net profit rate, which most concern investors as a management goal, this paper attempts to calculate and analyse the practicability of the MA expert system and the MACD expert system with a full sample size in the premise of funds management. Regarding the MA expert system, compared with the winning rate of one-time buy-in with all funds, that of step-by-step buy-in with 30% funds can merely increase 1%, but the annual rate of return and net profit margin will drop sharply to 26.36% with a reduction of 762.4 times in the number of annual transactions (see Figures 3 to 6). According to the market implication of this calculation result, step-by-step buy-in is meaningless for the MA system. Regarding the MACD expert system, compared with the winning rate of one-time buy-in with all funds, that of step-by-step buy-in with 30% funds can merely increase 6.31%, but its annual rate of return will see a significant decline to 50.93% with a sharp decrease in net profit margin of 50.95% and a reduction of 758.4 times in the number of annual transactions (see Figures 7 to 10). According to the market implication of this calculation result, step-by-step buy-in is also meaningless for the MACD system.

REFERENCES

Cao Yubo. (2010). Application of MACD index in securities investment, *Manager' Journal* (in Chinese), (4), 166–167.

http://en.wikipedia.org/wiki/Moving_average

http://forex-indicators.net/macd

http://wiki.mbalib.com/wiki/%E5%B9%B3%E6%BB%91%E5%BC%82%E5%90%8C%E7%A7%BB%E5%8A%A8%E5%B9%B3%E5%9D%87%E7%BA%BF

Liu, Y. Empirical study on senior managers and performances in companies of high-tech based on SPSS software regression analysis. Journal of Software, 7(7), 1569–1576.

Wang, Y. (1998). Application of moving average in the stock market, Zhe Jiang *Statistics (in Chinese)*, (2), 27–28.

Automotive, Mechanical and Electrical Engineering – Liu (Ed.)
© 2017 Taylor & Francis Group, London, ISBN 978-1-138-62951-6

Study of RSI and W%R reverse directional expert systems based on funds management

Haiping Huang & Pin Wang

Mathematics and Information Science Department, Guangxi College of Education, Nanning, Guangxi, China

ABSTRACT: Using mathematical modelling, this study takes authentic public securities data as the basis to test the RSI expert system and the W%R expert system. By taking the annual net profit margin, rate of return and winning rate as the management goal, the two reverse directional indexes are calculated and analysed, thus obtaining the RSI expert system and the W%R expert system. Compared with the winning rate of one-time purchase of all funds, that resulting from step-by-step purchase at 30% funds can merely increase 9.29% and 6.31% respectively. However, the annual rate of return and net profit margin drop significantly to 41.85% and 50.93%, and the number of annual transactions reduce by 84 and 758.4 times. According to the market implication of this calculation result, the method of funds management is meaningless for the RSI and W%R expert systems.

Keywords: RSI; W%R; Expert system; Funds management

1 INTRODUCTION

When carrying out technical analysis about investment in stocks, futures and precious metals, reverse directional indexes such as Relative Strength Index (RSI) and Williams%R (W%R) are commonly used. The basic investment principle states that no index or method has a success rate of 100%. Thus, how can we increase the yield rate, avoid investment risks, and prevent uncertainties when analysing indexes? Funds management, also known as risk management, is a common method adopted by many investors.

For investors who expect to survive in the investment market for the long run, especially those who intend to make a living by transactions, funds management is also a critical part in addition to excellent transaction skills. To a certain extent, funds management is even more important than transaction skills and technical analysis. Positive funds management policies may not help you gain profits immediately, but it enables your funds to increase steadily by accumulating long-term profits.

By taking A-share data from the Shanghai stock market of China from 26 August 2015 to 25 November 2016 as the basis, this paper attempts to test relevant data using mathematical modelling by taking the winning rate, annual rate of return and net profit margin as the management goal, thus conducting comparative analysis of the RSI system and the W%R system. The advantages and disadvantages of the two systems, as well as the optimal usage, are examined in order to provide an optimised investment scheme for investors.

A review of the relevant literature on the expert systems RSI and W%R has not been carried out here due to the limitation of the paper length.

The mathematic formula of RSI is:

$$RSI = \frac{100 \times RS}{1 + RS} \tag{1}$$

$$RS = \frac{\text{Average Rise Point in i days}}{\text{Average Dropped Point in i days}} (i = 1, 2, ..., n) \tag{2}$$

The W%R mathematical formula of Williams overbought/oversold (WMS%R) formula is:

$$\%R = \frac{high_{Ndays} - close_{today}}{high_{Ndays} - low_{Ndays}} \times (-100\%) \tag{3}$$

$$(close_{today} - low_{Ndays}) - (close_{today} - high_{Ndays}) = high_{Ndays} - low_{Ndays} \tag{4}$$

$$W\%R = \frac{C_n - H_n}{H_n - L_n} \times 100\% \tag{5}$$

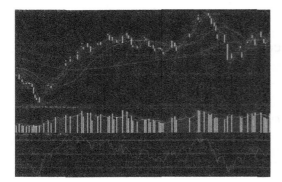

Figure 1. RSI expert system.

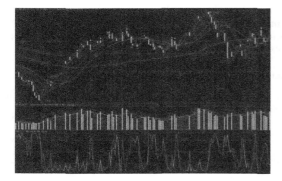

Figure 2. W%R expert system.

where n is the transaction period set by the dealer, C_n is the closing price in n day(s), L_n is the lowest price in n day(s), and H_n is the highest price in n day(s).

2 EMPIRICAL ANALYSIS OF RSI EXPERT SYSTEM AND W%R EXPERT SYSTEM

2.1 The experiment and results

1. Experimental procedure:
   The RSI expert system was developed according to the rule given by Welles Wilder, the transaction rule of which is: buy when $0 < RSI(14) < 20$ or $50 < RSI(14) < 80$, sell when $80 < RSI(14) < 100$, and wait when $20 < RSI(14) < 50$ (Wang Pin, 2012, 2011).

   Source code of the RSI expert system:
   ```
   N1 1 100 14
   LL 0 40 20
   ```

LH 60 100 80
```
LC: = REF(CLOSE,1);
RSI:SMA(MAX(CLOSE-LC,0),N1,1)/
SMA(ABS(CLOSE-LC),N1,1)*100,
colorwhite;
ENTERLONG:CROSS(RSI,LL);
EXITLONG:CROSS(LH,RSI)
```

The W%R expert system is designed in accordance with the Eastern philosophy principle that things will develop in the opposite direction when they become extreme. It is an index to measure overbought and oversold. By making use of the swing point to measure the overbought and oversold phenomenon in the share market, the W%R index can be used to predict high and low points in the circulation period (Li Yongchun, 2013).

When W%R is higher than 80, that is, it is in the oversold state, the stock market quotation is about to reach the bottom line, so buy-in should be taken into consideration.

When W%R is lower than 20, that is, it is in the overbought state, the stock market quotation is about to reach the top line, so sell-out should be taken into consideration.

When W%R reaches a high position, it is generally about to return. If the stock price continues to rise at this time, it will result in deviation, and the sell-out signal is given.

When W%R reaches a low position, it is generally about to rebound. If the stock price continues to drop at this time, it will result in deviation.

If W%R hits the top (bottom) continuously for several times with the local formation of double or multiple items (bottoms), then the sell-out (buy-in) signal is given.

Source code of the W%R expert system:
```
N 2.00 100.00 14.00
LL 0.00 100.00 20.00
LH 0.00 100.00 80.00
WR:=100*(HHV(HIGH,N)-CLOSE)/(HHV
(HIGH,N)-LLV(LOW,N));
ENTERLONG:CROSS(WR,LH);
EXITLONG:CROSS(WR,LL);
```

2. Experiment platform: V5.00 version of DZH securities information platform
3. Experimental parameters: one-time position opening or full position closing for funds meeting the conditions, and the transaction cost is 0.5%
4. Experimental samples: all A-share daily-line data from the Shanghai stock market (26 August 2015–24 November 2016)
5. Experimental process, time and results:

Table 1. The first test result of RSI expert system.

System test setting
Test method: Technical index – RSI (14)
Test time: 26/08/2015 – 24/11/2016 Calculating forced liquidation
Test share: 1,111 shares in total Initial input: 40,000.00 yuan
Buy-in conditions:
One of the following groups is tenable:
1. The following conditions are tenable simultaneously
1.1 Technical index: WP (14) index line WP cross and dncross [daily line]
When the condition is satisfied, according to the middle market value, the closing price is bought in with all funds
In case of continuous signal: no more buy-in
Sell-out condition: no sell-out condition
Conditions for closing a position: (according to the closing price)
Share selection by indexes: technical index: WP (14) index line WP dncross and cross [daily line]
System test report
Tested stock number: 1,111
Net profit: 9,543,976.00 yuan Net profit margin: 21.48%
Total profits: 10,353,684.00 yuan Total losses: -809,816.69 yuan
Transaction times: 1,122 Winning rate: 80.57%
Annual transaction times: 897.60 Profit/ loss transaction times: 904/218
Total transaction amount: 46,369,628.00 yuan Transaction fee: 33,692.84 yuan
Highest one-time profit: 524,012.50 yuan Highest one-time loss: -18,582.60 yuan
Average profit: 9,227.88 yuan Average loss: -721.76 yuan
Average profit: 8,506.22 Average profit / average loss: -1,278.52
Maximum continuous profit times: 41 Maximum continuous loss times: 4
Average transaction periodicity: 103.79
Average transaction periodicity with profits: 90.28 Average transaction periodicity with losses: 159.81
Profit coefficient: 0.85
Maximum floating profits: 53,951,048.00 yuan Maximum floating losses: 0.00 yuan
Maximum difference between floating profits and losses: 53,951,048.00 yuan
Total input: 44,440,000.00 yuan
----------------------------Statistics of buy-in signals------- ----------------------------
(Make statistics of the situation of all buy-in signal points, excluding signal deletion caused by funds and strategies in the transaction test)
Success rate: 78.74%
Signal number: 1,223 Annual average signal number: 978.40

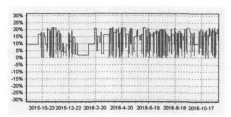

Figure 3. RSI earnings curve of one-time buy-in with all funds.

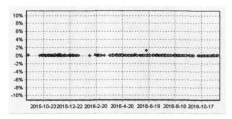

Figure 4. RSI buy-in signal distribution of one-time buy-in with all funds.

Table 2. The second test result of RSI expert system.

System test setting
Test method: Technical index – RSI (14)
Test time: 26/08/2015 – 24/11/2016 Calculating forced liquidation
Test share: 1,111 shares in total Initial input: 40,000.00 yuan
Buy-in conditions:
One of the following groups is tenable:
1. The following conditions are tenable simultaneously
1.1 Technical index: WP (14) index line WP cross and dncross [daily line]
When the condition is satisfied, according to the middle market value, the closing price is bought in with all funds
In case of continuous signal: buy-in on an equal basis
Sell-out condition: no sell-out condition
Conditions for closing a position: (according to the closing price)
Share selection by indexes: technical index: WP (14) index line WP dncross and cross [daily line]
System test report
Tested stock number: 1,111
Net profit: 3,993,072.00 yuan Net profit margin: 8.99%
Total profits: 4,182,184.25 yuan Total losses: -189,127.28 yuan
Transaction times: 1,017 Winning rate: 89.28%

(*Continued*)

Table 2. (*Continued*)

Annual transaction times: 813.60   Profit/ loss transaction times: 908/109

Total transaction amount: 18,416,068.00 yuan   Transaction fee: 38,637.22 yuan

Highest one-time profit: 254,867.52 yuan   Highest one-time loss: -10,933.96 yuan

Average profit: 4,112.28 yuan   Average loss: -185.97 yuan

Average profit: 3,926.32   Average profit / average loss: -2,211.31

Maximum continuous profit times: 76   Maximum continuous loss times: 3

Average transaction periodicity: 102.25

Average transaction periodicity with profits: 94.61   Average transaction periodicity with losses: 165.96

Profit coefficient: 0.91

Maximum floating profits: 48,395,508.00 yuan   Maximum floating losses: 0.00 yuan

Maximum difference between floating profits and losses: 48,395,508.00 yuan

Total input: 44,440,000.00 yuan

-------------------------------Statistics of buy-in signals-------------------------------

(Make statistics of the situation of all buy-in signal points, excluding signal deletion caused by funds and strategies in the transaction test)

Success rate: 82.34%

Signal number: 2,248   Annual average signal number: 1,798.40

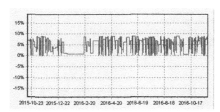

Figure 5. RSI earnings curve of step-by-step buy-in with 30% funds.

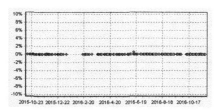

Figure 6. RSI buy-in signal distribution of step-by-step buy-in with 30% funds.

Table 3. The first test result of W%R expert system.

System test setting

Test method: Technical index – RSI (14)

Test time: 26/08/2015 – 24/11/2016   Calculating forced liquidation

(*Continued*)

Table 3. (*Continued*)

Test share: 1,111 shares in total   Initial input: 40,000.00 yuan

Buy-in conditions:

One of the following groups is tenable:

1. The following conditions are tenable simultaneously

1.1 Technical index: W&R (10,6) index line WR1 larger than 80 [daily line]

When the condition is satisfied, according to the middle market value, the closing price is bought in with all funds

In case of continuous signal: no more buy-in

Sell-out condition: no sell-out condition

Conditions for closing a position: (according to the closing price)

Share selection by indexes: technical index: W&R (10,6) index line WR1 smaller than 20 [daily line]

System test report

Tested stock number: 1,111

Net profit: 6,814,664.00 yuan   Net profit margin: 15.33%

Total profits: 19,752,260.00 yuan   Total losses: -12,937,755.00 yuan

Transaction times: 9,507   Winning rate: 67.52%

Annual transaction times: 7,605.60   Profit/ loss transaction times: 6,419/3,088

Total transaction amount: 400,031,968.00 yuan   Transaction fee: 303,485.59 yuan

Highest one-time profit: 88,710.94 yuan   Highest one-time loss: -59,017.63 yuan

Average profit: 2,077.65 yuan   Average loss: -1,360.87 yuan

Average profit: 716.80   Average profit / average loss: -152.67

Maximum continuous profit times: 17   Maximum continuous loss times: 7

Average transaction periodicity: 13.84

Average transaction periodicity with profits: 10.12   Average transaction periodicity with losses: 21.59

Profit coefficient: 0.21

Maximum floating profits: 51,224,712.00 yuan   Maximum floating losses: 0.00 yuan

Maximum difference between floating profits and losses: 51,224,712.00 yuan

Total input: 44,440,000.00 yuan

-------------------------------Statistics of buy-in signals-------------------------------

(Make statistics of the situation of all buy-in signal points, excluding signal deletion caused by funds and strategies in the transaction test)

Success rate: 64.95%

Signal number: 10,101   Annual average signal number: 8,080.80

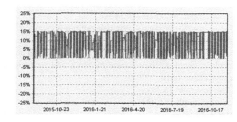

Figure 7. W%R earnings curve of one-time buy-in with all funds.

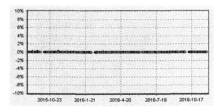

Figure 8. W%R buy-in signal distribution of one-time buy-in with all funds.

Table 4. The second test result of W%R expert system.

System test setting
Test method: Technical index – RSI (14)
Test time: 26/08/2015 – 24/11/2016   Calculating forced liquidation
Test share: 1,111 shares in total   Initial input: 40,000.00 yuan
Buy-in conditions:
　One of the following groups is tenable:
　1. The following conditions are tenable simultaneously
　1.1 Technical index: W&R (10,6) index line WR1 larger than 80 [daily line]
　　When the condition is satisfied, according to the middle market value, the closing price is bought in with 30.00% funds
　　In case of continuous signal: buy-in on equal basis
　Sell-out condition: no sell-out condition
　Conditions for closing a position: (according to the closing price)
　Share selection by indexes: technical index: W&R (10,6) index line WR1 smaller than 20 [daily line]

System test report
Tested stock number: 1,111
Net profit: 3,472,156.00 yuan   Net profit margin: 7.81%
Total profits: 8,921,514.00 yuan   Total losses: -5,449,293.00 yuan
Transaction times: 8,559   Winning rate: 73.83%
Annual transaction times: 6,847.20   Profit/ loss transaction times: 6,319/2,240
Total transaction amount: 201,667,840.00 yuan   Transaction fee: 510,531.06 yuan
Highest one-time profit: 23,975.38 yuan   Highest one-time loss: -16,991.85 yuan
Average profit: 1,042.35 yuan   Average loss: -636.67 yuan
Average profit: 405.67   Average profit / average loss: -163.72
Maximum continuous profit times: 21   Maximum continuous loss times: 5
Average transaction periodicity: 13.78
Average transaction periodicity with profits: 10.54
Average transaction periodicity with losses: 22.92
Profit coefficient:　0.24

(*Continued*)

Table 4. (*Continued*)

Maximum floating profits: 47,874,936.00 yuan   Maximum floating losses: 0.00 yuan
Maximum difference between floating profits and losses: 47,874,936.00 yuan
Total input: 44,440,000.00 yuan
----------------------------Statistics of buy-in signals-------
----
(Make statistics of the situation of all buy-in signal points, excluding signal deletion caused by funds and strategies in the transaction test)
Success rate: 59.34%
Signal number: 29,949   Annual average signal number: 23,959.20

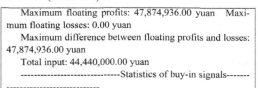

Figure 9. W%R earnings curve of step-by-step buy-in with 30% funds.

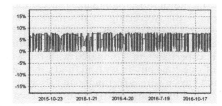

Figure 10. W%R buy-in signal distribution of step-by-step buy-in with 30% funds.

2.2 *Comparative analysis of results*

Table 5. Comparative analysis table.

	Winning rate	Annual rate of return	Net profit margin	Annual transaction times
One-time buy-in of RSI expert system with all funds	80.57	17.18	21.48	897.60
Step-by-step buy-in of RSI expert system with 30% funds	89.28	7.19	8.99	813.60
One-time buy-in of W%R expert system with all funds	67.52	12.27	15.33	7,605.60
Step-by-step buy-in of W%R expert system with 30% funds	73.83	6.25	7.81	6,847.20

3 CONCLUSION

By taking the winning rate, annual rate of return and net profit rate, which most concern investors as a management goal (Huang, 2013), this paper attempts to calculate and analyse the practicability of the RSI expert system and W%R expert system with a full sample size in the premise of funds management. Regarding the RSI expert system, compared with the winning rate of one-time buy-in with all funds, that of step-by-step buy-in with 30% funds can merely increase 9.29%, but the annual rate of return and net profit margin will drop sharply to 41.85% with a reduction of 84 times in the number of annual transactions (see Figures 3–6). According to the market implication of this calculation result, step-by-step buy-in is meaningless for the RSI system. Regarding the W%R expert system, compared with the winning rate of one-time buy-in with all funds, that of step-by-step buy-in with 30% funds can merely increase 6.31%, but its annual rate of return will see a significant decline to 50.93% with a sharp decrease in net profit margin of 50.95% and a reduction of 758.4 times in the number of annual transactions (see Figures 7–10). According to the market implication of this calculation result, step-by-step buy-in is also meaningless for the W%R system.

REFERENCES

http://en.wikipedia.org/wiki/Relative_strength_index

http://en.wikipedia.org/wiki/Williams_%25R

Huang, H-P. & Wang, P. (2013). Discussions on securities software expert system MA and RSI. *Advanced Materials Research*, 798–799, 757–760.

Li Yong-chun, & Cheng, H. (2013). Study on effectiveness of William overbought/oversold in technical analysis. *Commercial Times, 7*, 70–71.

Wang, P. (2011). Empirical analysis of RSI transaction system based on historical data. *Securities & Futures of China, 9*.

Wang, P. (2012). RSI investment decision-making based on consolidation market. *Journal of Quantitative Economics, 1*.

Automotive, Mechanical and Electrical Engineering – Liu (Ed.)
© 2017 Taylor & Francis Group, London, ISBN 978-1-138-62951-6

Weather routing based on modified genetic algorithm

Yanhui Wang, Xianming Zhu, Xiaoyu Li & Hongbo Wang
State Key Laboratory on Integrated Optoelectronics, College of Electronic Science and Engineering,
School of Jilin University, Changchun, Jilin Province, China

ABSTRACT: With the development of science and technology, using ships as the main transportation for the sea, route planning and weather routing become increasingly important. Ship weather routing provides the optimal routing for the ship based on predicted weather information, the state of the sea, and the ship's condition. This research uses wave data to achieve weather routing based on a modified genetic algorithm. Instead of using roulette wheel selection, the genetic algorithm uses a modified selection operator to achieve the goal of converging quickly and finding a better route. At the end, simulations are made to illustrate the result that the meteorological navigation time obtained by using a modified genetic algorithm is less than that using a traditional genetic algorithm.

Keywords: Weather routing; Modified genetic algorithm; Modified selection operator

1 INTRODUCTION

These days most of the world's trade is carried by ship, so the safety and economy of the ship's route are of considerable concern to the ocean transport companies. Through the loaded meteorological information, finding a safe and economic route is vital for sailing. This is because, during the ship's voyage, waves have the maximum impact on the ship's velocity, only considering the wave direction and significant height of combined waves and swell in this paper.

Genetic algorithm is a class of evolutionary laws of the evolution of the survival mechanism of the fittest. Genetic algorithm was first proposed by Professor J. Holland of the United States in 1975 (Holland, J. H., 1975), which had inherent implicit parallelism and better global optimisation ability. With the development of the genetic algorithm and its applications, there are more and more improvements to the genetic algorithm. D. Whitey (1991) (Michalewicz, Z. A., 1992) in his paper proposed a crossover operator based on adjacency based crossover. This operator is specific to the intersection of the individuals representing the gene with the sequence number, and applies it to the TSP problem. Jun Li of the University of Glasgow, UK (1992) (Gray, G.J., 1996), instructed Ph.D. students to extend genetic algorithms based on binary genes to genes such as Hexadecimal, Decimal, Integer and Float, so that the genetic algorithm can be more effectively applied to the direct optimisation of fuzzy parameters, system structure and so on. Jiang Lei (2005) (Chen & Jiang, 2003) proposed

parallel genetic algorithm to solve TSP problems, which made the algorithm get over obstacles to local convergence.

The modified selection operator proposed in this paper gets over the problems that the traditional genetic algorithm has, such as converging slowly and finding an optimal solution difficultly.

2 DATA

Meteorological data is loaded from ECMWF, which is an international organisation that includes 24 EU member states. It is vital to calculate the ship's speed in the ocean, to load the significant height of the combined waves and swell and the wave direction. The area studied lies from $0°N$ to $45°N$ and from $120°N$ to $180°E$ and they are divided by $1*1$ degrees latitude-longitude line. Because the ship's safety should be seriously considered in the ocean, when the wave height is higher than 4 meters the areas are set up as not navigable.

3 METHODOLOGY

In the course of route planning, our main considerations are route safety and economy. Nowadays, reducing navigation time means that large numbers will cost less to navigate. To achieve the purpose of route planning by finding a route with the shortest navigation time, a modified genetic algorithm is used.

3.1 Calculation of weight

The sea areas studied are divided into 1*1 degrees latitude-longitude line and 2,806 points. The point-to-point connection forms the route of the ship's voyage. The ship has eight sailing directions at each point, which are the north angle of 0 degrees, 45 degrees, 90 degrees, 135 degrees, 180 degrees, 225 degrees, 270 degrees and 315 degrees.

The weights to be calculated are the time of the point-to-point voyage (Panigrahi & Misra, 2010). The weights are expressed as:

$$T = L/V. \tag{1}$$

L is the distance between the point and the point of navigation. When the ship is sailing through the nodes, the wave height and the wave direction hinder the ship's speed. The ship's actual speed V of navigation is calculated by the empirical formula proposed by R.W. James (1957) (Haltiner & Hamilton, 1962). The formula is expressed as follows:

$$V = V_0 - \left[\alpha_1 + \alpha_2 * \cos(\beta - \alpha)\right] * H \tag{2}$$

where the angle 'α' gives the ship's course and 'β' is the wave direction. The V_0 is the ship's sailing speed in still water and H is the wave height. The ship parameters used in this paper are shown in the following table.

The α_1 and α_2 are constants and for the present ship type α_1 and α_2 are 0.30 and 0.15 respectively.

Using the formula (2), the speed of each point in 8 directions according to each point of the wave height and wave direction can be calculated. Since the distance between each point and the adjacent eight points is constant, then the weight can be calculated by formula (1).

3.2 Modified genetic algorithm

The flow chart of the genetic algorithm is shown below (Holland, J. H., 1975):

3.2.1 Initialise the chromosome

In this study, 300 chromosomes are used for genetic manipulation to find the optimal path and using the real number coding to initialise the chromosomes. Firstly, 2,806 points are numbered in the study area from bottom to top and from left to

Table 1. Ship parameters.

Type	Tanker
Loa	120 (m)
Dwt	1200 (ton)
Speed	18 (knots)

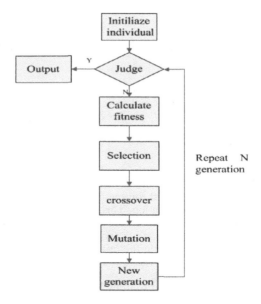

Figure 1. The flow chart of the genetic algorithm.

right, creating a Cartesian coordinate system with (0°N, 120°E) as the origin of the coordinate. Then, numbering each point by the formula as follows:

$$N = y + 46 * x \tag{3}$$

Here x is the value of the abscissa, y is the ordinate and N is the real number that will be used. In this study setting (20°N, 125°E) as the starting point and (45°N, 148°E) as the end point. From the starting point the ship has eight sailing directions heading for the next point until it reaches the specified end of the voyage. By this means initialising the chromosomes with a string of real strings.

3.2.2 Fitness evaluation

The selection of the individual fitness function directly affects the convergence speed of the genetic algorithm. The fitness function is calculated as follows:

$$Fitness = 1/sum. \tag{4}$$

Sum is the sum of the weights of adjacent real numbers in each chromosome.

3.2.3 Modified selection operator

In the traditional genetic algorithm, most of the selection operators are roulette selection method. Although the use of roulette selection is simple and practical, it has some problems that should be solved. First of all, at the beginning of the genetic algorithm, there may be a high degree of fitness of the individual, and then the probability that this individual is selected will be very large. This method will choose to copy a considerable number of individuals, which will result in the loss of diversity and hardly searching for a global optimal solution. At almost the end of the genetic algorithm phase, the fitness of each individual is very similar, causing the roulette selection to be useless.

In this paper, the author proposes a modified selection as follows: firstly, all the individuals are sorted by the bubble sorting method, and the individuals whose fitness ranked 1/6 after can then be eliminated. After that, make two copies of the individuals whose fitness ranked before 1/6 and select these individuals to enter the next generation. The remaining 2/3 individuals are all inherited to the next generation. When using this modified selection operator for genetic manipulation, there are several advantages as follows: (1) The individuals with very low fitness are directly eliminated, so that these individuals do not have the opportunity to enter the next generation and to improve convergence speed. (2) The individuals with better adaptability in the population can be increased rapidly, which makes the algorithm more efficient and practical. (3) Overcoming the shortcomings of the local convergence of the algorithm, and achieving global search to find the optimal solution.

3.2.4 *Crossover operator*
Selecting two parent individuals randomly and, when the crossover probability satisfies the condition, checking whether there are intersecting points between the two parental individuals. If there are intersecting points between the two parental individuals, the path exchanges after the intersection of two parent individual intersections.

3.2.5 *Mutation operator*
The mutation operator is considered to be the most effective method to get rid of the local convergence of the genetic algorithm. When a chromosome mutation probability satisfies the condition, one of the genes in the chromosome is mutated.

3.2.6 *Parameter selection*
Selecting a population size of 300, crossover probability of 0.7, and the mutation probability of 0.01. When the evolution to the 150 generation, the genetic algorithm terminates and outputs the optimal path.

4   RESULT

Respectively using the roulette selection operator and the modified selection operator of the genetic

algorithm for weather navigation. Setting (20°N, 125°E) as the starting point, and (45°N, 148°E) as the end point. Drawing two optimal paths of weather routing. The simulations are presented in the following:

The optimal route No. 1 is obtained by genetic algorithm with roulette selection operator and the optimal route No. 2 is obtained by genetic algorithm with modified selection operator. Comparing Figure 3 with Figure 2, the path obtained by genetic algorithm with modified selection operator has fewer inflection points and more smooth ones. Comparing Table 3 with Table 2, the modified genetic algorithm converges rapidly and evolutes a better path. It proves that using the modified

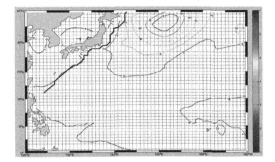

Figure 2.   The optimal route No. 1.

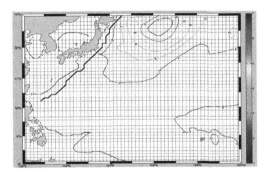

Figure 3.   The optimal route No. 2.

Table 2.   The total navigation time by roulette selection operator.

Generation	Total time (h)
1	63.2564
10	61.2664
40	61.0235
80	60.3987
120	59.8933
150	59.8933

Table 3. The total navigation time by modified operator.

Generation	Total time (h)
1	62.9771
10	61.0123
40	57.2522
80	57.2522
120	57.2522
150	57.2522

selection operator for genetic algorithm to find the optimal route is better than using the traditional genetic algorithm.

5 CONCLUSION

Now days ship as the main shipping tool at sea, with the scientific progress, ship's weather routing becomes more and more important. Because the genetic algorithm has the advantages of strong robustness and good flexibility, using the genetic algorithm to achieve the goal of weather routing is available. This study used a modified selection proposed by the author to achieve the goal of weather routing based on modified genetic algorithm and finding an optimal path. By using the modified genetic algorithm, finding the optimal route was quicker and more precise.

REFERENCES

Chen, X. F. & Jiang, L, (2003). Research on Holland Schemata Theorem [J] Proceeding of the Third International DCDIS Conference 247–251.

Gray, G.J. (1996). Nonlinear Model Structure Identification Using Genetic Programming, 33–42.

Haltiner, G.J. &.Hamilton, H.D. (1962). Minimal-time ship routing. *Journal of Applied Meteorology*, 1(1).

Holland, J. H. (1975a). Genetic algorithms and the optimal allocations of trials. *SIAM Journal of Computing*, 2(2), 88–105.

Holland, J. H. (1975b). *Adaptation in Natural and Artificial Systerm [M]*. Ann Arbor: The University of Michigan Press, 65–79.

Michalewicz, Z. A. (1992). Modified genetic algorithm for optimal control problems [J]. *Computatiom Math Application*, 23(12), 83–94.

Panigrahi, J.K. & Misra, S.K. (2010). Application of oceansat-1 MSMR analysed winds to marine navigation. *International Journal of Remote Sensing*, 31(10), 2623–2627.

Technologies in electrical and electronic control and automation

Automotive, Mechanical and Electrical Engineering – Liu (Ed.)
© 2017 Taylor & Francis Group, London, ISBN 978-1-138-62951-6

A full on-chip LDO regulator with a novel transient-response-enhanced circuit

Panpan Zhang, Peng Wang, Huaqun Meng, Siyuan Zhao & Fan Liu
Sichuan Institute of Solid-State Circuits, CETC, Chongqing, China

ABSTRACT: A full on-chip LDO regulator with a novel transient-response-enhanced circuit is presented in this paper. By adopting this new architecture, both load and line transient responses are apparently improved. Capacitive-coupling technology is used in the proposed circuit which has good stability and PSRR character. This new structure is finally implemented in HHNEC BCD-350 um technology. Simulation results show that 10.5 uA quiescent current is consumed and the maximum load current is 100 mA.

Keywords: Low-Dropout regulator (LDO); Transient-response-enhanced; Full on-chip

1 INTRODUCTION

As an important part in modern Very Large Scale Integration Circuit (VLSIC) designs, Low-Dropout (LDO) regulators are widely accepted and used in modern battery-powered portable devices. Owning the advantages of simple architecture, a fast-responding loop, much less output noise and lower production costs, the LDO regulator is an ideal choice for noise sensitive blocks.

To ensure stability, conventional LDO structures require an off-chip capacitor, which is always several microfarad and cannot be eliminated. The several microfarad large capacitor costs extra board space, and external pins are also needed. Furthermore, the parasitic resistor in the bulky large capacitor brings about large overshoots and undershoots at LDO output. Given the drawbacks of the off-chip capability, full on-chip LDO regulators have been studied by more and more researchers (Lau, Mok, and Leung, 2007). However, in this new area the frequency compensation strategy and achievement of a good transient response are still two main challenges in full on-chip LDO regulator design.

To achieve good line and load regulations, many researchers provided three-stage amplifier frequency compensation techniques in full on-chip LDO regulator design (Man, Mok and Chan, 2007). However, for some structures a requiring 100 uA minimum output current makes them unattractive. On the other hand, for power-saving impetus, the quiescent current should be minimised, however, low quiescent current leads to bad transient responses.

To overcome the disadvantages of previous architecture, this paper describe a full on-chip LDO regulator with novel transient-response-enhanced circuits. The proposed structure does not only improve the whole structure's transient response, but can be as an advanced compensation network. Thus, both good stability and better line and load regulations can be achieved in the full-output-current range.

The structure is shown as follows in this paper. Section 2 describes the structure and analyses the stability of the proposed LDO. In Section 3, the circuit design and transient-response-enhanced circuit are discussed. In Sections 4 and 5, the simulation results and conclusions are given, respectively.

2 PROPOSED LDO STRUCTURE AND STABILITY ANALYSIS

2.1 *Proposed LDO structure*

The proposed LDO is shown in Figure 1. This structure can be a three-stage amp including A_1, A_2 and a power MOS transistor (M_p). As Miller capacitor. C_m leads to poles splitting. Amplifier A_f and C_f constitute the transient-response-enhanced circuit, and can also be the Capacitive-Coupled Feedback stage (CCFB). Cm and the CCFB stage form the frequency compensation network, which

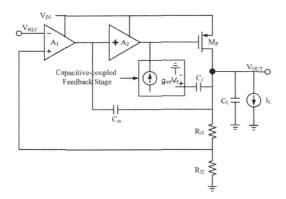

Figure 1. Structure of the LDO regulator in this paper.

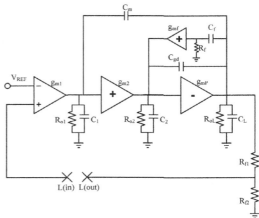

Figure 3. Topology of proposed LDO.

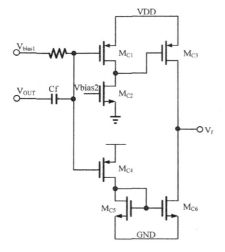

Figure 2. Schematic of transient-response-enhanced structure.

has been researched in numerous papers. R_{f1} and R_{f2} constitute feedback resistors, and C_L, I_L show the load condition.

The circuit of transient-response-enhanced structure used in this paper is given in Figure 2, consisting of M_{C1}~M_{C6}. V_{out} and V_f are connected to the output of the LDO regulator and the gate of M_p, respectively. This technology adopts push-pull movement to increase driving capability at the gate of M_p. A schematic of the proposed LDO with this technology is shown in Figure 5.

2.2 Stability analysis

Fig. 3 gives the topology of proposed LDO in this paper, and Fig. 4 gives the small-signal model of this new structure. $R_{O1\text{-}OL}$ is the output resistance,

and $C_{1\text{-}L}$ is the parasitic capacitance of each stage. C_{gd} represents the gate parasitic capacitor of the pass device.

According to Figures 3 and 4, formula (1) gives the small-signal transfer function of the proposed LDO regulator.

$$L(s) = \frac{A_{dc}F(1 - S/Z_1)(1 + S/Z_2)}{\left(1 + \dfrac{S}{P_0}\right)\left(1 + \dfrac{S}{P_1}\right)\left[\left(1 + S\left(\dfrac{1}{P_2} + \dfrac{1}{P_3}\right) + \dfrac{S^2}{P_2 P_3}\right)\right]}$$

(1)

where in (1), all symbols are defined by (2)–(9) and explained subsequently.

$$F = \frac{R_{f2}}{R_{f1} + R_{f2}} \tag{2}$$

$$A_{dc} = g_{m1}g_{m2}g_{mP} / R_{o1}R_{o2}R_{oL} \tag{3}$$

$$P_0 = R_{o1}R_{o2}R_{oL} / C_m g_{m2}g_{mP} \tag{4}$$

$$P_1 = \frac{g_{m2}}{(C_2 + C_{gd})(R_{oL}/g_{mP}) + C_{gd} + (g_{m2} + g_{mf})R_f C_f} \tag{5}$$

$$P_2 = \frac{g_{mP}\left[(C_2 + C_{gd})(R_{oL}/g_{mP}) + C_{gd} + (g_{m2} + g_{mf})R_f C_f\right]}{(C_2 + C_{gd})C_L} \tag{6}$$

$$P_3 = \frac{g_{mP}\left[(C_2 + C_{gd})(R_{oL}/g_{mP}) + C_{gd} + (g_{m2} + g_{mf})R_f C_f\right]}{(C_2 + C_{gd})R_{oL}R_f C_f} \tag{7}$$

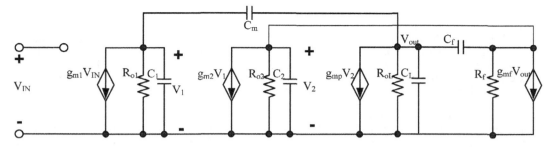

Figure 4. Small-signal modelling of proposed LDO.

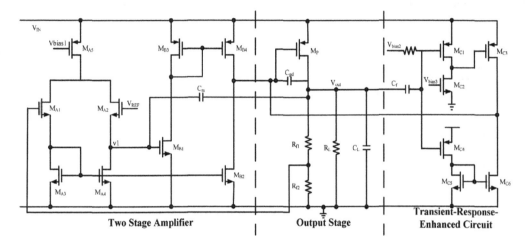

Figure 5. Schematic of the proposed LDO regulator.

$$Z_2 = 1/R_f C_f \qquad (8)$$

$$Z_1 = g_{m3}/C_{gd} \qquad (9)$$

F is the feedback factor. A_{dc} is the low-frequency gain. P_{0-3} represent poles of the system, and Z_{1-2} represent zeros. Since the trans-conductance g_{mp} and the output resistance r_{op} of the power PMOS-FET change with an increase or decrease in the load current, the stability of the proposed LDO will be studied at different load conditions. The stability of this circuit has been proved (Peng and Sansen, 2005; Leung and Mok, 2001).

3 CIRCUIT DESIGN AND TRANSIENT-RESPONSE-ENHANCED CIRCUIT DISCUSSION

3.1 The whole circuit

The whole circuit of the full on-chip LDO regulator in this paper is shown in Figure 5. $M_{A1} \sim M_{A4}$

compose the first stage amplifier, and M_{B1} and M_{B4} conform the second stage. This paper adopts pull-push architecture in the second stage to enhance the load transient response. M_p is a power transistor. $M_{C1} \sim M_{C6}$ constitute the pull-pushed transient-response-enhanced architecture. R_{f1} and R_{f2} constitute the feedback network. C_m is the Miller capacitor which is good for the LDO's stability.

3.2 Transient-response-enhanced circuit

The circuit of the new transient-response-enhanced structure is shown in Figure 5. For good driving capability, a push-pull structure is adopted (Man, Mok, and Chan, 2007). $M_{C1} \sim M_{C3}$ comprise the first channel to the gate of M_p, while $M_{C4} \sim M_{C6}$ compose the second channel. Once the voltage in the gates of M_{C1} and M_{C4} varies, the voltage in the drain of M_{C3} and M_{C6} changes in the same tendency because of the push-pull circuit channels.

Changes in the load current will lead to large output variations. Then, the voltage V_{out} is detected by C_f to pull the voltage in the gates of

M_{C1} and M_{C4} up (or down). When analysing the first circuit channel, the up (or down) voltage trend in gate M_{C1} causes the voltage in the drains of M_{C1} and M_{C2} to go down (or up). And then, transistor M_{C3} is turned on to charge (or discharge) the gate capacitance of the power transistor. Similarly, When analysing the second circuit channel, the up (or down) voltage trend in the gate of M_{C4} also leads transistor M_{C6} to charge (or discharge) the gate capacitance of the power transistor. Finally, the voltage of gate M_p is increased (or decreased), and then V_{out} will be decreased (or increased).

According to the analysis, transient response time is enhanced effectively by the dynamic push-pull scheme.

4 SIMULATION RESULTS

The proposed circuit is implemented in the HHNEC BCD_350 process, the quiescent current of this new structure is only 10.5 uA, while the maximum load current is 100 mA. In order to make a fair comparison on the transient performance, results with and without the transient response structure will be shown.

The simulation results of loop stability are shown in Figure 6. The comparison between structures with and without the transient-response-enhancement block is shown in Figures 7 and 8. Table 1 provides the performance summary of the proposed LDO regulator.

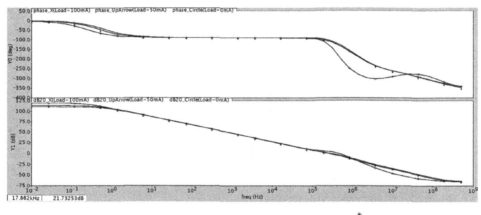

Figure 6. Loop stability at CL = 100 pF.

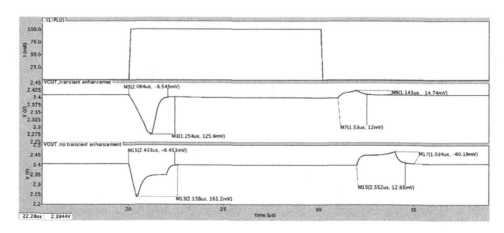

Figure 7. Comparison of load transient response of two LDOs with and without transient-response-enhanced technology.

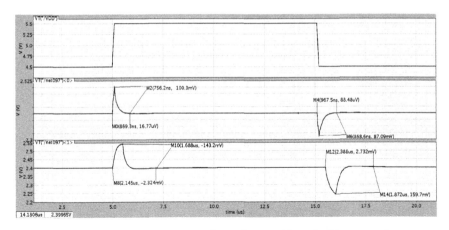

Figure 8. Comparison of line transient response of two LDOs with and without transient-response-enhanced technology.

Table 1. Performance summary of the proposed LDO.

V_{IN} (V)	5
V_{OUT} (V)	2.4
I_Q (uA)	10.5
I_{LOAD} (mA)	0–100
Load Reg. (mV/mA)	0.16
Line Reg. (mV/V)	3.7
PSRR (dB)	53
ΔV_{OUT} (mV)	125

5 CONCLUSION

In this paper, a full on-chip LDO regulator with a novel transient-response-enhanced circuit implemented in 0.35 um has been introduced. From the simulation results it can be seen that the proposed architecture improves the transient response greatly. Both undershoot and overshoot are greatly improved compared with the structure without transient-response-enhanced architecture.

Furthermore, this novel technology makes non-dominant poles locating at higher frequencies. Thus, the small compensation capacitor and full on-chip architecture lead to small-area LDO regulator.

REFERENCES

Leung, K. N. and Mok, P. K. T. (2001, Sep). Analysis of multistage amplifier-Frequency compensation. *IEEE Trans. Circuits Syst. I: Fund. Theory Appl.* 48(9), 1041–1056.

Lau, S.K., Mok, P.K.T. and Leung, K.N. (2007). A Low-Dropout Regulator for SoC with Q-Reduction. *IEEE J. Solid-state Circuits* 42, 658.

Man, T.Y., Mok, P.K.T. and Chan, M. (2007). A high slew-rate push-pull output amplifier for low-quiescent current low-dropout regulators with transient response improvement. *IEEE Trans, Circuits Syst. II, Exp. Briefs.* 54(9), 755–759.

Peng, X. and Sansen, W. (2005). Trans conductance with capacitances feed-back compensation for multistage amplifiers. *IEEE J. Solid-State Circuits.* 40, 1515.

Automotive, Mechanical and Electrical Engineering – Liu (Ed.)
© 2017 Taylor & Francis Group, London, ISBN 978-1-138-62951-6

A review on camera ego-motion estimation methods based on optical flow for robotics

Lingxi Lu
School of Mathematical Sciences, Peking University, Beijing, China

ABSTRACT: In the field of automotive engineering, the estimation of ego-motion plays an important role in robot control. In this review, we showed a general roadmap of camera ego-motion estimation, introduced the methods of optical flow estimation from image sequences, and compared three methods of ego-motion estimation from estimated optical flow in detail. We discussed the specialties of the mentioned methods and analyzed the error sources. We concluded that the optical flow-based ego-motion estimation methods can be used for velocity estimation but are not suitable for position or pose estimation. We encouraged researchers to focus on finding approaches to error elimination and finding less complicated methods.

Keywords: Optical flow; Ego-motion; Camera; Robot

1 INTRODUCTION

As robotics rapidly develops in the field of automotive engineering, ego-motion estimation becomes one of the primary tasks to achieve robot control. The ego-motion of a robot is defined as the robot's motion relative to the scene. Various types of sensors are invented for the purpose of ego-motion examination, and camera is the most common one among them.

The perception of ego-motion based on visual signals is first introduced by Von Helmholtz in 1950. Optical flow based ego-motion estimation is one of the most popular methods of camera ego-motion estimation. After American psychologist James J. Gibson introduces the concept of the Optical flow in the 1950s (Gibson, 1950), various methods of its determination are proposed, as well as the optimization constraints and techniques of ego-motion estimation from optical flow.

The rest of this paper is organized as follows: In Section 2, we define the problem of the camera ego-motion estimation. In Section 3, we introduce and compare three different methods of camera ego-motion estimation based on optical flow. In Section 4, we discuss about the

advantages and disadvantages of the optical flow-based methods. Finally, we draw a conclusion in Section 5.

2 PROBLEM DEFINITION

2.1 Pinhole camera model

To describe the projection of a point from the 3-D world coordinate system to the 2-D image plane, researchers have proposed several camera models, among which the pinhole model is the most popular.

Consider a point p in the world coordinate system $\{W : w_o, w_x, w_y, w_z\}$. The homogeneous coordinate of p in $\{W\}$ is $X_w = (x_w, y_w, z_w, 1)^T$. The pinhole camera model can be presented as (1).

$$m' = DK_0 M X_w$$
$$= \begin{pmatrix} \dfrac{f}{dx} & -\dfrac{f\cot\theta}{dx} & u_0 & 0 \\ 0 & \dfrac{f}{\sin\theta dy} & v_0 & 0 \\ 0 & 0 & 1 & 0 \end{pmatrix} \begin{pmatrix} R & T \\ 0 & 1 \end{pmatrix} X_w \quad (1)$$

where

$$K_0 = \begin{pmatrix} f & 0 & x_0 & 0 \\ 0 & f & y_0 & 0 \\ 0 & 0 & 1 & 0 \end{pmatrix}, \quad D = \begin{pmatrix} \dfrac{1}{dx} & -\dfrac{\cot\theta}{dx} & 0 \\ 0 & \dfrac{1}{\sin\theta dy} & 0 \\ 0 & 0 & 1 \end{pmatrix},$$

and $M = \begin{pmatrix} R & T \\ 0 & 1 \end{pmatrix}$.

The general steps of the pinhole model is shown in Figure 1. First, the matrix M transforms the coordinates of p in $\{W\}$ into the camera coordinate system. In M, R is a 3×3 matrix representing the camera's rotation and T is a 3×1 vector representing the camera's translation. Then K_0 projects point p to the image plane, where f is the camera focal length and (x_0, y_0) is the optic center on the camera image plane. Finally, D discretizes the 2-D coordinates in $\{I\}$, where (dx, dy) represents the size of each pixel in a CCD/CMOS image sensor with a intersection angle θ.

The matrix $K = DK_0$ contains all the intrinsic parameters of the camera, while the matrix M contains the camera ego-motions, which are called the extrinsic parameters. For a calibrated camera, we can estimate its extrinsic parameters if given a set of 3-D points in $\{W\}$ and their corresponding 2-D projections in the image. It's called a PnP (Perspective-n-Point) problem.

2.2 Motion field model

A disadvantage in solving PnP problems is that we have to acquire the 3-D coordinates of the points in the world coordinate system. Since Longuethig-

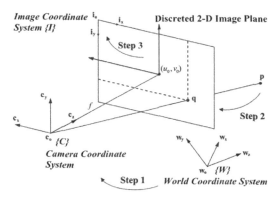

Figure 1. The general steps of the pinhole model. This figure is modified from (Armangu, 2003).

gins and Prazdny proposed the motion field model (Longuethiggins 1980), researchers have carried out camera ego-motion estimation methods based on optical flow, in which the 3-D coordinates of the points are not required.

Assume that a fixed point $P_i = (X_i, Y_i, Z_i)^T$ in the world coordinate and let $v = (v_x, v_y, v_z)^T$ and $\omega = (\omega_x, \omega_y, \omega_z)^T$ be the transitional velocity and the angular velocity of the camera respectively. The velocity of P_i relative to the camera satisfies (2).

$$\dot{P}_i = \left(\dot{X}_i, \dot{Y}_i, \dot{Z}_i \right)^T = -v - \omega \times P_i \qquad (2)$$

Consider the pinhole model and define the image of P_i as $p_i = (x_i, y_i)^T = \frac{f}{Z_i}(X_i, Y_i)^T$. Then we obtain

$$\dot{p}_i = \begin{pmatrix} \dot{x}_i \\ \dot{y}_i \end{pmatrix} = \frac{1}{Z_i} A_i \begin{pmatrix} v_x \\ v_y \\ v_z \end{pmatrix} + B_i \begin{pmatrix} \omega_x \\ \omega_y \\ \omega_z \end{pmatrix} \qquad (3)$$

where $B_i = \dfrac{1}{f} \begin{pmatrix} x_i y_i & -\left(f^2 + x_i^2\right) & fy_i \\ \left(f^2 + y_i^2\right) & -x_i y_i & -fx_i \end{pmatrix}$ and

$A_i = \begin{pmatrix} -f & 0 & x_i \\ 0 & -f & y_i \end{pmatrix}$. $\dot{p}_i = (\dot{x}_i, \dot{y}_i)^T$ is the optical flow at $(x_i, y_i)^T$. It can be regarded as the projection of 3-D motion vectors onto the 2-D image plane.

To estimate the ego-motion, five linearly independent data points are needed to solve the linear equation. If we acquire optical flows of more than five points, the camera ego-motion estimation problem will then turn into an optimization process, which is to find the best estimation of (Z_i, v, ω) such that $\{\dot{p}_i\}$ best fits $\{\dot{p}_i^*\}$, $i = 1, 2, \ldots, m$.

3 METHODS OF EGO-MOTION ESTIMATION

3.1 Optical flow estimation from image sequences

Due to the previous discussion, we have to acquire optical flows from an image sequence before estimating the camera ego-motion. The optical flow method considers the intensity of every voxel position and try to calculate the motion between image frames at time t and $t + \delta t$. The most popular method is based on the differential constraint. Consider a voxel at location (x, y, t) at time t will move to $(x + \delta x, y + \delta y, t + \delta t)$ in the next frame, where (x, y) is its position and t is the time. Let $I(x, y, t)$ be the intensity.

Assume that the intensity of each particular point is constant, then we have the intensity constancy constraint:

$$I(x,y,t) = I(x+\delta x, y+\delta y, t+\delta t) \qquad (4)$$

Assume that the movement is small. Use the Taylor series and we get (5).

$$I(x+\delta x, y+\delta y, t+\delta t) =$$
$$I(x,y,t) + \frac{\partial I}{\partial x}\delta x + \frac{\partial I}{\partial y}\delta y + \frac{\partial I}{\partial t}\delta t + \text{H.O.T} \qquad (5)$$

Combining (4) and (5), we get

$$\frac{\partial I}{\partial x}\frac{\delta x}{\delta t} + \frac{\partial I}{\partial y}\frac{\delta y}{\delta t} + \frac{\partial I}{\partial t}\frac{\delta t}{\delta t} = 0 \qquad (6)$$

or

$$I_x \dot{x} + I_y \dot{y} = -I_t \qquad (7)$$

where $(\dot{x}, \dot{y})$ is the optical flow of (x,y,t). Note that there are two unknowns in the equation, additional conditions are needed to estimate the flow, which draws various differential methods of optical flow estimation. The most well-known methods are Lucas-Kanade method (Lucas & Kanade, 1981) and Horn-Schunck method (Horn & Schunck, 1981). Other methods include Buxton-Buxton method (Wagemans, 1989), Black-Jepson method (Beauchemin & Barron, 1995), etc. There are also other methods estimating the optical flow without using the differential constraint.

3.2 Ego-motion estimation from optical flows

Once we acquired optical flows from an image sequence, we can use them for camera ego-motion estimation. The existed methods can be summarized as optical procedures. In this section, we mainly introduce three methods, while other methods vary with constraints and optimization techniques. For simplicity, we set the focal length $f = 1$.

3.2.1 The method of Zhang and Tomasi
According to (3), we cannot determine the absolute values of Z_i and v respectively as they only appear in the form $\frac{v}{Z_i}$. We add the constraint $\|v_z\| = 1$, without loss of generality.

Defining the residual as (8), Zhang and Tomasi regard the ego-motion estimation problem as to minimize the residual (Tong & Tomasi, 1999; Zhang & Tomasi, 2002). They develop an algorithm for the least square case as shown in (9).

$$r_i(Z_i, v, \omega) = \dot{p}_i^* - \dot{p}_i = \dot{p}_i^* - \left(\frac{1}{Z_i}A_i v + B_i \omega\right) \qquad (8)$$

$$(\hat{Z}_i, \hat{v}, \hat{\omega}) = \underset{Z_i, v, \omega}{\arg\min} \frac{1}{m}\sum_{i=1}^{m}\|r_i\|_2^2 \qquad (9)$$

The optimization problem could be solved by the Gauss-Newton update procedure, which determines a descent step at every iteration through linearization by Taylor series. m linear equations are formed:

$$\left(\frac{1}{Z_i}\right)^k A_i \Delta v^k + A_i v^k \Delta\left(\frac{1}{Z_i}\right)^k + B_i \Delta\omega^k = r_i^k \qquad (10)$$

with additional constraint $(t^k)^T \Delta t^k = 0$, where k is the iteration number.

The algorithm could be improved by observing that optimization parameter is separable as $(v, (\omega, Z_i))$. After v is solved, ω can be computed by solving a linear problem, and finally Z_i can be computed by the result of v and ω. Thus, the accuracy of the optimization problem could be improved by first computing Δv through Gauss-Newton method, and then calculate ω and Z_i respectively.

Note that a generalization of the optimization problem could be

$$(\hat{Z}_i, \hat{v}, \hat{\omega}) = \underset{Z_i, v, \omega}{\arg\min} \frac{1}{m}\sum_{i=1}^{m}f(r_i) \qquad (11)$$

where $f(r_i)$ could be $\|r_i\|_p^r$ or some more general loss functions.

3.2.2 Method of Zhuang et al
The method (Zhuang et al., 1988) assumes that not only optical flow of image points $\dot{p}_i^*$, but also their relative depth Z_i are given. In this case, depth information does not involve derivatives, and the optimization problem (Least Square Estimation) has only 6 unknowns:

$$(\hat{v}, \hat{\omega}) = \underset{v, \omega}{\arg\min} \frac{1}{m}\sum_{i=1}^{m}\|r_i\|_2^2 = \underset{v, \omega}{\arg\min} \frac{1}{m}\varepsilon^2(v, \omega) \qquad (12)$$

To get the minimal, we obtain the partial derivatives:

$$\frac{\partial \varepsilon^2}{\partial \omega_j} = 0, \quad \frac{\partial \varepsilon^2}{\partial v_j} = 0, \quad j = x, y, z \qquad (13)$$

Rearrange the functions we get the homogeneous linear equations

$$W(\omega_x, \omega_y, \omega_z, v_x, v_y, v_z)^T = b \qquad (14)$$

where $W = \sum_{i=1}^{m}(Z_i B_i, A_i)^T (Z_i B_i, A_i)$, and $b = \sum_{i=1}^{m}Z_i(Z_i B_i, A_i)\dot{p}_i^*$. W is positive symmetric.

Do the diagonalization to W we get the decomposition $W = U^T D U$, where U is orthonormal and D is diagonal. The solution of the Least Square estimation problem is given by

$$\left(\omega_x, \omega_y, \omega_z, v_x, v_y, v_z\right)^T = U^T D^{-1} U b \qquad (15)$$

3.2.3 Method of Raudies and Neumann

In this method (Raudies & Neumann, 2009), the constraint in (3) could be transformed into

$$0 = v\left(M_i - H_i \omega\right) \qquad (16)$$

where $M_i = \begin{pmatrix} f\ddot{x}_i^* \\ -f\ddot{x}_i^* \\ y_i \dot{x}_i^* - x_i \dot{y}_i^* \end{pmatrix}$, and

$$H_i = \begin{pmatrix} -\left(f^2 + y_i^2\right) & x_i y_i & fx_i \\ x_i y_i & -\left(f^2 + x_i^2\right) & fy_i \\ fx_i & fy_i & -\left(x_i^2 + y_i^2\right) \end{pmatrix}$$

Let $E_i^T = [-\left(f^2 + y_i^2\right), x_i y_i, fx_i, -\left(f^2 + x_i^2\right), fy_i, -\left(x_i^2 + y_i^2\right)]$, and $K^T = [v_x \omega_x, v_x \omega_y + v_y \omega_x, v_x \omega_z + v_z \omega_x, v_y \omega_y, v_y \omega_z + v_z \omega_y, v_z \omega_z]$. Regard K as $K(v)$, we get the linear optimization problem with respect to E and K:

$$\left(v, K(v)\right) = \arg\min_{v, K(v)} \sum_{i=1}^{m} \left(v^T M_i + K^T E_i\right)^2 \qquad (17)$$

Take partial derivatives of the optimization function with respect to K and v. Reform the equations and we obtain the homogeneous linear system of equations:

$$v^T C = 0 \qquad (18)$$

where

$$c_{jk} = \sum_{i=1}^{m} L_{ij} L_{kj},$$

$$L_{ij} = M_{ij} - \left(D_i E_i\right)^T \sum_{l=1}^{n} E_l M_{lj},$$

$$D_i = \sum_{l=1}^{n} E_l E_l^T \in R^{6 \times 6},$$

M_{ij} is the j-th component of M_i. The solution is given by the eigenvector corresponds to the smallest eigenvalue of matrix C.

4 DISCUSSIONS

The method of Zhang and Tomas minimized the standard residue by iteration. This minimization method has small variance and bias, and loss functions using p-norm will increase the robotness. Gauss-Newton method converges quickly so the calculation cost is also relatively small.

The method of Zhuang et al is a linearized method which transferred the nonlinear constraint into a linear minimization problem. The method requires less optical flow image point-depth pairs to do the estimation. Information from three, instead of eight points, are needed, so the method may be effective when eliable optical flow information are only provided in specific points.

Florian and Heiko's method has its superiority in its efficiency. The method has a very low computation complexity, so it has a faster performance than other optical flow estimation method. The method also shows numerical stability under noisy conditions.

These optical flow-based methods can be used for ego-motion parameters estimation, for example, the estimation of a quadrocopter's velocity. However, as errors occur in all the steps of the ego-motion estimation, these methods are not fit for position or pose estimation at present. The errors can be classified as the following three aspects. First, errors occur in the estimation of optical flow because the assumption of the intensity temporal constancy is not true in all situations. This kind of error is often caused by moving light sources or an additive Gaussian noise. Second, errors occur in estimation of ego-motion from optical flow due to additional sources besides ego-motion contributing to the optical flow. These additional sources, e.g. an independent moving object in sight, are usually called outlier noise. Third, in the integration from ego-motion to position or pose, the errors cumulate with time.

5 CONCLUSION

Ego-motion estimation is an important technology in automotive engineering. In this paper, we first give a general roadmap of the camera ego-motion estimation and summarize the estimation methods as various optimization procedures. Then, we introduce the methods of optical flow estimation from image sequences and methods of ego-motion estimation from estimated optical flow. Three ego-motion estimation methods are introduced in detail. After that, we show the differences of the three mentioned methods, and discuss the error sources of these optical flow-based methods.

Our review shows that the optical flow-based methods can be used for ego-motion parameters estimation but are not suitable for position or pose estimation as errors cumulate in all the steps. In addition, the complexity of the optical flow-based methods is high, and may have side effects on the robustness and real-time property. In the future, researchers need to focus on finding approaches to error elimination and less complicated methods to achieve camera ego-motion estimation.

REFERENCES

Armangué, X., Araújo, H., & Salvi, J. 2003. A review on egomotion by means of differential epipolar geometry applied to the movement of a mobile robot. Pattern Recognition, 36(12):2927–2944.

Beauchemin, S. S. & Barron, J. L. 1995. The computation of optical flow. Acm Computing Surveys, 27(3):433–466.

Burger, W. & Bhanu, B. 1990. Estimating 3d egomotion from perspective image sequence. Pattern Analysis & Machine Intelligence IEEE Transactions on, 12(11):1040–1058.

Gibson, J. J. 1950. The perception of visual surfaces. American Journal of Psychology, 63(3):367–84.

Horn, B. K. P. & Schunck, B. G. 1981. Determining optical flow. International Society for Optics and Photonics.

Longuethiggins, H. C. & Prazdny, K. 1980. The interpretation of a moving retinal image. Proceedings of the Royal Society of London, 208(1173):385–97.

Lucas, B. D. & Kanade, T. 1981. An iterative image registration technique with an application to stereo vision. In International Joint Conference on Artificial Intelligence, pages 285–289.

Raudies, F. & Neumann, H. 2009. An efficient linear method for the estimation of ego-motion from optical flow. In Dagm Symposium on Pattern Recognition, pages 21–28.

Tong, Z. & Tomasi, C. 1999. Fast, robust, and consistent camera motion estimation. In IEEE Conference on Computer Vision and Pattern Recognition, page 1164.

Wagemans, J. 1989. Visual cognition—computational, experimental, and neuropsychological perspectives—humphreys,gw, bruce,v. Physical Review, 82(1):98–99.

Zhang, T. & Tomasi, C. 2002. On the consistency of instantaneous rigid motion estimation. International Journal of Computer Vision, 46(1):51–79.

Zhuang, X., Haralick, R. M., & Zhao, Y. 1988. From depth and optical flow to rigid body motion. In Computer Vision and Pattern Recognition, 1988. Proceedings CVPR '88., Computer Society Conference on, pages 393–397.

Automotive, Mechanical and Electrical Engineering – Liu (Ed.)
© 2017 Taylor & Francis Group, London, ISBN 978-1-138-62951-6

A work charged car insulation system design and research

Ke Li, Lei Xia & Li Tan
State Grid Co. Electric Power Research Institute, Chongqing, China

Dahong Wang
The Smartech Institute, Shenzhen, China

ABSTRACT: For the high-voltage electric vehicle's arm-lifting position and pose problem in the process of operation in the power system, we put forward a prediction calculation method. On the basis of existing work of the high-voltage electric vehicle, a matrix method is used to construct the model of part of the lifting mechanism's position and pose of the high-voltage electric vehicle. Through calculation and analysis of the model, the prediction of the lifting mechanism's motion and position information is made, and this in turn make a reference to the unmanned remote operation control factors.

Keywords: Live working; Position and pose research

1 INTRODUCTION

Electricity as the main energy of modern society, plays an important role in the stability and development of modern society. High voltage power transmission and transformation system as power transmission channel. Its reliable working has a direct impact to the electric power system's safe and reliable operation. With the development of the economy and society, the interconnection, multipoint access, the increase of high-power and heavy load node, uncontrollable environmental factors affect the uncertainty of high-voltage power grid. In order to ensure the safe and reliable operation of the power grid, high-voltage electric inspection on a regular basis is widely used as a kind of effective means to provide power grid security in the routine maintenance. The high-voltage electric, vehicle as a kind of special equipment, arises at the historic moment. The high-voltage electric vehicle is a special kind of engineering vehicle. The traditional high-voltage electric vehicle has the following features: (1) manned aerial work, (2) complicated work environments, (3) timeliness demanding job tasks, (4) high objects bodily harm coefficient (Liu Cungen, 2015).

In this paper, on the basis of existing high-voltage electric vehicles, and considering the characteristics of the existing high-voltage electric vehicle, people bring the possibility of the risk to crew during work. To ensure the safety of people, there is intention to discuss the existing vehicle part of the lifting arm, and the movement characteristics of the structure. Building lifting arm terminal, insulated bucket movement model of and calculating the terminal spatial position and posture information of the lifting arm. Relevant technical accumulation and reference is provided for subsequent possible unmanned aerial electric vehicle research and development (Guo Yishen, 2009).

2 THE KINEMATICS MODEL

The function of the lifting mechanism is to transfer control instruction to control the position, including position control and pose control, to achieve the manipulator specified function. For a accurate kinematics model it is helpful to realise accurate posture adjustment, decoupling control and error compensation. This section studies its working mechanism and institution. According to the robot mechanism, we calculate its positive and inverse kinematics solution. To simplify the calculation, a part of the lifting mechanism and a part of the insulation arm are taken as a sample, as shown in Figure 1.

2.1 Institutions and work mechanism analysis

Figure 2 shows control process of the insulation arms. The device receives the control instruction, through system solution work out 1 #, 2 #, 3 # and 4 # hydraulic cylinder of the manipulated variable, by the multi-body movement, and is converted to hydraulic pole position (X, Y, Z) and attitude change (σ_X, σ_Y, σ_Z).

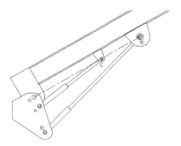

Figure 1. The structure diagram.

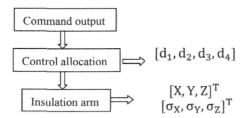

Figure 2. Insulation arms control process.

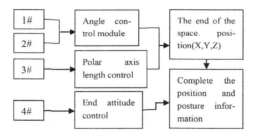

Figure 3. Insulation arm concrete implement process.

Figure 3 shows the details of the implementation process, 1# and 2# cylinder control angle changes, 3# polar axis cylinder control, and 4# control the pose angle at the end of the cylinder. The 1# and 2# cylinder can choose different work modes, alone or two cylinders together, to complete the control arm angle. 3# cylinder to complete the control arm 2 polar axis length. Therefore the triplex synergy can complete control arm space (x, y, z) at the end point of the manipulation. 4# stance at the end of the cylinder to complete the control arm manipulation.

Combining the movement mechanism, the control system can be divided into three modules: the angle control module, polar axis length control module and terminal pose control module.

2.2 *Manipulator arm kinematics analysis*

From the previous, the insulation arm control system can be divided into three parts: the angle

control module, polar axis length control module and the end of the attitude control modules. Kinematics modelling is applied to each module, and solves a full positive and inverse kinematics model manipulation.

Figure 4 shows angle control module movement mechanism and Figure 5 shows the physical prototype. As the figures show, the coordinates $O_0X_0Y_0Z_0$ are fixed on the body, and the origin is located in the rotation centre of insulation arm mechanism. Insulation O_0Y_0 axis points to the upper part, and O_0Z_0 goes along the normal position of the insulation arm plane. The coordinate and other axial the rotation angles follow the right-hand rule.

Three steering gears were respectively numbered 1–3 and located in $\varphi Lf = 0°$, 120°, 240°. Single for steering gear rotating centre is CLi. Steering gear and tilting device junction is zero ALi. Steering gear lever long respectively is PL. Steering gear lever variation is dLi, i = (1~3). Coordinates $O_aX_aY_aZ_a$, $O_bX_bY_bZ_b$, $O_cX_cY_cZ_c$, $O_dX_dY_dZ_d$ are located in 1 # and 2 # hydraulic cylinder rotation centre.

According to robot kinematics, assume λ_1 = A-Pi, Coordinate system $O_0X_0Y_0Z_0$ to $O_aX_aY_aZ_a$, $O_bX_bY_bZ_b$, $O_cX_cY_cZ_c$ and $O_dX_dY_dZ_d$. For the transformation of the matrix:

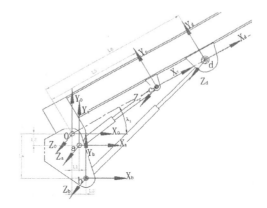

Figure 4. Angle control module movement mechanism.

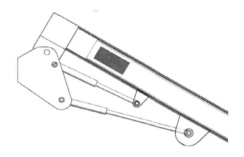

Figure 5. The entity prototype angle control module.

$$C_0^a = \begin{bmatrix} 1 & 0 & 0 & 0 \\ 0 & 1 & 0 & 0 \\ 0 & 0 & 1 & 0 \\ -L_1 & L_2 & 0 & 0 \end{bmatrix},$$

$$C_0^b = \begin{bmatrix} 1 & 0 & 0 & 0 \\ 0 & 1 & 0 & 0 \\ 0 & 0 & 1 & 0 \\ -L_3 & L_4 & 0 & 0 \end{bmatrix}$$

$$= \begin{bmatrix} 1 & 0 & 0 & 0 \\ 0 & 1 & 0 & 0 \\ 0 & 0 & 1 & 0 \\ -L_5\cos\lambda_1 & -L_5\sin\lambda_1 & 0 & 0 \end{bmatrix}$$
$$\cdot \begin{bmatrix} \cos\lambda_1 & -\sin\lambda_1 & 0 & 0 \\ \sin\lambda_1 & \cos\lambda_1 & 0 & 0 \\ 0 & 0 & 1 & 0 \\ 0 & 0 & 0 & 1 \end{bmatrix}$$

$$C_0^d = \begin{bmatrix} 1 & 0 & 0 & 0 \\ 0 & 1 & 0 & 0 \\ 0 & 0 & 1 & 0 \\ -L_6\cos\lambda_1 & -L_6\sin\lambda_1 & 0 & 0 \end{bmatrix}$$
$$\cdot \begin{bmatrix} \cos\lambda_1 & -\sin\lambda_1 & 0 & 0 \\ \sin\lambda_1 & \cos\lambda_1 & 0 & 0 \\ 0 & 0 & 1 & 0 \\ 0 & 0 & 0 & 1 \end{bmatrix}$$

$$SoC_0^a = [C_0^a]^{-1} = \begin{bmatrix} 1 & 0 & 0 & 0 \\ 0 & 1 & 0 & 0 \\ 0 & 0 & 1 & 0 \\ L_1 & -L_2 & 0 & 0 \end{bmatrix}$$

A point relative to the coordinates $O_aX_aY_aZ_a$ is $[0\ 0\ 0\ 1]^T$.

To launch the coordinates of a point relative to 0 coordinate system,

$$r_a^0 = \left[[r_a^a]^T \cdot C_a^0\right]^T = \begin{bmatrix} L_1 & -L_2 & 0 & 1 \end{bmatrix}^T$$

The same can be launched,

$$r_b^0 = \begin{bmatrix} L_3 & -L_4 & 0 & 1 \end{bmatrix}^T,$$

$$r_c^0 = \begin{bmatrix} L_5\cos\lambda_1 & L_5\sin\lambda_1 & 0 & 1 \end{bmatrix}^T,$$

$$r_d^0 = \begin{bmatrix} L_6\cos\lambda_1 & L_6\sin\lambda_1 & 0 & 1 \end{bmatrix}^T \quad (1)$$

Assuming $D_{ac} + \Delta D_{ac}$ is distance between a point and point c. $D_{bd} + \Delta D_{bd}$ is b and d, launch formula is:

$$\Delta D_{ac} = \sqrt{(L_1 - L_5\cos\lambda_1)^2 + (L_5\sin\lambda_1 + L_2)^2} - D_{ac}$$

$$\Delta D_{bd} = \sqrt{(L_3 - L_6\cos\lambda_1)^2 + (L_6\sin\lambda_1 + L_4)^2} - D_{bd}$$

Base $\lambda_1 = \alpha - \pi/6$, can calculate its angle control value and cylinder inverse kinematics relationship between the scales.

Figure 6 shows polar axis length control module movement mechanism and the prototype diagram shows that the location of the relationship between $O_dX_dY_dZ_d$ coordinate and $O_eX_eY_eZ_e$ coordinate system. Suppose $\lambda_2 = L_7 + \Delta d_3$, where L_7 is for preset values.

$$C_d^e = \begin{bmatrix} 1 & 0 & 0 & 0 \\ 0 & 1 & 0 & 0 \\ 0 & 0 & 1 & 0 \\ -\lambda_2 & -L_8 & 0 & 1 \end{bmatrix} \Rightarrow C_e^d = \begin{bmatrix} 1 & 0 & 0 & 0 \\ 0 & 1 & 0 & 0 \\ 0 & 0 & 1 & 0 \\ \lambda_2 & L_8 & 0 & 1 \end{bmatrix}$$

We can obtain,

$$[r_e^d]^T = [\lambda_2, L_8, 0, 1]^T$$

According to C_d^0, can launch

$$r_e^0 = \begin{bmatrix} \lambda_2\cos\lambda_1 - L_8\sin\lambda_1 + L_6\cos\lambda_1 \\ \lambda_2\sin\lambda_1 + L_8\cos\lambda_1 + L_6\sin\lambda_1 \\ 0 \\ 1 \end{bmatrix} \quad (2)$$

Comprehensive description above can introduce key points e spatial location.

According to the plane geometry relationship shown in Figure 7, if the input is 0, 4# cylinder is at the end of the attitude angle β:

$$\beta = \frac{\pi}{2} + \lambda_1 = \frac{1}{3}\pi + \alpha$$

The α variable value of 0 to turn the limit, in order to ensure the β unit in $\pi/2$, #4 cylinder to adjust operation and keep the balance.

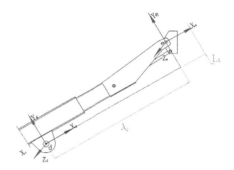

Figure 6. Polar axis length control module movement mechanism.

Figure 8 shows coordinate diagram established. For #4 cylinder coordinate system fixed end establishes f. Its coordinate direction remains the same as the e coordinate. G coordinate system, is set up to keep the end profile vertical upward; g is the coordinate system coordinates for a fixed direction. The origin position is as shown in Figure 7. O_gY_g is pointing vertically upward, and O_gY_g insulation arm is along the plane of the normal position. Coordinate other axial and corner also follow the right-hand rule. According to the Figure 8, the H point #4 cylinder activity, g point corresponds to the coordinate position for $(0, -L_{12}, 0)^T$, e point corresponds to the coordinate system of coordinates for $(L_{11}\cos(\lambda_3), L_{11}\sin(\lambda_3), 0)^T$, and f points corresponds to the coordinate system of coordinates for the $(-L_9, -L_{10}, 0)^T$. Therefore, it can be calculated:

$$C_e^g = \begin{bmatrix} 1 & 0 & 0 & 0 \\ 0 & 1 & 0 & 0 \\ 0 & 0 & 1 & 0 \\ -L_{11}\cos(\lambda_3) & -L_{11}\sin(\lambda_3) & 0 & 1 \end{bmatrix}$$

According to the above methods, it can be deduced:

$$C_g^e = \left[C_e^g\right]^{-1} = \begin{bmatrix} 1 & 0 & 0 & 0 \\ 0 & 1 & 0 & 0 \\ 0 & 0 & 1 & 0 \\ L_{11}\cos(\lambda_3) & L_{11}\sin(\lambda_3) & 0 & 1 \end{bmatrix};$$

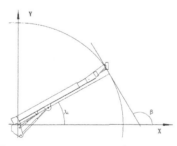

Figure 7. Insulation arm operation plane geometry relationship.

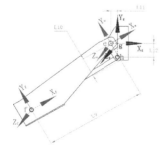

Figure 8. End attitude control module movement mechanism.

Thus, the coordinates of the position can be launched about h point relative to e point,

$$r_h^e = \left[\left[r_h^g\right]^T \quad c_g^e\right]^T$$
$$= \begin{bmatrix} (0, -L_{12}, 0, 1) \begin{bmatrix} 1 & 0 & 0 & 0 \\ 0 & 1 & 0 & 0 \\ 0 & 0 & 1 & 0 \\ L_{11}\cos(\lambda_3) & L_{11}\sin(\lambda_3) & 0 & 1 \end{bmatrix} \end{bmatrix}^T$$
$$= \left[L_{11}\cos(\lambda_3) \quad -L_{12} + L_{11}\sin(\lambda_3) \quad 0 \quad 1\right]^T$$

(3)

According to the distance formula, it can be calculated,

$$\Delta D_{fh} - D_{fh} =$$
$$\sqrt{\left(L_{11}\cos(\lambda_3) + L_9\right)^2 + \left(L_{10} - L_{12} + L_{11}\sin(\lambda_3)\right)^2}$$

To sum up, using $\lambda_3 = -\lambda_1$ can ensure to keep vertical upwards at the end, the manipulation of 4 # cylinder volume. Or according to ΔD_{fh} calculate λ_3 then, $\lambda_1 + \lambda_3$ can be obtained after the end of the execution.

3 CONCLUSION

As part of the lifting mechanism as a sample, this paper is committed to research on its kinematics model. First of all, its work mechanism is analysed, and we explain the control mode. And according to the movement mechanism we divide the control system into several modules. Secondly, combining movement mechanism, kinematics analysis was carried out on each module, its positive and inverse kinematics solution. (1) and (2) by simultaneous, can launch its spatial position and 1 #, 2 # and 3 # cylinder of the relationship between the output. Through simultaneous, (1) and (3) can be launched 4 # cylinder output and at the end of the relationship between the space position. After acquiring this relationship, we can accurately control the insulation work bucket at the end of the position and posture, providing better service for the staff.

REFERENCES

Chen L. (2001). Space manipulator posture and joint motion adaptive and robust control. *China Mechanical Engineering*, (05).

Guo, Yishen (2009) Free floating space manipulator posture, claw hand coordination movement at the end of the adaptive neural network control, *Mechanical Engineering Disciplines*, (07).

Liu, C. (2015) High voltage between the mechanical arm posture monitoring system research. Transmission and distribution equipment state assessment and fault diagnosis, journal (03).

Automotive, Mechanical and Electrical Engineering – Liu (Ed.)
© 2017 Taylor & Francis Group, London, ISBN 978-1-138-62951-6

Analysis of static impedance abnormity based on EMMI

Huaqun Meng, Rongqiang Li & Panpan Zhang
Sichuan Institute of Solid-State Circuits, Chongqing, China

Yanru Chen
Verakin High School of Chongqing, Chongqing, China

ABSTRACT: This paper presents an analytical approach for static impedance abnormity of the semiconductor integrated circuit. The design details and manufacturing process have been briefly described. The Emission Microscope (EMMI) is adopted in the analytical approach in order to locate the position that causes static impedance to be abnormally low. Combined with process analysis, the cause of static impedance abnormity will be precisely found.

Keywords: Static impedance; Emission Microscope; Process

1 INTRODUCTION

With the increasing integration ratio of the integrated circuit, the feature size of the component becomes smaller and smaller and the chip structure becomes more and more complicated. Locating the defects of the chip level becomes more difficult. In the presence of a leakage current, breakdown and the hot carrier effect of semiconductor devices, the failure point due to the process of electro luminescence brings about the production of a light-emitting phenomenon. The Emission Microscope (EMMI) takes advantage of this phenomenon to find the failure point of the device. The EMMI is sensitive to the defect of the doped region of the device, and can find out the abnormal point directly.

This paper introduces a failure analysis process of the semiconductor integrated circuit of the PIN switch driver, whose static impedance between the positive and negative power supply ports lowers during the using process. It will expound the PIN switch driver circuit principle, the method for locating the abnormal impedance using the EMMI and the analysis of the productive process to find out the reasons for the abnormality.

2 PROBLEM DESCRIPTION

The principle block diagram of the PIN switch driver is shown in Figure 1. The PIN switch driver has two ports: positive supplying with V_{CC} and negative supplying with V_{EE}. When the static impedance of the V_{CC} and V_{EE} ports is measured by the hand-held multimeter FLUKE 15B, the normal

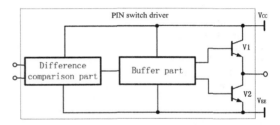

Figure 1. Principle block diagram of PIN switch driver.

device's resistance is at million ohm level while the abnormal device's resistance is at kilo ohm level.

2.1 *Preliminary analysis*

2.1.1 *Design analysis*
As is shown in Figure 1, the branches related to the impedance contain some resistors and transistors and are not pure resistance branches. When the hand-held multimeter FLUKE 15B applies a 0.23V voltage to measure the supply ports, the internal transistor in the open state can be equivalent to the static impedance at million ohm level. The output stage, having two power transistors (V1 and V2), is the same as the above.

Because there is no low impedance design, this can exclude the chip's low impedance being caused by design.

2.1.2 *Process analysis*
The PIN switch driver uses a complementary bipolar process. Because the circuit is composed of resistors and transistors, therefore the resistors

adopt the chrome silicon thin film resistor fabrication and have an insulating layer of silicon dioxide below them. Thus the resistors will not produce a leakage channel. Therefore, only the transistors have the leakage current channel, leading to the low static impedance of the device.

3 EMMI TEST

3.1 Sample impedance test

The samples' impedance measure results are shown in Table 1.

3.2 EMMI test result

When the normal power supply is applied on the V_{CC} and V_{EE} ports of the samples, the EMMI gets the images of the devices. The images are shown in Figure 2 and Figure 3.

3.3 EMMI image analysis

As is shown in Figure 2 and Figure 3, when comparing the image morphology of the normal sample and that of the abnormal sample, it is found that the abnormal sample's module includes not only all of the highlight spots of the normal sample's module, but also some others.

As is shown in Figure 3, the part marked out with the red circle contains the different highlight spots from Figure 2. According to the design lay-

Table 1. Sample impedance.

Sample no.	Test instrument	Test port's voltage	Impedance value	Device status
1#	FLUKE 15B	0.23 V	6 MΩ	Normal
2#			7.36 KΩ	Abnormal

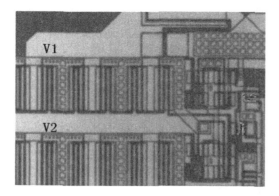

Figure 2. Image of normal sample.

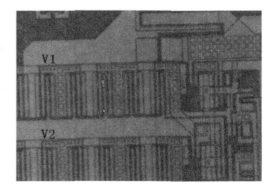

Figure 3. Image of abnormal sample.

out diagram, we can find that the corresponding component to the mark is the output transistor V1 of the PIN switch driver. Because the highlight spots are caused by the current, it can be determined that a leakage current exists in the positive and negative power supply branch of the output transistor of the static impedance abnormal circuit.

4 FAILURE MECHANISM ANALYSIS

According to the process of the transistor, the leakage current of the transistor is caused by either the PN junction defect or the oxide layer defects. There is a fault tree, as shown in Figure 4.

4.1 Oxide layer's leakage

If there is sodium ion of contamination in the oxide layer, it will cause the oxide layer to leak current. The leakage current of the oxide layer is caused by the ion contamination, and usually appears in a high temperature working environment and electrical ageing period, but can resume normal work in normal working conditions. It is not in accordance with the leakage current of the power supply ports of the PIN switch driver that are caused by the static low impedance phenomenon. Therefore, the possibility of the leakage of the oxide layer in the circuit of the PIN switch driver can be eliminated.

4.2 PN junction leakage current caused by defects

4.2.1 Material defects lead to PN junction leakage current

In the integrated circuit process, the crystal lattice defect of the silicon wafer is inevitable and in the epitaxial process the rate of the flowing gas, temperature, uniformity of the gas flow, etc. will also bring new lattice defects randomly distributed as point defects, line defects, planar defects and bulk. If PN junction is produced at lattice defect region by doping, there will be an accumulation of impurities and result in electric field localised in PN junction section to cause leakage current. Because the above-mentioned defects are random, and when the transistor area size becomes larger, the probability of the corresponding degree of the PN junction leakage current caused by the lattice defect is greater.

If the transistor CB junction and EB junction have some lattice defects, the transistor will exhibit the electrical characteristics of degeneration and CE leakage phenomenon. In the PIN switch driver circuit, if the defect falls on the output of the transistor, this will cause a phenomenon of low static impedance of the supply ports between V_{CC} and V_{EE}.

If the CS junction of substrate in the silicon wafer has some defects as above-mentioned, it will directly lead to the P type substrate between the NPN transistor V1 and V2 of the circuit of the PIN switch driver leaking current and causing the phenomenon of low static impedance.

Therefore, we cannot exclude the possibility of the material lattice defects causing the leakage current of the CE junction of the output transistor and the CS junction of the substrate, and resulting in a low value of static impedance between supply ports V_{CC} and V_{EE}.

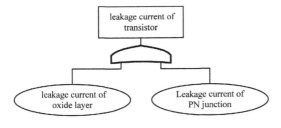

Figure 4. Fault tree.

4.2.2 *Lithography processing defects lead to CE junction leakage current*

In the process of forming the transistor, if a process parameter such as temperature is unstable, it will cause the base width of the transistor to become very thin, thus cause the soft breakdown of the transistor and CE junction leakage current of the transistor. Because the defects will be detected by the PCM process in wafer processing and the PCM test results of this batch of sample devices meet the requirements of the technical specifications, therefore, the possibility that the process parameter leads to the transistor base width being thin and causing a leakage current of the CE junction and low static impedance can be excluded.

In the process of forming the transistor, if the residual photo-resist or the adsorption of dust particles exists in the lithography process, it will not be possible to inject the impurity into the base area or cause the impurity density deficiency. The base width will become very thin, as shown in Figure 5. The above-mentioned defects will lead to CE junction leakage current of the transistor and low static impedance of the supply ports between V_{CC} and V_{EE}.

Although the process is carried out in an ultra-clean environment, dust particles in the air cannot be completely avoided. There will be a probability of a random attachment to the surface of the silicon wafer. According to the layout of the circuit of the PIN switch driver, if residual photo-resist or dust particles attach to the region of the output transistor, it will cause low static impedance of the supply ports between V_{CC} and V_{EE}.

5 VALIDATION

When comparing the I-V characteristic curves of the output transistor of the normal and the abnormal, as shown in Figure 6 and Figure 7, it can be found that the abnormal curves of the output

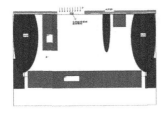
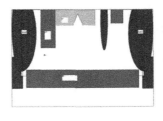
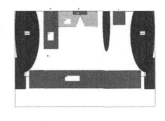

(a)   (b)   (c)

(a) Injection section after lithography
(b) Base area section after annealing process
(c) Transistor section after emitter area

Figure 5. Section of transistor process.

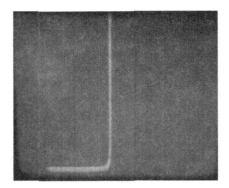

Figure 6. I-V curve of normal circuit.

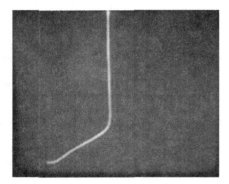

Figure 7. I-V curve of abnormal circuit.

transistor are resistive, which means that the PN junction leakage exists. Thus it can be determined that the reasons why the static impedance between the V_{CC} and V_{EE} power supply ports in the abnormal circuit is low are the lattice defects in the silicon semiconductor, which lead to PN junction leakage (CE leakage and CS leakage), and CE transistor leakage caused by process defects.

6 CONCLUSION

By EMMI analysis, the abnormal point is located accurately. The partly low static impedance between the V_{CC} and V_{EE} power supply ports in the device is due to material defects and the output transistor leakage current caused by the process.

ACKNOWLEDGEMENTS

This work was financially supported by Sichuan Institute of Solid-State Circuits, Chongqing.

REFERENCES

Chen, X., & Zhang, Q. (2005). *Transistor principle and design*. Publishing House of Electronics Industry.
Wang, Y. (1998). *Pulse and digital circuit*. Higher Education Press.

Automotive, Mechanical and Electrical Engineering – Liu (Ed.)
© 2017 Taylor & Francis Group, London, ISBN 978-1-138-62951-6

Application of an unmanned aerial vehicle in fire rescue

Bin Hu
Shanghai Fire Research Institute of Ministry of Public Security, Shanghai, China

Jinpeng Xing
Department of Mechanical and Electronic Engineering, Dongchang College of Liaocheng University, Shandong Liaocheng, China

ABSTRACT: With the rapid development of social economy and the expansion of the function of the fire brigade, firefighting and rescue operations of the public security forces are facing huge risks and challenges, especially in the face of complex and changeable large-scale fire scenes. The problem of how to access fire information accurately and in a timely manner, and the rapid and efficient implementation of a disaster relief strategy have become an urgent priority. In this paper, the research status of the Unmanned Aerial Vehicle (UAV), its specific application to fire rescue and the future development trend of the UAV will be examined.

Keywords: Firefighting; Firefighting and rescue; Large fire; Unmanned Aerial Vehicle

1   INTRODUCTION

An unmanned aircraft vehicle is a kind of non-manned aircraft operated by a radio remote control and a control device. It has the advantages of small size, low cost, ease of use, low environmen-tal requirements, and the integration of a wire-less image transmission system, with considerable autonomy, flight planning and image transmission capability (Eduard and Alex, 2016; Nijsure et al., 2016; KemperKoji et al., 2011). At present, commercial and military UAV systems can be divided

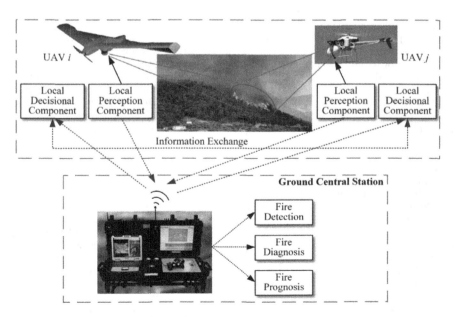

Figure 1.   The work principle diagram of UAV fire rescue.

into the following subsystems: ground command and control system, UAV flight platform, aerial video surveillance platform, wireless real-time image transmission system of microwave, GPS navigation and automatic driving system, high efficiency and high-capacity lithium polymer power supply unit, 3G transmission relay link etc. (Newaz et al., 2016; Jung et al., 2014; Lyu et al., 2016; Ortiz et al., 2013; Qian et al., 2013). Figure 1 shows the work principle diagram of UAV fire rescue (Yuan et al., 2015).

2 TECHNICAL ADVANTAGES OF UAV

Thanks to the advanced flight control platform and background video monitor system, with the perfect flight and ground support system, UAV can be implemented as a comprehensive long-term air monitoring for the ground. In order to achieve a lower overall cost of traditional means cannot be involved in the region for real-time monitoring and auxiliary rescue, the intelligence and advanced nature of UAV is reflected in such aspects as patrol path planning, intelligent analysis, fixed-point continuous monitoring, fire alarm, etc., and makes full use of technology in formulating contingency plans, establishing a rapid response mechanism, on-site fire archiving and forensics prevention means an important role (Marina et al., 2012; Namin et al., 2012). The UAV's technical advantages mainly comprise the following aspects:

1. Flexible. A small UAV is generally less than 100 kg, relying on flight control can be manipulated, only 1–2 people can complete such operations. In the case of poor roads, traffic disruption, the foot can be carried to the scene of disaster accidents, and take-off conditions is very simple, no requirements on the terrain, with unmanned aerial vehicles easy to carry, so the UAV has strong flexibility.
2. Vision comprehensive. The UAV uses broadband, data link technology can achieve over the horizon control, which has a very comprehensive view, according to the needs of the field, can be from different angles, different distances in different light conditions to work. The UAV can be realised at high altitude on the target of overall shooting. The distance and angle can also be adjusted, according to the decision-making needs.
3. Easy to operate. From a technical level, remote video transmission and control system of UAV and ground station access by the network interface, the operators of optical access network. Therefore, using only remote cameras and auxiliary equipment (camera, PTZ etc.), you can watch real-time video camera of the UAV directly.

4. Safe and reliable. Whether in the face of heavy rain, high temperature, typhoon, debris flow and other severe weather conditions or fire, explosions, collapse, toxic and other serious accident disaster sites, UAV technology can effectively avoid traditional fire rescue operations in the presence of short board, which can ensure the safety of the fire brigade.

3 POSSIBLE APPLICATIONS OF UNMANNED AERIAL VEHICLE IN FIRE FIGHTING FORCE

Combined with the actual needs of the fire fighting force, the use of UAV can solve the following four problems (Zeng et al., 2010):

1. Disaster investigation. When a disaster occurs, the use of UAV can ignore the disaster investigation, terrain and environment, and flexibility to carry out the investigation, especially some of the difficult and perilous disaster site, the investigation team to carry out investigation of the case, the UAV can expand the investigation quickly. The use of UAV detection can improve the efficiency of investigation effectively, and identify the key factors of disaster accidents in a timely manner, so that commanders can make the right decision, and can effectively avoid casualties, both to protect people from toxic, flammable, explosive and other dangerous environments, but also to make a comprehensive and detailed analysis of the situation on the ground.
2. Monitoring tracking. The role of UAV is not limited to disaster detection. In the process of handling the disaster, real-time monitoring and tracking by the UAV can provide an accurate view of the disaster. It is convenient for the headquarters at all levels to grasp the dynamic disaster situation in time, so as to make fast and accurate countermeasures to minimise disaster losses.
3. Assisted rescue. The use of unmanned aerial vehicle integration, or the flexibility to carry key equipment, can provide assistance in a variety of situations. First, the use of integrated voice, amplification module to convey instructions. The use of unmanned aerial vehicles to convey instructions to the ground propaganda or directive more effective, especially for high-altitude, high-level projects such as rescue, the UAV as the carrier, which can convey the key instructions effectively. The second is to open up a rescue channel.
4. Auxiliary supervision. The use of aerial photography for high-rise buildings to achieve full

real-time monitoring, timely detection of fire hazards, fire scene real-time control, building fire inspection or on-site fire image storage, air surveillance video access to other security or fire monitoring systems to support large-capacity long-term image storage and retrieval access to support by the intelligent terminal remote view and control some functions.

4 TECHNICAL REQUIREMENTS OF APPLICATION OF UNMANNED AERIAL VEHICLE IN FIRE RESCUE

In order for the fire forces to face the scene of the actual situation of disaster accidents, UAVs in the use of fire forces should also meet certain technical requirements.

1. Reliability. All kinds of fire forces facing a disaster site, especially a fire scene, in which the application environment is more complex, need to consider the wind, smoke, temperature, water and other environmental factors. Therefore, the UAV must be able to meet the harsh environment conditions.
2. Handling performance. The average speed of UAV cruise is 18 km/h, which can be visited in the range of 2 km radius, the endurance required to achieve single battery average more than 30 minutes, the route inspection requirements to achieve the continuous monitoring, and two inspections for clearance less than 15 minutes. In addition, the flight altitude, according to research, most of the disaster accident flight altitude of 200 metres below, taking into account the maximum height of the building, reaching 300 to 500 metres altitude should be able to meet the daily needs of the scene investigation.
3. Stability. The UAV image transmission should be clear and continuous, and the image of the wireless transceiver should have higher requirements. Wireless transmission can easily receive intereference; therefore, it is necessary to have some anti-interference ability, so as to achieve the image in the process of use.
4. Integrated compatibility. The application of a UAV in the fire force should achieve a lot of

expansion functions, which must have good integration and compatibility. Therefore, in full weight, consider the daily training and manoeuver requires the use of unmanned reconnaissance aircraft equipped with the required portable or car arrived at the scene near the individual into the mobile deployment site, emergency equipment development.
5. Rigour. If equipped with UAV, there is an inevitable need for quality training of the operators, so we must form a set of training evaluation mechanisms, the formation of non-daily management and use machine mechanisms and the operator will need training in these.

5 DOMESTIC RESEARCH STATUS OF UAV

Shenzhen City Corbett Aviation Technology Co. Ltd. is committed to the application of expertise in the UAV industry and research and development, specialising in multi-rotor UAV systems and their production, sales and service, with a number of UAV technology patents.

An example of a fire rescue UAV product is shown in Figure 2.

The F6 UAV is a fire and rescue six-rotor UAV, mainly used for city fire emergency rescue, forest fire prevention and routine inspections of various natural disasters and secondary disasters, on-site detection, search and rescue personnel, delivery of rescue supplies and other emergency rescue work.

The F6 fire UAV is mainly used for city fire emergency rescue work, and can be equipped with equipment to monitor the implementation of the fire, to find out the high temperature fire point, conduct a fire hazard survey, and look for survivors as part of the emergency rescue work, to reduce the loss of life and property.

Beichen Aerospace Science and Technology Co. Ltd. is a industry user UAV and application service providers, products used in police anti-terrorism, fire rescue, fire aerial surveillance, aerial video, agricultural plant protection, wiring and land resources survey in areas such as traction and power line.

BCA-Y4 is a vertical take-off and landing of four unmanned aerial rescue system for the fire depart-

Figure 2. RedKite F6 six-rotor fire rescue UAV.

Figure 3.　BCA-Y4 fire monitoring and rescue UAV system.

ment to perform accident investigation, search and rescue, and other tasks, and can be carried out in light of the amount of rescue equipment.

6　SUMMARY AND PROSPECT

Based on the investigation of UAV use in fire, monitoring and rescue functions, the UAV has shown obvious advantages in a large space, large areas of fire disaster reduction, based on its machine vision system and intelligent sensing system. With the continuous development of the social economy, the UAV will show increasing future advantages in fire fighting and fire protection in city buildings, public places, and rescue of forest parks.

ACKNOWLEDGEMENTS

This work was supported by Shanghai Science and Technology Commission (grant number 14dz12068 0 2).

REFERENCES

Eduard, B., & Alex, S.C. (2016). On the tradeoff between electrical power consumption and flight performance in fixed-wing UAV autopilots. *IEEE Transactions on Vehicular Technology, 65*(11), 8832–8840.

Jung, Y.S., You, J.Y., & Kwon, O.J. (2014). Numerical investigation of prop-rotor and tail-wing aerodynamic interference for a tilt-rotor UAV configuration. *Journal of Mechanical Science and Technology, 28*(7), 2609–2617.

KemperKoji, F.P., SuzukiJames, A.O., & Morrison, R. (2011). UAV consumable replenishment: Design concepts for automated service stations. *Journal of Intelligent & Robotic Systems, 61*(1), 369–397.

Lyu, Y., Pan, Q., Zhao, C., Zhang, Y., & Hu, J.W. (2016). Vision-based UAV collision avoidance with 2D dynamic safety envelope. *IEEE Aerospace and Electronic Systems Magazine, 31*(7), 16–26.

Marina, H.G.D., Pereda, F.J., Sierra, J.M.G., & Espinosa, F. (2012). UAV attitude estimation using unscented Kalman filter and TRIAD. *IEEE Transactions on Industrial Electronics, 59*(11), 4465–4474.

Namin, F., Petko, J.S., & Douglas, H. (2012). Analysis and design optimization of robust aperiodic micro-UAV swarm-based antenna arrays. *IEEE Transactions on Antennas and Propagation, 60*(5), 2295–2308.

Newaz, A.A.R, Jeong, S., & Lee, H. (2016). Hyejeong Ryu, Nak Young Chong. UAV-based multiple source localization and contour mapping of radiation fields. *Robotics and Autonomous Systems, 85*, 12–25.

Nijsure, Y.A., Georges, K., Nazih, K.M., Gagnon, G., & Gagnon, F. (2016). Cognitive chaotic UWB-MIMO detect-avoid radar for autonomous UAV navigation. *IEEE Transactions on Intelligent Transportation Systems, 17*(11), 3121–3131.

Ortiz, A., Kingston, D., & Langbort, C. (2013). Multi-UAV velocity and trajectory scheduling strategies for target classification by a single human operator. *Journal of Intelligent & Robotic Systems, 70*(1), 255–274.

Qian, M., Jiang, B., & Xu, D. (2012). Fault tolerant tracking control scheme for UAV using dynamic surface control technique. *Circuits, Systems, and Signal Processing, 31*(5), 1713–1729.

Yuan, C., Zhang, Y., & Liu, Z.X. (2015). A survey on technologies for automatic forest fire monitoring, detection, and fighting using unmanned aerial vehicles and remote sensing techniques. *Canadian Journal of Forest Research, 45*, 783–792.

Zeng, J., Yang, X.K., Yang, L.Y., & Shen, G.Z. (2010). Modeling for UAV resource scheduling under mission synchronization. *Journal of Systems Engineering and Electronics, 21*(5), 821–826.

Automotive, Mechanical and Electrical Engineering – Liu (Ed.)
© *2017 Taylor & Francis Group, London, ISBN 978-1-138-62951-6*

Attitude tracking of quadrotor UAV based on extended state observer

Yebin Liu, Xiangyang Xu & Zhonglin Yang
School of Automation, Beijing Institute of Technology, Beijing, China

ABSTRACT: This paper studies the attitude tracking control problem of the Quadrotor UAV. Considering that the quadrotor usually suffers external disturbances during practical missions, an Extended State Observer (ESO) is introduced to improve the accuracy and robustness of the attitude controller. And the stability analysis shows that the estimation error is convergent. Then the non-linear MIMO system was transformed into linear SISO system by the dynamic feedback. Finally, the effectiveness and the better characteristics of the strategy are illustrated by the simulation results, which lay a solid base for the engineering application.

Keywords: Quadrotor; Attitude control; Extended State Observer

1 INTRODUCTION

In recent years, the quadrotor has become increasingly popular for use in investigation and military fields as well as others. Accordingly, the study of the quadrotor control issues has become an intense area of research in engineering applications and automatic control fields. It is well known that the quadrotor is a typical non-linear system with under-actuated and coupling characteristics, which increases difficulty in the design of the controller. Besides this, an aircraft during flight suffers airflow, flapping, changing internal parameters and other unknown disturbances. These factors reduce the quadrotor flying qualities and could even cause loss of stability (Yebin Liu, 2016). The attitude controller, as the inner loop of the quadrotor control system, directly determines the performance of the mission. Therefore, the attitude controller design of providing the aircraft a good quality performance needs to be further developed.

Several strategies have been developed to solve this problem. (See Bouabdallah, 2004; Bouabdallah, 2005; and Wang, 2014). The PID and LQR have been proposed by Bouabdallah (2004), which assists hover control but has poor robustness. The sliding-mode control method (Bouabdallah, 2005) has a strong anti-interference capability, but the chattering is difficult to eliminate. The robust control method (Wang, 2014) is difficult to realize because the conservative estimation of the bound of uncertainties increases the burden of the controller.

In this paper, a backstepping-based control strategy with ESO compensation is proposed to control the attitude performance of the quadrotor, which can make the system more stable and eliminate the drawbacks produced by the external disturbances. With the ESO dynamic feedback compensation, the closed-loop has much stronger robustness than a conventional strategy, making it easier to be understood and realized. The control law and the observations are described in detail in this paper.

2 DYNAMICS OF THE QUADROTOR UAV

It is well known that the quadrotor is driven by four rotors. To describe the position and attitude of the quadrotor, the body co-ordinate frame is denoted by B-XYZ and the inertial co-ordinate frame is denoted by E-XYZ. The quadrotor has six degrees of freedom and four inputs, by varying the rotation rate ω_i of each rotor to provide corresponding thrust F_i and moment τ_i.

Assuming that the quality of the aircraft is uniformly distributed and its lift surface and centre of gravity are in the same plane, then according to the Newton-Euler equations, the mathematical mode of the system is as follows (Raffo, 2010):

$$\begin{cases} \sum f = m\ddot{\Gamma} \\ \sum \tau = I\dot{w} + w \times Iw \end{cases} \tag{1}$$

$\Gamma = [x\ y\ z]$ denotes the position of the aircraft under the inertial co-ordinate, $\sum f$ is the force vector except gravity, acting on the aircraft. $w = [\phi\ \theta\ \psi] \in R^3$ is the quadrotor Euler angle. $I = diag(I_x\ I_y\ I_z)$ is the inertial matrix. $\tau = [\tau_\phi\ \tau_\theta\ \tau_\psi]$ is the roll, pitch, and yaw control torque. The thrusts can be obtained by:

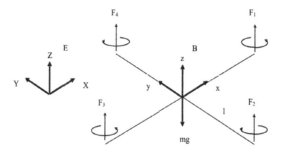

Figure 1. Co-ordinate frames of the quadrotor.

$$F_f = R \begin{bmatrix} 0 \\ 0 \\ U_1 \end{bmatrix} = R \begin{bmatrix} 0 \\ 0 \\ b(\Omega_1^2 + \Omega_2^2 + \Omega_3^2 + \Omega_4^2) \end{bmatrix} \quad (2)$$

$\Omega = [\Omega_1 \ \Omega_2 \ \Omega_3 \ \Omega_4]$ is the rotation rate of each rotor. The R denoting the transformation matrix of the body co-ordinate system to inertial frame is as follows:

$$R = \begin{bmatrix} c_\psi c_\theta & -s_\psi c_\phi + c_\psi s_\theta s_\phi & s_\psi s_\phi + c_\psi s_\theta c_\phi \\ s_\psi c_\theta & c_\psi c_\phi + s_\psi s_\theta s_\phi & -c_\psi s_\phi + s_\psi s_\theta c_\phi \\ -s_\theta & c_\theta s_\phi & c_\theta c_\phi \end{bmatrix} \quad (3)$$

where the notation has been adopted: $C_k = \cos k$, $S_k = \sin k$. The moment τ_i can be obtained by

$$\tau_i = \begin{bmatrix} U_2 \\ U_3 \\ U_4 \end{bmatrix} = \begin{bmatrix} bl(\Omega_4^2 - \Omega_2^2) \\ bl(\Omega_3^2 - \Omega_1^2) \\ d(\Omega_2^2 + \Omega_4^2 - \Omega_1^2 - \Omega_3^2) \end{bmatrix} \quad (4)$$

where b and d are assumed to be constant, the parameter l is the distance between the rotor and centre of the quadrotor. The Gyroscopic moment can be obtained by: $\tau_{gyro} = J_{TP} [-q \ p \ 0]^T \Omega$.

J_{TP} is the motor time constant, the Ω was adopted as: $\Omega = -\Omega_1 + \Omega_2 - \Omega_3 + \Omega_4$. By taking the external disturbances into consideration, the following is obtained:

$$\begin{cases} m\ddot{x} = (\sin\psi\sin\phi + \cos\psi\sin\theta\cos\phi)U_1 + md_x \\ m\ddot{y} = (-\cos\psi\sin\phi + \sin\psi\sin\theta\cos\phi)U_1 + md_y \\ m\ddot{z} = (\cos\theta\cos\phi)U_1 - mg + md_z \\ I_x\ddot{\phi} = (I_y - I_z)\dot{\theta}\dot{\psi} - J_{TP}\dot{\theta}\Omega + U_2 + I_x d_{\tau\phi} \\ I_y\ddot{\theta} = (I_z - I_x)\dot{\phi}\dot{\psi} - J_{TP}\dot{\phi}\Omega + U_3 + I_y d_{\tau\theta} \\ I_z\ddot{\psi} = (I_x - I_y)\dot{\phi}\dot{\theta} + U_4 + I_z d_{\tau\psi} \end{cases} \quad (5)$$

The $D = [d_x \ d_y \ d_z \ d_\phi \ d_\theta \ d_\psi]^T$ denotes the external interference, which is non-structural and bounded.

3 THE CONTROL SCHEME DESIGN

In this section, the control algorithms applied to the attitude sub-system of the quadrotor will be presented, including the backstepping control and the ESO design.

3.1 Backstepping control

Attitude control is the key of the whole control system, and this study adopted a backstepping controller for attitude without conventional linearization and identification online. Based on the above formula (5), the attitude sub-system can be expressed thus: $X = [x_1 \ x_2 \ x_3 \ x_4 \ x_5 \ x_6]^T = [\phi \ \dot{\phi} \ \theta \ \dot{\theta} \ \psi \ \dot{\psi}]^T$. Then the state space expression can be further described as:

$$\dot{X} = f(\mathbf{X}, \mathbf{U}) = \begin{cases} x_2 \\ x_4 x_6 a_1 + x_4 a_2 \Omega + b_1 U_2 \\ x_4 \\ x_2 x_6 a_3 + x_2 a_4 \Omega + b_2 U_3 \\ x_6 \\ x_4 x_2 a_5 + b_3 U_4 \end{cases} \quad (6)$$

where $a_1 = (I_y - I_z)/I_x$, $a_2 = -J_{TP}/I_x$, $a_3 = (I_z - I_x)/I_y$, $a_4 = J_{TP}/I_y$, $a_5 = (I_x - I_y)/I_z$, $b_1 = 1/I_x$, $b_2 = 1/I_y$, $b_3 = 1/I_z$.

As can be seen by the above formulae, every two state variables may constitute a group of sub-systems, meeting the requirements of backstepping for feedback. Use the roll angle as an example to illustrate the structure of backstepping strategy. Firstly, the tracking error: $z_1 = x_{1d} - x_1$, is considered, where x_{1d} is the expected value. Based on the Lyapunov stability theory, the system is stable while $V(z_1) > 0, V(\dot{z}_1) < 0$. Therefore, the positive definite Lyapunov function is chosen:

$$V(z_1) = \frac{1}{2}z_1^2 > 0 \ \underline{\underline{\square}} \ \dot{V}(z_1) = z_1(\dot{x}_{1d} - x_2) \leq 0 \quad (7)$$

In order to keep the stability, the virtual variable $a_1 = \dot{x}_{1d} + c_1 z_1$ was used instead of x_2 thus $\dot{V}(\mathbf{z}_1) = -c_1 z_1^2$.

Then the z_2 was adopted: $z_2 = x_2 - a_1$, The Lyapunov function and its derivative used are as follows:

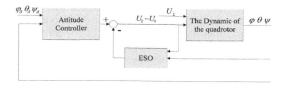

Figure 2. Structure of the control strategy.

$$V(z_1, z_2) = \frac{1}{2}z_1^2 + \frac{1}{2}z_2^2 \qquad (8)$$

$$\dot{V}(z_1, z_2) = z_2(x_4 x_6 a_1 + x_4 a_1 \Omega + b_1 U_2) \\ - z_2(\ddot{x}_{1d} - c_1(z_2 + c_1 z_1)) - z_1 z_2 - c_1 z_1^2 \qquad (9)$$

where $\dot{V}(z_1, z_2) < 0$, there is:

$$U_2 = (z_1 - x_4 x_6 a_1 - x_4 a_2 \Omega - c_1(z_2 + c_1 z_1) - c_2 z_2)/b_1 \qquad (10)$$

Using the same method, the control input U_3 and U_4 is obtained as follows:

$$U_3 = (z_3 - x_2 x_6 a_3 - x_2 a_4 \Omega - c_3(z_4 + c_3 z_3) - c_4 z_2)/b_2 \qquad (11)$$

$$U_4 = (z_5 - x_4 x_2 a_5 - c_5(z_6 + c_5 z_5) - c_6 z_2)/b_3 \qquad (12)$$

3.2 ESO design

In this sub-section, the design of the Extended State Observation (ESO), which is desired to compensate the external disturbances of the system, is considered. Use the roll angle as an example to illustrate the structure of the ESO (WEI Qingtong, 2016).

$$\begin{cases} \dot{x}_1 = \dot{\phi} \\ \dot{x}_2 = \ddot{\phi} = \{(I_y - I_z)\dot{\theta}\dot{\psi} - J_{TP} + U_2\}/I_x + n(t) \end{cases} \qquad (13)$$

where $n(t)$ is the disturbance and uncertainties, it can be written as:

$$\begin{cases} \dot{x}_1 = x_2 \\ \dot{x}_2 = f(x_1, x_2) + g(x_1, x_2)u + n(t) \end{cases} \qquad (14)$$

For this non-linear system, if $f(x_1, x_2)$ is derivable or the derivative is easy to estimate, the function to a new state of the original system can be extended, like: $x_3 = f(x, \dot{x})$. Based on the above analysis, a new ESO is designed as shown below.

As long as there is a reasonable choice of observer parameters, the state observer can guarantee a real-time observation.

$$\begin{cases} \dot{z}_1 = z_2 - \dfrac{a_1}{\varepsilon}(z_1 - x_1) \\ \dot{z}_2 = z_3 - \dfrac{a_2}{\varepsilon^2}(z_1 - x_1) + bu \\ \dot{z}_3 = -\dfrac{a_3}{\varepsilon^3}(z_1 - x_1) \end{cases} \qquad (15)$$

where $z_1 - x_1$ is the observation error and z_i is the observation state, then the observer error convergence is proved as below:

Firstly, the observation error is defined as $\eta = \begin{bmatrix} \eta_1 & \eta_2 & \eta_3 \end{bmatrix}^T$, where $\eta_1 = z_1 - x_1$, $\eta_2 = z_2 - x_2$, $\eta_3 = z_3 - x_3$. Then we have,

$$\begin{cases} \dot{\eta}_1 = -\dfrac{a_1 \eta_1}{\varepsilon} + \eta_2 \\ \dot{\eta}_2 = -\dfrac{a_2 \eta_1}{\varepsilon^2} + \eta_3 \\ \dot{\eta}_3 = -\dfrac{a_3 \eta_1}{\varepsilon^3} + \dot{n}(t) \end{cases} \qquad (16)$$

That is $\dot{\eta} = A\eta + B\dot{n}(t)$, in the formula $A = \begin{bmatrix} -\frac{a_1}{\varepsilon} & 1 & 0; & -\frac{a_2}{\varepsilon^2} & 0 & 1; & -\frac{a_3}{\varepsilon^3} & 0 & 0 \end{bmatrix}^T$, $B = \begin{bmatrix} 0 & 0 & 1 \end{bmatrix}^T$.

Make $\lambda^3 + a_1\lambda^2 + a_2\lambda + a_3 = \prod\limits_{}^{3}(\lambda + \lambda_i) = 0$, for matrix A, there are Vandermonde matrix Q contents:

$$A = Q diag\left\{-\dfrac{\lambda_1}{\varepsilon}, -\dfrac{\lambda_2}{\varepsilon}, -\dfrac{\lambda_3}{\varepsilon}\right\}Q^{-1} \qquad (17)$$

Define $\lambda_{min} = \min\{\lambda_1, \lambda_2, \lambda_3\}$, to get the results of (16):

$$\eta(t) = \exp(At)\eta(0) + \int_0^t \exp(A(t-\tau)) \cdot \dot{n}(t)d\tau B \qquad (18)$$

Supposing the upper bound of the $n(t)$ is $L(0 \leq L \leq +\infty)$, Then,

$$\|\eta(t)\| \leq \|\exp(At)\|\|\eta(0)\| + L\left\|\int_0^t \exp(A(t-\tau))\right\|\|B\|$$

$$\leq \left\|Q diag\left\{\exp\left(-\dfrac{\lambda_1}{\varepsilon}t\right), \exp\left(-\dfrac{\lambda_2}{\varepsilon}t\right), \exp\left(-\dfrac{\lambda_3}{\varepsilon}t\right)\right\}Q^{-1}\right\|\|\eta(0)\| + L\int_0^t \|\exp(A(t-\tau))\|\|d\tau B\|$$

$$\leq \|Q\|\|Q^{-1}\|\left\|\exp\left(-\dfrac{\lambda_{min}}{\varepsilon}t\right)\right\|\|\eta(0)\| + \|Q\|\|Q^{-1}\|L\int_0^t \exp\left(-\dfrac{\lambda_{min}(t-\tau)}{\varepsilon}\right)d\tau\|B\|$$

$$\leq \|Q\|\|Q^{-1}\|\left\|\exp\left(-\dfrac{\lambda_{min}}{\varepsilon}t\right)\right\|\|\eta(0)\| + L\dfrac{\varepsilon}{\lambda}\left(1 - \exp\left(-\dfrac{\lambda_{min}t}{\varepsilon}\right)\right)\|B\|$$

665

where $\|B\| = 1$, thus $\lim_{\varepsilon \to 0^+} \|\eta(t)\| = 0$. If appropriate parameters $\varepsilon, a_1, a_2, a_3$ are selected, the ESO states converge to system state and total disturbance in finite time. Thus, the control law $u = (v - z_3)/b$ is reached, the v means the pseudo-control signal (Yang Li-ben, 2015).

4 SIMULATION AND RESULTS

In this section, the effectiveness and the improved characteristics of the strategy are illustrated by the simulation results. The parameters of the quadrotor used in the simulation are shown in Table 1.

To verify the effectiveness of the new strategy in the attitude controller, different types of external disturbances are added to the three channels respectively. The disturbances are chosen as

$d_\varphi = 0.02 * \sin(0.8t + 50)$ $d_\theta = 0.1 * \cos(0.5t + 30)$, $d_\psi = 0.01 * \sin(0.2t + 50)$. To illustrate the manoeuvring ability of the attitude, the command chosen is:

$$[\varphi_d \ \ \theta_d \ \ \psi_d] = [0.2 * \sin(0.5t), 0.3 * \cos(0.5t), \\ - 0.2 * \sin(0.5t)] \text{ rad.}$$

Table 1. Quadrotor parameters.

Parameter	Value
Mass (m)	1 kg
Distance (l)	0.225 m
X-axis Inertias (I_x)	8.1×10^{-3} Nm $\cdot$ s²/rad
Y-axis Inertias (I_y)	8.1×10^{-3} Nm $\cdot$ s²/rad
Z-axis inertias (I_z)	1.42×10^{-2} Nm $\cdot$ s²/rad
J_{TP}	1.04×10^{-4} Nm $\cdot$ s²/rad

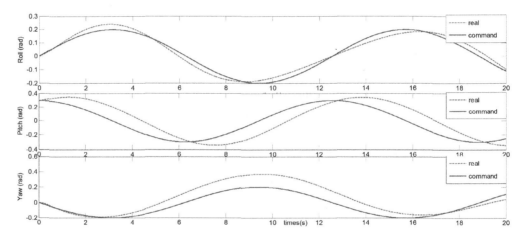

Figure 3. Comparison of the command and real angle with no ESO.

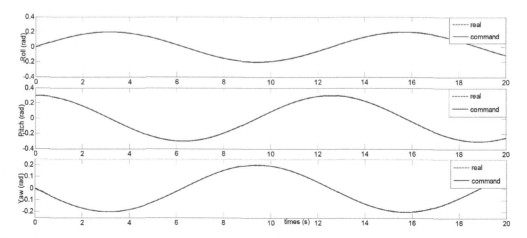

Figure 4. Comparison of the command and real angle with ESO.

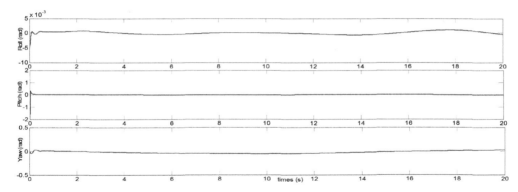

Figure 5.   Estimate errors of disturbance observers.

As is shown in Figure 3, the external disturbances will largely reduce the tracking performances of the attitude control system. The simulation results with ESO compensations are given in Figure 4, which indicate that the backstepping attitude controller has better characteristics and strong robustness. In Figure 5, the estimate errors of ESO are shown, proving that the introduced ESO has good disturbance observation.

5   CONCLUSION

In this paper, a backstepping control strategy with ESO compensations was proposed for the quadrotor UAV, which aims to improve the quality of performance in the presence of external disturbances. The ESO is designed to compensate for the dynamic system. The simulation results have demonstrated that the introduced strategy has good performance and strong robustness in exist of the external disturbances.

REFERENCES

Bouabdallah, S., Siegwart, R. (2005). Backstepping and sliding-mode techniques applied to an indoor micro quadrotor. *Robotics and Automation. 2005 Proceedings.* (pp. 2247–2252).

Han Jingqing. (2008). *Active Disturbance Rejection Control Technique.* Beijing: National Defence Industry Press.

Raffo, G.V., Ortega, M.G. and Rubio, F.R. (2010). An integral predictive non-linear control structure for a Quadrotor helicopter, automatic, 46(1), 29–39.

Bouabdallah, S., Noth, A. and Siegwart, R. (2004). PID vs LQ control techniques applied to an indoor micro quadrotor in *Proceedings of 2004 IEEE/RSJ International Conference*, 3, 2451–2456.

Wang Lu, Su Jian-bo. (2013). Attitude tracking of aircraft based on disturbance rejection control. *Control Theory & Applications.* 30(12).

Wang, Y.Q., Wu, Q.H. and Wang, Y. (2014). Distributed cooperative control for multiple quadrotor systems via dynamics surface control. *Non-linear Dynamics.* 75(3), pp. 513–527.

Wei Qingtong. Etc (2016). Backstepping-Based Attitude Control for a Quadrotor UAV Using Non-linear Disturbance Observer. *Proceedings of the 34th Chinese Control Conference*, pp. 771–776.

Yang Li-ben, Zhang Wei-guo, Huang De-gang. (2015). Robust trajectory tracking control for underactuated quadrotor UAV based on ESO. *Systems Engineering and Electronics,* 37 (9).

Yebin Liu, Xiangyang Xu. (2016). The Anti-interference Maneuver Control based on Extended Observer for Quadrotor UAV. Intelligent Human-Machine Systems and Cybernetics Proceedings, pp. 305–309.

Automotive, Mechanical and Electrical Engineering – Liu (Ed.)
© 2017 Taylor & Francis Group, London, ISBN 978-1-138-62951-6

Design of a high performance triple-mode oscillator

Zhuo Li, Suge Yue, Yantu Mo, Jia Wang & Fei Chu
Beijing Microelectronics Technology Institute, Beijing, China

ABSTRACT: A high performance triple-mode oscillator based on 0.25 um BCD process is presented in this paper. The conventional ring structure with mode control modules is adopted to produce a ramp signal and square wave. As the simulation results show, under the temperatures of −55°C–125°C, the output frequency varies from 485.9 KHz to 512 KHz in the master mode, with a maximum change rate of 2.82%. In the resistance adjustment mode, the adjustable range is 13.4 KHz–2.78 MHz. In the synchronisation mode, the synchronous frequency covers from 15 KHz to 4.2 MHz. An oscillator with great performance has thus been successfully applied in a DC-DC converter circuit.

Keywords: Triple-mode; resistance adjustment; synchronisation; high precision

1 INTRODUCTION

An oscillator is an important unit in many analogue systems which need a clock signal or switch control. A ramp signal oscillator, as a pulse, is a core module in DC-DC converter systems (Siu man et al., 2006; David Suo et al., 2010). The performance of the oscillator directly determines the characteristic of this kind of power system, so its stability is critical to the whole system (Zhao Chengguang et al., 2015). With changes in power voltage and environment temperature, the output frequency of an oscillator is required to be accurate and steady (Kaertner F X, 1989; Hajimiri A et al., 1998; Zhou Qianneng et al., 2013). Depending on the application condition, the internal oscillator may be required to provide clock signals of different frequency. However at present, oscillators usually have the disadvantages of fixed frequency, a small adjustable range and a simplex control mode.

In this paper, an oscillator with a triple-mode is proposed for a peak current buck DC-DC converter. The three schemas are the master mode, resistance adjustment mode and synchronisation mode. The output wave, ramp signal, large duty cycle square wave and standard square wave are respectively used for slope compensation, switch controlling and sub-class driving. Under the temperature of −55°C–125°C, the largest frequency range can cover 13.4 KHz–4.2 MHz. The circuit also has high frequency precision and great temperature stability.

2 CIRCUIT DESIGN

2.1 *Framework of the oscillator*

The framework of the oscillator is shown in Figure 1. The whole system is composed of mode detection, main oscillator, delay circuit, frequency phase detector, input and output controller, a total of five modules, implementing configuration and switch of the three modes. In the master mode, the RT port is floating while the delay circuit and phase frequency detector modules are closed. The capacitor in the main oscillator circuit is charged by the internal constant current reference to generate a stable ramp signal, RAMP, at the frequency of 500 KHz. The SYNC_CLK as an output port, exports a standard square wave. In the resistance adjustment mode, the port RT is connected to a certain resistance R_T with no synchronous clock forced into SYNC_CLK. The mode detection module opens but the delay circuit and phase frequency detector modules remain closed. The operating current of the oscillator is regulated by the value of resistance R_T. In the synchronisation mode, the RT port still needs a resistance, while the SYNC_CLK changes to be an input port. All the five module start to work. The frequency phase detector module generates a synchronisation

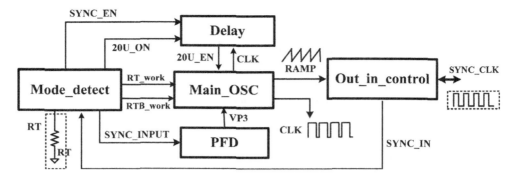

Figure 1. Framework of the oscillator.

control signal IP3 to modulate the charging current of the capacitance in order to achieve the function of synchronisation.

2.2 Mode detection circuit

The mode detection is used to detect the resistance R_T and the input external synchronous clock. It is also designed to export the control signal, SYNC_EN, 20U_ON, SYNC_INPUT, RT_work, in order to switch the oscillation modes. Figure 2 is the resistance recognition circuit.

As the circuit shows, when RT is floating, RT_float is low voltage and RT_float_B turns high. Otherwise, when a certain resistance R_T is contacted to the terminal R_T, a current path is formed along the C branch and GND. M3 and M4 have current flowing through which makes the node O_1 drop lower than the in-phase input of the comparator Amp2. Thus, the RT_float turns over and the signal RT_float_B is reduced to a low level. Both of the signals are used to enable the internal synchronous detection function.

Suppose the RT resistance value is RT. Since the node O2 is a constant 0.35 V voltage, I_C, the current floating in the path C is:

$$I_C = \frac{V_{O2}}{R_0 + R_T} = \frac{3.5V}{R_0 + R_T} \tag{1}$$

According to the formula, it can be seen that I_C is inversely proportional to the resistance R_T. Therefore, the current of the branch can be adjusted through resistance R_T, and then the gate voltage, Vbias_1 and Vbias_2, changes accordingly and synchronously.

As Figure 3 illustrates, while the terminal RT floating, RT_float and RT_float_B are low and high respectively, the transmission gates turn down and the node O_3 stays at high voltage due to the resistance R. As a result, the signal SYNC_EN

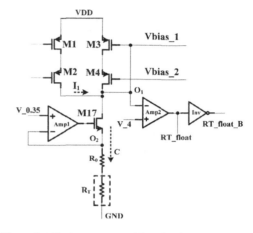

Figure 2. Resistance recognition circuit.

and 20U_ON keep the default low and high. In another scenario where there is R_T resistance but no synchronous clock, the transmission conducts and the node O_3 drops to a low level because of the low voltage from SYNC_IN. So the SYNC_EN and 20U_ON both decrease to GND. While in the synchronous mode, the O_3 changes along with the signal SYNC_IN and under the function of capacitance C1 and C2, both of the SYNC_EN and 20U_ON are in high voltage and act on the other synchronisation modules.

Figure 4 shows the modes control circuit in the mode detection module.

In the resistance adjustment mode, the module output SYNC_EN and 20U_ON are low and high respectively. The delay module is turned down and 20U_EN keeps the default low level. The two transmission gates are both open, in the meantime the output RTB_work is low and RT_work turns to a high level, which are made to select the operating current in the main oscillator.

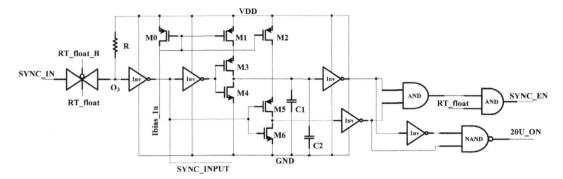

Figure 3. Circuit of synchronisation signal detection.

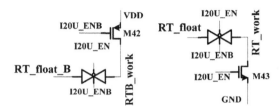

Figure 4. Circuit of mode control.

If operated in the synchronisation mode, the outputs of the module SYNC_EN and 20U_MODE are both set to a high voltage level which turns the delay module on. During the 20 us delay time, I20U_EN remains in high voltage while RTB_work is high and RT_work is the opposite. After the delay period, I20U_EN returns to low and the circuit turns back to the status of the resistance adjustment mode.

2.3 The master mode

Figure 5 is the schematic representation of the main oscillator. The power voltage VDD is 5V, while Vbias_1 and Vbias_2 are the gate voltages of the current mirror from previous class. RAMP is the output slope signal and CK is a large duty ratio square wave. According to the analysis of the mode detection circuit, in the master mode the terminal RT floats and M18 cuts off on account of the high voltage signal, RTB_work. And then the current path A is shut off. In the meantime, M20 is turned on and the current path B conducts which gives access to the internal constant reference Ibias_075 into the structure. If no synchronous clock is imported, the SYNC_EN's default is low. There is current flowing through M26-M28 and M29-M30. Hence, the current of M13-M14 is the sum of these two currents. By means of the mirror

M13-M16, a constant current Ion is produced to charge C1. On the basis of the operating principle of capacitance, the time of charging is:

$$I_{on} = C \times \frac{dU}{dt} = \frac{(V_2 - V_1)}{T_{on}} \cdot C_1 \qquad (2)$$

$$T_{on} = \frac{(V_2 - V_1)}{I_{In}} \cdot C_1 \qquad (3)$$

In a similar way, the discharging time of C1 is:

$$T_{off} = \frac{(V_2 - V_1)}{I_{off}} \cdot C_1 \qquad (4)$$

Therefore, the oscillation frequency is:

$$f = \frac{1}{T_{on} + T_{off}} \qquad (5)$$

In the circuit, the transistor M32 is designed in a large size so that I_{off} is a large current when M32 is operating. For this reason, the effects on frequency from T_{off} can be ignored and the oscillation is mainly decided by the constant current Ion.

2.4 Resistance adjustment mode and synchronisation mode

In the resistance frequency adjustment mode, with the analysis of the mode detection circuit, the values of Vbias_1 and Vbias_2 depend mainly on R_T. The RTB_work is low and RT_work is high in this situation. Therefore, the transistor M18 is on and M19 shuts down. So the pathway A is conductive and B is closed. The current I_A mirrored from previous classes charges C1. Under the control of Vbias_1 and Vbias_2, M5 and M6 constitute a mirror image structure with M3-M4 so that I_C

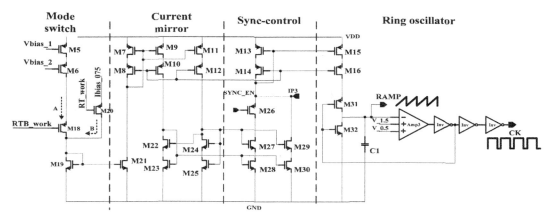

Figure 5. Main oscillator.

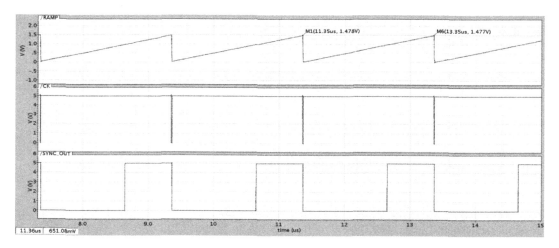

Figure 6. Wave of master mode.

as well as operating the current of the oscillator is subject to the R_T. Further, it is feasible to regulate the frequency by selecting proper resistance R_T.

In the synchronisation mode, based on the analysis above, during the 20 us delay time, I20U_EN is in high level and RT_work is opposite. It is operating in the resistance adjustment mode. After the delay period, RT_work starts to hop and the whole system switches to the resistance adjustment mode. At the same time, SYNC_EN jumps to a high voltage level to enable the PFD module. By way of comparison, the signal CLK and SYNC_IN, the synchronous conductor, IP3 is generated to modulate the current in M13-M14 in order to achieve the synchronisation function. The phase frequency detection circuit is designed according to a con-

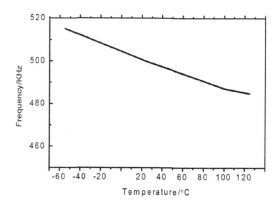

Figure 7. Frequency with temperature.

672

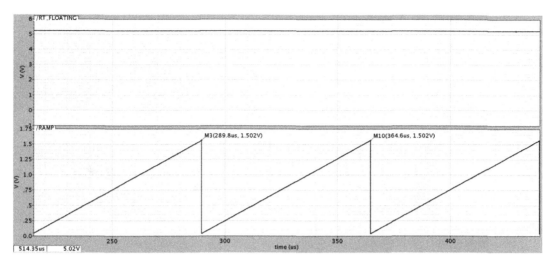

Figure 8.   Wave in resistance adjustment mode.

ventional structure based on the literature (Behzad Razavi, 2000).

3   SIMULATION RESULTS AND ANALYSIS

Based on a 0.25 um BCD process, the schematic is simulated and verified by Spectre embedded in the Cadence. The simulation results are shown as follows:

As shown in Figure 6, it is the output wave in master mode. Under the condition of normal temperature, the frequency of the oscillator is 499.7 KHz which deviates only by 0.06% from the desired value, 500 KHz. Thus it can be seen that the oscillating frequency has very high precision.

The frequency curve with temperature is illustrated in Figure 7. In the temperature range of −55°C–125°C, the frequency varies from 485.9 KHz to 512 KHz with a maximum offset ratio of 2.82% which means great stability of the whole system.

Figure 8 is the RAMP signal when RT is 343 KΩ. It can be seen that the RT_FLOT is a high voltage signal which is fully consistent with the design target. It represents that the resistance frequency adjustment function is correct. And the output frequency is 13.4 KHz in this condition.

Figure 9 illustrates the variable output frequency with the value of R_T resistance. The stable adjustment scope of frequency is 13.4 KHz–2.78 MHz in the resistance value of

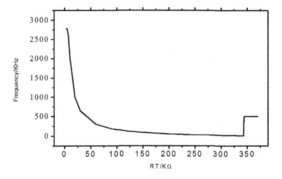

Figure 9.   Frequency with RT resistance.

3 KΩ–343 KΩ. The frequency decreases with increasing R_T resistance. When it is over 343 KΩ, the current is almost disconnected and the system returns back to the master mode. In the other status when R_T is too small to trigger effective changes in the current in the branch, the frequency will change very slowly.

Figure 10 shows the input synchronous clock SYNC_IN and output signal RAMP in the synchronisation mode. When no external signal comes in, the oscillation speed mainly depends on the resistance R_T. Otherwise, when there is a synchronous clock, after 20 us delay time, the RAMP goes to the synchronism and realises complete synchronisation at 80us. In this mode, the maximum synchronous scope is 15 KHz–4.2 MHz.

673

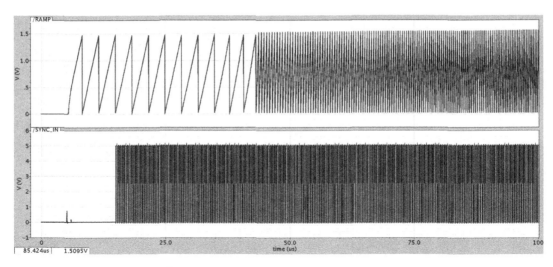

Figure 10. Wave of synchronisation mode.

4 CONCLUSION

In this paper, a high-precision, great stability triple-mode oscillator is proposed for a DC-DC converter. The oscillator is designed in a simple ring structure. The oscillator frequency is regulated by an adjustable current which is used to charge the capacitance. According to the simulation results, the oscillator be switched into three modes. In the master mode, the scope of the output frequency is 485.9 KHz–512 KHz during a temperature of −55°C–125°C with a maximum offset of 2.82%. In resistance adjustment mode, the frequency can be regulated from 13.4 KHz to 2.78 MHz. In the synchronisation mode, the external frequency can cover from 15 KHz to 4.2 MHz. The features of the circuit totally satisfy the demands of the DC-DC converter which have also been successfully applied in a DC-DC converter chip.

REFERENCES

Behzad Razavi. (2000). Design of analog CMOS integrated circuits. McGraw-Hill, Inc.

David Suo, Suo Zhouchao, Chiung A K. (2010). A buck converter with sawtooth voltage feed forward control. IEEE International Solid–State and Integrated Circuit Technology. Shanghai, China: IEEE Press, pp. 147–149.

Hajimiri A, Lee T. (1998). A general theory of phase noise in electrical oscillators. *IEEE Solid State Circuits, 33(2)*, 179–194.

Kaertner F X. (1989). Determination of the correlation spectrum of oscillators with low noise. *IEEE Trans Microw-Theory Tech., 37(1)*, 90–101.

Siu man, Mok Philip K T, Leung Ka Nang, et al. (2006). A voltage-mode PWM buck regulator with end-point prediction. *IEEE Trans Circuit Syst II, Expr Briefs, 53 (4)*, 2994–2998.

Zhao Chengguang, Feng Quanyuan, Wang Dan. (2015). Design of a low temperature drift and self-start function oscillator for AC-DC. 45(2), 213–216.

Zhou Qianneng, Wang Li, Li Hongjuan, Li Qi. (2013). Design of sawtooth-wave oscillator for DC-DC switching power. *Journal of Chongqing University of Posts and Telecommunications (Natural Science Edition), 25(4)*, 432–435.

Automotive, Mechanical and Electrical Engineering – Liu (Ed.)
© *2017 Taylor & Francis Group, London, ISBN 978-1-138-62951-6*

Electromagnetic compatibility design of power supply based on DSP industrial control system

Wei Liu
Logistics Group, Beijing Institute of Technology, Beijing, China

Xindi Zheng
School of Information and Electronics, Beijing Institute of Technology, Beijing, China

Zenan Lin
International School, Beijing University of Posts and Telecommunications, Beijing, China

Jian Xu
School of Automation, Beijing Institute of Technology, Beijing, China

ABSTRACT: With the rapid development of electronic technology, there is a need for higher requirements in electromagnetic compatibility of power supply systems in industrial control systems. This paper first puts forward the whole design of a high-frequency switching power supply system based on a Digital Signal Processor (DSP), and then designs the main controller, sampling circuit and drive circuit of the power system. It selects the important parameters of a high-frequency transformer and provides an improved design for an ElectroMagnetic Interference (EMI) filter circuit to filter out power electromagnetic interference. Simulation results show that the proposed DSP-based high-frequency switching power supply system in this paper can better filter out power electromagnetic interference and improve the effectiveness and reliability of high-frequency switching power supply systems.

Keywords: DSP; Switching power supply; ElectroMagnetic Interference; EMI

1 INTRODUCTION

In recent years, high-frequency switching power supply has been widely used in many industrial fields. The traditional switching power supply adopts the analogue component to regulate and control the output voltage, but this kind of regulate method has poor stability, and the testing and maintenance are more difficult. With the emergence of the Digital Signal Processor (DSP), DSP can be used in all aspects of the power system, which gives the switching power supply the advantages of good controllability, programmability and easy maintenance.

However, at present in DSP-based switching power supply systems, the interferences caused by the complex electromagnetic environment are large. These interferences will have a great impact on other devices (Zhou et al. 2004), and power electromagnetic compatibility has become the current research hotspot (Li et al. 2012; Yang, 2013; Salem et al., 2014; Wu et al., 2013; Chen et al., 2014; Yang et al., 2015). Li Min et al. (2012) analyse and discuss the EMI problems of on-board charging system switching power supply in the actual work, and adopt measures such as increasing the distance

between lines and serial termination to further reduce the system's ElectroMagnetic Interference (EMI). Yang Yubang (2013) analyses the causes of electromagnetic interference, and uses a filter circuit to suppress electromagnetic interference.

In order to filter out the electromagnetic interference of DSP-based high-frequency switching power supply, this paper firstly presents the overall design of high-frequency switching power supply system based on DSP, then designs the main controller, samples circuit and drives circuit of power system. It selects high-frequency transformer important parameters, and focuses on the EMI filter circuit design improvements.

2 SWITCHING POWER SUPPLY SYSTEM DESIGN

Switching power supply is equipment which converts the AC power supply to a stable voltage or current for the use of electrical devices (Brown et al. 2004). Its basic block diagram is shown in Figure 1. Switching power supply mainly consists of an EMI filter, rectifier filter, high-frequency transformer, high-frequency rectifier filter output circuit, control circuit, auxiliary power and other components. The basic working principle of switching power supply is: AC voltage input is converted into a DC voltage by the rectifier and filter. DC voltage is transformed into high-frequency AC voltage after high-frequency. After passing through a high-frequency rectifier filter, the high-frequency AC voltage is transformed into DC voltage and output.

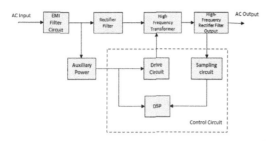

Figure 1. Block diagram of the basic power supply system.

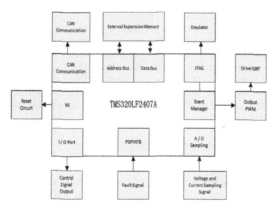

Figure 2. DSP control circuit basic structure.

2.1 Control circuit design

2.1.1 Primary controller

Switching power supply control circuit uses DSP. DSP has good programmability; its speed is far more than ordinary microprocessors. This paper takes TMS320 LF2407 chip as the main controller. The chip has a built-in Analogue-Digital Conversion (ADC) module, including two event management modules, EVA and EVB. The implementation of 40 million instructions only needs 1 second and the processing speed is fast. Figure 2 shows the DSP control circuit of the basic structure.

2.1.2 Sampling circuit

In this paper, the switching power supply circuit uses control on the voltage and current at the same time. The voltage sampling circuit diagram and current sampling circuit are shown in Figure 3 and Figure 4.

In the voltage sampling circuit, the linear optocoupler can be used to isolate the photo-electricity in order to prevent disturbance and guarantee the stable work of the system. Linear optocoupler HCNR201 takes an optical receiver circuit into the single-shot mode, and eliminates the nonlinear DC circuit to linear isolation by the nonlinear feedback circuit.

A current sampling circuit uses CHF-300E Hall sensor. The advantage of using Hall sensors is that the Hall sensors do not need to be connected in series in the process of current sampling. This non-contact test characteristic can isolate the high and low voltage circuits naturally. CHF-300E Hall sensor has great linearity, measurement accuracy and anti-interference ability.

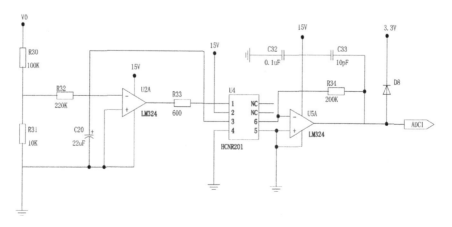

Figure 3. Voltage sampling circuit.

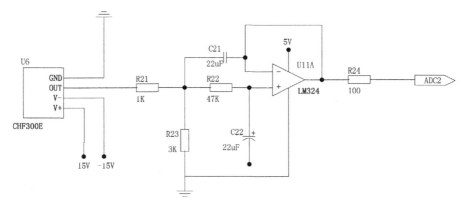

Figure 4. Current sampling circuit.

2.1.3 Drive circuit

MOSFET drive circuit is used to provide the required drive current and voltage for the MOS tube. As the PWM signal output current of DSP is only a few milliamperes, and the driving process of the capacitor charging instantaneous current, it needs to amplify the PWM voltage and current. As the switching power supply is at a higher frequency, so it needs to output the PWM signal and isolate the controlled object in order to make DSP work with stability. Based on this, the choice of PWM chip in this article is adjusted by the TLP250 second pin, the output by the sixth and seventh pin, and the output voltage regulator tube is adjusted by Z1. The driving circuit is shown in Figure 5.

2.2 High-frequency transformer

High-frequency transformer is an electromagnetic mutual induction, coupled device. In fact, it is a key component of the entire power circuit. It mainly consists of core and winding two parts. An important factor that affects the performance of the switching circuit is the high-frequency transformer. Therefore, the selection of appropriate high-frequency transformer parameters plays an important role in the performance of the switching circuit.

2.2.1 Magnetic core

Electromagnetic core should make the appropriate choice based of the output capacitor and the frequency of the output frequency. After selecting the appropriate electromagnetic core material, according to the characteristics of the core, working conditions, the highest temperature transformer and other factors, the maximum working flux density is calculated in order to achieve efficient use of the core and reduce losses. Most high-frequency transformers prefer an iron-containing material.

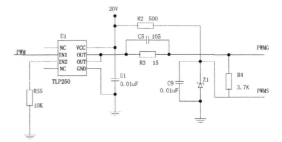

Figure 5. Drive circuit.

2.2.2 Transformer ratio

The ratio of the primary side turns to the secondary side turns (n) is the transformer ratio. In the calculation of transformer ratio, it needs to meet certain conditions, namely, when the input voltage is minimum; we can get the maximum voltage V_{0max} and the maximum duty cycle of the transformer D_{max}. In the minimum input voltage, the primary voltage and secondary voltage are the smallest, respectively V_{1min} and V_{2min}. The secondary voltage is

$$V_{2min} = (V_{0max} + V_f + V_x)/D_{max} \qquad (1)$$

where, V_f is the output rectifier voltage drop and V_x is the output winding on the pressure drop.

2.2.3 Winding turns

When both ends of the transformer are working, the primary side winding turns

$$N_1 = \frac{V_{1max}}{K_0 \times f \times A_e \times B_m} \qquad (2)$$

where, A_e is the cross-sectional area of the magnetic core, K_0 is the proportional coefficient, B_m is

677

the transformer flux density and f is the operating frequency of the transformer.

2.3 *EMI filter circuit design*

In the DSP industrial control system, the switching power supply provides power for various components of the control system. However, due to the high switching frequency of switching power supply, the switching power supply will form a strong electromagnetic interference in the high-frequency fast on and off operation. It will have an impact on other modules in the DSP industrial control system.

Switching power supply will not only interfere with other modules in the industrial control system, but also interfere other modules in the industrial control system. As shown in Figure 6, other modules in the industrial control system will produce differential-mode interference and common-mode interference on the switching power supply. At the same time, the switching power supply will produce radiation interference and noise interference to other modules in the industrial control system, including return noise and output noise. How to eliminate the industrial control system switching power supply to other modules of the interference and other modules of the switching power supply interference has been the main focus of the scholar's attention. The main current method is to add an EMI filter in the part of the switching power supply circuit. This filter can not only filter the interference signals generated by other modules in the industrial control system, but also filter out the radiation interference and noise interference of switching power supply to other modules in the industrial control system.

2.3.1 *Conventional EMI filtering*

Figure 7 is the traditional EMI filter circuit. As can be seen from Figure 7, the traditional EMI filter circuit is mainly constituted of three components of the differential-mode capacitor Cx, common-mode capacitor Cy and common-mode choke L. In the right part of the common-mode choke

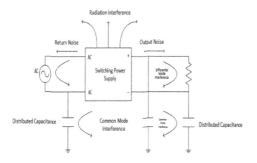

Figure 6. Interference and noise sources of switching power supply.

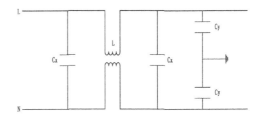

Figure 7. Traditional EMI filter circuit.

circuit, the two common-mode capacitors Cy are first in series and then in parallel with the differential-mode capacitor Cx, and then in parallel with the common-mode choke coil L. In the left part of the common-mode choke circuit, the differential-mode capacitor Cx is connected to the coil of the common-mode choke coil L. Since the upper and lower coils of the common-mode choke coil L are wound in opposite special structures, the magnetic fields generated by the currents flowing through the two coils cancel each other, and the common-mode current can be remarkably suppressed, and the differential-mode capacitance Cx. The common-mode capacitor Cy and the common-mode choke L can further improve the suppressing ability of the common-mode choke L to interfere with the common-mode current. However, when the differential current passes through the opposite coil, the magnetic fields cannot cancel each other. Therefore, the common-mode choke coil L does not have any inhibitory effect on the magnetic interference generated by the differential current.

In general, choose the metal film capacitor as differential-mode capacitor Cx with value 0.1–1 uF; select ceramic capacitor as common-mode capacitor Cy with value 2,200–6,800 pF; choose magnetic core with high permeability, and a common-mode choke coil L with good high-frequency performance.

2.3.2 *Improved EMI filtering*

Figure 8 shows the improved EMI filter circuit. As can be seen from Figure 8, the improvement is mainly the left part of the common-mode choke circuit. Add inductance in the differential-mode capacitor Cx and the upper and lower parts of common-mode capacitor Cy connected part and at the same time, also add inductance in the common-mode capacitor Cy and the output of the upper and lower parts. So, in the upper and lower parts, the new inductance and common-mode capacitor Cy constitute a T-type structure. The T-type circuit can not only improve the high-frequency performance, but also improve the resonant frequency and bypass effect. In addition, compared with the traditional EMI filter circuit, the improved EMI filter circuit has better loss performance.

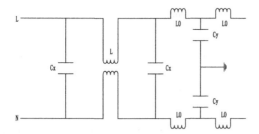

Figure 8. Improved EMI filter circuit.

3 SIMULATION

In the actual circuit design, the performance of the filter is impacted by a variety of factors. Therefore, the design of the various parameters of the EMI filter circuit is very difficult. However, the emergence of computer simulation technology greatly reduces the difficulty of designing EMI filter circuits, significantly shortens their design cycle, while providing test indicators of the design of the EMI filter circuit to obtain accurate simulation data for adjusting the EMI filter circuit. PSpice is a popular circuit design simulation tool. The simulation part of this section is based on the performance of the improved EMI filter circuit; analyses and improves the common-mode, differential-mode and inductance on the performance of EMI filter.

Figure 9 shows a traditional EMI filter circuit simulation model in which the input AC voltage is 220 V, the DC voltage is 0 V i.e. 220 V pure AC. Both differential-mode capacitors C1 and C2 are 0.1 uF, and both common-mode Capacitance C3 and C4 are 2,200 pF. The input impedance Rs1 is 50 ohms, the output impedance RL1 is also 50 ohms, and the common-mode choke inductance L is 5 mH. Figure 10 shows the improved EMI filter circuit simulation model. Compared with the traditional EMI filter circuit simulation model, the power section, input impedance, output impedance, common-mode capacitance, differential-mode capacitance, and common-mode choke inductance values are all the same. The inductance of L1 and L2 are both 0.36 mH.

Figure 11 shows the performance of the EMI filter circuit before and after the improvement. As can be seen from Figure 12, the loss of improved EMI filter circuit insertion is significantly lower than that of the traditional EMI filter circuit. When the frequency increases, the advantages of the improved EMI filter circuit performance are more obvious, and the difference between the two-insertion loss values becomes large.

Figure 12 shows the difference made by differential-mode capacitor Cx on the performance of the EMI filter circuit. The values of 0.1 uF, 0.22 uF and 0.47 uF are selected in the simulation. As can be seen from the figure, the resonant frequency of the

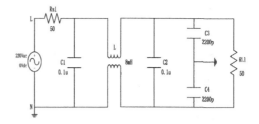

Figure 9. Traditional EMI filter circuit simulation model.

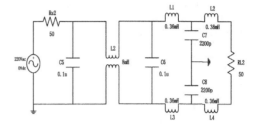

Figure 10. Improved EMI filter circuit simulation model.

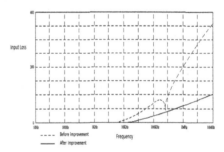

Figure 11. EMI filter circuit performance comparison before and after improvement.

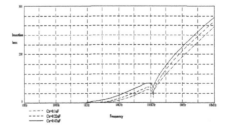

Figure 12. Effect of differential capacitance Cx on EMI filter circuit performance.

three curves are 150 kHz. That is, the differential-mode capacitor Cx improve the EMI filter circuit has no effect on the resonant frequency. In addition, 0.1 uF corresponding to the minimum insertion loss, 0.22 uF followed, 0.47 uF corresponding to the insertion loss is the largest. That is, with the increase of the differential-mode capacitance Cx, the insertion loss of the improved EMI filter circuit is increased and the performance is lowered.

Figure 13 shows the effect of the common-mode capacitor Cy on the performance of the EMI filter circuit. Simulation selects three differential-mode capacitor values of 2,200 pF, 3,300 pF and 4,700 pF. As can be seen from the Figure, 4,700 pF corresponding to the minimum resonant frequency, 3,300 pF as the second, 2,200 pF corresponding to the maximum resonant frequency. That is, with the common-mode capacitor Cy increasing, improved EMI filter circuit resonant frequency decreases. In addition, in the high-frequency region, 2,200 pF corresponding to the minimum insertion loss, 3,300 pF as the second, 4,700 pF corresponding to the maximum insertion loss. That is, in the high-frequency region, with the common-mode capacitance Cy increases, the insertion loss of the improved EMI filter circuit is increased and the performance is decreased.

Figure 14 shows the effect of the inductance L0 on the performance of the EMI filter circuit. In the simulation, three differential capacitance values of 0.03 mH, 0.36 mH and 0.72 mH are selected. As can be seen from the Figure, 0.72 mH corresponding to the minimum resonant frequency, 0.36 mH followed, and 0.03 mH corresponding to the maximum resonant frequency. That is, with the inductance L0 increasing, improved EMI filter circuit resonant frequency decreases. In addition, in the high-frequency part, 0.03 mH corresponds to the minimum insertion loss, 0.36 mH followed, and 0.72 mH cor-responding to the maximum insertion loss. That is, in the high-frequency region, the insertion loss of the improved EMI filter increases with the increase of the inductance L0 and the performance decreases.

4 CONCLUSION

This paper has the overall design to DSP-based high-frequency switching power supply system. It uses TMS320 LF2407 as the main control circuit controller, uses linear optocoupler HCNR201 to design voltage acquisition circuit and uses CHF-300E to design current acquisition circuit. High-frequency transformer important parameters are selected. To improve the design of EMI filter circuit, inductance is added to the upper and lower parts in the differential-mode capacitor and the common-mode capacitor connection part, while adding inductance to the output of the upper and lower parts of common-mode capacitor, making the loss of electromagnetic interference of EMI filter circuit performance is greatly improved, thereby improving the effectiveness and reliability of the DSP-based high-frequency switching power supply system.

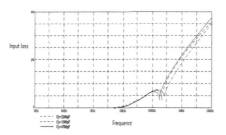

Figure 13. The impact of common-mode capacitor Cy on EMI filter circuit performance.

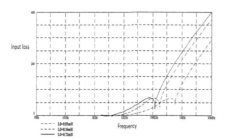

Figure 14. Effect of inductor L0 EMI filter circuit performance.

REFERENCES

Brown, M., Xu D. et al. (2004). *Switching power supply design guide (original version)*. Mechanical Industry Press.

Chen, Q., Zhang, Y., Lin, S. et al. (2014). Modeling and simulation of conducted EMI for front-end power supply. *International Power Electronics and Application Conference and Exposition, IEEE*, (pp. 1358–1362).

Li, M., Qiu, Z., Wang, X., et al. (2012). EMC simulation analysis of switching power supply for vehicle charging system. *Electronic Science and Technology, 25* (7), 125–129.

Li, Min, Qiu, Zijing, Wang Xiaoming, etc. (2012). EMC Simulation Analysis of Switching Power Supply for Vehicle Charging System [J]. Electronic Science and Technology, 25 (7): 125–129.

Salem, M., Hamouda, M. & Slama, J.B.H. (2014). Performance assessment of conventional modulation schemes in terms of conducted EMI generated by PWM inverters. *First International Conference on Green Energy ICGE*, IEEE, (pp. 207–212).

Wu, T.L., Buesink, F. & Canavero, F. (2013). Overview of signal integrity and EMC design technologies on PCB: Fundamentals and latest progress. *IEEE Transactions on Electromagnetic Compatibility, 55*(4), 624–638.

Yang, C.Y., Liu, Y.C., Tseng, P.J. et al. (2015). DSP-Based interleaved buck power factor corrector with adaptive slope compensation. *IEEE Transactions on Industrial Electronics, 62*(8), 4665–4677.

Yang, Y. (2013). High frequency switching power supply electromagnetic compatibility. *Electronic Quality*, (5): 72–74.

Zhou, C., Li, C., Zhu, F. et al. (2004). Study on electromagnetic compatibility of switching power supply. *Communications Power Technology, 21*(2), 16–19.

Automotive, Mechanical and Electrical Engineering – Liu (Ed.)
© 2017 Taylor & Francis Group, London, ISBN 978-1-138-62951-6

MAC design and verification based on PCIe 2.0

Zhiyong Lang, Tiejun Lu & Yu Zong

Beijing Microelectronics Technology Institute, Beijing, China

ABSTRACT: This article highlights the function of MAC layer which works in the Endpoint mode, in the Physical Layer based on the PCIe 2.0 protocol. After the study of the protocol, the forward design and verification is adopted. Simulation was performed using NC which is the Cadence's simulation tool, to communicate with the peer device in the 500 MHz. Two channels are used for data transmission in the design. The transmission rate is 5.0 G/second/lane/direction. In the end, we use SMIC 65 ns to synthesis. According to the simulation results, it is confirmed that the design works properly and meets the protocol requirements.

Keywords: PCIe; MAC; LTSSM; BUS

1 INTRODUCTION

The bus is a path through where data is transferred between different components of the computer. Using bus transmission can simplify the system structure and facilitate the modular design between the systems. The bus technology can greatly enhance the performance of the computer. With the development of computer hardware, bus technology has become the main bottleneck restricting the improvement of computer performance. According to the transmission mode, bus technology can be divided into parallel transmission and serial transmission. Parallel bus technology, which has the defects of ElectroMagnetic Interference (EMI), large number of pins and some others, is unable to meet the needs of high-speed data transmission. So serial transmission has gradually replaced the parallel transmission. High-speed serial transmission uses differential mode for data transmission, and it has a strong anti-interference ability. Compared with the parallel transmission mode, the serial transmission uses less signal lines. At present, the typical high-speed data bus includes PCI Express, USB, and SATA and so on.

PCI Express is the third generation of high-performance I/O bus, which is used to interconnect peripherals such as computing and communications platforms. It is primarily used for interconnection of peripheral I/O devices such as mobile devices, desktops, workstations, servers, embedded computing and communication platforms. PCIe bus has inherited powerful functions of the second generation, utilizing the new features of the computer architecture (Budruk R, 2004).

2 PCIE 2.0 PROTOCOL

PCIe enables interconnected devices to realize high-speed, high-performance and point-to-point data transmission between devices. PCIe 2.0 protocol provides a single link single-channel data transmission rate of 5.0 GB/s (PCI-SIG P C I. 2006), using package for data transmission. PCIe 2.0 protocol uses a layered architecture, which is divided into Transaction Layer, Data Link Layer and Physical Layer from top to bottom. The structure diagram is shown in Fig. 1.

The transaction layer is mainly used for generating outbound TLPs and receiving inbound TLPs. The primary function of the data link layer is to

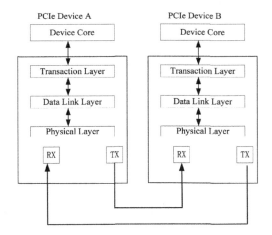

Figure 1. Layer of PCIe device.

ensure the integrity of the packets on the link. The transmitter of physical layer is responsible for receiving data packets from the Data Link Layer and then transmitting to the link, while the receiver is responsible for receiving packets from the link and transmitting the packets to the Data Link Layer. The Physical Layer mainly completes the link establishment, scrambling/descrambling, 8b/10bencoding/decoding and serial/parallel conversion. Two sub layers, namely logic Physical Layer and electrical Physical Layer, make up the Physical Layer. In order to promote and regulate the development of chips, the Intel specifically proposes the PHY Interface for the PCI Express Architecture for IC designers. The PIPE standard provides a uniform reference for circuit interfaces of PCI Express Physical Layer (Intel Corporation, 2008).

As is shown in Fig. 2, the PIPE standard further divides the Physical Layer into Media Access Layer (MAC), Physical Coding Sub layer (PCS), and Physical Media Attachment sub layer (PMA). The MAC is mainly used to initialize the link, byte striping/un-striping and byte scrambling/descrambling. 8b/10b encoding and decoding, elastic buffer and receiver detection are included in PCS module. PMA is mainly used for parallel to serial, clock recovery and PLL. PCS and PMA are digital analog hybrid circuits, usually called PHY. This article focuses on the functions and the verification implementation of the MAC layer.

In the whole architecture, the MAC connects the Data Link Layer and the PHY. It converts the upper control signal into the control signal of the PHY, so as to establish the link and transmit the data. The MAC mainly implements the functions of link establishment, frame header add/delete, byte striping/un-striping and data scrambling/descrambling (Wilen A, 2003).

2.1 LTSSM

After the reset, the MAC establishes the link through the Link Training And Status State Machine (LTSSM). The LTSSM consists of 11states and they can be divided into the following five sections: Link training states, Link retraining states, power management states, active power management and other status. The states of LTSSM are shown in Fig. 3 (Wang Qi. 2011).

The LTSSM is mainly responsible for the training and initialization of the link, working just after the reset. The normal state of transmitting packet is the L0 state. This process does not require software to participate in, as it is automatically initiated after reset. In the process of link training and initialization, the following tasks are performed:

1. Creating and setting the link width. The PCI Express device supports up to 32 channels of data transmission. The number of channels supported by the devices at both ends of the link may be different. Therefore, the link channel negotiation needs to be performed. The minimum link width is adopted to ensure that both ends of the device can transfer data.
2. Channel inversion is performed in multi-channel devices, if polarity inversion is required. So that signal crossing can be avoided at the time of connection.
3. Negotiating the data link rate. The transmission rate of the device is 2.5 GB/s at the beginning. If higher data rates are available, they are advertised during this process and, when the training is completed, devices will automatically go through a quick re-training to change to the highest commonly supported rate.
4. Bit lock. When link training begins, the receiver's clock is not synchronized with transmit clock of

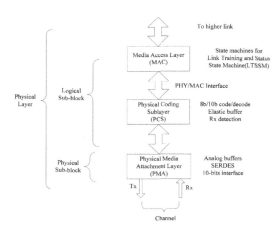

Figure 2.   Partitioning PHY layer for PCIe.

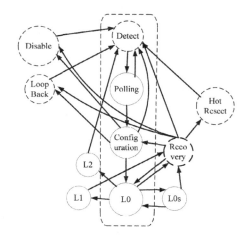

Figure 3.   The states of LTSSM.

682

incoming signal, and the receiver uses its CDR to recreate the clock by using the incoming bit steam as a clock reference, this is called bit lock.

5. Symbol lock. The receiver doesn't know where the boundaries of the 10-bit symbol are, hence, it searches for the COM Symbol to recognize the boundary of the symbol.

As is shown in Fig. 4, after the link is established, the MAC can receive the TLP/DLLP packets transmitted from the Data Link Layer. It adds STP or SDP to the frame header and adds END/EDB to the frame trailer. It also sends ordered sets to the multiplexer for transmission to the byte stripping. After scrambling, the packets will be sent to the 8b/10b encoder through the PIPE interface. All packets are serial-to-parallel. After the serial to parallel conversion, the packets will be transmitted to the link. The process of receiving data packets is the reverse process of sending data packets.

2.2 *Byte stripping*

The fundamental PCIe 2.0 agreement provides the rules of byte stripping. When there are multiple channels in the link, packets need to be split. The multiplexer selects the packets to be transmitted and then passes them to the byte striping. After the byte stripping, they will be transmitted to the scrambler for scrambling. Byte stripping rules in the protocol are as follows:

1. The total packet length of each packet (including the start and end characters) must be a multiple of 4 characters.
2. The TLP starts with STP and ends with the END/EDB character.
3. DLLP starts with SDP and ends with the END character.

4. When starting the transmission of a packet, the STP and SDP characters must be placed on channel 0.

2.3 *Byte scrambling*

The implementation of byte scrambling structure is shown in Fig. 5. Byte scrambling is achieved by a 16-bit Linear Feedback Shift Register (LFSR). PCIe 2.0 agreement generates a random sequence LFSR. The polynomial used for LFSR calculation has a coefficient expressed as 100 Bh.

3 MAC DESIGN

This paper is based on PCIe 2.0 protocol endpoint model to design the MAC layer, using Verilog language (Thomas D, 2008) for forward design. Two channels are used for data transmission between the devices, and the transmission rate is 5.0 G/second/lane/direction of raw bandwidth. According to the above functional description, MAC overall design diagram is shown in Fig. 5:

Os_build: Generating the data packets used in the link training and initialization. The channel designed by this module is two channels, and each channel is initialized separately. The data bit width of each channel is 8 bits, so the output bit width is 16 bits. The link number and lane number of TS1 and TS2 packets sent from this module are set to PAD, while Rate ID byte is set to 00 h (2.5 Gbps). If required, the processing is performed when the byte is split.

Os_div: In the link training and initialization phase, 8 bits of data transmitted by the os_build module simultaneously transfer to both channels. This module completes byte stripping of physical layer packet, and sets the corresponding data packet transmission completion signal back to LTSSM.

Timer_cnt: It is counting module, which starts counting the time at the beginning of data transmission, to see if there is a timeout. If time is out, this

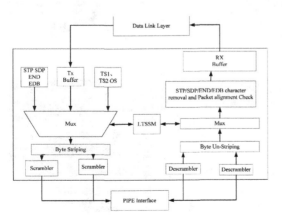

Figure 4. MAC layer block diagram.

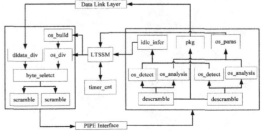

Figure 5. MAC layer diagram.

module will return the timeout signal to LTSSM to carry out the corresponding processing.

Dldata_div: This module strips the received TLP/DLLP. During byte stripping, it adds STP and END/EDB if the received packet is TLP. It also adds SDP and END if the packet is DLLP.

Byte_select: The module selects the packets to be sent based on the machine's state.

Scramble/descramble: This module scrambles/descrambles the packets according to the polynomials required by the protocol.

Os_detect: This module mainly judges the type of the received ordered set, and sets the corresponding flag to the os_analysis module. The module and os_analysis module work together to determine whether the received data meets the requirements. This module only needs to determine whether the received data packet is an ordered set or not. If it is an ordered set, it also needs to determine the type of the ordered set.

Os_analysis: The module is used to analyze the received TS1, TS2 ordered sets. It analyzes 1–5 bytes in all 16 bytes, and determines the link number and lane number. In this module, MAC negotiates the transmission speed.

Idle_infer: The module is used for electrical idle inference. If the corresponding ordered set is not detected for a certain period of time, it can be deduced that the link has entered the electrical idle state. This module sets the corresponding signal to the LTSSM module for the corresponding operation.

Pkg: This module is used to merge the TLP/DLLP packets from two channels. If the packet is TLP, STP and END/EDB are removed. If the packet is DLLP, SDP and END are removed. All packets will be transmitted to Data Link Layer.

4 SIMULATION WAVEFORM

After completing the code, we use Cadence's simulation tools (Cadence Design Systems, 2008) for function testing to check whether it meets the design requirements. The test results are as follows:

After the reset, Fig. 6 shows that the peer device is detected. Besides, the timing waveform of the verification structure is shown meeting the requirements of the PIPE protocol. Fig. 7 is a partial screenshot of TS1 packets sent during link establishment.

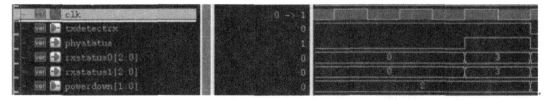

Figure 6.   Receiver detection.

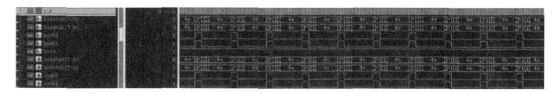

Figure 7.   Link establishment process.

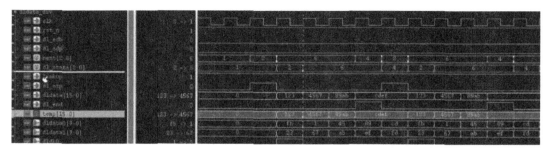

Figure 8.   The process of sending packets.

Fig. 8 shows the process of sending TLP packets. After receiving packets from the Data Link Layer, the packet is split into two channels, and STP characters on channel 0 and END/EDB characters on channel 1 are added respectively. In this example, the Data Link Layer indicates that the packet is right, so the END character is added to the end of the packet in channel 1. From the verification results, the design meets the protocol requirements.

5   CONCLUSION

This paper introduces the design and implementation of MAC sub layer in PCIe2.0 endpoint mode. Based on the design, the function simulation is carried out. Two channels are used for data transmission in the design. The transmission rate is 5.0 G/second/lane/direction. In this design, we use two channels to transmit packets, and the data width of single channel is 8 bits. Therefore, the clock frequency is 500 MHz in our design. We also use

SMIC 65 ns technology library to synthesis. The design can work normally under the 500 MHz clock, and furthermore achieves the goal that is expected.

REFERENCES

Budruk R, Anderson D, Shanley T. PCI express system architecture [M]. Addison-Wesley Professional, 2004.
Cadence Design Systems, Inc., "NC-Verilog Simulation Help," Product Version 8.2, Nov. 2008.
Intel Corporation. PHY Interface for the PCI Express Architecture, America: Intel Corporation, August, 2008.
PCI-SIG P C I. Express Base Specification Revision 2.0 [J]. 2006.
Thomas D, Moorby P. The Verilog® Hardware Description Language [M]. Springer Science & Business Media, 2008.
Wilen A, Thornburg R, Schade J P. Introduction to PCI Express [M]. Intel, 2003. [5] Wang Qi. PCI Express architecture guide [M], 1st. Beijing: China machine press, 2011.

Automotive, Mechanical and Electrical Engineering – Liu (Ed.)

Miniaturization of microstrip open-loop resonator with fragment-type loading

Ting Liu, Lu Wang & Gang Wang
School of University of Science and Technology of China, Hefei, China

ABSTRACT: Miniaturization of microstrip open-loop resonators is a key concern in design of microstrip components including an open loop. Fragment-type structures provide great flexibility in loading the open loop for miniaturization and other desired characteristics. With multi-objective optimization technique, miniaturized open loop can be readily designed, and specific requirements on characteristics such as parasitic resonant mode, quality factor, and loading density of the open loop may also be acquired with fragment-type loading.

Keywords: Fragment; Loading; Open loop; Optimization; Resonator

1 INTRODUCTION

Open-loop resonators have found applications in many RF/microwave designs (A. Gorur, 2002; X. B. Ma, 2013; J. R. Kelly, 2011; T. Li, 2012; J. Konpang, 2011). In addition to basic design concerns about resonant frequency, resonant mode, and quality factor, miniaturization of open-loop resonator is a key design concern in practice.

Loading the loop is the most common technique for miniaturization, and different loading structures have been proposed. In (P. Mondal and M. K. Mandal, 2008; R. K. Maharjan and N. Y. Kim, 2013; A. Gorur, 2003; R. Ghatak, 2012), stubs or its likes were proposed to load open loops for compact filters. In (X. Y. Zhang, 2007), capacitive loading structures were proposed to load the open loop for compact microstrip bandpass filter, which yielded 45% size reduction. In (Q. Zhao, 2014), fragment-type structures were proposed in loading open-loop resonator for compact bandpass filter. In (J. Konpang, 2009), stepped impedance patches were applied to load the open loop for compact diplexer. Other loading techniques such as loading with varactors and inductively have also been reported (X. G. Huang, 2014; L. Zhu and B. C. Tan, 2005).

It is still challenging to answer if there are more suitable loading structures for open loop to acquire better performance and smaller dimensions. Since any loading structure can be described by using fragment distribution, loading with fragments provides the potential to find a structure better than previously reported canonical loading structures.

In this letter, miniaturized microstrip open-loop resonator is designed with fragment-type loading.

Effects of fragment-type loading on performance of open-loop resonator are studied. With multi-objective optimization technique, miniaturized open loop can be readily designed, and specific requirements on characteristics such as parasitic resonant mode, quality factor, and loading density of the open loop may also be acquired with fragment-type loading.

2 MODELING AND DESIGN SCHEME OF OPEN-LOOP RESONATOR LOADED WITH FRAGMENTS

Geometry and design of an open-loop resonator with fragment-type loading is shown in Fig. 1. To acquire a fragment-type loading, design space inside the open loop is first gridded into cells, as shown in Fig. 1(a). The cells are then assigned with either "1" or "0", as shown in Fig. 1(b). By metalizing cells with "1", a fragment-type loading can be constructed as shown in Fig. 1(c).

In practice, certain simple structure is generally preset in the fragment-type loaded loop for fine

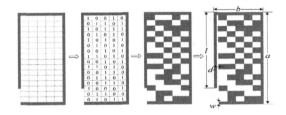

Figure 1. Loading an open-loop resonator with fragment-type structure.

frequency tuning. For example, part of the loop perimeter may be left free. As shown in Fig. 1(d), part of loop perimeter is separated from fragment loading by a slot of width d, which can be left to tune the resonant frequency by adjusting length l.

3 CHARACTERIZATION OF FRAGMENT-TYPE LOADING

For fragment-type loading in Fig. 1, "1" and "0" distributed in design space can be used to define a design matrix. The matrix can be constructed by recording "1" and "0" in any specified manner, if only the fragment distribution can be recovered in EM simulator. Once a design matrix is formed, design of fragment-type loading is to seek a proper design matrix to yield designated resonator performance.

In addition to design matrix, some specific indexes can also be defined such as density and barycenter of fragment loading. Density of fragment loading η can be defined as

$$\eta = \frac{number\ of\ metal\ cells}{number\ of\ total\ fragment\ cells} \qquad (1)$$

By dividing design space into several regions and assigning different loading densities to different regions, barycenter of fragment loading can be adjusted.

4 MULTI-OBJECTIVE OPTIMIZATION OF FRAGMENT-TYPE LOADING

Fragment-type loading provides great flexibility in seeking the overall best loading to meet multiple design objectives. The most suitable fragment-type loading structure can be found by optimization searching technique to determine which cells should be assigned with "1", as in Fig. 1(b). Optimization searching scheme such as multi-objective evolutionary algorithm based on decomposition combined with enhanced genetic operators (MOEA/D-GO) (D. W. Ding and G. Wang, 2013) can be applied.

Multi-objective optimization problem for design of fragment-type loading can be expressed as

$$\begin{aligned} minimize\ F(x) &= (f_1(x),\ f_2(x),\ ...,\ f_n(x)), \\ subject\ to\ &x \in \Omega\ , \end{aligned} \qquad (2)$$

where Ω is a decision space, and x is a decision variable that defines fragment-type loading, $f_i(x)\ (i=1,2,\cdots,n)$ represent n objectives for different design concerns.

For miniaturization of open-loop resonator, we may define the following objective function to restrict open loop size

$$f_1(x) = \max(L - L_{max}, 0), \qquad (3)$$

where L_{max} is the desired maximum perimeter and L is perimeter of an open loop.

For control of resonant frequency, we may define objective function

$$f_2(x) = \max(|f - f_0|, 0), \qquad (4)$$

where f_0 the desired resonant frequency of dominant mode is, f is the resonant frequency of open loop with fragment loading.

For control of resonant modes as required in bandwidth filter design, we may define objective function

$$f_3(x) = \max(f_p - f_{mode(k)}, 0), \qquad (5)$$

where f_p is desired frequency of parasitic resonant, and $f_{mode(k)}$ is resonant frequency of the kth resonant mode.

To restrict the loading density, we may define objective

$$f_4(x) = \max(\eta - \eta_0, 0), \qquad (6)$$

where η_0 is a desired loading density and η is loading density.

To control the quality factor, we may define objective

$$f_5(x) = \max(|Q_0 - Q|, 0), \qquad (7)$$

where Q_0 is the desired quality factor, Q is the quality factor of open loop with fragment loading under optimization, which can be obtained from Eigen Q by using EM simulation.

It should be remarked that ultimate characteristics can be pursued by setting unpractical desired values in objective functions. For example, $L_{max} = 0$ in object $f_1(x)$ will impose a searching for the minimum open loop, $f_0 = 0$ in object $f_2(x)$ will impose a searching for the lowest resonant frequency. For such unpractical desired values, optimization with MOEA/D-GO will implement an endless searching. In practice, we may check MOEA/D-GO's output stability for certain characteristics, and terminate the searching manually if acceptable characteristics have been achieved.

5 EFFECTS OF FRAGMENT-LOADING ON OPEN-LOOP RESONANT PERFORMANCE

Fragment-type loading may affect the open-loop resonator in different ways. To demonstrate the loading effects, we study the loading of an open loop of $a \times b = 6.4$ mm $\times$ 3.4 mm on 1.27 mm-thick Arlon AD600 PCB of relative dielectric constant of 6.15 and loss tangent of 0.003. Suppose the loop has $w = 0.2$ mm and $d = 0.1$ mm in Fig. 1(d), design space within the open loop can be discretized into 20×15 cells so that fragment cell has dimensions of 0.3 mm $\times$ 0.2 mm.

When loaded by fragment distribution of different fragment densities, the open-loop resonator will have different lowest resonant frequencies. The lowest resonant frequencies for different loading densities can be acquired by implementing MOEA/D-GO optimization searching with objective function $f_2(x)$ with $f_0 = 0$ and objective function $f_4(x)$ with several representative loading densities.

For each MOEA/D-GO searching, several optimized fragment-loaded open-loop resonators can be obtained. Since MOEA/D-GO belongs to stochastic optimization algorithm, it usually yields several optimized solutions approaching to Pareto-optimal set (D. W. Ding and G. Wang, 2013). By averaging the resonant frequencies for each density, we may obtain a characteristic resonant frequency.

Fig. 2 shows the distribution of resonant frequencies and corresponding quality factors for open loop with optimized loading of different densities. Fitting curves to the distributed resonant frequencies and Q values are also depicted. We find that the lowest resonant frequency can be achieved with loading density around 40%. In other words,

loading with fragment density around 40% will yield miniaturization of the open loop. It is also noted that Q value approaches to that of stepped impedance resonator for loading density larger than 80%.

6 CONCLUSION

Loading with fragments provides great flexibility and possibility to find the best performance of open-loop resonator. With optimization searching using MOEA/D-GO, open-loop resonator can be miniaturized by fragment-type loading. Other design concerns about parasitic resonant modes, quality factor, and loading density may also be considered in the multi-objective optimization.

ACKNOWLEDGMENTS

This work was supported in part by National Natural Science Foundation of China under Grant 61272471 and 61671421.

REFERENCES

Ding, D. W. and G. Wang, MOEA/D-GO for fragmented antenna design, Progr. Electromagn. Res. M, vol. 33, pp. 1–15, 2013.

Ghatak, R., M. Pal, P. Sarkar and D. R. Poddar, Dual-band bandpass filter using integrated open loop resonators with embedded ground slots, Microw. Opt. Tech. Lett., vol. 54, no. 9, pp. 2049–2052, Sept. 2012.

Gorur, A. A novel dual-mode bandpass filter with wide stopband using the properties of microstrip open-loop resonator, IEEE Microw. Wireless Compon. Lett. vol. 12, no. 10, pp. 386–388, Oct. 2002.

Gorur, A., C. Karpuz, M. Akpinar, A reduced-size dual-mode bandpass filter with capacitively loaded open-loop arms, IEEE Microw. Wireless Compon. Lett. vol. 13, no. 9, pp. 385–387, Sept. 2003.

Huang, X. G., Q. Y. Feng, Q. Y. Xiang, D. H. Jia, Constant absolute bandwidth tunable filter using varactor-loaded open-loop resonators, Microw. Opt. Tech. Lett., vol. 56, no. 5, pp. 1178–1181, May. 2014.

Konpang, J., A compact diplexer using square open loop with stepped impedance resonators, in IEEE Radio Wireless Symp. Dig., Jan. 2009, pp. 91–94.

Konpang, J., L. Thongnoi, A compact four-poles cross-coupled square open loop resonator diplexer, in Proc. German Microw. Conf. (GeMIC), Mar. 2011, pp.1–4.

Kelly, J. R., P. S. Hall, and P. Gardner, Band-notched UWB antenna incorporating a microstrip open-loop resonator, IEEE Trans. Antennas Propag., vol. 59, no. 8, pp. 3045–3048, Aug. 2011.

Li, T., H. Q. Zhai, L. Li, C. H. Liang, and Y. F. Han, Compact UWB antenna with tunable band-notched

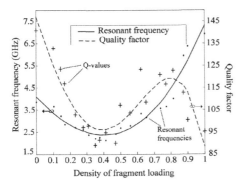

Figure 2. Frequency and Q change with density of optimized fragment loading.

characteristic based on microstrip open-loop resonator, IEEE Antennas Wireless Propag, Lett., vol. 11, pp. 1584–1587, 2012.

Ma, X. B., H. X. Zheng, Compact wideband bandpass filter using two open loop resonators, Wiley Microw. Opt. Tech. Lett., vol 55, no. 4, pp. 915–917, Apr. 2013.

Maharjan, R. K. and N. Y. Kim, Miniature stubs-loaded square open-loop bandpass filter with asymmetrical feeders, Microw. Opt. Tech. Lett., vol. 55, no. 2, pp. 329–332, Feb. 2013.

Mondal, P. and M. K. Mandal, Design of dual-band bandpass filters using stub-loaded open-loop resonators, IEEE Trans. Microw. Theory Tech., vol. 56, no. 1, pp. 150–155, 2008.

Zhang, X. Y., J. X. Chen, Q. Xue, Compact bandpass filter using open-loop resonators with capacitive loading, Microw. Opt. Tech. Lett., vol. 49, no. 1, pp. 83–84, Jan. 2007.

Zhu, L. and B. C. Tan, Miniaturized dual-mode bandpass filter using inductively loaded cross-slotted patch resonator, IEEE Microw. Wireless Compon. Lett. vol.15, no.1, pp.22–24, Jan. 2005.

Zhao, Q., G. Wang, D. Ding, Compact microstrip bandpass filter with fragment-loaded resonators, Microw. Opt. Tech. Lett. vol. 56, no. 12, pp. 2896–2899, Dec. 2014.

Automotive, Mechanical and Electrical Engineering – Liu (Ed.)
© 2017 Taylor & Francis Group, London, ISBN 978-1-138-62951-6

Power electronic digital controller based on real-time operating system

Lijuan Guo & Songmei Tao
Guangxi Electric Power Research Institute, Nanning, China

Lei Pei & Wensheng Gao
School of Electrical Engineering, Tsinghua University, Beijing, China

ABSTRACT: Typical power electronics has become a fundamental technology of power transformation in modern engineering. Researches on topology of power converter and related analogue circuit controlling strategy are difficult to be novel. On the other hand, research into new power semiconductor devices thrives and demand for higher power density leads to integrated circuits, which means controlling circuits and power device modules are integrated on the same semiconductor chip, and inventions on new circuit topology for higher-frequency switches. For converter controller parts, the core part of a main circuit, digital controlling platform and technology tends to replace the traditional analogous control model. This article introduces a real-time embedded system and applies into a boost circuit to realise DC-DC conversion.

Keywords: Real-time; Power electronic; Digital controller

1 INTRODUCTION

DSP/BIOS™ offers a low-level interface for application function. It supports management threads and external equipment. It also enables period and idle function to run in the background, and the system to get real-time analysis performance. Comparing programming with traditional assembly or C language, DSP/BIOS™ can control DSP hardware resources better and fit well to speed up the software development and debug processes. Two features must be noted in that all operations related to hardware have to be realised through the DSP/BIOS original function set. Developers should avoid control hardware directly, such as interrupt, timer, DMA controller, serial port and so on. Developers could configure not only by the graphing tool in CCS but also by calling API in the code. Second, programs with DSP/BIOS run differently to those without DSP/BIOS. The latter programs have total control of DSP while the former hand the control over to BIOS. Codes are not executed by sequence, but by both sequence and priority. Through this, we can realise very complicated and flexible multi-task functionality.

1.1 *API modules*

An application using DSP/BIOS calls API functions in the libraries to carry out specific task. There are different API modules, each with C-language calling interface. Main modules offered by CCS FOR C2000 are as follows:

- CLK module: set timer interrupt interval
- HST module: data exchange between mainframe and targeted system
- HWI module: set hardware interrupt service
- IDL module: manage idle function in the background, with the lowest priority
- LOG module: display event records
- MEM module: define memory usage of targeted system
- PIP module: manage data exchange between threads
- PRD module: realise period function
- RTDX module: real-time data exchange between mainframe and targeted system
- STS module: state statistics
- SWI module: manage software interrupt, with lower priority than HWI
- DEC module: interface to drive devices
- SIO module: realise data exchange between DEC and task or HWI
- MXB module: realise communications between tasks
- QUE module: queue management for tasks or threads.
- SEM module: synchronise tasks and threads.

2 SOFTWARE STRUCTURE

DSP/BIOS offers scalable real-time kernel and multi-threads management with priority. It is designed for applications that need real-time dispatch, synchronisation and communication. There are four main threads in DSP/BIOS kernel: IDL, TSK, SWI, HWI. Application starts from default point_c_int00 and runs in the flows below:

1. System initialisation, defined by term system in BIOS configuration, including register set and PLL double frequency clock.
2. Call user's main () function.
3. Call BIOS_start, execute HWI, SWI, and tasks in priority.
4. If there is no thread executed (no HWI or SWI), the application goes into IDL_F_loop to execute IDL thread.
5. If HWI comes, cpu deals with it immediately. When it finishes, if there is no SWI or tasks, cpu returns to IDL.

DSP/BIOS is a multi-task kernel, although it is not a strictly pre-emptive kernel. Only HWI can take cpu control from other types of threads. If there is no HWI, we must wait for a task or SWI to give up cpu control itself before other threads with higher priority can be executed. HWI can change sequence of threads by reorganising threads in priority. For example, if task A with priority 2 is being executed and there is a SWI B with priority 3 waiting in the queue, and now here comes a HWI. HWI takes cpu control and when it is finished, cpu will execute SWI first because it has higher priority.

Users can put their codes into any threads like HWI, SWI or TASK; the latter two are usual. They can set it by BIOS configuration or call API functions.

There are 150 API functions in the kernel, which offers HWI management, SWI start, tasks switch and synchronisation and communication between threads.

3 CIRCUIT SIMULATION

3.1 Boost circuit design

Step 1.   input = 5V, output = 15 V
Step 2.   DCM model: intermittence
Step 3.   switch frequency: TS = 10 us, f = 100 kHz
Step 4.   devices used: C1 = 100 uF, R1 = 147Ω. IRF840. MUR4100
Step 5.   Calculate inductance and duty ratio using formula in Chapter 2 of book *power electronics technology.*

$$M(D) = \frac{VOUT}{VIN} = \frac{1 + \sqrt{1 + \frac{4D^2}{K}}}{2} = 3 \tag{1}$$

$$D = \frac{k \times (VOUT - VIN)}{VIN} = \frac{2}{3}k \tag{2}$$

$$K = \frac{2 \times L}{R \times T_s} \tag{3}$$

$$k = D_1 + D_2$$
$$VOUT = \frac{D_1 + D_2}{D_2} \times VIN \tag{4}$$

$$D = D_1$$
$$= \frac{(VOUT - VIN)}{VIN} \times (D_1 + D_2)$$
$$= \frac{k \times (VOUT - VIN)}{VIN} \tag{5}$$

Applying designed circuit parameters, we can get:

$$L = \frac{1}{27}k^2 \times R \times T_S \tag{6}$$

$$D = \frac{2}{3}k \tag{7}$$

$$DCM: \quad I < \Delta i_L \Rightarrow L < 245 \times D(uH) \tag{8}$$

Plus,

$$T_{ON} = D \times T_S \tag{9}$$

Let L = 10 uH; we can get

$$k = \frac{3}{7}; \quad D = \frac{2}{7}; \quad T_{ON} = 2.86\,u \tag{10}$$

Step 6. Verifying circuit working mode:

$$I = \frac{VOUT}{R} \tag{11}$$

$$\Delta i_L = \frac{VIN}{2L} \times D \times T_S \tag{12}$$

The circuit works in DCM mode.

3.2 Boost circuit simulation

The program is designed to exercise close-loop control on the circuit to make the output voltage 15 V.

The closed-loop control structure is as follows:

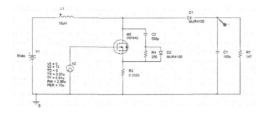

Figure 1.   Orcad simulation open-loop control.

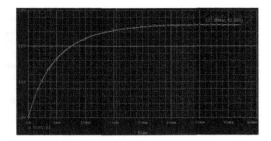

Figure 2.   VOUT.

4   PROGRAM DESIGN

The program is designed to exercise close-loop control on the circuit to make the output voltage 15 V.
The closed-loop control structure is as follows:
The program goes as follows:

1. Sysctrl. c: initialize the control system. You can also set it in DSP/BIOS configuration.
2. Main (): initialise clock/IO/PIE control, configure EPWM/ADC/IO, enable PIE interupt—PIE_INT1_6, enable CPU interrupt, clear wrong lock on EPwm1.
3. Call BIOS_start: detect and execute HWI sub-program, SWI sub-program and tasks according to the set priority.
4. PIE Interrupts—PIE_INT1_6.
5. CPU respond interrupts, goes into HW-ISR: ADC_isr ().
6. ADC_isr () start SWI—&SWI_ADC_Read and then exit, waiting for other HWI.
7. SWI_ADC_Read (PRIORITY = 3) start function ADC_Read (). ADC_Read () get Vout-Sample and ImosSample from ADC module, calculate Vout and Imos. And send signal: & ControlApplication_SEM.
8. & ControlApplication_SEM start task Control_TSK ().
9. Control_TSK () call function pi_calc () and calculate d for the next period.
10. pi_calc () sent output to EPWM module. EPWM module produces driving signal for real circuit.
11. BIOS calls function PRD20 ms periodly (&PRD_swi), and the object calls function But-

tonRead () to checks whether the button is on. If yes, it sends signal & StartButtonAction_SEM, which will start StartButton Action_TSK ().
12. Task StartButtonAction_TSK () checks the button again. If it is still closed, it enables EPWM module to drive TLP250.

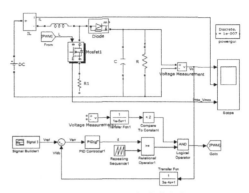

Figure 3.   Closed-loop control of digital control system.

5   EXPERIMENT RESULT

Figure 4.   Physical connection diagram.

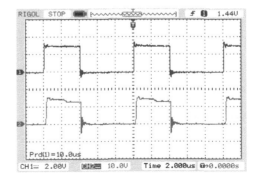

Figure 5.   Output of EPWM and TPL250.

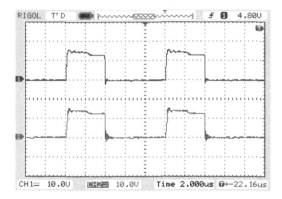

Figure 6.   Output of TPL250 and VG of MOSFET.

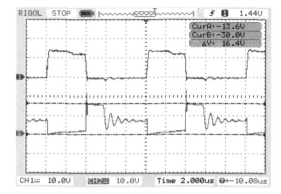

Figure 7.   VG and VD of MOSFET.

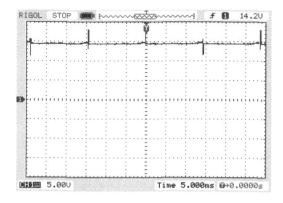

Figure 8.   VOUT = 15 V.

6   CONCLUSION

From the pictures above, we know that DSP output PWM wave with d = 0.4, amplitude = 3.3 V and f = 10 kHz; output of TPL250 is 0–15 V squarewave. MOSFET is properly driven and the circuit works in DCM mode. The output voltage is 15 V, which is the same as the original design goal.

REFERENCES

Buso, S. & Mattavelli, P. (2006). Digital control in power electronics: Synthesis lectures on power electronics (2ns ed.).
Dehong, Xu. (2010). Power Electronics. Science Press.
Jean, J. & Labrosse. (2007). *MicroC/OS-II: The real-time kernel* (2nd ed.). Beiging: Beijing University of Aeronautics and Astronautics Press.
Texas Instruments. (2012). TMS320F28X Data Manual. ZHCS889M.

Automotive, Mechanical and Electrical Engineering – Liu (Ed.)
© 2017 Taylor & Francis Group, London, ISBN 978-1-138-62951-6

Reactive power optimisation with small-signal stability constraints

Jiayun Xu
School of Electrical Engineering, Zhengzhou University, Zhengzhou, China

Xiaoming Wang
State Grid Zhejiang Haining Power Company, Haining, China

Kewen Wang
School of Electrical Engineering, Zhengzhou University, Zhengzhou, China

Congmei Zha
School of Electronic and Information Engineering, Zhongyuan University of Technology, Zhengzhou, China

ABSTRACT: Small-signal stability constraints should be considered in reactive power optimisation. A model of small-signal stability is constructed according to electromechanical oscillation modes. The objective function of reactive power optimisation is composed of system active power loss and small-signal stability index, with the former reduced and the latter enhanced by modifying related control parameters. Furthermore, improved Particle Swarm Optimisation (PSO) was taken advantage of when optimising the operative power. Testing examples reveals that the goal of reactive power optimisation can be efficiently reached from the proposed algorithm, and the power system will operate in a safe, economic and stable manner.

Keywords: Particle swarm optimisation; Reactive power optimisation; Power system; Small-signal stability; Parameters optimisation

1 INTRODUCTION

Reactive power optimisation is usually used in systems to improve voltage quality and reduce active power loss (Zhao Bo, 2005), by adjusting transformer taps, switchable capacitors, generator terminal voltages, etc.

After decades of research, the reactive power optimisation algorithm has developed adequately and various artificial intelligent methods are applied to reactive power optimisation, such as ant swarm optimisation (Ruan Ren-jun, 2010), genetic algorithm (Zhao Liang, 2010), immune algorithm (Luo Yi, 2013) and particle swarm algorithm (Liu Hong, 2009). The Particle Swarm Optimization (PSO) has advantages such as simple principle, stronger versatility, and is independent of the message. Even the memory characteristic for optimal message in PSO, classical PSO algorithm has the disadvantage of weaker capacity for local searching and for easily sinking into the local minimal solution (Wang Ling, 2008).

Current research on reactive power optimisation is mainly to improve the system stability index e.g. the voltage margin can be widely increased after optimisation by introducing voltage vague constraints (Li Ya-nan, 2001). In (Zhang An-an, 2010), load stability margin is used as voltage stability estimate index of integrated reactive power optimisation.

Because of the complexity of power systems, if only the steady state index is considered in the optimising process, the system small-signal stability may be decreased. For example, voltage quality can be improved by altering network parameters while the risk of system low frequency oscillation is increased. At present, reactive power optimisation with small-signal stability analysis has randomly been studied. When power system dispatch operation is researched, scholars put forward some methods for small-signal stability constrained optimal power flow, which provide reference for the study of reactive power optimisation with small-signal stability constraints. In (Li Pei-jie, 2013), the nonlinear semi-definite planning method is used to indicate the small-signal stability constraints with a series of equality. And inequality judging positive definite property of matrices are added into constraints in the interior point method optimal power flow. Differing from the above methods, the eigenvalues of system are calculated in (Sun Yuanzhang, 2005), and the minimum damping ratio of system electromechanical oscillation mode is used as the small-signal stability index which is linearized in optimisation. The objective function obtained by the above methods is the minimum cost of total power generation, which satisfies the optimal power flow with small-signal stability constraints.

In this paper, small-signal stability optimisation in systems is considered in optimising reactive power. The objective function of optimisation is composed of the network loss index and small-signal stability index. At the same time, the small-signal stability constraints are added to the constraint conditions to ensure the small-signal stability of the optimised system. The participation degree of small-signal stability in the objective can be adjusted by altering the weight coefficient in objective function. Examples show that the improved PSO algorithm can effectively solve the optimisation problem, and the reactive power optimisation method can provide safe, economical, and stabile system operation condition.

2 SMALL-SIGNAL STABILITY ANALYSIS METHOD

By adopting insert simulation technology into multi-machine systems, all machines and associated control systems can be derived using zero-order transfer blocks and first-order transfer blocks (Zhong Zhi-yong, 2000). After all the transfer blocks are given, connecting matrices can be obtained by using the port number of each transfer block. When zero-order transfer blocks are eliminated, the following state space equations of system are obtained:

$$\begin{cases} \dot{X} = AX + BR + E\dot{R} \\ Y = CX + DR \end{cases} \tag{1}$$

where A is state matrix. All the eigenvalue are given by QR algorithm.

Damping ratio gives an index to dynamic performance of system, and the damping ratio of conjugate complex eigenvalue $\lambda = \alpha + j\beta$ is:

$$\xi = -\alpha\sqrt{\alpha^2 + \beta^2} \tag{2}$$

Before reactive power optimisation, all eigenvalues of system are solved, small-signal stability index are formed by the maximum of real parts of eigenvalues and the minimum of damping ratios in electromechanical modes. α_0 is the maximum of real parts of eigenvalues before reactive power

optimisation. ξ_0 is the minimum damping ratio of electromechanical mode. When the electromechanical oscillation mode eigenvalue satisfies the following formulas, it indicates that small-signal stability is enhanced.

$$\alpha \leq \alpha_0 \quad \text{and} \quad \xi \geq \xi_0 \tag{3}$$

where α is the maximum real part of the electromechanical mode eigenvalue after reactive power optimisation; ξ is the minimum damping ratio of electromechanical mode.

3 REACTIVE POWER OPTIMISATION MODEL WITH SMALL-SIGNAL STABILITY

The goal of this paper is to minimise active power loss. Variable constraints for bus voltages and reactive powers are given in penalty function form. Equation constraints are nodal power balance equations, and the controlled variables are composed of the adjustable transformers taps, compensating capacitor and generator terminal voltages. The objective functions and relevant variable constraints are listed as:

$$\min F_Q = f_Q + f_C$$
$$f_Q = \sum_{k \in N_E} g_k(U_i^2 + U_j^2 - 2U_iU_j \cos\theta_{ij})$$
$$f_C = \sum_{n \in N_{PQ}} \lambda V_n(U_n - Un, \lim)^2 \tag{4}$$
$$+ \sum_{n \in (N_G + N_C)} \lambda_{Qn}(Q_n - Q_{n,\lim})^2$$

$$s.t. P_{Gi} - P_{Li} - U_i \sum_{j \in N_i} U_j(G_{ij}\cos\theta_{ij} + B_{ij}\sin\theta_{ij}) = 0$$
$$Q_{Gi} - Q_{Li} - U_i \sum_{j \in N_i} U_j(G_{ij}\sin\theta_{ij} + B_{ij}\cos\theta_{ij}) = 0$$
$$U_{i,\min} \leq U_i \leq U_{i,\max} \quad i \in N_B \tag{5}$$
$$T_{i,\min} \leq T_i \leq T_{i,\max} \quad i \in N_T$$
$$Q_{Gi,\min} \leq Q_{Gi} \leq Q_{Gi,\max} \quad i \in N_G$$
$$Q_{Ci,\min} \leq Q_{Ci} \leq Q_{Ci,\max} \quad i \in N_C$$

where $\lambda_{V_n}, \lambda_{Q_n}$ are penalty factors; $N_E, N_{PQ}, N_G, N_L, N_T$ and N_B are branch set, PQ node set, PV generator node set, compensating capacitor node set, transformer branch set and general node set; g_k is the conductance of branch k; U_i the voltage of node i; θ_{ij} the voltage angle difference between node i and j; G_{ij}, B_{ij} are coefficients of nodal admittance; Q_{Gi} is the reactive power of generators; T_i is the transformation tap of the transformers; Q_{Ci} is the reactive power compensation of the capacitors. T_i and Q_{Gi} are discrete variables which are properly disposed when being solved. $U_{n,\text{lim}}$, $Q_{n,\text{lim}}$ are given as:

$$\begin{cases} U_{n,\text{lim}} = U_{n,\min}; U_n < U_{n,\min} \\ U_{n,\text{lim}} = U_n; U_{n,\min} < U_n < U_{n,\max} \\ U_{n,\text{lim}} = U_{n,\max}; U_n > U_{n,\max} \end{cases}$$

(6)

$$\begin{cases} Q_{n,\text{lim}} = Q_{n,\min}; Q_n < Q_{n,\min} \\ Q_{n,\text{lim}} = Q_n; Q_{n,\min} < Q_n < Q_{n,\max} \\ Q_{n,\text{lim}} = Q_{n,\max}; Q_n > Q_{n,\max} \end{cases}$$

where $U_{n,\max}$, $U_{n,\text{lim}}$ are the upper limit and lower limit of nodal voltage; $Q_{n,\max}$, $Q_{n,\text{lim}}$ are the upper limit and lower limit of reactive power of generators or compensating capacitors.

When small-signal stability constraints are considered, the optimised small-signal stability is not expected to be reduced; it means that the eigenvalue of electromechanical oscillation mode should satisfy Equation 3. The controlled variable of small-signal stability optimisation is the same as that of reactive power optimisation. The objective function and constraint condition are listed as:

$$\min F_D = \begin{cases} 1 - f_{dt} \cdot \alpha \le \alpha_0 \quad and \quad \xi \ge \xi_0 \\ 1 + f_{dt} \cdot \alpha > \alpha_0 \quad and \quad \xi < \xi_0 \end{cases}$$
$$f_d = (\alpha - \alpha_0)^2 / (1 + \alpha_0^2) + (\xi - \xi_0)^2 / (1 + \xi_0^2)$$
$$s.t. \begin{cases} \alpha \le \alpha_0 \\ \xi \ge \xi_0 \end{cases}$$

(7)

where α_0 is the maximum real part of the eigenvalue of electromechanical oscillation modes before reactive power optimisation; ξ_0 the minimum damping ratio of electromechanical modes before reactive power optimisation; α the maximum real part of the eigenvalue of electromechanical modes after reactive power optimisation; ξ the minimum damping ratio of electromechanical modes before reactive power optimisation.

When reactive power optimisation and small-signal stability optimisation are both considered, reactive power optimisation is considered as control variables, and the constraint condition is composed of respective constraint conditions. To alter the weight of the optimisation result in small-signal stability optimisation, the objective functions in Eqation 4 and Equation 7 should be processed as follows, and the objective function of reactive power optimisation considering small-signal stability optimisation can be obtained:

$$\min F = \omega_2 F_R + \omega_2 F_D$$
$$F_R = F_Q / F^0 Q$$

(8)

where F_Q is the value in Equation 4 before reactive power optimisation; ω_1, ω_2 the loss of the system

and the weight coefficient in small-signal stability optimisation. When $\omega_1 = 0$ and $\omega_2 \ne 0$, the result of optimisation shows that the system losses are lower and the small-signal stability is enhanced.

4 REALISATION OF PSO ALGORITHM

PSO, put up forward by Kenmedy and Eberhart at 1995, is a kind of swarm intelligence stochastic optimisation algorithm. Compared with the evolutionary algorithms, the PSO algorithm retains the population-based global searching strategy. At the same time, the PSO algorithm is able to memory the best position of the particles (Wang Ling, 2008).

In this paper, we use the improved PSO algorithm. Suppose that at t time, the i-th particle location of n dimension search space is $X_i = [x_{i1}, x_{i2}, ..., x_{in}]$, with speed $V_i = [v_{i1}, v_{i2}, ..., v_{in}]$. First calculate the objective function of the particles, and determine the best position $L_i = [l_{i1}, l_{i2}, ..., l_{in}]$ of each particle and best position L_p reached by the swarm. Then use the following equations to update each particle's speed and position:

$$v_{ij}^{t+1} = u \cdot v_{ij}^t + c_1 r_1 (l_{ij}^t - x_{ij}^t) + c_2 r_2 (l_{pj}^t - x_{pj}^t)$$
$$x_{ij}^{t+1} = x_{ij}^t + v_{ij}^{t+1}$$
$$\mu = (\mu_1 - \mu_0) \cdot \arctan(1.56 \cdot (1 - (t/t_D)^m)) + \mu_0 \quad (9)$$
$$c_1 = 2.51 - 1.5 * t/t_D$$
$$c_2 = 1.51 + 1.5 * t/t_D$$

where, t means the t-th moment; t_D is the upper limit of iteration; c_1 and c_2 are nonnegative numbers, mean earning factors; r_1 and r_2 random Numbers in accordance with being uniformly distributed in interval [0,1]; μ inertia weight, and can harmonise the ability of the global and local optimisation of the particle swarm; μ_1 the upper limit of inertia weight, $\mu_1 = 0.9$; μ_0 the lower limit of inertia weight, $\mu_0 = 0.4$; the control factor $m = 0.4$.

From the updating formula of the velocity and position of particles, the particles in each iteration step use the individual optimal information and global optimal individual information. When the individual optimal information of particles dominates in the iteration process, the particle velocity will be smaller, and the particle swarm may fall into local optimum.

To avoid the above problems, Equation 8 formulates the objective function as the fitness of the particle after each iteration, and orders the particle according to the size of fitness, screening the excellent part of the individual directly into the next iteration, another part of the individual to initialise. Reinitialisation is equal to random variation, and the diversity of population can be guaranteed.

Then, steps of using PSO algorithm for reactive power optimisation are as follows:

Initialise: Set the maximum iterations number D, then assume that there are $M + N$ particles in the particle swarm ($M > N$), consider the variables of each particle with random value within the value range, and initialise the speed and position of each particle.

Evolution: Renew particle velocity and position, according to the iterative formula.

Sorting: Calculate the fitness of each particle, then sort the particles. If there is an individual that evolves to the target solution, it goes to (6), or (4).

Screening: The former M particles with higher fitness will directly enter into the next iteration, while the latter N particles with lower fitness will be reinitialszed.

Update information: Calculate the fitness of the N particles after being reinitialised, then renew the position of swarm. If no individuals evolve to the target solution or iterative times is less than D, it goes to (2), or (6).

Output: If the optimisation is successful, output the particle information that satisfy the requirements, otherwise output the information for optimisation failure.

5 EXAMPLES

Examples used in this paper are IEEE 3-generator 9-bus system and IEEE 8-generator 24-bus system, and the information such as system wiring diagram shows in (Tse T, 2010; Qiu Lei, 2011). In the example, the adjustable range of transformer ratio is 0.9~1.1 and the step length of adjustment is 0.025; There are ten compensating capacitor sets, and the magnitude of node voltage is 0.9~1.1. Relevant power data is expressed in per unit value, and benchmark capacity is 100 MVA.

The procedure of setting the related parameters in the PSO algorithm is as: $D = 100$, $M = 30$, $N = 10$. Adjust the parameters ω_1 and ω_2, then the proportion of that decreasing network loss and enhancing the system small-signal stability in the optimisation results is changed. Since using Power System Stabilizer (PSS) (Fu Hong-jun, 2013) is the main means of improving the system small-signal stability, the system is at the risk of low frequency oscillation. No matter adjusted, the system small-signal stability may not be effectively improved only by reducing the controlled quantity of network loss.

1. The example of 9-bus system:
All the active power generation of the 9-bus system reactive before power optimisation is $P_{GT} = 3.65234$, and the total loss $P_{loss} = 0.04008$. There are two electromechanical oscillation modes for the system, the minimum damping ratio $\xi_0 = 0.12249$: the maximum real part of eigenvalue $\alpha_0 = -1.11875$.

Table 1. The comparison of the optimisation result of 9-bus system when the weight coefficient is different.

Comparative item	$\omega_1 = 0.0$ $\omega_2 = 0.1$	$\omega_1 = 1.0$ $\omega_2 = 0.0$	$\omega_1 = 1.0$ $\omega_2 = 0.1$
P_{loss}/pu	0.05057	0.03951	0.03953
η_{save}/%	−26.17	1.42	1.37
t_d	75	80	75

The optimisation result of network loss P_{loss}, loss descent rate η_{save} and iterative times t_d is shown in Table 1. The optimisation result of electromechanical oscillation mode eigenvalue is shown in Form 2.

$$\eta_{save} = (P_{loss} - P_{loss}^0)/P_{loss}^0 \qquad (10)$$

where, P_{loss}^0 is the network loss before reactive power optimisation, and P_{loss} is the network loss after reactive power optimisation.

The change of steady state loss index of system is shown in Form 1 for different optimisation schemes. When $\omega_1 = 0.0$ and $\omega_2 = 0.1$, the optimisation objective is to enhance small-signal stability without considering the loss of system, and the loss after optimisation is larger; when $\omega_1 = 1.0$ and $\omega_2 = 0.0$, the optimisation objective is to decrease the loss of system without considering small-signal stability, and the loss after optimisation is the smallest; when $\omega_1 = 1.0$ and $\omega_2 = 0.1$, the objective of the optimisation is mainly to decrease the loss of system and slightly enhance small-signal stability, andhe loss after optimisation will decrease less.

The change of electromechanical oscillation mode eigenvalue of system is shown in Table 2 for different optimisation schemes. When $\omega_1 = 0.0$ and $\omega_2 = 0.1$, the optimisation objective is to enhance small-signal stability without considering the loss of system, and the minimum damping ratio of electromechanical oscillation mode after optimisation will decrease most; when $\omega_1 = 1.0$ and $\omega_2 = 0.0$, the optimisation objective is to decrease the loss of system without considering small-signal stability, and because it includes small-signal stability constraints, the small-signal stability of system changes little after optimisation. When $\omega_1 = 1.0$ and $\omega_2 = 0.1$, the optimisation objective is mainly to decrease the loss of system and slightly enhance small-signal stability, and the maximum real part of electromechanical mode eigenvalue after optimisation is between those in former two schemes.

2. The example of 24-bus system:
All the active power generation of the 9-bus system reactive before power optimisation is $P_{GT} = 27.79403$, and the total loss $P_{loss} = 0.35882$. There are two electromechanical oscillation modes

Table 2. The eigenvalue of electromechanical mode in 9-bus system when the weight coefficient is different.

$\omega_1 = 0.0$ and $\omega_2 = 0.1$		$\omega_1 = 1.0$ and $\omega_2 = 0.0$		$\omega_1 = 1.0$ and $\omega_2 = 0.1$	
Eigenvalue	Damping ratio	Eigenvalue	Damping ratio	Eigenvalue	Damping ratio
$-1.369 \pm j9.386$	0.144	$-1.123 \pm j9.074$	0.123	$-1.126 \pm j9.120$	0.123
$-15.207 \pm j24.795$	0.522	$-14.997 \pm j24.573$	0.520	$-14.980 \pm j24.553$	0.521

Table 3. The comparison of the optimisation result of 24-bus system when the weight coefficient is different.

Contract item	$\omega_1 = 0.0$ $\omega_2 = 0.1$	$\omega_1 = 1.0$ $\omega_2 = 0.0$	$\omega_1 = 1.0$ $\omega_2 = 0.1$
P_{loss}/pu	0.43442	0.23401	0.26414
η_{save}/%	-21.07	34.78	26.38
t_d	78	42	56

Table 4. The electromechanical oscillation mode eigenvalue of 24-bus system when the weight coefficient is different.

$\omega_1 = 0.0$ and $\omega_2 = 0.1$		$\omega_1 = 1.0$ and $\omega_2 = 0.0$		$\omega_1 = 1.0$ and $\omega_2 = 0.1$	
Eigenvalue	Damping ratio	Eigenvalue	Damping ratio	Eigenvalue	Damping ratio
$-3.010 \pm j16.248$	0.182	$-3.090 \pm j16.829$	0.181	$-2.524 \pm j17.808$	0.140
$-3.024 \pm j17.544$	0.169	$-2.187 \pm j19.187$	0.113	$-2.182 \pm j19.942$	0.108
$-3.143 \pm j22.780$	0.136	$-2.232 \pm j21.984$	0.101	$-2.214 \pm j21.992$	0.100
$-4.226 \pm j23.811$	0.174	$-2.160 \pm j23.021$	0.093	$-2.511 \pm j22.098$	0.112

for the system, the minimum damping ratio $\xi_0 = 0.09215$, and the maximum real part of eigenvalue $\alpha_0 = -2.14339$.

The optimisation result of network loss P_{loss}, loss descent rate η_{save} and iterative times t_d is shown in Table 3. The optimisation result of electromechanical oscillation mode eigenvalue is shown in Form 4.

As is shown in Table 3, the change regulation of the loss in 24-bus system is the same as that in 9-bus system. If only enhancing small-signal stability is considered, the loss will increase. If only considering decreasing the loss, the loss will decrease most. If mainly decreasing the loss and meanwhile optimising small-signal stability, the loss will decrease less.

As is shown in Table 4, the change regulation of electromechanical oscillation mode eigenvalue in 24-bus system is the same as that in 9-bus system. If only considering enhancing small-signal stability, small-signal stability index will change a lot and be the most steady. If only considering decreasing the loss, small-signal stability index will change little. If mainly decreasing the loss and meanwhile optimising small-signal stability, small-signal stability index will change more and be more steady.

6 CONCLUSION

In this paper, PSO algorithm is used to reactive power optimise the power system and introduce small-signal stability constraints into optimisation. The objective of reactive power optimisation is composed of system loss index that describes steady state characteristics and small-signal stability index that describes dynamic characteristics, which could decrease the loss and enhance small-signal stability. From the examples, based on the optimisation algorithm proposed in this paper, the goal of reactive power optimisation can be efficiently reached and the power system can be operated in a safe, economic and stable manner.

REFERENCES

Fu, H., Pan, L., Lin, T., Sun J. & Yu, G. (2013). Coordinatice optimization of PSS and DC-modulation based on improved PGSA. *Electric Power Automation Equipment, 33*(1), 75–80.

Li, P., Wei, H. & Bai, X. (2013). Small-signal stability constrained optimal power flow based on NLSDP. *Proceedings of the CSEE, 33*(7), 69–76.

Li, Y., Zhang, L. & Yang, Y. (2001). Reactive power optimization under voltage constraints margin. *Proceedings of the CSEE, 21*(9), 796–801.

Liu, H., Ge, S. & Li, H. (2009). Reactive power optimization based on improved particle swarm optimization algorithm with hard restriction regulation. *Journal of Tianjin University, 42*(9), 31–35.

Luo, Y. & Duo, J. (2013) Multi-objective reactive power optimization based on quantum immune colonial algorithm. *Electric Power Automation Equipment, 33*(9), 31–35.

Qiu, L., Wang, K. & Li, K. K. (2011). Construction design and parameter coordination of multi-band PSS. *Power System Protection and Control, 39*(5), 102–107.

Ruan, R., He, B., Kong, D. & Chen, L. (2010) Research on tournament-based ant colony algorithm for reactive power optimization. *Power System Protection and Control, 38*(12), 80–85.

Sun, Y., Yang, X. & Wang, H. (2005). Optimal power flow with transient stability constraints in power systems. *Automation of Electric Power Systems, 29*(16), 56–59.

Tse, T., Wang, K. W. & Chung, C. Y. (2000). Parameter optimization of robust power system stabilizers by probabilistic approach. *IEE Proceedings - Generation, Transmission and Distribution, 147*(2), 69–75.

Wang, L. & Li, B. (2008). Particle swarm and scheduling algorithm. *Tsinghua University Press* (pp. 1–10).

Zhang, A., Yang, H., He, X. & Yang, K. (2010). Parallel computing model for coordinated reactive power optimization considering load changing. *Power System Protection and Control, 38*(18), 13–18.

Zhao, B. & Cao Y. (2005). A multi-agent particle swarm optimization algorithm for reactive power optimization. *Proceedings of the CSEE, 25*(5), 1–7.

Zhao, L. & Lu, J. (2010). Multi-objective power optimization of wind farm based on improved genetic algorithm. *Electric Power Automation Equipment, 30*(10), 84–88.

Zhong, Z., Xie, H. & Wang, K. (2000). A novel modeling technique for modern power system dynamic studies *Proceedings of the CSEE, 20*(3), 30–33.

Zhou, X-J., Jiang, W-H. & Ma Li-li (2010). Reactive power optimization of power system based on improved genetic algorithm. *Power System Protection and Control, 38*(7), 37–41.

Automotive, Mechanical and Electrical Engineering – Liu (Ed.)
© 2017 Taylor & Francis Group, London, ISBN 978-1-138-62951-6

Research and design of beacon laser drive system

Dongya Xiao, Hongzuo Li & Huiying Zhang
School of Electronics and Information Engineering, Changchun University of Science and Technology, Changchun, China

ABSTRACT: This paper has designed drive circuits for beacon laser, mainly including the temperature control module and the power control module. In the temperature part, the nonlinear correction of thermistor, PID control compensation for ADN8830 and the output peripheral circuit are given. In the power part, the circuit for photocurrent detection and comparison is designed, and the power remains stable through an AVR single chip microcomputer which changes the intensity of the constant current source output. Also, the output power is tuneable through the digital potentiometer X9221. Stable output power, makes power is adjustable within the range indicator. This design can provide a temperature control precision of 0.02°C, and output power stability of about 0.25% – stable and tuneable.

Keywords: Drive circuit; PID control; Temperature control; Power control

1 INTRODUCTION

A semiconductor laser is also called a Laser Diode (LD), which has small volume, high efficiency, low power, easy modulation and other advantages. It is widely used in communications, medical, precision machining, and other fields. In laser communication transmitting systems, a beacon light source plays an important role, so the design of drive circuits is particularly important, related to the whole communication system. One feature of the LD is its high sensitivity to temperature. Working temperature can seriously make its output light performance; threshold current, output wavelength and quantum efficiency change along with the change of temperature. At the same time, as the light source in optical communication fields, it also needs stable power output to ensure the communication system operates normally. Therefore, it is necessary to adopt Automatic Temperature Control (ATC) and Automatic Power Control (APC) measures for laser, in order to make the laser work stably at the setting temperature and power.

2 DESIGN OF TEMPERATURE CONTROL

2.1 *Principle of temperature control*

The indicator of temperature control is 25 ± 0.02°C. Here we choose ADN8830 (Huang Qingchuan & Ma Jing., 2006) as the temperature control chip, which has the advantages such as high integration, quick response, high efficiency, high stability and

to drive TEC and keep the temperature constant. Its temperature drift voltage is less than 250 mV, which can make the setting temperature error control lower. An input voltage value of ADN8830 corresponds to a setting target temperature. TEC will heat or cool, when the current flows through it, which makes the surface temperature of LD remain stable near the setting temperature (Qin Xiqing et al., 2004). Also, the chip has the function of over-current protection.

The temperature variation of LD is detected by temperature sensor. Here we adopt a thermistor with Negative Temperature Coefficient (NTC). First the thermistor detects the temperature voltage value and compares with the set-point voltage. Then the controller outputs a control signal to adjust the direction and size of the current of TEC (Xia Jinbao et al., 2015), in order to realise the temperature control of the LD. The schematic of TEC temperature control is shown in Figure 1.

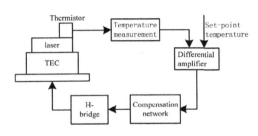

Figure 1. Schematic of TEC temperature control.

2.2 Nonlinear correction of thermistor

The relation between the value of thermal and temperature is not absolutely linear, which causes the bridge resistance circuit to also be unbalanced. To solve the nonlinear problem, here we adopt a way of linear compensation to achieve the largest linearisation output. The specific method is to put an ordinary resistance R_r and the thermistor R_T in parallel. When the value of resistance R_r equals the value of NTC resistance corresponding to the setting temperature, then the output is largely linear in the temperature range. The equation is given as follows:

$$U_{DB} + \Delta U_{DB} = V_{cc} \left(\frac{\frac{R_r(R_T + \Delta R_T)}{R_r + R_T + \Delta R_T}}{R_1 + \frac{R_r(R_T + \Delta R_T)}{R_r + R_T + \Delta R_T}} - \frac{R_3}{R_2 + R_3} \right) \quad (1)$$

When the temperature is at 25°C, thermistor resistance is approximately 10 kΩ, obtained by experiment. We should therefore choose a resistor R_r = 10 kΩ and in parallel with R_T, and then make $R_2 = R_3$ = 10 kΩ. We can get U_B = 2.5 V. When temperature is 21°C, thermistor corresponding about 12 kΩ, then make R_1 = (10 × 12)/(10 + 12) ≈ 5.45 kΩ. Now the corresponding output voltage U_{DB} = 0. This paper is based on temperature of 25°C, with the voltage of U_{DB} as the initial voltage. The resistance bridge circuit revised is shown in Figure 2.

From experimental test we obtain the curve shown as Figure 3. The relationship between output voltage U_{DB} and resistance is nearly linear, which means that the output voltage and temperature of the resistance bridge circuit revised have a linear relationship.

2.3 PID control compensation

PID is a linear controller, consisting of three control variables Proportional (P), Integral (I) and Differential (D) (Zhu Hongbin et al., 2009). It is a common feedback loop in industrial control applications. The PID controller parameter setting is the core content of TEC temperature control system design. Different PID controller parameters can realise different response speed and temperature stability of TEC.

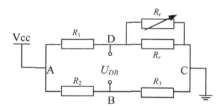

Figure 2. The resistance bridge circuit revised.

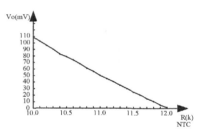

Figure 3. The relationship between output voltage and resistance.

We should adjust the parameters of the controller according to the relationship between the performance of dynamic-state and steady-state. The mathematical model of PID controller is:

$$u(t) = K_P \left[e(t) + \frac{1}{T_I} \int_0^t e(t)dt + T_D \cdot \frac{de(t)}{dt} \right] \quad (2)$$

The corresponding transfer function is:

$$G(s) = \frac{U(s)}{E(s)} = K_P \left[1 + \frac{1}{T_I s} + T_D s \right] \quad (3)$$

where K_P is proportional coefficient; T_I is integral time constant; T_D is differential time constant. Adjust the control parameters through Matlab simulation, to realise LD temperature control.

ADN8830 integrates low-noise amplifier which consists of PID compensation circuit with peripheral circuit among the TEMPCTL pin, COMPFB pin and COMPOUT pin, as shown in Figure 4. R_1 and R_3 control proportion coefficient, C_1 and R_3 control integral time constant, C_2 and R_1 control differential time constant, CF can increase the stability of the compensation circuit. Adjust the component parameters to get satisfactory control effect, and choose R_1 = 100 kΩ; R_2 = 1 MΩ; R_3 = 205 kΩ; C_1 = 10 μF; C_2 = 1 μF; CF = 330 pF.

2.4 Peripheral circuit design

ADN8830 using the way of differential output drives TEC. In order to provide the required current of TEC cooling, an H-bridge circuit is designed outside of ADN8830, as shown in Figure 5.

The pins of P1, P2, N1, N2, OUTA and OUTB respectively connect to the pins of P1, P2, N1, and N2, OUTA and OUTB of ADN8830. TEC driving mode uses asymmetric output, which can not only reduce the ripple effect, but also improve the efficiency of power source. P1 and N1 drive peripheral PMOS and NMOS working in switch mode in the way of PWM output, while P2 and N2 drive peripheral PMOS and NMOS in a linear way.

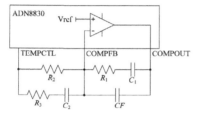

Figure 4. PID control compensation circuit.

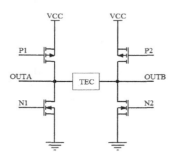

Figure 5. H-bridge circuit.

3 DESIGN OF POWER CONTROL

3.1 *Principle of power control*

The indicator of power control is 0~3 W, tunable, the stability of the output power is no more than 0.3%. The schematic diagram of power control is shown in Figure 6. First, the PhotoDetector (PD) detects the photocurrent of LD, and converts it to voltage signal through I/V conversion. The signal is converted by A/D conversion and then sent to an AVR Atmega16 controller and compared with the setting value. Next AVR outputs control signal and controls the constant current source through digital potentiometer (Lu Kai et al., 2012). The constant current source integrates operational amplifier, to ensure the drive current output is stable and avoids power fluctuations caused by the unstable current.

3.2 *Power detection and control circuit*

Power detection, which is converting the optical signal of the LD into electric signal, in order to compare with the control voltage signal. Then the controller outputs a command signal to change the size of the drive current. The process is repeated all the time until the output power remains stable. Power detection and automatic control circuit design is shown in Figure 7.

The PD works in reverse bias state, which can accelerate the movement of the electrons, reduce the probability of recombination with holes, and inhibit the effect of dark current. The response

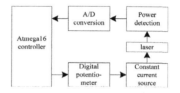

Figure 6. Schematic of power control.

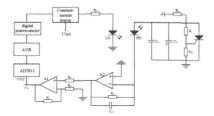

Figure 7. Power detection and control circuit.

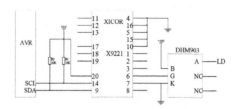

Figure 8. Constant current source DHM903 control circuit.

time can therefore be shortened, and the quantum efficiency and sensitivity improved. When PD detects optical power, the photon-generated carriers generated move directionally due to the electric field's action. The formation of current I_m is precisely because of the movement of electronic to N area. The relationship between current I_m and the incident light intensity P_{in} is:

$$I_m = RP_{in} \tag{4}$$

where R is the responsivity of PD.

The relation of output power of LD and current I_m is described as:

$$I_{in} = MP_0 \tag{5}$$

where M depends on the response of the PD. When LD and PD connect in positively driving mode, the relationship between the initial value of power voltage of setting system and output power of LD can be obtained as:

$$U_{SET} = \frac{R_6}{R_4} R_0 I_{in} = \frac{R_6}{R_4} R_0 M P_0 \tag{6}$$

3.3 Constant current source circuit

The output of digital potentiometer X9221 is controlled by AVR controller. X9221 is connected with K, G end of the constant current source DHM903. The circuit connection diagram is shown in Figure 8. Therefore if we adjust the value of digital potentiometer, the output current size of DHM903 will change, and the output power of LD will also change. They basically meet linear relationship. Finally, the power can realise a stable and adjustable output.

4 EXPERIMENT TESTS

4.1 Temperature test

The environment temperature is 20°C, and the setting temperature of LD is 25°C. The output voltage value of the thermistor after linear correction can be measured, then converting it into corresponding temperature value, we get the corresponding temperature curve of testing time for 10 min, as shown in Figure 9. As can be seen from the curve, the LD temperature is gradually stable when the temperature control lasting for 30 s, and then the temperature accuracy is within ±0.02°C. The experiment results show that the designed circuit can work fast and stably to control TEC, and ensure LD works at a constant temperature state.

4.2 Power test

The output power is set at 200 mW, test time is 60 min. The output power curve of actual measurement for LD is shown in Figure 10. It can be seen from the curve that the output power is stable, floating up and down slightly.

The power stability is defined as:

$$S = \frac{P_{max} - P_{min}}{P_{avg}} \qquad (7)$$

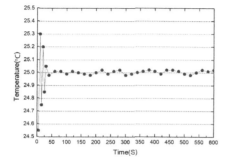

Figure 9.　Curve of LD working temperature in 10 min.

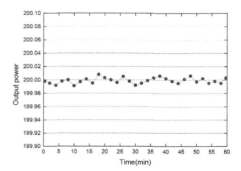

Figure 10.　Curve of LD output power in 60 min.

where S is power stability, P_{max} is the maximum of output power that is measured, P_{min} is the minimum of output power, P_{avg} is the average of the output power in 60 min. By calculation, we can get the stability of the output power is about 0.25%, less than 0.3%. The test results show that the designed circuits are feasible and stable, and can ensure the LD works at constant power state.

5 CONCLUSION

Drive circuits for beacon laser have been researched and designed, including temperature control circuits and power control circuits. Here adopted PID control compensation for temperature control and control accuracy can reach ±0.02°C, which will avoid the influence of temperature variation on the threshold current and the output wavelength. The constant current source with high precision is controlled by AVR single chip microcomputer, and 0.25% power stability is obtained. The designed drive circuits for beacon LD meet the design requirements and there may be some reference value to the drive circuit design and laser communication.

REFERENCES

Huang, Q. & Ma, J. (2006). Semiconductor laser temperature control circuit analysis and testing. *Optics & Optoelectronic Technology*, 4(3), 46–48.

Lu, K. et al. (2012). High power laser diode drive power supply. *Infrared and Laser Engineering*, 41(10), 2680–2684.

Qin, X. et al. (2004). A high performance TEC controller circuit based on ADN8830. *Optics & Optoelectronic Technology*, 2(1), 20–22.

Xia, J. et al. (2015). Design of semiconductor laser quick temperature control system. *Infrared and Laser Engineering*, 44(7), 1991–1995.

Zhu, H. et al. (2009). A design of high precision temperature controlling system for uncooled infrared focal plane. *Infrared Technology*, 31(3), 144–147.

Automotive, Mechanical and Electrical Engineering – Liu (Ed.)
© *2017 Taylor & Francis Group, London, ISBN 978-1-138-62951-6*

Research on a kind of intelligent garbage bin combined with solar street light

Yu Feng
School of Electrical and Electronic Engineering, North China Electric Power University, Beijing, China

ABSTRACT: At present, with the continuous improvement in distributed energy utilisation technology and people's awareness of energy saving and environmental protection, this paper presents a design scheme of an intelligent garbage bin combined with solar street light, and gives the principle of the combination of solar street lamp and intelligent garbage can. The design scheme for intelligent trash, includes garbage box cover automatic switching function, automatic voice prompt function, automatic garbage compression, garbage trunkful detection function and dynamic information acquisition and transmission function. The three principles of the combination of the trash can and the street lamp are: minimise the impact of their own original functions; in line with the surrounding landscape features; use modular design with free separation from each other.

Keywords: Solar energy; Street lamp; Intelligent trash can; ZigBee; Energy saving

1 INTRODUCTION

With the development of distributed energy technology and people's increasing awareness of energy saving and emission reduction, solar lights, as a new energy-efficient lighting tool, are used widely in city lighting systems (Yin, 2014).

Although an indispensable part of city life, there are many problems with trash and its usage process, including the difficulty of cleaning, low space utilisation, poor cleanliness, and so on. That these phenomena exist, fundamentally speaking, is because of the low level of intelligent trash.

However, the cost of improving the intelligence level of the trash can is extremely high. The combination of solar street lamps and trash cans, to a certain extent, can alleviate the above problems. On the one hand, the battery supplying energy for solar street lights and trash cans not only makes full use of solar energy, but also saves the cost of laying out the power grid for the trash can, reducing the operational cost; on the other hand, it may also save the city's layout area of trash can and street lamps.

In China's current cities, there are very few intelligent garbage bins combined with solar street lights (Zicheng, 2013). However, solar street lights and intelligent trash cans are very common; in other words, from a technical point of view, there is no problem to realise this thing. In addition to the technical aspects of the problem, the design is also very important when trash and street lights are combined. In consideration of the function of

the garbage can design, at the same time, a combination of these two methods should also be considered, paying attention to the coordination with the surrounding scenery, thus making people gain the feeling of beauty.

2 RESEARCH IDEAS

The research of this paper includes two parts: the research into the related function of the intelligent trash can, and the research into the method of combining the trash can and the street lamp. After considering the actual needs in real life, the combination of street-smart trash use should have five functions: automatic turning cover, automatic compression, voice prompt, garbage trunkful detection function, and dynamic collection and transfer function of information in the area.

For the combination of the trash can and the street lamp, this paper puts forward three principles: first, the minimum influence of their original function; two, the importance of keeping in line with the surrounding landscape features of artistic design; three, modular design, in that they can be freely separated from each other.

3 FUNCTION ANALYSIS OF INTELLIGENT GARBAGE CAN

As mentioned above, this paper argues that intelligent street trash should have five functions:

automatic turning cover, automatic compression, voice prompt, garbage trunkful detection function, dynamic collection and transfer function of information in the area.

For the dustbin cover automatic switching function, this paper describes a control starting circuit switch through the infrared induction system, which causes the transmission mechanism to automatically flip. When people leave, with the help of the delay circuit, the trash can automatically close the cover. For automatic compression function, the compression mechanism is used to compress the garbage. For the voice prompt function, first with the programming module entry into the memory, and then in the process of running through the speaker to achieve voice functions. The trunkful detection function is implemented through the pressure sensor. When garbage is compressed, if the measured force acting on the pressure sensor reaches the maximum value set previously, the sensor will pressure signals to SCM, and then trash flip mechanism to stop the operation (Qiang Zhou, 2016).

Finally, we consider the function of the dynamic acquisition and transmission of the information. It is achieved with the help of ZigBee technology. ZigBee is a new kind of wireless communication technology, which has the characteristics of close range, low difficulty, low cost, low power consumption and low speed. It is based on the wireless protocol of standard IEEE802.15.4, and can realise two-way communication by the mutual coordination between the wireless sensors. Through the use of ZigBee wireless sensor network technology, and using a series of sensors and chips to design the circuit, we can produce a smart garbage can (Wang, 2012; Hai Ma, 2010).

4 METHOD FOR REALISING INTELLIGENT GARBAGE CAN

4.1 Function of automatic opening and closing cover for garbage box

After access to information, we use thermal infrared sensors to achieve garbage automatic flip function (Huabin Wang, 2012).

The principle is as follows: when people get close to the trash to throw rubbish, the thermal infrared human body sensor can feel the people around, triggering the CC2430 chip and CC2430 chip, and the I/O port will drive the motor. The thermal infrared sensor induction principle is shown in Figure 1.

4.2 Automatic garbage compression function

The compression mechanism is connected to the garbage box cover—when the garbage can

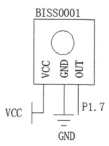

Figure 1. Schematic diagram of the infrared sensor circuit.

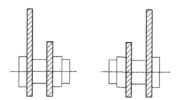

Figure 2. Trash liner structure.

automatically closes the cover, it drives the compression mechanism to compress the garbage. At the same time, due to the fact that the garbage bag is not easily separated from the tank of trash, an adjustable capacity trash liner has been designed to solve the problem.

The following illustrates the adjustable capacity trash liner. Due to the squeeze pressure, the original loose garbage moves close together, leading to relatively large attraction between the molecules; hence it is not easy to take the trash bag from the trash. Therefore, the structure of the screw nut is used to fix the nut and the movable baffle plate around the barrel wall, and the volume of the garbage bin can be varied. When the garbage can has been filled, and the rubbish needs to be taken out of the trash can, we can rotate the bolt and change the trash volume slightly, so that the garbage bag and the liner can be easily separated (Zhao, 2012). The trash liner structure is shown in Figure 2.

4.3 Voice prompt function

For the voice prompt module, this paper suggests using SYN6288 voice chip to achieve the conversion of text to speech.

4.4 Trunkful detection induction

This function is divided into two modules, weight monitoring module and volume monitoring module. First, for the weight monitoring module, this paper recommends the use of the TYC901 weight sensor. However, because the signal transmitted by the weight sensor is generally weak, it needs the signal to use a differential amplification circuit to amplify the weight sensor. Then, the volume monitoring module. In this paper, the idea is to use the laser sensor and photosensitive resistance. The components of the laser sensor consist of a laser, a laser detector, and a measuring circuit. By detecting the voltage of the photosensitive resistance, and combining with the weight of the trash can, we can determine whether the trash can is full. The circuit schematic diagram of the laser sensor and the photosensitive resistance are shown in Figures 3 and 4 respectively.

4.5 Dynamic collection and transfer function of information

The sensor nodes collect data through the integrated sensor module on a node, and these data are then transmitted over the network to the coordinator node. The coordinator node is responsible for collecting data transmitted over the network, and sending the data directly to the host computer through the serial port or through the network to achieve remote data transmission. Then the relevant departments can develop a simple data management centre software through the database technology. After receiving the data from the sensor network, the data centre is capable of completing the analysis and storage of the data (Hai Ma, 2010). After the data are processed by computer, the information platform will launch real-time information to the sanitation workers. With the help of this information, the sanitation workers can decide which trash needs cleaning up, and which does not need cleaning up, eliminating the trouble of performing a daily patrol and improving efficiency.

5 COMBINATION MODE OF GARBAGE CAN AND SOLAR ENERGY STREET LAMP

According to the actual situation in different countries, the garbage clean-up department and the road management department may not be the same. Therefore, the combination of the two may face a huge administrative reform. Besides, for the repair work of intelligent garbage cans or street lamps, because the work characteristic is different, the breakdown time is also very different; therefore, this article proposes three principles for combining them:

First, the minimum influence of their original function;

Two, in line with the surrounding landscape features of artistic design;

Three, modular design, in that they can be freely separated from each other.

A simple sketch map is shown below:

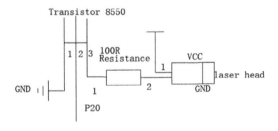

Figure 3. Schematic diagram of laser sensor circuit.

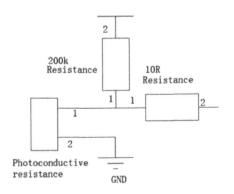

Figure 4. Circuit schematic diagram of photosensitive resistance.

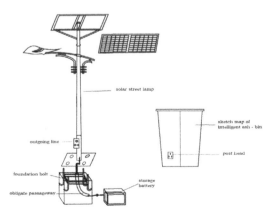

Figure 5. Schematic diagram of the combination of solar street lamp and intelligent garbage can.

6 CONCLUSION

It is an effective way of energy use to combine the intelligent garbage can with the solar street lamps, which can promote the progress of the social life. If the battery supplies energy for the solar street lights and the trash can at the same time, it not only makes full use of solar energy, but also saves the cost of laying out the power grid for the trash can and reduces the operation cost; on the other hand, it may also save city trash and street lamp layout area. Furthermore, through the ZigBee wireless communication technology, it can help improve the efficiency of street cleaning for cleaners, improve the efficiency of garbage disposal, and improve the efficiency of the urban operation. Therefore, this research is of value, and the intelligent garbage can combined with the solar street lamp has excellent development prospects.

ACKNOWLEDGEMENTS

This project was supported by North China Electric Power University students' innovation and entrepreneurship training programmes (2016193).

Author: Yu Feng (1996 -), male, Jiangsu Xuzhou, North China Electric Power University, undergraduate students, Beijing City, Changping District Huilongguan Beinong Road No. 2 North China Electric Power University, 10220618811362511, fengyuhm@163.com

REFERENCES

Ma, H. (2010). Design and implementation of remote data monitoring based on ZigBee wireless sensor network (masters dissertation). Wuhan University of Technology, China, 1–14.

Wang, H., Luo, Z., & Zeng, S. et al. (2012). Design of intelligent environmental sanitation garbage bin based on ZigBee. *Computing Technology and Automation, 31*(1), 48–50.

Yin Liu, Linlin Li, & Tang, J. (2014). Analysis on application mode of solar street lighting in urban lighting. *Zhao Ming Gong Cheng Xue Bao, 25*(5), 65–66.

Zhao, T., & Wang, H. (2012). Garbage can which automatically compress garbage. *China Science and Technology Information, 12*, 170–171.

Zhou, Q., Feng, G., Lin, L. et al. (2016). The design of a self-flip compressible multi-function intelligent trash. *New Product Development, 43*(5), 51–53.

Zicheng Luo, Min Tang, & Jun Xu. (2013). Study on a new type of energy saving and environmental protection solar lighting type luminous garbage bin. *Theoretical Research, 12*(1), 95–96.

Automotive, Mechanical and Electrical Engineering – Liu (Ed.)
© 2017 Taylor & Francis Group, London, ISBN 978-1-138-62951-6

Research on dual PWM converter for doubly fed motor generation system

Jianjun Xu, Lin Huang, Lianwei Zhang, Yina Zhou, Limei Yan & Hongyu Li
Department of Electrical Information Engineering, Northeast Petroleum University, Daqing, China

ABSTRACT: The control core of the doubly fed wind generation system is the dual PWM converter. Traditionally, the control method is used with the Proportional Integral (PI) controller. With the increase of wind turbine installed capacity, the system requirements for control precision and stability are higher. It is difficult to ensure that the PI control method can meet with control requirements in the case of an unbalanced power system. This paper adopts the method of Proportional Resonant (PR). The PR controller makes sure that the system's output is without static error. This method can also guarantee active power and reactive power of rotor side decoupling control at the same time, eliminating the coupling and feed-forward compensation to reduce the times of coordinate changes. This can reduce the influence of grid frequency deviation of the converter output inductor current, thus effectively improving the precision of the control algorithm and the electric energy's quality. The feasibility and correctness of the theory are verified by simulation experiment and result analysis.

Keywords: Double fed motor; Double PWM converter; PR controller; Power decoupling

1 INTRODUCTION

The Proportional Resonant (PR) controller contains proportional integral functions and the transfer function:

$$G_{PR}(s) = K_b + 2K_j \frac{s}{s^2 + \omega^2} \quad (1)$$

Compared with the PI controller, the PR controller can achieve zero steady state error, and has good anti-interference ability. However, there are the following problems in practice. In the circuit simulation data and components of precision limited, due to the large amount of computation the control algorithm cannot achieve real-time control; when there is power grid frequency fluctuation, it cannot effectively suppress the harmonic (Chen Wei, 2014; Ekanayake J. B. & Holdsworth L., 2003; Jin Yuanyuan, 2014). Therefore, with the advantages of PR control, the optimised PR control method is proposed. The transfer function is:

$$G(s) = K_b + 2K_j \frac{\omega_c s}{s^2 + 2\omega s + \omega^2} \quad (2)$$

The wind power generation system's units are becoming bigger. There is an urgent demand for improvement in the power grid's quality. Interference of large system is mainly 3,5,7,11 times low harmonic which can cause system noise and torque ripple. The resonant frequency ω in (2) is modified to be $h \cdot \omega$ ($h = 3,5,7$), which can be obtained as a function of the harmonic compensation term:

$$G_h(s) = \frac{K_h s}{s^2 + (h \cdot \omega)^2} \quad (3)$$

2 POWER GRID SIDE PR CONTROL SYSTEM

2.1 Establishment of mathematical model

Adopting controllable switch IGBT to control the power device of a power grid system can reduce harmonics of the system. Because of the high speed of the IGBT switch, low switching loss and large working range, the network side converter can realise its function better. The converter of the rotor side can be seen as a circuit load, and the main circuit structure of the network side converter is obtained, as shown in Figure 1.

When we are modelling, the electric potential of the power network is symmetrical, and the net side filter inductance is not linear and saturated. Thus, in order to better express the logic value of the power switch, when the S value is 1, the lower

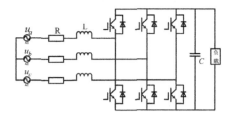

Figure 1. Main circuit structure of network side converter.

bridge arm is closed and the upper bridge arm is opened; when the S value is 0, the lower bridge arm is opened and the upper bridge arm is closed. According to Kirchhoff's voltage law, we can get (Morren J. & de Haan S. W. H., 2005; Teodorescu R. & Blaabjerg F., 2006; Xu Le, 2015):

$$L\frac{di_a}{dt} + Ri_a = u_a - \left(u_{dc}S_a + u_{NC}\right) \qquad (4)$$

Through Figure 1, we can get:

$$\begin{cases} u_a - Ri_a - L\dfrac{di_a}{dt} - S_a u_{dc} \\ = u_b - Ri_b - L\dfrac{di_b}{dt} - S_a u_{dc} \\ = u_c - Ri_c - L\dfrac{di_c}{dt} - S_c u_{dc} \\ c\dfrac{du_{dc}}{dt} = i_c - i_L = S_a i_a + S_b i_b + S_c i_c - i_L \end{cases} \qquad (5)$$

In the formula, u_a, u_b, u_c is three-phase voltage source; i_a, i_b, i_c is three-phase input current; i_L is load current; i_c is output current; u_{dc} is output voltage of DC side. In the system, the sum of three-phase current is zero and three-phase voltage is basic balanced, so we can get:

$$\begin{cases} i_a + i_b + i_c = 0 \\ u_a + u_b + u_c = 0 \end{cases} \qquad (6)$$

2.2 Control strategy

The control strategy of the dual PWM converter system's grid side is mainly to get good input characteristics, to maintain the stable output voltage of DC side, and to ensure that the input power factor is close to 1. The traditional control system adopts PI control mode. The structure is shown in Figure 2.

It can be concluded from Figure 3 that three-phase AC current in the grid side can be obtained by the coordinate transformation. Given the feedback voltage and the voltage comparison is

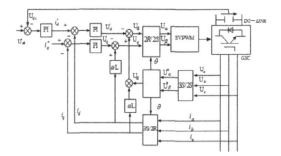

Figure 2. PI controller of the network side converter.

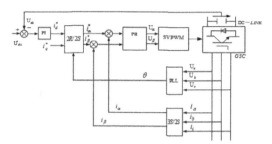

Figure 3. PI controller of the network side converter.

obtained after the reference current and through coordinate transformation to obtain the values in two-phase stationary coordinates. The system choose comparison of the current PR as control input and the output of PR controller as the input voltage SVPWM. SVPWM modulation is used to switch signal in order to control the main circuit of the converter. In this way we realise the network control strategy side converter.

3 PR CONTROL SYSTEM OF ROTER SIDE

3.1 Basic structure and function of rotor side converter

The rotor side converter is the hub of the doubly fed induction motor grid side and, and converter. The control of the wind power generation system is achieved by the rotor side converter. Through the rotor side PWM converter, we can control the doubly fed asynchronous motor conversion of wind energy utilisation and energy. The structure of the inverter circuit is similar, but the implementation is a two-way energy flow. The circuit is shown in Figure 4.

In the circuit, the power is divided into active power, reactive power and apparent power, which needs to be kept close to the unit power factor. The main function of the rotor side converter is to achieve the decoupling of active power and

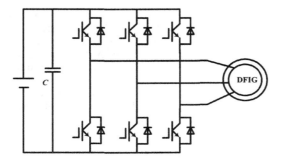

Figure 4. Structure of rotor side converter.

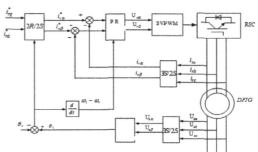

Figure 5. Control block diagram of PR control strategy.

reactive power through the control strategy. In the case of change of wind speed, the maximum wind energy capture can be achieved, and the utilisation rate of wind energy can be improved by (Yuan Guofeng & Chai Jianyun, 2015).

3.2 The mathematical model of three-phase static coordinate system of the double fed motor

Using a matrix to express three phase winding rotor to stator side's voltage, a differential operator is used to replace the differential symbol p:

$$
\begin{bmatrix} u_A \\ u_B \\ u_C \\ u_a \\ u_b \\ u_c \end{bmatrix} = \begin{bmatrix} -r_s & 0 & 0 & 0 & 0 & 0 \\ 0 & -r_s & 0 & 0 & 0 & 0 \\ 0 & 0 & -r_s & 0 & 0 & 0 \\ 0 & 0 & 0 & -r_r & 0 & 0 \\ 0 & 0 & 0 & 0 & -r_r & 0 \\ 0 & 0 & 0 & 0 & 0 & -r_r \end{bmatrix} \begin{bmatrix} i_A \\ i_B \\ i_C \\ i_a \\ i_b \\ i_c \end{bmatrix} + p \begin{bmatrix} \Psi_A \\ \Psi_B \\ \Psi_C \\ \Psi_a \\ \Psi_b \\ \Psi_c \end{bmatrix}
$$

$$(7)$$

1. The winding flux can be divided into self-inductance flux and transformer flux, so we can get the flux equation:

$$
\begin{bmatrix} \Psi_A \\ \Psi_B \\ \Psi_C \\ \Psi_a \\ \Psi_b \\ \Psi_c \end{bmatrix} = \begin{bmatrix} -L_{AA} & -L_{AB} & -L_{AC} & L_{Aa} & L_{Ab} & L_{Ac} \\ -L_{BA} & -L_{BB} & -L_{BC} & L_{Ba} & L_{Bb} & L_{Bc} \\ -L_{CA} & -L_{CB} & -L_{CC} & L_{Ca} & L_{Cb} & L_{Cc} \\ -L_{aA} & -L_{aB} & -L_{aC} & L_{aa} & L_{ab} & L_{ac} \\ -L_{bA} & -L_{bB} & -L_{bC} & L_{ba} & L_{bb} & L_{bc} \\ -L_{cA} & -L_{cB} & -L_{cC} & L_{ca} & L_{cb} & L_{cc} \end{bmatrix} \begin{bmatrix} i_A \\ i_B \\ i_C \\ i_a \\ i_b \\ i_c \end{bmatrix}
$$

$$(8)$$

2. The output torque equation is:

$$
T_e = \frac{1}{2} p \left(i_s^T \frac{dL_{sr}}{d\theta_r} i_r + r_r^T \frac{dL_{rs}}{d\theta_r} i_s \right)
$$

$$(9)$$

3.3 Control strategy

In the grid operation, vector control power is the output current of motor control through the excitation current of the rotor side, as long as rotor excitation current can realise control of active power and reactive power independently. Therefore, the control idea is: by controlling the excitation voltage, the frequency of the excitation current is scheduled to slip, and we can adjust the rotor current's active and reactive power independently.

Because wind speed is characteristically random and uncertain, so the frequency of rotor current is always changing, and the slip frequency is generated. The PR controller can adjust AC signals without static difference, so that the resonant frequency and the rotating difference frequency are equal, and the rotor current can be adjusted adaptively. Figure 5 shows the simulation block diagram of the rotor side converter under PR control strategy.

4 EXPERIMENTAL RESULTS AND ANALYSIS

Figure 6 shows the voltage and current waveform of the grid side converter when adopting the PR control strategy under sub-synchronous state and ultra-synchronous state respectively. Through the Figure we can get: grid side converter work in rectifying state when the system works in sub-synchronous state; when the system works in super-synchronous state, network side converter works in inverter and realises the two-way flow of energy. Observing the power network voltage and input current waveform shows that the grid side converter's input power factor is close to 1, the line current harmonic content is low, and close to the sine.

We can see that DC bus voltage did not change greatly when the given current is changed, and then the 0.02 s tends to be stable. In Figure 8, we can see the DC bus voltage is smaller than the PI

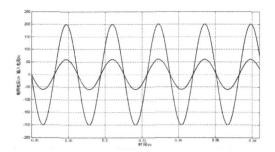

Figure 6. Voltage and input current waveforms of the grid side under the control of PR.

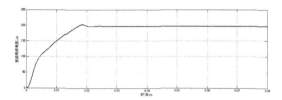

Figure 7. DC bus voltage under PI control strategy.

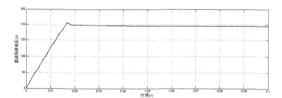

Figure 8. DC bus voltage under PR control strategy.

control voltage, and it has the function of harmonic suppression.

Increased and finally stabilised at 1,550 W. There is no disturbance of reactive power, which has always maintained a steady state.

5 CONCLUSION

The results under disturbance show that the motor can achieve the maximum wind energy capture, and also realise the adjustment of the speed. Changed the amplitude of stator current, stator but always work in 50 Hz and output waveform is always stable. This confirms that the system can realise variable speed with constant frequency. When the active power is under disturbance, the stator current response speed is fast, and the stator current reactive power is without disturbance. Tthe realisation process of active power and reactive power decoupling control, and ensure unit power factor operation system.

ACKNOWLEDGMENT

This work is supported by PetroChina Innovation Foundation (2016D-5007-0201).

This work is supported by 2015 Heilongjiang Province degree and graduate education reform research project [No.JGXM_HLJ_2015107]: "the research on the cultivation system of innovation ability of graduate students of electrical engineering degree". Corresponding Author is Yan Limei.

This work is supported by the graduate students' innovative research fund of Northeast Petroleum University (YJSCX2015–029 NEPU & YJSCX2016–027 NEPU).

REFERENCES

Chen, W. (2014). PI control of dual PWM converter in doubly fed wind generation System. *Proceedings of the CSEE, 29*(15), 2–3.
Ekanayake, J.B. & Holdsworth, L. (2003). Dynamic modeling of doubly fed induction generator wind turbines. *IEEE Transactions on Power Systems, 18*(2), 803–809.
Jin, Y. (2014). *Research of grid-connected inverter based on the PR control*. Zhejiang University, Hangzhou.
Morren, J. & de Haan, S.W.H. (2005). Ridethrough of wind turbines with doubly-fed induction generator during a voltage dip. *IEEE Transactions on Energy Conversion, 20*(2), 435–441.
Teodorescu, R. & Blaabjerg, F. (2006). Proportional-resonant controllers and filters for grid-connected voltage-source converters. *IEE Proceedings—Electric Power Applications, 153*(5), 750–762.
Xu, L. (2015). *The study of dual PWM converter for doubly-fed wind generator*. Hunan University, Changsha.
Yuan, G. & Chai, J. (2015). Study on excitation converter of variable speed constant frequency wind generation system. *Proceedings of the CSEE, 25*(8), 90–94.

Automotive, Mechanical and Electrical Engineering – Liu (Ed.)
© *2017 Taylor & Francis Group, London, ISBN 978-1-138-62951-6*

Research on single event effect based on micro-nano SRAM semiconductor devices

Shuai Wang

School of Physics, Nankai University, Tianjing, China

ABSTRACT: Along with the developing national aerospace industry and growing strategic consideration, the radiation resistance of semiconductor devices has been widely concerned Single event effect is the key factor which can affect the aerospace components' reliability, which raises wide attention. The semiconductor devices of the current mainstream technology are micro-nano scale, and sensitivity to single event effect is also increased. In this paper, a model is established to analyze the characteristics of the device, and single event effect of SRAM components is studied by simulation.

Keywords: Aerospace industry; Semiconductor devices; SRAM components

1 INTRODUCTION

In recent years, with the rapid development of science and technology, semiconductor integrated circuit, with its strong function, small volume, and low power consumption advantages has been widely used in various fields, especially in aerospace engineering and military applications. However, the harsh radiation environment put forward higher requirements for the reliability of semiconductor devices. The charged particles in the space environment can produce radiation damage to the spacecraft materials, integrated circuits and electronic components (EIA/JESD57 1996). The early detection of the radiation effect is mainly the total dose effect and displacement damage effect. However, it was first proposed by Binder in 1975 that the abnormal phenomena in the operation of spacecraft could be caused by the single event effect of the logic changes of the semiconductor devices caused by single heavy ions or high-energy protons (Binder et al 1980). During 1970–1980, some abnormal phenomena appeared in the satellite were systematically analyzed by Pickel and Blandford. Then it was confirmed that single event effect caused the problems (Pickel and Blandford 1980) Therefore, it is significant to study the effect of single event on the reliability of aerospace devices. The single event effect of micro-nano scale SRAM devices is one of the frontier research fields. In this paper, the mechanism of the single event effect, the device characteristics, and through the simulation and test verification, the single event effect on SRAM device is studied in these three aspects. As shown in Figure 1 is the article's structure.

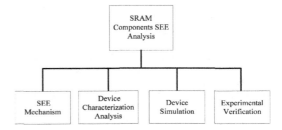

Figure 1. The overall structure of the article.

2 MECHANISM ANALYSIS OF SINGLE EVENT EFFECT

When a single high-energy charged, particle passes through the sensitive element of the electronic device, the energy is deposited, and the electron-hole pairs are generated. These charges are collected by the device electrodes. If these charges are greater than or equal to the critical charge, the function of the device is abnormal or the logic state is changed, it is single event effect. Figure 2 is the behavior of single event effect when heavy ion incident the device. As the formation process of single event effect is more complex, the action mechanism of different particles on different materials are different, but the results are the same. According to current international standards such as JEDEC89A (JEDEC 2001), EIA/JESD57 (EIA/JESD57 1996) and ASTM 1192 (Active Standard ASTM F1192 2006) etc., The main manifestations and classification include Single Event Upset, SEU, Single Event Latch-up, SEL, Single Event

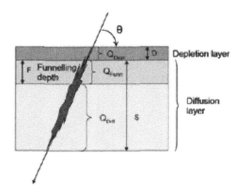

Figure 2. Heavy ion incident SRAM components.

Functional Interrupt, SEFI, Single Event Burnout, SEB, Single Event Gate Rupture, SEGR, Single Event Transient, SET. Among them, the soft errors refer to the change of the logic state and storage of the data, but the device is not damaged, such as SEU and SET, while hard errors could cause permanent damage to a device, such as SEB, SEGR.

With the introduction of new materials and new structures, as the CMOS technology keeps scale down, the research on the radiation effect of the devices is getting more and more attention. Especially, the research of single event effect has been widely developed. The research on the reliability of aerospace devices is very important for international sounding plan. The single event effect has been studied by Naval, Research Laboratory (Petersen et al 1992), Vanderbilt University (Petersen et al 2005) and Clemson University (Reed 1994) and other research institutions since the end of 70 s. Besides, as a result of the high-speed development of semiconductor technology, the single event effect research has received more attention as a new cross subject. The research on single event effect starts late in China, but in recent years many domestic institutions have made important achievements in the research of single event effect and hardening technology. For example, IMP and Atomic Energy Research Institute had carried out a single event simulation test in the 1990s. SEU and SET mechanism, modeling and numerical simulation were carried out by North west Institute of Nuclear Technology (NINT) and National University of Defense Technology (Chen et al 2008), and research on radiation effects of SOI devices in the institute of microelectronics of Chinese Academy of Sciences. The sensitivity of the total ionizing dose to the single event effect was studied by XinJiang Technical Institute of Physics and Chemistry Chinese Academy of Sciences, and the research on the test of single event effect and

the design of hardening was carried out in different degree in China Academy of Space Technology and Shanghai Academy of Spaceflight Technology.

Various types of single event effects have been found so far, including the SEU, SEL, SEFI, and SEB. However, the statistics show that the single event upset has a great influence on the operation and reliability of the spacecraft and the device. The most typical is the issue of the ITRS report has been defined the single event upset, which is one of the most important challenges in nano-technology (International Technology Roadmap for Semiconductors Reports 2009) integrated circuits. Moreover, its report pointed out that with the increase in device integration and decrease in cell size, the probability of the single event upset is constantly increasing. Soft error rate under 65 nm process from multiple upsets is 16% in 2007, and gradually increased to 64% of the 35 nm process in 2013 and in 2016, the data is 100% which process is 25 nm. Therefore, it is meaningful to study the single event effect on micro-nano scale SRAM device.

3 DEVICE CHARACTERISTICS INFLUENCE ON SINGLE EVENT EFFECT

The effect of device characteristics on the single event effect has been a hot research topic in the device level. This part is based on the RPP model to study the effect of device characteristics on the

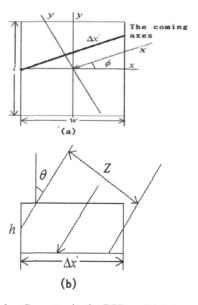

Figure 3. Geometry in the RPP model (a is top view and b is sectional view).

714

single event effect of SRAM components. Using LET value to calculate the upset rate under single event effect on orbit, thus the particle species and energy distribution in the space environment can be simplified as the LET curve, and the deposition of energy, as shown in the formula (1):

$$E_{\text{Deposition}} = L_{\text{Effective}} \times d \qquad (1)$$

The $L_{\text{Effective}}$ and D represent LET value and particle track length in the sensitive volume respectively. The valid value of LET is $L_{\text{Effective}} = L/cos\theta$, θ is included angle of incident direction and normal surface. A computational model for track length was first proposed by Pickel and Blandford (JEDEC 2001) in 1978, and then gradually developed into a classic RPP computation model. In the model, the sensitive areas are defined as a rectangular box (l, w, h), through a series of assumptions, so as to simplify the calculation of single event upset rate, as shown in Figure 3.

4 SIMULATION AND ANALYSIS OF SINGLE EVENT EFFECT OF SRAM DEVICE

At the present stage, there are two main research methods for single event effect. Space carrying experimental research, which has the advantages of intuitive and well-targeted research results, requires huge funding, and long research period. Another is Ground simulation experiment device, which has the advantage of flexibility in the selection of radiation sources, but the selection of the radiation source type is limited, however the funds are large. Therefore, the existing conditions are greatly restricted by the factors such as experiment, fund, and reuse conditions and so on. Simulation of single event effect by computer simulation is an effective method, and simulation results can feedback to the existing experimental conditions to evaluate the reliability of space SRAM device.

At the same time, with the development of the semiconductor simulation software, the simulation tool such as TCAD, which is an effective means to simulation single event effect of SRAM device. TCAD simulation, can reflect the regional energy deposition of SRAM device in detail and influence of various hardening technology of charge collection. And it can effectively shorten the design cycle, so as to reduce the cost. Specific simulation ideas such as Figure 4, using 3D model to replace SRAM device. Using the Sdevice, which is a tool in TCAD software, to carry on the single event effect simulation, and then production process is carried out after the simulation is completed. The chip goes through the single event test to verify the effectiveness of the simulation.

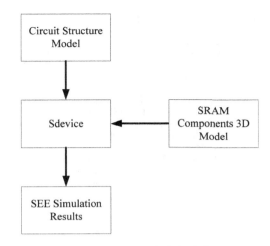

Figure 4. SRAM components SEE simulation procedure.

5 EXPERIMENTAL ANALYSIS AND VERIFICATION

SRAM components irradiation experiment was carried out with domestic test source. For example, using Chinese Academy of Sciences Heavy Ion Accelerator for single event effect test to verify the single event radiation resistance of the chip. The test results show that the simulation results are close to the experiment results.

6 SUMMARY

Commercial and military mainstream CMOS process of semiconductor manufacturing is getting smaller and smaller, and it has entered the micro-nano scale. Meanwhile, these devices usually have the characteristics of low power consumption, high performance and large capacity. In this context, comparing with large size devices, the single event effect as a key factor in the radiation hardening technology has emerged more new phenomena. For example, the multiple bit upset caused by single event effect, low energy proton direct ionization effect, etc. Therefore, it is necessary to study the single event effects of micro nano-scale SRAM device. In this paper, the theory, simulation and test methods are mainly carried out to study the single event effect of a SRAM component.

REFERENCES

Active Standard ASTM F1192. 2006. Standard Guide for the Measurement of Single Event Phenomena (SEP) Induced by Heavy Ion Irradiation of Semiconductor Devices.

Binder D., Smith E.C. & Holman A.B. 1975. Satellite Anomalies from Galactic Cosmic Rays. IEEE Trans Nucl Sci, 22(6): 2675–2680.

Chen S.M., Liang B., Liu B.W., et al. 2008. Temperature Dependence of Digital SET Pulse Width in Bulk and SOI Technologies. IEEE Trans Nucl Sci, 55(6): 2914–2920.

EIA/JESD57. 1996. Test Procedures for the Measurement of Single-Event Effects in Semiconductor Devices from Heavy Ion Irradiation.

International Technology Roadmap for Semiconductors Reports, 2009.

JEDEC STANDARD. 2001. Measurement and Reporting of Alpha Particle and Terrestrial Cosmic Ray-Induced Soft Errors in Semiconductor Devices.

Petersen E.L., Pickel J.C., Adams J.H., et al. 1992. Rate Predictions for Single Event Effects-Critique I. IEEE Trans Nucl Sci, 39(6): 1577–1599.

Petersen E.L., Pouget V., Massengill L.W., et al. 2005. Rate Predictions for Single Event Effects-Critique II. IEEE Trans Nucl Sci, 52(6): 2158–2167.

Pickel J.C. & Blandford J.T. 1980. Cosmic-Ray-Induced Errors in MOS Devices. IEEE Trans Nucl Sci, 27(2): 1006–1015.

Reed R.A. 1994. Prediction Proton-Induced Single Event Upsets Rates. Ph. D dissertation, Graduate School of Clemson University.

Weller R.A., Mendenhall M.H., Reed R.A., Schrimpf R.D., et al. 2010. Monte Carlo Simulation of Single Event Effects. IEEE Trans Nucl Sci, 57(4): 1726–1746.

Automotive, Mechanical and Electrical Engineering – Liu (Ed.)
© *2017 Taylor & Francis Group, London, ISBN 978-1-138-62951-6*

Design of an industrial robot system based on PLC control

Yingchen Dai

College of Mechanical and Electrical Engineering, Beijing Union University, Beijing, China

ABSTRACT: With the development of technology, industrial robots have become more widely applied in industrial production than ever, while at the same time the working environment of industrial robots is more complex. To ensure industrial robots' regular work in various electromagnetic environments, it is particularly important to carefully design the industrial robot system. This paper mainly introduces the structure, the Electromagnetic Compatibility (EMC) of the hardware circuit and a moving-control algorithm of an industrial robot's design.

Keywords: Robot system; PLC control

1 INTRODUCTION

In the field of industrial automation, Programmable Logic Controller (PLC), robots and CAD/CAM are the three technical backbones of automatic control, which play an important role in modern industry. So far, industrial robots and PLC has spread from design development to application, its application situation is an important symbol of the industrial automation level to a nation.

Industrial robots are electrical integrations of automatic mechanical equipment and systems. Through repeated programming and automatic control, industrial robots are able to complete certain operational tasks in the process of producing multifunctional and multiple degrees of mechanical freedom (Coiffet, 1980). In 1962, the first industrial robot "UNIMATE", by the company Unimation in the United States, was put into use in the United States company General Motors (GE), which marks the birth of the first generation of robots. From then on, industrial robot technology has continued to improve, and productions have been gradually updated and perfected, evolving into the first generation of programmable industrial robots with the control type of teaching and reappearing technology. In the 1980s, the manufacture of industrial robots formed the industry, which is widely used in industrial production. Examples are the PUMA robots of the USA, SMART robots of Italy, MOTOMAN robots of Japan, KUKA robots of Germany, IRB robots of ABB corporate, etc.

In China, industrial robots started in the early 1970s, after more than 20 years of development, and followed roughly three stages: germination stage in the 1970s, development stage in the 1980s and practical stage in the 1990s. With the development of technology, industrial robots can be designed very small, while their operating functions are more diverse. However, the working environment of industrial robots has become more complex, leading to industrial robots' failure to work. In order to make sure industrial robots can keep working regularly in a complex working environment, this paper discusses the structure design, hardware circuit design and motion control algorithm for the design of an industrial robot system based on PLC control.

2 THE STRUCTURE DESIGN OF INDUSTRIAL ROBOTS

There are many standards to judge the structure design of industrial robots. However, industrial robots can complete all kinds of tasks by programming. At the request of economy and practicality, the industrial robots' placement, size, number of joints, drive system and drive mode will vary according to different work tasks and environments. Therefore, the structure design of industrial robots must meet the basic parameters of working space, degree of freedom, payloads, motion precision and motion characteristic.

This paper presents an industrial robot with three joints, composed of lift arm, big arm and small arm, which do linear motion and rotation motion respectively. Figure 1 shows the structure diagram of the industrial robot, which is composed of base, lift arm, driving motor, big arm, small arm and end-effector.

In Figure 1, we can see that the base is an integral casting of aluminium, which is equipped with

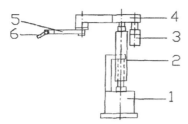

1-base 2-lift arm 3-drive motor 4-big arm
5-small arm 6-end-effector

Figure 1.   The structure diagram of an industrial robot.

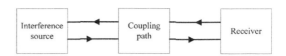

Figure 2.   The three-part basic composition of the EMC.

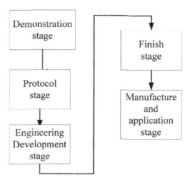

Figure 3.   Design flow of industrial robot's electromagnetic compatibility.

a drive motor of lift arm, reducer and silk pole. The lift arm is made from aluminium alloy and its cross section is a U shape, which is useful to improve rigidity and reduce the quality. The internal part of the lift arm is an installed rotary shaft of a big arm and bearing, bearing chock, drive motor and reducer. The big arm is made from thin-wall aluminium alloy, and its cross section is a U shape, which makes the arm certain strength, under load. It does not produce distortion and fracture. The big arm is equipped with a rotary shaft of small arm and bearing, bearing chock, drive motor and reducer, and the internal part of the big arm is installed with a tooth profile belt wheel transmission of the small arm. The small arm is made from aluminium alloy plate, which is installed with a wrist and gripper.

3   EMC DESIGN OF PLC HARDWARE CIRCUIT

Industrial robots commonly work in harsh and hazardous environments, such as in stamping, casting, heat treatment, welding, painting, plastic forming, simple mechanical processing and assembly process, and the atomic energy industry departments to complete harmful material handling or process operation; hence the design of the PLC hardware circuit is the key to an industrial robot (Matej & Milan, 2016). Therefore, electromagnetic compatibility design is a necessity to PLC hardware circuit design, and it includes interference source, coupling path and receiver, which is seen in Figure 2.

From Figure 2, we can see that the industrial robot is likely to be affected when under complex working conditions. Hence an electromagnetic compatibility design of the industrial robot through circuit design and shielding is needed. The overall electromagnetic compatibility design can be categorised into five steps: demonstration stage, protocol stage, engineering development stage, fin-

ish stage and manufacture and application stage, as shown in Figure 3.

In the electromagnetic compatibility design of the PLC hardware circuit, the power supply circuit electromagnetic compatibility design and the shielding method for the PLC hardware circuit are the key to an industrial robot, so we mainly focus on the power supply circuit design and shielding method design of the PLC hardware circuit.

3.1   EMC design of PLC power supply circuit

The PLC power supply circuit is not only affected by different kinds of electromagnetic interference noise, but also by surge currents. These interfere with the signal transmission to the PLC hardware circuit by the power supply circuit, which can interfere with the whole hardware circuit so that the industrial robot is unable to work normally. In order to prevent these interferences, we ensure a proper grounding design of the power supply circuit to conduct the surge current and other electromagnetic interference noise to the earth.

The purpose of a grounding design for the power circuit is the introduction of transient surge voltage into the earth. Another purpose of grounding design is to prevent interferences into the power supply, which then interfere with other load circuits. In order to conduct the transient surge through the case to avoid the transient surge of the components in the load circuit, the ground in the power circuit is connected with the ground of the industrial robot. The power circuit and the load circuit are carried out in many places. The connection between the ferrite beads and the PLC power supply circuit can then suppress electromagnetic interference noise. The grounding design of the PLC power supply circuit is shown in Figure 4.

3.2 *The shielding design on PLC hardware circuit*

Shielding is an important measure to improve the electronic system and electronic equipment of EMC. It can effectively prevent or reduce the radiation electromagnetic energy which is not needed and make sure that the equipment can work in the electromagnetic environment. Because an industrial robot is likely to be disturbed by electromagnetic noise, it is particularly important to effectively shield the PLC hardware circuit. The

factors of the shielding design that should be taken into consideration are shown in Figure 5.

In Figure 5, the electromagnetic environment of the industrial robot, including the type of electromagnetic field, the intensity of the field, the frequency and the distance between the control system and the source of the interference, should first be determined. Second, according to the electromagnetic environment, electromagnetic shielding requirements and the nature of the electromagnetic field, the appropriate material should be chosen. Third, after the determination of shielding materials, the shielding structure is designed. When the strong magnetic field is shielded, a multilayer shield structure can be used. The material's inner layer is made of low permeability material, the middle layer is made of medium permeability material, and the outer layer is made of high permeability material. Finally, the shield of the industrial robots should be well grounded.

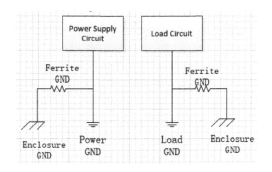

Figure 4. PLC power circuit GND design.

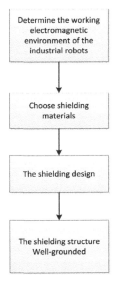

Figure 5. Shielding design of PLC hardware circuit.

4 MOVING-CONTROL ALGORITHM DESIGN

Moving-control is the control technology of position, speed, accelerated speed and force, which is the object in space and time. Moving-control of an industrial robot has three main aspects: high speed, high precision and wide range of control. An industrial robot not only needs higher location precision when a sudden stop happens in the industrial robot's high-speed movement, but also requires high precision to track the trajectory of time-varying speed and space. An industrial robot needs high precision control for acceleration and force, so it is very important to the design of an industrial robot's moving-control algorithm (Slivinschi et al., 2016).

In this paper, the design of the industrial robot's moving-control algorithm is planar linear interpolation algorithm, as shown in Figure 6. The starting position is Pi-1 and the end position is Pi, which are known. They can respectively use Pi-1 (Pxi-1, Pyi-1) and Pi (Pxi, Pyi), or Pi-1(θi-1, Φi-1)

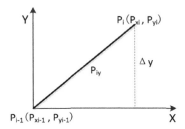

Figure 6. Planar linear interpolation algorithm.

and Pi(θi, Φi), or Pi-1(Pxi-1, Pyi-1) and Pi(Pxi, Pyi) to represent. In the first group of the above for the joint coordinate system of the pulse value, the second is angle value and the third is Card flute coordinate. The three groups of the above are equivalent.

5 SUMMARY

This paper discusses the structure and electromagnetic compatibility of a power supply circuit of hardware circuit and shielding of circuit, as well as a moving-control algorithm for industrial robot system design, which are essential to making industrial robots keep working regularly work in all kinds of complicated electromagnetic environments.

REFERENCES

Coiffet, P. (1980). Robots technology vol.1, modelling and control. Prentice-Hall, Inc.

Matej, K., & Milan, Š. (2016). Electromagnetic compatibility analysing of electrical equipment. *Diagnostic of Electrical Machines and Insulating Systems in Electrical Engineering (DEMISEE)*, 104–109.

Slivinschi, I., Mihai, S., & Clement, F. (2016). Algorithms and control structures for industrial robots TX90 model. *2016 IEEE International Conference on Automation, Quality and Testing, Robotics (AQTR)*, 1–4.

*Industrial production, manufacturing,
management and logistics*

Automotive, Mechanical and Electrical Engineering – Liu (Ed.)
© *2017 Taylor & Francis Group, London, ISBN 978-1-138-62951-6*

A method of electric power enterprise project plan audit based on full-text retrieval technology

Zijian Wang
State Grid Corporation of China, Beijing, China

Yunhui Chen, Hua Zhou & Fuxiang Li
State Grid Sichuan Electric Power Company, Chengdu, Sichuan, China

Qian Chen & Zhichao Ren
Research Institute of Economics and Technology of State Grid Sichuan Electric Power Company, Chengdu, Sichuan, China

Yan Wang & Hui Kang
College of Economics and Management, North China Electric Power University, Beijing, China

ABSTRACT: With the development of information technology, the State Grid Company urgently needs to change the present situation of the artificial project audit, to realise the automatic repeat audit. Therefore, this paper puts forward a project audit plan system for the electric power enterprises based on full-text retrieval technology. Firstly, the current situation of project audit in the State Grid Corporation and the research status of full-text retrieval technology are deeply analysed in this paper. Secondly, this paper simply introduces the concept of Lucene (a full-text search engine) and the full-text retrieval technology, detailing the process of full-text search and analysing the difference in Chinese analysis technology. Lastly, the text retrieval technology is applied in a project audit plan of the electric power enterprises and this paper provides a system design for the project audit plan of the electric power enterprises based on full-text retrieval technology.

Keywords: Full-text retrieval technology; Electric project audit plan; Lucene; Chinese analysis technology

1 INTRODUCTION

With three Intensification and five Systems completed, the State Grid Corporation of China is implementing a strategic transition from traditional to modern. And the lean management level enhances unceasingly. However, modern technology work support faces new challenges. Currently, a lot of information and project data are aggregated into the comprehensive planning system of the state grid corporation. But project audit of company still is given priority to with human review. Therefore, the company needs to move to automated audit in order to improve the audit efficiency. Specifically, this problem can be solved by applying full-text retrieval technology.

Full-text retrieval technology, invented in the 1950s, is a means of retrieval according to the content of the data information, taking all kinds of text information as the processing object.

With the development of full-text retrieval technology, more and more systems based on full-text retrieval technology are being developed in China, and the application field of full-text retrieval technology is widening more and more, such as geological survey data management, digital archives and enterprise electronic document retrieval.

In this paper, full-text retrieval technology based on Lucene is applied to establish the full-text index database of the project file in the database of the state grid company. Full-text retrieval technology is then used to retrieve relevant documents to realise project audit automation. There is no doubt that full-text retrieval technology has brought convenience to the company's project audit, and the labour cost is reduced, and work efficiency is improved. Accordingly, it is significant to research the project audit plan of the power enterprise based on full-text search.

2 FULL-TEXT RETRIEVAL TECHNOLOGY AND LUCENE SEARCH ENGINE

2.1 Basic concept

Full-text retrieval is a kind of retrieval method in which the computer indexing program builds an index for each word and indicates the number and location of the word in the article by scanning each word in the article. When the user enquires, the search procedure will be carried out to start the search according to the prior established index, and the results of the search will be fed back to the user.

Lucene is a sub-project of the Jakarta project group of the Apache Software Foundation. It is a full-text search engine tool kit based on Java open source. Lucene is not a complete full-text search engine, but a full-text search engine architecture, providing a complete query engine and index engine, part of the text analysis engine. It can be easily embedded into each application and be used as a full-text indexing engine of application background.

2.2 Full-text retrieval process

Full-text retrieval is divided into two processes, the index creation and the search index. Index creation is the process of extracting information from the data and creating the index. Search index is the process of searching for the previously created index after receiving the user's query request and returning the search results, as shown in Figure 1.

1. Indexing process for full-text retrieval. First of all, words analysis and language processing should be done to the files to be indexed, getting a series of words. Then, a dictionary and a reverse index table are formed by index creation.

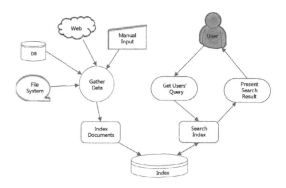

Figure 1. The process of Lucene full-text retrieval.

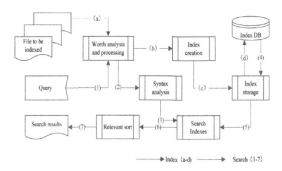

Figure 2. Full-text retrieval data flow chart.

Finally, these indexes are stored in the hard disk to form an index database.

2. Searching process for full-text retrieval. Firstly, using a syntax process to query requests for user input and getting a query tree by syntax analysis. Secondly, the query tree is used to search the index and the result of the search is obtained after the index is read into the memory. Next, sorting files from high to low according to the similarity of query requests. Finally, query results are fed back to the user, as shown in Figure 2.

2.3 Chinese words divided syncopation technology

Full-text search can be divided into index by Chinese character and index by word. Index by Chinese character is a simple method which searches index database for the words that have been broken, and then returns the retrieval result. For example, "dian wang", the retrieval system will firstly search "dian" and then the word "wang". However, for large amounts of data retrieval, this method has a considerable disadvantage in that every search will have tens of thousands of results, which will seriously affect the speed and accuracy of retrieval. Therefore, Chinese word segmentation is an effective method to improve the retrieval speed and accuracy.

The most important criterion to measure the Chinese word segmentation technology is the efficiency of word segmentation and the accuracy of word segmentation. Here are some of the existing Chinese word segmentation technologies:

The Lucene Chinese analyser mainly contains StandardAnalyser, ChineseAnalyser and JKAnalyser Chinese analyser. StandardAnalyser and ChineseAnalyser apply the principle of index by Chinese character, and JKAnalyser Chinese analyser adopts the bi-segment principle. However, no matter which of the above methods is used, it will have a large index redundancy and easily get the wrong result. Consequently, in the Chinese retrieval system, Lucene Chinese analyser is not up to the design requirements of the retrieval system.

At present, there are two major types of the third party Chinese analyser: JE analyser and Paoding analyser. Dictionary-based Forward Maximum Match Method is used to segment Chinese words in JE analyser, which supports dynamic extensions of dictionaries. The Paoding parser supports a user-defined dictionary, and is not limited to the name and catalogue of the dictionary. When the program is compiled, developers can visit a background thread for updates of the thesaurus and automatically compile the updated thesaurus into binary version and loading.

Learned through practice, the overall performance of the Paoding analyser is optimal considering the time, space and recall of three factors. It supports an unlimited number of easily added custom Thesauruses.

3 THE DESIGN AND IMPLEMENTATION OF PROJECT PLAN AUDIT SYSTEM IN POWER ENTERPRISE

3.1 System design requirements

The project plan audit system of the electric power enterprise shall meet the following requirements: a) when the project audit personnel input query statements into the full-text retrieval system, such as project name, project key indicators and other information, the system will return the results related to the query. Ultimately, this system is designed to shorten the audit time, reduce labour costs and achieve project audit automation. b) The system has the characteristics of fast search speed and high accuracy, which is helpful to improve the efficiency of project audit in the electric power enterprise.

3.2 System design and implementation

The design of a project plan audit system in an electric power enterprise is shown in Figure 3.

3.2.1 Index database creation

Firstly, load the existing project documents in the Power Grid Corp project database into the full-text search database, and define the structure of the project document. In fact, each incoming document is composed of a number of domains. For example, the project documents that need to be put in storage include the following information: the project name, the reporting unit, the declaration time, the project content, etc. The above information of each enterprise project can be used as

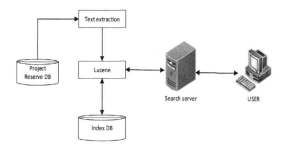

Figure 3. Design of project plan audit system for electric power enterprise.

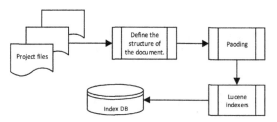

Figure 4. Lucene index database creation.

the domain of a document. Secondly, use Paoding analyser to segment the domain that needs to be segmented. Next, process the word after segmentation through the index device of Lucene. Finally, add the created index to the index database, and take responsibility for the management of data storage by storage of Lucene.

3.2.2 Search index

Firstly, input a search query into then system and then segment the query statement and parse it. Secondly, process the connection symbols in the input statement using the query analyser of Lucene. Thirdly, use "Paoding" analyser to segment the word that needed to be segmented. Finally, query device start search index database according to the segmentation results, and then get the results of the query and sort them by the correlation degree.

The key to the search index process is the calculation and ranking of document similarity. Lucene applies a vector space model to calculate and sort the similarity of documents. The principle of calculation is as follows:

Regard documents as a series of terms, and set a weight for each term, and then compute document correlation according to the weight of terms in the document.

Regard the weight of the total word in the document as a vector, expressed as follows:

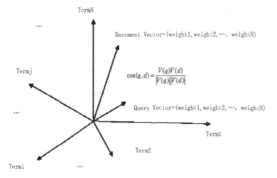

Figure 5.   N dimensional space.

$$Document = (term_1, term_2, \cdots, term_N)$$
$$Document\,Vector = (weight_1, weight_2, \cdots, weight_N)$$
$$\text{(1)}$$

Similarly, regard the query as a simple document, and use the vector to express it.

$$Query = (term_1, term_2, \cdots, term_N)$$
$$Query\,Vector = (weight_1, weight_2, \cdots, weight_N) \quad \text{(2)}$$

Put all the search out of the document vector and query the vector into an N dimensional space. Each term is one dimension, as shown in Figure 5.

Take the cosine of the included angle as the correlation of the score; the smaller the angle, the greater the cosine value, and the higher the score, the greater the correlation.

The correlation evaluation formula is as follows:

$$score(q,d) = \frac{V_q V_d}{|V_q||V_d|} = \frac{\sum_{i=1}^{N} QT_i * DT_i}{\sqrt{\sum_{i=1}^{N} QT_i^2} \sqrt{\sum_{i=1}^{N} DT_i^2}} \quad \text{(3)}$$

QT_i is the weight of the search item i of the query, DT_i is the weight of index term I of the document, N is the total number of items.

According to the score, the higher the score, the higher the relevance ranking.

3.3   System features

A power enterprise project planning audit system is a retrieval system based on full-text retrieval technology. According to the design requirements of the system, this system extends the related interface of Lucene to realise the corresponding system function. As Lucene is used as its technical support, the system has the advantages of good architecture and being Lucene object oriented. It is convenient to add or improve system functions by extending the related interface of Lucene. In addition, the default implementation query engine of Lucene is very powerful, which implements the following query operations: Boolean operations, group queries, fuzzy queries, etc. Therefore, it is quite simple to design the code. Furthermore, the retrieval accuracy and speed of the system meet the requirements of system design.

4   CONCLUSION

In this paper, full-text search technology is introduced into the research of the project plan of the electric power enterprise, which is conducive to realising the automation of the project audit of the electric power enterprise, and improve the efficiency and accuracy of the project audit. Lucene not only has outstanding object oriented features, but also has a very good structure. This system applies the Lucene search tool kit, so it is fully convenient to realise the system function through extending the Lucene-related interface.

ACKNOWLEDGEMENTS

The authors acknowledge the support of the "Key Technology and System Research of Company Project Plan Audit Analysis" project.

REFERENCES

Li, A. (2011). Design and implementation of the graduation thesis full-text retrieval system with Lucene. Ocean University of China.

Liu, Q. (2009). A solution to digital archives based on full-text search technology. Office Informatization.

Ye, L. (2013). Full-text retrieval of distributed geological survey data based on Lucene. Beijing: China University of Geosciences.

Yi, T. & Chen, Q. (2012). Comparison research of segmentation performance for Chinese analyzers based on Lucene. *Computing Engineering, 38*(22).

Zhao, S. (2015). Research and implementation of electronic document full-text retrieval system based on Lucene. Beijing Jiaotong University.

Automotive, Mechanical and Electrical Engineering – Liu (Ed.)
© 2017 Taylor & Francis Group, London, ISBN 978-1-138-62951-6

Advanced predictive quality adjustment strategy for aircraft parts

Fang Zhu
Department of Aircraft Maintenance and Engineering, Guangzhou Civil Aviation College, Guangzhou, China

Xiongfei Huang
Department of Marines, Naval Marine Academy, Guangzhou, China

Na Wei & Baoyu Ye
Department of Aircraft Maintenance and Engineering, Guangzhou Civil Aviation College, Guangzhou, China

ABSTRACT: Aircraft parts require a high degree of fabrication accuracy and multiple quality characteristics, which can be achieved in reality by a manufacturing process that consists of many fabrication steps to achieve specified precision of a part. Since manufacturing quality fluctuation is performed as composite variation, then quality variation prediction is becoming key to the quality assurance of aircraft parts. In this paper, aimed at the large dimension, complicated construction of aircraft parts used in large commercial airplanes, research actuality, existing problems and development trends of aircraft part fabrication were analysed. Air-craft parts fabrication technologies were discussed, including path and type of fabrication, fabrication process modelling, fabrication variation analysis, parameter optimisation based on fabrication stability modelling and quality variations adjustment and correcting, among others. Finally, a case study is performed to prove the proposed strategy. This work can provide a guideline for theoretical research and engineering application for aircraft parts manufacturing.

Keywords: Aircraft parts; Fabrication process; Quality variation; Quality correcting

1 INTRODUCTION

1.1 *Problem description*

Aircraft parts are products that have multiple quality characteristics, high fabrication precision, complex production technologies and many fabrication processes. Generally speaking, an aircraft part is usually produced through many fabrication stages. An aircraft part fabrication process normally involves more than one station or stage to fabricate a part to achieve the desired or designed high-precision and geometric-quality characteristics. The final part quality is greatly affected by the accumulation and transmission of variation resulting from process parameters and quality problems of the fabrication stages.

The field survey from the aircraft manufacturing factories indicates that the fabrication rejection rate of the multistage fabrication process is high. To get around this problem, factories have adopted a high-precision and high-cost measurement station after each of the fabrication stages, which is time consuming and expensive. The importance of the selection of the right fabrication parameters, especially for aircraft parts with complex structures, is also reflected in extensive literature that focuses on fabrication parameter optimisation (Liu et al., 2015). The complexity of an aircraft part fabrication process puts high demands on quality prediction to ensure the geometric quality of the final product. However, it depends largely on reliable historical data to generate references or benchmarks for effective process monitoring and the SPC method is not effective for root cause diagnosis and hence the quality variations correcting is left to quality control staff. Furthermore, it is not economic and practical to install a high-precision and high-cost measurement station after each of the fabrication stages in an aircraft parts fabrication process to monitor the geometric parameters of product quality characteristics.

The analysis of the complex interplay between geometric quality characteristics and fabrication process parameters in the different stages is the first and important problem for quality control and improvement in an aircraft part fabrication process. A great effort has been made in modelling of multistage fabrication processes (Wang et al., 2005, Loose et al., 2007) and fault diagnosis which takes into account both fixture and/or datum variations (Xie et al., 2010). Further research is needed to develop a systematic model analysis and quality prediction technology needs to be developed to correct the problem process parameter or parameters in an aircraft parts fabrication process.

1.2 *Related work*

Great efforts have already been conducted in earlier studies to describe deviation transmission modelling in the multistep fabrication process from quantitative engineering models. The popular engineering method is the application of a state space model for modeling and controlling quality of multistage fabrication or assembly process. Wang et al. (2005) proposed the equivalent fixture deviation concept and formulated the mathematic deviation transmission model. However, the research, as mentioned above, has concentrated on dimensional variation from kinematic sources, such as fixture support and locator variation, fabricate tool stack-up variation and part variation from the previous stage, etc. Less research has been conducted into the contribution of variation due to static deformation in a workstation, which can have a strong influence on fabrication precision in an aircraft part. Static deformation variations include part distortion and deflection due to clamping and cutting forces. When considering the static deformation during an aircraft part fabrication process, the modelling and prediction of the quality variation introduce new challenges.

There is some research on fabrication process deviation considered diagnosis has been reported, which based on product characteristics measurements scheme and data. Zhong & Hu (2006) proposed an engineering model to diagnose the fabrication process faults based on a 4-2-1 manufacturing process planning set-up. Xie et al. (2010) developed an effective and rapid rough set-based deviation source identification method. The fault diagnosis methods based on mathematic models and for single fault diagnosis cannot be reliable due to the absence of static variation sources, which could be important factors affecting the quality of a complex product. The diagnosis method based on traditional state space models (Wang et al., 2005) cannot ensure the correct fault detection if static variations are present. The earlier study has developed extensive and important methods of deviation source identification for a multistep fabrication process, but the usage of the earlier methods and strategy is limited, because all of those methods have deficiencies; for example, some methods are not considered the static deviation which caused by fixture pressure, only considering the relevance of fixture parameter deviation and quality characteristic data.

This paper presents a strategy and an experimental study of modelling and quality prediction for an aircraft part fabrication process. The sources of dimensional variation modelling include both kinematic and static variations. The quality prediction and adjustment method carries out engineering model and variation component statistical estimation, which has been developed for effectively solving the installation of real-time detecting equipment at some stages in the aircraft parts fabrication process. Therefore, the quality fault needs to be predicted by combining the detection and quality measurement data of an aircraft part. At the same time, the proposed quality prediction and adjustment strategy can solve the following problem: the conventional SPC techniques treat the multistage fabrication process as a whole and they lack the capability to discriminate among changes at different stages. On the other hand, the SPC methods are normally used to monitor quality problems in the manufacturing process and they do not provide the diagnostic capability to tell where the problems occur.

The rest of the paper is structured as follows. The engineering-based aircraft parts fabrication process model used in the adjustment strategy is developed and described in detail in Section 2. The quality prediction and adjustment method is presented in Section 3. An actual industrial case study is illustrated in Section 4. Finally, further discussion and conclusions are given in Section 5.

2 ENGINEERING MODEL

In the different aircraft manufacturing enterprises, classical aircraft parts fabrication processes involve many different fabrication stages where quality characteristic deviation can propagate from one fabrication stage to the next one and eventually accumulate to the final aircraft parts. Hence the engineering model of the aircraft part fabrication process is a very important part of quality characteristic prediction and adjustment of the aircraft parts fabrication process. In quality control applications, quality characteristic deviation and process error values are usually small. Hence, in order to facilitate the analysis of the problem, a linearisation of a non-linear system or a linear model is often acceptable for representation of the aircraft parts fabrication process. The final part geometric quality variation can be calculated in terms of fabrication and fixture variation, part variation, and fabrication/clamping force variation.

There are two types of process parameter deviations that are considered in aircraft parts fabrication process modelling. One type of deviation is the motorial deviations caused due to the fabrication faults in the set-up plan surfaces and fixture properties. Another type of deviation is the static deviation caused by the clamping and fabrication forces, which can cause elastic, linear fixture pressure deformation on the aircraft part being fabricated. In general, static-induced variations can be considered as one of the most critical process variables for quality prediction rather than others such as cutting-tool deflections, thermal deformation or geometric deviations of fabricate-tool axes, due to the fact that some variations in the fabrication process are partly controlled by the computer numerical control compensations.

In aircraft parts production, the fabrication and manufacturing errors are very small due to the application of advanced and high-precision fabricates or manufacturing facilities. Since the errors are small, the linearity of the modelling is held and the state space modelling method can be applied. An aircraft part requires much higher precision than a normal multistage fabrication or manufacturing process. Therefore, for an N-station aircraft part fabrication process, there should be N set-ups. It is possible to mathematically model the geometric dimensional expression of the fabrication characteristics for the set-up at the given station k as follows:

$$
\begin{aligned}
X_M(k) &= A_k X_M(k-1) + B_k X_P'(k) + w_k \\
Y(k) &= C_k X_M(k) + v_k, \quad k = 1, 2, \dots, N
\end{aligned} \tag{1}
$$

In Formula (1) above, $X_M(k)$ corresponds to the quality characteristic vector of the aircraft part with deviation at station k. $Y(k)$ corresponds to the measured data of the aircraft part quality property at station k. $X_M(k-1)$ corresponds to the quality characteristic vector with deviation at station $k-1$. $A_k X_M(k-1)$ corresponds to the transformation of quality characteristic from station $k-1$ to station k. $Xp'(k)$ is the latest fabricated quality characteristic vector at station k. $B_k Xp'(k)$ corresponds to the effect that the process deviations affect the aircraft part quality characteristic at station k. w_k and v_k are different types of noise. w_k corresponds to the noise affected on the aircraft part at station k. v_k corresponds to the measurement noise at station k.

The aircraft part quality characteristic vector transformation from station k to station $k+1$ can be represented as follows:

$$
X_{iM}^{k+1} = T_{MM}(k+1) X_{iM}^k \tag{2}
$$

In Formula (2) above, $T_M(k)$ is a 4×4 homogenous transformation matrix, corresponding to the position coordinate value of the machining tool in the fabrication. The formation of $T_M(k)$ accuracy of a fabrication tool is included both angular repeatability and axial linear, deriving from the performance or specifications of the fabricate tool. X_{iM}^{k+1} and X_{iM}^k correspond to the quality characteristic vector state of X_{iM} in station $k+1$ and station k respectively. $T_p(k)$ is also a 4×4 homogenous transformation matrix, corresponding to the position coordinate value of the fabrication fixture tool. The formation of $T_p(k)$ is dependent on the

repeatability and locator accuracy of the fabrication fixture, which are assumed to be available from fixture specifications or measurement. The differences between $T_M(k)$ and $T_p(k)$ are the deviations between station $k-1$ and station k. The 4×4 transformation matrix $T_{MM}(k+1)$ corresponds to the transformation between the station k and the station $k+1$, and can be expressed as follows:

$$
T_M(k+1) = T_{PM}(k+1) T_{PM}^{-1}(k) \tag{3}
$$

T_{PM} is a transformation matrix, corresponding to the transformation from the machine tool coordinate system to the part coordinate system. It can be deduced that $X_{iP} = T_{PM}^{-1}(k) X_{iM}^k$, $X_{iP} = T_{PM}^{-1}(k+1) X_{iM}^{k+1}$, the quality characteristic vector X_{iP} does not change; hence, from the equation $T_{PM}^{-1}(k) X_{iM}^k = T_{PM}^{-1}(k+1) X_{iM}^{k+1}$, it follows that $X_{iM}^{k+1} = T_{PM}^{-1}(k+1) T_{PM}^{-1}(k) X_{iM}^k$. $A(k+1)$ is the quality characteristics matrix that will not fabricated at station k, can be represented below:

$$
A_{k+1} = M_{k+1} T_{MM}(k+1) = M_{k+1} T_{PM}(k+1) T_{PM}^{-1}(k) \tag{4}
$$

The fabrication processes at station k will lead to static deviations and kinematic deviations. Z is the static deviation matrix vector, and it can be represented as $Z = E \cdot F$. E corresponds to the modulus of elasticity matrix, F corresponds to the fabrication force and fixture stress vector and, to calculate the static deviations, it is assumed that the Young's modulus, cutting and clamping forces, and the force deviations are known. Hence B_k can be represented as follows:

$$
B_k = N_k T_{PF}(k) T_M^{-1}(k) T_F^{-1}(k) T_{FM}^{-1}(k) + N_k Z(k) T_{PF}(k) \tag{5}
$$

N_k is the aircraft part quality characteristic that is fabricated at station k. The two terms on both sides of the plus sign (+) from left to right at the right-hand side of Equation (5) are caused by kinematic (fixture and set-up surface) deviations and static (fixture stress and fabrication forces) deviations respectively.

Based on the foregoing analysis, the aircraft part quality characteristic deviations $X_M(k)$ can be represented by:

$$
\begin{aligned}
\Delta X_M(k) &= A_k \Delta X_M(k-1) + B_k \Delta U(k) + w_k \\
\Delta Y(k) &= C_k \Delta X_M(k) + v_k \quad k = 1, 2, \dots, N
\end{aligned} \tag{6}
$$

In Formula (6) above, the input vector $\Delta U(k) = \left[\left[\Delta X_p'(k) \right] \left[\Delta Z(k) \right] \right]^T$ corresponds to the new deviations, which consist of the static deviations $\Delta Z(k)$ (including fixture stress and fabrication forces) and kinematic deviations (including set-up surface and fixture) $\Delta X_p'(k)$. It emerged from the fabrication process at station k.

729

3 QUALITY PREDICTION OF AIRCRAFT PARTS FABRICATION PROCESS

Based on the engineering model, a methodology to predict part geometric quality is discussed. According to the engineering model as discussed in the previous section, the mean shift and the change of variance or covariance of the measured or estimated data of the part geometric quality can be obtained. Under the different circumstances, it is very important to select the appropriate estimation method. The maximum likelihood estimation is effectively to estimation and maximises the likelihood function of the aircraft part quality characteristic measurement data. If matrix Y has n dimension and follows a multivariate normal distribution, $f(Y)$ corresponds to the probability density function, and can be represented as follows:

$$f(y) = (2\pi)^{-\frac{1}{2}} \left| \sum y \right|^{-\frac{1}{2}} e^{-\frac{1}{2}(Y-\mu_Y)\Sigma_Y^{-1}(Y-\mu_Y)} \qquad (7)$$

Given the aircraft part quality characteristic measurement data matrix Y and with the above Formula (7), the log-likelihood function of U and $\sigma_j^2, j = 1,2,...,P+Q+1$ can be derived as follows:

$$L(U, \sum_Y |Y) = \ln f(Y) = -\frac{n}{2}\ln 2\pi - \frac{1}{2}\ln\left|\sum Y\right|$$
$$-\frac{1}{2}(Y-\Gamma U)^T \sum_Y^{-1}(Y-\Gamma U) \qquad (8)$$

From Formula (8) above, the mean and covariance of U can be worked out as follows:

$$\frac{\partial(L(U, \sum_Y |Y))}{\partial(U)} = -\frac{1}{2}\cdot\sum_Y^{-1}\cdot\left[-Y^T\cdot\Gamma - (\Gamma)^T\cdot Y + (\Gamma)^T\cdot\Gamma\cdot U\right]$$
$$\frac{\partial(L(U, \sum_Y |Y))}{\partial(\sum_Y)} = -\frac{1}{2}\cdot\frac{1}{\sum_Y} + \frac{1}{2}(Y-\Gamma U)^T\frac{1}{(\sum_Y)^2}(Y-\Gamma U)$$
$$\sum_Y = (Y-\Gamma\cdot U)^T(Y-\Gamma\cdot U)$$
$$\qquad (9)$$

When the product quality measurement data Y and product and system design parameters Γ are known, we can estimate the U and $\sum_Y$ by Formulas (8) and (9).

4 INDUSTRIAL CASE STUDY

4.1 Description of the fabrication process

An industrial case study has been conducted to validate the methodology as presented in this paper.

The part as shown in Figure 1 is a cover board. With the datum features, the dimension specification is shown in Figure 1.

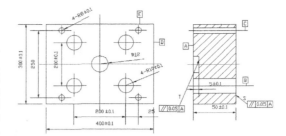

Figure 1. The dimension specifications of the aircraft part.

There are three fabrication stations in this part fabrication process as shown in Figure 1. After each of these fabrication stations or operations, the key quality characteristics of both the fabricated datum and the fabrication surfaces must be measured. Consequently, the three inspection stages are added to the end of each of the three fabrication stages respectively.

To validate the quality prediction method presented in Section 3, in this case study, consider two fabrication conditions: normal fabrication and faulted fabrication condition. When the fabrication process is normal, that means there are no fabrication process faults and the quality characteristic deviation is only caused by the natural deviation of the part fabrication process. When the fabrication process is abnormal, a fixture fault is introduced by rotating the fixture around pin $P_1(1)$ at the first station, and the clamping force will overload at the second station. Denote that φ is the rotating angular deviation of the fixture and the rotating angles of φ approximately follow a normal distribution, i.e. $\varphi \sim N(0.0683, 0.00411^2)$. For calculating the static-induced variation, we choose steel 20 as the material of the part and set the overload clamping force as 200 N.

Assume $\alpha = 0.05$, H_0 is rejected if $X^2 \le X_\alpha^2(n-1) = 9.49$ and H_0 is rejected if $|t| \ge t_{\alpha/2}(n-1) = 2.776$. With the faulted fabrication condition, there is overload at Z_2 and, at the same time, the mean variance and deviation of pin $P_1(1)$ has changed. The case study results coincide with the fabrication process in the test.

5 CONCLUSION

Quality prediction of the aircraft parts fabrication process is a critical aspect of quality control activities. However, the static-induced variation source has not been considered in previous research work. In this paper, a mathematic model-based quality prediction approach is proposed and applied to the aircraft parts fabrication process. A statistical estimation algorithm is used to calculate the improvement characteristics strategy. The case study validates that the proposed method could be a promising technique for quality prediction of an aircraft parts fabrication process.

REFERENCES

Liu, C.Q., Li, Y.G., Zhou, X., & Shen, W.M. (2015). Interim feature-based cutting parameter optimization for aircraft structural parts. *International Journal of Advanced Manufacturing Technology, 77*, 663–676.

Wang, H., Huang, Q., & Katz, R. (2005). Multi-operational machining processes modeling for sequential root cause identification and measurement reduction. *Journal of Manufacturing Science and Engineering, 127*(3), 512–521.

Xie, N., Chen, L., & Li, A.P. (2010). Fault diagnosis of multistage manufacturing systems based on rough set approach. *International Journal of Advanced Manufacturing Technology, 48*, 1239–1247.

Zantek, P.F., Wright, G.P., & Plante, R.D. (2006). A self-starting procedure for monitoring process quality in multistage manufacturing systems. *IIE Transactions, 38*(4), 293–308.

Zhong, W.P., & Hu, J.S. (2006). Modeling machining geometric variation in an N-2-1 fixturing scheme. *Journal of Manufacturing Science and Engineering, 128*, 213–219.

Automotive, Mechanical and Electrical Engineering – Liu (Ed.)
© *2017 Taylor & Francis Group, London, ISBN 978-1-138-62951-6*

Analysis on international fragmentation production of China's manufacturing industry

Yonghua Yang & Ying Chen

School of Economics and Management, Yunnan Normal University, Kunming, China

ABSTRACT: This paper constructs measure indexes of the international fragmentation production of the manufacturing industry based on the available data of the parts and components trade. With the use of the trade data of China's manufacturing goods by SITC Rev.3, the paper measures the international fragmentation of China's manufacturing production with Asia, Africa, Europe, Latin America, North America and Oceania, and so on. It also analyses China's product types, which participated significantly in the international fragmentation of production. The results show that the Chinese manufacturing industry has the highest degree of fragmentation in production with North America and Asia. The second highest is with Europe and Latin America, and the last is with Africa and Oceania. The method of China's fragmentation production with Asia, Europe, North America and Latin America is more dependent on importing parts and components. However, the method with Africa and Oceania is dependent on exporting parts and components. In addition, China's participation in international fragmentation of production is concentrated.

Keywords: International Fragmentation of Production; Parts and Components Trade; China's Manufacturing Industry

1 INTRODUCTION

Since the reform and opening up, China's export expensed rapidly. The rank of the total amount of China's exports rose from 28th in 1980 to 2nd in 2007. From 2009 to 2011, China remained the world's largest exporter. In recent years, the share of manufactured goods in total exports has remained at 95%, and exports of manufactured goods grew faster than the growth rate of the export of all goods. The basic phenomena of current products is that the different processes, sectors and components in the production process scattered to different countries and regions, thus the same final product was manufactured by different countries or regions' participation in the international fragmentation of production. Two or more countries or regions completed a final production through the FDI (Foreign Direct Investment) or by outsourcing activities to produce or assemble parts. A distinguishing feature is that the intermediate input circulated in nations, resulting in a large number of intermediate goods trade. Processing trade is one of the basic forms of the international fragmentation of production, as well as being China's main way to participate in the international fragmentation of production (Lufeng, 2004). It is valuable research to accurately measure the degree of international fragmentation of the Chinese manufacturing industry and comprehensively analyse the participation of China's manufacturing industries in the international fragmentation of production.

The input-output method has been commonly used in the literature[1] to measure the international degree of fragmentation in the production of our manufacturing industry (Zhao lei and Yang Yonghua, 2011). Due to the unavailability of the key variable, the intermediate input amount of data, the intermediate goods imports were therefore assumed, that is, the proportion of industry's total import of intermediate products is equal to the import of k industries. There are some deviations from the intermediate input in the actual situation of various industries. Therefore, using the input-output method to measure the true degree of international fragmentation of China's manufacturing production has its limitations. Therefore, this article will use the data of the parts and components trade, which is relatively easily available, to measure China's manufacturing industries international fragmentation of production.

---

[1]This approach is used by Feenstra and Hanson (1996), Hummels et al. (1998, 2001), Egger and Egger, (2003), Pincin et al. (2006) and Sheng Bin (2008).

2 MEASUREMENT OF INTERNATIONAL FRAGMENTATION OF PRODUCTION BASED ON PARTS AND COMPONENTS TRADE

It was Yeats that first used the measurement method of international fragmentation of production based on the parts and components trade. Since then, this method has been widely used. Some have used it to measure the degree of international fragmentation of production in certain countries or regions (Kaminski and Ng, 2001; Gaulier. etc, 2005). Others have used this method to measure an industrial level of international fragmentation (Lall. etc, 2004). The advantage of the measurement of international fragmentation of production based on the parts and components trade is the availability of data and the observability of the degree of international fragmentation in major countries.

The key to this method is the determination and recognition of components[2]. Referring to Athukorala's (2003) method, and according to the classification of SITC Rev. 3, we use the 225 species of five-bit code products, which belong to the classification of the SITC7 and SITC8 as trade indicators. Among them, 168 species of five-bit code products belong to the SITC7 sector, while 57 species belong to the SITC8 (see Appendix A). We can construct a metric index of international fragmentation of production based on the parts and components trade as follows:

$$\mathrm{F}P_{ij}^{T} = \frac{PC_{ij}}{T_{ij}}, \mathrm{F}P_{ij}^{X} = \frac{PCX_{ij}}{TX_{ij}}, \mathrm{F}P_{ij}^{M} = \frac{PCM_{ij}}{TM_{ij}} \quad (1)$$

Here, $\mathrm{F}P_{ij}^{T}$, $\mathrm{F}P_{ij}^{X}$ and $\mathrm{F}P_{ij}^{M}$ respectively represent the indicators of international fragmentation of manufacturing production measured from the perspective of trade import and export volume, export and import. "i" represents trading nation, and "j" represents some certain industry. PC_{ij} is the total import and export of the parts and components trade between country i and country j. T_{ij} is the total import and export of manufacturing industrial trade between country i and country j. PCX_{ij} is the total trade volume of parts and components that country i exports to country j. TX_{ij} is the total trade volume of manufacturing industrial trade that country i export to country j. PCM_{ij} is the total trade volume of parts and components that country i imports from country j. TM_{ij} is the total trade volume of manufacturing industrial trade that country i imports from country j.

[2]Some documents refer to BEC rev.3 as spare parts for capital goods (excluding transport equipment) with class code 42 and transport equipment spare parts with class code 53 as a measure of trade in components, such as Gaulier et al. (2006).

3 DATA SOURCE AND DESCRIPTION

The components selected in the equation (1) include 225 species of five-bit code products categorised by the SITC Rev.3, shown in Appendix A. Manufactures include all 1-bit code products classified by the SITC Rev.3 from SITC5 to SITC8. The data amount of imports, exports and the total imports and exports of parts and manufacturing are respectively the data of imports, exports and the total imports and exports of the 225 component products in Appendix A and all products from SITC5 to SITC8. From Asia, Africa, Europe, Latin America, North America and Oceania, we select 27 typical countries or regions[3] that China exports to most, including 10 countries or regions in Asia, 3 in Africa, 8 in Europe, 2 in Latin America and 2 in North America and Oceania. Based on the imports, exports and imports and exports of manufacturing between China and these representative countries, we analyse the international fragmentation of production of Chinese parts and manufacturing to Asia, Africa, Europe, Latin America, North America and Oceania. All the data of the import and export trade of the 225 component products and the SITC5 to SITC8 manufacturing products are obtained from the United Nations COMTRADE database.

4 CHINA'S PARTICIPATION IN INTERNATIONAL FRAGMENTATION OF MANUFACTURING PRODUCTION BASED ON PARTS AND COMPONENTS TRADE

According to equation (1) and using the data of imports, exports and import/export trades and manufacturing among China, Asia, Africa, Europe, Latin America, North America and Oceania countries or regions, we calculated and measured the international fragmentation of manufacturing

[3]These 27 countries were selected according to the export data of China, Asia, Africa, Europe, Latin America, North America and Oceania in 2001–2008, and the data of China's exports to continent countries for 9 consecutive years China, Hong Kong, Japan, South Korea, Singapore, India, Taiwan, the United Arab Emirates, Malaysia, Indonesia, Thailand; Africa selected three of the largest exporting countries and regions in Asia: China has the largest exporting countries: South Africa, Egypt, Nigeria; Europe selected eight of China's largest exporting countries: Germany, the Netherlands, Britain, Italy, Russia, France, Spain, Belgium, Latin America selected the two largest exporters of China: Mexico, Brazil; North America selected two of China's largest exporters: Canada, the United States; Oceania selected two of China's largest exporters: Australia, New Zealand.

Table 1. China's imports of parts and components from the world and the continent's share of manufacturing imports (unit: %).

Years	World	Asia	Europe	North America	Latin America	Africa	Oceania
1997	39.8	38.8	48.9	49.5	28.1	2.9	15.8
1998	41.9	39.3	53.1	51.3	30.3	7.4	17.0
1999	42.7	41.0	49.5	49.6	31.7	1.5	17.7
2000	44.4	43.5	44.9	53.0	57.2	3.6	13.1
2001	46.0	44.9	47.8	55.3	63.3	3.7	15.2
2002	46.9	47.5	45.4	53.0	58.8	1.9	13.6
2003	47.6	48.8	46.5	48.3	41.7	4.1	12.3
2004	48.4	48.5	47.7	51.1	42.9	4.2	14.9
2005	48.5	49.1	45.0	50.1	44.7	2.0	15.8
2006	51.0	50.5	47.4	54.9	50.3	2.1	14.2
2007	49.8	49.4	47.1	50.9	48.1	1.2	17.2
2008	49.0	48.4	46.3	49.0	49.1	2.2	18.5
Average value	46.4	45.8	47.5	51.3	45.5	3.1	15.4

Source: Author calculated from the United Nations COMTRADE database data.

Table 2. China's exports of parts and components from the world and the continent's share of manufacturing exports (unit: %).

Years	World	Asia	Europe	North America	Latin America	Africa	Oceania
1997	27.7	24.1	29.0	36.3	26.0	23.7	21.4
1998	29.3	26.7	27.9	35.3	35.3	24.3	23.0
1999	31.9	30.2	30.8	37.1	35.6	26.5	22.0
2000	34.1	33.6	33.6	37.9	31.1	27.4	23.4
2001	35.8	35.3	34.2	39.7	35.3	31.8	23.7
2002	36.7	37.9	33.5	38.9	32.7	33.8	24.8
2003	36.0	37.0	33.2	37.6	34.1	36.5	25.0
2004	36.7	37.9	35.5	37.5	35.2	35.6	25.8
2005	36.8	38.8	34.3	36.3	41.1	36.9	25.6
2006	36.9	40.0	33.1	36.5	40.8	35.9	27.9
2007	36.3	39.8	31.5	36.0	40.0	35.8	28.6
2008	36.9	40.7	32.7	35.2	41.3	39.8	28.9
Average value	34.6	35.2	32.4	37.0	35.7	32.3	25.0

Source: Calculated from the United Nations COMTRADE database data.

production from the point of import, export and import and export (See Table 1, Table 2 and Table 3).

Table 1 shows the share that China's import of components from the world and State account for the import of manufacturing. From the point of China's imports from the world, the international fragmentation of Chinese manufacturing production increased from 39.8% in 1997 to 51% in 2006, then declined, but the average degree of fragmentation is 46.4%. The degree of international fragmentation of China's imports from Asia also showed the same trends as the degree of China's imports from the world. However, the former data are slightly lower than the latter. The average degree of international fragmentation of China's imports from Asia (45.8%) is below the average degree of international fragmentation of China's imports from the world (46.4%). The degree of international fragmentation of China's imports from Europe, North America, Latin America, Africa and Oceania fluctuated from 1997 to 2008. The average degree of international fragmentation of China's imports from Europe is 45.8% and the average degree of international fragmentation of China's imports from North America is 51.3%, which are higher than the average degree of international fragmentation of China's imports from the world. The average degree of international fragmentation of China's imports from Latin America is 45.5% and is lower than the average

Table 3. China's share of imports and exports of parts and components from the world and continents (%).

Years	World	Asia	Europe	North America	Latin America	Africa	Oceania
1997	32.7	30.1	38.2	40.2	26.4	23.3	20.0
1998	34.5	32.2	38.7	39.8	34.7	24.2	21.6
1999	36.7	35.1	39.1	40.8	34.9	26.4	21.0
2000	38.6	38.1	38.4	42.1	47.2	23.7	20.9
2001	40.4	39.6	40.6	44.4	36.0	27.3	21.5
2002	41.3	42.3	38.9	42.5	40.8	28.4	22.0
2003	41.3	42.7	38.8	40.0	37.2	29.9	21.9
2004	41.9	42.9	40.3	40.5	37.5	28.6	23.5
2005	41.7	43.4	37.9	39.0	42.0	29.9	23.7
2006	42.5	44.5	37.7	40.0	42.6	30.6	25.4
2007	41.5	43.8	36.0	38.9	41.4	30.5	26.7
2008	41.3	43.8	36.7	37.9	42.4	34.5	27.7
Average value	39.5	39.9	38.4	40.5	38.6	28.1	23.0

Source: Calculated from the United Nations COMTRADE database data.

degree of international fragmentation of China's imports from the world. The average degree of international fragmentation of China's imports from Oceania and Africa are respectively 15.4% and 3.1%. Obviously the average degree of international fragmentation of China's imports from North America is the highest and the degree of international fragmentation of China's imports from Africa is the lowest from the point of import.

Table 2 reports the share of China's exports of parts and components from the world and continents to the manufacturing sector. From the perspective of exports, the degree of international fragmentation of production between China and the world and continents is on the rise. For example, the figures for China and the world and Asia are increasing between 1997 to 2006 and declining over the next two years. Between China and Europe, North America, Latin America and Africa, the international fragmentation of production showed a trend of increasing volatility, between China and Oceania, the international fragmentation of the production level is on the rise. Among them, the average degree of international fragmentation between China and Asia, North America and Latin America are respectively 35.2%, 37% and 35.7%, which are higher than that of China and the world. The average degree of international fragmentation between China and Europe, Africa and Oceania are respectively 32.4% 32.3%, 25%, which is lower than the average degree of international fragmentation of China and the world production level. Obviously, from the export point of view, the gap of international fragmentation degree from China and the continents is not large, but with North America it is still the

highest degree of international fragmentation. The degree of international fragmentation from China and Oceania is the lowest, and with Asia's degree of international fragmentation to improve the most. Compared with the degree of international fragmentation measured by the import angle in Table 1, the low degree of international fragmentation of China's exports also shows that China's participation in international fragmentation is more prominent in the form of imported parts.

Table 3 shows the share of imports and exports from the world and continents in 1997–2008. From the perspective of total imports and exports, the degree of international fragmentation production of China with the world, Asia, Latin America, Africa and Oceania is on the rise, while the degree of international fragmentation production from China and Europe are fluctuating, being 36.7% and 37.9% in 2008 respectively, lower than the 38.2% and 40.2% in 1997. From the overall and average figures, China and Asia and North America have a higher degree of international fragmentation, the average degree of international fragmentation from the two production levels are 39.9% and 40.5%, higher than the China and the world segmentation degree of production, which is 39.5%. From the changing trends, the degree of international fragmentation of China and Asia increases from 30.1% in 1997 to a high point of 44.5% in 2006. It showed a slight decline in the next two years, but remained at 43.8%, also shows that China is increasingly into the international fragment of Asia production system. In China and North America, the degree of international fragmentation production is in a relatively stable state, in 1997 it was 40.2% and higher than the continents, in 2007 it ranged at 40%, and then dropped by more than 37.9% in 2008. The average degree of international fragmentation from China and Europe and Latin America are respectively 38.4% and 38.6%, slightly lower than the average of international fragmentation production between China and the world; and China, Africa and Oceania international fragmentation production as a whole are growing, but the average levels are 28.1% and 23%, far lower than the average of international fragmentation production between China and the world.

The above content measures the international fragmentation production of manufacturing industry in our country from the point of view of imports, exports and import and export. From the perspective of import and export, the degree of international fragmentation from China, North America and Asia are at the highest degree, followed by China and Europe. Latin America's degree of segmentation, the gap of international fragment degree from China with them is not great, but China with Africa, Oceania fragment degree is low, with the front four there is a big gap. From the perspective of imports, the degree of international fragmentation from China and North America and Europe are of the highest degree, followed by the degree of international fragmentation from China and Asia and Latin America. The gap between these four is not particularly large, but China and Africa and Oceania are far lower than the previous four, and the gap is particularly large. From the perspective of exports, China and North America has the highest degree of international fragmentation, followed by China and Latin America and Asia, then come China and Europe and Africa's degree of international fragmentation, and finally comes Oceania. The gap among them is not great. It also shows the characteristics of China's manufacturing industry in international fragmentation production: on the whole this is through a large number of imports of intermediate goods, processing or assembly and then export. Of course, the division characteristics of production with the continents are also different. Compared with the export parts, in the international fragmentation mode of production, China and Asia, Europe, North America and Latin America are more dependent on imported components, but the international fragmentation from China and Africa and Oceania are just the opposite.

According to the method of Table A2 in Appendix A, the five-digit code corresponding to SITC Rev.3, which represents the component, is mapped to the three-digit horizontal code, which can be used to analyse the product types in China's manufacturing industry. Table 4 shows the import of China's top seven parts of the import share and the total can be seen from the data in the table.

Table 4. The import share and total of the top 7 products imported in China (unit: %).

Code Years	776	764	728	759	772	778	792	Total
1997	13.9	12.7	11.9	7.3	6.0	5.8	7.2	64.9
1998	17.1	15.6	9.3	8.1	6.2	5.4	6.5	68.1
1999	22.8	15.5	7.8	7.2	6.3	5.8	5.4	70.7
2000	27.0	15.6	7.5	7.6	6.9	6.1	2.8	73.5
2001	26.2	14.5	7.4	8.2	7.0	5.6	5.0	73.9
2002	30.7	12.1	7.1	8.8	6.9	5.9	3.5	75.0
2003	32.6	11.5	6.6	7.8	6.8	5.6	2.8	73.6
2004	34.7	10.8	7.2	7.0	7.0	5.7	2.3	74.7
2005	38.5	11.1	5.3	7.0	7.4	5.4	2.7	77.4
2006	39.6	10.6	4.7	6.6	7.6	3.6	78.1
2007	41.0	9.0	5.2	6.3	8.1	5.6	3.0	78.1
2008	39.5	8.8	5.1	5.5	8.1	5.7	2.7	75.3

Note: See Appendix A for product codes and classifications.
Source: Author calculated from the United Nations COMTRADE database data.

Table 5. China's exports of parts and components in the top seven export share of the product and the total (unit: %).

Code Years	764	759	851	776	778	821	848	Total
1997	12.8	5.7	19.4	4.4	7.1	5.7	7.2	62.4
1998	13.1	7.3	17.6	5.0	7.2	5.9	6.1	62.2
1999	14.3	7.4	15.6	6.6	8.0	6.2	4.9	63.0
2000	16.3	8.0	13.0	7.0	7.9	6.0	5.0	63.2
2001	18.0	10.2	11.8	5.8	7.7	5.9	5.0	64.3
2002	18.5	12.8	10.2	6.7	7.2	6.1	4.4	65.9
2003	19.2	13.2	8.9	7.2	7.0	6.2	4.4	66.1
2004	21.8	12.3	7.5	8.0	6.7	6.2	3.5	66.1
2005	23.8	11.3	7.3	7.8	6.3	6.3	3.2	66.0
2006	25.2	10.3	6.5	8.7	6.2	6.2	2.0	65.0
2007	24.5	9.2	6.0	8.5	6.5	6.4	1.6	62.7
2008	23.4	7.7	6.0	8.7	6.6	6.4	1.4	60.2

Note: See Appendix A for product codes and classifications.
Source: Author calculated from the United Nations COMTRADE database data.

China's participation in the international fragmentation production of products with a concentration, the share of imports of products which is China's imports of spare parts in the top seven accounted for about 70% of the total, and showed a growth trend, from 64.9% in 1997 increased to 78.1% in 2007, declined in 2008, but still accounted for 75.3%. Among them, the division of production up to 776, accounting for about 30%, and the growth trend is more obvious, from 13.9% in 1997

737

to 41% in 2007, 2008 also 39.5%. 776, 764, 728, 759, 772, 778, 792 international fragment is also more obvious.

Table 5 shows the export share of China's top seven exports of components and the total exports of China's exports in the top seven largest export share of the product accounted for more than 60%, of which the largest share is 764, accounting for 20%, an increase from 12.8% in 1997 to 25.2% in 2006, followed by a slight decline in 2007 and 2008, but still more than 23%, followed by the higher share of 851 and 759. In total, 851 products had a downward trend, and 759 products had an upward trend. The share of the gap among 776, 778, 821 and 848 products is not great. It can be concluded from the export of parts shown that the strongest participant in the degree of international fragmentation is 764, followed by 851 and 759, and finally 776, 778, 821 and 848 products.

5 CONCLUSION

The advantages of metrics based on the parts and components trade data to measure the degree of China's participation in the international fragmentation of production are the availability of data and the observability of the trade flow. Also it can compare the degree of international fragmentation of production between the major countries so as to comprehensively reflect the situation and the regional characteristics of China's participation in the international fragmentation of manufacturing production. Measurements based on the parts and components trade data show that the degree of international fragmentation of China's manufacturing production continues to increase. From the perspective of overall imports and exports, Chinese manufacturing industry has the highest degree of international fragmentation in its trade with North America and Asia. The second is with Europe and Latin America. The gap of the fragmentation degree is not very big among these four. However, the degree of international fragmentation of China's manufacturing trade with Africa and Oceania is still very low, which is different from the previous four. The degree of international fragmentation of China's importing of components from North America and Europe is more prominent, but it is less with Asia and Latin America. The gap of the fragmentation degree is not very big among North America, Europe, Asia and Latin America. Meanwhile, the degree of international fragmentation of China's importing from Africa and Oceania is still very low, which is different from the previous four. The degree of international fragmentation of China's exporting of components to North America is more prominent. However, it is less

with Latin America, Asia and Europe. The last is with Latin America. The gap of the fragmentation degree is not very big among China, North America, Europe, Asia, Latin America and Oceania. So we can see the characteristics of China's participation in the international fragmentation of manufacturing production. That is importing a large number of intermediate goods, then processing and assembling, finally exporting. China's participation in the international fragmentation of production is concentrated. The top seven imports of parts and components account for about 70% of China's total components imports, and the top seven exports of parts and components account for about 60% of China's total components exports. Those shares are continuously increasing.

Obviously, China's participation in the international fragmentation of manufacturing production has greatly accelerated the growth of manufacturing output and exports in China. Meanwhile, it also showed that China's manufacturing output and exports contained a large amount of the foreign investment. So, what kind of stage should Chinese manufacturing production sections are in the international division of production system, and how to strive for a higher degree of fragmentation in the production with high added value and more interest? This research direction is worthy of further investigation.

ACKNOWLEDGEMENTS

This research is supported by Natural Science Foundation of China (No.71163047) and Department of Science and Technology of Yunnan Province under grant number 2016FB118.

REFERENCES

Athukorala, Prema-chandra (2003). *Product Fragmentation and Trade Patterns in East Asia.* Departmental Working Papers 21, Australian National University, Arndt-Corden Department of Economics.

Egger, H. and Egger, P. (2003). Outsourcing and skill-specific employment in a small economy: Austria after the fall of the Iron Curtain. *Oxford Economic Papers*, 55(4), 625–643.

Feenstra, R.C. and Hanson, G.H. (1996). Globalization, Outsourcing, and Wage Inequality. *American Economic Review*, 86(2), 240–245.

Hummels, D., Ishii, J. and Yi, K.-M. (2001). The nature and growth of vertical specialization in world trade. *Journal of International Economics,* 54(1), 75–96.

Hummels, D., Rapoport, D. and Yi, K.-M. (1998). Vertical specialization and the changing nature of world trade. *Federal Reserve Bank of New York Economic Policy Review*, 4(2), 79–99.

APPENDIX A PART PRODUCT CODE

A1 Product code for the five-digit codes of the 225 SITC Rev.3 classifications of the representative parts included in SITC7 and SITC8.

71191	71192	71280	71319	71331	71332	71391	71392	71491
71499	71690	71819	71878	71899	72119	72129	72139	72198
72199	72392	72393	72399	72439	72449	72467	72468	72488
72491	72492	72591	72599	72689	72691	72699	72719	72729
72819	72839	72851	72852	72853	72855	73591	73595	73719
73729	73739	73749	74128	74135	74139	74149	74159	74172
74190	74291	74295	74380	74391	74395	74419	74491	74492
74493	74494	74519	74529	74539	74568	74591	74593	74595
74597	74699	74790	74839	74890	74991	74999	75910	75991
75993	75995	75997	75999	76491	76492	76493	76499	77129
77220	77231	77232	77233	77235	77238	77241	77242	77243
77244	77245	77249	77251	77252	77253	77254	77255	77257
77258	77259	77261	77262	77281	77282	77429	77549	77579
77589	77611	77612	77621	77623	77625	77627	77629	77631
77632	77633	77635	77637	77639	77641	77643	77645	77649
77681	77688	77689	77811	77812	77817	77819	77829	77833
77835	77848	77869	77883	77885	77889	78421	78425	78431
78432	78433	78434	78435	78436	78439	78535	78536	78537
78689	79199	79291	79293	79295	79297	81211	81219	81380
81391	81392	81399	82119	82180	84699	84848	85190	87119
87149	87240	87319	87329	87412	87414	87424	87426	87439
87449	87454	87456	87469	87479	87490	88114	88115	88123
88124	88134	88136	88422	88431	88432	88433	88439	88591
88592	88593	88597	88599	89124	89129	89191	89195	89199
89410	89935	89937	89949	89966	89984	89986	89996	89997

Source: Based on the classification of Athukorala (2003).

A2 SITC Rev.3 Classify the correspondence between the five-digit horizontal codes and the three-digit code levels.

Corresponds to the three-digit code			Others						
713	714	716	71192	71280	71819	71878	71899	72119	72129
723	724	728	72139	72198	72199	72439	72449	72467	72468
735	737	741	72488	72491	72492	72591	72599	72689	72691
742	744	745	72699	72719	72729	73719	73729	73739	73749
749	759	764	74380	74391	74395	74699	74839	74890	74991
771	772	776	74999	77429	79199	81211	81219	81380	81391
778	784	785	81392	81399	87119	87149	87240	87319	87329
792	821	846	88422	88431	88432	88433	88439	88591	88592
851	874	881	88593	88597	88599	89124	89129	89191	89195
891			89199	89410	89935	89937	89949	89966	89984
			89986	89996	89997				

Source: Based on the classification of Athukorala (2003).

Automotive, Mechanical and Electrical Engineering – Liu (Ed.)
© 2017 Taylor & Francis Group, London, ISBN 978-1-138-62951-6

Impact analysis of global production network on China's industrial chain upgrading

Yonghua Yang & Ying Chen
School of Economics and Management, Yunnan Normal University, Kunming, China

ABSTRACT: Based on the research of Humphrey and Schmitz (2000), this paper studies the industrial chain upgrading of the buyer-driven and producer-driven global production networks. It uses export sophistication to measure the degree of China's manufacturing industry chain upgrading. It then analyses the impact of the global production network on China's industrial chain upgrading. The paper puts forward a hypothesis, builds a model and analyses the conclusion.

Keywords: Industrial Chain Upgrading; International Production Network; Export Sophistication

1 INTRODUCTION

Chain upgrading is to create better products or product upgrading, use more efficient production methods or process upgrading. The aim is to expand the service range beyond the simple assembly of the value chain, including product designing, fabric purchasing, inventory management, production management purchasing or function upgrading, evolving from the global value chain of low-level, low value-added position to the high-tech, high value-added position. Humphrey and Schmitz (2000) summarise the four forms of industrial chain upgrading: (1) Process Upgrading: Through the restructuring of the production system or use of new technologies to improve input-output ratio, so that their production is more competitive than the competition; (2) Production Upgrading: Improve product quality and variety, the introduction of new features and new models to better quality, and use lower prices to compete with competitors; (3) Function Upgrading: From the production chain to design and marketing and other lucrative links across, such as from OEM to ODM to OBM conversion is often regarded as function upgrading path; (4) Intersectoral Upgrading: Refer to the value chain from one across to another value chain.

Research shows that the above four kinds of industrial chain upgrading methods are generally followed, from process upgrading to production upgrading then to function upgrading and finally intersectoral upgrading. However, the different types of global production networks have an impact on the speed of the industrial chain upgrading and the way of upgrading. In the buyer-driven

global production network, the value share of the value chain is mainly concentrated in the circulation areas, and the marginal value incremental rate from the production process to the circulation process is increasing. The upgrading path will generally be based on the trajectory: process → product → function → intersectoral upgrading. The upgrades also become more difficult as the sequence changes. Industrial clusters under the buyer-driven global production network from the product upgrading to the function upgrading process, the non-entity nature or the assets hollowness of the whole industry cluster will increase, that is, the cluster upgrading actually is a process that the original low value-added links continue to peel or separate.

In the producer-driven global production network, the value share of the value chain is mainly concentrated in the production areas, and the marginal value incremental rate from the production process to the circulation process is increasing. The upgrading path will generally be based on the trajectory: function upgrading → product upgrading → process upgrading → intersectoral upgrading. The upgrades also become more difficult as the sequence changes.

2 THE MEASURE OF INDUSTRIAL CHAIN UPGRADING

In the global production network system, a country's position in the global industrial chain will be reflected in the technical sophistication level of its production and export products. Therefore, this paper will use the technical sophistication of

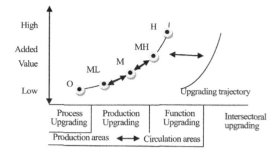

Figure 1. Buyer-driven global production network industry chain upgrading.

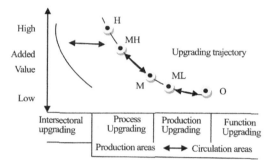

Figure 2. Producer-driven global production network industry chain upgrading.

export products to measure a country's industrial chain position. According to Hausman et al. (2007), each country's exports can be ordered at the productivity level corresponding to the product. However, since data on the productivity level of a product are difficult to obtain, the weighting reflects the dominant comparative advantage of a country on the exported product by weighting the per capita GDP of the country that exports a product. First of all, each country exports a product corresponding to the level of income or productivity indicators PRODY; second, the structure of a country's export-level industry-level measurement of the sophistication of indicators; finally, a country basket of export products corresponding to the level of income or productivity indicator EXPY. Using the method of Rodrik (2006), the measure of the sophistication of an arbitrary product i can be constructed as follows:

$$PRODY_i = \sum_j \frac{x_{ij} / X_j}{\sum_j \left(x_{ij} / X_j\right)} \bullet y_j \qquad (1)$$

$PRODY_i$ represents the sophistication of product i, x_{ij} represents the export value of country i,

and X_j is the total amount of all exports of country j. y_j denotes the per capita income level of country j, $PRODY_i$ is the weighted average of per capita income, and x_{ij}/X_j is the weight of product i in country j, reflecting the importance of product i in country j's export structure; $\sum_j \left(x_{ij} / X_j\right)$ is the sum of the product i's total exports in each country. The weighting variable reflects the importance of country i's exports in all countries that export such products, reflecting the comparative advantage of country i. $PRODY_i$ is the weighted average of the per capita income weighted by the dominant comparative advantage of the product i of the exporting country, and therefore the higher the explicit comparative advantage of the product i in the exporting country under the same other conditions, the higher the sophistication, the higher the sophistication of the product; if a product is densely exported by high-income countries, the sophistication of the product is higher, while a product-intensive low-income countries to export the product sophistication is low.

Product sophistication PRODY can be integrated into the national level, a country's export sophistication index EXPY can be defined as the weighted average of the sophistication of each country's export product PRODY, and the weight is the share of each product's export value to the country's total export value.

$$EXPY_j = \sum_i \frac{x_{ij}}{X_j} \bullet PRODY_i \qquad (2)$$

where $EXPY_j$ is the export sophistication of country j and x_{ij}/X_j is the share of exports of product i in total exports of country j.

This paper selected 118 countries and regions as the world's representative countries, using all countries from 2001–2011 export trade data and per capita GDP data. First, the export trade data are derived from the United Nations Trade Statistics Database (COMTRADE), which uses the three-digit code for product exports of the Standard International Trade Classification Third Edition (SITC Rev.3). Second, the per capita GDP of each country comes from PWT 8.1, which is the per capita GDP calculated at Purchasing Power Parity (PPP), as well as for each country. According to Formula (1), the export sophistication of 205 three-digit products is calculated from 2001 to 2011. Then, according to (2), we calculate the export sophistication of Chinese manufacturing industry from 2001 to 2011. In these 11 years, China's manufacturing export sophistication increased year by year. Therefore, China's manufacturing industry chain is escalating, as shown in Table 1.

Table 1. The degree of China's manufacturing industry chain upgrading (unit: US dollars).

Years	2001	2002	2003	2004	2005	2006
EXPY	17,840.34	18,467.89	19,469.93	19,730.86	20,265.65	20,921.59

Years	2007	2008	2009	2010	2011
EXPY	21,822.45	22,423.67	23,391.74	24,016.93	24,686.51

3 THE IMPACT OF GLOBAL PRODUCTION NETWORK ON CHINA'S INDUSTRIAL CHAIN UPGRADING

The international production network is organised in two ways: first, Foreign Direct Investment (FDI), and second, outsourcing, and China's mainly participation in international production network is introduction the FDI. On the other hand, the processing trade is one of the basic patterns of the international production network, and also one of the main ways for China to participate in the international production network. Therefore, this paper will analyse the main channels of international production network transfer and spillover technology—FDI and processing trade, indicating the international production network and the upgrading of the industrial chain.

3.1 Technical level

Of the many factors that affect the upgrading of the industrial chain, technology is the most important factor in industrial chain upgrading. The level of advanced technology in high-income countries is usually the main source of competitiveness. The techniques described here include not only product or process innovation based on research and development, but also the improvement of technology to achieve economies of scale, the effective organisation of supply or the ability to select an effective supply chain. The products' export sophistication reflect the technological level of the export enterprises (industry).

Thus, the following hypothesis can be obtained:

Hypothesis 1: China's industrial chain upgrading (export sophistication) depends on China's technology level.

3.2 The technical effect of FDI

The effect of FDI on export sophistication lies in the technical effect of FDI on the host country, that is, the technology transfer effect of FDI includes FDI's direct technology transfer effect and indirect technology transfer effect

According to Kojima (1987), a Japanese scholar, FDI should be understood to convey a package of capital, management skills, technical knowledge, and business resources to the recipient country. According to Dunning's compromise theory, FDI is only taken into account when transnational corporations (FDI carriers) have the advantages of monopoly, host country and internalisation. Multinational companies are both the owners of advanced technology and the innovators and transferors of advanced technology. Second, FDI has indirect effects on the technological progress of local firms in host countries, in addition to the direct transfer of technology to the host country's subsidiaries. In most cases, the technology spillover effects of FDI can be produced through the correlation effect between the multinational corporations and the local enterprises, the market competition and the demonstration effect, and the training and flowing of the human capital, which then improve the technical level of the host country, and enhance the sophistication of the host country's output and export products. Thus, the following hypothesis can be obtained:

Hypothesis 2: China's industrial chain upgrading (export sophistication) and China's FDI have a positive relationship.

3.3 OFDI reverse technology spillover effect

China's enterprises through foreign direct investment (OFDI) access to technology spillovers mainly through the following mechanisms: (1) Learning imitation effect. Foreign direct investment enterprises usually choose developed countries to establish transnational subsidiaries. On the one hand, companies directly access subsidiaries' technology through overseas acquisitions; on the other hand, OFDI firms greenfield investment in host countries which knowledge or technology spillovers, so that transnational subsidiaries can better track, imitate and learn from the host country enterprises and research institutions of technology research and development, which is conducive to investment in the country's technological innovation and technological progress. (2) Association effect. Multinational subsidiaries can absorb the technology spillover from the host country by the industry correlation effect, so that multinational subsidiaries have the opportunity to be embedded in the host country industrial chain and information channels, to better learn from affiliated enterprises, and to understand the latest technological developments in the industry. (3) Personnel movement effect. Multinational subsidiaries can employ local high-quality scientific research personnel to enhance their technological research and development capabilities, but also through the employment of local high-level technical personnel and management personnel to improve their production and management capabilities. (4) R&D cooperation. The main reason for the effectiveness of this mechanism is that through direct foreign investment they can carry out technical cooperation and share part of the cost of R&D with host governments and enterprises.

Hypothesis 3: China's industrial chain upgrading (export sophistication) degree and China's OFDI have a positive relationship.

3.4 *Processing trade*

The effect of international trade on export sophistication is reflected in the technical effect of international trade. The way processing trade through trade technology promotes product sophistication is: processing trade gives technologically backward countries the opportunity to imitate the technologies of the frontiers, while imitation is a process of improving technology; imports of new intermediate products can enhance the productivity of importing countries through input-output effects; import and export trade can avoid duplication of effort, improve the efficiency of research activities around the world, and so on. Processing trade facilitates the increase in output and export product sophistication of the participating countries.

Hypothesis 4: China's industrial chain upgrading (export sophistication) and China's processing trade have a positive relationship.

4 MEASUREMENT MODEL AND REGRESSION ANALYSIS

4.1 *Measurement model and variable selection*

According to the above theoretical hypothesis, the following measurement model is constructed:

$$PRODY_{it} = \alpha_i + \alpha_1 T_{it} + \alpha_2 FDI_{it} + \alpha_3 OFDI_{it} + \alpha_4 PT_{it} + u_{it} \quad (3)$$

where the dependent variable $PRODY_{it}$ is the export sophistication of country i in period t, T_{it} denotes the technological development level of the country i in the period t, mainly measured by the technological investment indicators and total expenditure of country i's technology activity; FDI_{it} represents the foreign direct investment that country i attracts in the period t; $OFDI_{it}$ represents the foreign direct investment of country i in period t; PT_{it} represents the processing trade of country i in period t, and is measured by the import volume of parts and components trade; α_i denotes the intercept term of country i; u_{it} indicates the error term.

4.2 *Data sources and processing*

The sample selected in this paper is Chinese manufacturing; the time span is 2003–2011. China's total expenditure data ST_{it} on science and technology activities come from the China Statistical Yearbook of Science and Technology (2004–2012); China's ASEAN FDI data, China's actual use of ASEAN FDI data and the exchange rate data come from the China Statistical Yearbook (2003–2012); China's import data of parts and components from the ASEAN come from the United Nations COMTRADE database.

4.3 *Measurement model regression*

Based on the above-mentioned processing, this paper estimates the model (1) using OLS under robust standard error and obtains the following preliminary results: the purpose of this paper is to analyse the different factors, especially the production network of China-ASEAN, on the industrial chain upgrading of our country's manufacturing industry as the dependent variable. The results

Table 2. Econometric analysis of the China–ASEAN production network on China's industrial chain upgrading.

Variable	Model 1	Model 2	Model 3
C	16,500.32	16,656.95	17,156.60
	(29.28492)*	(29.50019)*	(52.16783)*
FDI	4.102133	2.486972	
	(1.558947)**	(1.081070)	
OFDI	−4.483504	−4.032269	−3.983820
	(−4.914443)*	(−4.738074)*	(−4.623102)*
PT	−0.159496		
	(−1.153613)**		
RD	1.532049	1.447116	1.513705
	(8.374091)*	(8.367957)*	(9.238348)*
R²	0.995709	0.994282	0.992945
D.W	2.823247	2.026331	1.707677

of the different model tests are shown in Table 2. The insignificant variables from Model 1 to 3 are deleted in turn to estimate. The coefficient of RD in Model 1 is positive, which indicates that the investment of science and technology in our country has a significant impetus to the upgrading of the industrial chain in China. PT coefficient is negative, but not so significant, indicating that the processing trade imports on the impact of the industrial chain upgrading is not so obvious. FDI coefficient is positive, meaning that the use of ASEAN FDI from the impact on China's industrial chain upgrade is positive. OFDI coefficient is negative, t value is significant, and OFDI on the impact of China's industrial chain upgrading is negative. First, we consider removing the non-significant PT from the model, and obtain Model 2.

In Model 2, after the deletion of variable PT, the coefficient of RD is significantly positive, indicating that the investment of science and technology activities on China's industrial chain upgrading has promoted a significant role in science and technology activities, funding, for each additional unit, China's industrial chain to improve 1.447 units. OFDI coefficient is negative, and t value is significant; therefore, OFDI on the impact of China's industrial chain upgrading is negative. The coefficient of FDI is positive, but the significance is decreased, indicating that the FDI from ASEAN

to China's industrial chain upgrading is not significant. Therefore, we consider removing the variable FDI, and obtain Model 3.

In Model 3, the variable FDI is deleted, the t-value of RD regression coefficient is significantly positive and, for each additional unit of scientific and technological activities, China's industrial chain is increased by 1.5 units; China's FDI to ASEAN is negative on China's industrial chain.

ACKNOWLEDGEMENTS

This research is supported by Natural Science Foundation of China (No.71163047) and Department of Science and Technology of Yunnan Province under grant number 2016FB118.

REFERENCES

Hausmann, R., Hwang, J., & Rodrik, D. (2007). What you export matters. *Journal of Economic Growth, 12*, 1–25.

Humphrey, J., & Schmitz, H. (2002). How does insertion in global value chains affect upgrading in industrial dusters? *Re~onal Studies, 36*(9).

Lall, S. (2000). The technological structure and performance of developing country manufactured exports, 1985–1998. *Oxford Development Studies, 28.*

Automotive, Mechanical and Electrical Engineering – Liu (Ed.)
© 2017 Taylor & Francis Group, London, ISBN 978-1-138-62951-6

Logistics demand forecasting model applied in the Wuling mountain area

Huajun Chen & Ronglin Chen
Institute of Big Data, Tongren University, Guizhou, China

ABSTRACT: Logistics demand forecasting is one of the important factors to measure at a regional logistics development level. Logistics development in the Wuling area has rare opportunities and location advantages, but these factors are not equal to the reality of logistics demands; how to grasp the market information to predict decision-making, drive the scale of logistics demand and the high-speed development of the regional economy are of great significance. This paper studies the problems emerging in logistics industry development in the Wuling mountain areas, analyses the regional logistics demand in theory and its forecasting contents, finds reasonable logistics resource allocations used for the Wuling mountain area, builds the logistics demand forecasting model, and achieves the balance between logistics supply and demand to save costs.

Keywords: Logistics demand forecasting; Logistics demand; Wuling mountain areas

1 INTRODUCTION

Logistics activities are an important part of economic activity. In the rapid economic development of modern society, increasingly attention is being paid to this by the various countries around the world. The development level of the logistics industry is gradually being called one of the important indexes by which to measure a national modernisation level and comprehensive national strength. The logistics industry is becoming a new economic growth point, and has brought huge scale and economic benefits for the development of the local economy. But the development of the Wuling mountain area is relatively backward: it lacks the basis of analysing and forecasting logistics demand; has an imbalanced supply of and demand for logistics; has mismatched logistics quantity with regional resources; the logistics infrastructure is weak; small-scale logistics enterprises find it difficult to form large-scale operations; low information degree of logistics system; operating efficiency is not high. How to start from the demands of logistics for economic development, and the reasonable planning of regional logistics is a primary task to promote the development of the regional economy. But at present, the study of logistics planning takes qualitative analysis as its main principle, and there is a lack of quantitative analysis; also, few are for the rational planning of regional logistics development. This article aims to study this kind of situation.

The rest of this paper is organised as follows. In Section II, theory analysis and the forecasting contents of regional logistics demand are described. Section III provides the building of the logistics demand forecasting model, and then through the comparative study determines the effective demand forecasting method and selects the most appropriate method for the scale of logistics demand to make forecasts for the Wuling mountain area in a certain period of the future. Also, the technical route is provided in Section IV. Finally, the paper is terminated with a conclusion in Section V.

2 THEORY ANALYSIS AND FORECASTING CONTENTS OF REGIONAL LOGISTICS DEMAND

2.1 *The purpose of logistics demand analysis*

The purpose of logistics demand analysis is to ensure the relative balance between supply and demand and make sure that social logistics activities maintain high efficiencies and benefits. In a certain period of time, when the logistics capability for supply cannot meet the requirements, it will inhibit the effect of the demand; when it exceeds the demand, it inevitably leads to the supply of waste. Therefore, the logistics demand is the basis of the supply of logistics capability, which is its social and economic significance. By using qualitative and quantitative analysis methods, we understand the social economic activities for the demand intensity of logistics supply, and carry on effective demand management, so as to guide the social investment purposefully in logistics services, which will be conducive to reasonable planning, constructing a logistics infrastructure, and improving the logistics supply system.

2.2 The principle and characteristics of logistics demand analysis

From the perspective of the social and economic significance of logistics demand analysis, it is required that more attention should be paid to logistics demand analysis. Especially considering the condition that the current level of economic development in our country is still backward, and considering the condition that the logistics scale continues to grow, the quality of logistics services needs to improve, and the logistics demand structure keeps continuously developing. Logistics demand analysis should abide by the following principles.

Firstly, logistics enterprises should consciously strengthen their market investigation and research work and attach importance to market demand forecasts. Secondly, improve the ability of information management and utilisation, perfect the construction of the logistics information system, and attach great importance to the comprehensive analysis and evaluation of logistics demand information. Thirdly, pay attention to studying and drawing lessons from the successful experiences of logistics enterprises at home and abroad, and reduce the blindness and arbitrary behaviour of logistics demand analysis. Fourthly, emphasise the role of logistics consulting research institutions, lay emphasis on scientific prediction technology and combine this with expert analysis. Lastly, strengthen the analysis of the international and domestic economic situation, grasp and understand the characteristics of the whole social logistics demand, make explicit the requirements of the logistics enterprises services objectives in terms of the characteristics of quality and quantity, enhance the accuracy of logistics demand analysis, and find the market breakthrough point of logistics enterprises.

Logistics demand refers to the organisations or individuals suggested for products, services or information flow with the needs or the ability to pay. Analysis showed that either on the supply chain exists a lot of logistics demand, logistics demand exists in every link in the supply chain even throughout the supply chain, respectively to form intermediate or ultimate logistics demand of logistics demand. Compared with the demand for other commodities, logistics demand has its particularities, and these characteristics are related and influenced by each other. There are seven main characteristics, i.e. derivation, universality, variety, imbalance, part of the alternative, the qualitative specificity of time and space, and hierarchy.

2.3 The predictability and contents of regional logistics demand

Regional logistics demand is derived demand, it is brought about by the regional economic development itself, and therefore it is closely related to the development of the regional economy. Through the analysis of the theory of regional logistics demand, it can be seen that as the regional economy aggregates, the change of the industrial structure, resource distribution and regional logistics demand aggregates, and the quantity, structure and hierarchy of logistics demand also changes. Furthermore, with the development of the economy, the quantity of regional logistics demand presents the trend of obvious growth. In addition, through the analysis of the correlation data, it also shows that there exists a great correlation between the regional logistics demand and the regional economic level. Therefore, it can be thought that regional logistics demand has a certain predictability, and the indicators of the regional economy are available to forecast the regional logistics demand indicators.

Regional logistics demand forecasting aims to estimate and speculate the flow, source, flow direction, flow velocity and construction of goods that has not yet occurred within the area or is not clear at present, in order to study the size of the regional logistics demand and the demand hierarchy, for providing gist and decision in regional logistics planning. So, according to the characteristics and requirements of regional logistics services in regional logistics demand forecasting, the following can be predicted: the overall scale of the regional logistics demand; the flow distribution of regional logistics; the change of the main influencing factors.

3 BUILD THE LOGISTICS DEMAND FORECASTING MODEL

The basis of the neural network lies in neurons that are a biological model based on nerve cells in biological neural systems. People research biological neural systems, explore the mechanism of artificial intelligence and digitise the neurons, which produces the neuron model. A large number of the same neurons in this form are coupled together to form the nerve network, which is a highly nonlinear dynamic system. Although the structure and function of each neuron is not complex, the dynamic behaviour of neural networks is very complex. Therefore, a neural network can express different phenomena of the practical physical world.

3.1 BP neural network model

The BP network has many layers, and for the sake of simplicity of narrative, as an example it takes

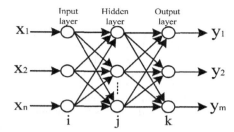

Figure 1. The schematic diagram of the BP neural network.

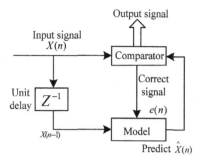

Figure 2. The schematic diagram of adaptive learning.

three layers to reduce computational formula. The schematic diagram of the BP neural network is shown in Figure 1.

Assuming the BP network has three layer networks, the input neurons use i as a serial number, and the input vector is $X = (x_1, x_2, x_3, \cdots, x_n)^T$, adding to $x_0 = -1$ as neurons of hidden layers introducing the threshold value. Neurons of hidden layers use j as a serial number, and its output vector is $O = (o_1, o_2, o_3, \cdots, o_l)^T$, adding to $O_0 = -1$ as neurons of output layers introducing the threshold value. Neurons of output layers use k as a serial number, its output vector is $Y = (y_1, y_2, y_3, \cdots, y_m)^T$, the expected output vector is $T = (t_1, t_2, t_3, \cdots, t_m)^T$, w_{ji} is the weight from the input layer to the hidden layer, v_{kj} is the weight from the hidden layer to the output layer, and the excited function is the function of logarithm Sigmoid:

$$f(x) = \frac{1}{1+e^{-x}} \tag{1}$$

Despite the BP network theory being solid, its physical concept being clear, and its versatility being strong, people in the process of using it find that the BP network has a relatively slow convergence speed with the shortcoming of convergence to local minimum points. For these defects, many improved methods have been proposed, for example, the Conjugate gradient method, the Newton method and the simulation annealing method. This paper aims for the following improvement: the adjustment to a self-adapted learning rate, which makes network of training automatically set the size of different learning rate in

Different stages. The norm of the adjustment learning rate is: check whether the revised value of the weight really reduces the errors function, if it does so, the value of the selected learning rate is small, and it can add a volume. If it does not do so, it produces over-regulation, and it should reduce the value of the learning rate and guarantee that the network can always accept the maximum learning rate for training. In Figure 2, the network is taken

as a predictor. Based on the previous time entered x (n−1) and model in (n−1) moment of parameter, it estimates n moment of output. Compared with the actual value x (n) (as the right answer), its differentials are called "new information". If "new information" e(n) equals zero, the parameters of the model are not amended, or should be amended to track environmental changes.

3.2 Support Vector Regression (SVR) model

Considering a given n learning sample (X_i, y_i), $X_i \in R^d, y_i \in R, i = 1, 2, \cdots, n$ the target of Multiple Linear Regression (MLR) is to resolve the regression function:

$$f(X) = (W \cdot X) + b \tag{2}$$

There, $W \in R^d, b \in R, (W \cdot X)$ is the inner product between W and X. In the previous learning algorithms, the optimising target makes the accumulation-$R_{emp}(f)$ of empirical risk minimised, namely sample loss function $L(X_i)$:

$$L(X_i) = g(y_i - f(X_i)) \tag{3}$$

For example, for the Least Square (LS) method, the resolved (W, b) should satisfy:

$$\min R_{emp}(f) = \sum_{i=1}^{n} (y_i - f(X_i))^2 \tag{4}$$

While statistical learning theory points out the minimum of empirical risk it does not ensure the minimum of expected risk. In the optimised target of Structural Risk Minimisation (SRM) minimisation, the parameters (W, b) of the above linear regression equations should meet:

$$\min Q(W, b) = \frac{1}{2}\|W\|^2 + C R_{emp}(f) \tag{5}$$

There into, $\|W\|^2$ reflects the generalisation ability of the regression function $f(x)$, and C is the penalty factor.

In the equation of (3), the common loss function $L(X_i)$ includes quadratic function, Laplace function, ε-non-sensitive function, etc.

ε-non-sensitive functions are as shown in Figure 3:

$$L_\varepsilon(X_i) = \begin{cases} 0, |y_i - f(X_i)| \le \varepsilon \\ |y_i - f(X_i)| - \varepsilon, |y_i - f(X_i)| > \varepsilon \end{cases} \quad (6)$$

There into, the regression error can be neglected in the scope of ε, which is more suitable for the data processing of the economic society system, because all the data targets of this system exist with higher error, often for many reasons.

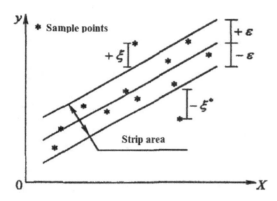

Figure 3. The schematic diagram of linear regression for nonsensitive function.

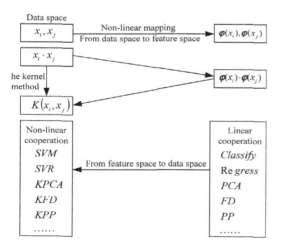

Figure 4. The framework diagram of the kernel method.

3.3 The kernel method

The kernel method is the generic term of a series of advanced and non-linear data processing technology, whose common feature is that these methods of data processing apply in kernel mapping. From the specific operation, the kernel method first uses the method of non-linear mapping to bring the original data from the data space mapping to the feature space, and then in the feature space corresponds to the linear operation, as shown in Figure 4. Because of using non-linear mapping, which is often very complex, it greatly enhances the capability of the non-linear data processing.

The kernel function realises the non-linear transformation between the data space and the feature space. Supposing x_i and x_j are the sampling points of the data space, the mapping function from the data space to the feature space is Φ, and the basis of the kernel method is to realise the inner product transformation of the vector:

$$(x_i, x_j) \to K(x_i, x_j) = \Phi(x_i) \cdot \Phi(x_j) \quad (7)$$

Generally, the non-linear transformation function $\Phi(\cdot)$ is quite complicated, while in the algorithm process the actual use of the kernel function $K(\cdot, \cdot)$ is relatively simple, which is the most charming place of the kernel method. In other words, when using the kernel method, you just need to consider how to choose an appropriate kernel function, and not care that the corresponding mapping $\Phi(\cdot)$ may have a complex expression and high dimension.

4 THE TECHNICAL ROUTE

The empirical research of logistics demand forecasting shows that, according to the above prediction models, it optimises to design the models, and

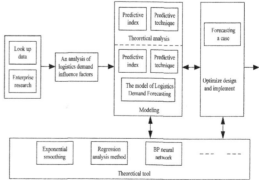

Figure 5. The technical route.

makes an improved new model to positively forecast the logistics quantity demanded for the Wuling mountain area for three years. Through the predicted results, it provides the theoretical basis on the development and plan of the logistics industry in the future for a given period. Figure 5 shows our adopted technical route.

5 CONCLUSION

In a word, this paper expounds the logistics demand forecasting models, the BP neural network model and the Support Vector Regression (SVR) model, then draws forth the technical route, and lastly concludes the full text. The disadvantage of this article is that it only considers the tangible logistics demand, and predicts the scale index of logistics demand, but does not think of the structural index of logistics demand, that is, the demand of logistics service quality, such as the aspects of intangible logistics demand, logistics efficiency, time, cost, etc. In the future, we will regard the above unconsidered sides, and further study them.

ACKNOWLEDGEMENTS

This work was supported by the Collaborative Fund Project of Science and Technology Agency in Guizhou Province Marked by the word LH on 7480 [2014].

REFERENCES

Bazaras, D., et al. (2016). Competence and capacity-building requirements in transport and logistics market. *Transport and Telecommunication Journal,* 17(1), 1–8.

Komori, O., et al. (2016). An asymmetric logistic regression model for ecological data. *Methods in Ecology and Evolution,* 7249–260.

Korczak, J., et al. (2016). Strategic aspects of an eco-logistic chain optimization. *Sustainability,* 277(8), doi: 10.3390 /su8040277.

Lukinskiy, V., & Dobromirov, V. (2016). Methods of evaluating transportation and logistics operations in supply chains. *Transport and Telecommunication,* 17(1), 55–59.

Škerlič, S., et al. (2016). A decision-making model for controlling logistics costs. *Model odlučivanja za kontrolu logističkih troškova; Tehnički vjesnik,* 23(1), 145–156.

Xu, S., & Xu, Y. (2012). Study on Shanghai logistics demand characteristics and influencing factors. *Modern Management,* 2:88–92. http://dx.doi.org/10.12677/mm. 2012. 22016.

Automotive, Mechanical and Electrical Engineering – Liu (Ed.)
© 2017 Taylor & Francis Group, London, ISBN 978-1-138-62951-6

Logistics equipment support capability evaluation

Guosong Zhen
Logistics College, Beijing, P.R. China

ABSTRACT: Logistics equipment support capability evaluation is a kind of cognitive activity that takes logistics equipment as the evaluation object, and takes the object attribute—logistics equipment support capability as the evaluation target. Its purpose is to evaluate the capability and the rationality of the current logistics equipment.

Keywords: Logistics equipment; Capability evaluation

1 INTRODUCTION

Logistics equipment support capability mainly embodies two kinds of capabilities. The first kind is the inherent support capability, that is, the equipment's own support capability when they are used under specified conditions. The second kind is the actual support capability. The inherent support capability is the base of the actual support capability, and the actual support capability is not larger than the inherent support capability. The two capabilities are close when the actual conditions and environment are consistent with the requirements and the personnel are skilled. The actual logistics equipment support capability can be measured by calculating the satisfaction degree of the inherent logistics equipment support capability. The actual support capability is usually expressed as a percentage; the closer it is to 100%, the stronger it is.

2 EVALUATION METHOD OF LOGISTICS EQUIPMENT SUPPORT CAPABILITY

Evaluation of logistics equipment support capability is divided into two levels. The first is the single-type logistics equipment support capability, including military and medical equipment; the second one is the systematic logistics equipment support capability. The method of evaluating the logistics equipment support capability is to calculate the degree of satisfaction of the inherent logistics equipment capability in order to measure the actual logistics equipment support capability. Next, evaluate the systematic logistics equipment support capability according to the single-type logistics equipment support capability. Based on the expert investigation method, this paper uses both, the Analytic Hierarchy Process (AHP) and

the multi index comprehensive evaluation method, to study the logistics equipment support capability.

3 EVALUATION INDEX SYSTEM FOR LOGISTICS EQUIPMENT SUPPORT CAPABILITY

The main elements that determine the logistics equipment support capability include personnel, logistics equipment, and the degree of personnel-equipment coordination. The evaluation index system settings for logistics equipment support capability should start from these three aspects, decompose, and subdivide them layer by layer. The specific index division is shown in Figure 1.

3.1 *Personnel satisfaction degree*

Personnel satisfaction degree includes three aspects—satisfaction degrees of logistics equipment operation, maintenance, and management personnel. The calculation method of each index is as follows. Operation personnel allocating rate equal to the actual operation personnel number/ the standard operation personnel number; operation personnel matching rate equal to the number of operation personnel in matched majors/the total number of operation personnel; operation personnel induction training rate equal to the number of operation personnel receiving induction training/ the total number of operation personnel; maintenance personnel allocating rate equal to the actual maintenance personnel number/the standard maintenance personnel number; maintenance personnel matching rate equal to the number of maintenance personnel in matched majors/the total number of maintenance personnel; maintenance personnel induction training rate equal to

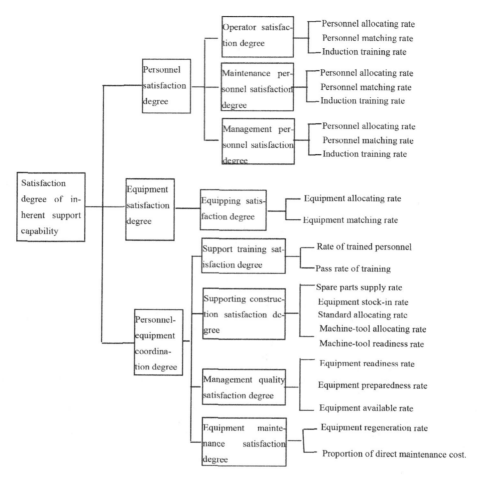

Figure 1. Evaluation index system for logistics equipment support capability.

the number of maintenance personnel receiving induction training/the total number of maintenance personnel; management personnel allocating rate equal to the actual maintenance personnel number/standard management personnel number; management personnel matching rate equal to the number of management personnel in matched majors/the total number of management personnel; management personnel induction training rate equal to the number of management personnel receiving induction training/the total number of management personnel.

3.2 Equipment satisfaction degree

The calculation method of each index is as follows. Equipment allocating rate equal to the actual number of logistics equipment/the stand number of logistics equipment; equipment matching rate

equal to (equipment matching rate of part A + equipment matching rate of part B +...equipment matching rate of part N) / N.

3.3 Personnel–equipment coordination degree

This can be divided into four aspects—training satisfaction degree, supporting construction satisfaction degree, management quality satisfaction degree, and equipment maintenance satisfaction degree.

3.4 Evaluation index synthesis model

In the index system of logistics equipment support capability evaluation, the indexes are mutually independent. At the same time, the linear weighting method is used to calculate the top index values to ensure that the main index well reflects each

754

sub-index in simple and intuitive results. Suppose that index Z consists of the following n sub-indexes.

$$x_1, x_2, \ldots\ldots, x_n, x_i \leq 1, i = 1, 2 \ldots\ldots, n$$

The synthesis model of logistics equipment support capability evaluation is,

$$A_i = \sum_{k}^{n} W_{ij} B_{ij}$$

Where Wij indicates the weight of index Bij for index Ai, and n indicates the number of index Bij contained by index Ai.

4 DEMONSTRATION OF LOGISTICS EQUIPMENT SUPPORT CAPABILITY

4.1 *Overall support capability evaluation for single-type logistics equipment*

The overall support capability of single-type logistics equipment is evaluated in the following steps.

Step 1: Determine the weight coefficients of evaluation indexes according to expert scoring.
Step 2: Calculate the values of bottom indexes in the system.
Step 3: Calculate the overall support capability of this type of logistics equipment.

The Delphy method was used in this study. 20 experts who have engaged in the logistics equipment for many years were invited to score the weight of each index in the index system. After repeated study and calculation, the bottom index values were finally determined, as listed in Table 1.

Based on formula, $A_i^* = \sum_k^n W_{ij} B_{ij}$, the support capability of this kind of logistics equipment can be calculated and was found to be 0.75627, indicating that the degree of satisfaction of the inherent support capability is 75.627% for this type of logistics equipment.

4.2 *Evaluation of system logistics equipment support capability*

The evaluation model used was: $p_g = \sum_{i=1}^{n} p_i \times \omega_i$ wherein i = 1, 2, 3, ...n; n indicates the number of divided logistics systems; Wi indicates the weight of the corresponding logistics system; pi indicates the degree of satisfaction of the inherent support capability of the corresponding logistics equipment; and pg indicates the satisfaction degree of the inherent support capability of the logistics equipment system.

The logistics equipment support capability evaluation results for all kinds of professional logistics service systems and the weight coefficients determined by expert investigation are shown in Table 2.

Table 1. Bottom index values.

Top index	Three level index	Index weight W_{ij}	Index value B_{ij}
Satisfaction degree of inherent support capability	Personnel allocating rate	0.0307	0.8
	Personnel matching rate	0.0171	0.7
	Induction training rate	0.0319	0.6
	Personnel allocating rate	0.0171	0.6
	Personnel matching rate	0.0095	0.6
	Induction training rate	0.0052	0.8
	Personnel allocating rate	0.0092	0.7
	Personnel matching rate	0.0052	0.9
	Induction training rate	0.0028	0.4
	Equipment allocating rate	0.1605	0.9
	Equipment matching rate	0.1000	0.7
	Rate of trained personnel	0.0946	0.8
	Pass rate of personal annual training	0.0771	0.8
	Pass rate of unit annual training	0.0912	0.9
	Rate of spare parts supply	0.0486	0.6
	Standard matching rate	0.0138	0.6
	Machine-tool allocating rate	0.0256	0.8
	Machine-tool readiness rate	0.0256	0.6
	Equipment readiness rate	0.0461	0.7
	Equipment preparedness rate	0.0245	0.6
	Equipment available rate	0.0461	0.7
	Proportion of direct equipment maintenance cost	0.0229	0.6
	Equipment regeneration rate	0.0689	0.6

Table 2. Weight coefficients.

Service	Satisfaction degree	Weight
Military supplies	0.8	0.12
Medical	0.9	0.08
Military transportation	0.8	0.30
Oil	0.8	0.25
Camp	0.6	0.1
Command	0.7	0.15

Plug them into the evaluation mode and obtain pg = 0.65728, indicating that the degree of satisfaction of the inherent support capability is 65.728% for this logistics equipment system.

5 CONCLUSION

Logistics equipment support capacity is related to multiple elements, including personnel, equipment, and the relation between personnel and equipment. In this study on the evaluation of logistics equipment support capability, the built evaluation index system needs constant improvement, and further research needs to be performed considering the combined influence of the logistics equipment theory and practicing researchers to better support logistics equipment construction.

REFERENCES

Hong Qingyin 2013. Research on deepening equipment support capability evaluation [J]. *Journal of Equipment Academy* 23(2): 12–14.
Li Fei 2015. Research on equipment support capability evaluation. *Academic Journal of the Chinese People's Armed Police Forces* 42(4): 35–39.
Liu Jun 2008. On the evaluation of military equipment support capability [J]. *Military Supplies Research* 47(4): 40–43.
Liu Junhu 2007. On the construction of evaluation index system for catering equipment support capability [J]. *Military Supplies Research* 35(2): 35–37.
Wang Ying 2014. AHP-based research on the equipment support capability of fire forces. *Academic Journal of the Chinese People's Armed Police Forces* 30(6): 50–54.

Automotive, Mechanical and Electrical Engineering – Liu (Ed.)
© 2017 Taylor & Francis Group, London, ISBN 978-1-138-62951-6

Optimisation and simulation of the production schedule for the missile general assembly process

Xujie Hu, Zhen Zhang & Yuanyuan Li

China Aerospace Construction Group Co. Ltd. Beijing, China

ABSTRACT: A key problem of production management in missile general assembly workshops is production scheduling. In this paper, the missile general assembly process was investigated. Network planning was introduced to describe the complex production process and express the logical relationship between working procedures and processing units. An optimising mathematical model was established in which the production sequence was the constraint, and the minimum flow time and waiting time of all units was the target. The genetic algorithm was applied to solve the model. Finally, the Plant-simulation software was used to simulate the production scheduling and to show the working procedure. The simulation demonstrated that the method proposed in this paper was helpful for making decisions about reasonable production arrangements and to enhance the productive efficiency.

Keywords: Missile General Assembly; Production Scheduling; Network Planning; Genetic Algorithm; Simulation

1 INTRODUCTION

As a key process in missile production, general assembly consists of a discrete and continuous process with a long production line, a great number of procedures and complex equipment. To a decomposable production task, the production scheduling is used to response production requirements, arrange manufacturing resources each operation uses, manufacturing sequence and manufacturing time each operation occupies. The aim of production scheduling is to optimise the production time and cost and at the same time to meet the production constraints. Therefore, the paper engaged in describing the method of the complex production process, and a production scheduling method that can meet the resources constraints was established based on the genetic algorithm. The method proposed aimed to flexibly optimise the production scheduling, to guarantee unobstructed logistics and effective manufacturing management, and to improve the execution efficiency and the total benefit of the production system.

2 NETWORK PLANNING MODELLING FOR WORKFLOW

Based on the character of the general assembly workflow, a dynamic network planning method for workflow analysis and modelling was proposed. The method described the massive dynamic character of missile general assembly workflow using simple network planning. This could supply workflow modelling and application with valuable references.

2.1 *The workflow of missile general assembly*

The workflow of missile general assembly is shown in Figure 1.

2.2 *Network planning modelling*

The network planning of general assembly workflow was defined as G = (V, E, W, S). The details are below:

V represents the set of all workstations and it is classified according to m working procedures. v is a specific workstation of a working procedure. It is a complete working unit containing main equipment and auxiliary equipment. For example, $\{v_{i1}, v_{i2}, ..., v_{ik}\}$ means there are k station sets of the same or similar function in the No. i working procedure.

$$V = \{V1, V2, ..., Vm\} = \{\{v_{11}, v_{12}, ..., v_{1k}\}, \{v_{21}, v_{22}, ..., v_{2k}\}, ..., \{v_{m1}, v_{m2}, ..., ...v_{mk}\}\}$$

E represents the matrix of relations between every two working procedures in the production process. It describes the transmission of main logistics among the procedures and workstations. E_{ij} is the logic matrix of relative workstations of all kinds of production tasks.

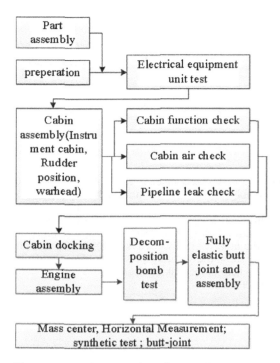

Figure 1. The workflow of missile general assembly.

E is expressed by Equation (1).

$$
E = \begin{array}{c} \\ V_1 \\ V_2 \\ \Gamma \\ V_m \end{array}
\begin{array}{cccc} V_1 & V_2 & L & V_m \end{array}
\left[\begin{array}{cccc}
E_{11} & E_{12} & L & E_{1m} \\
E_{21} & E_{22} & L & E_{2m} \\
\Gamma & \Gamma & \Gamma & \Gamma \\
E_{m1} & E_{m2} & L & E_{mm}
\end{array} \right]
\tag{1}
$$

Where,

$$
e_{i,j} = \begin{cases}
1 & V_i \ is \ e_{ij} 's & begining \ node \\
-1 & V_i \ is \ e_{ij} 's & ending \ node \\
0 & the \ others
\end{cases}
$$

W represents the set of all the production tasks in a workflow, which comes into the workflow network with the finish of each procedure in sequence. l is the number of production tasks.

$W = \{w_1, w_2, L, w_l\}$

S represents the composition of network planning and expresses its dynamics. S consists of the production task's process route matrix R, the production time matrix T, the transport time matrix Te and the workstation conditions matrix C.

In the network planning G = (V, E, W, S), V and E reflect the basic structure of the workflow network planning. The state variation of V and E influenced by W and S reflects the dynamics of the workflow network.

3 GENETIC ALGORITHM TO SOLVE THE MODEL

Considering constraints such as process requirement and production capacity of equipment, this paper established the general assembly production scheduling model, whose objective function was for the minimum production time and waiting time. The genetic algorithm simulated the mechanism of biological evolution in nature based on the natural selection principle. It was applied widely in many domains, such as combinatorial optimisation and machine study. Based on the model, the genetic algorithm, which used the processing paths of all the production units as chromosomes and the objective function as the fitness function, was designed to solve the model and to optimise the production scheduling.

3.1 Introduction of production scheduling

Production scheduling is used to distribute manufacturing resources reasonably in a limited period of time. The purpose is to balance the equipment's workload and to avoid production instability. The instability often occurs due to equipment failure caused by overload. Therefore, distributing resources reasonably in the physical constraints of production scheduling should be considered to improve production stability. Furthermore, workflow consists of a great number of processes with different characters. Each process can influence the production's regularity, stability and continuity. Production scheduling should be considered to balance the capacity, speed, degree of proficiency and stability between the former and latter processes. Grasping the target value and fluctuating value of each process is the precondition to execute production scheduling successfully.

3.2 Genetic algorithm for optimising

3.2.1 Coding strategy

For solving general assembly production scheduling, coding is very important for the genetic algorithm. Usually the process number corresponds with the chromosome encoding directly. This coding strategy may sometimes result in an illegal solution in the following genetic evolution. To solve this problem, the production routine of each equipment was used as a chromosome. Different chromosomes could be formed according to different routine combinations, as shown in Table 1.

Table 1.   Chromosome encoding.

One chromosome			
No. 1 production routine	No. 2 production routine	No. 3 production routine	...
Des1-Res1-Exv4-Bof1	Des2-Res2-Exv5-Bof2	Des1-Res2-Exv4-Bof6	

Before crossing
Father1 1141 2252 1246 ......
Father2 1252 2245 1155......

After crossing
Child1 1141 2245 1255......
Child2 1252 2252 1146......

Figure 2.   Example of chromosome crossing.

Before variation
Father1 1141 | 2252 | 1246 | ......
Feasible path set for gene value
(1,2)   (1,2,4,5,6)

After variation
Child1 1241 | 2252 | 1244 |......

Figure 3.   Example of Chromosome variations.

3.2.2   *Genetic operator design*

An optimal preservation strategy was used to choose the chromosomes. In this method the optimal production was preserved and the other chromosomes were selected according to fitness function's ratio in the whole chromosome fitness function. Progeny chromosomes could be generated through crossing and mutation.

Crossing was the key of the genetic algorithm. A PMX operator was built in this paper. Crossing operation is shown in Figure 2.

The aim of variation was to maintain population diversity. The variation operator was an auxiliary operator in the genetic algorithm due to its local search capability. Variation operation is shown in Figure 3.

4   REALISE GENETIC ALGORITHM OPTIMISING IN PLANT-SIMULATION

4.1   *Hybrid optimal diagram of genetic algorithm and simulation*

Plant-simulation software had a powerful engineering optimal model based on the genetic algorithm.

The model could establish seamless integration with the simulation model. The basic steps of forming the scheduling strategy are shown below:

1. Initialisation
For a fixed population size, the genetic algorithm parameter (variation and crossover operation), maximal cyclic iteration and initial population should be determined.
2. Integrating of the genetic algorithm model and the simulation model
Added the optimal function of the genetic algorithm to Plant-simulation, such as "GA Wizard", and "GA Sequence" task type.
3. Simulation optimising control
Input each chromosome of the population into the simulation model. The adaptation value of each chromosome could be obtained after the simulation. According to the adaptation value, the probability of coming into the next generation could be determined. The algorithm was executed circularly and stopped when requirements were met. The hybrid optimal diagram of the genetic algorithm and simulation is shown in Figure 4.

4.2   *Application example*

Aiming to optimise the production scheduling in a missile general assembly workshop, the genetic algorithm and Plant-simulation models were used.

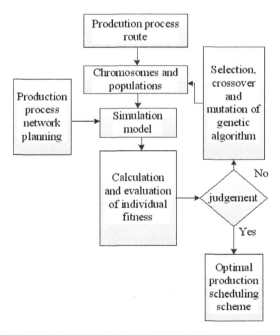

Figure 4.   Hybrid optimal diagram of genetic algorithm and simulation.

759

The production time of each workstation was input, and the procedure obeyed the rule of 'first-in, first-out'. The work site was controlled dynamically by the process control object. The simulation model is shown partly in Figure 5.

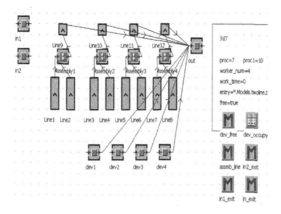

Figure 5.   Simulation model of general assembly.

Table 2.   Simulation result.

Project	Average production time for each workstation
Real production time	2 hours and 24 minutes
Simulation production time	1 hour and 55 minutes

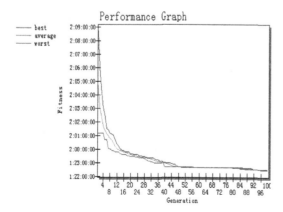

Figure 6.   Converge diagram of the genetic algorithm.

In the example, the crossing probability of the genetic algorithm model was set as 0.8. The variation probability was 0.1 and the adaptation function was minimum work time. The initial number of the population was set as 40 and evolution generation as 100. The simulation result is shown in Table 2.

The converge diagram of the genetic algorithm was shown in Figure 6.

The simulation result demonstrated that the simulation model integrating the genetic algorithm could solve production scheduling optimising more scientifically and reasonably. It could be applied to promote productivity.

5   CONCLUSION

Based on the analysis of the characters of missile general assembly, this paper established a mathematical optimal model for general assembly production.

Scheduling. The minimum production time was objective function, the process condition was constraints, and the production route of all work procedure was decision variation.

The simulation model, based on Plant-simulation integrating with the genetic algorithm, was explained and analysed. The network planning expression and modelling step was described in detail.

The simulation model was verified by a real assembly example. The result demonstrated the effectiveness of the method proposed in this paper.

REFERENCES

Holland, J. (1975). *Adaptation in Natural and Artificial Systems*. Ma: MIT Press.

Linn, R. & Zhang, W. (1999). Hybrid flow shop scheduling: A survey. *Computer & Industrial Engineering, 37,* 57–61.

Webster, S. T. (2008). The complexity of scheduling job families about a common due date. *Operations Research Letters,* 20(2), 65–74.

Xu, J. G. et al. (2004). The survey of production scheduling theory and method. *Computer Research and Development,* 41(2), 257–267.

Zheng, Z. et al. (2008). Simulation model of truck scheduling in workshop based on cellular automatic machine. *Systems Engineering Theory and Practice,* 2(2), 138–142.

Zhu, B. L. & Yu, H. B. (2010). The research on production model and algorithm. *Computer Integrated Manufacturing System,* 16(1), 34–36.

Automotive, Mechanical and Electrical Engineering – Liu (Ed.)
© 2017 Taylor & Francis Group, London, ISBN 978-1-138-62951-6

Research on bridge construction methods of power outage maintenance in distribution networks

Li Xu, Jie Niu, Dehua Zou, Li Tang, Lanlan Liu & Wen Li
Live Working Centre of State Grid Corporation of Hunan, Changsha, China

Yanyi Fu
State Grid Xiangxi Power Supply Company, Xiangxi, China

ABSTRACT: Live work is not the only way of implementation for complex live operation projects; the operation without power distribution line failure is another solution. The transition from live work to non-outage operation, and handling complex live operation projects with simple non-outage operation, is an effective way to reduce the security risk and improve the safety of live working. According to the characteristics of the power network and the requirement of live working, a most widely used non-outage maintenance method is explored, that is the "bridge construction method".

Keywords: Live working; Non-outage maintenance; Bridge construction method

1 INTRODUCTION

With the rapid development of economy and society, all sectors of society put forward higher and higher requirements for the reliability of power supply. In particular, some areas like the maintenance of electricity for major activities, power supplies in hot sensitive areas, and the construction of some key projects, have contributed to an even higher demand for uninterrupted power supply. Distribution network operation without power outage has become the most direct and effective means to improve the reliability and service quality of power supply. Aiming at the actual situation and existing problems of live work in the Beijing company, and in order to regulate the operation technology management, improve the level of non-outage equipment, and ensure the safety of site operation, it is necessary to do comprehensive research on the technology management system, operating method, and key equipment of the distribution network non-outage operation, and to carry out site promote and applied. Due to the increasing demand for continuous and reliable power supply and the requirement of Power Grid Corp to promote the live working in distribution network, the live working of distribution network has developed rapidly, and the conventional electrification projects have reached four types with a number more than 30. In addition, every region has also carried out a number of other live work projects according to the regional characteristics.

Parts of these projects involved a relatively high degree of complexity, and with the increase in complexity, the difficulty and risk also increased. This is evident from: (1) complicated working environment; (2) large labour intensity; (3) high level of difficulty; (4) more construction processes; (5) longer operation time; (6) outage working and live working happening alternately. So, compared with the risk of simple live working, the risk of complex live working increases significantly, and once the accident happens, irretrievable loss of life and property follow. From the lessons of the past, we can deeply realise that most accidents occurring during the construction of distribution lines are caused by personnel mistakes in the process of operations, and individual reliance on the technical safety requirements cannot fundamentally eliminate the occurrence of accidents.

2 OPERATION PRINCIPLE

Live work is not the only way of implementation for complex live operation projects; the operation without power distribution line failure is another solution. The transition from live work to non-outage operation, and dealing with complex live operation projects with the simple non-outage operation is an effective way to reduce the security risk and improve the safety of live working. According to the characteristics of the power network and the requirement of live working, Beijing region has explored a most widely used non-outage maintenance method called the "bridge construction method". Taking the replacement of the pole-mounted load-breaking switch with load as an example, simply explains the operation principle of the bridge construction method. The distribution line is equipped with equipment such as a large number of pole-mounted load-breaking switches (Figure 1) and isolation switches.

During the operation, the replacement of equipment with load is usually needed because of equipment failure. This work will be extremely complex if it is carried out with load on. On the one hand, the structure of the tension bar itself is complex, which makes the work difficult. On the other hand, to replace the pole-mounted load-breaking switch, not only six switch leads and between three and six arrester leads will be added, but also three divisional insulation lines. So by simply live working with load to replace such equipment, either the covering or the replacement of equipment, will all lead to the construction difficulty and time waste, which in turn brings high risk. If instead a bypass load switch and cable down lead are used to build a power outage area to implement the replacement of switch by the way of building a bridge, the above risks can be easily resolved as shown in Figure 2. Temporarily install a bypass load switch

in the appropriate position under the switch, and connect the bypass cable and the leading wire which connected the two sides of the switch by the conventional method of non-outage drainage line connecting, and the switch should be placed in the position of the break-brake when connecting.

When the connection is completed, put the operate switch in the position of the close-brake, and then make sure that the electricity—nuclear phase are all correct. At this time, the line load current will pass through the main line, bypass cable, and switching circuit respectively. Then use the hard insulated wire in the way of live work, as is shown in Figure 3. Break the leading line at both ends of the operating point. Because of the existence of the bypass cable, the line load current will be transferred quickly to the bypass cable; at this time the operation is only equivalent to a simple line of charged break.

Operating points at both ends of the leading line off after, a power cut (oval area) was built and completed, within this range, operators to perform good power outage, prove electricity, after the ground wire and other safety measures, can carry out the construction work, security is greatly improved; After construction, in reverse order, the connection of the main wire is restored by using a conducting wire in the reverse order (Figure 4). Remove the bypass cable and bypass switch, construction of the bridge construction method with load replacement switch is the end of the report.

Figure 3. Hard insulated wire.

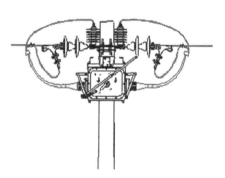

Figure1. Distribution line is equipped with a large number of column load switches.

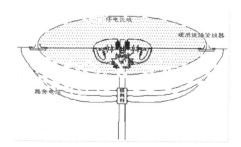

Figure 2. Install a bypass load switch.

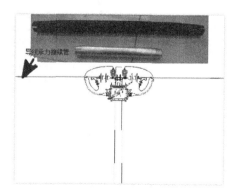

Figure 4. Main wire connection.

3 JOB STEPS

To implement the "bridge construction method", the work needs to arrange that two insulated bucket arm cars, a live working group and a power cut working group are in cooperation with each other. Among them, the live working group is responsible for the erection of the bypass load switch on the pole and the erection of the bypass cable and the connection of the cable on both sides of the operation point. The leading wire is cut off, and assistance is given to the power failure operation team to do the test, work on both sides of the ground to hang a good grounding and other work. The power cut operation team is responsible for the replacement of the switch on the column. The lifting of the switch on the column can be matched with an insulated bucket arm vehicle. A site survey of line construction is completed, a construction plan is made, a class meeting held, construction materials and construction tool prepared, field order, the arrangement of safety protection measures and other procedures in accordance with the provisions of the procedures (not repeated here). The main construction steps are described below: (1) install bypass load switch and detect; (2) bypass circuit construction; (3) break main wire, building blackout area; (4) carry out the power cut construction and replace the switch; (5) main wire bonding; (6) remove bypass switch and bypass cable; (7) end of construction and clean up the site.

4 EXTENDED APPLICATION-TEMPORARY CUT OFF METHOD

The hard insulating wire device which is introduced in the preceding paper can also realize the difficult problem of the special place to carry out the electric disconnection (Figure 3). We have a line terminal bar to change the live line work item (Figure 5). As an example, the method of using the non-power cut operation to solve the risk of complicated electricity operation is briefly described.

4.1 Task and risk analysis

Load extension, Need to change the line terminal to a pole. Such as the use of the terminal on the way of live operation, Because the rod is provided with a cross arm, user boundary load switch, arrester, cable and high-voltage cables and other equipment, carrying out the live work on the pole needs to be insulated from the charged body and the grounding body, mounting insulator, hanging wire and tightening wire. The working environment is extremely bad, job risk is huge, and almost impossible to implement.

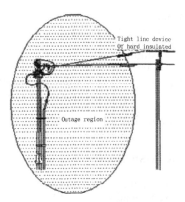

Figure 5. Blackout operation No.1.

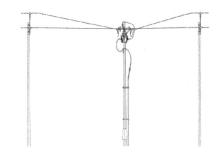

Figure 6. Blackout operation No. 2.

4.2 No power cut construction method—temporary cut-off-the-wire method

1. Live working: Pull the user demarcation load switch on the terminal pole, hard insulated wire device used for live group, the proper position of the power supply side of the terminal rod is slightly farther, temporary cut off wire using the aforementioned temporary cut wire, temporary construction of a power cut operation environment for the power cut operation team.
2. Power cut work: The terminal rod is arranged to install the anti-tension insulator in the blackout area by the power cut operation team, hanging wire and tightening wire, power outage operation.
3. Live working: Live teams and groups use the wire to bear force connection pipe, completion of the temporary cut off the main line of the connection, hard insulated wire. Easy and quick-to-use short-time power cut method to complete the terminal bar to change the pole of the non-power outage work (Figure 6).

4.3 Non-power outage construction effect analysis

This method is the use of short power outage, the implementation of a simple and effective implementation of the complex live. Although the short time to stop on the terminal connected to a power supply users, but the continuous supply of all power customers on the main line of the terminal bar is guaranteed (short-term power failure is also the goal of non-power failure operations), it can also effectively solve the security risks of carrying out live work on the terminal rod which is extremely poor in the environment.

5 CONCLUSION

The essence of the bridge construction method is to use a bypass load switch and bypass cable, commonly used in distribution line, using specially developed hard insulated wire, on both sides of the rod to build a temporary segment, the implementation of a small bypass operation in the range of this bar. Using the method of non-power failure operation, it is very dangerous and very complex to implement the simplest operation of the two charged off line, two time line and power cut off the implementation of the general power outage. Temporary cut-off-the-wire method involves complex live work to implement the simplest class of one charged off, the general power cut construction work carried out by the operation of the drain line and the power cut. Using the method of the construction of the non-power cut operation to solve the risk of complicated electricity operation, greatly reduces the risk of operations. Operating strength also decreases, fundamentally solving the problem of carrying out the personal safety risk of the high work brought by the complicated live work.

REFERENCES

Chinese Standards. *Technical guide for live working in distribution line* (Standard No. GB/T 18857–2008).

Hu, Y., ang L., Liu S. et al. The way of live working and safety protection. *High Voltage Technology*, 26(5), 34–35.

Li, G. et al. (2007). *Design and construction, operation and maintenance of power distribution lines*. Beijing: China Electric Power Press.

Liang, H. (2002). *Distribution line*. Beijing: China Electric Power Press.

Sun, C. et al. (2005). *Manual of power distribution lines*. Beijing: China Electric Power Press.

Automotive, Mechanical and Electrical Engineering – Liu (Ed.)
© 2017 Taylor & Francis Group, London, ISBN 978-1-138-62951-6

Research on logistics network site selection under imperfect information

Liu Yang, Xifu Wang & Jing Ye
School of Traffic and Transportation, Beijing Jiaotong University, Beijing, China

ABSTRACT: Differing from the traditional logistics network site selection problem, this paper aims at the optimisation of freight task, considering the unreliability of logistics network systems caused by collection-distribution centre failures, establishing a layout optimisation model of logistics networks based on imperfect information by extending the classical Uncapacitated Fixed-charge Location (UFL) model, thus effectively reducing loss caused by facilities' failures and improving the stability of logistics network systems.

Keywords: Logistics Network; Facilities Optimisation; Site Selection; Imperfect Information

1 INTRODUCTION

With the deepening regional cooperation in trade relations, the collaborative development of the Beijing-Tianjin-Hebei region has become central to China's major decisions and arrangements. As key elements of integrating Beijing-Tianjin-Hebei logistics, optimisation of logistics network nodes should take fully into account the construction of transport infrastructure networks, realising the integration of logistics network resources and improving the stability and efficiency of regional logistics network.

The site selection is an important component of logistics network optimisation, while facilities' failures are less considered in traditional site selection models. Some researchers analyse the site selection problem based on all facilities being in a fully reliable environment. In practice, however, destructive interference events (natural conditions, human factors, etc.) have occurred, resulting in facilities failures and immeasurable loss. Snyder (2003) and Ball (1993) consider the stochastic nature of logistics network and the reliability of facilities. On the basis of their ideas, this paper considers the condition of imperfect information in the site selection of collection-distribution centres.

2 ANALYSIS ON LOGISTICS NETWORK

In accordance with the existing logistics network facilities, a number of collection-distribution centres were chosen to meet the demand of goods collection in various regions of Beijing. In the latest administrative divisions, Beijing contains 16 districts and counties, and the distribution of road and freight railway networks in Beijing are shown in Figure 1.

Beijing's land transport logistics network is a radiation network, consisting of the highway, national and provincial highways, and freight railway lines, and the spatial distribution of traffic facilities in different districts and counties is uneven. In order to facilitate the data analysis, this paper extracts the actual land transportation logistic network into a network model from which alternative points of collection-distribution centres is determined by analysing the influence factors. These influence factors are: (1) The Intersection of Highway and Railway; (2) Relatively dense points of the road network; (3) Densely populated areas with large freight volume; (4) Uniformity of alternative points in the whole network. Finally, the alternative collection-distribution centres are given in Figure 2.

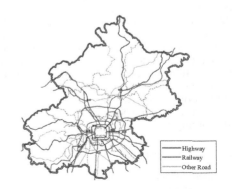

Figure 1. The logistics network of Beijing.

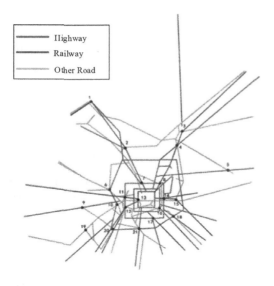

Legend:
━━ Highway
━━ Railway
━ Other Road

Figure 2. Alternative centres of Beijing logistics network.

3 SITE SELECTION MODEL BASED ON IMPERFECT INFORMATION

The goal of the site selection is to select a certain number of collection-distribution centres to meet the demand for goods collection in Beijing, and minimise the total cost. Based on this goal, this paper considers the site selection model based on imperfect information which means the status of the collection-distribution centres is unreliable (damaged or can be used). Freight task is carried out on the premise that the status of each collection-distribution centre is unknown, so the trucks have to follow a specific sequence of access to the collection-distribution centres, until the point can be used. When every point is damaged, the freight task must be surrendered and the penalty cost accepted.

In terms of cost, the conventional capacity-free location model only involves the cost of fixed infrastructure and transportation. Based on imperfect information, the additional cost of the additional transportation route should also be considered; this is called penalty cost. Therefore, the objective function of cost consists of four parts: fixed investment cost, transportation cost, storage cost and handling cost.

3.1 *Symbol description*

Describes the set of symbols associated with the model, variables, decision variables, and parameters as follows:

3.1.1 *Definition of related set*

I: Set of demand points i;

J: Set of alternative collection-distribution centres j;

$\bar{J}$: Set of alternative collection-distribution centres J with virtual collection-distribution centres j_0, $\bar{J} = J \cup \{J_0\}$

J^*: Set of built collection-distribution centres J;

J_i^*: Set of collection-distribution centres accessed from demand points i in any freight task;

$\bar{J}_i^*$: Set of collection-distribution centres and virtual collection-distribution centres accessed from demand point i in any freight task;

J_j^+; : Set of accessed points before alternative collection-distribution centres j, $J_j^+ = \begin{cases} J, j=j_0; \\ \bar{J} \backslash \{j\}, j \in J; \end{cases}$;

J_j^- : Set of accessed points after alternative collection-distribution centres, j, $J_j^- = \begin{cases} \{j_0\}, j=j_0; \\ \bar{J} \backslash \{j\}, j \in J; \end{cases}$ $j \in \bar{J}$;

3.1.2 *Definition of related variables*

h_i: Quantity of shipments from demand points i to any collection-distribution centres;

j_i^r: The r-th ($r = 2,...,|j_i^*|$) access to the collection-distribution centre from demand points i;

R: The number of times visiting actual collection-distribution centres and virtual collection-distribution centres, generally defined as large enough positive;

j_0: Virtual collection-distribution centre, $j_i^{|j_i^*|+1} = j_i^{|j_i^*|+2} = ... = j_i^R = j_0$;

d_{st}: Distance from s to t, define the distance from actual collection-distribution centres to virtual collection-distribution centres $d_{jj_0} = 0$;

d_{str}: Distance from s to t in a freight task from demand point i and the number of visiting times is r;

3.1.3 *Definition of decision variables*

$x_j = \begin{cases} 1, \text{collecting-distributing center } j \text{ is selected} \\ 0, \text{other} \end{cases}$;

$u_i = \begin{cases} 1, \text{Goods from demand point } i \text{ transported by road} \\ 0, \text{other} \end{cases}$;

$w_i = \begin{cases} 1, \text{Goods from demand point } i \text{ transported} \\ \text{by railway } 0, \text{other} \end{cases}$;

$y_{ij} = \begin{cases} 1, \text{Freight task from demand point } i \text{ to} \\ \text{the first visit collecting-distributing center } j; \\ 0, \text{other} \end{cases}$

$y_{ijj'r} =$

$$\begin{cases} 1, \text{Freight task from demand point i to the} \\ \quad (r-1)\text{-th visit collecting-distributing center} \\ \quad j, \text{the } r\text{-th visit is } j', \forall r = 2,\ldots,R \\ 0, \text{other} \end{cases};$$

$X = \{x_j\}_{j \in J}$, Set of decision variables x_j;

$U = \{u_i\}_{i \in I}$, , Set of decision variables u_i;

$W = \{w_i\}_{i \in I}$, Set of decision variables w_i;

$Y = \{y_{ij}\}_{i \in I, j \in \bar{J}}$, Set of decision variables y_{ij};

$Y' = \{yijj'r\}_{i \in I, j \in J, j' \in \bar{J}_{\bar{j}}, r=1,\ldots,R}$, Set of decision variables $y_{ijj'r}$.

3.1.4 Definition of related parameters

q: Damage probability of collection-distribution centre j (independent and identically distributed);

g_j: Fixed construction costs of collection-distribution centre j;

c_1: Unit cost of transportation by road;

c_2: Unit cost of transportation by railway;

δ: Unit penalty cost when every collection-distribution centre is damaged;

s_1: Unit cost of warehousing (road);

s_2: Unit cost of warehousing (railway);

v_1: Average speed of road transportation;

v_2: Average speed of railway transportation.

3.2 Model hypothesis

In order to build the model based on imperfect information, this paper proposes the following hypothesis.

(1) Damage probability of collection-distribution centre is independent and identically distributed; (2) The status of collection-distribution centres is unreliability (damaged or can be used) in every freight task; (3) When every collection-distribution centre is damaged, the penalty cost caused by the failure of the freight task has to be accepted; (4) The freight task between the collection-distribution centres is limited to only one mode of transport. Based on these hypotheses, the site selection model is built, considering fixed investment costs, transportation costs, storage costs and handling costs.

$$\underset{x,u,w,y,y'}{Min} \begin{pmatrix} \sum_{j \in I} g_i \cdot x_j \\ + \sum_{i \in I} h_i \sum_{j \in I} (c_1 \cdot u_j + c_2 \cdot w_j) \left(d_{ij} \cdot y_{ij} + \sum_{j \in I_j} \sum_{r=2}^{z} q^{r-1} \cdot d_{ij} y_{ij} \right) \\ + \sum_{i \in I} h_i \sum_{j \in I} \left(u_j \cdot \frac{s_1}{v_1} + w_j \cdot \frac{s_2}{v_2} \right) \left(d_{ij} \cdot y_{ij} + \sum_{j \in I_j} \sum_{r=2}^{z} q^{r-1} \cdot d_{ij} y_{ij} \right) \\ + \sum_{i \in I} h_i \sum_{j \in I} (u_j \cdot z_1 + w_j \cdot z_2) \end{pmatrix} \quad (1)$$

S.T.

$$\sum_{j \in \bar{J}} y_{ij} = 1, \forall i \in I \quad (2)$$

$$y_{ij} + \sum_{r=2}^{R} \sum_{j' \in J_{\bar{j}}} y_{ijj'r} \le x_i, \forall i \in I, j \in J \quad (3)$$

$$y_{ij} = \sum_{i' \in I_{\bar{i}}} y_{ij'2}, \forall i \in I, j \in \bar{J} \quad (4)$$

$$\sum_{j' \in J_{\bar{j}}^+} y_{ijj'(r-1)} = \sum_{j' \in J_{\bar{j}}} y_{ijj'r}, \forall i \in I, j \in \bar{J}, R = 3,\ldots,R \quad (5)$$

$$u_j + w_j = 1, \forall j \in J \quad (6)$$

$$\sum_{j \in J_{j_0}^+} y_{ijj_0 R} = 1, \forall i \in I \quad (7)$$

$$x_j \in \{0,1\}, \forall i \in I \quad (8)$$

$$x_j \in \{0,1\}, \forall i \in I \quad (9)$$

$$u_j \in \{0,1\}, \forall j \in J \quad (10)$$

$$w_j \in \{0,1\}, \forall j \in J \quad (11)$$

$$y_{ij} \in \{0,1\}, \forall i \in I, j \in \bar{J} \quad (12)$$

$$y_{ijj'r} \in \{0,1\}, \forall i \in I, j \in \bar{J}, j' \in J_{\bar{j}}, r = 2,\ldots,R \quad (13)$$

Based on this model, each freight task has to visit a collection-distribution centre (whether it is the actual or the virtual). In Eq. (2) the virtual centres are set so that the freight task can be finished in R times, (if none of the actual centres are available, the task will be finished in virtual centres); A centre can be visited only once in freight task (Eq. (3)); Access order must obey the rule from r-1 to r (Eqs. (4) and (5)); Only one type of transportation can be used between points (Eq. (6)); The last visit centre must be a virtual centre in a freight task (Eq. (7)).

4 ANALYTICAL SOLUTION

According to the real condition, some key data about the analysis are added.

4.1 Data complement

4.1.1 Demand points data

In this paper, demand points are divided into 16 parts according to the districts, freight volume of each points in Table 1 is set in accordance with the proportion of population. Further to this, Table 2 shows the location of each centre.

Table 1. Freight volume of each point in 2012.

District	Population (Million)	Freight Volume (Million tons)
Dongcheng	95.85	1211.601807
Xicheng	131.05	1656.551037
Chaoyang	376.67	4761.335971
Fengtai	223.72	2827.955726
Shijingshan	63.9	807.7345382
Haidian	348.43	4404.365339
Fangshan	98.65	1246.995496
Tongzhou	126.26	1596.002548
Shunyi	92.61	1170.646253
Changping	174.4	2204.521181
Daxing	143.53	1814.305763
Mentougou	30.65	387.4344851
Huairou	40.17	507.7730266
Pinggu	43.68	552.1415435
Miyun	47.4	599.1645871
Yanqing	32.33	408.6706983
Total	2069.3	26157.2

Table 2. Location of alternative centres and fixed construction costs.

Alternative centre	Longitude	Latitude	Fixed costs (Yuan)
1	115.9144	40.50884	3048000
2	116.2052	40.22605	2226000
3	116.6813	40.33467	4998000
4	116.6319	40.25716	3048000
5	117.0077	40.07009	3048000
6	116.0904	39.99113	4998000
7	116.3393	40.02506	3906000
8	116.454	40.01769	4998000
9	115.9603	39.82563	3048000
10	116.1468	39.88132	2226000
11	116.2113	39.92528	3906000
12	116.2247	39.84796	3048000
13	116.3105	39.90171	6276000
14	116.542	39.9185	3048000
15	116.5993	39.91103	2226000
16	116.4761	39.84263	4998000
17	116.4244	39.79022	1545000
18	116.6287	39.79626	1545000
19	115.9369	39.68469	3048000
20	116.0935	39.70794	1545000
21	116.3283	39.70364	1545000

4.1.2 Alternative points data

There are 21 alternative points, numbered 1–21, the land benchmark price is set as the fixed construction cost of each alternative point.

4.1.3 Distance of each point

Demand points and alternative points are both in latitude and longitude co-ordinates. So the co-ordinates of A and B are defined as A (LonA, LatA) and B (LonB, LatB) and d is the distance from A to B.

$$d = R * Arc\cos(C) * \frac{Pi}{180} \tag{14}$$

$$C = \begin{pmatrix} \sin(MLatA) \cdot \sin(MLatB) \cdot \cos(MLonA - \\ MLonB) + \cos(MLatA) \cdot \cos(MLatB) \end{pmatrix} \tag{15}$$

4.1.4 Other parameters

Penalty cost $\delta = 100,0000$ Y, and R = 6, q = 0.03; Average speed of road transportation is 75 km/h; Unit cost of transportation and warehousing is 5,000 Y/million tons*km; Average speed of road transportation is 85 km/h; Unit cost of transportation and warehousing is 2,500 Y/million tons*km, handling costs for railway is 50,0000 Y/million tons*km.

4.2 Result and discussion

According to the original data and parameters, the results obtained using Gurobi6.5.0 and Eclipse3.1 with java interface.

4.2.1 Result of site selection in Beijing

Using the basic data of 2012 and 2013, the results are identical for both years (2012 and 2013). Figure 3 presents the result for site selection, and 14 collection-distribution centres are selected from 21 alternative points.

Figure 3 marks the position of each collection-distribution centre (the black points and numbers show the alternative points, while red represents the demand points, the triangles are the selected collection-distribution centres, and the red lines are the path of freight task). 16 sets of demand points, corresponding to the 14 sets of collection-distribution centres, each of which is closest to the demand point. The demand points 2 and 3 correspond to the same alternative point 3 and the demand point 11 and 12 correspond to the same point 13.

4.2.2 The path to the collection-distribution centres in freight task

The number of collection-distribution centres allocated for each demand point is R = 6. According to Eq. (2), a virtual collection-distribution centre is the last point to visit in every freight task, so the 6th visit is marked as 0. Table 3 shows the order of the freight task access to centres for each demand point.

Table 3 shows the different path from the demand points to centres. Each collection-distribution centre can be obtained from demand point.

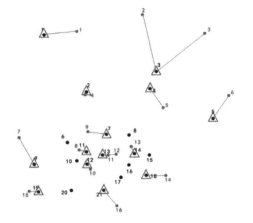

Figure 3.   Result of site selection.

Table 3.   The transportation routes from each demand region in Beijing to collection-distribution centres.

Demand point	1st	2nd	3rd	4th	5th	6th
1	1	2	7	13	11	0
2	3	4	7	13	11	0
3	3	4	7	13	11	0
4	2	7	13	11	12	0
5	4	3	5	12	13	0
6	5	13	12	21	14	0
7	9	19	12	11	13	0
8	11	12	13	7	14	0
9	7	13	11	12	21	0
10	12	11	13	7	14	0
11	13	11	12	21	18	0
12	13	11	12	21	18	0
13	14	18	21	12	11	0
14	18	14	13	11	12	0
15	19	9	12	11	13	0
16	21	12	11	13	7	0

Thus when some destructive interference events happen in Beijing and a part of collection-distribution centres failures occur, although imperfect understanding of collection-distribution centres exists, freight task can also be taken by other centres. So the complementarity between centres can maintain the stability of the logistics network in Beijing to some extent, and thus reduce the loss caused by the failures of the facilities.

5   CONCLUSION

Most of the traditional logistics network optimisation problems are based on the implicit assumption that the logistics node facilities will not be damaged; this paper considers collection-distribution centres in the logistics network of Beijing, which may be damaged in some circumstances. Based on the probability that failures of collection-distribution centres are independent and equal, and the penalty cost for failures of centres under imperfect information, with the aim of lowest cost, the optimal site selection scheme is obtained.

There are limitations in the research, to wit, not considering the influence of shipping and air transport modes in the freight task; and the possibility that collection-distribution centres failures may not be equal and independent in reality. Besides this, the probability of failures is linked with location, the status of the logistics network and the occurrence probability of unexpected events, all of which require further research.

REFERENCES

Ball, M. O, and Lin, F. L. (1991). A reliability model applied to emergency service vehicle location. *Operation Research.* 41(1), 18–36.

Elloymson, M.A. and Rya, T.L. (1966). Abrnach-nadbound algorithm for plant location. *Operations Research.* 14, 361–368.

Klose, A. and Drexl. A. (2001). *Lower bounds for the capacitated facility location problem based on column generation.* Technical Report. Universiat St. Gallen.

Snyder, L. V. (2003). Supply Chain Robustness and Reliability: Models and Algorithms. Evanston: Northwestern University.

ZHOU Ai-lian. (2006). Research on the method and application of nodes locations of enterprise logistics network. Southeast University.

Automotive, Mechanical and Electrical Engineering – Liu (Ed.)
© 2017 Taylor & Francis Group, London, ISBN 978-1-138-62951-6

Research on strategic development of smart logistics

Zhiguo Fan & Mengkun Ma
School of Management, Tianjin University of Technology, Tianjin, China

ABSTRACT: Smart logistics is based on an emerging technology, Internet of Things (IOT), whose idea is consistent with the automated, networked, visual, real-time, tracking and intelligent control development tendency of modern logistics industry. It can greatly reduce the cost of various industries and promote the upgrading of industry. It is the development tendency of future logistics. The transformation and upgrading from traditional logistics industry to "smart logistics" is an inevitable trend of social development. This paper is focused on the basic concept of IOT and smart logistics to analyse the main problems existing in the development of current smart logistics. Meanwhile, corresponding countermeasures and suggestions are proposed for such problem so as to better promote the development of smart logistics.

Keywords: Internet of Things; Smart logistics; Countermeasure

1 INTRODUCTION

In recent years, the logistics industry has become the most active emerging industry with fastest growth rate in the economic field of our country. However, compared with the developed countries, the development of the logistics industry in our country is still is very backward. According to the *Notification for National Logistics Operation in 2015* issued by National Development and Reform Commission, the ratio of total social logistics costs to GDP in 2015 is 16.0%, which is about 4.5% higher than the world average, equivalent to about twice that in the developed countries America and Japan. Too high logistics cost has become the important bottleneck restricting the logistics development of our country. Consequently, how to reduce the logistics cost and develop the modern logistics has become the vital subject for the economic development of our country. With the development of foreign trade and the deepening of international logistics communication, the concept of smart logistics has been known to various enterprises of our country extensively, including transportation, warehousing, production and marketing etc. Consequently, there is requirement for research and analysis of the present situation and existing problems of smart logistics, and for further discussion of its development strategy so as to guide enterprises to stride forward on the road of smart logistics.

2 INTERNET OF THINGS AND SMART LOGISTICS

Internet of Things (IOT) is also referred to as sensor network. Based on the Internet of computer, it is a network using Radio Frequency IDentification (RFID) technology, wireless communication technology, infrared sensor, global positioning system, laser scanner and other information sensing devices to connect anything with the Internet as agreed and to make information exchange so as to realise intelligent identification, positioning, tracking, monitoring and management. In this network, everything in the world ranged from tyre, toothbrush and house to facial tissue can actively make information to exchange through the Internet.

The system structure for IOT is composed of perception layer, communication layer and application layer, as shown in Figure 1. The perception layer is the foundation, the one joining between the physical world and information, world as well as being the source that IOT uses to identify things and gather information. The communication layer is mainly responsible for realising the transmission and communication of information, and transmitting and processing the information obtained by the perception layer. The application layer mainly provides the public service supporting environment for various specific applications. It is a platform to realise the identification and perception between things and things, and between humans and things, as well as to play the role of smart.

The concept of smart logistics was first proposed by Information Center of China Logistics Technology Association, Hua Xia Internet of Things and Editorial Office of Logistics Technology and Application together in December, 2009. Smart logistics refers that the advanced technologies in IOT, including RFID, sensor and global positioning system etc., which are applied in the basic links of the logistics industry like transportation, storage, distribution and express delivery etc. so as to realise the intelligent pattern and automated management of logistics

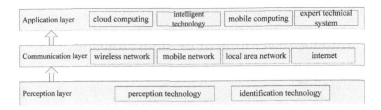

Figure 1. Structure of Internet of Things.

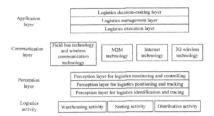

Figure 2. Technical structure of smart logistics.

industry, as well as the operation pattern of smart logistics with informatisation, intelligentisation and system automation. It mainly takes advantage of high technology and modern management means to realise the high-efficiency and low-cost intelligent operation of logistics distribution system.

The technical structure of smart logistics is based on the three-layer technical structure of IOT. According to the application requirement of modern logistics, each layer should be refined to provide a more effective technical support for modern logistics. Actually, the implementation process is composed of five steps. First, consider to realise the real-time sharing and mutual communication of basics IOT. Second, consider to perceive the information of each logistics activity, integrate the information of logistics and IOT, and guarantee the informatisation of logistics process of things. The third step is to take advantage of automation technology to realise the automation of handling process of things so as to reduce the strength of logistics handling and improve the accuracy of logistics. The fourth step is to use information technology to connect each logistics activity in series, so as to realise the integration of information in the whole logistics process. The last step is to make an in-depth analysis for each logistics activity, continually taking advantage of internet technology to deeply optimise the logistics activity and finally realise the smart logistics. Refer to Figure 2 for details.

3 RESTRICTION AND PROBLEM OF DEVELOPMENT FOR SMART LOGISTICS OF OUR COUNTRY

In general, many advanced modern logistics systems in our country have been provided with the advanced technical features like informatisation, digitisation, networking, integration, intelligentisation, flexibility, agility, visualisation and automation etc. However, compared with the developed countries like America and Japan etc., the smart logistics of our country still remains in the primary stage. There still are some prominent problems in the development of smart logistics.

3.1 Problems in government regulation system

First, there are conflicting policies from different departments. The development of smart logistics industry needs to integrate the latest achievements of multiple industries including products production, commerce circulation and information technology etc. The fuzzy definition of administration authority and the multi-administrative management pattern have set the multi-level administrative obstacles.

Second, government and business functions are mixed together. For a long time, the government power of our country is too concentrated. The administrative service divisions fail to realise the power of market in the resource allocation sufficiently. They adopt the method that the government directly control enterprises to enforce the development strategy of smart logistics industry made by the government. This tends to cause the mismatching in resource allocation of smart logistics industry development and reduce the allocation efficiency of smart logistics resource.

3.2 Problems in industrial technology

First, logistics enterprises lack the necessary management technology to implement smart logistics. At present, logistics enterprises have the problem to value the logistics facility and neglect the logistics software, which makes them have an insufficient modern logistics management philosophy. Second, logistics enterprises lack the necessary engineering technology to implement smart logistics. In consideration of saving the investment cost of logistics equipment at the initial stage, part of enterprises chooses the manpower to replace logistics equipment to rapidly promote the construction of product logistics network.

3.3 Logistics enterprise's own problem

Small scale, scattered layout and disordered management exist in logistics enterprises. At present, although many enterprises have begun to use IOT technology to build the smart logistics system, the scales of these enterprises are not generally large, and it is difficult to form the industrial cluster. Most of middle and small-sized enterprises have difficulties in logistics informatisation, where the management does not pay sufficient attention to the application of information technology due to lack of corresponding talents and capital. The supporting basic facilities still cannot be met even though the technology related to smart logistics has been introduced, which causes the enterprise benefit to not obviously improve.

3.4 Standardisation problem

Smart logistics is established based on the standardisation of logistics information, which requires realising the standardisation in the aspects related to codes like coding, file format, data interface, Electronic Data Interchange (EDI) and global positioning system (GPS) etc., so as to remove the information communication barriers among different enterprises. Compared with the developed countries, a lot of logistics information platforms and information systems all follow their own specification, which makes it is difficult to realise the information exchange and sharing among enterprises, platforms and organisations. It is hard for the whole electronic IOT to realise compatibility. It is difficult for data to exchange. The complete and smooth supply chain of commodities from production, circulation to consumption etc. is hard to form, which has seriously affected the management of logistics industry and the operation of e-commerce, and restricted the industrialisation of smart logistics products and the engineering of application.

3.5 Lack of professional talents

With the rapid development of the logistics industry, the problem of talent demand has been highlighted increasingly in our country. The scarcity of talent in smart logistics has become the bottleneck restricting the logistics development of our country. At present, our country needs the talents in logistics industry, at least 300,000. The majority of logistics enterprises lack the high-quality skilled talents at the front line of logistics and the high-level compound talents who are familiar with not only logistics management but also operation regulations of modern logistics informatisation.

4 COUNTERMEASURE AND SUGGESTION TO PROMOTE THE DEVELOPMENT OF SMART LOGISTICS IN OUR COUNTRY

4.1 Perfect the management system of smart logistics

First, government administrative departments should make the planning scheme of smart logistics from the perspective of supporting the social and economic development. In the macroscopic view, government administrative departments should integrate the administrative resources by changing the phenomenon that each department works for itself, so as to effectively promote the healthy development of smart logistics. In the microcosmic view, government administrative departments should break the industry monopoly by deactivating the mechanism that the executive power excessively intervenes with the market resource allocation so as to facilitate the multi-industry integration of logistics industry and information industry. Second, vigorously promote the construction of an authoritative industrial coordinating institution. The operation of the logistics industry is the product of comprehensive cooperation between multiple relevant industries. The industrial coordinating institution can take full advantage of its industrial organisational ability and information resource convergence ability to enhance the overall competitiveness and common interests of logistics enterprises and effectively control the internal bad enterprise behaviours of the logistics industry. Finally, take the smart logistics project as the key support project in the aspect of financial policy.

4.2 Optimise standard system of smart logistics industry

First, the standard system of smart logistics industry should form a unified standard in the transport formats of logistics code and logistics information, which is beneficial to removing the information interaction barrier among various enterprises in different regions, so as to provide a standardisation system guarantee for products to carry out logistics and sales activities in a wider range.

Second, establish the professional organisation supporting the standardisation of the smart logistics industry. The formulation and implementation for standards of the smart logistics industry is a systematic project. It needs to establish a unified and coordinated organisation specialised in the smart logistics industry to scientifically plan the technological development route of departments related to smart logistics business, and comprehensively allocate the limited logistics equipment resource and financial resource.

4.3 Make the planning of logistics development, and establish the sound relevant policies and regulations

For the problems in the logistics enterprises of our country (such aslarge quantity, small scale and backward management), it is necessary for the central government to plan the development of the logistics industry, introduce the development strategy of the logistics industry and cultivate a number of model enterprises in smart logistics industry with a high informationalised level and demonstration-promotion ability, so as to promote the development of the same industry and other enterprises. Perfect the relevant laws and regulations to create a fair and orderly market environment, remove the market barrier of each region and make production factors flow freely. Optimise resource allocation accordingly to form a unified, open and orderly market so as to better protect the legal interests of operators, safeguard the legitimate interests of logistics users, and realise the legislation, standardisation and systemisation of the logistics industry.

4.4 Speed up the construction of standard system for logistics informatisation

First, speed up the research and formulation of technology, code, safety, management and service standards in logistics information. Second, research and spread the application standard of bar code, radio frequency identification and other technologies in the business of storage, distribution, container and cold chain etc. Promote the standard system of logistics informatisation application in the key industries to gradually become perfect, including the automobile and its parts, food, drug, textile, agricultural materials and agricultural products etc. Finally, promote the linking of the logistics informatisation standard among data layer, application layer and exchange layer etc. to push the construction of the standard system in logistics informatisation.

4.5 Establish and continuously optimise logistics informatisation platform

One of the core problems for construction of smart logistics is the research and development for key technology. Consequently, it is necessary to concentrate on deploying a batch of major special projects in technology research and development, making the roadmap of technological development to research and develop the key technologies of modern logistics, and to reduce the universally applicable overall cost of key technologies.

The operation of smart logistics is inseparable from the support of the information platform. The information platform of smart logistics established with IOT, Cloud Computing, the Internet and other technologies provides the information service in real-time, including goods distribution, intelligent tracing and customer enquiry etc.

4.6 Speed up the development and cultivation of logistics talent resources

Talent resource is called the "soft power" of enterprise. For a series of problems existing in the development of smart logistics, it is a must to speed up the development and cultivation of logistics talent resource. First, actively introduce the excellent professional talent through favourable policy and high reward to enrich the talent team of enterprise. Second, strengthen the training for existing staff to improve their overall quality. Third, establish the effective evaluation and incentive mechanism to fully motivate the staff and strengthen their cohesion and working force. Fourth, establish the learning-based organisation to carry out the activities in rationalisation proposal, quality management and innovation, or carry out team learning etc. for the actual problems encountered, so as to exchange ideas, share experience and strengthen enterprise vitality.

5 CONCLUSION

As a new concept, smart logistics brings the new opportunity and challenge for the operation and management of modern logistics. Globalisation, intelligentization and professionalisation are the development tendencies of the modern logistics industry. The development of IOT has provided the advantageous technical guarantee for the realisation of smart logistics. With the vigorous development of IOT technology, logistics technology will be developed in the direction of intelligentization and visualisation. Smart logistics will be the next stage of informationalised logistics, which also is the general trend of modern logistics development.

REFERENCES

Jiang, Z. & Zhang, L. (2016). Research on intelligent logistics development model based on cloud computing and Internet of Things. *Management & Technology of SME,* (20), 124–125.

Wang, Z. (2014). "Smart logistics" is needed by urbanization. *China Business and Market,* (3), 4–8.

You, D. (2015). Discussion on the development of rural intelligent logistics under the background of new urbanization. *Journal of Commercial Economics,* (22), 40–41.

Zhong, B. (2012). the benign operation of the small and medium logistics enterprises in China. *China Business and Market,* (9), 29–33.

Automotive, Mechanical and Electrical Engineering – Liu (Ed.)
© 2017 Taylor & Francis Group, London, ISBN 978-1-138-62951-6

Virtual missile assembly system based customisation of plant simulation

Liangwen Ma, Zhen Zhang, Xiaoxu Su & Kezhen Guo
China Aerospace Construction Group Co. Ltd., Beijing, China

ABSTRACT: Because of the specialty using Plant Simulation to simulate missile assembly and not suitable for production designation, this paper presents a simulation template based on plant simulation. The input parameters and the solving process are templated, parameterised and standardised in the simulation template. Then, the template was used to simulate the missile assembly production process, which proved the validity of the template.

Keywords: Missile Assembly; Customisation; Simulation Template; Plant Simulation

1   INTRODUCTION

The aim of this paper is the virtual production of missile assembly. First, plant simulation was used to set up a logic model for the missile assembly factory. The soft can be visualised simulation of the production process, and output data curves, for example, equipment utilisation and power in the process of production. However, plant simulation is too professional to be promoted in the production design. The paper puts forward a method that encapsulates the whole simulation process effectively. Input conditions and output results are transparent for analysts, so that they can work well.

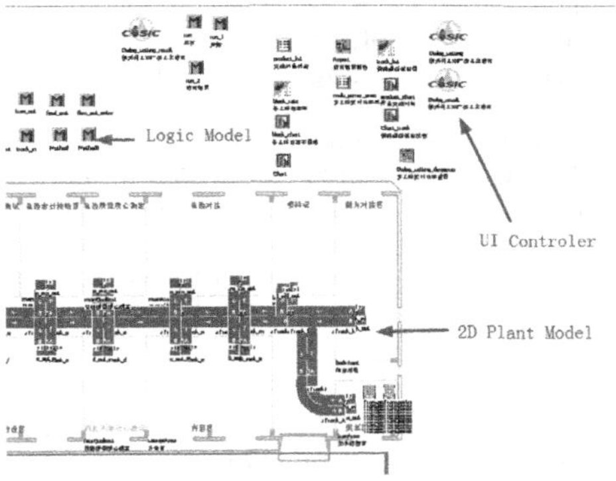

2   SCENARIO OVERVIEW

The whole model includes three parts, as shown in Figure 1: logic module, UI controls and visual model.

This simulation model can the simulate missile assembly process, and output production data in the process of simulation, such as blocking rate, logistics channel footprint, real-time power and so on. Finally, software is created by encapsulating this model. The simulation drives the model by modifying the number of missiles or production cycles. The software is based C# language in.net framework. It uses a lot of mainstream interface elements to improve the availability of the software. The interface is shown in Figure 2.

Figure 1.   Simplified elements model.

Figure 2.   Software interface.

3 BUILT-IN DEVELOPMENT MODE

The pretreatment is divided into three steps: plant modelling, modelling the logic relationship between modules, and solver control settings. The paper discusses two aspects from the user's perspective: UI modelling, and the writing and reading of parameters. The internal logic of the model is performed by the SimTalk programming language.

3.1 *UI modelling*

In plant simulation, you can use a dialogue control to receive the final user's input, as shown in Figure 3.

The Kind of dialogue control can meet basic requirements, but there are several limitations as described below:

1. The dialogue control cannot be used until the plant simulation is run. For user who are unfamiliar with the software, the use of this pattern is difficult, even leading to the wrong operation.
2. There is little UI control, few form and limited scalability, as shown in Figure 4.
3. The creation process is complex and there is no visual function. As shown in Figure 5, the coordinates need to be input manually for creating a simple textbox control, and the size of the textbox is non-adjustable.

3.2 *Interface callback function*

The response callback function of the interface in plant simulation is usually written by the SimTalk language, as shown in Figure 6.

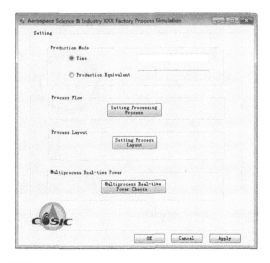

Figure 3. Dialogue control.

Figure 4. UI control sort.

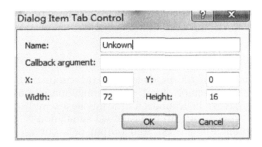

Figure 5. Static textbox control.

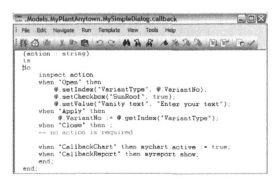

Figure 6. Callback function.

SimTalk is an interpreted simulation logic control language. Its grammar is similar to Basic. Plant Simulation provides a special debug window that speeds up the debug efficiency for users. Using a programming language interaction interface, although to a certain extent, improves the threshold of using the software, but greatly enhances the flexibility. Hence it provides a large operating space for custom development.

4 DEVELOPMENT KEY POINTS BASED PLANT SIMULATION COM INTERFACE

Other applications control plant simulation by calling COM control interface, for example, Windows applications, Office applications, Java applications, JS etc., and communicate with plant simulation by the interface. The interface defines the following methods, covering operations of the model and basic function input/output:

interfaceIRemoteControl: IDispatch

```
{
HRESULT NewModel();
HRESULT LoadModel(BSTR);
HRESULT SaveModel(BSTR);
HRESULT CloseModel();
HRESULT StartSimulation(BSTR);
HRESULT StopSimulation();
HRESULT ResetSimulation(BSTR);
HRESULT IsSimulationRunning ([out, retval]
VARIANT_BOOL*);
HRESULT SetPathContext(BSTR);
HRESULT ExecuteSimTalk(BSTR,[option al]
VARIANT,[out,retval]VARIANT*);
HRESULT GetValue(BSTR,[out,retval]VA RIA
NT*);
HRESULT SetValue(BSTR,VARIANT);
HRESULT Quit();
};
```

4.1 COM interface initialisation

This solution uses the Windows application by writing C# to invoke the COM interface. If you want to use the COM interface, you need to import the interface to the project of C#. The settings are as follows:

There are two key steps to load the model by initialisation code:

1. Statement at the beginning of the program, as follows:
2. Interface variable to be initialised and loaded missile assembly model created.

4.2 Modification of model parameters by COM interface

It is very convenient to modify the parameters by calling the SetValue function after the model is loaded. For example, to modify the missile assembly time and number:

```
eMPlant.SetValue(".Models.twoline.source
Area.num_or_time", true);
eMPlant.SetValue(".Models.twoline.source
Area.num", Decimal.ToInt32(knobControl2.Va
lue));
eMPlant.SetValue(".Models.twoline.source
Area.time_all", 0);
```

Name	Interop.eMPlantLib
Region	0
Copy Local	True
Parsed	True
Full Name	False
File Type	ActiveX
Logo	{CAF7631A-A39E-4141-956B-4B7B76359DF3}\1.0\0\tlbimp
version	1.0.0.0
Independent	False
Instruction	Tecnomatix Plant Simulation 9.0 Type Library
Path	D:\Seven\PlantSim\Projects\PlantSim\obj\Debug\Interop.eMPlantLib.dll

Figure 7. COM interface setting.

```
using System.Windows.Forms;
using System.Data;
using eMPlantLib;
using DevComponents.Instrumentation;
```

Figure 8. Statement library.

```
eMPlant = new RemoteControl();
eMPlant.LoadModel(@"d:\zongzhuang.spp");
```

Figure 9. Interface variable initialisation.

```
eMPlant.SetValue(".Models.twoline.EventCon-
troller. end", 0);
```

4.3 Simulation process control

The simulation process control is mainly composed of the following processes:

```
//Reset simulation
eMPlant.ResetSimulation(".Models.twoline.
EventControl ler");
//start simulation
eMPlant.StartSimulation(".Models.twoline.
EventControl
ler");
//stop simulation
eMPlant.StopSimulation();
```

4.4 Chart output

If the simulation is finished, we can active and view the chart through the COM interface. For example:

```
//Show Bolck Chart
eMPlant.ExecuteSimTalk("is    do.Models.two-
line.block_chart.active: = true;end;");
```

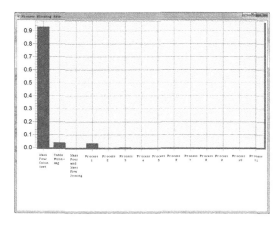

Figure 10. Blocking rate chart.

5 CONCLUSION AND PROSPECT

This paper includes two parts: plant simulation simulates the missile assembly production line and the development of a simulation template. The template can simulate various kinds of scenarios through changing the parameters on the application interface, and it greatly increases simulation efficiency. The template can be extended to other application scenarios. If reprocessing can automatically extract relevant output data, the template can cooperate iSight or ModelCenter software to make the whole process fitting, optimised or DOE. We can then find the correlation in the whole parameter space and find the best designation point.

REFERENCES

Guo, H. (2016). Optimization & design for job shop scheduling based on plant simulation. *Modern Manufacturing Engineering, 0*(2), 108–112.
Han, X. (2015). Simulation and analysis of engine testing line based on plant simulation. *Modular Machine Tool & Automatic Manufacturing Technique, 0*(11), 58–60.
Niu, Y. (2014). Cooperative process planning system for missile virtual assembly. *Ordnance Industry Automation, 8*, 5–7.
Zhou, J. (2011). *Production system simulation—Plant simulation application course.* Beijing: Electronics Industry Publisher.

Author index